International Exhibition & Conference for Power Electronics, Intelligent Motion, Renewable Energy and Energy Management (PCIM Europe 2024)

Nuremberg, Germany
11 – 13 June 2024

Volume 5 of 5

ISBN: 978-1-7138-9966-2

Printed from e-media with permission by:

Curran Associates, Inc.
57 Morehouse Lane
Red Hook, NY 12571

Some format issues inherent in the e-media version may also appear in this print version.

Copyright© (2024) by Mesago Messe Frankfurt GmbH
All rights reserved.

Printed with permission by Curran Associates, Inc. (2024)

For permission requests, please contact VDE VERLAG GMBH
at the address below.

VDE VERLAG GMBH
Bismarckstr. 33
P.O.B. 12 01 43
10625 Berlin, Germany

Phone: +49 30 34 80 01 - 0
Fax: +49 30 34 80 01 - 9088

kundenservice@vde-verlag.de

Additional copies of this publication are available from:

Curran Associates, Inc.
57 Morehouse Lane
Red Hook, NY 12571 USA
Phone: 845-758-0400
Fax: 845-758-2634
Email: curran@proceedings.com
Web: www.proceedings.com

TABLE OF CONTENTS

VOLUME 1

KEYNOTE

K01   AI BETWEEN HYPE AND INDUSTRIAL-GRADE - THE IMPACT OF AI ON THE
ENTIRE POWER ELECTRONICS LIFECYCLE.. 1
Rolf Hellinger

K02   INFRASTRUCTURE REQUIREMENTS FOR ELECTRIFIED HEAVY GOODS
TRANSPORT IN GERMANY AND THE EU.. 7
Martin Wietschel

K03   CHALLENGES AND SOLUTIONS TO POWER LATEST PROCESSOR
GENERATIONS FOR HYPER SCALE DATACENTERS 15
Gerald Deboy

GAN RUGGEDNESS

OP001   AN IMPROVED ULTRAFAST DESATURATION-BASED PROTECTION SCHEME
FOR GAN HEMT 19
Juncheng Lu

OP002   THE PERFORMANCE OF A GAN EMODE HEMT IN SURGE CURRENT
SCENARIOS SUCH AS THE ACTIVE SHORT CIRCUIT.. 24
Dominik Nehmer

OP003   GATE RESISTANCE EFFECT ON SHORT-CIRCUIT ROBUSTNESS OF P-GAN
HEMTS 34
Mohamed Lemine Dedew

ADVANCED PACKAGING TECHNOLOGIES

OP004   NEURAL NETWORK ASSISTED NUMERICAL SIMULATION BENCHMARKING
FOR ELECTRIC VEHICLE THERMAL MANAGEMENT SYSTEM 40
Ekin Alp Bicer

OP005   RELATIONSHIP BETWEEN POROSITY IN CU SINTERED BONDING AND
BONDING RELIABILITY... 49
Hideo Nakako

OP006   HIGH THERMAL DURABILITY OF THIN COPPER DIE-ATTACH LAYERS AND
FINITE ELEMENT MODEL SIMULATION.. 56
Takaaki Eyama

THERMAL CYCLING RELIABILITY

OP007   THERMAL SHOCK TEST LIFETIME IMPROVEMENT WITH OPTIMIZED
ADHESIVE STRENGTH BETWEEN EPOXY RESIN AND COPPER 62
He Kangjia

OP008   POWER CYCLING RELIABILITY AND FAILURE MODE ANALYSIS OF POL 67
Kenichi Koi

OP009   ACCELERATED POWER CYCLING OF GAN HEMTS USING SWITCHING LOSS
AND FAST TEMPERATURE MEASUREMENT 74
Wing Tai Leung

HIGH POWER CONVERTERS

OP010   CONTROL OF AN MMC-BASED HYBRID TRANSFORMER WITH STAR-POINT
VOLTAGE INJECTION 84
Rui Wang

OP011   PROTECTION AND CONTROL OF A DUAL MMC MEDIUM VOLTAGE SUPPLY 93
Max Dupont

OP012   STATION POWER ELECTRONICS CONVERTER WITH HIGH THERMAL
ENDURANCE TO POLE-TO-POLE SHORT CIRCUITS FOR LVDC DISTRIBUTION GRID 103
Frédéric Reymond-Laruina

GATE DRIVERS

OP013   SUPPRESSION OF OSCILLATIONS IN A SIC BRIDGE-LEG USING A CUSTOM
SINGLE-CHIP DIGITAL ACTIVE GATE DRIVER WITH 2×255 STRENGTH LEVELS 113
Qilei Wang

OP014   SIC MOSFET SHORT-CIRCUIT PROTECTION: A FASTER SOFT SHUT DOWN
METHOD FOR GATE DRIVERS 121
Julien Weckbrodt

OP015   PARAMETER IDENTIFICATION: GATE SENSOR FOR POWER TRANSISTOR
TOLERANCE COMPENSATION IN ADVANCED GATE DRIVER ICS 128
Christopher Wille

ADVANCED CONTROL TECHNIQUES ON ELECTRICAL DRIVES I

OP016   AN INNOVATIVE HIGH-SPEED TRACK RANGE RESTART STRATEGY FOR
PERMANENT MAGNET SYNCHRONOUS MOTOR 135
Anna Corbitt

OP017   STEADY-STATE ERROR REDUCTION OF REINFORCEMENT LEARNING BASED
INDIRECT CURRENT CONTROL OF PERMANENT MAGNET SYNCHRONOUS
MACHINES 140
Tobias Schindler

OP018   PERFORMANCE COMPARISON OF USING SHUNT-BASED AND INTEGRATED
CURRENT SENSING FOR SENSORLESS FIELD-ORIENTED CONTROL 150
John Emmanuel Tan

GAN CONVERTERS

OP019   DESIGN OF HIGH-POWER INVERTER WITH 12 PARALLEL GAN DEVICES 161
Takashi Sawada

OP020   OVER 99.7% EFFICIENT GAN-BASED 6-LEVEL CAPACITIVE-LOAD POWER
CONVERTER 167
Stefan Mönch

OP021   CASCADED PRIMARY-SIDE-ONLY CONTROL OF A COMPACT 2 MHZ 500 W
WIRELESS POWER TRANSFER SYSTEM 174
Tim Krigar

ADVANCED MATERIALS AND TECHNOLOGIES

OP022   POWER MODULE EVALUATION USING ULTRA HIGH HEAT DISSIPATION AND
HIGH HEAT RESISTANCE RESIN SHEET CONTAINING CARD HOUSE TYPE BORON
NITRIDE FILLER 180
Ayano Imai

OP023   INVESTIGATING TEMPERATURE DEPENDENT WARPAGE IN METAL CERAMIC
SUBSTRATES FOR POWER ELECTRONICS DEVICES 190
Benjamin Fabian

OP024   DEGRADATION MODE ANALYSIS OF DIFFERENT BONDING TECHNOLOGIES
OF SIC POWER SEMICONDUCTORS STRESSED BY ACTIVE POWER CYCLING 197
Rasched Sankari

CHARGING STATION TECHNOLOGY

OP025   IMPLEMENTATION AND VERIFICATION OF A 50KW OPPORTUNITY WIRELESS
CHARGER DESIGN 205
Carlos Costas Sos

OP026   PERFORMANCE EVALUATION OF SILICON-BASED 3-LEVEL VIENNA
RECTIFIER IN ISOPLUS SMPD PACKAGE 214
Karsten Haehre

OP027   PERFORMANCE ANALYSIS OF A 25-KW SIC-BASED DUAL ACTIVE BRIDGE
CONVERTER BASED ON PARALLEL-CONNECTED DEVICES 222
Francesco Porpora

MODELLING AND MONITORING

OP028   SEMICONDUCTOR CHIP MODELS ARE THE KEY FOR ENABLING VIRTUAL
DESIGN AND OPTIMIZATION WORKFLOWS OF POWER ELECTRONIC SYSTEMS 230
Stefan Haensel

OP029   IMPROVED RESONANT FREQUENCY-BASED PARASITIC INDUCTANCE
ESTIMATION METHOD FOR SIC MOSFET HALF-BRIDGE CIRCUIT 238
Hongpeng Zhang

OP030   FAST SIMULATOR WITH INVERTER TEMPERATURE ESTIMATION FOR
TRACTION EDRIVES IN VEHICLES SUBJECTED TO DRIVING CYCLES 248
Simone Giuffrida

SOLID STATE TRANSFORMERS

OP031 A NEW FAMILY OF THREE-PHASE-UNFOLDER-BASED MVAC-LVDC SOLID-STATE TRANSFORMERS 254
Jonas Huber

OP032 VOLTAGE BALANCING OF A SPLIT-CAPACITOR IGCT 3L-NPC LEG FOR THE RESONANT DC TRANSFORMER 264
Renan Pillon Barcelos

OP033 COMPARATIVE ANALYSIS OF UNIDIRECTIONAL HIGH STEP-UP CONVERTERS FOR MEDIUM VOLTAGE APPLICATIONS 274
Stefan Subotic

ADVANCED CONTROL TECHNIQUES ON ELECTRICAL DRIVES II

OP034 STARTUP BEHAVIOR OF HARMONIC SUPPRESSION IN ELECTRICAL MACHINES USING ITERATIVE LEARNING CONTROL AND NEURAL NETWORKS 284
Annette Mai

OP035 ANALYTICAL APPROACH OF THE VECTOR CURRENT CONTROL FLUX-WEAKENING STRATEGY FOR PERMANENT MAGNET SYNCHRONOUS MACHINES 290
Oriol Subirats Rillo

POWER ELECTRONICS FOR E-MOBILITY

OP036 INVESTIGATION ON DIRECT LIQUID COOLING DESIGN OF POWER MODULES WITH FLAT BASEPLATE FOR AUTOMOTIVE APPLICATION 298
Nobuhide Arai

OP037 A NOVEL APPROACH FOR AFFORDABLE ELECTRIC VEHICLES BASED ON DUAL 48V BATTERY SYSTEM WITH MULTI-FUNCTIONAL 3-LEVEL CONVERTER 305
Radovan Vuletic

OP038 AN INNOVATIVE 3-LEVEL SOLUTION FOR AUTOMOTIVE APPLICATIONS: EMPACK 315
Pranav Panchal

OP039 GATED RECURRENT UNITS-ASSISTED STATE-SPACE MODELING FOR ELECTRIC VEHICLE TEMPERATURE PREDICTION 322
Xinyuan Liao

OP040 NOVEL BIDIRECTIONAL SINGLE-STAGE ISOLATED 600-V GAN M-BDSBASED SINGLE/THREE-PHASE-OPERABLE EV ON-BOARD CHARGER 330
Sven Weihe

ENCAPSULATION MATERIALS

OP041 APPLICATION-SPECIFIC INVESTIGATION OF INORGANIC POTTING MATERIAL IN DRIVE TRAINS 338
Soenke Fleck

OP042   THE INFLUENCE OF THE GLASS TRANSITION TEMPERATURE OF EPOXY
MOLD COMPOUNDS ON THE RELIABILITY OF A SEMICONDUCTOR DEVICE 343
   Stefan Schwab

OP043   CORROSION RESISTANT PACKAGING FOR POWER SEMICONDUCTOR
MODULES - MODIFIED INSULATION MATERIALS FOR CONTAMINATED
ENVIRONMENTS 351
   Michael Hanf

OP044   INVESTIGATION OF INORGANIC ENCAPSULATION MATERIALS IN POWER
ELECTRONIC SYSTEMS FOR HIGH POWER DENSITY APPLICATIONS 361
   Stefan Behrendt

OP045   CHARACTERIZATION OF THERMALLY AGED SILICONE GELS FOR POWER
SEMICONDUCTOR MODULES... 369
   Sonja Madloch

POWER QUALITY

OP046   A COORDINATED CONTROL OF HYBRID SINGLE-PHASE AC/DC MICROGRIDS
BASED ON THE NATURAL HARMONIC INJECTION CONCEPT 378
   Mehdi Baharizadeh

OP047   A HIGH-POWER DENSITY SIC BASED TP PFC WITH HIGH-FREQUENCY RIPPLE
CANCELLATION LEG.. 383
   Serkan Dusmez

OP048   HIGH FREQUENCY ACTIVE FILTER FOR AC-DC HIGH POWER CONVERTERS 390
   Sarah Sifoune

OP049   LABORATORY SETUP FOR ACCURACY INVESTIGATION OF ELECTRICITY
METERS AND MONITORS UNDER INDUSTRY-TYPICAL OPERATING CONDITIONS 397
   Matthias Schmidt

GRID CONNECTED CONVERTERS

OP050   REAL-TIME EVALUATION OF WEIGHTING FACTORLESS PREDICTIVE
CONTROL OF LCL FILTER EQUIPPED GRID-SIDE CONVERTERS USING SORTING
NETWORKS.. 403
   Kristóf Bándy

OP051   RELAXED ROBUST CONTROL WITH PRAGMATIC SHORTAGE OF PASSIVITY
FOR WIND, STORAGE AND PV POWER CONVERTERS 411
   Sergio De Lopez Diz

OP052   AN EFFECTIVE DC VOLTAGE REGULATION OF ACTIVE FRONT-END
RECTIFIER THROUGH MODEL PREDICTIVE CONTROL.. 419
   Mobina Pouresmaeil

OP053   BI-DIRECTIONAL 11KW MULTI-LEVEL ACTIVE-NEUTRAL-POINT-CLAMPED
AC-DC CONVERTER USING 600V/750V SI SUPER-JUNCTION AND SIC MOSFETS FOR
HIGH-EFFICIENCY AND HIGH-DENSITY APPLICATIONS.. 424
   Mengxing Chen

OP054 A STUDY OF GRID-FORMING INVERTER CONTROL STRATEGY FOR FAULT-RIDE-THROUGH CAPABILITY.. 433
Hirofumi Uemura

PASSIVE COMPONENTS

OP055 FILM CAPACITORS FOR HIGH TEMPERATURE AC-DC INVERTER APPLICATIONS 440
Adel Bastawros

OP056 LOSS REDUCTION IN HF-TRANSFORMERS USING LAMINATED FERRITE E-CORES 447
Lukas Reißenweber

OP057 MULTIGAP TOROIDAL TRANSFORMER AND INDUCTORS FOR OVERCOMING FRINGING LOSSES IN HIGH FREQUENCY CONVERTERS 456
Pau Colomer

OP058 STUDY ON SAMPLE GEOMETRIES FOR FERRITE CHARACTERISATION IN THE MHZ RANGE 463
Till Piepenbrock

OP059 FEM-SUPPORTED AND NON-DESTRUCTIVE MAGNETIC CHARACTERIZATION METHOD FOR NON-LAMINATED STEEL 472
Stefan Tobler

DRIVES FOR HIGH DEMANDING APPLICATIONS

OP060 HIGHLY-COMPACT BEARINGLESS AXIAL-FLUX MOTOR FOR A PEDIATRIC IMPLANTABLE FONTAN BLOOD PUMP 480
Andreas Horat

OP061 A NOVEL PERMANENT MAGNET SYNCHRONOUS MOTOR DRIVE FOR REACTION WHEELS IN SATELLITES 490
Baris Colak

OP062 EXPLORING HIGH FREQUENCY OPERATION OF MOTOR DRIVES: PRACTICAL INSIGHTS ON EFFICIENCY AND LOSS 497
Asantha Kempitiya

OP063 HIGH POWER DENSITY SYSTEM DESIGN FOR GAN-BASED LV MOTOR DRIVES 502
Marco Cannone

OP064 DESIGN OF GAN TRANSISTOR BASED VARIABLE SPEED DRIVE INVERTER WITH OUTPUT VOLTAGE FILTERING.. 510
Kaspars Kroics

IGBT

OP065 THE 8TH GENERATION LV100 IGBT MODULE WITH HIGHER CURRENT RATING 518
Daichi Otori

LOSS REDUCTION BY LAMINATIGN FERRITE E CORES.. 525
Lukas Reißenweber

OP066   NEW PLANAR 4.5 KV SPLIT-GATE (SG) SI-IGBT DEVICE FOR IMPROVED
SWITCHING CHARACTERISTICS AND HIGH FREQUENCY OPERATION 534
Gaurav Gupta

OP067   4.5 KV DOUBLE-GATE REVERSE-CONDUCTING PRESS-PACK IEGT 543
Satoshi Yoshida

DEVICE CONCEPTS

OP068   EVALUATION OF A 3 KV POLARIZATION SUPERJUNCTION GAN HEMT........................... 549
Alireza Sheikhan

OP069   MORE THAN 1200 V BREAKDOWN AND LOW AREA-SPECIFIC ON STATE
RESISTANCES BY PROGRESS IN LATERAL GAN-ON-SI AND GAN-ON-INSULATOR
TECHNOLOGIES.. 557
Richard Reiner

OP070   NOVEL 200 V MOSFET TECHNOLOGY PUSHES MOTOR DRIVE INVERTER
EFFICIENCY TO AN UNPRECEDENTED LEVEL 564
Mark Thomas

DEGRADATION MECHANISMS

OP071   MOISTURE ROBUST CHIP DESIGN - IMPROVED EDGE-TERMINATIONS FOR
HIGH LIFETIME UNDER HIGH HUMID CONDITIONS... 571
Michael Hanf

OP072   METHOD FOR MEASURING THE INITIAL STATE OF A SOLDER JOINT
DELAMINATION IN A 3D PCB INTEGRATION ASSEMBLY OF SIC MOSFETS................................. 581
Souhila Bouzerd

OP073   GENERIC LIFETIME MODEL FOR WIRE BONDS DEGRADATION IN IGBT
MODULES BASED ON A FRACTURE MECHANICS PARAMETER... 589
Merouane Ouhab

ADVANCED CONVERSION CONCEPTS

OP074   MODULAR COAXIAL POWER CONVERTER FOR HIGH-DENSITY INTEGRATION
INTO MEDIUM-VOLTAGE CABLES 599
Mark Cairnie

OP075   CONTROLLED INDUCTOR BASED BCM BUCK CONVERTERS 608
Ziv Gellman

OP076   INFLUENCE OF VARYING COMMON MODE CHOKE SIZES ON THE
PERFORMANCE AND STABILITY OF AN ACTIVE EMI FILTER.. 615
Patrick Körner

PHOTOVOLTAIC SYSTEMS

OP077   A HIGH EFFICIENCY BATTERY CHARGER WITH MAXIMUM POWER POINT TRACKING FOR MAGNETIC ENERGY HARVESTERS 625
Antonio Miguel Munoz Gomez

OP078   SYMMETRIC FLYING-CAPACITOR BOOST CONVERTER FOR MEDIUM-VOLTAGE PHOTOVOLTAIC APPLICATIONS 635
Luis Alves Rodrigues

OP079   COMPARISON OF SI IGBT, SIC MOSFET AND ADJUSTABLE HYBRID SWITCH PV INVERTERS FOR DIFFERENT GEOGRAPHICAL LOCATIONS 645
Tanya Thekemuriyil

MODEL BASED SYSTEM ANALYSIS

OP080   OPTIMISING A POWER MODULE FOR ELECTRICAL AND THERMAL PERFORMANCE AND SYMMETRY USING EDA TOOLS 655
Wilfried Wessel

OP081   CONDUCTOR-BASED MODELING OF VOLTAGE DISTRIBUTION ALONG A SINGLE-TOOTH WINDING OF ELECTRICAL MACHINES 665
Hujun Peng

OP082   REDUCTION OF PWM HARMONICS WITH CARRIER PHASE SHIFTING IN A DUAL-STATOR PMSM WITH MAGNETIC COUPLED WINDINGS 672
Bünyamin Tekir

VOLUME 2

SIC DEVICES

OP083   THE NEW COOLSIC MOSFET 1200 V G2: ELECTRICAL PERFORMANCE AND COMPACT MODELLING 681
Andreas Huerner

OP084   PARALLELING SIC-POWER-MOSFET BODY DIODES UNDER HARSH SWITCHING CONDITIONS 690
Michael Rauh

OP085   3.3KV SBD-EMBEDDED SIC-MOSFET MODULE FOR TRACTION USE 699
Yoichi Hironaka

OP086   DEAD TIME OPTIMIZATION FOR HIGH POWER SIC MOSFET MODULE IN CONSIDERATION OF PARASITIC COMPONENTS 707
Pham Ha Trieu To

WBG RELIABILITY

OP087   PERFORMANCE INSTABILITY OF 650 V P-GAN GATE HEMT DEVICE UNDER TEMPERATURE-RELATED POSITIVE GATE BIAS STRESSES 717
Renze Yu

OP088 GATE OXIDE RELIABILITY OF CURRENT GENERATION 1.2 KV SIC MOSFETS UNDER STEP-WISE INCREASED GATE VOLTAGE... 723
  Roman Boldyrjew-Mast

OP089 AN ACCELERATED DYNAMIC GATE SWITCHING STRESS TEST CONCEPT OF SIC MOSFETS AT HIGH DRAIN-SOURCE VOLTAGE (HV-GSS) 731
  Clemens Herrmann

OP090 SILICON CARBIDE POWER DEVICE USE IN SPACECRAFT AND AIRCRAFT 739
  Akin Akturk

POWER ELECTRONICS FOR E-MOBILITY/ CONTROL

OP091 CURRENT RIPPLE REDUCTION BY COMBINATION OF SI IGBT AND SIC MOSFETS IN HEAVY DUTY FUEL CELL TRUCKS... 745
  Yavuz Gürlek

OP092 EVALUATION OF ACTIVE GATE DRIVERS WITH SWITCHABLE GATE RESISTORS AND INTERMEDIATE VOLTAGE LEVELS FOR SIC MOSFETS IN WLTC 754
  Michael Frank

OP093 PERFORMANCE EVALUATION OF TCM-BASED, ZERO-VOLTAGE SWITCHING (ZVS) THREE-PHASE INVERTER FOR ELECTRIC VEHICLE DRIVE SYSTEMS 764
  Khizra Abbas

OP094 A PARTIAL LOAD THREE-PHASE TRIANGULAR CURRENT MODE MODULATION CONCEPT WITH AN OPTIMIZED FILTER INDUCTOR FOR HIGH EFFICIENCY TRACTION DRIVES 774
  Bhaskar Chatterjee

DC-DC CONVERTERS I

OP095 GAN VS SI SYNCHRONOUS RECTIFIER FOR LLC CONVERTER 784
  Gokhan Sen

OP096 CO-SIMULATION DESIGN OF A GAN-BASED THREE-PHASE LLC CONVERTER WITH INTEGRATED THREE-PHASE MAGNETICS 791
  Jhih-Cheng Hu

OP097 SWITCHING ASSISTING CIRCUIT IMPROVING THE EFFICIENCY OF DC-DC CONVERTERS BASED ON PIEZOELECTRIC RESONATORS... 797
  Ghislain Despesse

OP098 TRANSFORMER-BASED FIXED-RATIO RESONANT DC-DC CONVERTERS FOR 48V DATA CENTERS 803
  Xufu Ren

PFC CONVERTERS

OP099 HIGH-DENSITY 3.3 KW GAN RECTIFIER FOR SERVER APPLICATIONS COMPRISING A 130 KHZ TOTEM-POLE PFC AND A 500 KHZ LLC....................................... 812
  Manuel Escudero Rodriguez

OP100 ADDRESSING POWER SWITCH TECHNOLOGY SELECTION SI/SIC/GAN IN HIGH EFFICIENCY ZVS-PFC RESONANT CONVERTERS 822
Marco Torrisi

OP101 BUCK-TYPE CURRENT UNFOLDING CONVERTER WITH DISCONTINUOUS CONDUCTION MODE IN ULTRA-LOW POWER-FACTOR OPERATION 831
Tomoyuki Mannen

OP102 GAN BASED BI-DIRECTIONAL 6.6KW INTERLEAVED TOTEM-POLE PFC WITH 13KW/L POWER DENSITY AND HIGH EFFICIENCY 837
Juncheng Lu

SIC MODULES

OP103 THE DESIGN OF A 2KV 1700A SIC MOSFET DUAL MODULE 843
Jorge Mari

OP104 TECHNOLOGICAL APPROACHES TO HIGH-POWER DENSITY SIC POWER MODULE FOR AUTOMOTIVE 849
Takeshi Tokorozuki

OP105 EXTREMELY COMPACT SIC POWER MODULE FOR EV TRACTION INVERTERS IN THE 250 KW CLASS 855
Raffael Schnell

OP106 BENEFITS OF .XT INTERCONNECTION TECHNOLOGY FOR 3.3 KV XHP 2 MODULE WITH 3.3 KV COOLSIC MOSFET 863
Matthias Bürger

ADVANCED COOLING

OP107 LARGE-AREA BONDING WITH LMEE: SUPPRESSION OF THE DEGRADATION OF THE JUNCTION-TO-WATER THERMAL RESISTANCE IN POWER MODULES 870
Yo Mochizuki

OP108 ACTIVE THERMAL CONTROL OF SIC MOSFETS UTILIZING TRANSIENT THERMAL CHARACTERIZATION 875
Varaha Satya Bharath Kurukuru

OP109 THERMAL MANAGEMENT SOLUTIONS BY ADDITIVE MANUFACTURING – POWDER BED FUSION AND DIFFUSION BONDING 883
Simon Jahn

OP110 ADVANCED PUMPED TWO-PHASE COLD PLATE FOR COOLING POWER ELECTRONICS 888
Elizabeth Seber

DC-DC CONVERTERS II

OP111 FEASIBILITY STUDY OF HIGH-POWER DENSITY ISOLATED CLLC DC-DC INTERFACE WITH WIDE RANGE OF VOLTAGE/CURRENT REGULATION 893
Oleksandr Husev

OP112   DC-BIAS REDUCTION IN HIGH-FREQUENCY DUAL ACTIVE BRIDGE DC-DC
CONVERTERS THROUGH SLOW DC MEASUREMENTS 903
Patrick Lenzen

OP113   OPTIMIZED CURRENT SHARING TECHNIQUE FOR INTERLEAVED CLLC
CONVERTERS FOR MINIMAL OUTPUT CURRENT DISTORTION 909
Martin Gendrin

OP114   PRIMARY-SIDE OUTPUT REGULATION PRINCIPLES IN DYNAMIC MULTI-MHZ
INDUCTIVE POWER TRANSFER SYSTEMS AND ISOLATED DC/DC CONVERTERS 916
Ioannis Nikiforidis

SMART GRID

OP115   LOW VOLTAGE DC-GRIDS WITH GALVANIC ISOLATION: SYSTEM
DISCUSSION, EFFICIENCY AND PERFORMANCE COMPARISON TO AC-FEEDING....................... 926
Lukas Fräger

OP116   IMPLEMENTATION AND EXPERIMENTAL EVALUATION OF AN ADAPTIVE DC
GRID CONTROLLER FOR DECENTRALISED GRID CONTROL 933
Steffen Menzel

OP117   DEMONSTRATING THE EFFECTIVENESS OF A DC SOLID-STATE CIRCUIT
BREAKER'S FAST RESPONSE TIME.. 942
Ehab Tarmoom

OP118   MODELLING AND SIZING SENSITIVITY ANALYSIS OF A FULLY RENEWABLE
ENERGY-BASED ELECTRIC VEHICLE CHARGING STATION MICROGRID 949
David A. Stone

MEASUREMENT TECHNIQUES AND METHODS

OP119   LED POWERED ROTOR TELEMETRY SYSTEM... 958
Raphael Beyerle

OP120   'INFINITY GATE SENSOR': A DIFFERENTIAL MAGNETIC FIELD SENSOR FOR
MEASURING GATE CURRENT OF SIC POWER TRANSISTORS.. 966
Yushi Wang

OP121   CHARACTERISING WIDE BANDGAP POWER MODULES: VALIDATING THE M-
SHUNT CONCEPT FOR HIGH-POWER APPLICATIONS IN THE KILOAMPERE RANGE 976
Hauke Lutzen

OP122   CHARACTERIZATION OF POWER-MODULE PARASITICS: SUB-NANOSECOND
LARGE SIGNAL PULSING VS. DOUBLE-PULSE TESTING 986
Gerhard Groos

STATISTICAL VARIATIONS IN THE PARASITIC CAPACITANCE OF A COIL.................................... 997
Kevin Talits

HIGH VOLTAGE SWITCHES

PP001   A 4.5 KV FAST RECOVERY DIODE PLATFORM FOR HIGH-CURRENT IGBTS 1002
Jan Vobecky

PP002   6.5 KV INNOVATIVE SILICON POWER DEVICE (I-SI) MODULE WITH HIGH POWER DENSITY AND LOW LOSS BY STORED CARRIER CONTROL 1007
Takashi Hirao

PP003   HIGH CURRENT DENSITY 4.5KV PRESSPACK IGBTS PUSH SOA LIMITS 1013
Hossein Davoodi

PP004   2.5KV IGBT MODULE WITH HIGH RELIABILITY FOR RENEWABLE APPLICATIONS 1018
Akiyoshi Masuda

PP005   NEW GENERATION 4.5KV IGCT AND FAST RECOVERY DIODE FOR RAILWAY POWER SUPPLY APPLICATIONS 1025
Umamaheswara Reddy Vemulapati

PP006   NEXT GENERATION 4.5 KV IGBT-ONLY STAKPAK MODULE WITH REDUCED LOSSES AND HIGH TEMPERATURE CAPABILITY 1031
Jeremy Jones

THERMAL MODELLING AND SIMULATIONS

PP007   FINITE ELEMENT ANALYSIS OF THE UPSCALING OF WARPAGE AND BIFURCATION HYSTERESIS LOOPS: FROM CU/SI DIE TO LARGE WAFERS 1039
Vincenzo Vinciguerra

PP009   MAXIMUM JUNCTION TEMPERATURE SIMULATION AND VALIDATION FOR THE HOT SPOT IN MULTI-CHIP SIC POWER MODULE 1046
Wonjin Dylan Cho

PP010   INTEGRATION OF CFD-SIMULATION RESULTS IN PLECS USING LOOKUP TABLES 1051
Simon Cepin

PP011   PCB ONLY THERMAL MANAGEMENT TECHNIQUES FOR EGAN FETS IN A HALF-BRIDGE CONFIGURATION 1057
Adolfo Herrera

HIGH POWER DENSITY DESIGNS

PP013   FROM 4X TO 3X STPAK – OPTIMIZATION FOR A MORE COMPACT EV TRACTION INVERTER SOLUTION 1065
Vittorio Giuffrida

PP014   A MULTI-OBJECTIVE STRUCTURAL OPTIMIZATION METHOD BASED ON MULTI-PHYSICS SIMULATIONS FOR POWER MODULE 1072
Baihan Liu

PP015   HOLISTIC APPROACH TO MAXIMIZE LIFETIME AND POWER DENSITY IN HIGH POWER SEMICONDUCTOR MODULES 1077
Martin Schulz

PP016   REGULATED HIGH DENSITY SWITCH CAPACITOR TOPOLOGY 1082
Pierrick Ausseresse

PP017   SILICON INTERPOSER AS A SUBSTRATE FOR POWER MODULES WITH HIGH
POWER DENSITY AND SUPERIOR THERMAL PERFORMANCE 1087
  Ahmed Ammar

SPECIAL CONVERTER APPLICATIONS

PP018   ANALYTICAL MODELING AND STABILITY CHARACTERIZATION OF A
DAMPED VSCC CM ACTIVE EMI FILTER FOR SINGLE- AND THREE-PHASE AC-DC
APPLICATIONS 1092
  Timothy Hegarty

PP020   A REPETITIVE HIGH VOLTAGE NANOSECOND PULSE GENERATOR: FIRST
PROTOTYPE DESIGN AND TEST RESULTS 1101
  Serge Gavin

PP021   FREQUENCY SHIFT KEYED DUAL SIDE CONTROL OF INDUCTIVE POWER
TRANSFER: AN APPLICATION OF TALKATIVE POWER CONVERSION 1105
  Hamzeh Beiranvand

PP022   STUDY OF A MULTI-ACTIVE BRIDGE CONVERTER FOR A DOMESTIC
ELECTRICAL GRID 1113
  Abdennour Merrouche

INTEGRATION TECHNOLOGIES AND RELIABILITY DESIGN

PP023   FABRICATION DEVELOPMENT FOR GATE DRIVER EMBEDDED DOUBLE-
SIDED COOLING SIC POWER MODULE FOR ELECTRIC VEHICLE APPLICATION 1123
  Anna Corbitt

PP024   PRINTED CIRCUIT EMBEDDING OF PREPACKAGED 150V POWER MOSFETS IN
A PORTABLE WELDING APPLICATION 1128
  Thomas Gebhard

PP025   PROCESS CHALLENGES AND PROGRESS TOWARDS DIRECT CONNECTION OF
AUTOMOTIVE POWER MODULES (TMM) TO HEATSINK 1133
  Indrajit Paul

PP026   OPTIMIZING PCB STACKUPS FOR ENHANCED GAN TRANSISTOR
PERFORMANCE IN HIGH-POWER APPLICATIONS 1139
  Philipp Czerwenka

PP027   NEW GENERATION CERAMIC SUBSTRATES – KEY COMPONENTS FOR POWER
ELECTRONIC APPLICATIONS: PROCESSING AND CHARACTERIZATION 1147
  Stefanie Schindler

PP028   AI-ENHANCED VACUUM REFLOW OVEN: PRECISION CONTROL FOR
RELIABLE LARGE-AREA SOLDERING 1152
  Chih Hui Lee

PP030   CORROSION-COMPATIBLE DRIVE ELECTRONICS FOR ELECTRIC VEHICLES
AND INDUSTRIAL POWER MODULES 1158
  Tom Petzold

PP031 EVALUATING THE SAFETY ISOLATION OF THE PACKAGE IN AN INTEGRATED POWER DEVICE 1168
Thomas Anthony Capobianco

CONTROL METHODS I

PP032 FLEXIBLE CONTROL SYSTEM FOR MODULAR ONE-PHASE INTERLEAVED GAN-BASED TOTEM POLE PFC USING REAL-TIME HARDWARE 1174
Oleksandr Solomakha

PP033 A PEAK CURRENT MODE CONTROL METHOD FOR PFC 1180
Sean Yu

PP034 ADAPTIVE RESONANT CONTROLLER FOR A THREE-PHASE PFC CONVERTER FOR AN ON-BOARD CHARGE APPLICATION 1185
Rami Troudi

PP035 SYNTHESIS OF A FIELD ORIENTED CONTROL ALGORITHM BY USING TWO DIFFERENT POLE-ZERO COMPENSATION APPROACHES..1192
Marco Denk

PP037 AVERAGE CURRENT MODE CONTROL AND ITS LOOP DESIGN 1200
Niklas Schwarz

PP038 NOVEL POWER FEED-FORWARD REGULATION FOR DUAL STAGE PFC+DCDC CONVERTERS 1207
Alfredo Medina-Garcia

HIGH POWER AC-DC AND DC-AC CONVERTER

PP039 22 KW BI-DIRECTIONAL WALL-BOX CHARGER WITH 1200 V SIC MOSFET.................. 1212
Sanbao Shi

PP040 DYNAMIC SWITCHING FREQUENCY SELECTION FOR EFFICIENCY OPTIMIZATION IN ON-BOARD CHARGER PFC STAGE BASED ON NOVEL SIC MOSFET POWER MODULE.. 1217
Giuseppe Aiello

PP041 DESIGN AND OPTIMIZATION OF SIC-BASED 11KW MOTOR DRIVE WITH HIGH EFFICIENCY 1222
Iris Liu

PP042 MODEL DESIGN DEVELOPMENT FOR FALSE TURN-ON CHARACTERIZATION IN SIC-BASED ACTIVE T-TYPE CONVERTER CONSIDERING ALL PARASITICS 1227
Amir Babaki

PP043 EFFICIENCY INVESTIGATIONS OF AN AUXILIARY RESONANT COMMUTATED POLE INVERTER.. 1233
Markus Zocher

PP044 A NOVEL HYBRID TWO-STAGE AC-DC CONVERTER WITH SOFT-SWITCHED CCM PFC STAGE FOR EVS CHARGING APPLICATIONS.. 1242
Lei Wang

PP045   A METHOD FOR TUNING LEAKAGE INDUCTANCE IN TRANSFORMERS 1249
Rosemary O'Keeffe

PP046   LOW COST HIGH DENSITY 300W/20V AC-DC CONVERTER ENABLED BY GAN
POWER ICS.. 1254
Tom Ribarich

PP047   25KVA GRID-TIED BI-DIRECTIONAL T-TYPE INVERTER WITH HIGH-
EFFICIENCY AND HIGH-POWER DENSITY USING SIC MOSFETS.. 1259
Tamanna Bhatia

PP048   COST-EFFECTIVE EFFICIENCY ENHANCEMENT IN AC-DC CONVERTERS: A
STUDY ACROSS THE FULL LOAD CYCLE 1264
Sebastian Gick

E-MOBILITY TRACTION I

PP049   NEXT GENERATION POWER MODULE WITH PARALLEL CONNECTED SIC
MOSFETS FOR BEV TRACTION INVERTERS.. 1272
Kohei Tanikawa

PP051   INVESTIGATION OF COMMON SOURCE FEEDBACK IN SIC POWER MODULES
REGARDING PERFORMANCE AND SHORT CIRCUIT ROBUSTNESS 1277
Dominik Ruoff

PP052   HYBRIDPACK DRIVE POWER MODULES WITH SIC-MOSFET'S AND
MONOLITHIC RC- SNUBBER CHIPS FOR OPTIMIZED POWER DENSITY........................ 1283
Andre Uhlemann

PP053   ROBUST AUXILIARY POWER SUPPLY FOR EVS BASED ON INNOVATIVE
STI2GAN 650V IC... 1289
Federica Cammarata

PP054   IMPACT OF VARIOUS SILICON DIODES ON THE HYBRID SWITCH INVERTER 1297
Michael Walter

PP055   ADVANCED PULSE SEQUENCE FOR SALIENCY-BASED HIGH-ACCURATE
ROTOR POSITION ESTIMATION OF RAILWAY TRACTION LOCOMOTIVE MOTORS 1307
Markus Vogelsberger

CONTROL TECHNIQUES

PP056   OPTIMIZED HALF-BRIDGE GATE-DRIVE WITH LOW TIME-SKEW FOR RC-
IGBTS AND SIC-MOSFET DEAD-TIME CONTROL 1315
Jan Fuhrmann

PP057   DESIGN OF A TRACTION INVERTER BASED ON PCB-EMBEDDED GAN
DEVICES 1322
Maurizio Tranchero

PP058   OPTIMIZING ELECTRIC VEHICLE PERFORMANCE WITH GAN DESIGN........................ 1330
Andrew Patterson

PP059   FAST ANALYTICAL CALCULATION OF THE MAGNETIC FIELD IN PERMANENT
MAGNET SYNCHRONOUS MACHINES WITH FLUX BARRIERS INCLUDING
SATURATION 1336
     Martin Ackermann

PP060   MODELING AND CONTROL OF LCL FILTERED 3L-VSCS IN INTERLEAVED
TOPOLOGY 1346
     Adeel Jamal

PP062   ENHANCING SAFETY AND EFFICIENCY FOR ISOLATED PLC I/O DESIGNS
WITH SPI DAISY CHAIN 1352
     Travis Lenz

VOLUME 3

PP063   COST-EFFECTIVE METHOD TO DISCHARGE DC LINK CAPACITORS WITH SIC
POWER MODULES 1361
     Paul Kanatzar

POWER QUALITY

PP064   A STUDY ON CIRCULATION CURRENT IN PARALLEL OPERATION OF
TRANSFORMER LESS UPS 1368
     Koji Kato

PP065   DESIGN CHALLENGES AND CONSIDERATIONS FOR GATE DRIVERS OF SIC
MOSFETS AND THEIR TESTING 1374
     Niranjan Hegde

PP066   A PORTABLE EFFICIENCY CHARACTERIZATION SETUP FOR TECHNOLOGY
DEMONSTRATION OF POWER MODULES 1380
     Sebastian Tengvall

PP067   FAST EME CHARACTERIZATION OF BARE-DIE SIC MOSFETS 1385
     Robert Kragl

PP068   THEORETICAL COMPARISON OF COMPONENT-RELATED MEASUREMENT
METHODS OF PHOTOVOLTAIC INVERTERS FOR LONG-TERM TESTING 1393
     Niclas Reitz

DYNAMIC TRANSIENTS AND RELIABILITY OF HIGH-VOLTAGE SILICON & 4H-SIC
BIPOLAR JUNCTION TRANSISTORS UNDER AVALANCE AND SHORT-CIRCUITS 1402
     Mana Hosseinzadehlish

PP069   POWER CYCLING TEST OPTIMIZATION TOWARD RELIABILITY ASSESSMENT
OF SINTERED POWER MODULES 1410
     Robert Graham

PP070   REAL-TIME ESTIMATION AND SENSITIVITY ANALYSIS OF PARASITIC
CAPACITANCES IN ELECTRIC DRIVE SYSTEMS 1418
     Mohammadreza Bagheribavaryani

MODELLING AND TESTING

PP071 PARASITIC COMPONENT EFFECTS OF INTERNAL AND EXTERNAL PACKAGE
LEVEL ON SWITCHING PERFORMANCE OF SIC POWER MODULE 1428
Nguyen Nghia Do

PP072 A MULTI-PHYSICS ITERATIVE APPROACH FOR TEMPERATURE ESTIMATION
IN SIC POWER MODULE FOR ELECTRIC VEHICLE 1434
Stefano Orlando

PP073 VOLTAGE BALANCING METHOD FOR SERIES CONNECTION OF 50 SIC
MOSFETS 1441
Antoine Philippe

VOLTAGE BALANCING METHOD FOR SERIES-CONNECTION OF 50 SIC MOSFETS 1449
Antoine Philippe

PP074 A LABORATORY-SCALE MMC-BASED DC SYSTEM WITH RCP AND PHIL
SIMULATION CAPABILITIES 1457
Marc René Lotz

PP075 FILM CAPACITOR STANDARD SERIES DIGITALIZATION: ELECTROMAGNETIC
& THERMAL MODELLING IMPLEMENTATION IN CLARA WEB TOOL 1467
Fernando Aunon

PP076 ACCURACY EVALUATION AND PROPOSED DYNAMIC TUNING PROCEDURE
OF A COMPACT SIC SPICE MODEL 1475
Austin Curbow

PP077 INVESTIGATION OF USE-CASE-DEPENDENT MODELING APPROACH FOR
SWITCHED-MODE POWER CONVERTER FOR LVDC GRID EVALUATION 1485
Melanie Lavery

PP078 AVERAGED MODEL WITH BLOCKING CAPABILITY FOR SOLID-STATE
TRANSFORMERS 1495
Ahmed Meligy

ADVANCED COMPONENTS

PP080 SURFACANT-MODIFIED NANOCOMPOSITE THIN-FILM CAPACITORS 1504
Bartosz Gackowski

PP081 INCREASING ENERGY STORAGE CAPABILITIES OF POWDER CORES BY
ADAPTING THE WINDING AND THE USE OF FRINGING FLUX 1511
Paul Winkler

PP082 PEEC-BASED THERMAL MODELING OF PASSIVE COMPONENTS 1516
Sascha Langfermann

PP083 GALVANICALLY ISOLATED POWER SUPPLY FOR GATE DRIVERS IN HIGH
VOLTAGE APPLICATIONS 1523
Priyanka Ghosh

PP084 FABRICATION TECHNIQUE FOR NOVEL NANOCRYSTALLINE CORES WITH
HIGH SATURATION POLARIZATION AND LOW LOSSES 1532
Merlin Thamm

PP085 EXCITATION-DEPENDENT TEMPERATURE BEHAVIOR OF THE QUASI-STATIC
HYSTERESIS LOSS ENERGY DENSITY OF N87 FERRITE MATERIAL 1538
Jeremias Kaiser

PP087 PASSIVE METHODS LIMITING LEAKAGE CURRENT IN METAL-OXIDE
VARISTOR AS VOLTAGE CLAMPING DEVICE USED DC LOW VOLTAGE POWER
ELECTRONICS-BASED CIRCUIT BREAKERS 1545
Kenan Askan

GAN DEVICES AND APPLICATIONS

PP088 ESD SOLUTIONS FOR 650V NORMALLY-OFF ALGAN/GAN HEMTS 1555
Thanh Hai Phung

PP089 A SIMULATIVE STUDY OF MEASUREMENT ERRORS DURING DOUBLE PULSE
TESTING OF GAN DEVICES 1561
Severin Klever

PP090 PARALLEL CONNECTION OF GAN FETS: AN EXPERIMENTAL INVESTIGATION
APPROACH 1568
Marco Palma

PP091 REPETITIVE SHORT CIRCUITS ON 650 V GAN 1574
Adrien Lambert

PP092 COMPARISON OF SWITCHING LOSSES AND DYNAMIC ON RESISTANCE OF
600 V-CLASS GAN HEMTS 1584
André Thönnessen

PP093 PERFORMANCE EVALUATION OF DEADTIME AND GATE RESISTANCE FOR
PARALLEL CONNECTED GAN HEMTS 1590
Junhyeok Jegal

PP094 REACHING BEYOND 1200V: LATERAL GAN HEMTS FOR HIGH-RELIABILITY
EV AND INDUSTRIAL APPLICATIONS 1598
Kamal Varadarajan

SIC DEVICES AND TECHNOLOGIES

PP095 SMARTSIC 150 & 200MM ENGINEERED SUBSTRATE: INCREASING SIC POWER
DEVICE CURRENT DENSITY UP TO 30% 1604
Eric Guiot

PP096 DYNAMIC TRANSIENTS IN HIGH-VOLTAGE SILICON AND 4H-SIC NPN
BIPOLAR JUNCTION TRANSISTORS 1610
Mana Hosseinzadehlish

PP097 AN ADVANCED MULTI-ASPECT PERFORMANCE ANALYSIS OF PLANAR-GATE
1.2 KV SIC POWER MOSFETS 1613
Anja Katerina Brandl

PP098   SIC MOSFET DIE SORTING AND PARALLEL FOR OPTIMAL MODULE DESIGN 1621
Zhong Ye

PP099   SIMULATION APPROACH FOR RADIATED ELECTRO-MAGNETIC FIELDS
ESTIMATION ON ACEPACK DRIVE SIC POWER MODULE 1627
Andrea Cusumano

CONTROL METHODS II

PP100   EXACT ANALYSIS OF CONTROL-TO-OUTPUT TRANSFER FUNCTIONS OF
PWM-CONVERTERS - A COMPARISON OF TWO METHODS... 1634
Daniel Breidenstein

PP101   3-LEVEL FLYING CAPACITOR MULTILEVEL TOPOLOGY WITH DELTA-SIGMA
MODULATION 1642
Jannik Maier

PP102   MODEL BASED CONTROLLED POWER CONVERTER TEST PLATFORM.......................... 1651
Dawid Koczy

PP103   EDUCATIONAL HARDWARE TRAINER FOR TEACHING THE DUAL ACTIVE
BRIDGE IN A DC GRID 1658
Peter Van Duijsen

PP104   STUDY OF THE OPERATING PERFORMANCE OF A FCS-MPC-CONTROLLED
MATRIX-CONVERTER FOR PMSM AT DIFFERENT FREQUENCY RATIOS 1664
Robert Zipprich

PP105   ENHANCING REACTIVE POWER CAPACITY IN BATTERY-FED POWER
CONDITIONING SYSTEMS 1673
Lucas Araujo

PP106   PULSE SHARING: ACHIEVING HIGH EFFICIENCY AND EXCELLENT REGULA-
TION IN MULTI-OUTPUT FLYBACK POWER SUPPLIES 1680
Xingda Yan

PP107   RELIABILITY-OPTIMIZED SPACE VECTOR MODULATION (RO-SVM) FOR
SEMICONDUCTORS LIFETIME ENHANCEMENT 1686
Amin Rezaeizadeh

INTELLIGENT POWER MODULES

PP108   ANALYSIS AND OPTIMIZATION OF INTERNAL COUPLING INTERFERENCE IN
INTEGRATED SIC POWER MODULE BASED ON DBC 1693
Chenhang Zeng

PP109   MULTISPECTRAL ELECTROLUMINESCENCE SENSING OF SIC MOSFETS FOR
JUNCTION TEMPERATURE AND CURRENT EXTRACTION 1703
Lukas Ruppert

PP110   SIC-IPM FOR COMPACT AND ENERGY EFFICIENT LOW-POWER MOTOR
DRIVES 1712
Jongmu Lee

PP111 CONCEPT FOR A GAN-BASED INTELLIGENT MOTOR CONTROLLER WITH
INTEGRATED FAILURE PREDICTION FOR THE INVERTER AND THE DRIVE 1717
Christoph Blechinger

PP112 INTRODUCING THE NEW 1200 V CIPOS MAXI IM817 INTELLIGENT POWER
MODULE FOR MOTOR DRIVE APPLICATIONS 1724
Kihyun Lee

PP113 THERMAL PERFORMANCE OF INFINEON'S NEW 600 V CIPOSTM MICRO IM241
IPM FOR LOW POWER MOTOR DRIVE SYSTEMS WITHOUT HEATSINK 1732
David Jo

INTRODUCING THE NEEW 1200 V CIPOSTM MAXI IM12BXXXC1 INTELLIGENT POWER
MODULE FOR MOTOR DRIVE APPLICATIONS 1737
Kihyun Lee

INTELLIGENT GATE DRIVE UNITS

PP114 AN ADAPTIVE DEAD TIME CONTROL BASED ON SWITCH NODE VOLTAGE
DERIVATIVE.. 1745
Lukas Knappstein

PP115 COUPLING COIL DESIGN AND POSITIONING OPTIMIZATION ON NEW HIGH
POWER SEMICONDUCTOR MODULE FOR FAST SHORT CIRCUIT DETECTION 1751
Yannick Dumollard

PP116 ENABLING ACTIVE THERMAL CONTROL VIA AN ADAPTIVE MULTI-VOLTAGE
GATE DRIVER.. 1759
Tianlong Albert

PP117 INNOVATIVE GATE DRIVE METHOD TRIC3 FOR MOTOR.................................... 1765
Hisashi Sugie

PP118 A NEW CLASS OF SOLID STATE ISOLATORS ENHANCES THE RELIABILITY OF
SOLID STATE RELAYS 1770
Wolfgang Frank

PP119 A SELF-DRIVING 3-LEVEL ACTIVE GATE DRIVER NETWORK TO CONTROL
THE SWITCHING SLEW RATE FOR SIC MOSFETS 1775
Vin Loong Choo

E-MOBILITY TRACTION II

PP121 ANALYSIS OF LONG-TERM RELIABILITY OF SIC IN TRACTION INVERTER
CONSIDERING VTH INSTABILITY 1781
Chi Zhang

PP122 EFFICIENT MAPPING OF ON-DEMAND DRIVE LOAD PROFILES ON INVERTER
STRESS.. 1788
Zlatko Bosnjic

PP123 EV TRACTION INVERTER OPTIMAL DESIGN IS DOMINATED BY 3-LEVEL
ANPC 1797
Timothé Delaforge

PP124   INTRODUCTION OF POWER SEMICONDUCTOR OPTIONS FOR AN EXCITER OF
ELECTRICALLY EXCITED SYNCHRONOUS MOTOR 1804
   Yeriel Bai

PP125   A NOVEL HIGH POWER DENSITY THREE PHASE TRACTION INVERTER
ARCHITECTURE FOR ELECTRIC VEHICLE (EV) APPLICATIONS.. 1809
   Yiyang Yan

PP126   A MODULAR DC-LINK CAPACITOR SOLUTION FOR THE MAIN POWERTRAIN
INVERTER OF XEV 1814
   David Olalla

PP127   FAULT IDENTIFICATION TESTING METHODS FOR A COMMERCIAL TRACTION
INVERTER 1821
   Anna Corbitt

PP128   SHORT CIRCUIT ROBUSTNESS FOR TRACTION INVERTERS FROM AN
APPLICATION POINT OF VIEW 1828
   Karl Oberdieck

INVESTIGATIONS OF PARTICULAR SIC DEVICE PHENOMENON

PP129   THE IMPACT OF THE DEADTIME ON THE STABILITY OF 1.2KV SIC MOSFET
BODY DIODE UNDER HARD SWITCHING WITH SYNCHRONOUS RECTIFICATION.................... 1835
   Mohammed Amer Karout

PP130   RC-DC SNUBBER IMPLEMENTATION FOR SUPPRESSION OF DIODE VOLTAGE
PEAK AND RINGING IN A FULL SIC HALF-BRIDGE POWER MODULE 1844
   Emanuela Alfonzetti

PP131   SUB-5 SECOND WIDE-BANDGAP POWER DEVICE CALORIMETRIC
MEASUREMENTS UTILZIING OPTICAL SENSORS AND PELTIER ELEMENTS 1851
   Ruben Schnitzler

PP132   SIC TRENCH MOSFETS IN AVALANCHE MODE WITH RC SNUBBER CIRCUIT............... 1858
   Sebnem Tuncay

PP133   HIGH-FREQUENCY OSCILLATIONS IN SIC MOSFET POWER MODULES
DURING TURN-ON SWITCHING TRANSIENT – ANALYSIS BASED ON SIMULATIONS
AND MITIGATION METHODS.. 1865
   Rajani Kumar Thirukoluri

PP134   A DYNAMIC CURRENT BALANCING METHOD USING FULL-COUPLED
INDUCTORS IN PARALLELED GATE BRANCHES.. 1872
   Jianwei Lv

PP135   QUANTITATIVE PERFORMANCE COMPARISON OF LARGE-FORMAT SIC
MOSFET AND SI IGBT MODULES 1878
   Arthur Boutry

THERMAL MANAGEMENT AND ADVANCED COOLING

PP136   SOLDER PREFORM TECHNOLOGY FOR IMPROVED THERMOMECHANICAL
PERFORMANCE IN MOLDED POWER MODULE PACKAGE-ATTACH 1886
   Joseph Hertline

PP138   EFFECT OF FLIP-CHIP DIE-ATTACH ON THE THERMAL BEHAVIOR OF POWER GAAS DIODES 1891
Felix Steiner

PP139   INFLUENCES OF SOLDER DELAMINATION ON THE THERMAL PERFORMANCE IN AUTOMOTIVE TRACTION MODULE 1896
Hansol Seo

PP141   DEVELOPMENT OF A PASSIVE CAPILLARY-PUMPED COOLING SYSTEM FOR HIGH-PERFORMANCE ELECTRONICS 1902
Justin Fey

PP143   ADVANCED COOLING OF POWER ELECTRONICS WITH COPPER COLD SPRAYED ALUMINIUM HEATSINKS & BUSBARS 1907
Michael Dasch

PP144   COLD PLATE DESIGN FOR COOLING LV100 SILICON CARBIDE POWER MODULE PACKAGING 1910
Wahid Cherief

PP145   AN IMPROVED DOUBLE-LAYER SPACER IN DOUBLE-SIDED COOLING POWER MODULE 1917
Linhao Ren

RELIABILITY TESTING

PP146   POWER CYCLING OF 1.7KV MULTI-CHIP POWER MODULES – SIC MOSFETS VS SILICON IGBTS 1923
Nick Baker

PP147   POWER CYCLING CAPABILITY OF DISCRETE SIC MOSFET DEVICES WITH DIFFERENT DESIGNS 1930
Luhong Xie

PP148   MODEL-BASED PARAMETER TUNING OF SEMICONDUCTOR DEVICES IN DC POWER CYCLING TEST 1936
Yi Zhang

PP149   INFLUENCE OF TRANSFER MOLDING ON THE RELIABILITY OF DCM SIC POW-ER MODULES 1942
Jacek Rudzki

PP150   DAMP HEAT BEHAVIOR OF HIGH HEAT CAPACITORS FOR APPLICATIONS IN ELECTRIC VEHICLES 1951
Adel Bastawros

PP151   INFLUENCE OF THE GATE VOLTAGE DURING ON-TIME ON THE POWER CYCLING CAPABILITY OF SIC MOSFETS 1955
Patrick Heimler

PP152   INVESTIGATION OF THE TEMPERATURE MEASUREMENT VIA VSD(T)-METHOD APPLIED TO PARALLELED SIC MOSFET CHIPS DURING POWER CYCLING 1964
Kevin Ladentin

PP153   APPROACHES OF TSEP MEASUREMENTS FOR POWER SEMICONDUCTORS 1969
Philipp Hauenschild

PP154 REALTIME JUNCTION TEMPERATURE ESTIMATION IN SIC POWER MODULES
BASED ON MULTIPLE TSEP ACQUISITION 1978
Kevin Muñoz Barón

HIGH VOLTAGE WBG DEVICES

PP155 ENHANCED CURRENT MEASUREMENT APPROACH FOR NON-ISOLATED 6.5
KV SILICON CARBIDE MOSFETS 1987
Xinyuan Du

PP156 NEW 2KV SIC-MOS TECHNOLOGY FOR APPLICATION FIELDS IN THE
INDUSTRIAL LANDSCAPE 1991
Igor Kasko

PP157 HIGH TEMPERATURE EXPERIMENTAL CHARACTERIZATIONS OF COSS OF 3.3
KV SIC MOSFET FOR MEDIUM VOLTAGE PV APPLICATIONS... 1999
Paul Schmidt

PP158 IMPACT OF GATE CONTROL ON THE SWITCHING PERFORMANCE OF 3.3KV
SBD-EMBEDDED SIC-MOSFET... 2006
Junya Sakai

PP159 COMPARATIVE ASSESSMENT OF OVERLOADABILITY POTENTIAL OF 3.3 KV
SI-IGBTS AND SIC-MOSFET POWER MODULES 2013
Muhammad Nawaz

PP160 IMPROVED RELIABILITY OF A 2200 V SIC MOSFET MODULE WITH AN EPOXY-
ENCAPSULATED INSULATED METAL SUBSTRATE... 2022
Hiroshi Kono

PP161 PARALLELING 3.3-KV/800-A RATED SIC-MOSFET MODULES – AN
OPTIMIZATION METHOD.. 2028
Hiroyuki Irifune

PP162 PERFORMANCE ASSESSMENT OF 10 KV SIC MOSFET AND PIN DIODE IN 3L-
NPC CONVERTER TOPOLOGY 2036
Renato Amaral Minamisawa

VOLUME 4

PP163 PERFORMANCE EVALUATION OF COOLSIC 2 KV SIC MOSFET DISCRETE IN
1500 V DC LINK SYSTEMS 2041
Ajith Kumar Sekar

PP164 A NEW 2.3 KV RATED SIC MOSFET MODULE WITH LOW-INDUCTANCE HIGH-
POWER PACKAGE HPNC FOR 1500 VDC APPLICATIONS 2049
Junya Kawabata

PACKAGING AND INTERCONNECTION MATERIALS

PP166 MECHANISM FOR IMPROVING THE HEAT-RESISTANCE OF ADHESIVE
INTERFACE IN FLEXIBLE PRINTED CIRCUITS.. 2053
Keita Suzuki

PP167   A SYSTEMATIC COMPARISON STUDY OF DIFFERENT BONDING
TECHNOLOGIES FOR SUBSTRATE ATTACHMENT OF POWER ELECTRONICS............................ 2060
   Lisheng Wang

PP168   STABILITY OF PRESSURE SINTERED INTERCONNECTS AS A FUNCTION OF
TEMPERATURE AND ENVIRONMENTAL CONDITIONS.. 2067
   Kentaro Yoshioka

PP169   THE EFFECT OF NANO-CU INTERCONNECTION MATERIALS ON THE
THERMOMECHANICAL PROPERTIES OF SIC DOUBLE-SIDED POWER MODULES 2074
   Suhang Wei

PP170   ALL-IN-ONE-SINTERING: DIE-ATTACH AND SUBSTRATE-ATTACH ON BARE
COPPER IN A PRESSURE ASSISTED SINTERING ONE-STEP PROCESS.. 2082
   Battist Rabay

PP171   SEQUENTIAL MANUFACTURING OF HIGHLY FUNCTIONALIZED THREE-
DIMENSIONAL CERAMIC COMPONENTS FOR POWER ELECTRONICS... 2088
   Lars Rebenklau

PP173   PARAMETRIC STUDY OF DAMAGE EVOLUTION IN SILVER SINTERED
LAYERS OF DOUBLE SIDED POWER ELECTRONICS MODULES OF ELECTRICAL
VEHICLES... 2094
   Saeed Akbari

DC-DC CONVERTER I

PP174   TRISTATE MODIFIED BOOST CONVERTER.. 2104
   Johannes Gragger

PP175   COMPARATIVE EVALUATION OF THE CENTER TAPPED BOOST CONVERTER
TOPOLOGY 2112
   Bryan Radix

PP176   COMPARISON OF MULTI-LEVEL TOPOLOGIES TO REDUCE THE
COMPONENTS VOLTAGE STRESSES WHEN POWERED FROM INDUSTRIAL DC GRIDS.............2119
   Katharina Machtinger

PP177   HARD-SWITCHING HIGH-FREQUENCY GAN-BASED DC-DC CONVERTERS
WITH CONCOMITANT DATA TRANSMISSION FUNCTIONALITY 2128
   Abdelmoumin Allioua

PP178   EFFICIENT DESIGN OF HIGH-CURRENT, LOW-OUTPUT VOLTAGE DC-DC
CONVERTERS USING ARTIFICIAL INTELLIGENCE-BASED TOPOLOGY SELECTION
AND OPTIMIZATION 2138
   Thomas Harmand

HIGH POWER DC-DC CONVERTER I

PP180   A SIC BASED 60KW LLC CONVERTER WITH NOVEL TRANSFORMER DESIGN
FOR IMPROVING VOLTAGE BALANCE.. 2146
   Frank Wei

PP181 ANALYSIS OF INVERTER OPERATION MODES OF AN IGBT-BASED ZCS LLC CONVERTER FOR A 2 KW AUTOMOTIVE ON-BOARD DC-DC 2152
Daniel Urbaneck

PP182 DUAL OUTPUT HYBRID CONVERTER FOR 48 V DATA CENTERS: M-HSC.................... 2162
Simone Mazzer

PP183 3.6KW HIGH EFFICIENCY SIC-BASED HV/LV DC-DC CONVERTER FOR EVS 2167
Veera Bharath Chandra Reddy Gandluru

PP184 BIDIRECTIONAL DC-DC TOPOLOGIES COMPARISON FOR 800 V AUTOMOTIVE APPLICATIONS INTEGRATING 650 V GAN-ON-SI DEVICES.. 2175
Ilias Chorfi

PP185 ANALYSIS OF PHASE SHIELDING METHOD BASED ON ?-CR-Y THREE-PHASE INTERLEAVED LLC CONVERTER.. 2182
Jin Wen

PP186 22KW IMS-BASED BIDIRECTIONAL DC-DC CONVERTER USING SURFACE MOUNT SIC MOSFETS FOR OBCS 2185
Hamlin Wang

PP187 COMPARATIVE ANALYSIS OF DC-DC CONVERTERS FOR ELECTROLYZERS USING GEOMETRIC PROGRAMMING 2190
Tim McRae

PP188 DESIGN CONSIDERATION OF BI-DIRECTIONAL CLLLC RESONANT CONVERTER IN ENERGY STORAGE SYSTEMS 2200
Sheng-Yang Yu

SMART-GRID TECHNOLOGIES

PP189 ADAPTIVE FAST CHARGING SYSTEM WITH SECOND LIFE BATTERIES - AN OVERVIEW OF A RESEARCH PROJECT 2208
Lukas Böhning

PP190 PARALLEL OPERATION AND SYNCHRONIZATION OF MICROGRIDS BY USING THE THEVENIN THEOREM 2217
Marius Block

PP192 21 KA SOLID STATE DC BREAKER FOR SUPERGRID INSTITUTE'S HIGH POWER TEST FACILITY 2227
Christophe Conilh

PP193 DESIGN AND ANALYSIS OF A 50KW SIC-BASED ACTIVE FRONT END WITH A VERY SMALL LINE CHOKE FOR DC-GRIDS 2234
Raphael Otte

PP194 INVESTIGATION OF LOAD TRANSITIONS BETWEEN LOADED AND LOAD FREE CONDUCTOR SEGMENTS IN INDUSTRIAL CONDUCTOR SYSTEMS 2240
Jan-Niklas Koch

PP195 A METHOD TO CONTROL VOLTAGE AND POWER FLOW IN A DC GRID 2248
Peter Van Duijsen

ENERGY STORAGE SYSTEMS

PP196 CONSIDERATIONS ON A HIGH-CELL-COUNT CONVERTER-BASED BATTERY STORAGE SYSTEM WITH REDUCED COMMUNICATION EFFORT 2258
Paul Aspalter

PP197 STUDYING CONVERTORS FOR VOLTAGE EQUALIZATION IN ENERGY STORAGE SYSTEM WITH ACTIVE BMS 2268
Dimitar Arnaudov

PP198 CHALLENGES OF HIGH SIDE GATE DRIVER AND DISCONNECT MOSFET FOR BATTERY PROTECTION UNIT DURING START-UP, TURN-OFF AND OVER CURRENT EVENTS.. 2273
Niranjan Suravarapu Reddy

PP199 ELECTRIC INSULATION COORDINATION TO PREVENT ELECTRIC ARCS IN LITHIUMION BATTERIES 2278
Daniel Chatroux

PP201 BATTERY CHARGER WITH IMPEDANCE SPECTROSCOPY CAPABILITY FOR LI-ION CELLS.. 2286
Christian Branas

EMC

PP202 EFFICIENCY, VOLUME AND CO2 EMISSIONS IMPACT IN A PFC CONVERTER WITH AN ACTIVE FILTER SOLUTION FOR OBC APPLICATION.. 2294
Kelly Ribeiro

PP203 ANALYTICAL AND EXPERIMENTAL VALIDATION COMMON MODE FEEDBACK LOOP FOR A THREE-PHASE_LEVEL VIENNA RECTIFIER............................ 2303
Daniel San Laureano Igartuburu

PP204 ROBUSTNESS OF FREQUENCY-DOMAIN TERMINAL MODELING OF ELECTROMAGNETIC INTERFERENCES IN STATIC CONVERTERS 2309
Mehyeddine Singer

PP205 STUDY OF EMI BEHAVIOR OF A 2-LEVEL GAN-INVERTER – SIMULATION AND MEASUREMENT... 2316
Benedikt Kohlhepp

COMMON MODE CURRENTS IN RESONANT CIRCUITS GENERATED WITH A DELTA-SIGMA MODULATED VOLTAGE SOURCE INVERTER.. 2326
Tobias Haas

PP206 ANALYSIS OF COMMON-MODE NOISE GENERATED DUE TO FAST-SWITCHING GAN DEVICES IN TOTEM-POLE PFCS 2334
Serkan Dusmez

PP207 CONDUCTED EMI FROM GAN-BASED 48V TO 12V DC-DC-CONVERTERS FOR AUTOMOTIVE APPLICATIONS 2342
Erik Kampert

ADVANCED DESIGN

PP208 APPLIED DESIGN AUTOMATION FOR FINDING FEASIBLE DESIGNS FOR HIGH-FREQUENCY PLANAR TRANSFORMERS 2350
Rando Raßmann

PP209 FREQUENCY DEPENDENT AREA PRODUCT METHOD 2359
Alfonso Martínez

HIGH RESOLUTION MIXED-SIGNAL PULSE WIDTH MODULATOR FOR HIGH-FREQUENCY DC-DC CONVERTERS 2364
Tim McRae

PP210 DESIGNING A CONTROL LIBRARY FOR GRID-FOLLOWING AND GRID-FORMING POWER INVERTERS 2370
Lars Lindner

PP211 INTELLIGENT OPTIMISATION OF A WIND TURBINE DIGITAL TWIN MODEL 2377
René Reimann

PP212 THERMAL TRANSIENT DIGITAL TWIN MODELLING FOR POWER CONVERTERS 2386
Xianghao Mo

PP213 A DIGITAL TWIN APPROACH TOWARD LIFETIME ANALYSIS AND PREDICTIVE MAINTENANCE OF POWER SEMICONDUCTORS FOR RAILWAY APPLICATION 2394
Emmanuel Batista

INDUCTORS

PP214 SATURABLE FERRITE CORE INDUCTORS IN LCL FILTERS OF THREE-PHASE VOLTAGE SOURCE INVERTERS 2400
Marius Kaufmann-Bühler

PP215 2D COPPER LOSS ANALYTICAL MODEL FOR PLANAR INDUCTOR COMBINING HIGH AND LOW PERMEABILITY MATERIALS 2408
Idriss Nachete

PP216 CNC-MANUFACTURED POWER INDUCTORS WITH EXCELLENT BANDWIDTH FOR MULTI-MEGAWATT CONVERTERS 2416
Thomas Kreppel

PP217 ANALYTICAL EVALUATION OF DIFFERENTIAL MODEL DC EMI FILTER INDUCTORS USING MATERIAL SATURATION COEFFICIENT 2425
Lukas Mueller

PP218 DESIGN AND PERFORMANCE EVALUATION OF AIR CORE INDUCTORS FOR VERY HIGH FREQUENCY POWER CONVERSION 2431
Florentin Salomez

PP220 IMPROVING MULTI-PHASE FERRITE MAGNETICS BY COUPLING FOR MV AND UPS CONVERTERS 2438
Michael Schmidhuber

E-MOBILITY CHARGING

PP221   22-KW BIDIRECTIONAL SINGLE-STAGE DIRECT-AC-AC POWER CONVERSION ON-BOARD CHARGER WITH HIGH-POWER-DENSITY IMPLEMENTATION....................................2448
Oscar Lucia

PP222   BENCHMARKING DC FAST CHARGERS: A COMPARATIVE ANALYSIS OF POWER CONVERTER STRUCTURES FOR WIDE VOLTAGE RANGE 2453
Sadik Cinik

PP223   PERFORMANCE OPTIMIZATION OF SINGLE-PHASE ON-BOARD CHARGERS WITH RIPPLE PORT 2461
Davide Gottardo

PP224   A REDUCED-SENSOR MODULAR DUAL ACTIVE BRIDGE-BASED BATTERY CHARGING SYSTEM FOR ELECTRIC VEHICLES USING AN IMPROVED LINEAR EXTENDED STATE OBSERVER..2469
Armel Asongu Nkembi

PP225   BIDIRECTIONAL NON-ISOLATED THREE-PHASE ONBOARD CHARGER WITH A LOW-VOLTAGE LOWER-PHASE OPERATION MODE 2478
Steffen Frei

PP226   CONTROL OF A THREE-PHASE INDUCTIVE POWER TRANSFER SYSTEM BASED ON DD²Q COIL TOPOLOGY 2488
Nikola Mirkovic

PP227   COMPARISON OF TWO BIDIRECTIONAL 11KW 400V CLLC AND CLLLC RESONANT CONVERTERS FOR EV APPLICATIONS 2494
Hasan Mousavi Somarin

PP228   DYNAMIC WIRELESS CHARGING SYSTEM DESIGN FOR EXTRA-URBAN AREAS BASED ON RESONANT INDUCTIVE POWER TRANSFER 2503
Irene Maria Torres Alfonso

PP229   BIDIRECTIONAL ISOLATED 400-12V DC-DC CONVERTER WITH IMPROVED POWER DENSITY AND FULL-RANGE OPERAION FOR EV APPLICATIONS 2513
Oscar Lucia

HIGH POWER DC-DC CONVERTER II

PP230   GAIN OPTIMIZATION CONTROL METHOD FOR CLLLC RESONANT CONVERTERS UNDER PHASE SHIFT MODE 2518
Sean Yu

PP231   ANALYSIS OF COMMON AND SPLIT DC-BUS INTERLEAVED H-BRIDGE CONVERTERS FOR HIGH-CURRENT LOW-RIPPLE APPLICATIONS...2524
Bhavana Gudala

PP232   OPTIMAL FREQUENCY OPERATING POINTS FOR HYBRID SWITCHED CAPACITOR CONVERTERS AND LOSSLESS CURRENT SENSE METHOD 2532
Simone Mazzer

PP233   DESIGN AND TESTING OF A 250 KW 50 KHZ SIC-BASED HALF-BRIDGE-SERIES-RESONANT-CONVERTER 2538
Daniel Haake

PP234   30KW - 97% EFFICIENCY ISOLATED DC-DC CONVERTER WITH LARGE INPUT VOLTAGE RANGE BASED ON A BOOST DAB ASSOCIATION 2547
Jean-Jacques Huselstein

PP235   ANALYSIS OF A FULL-BRIDGE PUSH-PULL FORWARD DUAL ACTIVE BRIDGE DC-DC CONVERTER 2557
Gean Sousa

DC-DC CONVERTER II

PP236   SYMMETRICAL OPERATION OF FOUR CHANNEL RESONANT BOOST DC-DC CONVERTERS IN CONTINUOUS CONDUCTION MODE 2566
Kristóf Bándy

PP237   IMPACT OF MAGNETICS TOLERANCE ON THE POWER SHARING OF PARALLEL DUAL-OUTPUT PHASE-SHIFT FULL-BRIDGE CONVERTERS 2576
Riccardo Mandrioli

PP238   A BALANCING CONVERTER WITH SERIES CONNECTED MOSFETS FOR +/-700V BIPOLAR DC GRIDS 2583
Sachin Yadav

PP239   OPTIMIZATION AND DESIGN OF LOW-VOLTAGE AND HIGH-CURRENT POINT-OF-LOAD CONVERTER UNDER 48V BUS ARCHITECTURE 2591
Jiajia Guan

PP240   INTERLEAVED BOOST CONVERTER EFFICIENCY AND POWER DENSITY MODEL FOR ACTIVE AND PASSIVE COMPONENT DESIGN 2596
Damien Lemaitre

NOVEL AND ADVANCED SEMICONDUCTOR DEVICES

PP241   EVALUATION OF A HYBRID POWER SWITCH BASED ON TRENCH CLUSTERED IGBT AND SIC MOSFET 2606
Alireza Sheikhan

PP242   CONTRIBUTIONS FOR BUILDING BLOCKS FOR NORMALLY-OFF 650V GAN-ON-SI POWER INTEGRATED CIRCUITS 2612
Thanh Hai Phung

PP243   NEW BIDIRECTIONAL ASYMMETRIC HIGH VOLTAGE TVS (TRANSIENT VOLTAGE SUPPRESSOR) DIODE 2620
Boris Rosensaft

PP244   ISO247: HIGH PERFORMANCE CERAMIC BASED ADVANCED ISOLATED DISCRETE PACKAGE TO FULLY EXPLOIT THE ADVANTAGES OF SIC MOSFET 2627
Sachin Shridhar Paradkar

PP245   IMPACT OF CURRENT RIPPLE REDUCTION USING HIGH SWITCHING FREQUENCIES ON PMSM EFFICIENCY 2632
Jannik Fuchs-Gade

PP246 MAXIMIZING COST-EFFICIENCY IN ELECTRIC DRIVETRAINS: A SIC/SI
FUSION SWITCH APPROACH 2638
   Matthias Ippisch

ADVANCED CONTROL

PP247 CONCISE AND RELIABLE SIC MOSFET DRIVER CIRCUITS 2646
   Zhong Ye

PP248 ARTIFICIAL INTELLIGENCE ENHANCED RESOLVER SYSTEM FOR
AUTOMOTIVE TRACTION INVERTER APPLICATIONS BASED ON AURIX TC4X.......................... 2651
   David Zipperstein

PP250 MULTIFUNCTIONAL GRID MANAGER TOPOLOGY WITH CONFIGURABLE
OUTPUT 2657
   Peter Van Duijsen

PP252 CO2 FOOTPRINT OF MEDIUM VOLTAGE DC SOLID STATE
TRANSFORMER 2663
   Adriana Campos

SIC MOSFET

PP253 THERMO-ELECTRICAL ANALYSIS AND PERFORMANCE: A COMPARATIVE
STUDY BETWEEN MODULAR AND DISCRETE APPROACHES... 2673
   Stefano Orlando

PP254 IMPACT OF PARAMETER SPREAD IN PARALLEL-OPERATED SIC MOSFETS
FOR HARD-SWITCHING CONVERSION.. 2680
   Andrea Piccioni

PP255 ASSESSMENT OF THE RDS,ON OF SIC MOSFET DIES THROUGH KELVIN WIRE
CONNECTION 2686
   Philipp Rehlaender

PP256 CHALLENGES IN SCALING SIC SINGLE-CHIP MEASUREMENTS TO
CORRESPONDING POWER MODULES 2693
   Hao Wang

PP257 SWITCHING PERFORMANCE EVALUATION OF HIGH-POWER 1.7 KV SIC
MOSFET MODULES USING A COMMON BUSBAR DESIGN.. 2700
   Sebastian Neira

PP258 CHARACTERIZING THE SWITCHING BEHAVIOR OF A 1.2 KV MIXED SIC JFET
AND MOSFET HALF BRIDGE... 2708
   Tim Ringelmann

VOLUME 5

WBG HIGH FREQUENCY APPLICATION

PP259 PERFORMANCE EVALUATION OF THE PACKAGING OF SIC DIODES IN A 6.78
MHZ WIRELESS POWER TRANSFER SYSTEM.. 2718
   Ioannis Nikiforidis

PP260   VOLTAGE WAVEFORM GENERATION FOR SAWYER-TOWER COSS LOSS
MEASUREMENTS USING A HYBRID POWER CONVERTER 2724
Malachi Hornbuckle

PP261   EVALUATION OF SIC DEVICES FOR OVER 500KHZ APPLICATION BASED ON
BUCK CIRCUIT 2730
Minli Jia

PP262   LINEARIZATION OF DRAIN-SOURCE CAPACITANCES FOR ANTISERIAL
CONFIGURATED SIC MOSFETS IN HIGH FREQUENCY SOLID STATE SWITCHES 2737
Lars Dresel

SIC RUGGEDNESS

PP263   EFFECTS OF NON-KILLER DEFECTS ON SIC MOSFET SHORT-CIRCUIT
RUGGEDNESS AND RELIABILITY 2745
Sara Kuzmanoska

PP264   DYNAMIC REVERSE BIAS TEST: ELECTRO-THERMAL CHARACTERIZATION
OF SIC MOSFETS 2751
Giuseppe Mauromicale

PP266   RADIATION HARDNESS OF SIC BASED INVERTERS BASED ON AN EV
MISSION PROFILE 2758
Hadiuzzaman Syed

PP267   RAPID SHORT CIRCUIT PROTECTION USING DIDT DETECTION FOR SIC
POWER MODULES 2764
Koki Samura

PP268   COMPARISON OF DYNAMIC GATE STRESS TEST RESULTS OF SIC MOSFETS 2769
Mathias Gebhardt

PP279   EXTENDING SIC MOSFET SHORT-CIRCUIT WITHSTANDING TIME BY TWO-
LEVEL TURN-OFF GATE DRIVING 2778
Kwokwai Ma

PP270   EXPERIMENTAL INVESTIGATIONS ON PARASITIC TURN-ON OF 1.2KV SIC
MOSFET DISCRETE DEVICES 2786
Thanh-Toan Pham

PP271   BEHAVIOR MODELLING THE SHORT CIRCUIT CHARACTERISTICS OF SIC
MOSFETS USING COMPACT MODELS 2791
Qing Sun

THERMAL CHARACTERIZATION

PP273   THERMAL ANALYSIS AND MODELLING OF CHARGING STATIONS FOR
ELECTRIC VEHICLES 2796
Ruben Kopischke

PP274   JUNCTION TEMPERATURE MEASUREMENT OF A 3.3 KV SILICON CARBIDE
MOSFET POWER MODULE 2803
Michael Gleissner

PP275   INNOVATIVE 3D POWER MODULE DEFAULTS DETECTION VIA THERMAL
IMPEDANCE ANALYSIS AND SIMULATIONS...2811
   Louis Alauzet

PP276   THERMAL CHARACTERIZATION OF AN AIR-COOLED PEBB BASED ON SIC
MOSFET POWER MODULES 2819
   Alexandre Marie

PP277   THERMAL BEHAVIOUR OF SIC MOSFET WITH PLANAR PACKAGING
TECHNOLOGY 2826
   Yijun Ye

RELIABILITY AND AVAILABILITY

PP279   IMPLEMENTING MODULE HEALTH MONITORING IN EV TRACTION
INVERTERS 2831
   Karol Rendek

PP280   RELIABILITY TESTS OF COPPER THICK-FILM SUBSTRATES FOR POWER
ELECTRONIC APPLICATIONS.. 2838
   Henry Barth

PP281   POWER MODULE SOLUTIONS WITH IMPROVED RELIABILITY FOR ELEVATOR
DRIVE APPLICATIONS 2843
   Tiago Jappe

PP282   FAIL-OPERATIONAL LLC TOPOLOGIES WITH FAULT-TOLERANCE
INTEGRATED REDUNDANT CAPABILITIES 2850
   Aswathy M. Prince

PP283   THERMAL AND RELIABILITY OPTIMIZATION OF CLIPS IN SIC MOSFET
POWER MODULES... 2860
   Zexiang Zheng

PP284   CONDITION MONITORING OF A GAN FULL-BRIDGE BY MEANS OF FORWARD
VOLTAGE IN CONTINUOUS OPERATION.. 2866
   Michael Vogt

PP285   A SIMPLE AND LOW COST OVERCURRENT PROTECTION SYSTEM BASED ON
COMMERCIAL SHUNT FOR WIDE-BANDGAP DEVICES 2874
   Emanuele Martano

PP286   SVM-BASED FAULT-TOLERANT CONTROL FOR A CASCADED H-BRIDGE
MULTILEVEL CONVERTER UNDER MULTIPLE OPEN-CIRCUIT SWITCH FAULTS........................ 2880
   Dong Xie

PP287   REVOLUTIONIZING MOBILITY: THE SECOND LIFE OF ONBOARD CHARGING
SYSTEMS IN COMMERCIAL VEHICLES 2886
   Ajay Krishna Voppu Muralikrishna

LOW VOLTAGE SWITCHES

PP288   A BEHAVIORAL TRANSIENT MODEL FOR IGBT DEVICE WITH ANTI PARALLEL
FREEWHEELING DIODE.. 2893
   Shiwu Zhu

PP289   PARAMETER EXTRACTION FOR AN ANN-ASSISTED IGBT MODEL IN
TRANSIENT SIMULATIONS 2901
Huaiyuan Zhang

PP290   FABRICATION OF 600V RC-IGBT USING 300MM WAFER 2909
Masaki Ueno

PP291   NEXT LEVEL OF POWER MODULE SOLUTION FOR PV C&I STRING INVERTER
WITH 1200V H7 TECHNOLOGY IN EASY3B PACKAGE 2914
Tilo Poller

PP292   ANALYSIS OF MOSFET SWITCHING LOSSES IN RESONANT CONVERTERS
USING ELECTRICAL AND THERMAL MEASUREMENTS AND LOSS TRENDS WITH
MOSFET SIZE VARIATION 2921
Alfio Scuto

PP293   OPTIMOS 6 135V FOR HIGH POWER MOTOR DRIVES 2930
Kunal Jha

PP294   AUTO POWER-SOI: SHAPING THE FUTURE OF BATTERY MONITORING
TECHNOLOGY... 2937
Alex Lim

LIFETIME MODELLING AND CONDITION MONITORING

PP295   UNDERSTANDING THE IMPACT OF IEC60747-17 ON CAPACITIVE AND
MAGNETIC COUPLERS... 2942
Shu Ee Ong

PP296   PARIS LAW APPLIED TO WIRE BONDS DEGRADATION USING CRACK
GROWTH MEASUREMENT 2948
Merouane Ouhab

PP297   CONDITION MONITORING TECHNIQUE OF POWER ELECTRONIC MODULES
VIA SQUARE-WAVE GATE SIGNAL EXCITATION 2956
Isabel Austrup

PP298   STATISTICS-BASED LIFETIME SIMULATION ENVIRONMENT FOR POWER
MODULES INCORPORATING DEGRADATION MODELS 2963
Karthik Debbadi

PP299   POWER CYCLING RESULTS FOR RELIABILITY STUDIES OF SIC-INVERTERS 2972
Robert Keilmann

PP300   GAN CASCODE IN HIGH SPEED DRIVEN AIR COMPRESSORS FOR
AUTOMOTIVE FUEL CELLS.. 2981
Florian Lippold

PP301   PROGNOSTIC ANALYSIS OF IGBT HEALTH: REAL-TIME ON-STATE VOLTAGE
PREDICTION THROUGH MACHINE LEARNING 2986
Tanya Thekemuriyil

PP302   ROBUSTNESS ANALYSIS OF TEMPERATURE-SENSITIVE ELECTRICAL
PARAMETERS OF IGBTS 2995
Laurids Schmitz

PP303   OBSERVATION OF THERMAL-RESISTANCE INCREASE OF DEGRADED IGBT
MODULES BY VCE (SAT) MEASUREMENT IN A CHOPPER CIRCUIT 3002
     Kazunori Hasegawa

PULSE WITH MODULATION METHODS

PP304   MODULATION TECHNIQUE FOR REDUCED AC CONTENT OF THE DC LINK
CURRENT IN THREE-PHASE TWO-LEVEL INVERTERS 3007
     Steffen Frei

PP305   COMMON MODE CURRENTS IN RESONANT CIRCUITS GENERATED WITH A
DELTA-SIGMA MODULATED VOLTAGE SOURCE INVERTER 3017
     Tobias Haas

PP306   EVALUATION OF NEW MODULATION SCHEME FOR 3L-ANPC USING BOTH
CURRENT PATHS IN ZERO STATE 3020
     Felix Eichler

PP307   AN INNOVATIVE SYNCHRONOUS RECTIFICATION METHOD FOR 11KW CLLC
CONVERTER 3029
     Sanbao Shi

PP308   INTERLEAVED ASYNCHRONOUS DELTA-SIGMA MODULATION CONCEPT FOR
DYNAMIC POWER CONVERTERS 3034
     Philipp Czerwenka

PP309   HIGH RESOLUTION MIXED-SIGNAL PULSE WIDTH MODULATOR FOR HIGH-
FREQUENCY DC-DC CONVERTERS 3042
     Tim McRae

PP310   IMPLEMENTATION AND CONTROL OF OPTIMIZED PULSE PATTERNS FOR
SALIENT PERMANENT MAGNET SYNCHRONOUS MACHINES IN ELECTRIC VEHICLES 3045
     Maximilian Hepp

PP311   A 3-LEG INTERLEAVED TP PFC WITH A 90° PHASE-SHIFTED ASYMMETRIC
LEG FOR REDUCED MAGNETICS 3060
     Serkan Dusmez

PP312   FAULT-TOLERANT OPERATION ANALYSIS OF A FIVE-PHASE THREE-LEVEL
TNPC INVERTER FOR ELECTRIC AIRCRAFT PROPULSION SYSTEMS 3067
     Chanuch Chaisakdanugull

AC-DC AND DC-AC CONVERTER

PP313   CCM TOTEM-POLE PFC FOR ULTRA-HIGH POWER DENSITY USB-PD
CHARGERS 3077
     Manuel Escudero Rodruigez

PP314   COMPARISON OF HYBRID SI/SIC AND SIC TWO-LEVEL AND THREE-LEVEL
CONVERTERS FOR LOW-VOLTAGE LOW-POWER APPLICATIONS 3086
     Tim Augustin

PP315   ANALYSIS OF ANALOGUE CURRENT AND FLUX BALANCING FOR THE DUAL-
ACTIVE-BRIDGE CONVERTER 3096
     Christophe Basso

PP316 DESIGN AND OPTIMIZATION OF A SINGLE-STAGE PHOTOVOLTAIC
MICROINVERTER WITH INTEGRATED MAGNETICS 3103
Jin Wen

PP317 EXPERIMENTAL INVESTIGATION OF CLASS F INVERTER UNDER VARIOUS
LOAD CONDITIONS.. 3110
Baptiste Daire

PP318 ANALYSIS, MODELING, DESIGN, AND LIMITATIONS OF CURRENT INJECTION
BASED UPF RECTIFIER WITH SMALL DC-LINK CAPACITOR... 3118
Ramkrishan Maheshwari

PP319 HIGH-EFFICIENT ISOLATED AC-DC CONVERTER WITH CIRCULATING
CURRENT REDUCTION FOR AC ADAPTERS 3125
Hiroki Watanabe

PP320 A PHASE-LOCKED LOOP (PLL) BASED STRATEGY FOR ACCURATE BLANKING
TIMES IN BRIDGELESS TOTEM-POLE PFCS.. 3130
Sandu Tigira Tigira

PP321 CIRCULATING CURRENTS IN COUPLED MULTI-TERMINAL HYBRID AC-DC
GRIDS... 3136
Fabian Herzog

ADVANCED CONVERTER TOPOLOGIES

PP322 COMPARISON OF 4500V STATE-OF-THE-ART XHP3 IGBT AND CONVENTIONAL
IHV IGBT FOR 3300V 3-LEVEL ANPC MEDIUM VOLTAGE DRIVES 3142
Martin Knecht

PP323 GENERALIZED SWITCHING SEQUENCE FOR VOLTAGE BALANCING IN A
FLYING CAPACITOR DC-DC CONVERTER WITH QUASI-2-LEVEL MODULATION 3150
Jose Andres Aguilar Croston

PP324 OPTIMIZATION-BASED SIZING OF A MODULAR MULTILEVEL CONVERTER
BASED ON 650 V GAN MODULES FOR NEW LVDC/MVDC GRIDS...................................... 3160
Gregoire Le Goff

PP325 A NOVEL THREE-PHASE LOW-SWITCH-COUNT AC-DC GRID CONVERTER
TOPOLOGY WITH GALVANIC ISOLATION... 3169
Liska Steenbock

PP326 SINGLE-STAGE LED DRIVER BASED ON COUPLED INDUCTOR POWER
FACTOR CORRECTION AND LLC CONVERTER.. 3175
Alireza Ramezan Ghanbari

PP327 A INVERSE COUPLED DC-DC BOOST INDUCTOR WITH 2-KV SIC MOSFET
MODULE FOR 1500V SOLAR INVERTER MPPT... 3181
Yusi Liu

PP328 ENVIRONMENTAL IMPACT OF MODULAR POWER ELECTRONICS SYSTEMS
CONSIDERING DIAGNOSTIC-DRIVEN UNIT REPLACEMENT 3187
Briac Baudais

POWER ELECTRONICS FOR RAILWAY APPLICATIONS

PP329   SWITCHING PERFORMANCE COMPARISON OF 3.3 KV SIC MOSFET AND SI IGBT POWER MODULES FOR RAILWAY TRACTION SYSTEMS 3197
Yue Zhao

PP330   COMPARISON OF THREE-LEVEL INVERTER TOPOLOGIES FOR MVDC REVERSIBLE RAILWAY SUBSTATIONS 3206
Luc Bimmel

PP331   CONTROL OF BIDIRECTIONAL POWER FLOW IN RAILWAY CATENARY OVERHEAD LINES.. 3213
Peter Van Duijsen

PP332   A RAIL TRACTION CONVERTER PLATFORM BASED ON POWER MODULE IMPLEMENTATIONS WITH 450 A, 600 A AND 800 A 3.3 KV IGBT MODULES 3221
Ekrem R. Gunes

PP333   COMPARISON OF SELECTED MEGAWATT-LEVEL TRACTION CONVERTER POWER MODULE IMPLEMENTATIONS IN TERMS OF COMMUTATION INDUCTANCE AND PRACTICALITY.. 3229
Abdulkerim Ugur

CURRENT RELATED TESTING

PP334   PITFALLS AND THEIR AVOIDABILITY IN THE DOUBLE-PULSE TEST 3237
Nikolas Förster

PP335   MODELING AND SIMULATION OF FLUXGATE BASED CURRENT SENSOR 3247
Yunus Çay

PP336   SIGMA-DELTA BASED CURRENT ACQUISITION WITH REDUCED SETTLING TIME 3256
Joschka Randerath

PP337   CHARACTERISATION OF WIDE-BANDGAP SEMICONDUCTORS IN DOUBLE PULSE TESTING USING OPTICALLY ISOLATED PROBES... 3264
Lennart Hoffmann

PP338   NON-INVASIVE BATTERY CONDITION TESTING USING ELECTRICAL SIGNALS AND OSCILLOSCOPES... 3269
Srikrishna N. H

PP339   INSTRUMENTATION REQUIREMENTS FOR FAST 130 V/NS SWITCHING OF 1700 V, 35 M? SIC MOSFETS 3276
Matthew Appleby

POWER ELECTRONICS FOR AEROSPACE APPLICATIONS

PP340   CONCEPTUALIZATION AND EXPERIMENTAL ASSESSMENT OF DESIGN ASPECTS FOR 3-LEVEL ANPC INVERTERS 3286
Lukas Radomsky

PP341 DESIGN OF A HIGH POWER DENSITY INVERTER AND FOC IMPLEMENTATION FOR UAVS 3296
Matthias Neuner

PP342 HIGHLY-INTEGRATED, FLEXIBLE POWER SOLUTION FOR AEROSPACE 5KVA – 20 KVA MOTOR DRIVE APPLICATIONS 3305
Alain Calmels

PP343 DATABASE-SUPPORTED PRELIMINARY DESIGN, SIMULATION AND EVALUATION OF POWER CONVERTERS IN ELECTRIC AIRCRAFT PROPULSION SYSTEMS 3315
Jeff Kugener

PP344 DESIGN AND ANALYSIS OF GATE-DRIVER FOR SIC-BASED INVERTER FOR MEGAWATT SCALE ALL ELECTRIC AIRCRAFT 3318
Jeff Kugener

MEASUREMENT TECHNIQUES AND METHODS

PP345 ADDRESSING TESTING CHALLENGES FOR POWER MODULES AND THREE-LEVEL INVERTERS 3328
Oleg Fotteler

PP346 CHARACTERIZATION OF THE BONDING QUALITY OF SILVER SINTERED COMPOUNDS BY MEANS OF LASER-INDUCED BREAKDOWN SPECTROSCOPY 3334
Yannick Bockholt

PP347 INVERTER-INTEGRATED MEASUREMENT OF THE FREQUENCY-DEPENDENT WINDING IMPEDANCE OF ELECTRIC MACHINES 3340
Christian Mühlfeld

PP348 COMPENSATION TECHNIQUES FOR BANDWIDTH-DISTORTED MEASUREMENTS OF FAST TRANSIENTS IN DOUBLE PULSE TESTS 3347
Christian Lottis

PP349 AN AERODYNAMIC LOAD MEASUREMENT TECHNIQUE FOR AUTONOMOUS AERIAL VEHICLES 3353
Mehmet Oguz Girgin

COMPENSATION TECHNIQUES FOR BANDWIDTH-DISTORTED MEASUREMENTS OF FAST TRANSIENTS IN DOUBLE PULSE TESTS 3358
Christian Lottis

PP350 A HIGH-BANDWIDTH MULTILEVEL COUNTER CIRCUIT FOR BEARING CURRENT EVALUATION 3364
Felix Schulte

TRANSFORMERS

PP351 CORE LOSS MODEL FOR CONSIDERING ANISOTROPY AND TEMPERATURE EFFECTS ON ELECTRICAL STEEL UNDER POWER ELECTRONIC CONDITIONS 3371
Michael Owzareck

PP353   CIRCULAR ECONOMY ORIENTED AND RECONFIGURABLE PLANAR
TRANSFORMER DESIGN FOR ISOLATED DC-DC CONVERTERS 3380
    Fabian Groon

PP354   CONTROLLABLE MAGNETICS: VARIABLE TRANSFORMERS AND VARIABLE
INDUCTORS, THEORY – PRODUCTION – APPLICATION 3390
    Florian Fenske

PP355   A THREE-PHASE INTERLEAVED LLC INTEGRATED TRANSFORMER USING
PCB WINDINGS FOR FUEL CELL DCDC CONVERTERS 3395
    Jiajia Guan

PP356   TESTING THE PRIMARY-SECONDARY COIL COUPLING OF HIGH-FREQUENCY
TRANSFORMER IMPLEMENTED ON ETD AND TOROIDAL CORES 3400
    Alexis Gioda

Author Index

PCIM Europe 2024, 11– 13 June 2024, Nuremberg   DOI: 10.30420/566262382

Performance Evaluation of the Packaging of SiC Diodes in a 6.78 MHz Wireless Power Transfer System

Prateek Wagle[1], Ioannis Nikiforidis[1,2], Masashi Fukai[3], Shu Takeuchi[3], Kengo Tashiro[3], Paul D. Mitcheson[1,2]

[1] Imperial College London, United Kingdom
[2] Bumblebee Power, United Kingdom
[3] Sansha Electric MFG Co., Finland

Corresponding author: Ioannis Nikiforidis, i.nikiforidis@imperial.ac.uk
Speaker: Ioannis Nikiforidis, i.nikiforidis@imperial.ac.uk

Abstract

This paper evaluates the performance of three 1.2 kV Silicon Carbide diodes for high frequency applications, including a custom packaged 20 A rated device by SanRex. The diodes were tested in a high voltage half-bridge Class DE rectifier design in an inductive power transfer system operating at 6.78 MHz. The operating conditions were matched for all the devices and our measurements confirmed that the reduced parasitic inductance of the custom packaged diode from SanRex had significantly less ringing which resulted in lower power losses in the devices, as well as lower harmonic content on the secondary coil current and the output current.

1 Introduction

The recent advances in wide band-gap technology have enabled switching power converters to operate efficiently at multi-MHz frequencies [1]. Apart from the increased power density, high frequency is a key element for lightweight Inductive Power Transfer (IPT) applications over large air gaps [2]. However, at such high frequency, the parasitic elements of the devices have a significant impact on the operation and tuning of the converter [3].

Class E type converters, which are typically used at high frequency because of their soft-switching capabilities, can effectively absorb the non-linear junction capacitance of the semiconductors [4]. However, the ringing produced by the package inductance is still something that has to be minimised, and can be the limiting factor in further increasing the operating frequency. One such example is our recent work on a 3 kW IPT system, where the operating frequency was reduced to 3.39 MHz because of the ringing caused by the diodes on the rectifier [5].

Some of the negative effects of excessive ringing include reduced power capability of the converter, increased power losses on the device, electromag-

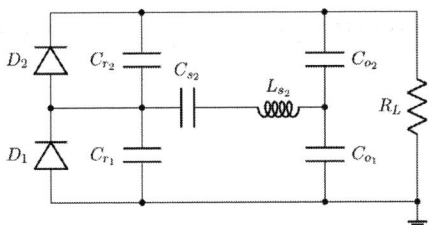

Fig. 1: Half-bridge Class DE rectifier circuit diagram, driven by the secondary coil L_{s_2}, which is fully resonant with the inclusion of the series capacitor C_{s_2}. The voltage across the load R_L is stabilised by the output capacitance C_o.

netic compatibility issues, and increased stresses on the filtering components [5]. In this work, we built Silicon Carbide (SiC) diode samples using a custom package specifically for high power, high frequency applications, and compared their performance over other similar off-the-shelf devices.

2 Motivation and Approach

This paper primarily focuses on investigating and minimising the effects of the parasitic package inductance. Here, the ringing caused by the resonance of this parasitic inductance with the junction capacitance is studied in the context of the passive Class DE rectifier [6] in a multi-MHz IPT system.

2718

PCIM Europe 2024, 11– 13 June 2024, Nuremberg   DOI: 10.30420/566262382

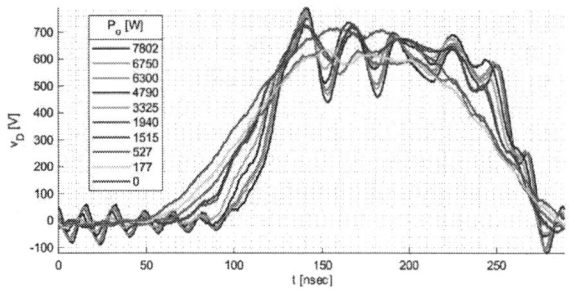

(a) Diode voltage

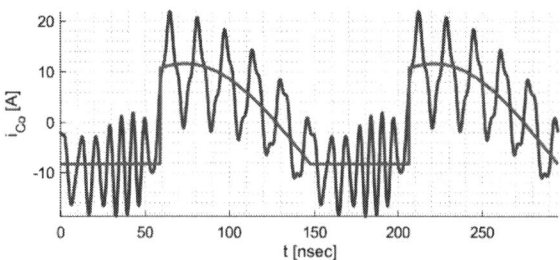

(b) Output capacitor current

Fig. 2: Waveforms of the voltage across the diode (top) and current through the output capacitor (bottom) in a Class DE rectifier. The voltage is measured across on the entire range of the load and the current is simulated with ideal diodes (orange) and the actual model of the device.

The Class DE rectifier (Fig. 1) is effectively the high-frequency version of the current-driven Class D rectifier [7], because of non-negligible parasitic junction capacitance of the diode. The duration of the dead-time can be controlled with the shunt capacitance C_r.

2.1 Effects of Parasitic Inductance in multi-MHz Applications

The measured voltage across one of the diodes in a 3.47 MHz, 8 kW full-bridge Class DE rectifier across the entire load range is shown in Fig. 2a. As the load increases, so does the current through the devices, which in turn increases the effect of the resonance of the parasitic inductance of the TO-247-2 package with the junction capacitance and the external shunt capacitor. This results in increased voltage stresses and power consumption on the device.

In addition, the parasitic inductance also significantly increases the stress on the output filter capacitor, which needs to act as a low impedance path for all the high frequency currents. This is shown in the simulated waveforms in Fig. 2b in LT-

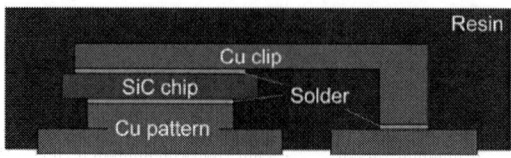

(a) Side view cross-section

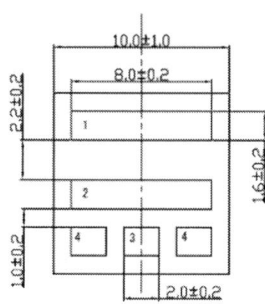

(b) Low coupling

Fig. 3: Cross-section of the clip-bonding process in the custom SiC diode (a), along with the dimensions of the pads at the bottom of the surface mount packaging (b).

spice, where the ideal current is substantially less than the case with the real device model.

2.2 Custom SiC Diode Packaging

For optimal operation at multi-MHz frequencies, the device itself needs to withstand high voltages and have low conduction losses, and the packaging needs low parasitic inductance. SiC intrinsically satisfies the device requirements; however packaging with traditional wire bonding would require many wires against a small source pad to reduce the inductance, which is not ideal. Therefore, in this work we tried the clip-bonding process applied to a surface mount package shown in Fig. 3, which optimises the usage of the limited pad area while keeping the inductance minimal. Additionally, we can prevent a thermal runaway on a spot and have more ways for heat dissipation.

3 Experimental Setup

To test the high-frequency performance of various diode packages we developed the 6.78 MHz, 1 kW IPT system shown in Fig. 4. The magnetic link consist of two two-turn planar square coils made of copper pipe with 10 mm diameter, with a measured inductance of around 2.5 µH. The coils have roughly 30 cm outer side length and are separated by a 15 cm air gap. The switching networks of the

2719

PCIM Europe 2024, 11– 13 June 2024, Nuremberg   DOI: 10.30420/566262382

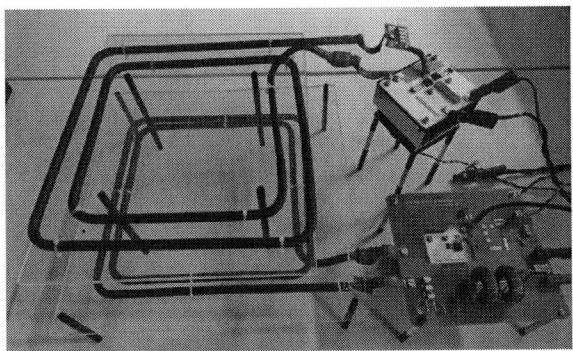

Fig. 4: Photo of the IPT experimental setup for evaluation the diode performance.

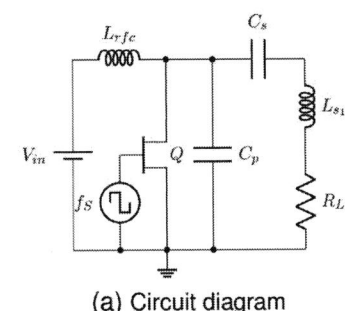

(a) Circuit diagram

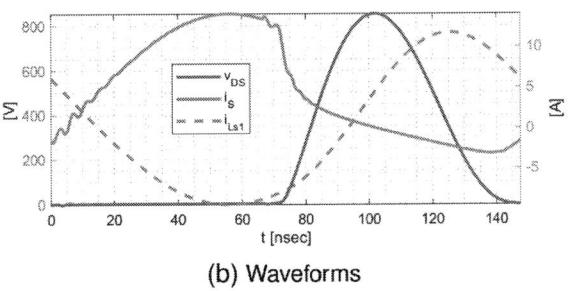

(b) Waveforms

Fig. 5: Circuit diagram and key waveforms of the single-ended Class E inverter in Fig. 4. In blue is the drain voltage, in solid orange is the drain current, and in dashed orange is the primary coil current. The primary coil L_{s_1} is inductively coupled to the secondary coil L_{s_2} in Fig. 1.

rectifier and inverter boards (top right and bottom right in Fig. 4, respectively) are on insulated metal substrate (IMS) printed circuit boards (PCB) with the Ventec VT-4B7 SP for insulation, which provides better thermal performance for surface mount components compared to FR4.

3.1 Inverter Design

The primary coil driver is a single-ended Class E inverter with an infinite input choke, shown in Fig. 5a. The inverter is based on the Cree C3M0065100J SiC MOSFET and designed based on the method

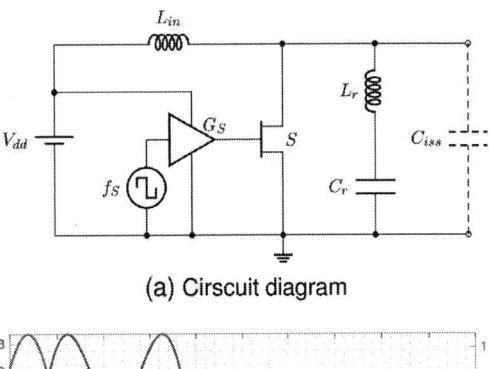

(a) Cirscuit diagram

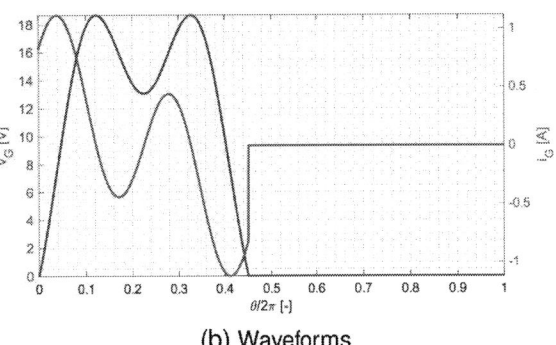

(b) Waveforms

Fig. 6: Circuit diagram and gate voltage and current waveforms of the Class Φ_2 gate driver. The resonant tank formed by L_r and C_r is resonating close to the second harmonic to achieve the voltage shape, and L_{in} provides ZVS.

explained in detail in [8] to match the primary coil with a 45 % duty cycle. The simulated drain and output waveforms of the inverter with 225 V supply voltage are given in Fig. 5b, where the MOSFET has both zero voltage switching (ZVS) and zero voltage derivative switching (ZVDS).

3.2 Gate Driver Design

Driving the SiC transistor of the primary (Cree C3M0065100J) efficiently at 6.78 MHz is no trivial task, and hard-switching half bridge drivers can be very inefficient [9], [10]. For this reason we also designed a Gallium Nitride (GaN Systems GS66504B) based Class Φ resonant gate driver, shown in (Fig. 6a). The design method for the gate driver is based on the work in [11] and further optimised using the methods in [4]. The gate driver is designed to match the input capacitance C_{iss} of the SiC power MOSFET at the desired duty cycle, while being supplied with the same voltage as the GaN gate driver (UCC27512 from TI). The ideal gate voltage and current waveforms are shown in Fig. 6b, where the gate voltage always remains within the optimal switching value range.

2720

PCIM Europe 2024, 11–13 June 2024, Nuremberg   DOI: 10.30420/566262382

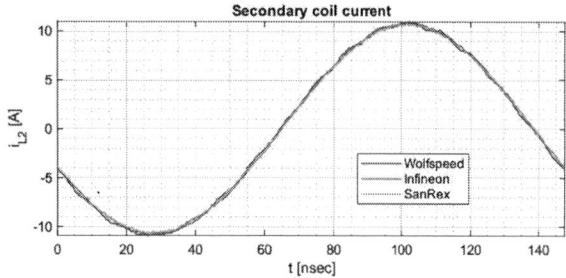

Fig. 7: Caption

3.3 Rectifier Design

The main design equations for the Class DE rectifier for a given load R_L can be summarised to [12]:

$$\overline{Z}_{ac} = \frac{\sin^2 \phi + \mathrm{j}\,(\sin \phi \cos \phi - \phi)}{\pi \omega C_r}, \tag{1}$$

where $\overline{Z}_{ac}$ is first order input impedance, j is the imaginary unit, ω is the angular frequency, C_r is the shunt capacitance, and

$$\phi = \tan^{-1}\left(2\frac{\sqrt{\pi \omega C_r R_L}}{\pi - \omega C_r R_L}\right). \tag{2}$$

Subsequently, the reflected impedance to the primary becomes

$$\overline{Z}_{eq} = k^2 X_{L_{s_1}} X_{L_{s_2}} \cdot$$
$$\cdot \frac{R_{ac} - \mathrm{j}\,\left(X_{C_{s_2}} + X_{C_{ac}} + X_{L_{s_2}}\right)}{R_{ac}^2 + \left(X_{C_{s_2}} + X_{C_{ac}} + X_{L_{s_2}}\right)^2}, \tag{3}$$

where k is the coupling factor of the magnetic link, $X_{L_{s_1}}$ is the reactance of the primary coil, $X_{L_{s_2}}$ is the reactance of the secondary coil, R_{ac} is the input resistance of the rectifier, $X_{C_{s_2}}$ is the secondary series resonant capacitor reactance, and X_{ac} is the input reactance of the rectifier.

An effective low duty cycle of around 25 % (large C_r) was chosen for maximum power capability [13] and lower ringing, due to the slower transition of the diode voltage. The secondary coil current was set at 11 A peak and the output voltage at full load was 540 V. After Eq. (1) is calculated, Eq. (3) can be determined, which is used for the optimal tuning of the inverter. Also, we used high Q C0G ceramic capacitors both the the resonant and the bulk capacitors to maximise performance and reduce deviation due to temperature and bias.

We used three different 1.2 kV diodes on the rectifier side: the SanRex prototype mentioned in section 2.2 rated for 20 A, the Wolfspeed C4D20120H

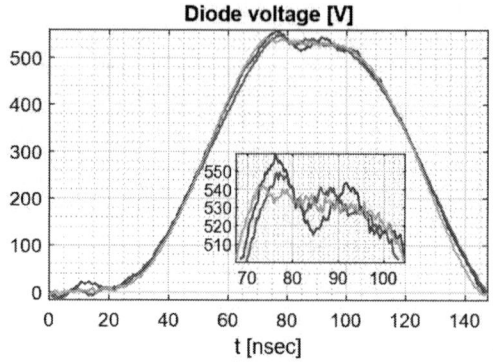

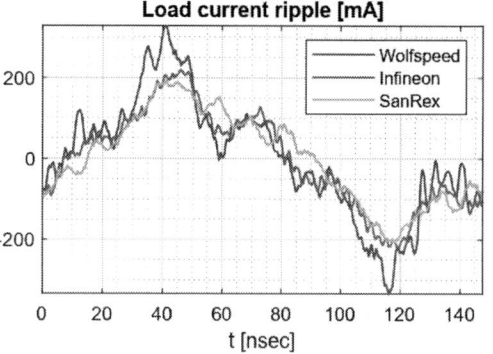

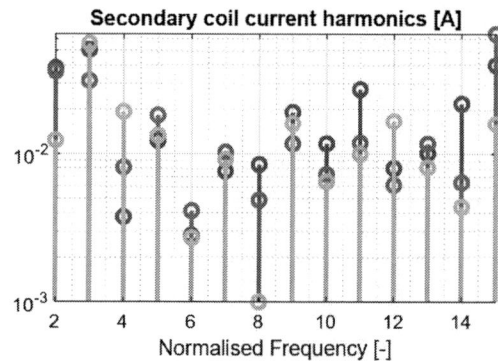

Fig. 8: Diode voltage with detail at peak, ripple of the output current, and harmonic content of the secondary coil current (logarithmic scale) for the three instances of the rectifier.

(also used in [5]), and the Infineon IDM10G120C5. The C4D20120H has the advantage of the highest current rating but in a TO-247-2 package, and the IDM10G120C5 was the only available diode in a surface mount package (PG-TO252-2) capable of managing the power level of the experiment.

4 Performance Results

All rectifiers had exactly the same load and series resonant capacitor during the experiments. Since all diodes have different junction capacitance, C_r

2721

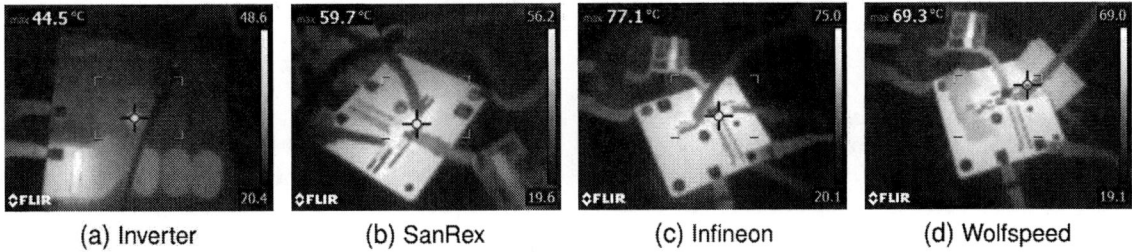

| (a) Inverter | (b) SanRex | (c) Infineon | (d) Wolfspeed |

Fig. 9: Thermal camera images of the rectifiers and the inverter after 15 min of operation at full load. There was no forced cooling and no heatsink attached to the rectifier IMS board for consistent results.

had to be adjusted accordingly in order to achieve the same tuning for each individual case. Also, the PCB layout in all three rectifier IMS boards is also identical. The secondary coil current was measured with the Pearson 6585 current monitor, the output current was measured with the Keysight N2783B current probe, and the voltage across the diodes was measured with the Tektronix TIVH08 probe to ensure high accuracy and minimal loading to the circuit.

A strong proof that all rectifiers had the same exact tuning and hence the comparison is fair is the waveform of the secondary coil current in Fig. 7, which has the same amplitude for all instances for the same DC load. In addition, the shapes of the switching waveform of the diodes in Fig. 8a also match. Since all the passive component and the primary driver are the same for all rectifier boards, any differences in performance can only be attributed to the diodes.

The custom clip-bonded package of the SanRex diode achieved the lowest ringing on the diode voltage according to Fig. 8a), which in part resulted in lower output current ripple in Fig. 8b. The peak ringing in Fig. 8b of the through-hole package is double of that of the surface mount devices. Also, the clip-bonded package had 19 % lower total harmonic distortion on the secondary coil current (0.66 % compared to 0.81 % in Infineon and 0.81 % in Wolfspeed). The total harmonic content of the current through the secondary coil is shown in detail in Fig. 8c.

The effect of the lower inductance and hence lower ringing can also be demonstarted from the thermal performance of the diodes (Fig. 9), where again the SanRex prototype was superior. Even though the through-hole device has the biggest current rating, it was still almost 8° hotter than the clip-bonded one. It must be pointed out that the current rating of the

Infineon device was the lowest one, which definitely contributed negatively in the thermal performance.

5 Conclusion

Motivated by our previous experience with traditional SiC diodes packaging for MHz IPT systems, we proposed the clip-bonding process in this paper. A 1 kW, 6.78 MHz test rig was designed and developed for comparison and performance evaluation between various package technologies. The performance of our prototype was noticeably better compared to the rest of the devices, both in signal integrity and temperature, suggesting a significantly lower parasitic inductance.

6 Acknowledgements

This work was supported in part by EPSRC Safe Power Delivery Using a Reconfigurable Mesh of Inductive Transceivers, under Grant EP/X020606/1, and in part by SanRex.

References

[1] G. Meneghesso, M. Meneghini, and E. Zanoni, *Gallium Nitride-enabled High Frequency and High E ciency Power Conversion.* Springer, 2018.

[2] J. M. Arteaga, S. Aldhaher, G. Kkelis, D. C. Yates, and P. D. Mitcheson, "Multi-MHz IPT systems for variable coupling," *IEEE Transactions on Power Electronics*, vol. 33, no. 9, pp. 7744–7758, 2018. DOI: 10.1109/TPEL.2017.2768244.

[3] S. Lu, T. Zhao, Z. Zhang, K. D. T. Ngo, R. Burgos, and G.-Q. Lu, "Low parasitic-inductance packaging of a 650 v/150 a half-bridge module using enhancement-mode gallium-nitride high electron mobility transistors," *IEEE Transactions on Industrial Electronics*, vol. 70, no. 1, pp. 344–351, 2023. DOI: 10.1109/TIE.2022.3148750.

[4] I. Nikiforidis, J. M. Arteaga, C. H. Kwan, N. Pucci, D. C. Yates, and P. D. Mitcheson, "Generalized multistage modeling and tuning algorithm for Class EF and Class Φ inverters to eliminate iterative retuning," *IEEE Transactions on Power Electronics*, vol. 37, no. 10, pp. 12 877–12 900, 2022. DOI: 10.1109/TPEL.2022.3176391.

[5] I. Nikiforidis, C. H. Kwan, D. C. Yates, K. Bampouras, J. Gawith, *et al.*, "A 3 kW 3.39 MHz DC/DC inductive power transfer system with power combining converters," in *2023 IEEE Wireless Power Technology Conference and Expo (WPTCE)*, 2023, pp. 1–6. DOI: 10.1109/ WPTCE56855.2023.10215793.

[6] D. Hamill, "Class DE inverters and rectifiers for DC-DC conversion," in *PESC Record. 27th Annual IEEE Power Electronics Specialists Conference*, vol. 1, 1996, 854–860 vol.1. DOI: 10.1109/PESC. 1996.548681.

[7] M. Kazimierczuk, "Class d current-driven rectifiers for resonant dc/dc converter applications," *IEEE Transactions on Industrial Electronics*, vol. 38, no. 5, pp. 344–354, 1991. DOI: 10.1109/41. 97554.

[8] K. Van Schuylenbergh and R. Puers, *Inductive powering: basic theory and application to biomedical systems*. Springer, 2009.

[9] J. Xu, Z. Tong, and J. Rivas-Davila, "1 kw mhz wideband class e power amplifier," *IEEE Open Journal of Power Electronics*, vol. 3, pp. 84–92, 2022. DOI: 10.1109/OJPEL.2022.3146835.

[10] Z. Ye, Z. Tong, L. Gu, and J. Rivas-Davila, "A high frequency resonant gate driver for sic mosfets," in *2021 IEEE 22nd Workshop on Control and Modelling of Power Electronics (COMPEL)*, 2021, pp. 1–5. DOI: 10.1109/COMPEL52922.2021. 9645982.

[11] H. Jedi, T. Salvatierra, A. Ayachit, and M. K. Kazimierczuk, "High-frequency single-switch zvs gate driver based on a class Φ_2 resonant inverter," *IEEE Transactions on Industrial Electronics*, vol. 67, no. 6, pp. 4527–4535, 2020. DOI: 10.1109/TIE.2019.2927192.

[12] H. Sekiya, X. Wei, T. Nagashima, and M. K. Kazimierczuk, "Steady-state analysis and design of class-de inverter at any duty ratio," *IEEE Transactions on Power Electronics*, vol. 30, no. 7, pp. 3685–3694, 2015. DOI: 10.1109/TPEL.2014. 2339355.

[13] M. K. Kazimierczuk and D. Czarkowski, *Resonant power converters*. John Wiley & Sons, 2012.

PCIM Europe 2024, 11– 13 June 2024, Nuremberg   DOI: 10.30420/566262383

Voltage Waveform Generation for Sawyer-Tower C_{OSS} Loss Measurements Using a Hybrid Power Converter

Malachi Hornbuckle[1], Steven Abrego[1], Katherine Liang[1], Sara Davidova[1], Zikang Tong[1], Juan Rivas-Davila[1]

[1] Stanford University, USA

Corresponding author: Malachi Hornbuckle, malachih@stanford.edu
Speaker: Malachi Hornbuckle, malachih@stanford.edu

Abstract

This paper presents the design and implementation of a converter topology to generate voltage excitations for C_{OSS} loss characterizations in MHz-frequency power switching devices. The assembled converter can drive a Sawyer Tower circuit with trapezoidal and sinusoidal voltage waveforms, up to 1.2 kV$_{PP}$ at MHz frequencies. Our work is intended to reduce the size and cost of existing characterization systems, improving access to C_{OSS} data.

1 Introduction

Zero-voltage switching (ZVS) plays an important role in enabling the efficient design of switching power converters at MHz frequencies, mitigating the frequency dependent energy losses that arise from turning on a switching device while its output capacitance ($C_{OSS} = C_{GD} + C_{DS}$) is charged. To utilize ZVS, converters can be designed such that the drain-source voltage (V_{DS}) of a switching device automatically transitions to zero before a conductive drain-source channel is formed as the device is turned on. Figure 1 includes a small selection of topologies that take advantage of ZVS. Traditional analysis of converters utilizing ZVS assumes that the resonant charging and discharging of C_{OSS} is lossless, and typical device datasheets characterize C_{OSS} as a small-signal lossless capacitance whose value depends on bias voltage. However, parasitics in the output capacitance of devices can lead to significant losses even when ZVS is utilized, reducing the efficiency of converters below predicted values [1]. Losses arising from the charging and discharging of C_{OSS} have been observed in Si, SiC, and GaN devices, motivating efforts to characterize its large-signal behavior. While multiple large-signal methods for measuring C_{OSS} losses exist [2], [3], the Sawyer Tower circuit shown in Fig. 2 has become a popular way to test devices in

both Si and wide-bandgap materials [4]–[6]. The results of these tests show that C_{OSS} losses should be considered when choosing devices for high-frequency power conversion [7].

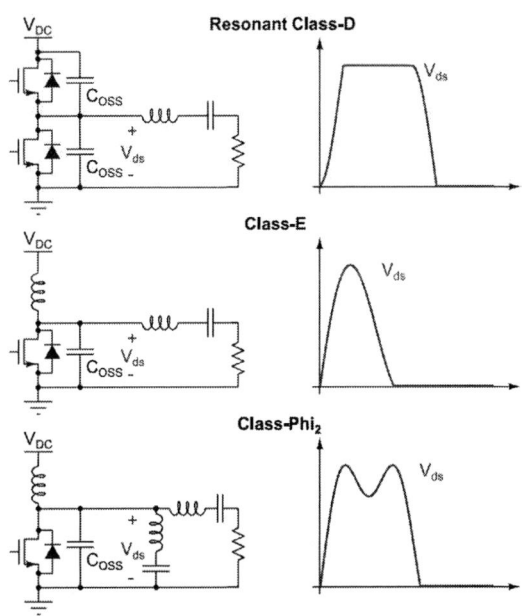

Fig. 1: Examples of converter topologies that use ZVS to reduce switching losses.

Sawyer Tower measurements require the generation of a high-voltage, high-frequency waveform mimicking the $V_{DS}(t)$ that a device would be subjected to in a power conversion topology. This is

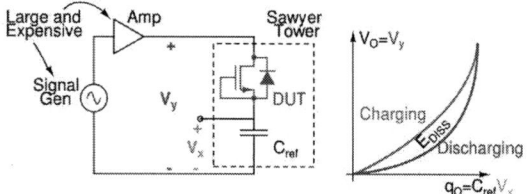

Fig. 2: The Sawyer Tower method for measuring C_{OSS} hysteresis losses.

typically accomplished with a large and expensive broadband power amplifier, reducing the accessibility of Sawyer Tower C_{OSS} loss measurements. In this paper we demonstrate an alternative waveform generator: a small, low-cost "hybrid converter" intended to reduce the cost of driving Sawyer Tower C_{OSS} loss measurement systems.

2 Sawyer Tower Measurement Method

The Sawyer Tower circuit, originally used to characterize Rochelle salt for use in capacitors [8], allows measurement of the large-signal C_{OSS} hysteresis that leads to energy loss. Shown in Fig. 2, the Sawyer Tower consists of a device under test (DUT) in series with a reference capacitor C_{ref}. The entire structure is subjected to a voltage waveform, and a highly ideal C_{ref} can be chosen such that the DUT is subjected to a desired fraction of this waveform's total voltage swing [5]. By probing V_x and V_y as shown in Fig. 2, the charge (Q_{OSS}) and voltage (V_{DS}) of C_{OSS} on the DUT can be determined:

$$Q_{OSS} = C_{ref} \cdot V_x \; ; \; V_{DS} = V_y - V_x \qquad (1)$$

To calculate E_{DISS}, the energy dissipated as C_{OSS} is charged and discharged, V_{DS} can be integrated with respect to Q_{OSS} over a cycle of the voltage waveform:

$$E_{DISS} = \oint V_{DS} \cdot dQ_{OSS} \qquad (2)$$

This energy loss can be interpreted graphically as a hysteresis in the Q-V plot of the output capacitance, resulting in more energy being consumed during charging than returned during discharging (Fig. 2).

A useful feature of the Sawyer Tower test setup is flexibility in the applied voltage waveform. For example, in [1] it is used to generate the Φ_2 waveform seen in Fig. 1, allowing characterization for a specific circuit application. For general device characterizations, it is useful to vary the applied waveform

to reveal trends in E_{DISS}. Our goal in designing a converter for Sawyer Tower measurements was to enable the generation of voltage waveforms that will be most useful for C_{OSS} loss characterizations.

3 Hybrid Converter Design

We designed a "hybrid converter" with the ability to operate in two separate modes: one generating trapezoidal voltage waveforms and one generating sinusoidal voltage waveforms. Trapezoidal waveforms are chosen to allow dV/dt to vary without changing overall waveform amplitude or frequency, enabling direct characterization of E_{DISS} with regard to voltage slew rate. Sinusoidal waveforms allow the excitation amplitude and frequency to be varied, and allow comparison to E_{DISS} values from prior works [1], [5]. Another factor influencing the waveforms chosen was the feasibility of generating them given a range of DUTs and C_{ref} values that may make up the Sawyer Tower.

Figure 3 shows the converter topology we have assembled. It is a combination of a full bridge voltage mode Class-D amplifier [9] and a current mode Class-D converter [10], with "enable" transistors to toggle between sub-topologies. Variable inductors (L_R) tune the output waveform to the shape desired for characterization. In voltage mode operation (enable channel off; Fig. 4 shows a simplified schematic), trapezoidal voltage waveforms are generated. Current mode operation (enable channel on; Fig. 5 shows a simplified schematic) is used to generate a sinusoidal voltage waveform for DUT characterization. In both modes of operation ZVS is utilized, reducing the risk of the switching devices overheating. Additionally, the switch devices all initially conduct current in the source to drain direction when turned on, so their reverse-body diodes allow some flexibility in the gating waveforms shown in Fig. 4 and Fig. 5.

A few modifications from the traditional topologies in [9], [10] have been made to accommodate dual operation. Reverse blocking diodes (D_{RB}) and buffer capacitors (C_{buf}) are added above the full bridge high-side switches to stabilize the voltage at this node and keep the switching node voltage from being pinned to V_{DC} during sinusoidal operation. Buffer capacitors have also been added to stabilize the drain voltage of the enable transistor and prevent any interaction with L_{choke} during trapezoidal

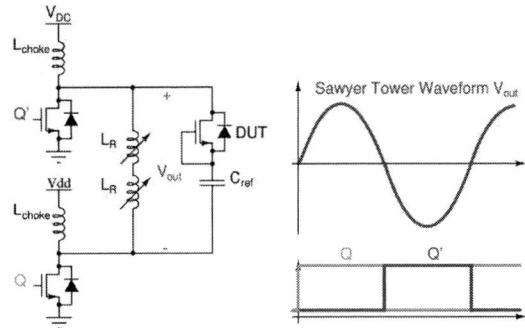

Fig. 5: Simplified schematic for current mode operation (sinusoidal waveform).

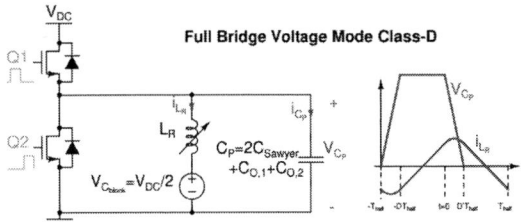

Fig. 6: AC equivalent half circuit for voltage mode operation (trapezoidal waveform).

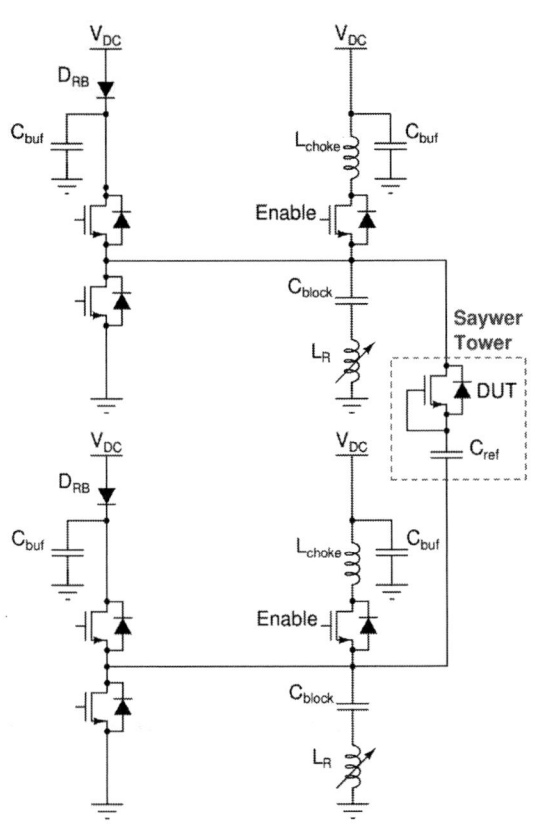

Fig. 3: (a) Full converter topology, plus simplified topologies and gate drive patterns in (b) voltage mode and (c) current mode.

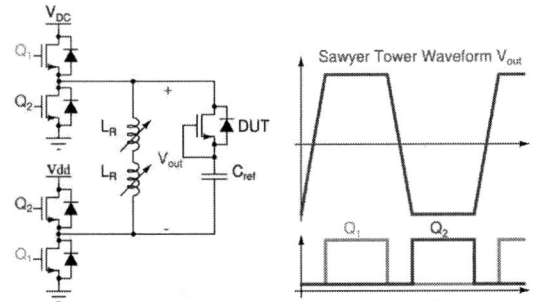

Fig. 4: Simplified schematic for voltage mode operation (trapezoidal waveform).

operation. Additionally, the chassis of the variable inductors need to be ground-referenced, so DC blocking capacitors C_{block} are added in series.

4 Full Bridge Class-D Operation: Trapezoidal Waveform Generation

To analyze the converter operating in trapezoidal mode, the equivalent AC half-circuit in Fig. 6 is obtained by assuming the capacitances connected to the switching node can be lumped together into a single capacitance C_P. This includes $C_{O,1}$ and $C_{O,2}$ from the switching devices, along with $2C_{Sawyer}$ where C_{Sawyer} is the total capacitance of $C_{OSS,DUT}$ in series with C_{ref}. To simplify the analysis, we will assume that C_P is linear. In periodic steady state, the average voltage of the switch node and C_{block} must be equal at half of V_{DC}. As shown in Fig. 6, we define T_{half} as half the period of the switching frequency, D as the fraction of T_{half} for which Q_1 or Q_2 is on, and D' = (1 - D) as the dead time where the switching node voltage V_{C_P} transitions between V_{DC} and ground. Defining t=0 at the beginning of the dead time, the differential equation and initial conditions in Eq. 3 determine how the circuit behaves until $D'T_{half}$:

$$LR C_P \frac{d^2 V_{C_P}}{dt^2} + V_{C_P} - \frac{V_{DC}}{2} = 0$$
$$V_{C_P}(t=0) = V_{DC} \qquad (3)$$
$$I_{L_R}(t=0) = C_P \frac{dV_{C_P}}{dt}(t=0) = \frac{V_{DC} D T_{half}}{4 L_R}$$

The solution to Eq. 3 is a sinusoidal oscillation of V_{C_P} around $V_{DC}/2$:

$$V_{C_P}(t) = \frac{V_{DC}}{2} cos\left(\frac{t}{\sqrt{L_R C_P}}\right)$$
$$- \frac{D T_{half} V_{DC}}{4\sqrt{L_R C_P}} sin\left(\frac{t}{\sqrt{L_R C_P}}\right) \qquad (4)$$
$$+ \frac{V_{DC}}{2}$$

For trapezoidal waveforms, we are interested in cases where V_{C_P} is roughly linear in time, so we take its Maclaurin expansion:

$$V_{C_P}(t) \approx V_{DC} - \frac{D T_{half} V_{DC}}{4 L_R C_P} t \qquad (5)$$

Constraining Eq. 5 to force $V_{C_P}=0$ at the end of the dead time yields Eq. 6, which gives an estimate for L_R to achieve a given dead time:

$$L_R \approx \frac{D D' T_{half}^2}{4 C_P} \qquad (6)$$

Equation 5 cannot be used to immediately set the variable inductors L_R to the perfect position for a desired waveform due to the assumptions in its derivation, as well as the general difficulty of setting variable inductors to a specific value. Rather, it provides a starting guess from which the inductors can be tuned to an ideal operating point. As a result, the hybrid converter will be able to apply differential trapezoid waveforms to the Sawyer Tower circuit with variable dV/dt controlled by tuning its variable inductors.

5 Current Mode Class-D Operation: Sinusoidal Waveform Generation

In sinusoidal mode, the equivalent AC half-circuit in Fig. 7 can be used to analyze behavior. In contrast to the trapezoidal mode analysis, T is defined as the period of the switching frequency and the switch is assumed to be on for half of T. Due to the existence of the choke inductor, the average voltage of C_{block} and the switching node must be V_{DC} in periodic steady state. When the switch is off, Eq. 7 determines the circuit behavior:

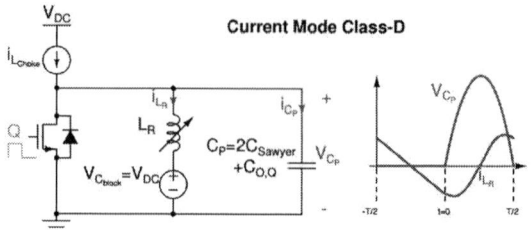

Fig. 7: AC equivalent half circuit for current mode operation (sinusoidal waveform).

$$L_R C_P \frac{d^2 V_{C_P}}{dt^2} + V_{C_P} - V_{DC} = 0$$
$$V_{C_P}(t=0) = 0 \qquad (7)$$
$$I_{L_R}(t=0) = C_P \frac{dV_{C_P}}{dt}(t=0) = \frac{T V_{DC}}{4 L_R}$$

We can see that V_{C_P} will ring sinusoidally, so we will choose the resonant frequency of L_R and C_P to roughly match the switching frequency:

$$2\pi f_{SW} = \frac{1}{\sqrt{L_R C_P}} \Rightarrow L_R = \frac{T^2}{4\pi^2 C_P} \qquad (8)$$

Equation 8 does not yield the exact value for L_R such that the solution of Eq. 7 reaches zero exactly when t = T/2. Rather, it gives a good starting point from which the converters L_R variable inductors can be tuned to achieve a good differential sinusoid approximation across the Sawyer Tower circuit. Due to the nonlinearity of C_P and the difficulty of accurately setting variable inductor values, tuning would still be necessary even if Eq. 8 was more accurate.

6 Converter Implementation

The assembled converter is shown in Fig. 8 compared to the much larger power amplifier in our lab which it is intended replace. Gating signals for each switching device, as well as the enable transistors, are sent to the board over a ribbon cable. Utilizing 600 V GaN power transistors (Gs66504b), the converter can produce output voltages up to 1.2 kV$_{PP}$ at MHz frequencies. SiC diodes were added antiparallel to the GaN devices to improve their reverse conduction, and the variable inductances are implemented by Harris roller inductors (755017A4530) along with stepper motors controlled by an Arduino Uno microcontroller. The relationship between inductance and position has been characterized for each roller inductor, with Eqs. 6 and 8 determining their initial position guesses. After setting the roller inductors to their initial positions, manual tuning is used to refine the Sawyer

PCIM Europe 2024, 11– 13 June 2024, Nuremberg   DOI: 10.30420/566262383

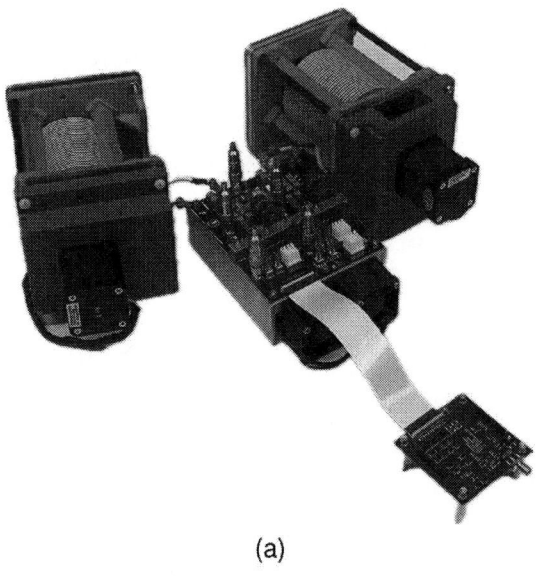

(a)

(b)

(c)

Fig. 8: (a) Fully assembled converter (b) Main converter board (c) Power amplifier previously used for C_{OSS} loss measurements.

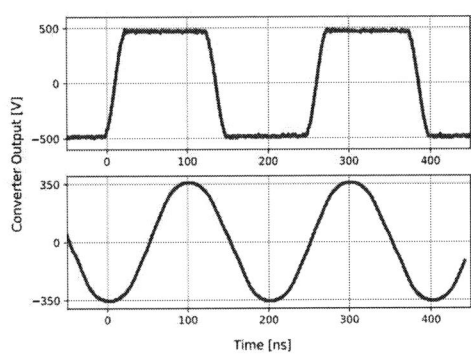

Fig. 9: Measured converter voltage waveforms in trapezoidal mode (1 kV$_{PP}$, 4 MHz) and sinusoidal mode (700 V$_{PP}$, 5 MHz).

Tower waveform. The board includes mounting points for small DUT and C_{ref} boards, making these components easily interchangeable. Sample waveforms from initial testing of the converter in both modes of operation are shown in Fig. 9.

7 Conclusion

In this paper, we present the design and implementation of a converter to generate voltage waveforms for Sawyer-Tower C_{OSS} loss characterizations. Unloaded, our converter is able to generate sinusoidal and trapezoidal voltage waveforms up to 1.2 kV$_{PP}$ at switching frequencies from 1 MHz to 30 MHz. By generating waveforms that are of interest for Sawyer Tower characterizations of high-frequency switching device output capacitances, this converter presents an alternative to the large and expensive broadband power amplifiers that would otherwise be needed for such measurements. Increased access to quality C_{OSS} loss measurement systems will enable circuit designers to build more efficient converters and manufacturers to reduce soft switching losses in future devices.

Acknowledgement

The authors would like to thank PowerAmerica for supporting this work.

2728

References

[1] G. Zulauf, S. Park, W. Liang, K. N. Surakit-bovorn, and J. Rivas-Davila, "Coss losses in 600 V GaN power semiconductors in soft-switched, high- and very-high-frequency power converters," *IEEE Transactions on Power Electronics*, vol. 33, no. 12, pp. 10 748–10 763, 2018. DOI: 10.1109/TPEL.2018.2800533.

[2] Q. Song, R. Zhang, Q. Li, and Y. Zhang, "Origin of soft-switching output capacitance loss in cascode gan hemts at high frequencies," *IEEE Transactions on Power Electronics*, vol. 38, no. 11, pp. 13 561–13 566, 2023. DOI: 10.1109/TPEL.2023.3299977.

[3] M. Samizadeh Nikoo, A. Jafari, N. Perera, and E. Matioli, "Measurement of large-signal coss and coss losses of transistors based on nonlinear resonance," *IEEE Transactions on Power Electronics*, vol. 35, no. 3, pp. 2242–2246, 2020. DOI: 10.1109/TPEL.2019.2938922.

[4] J. Fedison and M. Harrison, "Coss hysteresis in advanced superjunction mosfets," in *2016 IEEE Applied Power Electronics Conference and Exposition (APEC)*, 2016, pp. 247–252. DOI: 10.1109/APEC.2016.7467880.

[5] N. Perera, G. Kampitsis, R. van Erp, J. Ançay, A. Jafari, *et al.*, "Analysis of large-signal output capacitance of transistors using sawyer–tower circuit," *IEEE Journal of Emerging and Selected Topics in Power Electronics*, vol. 9, no. 3, pp. 3647–3656, 2021. DOI: 10.1109/JESTPE.2020.2992946.

[6] D. Bura, T. Plum, J. Baringhaus, and R. W. De Doncker, "Hysteresis losses in the output capacitance of wide bandgap and superjunction transistors," in *2018 20th European Conference on Power Electronics and Applications (EPE'18 ECCE Europe)*, 2018, P.1–P.9.

[7] G. Zulauf, Z. Tong, J. D. Plummer, and J. M. Rivas-Davila, "Active power device selection in high- and very-high-frequency power converters," *IEEE Transactions on Power Electronics*, vol. 34, no. 7, pp. 6818–6833, 2019. DOI: 10.1109/TPEL.2018.2874420.

[8] C. B. Sawyer and C. H. Tower, "Rochelle salt as a dielectric," *Phys. Rev.*, vol. 35, pp. 269–273, 3 Feb. 1930. DOI: 10.1103/PhysRev.35.269.

[9] D. Hamill, "Class DE inverters and rectifiers for dc-dc conversion," in *PESC Record. 27th Annual IEEE Power Electronics Specialists Conference*, vol. 1, 1996, pp. 854–860. DOI: 10.1109/PESC.1996.548681.

[10] H. Kobayashi, J. Hinrichs, and P. Asbeck, "Current mode Class-D power amplifiers for high efficiency RF applications," in *IEEE MTT-S IMS Digest*, vol. 2, 2001, pp. 939–942. DOI: 10.1109/MWSYM.2001.967047.

PCIM Europe 2024, 11– 13 June 2024, Nuremberg   DOI: 10.30420/566262384

Evaluation of SiC Devices for Over 500kHz Application Based on ZVS Buck Circuit

Minli Jia[1], Hao Sun[1], Kang Liu[1], Sicheng Gong[1], Zhen Zhou[1]

[1] Navitas Semiconductor, Shanghai EV Team, China

Corresponding author: Minli Jia, minli.jia@navitassemi.com
Speaker: Minli Jia, minli.jia@navitassemi.com

Abstract

SiC devices are used in soft-switching topologies, where the switching frequency can be further increased, thus reducing the size of the passive devices, and increasing the power density of the product. Planar and trench gate are two main structures of SiC MOSFETs development technology. Despite the same technology, different manufacturers in the switching and thermal characteristics differ from each other, so it is necessary to compare and analyze the characteristics of different manufacturers of the devices to guide the engineers to make application selection. This paper is based on the design of a 600 kHz, 3.6 kW soft-switching ZVS Buck converter. First of all, the drive circuit for SiC devices high-frequency applications is optimized to improve its drive reliability. Moreover, the key parameters affecting the high-frequency applications of SiC devices are revealed by power loss breakdown. In addition, SiC devices from three manufacturers, with different structures and similar parameters, were selected for experimental comparison. The results verify the correctness of the analysis and provide a reference design for the high-frequency application of SiC devices.

1 Introduction

With the development of electric vehicles (EV), energy storage technology and photovoltaic power generation, SiC, as the third generation of wide bandgap semiconductor power devices, is more and more widely used to reduce the size and improve the power density of the products.

For the switching characteristics of SiC, Literature [1] analyzes in detail and compares it with Si devices in terms of static characteristics, switching characteristics, temperature characteristics and losses, showing the excellent conduction and switching performance of SiC devices. Literature [2] based on the application of soft-switching inverter, the switching frequency is controlled at 500 kHz, and the power density of the product is greatly improved. Literature [3] used the topology of CLLC to compare the design of constant and variable bus voltage and proved the effect of the turn-off loss on the efficiency under high frequency conditions.

SiC, as the third generation of wide-bandgap semiconductor devices in product applications, has always been a research hotspot. For this reason, this paper will be based on the high-frequency soft-switching Buck circuit topology, selecting the three manufacturers of different SiC technologies for experimental comparison and verification.

2 SiC Technology and Features

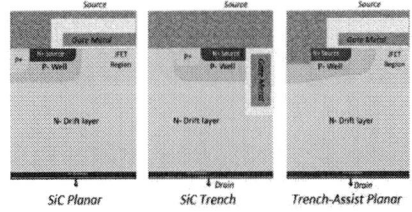

Fig. 1   Different SiC MOSFET technologies.

Planar and trench are the two main technologies of SiC MOSFETs, and many complementary planar and trench structures have been developed to optimize the switching performance of the devices. Three typical structures of planar, trench and trench-assisted planar, are shown in Fig. 1, and detailed comparisons of different technologies are described in the literature [4, 5]. Structurally, the most obvious feature of the planar type is that the gate and the source are on the same level, showing a "planar" distribution, and the trench is parallel to the substrate. On the other hand, the gate of the trench type is located below the source, forming a "trench" in the semiconductor material, and the trench and gate are perpendicular to the substrate.

Planar gate process is relatively simple, easy to achieve a better quality of the gate oxide layer, which has a strong resistance to voltage shocks. The practical application of higher reliability in the overload conditions is not easy to be damaged. However, the planar gate structure limits the reduction of the metric spacing to some extent due to the gate being transverse, which is unfavorable to the on-resistance and switching speed of the device.

Trench gate structure can make the power chip cell space greatly reduced, in the performance than the planar gate SiC MOSFET and show lower on-resistance, stronger switching performance, etc. But due to the reduction of wafer area, it will cause the increase of its own thermal resistance, and heat dissipation is unfavorable. In addition, the structure of the trench gate needs to dig out the trench on the substrate; the gate will be buried in the formation of vertical channel; the process is obviously more complex compared with the structure of the planar gate. Thus, unit consistency is poor. At the same time, the trench gate in the silica gate to withstand the electric field strength than in the silicon based MOSFET is much higher, so the reliability of the gate oxide layer will have some problems. It needs to be solved by improving the gate oxidation process, or by improving the electric field concentration at the bottom of the gate through different structural designs.

The trench assisted planar SiC absorbs the advantages of the planar and trench structures and it has the characteristics of high reliability of the gate oxide layer, strong resistance to voltage shocks, fast switching speed and low thermal resistance.

3 Buck Circuit Main Parameters Design

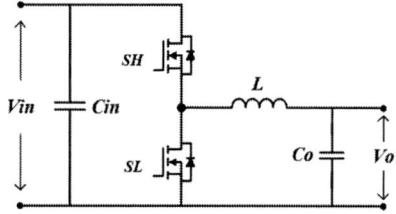

Fig. 2 Topology of the Buck circuit.

Fig. 2 shows the half-bridge Buck circuit topology, *SH* and *SL* are the upper and lower SiC devices of the bridge arm respectively. *L* is the output inductance. *Cin* is the input and *Co* is the output capacitance for current ripple filter.

3.1 Output Inductor Design

Continuous Current Mode (CCM), Boundary Current Mode (BCM) and Synchronized Continuous Current Mode (SR-CCM) (Fig. 3) are the three working modes of Buck converter, and each mode corresponds to different inductor values. In this paper, we focus on the high-frequency operation of SiC devices under soft-switching conditions, so SR-CCM is selected for design.

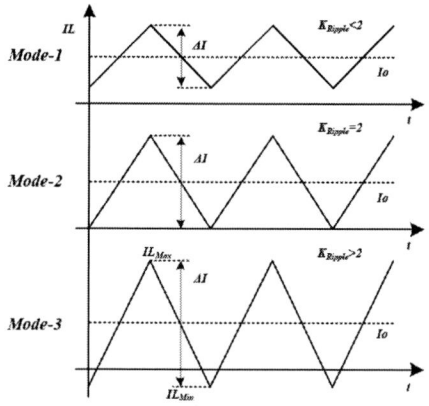

Fig. 3 Three modes of Buck circuit control.

The maximum and minimum value of the inductor current is defined as IL_{Max} and IL_{Min}, respectively, and the difference $\Delta IL = IL_{Max} - IL_{Min}$. The output current is defined as I_O, and the expression of the inductor ripple coefficient is:

$$K_{Ripple} = \frac{\Delta IL}{I_O} \qquad (1)$$

In Fig. 3, the three modes correspond to the ripple coefficients $K_{Ripple} < 2$, $K_{Ripple} = 2$ and $K_{Ripple} > 2$, respectively, and combined with equation (1), the expression for the inductance at CCM mode is as follows:

$$L = \frac{(Vin - Vo) \cdot D}{2 \cdot fs \cdot K_{Ripple} \cdot I_o} \qquad (2)$$

where *fs* is the switching frequency, *D* is the duty cycle, *I$_o$* is the output current, and the value of the IL_{Min} determines the value of K_{Ripple}.

Due to the existence of capacitance Coss of SiC device, IL_{Min} must be large enough to realize the ZVS turn-on during the current exchange, otherwise it will enter the hard switching mode. According to the inductor current and capacitor voltage energy relationship $L \cdot IL_{Min\_ZVS}^2 = Coss \cdot Vin^2$, combined with the equation (2), meeting the soft switching of the minimum current expression can be obtained as follows:

$$IL_{Min\_ZVS} = Vin\sqrt{\frac{Coss}{L}} \qquad (3)$$

Based on equation (3), Table 1 gives the calculated values of inductance at 800 V input, 400 V output, and 3600 W power output, corresponding to fs=300 kHz, fs=600 kHz, and fs=800 kHz, respectively.

fs(kHz)	D	IL$_{Min}$(A)	L(µH)
300	1	5A	22.2
600	1	5A	11.1
800	1	5A	8.3

Table 1 *L* vs *fs* in SR-CCM mode

Taking GeneSiC's G3R75MT12J (1200 V / 75 mΩ) SiC MOSFET as an example, its junction capacitance Coss=47 pF. Combined with equation (3) and Table 1, the minimum current to meet the soft switch operation can be obtained as: $IL_{Min\_ZVS} = 1.65A$, and the actual design current is $IL_{Min} = 5A > IL_{Min\_ZVS}$, which meets the design requirements of the soft switching.

3.2 Driver Circuit Design

The gate threshold voltage of SiC devices is usually between 1.5 V and 5 V. Fig. 4 shows a commonly used circuit with negative voltage drive, characterized by simplicity and reliability.

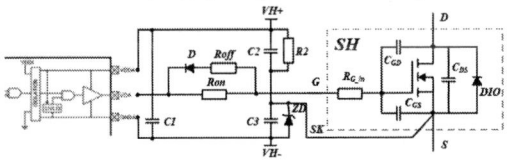

Fig. 4 Typical drive circuit for SiC devices.

Due to the Cgs capacitance of the SiC device, with the increase of the switching frequency, the loss of the driver IC increases, the temperature rises, and in the absence of heat dissipation measures, the driver chip will fail. Fig. 5 shows the derating curve given by the SILICON LABS Si823X series specifications. When its case temperature reaches to 150 ° C, the drive current will be limited to zero.

The expression for calculating the driver loss is as follows (Equation (4)), and it is mainly proportional to the switching frequency when the device parameters are determined.

$$P_{Loss\_dr} = V_{dr} \cdot Q_g \cdot f_s \qquad (4)$$

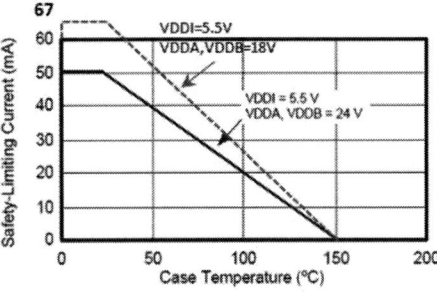

Fig. 5 Current Derating VS Case Temperature.

The Q_g of G3R75MT12J is 47 nC. When the driving voltage is 18V, the drive losses corresponding to different frequencies calculated based on equation (4) are shown in Table 2.

fs(kHz)	Qg (nc)	Ploss_dr (W)	Current (A)
300	47	0.085*2=0.17	9
600	47	0.508*2=1.016	56
1000	47	0.846*2=1.692	94

Table 2 Drive loss vs *fs*

The derating curve under the 18V drive voltage is shown as the dashed line in Fig. 5. When the switching frequency is 600kHz, the loss of the driver IC chip has exceeded its rated curve. Therefore, this paper presents the solution as shown in Fig. 6, which combines the driver IC with the totem pole circuit to solve the driver IC heating problem.

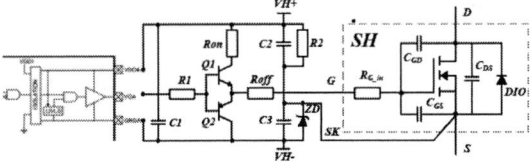

Fig. 6 Totem-pole based driver circuit design.

3.3 Loss Calculation

Loss breakdown can help us clearly understand the risk of devices in high frequency applications and give the optimal design direction. In addition to the drive loss (Equation (4)), there are the following loss components:

Conduction loss:

$$P_{con} = I_{DS\_RMS}^2 \cdot R_{DS\_On} \qquad (5)$$

Switching on loss:

$$P_{ON\_Loss} = E_{on} \cdot f_s \qquad (6)$$

Switching off loss:

$$P_{OFF\_Loss} = E_{off} \cdot f_s \qquad (7)$$

Body diode conduction loss:

$$P_{Dio\_Loss} = V_{sd} \cdot I_{sd\_avg} \qquad (8)$$

Body diode reverse recovery loss:

$$P_{rr} = V_{in} \cdot Q_{rr} \cdot f_s \qquad (9)$$

For soft switching applications, the turn-on loss of the device is zero; The turn-off loss does not include Eoss energy, which is released at the on-time and not consumed through the internal resistance of the device.

4 Simulation and Experimentation

4.1 Simulation Analysis

Parameters	Value
Vin	800V
Vo	400V
Po	3.6kW
fs	600kHz
L	11uH

Table 3 Main design parameters

The main parameters of the Buck converter designed in this paper are shown in table 3, the input and output voltage are 800 V and 400 V, respectively. The output power is 3.6kW and switching frequency is 600 kHz. The output inductance is 11uH. The simulation model is built based on SIMetrix/SIMPLIS software as shown in Fig. 7.

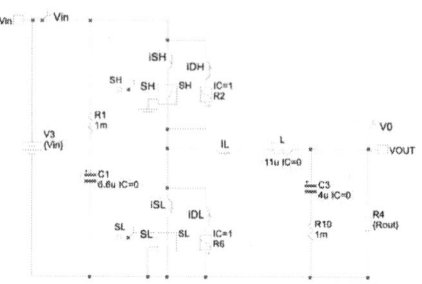

Fig. 7 Buck circuit Simulation mode

Fig. 8 shows the simulated waveforms, under the SR-CCM control mode. The upper side is active device, the conduction RMS current is 8.8A, the ZVS-ON current is 4.3A, and the turn-off current is 24.5A. The lower is synchronous rectifier device, and the conduction RMS current is 5.79A.

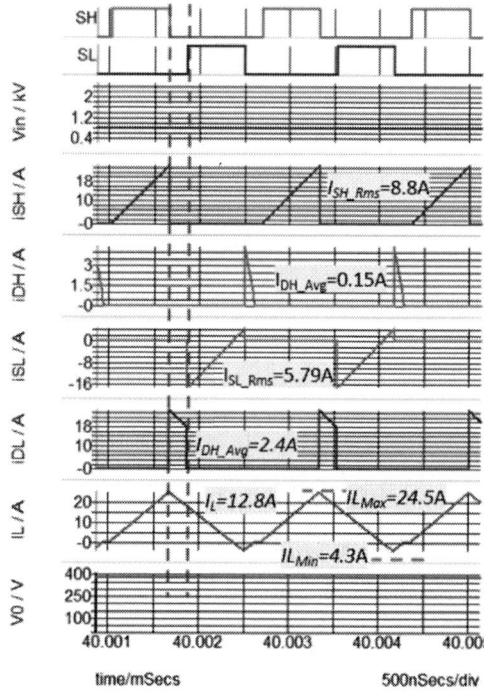

Fig. 8 Main Simulation Waveforms

4.2 SiC Performance Comparison

Different SiC technologies have different device characteristics. In this paper, three representative 1200V TO-263-7 packaged SiC devices with similar specifications are selected for comparative analysis. They are *0075120* from competitor A using planar technology, *120R060* from competitor B using trench technology and G3R75MT12J from GeneSiC (Navitas semiconductor) using trench-auxiliary planar technology.

Items	Competitor A	Competitor B	G3R75MT12J
SiC	Planar	Trench	Trench-Assist Planar
IDS	26A@100°C	19.7A@100°C	27A@100°C
Vgs	-5V ~18V	-4V ~15V	-5V ~18V
Rds_on	MAX=83mΩ	MAX=90mΩ	MAX=85mΩ
Qg	34nC	48nC	47nC
Eoss	22uJ	33uJ	18uJ
Vsd	3.9V(175°C)	4.0V(175°C)	4.4v(175°C)
Rg(int)	6 Ω	9 Ω	1.3 Ω
Rjc	Max=0.83°C	Max=1.1°C/w	Max=0.77°C/w

Table 4   Key Parameters Comparison

Table 4 lists some of the key comparison parameters. In terms of on-state current IDS, drive voltage Vgs, Rds_on and Eoss, the parameters of competitor A and GeneSiC differ little, thanks to the fact that both are planar technology, while competitor B is trench-technology. In terms of driving energy Qg, competitor A has the smallest value, while competitor B and GeneSiC have the same value. There is little difference in the voltage drop of body diode. In terms of the internal resistance Rg(int) of the gate, GeneSiC is the smallest, which can achieve faster switching speed. In terms of thermal resistance Rjc, despite the advanced heat treatment process adopted by competitor B, the Rjc is still large, while GeneSiC is the least.

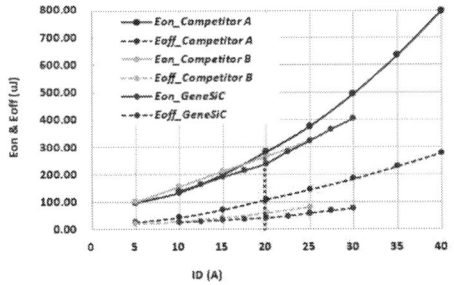

Fig. 9   Eon and Eoff vs ID

Same conditions: Vbus=800 V, Rg_ext= 2 Ω, Vg=-5/18 V(competitor A is -5/15 V), Tj=150 ℃. FIG. 9 shows the comparison curve of Eon and Eoff with current ID (the data is converted according to manufacturer's specifications). Under the same 20 A turn-on and turn-off current, G3R75MT12J has the smallest switching energy, while competitor B is slightly larger, and competitor A has the largest.

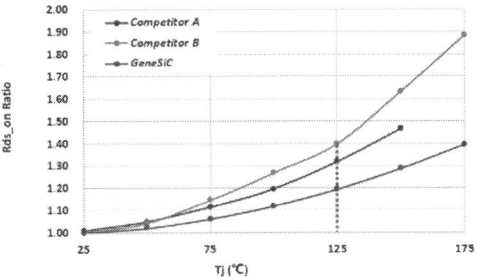

Fig. 10   Rds_on vs Tj

FIG. 10 shows the variation curve of Rds_on of three devices with Tj. at 125℃ junction temperature, GeneSiC's SiC has a low conduction loss, and its resistance change rate is 1.2, while competitor A and competitor B is 1.31 and 1.4, respectively.

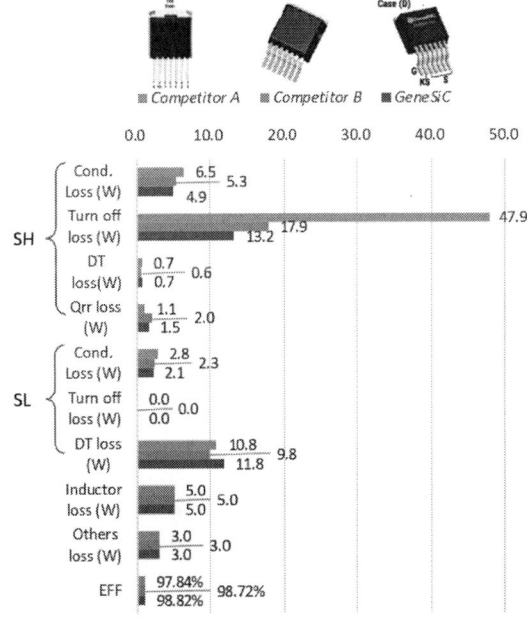

Fig. 11   ZVS SR-CCM buck circuit loss breakdown

Based on the above analysis, the loss breakdown under the condition of 600 kHz switching frequency with reference to the data specification in Table 3 is shown in Fig. 11. Since the upper devices SH is ZVS switching, its turn-on loss is zero;

the lower device SL is synchronous rectification, its switching losses is zero. A comparison of the data shows that GeneSiC and Competitor B's SiC devices turn off fast with low turn-off loss, and competitor A has the highest turn-off loss. In terms of conduction loss, GeneSiC's SiC has a small rate of change of resistance and has the lowest conduction loss.

4.3 Experimental Validation Analysis

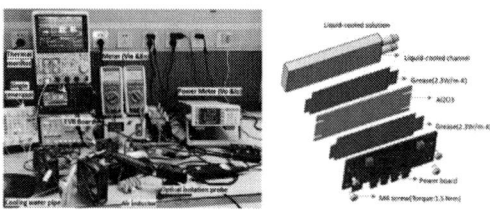

Fig. 12 Experimental platform and thermal solution

Fig. 12 shows the experimental platform and heat dissipation design. The output filter adopts air inductance to eliminate the influence of magnetic core on efficiency. Optical isolation probes are used to improve measurement accuracy. The system adopts water-cooled heat dissipation, PCB adopts copper-inlay technology, AL2O3 board provides insulation and efficient heat treatment of the device, and the thermal resistance of the system is close to 2.0 ℃/W.

	Ron（Ω）	Roff（Ω）
Competitor A	2	1
Competitor B	3	1
GeneSiC	6.9	2.2

Table 5 Drive Resistor Configuration

To ensure that each SiC device is in optimal driving condition, Table 5 lists the corresponding turn-on and turn-off resistance values. Fig. 13, Fig. 14 and Fig. 15 correspond to the measurement waveform and thermal test results of each device respectively. The dv/dt of competitor A, competitor B and G3R75MT12J are 54 V/ns, 87 V/ns and 82.9 V/ns respectively. Obviously, the latter two have fast turn-off speed and small turn-off loss, and competitor A has slow turn-off speed, the largest turn-off loss, and the highest thermal imaging temperature, which has reached 101 ℃. The experimental data are consistent with the analysis.

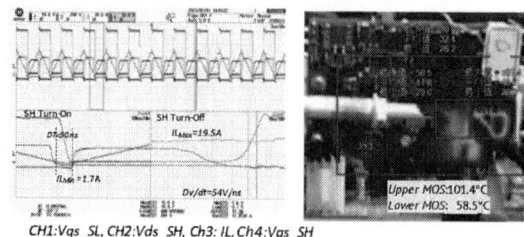

CH1:Vgs_SL, CH2:Vds_SH, Ch3: IL, Ch4:Vgs_SH

Fig. 13 Switching Waveforms and Thermal of Competitor A Devices

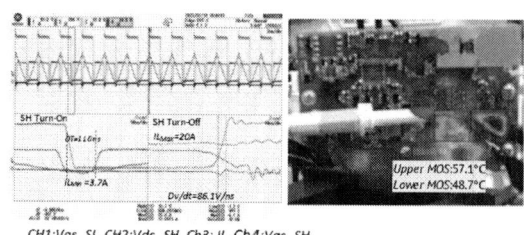

CH1:Vgs_SL, CH2:Vds_SH, Ch3: IL, Ch4:Vgs_SH

Fig. 14 Switching Waveforms and Thermal of Competitor B Devices

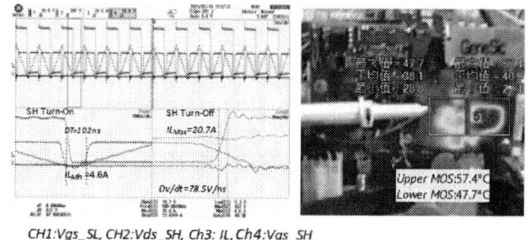

CH1:Vgs_SL, CH2:Vds_SH, Ch3: IL, Ch4:Vgs_SH

Fig. 15 Switching Waveforms and Thermal of G3R75MT12J Devices

Fig. 16 shows the comparison between calculated and measured efficiency. GeneSiC's device has the smallest on-resistance and fast switching characteristics, and the overall efficiency is up to 98.8% at 600kHz switching frequency, which is more suitable for high-frequency applications.

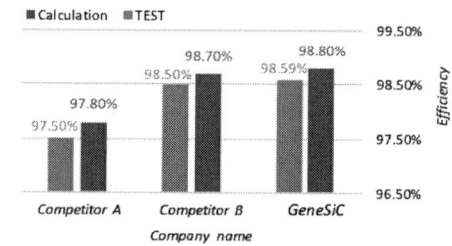

Fig. 16 Calculated vs Tested Efficiency

5 Conclusion

In this paper, a comparative study based on the ZVS SR-CCM Buck circuit is carried out to verify the feasibility of SiC in high-frequency applications, which is summarized as follows:

i. Drive loss is proportional to the switching frequency. Because SiC Qg is still relatively high, for high-frequency applications, the driver IC's own loss and heat dissipation must be considered.

ii. Switching turn-off loss is a key parameter that restricts the high-frequency application of SiC, which is directly proportional to the switching frequency. When designing, try to select the device specifications Eoff small. Engineering experience, in the device 30W total loss, turn-off loss should be controlled at about 15 W. The turn-off energy (excluding Eoss) at 600 kHz should meet Eoff<25 uJ.

iii. Thermal resistance RjC is also an important parameter of device selection, the same Rds_on and switching loss, small Rjc can meet the SiC device of higher power applications, and thus improve product reliability.

Based on the above analysis, SiC can also be applied to the design of high-power high-frequency soft-switching applications above 500kHz.

References

[1] Chen, Zheng, D. Boroyevich , and J. Li . "Behavioral comparison of Si and SiC power MOSFETs for high-frequency applications." Applied Power Electronics Conference & Exposition IEEE, 2013.

[2] Li Y, Ye Z, He N, et al. Efficiency improvement of a SiC-MOSFET 500 kHz ZVS inverter[C]//2016 IEEE 7th International Symposium on Power Electronics for Distributed Generation Systems (PEDG). IEEE, 2016: 1-8.

[3] Li, Bin, et al. "Bi-directional on-board charger architecture and control for achieving ultra-high efficiency with wide battery voltage range." 2017 IEEE Applied Power Electronics Conference and Exposition (APEC). IEEE, 2017.

[4] Chaturvedi, Mayank, et al. "Comparison of commercial planar and trench SiC MOSFETs by electrical characterization of performance-degrading near-interface traps." IEEE Transactions on Electron Devices 69.11 (2022): 6225-6230.

[5] Wei, Jiaxing, et al. "Review on the Reliability Mechanisms of SiC Power MOSFETs: A Comparison between Planar-Gate and Trench-Gate Structures." IEEE Transactions on Power Electronics (2023).

PCIM Europe 2024, 11– 13 June 2024, Nuremberg   DOI: 10.30420/566262385

Linearization of Drain-Source Capacitances for Antiserial Configurated SiC MOSFETs in High Frequency Solid State Switches

Lars Dresel[1], Vefa Karakasli[1], Gerd Griepentrog[1]

[1] Technical University of Darmstadt, Germany

Corresponding author: Lars Dresel, lars.dresel@lea.tu-darmstadt.de
Speaker: Lars Dresel, lars.dresel@lea.tu-darmstadt.de

Abstract

This paper introduces a simulation model for SiC MOSFETs with nonlinear output capacitance C_{ds} to forecast capacitive current in antiserial topologies. In applications like ac-ac converters for solid-state transformers, hybrid 4-quadrant switches for ac power conversion, or ac switches for particle accelerator cavities, the necessity arises to block ac voltages up to several kV with frequencies up to $10, MHz$. The paper presents a PLECS model where two identical SiC MOSFETs are arranged in an antiserial configuration. As SiC MOSFETs exhibit voltage-dependent behavior in their output capacitances $C_{ds} \sim \frac{1}{\sqrt{v_{ds}}}$, this characteristic is incorporated into the model. Consequently, this unique configuration leads to a physical linearization of the drain-source capacitance.

1 Introduction

Over the past few years, the role of semiconductor devices has expanded, particularly in the utilization of semiconductors as solid-state switches in DC and drive systems for controlling electric motors and general converters. For high voltage applications such as grid systems and particle accelerators, mechanical switches have been traditionally favored. However, with the recent developments in wide band-gap semiconductors, such as SiC MOSFETs, semiconductors are increasingly considered for these applications due to their enhanced switching frequency, power handling, thermal efficiency, and extended lifespan [1], [2]. These benefits allow to develop compact high-efficiency solid-state switches for high voltage applications.

The operational lifespan of vacuum relays in rigorous settings is limited, which averages six months. In such harsh environment they need to deal with high requirements like robustness against neutron radiation, high operating frequencies and ionizing particles. The mechanical components of vacuum relays have sparked inquiries into their potential replacement with semiconductor alternatives such as Si IGBTs or SiC MOSFETs. Especially SiC MOSFETs seem to have a good performance and higher robustness in particle accelerators due to

their small chip surface, which reduces the probability to be hit by stray neutrons [3]. To block AC voltages with semiconductors, several topologies have been proposed [4], [5]. A traditional approach in [4] involves configuring four diodes in a full-bridge arrangement parallel to a single transistor. This approach allows AC current flow through two diodes and the transistor during each half-wave. Another approach is to connect two transistors in antiserial configuration, as shown in Fig. 1, which allows blocking both negative and positive voltages.

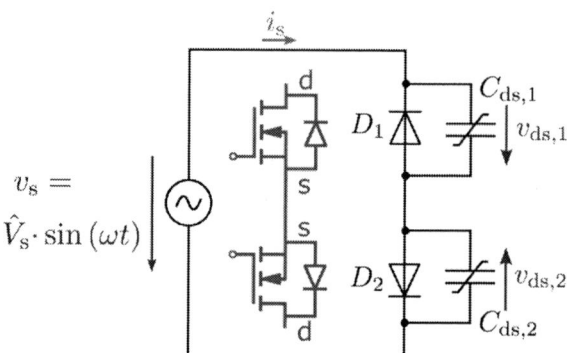

Fig. 1: Replacement circuit of antiserial connection of two MOSFETs

One significant advantage of the antiserial transistor configuration is the reduction in conduction losses. However, it requires precise control over the conduction sequence of transistors in both di-

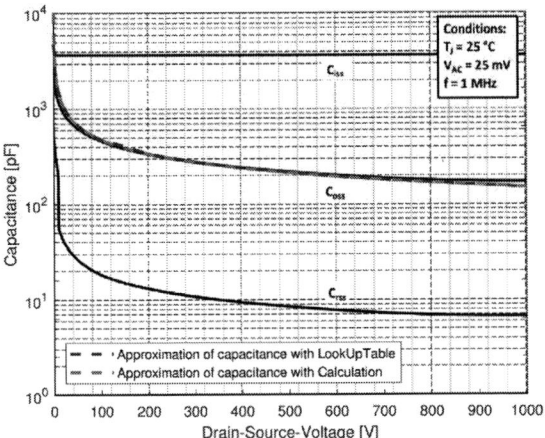

Fig. 2: Approximation of nonlinear output capacitance C_{ds} [8]

rections, a process that is notably complex, as discussed in [4]. The capacitive current i_s is capable in flowing in both directions through the parallel parasitic output capacitance, which is approximately equal to the drain-source capacitance $C_{oss} \approx C_{ds}$. In evaluating SiC MOSFETs with respect to their output and reverse transfer capacitances, C_{oss} and C_{rss} respectively, it is observed that these capacitances exhibit nonlinear behavior, as discussed in [6] and [7]. This behavior is characterized by a decay with an increase in drain-source voltage v_{ds}, as illustrated in Fig. 2. At high frequencies, a high capacitive current is observed, if the MOSFETs are in OFF-state. Due to the antiserial configuration of the MOSFETs, the current follows a linear relationship in response to a sinusoidal voltage v_{ds}, despite the anticipated nonlinearity of the output capacitance C_{oss}. To grasp the characteristics of SiC MOSFETs in design, a comprehensive nonlinear MOSFET model is indispensable.

The initial part of the paper explains the theoretical aspects related to the phenomenon of nonlinear capacitances and a calculation of the linearization for antiserial connection. In Section 3, the calculation is verified in a simulation. Moreover, a setup is presented and a measurement is taken. The comparison between simulation and measurement results is provided in Section 4, which verifies the analysis and modeling. Finally, the conclusion is drawn in Section 5.

2 Phenomenon of Nonlinearities in SiC MOSFETs and calculation of the linearization

Si and SiC MOSFETs output capacitances C_{oss} have a nonlinear characteristic, which varies with the drain-source voltage v_{ds}. However, comparing Si and SiC MOSFETs, the phenomenon is more clearly visible for SiC MOSFET. Different layers of doped regions cause parasitic capacitances in the MOSFET junctions, such as the gate-drain capacitiance (MILLER-Capacitance) C_{gd}, the gate-source capacitance C_{gs} and the drain-source capacitance C_{ds}. The output and the MILLER capacitances C_{oss} and C_{rss} are strongly dependent on the drain-source voltage v_{ds}. Since the drain-source current i_s is mainly caused by C_{ds}, the focus is only on this capacitance. As explained in [2], C_{ds} is caused by the n-doped region and the source connection and varies with the depletion region width W_{dsj} of the drain body junction. In the realm of capacitance, there exist various definitions that are utilized in different manners, occasionally leading to misapplication, particularly when describing nonlinear capacitances. The Differential Capacitance $C_{ds,I}$ is often associated with "small signal behavior," whereas the magnitude or amplitude capacitance $C_{ds,Q}$ is employed to establish the connection between voltage, stored charge, and capacitance. The differential capacitance $C_{ds,I}$ can be assumed as followed:

$$C_{ds,I} = \frac{A_{dsj} \cdot \varepsilon_s}{W_{dsj}} \tag{1}$$

$$W_{dsj} = \sqrt{\frac{2\varepsilon_s \left(v_{ds} + V_{bi}\right)}{qN_b}} \tag{2}$$

where A_{dsj} is the area of the drain-body junction and ε_s the semiconductors dielectric constant. W_{dsj} itself depends on the applied drain-source voltage v_{ds}, where V_{bi} is the built-in potential, q the fundamental electronic charge and N_b the base dopant density. For $v_{ds} = 0$, C_0 is calculated in Eq. (3). Moreover, for very high voltages $v_{ds} >> V_{bi}$, $C_{ds,I}$ can be simplified as given in Eq. (4):

$$C_0 = A_{dsj}\sqrt{\frac{qN_b\varepsilon_s}{2V_{bi}}} \tag{3}$$

$$C_{ds,I} = \frac{C_0}{\sqrt{1 + \frac{v_{ds}}{V_{bi}}}} \approx C_0 \cdot \sqrt{\frac{V_{bi}}{v_{ds}}} \tag{4}$$

As explained in [1] the electron density of the region away from the gate oxide decreases, causing a smaller value of N_b. In total the effective value of $C_{ds,I}$ is mainly caused by the smaller area A_{dsj} and N_b, when the MOSFET is turned off, leading to its characteristic nonlinearity.

The nonlinear differential capacitance data of the SiC MOSFET C2M0045170D by the manufacturer Cree Wolfspeed, is depicted in the Fig. 2. All those capacitances are measured by the company with a "small signal approach" and are therefore differential capacitances, which means:

$$C_{ds,I}(t) = \frac{dQ_{ds}(t)}{dv_{ds}} \quad (5)$$

From that, the charge Q_{ds} being stored in the capacitor can be derived by taking Eq. 4:

$$\begin{aligned}
Q_{ds}(t) &= \int C_{ds,I}(t) \cdot dv_{ds} \\
&= \int C_0 \sqrt{V_{bi}} \cdot v_{ds}^{-\frac{1}{2}} \cdot dv_{ds} \\
&= 2 \cdot C_0 \cdot \sqrt{V_{bi} \cdot v_{ds}} \quad (6)
\end{aligned}$$

This can be used to calculate the drain-source voltage v_{ds}, which depends on the charge Q_{ds}:

$$v_{ds} = \frac{1}{V_{bi}} \cdot \left(\frac{Q_{ds}(t)}{2 \cdot C_0} \right)^2 \quad (7)$$

Using Eq. (6), the magnitude capacitance can be expressed as:

$$\begin{aligned}
C_{ds,Q}(v_{ds}) &= \frac{Q_{ds}}{v_{ds}} = \frac{2 \cdot C_0 \cdot \sqrt{V_{bi} \cdot v_{ds}}}{v_{ds}} \\
&= 2 \cdot C_0 \cdot \sqrt{\frac{V_{bi}}{v_{ds}}} = 2 \cdot C_{ds,I} \quad (8)
\end{aligned}$$

This means the magnitude capacitance for this case is twice as much as the differential capacitance. It is essential to note that this consideration is applicable exclusively to capacitances characterized by a $\frac{1}{\sqrt{v_{ds}}}$ dependency.

In the next step, a sinusoidal voltage v_s is applied to two antiserial configurated SiC MOSFETs, as indicated in Fig. 1. The voltage can be considered as $v_s = v_{ds,1} - v_{ds,2}$. Due to the body diodes of the MOSFETs there will be a clamping of the drain-source voltages, which cannot be negative:

$$v_{ds,1} \geq 0 \qquad v_{ds,2} \geq 0$$

Upon the initial increase of v_s, one of the previously uncharged capacitors, for instance

$C_{ds,I,1}$, is charged through the counteracting body diode, specifically D_2. Following this initial charge, a continuous charge transfer occurs between the two capacitors, leading to both diodes becoming reverse-biased due to the positivity of $v_{ds,1}$ and $v_{ds,2}$. For steady state, the current is:

$$i_s = \frac{dQ_{ds,1}}{dt} = -\frac{dQ_{ds,2}}{dt} \quad (9)$$

At steady state, the rates of charging and discharging between the capacitors become equivalent, leading to a constant total charge across the antiserial configuration of two MOSFETs, as given:

$$Q_{ds,1} + Q_{ds,2} = Q_{ds,max} = \text{const.} \quad (10)$$

The maximum charge $Q_{ds,max}$ can be calculated by using Eq. (6) and is reached when the maximum voltage is attained. This occurs, whenever v_{ds} reaches the peak amplitude of v_s at $v_{ds} = \hat{V}_s$:

$$Q_{ds,max}(v_{ds} = \hat{V}_s) = 2 \cdot C_0 \cdot \sqrt{V_{bi} \cdot \hat{V}_s} \quad (11)$$

If the applied voltage v_s is zero, each capacitor stores the same charge $\frac{Q_{ds,max}}{2}$, which results in an equal voltage across both capacitors. Equation (7) is now used to calculate both drain-source voltages $v_{ds,1}$ and $v_{ds,2}$ for $v_s = 0$:

$$\begin{aligned}
v_{ds,1}(v_s = 0) &= v_{ds,2}(v_s = 0) \\
&= \frac{1}{V_{bi}} \left(\frac{Q_{ds}}{2 \cdot C_0} \right)^2 \\
&= \frac{1}{V_{bi}} \left(\frac{Q_{ds,max}}{2 \cdot 2 \cdot C_0} \right)^2 \\
&= \frac{1}{V_{bi}} \left(\frac{2 \cdot C_0 \cdot \sqrt{V_{bi} \cdot \hat{V}_s}}{2 \cdot 2 \cdot C_0} \right)^2 \\
&= \frac{\hat{V}_s}{4} \quad (12)
\end{aligned}$$

Thus, at $v_s = 0$, the charges have to be equal in both capacitors, as shown in Table 1:

v_s	$v_{ds,1}$	$Q_{ds,1}$	$v_{ds,2}$	$Q_{ds,2}$
$+\hat{V}_s$	$\hat{V}_s$	$Q_{ds,max}$	0	0
$-\hat{V}_s$	0	0	$\hat{V}_s$	$Q_{ds,max}$
0	$\frac{\hat{V}_s}{4}$	$\frac{Q_{ds,max}}{2}$	$\frac{\hat{V}_s}{4}$	$\frac{Q_{ds,max}}{2}$

Tab. 1: Drain-Source charges depending on applied voltage

In order to derive general equations for v_s, the charges are substituted into the voltages $v_\mathrm{s} = v_{\mathrm{ds},1} - v_{\mathrm{ds},2}$:

$$
\begin{aligned}
v_\mathrm{s} = v_{\mathrm{ds},1} - v_{\mathrm{ds},2} &= \frac{1}{V_\mathrm{bi}} \left(\frac{Q_{\mathrm{ds},1}}{2 \cdot C_0} \right)^2 - \frac{1}{V_\mathrm{bi}} \left(\frac{Q_{\mathrm{ds},2}}{2 \cdot C_0} \right)^2 \\
&= \frac{Q_{\mathrm{ds},1}^2 - Q_{\mathrm{ds},2}^2}{V_\mathrm{bi} \cdot (2 \cdot C_0)^2}
\end{aligned} \tag{13}
$$

With Eq. (10) and (11) a voltage equation depending on the charges $Q_{\mathrm{ds},1}$ or $Q_{\mathrm{ds},2}$ is derived:

$$
\begin{aligned}
v_\mathrm{s} &= \frac{Q_{\mathrm{ds},1}^2 - (Q_{\mathrm{ds,max}} - Q_{ds,1})^2}{V_\mathrm{bi} \cdot (2 \cdot C_0)^2} \\
&= \frac{Q_{\mathrm{ds},1}^2 - (Q_{\mathrm{ds,max}}^2 - 2 \cdot Q_{\mathrm{ds},1} \cdot Q_{\mathrm{ds,max}} + Q_{\mathrm{ds},1}^2)}{V_\mathrm{bi} \cdot (2 \cdot C_0)^2} \\
&= \frac{Q_{\mathrm{ds},1}^2 - Q_{\mathrm{ds,max}}^2 + 2 \cdot Q_{\mathrm{ds},1} \cdot Q_{\mathrm{ds,max}} - Q_{\mathrm{ds},1}^2}{V_\mathrm{bi} \cdot (2 \cdot C_0)^2} \\
&= \frac{2 \cdot Q_{\mathrm{ds},1} \cdot Q_{\mathrm{ds,max}} - Q_{\mathrm{ds,max}}^2}{V_\mathrm{bi} \cdot (2 \cdot C_0)^2} \\
&= \frac{Q_{\mathrm{ds,max}}^2}{V_\mathrm{bi} \cdot (2 \cdot C_0)^2} \left(2 \cdot \frac{Q_{\mathrm{ds},1}}{Q_{\mathrm{ds,max}}} - 1 \right)
\end{aligned} \tag{14}
$$

The first part of Eq. (14) is equal to $\hat{V}_\mathrm{s}$, which is extracted from Eq. (11) as $\hat{V}_\mathrm{s} = \frac{Q_{\mathrm{ds,max}}^2}{V_\mathrm{bi} \cdot (2 \cdot C_0)^2}$. Thus, it can be further simplified to determine the voltage v_s depending on the charges $Q_{\mathrm{ds},1}$ or $Q_{\mathrm{ds},2}$:

$$
v_\mathrm{s} = \hat{V}_\mathrm{s} \left(2 \cdot \frac{Q_{\mathrm{ds},1}}{Q_{\mathrm{ds,max}}} - 1 \right) \tag{15}
$$

$$
Q_{\mathrm{ds},1} = \frac{Q_{\mathrm{ds,max}}}{2} \left(1 + \frac{v_\mathrm{s}}{\hat{V}_\mathrm{s}} \right) \tag{16}
$$

$$
\begin{aligned}
Q_{\mathrm{ds},2} &= Q_{\mathrm{ds,max}} - Q_{\mathrm{ds},1} \\
&= \frac{Q_{\mathrm{ds,max}}}{2} \left(1 - \frac{v_\mathrm{s}}{\hat{V}_\mathrm{s}} \right)
\end{aligned} \tag{17}
$$

As shown in these equations, the charges are directly proportional to v_s. The relationship between the current i_s and the voltage v_s is determined by substituting Eq. (16) or (17) into Eq. (9) as given:

$$
\begin{aligned}
i_\mathrm{s} &= \frac{\mathrm{d}Q_{\mathrm{ds},1}}{\mathrm{d}t} = -\frac{\mathrm{d}Q_{\mathrm{ds},2}}{\mathrm{d}t} \\
&= \frac{Q_{\mathrm{ds,max}}}{2 \cdot \hat{V}_\mathrm{s}} \cdot \frac{\mathrm{d}v_\mathrm{s}}{\mathrm{d}t} = C_0 \cdot \sqrt{\frac{V_\mathrm{bi}}{\hat{V}_\mathrm{s}}} \cdot \frac{\mathrm{d}v_\mathrm{s}}{\mathrm{d}t}
\end{aligned}
$$

$$
\boxed{i_\mathrm{s} = C_{\mathrm{ds,I}}(\hat{V}_\mathrm{s}) \cdot \frac{\mathrm{d}v_\mathrm{s}}{\mathrm{d}t}} \tag{18}
$$

The total capacitance is only dependent on the applied voltage amplitude $\hat{V}_\mathrm{s}$, which stays constant for sinusoidal voltages. It is evident, that the current indicates a linear capacitive behavior. Thus, the current amplitude $\hat{I}_\mathrm{s}$ $\left(i_\mathrm{s} = \hat{I}_\mathrm{s} \cdot \sin(\omega t - 90°) \right)$ is constant.

Substituting the differential capacitance with Eq. (4), the current i_s is:

$$
\begin{aligned}
i_\mathrm{s} &= C_0 \cdot \sqrt{\frac{V_\mathrm{bi}}{\hat{V}_\mathrm{s}}} \cdot \frac{\mathrm{d}v_\mathrm{s}}{\mathrm{d}t} \\
&= C_0 \cdot \sqrt{\frac{V_\mathrm{bi}}{\hat{V}_\mathrm{s}}} \cdot \omega \cdot \hat{V}_\mathrm{s} \cdot \cos(\omega t) \\
&= C_0 \cdot \sqrt{V_\mathrm{bi} \hat{V}_\mathrm{s}} \cdot \omega \cdot \cos(\omega t)
\end{aligned} \tag{19}
$$

The general expressions for the voltages $v_{\mathrm{ds},1}$ and $v_{\mathrm{ds},2}$ are hereby defined as:

$$
\begin{aligned}
v_{\mathrm{ds},1} &= \frac{1}{V_\mathrm{bi}} \left(\frac{Q_{\mathrm{ds},1}}{2 \cdot C_0} \right)^2 \\
&= \frac{1}{V_\mathrm{bi}} \left[\frac{\frac{Q_{\mathrm{ds,max}}}{2} \left(1 + \frac{v_\mathrm{s}}{\hat{V}_\mathrm{s}} \right)}{2 \cdot C_0} \right]^2 \\
&= \frac{1}{V_\mathrm{bi}} \left[\frac{\sqrt{V_\mathrm{bi} \cdot \hat{V}_\mathrm{s}} \left(1 + \frac{v_\mathrm{s}}{\hat{V}_\mathrm{s}} \right)}{2} \right]^2 \\
&= \frac{\hat{V}_\mathrm{s}}{4} \left(1 + \frac{v_\mathrm{s}}{\hat{V}_\mathrm{s}} \right)^2
\end{aligned} \tag{20}
$$

$$
v_{\mathrm{ds},2} = \frac{\hat{V}_\mathrm{s}}{4} \left(1 - \frac{v_\mathrm{s}}{\hat{V}_\mathrm{s}} \right)^2 \tag{21}
$$

The voltages, $v_{\mathrm{ds},1}$ and $v_{\mathrm{ds},2}$, consistently maintain positivity owing to the presence of opposing diodes. They signify a quadratic relationship with v_s attaining their zenith whenever v_s reaches its maximum or minimum peak amplitude $v_\mathrm{s} = \pm \hat{V}_\mathrm{s}$.

3 PLECS model

To validate the analytical findings, a non-linear MOSFET model is proposed within PLECS. According to the schematic in Fig. 1 the drain-source capacitor is voltage dependent as shown in the datasheet in Fig. 2. Figure 3 shows the submodel of a nonlinear capacitor which approximates its capacitance.

In the proposed model, the voltage v_ds is utilized as input for a 1-dimensional look-up table (1D-LUT). The capacitor values for the LUT are predetermined and can be either acquired from the datasheet or calculated using Eq. (4). The result of the LUT,

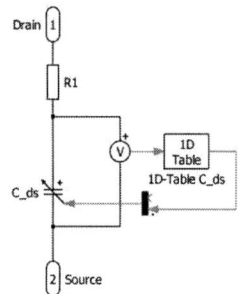

Fig. 3: Submodel of nonlinear capacitor

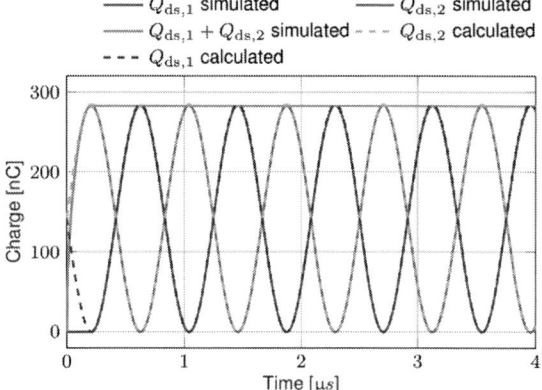

Fig. 4: Simulated and calculated charges

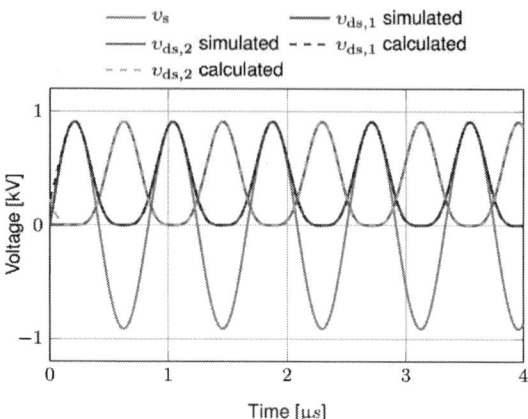

Fig. 5: Simulated and calculated drain-source voltages

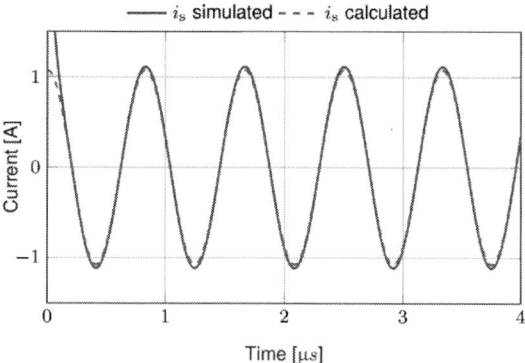

Fig. 6: Simulated and calculated current

obtained via these two approaches, are depicted in Fig. 2. Afterwards the value is given to a variable capacitor. A series resistor with a very low value of $1\,\text{m}\Omega$ is connected to the capacitor in order to solve the differential equation. It has been observed that simulations require a longer duration when the LUT values are determined analytically. As a result, the LUT derived from datasheet values is preferred for use in simulations. A simulation is run with $\hat{V}_s = 0.91\,\text{kV}$, $V_{\text{bi}} = 1\,\text{V}$ and $f = 1.2\,\text{MHz}$. Using Eq. (4), (11) and (18) the differential capacitance, the maximum charge and the current amplitude can be calculated:

$$Q_{\text{ds,max}} = 2 \cdot 4.72\,\text{nF} \cdot \sqrt{1\,\text{V} \cdot 910\,\text{V}} = 285\,\text{nC}$$

$$C_{\text{ds,I}}(\hat{V}_s = 910\,\text{V}) = \frac{4.72\,\text{nF}}{\sqrt{\frac{910\,\text{V}}{1\,\text{V}}}} = 156\,\text{pF}$$

$$\hat{I}_s = 2 \cdot \pi \cdot 1.2\,\text{MHz} \cdot 156\,\text{pF} \cdot 910\,\text{V} = 1.07\,\text{A}$$

The analytical and simulated outcomes, corresponding to the circuit depicted in Fig.1 are shown in Fig. 4, 5 and 6.

Figure 4 shows the exchange of the charges $Q_{\text{ds,1}}$ and $Q_{\text{ds,2}}$. When one of the capacitor is discharged, the other capacitor is charged. Also, the plot shows that the sum of the charges stays constant, as derived in in Eq. (11). A comparison between the simulated and analytically derived charges reveals a slight mismatch. Due to small interpolations of the values of the capacitance, some inaccuracies occur, which cause a small shift of the charges. Figure 5 displays the calculated and simulated drain-source voltages, consistently remaining positive as presumed in the preceding section. The entire voltage v_s is consistently concentrated on a single capacitor. In steady state, it is evident that the analytically calculated and simulated voltages align perfectly with each other. Figure 6 shows the calculated and simulated source currents, which match each other in steady state. At the beginning of the simulation, the current is very high due to transients, which are neglected in the calculations. As defined in Eq. (18), the current indicates a linear dependency on the voltage amplitude $\hat{V}_s$ for two nonlinear capacitors connected in series.

PCIM Europe 2024, 11– 13 June 2024, Nuremberg   DOI: 10.30420/566262385

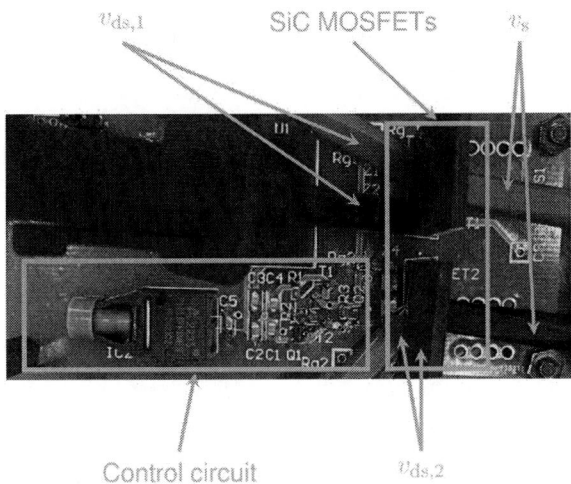

Fig. 7: Measurement Setup

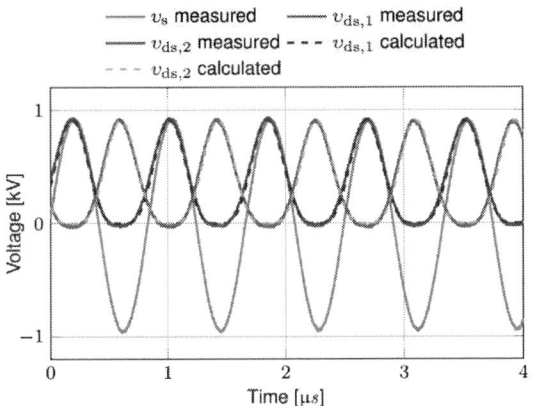

Fig. 8: Measured and calculated voltages, $\hat{V}_s = 910\,\text{V}$

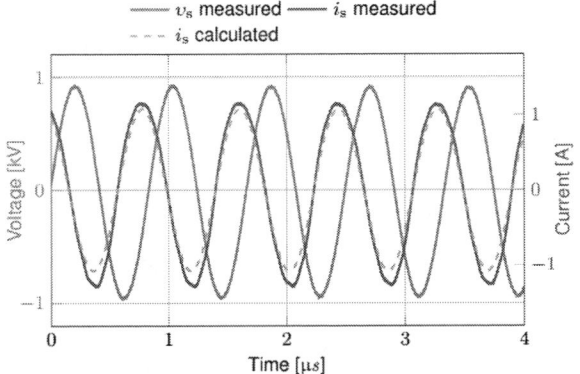

Fig. 9: Measured and calculated current, $\hat{V}_s = 910\,\text{V}$

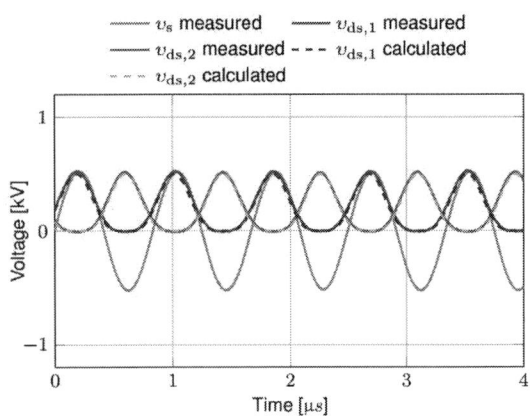

Fig. 10: Measured and calculated voltages, $\hat{V}_s = 500\,\text{V}$

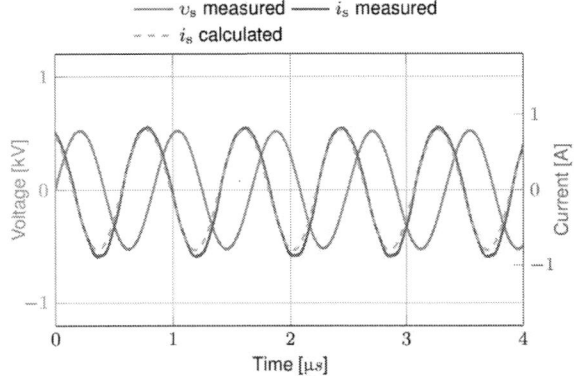

Fig. 11: Measured and calculated current, $\hat{V}_s = 500\,\text{V}$

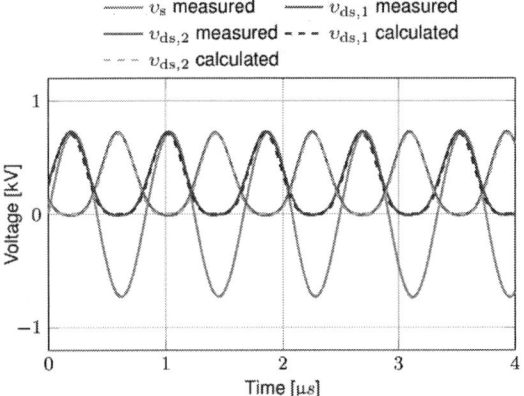

Fig. 12: Measured and calculated voltages, $\hat{V}_s = 700\,\text{V}$

4 High voltage/High Freqeuncy measurement

The verification of the analytical model and non-linear MOSFET model necessitates the construc-tion of a test setup, detailed in Fig. 7. The measure-ment results are only compared with the analytical model as the previous section demonstrates a neg-ligible difference between the analytical results and simulation results. To generate the high-frequency voltage source, a resonant circuit from a particle

2742

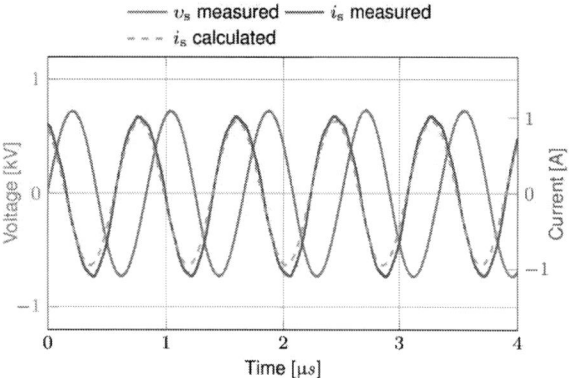

Fig. 13: Measured and calculated current, $\hat{V}_s = 700\,\text{V}$

accelerator was employed.

In this experimental configuration, a sinusoidal voltage v_s is applied to the drain-pins of the MOSFETs. Measurements of the drain-to-source voltages, $v_{\text{ds},1}$ and $v_{\text{ds},2}$, are taken from the corresponding pins of the MOSFETs when they are in OFF-state. The current i_s is measured in the resonant circuit. The experimental and analytical model results, illustrated in Fig. 8 and 9 is conducted with $\hat{V}_s = 0.91\,\text{kV}$, $V_{\text{bi}} = 1\,\text{V}$ and $f = 1.2\,\text{MHz}$. To observe variations in i_s, the applied voltages were changed to $\hat{V}_s = 500\,\text{V}$ and $\hat{V}_s = 700\,\text{V}$ for the same frequency, as depicted in Fig. 10 to Fig. 13.

It is evident that the analytical and MOSFET models are in close agreement with the experimental results, accurately predicting the expected current. Moreover, the findings depicted in Fig. 9, 11 and 13 validate the relationship delineated in Eq. (19), demonstrating the proportionality of the current to the square root of the voltage amplitude $\hat{V}_s$.

5 Conclusion

In this paper, the linearization phenomenon of two antiserially configured SiC MOSFETs with nonlinear drain-source capacitance is explored. A mathematical derivation that predicts the expected current i_s for sinusoidal voltages is presented. Subsequently, a PLECS model is constructed to validate these calculations. Both simulation and experimental results demonstrate the effectiveness of the calculations in estimating capacitance, stored charge, voltages, and current within this topology.

This configuration finds application in multistage solid-state switches or in [9], as depicted in Fig. 14, where the mathematical approaches aid in estimating capacitance and current. These derivations

offer a streamlined method for calculating circuits employing antiserially connected MOSFETs. However, it is crucial to note that these calculations hold true only in the absence of additional parasitic capacitances. Measurement data indicates that such parasitics can affect voltage balance across MOSFETs, thus altering voltage distribution.

Furthermore, the assumption of linearized capacitance applies solely to MOSFETs featuring a $\propto \frac{1}{\sqrt{v_{\text{ds}}}}$ drain-source capacitance. MOSFETs with different output capacitances might produce sinusoidal currents with harmonics. Nevertheless, this mathematical model remains valuable for high voltage applications, enabling the development of simulation models and prediction of capacitive currents.

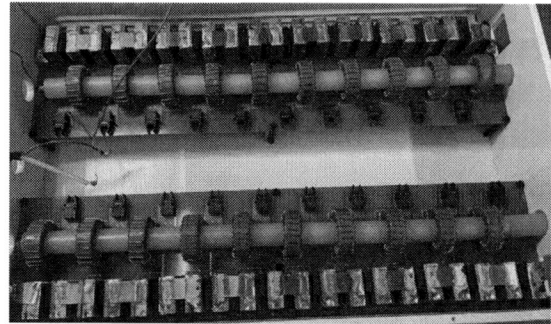

Fig. 14: Design of a High Voltage/High Frequency Solid State Switch

References

[1] K. Chen, Z. Zhao, L. Yuan, T. Lu, and F. He, "The impact of nonlinear junction capacitance on switching transient and its modeling for sic mosfet," *IEEE Transactions on Electron Devices*, vol. 62, no. 2, pp. 333–338, 2015. DOI: 10.1109/TED.2014.2362657.

[2] B. J. Baliga, *Gallium nitride and silicon carbide power devices*, New Jersey, 2016.

[3] M. Khani, G. Griepentrog, A. Sokolov, and E. Kozlova, "Radiation hardness of si- and sic-power-mosfets in particle accelerator environments," in *PCIM Europe 2022; International Exhibition and Conference for Power Electronics, Intelligent Motion, Renewable Energy and Energy Management*, 2022, pp. 1–7. DOI: 10.30420/565822163.

[4] G. Venkataramanan and N. Kogalur, "A hybrid 4-quadrant switch for ac power conversion," in *2019 IEEE Energy Conversion Congress and Exposition (ECCE)*, 2019, pp. 5487–5493. DOI: 10.1109/ECCE.2019.8912631.

[5] W. Khan, J. A. Dar, K. S. Parihar, and M. K. Pathak, "Single-stage isolated ac/ac converter for solid-state transformer," in *2022 IEEE Global Conference on Computing, Power and Communication Technologies (GlobConPT)*, 2022, pp. 1–6. DOI: 10.1109/GlobConPT57482.2022.9938301.

[6] M. Hayati, S. Roshani, M. K. Kazimierczuk, and H. Sekiya, "A class-e power amplifier design considering mosfet nonlinear drain-to-source and non-linear gate-to-drain capacitances at any grading coefficient," *IEEE Transactions on Power Electronics*, vol. 31, no. 11, pp. 7770–7779, 2016. DOI: 10.1109/TPEL.2015.2512928.

[7] K. Yonekura, X. Wei, T. Nagashima, H. Sekiya, and T. Suetsugu, "Class-d inverter with mosfet nonlinear parasitic capacitance," in *2015 International Conference on Renewable Energy Research and Applications (ICRERA)*, 2015, pp. 944–948. DOI: 10.1109/ICRERA.2015.7418549.

[8] *Silicon carbide power mosfet, c2m mosfet technology, n-channel enhancement mode.* C2M0045170D, CREE Wolfspeed, 2016.

[9] W. Xu and A. Q. Huang, "15kv/50a sic ac switch based on series connection of 1.7kv mosfets," in *2022 IEEE Energy Conversion Congress and Exposition (ECCE)*, 2022, pp. 1–6. DOI: 10.1109/ECCE50734.2022.9947463.

PCIM Europe 2024, 11– 13 June 2024, Nuremberg   DOI: 10.30420/566262386

Effects of Non-killer Defects on SiC MOSFET Short-Circuit Ruggedness and Reliability

Sara Kuzmanoska[1], Prajeesh Karimbankara [1], Kaone Bogopa[1], Philipp Rehlaender[1], Swapna Sunkari[2], Hrishikesh Das[2]

[1] onsemi, Germany
[2] onsemi, USA

Corresponding author: Sara Kuzmanoska, sara.kochoska@onsemi.com
Speaker: Sara Kuzmanoska, sara.kochoska@onsemi.com

Abstract

The manufacturing process of Silicon Carbide (SiC) devices, from the substrate to the epitaxial growth and subsequent fabrication stages, introduces various defects [1-2]. These defects are classified into killer and non-killer defects (NKDs). NKDs deviate from the perfect crystal structure and consequently impact the performance and reliability of the device [3]. This paper concentrates on the different types of NKD defects found in 900 V SiC MOSFETs, including basal plane dislocations (BPDs), stacking faults (SFs), partials and micropipes (MPs). Furthermore, a set of devices that do not contain any NKDs (considered as good) are also included for the comparison purpose. The devices undergo electrical characterization before and after being subjected to single and repetitive short-circuit (SC) testing. For the first time, this study establishes a link between the existence of non-killer defects (NKDs) in SiC devices and their influence on electrical performance, short-circuit (SC) ruggedness and reliability. The critical short-circuit energy (E_{SC}) for the specific NKDs tends to be higher on average and exhibits a narrower variation compared to the devices without NKDs. On the other hand, the group labeled as 'SFs' shows higher short-circuit peak currents (I_{Dmax}). Additionally, there is a noticeable variability in the short-circuit withstand time (t_{SCWT}) for both groups 'SFs' and 'good'.

1   Introduction

The performance of Silicon Carbide (SiC) MOSFETs under harsh conditions is critical for their widespread adoption [4-6]. A key factor is their ability to withstand short-circuit conditions as SiC technology provides a significantly shorter short-circuit withstand time (t_{SCWT}) compared to silicon (Si) technology [7]. Any significant variation or outliers in the short-circuit withstand time poses a challenge in ensuring reliable device limits. The short-circuit (SC) ruggedness of SiC devices is influenced not only by certain device parameters like on-state resistance ($R_{DS,on}$) and threshold voltage (V_{th}), but also by setup parameters such as stray inductance (L_{stray}) and the external gate resistance $R_{G,ext}$ [8,9].

Despite extensive efforts to reduce defects, SiC devices still contain numerous defects. 'Killer defects', as they are known, cause the devices to fail the electrical tests. However, the SiC devices with non-killer defects pass all electrical and reliability tests such as unclamped inductive switching (UIS) or the 1000-hour high-temperature reverse bias

(HTRB) test [1]. As identified in [2], some defects, such as dislocation and stacking faults, can create leakage paths or decrease conductivity in the drift region. The exposure of SiC devices under the rigorous short-circuit (SC) conditions with very high current and voltage, as well as high temperature, may trigger some of these defects and impact device SC reliability.

In this study, an electrical characterization and SC reliability testing is conducted on 900 V SiC MOSFETs. These MOSFETs, packaged in TO-247-3L package and manufactured under identical process conditions are divided into six distinct groups based on the presence of NKDs. The presence of the NKDs in the epitaxial layer was detected by use of a SICA88. The first group, referred to as 'good' devices contains devices with no NKDs. The remaining groups are named after the presence of NKDs, such as group with basal plane dislocations (BPDs) being called 'BPDs', the group with stacking faults (SFs) being called 'SFs', the group with micropipes (MPs) being called 'MPs', and the group with partial stacking fault defects being called 'Partials'.

The process of electrical characterization is carried out using the B1505A power device analyzer from Keysight. The DC characterization for each group encompasses parameters such as the on-state resistance ($R_{DS,on}$), the threshold voltage (V_{th}), the breakdown voltage (BV), the forward voltage (V_F) and the reverse leakage current (I_{DSS}). These parameters can be linked to the existing defects in the devices.

In addition, devices that have been pre-characterized undergo testing under a single short-circuit (SC) test, type I, and repetitive SC. For these SC tests, an analysis is conducted on more than 80 samples, and the SC results are associated with the existing defect. Table 1 displays the total number of devices tested for each individual group and test.

Device Group	DC Characterization	Single SC Test	Repetitive SC Test
Good	30	10	5
BPDs	30	10	5
SFs	30	10	5
Partials	30	10	2
MPs	10	0	0

Table 1. Total number of tested devices per group

2 NKD Effects on SiC Devices

2.1 NKDs impact on SiC MOSFETs static parameters and performance

In Table 2 are elaborated the test conditions and limits for the *I-V* sweeps that are utilized for static characterization. Individual data points are derived for the correlation between each NKD group and static performance. Figure 1 illustrates the initial DC results extracted for each group, including the on-state resistance $R_{DS,on}$, threshold voltage V_{th}, body diode forward voltage V_F and reversed current leakage I_{DSS}.

Parameters	DC Test Conditions	Unit
I_{GSS}	V_{GS} = 20	(V)
V_{th}	V_{DS} = {0.1, 4, 10}	(V)
I_D-V_{DS}	V_{GS} = {15, 18, 20}	(V)
	I_D = 80	(A)
I_F-V_F	V_{GS} = {0, -3, -5}	(V)
BV	V_{GS} = -5	(V)
	I_D = 4	(mA)
I_{DSS}	V_{DS} = 900	(V)
	V_{GS} = -5	(V)

Table 2. Test conditions applied for DC characterization

The devices categorized as 'good', 'SFs' and 'BPDs' exhibit no difference in average on-resistance $R_{DS,on}$ = 8.45 mΩ. In contrast, the 'partials' and 'MPs' groups demonstrate 3% higher values. A similar pattern is observed for the V_{th}, where 'partials' and 'MPs' groups demonstrating 2-3% elevated threshold voltage values. Although micropipes have been documented in literature [10, 11] as harmful or as killer defects causing reliability problems for SiC devices, this study were not observe any specific issues in the devise containing micropipes. The presence of the NKDs does not significantly influence the electrical characterization parameters, nor can any of the defects be correlated to a device-to-device variation.

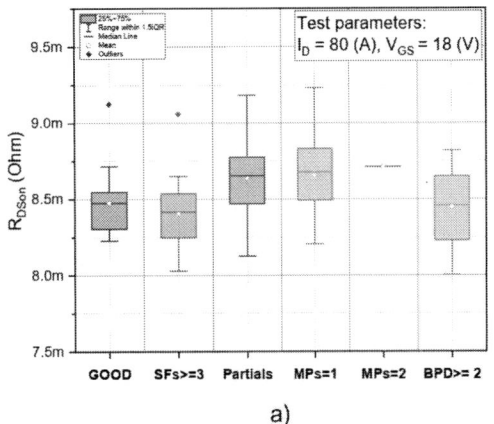

a)

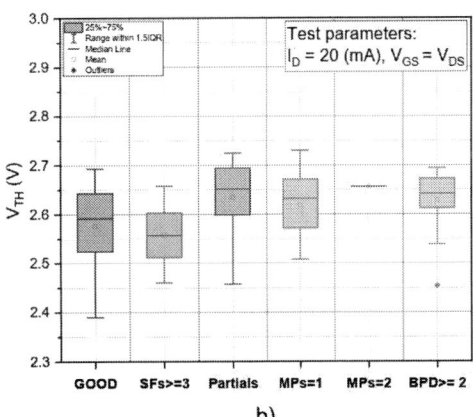

b)

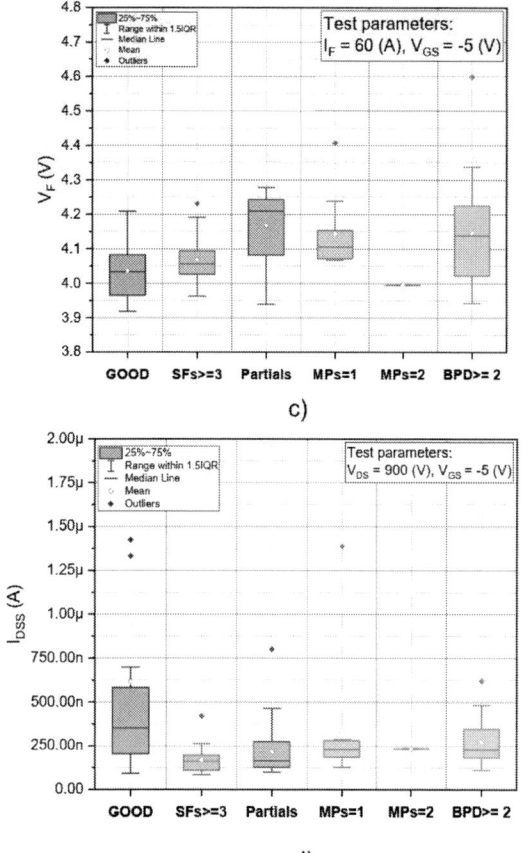

c)

d)

Fig. 1: DC characterization of six groups with and without NKD at room temperature: (a) on-state resistance, (b) threshold voltage, (c) body diode forward voltage and (d) reverse leakage current

2.2 NKDs impact on SiC MOSFETs short-circuit performance.

SiC devices from all groups, as indicated in Table 3 undergo testing under short-circuit conditions, specifically single SC Type I and repetitive SC. The testing methodology is outlined in Figure 2 and the conditions applied for the short-circuit testing are detailed in Table 3.

Parameters	SC Test Conditions	Unit
V_{DS}	500	(V)
V_{GS}	18	(V)
$R_{G\ (on/off)}$	10	(Ohm)
Temp	Ambient ~25	(°C)

Table 3. Test conditions applied for SC testing, applicable both single and repetitive SC test.

2.2.1 Single Short-Circuit Ruggedness

In the single SC test, the procedure commences with a pulse width of 1 us, which than incrementally is increased by 0.1 us steps until the point of failure. A pause of 20 seconds is applied between each SC pulse. The current and voltage conditions remain consistent with those specified in Table 3.

During the short circuit tests, the average SC peak current for the 'good', 'BPDs' and 'Partials' group is approximately 1.6 kA, whereas the 'SFs' group exhibits an average peak current that is roughly 5% higher.

When considering t_{SCWT}, the 'BPDs' and 'Partials' groups display longer durations compared to the 'SFs' and 'good' groups, with an average difference of approximately 10%.

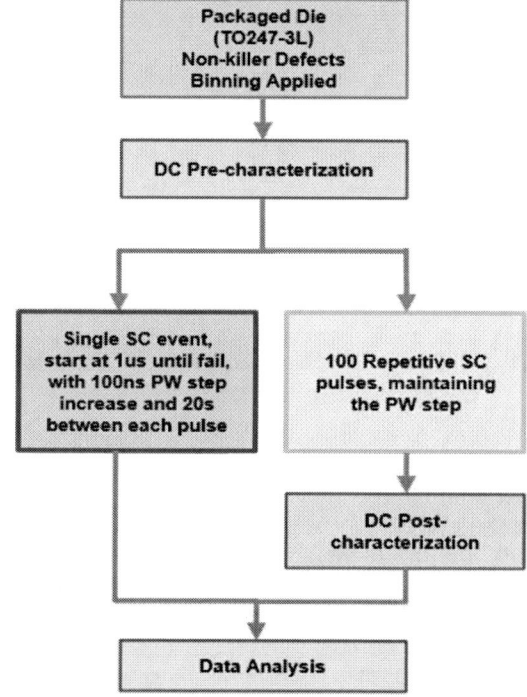

Fig. 2 The step-by-step testing procedure used for the single and the repetitive short-circuit tests

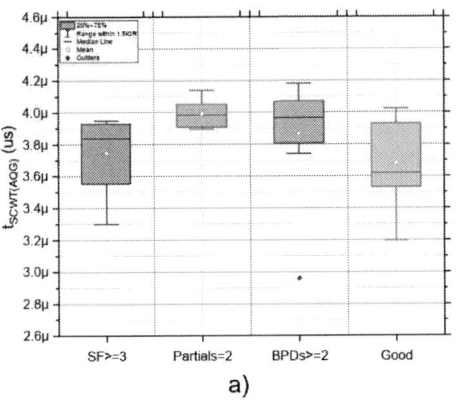

a)

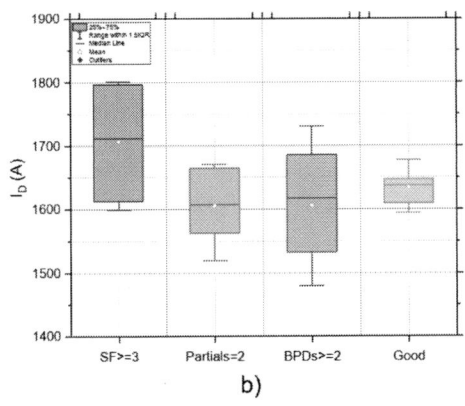

b)

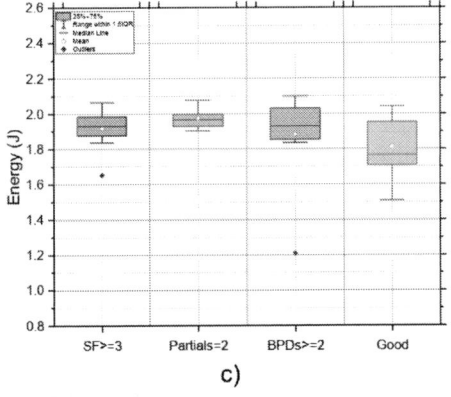

c)

Fig. 3 Single short-circuit results obtained on a group of SiC MOSFETs without NKD and three groups with NKDs respectively: (a) short-circuit withstand time (b) peak drain current during short-circuit test (c) maximum energy withstand capability during short-circuit

Moreover, devices in the 'good' group demonstrate an average short-circuit withstand energy (E_{SC}) of 1.8 J, along with a significant device-to-device variation. In contrast, the other NKD groups, namely 'BPDs', 'SFs' and 'Partials', exhibit an E_{SC} that is approximately 5% to 10% higher

compared to the 'good' devices. The experimental results reveal a substantial variation between devices, particularly in terms of SC peak current I_{D-max}, t_{SCWT} and a few outliers. The existence of NKDs does not seem to have a significant correlation with the short-circuit behavior of the SiC devices, which could potentially be due to the number of defects present. Although the 'good' group was specifically chosen for having no NKDs, there is still significant variability in the results. This variation could be attributed to the way the samples were selected. Further investigation is needed to determine whether this variation is systematic or random.

2.2.2 Repetitive Short-Circuit Reliability

Throughout the repetitive short-circuit testing, the approach and testing conditions remain consistent, as outlined in Fig. 2 and Table 3 respectively. Each group is subjected to 100 cycles of testing at energy levels around 1 J and 1.5 J, given that the average withstand Esc during the single SC is 1.8 J. The goal of the post-characterization is to detect any changes in static parameters, offering valuable insights into how the device behaves under sustained and repeated short-circuit conditions.

Repetitive short-circuits can result in permanent damage to SiC MOSFETs and cause a reduction in capacity due to physical and chemical alterations in the energy storage devices. An example waveform of the repetitive short-circuit testing is provided in Fig. 4.

In the results from the repetitive SC tests, there was no noticeable degradation in any of the static parameters. As depicted in Fig. 5, no degradation is observed for any of the devices subject to 100 cycles at SC energy levels of 1 J and 1.5 J.

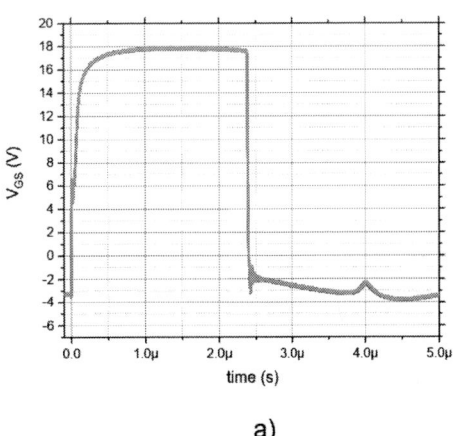

a)

PCIM Europe 2024, 11– 13 June 2024, Nuremberg DOI: 10.30420/566262386

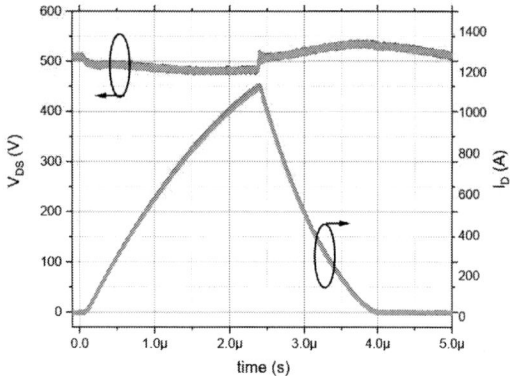

b)

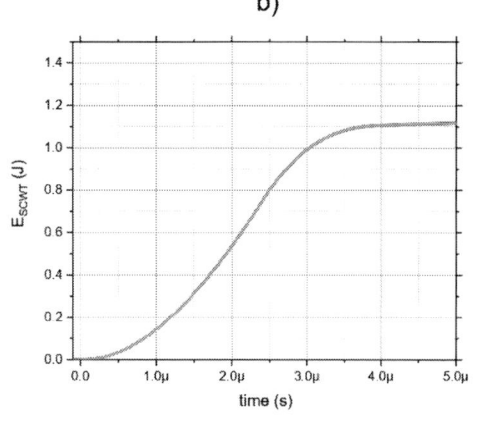

c)

Fig. 4 An example waveforms overlapped on top of each other from 100 repetitive SC cycle tests for: a) V_{GS}, b) V_{DS} and I_D and c) E_{SC}

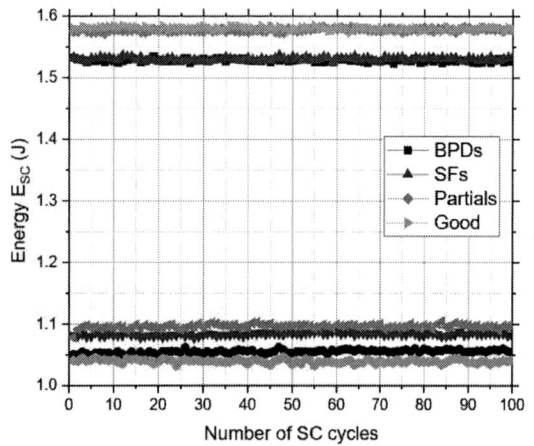

Fig. 5 Repetitive short-circuit results obtained on a group of SiC MOSFETs without NKD and three groups with NKDs respectively

3 Conclusion

During the manufacturing procedure, SiC MOSFETs with various NKDs were initially mapped. These SiC MOSFETs were classified based on the present defect into groups such as basal plane dislocations (BPD), stacking faults (SFs), micropipes (MPs), partial SFs (Partials) and (good) group containing no NKDs. Multiple samples from each group underwent measurement testing to assess the influence of the NKDs on the device's robustness and reliability.

The initial analysis involved the static characterization of selected SiC MOSFETs to evaluate changes in electrical parameters. During static characterization, the electrical properties show minimal variation compared to the reference 'good' MOSFETs.

For most of the electric parameters, a minor difference of 2-3% is observed, which is not significant enough to identify the presence of any defects in the devices. This observation is based on the number of defects present in the devices under study.

Single short circuit testing results indicate that the E_{SC} energy is approximately 5% to 10% higher on average for the groups with NKDs compared to the devices without any NKDs. Moreover, the groups with NKDs exhibit a narrower E_{SC} variation.

A wide variation is observed in peak current during the short-circuit for all groups. However, the groups adhere to the anticipated trend, with those experiencing higher I_{Dpeak} demonstrating a shorter t_{SCWT}. Specifically, the SF group exhibit higher peak currents and shorter t_{SCWT}, while the BPDs and Partials groups display a longer t_{SCWT}.

In the analysis of repetitive short circuit, all groups exhibit practically no degradation, suggesting that their robust performance under these conditions is not impacted by the presence of these specific NKDs.

References

[1] H. Das, et al. "Statistical Analysis of Killer and Non-Killer Defects in SiC and the Impacts to Device Performance", Material Science Forum Vol. 1004, pp 458-463, (2020)

[2] L. Li, at al. "Effect of wafer defects on electrical properties and yields of SiC Devices", 2021 Journal of Physics: Conf. Ser. 2033 012095

[3] T. Neyer, et al. "Is there a perfect SiC MosFETs Device on an imperfect crystal", 2021 IEEE International Reliability Physics Symposium (IRPS)

2749

[4] B. Shi, et al. "A review of silicon carbide MOSFETs in electrified vehicles: Application, challenges, and future development", IET Power Electronics, 2023

[5] T. Kimoto "Fundamentals, Commercialization, and Future Challenges of SiC Power Devices", IEEE International Meeting for Future of Electron Devices 2023

[6] J. An, et al. "Methodology for Enhanced Short-Circuit Capability of SiC MOSFETs", Proceedings of the 30th International Symposium on Power Semiconductor Devices & ICs, 2018

[7] J. O. Gonzalez, et al. "Benchmarking the robustness of Si and SiC MOSFETs: Unclamped inductive switching and short-circuit performance", Microelectronics Reliability Vol. 138 (2022)

[8] S. Kochoska, et al. "The Impact of Different Test Methodologies on Short-Circuit Ruggedness of SiC MOSFETs", PCIM Europe 2023, 09-11 May 2023

[9] C. Unger, et al. "Particularities of the Short-Circuit Operation and Failure Modes of SiC-MOSFETs", IEEE Journal of Emerging and Selected Topics in Power Electronics, Vol. 9, NO. 5, October 2021

[10] H. Wang, et al. "Micropipes in SiC Single Crystal Observed by Molten KOH Etching", Special Issue Wide Bandgap Semiconductor Materials and Devices, October 2021

[11] J.L. Liu, et al. "Methods for the reduction of the micropipe density in SiC single crystals", Journal of Material Science, Vol. 42 pages 6148–6152, 2007

PCIM Europe 2024, 11– 13 June 2024, Nuremberg   DOI: 10.30420/566262387

Dynamic Reverse Bias Test: Electro-Thermal Characterization of SiC MOSFETs

Giuseppe Mauromicale[1], Alessandro Sitta[1], Luciano Salvo[1], Michele Calabretta[1]

[1] STMicroelectronics, Catania, Italy

Corresponding author: Giuseppe Mauromicale, giuseppe.mauromicale@st.com
Speaker: Giuseppe Mauromicale, giuseppe.mauromicale@st.com

Abstract

Beyond the traditional static biased tests, novel dynamic electric assessments are being developed for wide band-gap devices and package solutions, to assess their reliability behaviour in the automotive environment. In the dynamic reverse bias (DRB) test, the power devices are dynamically stressed with very high slew rates of drain-source voltage for one thousand hours. The test evaluates the reliability at applicative-like conditions in terms of switching frequency and bus voltage. This paper presents a DRB test characterization: more in details, all the most relevant electrical and thermal aspects are both characterized. Hence, the voltage waveforms are assessed to ensure the compliance with guideline and the alignment with the application. Main test parameters, such as frequency, temperature and voltage, are correlated by dedicated analysis. Moreover, power losses due to device's capacitances are calculated by a straightforward approach. A thermal experimental method is developed to perform a benchmark.

1 Introduction

In automotive power electronic environment, harsh conditions can be possible, both in terms of ambient conditions (high temperature and/or humidity values) and electric parameters. Hence, reliability tests are of paramount importance to assess the behaviour of the power semiconductor devices after accelerated and/or stressful trials. Wide band-gap materials-based power devices, such as those based on silicon carbide (SiC), have superior electric and thermal properties in comparison to the ones based on silicon [1], [2]. For this reason, in recent years the power semiconductor devices based on SiC have been used in automotive field [3]–[5]. However, some issues can arise due to very high speed switching transients in this kind of devices [6]. For the above mentioned reasons, dedicated reliability tests are needed [7].

Among the others, there are the dynamic reverse bias (DRB) and dynamic gate stress (DGS) tests, both present in the AQG 324 guideline from the European Center for Power Electronics (ECPE) [8]. While DGS is related to a dynamic stress on the gate structure [9], the DRB test involves the time-varying drain-source high voltage. DRB in-

vestigates mainly the effect of electric quantities during a 1000 h stress. Additionally, the dynamic high humidity, high temperature reverse bias (dyn-H3TRB) triggers some effects, like corrosion and the dendritic growth, thanks to the combination of voltage, humidity and temperature. DRB is executed at discrete package level, with room values of temperature and humidity. This is a no-load test: the currents involved in the trial are due to charging and discharging of internal device structures [10]. Self-heating is due to the switching losses occurring at application-like switching frequency, due to the non-idealities of the device and of the circuit. DRB would be related to the blocking voltage-related robustness of the power device.

In this paper, a comprehensive electric and thermal characterization method is developed, in order to fully understand the proper setup of a DRB test on discrete packages. This is mandatory to guarantee that slew rate and peak voltages are compliant with AQG 324 and final applications. The paper is organized in the following way. Section 2 introduces the possible issues related to high dv/dt in SiC devices. In Section 3, the test vehicle is firstly described, then the experimental and theoretical methods used in the study are outlined. Section

4 reports the results and comments of the study, while in Section 5 the conclusions are drawn.

2 Issues related to high dv/dt in SiC power devices

This Section presents a brief outline of possible issues in SiC power devices. Robustness to dv/dt fast transients has already been evaluated in the past [11], [12] for example on SiC diodes and rectifiers, but on old technologies, or with bus voltages beyond the datasheet values. There is a lack of papers for the long-term reliability evaluation on SiC MOSFETs. In paper [13], this test is executed, but few details are reported, and the stress is passive. Possible phenomena involved in this kind of reliability assessment, related to previous literature, are briefly outlined in the following. However, it must be remarked that in most cases no reliability studies have been done. The first one is the incomplete ionization of dopants in SiC materials [14], especially when very fast rise time occurs, in comparison with ionization time constants [15]–[17]. In this way, the p-type dopant would be not ionized in time, resulting in a wider depletion region's spread [15]. In another recent paper [18], electron hopping along the defects is supposed to be the phenomenon related to the degradation of devices when high bias switching events are employed. In general, an appropriate design of edge terminations is a must for SiC devices, where higher electrical fields are present [15], [18]. DRB can be the most appropriated test in this regards [13], [19].

3 Materials and Methods

3.1 Materials

SiC Gen3 MOSFETs from STMicroelectronics, housed in discrete four-leaded TO-247 package (with Kelvin source pin), are considered in the experimental assessment of this paper. This technology has a nominal breakdown voltage equal to 1200 V. This die is used in power modules, with a two-level inverter topology, to feed the electric motor in full-electric and hybrid vehicles. The equipment used to stress the devices is from NI-SET company: it is capable of doing both DRB and dyn-H3TRB tests. In this machine, the devices are arranged in many half-bridge topology, in such a way that each device is switched in a complementary way with the other one of the half-bridge, with a small dead time. A picture of experimental environment has been depicted in Fig. 1. It can be noted that the power

Fig. 1: DRB equipment, with TO-247 packages and differential probe

devices are mounted to the equipment through a board, without any heatsink.

MSO54B Tektronix oscilloscope has been employed to characterize drain-to-source - V_{DS} - and gate-to-source - V_{GS} - voltages waveform, using a differential (THDP0200) and two passive (TPP0500B) probes. The B2902A sensing measurement unit (SMU) from Keysight is used for the thermal characterization, in order to force a fixed DC power. Negative temperature coefficient (NTC) sensor B57541G1104F from EPCOS/TDK has been employed. It has a resistance at 25 °C equal to $100 \, k\Omega$ and exponential derating factor $B_{25/85}$ equal to 4072 K. This NTC sensor is attached on the package's molding case to evaluate the case temperature T_c.

3.2 Methods

DRB test

DRB test consists of switching two devices arranged as half-bridge from on-state to off-state and vice versa, not applying any load current. As previously discussed, this trial assesses the capability of edge termination to withstand high voltage slew rate during turn-off. $V_{GS,on}$ is set to 18 V when the device is in on-state, while $V_{GS,off}$ = - 5 V when the switch is turned off and it should block bus voltage V_{DC}. Negative V_{GS} has been considered to avoid the Miller false turn on phenomenon, also referred to as spurious turn-on. Test schematic and probes'

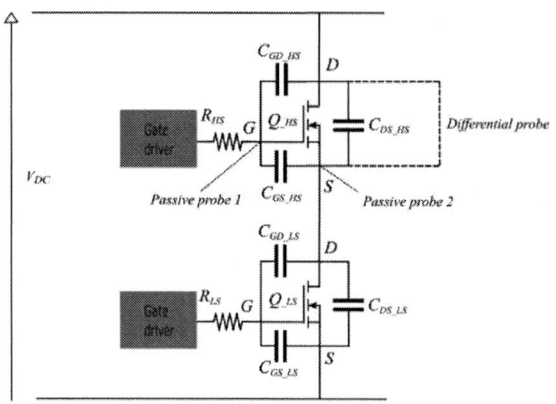

Fig. 2: DRB test equivalent circuit, with probe placement

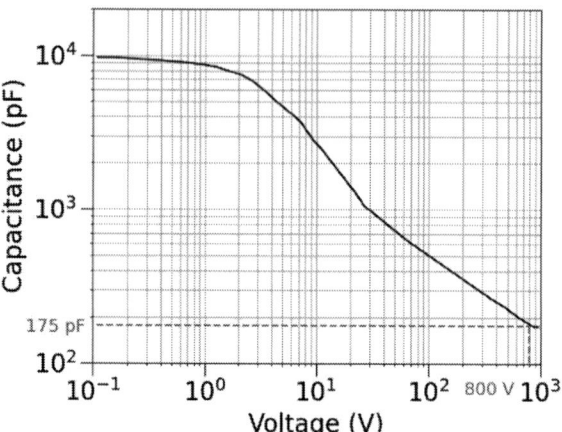

Fig. 3: Plot of output capacitance at different voltages for a SiC MOSFET

positions are shown in Fig. 2.

Different test conditions are explored to investigate the relation among several parameters, such as dissipated power, frequency and temperature. In general, many analytical models of power MOS-FETs, either in silicon and in SiC, have been developed in recent years [20]. The role of internal capacitances inside a SiC MOSFET is studied in [21], in which the authors consider the no-load case to understand the switching limits of the devices, although they consider only the turn on case. The proposed experiment is complementary because it is focused on turn-off behavior.

Moreover, deep analysis is required to ensure the compliance with the AQG 324 guidelines. It is requested a minimum slew rate dv/dt of 50 V/ns, drain-source peak voltage $V_{DS,peak}$ between 80% (i.e., 960 V) and 95% (i.e., 1140 V) of nominal breakdown voltage, and minimum frequency of 25 kHz. A trade-off of such aspects should be found, i.e., higher V_{DC} induces higher dV/dt, but the $V_{DS,peak}$ high limit should be accounted. Characterization is performed considering V_{BUS} = 800 V, and switching frequencies f = 1, 3, 10, 15 and 25 kHz.

Power losses estimation

This section will present an explanation of the measurement methodologies and the formulation used in the paper to compute power losses. Preliminary temperature measurement has assessed that device self-heating occurs during DRB. Hence, this phenomenon is investigated relating to the DRB power losses. As first instance, it is considered that power losses are induced by the energy stored in internal capacitance. It is remarked that the con-

duction losses are not present, due to the absence of load. Such energy, named as E, depends on capacity C and voltage V, as showed in the following formula (no load conditions):

$$E = \int_{V_0}^{V_1} C(V) \cdot V \, dV \qquad (1)$$

where capacitance $C(V)$ is the output capacitance of the MOSFET C_{oss}, made up by two contributions, i.e. C_{GD} and C_{DS}, and it depends from voltage V, as shown in Fig. 3. In fact, E is calculated integrating from V_0 to V_1, which represent V at the start (V_0) and at the end (V_1) of the turn-off transition, respectively.

Switching power loss P_{calc} is calculated from switching energy E multiplied by the frequency f:

$$P_{calc} = 2 \cdot f \cdot E = 2 \cdot f \cdot \int_{V_0}^{V_1} C(V) \cdot V \, dV \qquad (2)$$

It should be noted that the energy due to C_{oss} in one switching cycle considers two times the term $\int_{V_0}^{V_1} C(V) \cdot V \, dV$, related to the energy stored by C_{oss}. To clarify such an aspect, it is assumed that power losses are calculated for the high side (HS) device of half bridge. During turn-on, HS device's C_{oss} is discharged during its turn-on, causing an heating due to internal resistive paths. Moreover, low side (LS) device induces on the HS device the same contribution in terms of energy when LS is turned on, which occurs at the same time of HS turn-off [10], [20], [22].

To validate the outcome of Eq. (2), the power losses are estimated using T_c. During DRB test at different

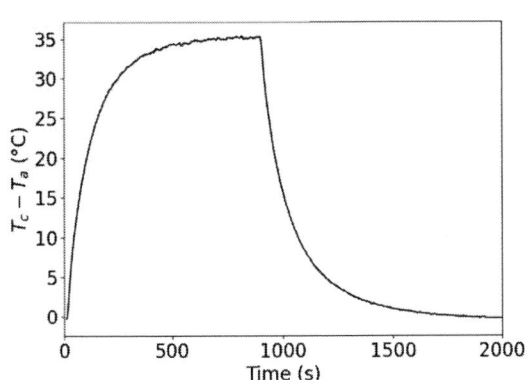

Fig. 4: Measured T_c, with respect to T_a, during heating and cooling. Resulting $Rth, c - a$ is 34 K/W.

switching frequency values, the device temperature T_j is measured at steady-state conditions (i.e., after 20 minutes). Assuming thermal linear behavior during the experiment, T_c is related with the experimental power losses P_{exp} by means of case-ambient thermal resistance $R_{th,c-a}$:

$$P_{exp} = \frac{(T_c - T_a)}{R_{th,c-a}} \tag{3}$$

Power resulting from Eq. 2 should be compared with Eq. (3) outcome to validate the assumption related to power losses sources, which is the energy discharged by device C_{oss}. Accurate estimation of $R_{th,c-a}$ is requested to perform such a comparison.

Thermal resistance measurement

For $R_{th,c-a}$ estimation, device's terminals are connected with the the SMU channel, in the same thermal environment of DRB test ($T_a = 23°C$). Current I_{SD} is set equal to 0.4 A, generating a dissipated power of 1.028 W. NTC sensor for T_c estimation is attached with conductive epoxy glue. Signal is sampled at 1 Hz during the heating forcing, till the steady-state condition, and during cooling down. Measured (T_c) with respect to T_a is depicted in Fig. 4, in which it is assessed that steady stead is achieved after approximately 600 s. Normalizing by the dissipated power (1.028 W), $Rth, c - a$ results 34.4 K/W.

4 Results and Comments

The experimental V_{GS} and V_{DS} waveforms, obtained at 800 V and 25 kHz, are reported in Fig. 5. Comparing V_{DS} and V_{GS} curves in Fig. 5-(a), dead time can be estimated in the order of 200 ns. During turn-off, V_{GS} rings between − 13.7 V and

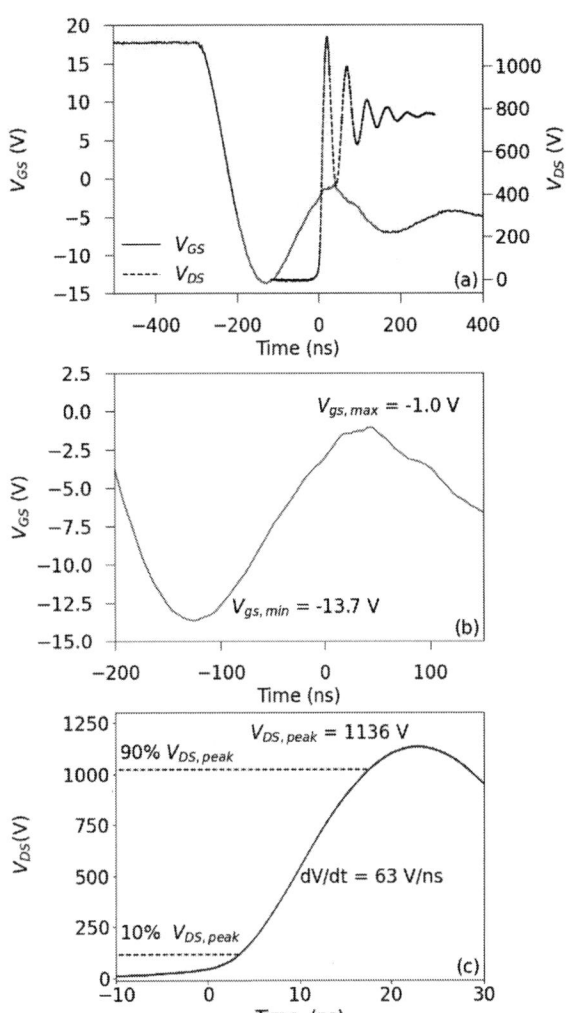

Fig. 5: Waveform of V_{DS} and V_{GS} during turn-off (a), with focus on V_{GS} peak (b) and V_{DS} rise (c).

-1 V, as highlighted in Fig. 5-(b). Considering that threshold voltage V_{th} is approximately 3 V, spurious turn-on may be excluded. From Fig. 5(c), dv/dt, which results equal to 63 V/ns, is computed as the ratio between the 90% and 10% of $V_{DS,peak}$ and the related rising time, which is 14 ns. Furthermore, $V_{DS,peak}$ is 1136 V, resulting in a overshoot of 336 V with respect to V_{DC}, equal to 800 V. In last analysis, these waveforrms are fully compliant with AQG 324 requirements.

Table 1 shows comparison among T_c, P_{exp} and P_{calc} at different f. P_{exp} and P_{calc} are quite aligned when frequency is between 1 to 15 kHz. Considering f = 25 kHz, Eq. (2) overestimates P_{exp} of approximately 30%. This result can be explained by the fact that $R_{th,c-a}$ is measured between 23 °C and

Tab. 1: Measured T_c, experimental P_{exp} and calculated P_{calc} at different frequency

f (kHz)	T_c (°C)	P_{exp} (W)	P_{calc} (W)
1	27.8	0.150	0.153
3	40.0	0.522	0.460
10	64.4	1.392	1.532
15	91.0	2.081	2.299
25	123.0	3.059	3.831

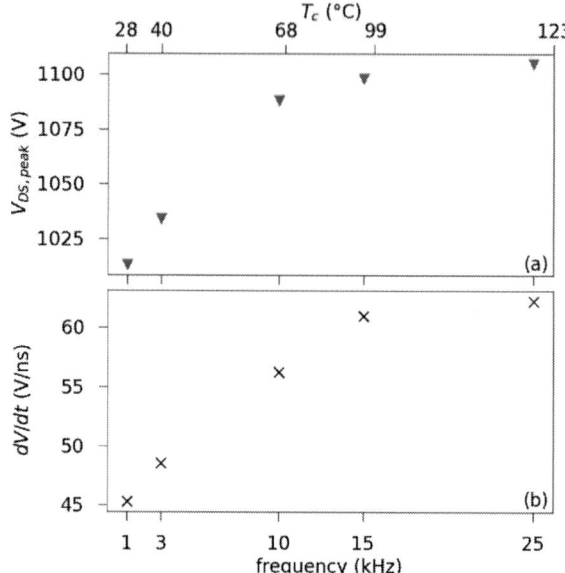

Fig. 6: $V_{DS,peak}$ (a) and dv/dt (b) at different f

56 °C, while T_c at 25 kHz is 123 °C. Indeed, temperature is 66 °C higher during DRB than in thermal resistance assessment. It is expected that the heat transfer coefficient increases with temperature due to a possible spark of turbulence flow triggered by the higher temperature gradient. Moreover, in the case study, device is in still air environment and without any heatsink. Radiation, which cannot be neglected, is enhanced by increasing temperature in this context [23]. Literature experimental assessments have confirmed that R_{th} decreases when dissipated power, thus device temperature, increases [24]. This trend is enhanced in case of small-size heatsink, which is similar to the presented case study, where the heasink is not employed. An experimental study has quantified that R_{th} can decrease of 10-20%, when dissipation power growths [25]. As a result, the actual $R_{th,c-a}$ during DRB at 25 kHz is lower than 34.4 K/W due to the non-linear convective and radiated fluxes involved in device cooling. Those are because of temperature difference between thermal resistance and DRB assessments. Considering Eq. (3), in which $R_{th,c-a}$ is at denominator, the misalignment between P_{calc} and P_{exp} is, almost partially, explained.

Finally, during the experimental activity, tests are executed to explore the dependence of dV/dt and V_{peak} parameters on the switching frequency. V_{peak} and dV/dt values increase at higher frequency and, therefore, at higher temperature, as depicted in Fig. 6 and recalling the positive correlation between frequency and temperature correlation from Table 1. Temperature, due to self-heating phenomenon, can cause a change in device's parameters and, in turn, in dV/dt and V_{peak}. More specifically, it must be recalled that higher temperature results in lower threshold voltage. In this case, higher frequency implies both higher temperature, indeed lower threshold voltage, and higher dv/dt. In literature, analytic formulations for turn-on and turn-off

of SiC MOSFETs have been proposed, correlating dv/dt with parameters such as threshold voltage and transconductance [21], [26]. Considering that these two parameters decrease as the temperature increases and are as ratio in the formulation, trend for dv/dt should be defined case by case. According to these approaches, it is not possible to establish if the decrease of transconductance and threshold voltage result in an increase or decrease of dv/dt.

5 Conclusions

In this paper, a deep electro-thermal characterization for dynamic reverse bias test has been presented. After a brief outline on possible phenomena related to this reliability test, experimental and easy-to-implement analytical approaches have been presented and discussed. After methodologies' explanation, the acquired waveforms have demonstrated that implemented DRB test is compliant with ECPE AQG 324 guidelines. Moreover, power losses have been estimated both experimentally and by some simple formulas, finding a good agreement till 15 kHz. The easy-to-implement formulation overestimates the experimental power losses at the highest frequency level (25 kHz). This may be due to the $R_{th,c-a}$ non-linear behavior with the temperature. Correlation among test parameters has highlighted the link between frequency, drain-source voltage waveform and threshold voltage, according to other literature findings. Future papers will dealt with the execution and data analysis of long reliability trials, and with a thermal resistance estimation at temperature close to DRB test.

References

[1] A. Chvála, J. Marek, J. Drobnỳ, L. Stuchlíková, A. A. Messina, et al., "Characterization and evaluation of current transport properties of power SiC Schottky diode," en, Materials Today: Proceedings, vol. 53, pp. 285–288, 2022. DOI: 10.1016/j.matpr.2021.06.150.

[2] B. J. Baliga, "Silicon Carbide Power Devices: Progress and Future Outlook," IEEE Journal of Emerging and Selected Topics in Power Electronics, vol. 11, no. 3, pp. 2400–2411, Jun. 2023. DOI: 10.1109/JESTPE.2023.3258344.

[3] G. Mauromicale, A. Cascio, M. Papaserio, D. G. Cavallaro, G. Bazzano, et al., "Directly Cooled Silicon Carbide Power Modules: Thermal Model and Experimental Characterization," in PCIM Europe digital days 2021; International Exhibition and Conference for Power Electronics, Intelligent Motion, Renewable Energy and Energy Management, VDE, 2021, pp. 1–5.

[4] L. Anoldo, C. Triolo, S. Panarello, F. Garescì, S. Russo, et al., "Study of the Thermomechanical Strain Induced by Current Pulses in SiC-based power MOSFET," IEEE Electron Device Letters, vol. 42, no. 7, pp. 1089–1092, 2021.

[5] B. Shi, A. I. Ramones, Y. Liu, H. Wang, Y. Li, et al., "A Review of Silicon Carbide MOSFETs in Electrified Vehicles: Application, Challenges, and Future Development," en, IET Power Electronics, vol. 16, no. 12, pp. 2103–2120, Sep. 2023. DOI: 10.1049/pel2.12524.

[6] H. Chen and D. Divan, "High speed switching issues of high power rated silicon-carbide devices and the mitigation methods," in 2015 IEEE Energy Conversion Congress and Exposition (ECCE), Montreal, QC, Canada: IEEE, Sep. 2015, pp. 2254–2260. DOI: 10.1109/ECCE.2015.7309977.

[7] N. Kaminski, S. Rugen, and F. Hoffmann, "Gaining Confidence - A Review of Silicon Carbide's Reliability Status," in 2019 IEEE International Reliability Physics Symposium (IRPS), Monterey, CA, USA: IEEE, Mar. 2019, pp. 1–7. DOI: 10.1109/IRPS.2019.8720578.

[8] "AQG 324 ECPE Guideline - Qualification of Power Modules for Use in Power Electronics Converter Units in Motor Vehicles," ECPE European Center for Power Electronics, Nuremberg, DE, Standard, May 2021.

[9] A. A. Gómez Gómez, J. R. García Meré, A. Rodríguez Alonso, J. Rodríguez Méndez, C. Jiménez, J. Roig, et al., "Deep Investigation on SiC MOSFET Degradation under Gate Switching Stress and Application Switching Stress."

[10] N. Perera, A. Jafari, R. Soleimanzadeh, N. Bollier, S. G. Abeyratne, and E. Matioli, "Hard-Switching Losses in Power FETs: The Role of Output Capacitance," IEEE Transactions on Power Electronics, vol. 37, no. 7, pp. 7604–7616, Jul. 2022. DOI: 10.1109/TPEL.2021.3130831.

[11] L. Z. P. A. Losee, "Degraded Blocking Performance of 4H-SiC Rectifiers under High dV/dt Conditions," in Proceedings. ISPSD '05. The 17th International Symposium on Power Semiconductor Devices and ICs, 2005., Santa Barbara, CA, USA: IEEE, 2005, pp. 219–222. DOI: 10.1109/ISPSD.2005.1487990.

[12] E. Van Brunt, G. Wang, J. Liu, V. Pala, B. Hull, et al., "Operation of 4H-SiC Schottky diodes at dV/dt values over 700 kV/µs," in 2016 28th International Symposium on Power Semiconductor Devices and ICs (ISPSD), Prague, Czech Republic: IEEE, Jun. 2016, pp. 67–70. DOI: 10.1109/ISPSD.2016.7520779.

[13] E. Mengotti, E. Bianda, D. Baumann, G. Schlottig, and F. Canales, "Industrial approach to the chip and package reliability of SiC MOSFETs (Invited)," in 2023 IEEE International Reliability Physics Symposium (IRPS), Monterey, CA, USA: IEEE, Mar. 2023, pp. 1–6. DOI: 10.1109/IRPS48203.2023.10118084.

[14] N. Donato and F. Udrea, "Static and Dynamic Effects of the Incomplete Ionization in Superjunction Devices," IEEE Transactions on Electron Devices, vol. 65, no. 10, pp. 4469–4475, Oct. 2018. DOI: 10.1109/TED.2018.2867058.

[15] Z. Tong, J. Roig-Guitart, T. Neyer, J. D. Plummer, and J. M. Rivas-Davila, "Origins of Soft-Switching C_{oss} Losses in SiC Power MOSFETs and Diodes for Resonant Converter Applications," IEEE Journal of Emerging and Selected Topics in Power Electronics, vol. 9, no. 4, pp. 4082–4095, Aug. 2021. DOI: 10.1109/JESTPE.2020.3034345.

[16] P. G. Neudeck and C. Fazi, "High-field fast-risetime pulse failures in 4H- and 6H-SiC pn junction diodes," en, Journal of Applied Physics, vol. 80, no. 2, pp. 1219–1225, Jul. 1996. DOI: 10.1063/1.362922.

[17] M Lades, W Kaindl, N Kaminski, E Niemann, and G Wachutka, "Dynamics of incomplete ionized dopants and their impact on 4H/6H-SiC devices," IEEE Transactions on Electron Devices, vol. 46, no. 3, pp. 598–604, 1999.

[18] J. P. Kozak, R. Zhang, J. Liu, K. D. Ngo, and Y. Zhang, "Degradation of SiC MOSFETs under high-bias switching events," *IEEE Journal of Emerging and Selected Topics in Power Electronics*, vol. 10, no. 5, pp. 5027–5038, 2021.

[19] P. Godignon, J. Montserrat, J. Rebollo, and D. Planson, "Edge Terminations for 4H-SiC Power Devices: A Critical Issue," *Materials Science Forum*, vol. 1062, pp. 570–575, May 2022. DOI: 10.4028/p-lom714.

[20] D. Christen and J. Biela, "Analytical Switching Loss Modeling Based on Datasheet Parameters for mosfets in a Half-Bridge," *IEEE Transactions on Power Electronics*, vol. 34, no. 4, pp. 3700–3710, Apr. 2019. DOI: 10.1109/TPEL.2018.2851068.

[21] D. Cittanti, F. Iannuzzo, E. Hoene, and K. Klein, "Role of parasitic capacitances in power MOSFET turn-on switching speed limits: A SiC case study," in *2017 IEEE Energy Conversion Congress and Exposition (ECCE)*, Cincinnati, OH, USA: IEEE, Oct. 2017, pp. 1387–1394. DOI: 10.1109/ECCE.2017.8095952.

[22] J. Gareau, R. Hou, and A. Emadi, "Review of Loss Distribution, Analysis, and Measurement Techniques for GaN HEMTs," *IEEE Transactions on Power Electronics*, vol. 35, no. 7, pp. 7405–7418, Jul. 2020. DOI: 10.1109/TPEL.2019.2954819.

[23] W. Zhang and G. Feng, "A Quick PCB Thermal Calculation for Power Electronic Devices with Exposed Pad Packages," in *PCIM Asia 2019; International Exhibition and Conference for Power Electronics, Intelligent Motion, Renewable Energy and Energy Management*, Jun. 2019, pp. 1–8.

[24] K. Górecki and K. Posobkiewicz, "Selected Problems of Power MOSFETs Thermal Parameters Measurements," en, *Energies*, vol. 14, no. 24, p. 8353, Jan. 2021. DOI: 10.3390/en14248353.

[25] K. Górecki and J. Zarębski, "Modeling the Influence of Selected Factors on Thermal Resistance of Semiconductor Devices," *IEEE Transactions on Components, Packaging and Manufacturing Technology*, vol. 4, no. 3, pp. 421–428, Mar. 2014. DOI: 10.1109/TCPMT.2013.2290743.

[26] A. Hu and J. Biela, "Evaluation of the I_{max}-f_{sw}-dv/dt Trade-off of High Voltage SiC MOSFETs Based on an Analytical Switching Loss Model," in *2020 22nd European Conference on Power Electronics and Applications (EPE'20 ECCE Europe)*, Lyon, France: IEEE, Sep. 2020, P.1–P.11. DOI: 10.23919/EPE20ECCEEurope43536.2020.9215911.

PCIM Europe 2024, 11– 13 June 2024, Nuremberg   DOI: 10.30420/566262389

Radiation Hardness of SiC Based Inverters Based on an EV Mission Profile

Hadiuzzaman Syed[1], Stephan Schwaiger[1], Sudhanshu Goel[1], Alberto Martinez-Limia[1], Klaus Heyers[1]

[1] Robert Bosch GmbH, Mobility Electronics, Germany

Corresponding author: Hadiuzzaman Syed, hadiuzzaman.syed@de.bosch.com
Speaker: Hadiuzzaman Syed, hadiuzzaman.syed@de.bosch.com

Abstract

Cosmic radiation (CR) ruggedness is one of the main design features of the SiC MOSFETs used in the EV inverters. Over its entire lifetime the EV inverter is exposed to cosmic radiation and prone to single-event-burnout failures. The rate of failure or CR FIT (failure in time) is a commonly agreed parameter that defines the radiation ruggedness of the SiC. CR FIT rate is also dependent on the mission profile that is particular to a certain inverter application. Experimental results combined with the application specific mission profile eases the design for a SiC technology optimized for the use in the real-life scenarios.

1   Introduction

Silicon carbide (SiC) devices are widely used in the automotive grade inverters [6 ,7, 8, 9]. Compared to the silicon devices silicon carbide-based switches have three times the bandgap energy, thermal conductivity, and substantially larger breakdown field. These material properties enable a thinner drift region with similar voltage blocking capability. Thinner drift region along with higher doping result in a low resistance device having the same or higher blocking voltage capability than that of the silicon.

Silicon carbide devices normally experience high electric field during their operation in the inverter as the battery voltage is reaching 800 V and higher. The high electric field during operation enhances the chance of a device being failed under background neutron radiation. Neutron radiation is one of the major sources for extrinsic device failures [1].

The probability of the failure becomes higher once the inverter is operating in more rugged terrains i.e., high altitudes.

To understand the device failure under terrestrial neutron radiation, accelerated test needs to be performed with neutron sources that have very similar spectrum as the terrestrial neutron. We have performed these accelerated tests under the Chiplr beamline at the ISIS spallation source, Rutherford Appleton Laboratory, UK. The Chiplr beamline has very similar spectrum as the terrestrial neutron [2]. The accelerated tests resulted a CR FIT rate as a function of the drain-source blocking voltage curve at sea-level. This CR FIT curve and the failures in the devices were analyzed and further device level optimization was performed to have a radiation rugged silicon carbide device for the automotive inverter. The optimized devices were tested at the same beamline, and they showed superior performance in terms of

CR FIT compared to the previous generation of devices.

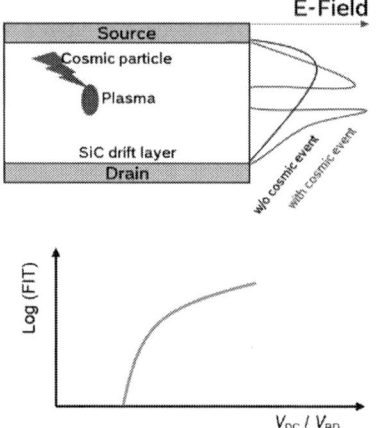

Figure 1: Sketch of the electrical field in a SiC MOSFET with and without interaction with a cosmic particle and theoretical failure rate as a function of relative voltage (normalized to breakdown voltage).

2 Device Optimization

Cosmic particles with high energies can interact with a SiC atom in the drift zone of a high voltage power devices as shown in Figure 1.

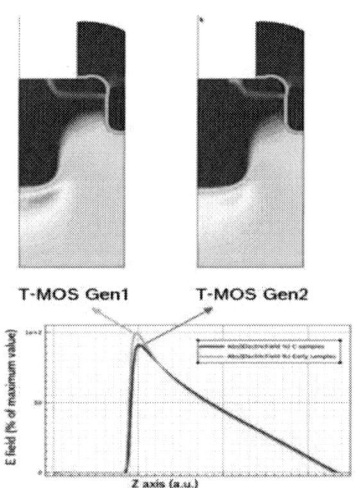

Figure 2: Electric field distribution of Bosch T-MOS Gen1 and T-MOS Gen2 device in the blocking state.

Due to this interaction an electron-hole plasma is created resulting in the local collapse of the electric field. Operating the device near the avalanche

voltage may result in catastrophic failure even at the sea level.

A straightforward approach to have less cosmic ray related failure is to increase the breakdown voltage of the device, so that the DC link versus breakdown voltage ratio (V_{DC}/V_{BD}) decreases (as shown in Fig. 1). But increasing the breakdown voltage increases device on state resistance as the drift region becomes thicker.

In Bosch T-MOS Gen2 devices the on-state resistance was a trade off with the breakdown voltage. To achieve a high degree of cosmic radiation ruggedness an additional optimization in terms of the electric field distribution in the device was implemented.

As reported previously [3], the neutron induced local failure rate (P_{lf}) in a silicon carbide MOSFET depends exponentially on the local electric field ($E(x)$) and is integrated over the entire device volume to get the total failure (P_f) probability:

$$P_{lf}\big(E(x)\big) = v_0 \exp\left(-\frac{E_b}{E(x)}\right)$$

$$P_f = \int_{\Omega} P_{lf}(E)\mathrm{d}\Omega$$

The parameters v_0 and E_b are fitting parameters for SiC. The device cross section and the electric field distribution of the 1st generation (T-MOS Gen1) and 2nd generation (T-MOS Gen2) samples are shown in the Figure 2. As shown in Figure 2 the electric field at the junction of the 2nd generation device is smaller than that of the 1st generation during the blocking state. Reduction of this electric field reduces the total failure probability significantly.

Reduction of the electric field at the junction during the blocking state is achieved by increasing the width of the P+ region in the trench MOSFET and the larger radius of the p-n junction at the corner.

The reduction of the electric field along with optimization in the breakdown voltage significantly improves the cosmic radiation related failure rate of the Bosch T-MOS Gen2 device with an excellent breakdown voltage versus on state resistance trade-off.

3 Measurement

Bosch T-MOS Gen1 and Gen2 devices with the rated voltage of 1200 V were measured at the ChipIR beamline. For each single measurement run 200 devices had been radiated and the blocking voltage was swept from low to high. For each DC-bias voltage at least 30 devices or more have been considered at the start i.e., before the radiation and the radiation dosage was stopped after there had been minimum of 10 failed devices. The equivalent time for non-accelerated condition was calculated from the reference flux as given in

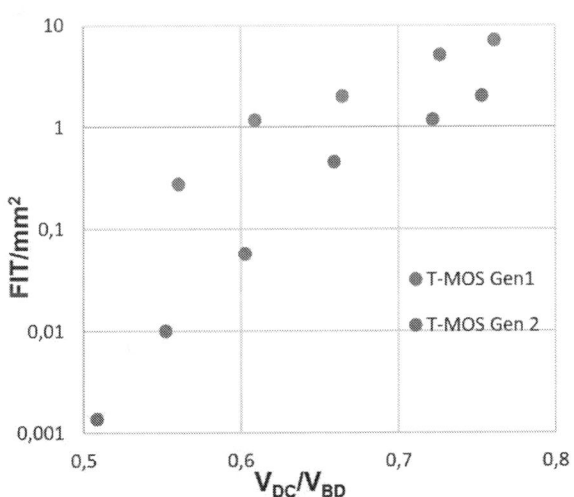

Figure 3: Measured FIT Rate curve for T-MOS Gen1 and T-MOS Gen2.

JESD89A and the devices at the beam stop and the number of failed devices had been considered to calculate the FIT rate for each of the DC bias condition [4]. The test procedure followed JEP151A guideline [5]. The worst-case FIT value

was computed with a 95 % confidence level using a chi-square function as recommended by the JEP151A.

In Figure 3 the FIT/mm^2 rate as a function of the drain-source blocking voltage curve is plotted for Bosch T-MOS Gen1 and T-MOS Gen2 at sea level. The device breakdown voltage is increased for T-MOS Gen2 compared to T-MOS Gen1 [8]. The blue and orange points are the measured FIT/mm^2 values for T-MOS Gen1 and Gen2 respectively at different V_{DC}/V_{BD} ratio. It can be seen from the measurement results that the T-MOS Gen2 devices have ~ 5 times lower FIT/mm^2 value than the T-MOS Gen1 devices $V_{DC}/V_{BD} > 0.7$. The improvement is more than an order of magnitude when $V_{DC}/V_{BD} < 0.65$.

4 Module FIT /PPM Rate based on EV Mission Profile

Table 1 shows an example EV mission profile with voltage stress on each of the logical switch of an automotive half-bridge inverter and the corresponding stress duration over the lifetime. The charging time is significantly longer than the driving operation.

Driving		Charging			
VDS	Time	VDS_HS	Time	VDS_LS	Time
[V]	[h]	[V]	[h]	[V]	[h]
672	0	288	0	192	0
696	100	312	150	208	150
720	400	336	2000	224	2000
744	1200	360	5800	240	5800
768	3100	384	6100	256	6100
792	2600	408	6200	272	6200
816	450	432	6000	288	6000
840	100	456	3500	304	3500
864	50	480	250	320	250

Table 1: EV mission profile for FIT and PPM calculation.

The device during the driving is mostly stressed with a voltage range of around 720 V to 800 V. The voltage during charging can be either a worst-case approximation, with the full DC-link voltage over one switch, or a voltage balancing can be assumed. The voltage balancing is either depending on the connected load like the resistance of the electrical machine or on the other hand is explicitly implemented using a balancing technique in the inverter design. This can be done either by two resistances in series to divide the DC-link voltage by half or with sophisticated gate-driver technique using the SiC MOSFET in the sub-threshold region. In approach with two additional resistances, the tolerance of the resistance needs to be considered, and the voltage was assumed to be divided between the top (V_{DS_HS}) and bottom (V_{DS_LS}) switches by 60%/40% respectively. As an example, a DC-link voltage of 800 V with the resistance tolerance into consideration, top switch is stressed with 480 V and the bottom switch has a voltage of 320 V.

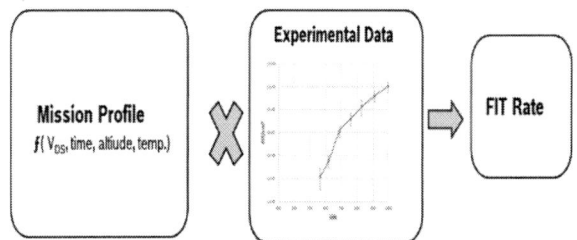

Figure 4: Convolution of CR FIT rate based on mission profile and experimental data.

The altitude profile in Table 2 is also included in the EV mission profile. The altitude profile is composed of two parameters. Height and the corresponding percentage of the life time the EV drives in that altitude is shown.

The altitude profile is useful to calculate the altitude scaling factor as the experimental FIT versus blocking voltage curve is given only for sea level.

As shown in Figure 4 the blocking voltage (V_{DS}), corresponding time, altitude for a certain temperature is convoluted with the FIT vs V_{DS} curve from the accelerated test in the ChipIr beamline. The experimental curve was interpolated between the shown measurement points in Figure 3 in 1 V interval to cover all the voltages in the mission profile. The FIT for the T-MOS Gen2 is relatively low when convoluted with the EV mission profile as the switches are mostly stressed with the DC link voltage in the range of 720 V to 800 V during the driving mode. The contribution from this voltage range to the FIT rate of the device is quite insignificant for the T-MOS Gen2 device. Because of the balancing resistors the voltage stress in the charging mode in the worst case is below 500 V. The improved CR FIT rate with respect to V_{DS} characteristics due to the optimized device design and the balancing resistor network for the charging mode results in a very low FIT rate for this exemplary EV mission profile.

The PPM rate for the module can also be calculated from the chip FIT rate in the following way:

$$PPM_{\text{Rate}} = (FIT_{Rate} * hour_{\text{Total}})/10^3$$

The power module shown in in Figure 5 has eight

Altitude Pofile	
Height [m]	Time [%]
250	68
1000	22
2000	6
3000	3.5
4000	0.3
5000	0.2

Table 2: Altitude profile for EV over the lifetime.

1200 V rated SiC chips in total. The module with T-MOS Gen2 devices achieve less than 1 PPM failure rate related to the cosmic radiation as shown in Figure 5.

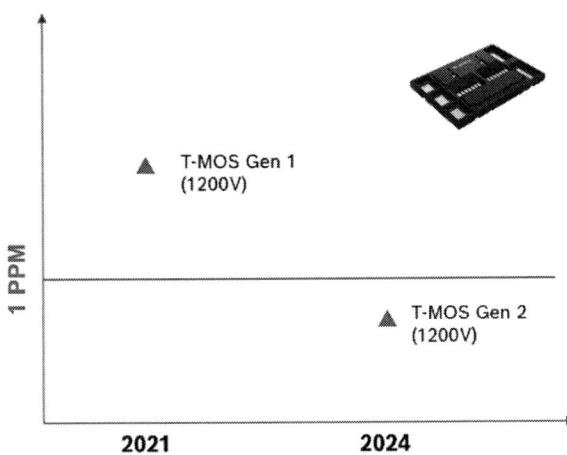

Figure 5: *T-MOS Gen1 and T-MOS Gen2 PPM rate comparison.*

5 Conclusion

1200 V rated Bosch SiC semiconductors were tested under accelerated condition to understand the neutron induced failure rate in the device. Based on the measurement results of the first generation (T-MOS Gen1) device second generation (T-MOS Gen2) device design was optimized. Device electric field during the blocking state was optimized along with an increase in the breakdown voltage in the T-MOS Gen2 device.

The reason for this two-degree optimization is to have a better on state resistance with significant improvement in the cosmic radiation ruggedness. The two-degree optimization in the device level produced a more radiation rugged device which is confirmed by the measurement. On the other hand, the $R_{DSon}*A$ @ 175°C was reduced by 26 % from Bosch SiC T-MOS Gen1 to Gen2 devices.

The radiation beam measurement result was convoluted with an EV mission profile to calculate the FIT for both generation of devices were compared.

Furthermore, PPM rate of an eight-chip module was calculated with both semiconductor generations. Bosch T-MOS Gen2 devices in the module showed PPM rate <1 in terms of cosmic radiation.

Acknowledgement

This project has received funding from the Key Digital Technologies Joint Undertaking (KDT JU) under Grant Agreement No101007237. The JU receives support from the European Union's Horizon 2020 research and innovation program and Germany, France, Italy, Sweden, Austria, Czech Republic, Spain.

The authors would also like to thank the Chiplr colleagues for their excellent support during the accelerated test.

References

[1] J. F. Ziegler, "Terrestrial cosmic ray intensities," IBM J. Res. Dev.,vol. 42, no. 1, Jan. 1998

[2] Carlo Cazzaniga, Simon P. Platt and Christopher D. Frost, "Preliminary Results of Chiplr, a new Atmospheric-like Neutron Beamline for the Irradiation of Microelectronics", proceeding of the SELSE-13: "The 13th Workshop on Silicon Errors in Logic – System Effects",21-22 March 2017, Northeastern University, Boston, Massachusetts (USA).

[3] A. Akturk et al., "Predicting Cosmic Ray-Induced Failures in Silicon Carbide Power Devices," in IEEE Transactions on Nuclear Science, vol. 66, no. 7, July 2019, doi: 10.1109/TNS.2019.2919334.

[4] C. Felgemacher, S. V. Araújo, P. Zacharias, K. Nesemann and A. Gruber, "Cosmic radiation ruggedness of Si and SiC power semiconductors," *2016 28th International Symposium on Power Semiconductor Devices and ICs (ISPSD)*, Prague, Czech Republic, 2016, doi: 10.1109/ISPSD.2016.7520775

[5] Test Procedure for the Measurement of Terrestrial Cosmic Ray Induced Destructive Effects in Power Semiconductor Devices, JEP151A, Jan. 1, 2022

[6] Klaus Heyers, Stephan Schwaiger, Christian Banzhaf, Michael Grieb, „SiC-Trench-MOSFETs for automotive drive application", in Bauelemente der Leistungselektronik und ihre Anwendungen 2017 - 7. ETG-Fachtagung.

[7] M. Boesing and D. Schweiker, "Design Aspects in SiC MOSFET based High Performance Automotive and Commercial Vehicle Inverters," PCIM Europe 2023; International Exhibition and Conference for Power Electronics, Intelligent Motion, Renewable Energy and Energy Management, Nuremberg, Germany, 2023, doi: 10.30420/566091063.

[8] S. Schwaiger, K. Heyers, A. Martinez-Limia, K. Oberdieck and C. Foerster, "Advanced SiC Trench-MOS Technology for Automotive Application," PCIM Europe 2023; International Exhibition and Conference for Power Electronics, Intelligent Motion, Renewable Energy and Energy Management, Nuremberg, Germany, 2023, doi: 10.30420/566091096.

[9] Robert Bosch GmbH, SiC Technology, https://www.bosch-semiconductors.com/power-se miconductors-and-modules/

PCIM Europe 2024, 11– 13 June 2024, Nuremberg   DOI: 10.30420/566262390

Rapid Short Circuit Protection Using di/dt Detection for SiC Power Modules

Koki Samura[1], Kentaro Yoshida[1], Kakeru Iwashita[1], Seiichiro Inokuchi[1]

[1] Power Device Works, Mitsubishi Electric Co., Japan

Corresponding author: Koki Samura, Samura.Koki@cj.MitsubishiElectric.co.jp
Speaker: Koki Samura, Samura.Koki@cj.MitsubishiElectric.co.jp

Abstract

A new generation of Mitsubishi automotive power module (J3-Series) is currently close to mass production. This paper presents a new short-circuit protection for SiC J3-T-PM, a member of J3-Series. The J3-T-PM requires a high-speed protection circuit due to achieving smaller chip size and lower on-state resistance than IGBT modules. The short-circuit protection method uses Desaturation (DESAT) and di/dt detection for the automotive SiC power module.
First, to verify the operation of the proposed circuit and study its parameters setting, short-circuit tests are conducted by applying it to the Si module J1 Series(700A/650V), and compared to current sensor method, DESAT method, and the proposed protection circuit. Finally, the proposed protection circuit is applied to SiC J3-T-PM and a short-circuit test is conducted. The effectiveness and robustness of the circuit are verified at -40,25 and 175°C, which contributes to low loss and miniaturization of J3-T-PM.

1   Introduction

In recent years, there has been a growing interest in environmental issues and energy sustainability, and the automotive industry is also required to develop environmentally friendly technologies. Under these circumstances, electric vehicles (EV) are gaining attention as a means of achieving zero emissions and sustainable transportation. On the other hand, one of the main problems hindering the shift to EV is that the cruising range is shorter than that of engine vehicles. SiC power modules are attracting attention as a device to solve this problem due to their small size and low loss. Mitsubishi Electric has been developing SiC devices since the early 1990s, and in 2010, we released the world's first SiC products [1]. By using our device technologies, a new automotive SiC power module J3-T-PM (**Fig. 1**) which is compact and low loss is now being developed.
One of the key points in designing a compact SiC module is the trade-off between short-circuit withstand capacity and on-state resistance. Since SiC modules have a smaller chip size and lower short-circuit withstand capacity than IGBT modules, rapid protection circuits are required. Various short-circuit protection circuits have been proposed to improve the trade-off [2]. This paper introduces a new short-circuit protection SCM (Short Circuit Monitor) circuit.

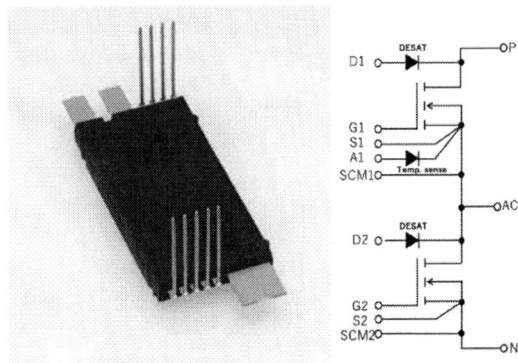

Fig. 1   Mitsubishi J3-T-PM module

This feature can be used in combination with the conventional DESAT method to speed up the short-circuit protection circuit. Currently, there are two methods of short-circuit protection circuits widely used for automotive power modules, on-chip current sensor and DESAT protection. In general, on-chip current sensor method enables high-speed protection. However, it is not reasonable to apply this method to expensive SiC devices because it utilizes a portion of the effective area of the chip as a sensor and leads to higher cost. Therefore, DESAT method is commonly used for SiC. In DESAT method, to prevent malfunctions, a blanking time when the protection circuit cannot react is needed, making it difficult to implement the fast protection circuit required for SiC devices.

2764

2 Proposed SC protection

The circuit diagram of the proposed SCM circuit is shown in Fig. 2. The SCM circuit can directly detects the di/dt of short-circuit current by the parasitic inductance of the package to achieve high-speed protection. It is composed of a couple of discrete components including diode, NPN transistor, resistor, and capacitor, which are connected to the parasitic inductance L_{PKG} between the Kelvin source (KS) terminal and the main current terminal. The SCM circuit uses the main current terminal signal as SCM signal. Fig. 3 shows the timing chart of the proposed circuit. The solid and dashed lines show the waveforms with and without the SCM circuit, respectively. When a short-circuit occurs, a large di/dt generates a positive voltage between KS and SCM terminal during period ①. This positive voltage charges C1 through D1 and causes Q1 to turn on when the base-emitter voltage of Q1 exceeds a certain value. When Q1 turns on during period ②, the gate voltage is clamped by D2 and R3, resulting in suppressing the short-circuit peak current. The gate is then safely shut down by the conventional DESAT protection circuit during period ③. D3 is required to protect the SCM circuit from the negative voltage generated during soft turn-off of period ③. Since di/dt is lower during a normal turn-on period ④ than during a short-circuit, the gate voltage can be suppressed only during a short circuit by setting the appropriate threshold.

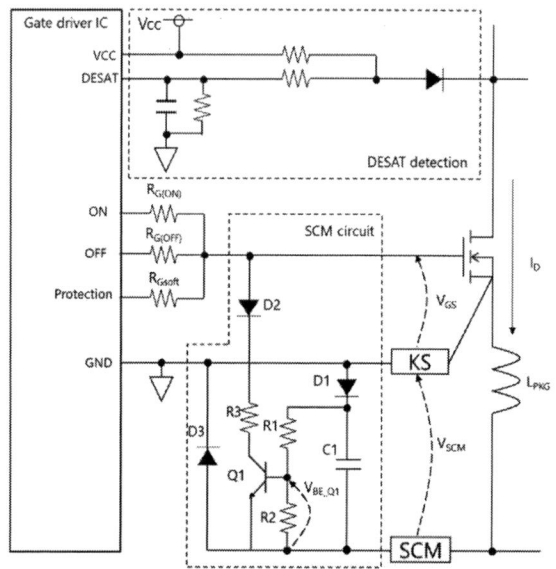

Fig. 2 Circuit diagram of the proposed protection

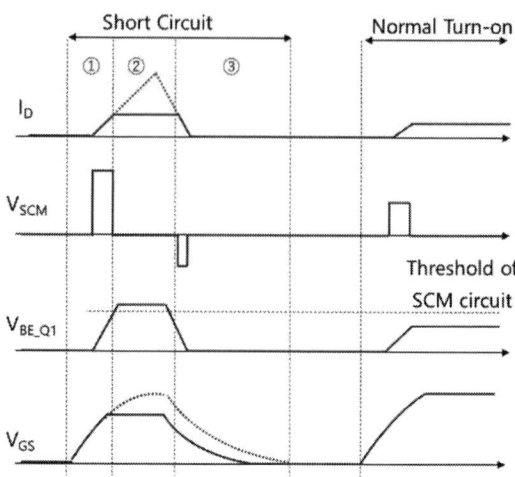

Fig. 3 Timing chart of the proposed circuit

3 Experimental result

3.1 Verification with a Si module

In order to verify the operation of the SCM circuit and to study the parameter settings, we applied it to a Si module Mitsubishi J1 Series (Fig. 4, 700A/650V) and performed short-circuit tests. Fig. 5 shows the short-circuit waveforms with several SCM circuit constants. As shown in Fig. 5 (a), the operation timing and duration of the SCM circuit can be adjusted by changing R1 and C1, and as shown in Fig. 5 (b), by changing R3, the voltage value at which the gate voltage is clamped can be adjusted. As explained in Chapter 2, it is necessary to adjust these constants so that the SCM circuit is activated only in the event of a short-circuit without malfunction during normal operation. These results suggests that this circuit can easily adjust the threshold and the timing of its operation and can be applicable to various power module with different characteristics.

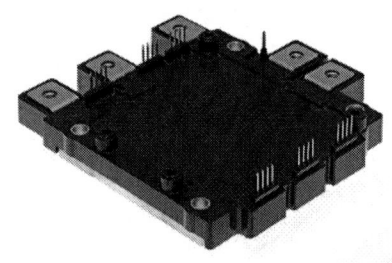

Fig. 4 J1 Series module

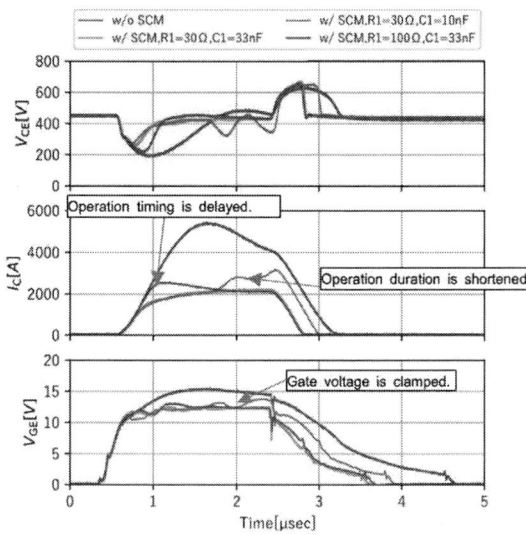

(a) Parametric study of R1 and C1

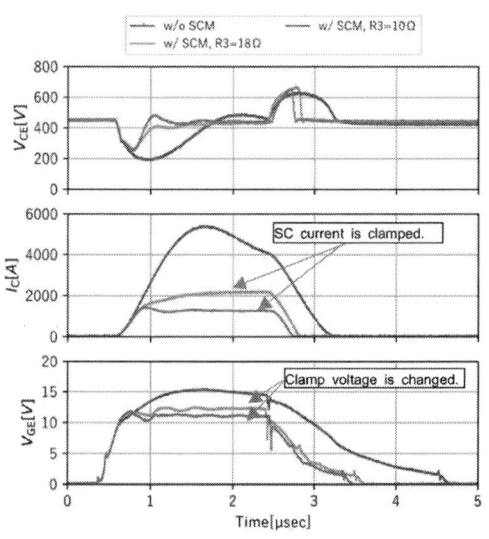

(b) Parametric study of R3

Fig. 5 SC waveform of SCM circuit with different settings (VCC=450V, Tvj=25°C)

3.2 Comparison of protection method using a Si module

To compare each protection method including on-chip current sensor method, a short-circuit test was performed on the Si module. J1 series has on chip current sensor and enable to implement fast protection circuit. Fig. 6 and Table 1 shows the waveform and characteristics of each protection method. The results show that the SCM circuit has best characteristics among the three protection

methods. The SCM circuit immediately clamps the gate voltage in the event of a short circuit, reducing short-circuit loss E_{SC} by 54% compared to DESAT method.

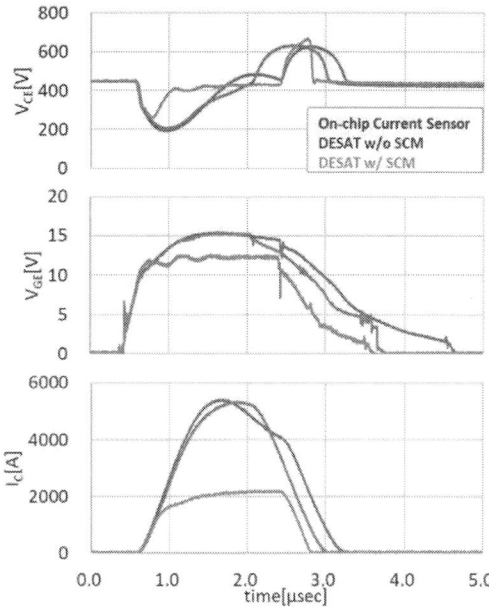

Fig. 6 SC waveform of each protection method with Si module

Method	Delay [μsec]	I_{Cpeak}[A]	E_{SC}[J]
On-chip current sensor	1.68	5440	3.2
DESAT	2.05	5480	3.6
DESAT with SCM	2.05	2320	1.7

Table 1 SC characteristics of each protection method with Si module

3.3 Short circuit test with SiC module

We applied the SCM circuit to J3-T-PM shown in Fig. 1 and performed short-circuit tests to verify the effectiveness of the circuit in SiC modules. To facilitate the application of the SCM circuit, the J3-T-PM has a signal terminal for SCM that takes out the potential of the main terminal. The results of the short-circuit test are shown in Fig. 7. The results show that the SCM circuit works as designed also in the SiC module, and the short-circuit loss is significantly reduced by 57%. Compared to the SC waveform of VGS=15V, the waveform with the SCM circuit has lower loss. The circuit enables SiC to be driven with a high gate voltage of 20V while maintaining short-circuit durability. Fig. 8 shows the $R_{DS(on)}$ improvement of

J3-T-PM when increasing the gate voltage. Changing the gate voltage from 15V to 20V in J3-T-PM improves $R_{DS(on)}$ by 32% at 25°C, contributing to the miniaturization of and low loss of the module.

To verify the SCM circuit operation in the worst-case scenario, short-circuit tests were conducted at different temperatures (-40, 25, and 175°C) and VDD=870V. The results are shown in Fig. 9 and Table 3. The SCM circuit reduces I_{SC} and E_{SC} at all temperatures and was proven to be highly robust protection circuit.

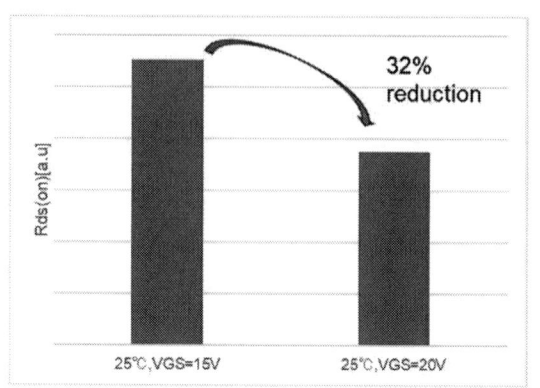

Fig. 8 Comparison of $R_{DS(on)}$ of J3-T-PM with different gate voltage

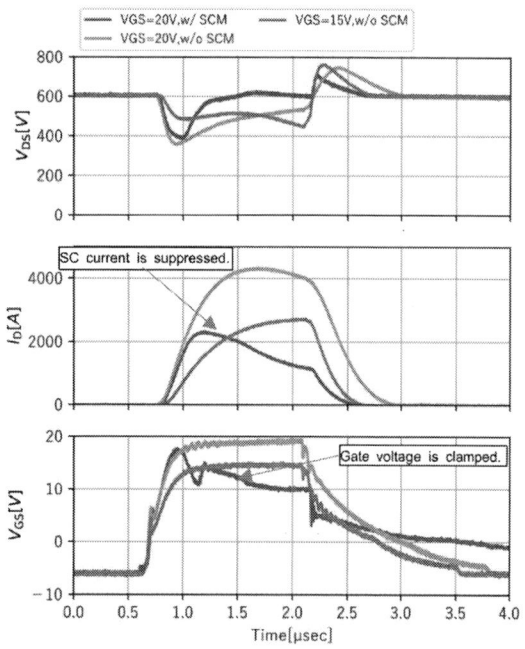

Fig. 7 SC waveform of SiC J3-T-PM (VDD=600V, Tvj=25°C)

Condition	I_{SCpeak}[A]	E_{SC}[J]
VGS=20V with SCM	2320	1.3
VGS=20V without SCM	4330	3.0
VGS=15V without SCM	2740	1.6

Table 2 SC characteristics of SiC J3-T-PM (VDD=600V,Tvj=25°C)

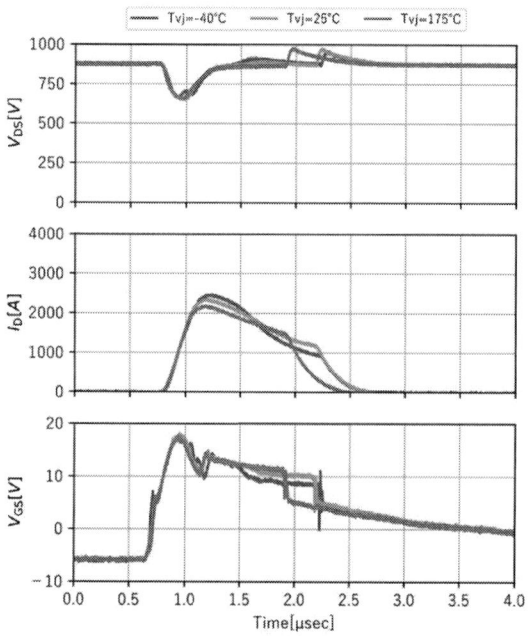

Fig. 9 SC waveform with SCM circuit at different junction temperature (VDD=870V)

Condition	I_{SCpeak}[A]	E_{SC}[J]
Tvj=-40°C	2450	2.0
Tvj=25°C	2350	2.0
Tvj=175°C	2190	1.7

Table 3 SC characteristics with SCM circuit at different junction temperature (VDD=870V)

4 Conclusion

In this paper, the rapid short-circuit protection method, SCM circuit, is proposed to take counter-measures against short-circuits of SiC. The SCM circuit starts to operate by detecting the di/dt of the short-circuit current during a short-circuit and using the parasitic inductance of the package effectively. When the SCM operates, it can suppress the gate voltage during a short-circuit, and by combining it with the conventional DESAT protection method, rapid short-circuit protection can be achieved.

When compared to three short-circuit protection methods, the method combining the proposed SCM circuit and DESAT protection in this paper improved losses by 54% compared to the conventional DESAT protection method, demonstrating the best characteristics.

When the SCM circuit was applied to the SiC module J3-T-PM, the losses were reduced by 57% compared to the DESAT protection method. By applying the SCM circuit to the SiC module J3-T-PM, it is possible to achieve both short-circuit durability and high gate voltage drive of 20V, which result in loss and minituarization of the module.

Through this technology, we will promote carbon-neutral efforts and contribute to the protection of the global environment.

References

[1] S.Nakata, Y.Nakaki, N.Miura: SiC Power Device Technology and Application, Mitsubishi Denki Giho, 84, No.4, 215–218 (2010)

[2] F.Sommer, N.Soltau, F.Stamer,et al.: Mirror Source based Overcurrent and Short Circuit Protection Method for High Power SiC MOSFETs, PCIM Europe 2021, 1712-1728(2021)

[3] K.Hideo, R.Yoneyama, T.Tokorozuki, Y.Higashi: Mitsubishi Denki Giho, 98, No.3, 3-01-3-05 (2024)

PCIM Europe 2024, 11–13 June 2024, Nuremberg DOI: 10.30420/566262391

Comparison of Dynamic Gate Stress Test Results of SiC MOSFETs

Mathias Gebhardt [1], Gabriel Lieser[1]

[1] SET GmbH, NI, Germany

Corresponding author: Mathias Gebhardt, mathias.gebhardt@ni.com
Speaker: Mathias Gebhardt, mathias.gebhardt@ni.com

Abstract

Silicon MOSFETs are continuously replaced by silicon carbide MOSFETs in several applications such as automotive, railway, or renewables. While utilizing the advantages of silicon carbide, such as higher junction temperatures, switching speeds, or efficiency, reliability must be equal or better. Switching a silicon carbide MOSFET using a bipolar gate-source voltage causes drift of the gate-source threshold voltage and the drain-source on resistance. This paper describes the setup used for dynamic gate stress tests of discrete silicon carbide MOSFETs of five manufacturers. It aims to show whether a static voltage below minimum gate-source voltage, an undershoot, or the difference between gate-source on and off state voltage affect the gate-source threshold voltage drift. Therefore, gate switching waveforms as well as observed drift are provided and compared.

1 Introduction

Dynamic gate stress (DGS) according to AQG 324 [1] as well as gate switching stress (GSS) according to JEP195 [2] evaluate the effect of gate switching pulses on silicon carbide (SiC) MOSFETs. Gate switching can cause a drift of gate-source threshold voltage ($V_{\mathrm{GS,th}}$) and drain-source on resistance ($R_{\mathrm{DS,on}}$) [3]. $V_{\mathrm{GS,th}}$ drift observed during test mainly depends on the number of switching cycles. In addition, a minimum gate-source voltage slope ($\mathrm{d}V_{\mathrm{GS}}/\mathrm{d}t$) is required, and a higher $V_{\mathrm{GS,th}}$ drift is expected when gate-source off state voltage ($V_{\mathrm{GS,off}}$) is further decreased.

While [4] provides results on stress and ambient temperature as well as frequency dependency this work focuses on the effect of preconditioning, $V_{\mathrm{GS,off}}$, and gate resistor (R_{G}) on $V_{\mathrm{GS,th}}$ drift. First, the transition from stress to readout and the pass-through from last stress state to measured $V_{\mathrm{GS,th}}$ is shown. Afterwards, the effect of $V_{\mathrm{GS,off}}$ with constant gate-source on state voltage ($V_{\mathrm{GS,on}}$) as well as with constant difference between gate-source on and off state voltage ($\Delta V_{\mathrm{GS,on-off}}$) on $V_{\mathrm{GS,th}}$ drift of five commercially available 1200 V, 80 mΩ, TO 247-4L devices is investigated. Finally, the influence of R_{G} on $V_{\mathrm{GS,th}}$ drift is analyzed. The aim is to show whether the increase in $V_{\mathrm{GS,th}}$ drift is caused by either lower $V_{\mathrm{GS,off}}$, higher $\Delta V_{\mathrm{GS,on-off}}$, or under-

shoots caused by lower R_{G}.

The manufacturers and devices match those of the reported results on DGS test with and without drain-source voltage (V_{DS}) bias [5]. All $\mathrm{d}V_{\mathrm{GS}}/\mathrm{d}t$ values given herein are measured from 40 % to 60 %, see JEP190 [6, Table 1].

2 Test setup

Fig. 1 shows the block diagram of the used DGS test system. There is a PWM unit and a power stage for stress generation. Switching frequency has been set to 500 kHz with a duty cycle of 50 %. $V_{\mathrm{GS,on}}$ and $V_{\mathrm{GS,off}}$ of the power stage can be configured to be in the range of –35 V to 35 V. To configure $\mathrm{d}V_{\mathrm{GS}}/\mathrm{d}t$ and control the over- and undershoot R_{G} can be selected out of 0 Ω, 1.2 Ω, 3.9 Ω, 6.8 Ω and 10 Ω. Drain has been shorted to source during stress as all testes have been performed without V_{DS} bias. The case temperature (T_c) is actively controlled to 25 ℃ ± 1 ℃ during stress as well as in-situ readout. Maintaining a constant temperature during readout is imported as the thermal drift of $V_{\mathrm{GS,th}}$ is expected to be in the same order of magnitude as the $V_{\mathrm{GS,th}}$ drift caused by DGS [4, Fig. 6].

The in-situ readout circuit measures $V_{\mathrm{GS,th}}$ and $R_{\mathrm{DS,on}}$ afterwards. The transition from stress to readout as well as the readout sequence itself is controlled by an FPGA. This ensures repeatable timings for every measurement. The stress

PCIM Europe 2024, 11– 13 June 2024, Nuremberg   DOI: 10.30420/566262391

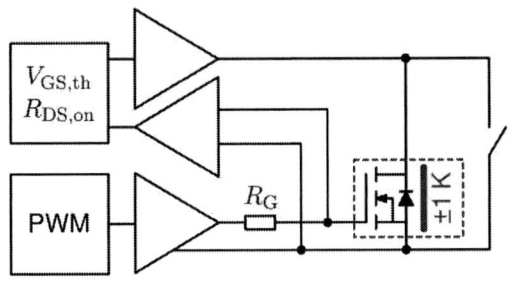

Fig. 1: Block diagram of DGS system

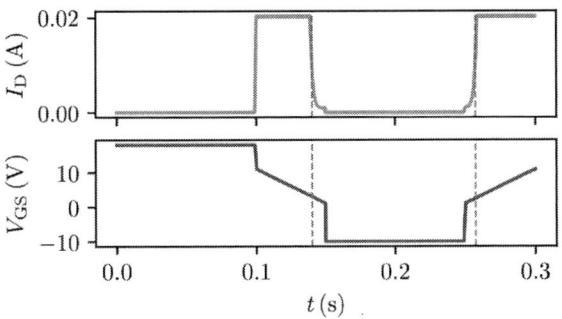

Fig. 2: Exemplary I_D and V_GS waveform of a $V_\mathrm{GS,th}$ measurement

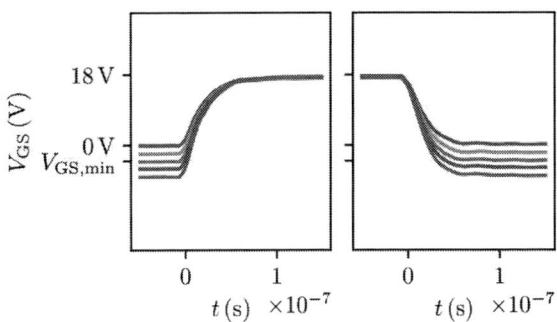

Fig. 3: V_GS switching edges of manufacturer A with constant $V_\mathrm{GS,on}$ of 18 V

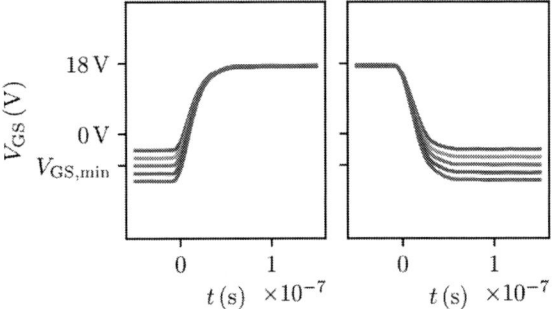

Fig. 4: V_GS switching edges of manufacturer B with constant $V_\mathrm{GS,on}$ of 18 V

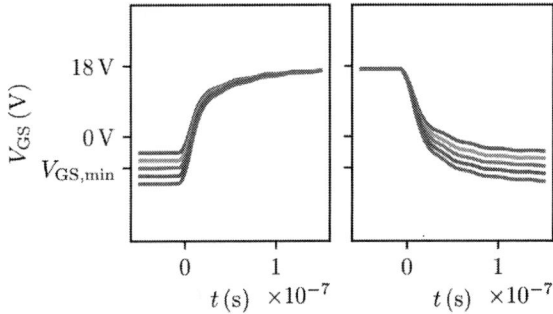

Fig. 5: V_GS switching edges of manufacturer C with constant $V_\mathrm{GS,on}$ of 18 V

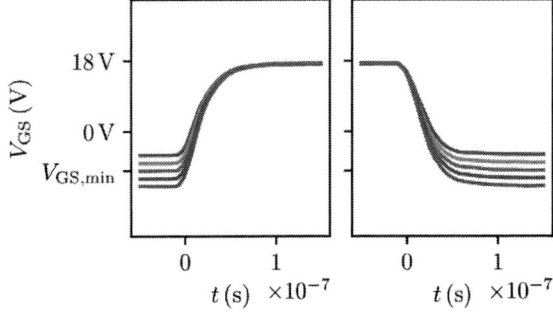

Fig. 6: V_GS switching edges of manufacturer D with constant $V_\mathrm{GS,on}$ of 18 V

is stopped after the off phase right before the next rising edge. $V_\mathrm{GS,th}$ has been measured according to [7] by first a 100 ms positive preconditioning pulse at $V_\mathrm{GS,on}$. Followed by a down slope from 11 V to 1 V to determine down-slope gate-source threshold voltage ($V_\mathrm{GS,th-}$) when drain current (I_D) equals 10 mA. Afterwards, a 100 ms negative preconditioning pulse at $V_\mathrm{GS,off}$ has been performed. Finally, an up slope from 1 V to 11 V determines up-slope gate-source threshold voltage ($V_\mathrm{GS,th+}$) when I_D equals 10 mA. An exemplary I_D and gate-source voltage (V_GS) waveform is shown in Fig. 2. Both green vertical lines mark the 10 mA crossings where the respective $V_\mathrm{GS,th}$ is measured.

The following figures show the rising and falling edges of a gate switching event. All figures share the same x- and y-axis scaling. Minimum gate-source voltage ($V_\mathrm{GS,min}$) is marked according to manufacturer's respective data sheet rating. Figs. 3 to 7 show the resulting switching edges for each manufacturer for $V_\mathrm{GS,on}$ of 18 V and $V_\mathrm{GS,off}$ of –4 V, –2 V, +0 V, +2 V and +4 V relative to manufacturer's respective $V_\mathrm{GS,min}$. R_G has been set to 10 Ω to avoid any over- or undershoot.

Figs. 8 to 12 show the resulting switching edges for each manufacturer while $V_\mathrm{GS,off}$ has been set

2770

PCIM Europe 2024, 11– 13 June 2024, Nuremberg   DOI: 10.30420/566262391

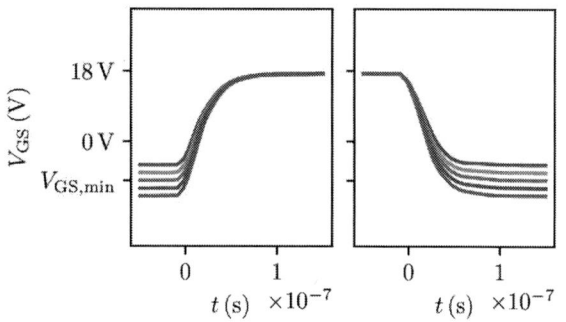

Fig. 7: V_{GS} switching edges of manufacturer G with constant $V_{GS,on}$ of 18 V

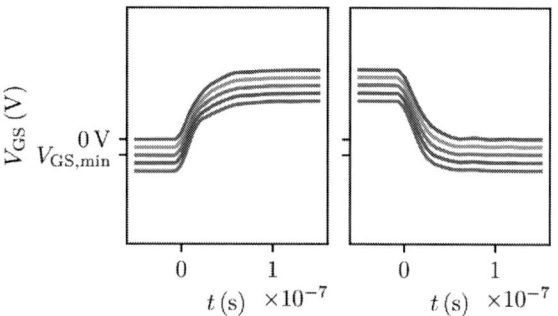

Fig. 8: V_{GS} switching edges of manufacturer A with constant $\Delta V_{GS,on-off}$ of 18 V

to $-4\,\text{V}$, $-2\,\text{V}$, $+0\,\text{V}$, $+2\,\text{V}$ and $+4\,\text{V}$ relative to manufacturer's respective $V_{GS,min}$ as well but $\Delta V_{GS,on-off}$ has been set to 18 V. R_G has been kept at $10\,\Omega$ to avoid any over- or undershoot.

Figs. 13 to 17 show the resulting switching edges for each manufacturer where $V_{GS,off}$ has been set to manufacturer's respective $V_{GS,min}$ and $V_{GS,on}$ of 18 V. The horizontal dashed lines mark manufacturer's respective $V_{GS,min}$ minus 4 V. R_G has been set to $0\,\Omega$, $1.2\,\Omega$, $3.9\,\Omega$, $6.8\,\Omega$ and $10\,\Omega$ respectively. It can be observed, that R_G of $10\,\Omega$ does not trigger

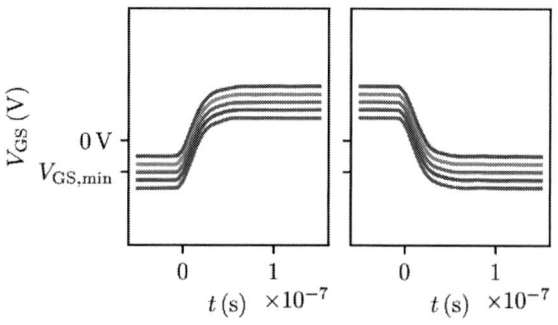

Fig. 9: V_{GS} switching edges of manufacturer B with constant $\Delta V_{GS,on-off}$ of 18 V

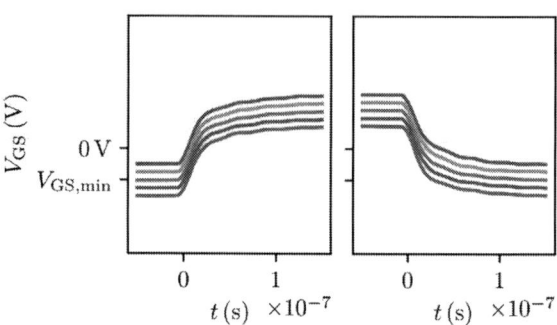

Fig. 10: V_{GS} switching edges of manufacturer C with constant $\Delta V_{GS,on-off}$ of 18 V

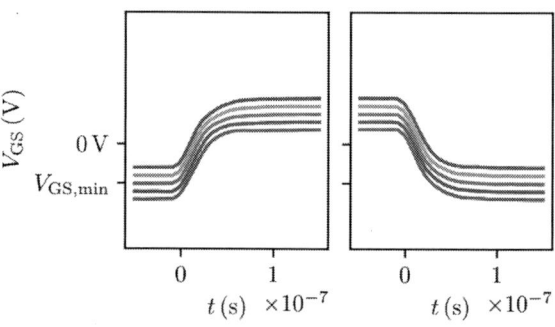

Fig. 11: V_{GS} switching edges of manufacturer D with constant $\Delta V_{GS,on-off}$ of 18 V

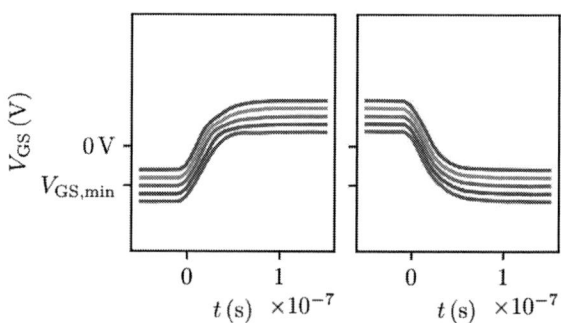

Fig. 12: V_{GS} switching edges of manufacturer G with constant $\Delta V_{GS,on-off}$ of 18 V

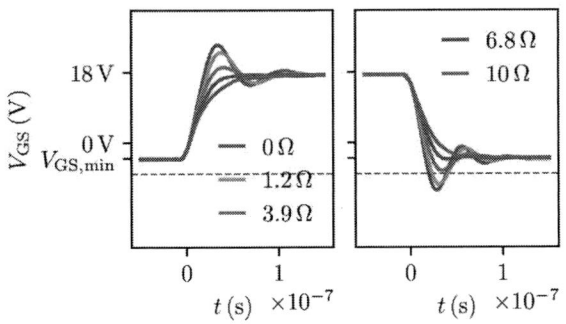

Fig. 13: V_{GS} switching edges of manufacturer A for different R_G

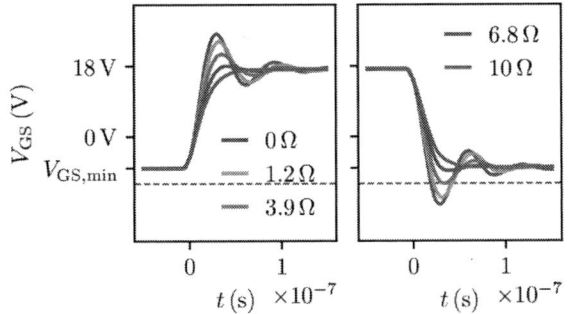

Fig. 14: V_{GS} switching edges of manufacturer B for different R_G

any over- or undershoot while R_G of 6.8 Ω already triggers small over- and undershoots for some manufacturers. R_G equal or below 3.9 Ω are causing under- and overshoot for all manufacturers.

Table 1 compare the measured dV_{GS}/dt from 40 % to 60 % in V/ns across all manufacturers and R_G for $V_{GS,on}$ of 18 V and $V_{GS,off}$ of manufacturer's respective $V_{GS,min}$.

A comparison of dV_{GS}/dt from 10 % to 90 % and dV_{GS}/dt from 40 % to 60 % can be seen in table 2. The right column shows the difference between

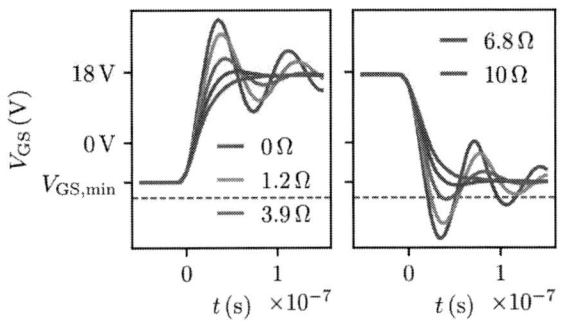

Fig. 16: V_{GS} switching edges of manufacturer D for different R_G

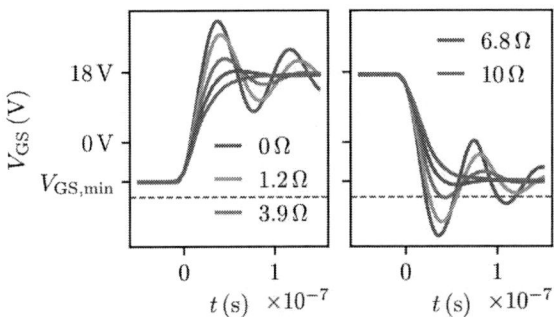

Fig. 17: V_{GS} switching edges of manufacturer G for different R_G

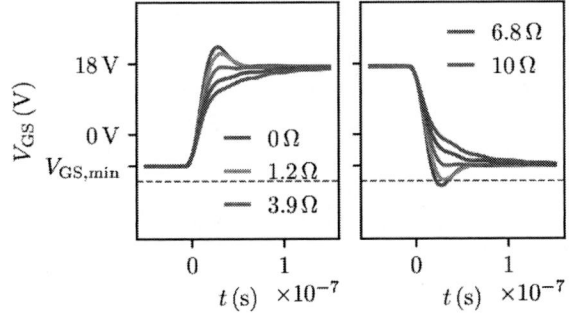

Fig. 15: V_{GS} switching edges of manufacturer C for different R_G

	0 Ω	1.2 Ω	3.9 Ω	6.8 Ω	10 Ω
A	1.19	0.98	0.80	0.66	0.48
B	1.69	1.43	1.16	0.97	0.79
C	1.69	1.49	1.19	0.92	0.68
D	1.41	1.26	1.09	0.87	0.67
G	1.33	1.19	1.00	0.83	0.63

Tab. 1: Rising edge dV_{GS}/dt in V/ns

Tab. 2: dV_{GS}/dt for R_G of $10\,\Omega$ in V/ns

	10 % to 90 %	40 % to 60 %	%
A	0.36	0.48	−25
B	0.59	0.79	−25
C	0.27	0.68	−60
D	0.47	0.67	−29
G	0.45	0.63	−28

both values in % relative to dV_{GS}/dt from 40 % to 60 %. Using dV_{GS}/dt from 10 % to 90 % will give an about 30 % smaller value when setting R_G in a way that there are no under- and overshoots.

3 Test results

The following figures, except Fig. 18, show the absolute $V_{GS,th-}$ change referred to the initial reading before stress over cycles of each manufacture in a double-logarithmic plot sharing common x- and y-axis scales.

3.1 Transition from stress to readout

When stress is stopped and the device is connected to readout logic timing must be repeatable to measure $V_{GS,th}$ correctly [7]. Therefore, FPGA based timing has already been implemented into the first lab-based demonstrator setups. In addition, to switch-over time dependencies the last stress state can be observed in $V_{GS,th}$ reading as well. Fig. 18 shows the $V_{GS,th}$ reading on the left side and an exemplary V_{GS} waveforms during transition time for each group. These measurements have been performed using the first lab-based demonstrator. Although all timings have already been implemented on an FPGA there was no synchronization between the transition from stress to readout and the actual stress state. In addition, $V_{GS,th}$ measurement has been performed by first, a 100 ms negative preconditioning, followed by an up slope to determine $V_{GS,th+}$ and, second, a 100 ms positive preconditioning, followed by a down slope to determine $V_{GS,th-}$.

This has resulted into $V_{GS,th}$ measurements that can be put into three groups. First, those where stress has been stopped during the on-phase, depicted by the top right waveform. Second, those where stress has been stopped during the off-phase, depicted by the bottom right waveform. Third, transitions to readout during the on to off transition as shown by the middle right waveform.

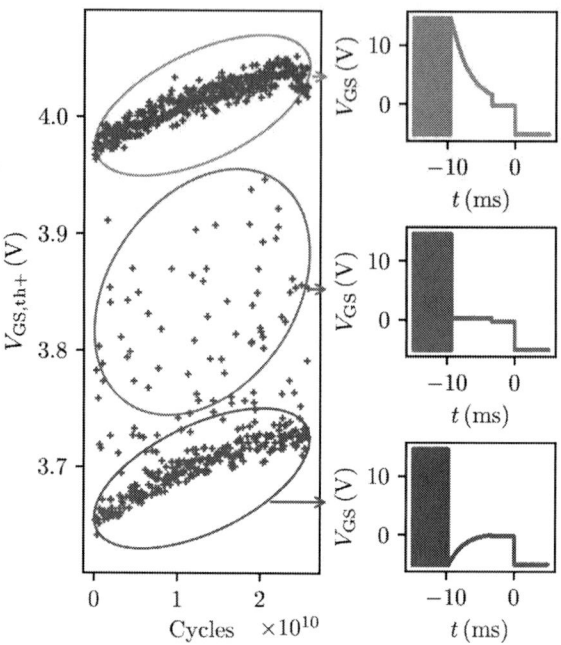

Fig. 18: $V_{GS,th-}$ drift over cycles of manufacturer A with constant $V_{GS,on}$ of 18 V

The pass-through of the last stress state could only be observed for $V_{GS,th+}$ that is determined first. There could be no significant influence on $V_{GS,th-}$ observed that is determined second.

By taking these results into account the questions arises why a 5 µs on- or off-time passes through a 100 ms preconditioning time but could not be observed during the second $V_{GS,th}$ measurement slope anymore. One explanation could be that the last switching event causes a shift of the first $V_{GS,th}$ reading rather than the on- or off-state. Therefore, it is important that a DGS test system not only provides a repeatable timing but also always transitions form the same stress state to readout sequence to gain repeatable $V_{GS,th}$ readings.

3.2 Influence of static gate-source off state voltage

While all manufacturers share same $V_{GS,on}$ and R_G, $V_{GS,off}$ has been set to −4 V, −2 V, +0 V, +2 V and +4 V relative to manufacturer's respective $V_{GS,min}$. It is expected that lower $V_{GS,off}$ causes a higher $V_{GS,th}$ drift. Figs. 19 to 23 show the observed $V_{GS,th-}$ drift over cycles. Manufacturer A has shown a huge increase in $V_{GS,th}$ drift for $V_{GS,min}$ −2 V and −4 V, as can be seen Fig. 19. $V_{GS,th}$ drift for $V_{GS,min}$ −2 V and +4 V of Manufacturer C, see Fig. 21, has started negative resulting in parts of the curve not

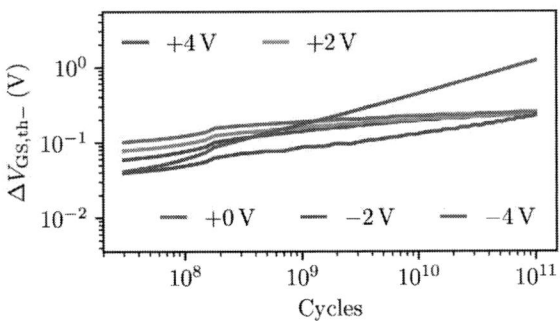

Fig. 19: $V_{\mathrm{GS,th-}}$ drift over cycles of manufacturer A with constant $V_{\mathrm{GS,on}}$ of 18 V

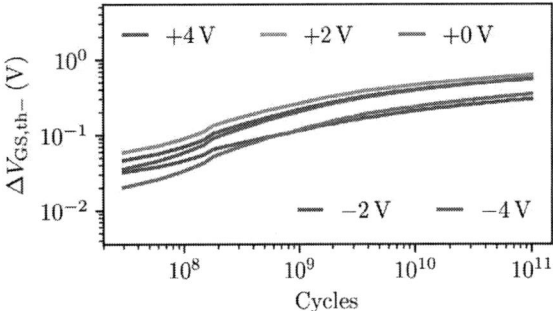

Fig. 22: $V_{\mathrm{GS,th-}}$ drift over cycles of manufacturer D with constant $V_{\mathrm{GS,on}}$ of 18 V

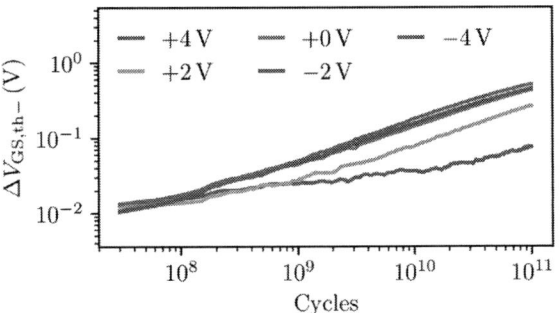

Fig. 20: $V_{\mathrm{GS,th-}}$ drift over cycles of manufacturer B with constant $V_{\mathrm{GS,on}}$ of 18 V

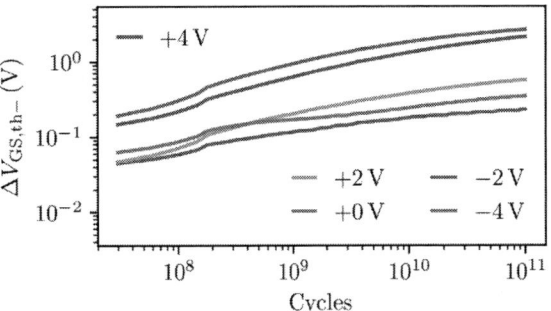

Fig. 23: $V_{\mathrm{GS,th-}}$ drift over cycles of manufacturer G with constant $V_{\mathrm{GS,on}}$ of 18 V

shown in the double logarithmic plot. For manufacturer D no clear dependency on $V_{\mathrm{GS,off}}$ could be observed, see Fig. 22. Manufacturers B and G, see Figs. 20 and 23, show the expected behavior.

3.3 Influence of constant difference between gate-source on and off state voltage

In contrast to the previous test $V_{\mathrm{GS,on}}$ has been set to $V_{\mathrm{GS,off}}$ plus 18 V. $V_{\mathrm{GS,off}}$ has been set to $V_{\mathrm{GS,min}}$ −4 V, −2 V, +0 V, +2 V and +4 V relative to manufac-

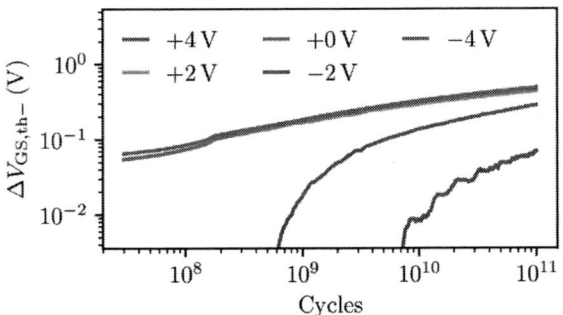

Fig. 21: $V_{\mathrm{GS,th-}}$ drift over cycles of manufacturer C with constant $V_{\mathrm{GS,on}}$ of 18 V

turer's respective $V_{\mathrm{GS,min}}$. It is expected that lower $V_{\mathrm{GS,off}}$ causes a higher $V_{\mathrm{GS,th}}$ drift. Figs. 24 to 28 show the observed $V_{\mathrm{GS,th-}}$ drift over cycles. Manufacturer G clearly shows the opposite, see Fig. 39, where $V_{\mathrm{GS,th}}$ drift decreases with lower $V_{\mathrm{GS,off}}$. A first, negative drift can be observed for $V_{\mathrm{GS,min}}$ −4 V of manufactures B, see Fig. 25, and D, see Fig. 27. As can bee seen in Fig. 26, manufacturer C shows no significant sensitivity.

Comparing the $V_{\mathrm{GS,th}}$ drift of constant $V_{\mathrm{GS,on}}$ to constant $\Delta V_{\mathrm{GS,on-off}}$ less $V_{\mathrm{GS,th}}$ drift for constant $\Delta V_{\mathrm{GS,on-off}}$ can be observed for manufacturer A, see Figs. 19 and 24. As can be seen in Figs. 21 and 26, manufacturer C shows no significant sensitivity, neither for constant $V_{\mathrm{GS,on}}$ nor constant $\Delta V_{\mathrm{GS,on-off}}$.

3.4 Influence of undershoot and gate resistor

To prove a possible undershoot sensitivity R_{G} has been varied while $V_{\mathrm{GS,on}}$ has been set to 18 V and $V_{\mathrm{GS,off}}$ to manufacturer's respective $V_{\mathrm{GS,min}}$. It is expected to see an increase in $V_{\mathrm{GS,th}}$ drift for higher undershoots and therefore lower R_{G}. Figs. 29 to 33 show the observed $V_{\mathrm{GS,th-}}$ drift over cycles. Manu-

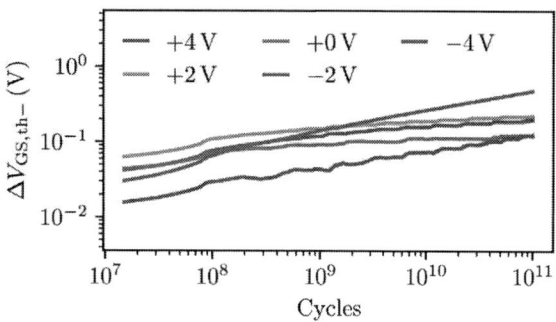

Fig. 24: $V_{GS,th-}$ drift over cycles of manufacturer A with constant $\Delta V_{GS,on-off}$ of 18 V

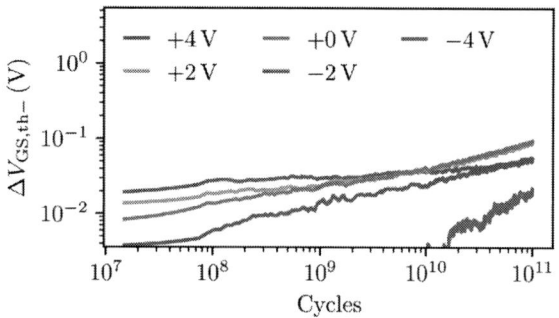

Fig. 25: $V_{GS,th-}$ drift over cycles of manufacturer B with constant $\Delta V_{GS,on-off}$ of 18 V

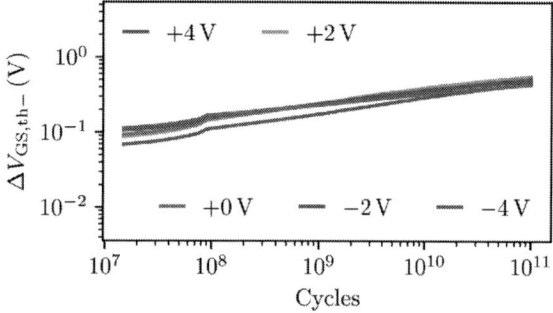

Fig. 26: $V_{GS,th-}$ drift over cycles of manufacturer C with constant $\Delta V_{GS,on-off}$ of 18 V

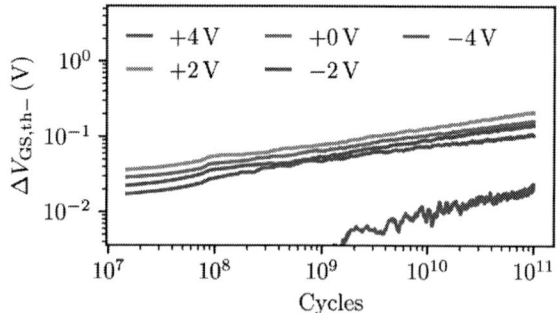

Fig. 27: $V_{GS,th-}$ drift over cycles of manufacturer D with constant $\Delta V_{GS,on-off}$ of 18 V

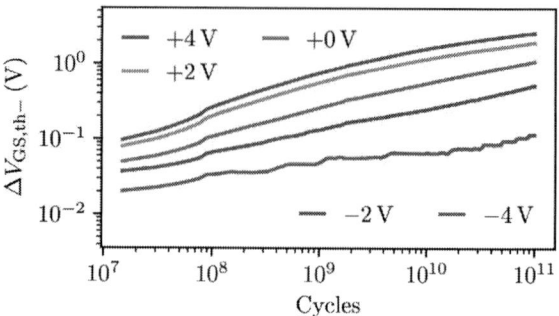

Fig. 28: $V_{GS,th-}$ drift over cycles of manufacturer G with constant $\Delta V_{GS,on-off}$ of 18 V

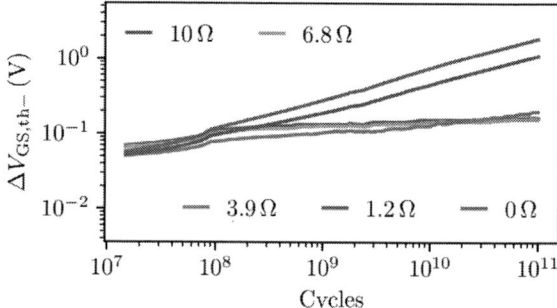

Fig. 29: $V_{GS,th-}$ drift over cycles of manufacturer A for different R_G

facturer A meet the expectation, see Fig. 29. While manufacturer B shows no sensitivity during first 10^{10} cycles but a decrease in $V_{GS,th}$ drift slope can be observed for R_G of 0 Ω and 1.2 Ω afterwards, see Fig. 30. As can be seen in Fig. 31, manufacturer C does not show any significant sensitivity. In contrast to what is expected, lower R_G values and therefor higher undershoots cause a more negative $V_{GS,th}$ drift in the first stage of manufacture D. As the double-logarithmic plot, see Fig. 32, is not able to show the negative $V_{GS,th}$ drift Fig. 34 provides a semi-logarithmic plot of manufacturer D in addition. V_{GS} drift of manufacturer G decreases with lower R_G.

When comparing these results to the two previous test, manufacturer C shows no significant sensitivity at all. Manufacturers D and G continued to show less $V_{GS,th}$ drift when R_G is decreased and the undershoot as well as the overshoot is increased therefor.

4 Conclusion

When performing in-situ measurements to track $V_{GS,th}$ drift over cycles both repeatable timing and a well-defined transition from stress to readout are

PCIM Europe 2024, 11– 13 June 2024, Nuremberg   DOI: 10.30420/566262391

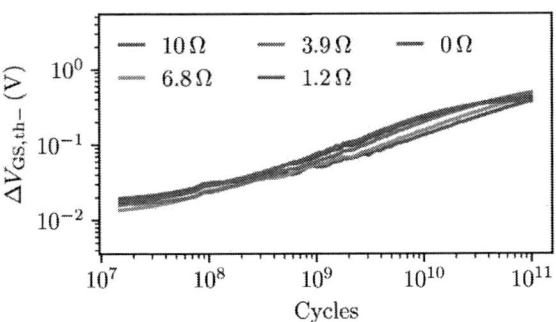

Fig. 30: $V_{\mathrm{GS,th-}}$ drift over cycles of manufacturer B for different R_{G}

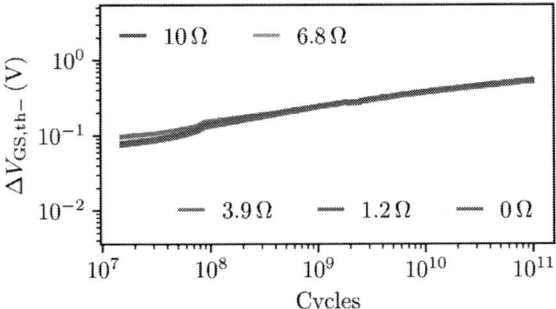

Fig. 31: $V_{\mathrm{GS,th-}}$ drift over cycles of manufacturer C for different R_{G}

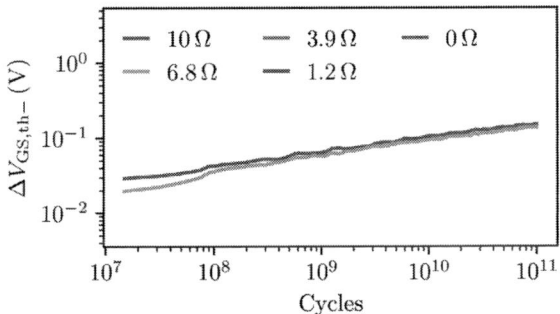

Fig. 32: $V_{\mathrm{GS,th-}}$ drift over cycles of manufacturer D for different R_{G}

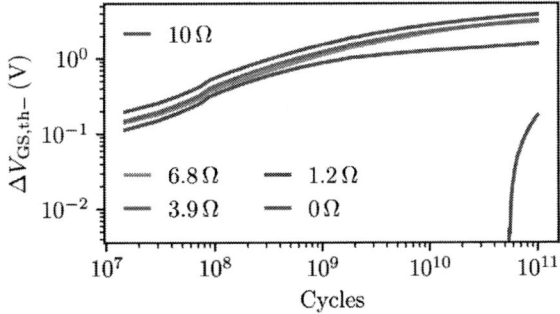

Fig. 33: $V_{\mathrm{GS,th-}}$ drift over cycles of manufacturer G for different R_{G}

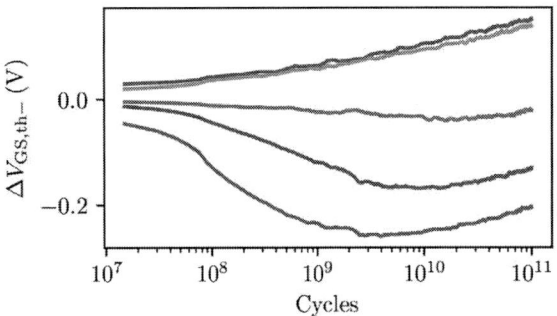

Fig. 34: $V_{\mathrm{GS,th-}}$ drift over cycles of manufacturer D for different R_{G} in a semi-logarithmic plot

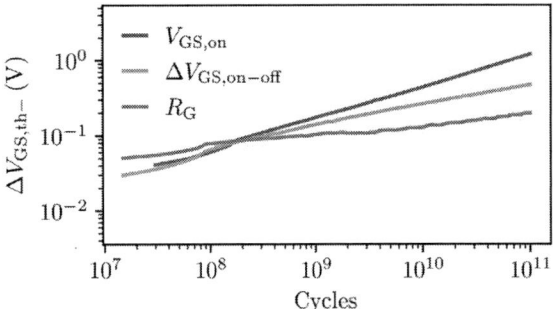

Fig. 35: $V_{\mathrm{GS,th-}}$ drift over cycles of manufacturer A for $V_{\mathrm{GS,off}}$ of $V_{\mathrm{GS,min}}$ –4 V

required. It has to be ensured that stress is always left at the same gate state and time between end of stress and start of preconditioning is constant.

To compare the $V_{\mathrm{GS,th}}$ drift of static $V_{\mathrm{GS,off}}$, constant $\Delta V_{\mathrm{GS,on-off}}$, and undershoot by R_{G} Figs. 35 to 39 show the $V_{\mathrm{GS,th-}}$ drift over cycles for each manufacture. For $V_{\mathrm{GS,off}}$ and $\Delta V_{\mathrm{GS,on-off}}$ $V_{\mathrm{GS,min}}$ –4 V has been chosen. R_{G} has been selected such as the undershoot equals $V_{\mathrm{GS,min}}$ –4 V as well. Therefore, 3.9 Ω has been selected for manufacturers A, B, D, and G while for manufacturer C 1.2 Ω has been selected.

Manufacturer A, see Fig. 35, shows the highest sensitivity to static $V_{\mathrm{GS,off}}$. Reducing $V_{\mathrm{GS,on}}$ to gain a constant $\Delta V_{\mathrm{GS,on-off}}$ reduces the $V_{\mathrm{GS,th}}$ drift as well. Short time undershoots cause less $V_{\mathrm{GS,th}}$ drift compared to constant $V_{\mathrm{GS,off}}$.

Constant $V_{\mathrm{GS,off}}$ as well as short term undershoots result in an equal $V_{\mathrm{GS,th}}$ drift behavior of manufacturer B, see Fig. 36. Reducing $V_{\mathrm{GS,on}}$ causes a negative $V_{\mathrm{GS,th}}$ drift for the first 10^{10} cycles and therefore significantly lowers the overall $V_{\mathrm{GS,th}}$ drift at 10^{11} cycles.

No significant sensitivity to any change in stress

2776

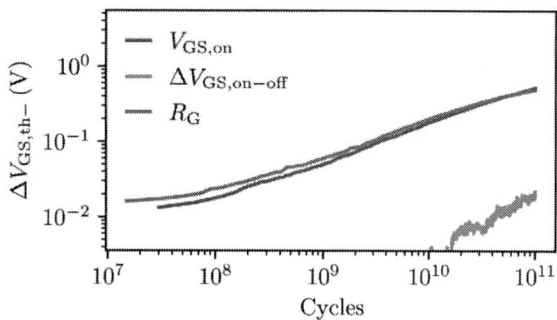

Fig. 36: $V_{GS,th-}$ drift over cycles of manufacturer B for $V_{GS,off}$ of $V_{GS,min}$ −4 V

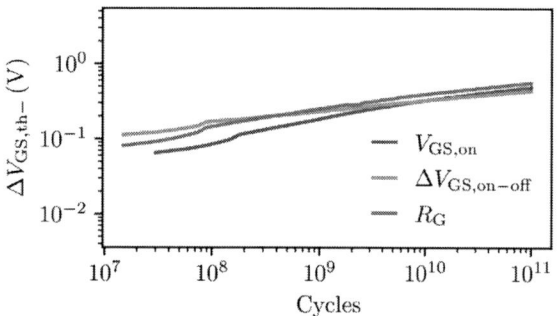

Fig. 37: $V_{GS,th-}$ drift over cycles of manufacturer C for $V_{GS,off}$ of $V_{GS,min}$ −4 V

could be observed for manufacturer C, see Fig. 37. As can be seen in Fig. 38, decreasing $V_{GS,on}$ as well as short time undershoots cause a negative $V_{GS,th}$ drift in the first stage. A significantly lower $V_{GS,th}$ drift can be observed compared to static $V_{GS,off}$. Manufacturer G shows a comparable $V_{GS,th}$ drift for both static $V_{GS,off}$ and short term undershoots, see Fig. 39. $V_{GS,th}$ drift of reduced $V_{GS,on}$ is significantly lower.

In conclusion, the sensitivity to static $V_{GS,off}$, short term undershoots, and constant $\Delta V_{GS,on-off}$ is de-

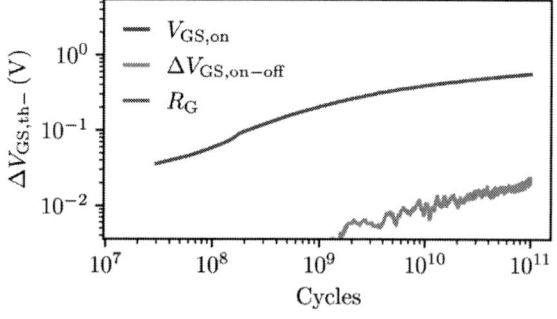

Fig. 38: $V_{GS,th-}$ drift over cycles of manufacturer D for $V_{GS,off}$ of $V_{GS,min}$ −4 V

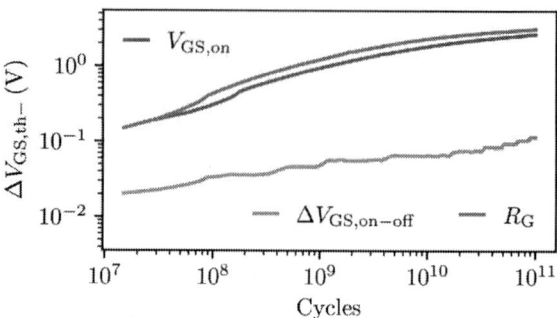

Fig. 39: $V_{GS,th-}$ drift over cycles of manufacturer G for $V_{GS,off}$ of $V_{GS,min}$ −4 V

vice dependent. There are manufactures that show no sensitivity at all. Some meet the expectation that lower $V_{GS,off}$ or higher undershoots cause a higher $V_{GS,th}$ drift. Others show an initial negative $V_{GS,th}$ drift.

References

[1] "ECPE guideline AQG 324, qualification of power modules for use in power electronics converter units in motor vehicles," ECPE, AQG 324, 2021.

[2] "Guideline for evaluating gate switching instability of silicon carbide metal-oxide-semiconductor devices for power electronic conversion," JEDEC, JEP195, 2023.

[3] P. Salmen, M. W. Feil, K. Waschneck, H. Reisinger, G. Rescher, and T. Aichinger, "A new test procedure to realistically estimate end-of-life electrical parameter stability of sic mosfets in switching operation," in *2021 IEEE International Reliability Physics Symposium (IRPS)*, 2021, pp. 1–7. DOI: 10.1109/IRPS46558.2021.9405207.

[4] F. Heidemann, M. Gebhardt, and S. Strasser, "Reliability challenges in SiC components – performing dynamic test methodologies like DGS for meaningful measurement results," in *2024 China Semiconductor Technology International Conference (CSTIC)*, in press, 2024.

[5] M. Gebhardt and G. Lieser, "Influence of VDS bias on sic mosfet dynamic gate stress," in *2024 13th International Conference on Integrated Power Electronics Systems (CIPS)*, 2024, pp. 228–232.

[6] "Guideline for evaluating dV/dt robustness of SiC power devices," JEDEC, JEP190, 2022.

[7] "Guidelines for measuring the threshold voltage (VT) of SiC MOSFETs," JEDEC, JEP183A, 2023.

PCIM Europe 2024, 11– 13 June 2024, Nuremberg   DOI: 10.30420/566262392

Extending SiC MOSFET Short-Circuit Withstanding Time by Two-Level Turn-Off Gate Driving

Dinesh Palaniappan, Kwok Wai Ma

Infineon Technologies Asia Pacific Pte Ltd, Singapore

Corresponding author: Kwok Wai Ma, Kwokwai.ma@infineon.com

Abstract

Short-circuit withstanding time t_{SC} requirements in industrial motor drive applications are often difficult to meet for commercially available silicon carbide (SiC) MOSFETs without compromising $R_{DS(ON)}A$ performance. This paper proposes using a two-level turn-off (TLTO) mechanism to slow down the short-circuit energy rise in the SiC MOSFET, and thus extend its t_{SC}. The intrinsic t_{SC} of SiC MOSFETs at different V_{GS} and V_{DS} is first characterized by destructive tests, as the basis for selecting key TLTO parameters. Using a digital gate driver IC for desaturation (DESAT) detection with TLTO, the extension of the short-circuit withstanding time to 10 μs and beyond is demonstrated under different short-circuit impedances.

1 Introduction

Silicon carbide (SiC) power MOSFETs have lower switching losses and conduction losses than traditional power devices such as silicon IGBTs. Thus, they are becoming more widely accepted in recent years in several industrial power applications. They contribute to the growing trend in the design of power-electronic converters to improve efficiency and power density.

In certain industrial power applications, especially motor drives, short-circuit withstanding capability is a key requirement for the robustness of power semiconductor devices to protect them from possible fault patterns such as inverter shoot-through or phase-to-phase short circuits. This capability is often specified in datasheets as the short-circuit withstanding time t_{SC} of power semiconductors, which is expected to be longer than the response time required for system protection. Traditionally, IGBTs used in industrial applications have been specified with a t_{SC} of 10 μs [1]. Newer IGBT generations targeting motor-drive applications are being released with the t_{SC} lowered to 8 μs or less to reduce conduction losses. Some new IGBT families targeting power supply applications are being released without any t_{SC} rating as a trade-off for achieving better switching and conduction losses. Similar design concept is also seen in the SiC power MOSFETs.

One major design target of the SiC power MOSFET is to reduce the specific on-resistance $R_{DS(ON)}A$ to reduce the conduction losses of the device. Most of the commercially available SiC power MOSFETs in the market today have no t_{SC} ratings, or only very short ones specified e.g. 2-3 μs to achieve the lowest $R_{DS(ON)}A$. It is possible to design and specify SiC MOSFETs with a longer t_{SC}, however this would result in lower channel conductivity and thus higher $R_{DS(ON)}A$. Optimizing the design of SiC power MOSFETs for better performance (lower $R_{DS(ON)}$) instead of short-circuit robustness (longer t_{SC}) is generally acceptable for power supply applications. However, using such SiC power MOSFETs without or with short t_{SC} in industrial motor-drive applications is a challenge for the protection circuitry of the system to provide fast and reliable responses to fault [2]. Extending the t_{SC} of SiC power MOSFETs without sacrificing $R_{DS(ON)}A$ enables the use of SiC MOSFETs in industrial power applications that require robust fault protection.

This paper proposes an application technique that extends the t_{SC} of a SiC power MOSFET without compromising its $R_{DS(ON)}$. The proposed

2778

concept makes use of configurable desaturation short-circuit protection together with configurable two-level turn-off gate drive features that are available in modern digital gate driver ICs. The operation principle, selection of key parameters, hardware setup and the measurement results are explained in the following sections.

2 Extend short-circuit capability by TLTO

During overcurrent and short-circuit faults in power semiconductors, such as Si IGBTs and SiC MOSFETs, voltage overshoots and severe gate oscillations at turn-off are quite typical. These are due to the very high negative turn-off current gradient in conjunction with commutation loop inductance. This poses a challenge to designers, as the peak overshoot voltage needs to be kept within the maximum voltage limit of the power switch for safe operation. This is even more important in wide bandgap devices like SiC MOSFETs because of their inherent high-speed switching capabilities with very high di/dt values. The DC link bus voltage may have to be reduced or the turn-off speed reduced accordingly.

To suppress such voltage overshoot during overcurrent and short circuit events, two-level turn-off (TLTO) concept is widely used. It is a well-known feature in many commercially available gate driver ICs. When TLTO is used during such fault turn-off conditions, the gate voltage instead of going down directly to the turn-off gate voltage is reduced and maintained at an intermediate level (typically in the range of 9 to 12 V) for a desired duration before initiating the final turn-off. The intermediate gate voltage will reduce the fault current prior to final turn-off, and thus reduce the voltage overshoot and oscillations [3-5].

In this paper, we propose a new use case for TLTO to extend the short-circuit withstanding capability of SiC MOSFETs.

For SiC MOSFETs, the saturation current is directly controlled by its turn-on gate voltage, while the short-circuit withstanding capability is determined by short-circuit energy, which is the time integral of saturation current and blocking voltage during a short-circuit event.

With TLTO operation during a SiC MOSFET short circuit, the gate voltage will be maintained at an intermediate level that is lower than the normal operating gate voltage, but still above gate threshold voltage V_{TH}. The short-circuit fault current will be brought down considerably from the original peak current magnitude, and thereby reduce the instantaneous power dissipation as well as the rate of rise of short-circuit energy E_{SC} within the device.

With the TLTO short-circuit protection scheme applied to the SiC MOSFET, a high gate voltage $V_{GS,H}$ will be used during normal operation to give low $R_{DS(ON)}$ for best conduction performance. When short circuits occur, the gate voltage will still be maintained at $V_{GS,H}$ for a period t_H. Afterwards, the gate voltage will be reduced to an intermediate value $V_{GS,L}$ lower than $V_{GS,H}$ to reduce short-circuit current, and will be maintained at this level for a period t_L before final shut-down. The new short-circuit withstanding time with TLTO will now be $t_H + t_L$. Since the rate of rise of short-circuit energy E_{SC} is lower when gate voltage is reduced, $t_H + t_L$ will be longer than the t_{SC} value specified at $V_{GS,H}$. This implementation thus enables t_{SC} to be extended without compromising conduction loss performance.

3 Test setup with digital gate driver IC

A typical double-pulse characterization setup was used, as indicated in Fig. 1; the simplified schematic of the setup is shown in Fig. 2. The low-side switch of the half bridge is used as device under test (DUT), with the top-side switch position being replaced by low-impedance short circuit. The evaluation board of a digital gate driver IC 1ED3890MC12M (area highlighted in red), together with its digital configuration board from Infineon (area highlighted in yellow), were used to perform short-circuit measurements on a 30 mΩ 1200 V SiC MOSFET IMW120R030M1H from Infineon. An external auxiliary power supply board (area highlighted in green) was used to vary the turn-on gate voltage applied to assess the SiC MOSFET short-circuit capability under different conditions, with the turn-off gate voltage always kept at 0 V. Default gate resistors in the evaluation board of 22 Ω and 10 Ω were used for gate turn-on and turn-off, respectively.

PCIM Europe 2024, 11– 13 June 2024, Nuremberg   DOI: 10.30420/566262392

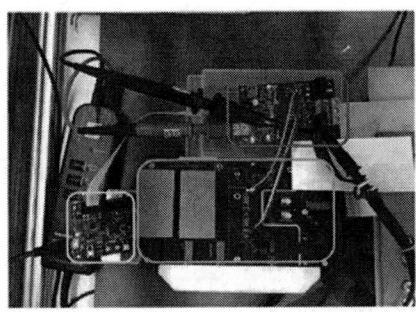

Fig. 1 Double-pulse test setup with SiC MOSFET IMW120R030M1H and digital gate driver IC 1ED3890MC12M from Infineon

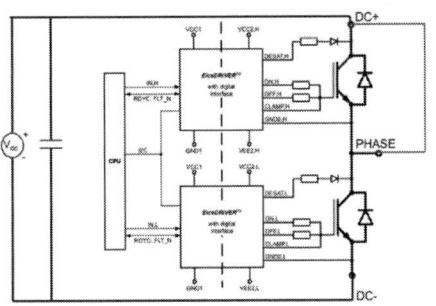

Fig. 2 Simplified schematic diagram of the double-pulse test setup

Additional measurements at selected conditions were carried out with lower turn-on gate resistance of 2.2 Ω as well to study its impact on t_{SC}. All measurements were carried out at room temperature.

A distinctive advantage of the digital gate driver IC 1ED3890MC12M is that many of the integrated protection features like DESAT, TLTO, soft turn-off, UVLO, etc. are digitally configurable. This feature enables the proposed concept to be implemented in real application board development without adding external circuitry to the gate drive.

4 Characterize t_{SC} at different V_{GS} and V_{DS} by destructive testing

To emulate a hard-switching fault scenario with nearly zero impedance in the fault path, DC+ and phase output terminals of the half bridge in the test setup were shorted directly using a short cable. Understanding the effects of gate and drain voltages on t_{SC} is a pre-requisite to implement the t_{SC} extension using the TLTO mechanism.

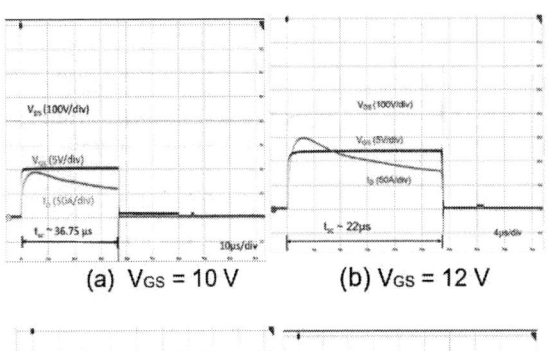

Fig. 3 Destructive testing to measure t_{SC} at V_{DS} of 400 V under different V_{GS} of (a) 10 V, (b) 12 V, (c) 15 V, and (d) 18 V

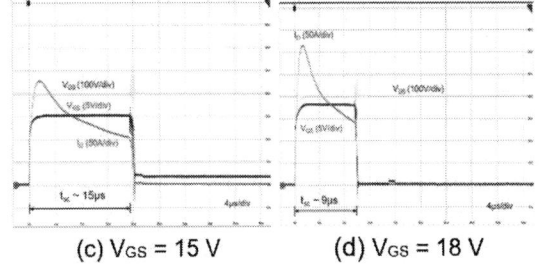

Fig. 4 Destructive testing to measure t_{SC} at V_{DS} of 600 V and V_{GS} of 18 V

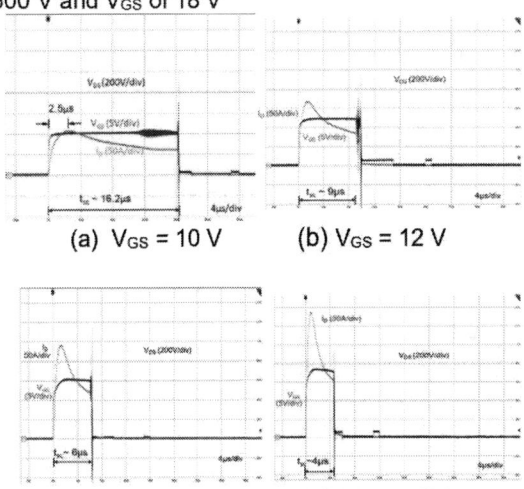

Fig. 5 Destructive testing to measure t_{SC} at V_{DS} of 800 V under different V_{GS} of (a) 10 V, (b) 12 V, (c) 15 V, and (d) 18 V

2780

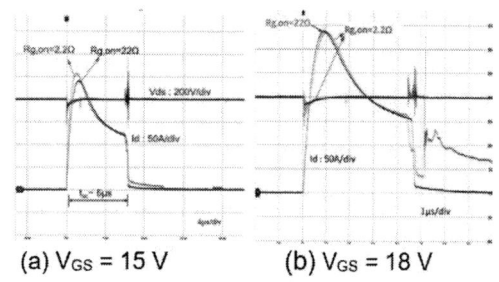

(a) V_{GS} = 15 V (b) V_{GS} = 18 V

Fig. 6 Effect of R_{GON} on t_{sc} at V_{DS} of 800 V and V_{GS} of (a) 15 V and (b) 18 V

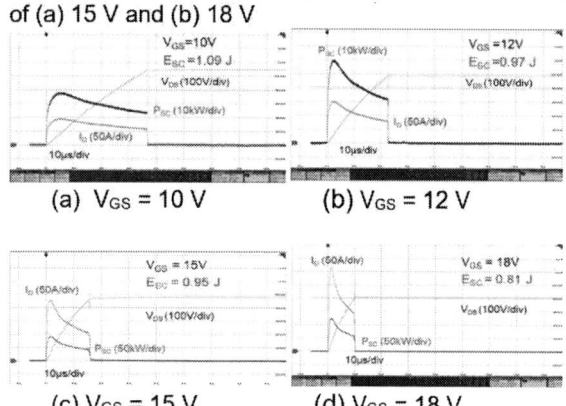

(a) V_{GS} = 10 V (b) V_{GS} = 12 V

(c) V_{GS} = 15 V (d) V_{GS} = 18 V

Fig. 7 Measured short-circuit withstanding energy E_{SC} at V_{DS} of 400 V with gate voltages V_{GS} of (a) 10 V, (b) 12 V, (c) 15 V, and (d) 18 V

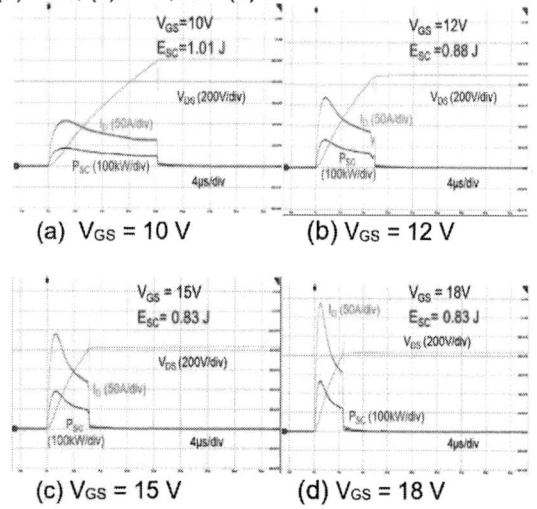

(a) V_{GS} = 10 V (b) V_{GS} = 12 V

(c) V_{GS} = 15 V (d) V_{GS} = 18 V

Fig. 8 Measured short-circuit withstanding energy E_{SC} at V_{DS} of 800 V with gate voltages V_{GS} of (a) 10 V, (b) 12 V, (c) 15 V, and (d) 18 V

To characterize the short-circuit withstanding capability of the SiC MOSFET, destructive tests were performed on the DUTs at the selected turn-on gate voltages of 10 V, 12 V, 15 V and 18 V,

Short-circuit withstanding time t_{SC} (μs)

$R_{G,ON}$ 22Ω		V_{GS} (V)			
		10	**12**	**15**	**18**
V_{DS} (V)	**400**	36.8	22.8	15.4	9.7
	600				6.2
	800	16.2	9.1	6.1	4.6

$R_{G,ON}$ 2.2Ω		**10**	**12**	**15**	**18**
V_{DS} (V)	**400**				8.9
	800		9.25	6	4.4

Short-circuit withstanding energy E_{SC} (J)

$R_{G,ON}$ 22Ω		V_{GS} (V)			
		10	**12**	**15**	**18**
V_{DS} (V)	**400**	1.09	0.97	0.95	0.81
	800	1.01	0.88	0.82	0.82

$R_{G,ON}$ 2.2Ω	**10**	**12**	**15**	**18**
V_{DS} (V) 800		0.89	0.84	0.82

Table 1 Short-circuit withstanding time t_{SC} and short-circuit withstanding energy E_{SC}, measured by destructive testing at different V_{GS}, V_{DS} and $R_{G,ON}$.

and drain voltages of 400 V, 600 V and 800 V, to ascertain the relationship of t_{SC} with V_{GS} and V_{DS}. Gate voltages at 15 V and 18 V are considered as high gate voltage $V_{GS,H}$, as they correspond to operation at normal conditions, while gate voltages at 10 V and 12 V are considered as low gate voltage $V_{GS,L}$, as they correspond to the intermediate voltage for TLTO during a short-circuit fault scenario. Figs. 3 - 5 show the effect of gate voltage V_{GS} and drain-source voltage V_{DS} on the peak fault current and the short-circuit withstanding time t_{SC}, measured by destructive tests at V_{GS} of 10 V, 12 V, 15 V, 18 V and V_{DS} = 400 V, 600 V and 800 V.

Measurements with lower R_{GON} of 2.2 Ω were performed and compared with higher R_{GON} of 22 Ω and the results are shown in Fig. 6. Although the initial rate of current rise is slightly faster with lower R_{GON}, the impact of R_{GON} on t_{SC} is negligible.

The measured short-circuit withstanding energy E_{SC} at different gate voltages V_{GS} with V_{DS} = 400 V and 800 V are shown in Figs. 7 and 8, respectively. E_{SC} values at different gate voltages V_{GS} are used to determine TLTO parameters.

The measurement results of t_{SC} and E_{SC} by destructive testing at different V_{DS}, V_{GS} and $R_{G,ON}$ are summarized in Table 1.

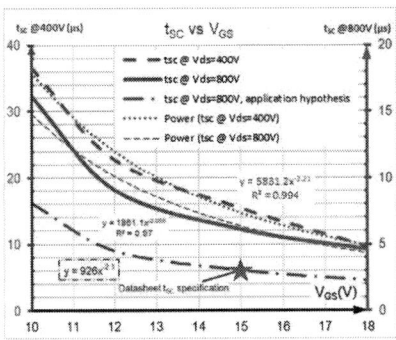

Fig. 9 Effect of V_{GS} on short-circuit withstanding time at V_{DS} = 400 V (broken line) and 800 V (solid line)

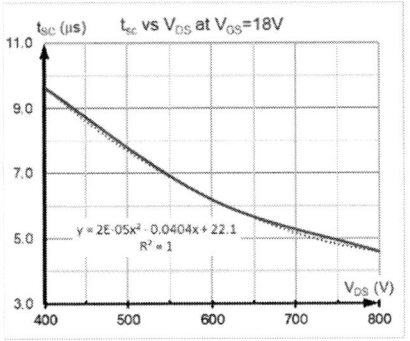

Fig. 10 Effect of V_{DS} on short-circuit withstanding time at V_{GS} = 18 V

The relationship of t_{SC} with V_{GS} and V_{DS} at $R_{G,ON}$ of 22 Ω are plotted in Figs. 9 and 10, respectively. It can be observed that t_{SC} at V_{DS} =800 V is roughly about 0.4 times that of t_{SC} at V_{DS}= 400 V.

From Fig. 9, the t_{SC} destructive measurement result of 6 μs is found to be double that of the datasheet value of 3 μs at the same conditions of V_{DS} of 800 V and V_{GS} of 15 V. Assuming the same derating factor between the measurement results and datasheet specification, a relationship of t_{SC} against V_{GS} at V_{DS} of 800 V can be extrapolated from the solid curve (measurement results) and plotted as the lowest dotted curve in the graph, as a hypothetical specification for further verification.

However, it should be noted that these measurement results were obtained from a limited number of test samples only. Before the test results can be used as a product specification, a significant design margin must be added to account for factors such as high-volume production variation, temperature and repetitive short-circuit capability degradation. These topics are beyond the scope of this study.

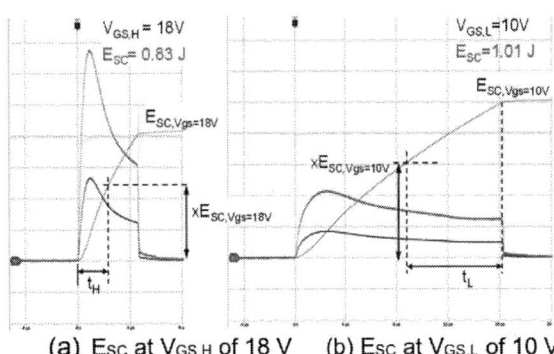

(a) E_{SC} at $V_{GS,H}$ of 18 V   (b) E_{SC} at $V_{GS,L}$ of 10 V

Fig. 11 Estimation of t_L from t_H, based on E_{SC} consumed at different stages of short circuit, with gate voltages first at (a) $V_{GS,H}$ and then (b) $V_{GS,L}$

With the characteristics of SiC MOSFET t_{SC} and E_{SC} versus V_{GS} now known, TLTO parameters can then be set to extend t_{SC} for two different short-circuit scenarios, namely short-circuit type I or hard switching fault (HSF) and short-circuit type II or fault under load (FUL).

5 t_{SC} extension using TLTO under HSF

Key parameters in the 1ED3890MC12M digital driver IC for TLTO protection can now be set depending on the specific application requirements. Selecting V_{DS}, $V_{GS,H}$ and $V_{GS,L}$ as typical operating conditions at 800 V, 18 V and 10 V, respectively, the parameters t_H (DUT operation duration at normal gate voltage $V_{GS,H}$ at the onset of short-circuit) and t_L (DUT operation duration at low gate voltage $V_{GS,L}$ prior to full shut-down) can be determined based on E_{SC} consumed during t_H and t_L, with gate voltages at $V_{GS,H}$ and $V_{GS,L}$ respectively, as shown in Fig. 11.

5.1   Choose t_H and t_L based on E_{SC} consumption

The parameter t_H should initially be chosen as a small percentage of t_{SC} at V_{DS} of 800 V. The value of t_{SC} (V_{GS}) at V_{DS} of 800 V can be read from the solid curve in Fig. 9 or estimated using the empirical relationship of $t_{SC} = 1861\,V_{GS}^{-2.1}$ from curve-fitting.

At the chosen t_H, the percentage x of short-circuit energy consumed from the E_{SC} limit when $V_{GS}=V_{GS,H}$ can be read from Fig. 11(a). This percentage x can then be used to estimate the short-circuit time left to reach the remaining E_{SC} limit, with V_{GS} already changed from $V_{GS,H}$ to $V_{GS,L}$ during TLTO, using Fig. 11(b). This remaining time will be the maximum limit of t_L before device failure at $V_{GS}=V_{GS,L}$.

Since the short-circuit withstanding time will increase as the gate voltage reduces, a shorter t_H can be offset more by a longer t_L. This would lead to the total short-circuit withstanding time with TLTO of $t_H + t_L$ being longer than the intrinsic device short-circuit withstanding time t_{SC}. The maximum extension of t_{SC} with TLTO is achieved by minimizing t_H and maximizing t_L correspondingly.

5.2 Verification of t_{SC} extension using TLTO

The performance of SiC MOSFET short-circuit protection with t_{SC} extension by TLTO, realized using digital gate driver IC 1ED3890MC12M, has been experimentally verified with different t_H by adjusting the driver IC DESAT filter time settings until the SiC MOSFET reaches its destruction limit and fails.

SiC MOSFET short-circuit fault is detected by the desaturation protection of the driver IC, with driver IC parameter settings of:

- DESAT threshold voltage = 6.12 V
- Leading edge blanking time = 200 ns
- Filter time from 0.075 µs to 2.175 µs
- Fault turn-off set to TLTO mode with $V_{GS,L}$ =10 V and t_L = 7.75 µs

The DESAT filter time settings, total short-circuit time (=$t_H + t_L$) and the protection results from different tests are tabulated in Table 2. The measurement waveforms are shown in Fig. 12 (a)-(d).

With fixed V_{GS} of 18 V and V_{DS} of 800 V, the SiC MOSFET destruction limit is 4.6 µs, as previously shown in Fig. 5(d). Using TLTO, the SiC MOSFET is now safely turned off with a total short-circuit withstanding time of 10 µs, as shown in Fig 12 (c).

The SiC MOSFET failure in Fig. 12(d) can be predicted based on the principle of E_{SC} consumption described in the previous section. With t_{SC} = 4.6 µs at V_{DS} = 800 V, and $V_{GS,H}$ =18 V, and t_H chosen as 2.8 µs, the E_{SC} consumption percentage x can be read from Fig. 11(a) to be 69%.

Test	DESAT filter time	TLTO duration (t_L)	$t_H + t_L$	Result
(a)	0.075 µs	7.75 µs	8.4 µs	Safe turn-off
(b)	1.075 µs	7.75 µs	9.5 µs	Safe turn-off
(c)	1.575 µs	7.75 µs	10 µs	Safe turn-off
(d)	2.075 µs	7.75 µs	9 µs	Device failed

Table 2 Results of SiC MOSFET desaturation short-circuit protection with TLTO at different DESAT filter time settings

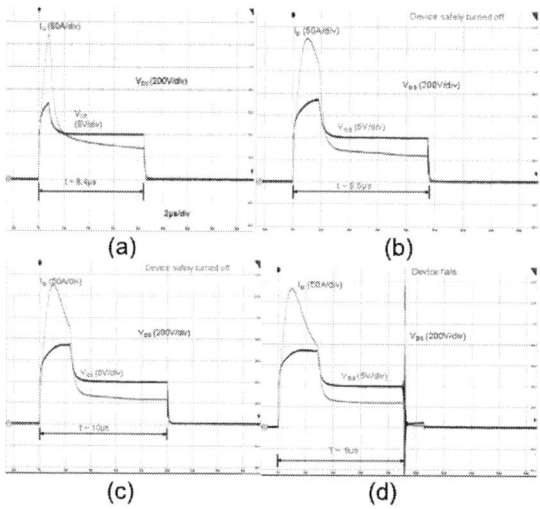

Fig. 12 DESAT protection waveforms of SiC MOSFET IMW120R030M1H with TLTO at V_{DS} of 800 V, $V_{GS,H}$ of 18 V and $V_{GS,L}$ of 10 V, at different DESAT filter time settings of (a) 0.075 µs, (b) 1.075 µs, (c) 1.575 µs and (d) 2.175 µs.

From Fig. 11(b), the maximum remaining short-circuit time t_L with $V_{GS,L}$ =10 V at x = 69 % is 5.7 µs. The total short-circuit time with TLTO (= $t_H + t_L$) is thus estimated to be 8.5 µs, which matches well with the experimental results.

It can be inferred from the failure in Fig. 12(d) that the short-circuit energy consumption percentage x with applied $V_{GS, H}$ increases nonlinearly with DESAT filter time, and thus eventually limits the maximum t_L to extend SiC MOSFET short-circuit capability. This demonstrates the need for fast short-circuit DESAT detection to fully utilize the benefits of t_{SC} extension by TLTO.

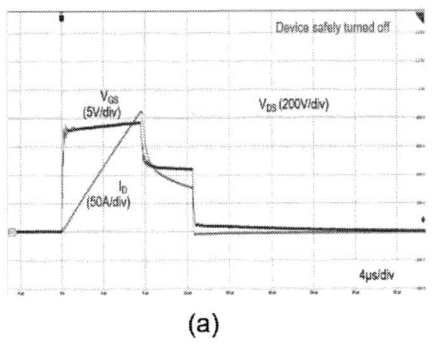

(a)

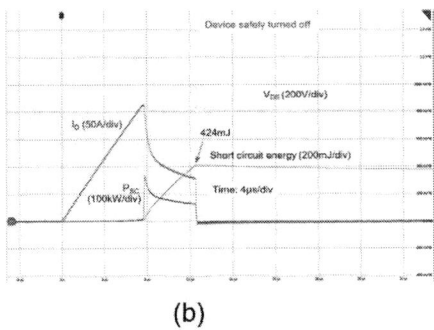

(b)

Fig. 13 DESAT protection of SiC MOSFET IMW120R030M1H with TLTO turn-off during FUL. (a) turn-off waveforms, (b) short-circuit energy

6 t_{SC} extension using TLTO under FUL

To emulate a fault under load (FUL or short-circuit type II) scenario with relatively high impedance in the fault path, the low impedance short-circuit connection between DC+ and the phase output terminals of the half bridge in the setup shown in Fig. 2 was replaced by a 25 µH air core inductor. With V_{DS} = 800 V and $V_{GS,H}$ = 18 V, the new parameter settings of the digital driver IC are:

- DESAT protection was enabled
- DESAT threshold voltage = 9.18 V
- Leading edge blanking time = 200 ns
- Filter time = 2.175 µs
- Fault turn-off set to TLTO mode with $V_{GS,L}$ =10 V and t_L = 5 µs

The results of the short-circuit protection of the SiC MOSFET, with t_{SC} extension by TLTO under FUL, is shown in Fig. 13(a). The SiC MOSFET was safely turn-off after 12.6 µs. Energy dissipation during FUL as shown in Fig. 13(b) of 424 mJ is less than half of 900 mJ in HSF, due to

a slower rate of rise of current limited by fault path impedance.

The longer t_{SC} and lower E_{SC} during FUL shows that FUL in a SiC MOSFET is less critical than in HSF. The results here demonstrate that the principle of t_{SC} extension by TLTO is applicable to both HSF and FUL.

7 Conclusion

SiC power MOSFETs are often optimized nowadays for low $R_{DS(ON)}$ with limited short-circuit withstanding capability. It is challenging for such SiC MOSFETs to give robust fault protection in industrial motor drives applications. To address this problem, this paper proposes to use TLTO to extend the short-circuit capability of SiC MOSFETs. Destructive double-pulse testing is used to characterize SiC MOSFET t_{SC} and E_{SC} at different V_{DS} and V_{GS}. Gate driver ICs with DESAT and TLTO parameters that are digitally configurable are used to demonstrate t_{SC} extension using TLTO under HSF. t_H and t_L are chosen based on the E_{SC} consumed at different short-circuit stages.

The performance of t_{SC} extension with TLTO is verified experimentally with a wide range of filter times until the SiC MOSFET eventually fails, and has demonstrated a safe turn-off of the SiC MOSFET with a total short-circuit time of 10 µs, compared to the destruction limit of 4.6 µs at V_{GS} of 18 V and V_{DS} of 800 V. To achieve the full benefit of t_{SC} extension by TLTO, fast short-circuit DESAT detection is needed. The concept is also demonstrated for protection under FUL, with the SiC MOSFET being safely turned off after 12.6 µs, and energy dissipation being less than half of that in HSF. The principle of t_{SC} extension by TLTO is thus applicable to both HSF and FUL.

References

[1] R Rahul Chokhawala, Jamie Catt, and Laszlo Kiraly. "A Discussion on IGBT Short-Circuit Behavior and Fault Protection Schemes," IEEE Transactions on Industry Applications, Vol 31, No 2, 1995.

[2] Diane-Perle Sadik, Juan Colmenares, Georg Tolstoy, Dimosthenis Peftitsis, Mietek

Bakowski, Jacek Rabkowski and Hans-Peter Nee, "Short-Circuit Protection Circuits for Silicon-Carbide Power Transistors," IEEE Transactions on Industrial Electronics, Vol. 63, No. 4, April 2016.

[3] X. She, A. Q. Huang, O. Lucia, and B. Ozpineci, "Review of silicon carbide power devices and their applications," IEEE Trans. Ind. Electron., vol. 64, no. 10, pp. 8193–8205, 2017.

[4] Jiangui Chen, Yan Li , Mei Liang, R. Kennel, Jiayu Liu and Haobo Guo, "A Novel Gate Driver for Suppressing Overcurrent and Overvoltage of SiC MOSFET," 10th International Conference on Power Electronics-ECCE Asia, May 27 - 30, 2019, Busan, Korea.

[5] Minseob Shim, Kyoungho Lee, Jonghyun Kim and Kihyun Kim, "Multistep Soft Turn-Off Time Control to Suppress the Overvoltage of SiC MOSFETs in Short-Circuit State," IEEE Access, V10, pp: 46408 - 46417, 22 April 2022. DOI: 10.1109/ACCESS.2022.3169764

PCIM Europe 2024, 11– 13 June 2024, Nuremberg   DOI: 10.30420/566262393

Experimental Investigations on Parasitic Turn-on of 1.2kV SiC MOSFET Discrete Devices

Thanh-Toan Pham[1], Kwangwon Lee[2], Jimmy Franchi[1], Sara Kuzmanoska[3], Martin Domeij[1]

[1] onsemi, Sweden

[2] onsemi, South Korea

[3] onsemi, Germany

Corresponding author: Thanh-Toan Pham, thanh-toan.pham@onsemi.com
Speaker: Thanh-Toan Pham, thanh-toan.pham@onsemi.com

Abstract

The paper presents detailed experimental investigations on Parasitic Turn-on (PTO) of 1.2 kV SiC MOSFETs discrete devices in TO247-4L package. PTO correlation to capacitance profiles is shown using two devices with different Ciss/Crss ratios. Key focuses are on the devices that are highly susceptible to PTO for better understanding the device behavior under PTO. Investigations include the approach for measuring the non-trivial HS V_{GS} pull-up signal during BD turn-off and its correlation to measured current. PTO dependent on different Rg_ext will be shown to demonstrate the effectiveness of this method to evaluate device's PTO ruggedness. PTO measurements at high temperature will also be presented and compared to PTO at RT. Interesting effect of reduced PTO at high temperature compared to RT is presented.

1   Introduction

PTO is described as an effect of unwanted pull-up of the High Side (HS) device V_{GS} over VTH thereby inducing channel current during the Body Diode (BD) turn-off process. Previous reports show that PTO is strongly dependent on C_{ISS}/C_{RSS} ratio and less dependent on device's VTH [1-2]. Recently, we demonstrated a highly robust device against PTO can be achieved using optimized capacitance profile [2]. However, the investigation also showed that different devices on the market may experience different degrees of PTO [2]. It is therefore important to study PTO in deeper detail, especially with the devices that are highly susceptible to PTO. One of important challenges when investigating PTO is to measure the miller capacitance transient induced V_{GS} pulled-up on HS device. Here, due to high common mode potential (e.g. V_{DD}=800V), the induced differential potential HS V_{GS} could not be measured using low voltage passive probe due to the lack of grounding potential. Also, due to the relatively low range of induced HS V_{GS}, high voltage differential probes are not suitable for these measurements. In previous works [1-2], only measured current overshoots with different HS R_{G_EXT} were shown during PTO. This point will be addressed

with details board design and measurements set-up.

Thanks to this feasibility, we will again demonstrate the PTO correlation to devices capacitance profile using devices with the same die size and significant differences in C_{ISS}/C_{RSS} ratio. After that, measurements will be focused on the device that is highly susceptible to PTO, target a better understanding of device behaviors under strong PTO. The investigations included the correlation of measured current overshoot and its correlation to HS V_{GS} pulled-up signal with HS R_{G_EXT} variation for PTO evaluation. These measurements confirm the effectiveness of PTO evaluation by varying HS R_{G_EXT}.

Also, previous investigations on PTO indicate that devices could become more susceptible against PTO with increased temperature [4], and this is believed due to device VTH reduction at elevated temperature [4]. It is therefore generally believed that PTO can become a serious issue when operating SiC MOSFETs under high temperature conditions. In recent work, we demonstrated that with proper capacitance profile, even under high temperatures where device VTH is reduced, devices can still be highly rugged against PTO [2]. In this work, we examine the device's behaviors at high temperature when it is exposed to strong PTO events. An interesting observation is that, when the device is in strong PTO, its PTO behavior is indeed becoming softer at high temperature compared to RT.

2 Results and Discussion

2.1 Experimental Details and HS V_{GS} Measurements

Fig.1a depicts the actual DPT internal evaluation board and the NCP51705 gate driver. For switching current measurements, Pearson probes with 100MHz bandwidth were used to measure current on HS and LS devices. Differential high voltage probes THDP0200 with 200 MHz bandwidth were used to measure V_{DS} on the HS and LS device. A galvanically isolated high impedance differential probe TIVP02 (IsoVu) with 200 MHz bandwidth was used to measure induced V_{GS} on the HS device during PTO event.

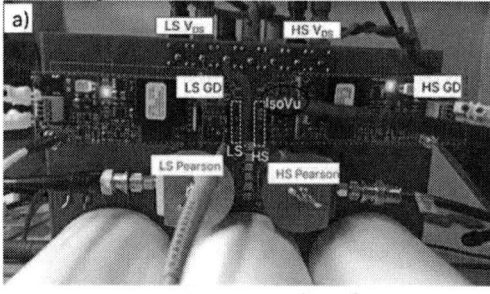

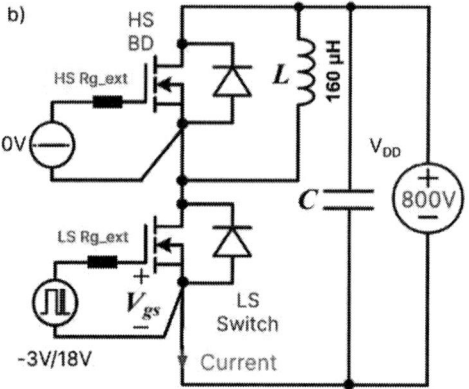

Fig.1a: DPT evaluation board equipped with its probes and NCP51705 gate driver. Fig.2b: Simplified circuit diagram of a half-bridge configuration in double pulse test.

Fig1.b represents the simplified schematics of a basic half-bridge configuration in double pulse test. For PTO evaluation, a similar approach compared to previous work [2] was employed with a fixed LS device to obtain a fixed di/dt during active switch turn-on, i.e., BD turn-off. Complementary device, i.e., HS BD (DUT) condition was varied in devices,

HS R_{G_EXT} and temperatures to evaluate PTO, its dependent on die design and the PTO temperature dependence.

2.2 Capacitance Profile and PTO

Fig.2a represents C(V) characteristics of the devices considered in this work. The devices with same die size ($25mm^2$) with different designs at die level intended to obtain comparable VTH and different capacitance profile. Device #2 shows remarkable higher C_{RSS} compared to device #1 for the entire V_{DS} range. At the same time, device's C_{ISS} are comparable between two devices. Fig.2b shows the area independent C_{ISS}/C_{RSS} ratio of the two devices. It is obvious that device #1 C_{ISS}/C_{RSS} ratio is significantly higher compared to device #2. As discussed previously, the C_{ISS}/C_{RSS} ratio is believed to be the main factor determining the susceptibility to PTO [2].

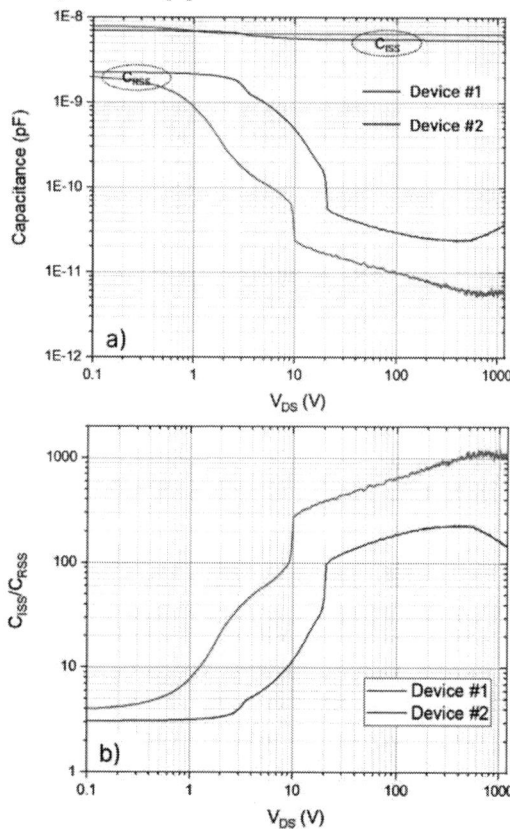

Fig.2a: Measured C(V) Characteristics of Device #1 and Device #2 of the same die size with different die designs. Fig.2b: Area independent C_{ISS}/C_{RSS} ratio from device #1 and device #2

Fig.3 shows PTO ruggedness of two devices when they were exposed into the same stressing condition. It is showing that under V_{DD}=800V, in a fast turn-on event, e.g., di/dt≈5A/ns, with a typical external R_{G_EXT}=10 Ω, significant differences can be

2787

observed when using devices with different capacitance profiles. Device #1 with optimized C_{ISS}/C_{RSS} ratio over the entire V_{DS} range shows typical transient characteristics of a BD during its turn-off process. Measured current experienced reverse recovery phase with the HS V_{DS} rising to V_{DD}. Voltage transient induced a voltage overshoot on its V_{GS} that is rapidly decaying when the V_{DS} transient has completed. Device #1 PTO ruggedness at different current densities at high V_{DD}, fast switching, high R_{G_EXT} and high temperature were demonstrated previously [2].

strong current overshot during the reverse recovery phase is well correlated to the measured pulled-up HS V_{GS} signal. Here, the HS V_{GS} signal was strongly pulled-up during the HS V_{DS} transient phase. The measurements show the usefulness of an optical isolated probe for PTO characterization. Measured V_{DS} on the HS device showing that PTO strongly slowing down dV/dt of the HS BD. Together with long duration current overshoot, this slower V_{DS} turn-off speed induces a high-power switching loss during BD turn-off and causes unnecessary thermal stress on the device when it experiences the PTO.

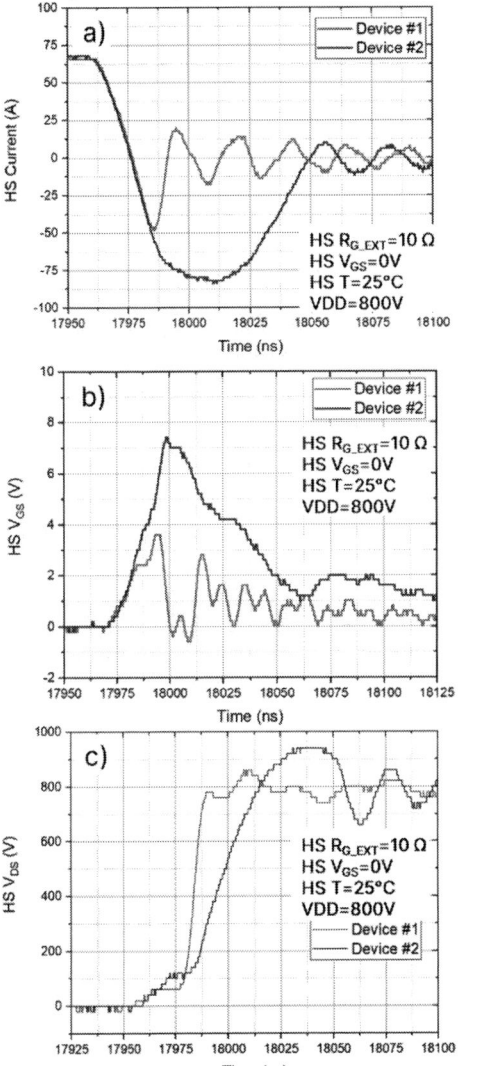

Fig.3: Measured BD characteristics with device # 1 and device #2 as DUT during its turn-off transient, included: a) BD current; b) BD V_{GS}; c) BD V_{DS}.

On the other hand, device #2 with much lower C_{ISS}/C_{RSS} ratio show peculiar BD turn-off characteristics. A very strong current overshoot is observed during its reverse recovery phase. The

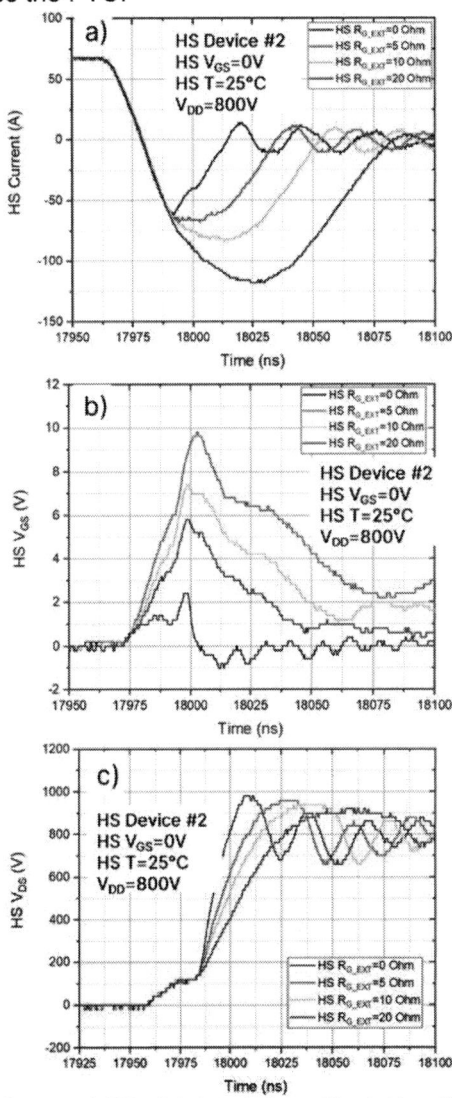

Fig.4: Measured BD characteristics with device # 2 as DUT with different HS R_{G_EXT} at RT during its turn-off transient, included: a) BD current; b) BD V_{GS}; c) BD V_{DS}.

Fig.4 represents detailed PTO measurements using device #2 as DUT at 25°C with varying HS R_{G_EXT}. The measured HS current is correlated to

HS V_{GS} overshoot and measured HS V_{DS}. The measurements further confirmed the effectiveness of PTO evaluation by varying HS R_{G_EXT}. The measurements also indicate that under a strong PTO event, significant current can be induced. This is therefore important to design a device that is highly rugged against PTO.

2.3 High Temperature PTO

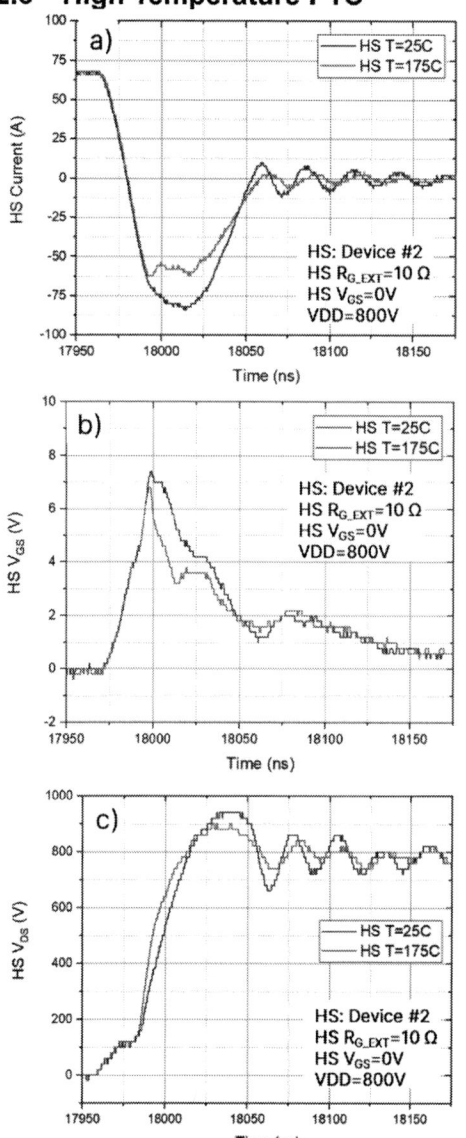

Fig. 5: Measured BD turn off characteristics with device #2 as DUT under RT and 175°C, included: a) BD current; b) BD V_{GS}; c) BD V_{DS}.

Fig.5 represents PTO evaluation with device #2 as the DUT when stressing device at room temperatures and high temperature. It is showing that under substantial stressing condition, e.g., V_{DD}=800V, di/dt≈5A/ns with R_{G_EXT}=10 Ω, unexpected effect is observed with PTO at 175°C appears to be smaller

compared to PTO at 25°C. Here, measured HS V_{GS} and measured current overshoot are both lower at 175°C compared to its counterparts measured at RT. It is believed that the observed effects can be related to electron mobility reduction in the SiC drift region as well as JFET region at 175°C. Miller feedback also became slightly weaker, as shown in Fig. 5b. The effects were mentioned also in previous work [4]. Further investigations with TCAD calibration and simulations may be needed for a concrete conclusion.

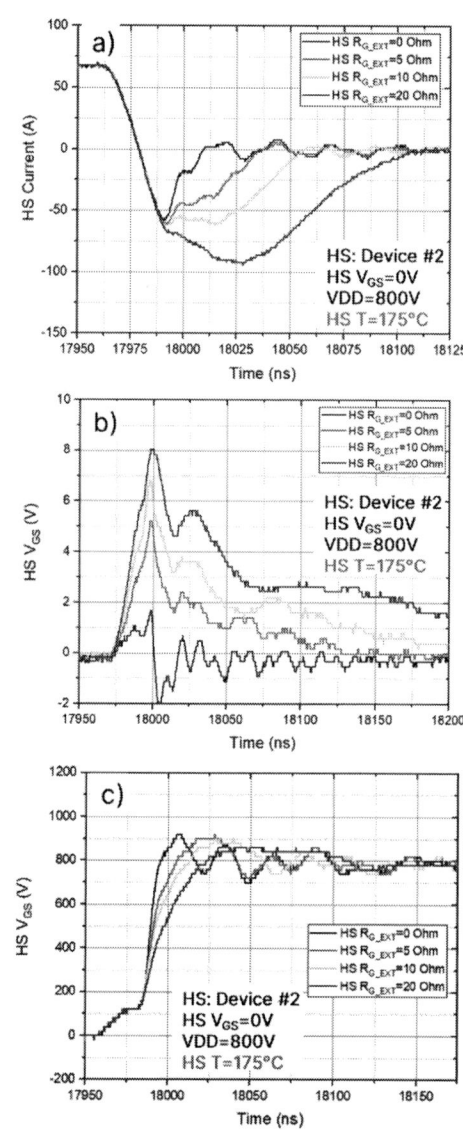

Fig.4: Measured BD characteristics with device #2 as DUT with different HS R_{G_EXT} at 175°C during its turn-off transient, included: a) BD current; b) BD V_{GS}; c) BD V_{DS}.

Fig. 6 represents further detailed PTO measurements with HS Temperature at 175°C. Here, the measured signal was plotted on the same scale

2789

compared to Fig. 4. The measurements further confirmed that for a device that is highly susceptible to PTO, its PTO behavior is improving at HT compared to RT.

3 Summary

The work presents experimental investigations of PTO behavior of 1.2 kV SiC MOSFET discrete devices in TO247-4L packages. Clear correlation of PTO ruggedness against device's capacitances characteristics is observed. When device experienced PTO, measured current overshoot is correlated to its measured Miller induced HS V_{GS} thanks to IsoVu probe implementation. The measurements confirmed the effectiveness of PTO evaluation by varying HS R_{G_EXT}. PTO measurements at 175°C appear softer than PTO behavior measured at RT.

References

[1] T. -T. Pham, J. Franchi, K. Lee and M. Domeij, "1.2 kV SiC MOSFET Body Diode Turn-Off in Fast Switching: Channel Conduction, Carrier Plasma and Parasitic Turn-On," in *35th International Symposium on Power Semiconductor Devices and ICs (ISPSD), Hong Kong*, pp. 334-337 (2023).

[2] T.-T. Pham, J. Franchi, S. Kang, KS. Park, D. Choi, M. Domeij, "1.2 kV SiC MOSFET with Low Specific ON-Resistance and High immunity to Parasitic Turn-On," in *International Conference on Silicon Carbide and Related Material (ICSCRM)* 2023.

[3] https://www.tek.com/en/products/oscilloscopes/oscilloscope-probes/isovu-isolated-probes

[4] K. Sobe, T. Basler, and B. Klobucar, "Characterization of the parasitic turn-on behavior of discrete CoolSiC™ MOSFETs," in *PCIM Europe*, pp. 1-7, 2019.

PCIM Europe 2024, 11– 13 June 2024, Nuremberg   DOI: 10.30420/566262394

Behavior Modelling the Short Circuit Characteristics of SiC MOSFETs using Compact Models

Qing Sun[1], Andreas Huerner[1], Rudolf Elpelt[1]

[1] Infineon Technologies AG, Germany

Corresponding author: Qing Sun, qing.sun@infineon.com
Speaker: Qing Sun, qing.sun@infineon.com

Abstract

An electro-thermal compact model of SiC MOSFETs, capable of accurately estimating short circuit behavior of SiC MOSFETs, is presented in this paper. The model has been validated for standard static and dynamic characteristics under different temperatures, and calibrated through high-voltage I-V output measurements to ensure that the drain-current in the saturation region is depicted accurately. Thanks to an extended thermal impedance network, the simulation results show an excellent agreement with the non-destructive short circuit measurements over a wide range of gate voltages and short-circuit pulse time. The accurate estimation of short circuit power, turn-off overvoltage, and critical temperature provides beneficial information for assessing reliability and analyzing the fault mode in virtual prototyping. Ultimately, these models can help electrical engineers protect power electronic systems by detecting short circuit events precisely.

1 Introduction

Compact models are powerful tools that assist electrical engineers investigate and predict the static and dynamic characteristics of power devices during the development phase. This can, potentially, lead to a significant reduction in time and cost. In compact models, electrical and thermal properties of SiC MOSFETs are often described by specific elements through continuous and differentiable equations, resulting in fast and robust simulations in the SPICE simulators [1] [2]. They are less complicated than compact models based on dimensions and technological process parameters of the real device. The reduced complexity and robustness of the compact models discussed in [1] [2], do not necessarily hinder the potential for more specialized features, such as estimating short circuit behavior of SiC MOSFETs.

In the past, several publications have proposed various complicated physics-based [3] [4] and analytical [5] – [7] models developed for short circuit simulations. These models have demonstrated a good agreement between simulation and measurement results, but only under limited testing conditions.

This paper demonstrates that the SiC MOSFET model introduced in [1] [2] is capable of insightfully estimating the short circuit behavior under a wide range of testing conditions, with the help of well-calibrated drain current in the high drain voltage region and an extended six-element Cauer thermal network.

2 Compact Model Requirements

The compact model of a 1200 V SiC MOSFET in a TO247-4 package is discussed in this paper. Figure 1 shows the structure of this model, including the main chip model, thermal network model, and the package parasitic model. Details of the model have been discussed in [1] [2].

2.1 Thermal Network

To simulate the heat path from the SiC MOSFET to the back of the lead frame or a heatsink, the heat flow, i.e., dissipated power, is represented by the electrical power source, P_j, in the thermal networks shown in Fig. 1 and Fig. 2. The interactive coupling of thermal and electrical description of these models is enabled by simultaneously feeding a current proportional to the dissipated power into the thermal network. The virtual junction temperature is represented by the voltage at the node, T_j, and interacts in return with the temperature-dependent parameters of the MOSFET, such as threshold voltage and drain-to-source resistance.

Normally, thermal networks published in datasheets and compact models consist of up to four elements. To capture the instant thermal response in the MOSFET channel, the thermal network in this compact model has been expanded from four to six resistor-capacitor elements (shown in Fig. 2). By representing the thermal behavior for time constants smaller than 10 µs, the two newly added elements help describe the short circuit behavior accurately.

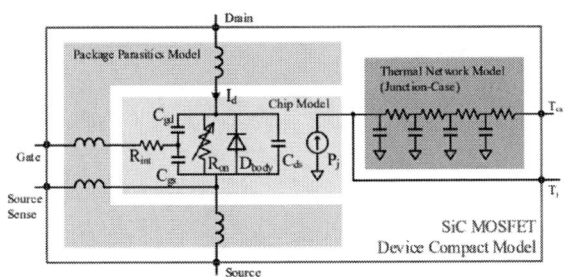

Fig. 1 Simplified electrical structure of a compact model for SiC MOSFETs in a TO247-4 package.

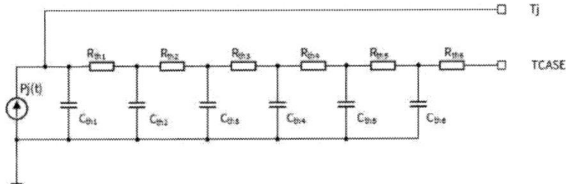

Fig. 2 Six-element thermal network in a compact model for SiC MOSFETs.

2.2 MOSFET Current Source

A precise description of the I-V output characteristics at the low drain-source voltage is important to accurately determine the on-state static characteristics of the device. For dynamic behaviors such as fast switching and short circuit events, well simulated I-V output characteristics at high drain-source voltage (of several hundred volts up to DC link voltage) becomes crucial.

Figure 3 presents the measured output-characteristic for drain voltages up to 800 V. This specific method guarantees the accurate estimation of

drain current in the high voltage region – vital for accurately describing the short circuit current.

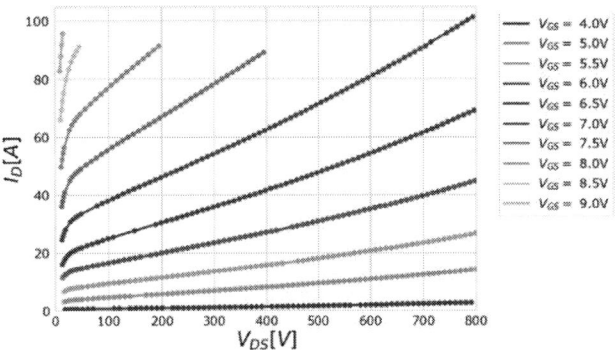

Fig. 3 High voltage output-characteristic at various gate voltages.

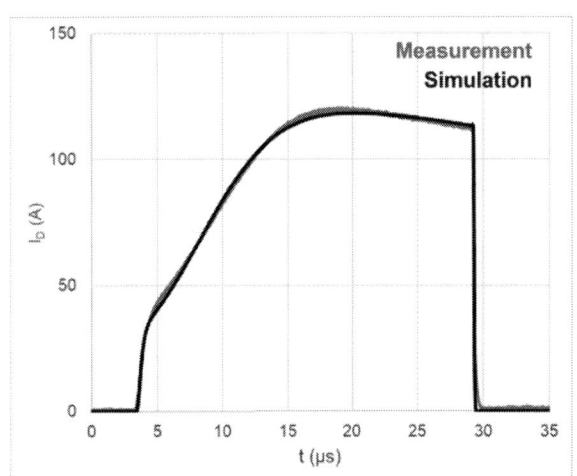

Fig. 4 Demonstration of a low-power short circuit pulse.

In addition, threshold voltage and transconductance help define the initial stages of the rapid drain current rise observed in Fig. 4. The gradual increase in current during the second phase can be described by the threshold voltage's dependence on temperature in the model. After reaching the peak value, the drain current decreases slightly. This can be linked to the increase in the drain-to-source resistance arising from temperature rise.

Details of the current source in the model have been provided in [1] and more recently emphasized in [2], which has also proposed a possible mathematical representation of the drain current in compact models.

It is important to note that the standard models available on the homepage of Infineon's website are not suitable for short circuit simulation without modifying the thermal networks.

3 Short Circuit Characteristics in Compact Modelling

Figure 5 shows a direct comparison between the simulated and measured non-destructive type-I short circuit waveforms with a gate voltage of 18 V for a short-circuit pulse time of 1.8 µs. For the measurement, the external gate resistance was set to 1.8 Ohms. To minimize the impact of discrepancies caused by parasitic elements in the simulation circuitry and measurement setup, the components of DC link voltage supply and gate voltage source in the simulation were defined using the respective measurement results as text input files. Consequently, the simulated (red) and measured (blue) waveforms of drain voltage and gate voltage closely mirror each other.

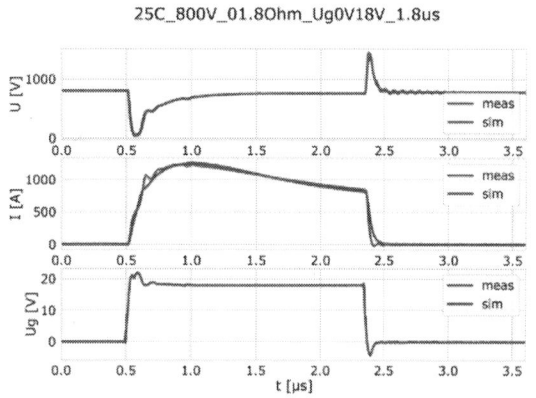

25C_800V_01.8Ohm_Ug0V18V_1.8us

Fig. 5 Simulated (red) and measured (blue) drain voltage, drain current, and gate voltage waveforms at non-destructive short circuit events. 1.8 µs short-circuit pulse time at a gate voltage of 18 V. Conditions: $R_{G,ext}$ = 1.8 Ohm, V_{DC} = 800 V, T = 25°C.

At the onset of the simulated current's curve shown in Fig. 5, several oscillations can be observed. These oscillations are a direct result of the oscillations present in the input drain and gate signals. To mitigate this influence and improve the accuracy of the results, a larger external gate resistance of 48 Ohms was selected for subsequent experiments. This change allowed for a more ac-

curate representation of the short circuit waveforms in the simulation, as demonstrated in Fig. 6. The simulated short circuit waveforms show a good agreement with the observed measurements, at a gate voltage of 18 V and a larger external gate resistance of 48 Ohms, for a short-circuit pulse time of 2 µs.

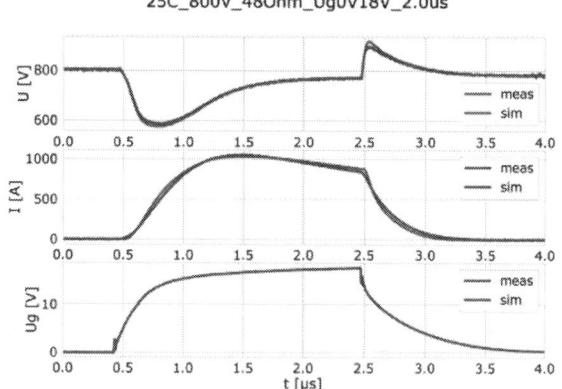

25C_800V_48Ohm_Ug0V18V_2.0us

Fig. 6 Simulated (red) and measured (blue) drain voltage, drain current, and gate voltage waveforms at non-destructive short circuit events. 2 µs short-circuit pulse time at a gate voltage of 18 V. Conditions: $R_{G,ext}$ = 48 Ohm, V_{DC} = 800 V, T = 25°C.

The simulation accurately reflected the measured drain current, thereby confirming that the compact model successfully replicates the short circuit behavior. To validate the model's applicability under various conditions, additional tests were carried out for lower and higher gate voltages. These tests aimed to demonstrate the versatility and robustness of the compact models in simulating short circuit behaviors under various scenarios.

In line with this, Figure 7 and Figure 8 show a direct comparison between the simulated and measured non-destructive short circuit waveforms under two completely different conditions. Figure 7 presents a longer short-circuit pulse time of 30 µs with a lower gate voltage of 6 V, while Figure 8 shows a shorter pulse time of 1.5 µs with a gate voltage of 21 V. Although slight mismatches can be seen, the overall agreement between the simulated and measured waveforms is satisfactory.

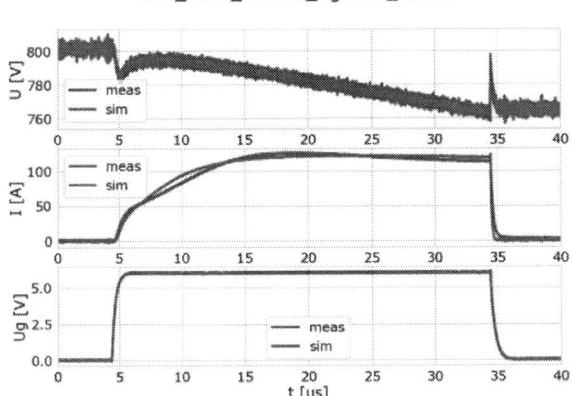

25C_800V_48Ohm_Ug0V6V_30.0us

Fig. 7 Simulated (red) and measured (blue) drain voltage, drain current, and gate voltage waveforms at non-destructive short circuit events. 30 μs short-circuit pulse time at a gate voltage of 6 V. Conditions: $R_{G,ext}$ = 48 Ohm, V_{DC} = 800 V, T = 25°C.

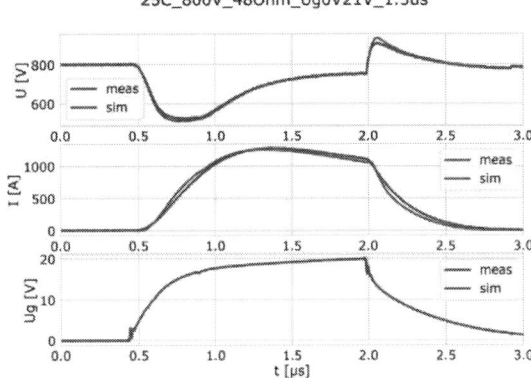

25C_800V_48Ohm_Ug0V21V_1.5us

Fig. 8 Simulated (red) and measured (blue) drain voltage, drain current, and gate voltage waveforms at non-destructive short circuit events. 1.5 μs short-circuit pulse time at a gate voltage of 21 V. Conditions: $R_{G,ext}$ = 48 Ohm, V_{DC} = 800 V, T = 25°C.

In addition, further statistical analyses were performed to assess the accuracy and reliability of the model for short circuit simulation. A total of 124 measurements were taken with gate voltages ranging from 6 V to 21 V and pulse length from 1 μs to 36 μs. Subsequently, simulations were also carried out to evaluate the model's performance across the same range of conditions.

The results of these analyses are shown in Fig. 9. The bar graphs illustrate the relative deviation percentages (x-axis) of the simulated short circuit energy and peak short-circuit drain current compared to the measured values. The x-axis displays the relative deviations, while the y-axis presents the measurement count. The maximum deviation observed is less than 20%, and notably, 90% of the simulation values lie below 10% deviation for short circuit energy.

This indicates a high degree of accuracy in the simulations further reinforcing the reliability and versatility of compact models in replicating the short circuit behavior across various experimental conditions.

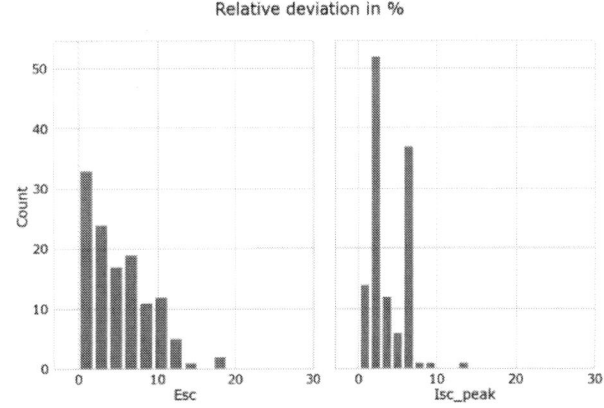

Fig. 9 Relative deviation percentages of (a) simulated short circuit energy and (b) simulated maximum short circuit current compared to measured values.

4   Conclusion

This paper presented a newly developed compact model of SiC MOSFETs that can replicate type-I short circuit events. The model's performance has been demonstrated through a wide range of simulations that showed a good agreement with the measurement waveforms obtained during non-destructive short circuit events. The consistency between the simulation and measurement results confirms the accuracy and reliability of these compact models.

Thanks to the compact model's ability to accurately simulate short circuit events across a wide range of conditions, it can serve as a valuable resource for electrical engineers to understand and carry out failure analysis in sophisticated power electronic applications. Furthermore, it helps the engineers design advanced and resilient protection mechanisms, reducing the likelihood of costly failures and preventing potential damage.

5 References

[1] P. Sochor, A. Huerner, and R. Elpelt, "A Fast and Accurate SiC MOSFET Compact Model for Virtual Prototyping of Power Electronic Circuits," PCIM Europe 2019; International Exhibition and Conference for Power Electronics, Intelligent Motion, Renewable Energy and Energy Management, Nuremberg, Germany, 2019, pp. 1–8.

[2] P. Sochor, A. Huerner, Q. Sun, and R. Elpelt, "Characteristics of SiC MOSFET Compact Models Suitable for Virtual Prototyping of Power Electronic Circuits," 2023 11th International Conference on Power Electronics and ECCE Asia (ICPE 2023 – ECCE Asia), Jeju Island, Republic of Korea, 2023, pp. 112–119.

[3] L. Ceccarelli, P. Diaz Reigosa, A. S. Bahman, F. Iannuzzo, and F. Blaabjerg, "Compact electro-thermal modeling of a SiC MOSFET power module under short-circuit conditions," IECON 2017 – 43rd Annual Conference of the IEEE Industrial Electronics Society, Beijing, China, 2017, pp. 4879–4884.

[4] M. M. Hossain, L. Ceccarelli, A. U. Rashid, R. M. Kotecha, and H. A. Mantooth, "An Improved Physics-based LTSpice Compact Electro-Thermal Model for a SiC Power MOSFET with Experimental Validation," IECON 2018 – 44th Annual Conference of the IEEE Industrial Electronics Society, Washington, DC, USA, 2018, pp. 1011–1016.

[5] X. Zhao, H. Li, Y. Wang, Z. Zhou, K. Sun, and Z. Zhao, "A Temperature-dependent PSpice Short-circuit Model of SiC MOSFET," 2019 IEEE Workshop on Wide Bandgap Power Devices and Applications in Asia (WiPDA Asia), Taipei, Taiwan, 2019, pp. 1–5.

[6] H. Li, Y. Wang, X. Zhao, K. Sun, Z. Zhou, and Y. Xu, "A Junction Temperature-based PSpice Short-circuit Model of SiC MOSFET Considering Leakage Current," IECON 2019 – 45th Annual Conference of the IEEE Industrial Electronics Society, Lisbon, Portugal, 2019, pp. 5095–5100.

[7] M. Riccio, V. d'Alessandro, G. Romano, L. Maresca, G. Breglio, and A. Irace, "A Temperature-Dependent SPICE Model of SiC Power MOSFETs for Within and Out-of-SOA Simulations," in IEEE Transactions on Power Electronics, 2018, Vol. 33, No. 9, pp. 8020–8029.

PCIM Europe 2024, 11– 13 June 2024, Nuremberg   DOI: 10.30420/566262396

Thermal Analysis and Modelling of Charging Stations for Electric Vehicles

Ruben Kopischke, Christian Koppe, Tim Schmidt, Mohamed Ayeb, Ludwig Brabetz

Universität Kassel, Germany

Corresponding author: Ruben Kopischke, kopischke@uni-kassel.de
Speaker: Ruben Kopischke, kopischke@uni-kassel.de

Abstract

This paper presents a concept to investigate the thermal behavior of charging stations for electric vehicles in the power range of 100 kW. Based on experimental data of a dummy charging station, thermal parameters are determined to allow a mathematical description of the thermal processes. The resulting model enables continuous monitoring and better control of the thermal variables during the charging process, allowing the operation to be optimized through appropriate thermal monitoring.

1   Introduction

The charging behavior of today's electric vehicles is influenced by many different factors and varies depending on the vehicle. In automotive batteries, different electrochemical compositions such as NMC (nickel manganese cobalt), NCA (nickel cobalt aluminum) and LFP (lithium iron phosphate) are commonly used [1]. Furthermore, different charging strategies such as CC-CV (constant current - constant voltage) and MS-CC (multistep constant current) are applied in practice [2]. The increasing use of electric vehicles and the driving behavior of their users often lead to peak load demands on charging stations, causing additional strain on the infrastructure [3]. This, combined with the variability in electric vehicle usage profiles, results in charging stations experiencing a wide range of load conditions.

Charging strategies used in practice for electric vehicles typically start the charging process with a high charging current, especially at a low state of charge, to ensure a fast initial charge. As the battery continues to charge, the charging current is adjusted to maintain the battery's lifespan.

With the CC-CV strategy, the charging current remains constant at a high level at the beginning to enable rapid charging. As soon as a certain voltage threshold is reached, the process switches to the constant voltage phase (CV), in which the current is continuously reduced to prevent overcharging and

to charge the battery optimally [4].

In contrast, multi-stage constant current charging strategies (MS-CC) implement an adjustment of the charging current by gradually switching between different charging stages based on defined switching conditions, such as reaching certain SoC levels [5] or voltage limits [6]. These targeted transitions between the charging stages aim to maximize charging efficiency while avoiding overcharging the battery.

Within the charging station, the charging current leads to high thermal losses in the installed power electronics and passive components, which can lead to overheating of certain components and consequently to a reduction in charging power, commonly known as derating. An early assessment of these risks is essential in the development phase but is also difficult to describe mathematically due to the complex thermal relationships.

In this paper, an approach is presented that makes it possible to estimate the expected thermal conditions at an early stage of development. This is done with the help of a simple to set up dummy charging station. In addition to describing the structure of the dummy station, this paper provides insights into the cooling performance of the station and a modeling concept for the thermal behavior of a charging station. The elaborated model represents the central approach for estimating the expected thermal conditions, with the results being presented and discussed.

2 Design of the dummy charging station

The design of the dummy charging station is based on a real charging station that is being investigated as part of the BMWK joint research project PVtec-Charger. The dummy station was developed from the CAD data of the charging station, which reflects the structure of the real charging station. The thermal active areas were identified, and the power losses of the individual components were realized using heating mats. In the current setup of the dummy charging station, heating mats with a total output of $P_{tot} \approx 4\,kW$ are deployed to emulate the power losses of various components of the real station. In addition, the internal structure of the components was taken into account using metal plates in order to reflect the thermal capacities of the real station. The dummy station as it is set up in the laboratory is shown in Fig. 1.

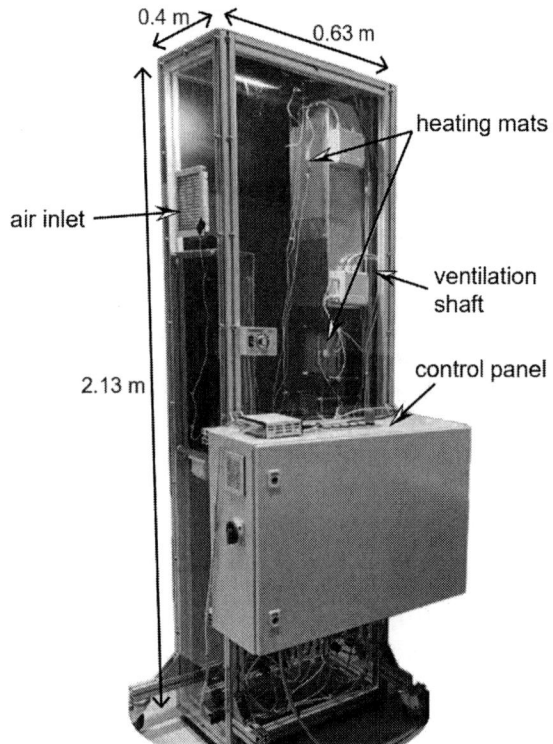

Fig. 1: Dummy station as set up in the laboratory.

The image shows the dimensions of the dummy station with a height of approx. 2.1 m, a width of 0.4 m and a depth of approx. 0.6 m. The air inlet can be seen at the front, the ventilation shaft, which can force convection through the fan, is located at the rear. The use of transparent polycarbonate

(Makrolon), which has a high thermal stability, allows a view of the inside of the station, such as some built-in heating mats and metal panels. They can be replaced with metal panels to better represent the thermal behavior within the station.

2.1 Customized power loss areas

The locations of all the installed sensors and the arrangement of the heating mats, fan and air inlet and outlet are shown to scale in Fig. 2.

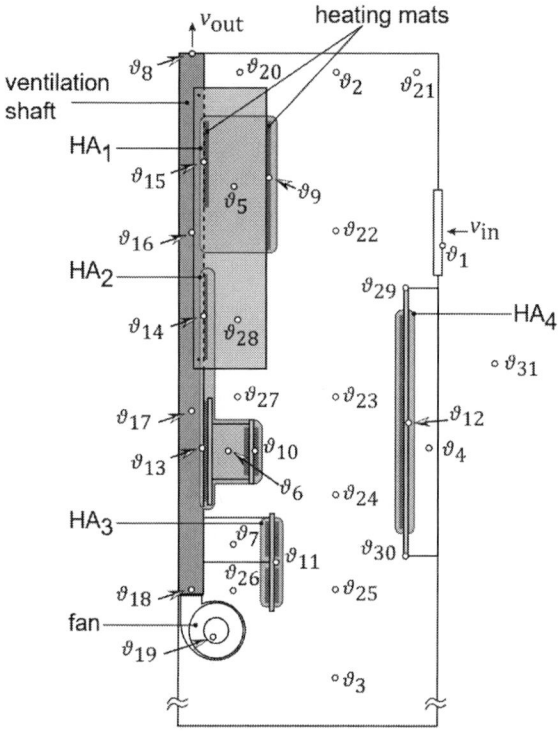

Fig. 2: Positioning of Sensors ϑ_n and heating areas HA_m.

The heating mats are assigned to four heating areas HA_m, $m = 1, ..., 4$. Their dissipated power heats are regulated using a control panel. They represent the thermally active areas which are specially designed to match the components used in the real charging station. Essentially, these are HA_1 for the power unit and the output module, HA_2 for the DC/DC converter transformer, HA_3 for the grid-side inductance and the inverter-side inductance of the AC/DC power unit and HA_4 for the controller module. Each area is modeled on the corresponding real components in terms of size and dimension in order to enable an accurate simulation of the thermal conditions in the dummy charging station.

2.2 Controllability and modular design

The modular design of the dummy station allows for easy expansion and modification of the system, especially regarding the flexible spatial arrangement of the individual components. The control of each heating zone is individually configurable, so that a targeted control and adjustment of the thermal conditions is possible. In addition, the station enables variable control of the air flow, adjustable power setting and sensitive temperature control to ensure optimal adaptation to different requirements and conditions. Furthermore, the system design allows for any number of sensors and positioning.

3 Cooling System

This section presents the structure of the cooling system and the results of the cooling performance of the fan for the individual heating areas. In particular, the cooling performance is examined with regard to the heat flow that can be dissipated in the individual heating zones through the use of the fan.

3.1 Setup of the cooling system

The cooling path is arranged so that air is supplied through the air inlet at the front of the station. On the opposite side, at the bottom of the station, the fan draws in air and blows it into a dedicated ventilation shaft that removes heat through the top of the station.

The air inlet and outlet are indicated in Fig. 2 by v_{in} at the front of the dummy station and v_{out} at the top of the shaft. When the fan is at maximum speed, the air reaches an average velocity of around $v_{out} \approx 4.5\,\mathrm{ms}^{-1}$ which was measured at an average temperature of around 30.4 °C in the ventilation shaft.

3.2 Cooling Performance

In the current setup of the dummy station, the heat dissipation by free convection $P_{frc,m}$ and forced convection $P_{foc,m}$ can be well determined by controlling setpoint temperatures. Figure 3 shows the determination of free and forced convection by the fan in an experiment using HA_1 as an example.

The power measured at HA_1, which corresponds to the power loss, is plotted on the left axis. The temperature is shown on the right axis. A simple PI controller is used to regulate the temperature by adjusting the power loss until the system reaches a temperature of 80 °C in steady state with the fan switched off. The fan is then switched on at nominal speed and the controller regulates the temperature

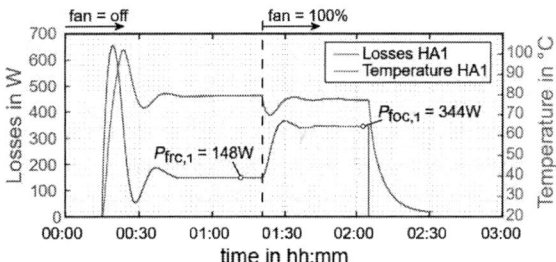

Fig. 3: Temperature and power loss curve for the first heating area (HA_1). The temperature is controlled at 80 ° C.

back to 80 °C under these new conditions. The powers produced here correspond to those of free and forced convection. Using Eq. 1 the forced dissipated heat output $P_{air,m}$ is determined for a specific heating area m.

$$P_{air,m} = P_{foc,m} - P_{frc,m} \qquad (1)$$

The results of the conducted tests show that the forced convection, induced by the cooling fan, enhance the cooling performance by 65 % to 132 % compared to a free convection. The mentioned figures are calculated for each heating area using

$$\eta_m = \frac{P_{air,m} \cdot 100\,\%}{P_{frc,m}}. \qquad (2)$$

Figure 4 depicts an exemplary heat map of the charging station during a test where all areas were heated to illustrate the temperature distribution inside the station.

4 Thermal Model

An analytical description of the thermal relationships in a complex 3D geometry, whereby all three heat transfer mechanisms, heat conduction, heat radiation as well as free and forced convection, must be taken into account, is hardly feasible. An FEM calculation would be an alternative, although a very high effort in parameterization and calculation is to be expected here. In this section, an alternative, data based generic approach is adopted and the first results for a stationary thermal model are presented.

4.1 Generic structure of the thermal model

The developed thermal model aims to predict the temperature increase, induced by the heat dissipation under different load conditions, at specific N

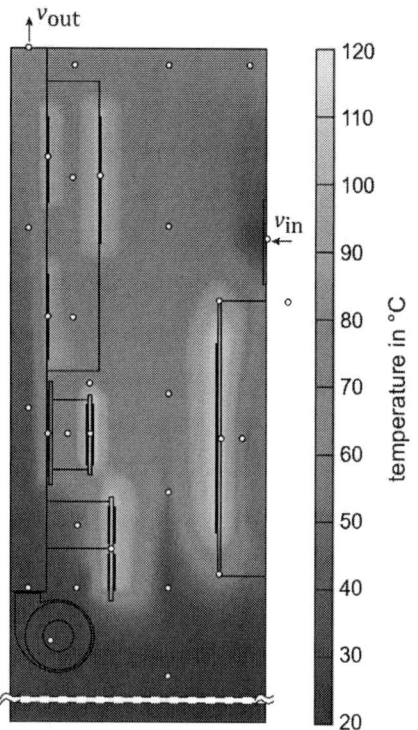

Fig. 4: Heat map of the dummy station during an experiment caused by free convection.

locations inside the charging stations. This temperature increase $\Delta\vartheta_n$, $n = 1, ..., N$ is first captured using dedicated temperature sensors at the chosen locations under representative heat dissipation conditions. The collected temperature increase data is then used to parametrize a Hammerstein model structure which consists of a static nonlinear map followed by a linear dynamic transfer function as depicted in Fig. 5. The Hammerstein model represents a widely spread approach to represent nonlinear dynamic behavior [7].

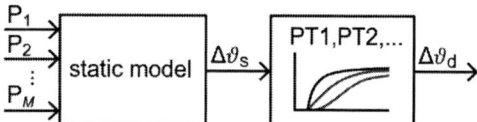

Fig. 5: Structure of the model based on [7].

The first block contains the static model, which calculates the temperature change as a function of the applied power P_m with $m = 1, ..., M$. The result of the thermal computation in the steady state is $\Delta\vartheta_s$. With the help of a dynamic transfer element, whereby a first or a second order transfer function should already capture the main transient behavior

in our case, the model can be extended to calculate dynamic temperature changes $\Delta\vartheta_d$, which is represented in the second block of the generic model structure.

4.2 Static thermal model

If the heat transfer mechanisms were linear, then

$$\Delta\vartheta_n = \sum_{m=1}^{M} \alpha_{n,m} \cdot P_m \qquad (3)$$

applies, whereby the alpha coefficients are constant. In order to determine them, the stationary temperature values must be captured under representative heat dissipation conditions in all heating areas. After the conduction of the measurements under S different load conditions, the temperature measurement data obtained is gathered in a temperature vector

$$Y = \begin{pmatrix} \Delta\vartheta_{n,1} \\ \Delta\vartheta_{n,2} \\ \vdots \\ \Delta\vartheta_{n,S} \end{pmatrix}, \qquad (4)$$

where $\Delta\vartheta_{n,s}$ represents the stationary temperature increase at a position n in the dummy station under the respective load condition $s = 1, ..., S$. In addition, the dissipated heating powers are gathered in a matrix:

$$F = \begin{pmatrix} P_{1,1} & P_{1,2} & \cdots & P_{1,M} \\ P_{2,1} & P_{2,2} & \cdots & P_{2,M} \\ \vdots & \vdots & \ddots & \vdots \\ P_{S,1} & P_{S,2} & \cdots & P_{S,M} \end{pmatrix}. \qquad (5)$$

The columns of F represent a specific heating area m within the charging station for the respective measurement s. Therefore, we obtain from Eq. 3 the matrix equation

$$Y = F\alpha_n, \qquad (6)$$

with α_n a vektor composed by the coefficients $\alpha_{n,m}$ of the linear model from Eq. 3. Equation 6 can be solved for α_n using the least square method:

$$\alpha_n = \begin{pmatrix} \alpha_{n,1} \\ \alpha_{n,2} \\ \vdots \\ \alpha_{n,M} \end{pmatrix} = (F^T F)^{-1} F^T Y. \qquad (7)$$

Through a rich variation of the dissipated powers in the experimental design, the invertibility of $(F^T F)$ is ensured.

However, due to the non-linearity of the heat transfer mechanisms, in particular through convection and thermal radiation, the alpha values themselves must be represented as functions of the temperatures and the forced air flow $\dot{Q}_{\text{fan}}$. Therefore the linear model (Eq. 3) is extended to a non-linear model:

$$\Delta\vartheta_n = \sum_{m=1}^{M} \alpha_{n,m}\left(\dot{Q}_{\text{fan}}, \Delta\vartheta_{n=1,\cdots,N}\right) P_m. \quad (8)$$

However, as within the deployed experiment design, the dissipated powers and the fan speed, rather than the temperatures and the air flow, are tuned. Thus, the model is further modified to an equivalent form:

$$\Delta\vartheta_n = \sum_{m=1}^{M} \alpha_{n,m}\left(v_{\text{fan}}, P_{m=1,...,M}\right) P_m, \quad (9)$$

which is more appropriate for parametrizing the model as well as for its deployment to predict the thermal behavior under different load conditions, where the dissipated powers and the fan speed are given.

The analysis of the measurement data showed that a simple linear relationship between the alpha coefficients and the dissipated powers is appropriate. The initial design matrix F from Eq. 5 is therefore extended by additional columns with quadratic and the twofold interaction terms of the dissipated powers. The calculation of the alpha coefficients can further be carried out according to Eq. 7 while using the extended design matrix F.

4.3 Data Basis

In order to obtain a rich data base for the parametrization of the model, a three levels factorial experimental design to set the heating power for each heating area is deployed. The three power levels are defined individually for each heat area as P_{100}, $1/2 \cdot P_{100}$ and zero, whereas P_{100} is the needed power to heat up the respective zone to 100 °C. The measurement plan created for the model currently comprises 240 measurement points, including the measurement points related to the variation of the forced convection by the fan.

4.4 Measurement Results

For the static thermal model, the final stationary temperature values are of particular importance, with the temperature increase is calculated using the ambient temperature as a reference. These steady-state end values are reached at various times depending on the position of the sensors. The measurement time is therefore determined by the sensor that requires the longest time to reach the steady state.

Figure 6 shows the temperature rise for an exemplary series of measurements, whereby only the heating zone HA_1 was subjected to a power dissipation of $1/2 \cdot P_{100}$ while no power was dissipated in the other areas.

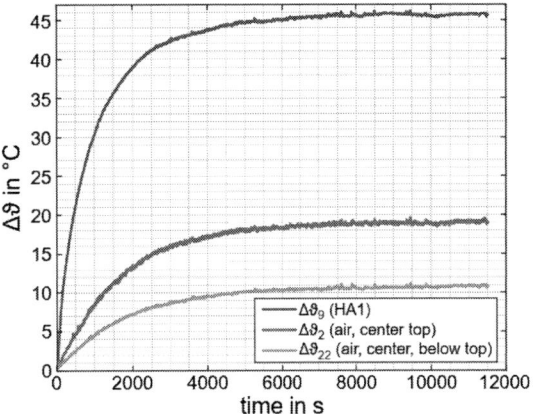

Fig. 6: Sample measurement, with $HA_1 = 1/2 \cdot P_{100}$ and $HA_{2-4} = $ zero.

The temperature increases with regard to the ambient temperature are depicted for three different locations in Fig. 6. The ambient temperature for this measurement was approximately $\vartheta_{\text{amb}} = 21\,°\text{C}$ and the stationary end value at the position of sensor ϑ_9 was around 67 °C. The results of another experiment with overall higher load condition are given in Table 1.

Tab. 1: Settings and sample results for a measurement.

	Setting	P_{loss} in W	$\Delta\vartheta_{\text{s,n}}$ in °C
HA_1	$1/2 \cdot P_{100}$	164,29	68,46 $(n=9)$
HA_2	P_{100}	212,61	93,86 $(n=10)$
HA_3	P_{100}	103,04	87,61 $(n=11)$
HA_4	$1/2 \cdot P_{100}$	145,67	67,73 $(n=12)$

The table contains information about the settings for this measurement and the results for example sensors ϑ_{9-12}, which are located within the heating areas. In this measurement, HA_1 and HA_4 were loaded simultaneously with $1/2 \cdot P_{100}$ and HA_2 and HA_3 with the full power P_{100}. Taking into account the ambient temperature, which was approximately $\vartheta_{\text{amb}} = 19\,°\text{C}$ during this measurement, a temperature of approximately $112\,°\mathrm{C}$ is reached at ϑ_{10}. In another measurement, in which all heating ranges are set to P_{100}, the maximum absolute temperature was $\vartheta_{\mathrm{s},10} = 122\,°\mathrm{C}$.

4.5 Results for the stationary thermal model

This section shows the results of the steady-state thermal model for the dummy station. The modeling results for sensor ϑ_{23}, which is positioned in the middle of the station as an air temperature sensor, are shown as an example in Figure 7.

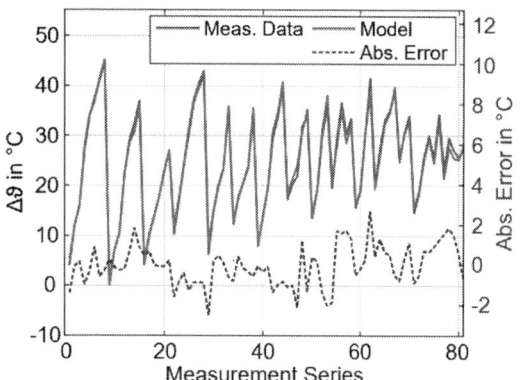

Fig. 7: Model results for sensor ϑ_{23} with measured and predicted data on the left axis and the abs. error on the right axis.

The series of measurements are plotted on the abscissa, with the blue curve representing the real measured values and the orange curve the results of the model. For the shown sensor ϑ_{23} the absolute error is in the range of approx. –2.5 °C to 2.5 °C and the root mean square error (f_{rms}) reaches a value of 0.94 °C, which indicates a good fit of the simple model with the measured data.

To validate the steady-state thermal model, the model was only trained with every other measurement. The model was then used to predict the remaining measurements. This methodology made it possible to verify the performance of the model under different conditions. The result of the validation showed an accurate curve for all sensors, very similar to the one shown in Fig. 7. This consistency between the model results and the real data underlines the validity of the proposed modeling approach.

The next step consists of the parametrization of the dynamic extension of the model using first and second order linear dynamics depending on the transient behavior captured by each deployed temperatur sensor.

5 Conclusion

In this paper, a concept for investigating the thermal behaviour of charging stations for electric vehicles in the power range of 100 kW was described and a method for creating a data-based thermal model was demonstrated. A data set was generated with which the developed stationary thermal model could be parametrized. The validation of the model has shown that the prediction of the thermal behavior of charging stations for electric vehicles is feasible with high accuracy. This confirms the suitability of the developed data-based approach for the investigation and optimization of thermal management strategies. The findings offer a solid basis for future investigations and optimizations in the field of electric vehicle charging technology and underline the viability of considering thermal aspects in the early stages of development. Appropriate thermal monitoring can prevent inadmissible temperatures in the charging stations to ensure controlled and safe operation and thus protect the power electronics, filters and control elements in the charging stations.

Acknowledgements

The work has been conducted in the frame of the collaborative research project PVtec-Charger (FKZ: 03ETE037E), which is founded by the German Federal Ministry for Economic Affairs and Climate Action (BMWK) on the basis of a decision by the German Bundestag.

References

[1] S. Hasselwander, M. Meyer, and I. Österle, "Techno-economic analysis of different battery cell chemistries for the passenger vehicle market," *Batteries*, vol. 9, no. 7, p. 379, 2023. DOI: 10.3390/batteries9070379.

[2] M. Brenna, F. Foiadelli, C. Leone, and M. Longo, "Electric vehicles charging technology review and optimal size estimation," *Journal of Electrical Engineering & Technology*, vol. 15, no. 6, pp. 2539–2552, 2020. DOI: 10.1007/s42835-020-00547-x.

[3] T. Schmidt, B. Löwer, M. Ayeb, and L. Brabetz, "Entwicklung und aufbau einer simulationsumgebung zum abbilden einer ladeinfrastruktur für elektrofahrzeuge," in *Symposium Hybrid- und Elektrofahrzeuge*, Oct. 2022.

[4] J. M. Amanor-Boadu and A. Guiseppi-Elie, "Improved performance of li-ion polymer batteries through improved pulse charging algorithm," *Applied Sciences*, vol. 10, no. 3, p. 895, 2020. DOI: 10.3390/app10030895.

[5] F. An, R. Zhang, Z. Wei, and P. Li, "Multistage constant-current charging protocol for a high-energy-density pouch cell based on a 622ncm/graphite system," *RSC Advances*, vol. 9, no. 37, pp. 21 498–21 506, 2019. DOI: 10.1039/c9ra03629f.

[6] R. Rai, M. Gaglani, S. Das, and T. Panigrahi, "Multi-level constant current based fast li-ion battery charging scheme with lms based online state of charge estimation," in *2020 IEEE Kansas Power and Energy Conference (KPEC)*, 2020, pp. 1–6. DOI: 10.1109/KPEC47870.2020.9167541.

[7] M. Ayeb, L. Brabetz, G. Jilwan, and P. Graebel, "A generic modeling approach for automotive power net consumers," in *SAE Technical Paper Series*, ser. SAE Technical Paper Series, SAE International400 Commonwealth Drive, Warrendale, PA, United States, 2012. DOI: 10.4271/2012-01-0924.

PCIM Europe 2024, 11– 13 June 2024, Nuremberg   DOI: 10.30420/566262397

Junction Temperature Measurement of a 3.3 kV Silicon Carbide MOSFET Power Module

Michael Gleissner [1], Matthias Bürger[2], Mark-M. Bakran [1]

[1] University of Bayreuth, Department of Mechatronics, Center of Energy Technology, Germany
[2] Infineon Technologies AG, Warstein, Germany

Corresponding author: Michael Gleissner, michael.gleissner@uni-bayreuth.de
Speaker: Michael Gleissner, michael.gleissner@uni-bayreuth.de

Abstract

The static temperature distribution in a 3.3 kV SiC MOSFET power module with several parallel chips depending on power loss is examined using an infrared temperature camera. Moreover, the dynamic infrared temperature curves are compared with the junction temperature measurement based on the internal gate resistance and forward voltage of the body diode as temperature-sensitive electrical parameters. The thermal impedance measurements with these three different temperature sensors are compared. Moreover, the influence of different chip temperatures on lifetime simulation is investigated.

1 Introduction

Typically, the junction temperature is determined by measuring the forward voltage. This approach yields a virtual junction temperature, representing an average value applicable to single or multiple parallel chips spanning the entire chip area [1]. However, this method does not allow for the recording of temperature distributions or extreme values. To obtain precise insight into the temperature distribution within a power module, it must be prepared accordingly, e.g. by opening the module and employing an infrared camera for temperature measurement, as discussed by [2]. Numerous publications delve into this subject matter for IGBT power modules [3]–[9]. This paper analyzes a 3.3 kV SiC MOSFET module mounted on a water cooler [10]–[12]. The temperature distribution of the chips depending on the current is determined by an infrared camera in section 2. Dynamic temperature measurements based on the internal gate resistance and infrared are compared in section 3. The thermal impedance curve is determined in section 4 based on forward voltage measurement of the body diode at low current, internal gate resistance and infrared. The thermal impedance measurement results for the hottest and coldest chip are compared with the mean value in a life time simulation in section 5. In section 6, the influence of varying diode output characteristic on temperature measurement based on the forward voltage is analyzed.

2 Temperature distribution of paralleled chips within the module

A 3.3 kV SiC MOSFET half-bridge module with 32 parallel chips per switch is modified by omitting the silicone gel and blackening for temperature measurements with an infrared camera (see Fig. 1).

Fig. 1: Measurement setup with prepared module, IR camera and junction temperature monitor

The sampling rate of the Optris PI 640i camera is 32 Hz. The overall resolution is 640x460 pixels. To record all chips, there is a resolution of 43x32 pixels per chip area. The rectangular measurement areas for the mean temperature of each chip are depicted in Fig. 2a.

2803

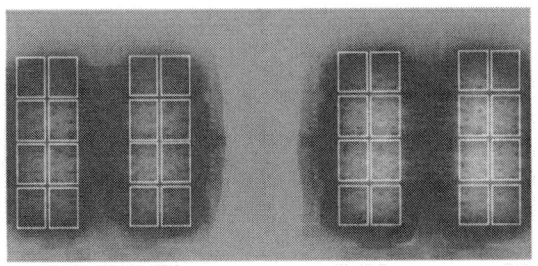

(a) Infrared view with marked chip areas

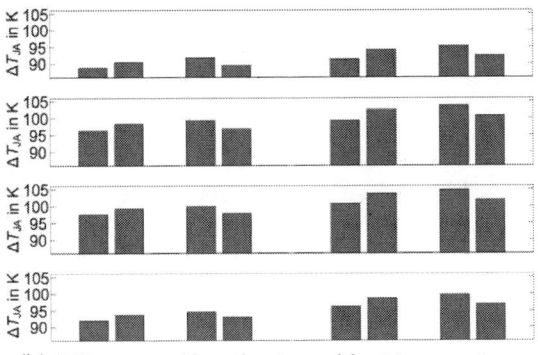

(b) Difference of junction to ambient temperature

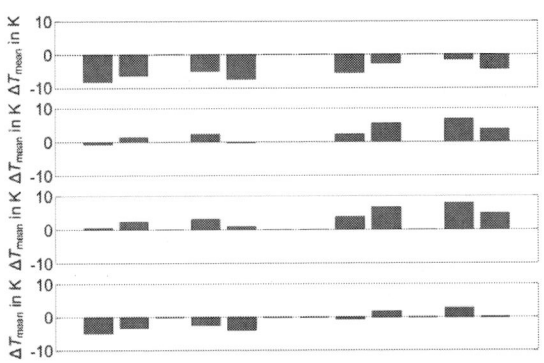

(c) Deviation from mean temperature of all chips

Fig. 2: Static temperature distribution of 32 chips at 500 A heating current. The arrangement of the bars is according to the chip position within the module.

The switch is permanently turned on with 15 V at the gate for all measurements. A current source sets the current for the conduction losses. The module is mounted on a water cooling system with a water temperature of 21 °C. Figure 2b shows the difference between chip and ambient temperature of the chips in the upper range from 89 °C to 105 °C at 500 A heating current. The associated deviation from the mean value of all chips 97 °C is also shown in Fig. 2c. In these bar charts, one bar equals one chip. The upper bars correspond to the upper 8 chips, and the lower bars to the lower 8 chips according to the arrangement of the chips in the module. The coolant flows downwards on the left side and upwards on the right side. This flow direction is visible in the chip temperatures, as the chips on the right side are warmer than their equivalents on the left side. The mean and individual chip temperatures depending on the heating current are illustrated in Fig. 3a. Figure 3b depicts the difference between hottest and coldest chip depending on current. As expected, these values increase with the current.

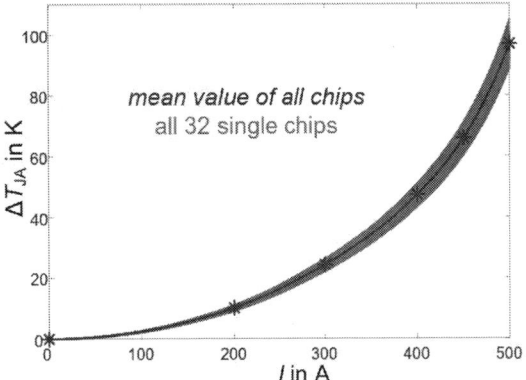

(a) Difference of junction to ambient temperature

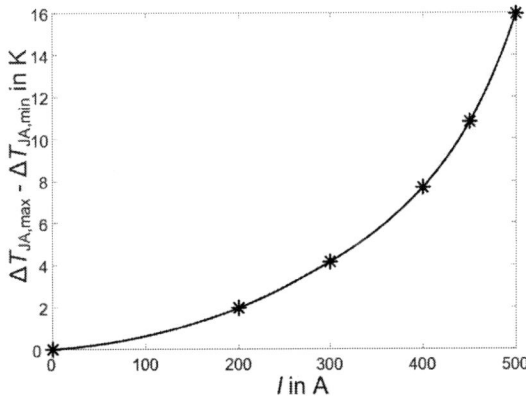

(b) Difference between hottest and coldest chip

Fig. 3: Temperature distribution of chips depending on current based on IR measurement

3 Dynamic junction temperature measurement based on the internal gate resistance

The junction temperature monitor, which is based on the internal gate resistance as temperature-sensitive parameter and described in detail in [13], [14], has been applied to measure the junction temperature as an average of all 32 chips, because only a common gate connection of the total module is available. The gate impedance of the switch measured with a vector network analyzer Bode 100 from OMICRON Lab is depicted in Fig. 4. At resonance frequency of about 1 MHz, only the real part of the impedance is measured and it depends on the temperature of the chip.

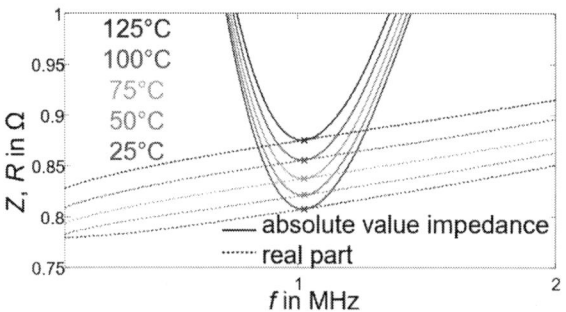

Fig. 4: Absolute value and real part of gate impedance of SiC MOSFET module depending on frequency and temperature measured with a vector network analyzer

The slight non-linearity, which is noticeable with the gate impedance curves of the network analyzer as well as with the junction temperature monitor, (see Fig. 5) is taken into account when converting the resistance value to temperature.

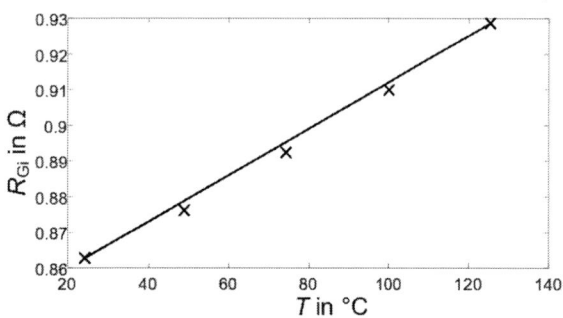

Fig. 5: Calibration curve internal gate resistance and temperature of junction temperature monitor with slight non-linearity

The chip temperature is adjusted by the heating current according to the circuit diagram illustrated in Fig. 6. The dynamic temperature measurement in Fig. 7 shows, that the mean temperature of all chips is close to the temperature measured by the junction temperature monitor. A complete match is not likely because most of the internal gate resistance is located in the center of the chips and

the IR camera averages the entire chip area. The temperature distribution over the chip surface is not homogeneous and the temperature is slightly lower in the center, because there are no active semiconductor cells in the area of the gate resistor and generating losses.

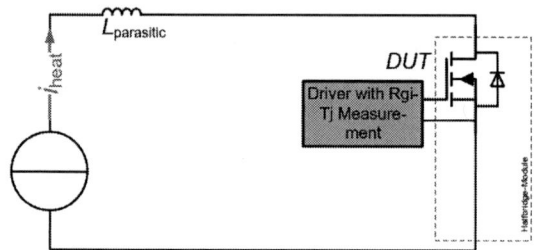

Fig. 6: Circuit diagram of heating current source and temperature monitor configured for measurement during permanent on state of the DUT.

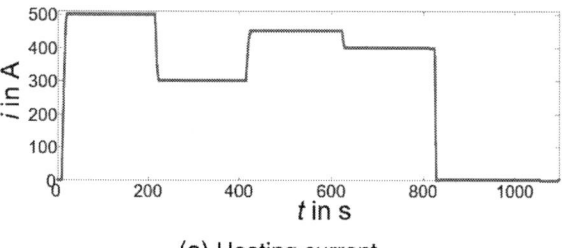

(a) Heating current

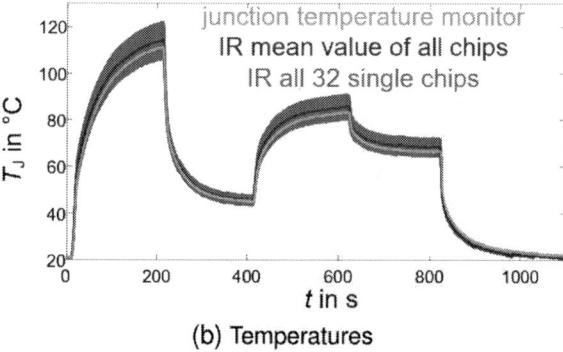

(b) Temperatures

Fig. 7: Dynamic temperature measurement with IR camera and junction temperature monitor

A measurement was only carried out in permanent on mode because the removal of the insulation gel means that the module is no longer suitable for operation with a nominal voltage in the 2 kV range. In principle, measurements can also be carried out with the junction temperature monitor in switching operation of the converter. The measurement phase of the junction temperature monitor can be configured in on- or off-state of the DUT.

4 Determination of the thermal impedance with three different temperature sensors

Another temperature-sensitive parameter is the forward voltage of the body diode of the SiC MOSFET when a small measuring current is applied. For this purpose, the measurement setup is supplemented by a 100 mA current source as well as a parallel and serial switch to the test object. They enable a fast commutation of the heating current and thus an immediate temperature measurement based on the forward voltage (see Fig. 8). Additionally, the serial switch ensures that the small sense current is blocked from the parallel path. The additional serial and parallel switch are of the same semiconductor type like DUT. The junction temperature monitor measures the internal gate resistance during the off-state of the DUT at a negative gate voltage of –5 V so that the body diode is effective.

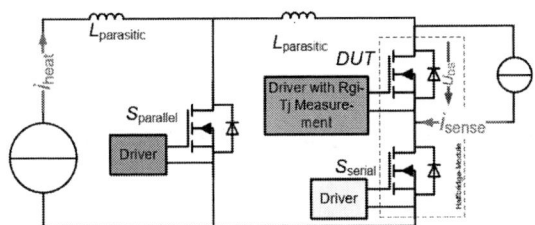

Fig. 8: Circuit diagram of heating current source, temperature monitor and forward voltage measurement at a small current. The serial and parallel switch are added for a fast commutation of the load current and decoupling.

Figure 9 illustrates the temperature measurement during the cooling phase with three different sensors. For the IR camera only the mean value of all 32 chips is shown. Small jumps in the IR signal can be seen due to the camera's calibration flag. Due to the thermal drift of bolometers, all measuring IR cameras require an offset correction every few minutes. This correction is done by a motor-driven movement of a blackened piece of metal on the front of the image sensor. The temperature measurement based on IR, internal gate resistance and forward voltage measurement are in good agreement for the long time range (see Fig. 9a). The zoom for the first 5 ms is depicted in Fig. 9b. The IR signal is here constant due to the very low sampling rate of the camera. The junction temperature monitor sends the first value about 20 µs after the driver turn-off. The forward voltage signal is shown

in three variants and has been recorded with an oscilloscope. The signal for the long time range of 500 s is sampled too small. Therefore, an additional measurement with a shorter time range and thus higher sampling rate has been done. Due to the noise, this signal is shown with and without a moving average filter. When measuring the forward voltage at short-time range, it can be seen that the temperature rises within the first 200 µs, which is not plausible because the heating was previously switched off.

(a) Total time range

(b) Zoom of the first 5 ms

Fig. 9: Measurement of cooling phase with various temperature sensors

The cause lies in the calibration of the probe as can be seen in Fig. 10. The probe was calibrated with a slight overshoot or undershoot at the oscilloscopes calibration pins. The forward voltage of the SiC MOSFET is positive when the heating current is applied and negative during the sense current phase. There is a slight overshoot due the parasitic inductance in the setup (see Fig. 10a). A clipping of the input of the oscilloscope has to be avoided. The temperature sensitivity of the forward voltage at 100 mA is only –2.5 mV K^{-1}. This represents a special challenge for the measuring system. Depending on the calibration of the probe, the temperature rises or falls in the first microseconds (see Fig. 10b).

PCIM Europe 2024, 11– 13 June 2024, Nuremberg   DOI: 10.30420/566262397

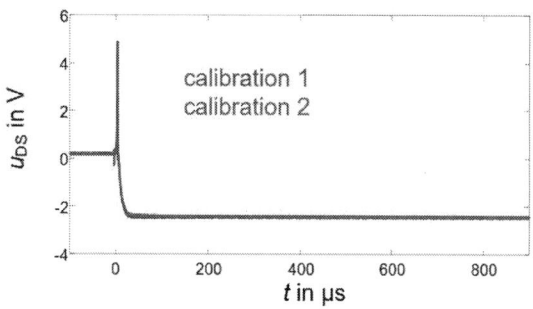

(a) Total view of SiC MOSFET forward voltage measurement

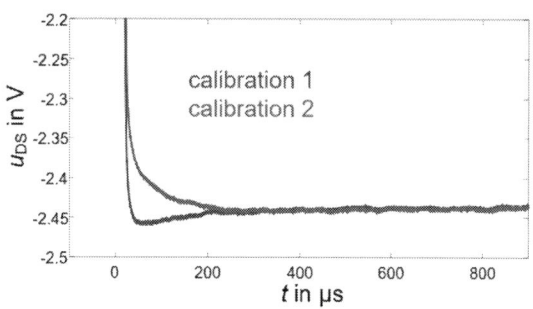

(b) Zoom view of SiC MOSFET forward voltage measurement

Fig. 10: Influence of probe calibration on forward voltage based temperature signal for the first 200 µs

This makes it difficult to obtain a correct temperature measurement based on the forward voltage for the first microsecond time range. A $\sqrt{t}$-extrapolation can be applied to determine the initial temperature for the forward voltage method [15]. The thermal impedance calculated from the cooling curve is plotted double logarithmically in Fig. 11 for the three temperature sensors.

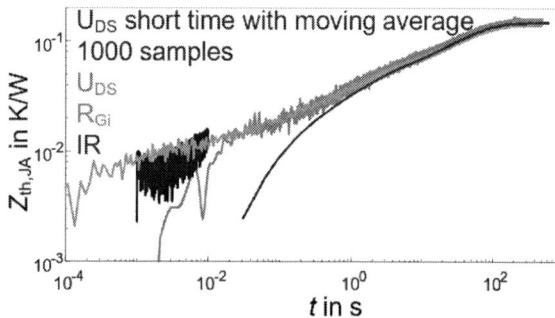

Fig. 11: Comparison of thermal impedance measurement based on three different temperature sensors

5 Lifetime simulation based on various thermal impedances

The IR camera measurement enables the determination of the thermal impedance of the hottest and coldest chip. They are merged with the thermal impedance curve based on the internal gate resistance temperature measurement. A fitting with three thermal Cauer network elements has been performed and the results for the maximum, mean and minimum thermal impedance are illustrated in Fig. 12. The stationary thermal resistance of the hottest chip is 9 % higher than the mean value and for the coldest chip it is 9 % smaller. This deviation also depends on the cooling system and can be reduced by a more complex cooling system.

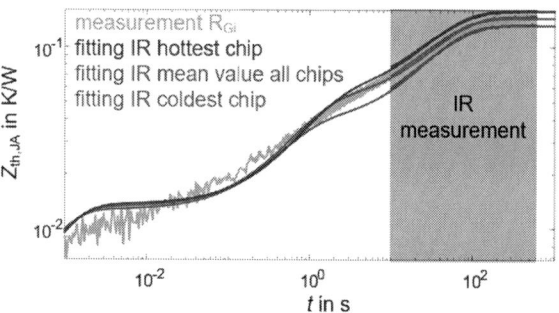

Fig. 12: Measurement and fitting of thermal impedances for lifetime simulation

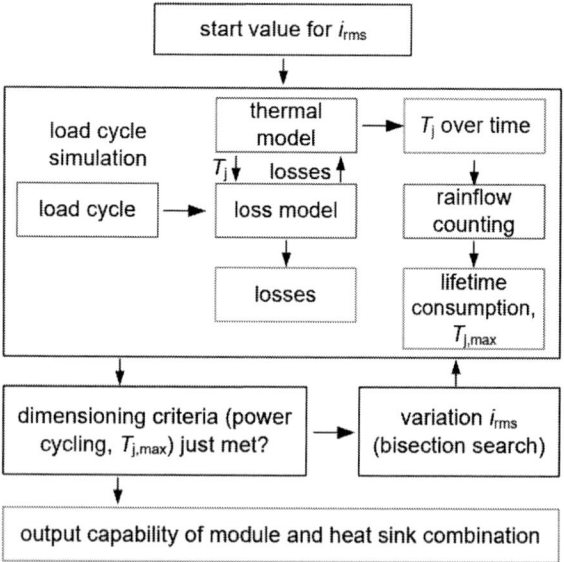

Fig. 13: Overview of the load cycle simulation and determination of the output capability regarding current or lifetime [9], [16]

2807

These three thermal impedance curves are applied in a load cycle simulation with MATLAB Simulink (see [9], [16]) to investigate the influence on lifetime consumption and performance (see Fig. 13). A steady-state condition is determined depending on the rms current, load cycle profile, thermal, and loss model. The loss and thermal models influence each other. Based on the steady-state temperature profile, the lifetime is determined with a rain flow counting algorithm (see [17]).

The load cycle profile is illustrated in Fig. 14a, and the losses are extracted from the data sheet of the module. Fig. 14b shows an example of the simulated junction temperatures at 45 °C ambient temperature for the three thermal impedances at identical rms current. The largest temperature ripple for the mean thermal impedance is about 44 K. For the hottest chip thermal impedance simulation the largest temperature ripple is 6 K more and for the coldest chip thermal impedance simulation 5 K less. The lifetime for this hypothetical simulation with the thermal impedance value of the hottest and coldest chip with identical rms current decreases or increases according to the lifetime curve. These results are very similar to the identical approach with a water cooled IGBT module in [9].

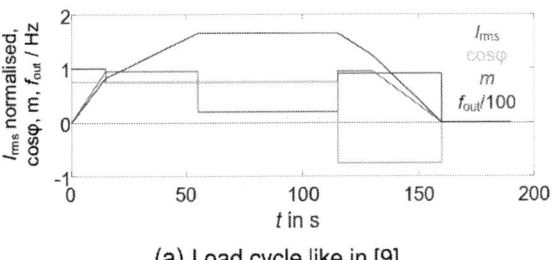

(a) Load cycle like in [9]

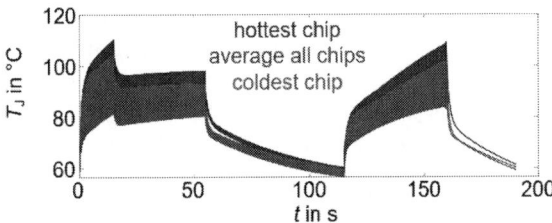

(b) Simulated junction temperature at 45 °C ambient temperature and for different thermal impedances with identical maximum rms current.

Fig. 14: Load cycle simulation

With the identical lifetime as the boundary condition for the simulation, the thermal impedance curve of the coldest chip achieves a 5 % higher and that of the hottest chip a 4 % lower output power than the simulation with the mean thermal impedance.

6 Influence of inhomogeneous diode output characteristic on U_{SD} temperature measurement

The temperature measurement based on the forward voltage provides an average value for parallel chips. This section examines the effects on the temperature measurement if the diodes have very different output characteristics. The chips in the 3.3 kV module cannot be measured individually. Therefore, three commercial 1.2 kV SiC MOSFETs of the type C3M0016120K from Wolfspeed are used. The output characteristic curves of the body diodes were determined at temperatures from 25 °C to 125 °C in 25 °C steps and at a gate voltage of –5 V with a Keysight Technologies power device analyzer B1506A (see Fig. 15).

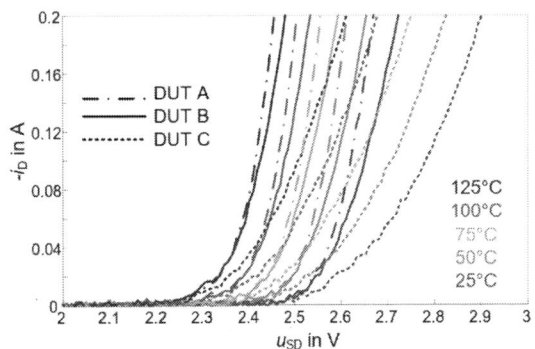

Fig. 15: Measured output characteristic of C3M0016120K SiC MOSFET body diodes at various temperatures for three devices with strongly varying characteristic

These measured output characteristics can be combined with each other to determine the forward voltage in a parallel connection of two devices at a given sense current of 100 mA. Therefore, the points must be found for two output curves at which the total current corresponds to the sense current at identical voltage. DUT A represents the device with the best output characteristic, which means lowest forward voltage at a given output current, and DUT C the worst output characteristic. DUT B is between the two extremes. Normally, the chips in a module have a very similar output characteristic. The temperature of the chips can vary due to a different thermal arrangement as shown in section 2 and [9].

Figure 16 depicts the calculated results for three scenarios with varying temperatures and each two devices in parallel:

- DUT B warmer and with identical output characteristic combined with DUT B

- DUT A warmer and with better output characteristic combined with DUT C

- DUT A colder and with better output characteristic combined with DUT C

For each scenario, a calibration characteristic is assumed in which both DUTs have the same temperature. Figure 16a shows the deviation of the calculated value from the hottest chip and Fig. 16b shows the deviation from the mean temperature of both chips. The actual temperature difference between the two chips is indicated on the x-coordinate. For chips with identical output characteristic but different temperatures, the calculated value is about 3 K higher than the real mean value at a given temperature difference of 25 K. This is in accordance with [18]. In the combinations of the two very different output characteristics, the temperature of DUT A with the better output characteristic determines the temperature measurement based on the forward voltage. At a given temperature difference of 25 K, the calculated mean value is about 8 K higher than the real mean value, when the DUT with better output characteristic is warmer. If the DUT with the better output characteristic is 25 K colder, than the calculated mean value is 8 K below the real mean value. For arrangements with more than two parallel chips, it depends on the distribution of output characteristics as well as temperature.

7 Conclusion

This article analyzes the temperature distribution of several paralleled chips within a 3.3 kV SiC MOSFET module. The temperature spread is very homogeneous and depends on the current and the cooling system. The influence of the hottest and coldest chip on the output performance at identical lifetime is also low as has been shown with a simulation. The junction temperature can be measured dynamically very well using the junction temperature monitor based on the internal gate resistance. This method is superior to the forward voltage method, as it can be applied in converter

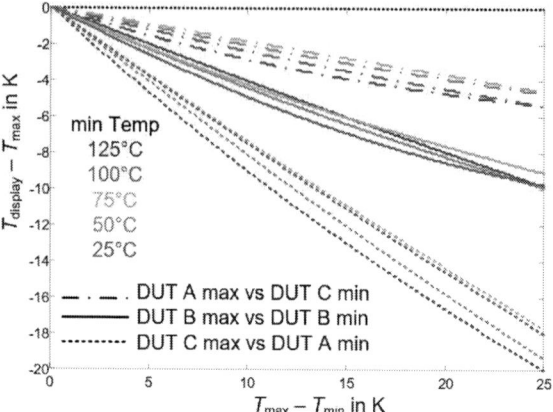

(a) Difference of calculated and actual maximum value

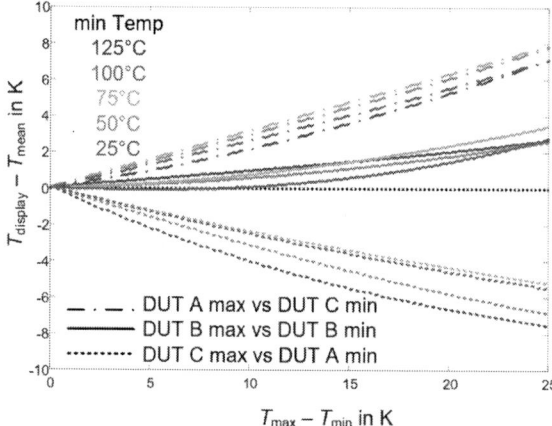

(b) Difference of calculated and actual mean value

Fig. 16: Difference of calculated values depending on actual temperature difference of hottest and coldest chip as well as diode output characteristic for two parallel devices

operation. Moreover, the time range of the first microseconds after heating current turn-off can be measured directly. Furthermore, the influence of a output characteristic spread of the diodes on the forward voltage temperature measurement has been investigated.

References

[1] U. Scheuermann and R. Schmidt, "Investigations on the VCE(T)-Method to Determine the Junction Temperature by Using the Chip Itself as Sensor," in *Proceedings of the International Exhibition and Conference for Power Electronics, Intelligent Motion and Power Quality (PCIM Europe)*, Berlin [u.a.]: VDE VERLAG, 2009.

[2] D. L. Blackburn, "Temperature measurements of semiconductor devices - a review," in *Twentieth Annual IEEE Semiconductor Thermal Measurement and Management Symposium (IEEE Cat. No.04CH37545)*, 2004, pp. 70–80. DOI: 10.1109/STHERM.2004.1291304.

[3] L. Dupont, Y. Avenas, and P.-O. Jeannin, "Comparison of Junction Temperature Evaluations in a Power IGBT Module Using an IR Camera and Three Thermosensitive Electrical Parameters," *IEEE Transactions on Industry Applications*, vol. 49, no. 4, pp. 1599–1608, 2013. DOI: 10.1109/TIA.2013.2255852.

[4] C. Chen, V. Pickert, B. Ji, C. Jia, A. C. Knoll, and C. Ng, "Comparison of TSEP Performances Operating at Homogeneous and Inhomogeneous Temperature Distribution in Multichip IGBT Power Modules," *IEEE Journal of Emerging and Selected Topics in Power Electronics*, vol. 9, no. 5, pp. 6282–6292, 2021. DOI: 10.1109/JESTPE.2020.3047738.

[5] C. Fang, T. An, F. Qin, X. Bie, and J. Zhao, "Study on temperature distribution of IGBT module," in *2017 18th International Conference on Electronic Packaging Technology (ICEPT)*, 2017, pp. 1314–1318. DOI: 10.1109/ICEPT.2017.8046680.

[6] J. Chen, E. Deng, L. Xie, X. Ying, and Y. Huang, "Investigations on Averaging Mechanisms of Virtual Junction Temperature Determined by VCE (T) Method for IGBTs," *IEEE Transactions on Electron Devices*, vol. 67, no. 3, pp. 1106–1112, 2020. DOI: 10.1109/TED.2020.2969727.

[7] F. Nehr and U. Scheuermann, "Consequences of Temperature Imbalance for the Interpretation of Virtual Junction Temperature Provided by the VCE(T)-Method," in *Proceedings of the International Exhibition and Conference for Power Electronics, Intelligent Motion, Renewable Energy and Energy Management (PCIM Europe)*, 2022. DOI: 10.30420/565822086.

[8] Di Zhao, C. Guo, Y. Li, S. Pan, S. Feng, and H. Zhu, "Study of the temperature distribution in insulated gate bipolar transistor module under different test conditions," *Microelectronics Reliability*, vol. 140, 2023. DOI: 10.1016/j.microrel.2022.114880.

[9] M. Gleissner and M.-M. Bakran, "Analysis of inhomogeneous temperature distribution in power modules for different cooling systems and the influence on lifetime consumption," in *Proceedings of the 25th European Conference on Power Electronics and Applications (EPE ECCE Europe)*, 2023. DOI: 10.23919/EPE23ECCEEurope58414.2023.10264638.

[10] V. Jadhav, J. Czichon, and M. Buerger, "Efficient and Optimized Traction Converter System Enabled by the New 3.3kV CoolSiC .XT Mosfet in XHP 2 Package," in *Proceedings of the International Exhibition and Conference for Power Electronics, Intelligent Motion, Renewable Energy and Energy Management (PCIM Europe)*, 2023. DOI: 10.30420/566091119.

[11] M. Buerger, K.-H. Hoppe, and K. Schraml, "New XHP 2 Module using 3.3kV CoolSiC MOSFET and .XT Technology," in *Proceedings of the International Exhibition and Conference for Power Electronics, Intelligent Motion, Renewable Energy and Energy Management (PCIM Europe)*, 2023. DOI: 10.30420/566091115.

[12] J. Czichon, U. Schwarzer, V. Jadhav, and N. Hoffmann, "Performance Analysis of a PEBB Demonstrator with High Power 3.3 kV CoolSiC in XHP2 for Modern Railway Traction Systems," in *Proceedings of the International Exhibition and Conference for Power Electronics, Intelligent Motion, Renewable Energy and Energy Management (PCIM Europe)*, 2023. DOI: 10.30420/566091034.

[13] T. Kestler and M.-M. Bakran, "A plug and measure device for junction temperature monitoring in real converter environment," in *Proceedings of the International Exhibition and Conference for Power Electronics, Intelligent Motion, Renewable Energy and Energy Management (PCIM Europe)*, 2019.

[14] M. Gleissner, D. Nehmer, and M.-M. Bakran, "Junction Temperature Measurement Based on the Internal Gate Resistance for a Wide Range of Power Semiconductors," *IEEE Open Journal of Power Electronics*, vol. 4, pp. 293–305, 2023. DOI: 10.1109/OJPEL.2023.3265850.

[15] C. Herold, M. Beier, J. Lutz, and A. Hensler, "Improving the accuracy of junction temperature measurement with the square-root-t method," in *19th International Workshop on Thermal Investigations of ICs and Systems (THERMINIC)*, 2013, pp. 92–94. DOI: 10.1109/THERMINIC.2013.6675204.

[16] T. Kestler, V. Damec, and M.-M. Bakran, "Differences in Dimensioning SiC MOSFETs and Si IGBTs for Traction Inverters," in *Proceedings of the 20th European Conference on Power Electronics and Applications (EPE ECCE Europe)*, 2018.

[17] Infineon Technologies AG, "AN2019-05: PC and TC Diagrams," *Application Note*, no. Revision 2.1, 2021-02-1919.

[18] M. Denk and M.-M. Bakran, "Comparison of UCE- and RGi-based Junction Temperature Measurement of Multichip IGBT Power Modules," in *Proceedings of the 17th European Conference on Power Electronics and Applications (EPE ECCE Europe)*, 2015. DOI: 10.1109/EPE.2015.7309067.

Innovative 3D Power Module Defaults Detection via Thermal Impedance Analysis and Simulations

Louis Alauzet[1,2], Anne Castelan[1,3], Sophie Regnier[1], Jean-Pierre Fradin[1], Alexandre Dezalay[1], Patrick Tounsi[2]

[1] Icam School of Engineering, Toulouse campus, France
[2] LAAS-CNRS, Univ. De Toulouse, INSA-Toulouse, France
[3] Laboratory of Plasma and Energy Conversion (LAPLACE), University of Toulouse, France

Corresponding author: Louis Alauzet, louis.alauzet@icam.fr
Speaker: Louis Alauzet, louis.alauzet@icam.fr

Abstract

This article presents a method for detecting physical defects in an innovative 3D Silicon (Si) power module by measuring thermal impedance (Zth). A non-invasive method for measuring the junction temperature of dies is introduced (voltage drop across a diode under constant current), and a comparison is made with simulation results (CELSIUS EC SOLVER software). The observed differences are analyzed to propose hypotheses for defects origins and location. A scanning electron microscopy (SEM) observation is also used to validate these hypotheses. The results confirm that the comparison between experimental and simulation data of Zth is a precise method for determining the areas of physical defects.

Terms and definitions

α: thermal diffusivity [m^2/s]

λ: thermal conductivity [W/m/K]

P: Power dissipated by the die [W]

$Tj_c(t)$: Junction temperature of a die acquired after cooling during a time t [°C]

$Tj_h(t)$: Junction temperature of a die acquired after heating during a time t [°C].

T_{ref} : Ambient temperature [°C]

1   Introduction

The electrification of transportation systems (aircrafts, automotive, trains), along with current environmental challenges, have led to an increased demand for higher performance in power electronic modules. It is crucial to reduce electrical and thermal losses. Packaging plays a significant role in this task: it allows efficient heat dissipation from the dies and enables an expansion of the Safe Operating Area (SOA). The need for improved performance has driven innovation in packaging technology.

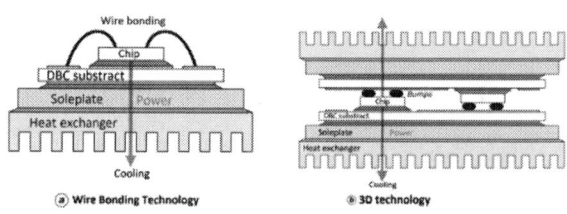

Fig. 1   bonding 2D and bump 3D architectures

Traditional 2D architectures evolves into 3D architectures, as shown in Fig.1 where bonding wires are replaced by a second substrate and cooling system, with an electrical connection made of bumps [1]. As it can be seen on the same figure, dies are also staggered (alternate between the first and second substrate) to improve cooling efficiency.

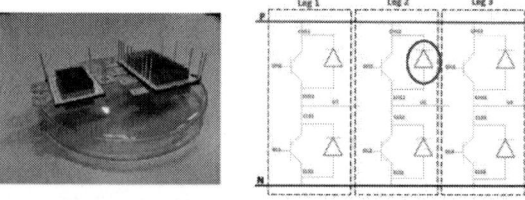

Fig. 2   Monobox module and its electrical composition

a*PSI*[3D] startup offers a 3D module, named Mono-box, with low theoretical thermal resistance (Rth) of 0.35°C/W for its diodes.

The module is a half-bridge three-phase inverter designed for the automotive industry, as shown in Fig.2. Each switching cell consists of two IGBTs and two diodes in anti-parallel configuration. The dies are directly sintered onto a substrate (Back-side), and the connection on the other side is achieved by sintered bumps (Frontside). The substrates use DBC (Direct Bonded Copper) technology. Heatsinks are directly brazed onto them.

The main advantages of this architecture are a reduction in parasitic inductance (caused by wire bonding) and a decrease in Rth (0.75 °C/W for the Acepack 2 sixpack topology module, which is the 2D equivalent of the 3D Monobox (0.35°C/W) by STMicroelectronics). The innovation of this 3D module is the fact that there are no thermal interface materials with high thermal conductivity: everything is brazed or sintered. That's why the Mono-box power density theoretically reach 100kW/L, where best commercialized modules struggle to reach 60kW/L (Fig.3).

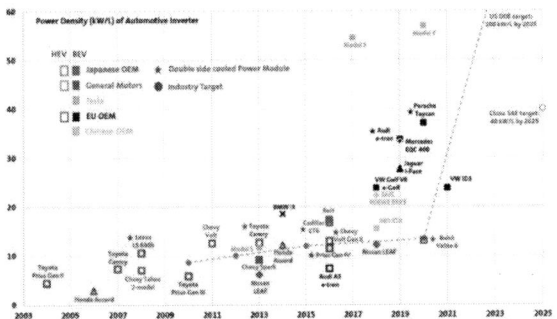

Fig. 3 Power density of traction inverters from various electrical vehicles [1]

While very promising for the future, the use of this architecture still faces a major struggle: the technological maturity of the module assembly processes. In fact, this kind of architecture has been on the market since 2008 [1], but mostly with one phase inverters. This module being a three-phase inverter, manufacturing is a greater challenge: the mechanical and thermal constraints currently encountered during the manufacturing process and the use of the module can lead to delamination and fractures between the various materials inside the module.

The objective of this article is therefore to detail a method for the thermal characterization of a power module through comparison between thermal impedance measurements and simulations, to perform a design validation.

2 Methods of measurement

2.1 Cooling and pulsed heating Zth

The thermal impedance can be obtained by two distinct processes depending on the curve studied: Cooling or heating. In both case it describes the thermal comportment of a module to a particular pulsed power [2].

The heating curve Zth expression is:

$$Z_{th_h} = \frac{Tj_c(t) - \mathrm{T_{ref}}}{P} \tag{1}$$

The cooling curve Zth expression is:

$$Z_{th_c} = \frac{Tj(t=0) - Tj_c(t)}{P} \tag{2}$$

With the cooling method, a power is applied to the device until it is thermalized. Then the power is shut down and the junction temperature is acquired until the device is again thermalized. With this method a curve is obtained, depending on the frequency of acquisition of the temperature.

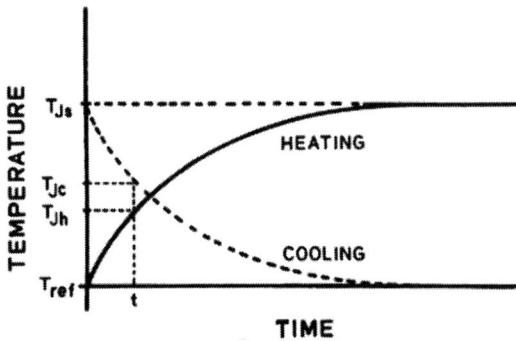

Fig. 4 Comparison of cooling and heating curves for a semiconductor device

With the heating method, a power is applied for a known time, then at the end of the injection the junction temperature is measured. It is repeated with increasingly long injection times, until thermalization time is obtained. With this method, only few points are obtained, corresponding to power injection times (from 100µs to 100s, with 5 points per decade in this article)

In theory, if the thermal diffusivity and conductivity of the tested device are not dependent of the temperature, the two temperature curves should be conjugate (Fig.4). As the thermal conductivity of the Silicon is temperature dependent, it should be expected that results will differ depending on which method is used.

2.2 Extrapolation of junction temperature

Whatever method is chosen to get the Zth, measuring the junction temperature is needed. To do so, many solutions exist: Thermocouple, IR cameras and optical fibers to name but a few. But as all of those solutions need a direct access to the die tested, or at least a visual access, they are not suited to obtain junction temperature of 3D modules.

Another well-known solution is the thermally sensitive electrical parameter (TSEP). It's a non-intrusive solution since only an electrical parameter is measured. To determine the thermal characterization of the Monobox, it has been decided that the Zth studied would be the diode's Zth (easier to measure than the IGBT with TSEPs) [3]. Hence the voltage drop of the diode under low continuous current (Vf) will be the TSEP used [3].

This TSEP shows good linearity and is quite sensible to thermal variation (2.39mV/°C) (see Fig.5). This curve has been obtained by heating the diodes with a modification of the temperature of the cooling system, between 20°C to 90°C, and measuring the voltage drop with a 10mA injection. Moreover, with 10mA forward current, the power dissipation of the measure is negligible (under 5mW at 20°C), meaning it's a truly non-invasive solution that won't self-heat diodes during measure.

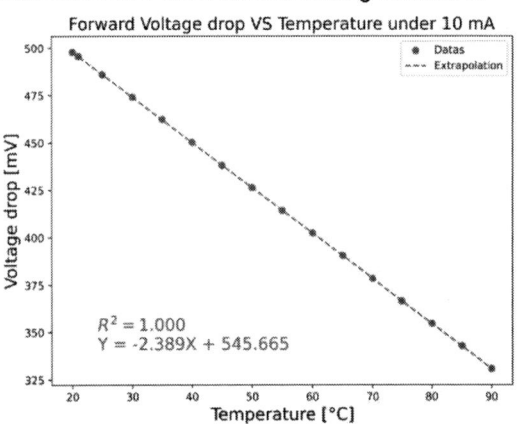

Fig. 5 Temperature VS Voltage drop under 10mA for Monobox diodes.

A drawback of the use of Vf as a TSEP is the fact that the measure of the junction temperature cannot be obtained during the self-heating process, nor after few hundreds of microseconds after it's shutdown. This is due to non-thermal electrical transient variations. Its origin is still unknown, but certainly due to parasite inductances, resistances, and capacitances in the circuit or module [4]. [4] and [3] propose the solution of the extrapolation of the junction temperature: In a semi-infinite solid, if

the dissipation of power is surfaced based, the evolution of the temperature is linear with the root of the time (cf. Fig.6 and equation 4).

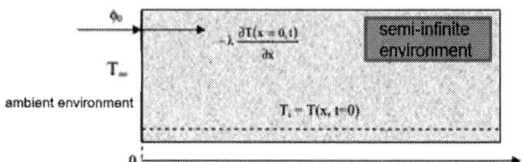

Fig. 6 Semi-infinite slab scheme with imposed surface flux

$$\bar{T}(x,t) = \text{ierfc}\left(\frac{x}{2\sqrt{\alpha t}}\right)\frac{2P}{\lambda}\sqrt{\alpha t} \tag{3}$$

$$\Rightarrow T_j(0,t) = \frac{2P}{\lambda}\sqrt{\alpha t} + T_a \tag{4}$$

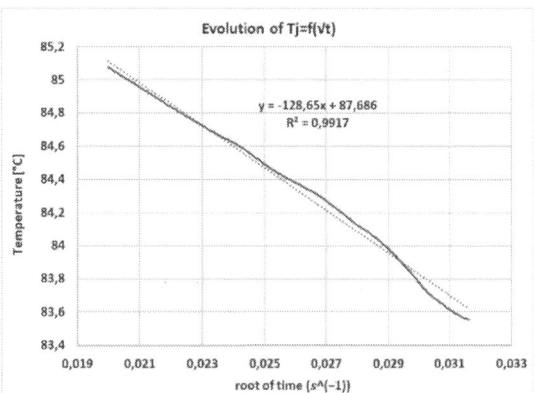

Fig. 7 Typical form of $T_j = f(\sqrt{t})$ during tests

It should be noted that the power dissipation of a diode is not surface based, but volumic [4]. Despite this approximation, extrapolation has proven to give precise results when compared to other means of temperature measurement. [4]. The Fig.7. shows good linearity of the temperature with the root of time, using the diodes of the module.

Hence even with non-thermal electrical transient, extrapolation allows the use of the voltage drop of the diode under low continuous current to get junctions temperatures, for cooling and pulsed heating measures.

2.3 First results: comparison between heating and cooling curves

The two possible tests were conducted to determine if a noticeable difference could be found between them.

PCIM Europe 2024, 11– 13 June 2024, Nuremberg   DOI: 10.30420/566262398

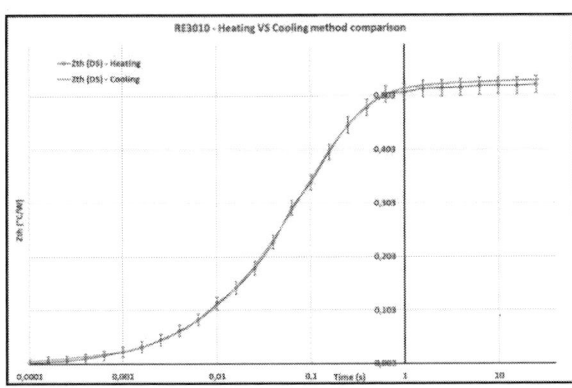

Fig. 8 Comparison on the RE3010 module (double sided cooling) between Zth curves using cooling and heating method.

For the Rth value, the relative deviation is less than 1.7%, (Fig.8) and under the error margin. A small difference can be observed for the smallest times, which can be explained by the difference of conductivity of Si depending on the temperature. Hence for this module it is shown that the two methods give about the same results. However, the constraints to be respected during testing are different: for the heating curves, each point requires the module to be heated. It's not a problem for times under 1 second, but the fact that each point is tested multiple times to reduce dispersion of the results could lead to premature aging of the module. Where more than 40 temperature cycling are needed to get a heating curve, only 4 are done to get the cooling one.

Since it has been noticed during tests that the module ages after multiple cycles, the cooling method will be preferred.

3 Simulation analysis

3.1 Power module presentation

To compare results of said TSEP temperature measurement and hence to conclude on the presence of defects, a simulated functioning module was made.

The 3D thermofluidic finite element volume software used for this study is CELSIUS EC SOLVER. Numerical simulation aims to precisely reproduce the experimental conditions of the tests.

The cooling fluid is a 50% glycoled water, at 20°, and 4L/min flow rate in each cooling duct.

To precisely reproduces experimental tests, three cooling solutions are available on this module (one or double side cooling), that are defined as Front side cooling (FS), Back Side cooling (BS) and Double side cooling (DS) (Fig.11).

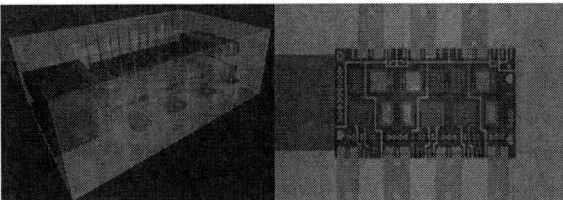

Fig. 9 Isometric view of power module (left) and top view of one's power module side (right)

The direction of cooling is based on the position of the active cooling duct with the heating component (Fig.11).

To be close to experimental test conditions, when a one side cooling simulation is done, the cooling duct associated to the turned off pump is still full of glycoled water, with no flow rate.

As the power module geometry is complex (strong scale factor), define a precise mesh is a key factor for an accurate simulation result.

Bumps are doing thermal and electrical contact, so their mesh shape must be well defined. Final non conform parallelepiped volumic mesh provides a good definition of the assembly, as shown in Fig.10.

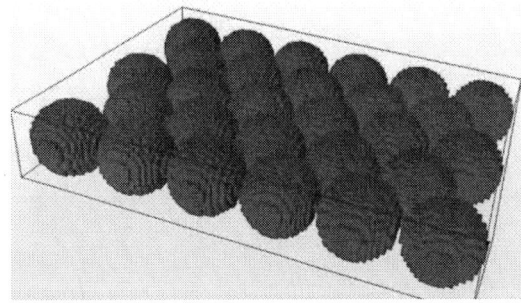

Fig. 10 Mesh of thermal bumps inside power module

It should be noted that the thermal conductivity of the Silicon of the dies is dependent of its temperature (Fig.12). Other material properties are summarized in Table 1.

2814

	Conductivity (W/m.K)	Specific heat (J/kg.K)
PEI-Plastic	0.25	1260
Al2O3 Curamik	24	900
Aluminum metallization	235	921
Porous silver	100	232
Copper pure	386	380
Glue joitns	0.1	700
Underfill	0.7	3000
Sumitomo braze	20	1005

Table. 1 3D power module material properties

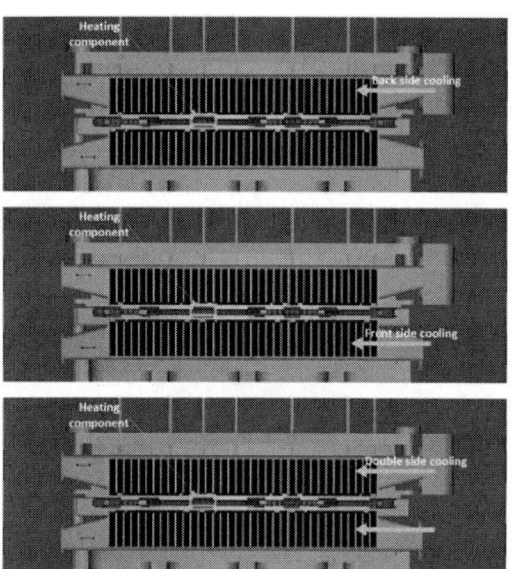

Fig. 11 Three types of possible cooling configuration for a heating component inside the module: BackSide, FrontSide, and double side.

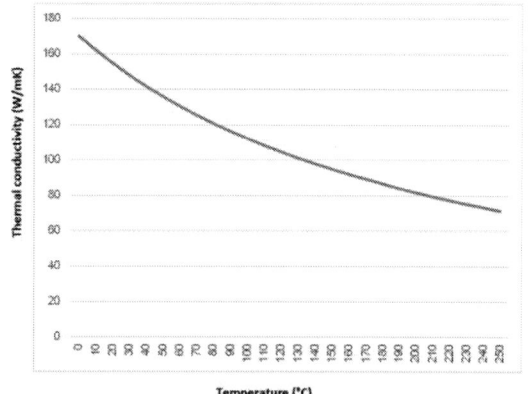

Fig. 12 Thermal conductivity of Silicon

3.2 Simulated thermal impedances on 3D module and classic 2D module.

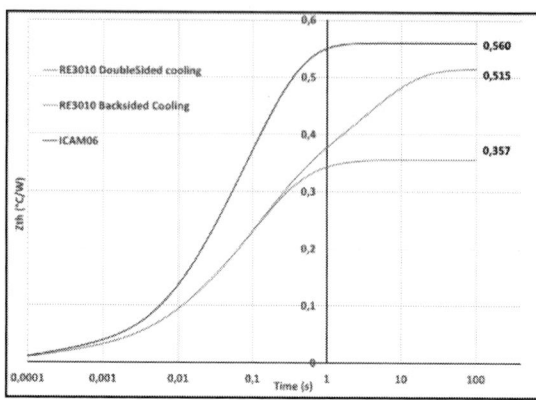

Fig. 13 Thermal impendences curves (Zth) of RE3010 with doubleside and backside cooling and Icam06 modules

To prove the interest of such 3D modules of classic ones, a simulation of the 3D RE3010 module and the ICAM06 one (a classic 2D module made by $aPSI^{3D}$, using only one ceramic substrate with bonding dies) has been done (Fig.13).

When the RE3010 is backside cooled, part of the flux passes through the bumps to the opposite ceramic substrate before using other dies to transfer the heat flux. This contributes to die cooling and explains the difference between Icam06 Rth and backside cooling RE3010 Rth.

This study shows the theorical 3D architecture is a huge improvement (diminution of the Rth by 65%) compared to a classic module.

4 Experimental setup and Tests

Tests have been conducted on the DepTH-LAB [6] platform at Icam Toulouse.

The electrical part of the test bench consists of two current circuits. The first circuit comprises a power current source with an inductance (L) of 8mH, ensuring a constant current flow even with minimal injection times. This circuit (Fig.14) utilizes two Mosfets (T1 and T2) to control the power current (Ip) through the diode. Precise switching of the Mosfets is essential to ensure accurate current injection timing and prevent the circuit from remaining open with the inductance still charged.

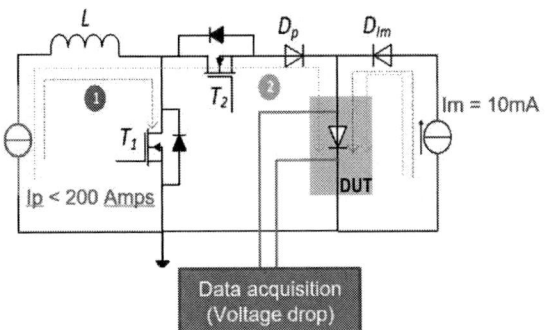

Fig. 14 Electrical schematic of the test bench

The second circuit consists solely of a 10mAmps current source (Im) that remains continuously active, providing temperature readings when the power current (Ip) is deactivated.

Voltage drop is continuously monitored to extrapolate the temperature when the power current is off and to determine the power dissipation of the diode when the power current is on.

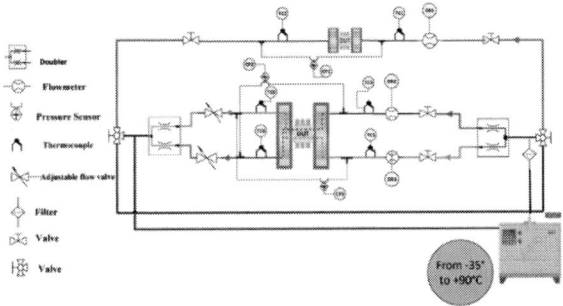

Fig. 15 hydraulic schematic of the test bench

The thermal part of the test bench (Fig.15) has been constructed to allow such control on the cooling process. A cryostat is used to precisely set the temperature of the cooling fluid (50% glycoled water, at 20°C).

Splitters are employed to enable cooling of the module on both sides, and two valves can be used to fine-tune the flow rates independently. Temperature of the liquid can be acquired by thermocouples if needed. The temperature of the liquid used for subsequent tests is 20°C unless otherwise specified.

5 Comparison between tests and simulations for Zth Curves

A first steady state analysis is done, using both simulation software and experimental setup previously discussed.

The first studied module, named RE3010 is a complete 3D module, with radiators. In each case the diode under test will be the diode HS2 (visible in red circle in Fig.2).

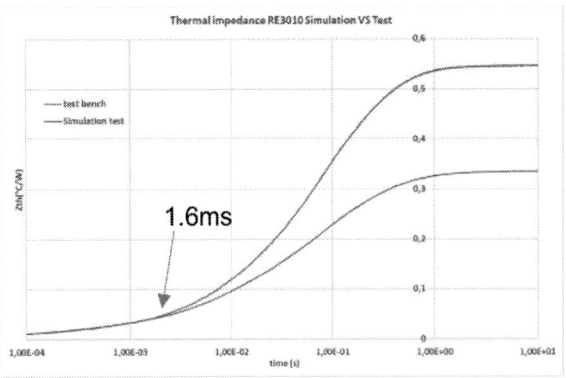

Fig. 16 Simulated and tested Zth on RE3010 double sided cooling

The differences between experiments and simulations show that the thermal resistance (Rth) of real modules is much higher than it should be (relative deviation of 63%).

This indicates that defects are present. By studying separation moment between simulated and test curves and by linking it with the thermal diffusivity and thermal transit time of each material of the module a location where a defect is present can be extrapolated. It should be added that since cooling takes place in a true 3D environment, with thermal loops through the two substrates that occur due to other unheated dies, the study of the structure functions usually used [5] is of little relevance here and the deduction of the delamination site is rendered more difficult. These Zth results are shown in Fig.16. The comparison between the two curves of the figure indicates a separation at approximately 1.6ms.

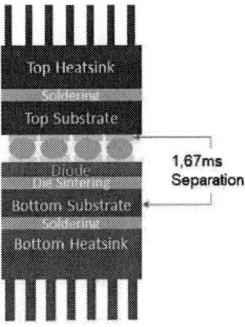

Fig. 17 Trench View of the module with proposed sites of defect

Then, thanks a study of the different thermal diffusivity and transit times of materials, a conclusion can be drawn: defects might be present either in the upper bump bonding or in the bottom substrate of the die (cf. Fig.17).

To investigate the two possible defects, it has been decided to manufacture a new module, named ICAM06. It's a half-module, meaning that the bumps are replaced for classical bonding. As there isn't top substrate nor top heatsink the cooling process can only be achieved by its backside.

A special simulation model is created for this module.

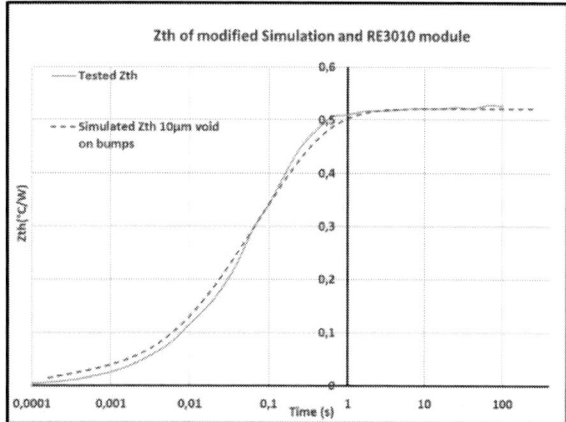

Fig. 18 Simulated and tested Zth on ICAM06

It appears with those curves (Fig.18) that there is no significant difference between simulations and tests. Meaning that there is no delamination in substrates.

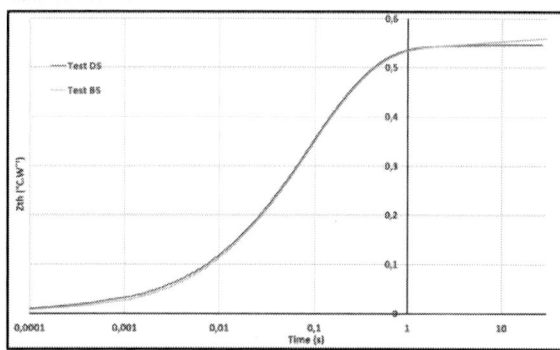

Fig. 19 Comparison of RE3010 double and backside cooling Zth curves

The Fig.19 seems to indicate that there is no difference between double sided cooling and backside cooling on the RE3010 module, which could mean that the bumps play no role in the cooling process. To test the possible delamination of the bumps, simulations have been conducted with air gaps between bumps and their sintering, to be

compared with experimental results. The obtained curves in Fig.20 demonstrates that the addition of these imperfections enables the simulation to closely match the experimental results. The differences in Rth do not exceed 6% (maximum deviation observed at the backside).

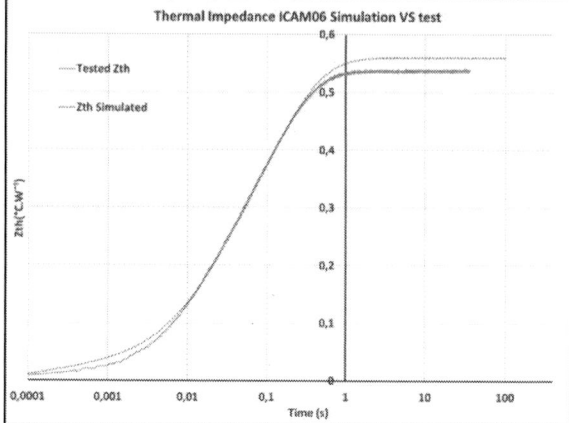

Fig. 20 Simulated and tested back side Zth on RE3010 with bumps defects simulated.

As simulations match tests, it is considered that defects indeed come from the bumps, or its sintering.

6 SEM analysis

Some interfaces between small Cu bumps / Sintered Ag / Cu substrate are observed using SEM (Scanning Electron Microscope Hitachi TM4000Plus) to conclude about delamination of bumps.

6.1 Preliminary preparations

Before observations, the module has been cut following the Fig.21, polished using SiC papers (from grade P600 to P1200) and finally polished with diamond suspension up to a size of 1μm.

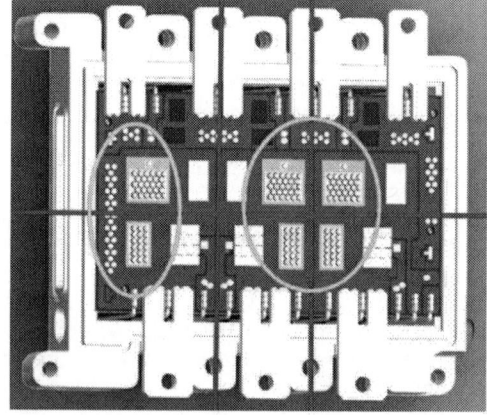

Fig. 21 Position of cutting areas (red lines: Limit of the cutting – green circles, planned observation areas)

6.2 SEM observations

Some of the observed samples indicate sintered Silver as waited, showing sintered silver without any microcracks nor cracks (Fig.22-left); but most of the observations indicate some defaults as:

- Inefficient sintering (fig.22-right)
- Microcracks (fig.23-left)
- Total delamination between sintered silver and bumps due to a frequent lack of diffusion between bumps and silver powder (Fig.23-right).
- Compression areas in the substrate leading to cracks (Fig.22-right).

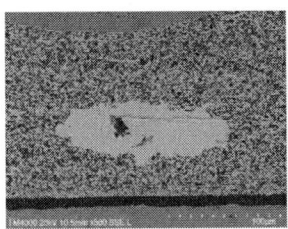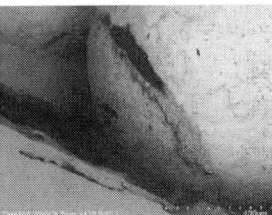

Fig. 22 Left: Well sintered Silver – Right: Microcracks of the substrate and inefficient sintering

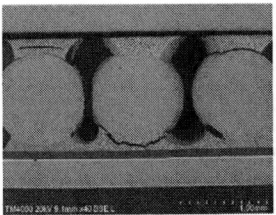

Fig. 23 Left: Microcracks – Right: Total delamination between bump and silver.

Those observations conclude that the defects of the module were indeed situated at the level of the bumps. The advantage of the SEM observation remains the capability to conclude on the origin of the defect (bad quality of sintering), where the thermal curve analysis could only describe the place where the defect occurred (but is non-invasive and doesn't necessitate the destruction of said module). Advice was given to the manufacturer (aPSI3D) to resolve those defects: More Sintering should be used, of better quality.

7 Conclusions

A non-intrusive test has been developed to acquire diodes junction temperature of 3D inverter module, using the voltage drop of diodes under low current, and the extrapolation of temperature with the surface-based power dissipation hypothesis.

A perfect module was simulated using the CELSIUS EC SOLVER software. The comparison between simulations and tests, linked with thermal diffusivity, has made possible to formulate two hypotheses on the defects.

The possible delamination in the substrate was not observed on a one substrate module (ICAM06). Modified simulations of the module with air gap between the bumps and the substrate proved to obtain Zth curves very similar to tested ones.

SEM observations tend to demonstrate that bumps and their sintering show a large disparity of quality, which comfort the fact that the defects of 3D modules can be found by thermal impedance curves analysis and comparisons.

References

[1] M. Liu, A. Coppola, M. Alvi, and M. Anwar, "Comprehensive review and state of development of double-sided cooled package technology for automotive power modules," IEEE Open Journal of Power Electronics, vol. 3, pp. 271–289, 2022.

DOI: 10.1109/ojpel.2022.3166684.

[2] D. L. Blackburn et F. F. Oettinger, « Transient thermal response measurements of power transistors », IEEE Transactions on Industrial Electronics and Control Instrumentation, no 2, p. 134-141, 1975.

[3] Q.C.Nguyen, "développement d'outils electrothermiques pour la localisation de defauts et pour l'optimisation de la performance de modules mecatroniques de puissance SiC," Theses, INSA de Toulouse, Sep. 2021..

[4] B. Thollin, « Outils et méthodologies de caractérisation électrothermique pour l'analyse des technologies d'interconnexion de l'électronique de puissance », Université de Grenoble, 2013.

[5] V. Szekely, "A new evaluation method of thermal transient measurement results," ´ Microelectronics Journal, vol. 28, no. 3, pp. 277–292, Mar. 1997. DOI: 10.1016/s0026-2692(96)00031-6.

[6] https://en.icam.fr/research/depth-lab/

PCIM Europe 2024, 11– 13 June 2024, Nuremberg   DOI: 10.30420/566262399

Thermal characterization of an air-cooled PEBB based on SiC MOSFET Power Modules

Alexandre Marie[1], Maria Alejandra Castellanos Taita[1,2], Benjamin Vieillefosse[1], Joseph Fabre[2], Philippe Ladoux[3], Jean-Pierre Fradin[1]

[1] Icam School of Engineering, France

[2] SCLE-SFE, France

[3] Laboratory of Plasma and Energy Conversion (LAPLACE), France

Corresponding author: Alexandre Marie, alexandre.marie@icam.fr
Speaker: Alexandre Marie, alexandre.marie@icam.fr

Abstract

This paper presents the thermal characterization of an air-cooled Power Electronic Building Block (PEBB) based on two SiC MOSFET Power Modules. The experimental thermal transient response of the assembly is compared to a thermofluidic 3D model. The results show good agreement between the predicted and measured values and allow to validate the possibility to use conventional air-cooling technology for this high dissipation application. In the case of a power step, the transient thermal response of PT100 temperature sensors placed within the heatsink and their relevance to evaluate the converter operating temperature are also discussed. Finally, the impact of a partial loss of the cooling fans on the thermal performance of the assembly is experimentally evaluated.

1 Introduction

The thermal management of high voltage power converters in railway applications is now a common problematic [1]–[3]. Recently, the development of SiC DC/DC converters for trackside energy storage systems have been considered by Fabre et al. [4]. In this application, the authors proposed in particular the utilization of high voltage power MOSFET modules within a resonant dual-bridge converter stage [5]. The power module considered, developed by Mitsubishi, is based on a new packaging technology allowing in particular to reduce the number of interface layers between the dies and the baseplate cooling surface [6]. However, experimental data regarding the thermal management of this recent packaging technology remain scarce, in particular for such high power and high tension SiC transistors. Considering this, a test bench allowing the thermal characterization of this type of power module in a configuration representative of the actual converter implementation has been developed within the research platform DepTH-LAB [7]. As planned in the industrial application, the cooling of the assembly is ensured by a forced air convection

heatsink. The experimental results are compared to a 3D thermofluidic model to evaluate the possibility to predict the thermal performance of the power converter. Then, the relevance of classical PT100 temperature measurements to monitor the assembly transient thermal response is discussed thanks to their comparison to junction temperature measurements obtained by temperature-sensitive electrical parameter (TSEP) methods. Finally, the cooling efficiency reduction associated to the failure of some of the cooling fans is experimentally evaluated.

2 Experimental setup

2.1 Power modules cooling system and electrical management

In the work presented here, two Mitsubishi LV100 MOSFET power modules (FMF750DC-66A, rated voltage $3300\ V$, rated current $750\ A$ [6][8]) are attached to an aluminium heatsink. Each module is based on a half-bridge topology and is therefore comprised of two independant switches. A layer of thermal interface material (Electrolube HTSP50T, $\lambda \approx 3\ W.m^{-1}.K^{-1}$) approximately $100\mu m$ thick is applied between the modules and the aluminium heat sink. Each module/heatsink mechanical contact is ensured by 4 screws placed in the corresponding fixation points and tightened with a torque of $5\ N.m$. The forced convection cooling air flow is ensured through the heatsink thanks to three fans (Sunon PMD2412PMB1-A2) attached upstream of a $200\ mm$ inlet duct. Additionally, eight heatpipes are embedded within the heatsink in order to enhance the heat flow spreading along the fins direction. A picture of the corresponding assembly is presented in Fig.1.

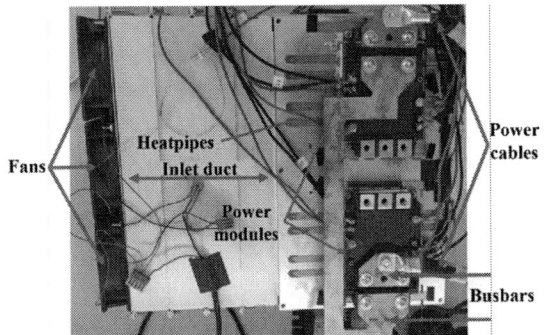

Fig. 1: Picture of the power module thermal characterization test bench

The two modules are then wired in series thanks to copper busbars and connected to a $340\ A$ DC power supply (EA-PSI 9040-340 3U). Finally, in order to turn-on the transistors, each switch gate voltage is set to $+17\ V$ thanks to 4 independant DC power supply. A schematic of this electrical circuit can be found in Fig. 2.

2.2 Thermal instrumentation

To measure the different thermal resistances of the assembly as well as its transient thermal response, three independant temperature measurements types are implemented within the experimental setup. First, two $0.5\ mm$ sheathed PT100 (TC DIRECT, ref 515-110) are placed in two $1 \times 1\ mm$ rectangular grooves machined directly below the hot spots of each module and held in place with the help of cyanoacrylate glue. Then, two surface mounted PT100 sensors (cylindrical, diameter $16\ mm$) are placed on the heatsink exposed surface, close to the power modules. A visualisation of these two measurements is represented in Fig. 3. Finally, the junction temperatures of each of the four switches of the assembly are evaluated thanks to the corresponding on-state resistance temperature dependance. To this end, a prior characterization

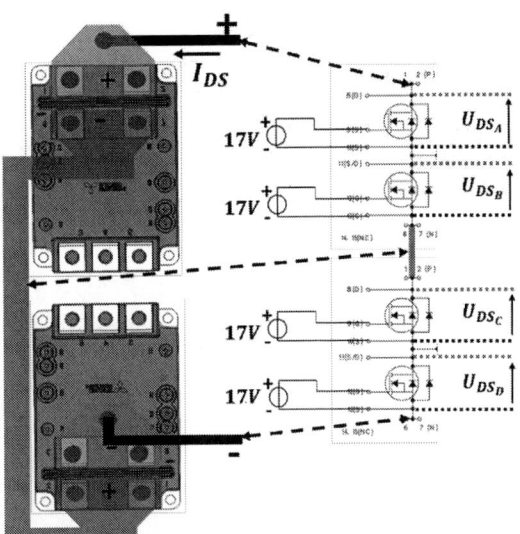

Fig. 2: Schematic of the modules electrical connections

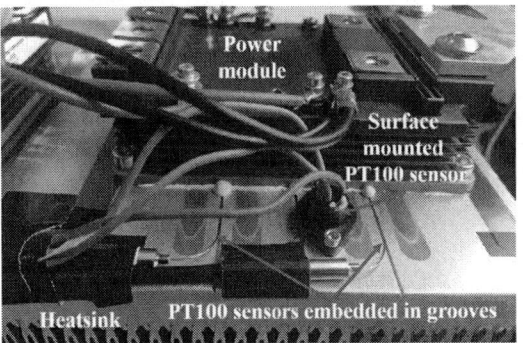

Fig. 3: Physical implementation of PT100 sensors on the heatsink

of the curves $(I_{DS}, U_{DS_{A/B/C/D}})$ (see Fig. 2) has been performed in a thermal oven with the help of a Keysight B1505 power device analyzer/curve tracer. At each measurement point, $200\ \mu s$ long I_{DS} current pulses have been used in order to ensure that self-heating of the dies was negligible during the characterization. An example of thermosensitive calibration curve $R_{DS,on} = \frac{U_{DS}}{I_{DS}} = f(T)$ is presented in Fig. 4.

For each of the temperature measurements defined above, the associated thermal impedance is defined as :

$$Z_{th} = \frac{T - T_{ref}}{\Phi} \quad (1)$$

In this expression, T_{ref} represents the air temperature at the inlet of the heat sink, Φ the power dissipated by the considered switch (in permanent regime) and T is either the junction temperature

PCIM Europe 2024, 11– 13 June 2024, Nuremberg   DOI: 10.30420/566262399

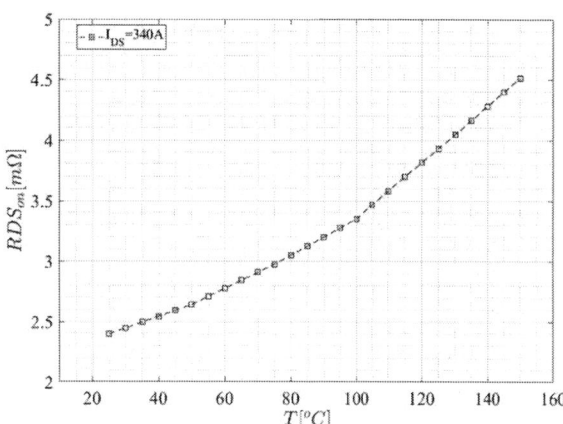

Fig. 4: Example of calibration curve $R_{DS,on} = f(T_j)$, $I_{DS} = 340\ A$

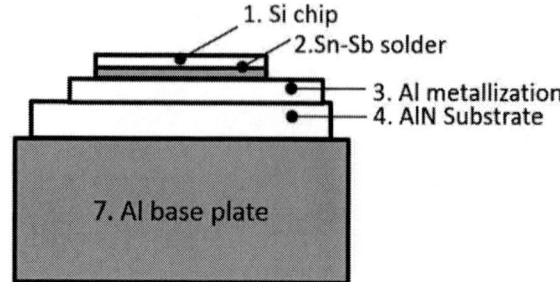

Fig. 5: Schematic of the internal power module mechanical assembly, reproduced from [6]

T_j, the temperature measured by the sheathed PT100 sensors T_{PT100} or the temperature $T_{S-PT100}$ measured by the surface mounted PT100 sensors. For each of these thermal impedances, the corresponding thermal resistance R_{th} is the thermal impedance value achieved in permanent regime. Finally, all thermal and electrical measurement are acquired thanks to a AOIP FD5 acquisition system with a sampling frequency of $1\ Hz$.

3 Thermofluidic modelisation of the assembly

Concurrently to the experimental setup presented above, a complete thermofluidic model of the assembly has been developed to be compared to the experiment. The inner modelisation of the power modules is briefly presented and the complete model is then detailed.

3.1 Thermal model of the power module

The 3D geometrical modelling of the FMF750DC-66A module was based on the decomposition in the different layers of the module detailed in the publication of Stumpf et al. [6] and represented in Fig. 5. The internal position of the dies placed on each substrate has been provided by the manufacturer. Then, this 3D geometry has been implemented in the finite volume method thermal solver 6SigmaET v16.3. In order to compare the model to the module transient thermal response provided by the manufacturer datasheet [8], a constant temperature has been imposed on the external cooling surface of the module and a constant dissipation $\Phi = 50\ W$ is dissipated in the active area of each MOSFET die

of the module. Then, independancy to the meshing refinement has been ensured and the different layers thicknesses and thermal properties have been adjusted to ensure the closest possible match between the model and the manufacturers impedance curve defined in eq. (1). The comparison between the adjusted model and the manufacturers data is presented in Fig. 6. The normalized thermal impedance displayed in this figure is defined as :

$$\overline{Z_{th}}(t) = \frac{Z_{th}(t)}{Z_{th}(t = 10s)} \qquad (2)$$

Apart from a slight difference for $t < 0.1s$, which

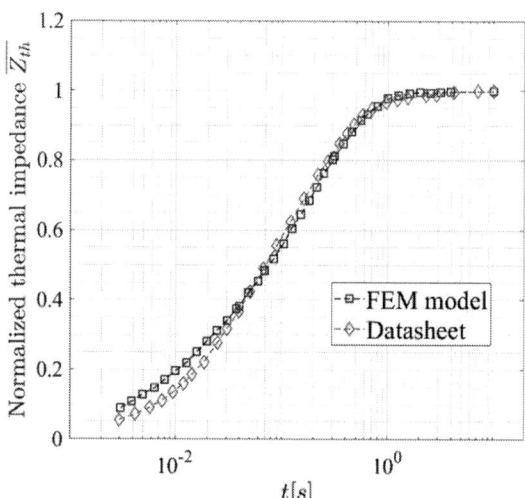

Fig. 6: Comparison of the module thermal impedance to the curve given by the manufacturer

can be explained by the absence of the thermal mass associated to bonding wires (not taken into account in the numerical model), the agreement between the two curves is very satisfactory. Considering that these small time scales will not be

2821

considered in the study presented therafter, this numerical model of the module is therefore satisfactory and will be used in the complete modelisation of the assembly presented in the following section.

3.2 Complete thermofluidic model

A complete thermofluidic model, schematically represented on Fig. 7, has been developped on 6SigmaET v16.3. In this model, all the components of the assembly have been taken into account. In particular, the sheathed PT100 sensors are represented in the model as well as the cyanoacrylate glue holding them in the grooves. Finally, the flow rate induced by the fans is deduced from the fans characteristic curve and from the pressure drop induced by the airflow along the heat sink fins calculated by the solver. A summary of the main ma-

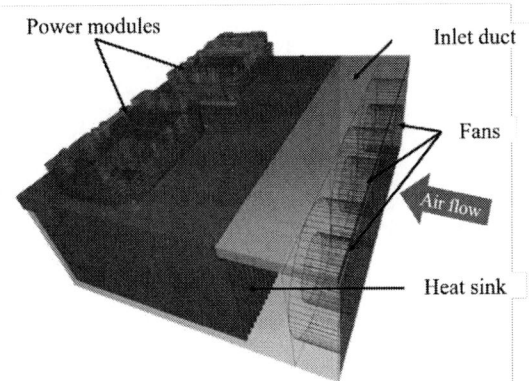

Fig. 7: Schematic view of the complete thermofluidic model of the assembly

terials properties used in the model is detailed in Table 1.

An example of temperature field given by the model in a transverse cross section passing through some of the active dies is presented in Fig. 8. It can be seen that the thermal gradient is mainly located in the heatsink. Indeed, conduction within the module only account for $\approx 25\%$ of the total junction overheat and the thermal interface for less than 10%.

4 Experimental results and comparison to simulations

In this section, the measured thermal response will be compared to the temperature variations predicted by the model. In all experimental results presented thereafter, a constant current $I_{DS} = 340\ A$ is forced through the two modules in series at at

the initial time $t = 0$. This induces, in steady state, a power dissipation of around $650\ W$ per module.

4.1 Junction to air thermal impedance

The measurement of the global thermal impedance $Z_{th,j->air}$ (evaluated from the temperature difference $T_j - T_{air}^{inlet}$) is compared to the curve predicted by the model in Fig. 9. For the sake of clarity and given that no significant differences are observed between the 4 switches, the curves presented here represent an average value for the 4 switches considered in the 2 modules setup. Overall, despite slight inaccuracies for $t < 40s$ and $t > 200s$, the model seems to capture both the assembly thermal dynamic and permanent regime thermal resistance satisfactorily.

At this stage, it is important to note that the on-state current, and thus the power dissipation, was limited in the test bench by the DC power supply. Indeed, in the final application a higher power dissipation of $1300\ W$ per module (i.e. $650\ W$ per switch) needs to be managed by the cooling system. In addition, a worst case scenario of 45^oC inlet air temperature needs to be considered, which was also not achievable in the laboratory. However, considering the good consistency between experiments and simulations, it appears possible to extrapolate from the measured thermal resistance $R_{th,j->air}^{exp} = 123^oC/kW$ the junction temperature $T_j^{operationnal}$ which would be achieved under these more constraining operationnal parameters :

$$T_j^{operationnal} = 45^oC + 0.650 \times R_{th,j->air}^{exp} = 131.5^oC$$
(3)

Hence, given the maximum operating temperature T_j given by the manufacturer, the forced air-cooling system considered here seems to be well-adapted to the thermal management of the modules in this railway power converter application, despite the relatively high power dissipation.

4.2 Physical sensors measurement and associated response time

Given the experimental parameters considered above, a similar experiment/simulation comparison is conducted here for the thermal response of the sheathed sensors placed within the heat sink grooves below the hot spots of the modules. The comparison of the corresponding thermal impedances curves is represented in Fig. 10. Once again, the dynamic predicted by the model underestimates slightly the measured temperatures at

PCIM Europe 2024, 11– 13 June 2024, Nuremberg — DOI: 10.30420/566262399

Material	Thermal conductivity $[W.m^{-1}.K^{-1}]$	Density $[kg/m^3]$	Specific heat $[J.kg^{-1}.K^{-1}]$
Air	0.026	1.19	1005
AlN substrate	180	3250	780
Aluminium (module baseplate)	116	2800	880
Aluminium (heatsink)	209	2700	900
Thermal interface material	3	1000	1000
Cyanoacrylate glue	0.2	1070	1420
PT100 sheathing	50	3600	930

Tab. 1: main materials thermophysical properties used in the 3D model

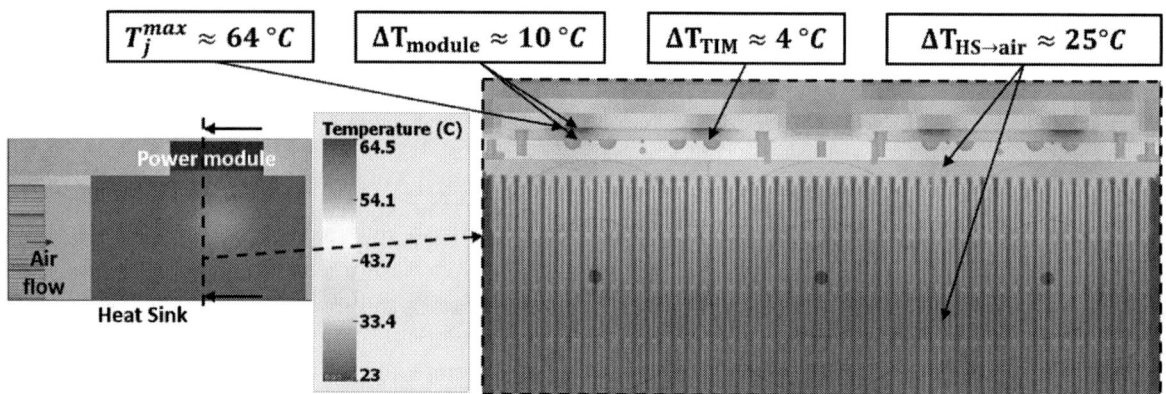

Fig. 8: Exemple of a numerical temperature field calculated with 6SigmaET, heat dissipation $\Phi = 315\ W$ per switch in permanent regime, $T_{ref} = 23°C$

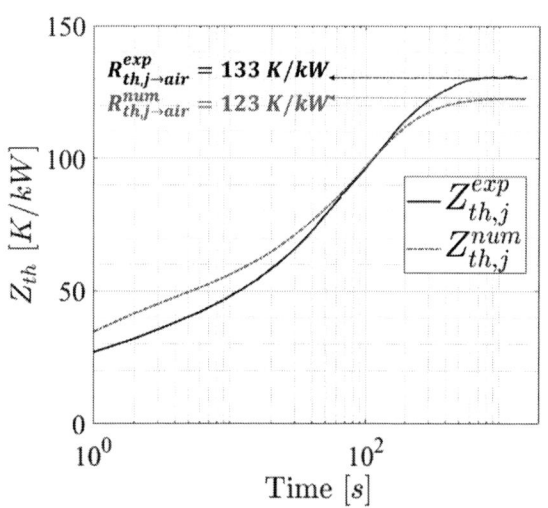

Fig. 9: Comparison of the numerical and experimental junction to heatsink thermal impedances, $I_{DS} = 340\ A$, $T_{air}^{inlet} = 24°C$

$t > 100s$ but overall matches relatively well the experiment.

In addition, the experimental curves of the surface mounted sensors, represented on the same graph, allow to highlight the significant response time associated to this type of sensors. Such a delayed measurement was indeed expected, considering the relatively high thermal mass of the surface mounted sensors as well as their placement on the heatsink relatively far from the hot spot of the modules.

Considering these results and in order to evaluate more precisely the response time and relevancy of each measurement, the experimental response time τ and attenuation factor K have been defined for each type of measurement according to :

$$\frac{T(\tau) - T_{ref}}{T(t \rightarrow \infty) - T_{ref}} = 0.63 \quad (4)$$

$$K = \frac{T(t \rightarrow \infty) - T_{ref}}{T_j(t \rightarrow \infty) - T_{ref}} \quad (5)$$

In other words, τ represents the time necessary to measure 63% of the permanent regime overheat after the application of a power step and η evaluates the underestimation of the junction temperature

2823

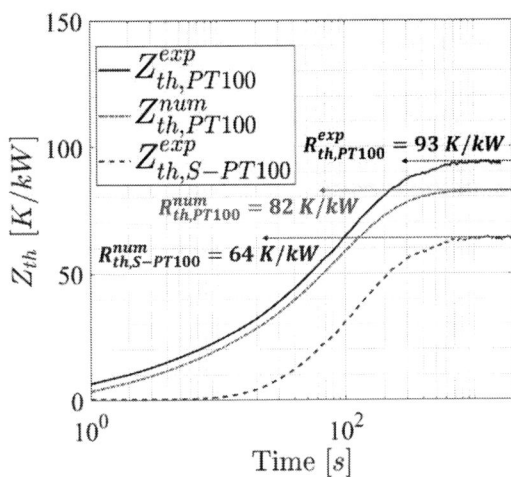

Fig. 10: Aluminium baseplate to heatsink pseudo-impedance, $I_{DS} = 340\ A$, $T_{air}^{inlet} = 24C$

made when considering the corresponding physical sensor. The computed values of these parameters are synthetized in Table 2.

Measurement type	Response time τ	Attenuation factor K
Junction	$57\ s$	1
PT100 (grooves)	$81\ s$	0.72
PT100 (surface mounted)	$144\ s$	0.49

Tab. 2: Synthesis of the measured response time and attenuation factor for each type of measurements implemented in the test bench

The calculated values of τ confirm that the surface-mounted sensor is significantly delayed from the intrinsic response of the modules (given by the junction temperature measurement). It is however to be noted that the sheathed PT100 sensors placed below the modules induce a response time only 40% higher than the intrinsic response time of the modules and are also able to capture almost 75% of the actual junction overheat.

4.3 Effect of fan failure

In order to evaluate the impact of the failure of one or two of the three fans on the thermal performance of the assembly, additional tests in which only one or two of the fans are turned on have been conducted. Apart from the turned off fans, the experimental protocol remains the same as the one

detailed in section 2. The three fans are thereafter denoted F_l, F_c, F_r, for the fans on the left-hand side, center and right-hand side of the air flow respectively (see Fig. 1). A synthesis of the measured junction to air thermal resistances is presented in Table 3. Given the observed differences between the 4 switches in some fan failures configurations (in particular the one where only one lateral fan is on), the lowest and highest switches thermal resistances are indicated for each configuration.

Number of active fans	F_l	F_c	F_r	$R_{th,j->air}^{exp}$ $[K/kW]$
3	ON	ON	ON	$126 - 135$
2	ON	OFF	ON	$140 - 147$
2	ON	ON	OFF	$140 - 154$
1	OFF	ON	OFF	$170 - 176$
1	ON	OFF	OFF	$166 - 200$

Tab. 3: Synthesis of the measured junction to air thermal resistances for the different fan failures configurations

Overall, the loss of one fan seems to have a limited impact on the cooling system performance, in particular when the turned off fan is the one placed in the middle (F_c). However, the increase in thermal resistance in much more visible when only one fan is operating. In particular, when only the left-hand side fan (F_l) is turned on, the switch on the right-hand side of the airflow exhibits a significantly higher thermal resistance ($200\ K/kW$ instead of $140\ K/kW$ in the reference case).

5 Conclusion

In this paper, the thermal characterization of an air-cooled assembly based on LV100 Mitsubishi MOSFET power modules has been presented. Overall, it is important to note that, even with the consideration of heat dissipation of $1.3\ kW$ per module (i.e. a mean power density of $\approx 10W/cm^2$ at the module baseplate), the results presented here show that this power module technology can be well thermally managed with conventional air-cooled technology. In addition, the comparison to a 3D thermofluidic model showed that it was possible to predict, with a traditionnal FEM modeling approach, both the transient thermal response and permanent regime thermal resistance of the system with a reasonable accuracy. Finally, miniature PT100 sensors placed in the heatsink directly below the modules have been shown to be able to provide a measurement response time of the same order of magnitude as to the one given by a junction temperature measurement. On the contrary, surface mounted sensors exhibit much longer response time and a significant underestimation of the maximum temperature of the power converter.

Acknowledgment

This work was funded by the French government as part of the France 2030 program (RACCOR-D project)

References

[1] A. Rujas, V. M. Lopez, I. Villar, T. Nieva, and I. Larzabal, "Sic-hybrid based railway inverter for metro application with 3.3 kv low inductance power modules," in *2019 IEEE Energy Conversion Congress and Exposition (ECCE)*, IEEE, 2019, pp. 1992–1997.

[2] H. Ke, G. Chang, W. Zhou, C. Li, Y. Peng, and X. Dai, "3.3 kv/500a sic power module for railway traction application," in *PCIM Asia 2018; International Exhibition and Conference for Power Electronics, Intelligent Motion, Renewable Energy and Energy Management*, VDE, 2018, pp. 1–3.

[3] X. Li, D. Li, G. Chang, W. Gong, M. Packwood, *et al.*, "High-voltage hybrid igbt power modules for miniaturization of rolling stock traction inverters,"

IEEE Transactions on Industrial Electronics, vol. 69, no. 2, pp. 1266–1275, 2021.

[4] J. Fabre, P. Ladoux, and H. Caron, "Fixed (trackside) energy storage system for dc electric railways based on full-sic isolated dc-dc converters," *Electronics*, vol. 12, no. 7, p. 1675, 2023. DOI: 10.3390/electronics12071675.

[5] J. Fabre, P. Ladoux, H. Caron, A. Verdicchio, J.-M. Blaquière, *et al.*, "Characterization and implementation of resonant isolated dc/dc converters for future mvdc railway electrification systems," *IEEE Transactions on Transportation Electrification*, vol. 7, no. 2, pp. 854–869, 2020.

[6] E. Stumpf, E. Wiesner, H. Uemura, and S. Iura, "Lv100-a dual power module for the next generation railway inverters," *Bodo's Power Syst*, vol. 2, pp. 32–34, 2017.

[7] Icam Toulouse. "Depth-lab, icam toulouse campus research platform." (2024), [Online]. Available: https://en.icam.fr/research/depth-lab/ (visited on 03/19/2024).

[8] Mitsubishi Electric corporation. "Fmf750dc-66a datasheet." (Nov. 2023), [Online]. Available: https://www.mitsubishielectric.com/semiconductors/powerdevices/topics/sic/fmf750dc-66a.html (visited on 04/05/2024).

PCIM Europe 2024, 11– 13 June 2024, Nuremberg   DOI: 10.30420/566262400

Thermal Behavior of SiC MOSFET with Planar Packaging Technology

Yijun Ye[1]; Alexander Hensler[1]; Thomas Bigl[1]; Pietro Botazzoli[1]; Ralf Schmidt[1]; Thomas Basler[2]; Josef Lutz[2]
[1] Siemens AG, Germany.
[2] Chemnitz University of Technology, Germany.

Corresponding author and speaker: Yijun Ye, yijun.ye@siemens.com

Abstract

This paper discusses the thermal characteristics of SiC MOSFET power modules using different planar packaging concepts. The thermal performance is validated both with thermal impedance measurements and transient thermal simulations. Finally, the results are compared with a standard TO-247-4 package. Simulation and experimental findings demonstrate that planar packaging technology significantly enhances both transient and steady-state thermal performance, leading to a reduced junction temperature for SiC MOSFETs both in regular inverter application, in overload situations and especially during surge current events.

1   Motivation

SiC MOSFETs are increasingly gaining prominence in automotive and industry applications because of their superior electrical properties with low switching losses and possible high-power density. With advancements in SiC MOSFETs, the inverter efficiency can reach values well above 99% [1]. However, there remains a challenge for the cooling of SiC power modules related to the increased power density in the comparably small dies. Therefore, the power module design should also be optimized for thermal management.

In this paper, SiC MOSFETs were assembled with different planar packaging concepts. The performance of different packages was validated in thermal impedance measurement and transient thermal simulation. Finally, the thermal measurement results for the different planar packages are discussed and compared with a standard TO-247.

2   Package Concepts and Test Approach

2.1   Planar Packaging

The investigated 1200 V SiC MOSFETs had a total chip area of 0.25 cm^2 and a specific R_{DSon} of 4.25 mΩcm^2 at 25°C. The same chip type was assembled into three different packages.
The cross-sectional view of these SiC MOSFET package types is shown in Fig. 1 (a, b, c). The TO-247 with the same chip was used as a reference sample, see Fig. 1 (d).

a)   SiC with planar concept A (AMB)

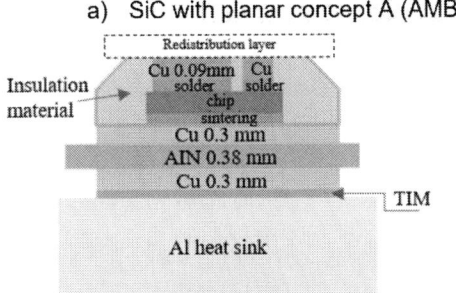

b)   SiC with planar concept B (IMS)

c)   SiC with planar concept C (Cu-lead frame)

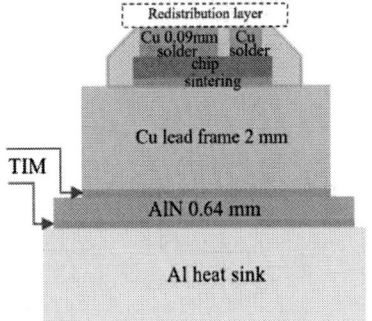

2826

d) SiC with bond wires
(Standard discrete package)

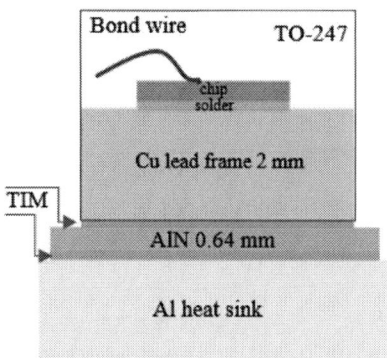

Fig. 1: Three types of planar package concepts for SiC MOSFET and standard discrete package TO-247. TIM: thermal interface material.

- **Planar concept A with Active Metal Brazing (AMB):**

  Four SiC MOSFET chips were Ag-sintered to AMB substrates, and the top-side source and gate areas were soldered to a redistribution layer, whose coefficient of thermal expansion (CTE) is 17 ppm/K. The insulation material around the chip is epoxy based, which has a low thermal conductivity of < 0.5 W/(m·K). Therefore, almost no heat can be dissipated through the insulation material.

- **Planar concept B with Insulated Metal Substrate (IMS):**

  The full chip top area of four parallel chips was soldered to a redistribution layer. The die bottom side was Ag-sintered on a thick copper layer, which was isolated from the bottom copper layer using a dielectric layer. The 2 mm thick top core copper layer, 0.18 mm dielectric layer and 0.3 mm bottom layer form the IMS substrate. More literature about insulated metal substrates can be found in [2-4], the comparison about IMS and DBC can be found in [5].

- **Planar concept C with Cu lead frame:**

  Four chips were sintered on a Cu lead frame of 2 mm thickness and the chip top area was soldered to a redistribution layer. A 0.64 mm AlN layer was used between the 2 mm Cu lead frame and the heatsink to provide electrical isolation, which was also utilized for devices with the TO package. The thermal interface material was applied on both sides of the AlN ceramic.

- **TO-247 package (reference sample):**

  The investigated TO packages contain only one SiC die that is soldered onto a 2 mm Cu lead frame. The electrical connection to the die topside is realized with Al bond wires Again a 640 µm AlN ceramic was used as isolation layer.

2.2 Test Approach

All three types of investigated planar packaging concepts (type A, B, C) were realized in half-bridge topology with four chips in parallel per switch. During application, the top and bottom SiC MOSFETs are switched at high frequency and the thermal cross-coupling effect between top and bottom switches is not negligible. Therefore, the load current was fed through both top and bottom SiC MOSFETs during thermal impedance measurement. Accordingly, the measurement current for T_{vj}-sensing also flowed through all body diodes. In this manner, not only the self-heating effect but also the cross-coupling effect was measured.

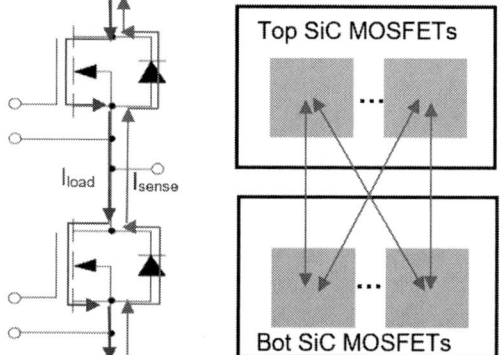

Fig. 2: Thermal impedance measurement circuit and cross coupling effect between top and bottom switches.

Since the reference sample TO-247 (d) contains only one chip inside the package, the planar packages (concept A and concept C shown in Fig. 1) were also assembled with only one chip for direct comparison. The package was heated up at 50% nominal current for 10 s and cooled down for 10 s, resulting in a power density of 350 W/cm².

The TO-247 was screwed with 3 Nm torque onto the same Al heat sink, with a thermal interface material in between. Additional force was applied on the TO packaging with 2 Nm torque for a better contact to the heatsink. An AlN ceramic layer (640 µm thick) was used between the TO package and heatsink to provide an isolation layer, which was also utilized for devices with the planar package concept C using 2 mm Cu lead frame.

3 Test and Simulation Results

3.1 Thermal Impedance Measurement

In 3, the thermal impedance of planar concept B and C (IMS and Cu) is compared with the planar concept A with AMB substrate. Both concepts B and C exhibit much better thermal performance with the improvement starting from 4 ms.

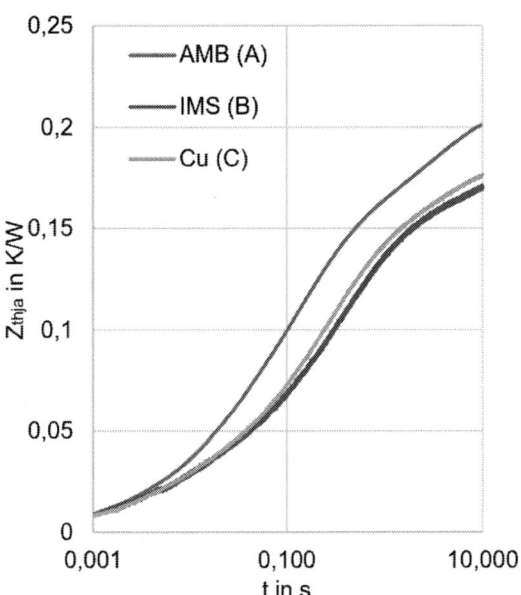

Fig. 3: Thermal impedance measurement results of a half-bridge with four chips in parallel. Test condition: t_{on} /t_{off} = 10 s, power density: 220 W/cm^2, chip size: 0.25 cm^2.

This is particularly advantageous for fast overload events, such as surge currents. The reason is due to the thicker a copper layer (2 mm instead of 300 µm) close to the chip, which acts as a thermal capacity in the ms time range. The increased thickness of the Cu layer allows the heat to spread laterally and pass through the AlN ceramic with a larger area. Consequently, the thermal impedance of concepts B and C were lower than concept A.

In the targeted inverter application, the overload capability is defined by two standard load duty cycles: low overload (lasting 3 s) and high overload (lasting 7 s). The thermal impedance at both load cycles was compared and summarized in Table 1.

At t = 3 s, the thick-Cu concepts (Cu and IMS) exhibited superior thermal performance, with a 13% and 14% reduction in the Z_{thja} value compared to the package with AMB. At t = 7 s, the difference widened slightly, showing a 15% and 17% improvement for the thick-Cu designs re-

spectively. The planar concept with the IMS exhibited comparable thermal impedance to the packaging concept C with Cu during the initial 0.1 s. This similarity arises because the SiC chips in both packages were sintered onto a 2 mm Cu layer, allowing efficient lateral heat spreading within the thick Cu layer.

t in s	3 s overload cycle	7 s overload cycle
Z_{thja} of AMB (A)	0.183 K/W	0.197 K/W
Z_{thja} of IMS (B)	0.160 K/W	0.168 K/W
Z_{thja} of Cu (C)	0.162 K/W	0.173 K/W

Table 1 The thermal impedance of three types packaging with four parallel SiC chips in half-bridge.

After 0.1 s, the IMS concept demonstrated slightly better thermal performance than the planar concept with Cu. One reason for this behaviour is the additional thicker isolation layer (AlN ceramic) for the planar concept C with Cu to insulate the backside potential from the heat sink.

Despite the significantly higher thermal conductivity (λ) of AlN compared to the dielectric layer in IMS (as indicated in Table 2), the necessity for AlN ceramic introduced a trade-off. Specifically, it resulted in two TIM layers on both side of the ceramic, rather than a single TIM layer for the IMS design.

Material	AlN in AMB (A)	Dielectric in IMS (B)	AlN for Cu lead frame (C)
thickness in mm	0.38	0.18	0.64
Breakdown voltage in kV	>5	>5	>5
λ in W/(mK)	170	14	170
R_{th} in K/W	0.09	0.51	0.15

Table 2 Layer thickness and fundamental properties.

The thermal impedance of the TO-247 (single chip) is measured as reference, the setup was described in 2.2. Two types of planar concepts were tested in the single chip version, both were mounted on the same Al heatsink and subjected to identical load current. The comparison results are shown in Fig. 4.

At short time intervals (millisecond range), both planar concepts exhibited superior thermal behaviour compared to the TO-247 package. Specifically, at 3 ms, approximately a 40% reduction in Z_{thja} value was observed with planar concepts. Further at 10 ms, the planar package concept A (AMB) with a thermal impedance of 0.12 K/W, outperforms the TO-247 package (0.163 K/W) by approximately 20%. Similarly, the planar concept C with thick Cu demonstrates a remarkable 33% reduction in Z_{thja} value compared to the TO-247, see Table 3.

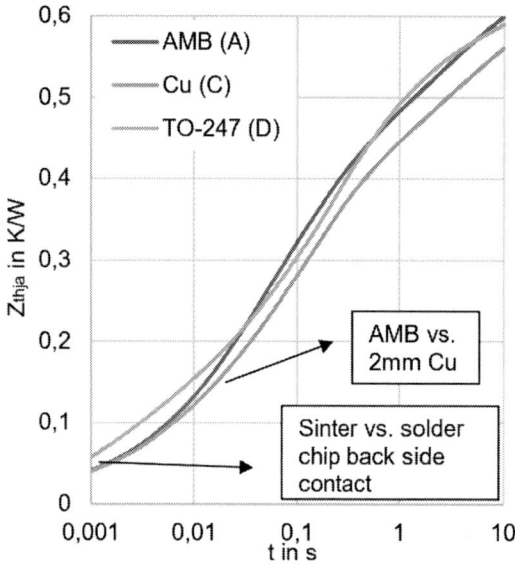

Fig. 4: Thermal impedance measurement result of planar concepts (type A, C) vs. TO-247 (type D) with single SiC chip. Test condition: t_{on}/t_{off} = 10 s, power density 350 W/cm².

Z_{thja} in K/W	3 ms	10 ms	3 s	7 s
planar Cu (C)	0.07	0.12	0.5	0.55
planar AMB (A)	0.073	0.13	0.54	0.59
TO-247 (D)	0.1	0.16	0.55	0.59
$\Delta Z_{thja}(t_{on})$ TO vs. planar Cu (C)	40%	33%	10%	8%

Table 3 The thermal impedance of three types of packaging with single SiC chip (25 mm² die area).

The improved thermal performance of the planar package concept at short time interval can be attributed to its sintered chip design. Additionally, the chip front-side connection via soldered copper to a redistribution layer also contributes to its

superior behaviour, especially for short pulse times.

Notably, this difference in the thermal impedance results to a remarkable improvement in short-term failure cases, particularly the surge current capability of a planar packaging concept. For detailed information on surge current capability, refer to [6]. At 0.5 s, the planar concept with AMB began to exhibit a tendency of lower thermal impedance than the TO-247.

This difference arises because the TO-247 package includes an additional TIM layer between the TO package and ceramic substrate. The comprehensive comparison between planar AJTs and TO-package is summarized in Table 3.

3.2 Thermal Simulation Results

The thermal impedance of TO-247 package and planar Cu package were investigated using Siemens 3D thermal simulation tool (Simcenter 3D). The power loss was concentrated at the top surface of the SiC MOSFET, because most of the energy dissipation primarily occurs in the ca. 10 μm active layer (highlighted in yellow in Fig. 5). The green area corresponds to the junction termination area.

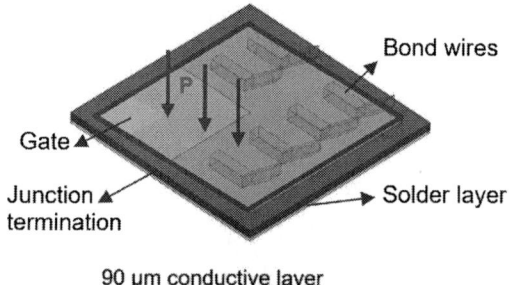

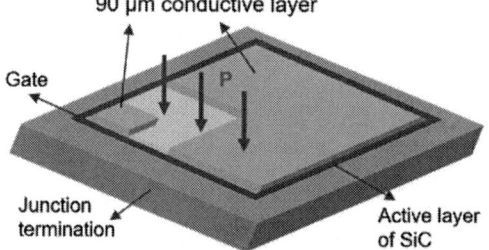

Fig. 5: simulation model of TO-247 with bond wires and planar concept with Cu conductive layer (only SiC die is shown).

The thermal conductivity of SiC decreases with increasing temperature. Therefore, the temperature dependence of thermal conductivity and capacity was taken into consideration in this simulation. The temperature dependent curves of the different materials could be found in [7].

Fig. 6 presents the comparison between the measured thermal impedance results (solid line) over a 10 s interval and the simulation results (dashed line) over a 1 s interval. The planar packaging with Cu consistently exhibits lower thermal impedance than the TO-247 during this time interval. Specifically, at 10 ms, the planar package C with Cu achieved with Z_{th}= 0.12 K/W an approximate 35% reduction in the thermal impedance compared to the TO-247 (with a thermal impedance of 0.163 K/W).

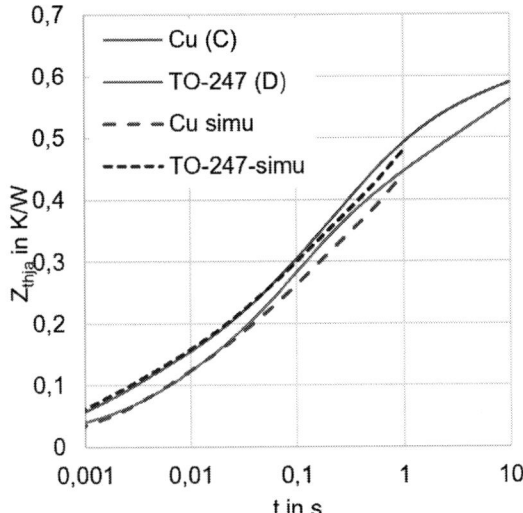

Fig. 6 Thermal impedance measurement and simulation result of planar Cu (C) shows lower thermal impedance value than TO-247 (D).

The blue dashed curve represents the simulation result for the TO package, which fits well to the measured thermal impedance curve up to 0.1 s. The red dashed curve is the simulation result of the planar Cu package, which also lies close to the measured Z_{th} curve (red solid line) in the first 0.05 s interval. However, beyond 0.05 s, simulation result and measurement slightly differ, which can be explained with the simplified simulation model consisting of co-planar layers. The simulation cannot represent the TIM layer very well, for example the thickness might not be homogonous, and thermal conductivity may differ from data sheet value with varying bond line thickness and applied pressure.

4   Summary

Superior thermal behaviour was demonstrated for the planar packaging concepts in comparison to traditional TO-247, particularly within short time ranges. At 10 ms, the planar package with Cu exhibits a 33% reduction in thermal impedance, leading to significant improvements of short-time events like surge current. The key lies in a good top-side contact, which provides additional thermal capacitance. Furthermore, under overload conditions, the planar AJTs consistently outperformed the TO-247 package.

Among three types of planar concept, the thermal behaviour of planar concept B and C (IMS and Cu) outperforms that of the planar concept with AMB. In summary, the planar package concepts show a clear potential to enhance thermal performance.

5   Literature

[1] A. Hensler; T. Bigl; S. Neugebauer; S. Pfefferlein.: Air Cooled SiC Three Level Inverter with High Power Density for Industrial Applications, International Exhibition and Conference for Power Electronics, Nürnberg, PCIM 2017.

[2] S. Idaka: New Packaging Concepts: Bridging Devices and Applications, International Seminar on Power Semiconductors Proceedings, Prague, 2021.

[3] Tokuyama, T., Kusukawa, J., and Nakatsu, K., 2014, "Insulated Metal Substrate for Power Modules Using Anodic Oxide Film of Aluminum," 2014 International Power Electronics Conference (IPEC-Hiroshima 2014—ECCE ASIA), Hiroshima, Japan, May 18–21, pp. 2904–2909.

[4] M. Correvon; J. Nagashima; R. Apter.: Power Modules with IMS Substrates for Automotive Applications. IEEE Vehicular Technology Conference, 2002.

[5] E. Gurpinar; B. Ozpineci; S. Chowdhury.: Design, analysis, comparison, and experimental validation of insulated metal substrates for high-power wide-bandgap power modules, Transactions of the ASME. Journal of Electronic Packaging, 2020.

[6] Y. Ye; A. Hensler; T. Bigl; P. Botazzoli; R. Schmidt; T. Basler; J. Lutz.: Surge Current Test of SiC MOSFET with Planar Assembling and Joining Technology, 13th international conference on integrated power electronics systems, Düsseldorf, 2024.

[7] Lutz, J., Schlangenotto, H., Scheuermann, U., D., and Doncker, R., 2011, Semiconductor Power Devices: Physics, Characteristics, Reliability, Springer, Berlin/Heidelberg.

PCIM Europe 2024, 11– 13 June 2024, Nuremberg   DOI: 10.30420/566262402

Implementing Module Health Monitoring in EV Traction Inverters

Karol Rendek[1], Adam Matajs[1]

[1] onsemi Piešťany, Slovakia

Corresponding author: Karol Rendek, karol.rendek@onsemi.com
Speaker: Karol Rendek, karol.rendek@onsemi.com

Abstract

The traction inverter in an electric vehicle (EV) is a critical component and its unexpected failure during operation can result in a dangerous situation for the occupants. This paper describes a novel approach to add accurate Lifetime Prediction Functionality (LPF) to traction drives based on actively monitoring various device parameters through on-chip sensors.

An enhanced method for predicting the remaining useful life (RUL) of a system involves monitoring the percentage shift of various health parameters, such as R_{ds-on}, V_{th}, and T_j, and analyzing them using principal component analysis (PCA). This approach improves the accuracy and reliability of the lifetime prediction model, building upon the traditional Coffin-Manson cycle-based calculation. It is important to note that relying on a single method for calculating lifetime usage is inadequate. To ensure effective online power switch health monitoring in the future, a combination of multiple methods is necessary. The LPF requires the semiconductor manufacturer to construct a database of lifetime parameters. Data is collected from various stages of the qualification process, including power module qualification, system level qualification, motor emulator testing, and the final phase where the LPF is tested and fine-tuned for a specific powertrain application.

1   Introduction

The expanding traction xEV market is looking to enhance the design of power electronics to lower the cost of the final product BOM, while also increasing the power/liter ratio to support the production of EVs for the mass market. One of the key factors that can contribute to advancements is the reduction in SiC technology prices, along with the integration of LPF into traction inverters. The LPF allows for the optimization of BOM costs in traction inverters for electric vehicles in the medium and low-price range, serving the mass market. In this work, the goal is to identify any potential failures and track the lifetime usage using T_j cycles. [1]

This feature offers a convenient solution for end users by monitoring the traction inverter's power module health and notifying of potential failures or when it has reached its lifetime usage based on cycles. By implementing effective measures, the likelihood of an EV traction inverter malfunctioning in real-world conditions is minimized, ensuring that EV drivers can enjoy the necessary dependability and convenience throughout the lifespan of their vehicle.

2   Overview of Lifetime Prediction Models

Various approaches can be utilized in lifetime prediction models for power components in electric vehicle traction inverters for component durability and predict failure assessment. These models often involve Remaining Useful Life (RUL) and thermal cycle-based models, among others. Thermal cycle-based models, such as the Coffin-Manson model, are designed specifically to evaluate the fatigue life of components subjected to recurrent thermal stress.

Equation (1) is used in the Coffin-Manson model to predict the number of cycles to failure (N_f), which is calculated using thermal stress cycles. These cycles are generated by fluctuations in load current, that affects die temperature, resulting in material deterioration and final failure [2]. Equation (1) provides a quantitative description of this relationship, allowing us to calculate Nf and gain insights into the expected operational lifespan of the component.

2831

$$N_f = K * \Delta T^{-\beta_1} * e^{\frac{\beta_2}{(273+T_{j,max})}} * t_{on}^{-\beta_3} \qquad (1)$$

Applying the Palmgren-Miner rule allows for a more concrete expression of the results. The DUT is considered to have reached the end of its life when the value one is attained, according to this regulation, that assigns a value between zero and one. [2]

$$D = \sum_{i=1}^{j} \frac{n[i]}{N_f[i]} \qquad (2)$$

Equation (2) is summary of the partial damage caused by each thermal cycle.
Where:

j – is the total number of thermal cycles used in the Rainflow algorithm.

n_i – is the number of full cycles (1) or half cycles (0.5) of thermal cycles.

N_f – is the number of cycles until failure [7].
The Rainflow algorithm works by analyzing the local extrema of the provided waveform. Once all extrema points have been identified, a three-point comparison rule is employed to detect cycles. This method evaluates three points and determines whether a cycle is detected. Once the cycle is detected, the initial starting point is removed, leaving only the remaining two points for the subsequent comparison. This process continues until all the extreme points have been utilized. It generates the cycles marked as 1 and the half cycles marked as 0.5. [7], [9]
Another indicator of lifetime prediction is the percentage parameter shift, which represents the shift from the initial value. When there is a parameter shift, the initial nominal values undergo changes, but the power module does not fail right away. A monitoring algorithm is used to collect and calculate the initial values separately for each power switch. Power module wear may be indicated by a shift in the percentage parameter. A shift like this is also part of the AGQ324 guideline. [3]
RUL-based models can predict the time remaining before a power module fails, allowing for proactive measures to prevent such failures in the final application. There are three main groups of RUL models [4]:

- Analytical-based - predicts lifetime based on equations describing the system.

- Data-driven - predicts using a historical database of measured devices.

- Hybrid models - combine different approaches.

Ensuring the accuracy and reliability of the collected data set is crucial for accurately predicting the remaining lifetime of power modules. The data set is analyzed using statistical methods to ensure its accuracy. Calculations are performed to determine monotonicity, prognosability, and trendability. [5]
The concept of monotonicity refers to the direction of the trend, whether it is positive or negative, for certain parameters related to mechanical devices, electric devices, or electric parts. It indicates whether the device returns completely or partially to its previous state. As an example, the concept of mono-tonicity does not factor into the estimation of battery lifespan. [8]

$$Monotonicity = \left| \frac{no.\ of\ d/dx>0}{n-1} - \frac{no.\ of\ d/dx<0}{n-1} \right| \qquad (3)$$

Where 'n' represents the number of observations. The variance in the critical failure point is defined by prognosability. When there is a high level of variability in the critical point of prognosability, the outcome of extrapolating a predictive parameter to failure becomes increasingly unfavorable. [8]

$$prognosability = \exp\left(\frac{-std(p_{fail})}{mean|p_{start}-p_{fail}|} \right) \qquad (4)$$

Where 'p' determines the value of the predictive parameter.
Trendability refers to the extent to which various parameters in different DUTs exhibit the same or similar function that can be described using the same functional form. [8]

$$t_i = \frac{no.\ of\ d/dx > 0}{n-1} - \frac{no.\ of\ d^2/dx^2 > 2}{n-2} \qquad (5)$$

$$Trendability = 1 - std(t_i) \qquad (6)$$

Where 'n' represents the number of observations in the ith parameter.
RUL estimation employs a variety of techniques, including neural networks, fuzzy logic, and analytical equations that model system dynamics.

3 LPM Model Data Collection

The methodology used for lifetime prediction model (LPM) data collection and training are summarized in Table 1.

During the initial data collection phase, the AQG324 reliability testing will be running concurrently. This testing will specifically target any packaging-related issues that may arise. The information gathered from the AQG324 testing is utilized to create a model for the initial LPM. The first version of LPM undergoes testing simultaneously with T_j cycling measurements in system level traction inverter mode. Fig. 1 illustrates the parameters measured during the long-term cycling test and their variations in the selected IGBT module.

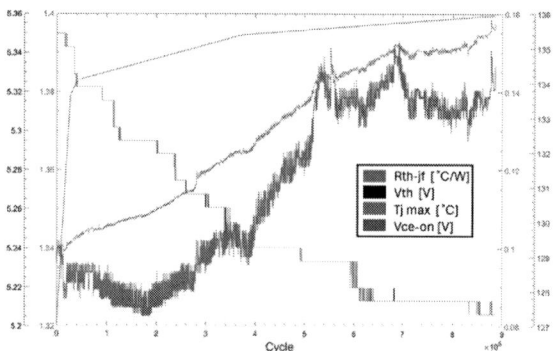

Fig. 1 Measured $R_{th\text{-}jf}$, V_{th}, $V_{CE\text{-}ON}$ and $T_{j\text{-}max}$ from the AGQ324 power cycling test for selected IGBT

Table 1 Phases of LPM model database data collection and model training.

Phase no.	Test Method	LPM model parameters	Test description, conditions
1.	AQG324 Packaging Level Testing	V_{th} – Gate threshold voltage T_j – Junction temperature $R_{DS\text{-}ON}$ – Power switch ON resistance	DC current
2.	Inverter Mode Inductive Load Testing		Inv. Mode – PWM drive, inductive load
3.	Motor Emulator Testing		EV motor load, PWM driving, driving profile
4.	Product Qualification Testing		Product qualification testing

In the second phase, the system level testing for the inverter mode application was conducted. This phase involves simulating real-world scenarios while running in a test-bench. The data collected in the second phase can be utilized to expand the LPM model, and the verification is conducted in phase 3 through the Motor Emulator test. The data

collected in phase 3 can be used to expand the LPM. During phase 3, the accuracy of lifetime prediction for LPM is evaluated and expanded using data of various drive-cycles. Additionally, phase 3 testing can be performed with the target motor of the application which increases the applicability of the results.

During phase 4, which is the final development stage of the LPM, thorough testing is carried out to ensure that the model meets all the final product's specifications.

Final testing and predictive model training are carried out in the finished product during the qualification phase, where final testing and qualification of the LPF is performed in parallel with demonstrated useful lifetime testing.

4 Experimental Testing

During LPM training, we collected data from a variety of tests, including the AQG324 power cycling phase and the system level inverter mode test. Measured data includes V_G, $V_{CE\text{-}ON}$/$V_{DS\text{-}ON}$, $I_{CE\text{-}ON}$/$I_{DS\text{-}ON}$, V_{th} (measured during off cycles, not switching mode), and T_j. Additionally, Rth and $R_{DS\text{-}on}$ are calculated. This measured and calculated data serves as lifetime parameters. [6]

These values are measured and collected by a lifetime monitoring system in parallel with the test. Other lifetime parameters such as I_{gss}, T_{on}/T_{off} time and Gate Noise measurement [6] are not used in this methodology since the integration of the measurement into the end application is complex and the listed parameters are sufficient for our purpose.

Ensuring the reliability of the lifetime prediction model (LPM) is crucial during the power module qualification phase, system level qualification, and finished good application. To achieve this, our developed methodology gathers data from the AQG324 power cycling test, which encompasses the module qualification phase and system level qualification in inverter mode. During phase 1 and phase 2, the controlled parameter T_j (die junction temperature) undergoes periodic cycling, resulting in accelerated module degradation. The data collection methodology, beginning with the module qualification phase and concluding with the final product development and qualification, ensures the collection of adequate data, which is crucial for the reliability of the lifetime model.

It is crucial to maintain reproducibility in all measurements. The focus should be on preventing any interference from the test setup that could impact the accuracy of the measured modules. This is a fundamental consideration when it comes to measurement methodology. Reproducibility of measurements ensures that changes in monitored data are only due to power module degradation and not due to test setup changes which would lead to incorrect lifetime prediction model training. The measurement system developed collects data from the AQG324 power cycling test phase and updates the module database with all lifetime parameter records. In the next phase, which is the system-level inverter mode qualification, the collected data is uploaded to the same database. The data acquisition aspect of inverter mode testing requires a more rigorous approach, which is why a measurement board with an FPGA triggering ADC scheme is utilized. For accurate system level measurement in inverter mode, it is essential to synchronize the measurement with the motor control PWM. Additionally, the data acquisition needs to be performed in the microsecond range, taking into consideration that the Vds-on time span is only a few microseconds. This is assuming a PWM motor control frequency of approximately 10kHz. An example of monitored data in AQG324 power cycling is shown in Fig.2, a). The figure is from a developed lifetime parameter monitoring system. The system offers initial and allows for changes in data, either in absolute terms or as a percentage. Fig.2, b) displays the data collected, illustrating the variation in response to the number of cycles. The dashed line represents the initial value of the lifetime parameter, while the x-axis represents the approximate change in data over a one-month period of power cycling testing.

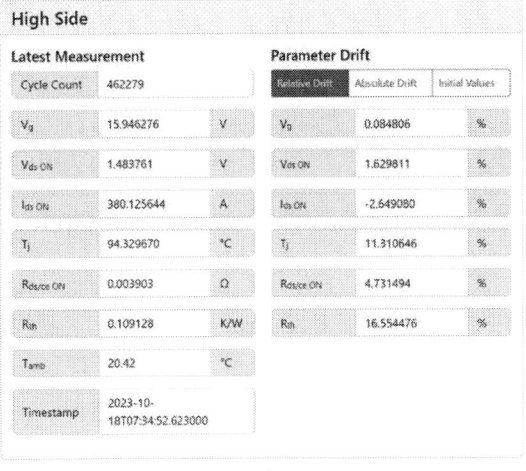

a)

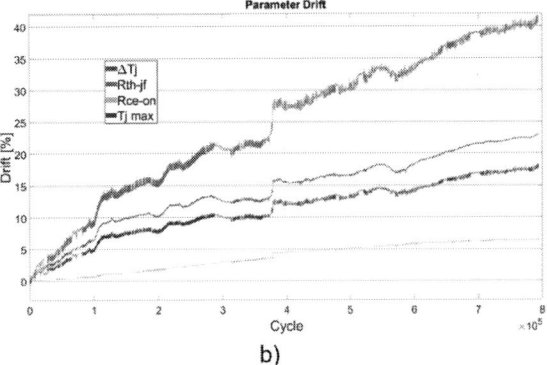

b)

Fig. 2 a) Parameter drift in AQG324 test, b) Long- term data tendency and shift from the initial value. Example measurement of an IGBT transistor.

5 RUL Model Development and Training

In the context of LPM with RUL model, the decision was made to use a data-driven type of RUL model. Understanding the behavior of transistors and accounting for various external factors such as heat dissipation, cooling, and EMC can be quite challenging when modeling different DUTs. An alternative method is to establish a database containing verified DUTs and then utilize the data to train the predictive model. We utilize an advanced algorithm that analyzes online data from monitored power switches and cross-references it with our database to accurately forecast the remaining useful lifetime of the devices.

When developing a lifetime prediction model, it is essential to incorporate lifetime prediction parameters into the predictive models. Data pre-processing is essential for standardizing the format, such as applying techniques like data smoothing and filtering. Following these steps, the pre-processed data undergoes testing to determine the optimal parameters for the RUL model's performance. Monotonicity, prognosability and trendability are used. The output values of these three indicators of the prediction model's quality are presented on a scale ranging from zero to one. The lowest quality is represented by zero, while the optimal value to achieve the most suitable parameter is one. The selected parameters cannot be correlated.

After completing the analysis, the next step is Principal Component Analysis (PCA). PCA is a valuable technique for reducing the size of a data set while preserving its important features that contribute to the prediction model. When performing PCA analysis, it is necessary to process the data to account for potential variations in the range of values. This is done through a technique called normalization, which ensures that all values are made equal. For example, when considering I_{CE}/I_{DS} parameter across hundreds of amps and a wide range of T_j from 20 ℃ up to 175 ℃, it can have a substantial impact on subsequent data analysis and predictive model testing. Before proceeding with data normalization, it is essential to calculate the mean and standard deviation of the input data. The mean and standard deviation are calculated using the raw ADC data from the database.

$$Normilize = \frac{(original\ data - Mean\ value)}{Std\ value} \quad (7)$$

The PCA analysis involves processing of normalized data. The output of the PCA analysis is used in the RUL prediction model.

As previously stated, data-based models analyze collected lifetime data in comparison to data from historical databases. Predictive models in Matlab libraries were used to develop RUL predictions. In this case, the Pairwise Similarity Model was employed, to compare the degradation profile of the tested DUT with DUTs from the database. Setting the parameters and testing the accuracy of the predictive model with lifetime usage data is a crucial step in the development and training of the RUL model. There are multiple choices available in Matlab for configuring predictive models. Given the versatility of the RUL model's settings, a comprehensive testing process was employed to identify an optimal predictive maintenance configuration.

Table 2 displays the RUL model settings that were tested using the automated script to select optimal settings. The "estRUL_output" column contains the simulated data for the remaining life thermal cycles to failure, which is used later in the evaluation process.

The performance of the prediction model was quantified by testing the RUL model with known percentages of real data. The percentage prediction error was calculated. The results were summarized in Fig. 3. The data plotted reveals a clear pattern of repeating prediction errors. The graph provides a clear representation of the factors that influence predicted outcomes, distinguishing between those that have an impact and those that do not. This information facilitates the configuration of the prediction model and reduces the testing time. To build a reliable RUL model, it is important to thoroughly test the prediction model across various usage levels of the DUTs under examination.

Table 2 Example of RUL model settings and corresponding estimated remaining lifetime.

Trial	Method	Neighbor	History	Range	estRUL_output
1	correlation	10	10	0	272004
2	dtw	10	10	0	307829
3	correlation	20	10	0	272004
4	dtw	20	10	0	327268
21	correlation	10	20	0	409805
22	dtw	10	20	0	307747
23	correlation	20	20	0	409805
24	dtw	20	20	0	327224
41	correlation	10	30	0	418288
42	dtw	10	30	0	307612
43	correlation	20	30	0	418288
44	dtw	20	30	0	327147
61	correlation	10	40	0	414264
62	dtw	10	40	0	307612
63	correlation	20	40	0	431214
64	dtw	20	40	0	327147
81	correlation	10	50	0	401918
82	dtw	10	50	0	307653
83	correlation	20	50	0	439710
84	dtw	20	50	0	327166
201	correlation	10	10	0.1	272004
202	dtw	10	10	0.1	307829
203	correlation	20	10	0.1	272004
204	dtw	20	10	0.1	327268
221	correlation	10	20	0.1	409805
222	dtw	10	20	0.1	307747
223	correlation	20	20	0.1	409805
224	dtw	20	20	0.1	327224
601	correlation	10	10	0.3	272004
602	dtw	10	10	0.3	307829
603	correlation	20	10	0.3	272004
604	dtw	20	10	0.3	327268
2197	correlation	90	100	1	391149
2198	dtw	90	100	1	381669
2199	correlation	100	100	1	391149
2200	dtw	100	100	1	381669

Simulation of different stages of lifetime usage can be achieved by using different percentage amount of original data, such as 50%,70% and 90%. After completing all of the tests that are required and evaluating all of the exported test data, the following conclusion can be reached. The "Neighbor" and "History" options, as well as when "Method" is set to "correlation" or "dtw", create the largest difference in the RUL model parameters in the result. Settings with a percentage error in the range of -5.5% to 5.5% were chosen. A summary of the results in Table 3, indicates that different settings have different prediction percentage errors, some below 1% and some greater than 1%. Table 3 shows that the RUL training process should accept percentual error of around 1.5% higher than aiming for any prediction below 1%.

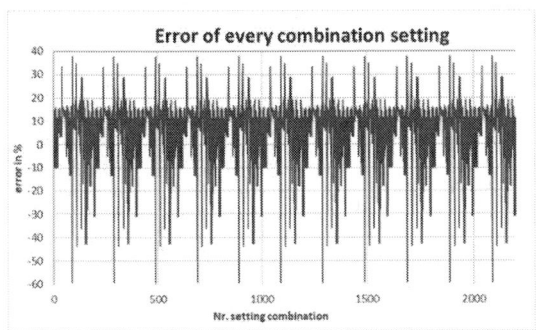

Fig. 3 Percentage error of prediction result

This RUL model setting compromise allows us to set predictive model with a setting with acceptable results in 1.5%, assuming used lifetime parameter dataset.

Table 3 Example of the best results from the prediction model

DUT 50%	Trial	Method	Distance	Neighbor	History	Estimated RUL	Real remaining RUL	Percentage error
	81	correlation	euclidean	10	50	429614	431661	-0.474
	101	correlation	euclidean	10	60	431387	431661	-0.063
6	129	correlation	euclidean	50	70	426633	431661	-1.165
	131	correlation	euclidean	60	70	428761	431661	-0.672
	183	correlation	euclidean	20	100	426229	431661	-1.258
	81	correlation	euclidean	10	50	436909	442767	-1.323
	151	correlation	euclidean	60	80	437532	442767	-1.182
11	163	correlation	euclidean	20	90	439208	442767	-0.804
	165	correlation	euclidean	30	90	439289	442767	-0.785
	167	correlation	euclidean	40	90	436452	442767	-1.426
	169	correlation	euclidean	50	90	439667	442767	-0.700

DUT 70%	Trial	Method	Distance	Neighbor	History	Estimated RUL	Real remaining RUL	Percentage error
1	101	correlation	euclidean	10	60	286194	289530	-1.152
6	88	dtw	euclidean	40	50	258487	258997	-0.197
6	188	dtw	euclidean	40	100	258550	258997	-0.173
11	189	correlation	euclidean	50	100	265569	265660	-0.034

DUT 90%	Trial	Method	Distance	Neighbor	History	Estimated RUL	Real remaining RUL	Percentage error
1	65	correlation	euclidean	30	40	93148	96510	-3.484
6	81	correlation	euclidean	10	50	85552	86332	-0.903
	171	correlation	euclidean	60	90	87834	88553	-0.812
	173	correlation	euclidean	70	90	87535	88553	-1.150
11	175	correlation	euclidean	80	90	87535	88553	-1.150
	177	correlation	euclidean	90	90	87535	88553	-1.150
	179	correlation	euclidean	100	90	87535	88553	-1.150

The ramification of the results in Table 3 reveals the dependency of the prediction model settings. Modules with 50% or 90% lifetime usage require different RUL model settings. This requires the implementation of an adaptive RUL model setting in the application that varies based on lifetime usage. The RUL model should be initialized using original "pristine" sample settings. Adaptive model prediction setting will alter with lifetime usage, making prediction more accurate as module lifetime usage increases.

Figure 4 presents an example of RUL model training. The red curve is data from PCA analysis tested with available RUL data models of one type of traction power module.

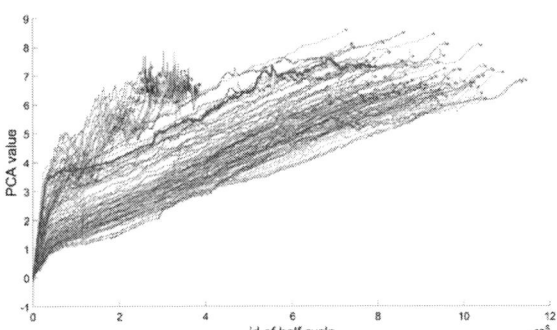

Fig. 4 Tested RUL model, red curve, of one type of traction power module.

6 LPM Implementation in Traction Inverter

The whole process of LPM development and implementation in traction inverters can be summarized as the collection of data in all four phases outlines in section 2. The data is processed using a Matlab based application and the lifetime prediction model is developed and trained using all of the data collected in the lifetime database. To enhance the development of the lifetime model, it is necessary to establish a connection between the measurement system database and the Matlab RUL prediction application. The LPM includes three methods for predicting the lifetime of a system: the RUL model, the percentage shift model and the T_j cycle-based lifetime model. The Matlab model can be exported as C or HDL code as required by the final application.

An example of the proposed combined lifetime prediction model embedded in a traction inverter application controller or dedicated MCU/FPGA is shown in Fig.5. The arrows depicted in the figure provide a clear illustration of the flow of data processing. The figure displays the data acquisition part and the block related to lifetime prediction. However, it does not include the power stage, gate drive, or any other blocks.

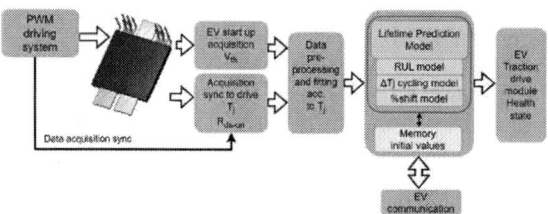

Fig. 5 LPF Data acquisition flow and processing with Remaining Lifetime Prediction model

7 Conclusion

The development of an LPM and its incorporation into control electronics is a possibility. An effective LPM implementation requires the collecting of defined data into a database. This must be done on the side of the semiconductor supplier, and it must be incorporated into the development process beginning with the packaging qualification phase. With the use of a lifetime prediction model that monitors the percentage shift of uncorrelated health parameters such as Rds-on (SiC MOSFET), Vth, and Tj parallel processing in PCA analysis, the remaining useful lifetime model is made more robust. Another parallel processing loop to this model is the Coffin-Manson cycle-based lifetime consumption calculation. It is not sufficient to calculate the lifetime using a single method, and to monitor the health of power switches in the future, it will be necessary to combine multiple different lifetime prediction methods.

Tests of the RUL model was presented utilizing real measured data and key settings were identified by an automated RUL model estimation accuracy test. The test results of the trained prediction model show percentage error of less than 3.5%. Expanding the database of collected data and further LPM training makes the lifetime prediction model more robust and reduces the percentage prediction error.

References

[1] M. Witczak, M. Mrugalski, and B. Lipiec, "Remaining Useful Life Prediction of MOSFETs via the Takagi–Sugeno Framework," Energies, vol. 14, no. 8, p. 2135, Jan. 2021, doi: https://doi.org/10.3390/en14082135.

[2] R. Bayerer, T. Herrmann, T. Licht, J. Lutz and M. Feller, "Model for Power Cycling lifetime of IGBT Modules - various factors influencing lifetime," 5th International Conference on Integrated Power Electronics Systems, Nuremberg, Germany, 2008, pp. 1-6.

[3] S.Kim, N.H.Kim, J.Choi, "Prediction of remaining useful life by data augmentation technique based on dynamic time wrapping," Mechanical Systems and Signal Processing,vol. 136, 2020, DOI: https://doi.org/10.1016/j.ymssp.2019.106486.

[4] K. Goebel and P. Bonissone, "Prognostic information fusion for constant load systems," *2005 7th International Conference on Information Fusion*, Philadelphia, PA, USA, 2005, pp. 9 pp., doi: 10.1109/ICIF.2005.1592000.

[5] MathWorks, "Feature Selection for Remaining Useful Life Prediction". Available at: https://www.mathworks.com/help/predmaint/ug/feature-selection-for-remaining-useful-life-prediction.html (2024,March 4)

[6] Z. Ni, X. Lyu, O. P. Yadav, B. N. Singh, S. Zheng and D. Cao, "Overview of Real-Time Lifetime Prediction and Extension for SiC Power Converters," in *IEEE Transactions on Power Electronics*, vol. 35, no. 8, pp. 7765-7794, Aug. 2020, doi: 10.1109/TPEL.2019.2962503.

[7] A. Nieslony, "Determination of fragments of multiaxial service loading strongly influencing the fatigue of machine components," Mechanical Systems and Signal Processing, pp. 2712-2721, 2009.

[8] Coble, Jamie & Hines, J. "Identifying optimal prognostic parameters from data: A genetic algorithms approach," Annual Conference of the Prognostics and Health Management Society, San Diego, CA, September 2009.

[9] ASTM international, "Standard Practices for Cycle Counting in Fatigue Analysis," Available at: https://tajhizkala.ir/doc/ASTM/E1049-85%20(Reapproved%202011)e1.pdf (2024, February 14)

PCIM Europe 2024, 11– 13 June 2024, Nuremberg   DOI: 10.30420/566262403

Reliability Tests of Copper Thick-Film Substrates for Power Electronic Applications

Henry Barth[1] , Sebastian Letz[2], Lars Rebenklau[1]

[1] Fraunhofer-Institut für Keramische Technologien und Systeme IKTS, Germany

[2] Fraunhofer-Institut für Integrierte Systeme und Bauelementetechnologie IISB, Germany

Corresponding author:	Henry Barth, henry.barth@ikts.fraunhofer.de
Speakers:	Henry Barth, henry.barth@ikts.fraunhofer.de
	Sebastian Letz, sebastian.letz@infineon.com

Abstract

Copper thick-film substrates were produced at Fraunhofer IKTS with a commercial copper thick-film paste system and an IKTS proprietary copper paste to compare them with commercial DBC substrates. Thermal shock testing at IKTS resulted in a total failure of the DBC substrates after only 50 cycles. The copper thick-film substrates survived the planned 1000 temperature cycles without damage, though. Power cycling at Fraunhofer IISB showed a significantly lower reliability for copper thick-film substrates compared to the DBC reference. However, a 25 % higher characteristic lifetime was observed for the thick-film substrates compared to the DBC reference.

1   Introduction

Industrialized countries are facing major challenges in transforming the industrial and mobility sectors toward climate neutrality. In the future, more electrical energy will have to be generated, transported and converted. Reliable, energy efficient and cost-effective power electronics will play a decisive role in this change. In e-mobility especially, the ambient and operating conditions pose major challenges for power electronics.

Power electronics modules are used in the electric powertrain. The necessary power electronic devices (IGBTs, SiC MOSFETs and free-wheeling diodes) are mounted on a circuit carrier consisting of a metal-ceramic composite. State of the art is direct bonded copper (DBC). Two copper foils are bonded to both sides of an Al_2O_3 ceramic at just over 1000 °C [1]. In high-end applications, the copper foils are brazed to a Si_3N_4 ceramic with active brazing at around 900°C (Active Metal Brazing; AMB) [2]. Subsequently, the required structures are etched in both technologies. As an alternative to these, there is the copper thick-film technology. Using the screen-printing process, copper pastes are deposited in the desired structure on the ceramic right away. After leveling and drying the copper layer is sintered at approx. 900°C, as well [3]. In this work, the copper thick-film substrates are compared with the standard DBC solution in terms of long-term stability.

2   Sample Preparation for Reliability Testing

A key objective of the internal Fraunhofer research project was to compare the electrical, thermal and mechanical properties of a thick-film copper substrate in the context of the production technology with standard DCB substrates on the market. Three types of samples with identical layout on the top side were prepared. The test layout from the previous EU-funded project HOPE (FP6-019848) was used, due to the following reasons. On the one hand, the layout of the circuit carrier must have a certain adaptability for the planned investigations on active load cycling resistance with the equipment available at the IISB. On the other hand, it should have a low degree of complexity for efficient manufacturability of test specimens and clear interpretation of damage patterns after the ageing tests.

As a reference system, DBC substrates with 0.38 mm thick Al_2O_3 ceramic and 0.3 mm thick copper on both sides without dimples were obtained in commercial quality from Rogers Germany GmbH.

At IKTS two types of copper thick-film substrates with identical layout were manufactured by using two different copper thick-film pastes. The first one is the commercially available paste system C740x from Heraeus, which serves as a thick-film reference, and the second one is a development of the IKTS. The thick-film process for thick copper layers involves printing the copper thick-film paste, drying and sintering it in a nitrogen atmosphere. An essential part of the research project at IKTS was the determination of screen, stencil and other printing parameters for thick copper layers with low residual porosity. One way of producing copper thick-film layers of approx. 100 μm is to combine screen and stencil printing. A microscope image of the surface of such a sample is shown in the following figure.

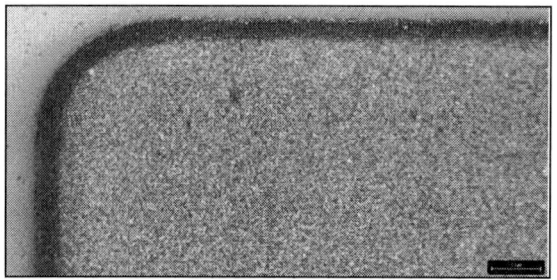

Fig. 1 Microscope image of copper thick-film surface

First, a 10 μm thick layer of copper was applied as an adhesion layer between the metal layer and the ceramic using a thick-film process. A screen with a high mesh count was used for this purpose. In the second step, the remaining 90 μm were processed using stencil printing. The cross-section in Fig. 2 shows a copper layer with the porosity typical of a thick-film.

The final printed copper thick-film thickness was around 100 μm on one side only. This was done on purpose to address a possible application,

namely liquid-cooled substrates. An Al_2O_3 ceramic with 0.38 mm thickness was used as insulator, too.

Fig. 2 Cross section of 100 μm thick copper thick-film on top of Alumina (Al_2O_3 96%)

Thermal characterization of the samples using direct cooling on the bottom of the substrate at a coolant temperature of 35°C resulted in a thermal resistance of 0.46 K/W for the DBC substrate, 0.64 K/W for the thick-film substrate with IKTS paste and 0.67 K/W for the thick-film substrate with the reference paste. The DBC substrate is superior to the thick-film variants. For comparison, the thermal resistance of a DBC substrate with thermal paste between the heat sink and substrate was determined to be 0.74 K/W. This represents the state of the art and shows that a thermal improvement is possible with thick-film substrates, provided that direct cooling concepts are used.

For the power cycling experiments, semiconductor chips from Infineon of the type IGBT3 SIGC100T65R3E were soldered with a SAC305 solder onto the substrates, see Fig. 3.

Fig. 3 In project produced sample with one-sided copper thick-film HOPE Layout for aging by active power cycling

The IGBT was contacted with 144 wire bonds. A total amount of 85 samples were prepared, whereby 30 samples were made with IKTS thick-film paste, 30 samples with the commercial paste and 25 samples with commercial DBC-Substrates.

A sample produced in the project intended for power cycling with an IGBT mounted on a thick-film copper layer is shown in Fig. 3.

3 Ageing with Thermal Shock

The passive temperature cycling tests at IKTS were performed in a two-chamber system according to DIN EN 60749-25 and ECPE Guideline AQG 324 [4].
The temperature range was between - 40 °C and + 125 °C. The basket with the samples was transferred from the hot to the cold chamber and vice versa within 30 seconds and remained there for more than 15 minutes. An initial DBC substrate is shown in below.

Fig. 4   Initial state of a DBC-sample.

The DBC substrates failed after max. 50 cycles with the typical conchoidal fracture, see figure below.

Fig. 5   Conchoidal fracture of DBC after 50 temperature shock cycles.

Even after 1000 thermal shock cycles, the thick-film samples showed no signs of delamination, see following figure.

Fig. 6   Copper thick-film sample after 1000 temperature shock cycles.

The results of the temperature shock test show an outstanding performance of the thick film substrates. In future work, the influence of the following aspects must be investigated:

- the thickness of the thick film
- the coefficient of thermal expansion of the paste
- the copper layer on only one side of the ceramic substrate

4 Ageing with Power Cycling

For the power cycling test, a minimum temperature of T_{min} = 40 °C, a temperature swing of ΔT = 100 K, an on-duration of t_{on} = 3 s, an off-duration of t_{off} = 9 s and a load current I_L = 114 A for the DBC samples and 82 A for the thick-film substrate samples was applied. A liquid cooling system was used, onto which the substrates were thermally contacted by using KeraTherm (white) TIM foils. The electric contacts were realized by using spring contact probes.

The failure criteria according to AQG324 [4] determined the time of a test specimen failure. The results of the power cycling test are evaluated and presented using the Weibull distribution, shown here in the form of the cumulative failure probability F:

$$F = 1 - e^{-(n/\alpha)^{\beta}} \qquad (1)$$

Therein, α is the scale parameter, which describes the lifetime at a failure probability of 63,2 % and β is the shape parameter, which characterizes the steepness of the cumulative distribution and is larger than 1 for ageing processes (wear out). The variable n is the evaluation time or number of cycles.

The failure times of all samples are depicted in the typical Weibull failure probability plot in the following figure.

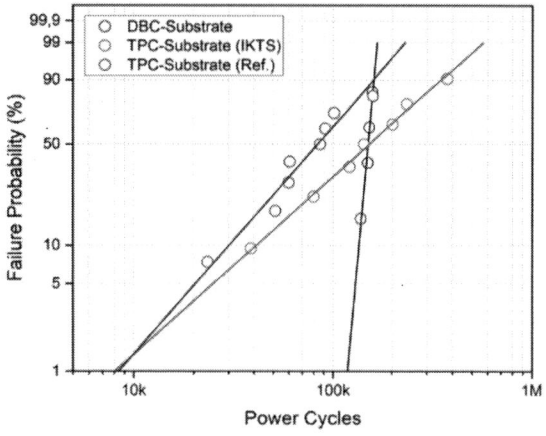

Fig. 7 Weibull plot to visualize the results of the power cycling experiment

The failure data is further fitted for a two parameter Weibull function with maximum-likelihood estimator [5]. The DBC samples have a significantly higher reliability compared to both thick-film substrate systems, which is indicated by the higher slope of the DBC line-fit. In case of the IKTS substrate system, some samples exist, which have a visibly higher lifetime compared to the DBC reference. One thick-film sample survived 761.329 cycles, at which the test was stopped, and is treated as right-censored.

The highest characteristic lifetime (63% failure probability) is obtained with the IKTS thick-film substrate system at 196.617 cycles. It approximately doubles the lifetime of the reference thick-film technology with 101.928 cycles. The DBC substrates show with 153.643 cycles only around 75 % of the characteristic lifetime of the IKTS thick-film technology. However, it must be mentioned that the DBC substrate result has lower statistical significance, since it is based on a sample number of only five samples in total.

The reliability is rather small and comparable for both thick-film systems. It is 1.4 and 1.8 for the IKTS and the reference system, respectively. An around 8.5 times larger value of 17.3 is obtained with the DBC substrates. This is expectable, since the thick-film samples are manufactured in a laboratory process and are compared to commercial products, which have a significant higher degree of maturity.

All three sample types showed a mixture of heating voltage and thermal resistance increase as failure indicators. A physical failure analysis showed for most samples solder fatigue, as can be seen in Fig. 8. Also some bond wire lift-offs and heel cracks were observed. An allocation to the lifetime or other aspects, like copper layer thickness, position on the test bench, etc. showed no systematic trends.

The larger scattering of the thick-film substrates lifetime in power cycling is attributed to the larger variations in the laboratory processing compared to the series production process of the DBC substrates.

It is expected that the thick-film samples with the lower lifetimes are suffering from stronger bending of these substrates during the power cycling tests. However, a clear correlation with the copper metallization thickness or other externally observable parameters could not be identified. Further investigations on the thick copper thick-film must therefore be done in the future.

5 Conclusion

Copper thick-film technology-based substrates for power electronics devices have been successfully developed and prepared for benchmarking their reliability against commercially available DBC substrates in temperature shock and active power cycling tests. The thick-film technology showed superior reliability in the thermal shock tests without delamination after 1000 cycles, whereas the DBC substrates (without dimples) failed after around 50 cycles with the typical conchoidal fracture. In the power cycling experiments, the thick-film substrates showed a 25 % higher characteristic lifetime compared to the DBC substrates but at the cost of a significantly lower reliability (ca. 8.5 times), which must be improved in the future.

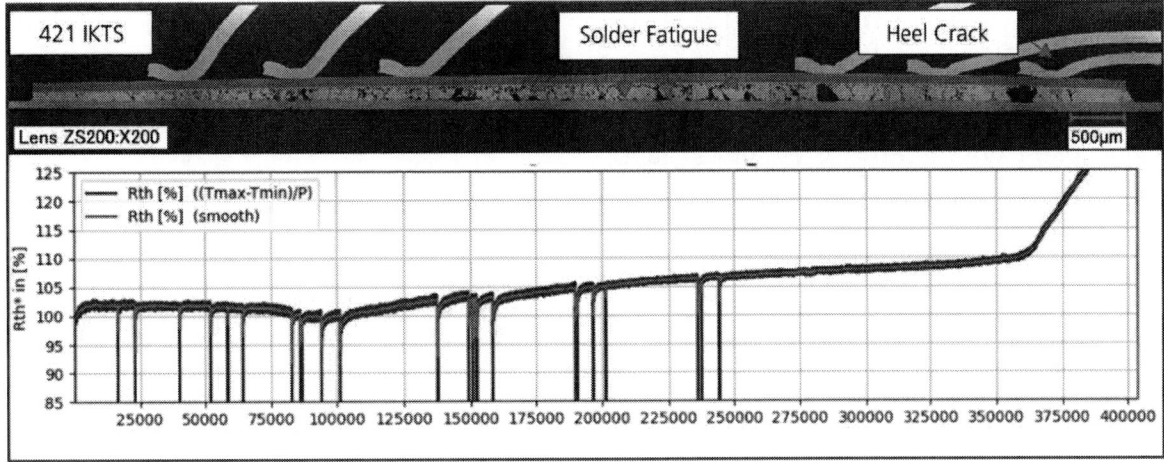

Fig. 8 Physical defect analysis of selected samples after the power cycling test. Top: Metallographic cross-section with solder fatigue. Bottom: Increase in the thermal resistance of a thick-film sample as a function of the power cycles

With a higher degree of maturity for the manufacturing process, the copper thick-film technology may become a promising candidate as reliable circuit carrier technology for power electronics circuits.

References

[1] J. Schulz-Harder, Hg., *DBC substrates as a base for power MCM's.* Proceedings of 3rd Electronics Packaging Technology Conference (EPTC 2000) (Cat. No.00EX456), 2000.

[2] A. Pönicke, J. Schilm, A. Triebert, K. S empf, T. Gestrich, H.-P. Martin, G. Böh m, D.Schnee, *Aktivlöten von Kupfer mit Aluminiumnitrid- und Siliziumni-tridkeramik (Active Metal Brazing of Copper with Aluminium Nitride and Silicon Nitride Ceramics).* Keramische Zeitschrift. 63. 334-342.

[3] K. Reinhardt *et al., Hg., Lead-oxide-free copper thick-film paste for alumina substrates.* 33rd International Spring Seminar on Electronics Technology, ISSE 2010, 2010.

[4] AQG 324: ECPE Guideline, *Qualification of Power Modules for Use in Power Electronics Converter Units in Motor Vehicles.* Release no.: 03.1/2021, 31.05.2021.

[5] M. Modarres, M. Amiri, C. Jackson *Probabilistic Physics of Failure Approach to Reliability: Modeling, Accelerated Testing, Prognosis and Reliability Assessment,* DOI:10.1002/9781119388692, pp. 136-168, 2017.

PCIM Europe 2024, 11– 13 June 2024, Nuremberg   DOI: 10.30420/566262404

Power Module Solutions with Improved Reliability for Elevator Drive Applications

Ábel Tőkés[1], Tiago Jappe ©[2], Matthias Tauer[2]

[1] VINCOTECH, Hungary
[2] VINCOTECH, Germany

Corresponding author: Tiago Jappe, tiago.jappe@vincotech.com
Speaker: Tiago Jappe, tiago.jappe@vincotech.com

Abstract

Elevator drive applications impose very stringent reliability requirements throughout their system-level lifetime. Focusing on the reliability of the power modules used in these applications, several stressors can lead to failures. The paper aims to benchmark different power module thermal technologies against an elevator drive's standard mission profile. Using an empirical model calibrated with experimental measurements, baseplate power modules were shown to achieve higher predicted lifetime compared to the baseplate–less variants. Nonetheless, the predicted lifetime of the more cost-competitive baseplate–less modules still falls well within the required lifetime for elevator drive applications.

1 Introduction

Elevator drive applications have stringent system-level requirements – not least in terms of power, efficiency, robustness, reliability, cost, size, and safety. These requirements significantly impact technological innovation in the overall design of power conversion systems. Reliability is a preponderant factor in power module selection [1], [2]. Here, reliability, defined as a function of time, $R(t)$, refers to the probability that the system continues to perform its intended function without failure after operating for a given time, (t). Various stressors, such as temperature variations, humidity, vibration, and radiation, can impact power module operation and increase the probability of failure.

The most prominent and well understood ageing mechanism acting on power modules is thermo-mechanical stress caused by temperature swings across material layers (metals, semiconductors, and ceramics) with mismatching coefficients of thermal expansion (CTE). Over time, these differentials in thermal expansion can lead, first to minor, then growing, structural degradation and, ultimately, premature failure, reducing the module's lifespan [3]. Power module architectures can be categorized based on the presence or absence of a baseplate. In baseplate–less power modules, the backside of

the direct bonded copper (DBC) substrate is directly mounted onto the heatsink via a thermal interface material (TIM), with a relatively high thermal resistance. In baseplate power modules, by contrast, the baseplate is soldered to the DBC substrate, facilitating heat spreading and reducing temperature peaks. Consequently, baseplate power modules tend to exhibit lower overall thermal resistance. However, the addition of a Cu or AlSiC baseplate considerably increases module cost.

Several alternatives have been proposed to overcome the thermal performance limitations of baseplate–less power modules. VINcoPress technology [4] offers an innovative solution that relies on pressing silicone potting gel between the substrate

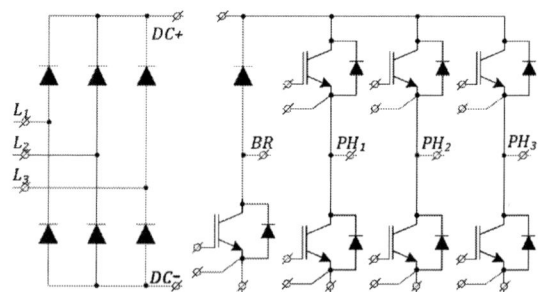

Fig. 1: Power Integrated Module (PIM) topology combined by: a three-phase passive rectifier, a three-phase motor inverter and a brake chopper.

and the heatsink with a uniformly distributed pressure. This prevents uncontrolled forces from acting on the substrate and minimizes cracking risks during assembly.

The proposed investigation benchmarks different power module thermal technologies, focusing on power modules with and without a baseplate. The selected topology is the standard PIM, composed of (i) a three-phase passive rectifier, (ii) a three-phase motor inverter, and (iii) a brake chopper, as demonstrated in Figure 1.

Both power module solutions share the same specifications, chipset technology, and number of wires for bare-die bonding. The power module solutions are: (i) B0-SP12PMA100M7-LQ99A78T – a baseplate–less power module with improved gel press technology (VINcoPress) [4]; (ii) 30-F212PMA100M7-L880A79 – a power module with a baseplate.

2 Elevator Drive Load Mission Profile

The load mission profile captures an average elevator's operation in a standard office environment, based on the following specifications :

- 200 000 rides up and down per year;

- 1 s hold time before and after each ride with 30 A_{DC} covering actuation of the mechanical brakes;

- 0.5 s jerks before and after each acceleration and deceleration to smoothen the ride;

- 8 kHz switching frequency, f_{sw}.

Virtual junction temperatures are predicted using simulation approach by VINcoSIM, available on the Vincotech website. Because the inverter switches

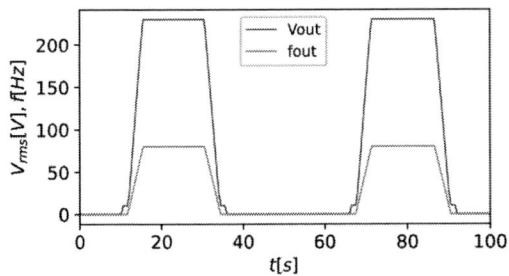

Fig. 2: Elevator mission profile: Voltage and motor frequency during a ride up and down.

and diodes are the only components that are subject to considerable heating, they are identified as the bottleneck components. As a result, the lifetime calculations are limited to them.

The motor's operational current, voltage, and frequency during a ride up and down are shown in Figure 2. It defines the load mission profile used as reference for the proposed investigation.

Figure 3, highlights features specific to the slow frequency and hold phases. It demonstrates that motor currents behavior during the selected load mission profile, focusing during elevator (i) upwards acceleration and (ii) downwards deceleration operations.

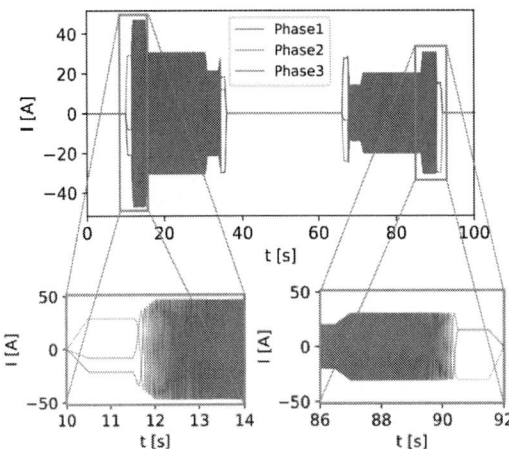

Fig. 3: Mission profile of the three current phases, zooming in on upwards acceleration and downwards deceleration to highlight features of the low frequency and the hold phases.

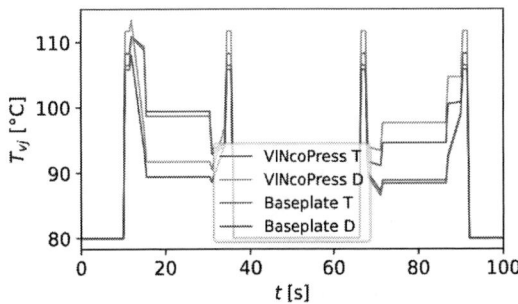

Fig. 4: Virtual junction temperature changes in the elevator mission profile.

These inputs were entered phase-by-phase into VINcoSIM [5] for both modules, resulting in the junction temperatures for all devices, depicted in Figure 4. It demonstrates the worst case scenario for the three phases during the hold time. When the motor operates at over 1 Hz, all the three phases experience roughly the same temperature fluctuations, just in different phases. During the hold times when the elevator is not in motion, the output frequency is zero and all phases carry direct current for one second. During this time, one phase carries as much current as the other two combined. This causes it to heat to the highest temperature and experience two times higher losses.

When the elevator travels upwards, temperatures are highest during acceleration across all phases, making it unnecessary to consider the impact of the subsequent hold time. During downwards travel, however, the maximum temperature attained during deceleration is inferior to the hottest phase during the subsequent hold time but higher than the other two phases, as show in Figure 5. The lifetime calculation must, therefore, consider the hold temperature that experiences the highest temperature swing, i.e., one phase, switch, and diode, or one-sixth of the sixpack. For the remaining five-sixths of the components, the highest temperature is reached at the end of deceleration.

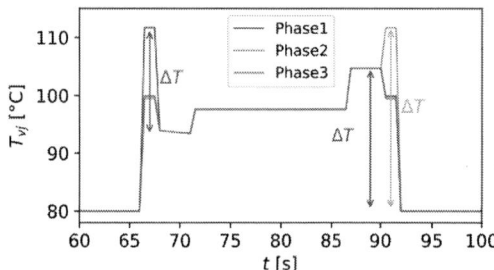

Fig. 5: Temperature differences of the three phases during hold time and downwards travel, VINcoPress inverter diode – arrows show the ΔT of the temperature swings.

Additionally, for upwards and downwards travel, an extra cycle must be added between the lowest temperature point of deceleration/acceleration and the highest temperature phase of the hold time. This cycle affects only 1/6 of the components; for the remaining 5/6, this additional temperature swing can be neglected when calculating ageing.

As a result, the following temperature cycles are used as an input for the lifetime calculation as demonstrated in Table 1.

The highest junction temperatures and temperature swings occur during the acceleration jerk when travelling upwards. In both modules, this cycle is

Tab. 1: Mission profile cycle parameters.

| Mission Profile | | Temperature ΔT [°C] | | | |
| | | VINcoPress | | Baseplate | |
Phase	N_{cyc}	IGBT	Diode	IGBT	Diode
up, acceleration jerk	200 000	30.9	33.4	30.9	28.1
up, end hold	33 000	14.7	21.2	13.6	17.2
down, deceleration	166 667	19.1	24.7	18.8	20.8
down, end hold	33 333	28.3	31.7	26.5	25.8
down, start hold	33 333	21.1	18.3	19.9	14.7

N_{cyc} represents the number of cycles per year.

Tab. 2: Top temperatures and R_{th} of bare-dies.

| Module | VINcoPress | | Baselate | |
Chip	IGBT	Diode	IGBT	Diode
T_h	110.9°C	113.4°C	110.9°C	108.1°C
R_{th}	0.46 K/W	0.68 K/W	0.43 K/W	0.58 K/W

responsible for roughly 2/3 of the ageing stress. As shown in the Table 2, the temperature is higher on diode considering the baseplate–less power module architecture. This is due to a particularity in its power module layout.

3 Power Cycling Model

3.1 Power Cycling Tests

Power cycling (PC) tests trigger the same ageing mechanisms that normally cause a power module to fail in the field. The reason is obvious: During operation, the module's semiconductor heats up. When it is switched off or used with less power, it cools back down. This results in a series of temperature cycles characterized by varying temperature swings (ΔT), maximum temperatures (T_{max}), and heating times (t_{on}). Because power modules are made of materials with different coefficients of thermal expansion, these repeated cycles lead to mechanical stresses that are most pronounced near the heat source – the junction. As a result, the bond wire and die solder experience the highest temperature swings.

To accelerate ageing, PC tests subject samples to temperature swings and maximum temperatures exceeding those observed in the field with shortened cycle lengths. The PC parameters used here are: ΔT=100 K, T_{max}=150°C, t_{on}=0.5 s. Tests were carried out following a constant current strategy, as this best characterises the application conditions [6]: Neither the current nor any other parameters were actively changed, allowing device temperatures to increase as solder ageing increases R_{th}. While PC tests were carried out on IGBTs and diodes, and all four test results are used in the lifetime calculations, only the results of the IGBTs are demonstrated.

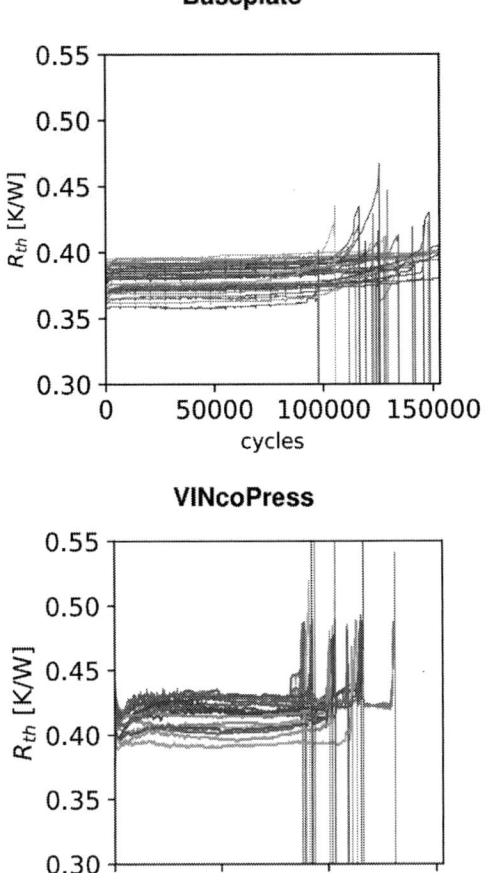

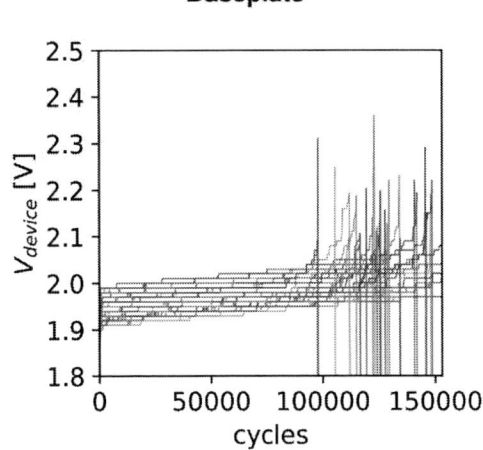

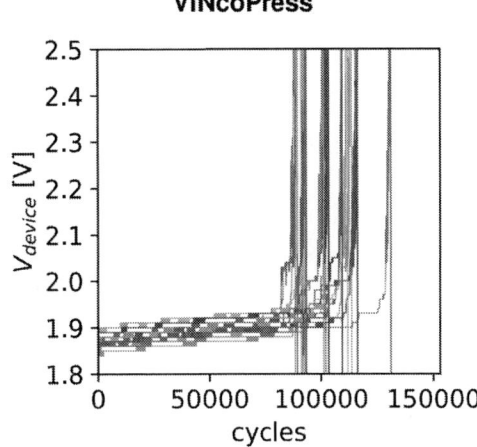

Fig. 6: Voltage V_{device} change during the PC test, indicating wire bond lift-off failures.

Fig. 7: Values of R_{th} change during the PC tests, indicating die attach solder and substrate ageing.

Values from R_{th} and V_{device} are constantly monitored throughout the PC tests, allowing to distinguish between the two leading failure causes: solder and wire bond ageing (see Figure 6 and Figure 7). Both tests results demonstrate that they are dominated by V_d failures, indicating that lifetime is limited by the wire bond and not the solder [3]. This is partly because Vincotech's improved solder material extends power cycling lifetime. When older solder materials, still prevalent in the industry, are used, solder-bond durability is reduced by half, making solder ageing a frequent bottleneck for longevity.

The Weibull distribution is fitted on the power cycling failures to facilitate further calculation:

$$F(t) = 1 - e^{-\left(\frac{t}{\eta}\right)^k} \tag{1}$$

where $F(t)$ is the unreliability or failure cumulative distribution function for a chip, t is time in cycles, k is the shape parameter, and η is the scale parameter.

Figure 8 shows the Weibull graphs for the PC tests on the IGBTs in both power module modalities.

As can be seen in the figures, the baseplate module offers a longer lifetime. Because of its higher scale parameter and lower shape parameter, the difference in lifetime depends on the desired reliability. At above 10% failure, this difference becomes significant and the confidence ranges separate. Both tests used the same chip size, test parameters, and wire bond number, type, and thickness. While both modules had to be slightly overloaded with current to reach the target ΔT, the baseplate module required slightly higher current (124% of the nominal current) due to its lower thermal resistance,

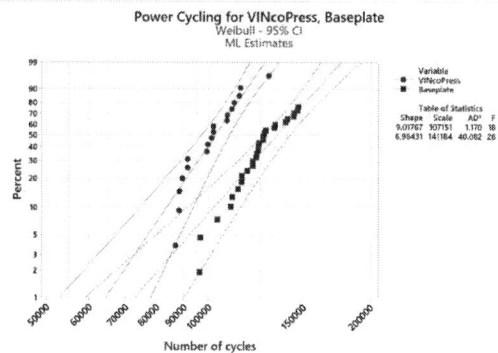

Fig. 8: Weibull diagram comparing power cycling test results for the switches in both module technologies with improved solder.

R_{th}. For comparison, the VINcoPress only needed only 113%. It should be noted that this difference does not explain the baseplate module's increased longevity.

An explanation for its increased longevity is that the baseplate – a thick layer of copper with high CTE – enhances the silicon chip's effective thermal expansion. This increases the stress on the die attach solder while decreasing that on the wire bond. Thanks to the increased lifetime of the improved solder, wire bond aging continues to be the leading failure cause, leading to improved overall lifetime.

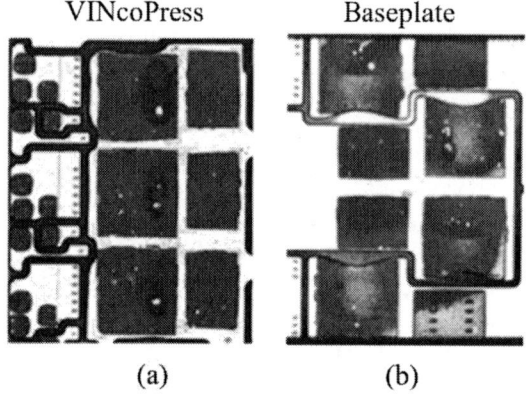

Fig. 9: C-SAM images of solder ageing after PC.

Figure 9 shows C-SAM images of the die attach solder layer of the samples after the PC test. The VINcoPress solders present only minor signs of ageing, visible as a small, slightly darker spot. While ageing is more pronounced in the baseplate solders, they remain below the failure criterion of a 20% increase in R_{th}. This supports the idea that the baseplate increases wire bond lifetime. Ageing of

the ceramic substrates can also be observed. It, too, is more pronounced on the baseplate module, but only where the chip is very close to the substrate trench. This substrate and solder ageing contributes both to the increase in R_{th}. Nevertheless, wire-bond lift-off continues to be the main cause of failure [3].

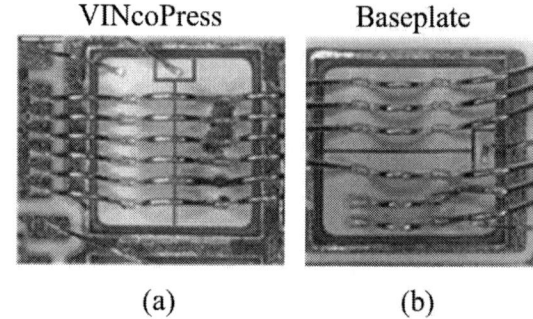

Fig. 10: Optical image of a typical chip's wire bonds after PC testing.

Optical images such as Figure 10.(a) reveal some more details about the failure mechanisms: all wire bond pads are lifted off, and on the left side, the damage is typically limited on it. These were probably the first bond pads to fail, as the current flowed from the left, and these bonds were the hottest in the beginning. On the right side, there is more damage, especially on the last one or two wires, as in the end, these wires had to carry more than 100 A. Even the chip top metallization is damaged in some cases. This also explains why the VINcoPress chips show some sign of solder ageing only on the right side, under the wire bonds: this area of the chip must have been overheated in the last part of the test.

Figure 10.(b) shows the top side of the same component as the top IGBT in Figure 9.(b). While the C-SAM is taken from the back, the optical image is from the front, and thus mirrored. These images reveal that the wires lifted off early on the top half of the chip, likely due to the voids that locally increased the thermal resistance R_{th} beneath that half, resulting in a higher temperature. Subsequently, the bottom half had to carry most of the load alone, leading to more severe solder ageing and, ultimately, molten wires. Visual examination further reveals that the chips have lost their shininess and taken on a grey hue, a phenomenon commonly referred to as aluminium reconstruction.

3.2 Lifetime Model and Calculation

The lifetime model that connects the PC results and the mission profile – representing the module's life cycle in the field – is inspired by the LESIT and models demonstrated in [2]. Hence, the lifetime model is given by:

$$N_f = A \cdot \Delta T^{\gamma} \cdot e^{\left(\frac{E_{\alpha}}{\kappa_B \cdot T_{mean}}\right)} \cdot t_{on}^{\beta} \quad (2)$$

The parameters that define lifetime dependence on the cycle parameters γ, E_{α}, and β, are determined by a series of PC tests under varying conditions. These parameters are specific to Vincotech's power module manufacturing process and technology, with different values than published by others, but in the same order of magnitude. Performing the lifetime calculation first requires determining the parameter A in equation (2). It is by using the scale parameter of the Weibull distribution as characteristic N_f as well as the IGBTs' and diodes' PC parameters for both module types. Next, the cycles from the mission profile in Table 1 are taken and the contribution of every cycle to ageing is calculated:

$$C_{cyc} = \frac{N_{cyc}}{A_{chip} \cdot \Delta T_{cyc}^{\gamma} \cdot e^{\left(\frac{E_a}{\kappa_B \cdot T_{meancyc}}\right)} \cdot t_{on.cyc}^{\beta}} \quad (3)$$

These contributions are then added up according to the Palmgren–Miner rule and entered into the Weibull–distribution to get the CDF of failure of a single chip:

$$F_{chip(T\ or\ D)} = 1 - e^{-t(\sum C_{cyc})^k}, \quad (4)$$

where k is the Weibull shape parameter and the scale parameter, η, is included in C_{cyc} through A. As all six-six IGBTs and diodes in the inverter have the same independent probability to fail, the reliability of the module, (R_m), at a given time, t, is given by the product of reliability of the chips, the six "T" switches, and the six "D" diodes:

$$1 - F_m = R_m$$
$$R_m = R_T^6 \cdot R_D^6 = (1 - F_T)^6 \cdot (1 - F_D)^6 \quad (5)$$

4 Results and Discussion

Figure 11 shows the results of the predicted lifetime for different thermal technologies for power modules. The results are also demonstrated in Table 3. The predicted lifetime for a baseplate module is somewhat higher than the baseplate–less module due to (i) the lower junction temperature experienced by chips for a given load mission profile and (ii) their higher power cycling lifetime.

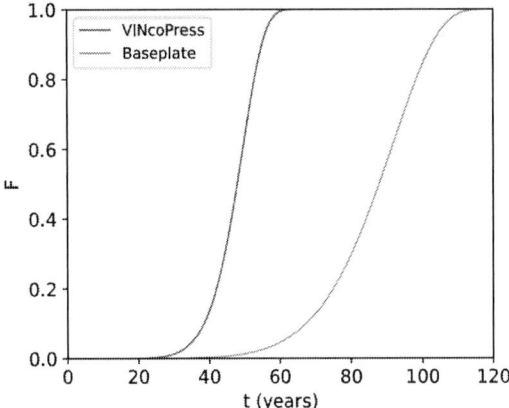

Fig. 11: Unreliability function of both power module thermal technologies.

Tab. 3: Predicted lifetime results.

Technology	VINcoPRESS	Baseplate
B1 lifetime	29.4 years	48.1 years

Nevertheless, the more cost-effective baseplate–less power modules still effectively provide the long lifetimes demanded by application requirements. In conclusion, the remarkable thermal management capabilities of baseplate–less power modules offered by VINcoPress technology make this configuration a recommended solution for demanding reliability-critical applications (more than 25 years), such as elevator drives.

4.1 Comparison to SAC

Most power module manufacturers use an SAC-type solder material, which consists of tin, 3-4% silver, and 0.5-0.7% copper. Because this solder has its limitations in terms of its power cycling capability, Vincotech has for some years been using an improved solder material that roughly double the power cycling lifetime. This gives an advantage both against previous Vincotech modules and competitors that still use SAC-type die attach solders. Figure 12 benchmarks their performance against each other. It demonstrates the power cycling lifetime extension provided by improved solder material combined with each power module thermal technology.

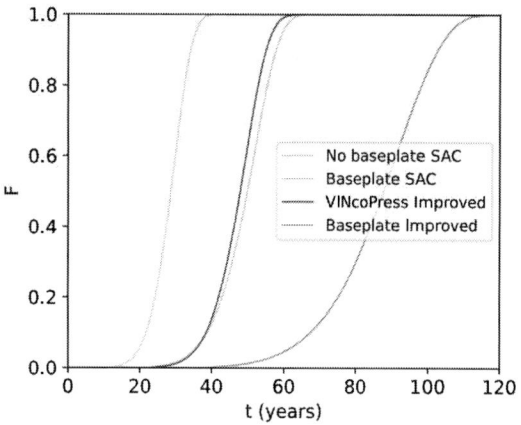

Fig. 12: Unreliability function for both module thermal technologies with traditional SAC-type and improved solder alloys.

5 Conclusion

In conclusion, while the power cycling lifetime was found to be slightly lower in baseplate-less modules, VINcoPress power modules remain a viable choice for meeting the stringent performance requirements of elevator drive applications. Moreover, considering their lower cost, this study positions power modules using VINcoPress technology as the solution of choice for reliability-critical applications such as the one investigated here.

References

[1] J. Lutz, "Packaging and reliability of power modules," in *8th International Conference on Integrated Power Electronics Systems–CIPS 2014*, 2014, pp. 1–8.

[2] R. Bayerer, T. Herrmann, T. Licht, J. Lutz, and M. Feller, "Model for power cycling lifetime of igbt modules - various factors influencing lifetime," in *5th International Conference on Integrated Power Electronics Systems – CIPS 2008*, 2008, pp. 1–6.

[3] S. Yang, D. Xiang, A. Bryant, P. Mawby, L. Ran, and P. Tavner, "Condition monitoring for device reliability in power electronic converters: A review," *IEEE Transactions on Power Electronics*, vol. 25, no. 11, pp. 2734–2752, 2010.

[4] M. Tauer, T. Jappe, T. Panyik, Z. Gyimothy, and M. Buza, "Power module packaging comprising direct pressed substrate with superior thermal performance and reliability," in *12th International Conference on Integrated Power Electronics Systems – CIPS 2022*, 2022, pp. 1–5.

[5] Vincotech, *VINcoSIM – The Integrated Simulation Environment*, https://www.vincotech.com/support-and-documents/simulation-software.html, 2021.

[6] S. Schuler and U. Scheuermann, "Impact of test control strategy on power cycling lifetime," in *PCIM Europe 2010; International Exhibition and Conference for Power Electronics, Intelligent Motion, Renewable Energy and Energy Management*, 2010, pp. 1–6.

PCIM Europe 2024, 11– 13 June 2024, Nuremberg   DOI: 10.30420/566262405

Fail-Operational LLC Topologies with Fault-Tolerance Integrated Redundant Capabilities

Aswathy M. Prince, Ayman Ayad
Vitesco Technologies GmbH, Germany

Corresponding author: Aswathy M. Prince, aswathy.marangattu.prince@vitesco.com
Speaker: Aswathy M. Prince, aswathy.marangattu.prince@vitesco.com

Abstract

The paper presents a novel fail-operational (FO) strategy designed for HV/LV DC-DC converters in Autonomous Drives (AD). Its primary objective is to ensure high availability through standalone fault-tolerant capability as well as redundancy for maintaining compliance with the Automotive Safety Integrity Level (ASIL) D, the highest safety standard. The proposed approach is implemented in Input-Parallel Output-Parallel (IPOP) LLC converters, where one of the two modules possesses inherent fault-tolerant capability. Each LLC module is designed to operate at 400 / 12 V and 1.8 kW, with an input voltage range spanning from 220 V to 440 V. Results indicate consistent power supply, soft switching capabilities both pre- and post-fault occurrence, and efficiencies exceeding 96 %.

1 Introduction

In response to escalating concerns regarding global warming and air pollution, collaborative efforts between governmental bodies and industries have aimed to curtail carbon dioxide (CO_2) emissions. Consequently, electro-mobility (E-mobility) has emerged as a focal point owing to its notably low CO_2 emissions, finding extensive application within the transportation sector. However, the deployment of E-mobility applications, encompassing more electric aircraft, ships, vehicles, and propulsion spacecraft, poses significant risks to human lives and property in the event of electrical power system failures. Hence, ensuring the reliability of E-mobility systems stands as a paramount concern, mandating swift remedial measures in case of failure [1]-[2]. This research focuses on enhancing the efficiency, safety, and reliability of battery electric vehicles (BEVs) to increase their appeal to potential customers. Technological advancements in the automotive sector are driving the development of more compact and efficient DC-DC converters, which are crucial for transforming high voltage (HV) DC power from the vehicle's battery into the lower voltage (LV) required by various electrical components. These components include auxiliary loads like infotainment systems, lighting, sensors, air conditioning, and various driver assistance functions, highlighting the importance of improving BEV technology for broader market acceptance. [3].

This study focuses on identifying the primary sources of failure in HV primary-side semiconductors and resonant capacitors within converters, which together account for 72% of total failures in this converter type. Semiconductor faults on the primary side, particularly open circuit (OCFs) or short-circuit (SCFs) faults in Si/SiC-based devices, contribute significantly to converter damage. The study specifically investigates faults in HV side switches due to their fivefold higher probability of failure compared to secondary side switches. Fault-tolerant strategies are crucial to ensure reliability, including expedited fault detection to minimize shoot-through durations and prompt deactivation of both high-side and low-side switches upon SCF detection. Additionally, proper remedial action, such as isolating faulty switches and reconstructing the converter, is essential to maintain stable output voltage and enable normal operation during unforeseen SCFs, thereby averting system failures [5].

Numerous articles have proposed solutions to address reliability concerns, categorized into additional switch [6]–[8], redundant converter [9], [10], and reconfiguration [11]–[14] strategies. In situations where SCFs occur in the primary switch of a converter, output voltage regulation is maintained by incorporating an additional switch. This approach, discussed in reference [6], involves adding a switch and TRIAC in parallel with the original switch, while references [7] and [8] propose integrating a switch and fuse into a buck converter. These methods use extra switches to transform the converter into a buck-boost configuration after a switch failure, thus ensuring single fault-tolerant capability. However, the reliance on TRIACs or fuses is essential for isolating faulty switches. Redundant converters [9]-[10], favored for their robust fault-tolerant features, can be configured with "n+k" redundancy, where "n" represents active converters for power transfer and "k" denotes redundant converters for fault tolerance. They operate in either cold or hot redundancy modes, with cold redundancy involving inactive redundant converters during normal operation and hot redundancy ensuring continuous operation even after a failure. While robust designs often employ all converters simultaneously to simplify control, high-efficiency designs may optimize efficiency by activating converters based on load currents. While utilizing redundant converters enhances fault tolerance and availability, there's a trade-off concerning the number employed: increasing redundancy for better availability can reduce power density and increase costs.

A reconfiguration approach has been proposed for full-bridge (FB) converters to overcome limitations associated with redundant converters [11]–[14]. In the event of a SCF, the FB cell transforms into a half-bridge configuration by deactivating the complementary switch on the faulty leg. While this approach achieves single fault-tolerant capability with only one converter, it may introduce additional losses and increased current stress under fault conditions. To address the need for high reliability in E-mobility applications, an LLC converter with double fault-tolerant capabilities is proposed, incorporating two diodes on the secondary side to withstand double SCFs and potentially eliminating the need for redundant converters [14].

Consequently, reconfiguration offers high power density and low cost in [11]–[14]. Nevertheless, the equivalent circuit of the converters undergoes changes following a fault, leading to reduced voltage gain or a narrower operating frequency range under fault conditions. Consequently, the efficiency and dynamic performance of the converter are compromised in [11]–[14]. The concepts outlined in [15] discuss two primary solutions: redundant converters and stand-alone fault-tolerant converters. Both strategies play a crucial role in

enhancing system availability, especially in applications intolerant to power interruptions.

This paper proposes fault-tolerance integrated redundant solutions to combine the benefits of standalone fault-tolerance mechanisms, such as rectifier or transformer reconfiguration, with redundancy to ensure high availability of power supply, crucial for safety-critical applications. In an IPOP converter featuring two redundant LLC modules, one functions as a standalone fault-tolerant module, while the other serves as either a standard FB LLC module or a diverse HB LLC module. The analysis focuses on cold redundancy, where the second module activates only after the complete failure of the first redundant module with standalone fault-tolerant capability. Section 2 provides a brief overview of FB LLC converter, while Section 3 discusses IPOP fault-tolerant LLC with rectifier reconfiguration and a generic redundant second module, as well as IPOP fault-tolerant LLC with transformer reconfiguration and a diverse redundant second module, including the determination of switching frequency. Simulation results are presented in Section 4, followed by the conclusion in Section 5.

2 Normal Full-Bridge LLC

The LLC converter, comprising a series inductor, a shunt parallel inductor, and a series capacitor, is classified under the series-parallel resonance converters (SPRC) family, offering various advantageous features such as soft switching, galvanic-isolation, high power density and high efficiency.

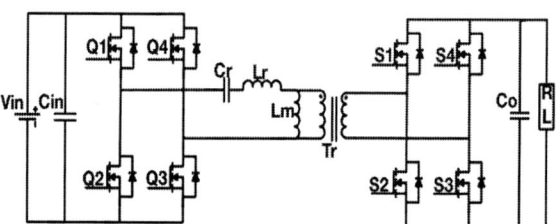

Fig. 2   Full-Bridge LLC converter.

A typical schematic of an LLC converter is depicted in Fig. 1. Unlike traditional pulse width modulation (PWM) based converters, LLC converters are governed by frequency modulated PWM signals, enabling output voltage regulation solely by adjusting the switching frequency (fsw). The switches Q1, Q3, and Q2, Q4 operate in a complementary manner to generate a square-wave voltage waveform, with a standard fixed 50 % duty cycle. A minimal dead-time is typically introduced between the activation and deactivation of the complementary switching pair to prevent cross-conduction and facilitate ZVS and ZCS. The resonant tank comprises the resonant capacitance (Cr) and two inductances—the series resonant inductance (Lr) and the transformer's magnetizing inductance (Lm). Upon applying a square waveform from the switching H-bridge to the LLC resonant tank, the tank generates a resonant sinusoidal current, scaled by the transformer. Subsequently, the synchronous rectifier circuit rectifies the AC signal, and the output capacitor filters the rectified signal to produce a DC voltage. Due to the presence of three distinct reactive elements Lr, Lm, and Cr - the LLC circuit can resonate at two different frequencies. However, optimal efficiency necessitates operation near the main resonance frequency (fr) formed by Lr and Cr. Depending on the load condition or input voltage variation, the converter can operate at, below, or above the Fr frequency to meet gain requirements while maintaining ZVS.

Analyzing the LLC converter yields nonlinear transfer functions from the physical model. To simplify the design, a common approach [16] known as the first harmonic approximation (FHA) technique is utilized, assuming that power transfer from the source to the load correlates entirely with the fundamental component of voltage and current. Consequently, all current and voltage waveforms are assumed to be purely sinusoidal, with FHA also providing the necessary conditions for achieving ZVS on the primary side MOSFETs and ZCS on the LV side MOSFETs. The normalized voltage gain function can be determined by Eq. (1):

$$G\left(Q_e, L_n, f_n\right) = \left| \frac{L_n * f_n^2}{\left[\left(L_n+1\right) * f_n^2 - 1\right] + j\left[\left(f_n^2 - 1\right) * Q_e * L_n * f_n\right]} \right| (1)$$

$$Q_e = \frac{\sqrt{\frac{L_r}{C_r}}}{R_e} \qquad L_n = \frac{L_m}{L_r} \qquad f_n = \frac{f_{sw}}{f_r} \quad (2)$$

where G (Qe, Ln, fn): Normalized voltage gain function of the resonant tank; Qe: Quality factor; Re: Equivalent load resistance of the resonant tank; Ln: Inductance ratio; fn: Normalised frequency; fsw: Switching frequency; fr: Series resonant frequency [16]. The normalized voltage-gain function, as described by Eq. (1), provides insight into the behavior of G in relation to the factors fn, Ln, and Qe. While fn serves as the control variable in the gain function, Ln and Qe act as fixed param-

eters once their physical characteristics are determined during design. G is subsequently adjusted by fn post-design completion. It's imperative to note that the value of G is not less than zero, evident from its derivation via the modulus operator, which encompasses both real and imaginary numbers. However, in this context, only the magnitude of these numbers is relevant [16].

3 Proposed Fail-Operational LLC

The proposed strategy, implemented in IPOP LLC converters as shown in Fig. 3 and Fig. 5, integrates fault-tolerant capability within one of the two modules. The first module, equipped with standalone fault-tolerant capability, utilizes the failed switches to reconfigure the converter's primary side into a HB configuration upon detecting a fault. This reconfigured mode allows the converter to continue supplying nominal power to loads without compromising its core functionality. The other module operates as a standard FB LLC converter. Both modules are interconnected in an IPOP configuration, offering fault tolerance in either hot or cold redundancy modes. The minimum number of redundant converters required is determined either by compliance with reliability standards or by considering factors such as power density and cost. The proposed topology integrates multiple fault-tolerant capabilities or FO mechanism through standalone fault tolerance and redundancy, ensuring maximum availability.

3.1 IPOP Fault-Tolerant LLC with Rectifier Reconfiguration

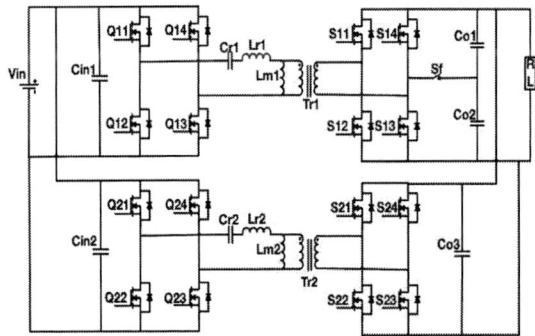

Fig. 3 Proposed IPOP fault-tolerant LLC with rectifier reconfiguration

Figure 3 shows the proposed IPOP fault-tolerant LLC with rectifier reconfiguration. The proposed converter consists of two modules, with the first module resembling a FB LLC converter structure but featuring a distinct secondary side. In this module, the output capacitor comprises a split capacitor bank, capable of functioning as a voltage doubler if necessary. This design renders the first module resilient to both OC and SC faults, making it suitable for various semiconductor faults on the primary side. The fundamental concept of this topology is to maintain converter operation following a fault, such as SC or OC of a semiconductor, by reconfiguring it into an HB-LLC instead of a FB-LLC, thereby isolating the faulty leg. The damaged device serves as a circuit path, converting the FB-LLC into an HB-LLC, as depicted in Fig. 4.

Upon reconfiguration, the output voltage decreases by half, necessitating a reconfigurable rectifier to maintain a constant output voltage. The analysis in this paper focuses on cold redundancy, wherein if the first module with standalone fault-tolerant capability fails, the standby LLC can be utilized, employing cold redundancy. This FO mechanism, leveraging standalone fault tolerance and redundancy, has no adverse effects on the converter's efficiency. The current paths of both the primary and secondary sides in this configuration during positive and negative excitation cycles are illustrated in Fig. 4(a) and Fig. 4(b), respectively.

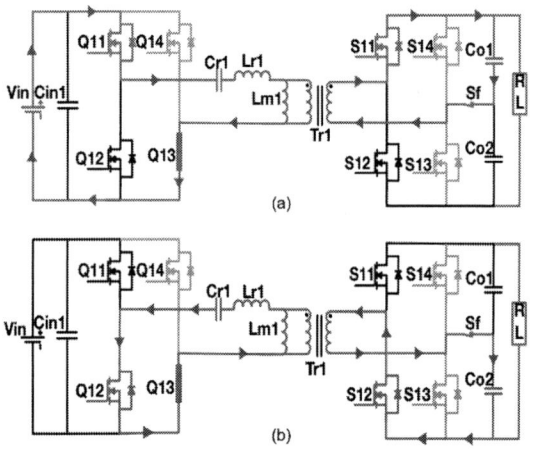

Fig. 4 (a) Current flow direction during positive excitation voltage, (b) during negative excitation voltage.

To illustrate the converter's operation, a hypothetical fault scenario is considered where switch Q13 in Fig. 4 experiences permanent damage, resulting in a detected SC fault. Consequently, the converter cannot function normally as the H-bridge functionality is compromised. In response, the reconfiguration strategy is activated. The failed switch Q13 is repurposed as a passive uncontrollable path that remains closed within the converter. To transition the primary side of the converter into

a HB configuration, switch Q14 is permanently kept open during this mode of operation. Switches Q11 and Q12 are then controlled with complementary PWM signals, effectively forming a HB circuit. To restore the converter's output power, the capacitive voltage divider is activated. During the rectification cycle, the upper capacitor of the voltage doubler circuit charges during the positive half, while the lower capacitor charges during the negative half. Switches S13 and S14 remain unutilized and ungated in this mode of operation. Consequently, this configuration transforms the FB rectifier into a full-wave rectifier by effectively bypassing switches S13 and S14.

In the event of a fault detection on the HV side, switch Sf remains in a closed or conducting state. Notably, Sf is configured as a bidirectional power switch - a back-to-back connected N-MOSFETs in common-drain configuration - which remains inactive under normal operating conditions. It is important to highlight that the fault in switch Q13 is merely an example, and in the occurrence of an SC fault at any other switch in the primary side H-bridge, a different switch must be opened to facilitate reconfiguration to an HB. Furthermore, in case of an OC fault detected in any of the switches, the other switch in the same branch must be continuously maintained in an on-state to enable the transition from a FB to an HB configuration.

3.2 IPOP Fault-Tolerant LLC with Transformer Reconfiguration

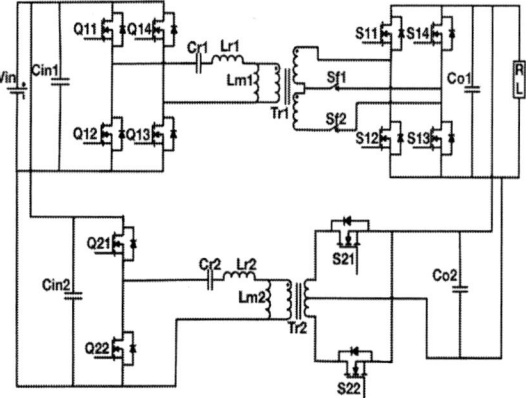

Fig. 5 Proposed IPOP fault-tolerant LLC with transformer reconfiguration.

Figure 5 depicts the proposed IPOP fault-tolerant LLC with transformer reconfiguration. In this topology, the first module comprises a FBLLC with a FB synchronous rectifier, while the second module features a HB LLC with a center-tapped full-wave synchronous rectifier. This arrangement introduces redundancy with diversity compared to the previous topology depicted in Fig. 3, which employed generic redundancy. As previously stated, the choice of redundancy is contingent upon specific requirements and applications.

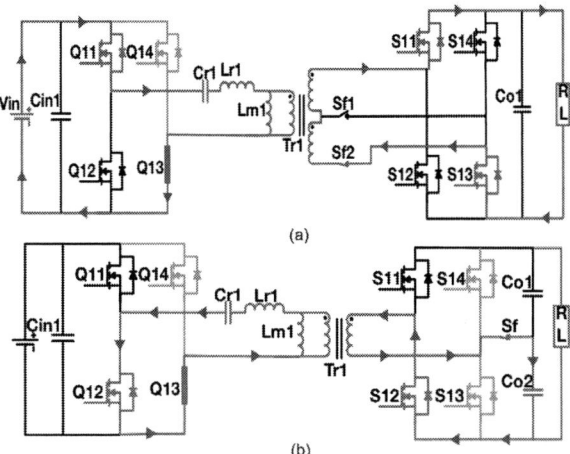

Fig. 6 (a) Current flow direction during positive excitation voltage, (b) during negative excitation voltage.

Figure 6 illustrates the current path of the primary and secondary sides in the HB configuration during both the positive and negative halves of the cycle. During normal operation, Sf1 remains closed or in a conducting state, receiving the necessary gating, while Sf2 is in an open or non-conducting state. In the event of a fault, assuming reconfiguration of the primary side from a FB to a HB configuration has occurred in module 1 of the LLC, a corresponding reconfiguration of the secondary bridge is necessary to achieve fault tolerance. In this strategy, Sf2 connects the bottom node of the transformer (Tr1) to the middle node of the second leg on the secondary or LV side, while Sf1 remains open or ungated. This configuration ensures that Sf2 is in a closed or conducting state, thereby enabling gating.

Due to the inherent characteristics of the topology, an HB configuration can only generate half of the AC square-wave voltage compared to an FB configuration, resulting in a halved output voltage from the rectifier. To address this limitation, the transformer employed in the converter is utilized. In the reconfiguration mode, both the secondary and tertiary windings of the transformer are energized and utilized simultaneously. This effectively increases the effective turns ratio of the transformer from n:1 to n:2 (where n is the turns ratio), thereby

doubling the AC voltage output from the secondary side of the transformer to the rectifier stage. Consequently, after reconfiguration, the nominal voltage and power can be restored at the converter output.

3.3 Determination of Frequency Range

The system specifications, encompassing parameters like minimum and maximum input voltage, output voltage, and output current, along with the resonant tank parameters designed through optimization techniques based on these input parameters, are summarized in Table 1.

	Parameter	Value	Unit
Input	Voltage (nom.)	400	V
	Voltage range	240-440	V
	Switching frequency range	52-242	kHz
	Resonant frequency	100	kHz
Output	Voltage (nom.)	12	V
	Voltage range	10-14	V
	Power (nom.)	1.8	kW
	Current	150	A
	Voltage ripple	0.1	V
Resonant tank	Inductance	7.7	µH
	Magnetizing Inductance	54.3	µH
	Capacitance	328	nF
	Minimum gain	0.8	
	Maximum gain	1.4	

Table 1 Design parameters.

It is noteworthy that the overall system gain is not same as the desired resonant tank gain. Consequently, following the determination of the number of turns ratio and the minimum and maximum gain of the resonant tank derived from the system specifications, the subsequent step involves plotting the gain against the normalized switching frequency while considering various Q values. This analysis aids in identifying the optimal Qmax value that meets the gain requirement. Furthermore, careful selection of the operating frequency range is essential to meet power density targets, ensure compatibility with cooling technologies, and maintain EMC system compatibility.

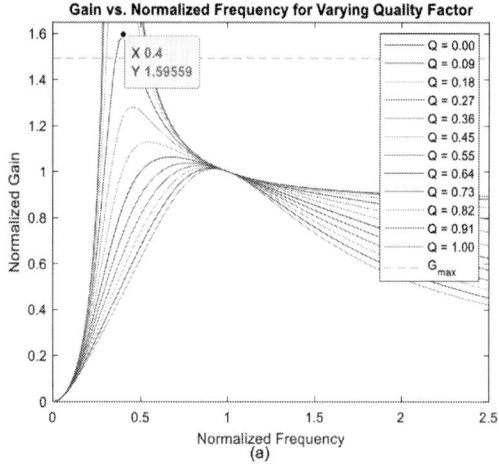

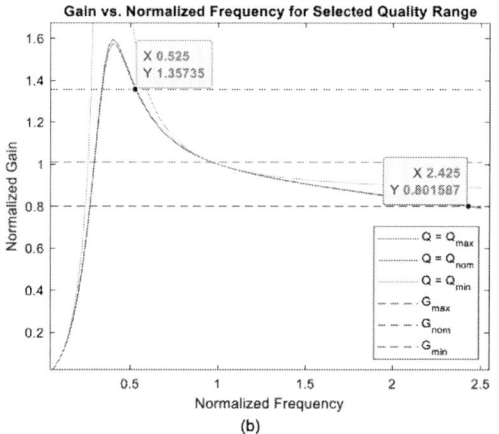

Fig. 7 (a) Gain vs. normalized frequency for varying quality factor, (b) Gain vs. normalized frequency for selected quality range

Figure 7(a) illustrates the gain versus normalized frequency for varying quality factors, aiding in the selection of an appropriate Qe for a fixed Ln value of 7. The chosen Qe should yield a gain curve where the peak just exceeds the Gmax line and here it is 0.27. Additionally, Fig. 7(b) depicts the plot of gain versus normalized frequency for a selected quality range, clearly indicating the minimum and maximum switching frequencies. The

minimum switching frequency occurs when the Gmax line intersects the right side of the Qmax curve, while the maximum switching frequency occurs when the Gmin line touches the right side of the Qmax curve.

The LLC converter is associated with two main frequencies: the resonant frequency, fr, formed by Lr and Cr alone and is set at 100 kHz, and the pole frequency, fp, formed by Lr, Lm, and Cr. Due to the inclusion of Lm, the converter's switching frequency at peak resonance (fsw) becomes load-dependent, ranging within fp ≤ fsw ≤ fr as the load varies. While Eqn. (3) shows the fr and it holds true regardless of the load, Eqn. (4) describes fp accurately only under no-load conditions. Below fp, the converter loses its ZVS capabilities as it transitions from the inductive to the capacitive region. Irrespective of the combination of Ln and Qe employed, all curves converge at the point (fn, G) = (1, 1). This point corresponds to fn = 1, or fsw = fr, denoting series resonance where XLr − XCr = 0 at fr. Consequently, the voltage drops across Lr and Cr be-comes zero, allowing the input voltage to be directly applied to the output load, resulting in a unity voltage gain of G = 1.

$$fr = \frac{1}{2\pi\sqrt{LrCr}} \tag{3}$$

$$fp = \frac{1}{2\pi\sqrt{(Lr + Lm)Cr}} \tag{4}$$

$$fp = fn \times fr = 0.4 \times (100 \times 10^3) = 40kHz \tag{5}$$

$$fsw_{min} = fn \times fr = 0.525 \times (100 \times 10^3) = 52.5kHz \tag{6}$$

$$fsw_{max} = fn \times fr = 2.425 \times (100 \times 10^3) = 242.5kHz \tag{7}$$

Equations (5), (6), and (7) specify fp, the minimum switching frequency, and the maximum switching frequency, respectively, for the FO LLC converter with each module rated at 1.8 kW and 400/12 V. The desired inductive operation region, where the LLC can regulate output over wide line and load variations with minimal switching frequency variation and excellent efficiency, lies between these minimum and maximum switching frequencies. In typical power-supply converter design, primary considerations include line regulation, load regulation, and efficiency. Both types of regulation are achieved through voltage-gain adjustment, typically accomplished via frequency modulation in an LLC converter. Efficiency is a significant ad-vantage of LLC converters, achieved by minimizing switching losses through maintaining primary-side

ZVS and secondary-side ZCS throughout the operating range. Ensuring operation within the recommended design area, typically on the right side of the resonant gain curves region, facilitates ZVS. Selecting smaller Ln values increases peak gain, keeping the design out of the capacitive region, although this may elevate conduction losses due to higher magnetizing current. Optimal Ln values are typically around 5, coupled with Qe values around 0.5, to achieve moderate gain curve slopes without sacrificing efficiency. Iterative refinement is often necessary to fine-tune these parameters.

4 Simulation Results

The FO LLC topologies are simulated using PLECS Blockset 4.7.5, incorporating a soft start mechanism and focusing on cold redundancy. Each LLC converter module is configured for 1.8 kW and 400/12 V operation, with a fr set at 100 kHz. The duration of the dead time is set to 300 ns, and the turns ratio of the transformer is selected as 32. Variable frequency control is achieved through a voltage-controlled oscillator, allowing precise frequency adjustment in response to voltage fluctuations to ensure stable and efficient system operation.

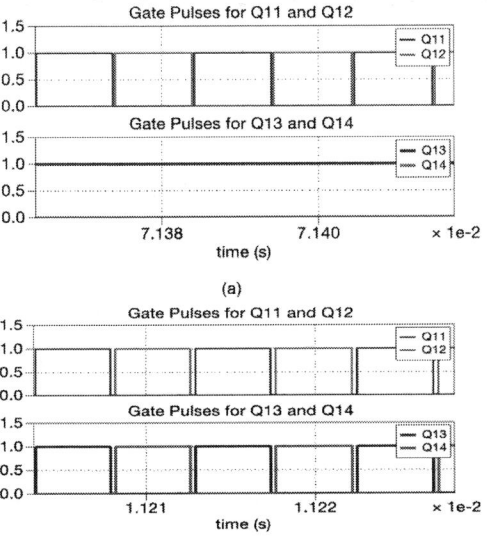

Fig. 8 Gate pulses (a) for fault-free, and (b) under fault operation.

The switches Q11, Q13, Q12, and Q14 are controlled by complementary PWM signals with a fixed 50 % duty cycle. The simulation duration is set at 0.1 seconds (100 ms), with a simulated SC fault introduced at 0.05 s on Q13. The main objective of the study is to conduct FO analysis, as-

sessing the impacts of this fault on both the primary and secondary sides of the standalone fault-tolerant LLC topologies with rectifier and transformer reconfiguration.

Figures 8(a) and 8(b) illustrate the gate pulses for the HV side switches Q11, Q12, Q13, and Q14. At 50 ms, a SC fault is induced in Q13. To reconfigure the primary side of the converter into a HB, Q14 is permanently open-circuited. With Q13 now in a permanent short-circuit state, it functions as part of the current commutation path, serving as a closed link. Consequently, Q11 and Q12 are gated with complementary PWM control signals to form the HB circuit.

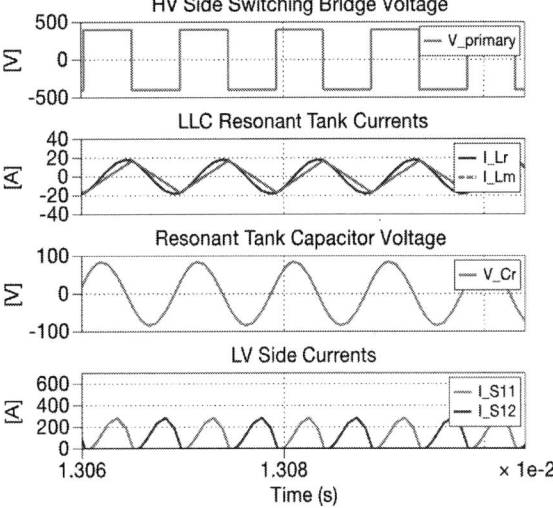

Fig. 9 Simulation waveforms for fault-free operation.

Figures 9 and 10 illustrate the crucial switching diagrams for fault-free and fault-induced operations. In Figure 9, the initial waveform depicts the bidirectional voltage of 400 V on the HV side. Subsequently, the resonant tank currents are presented, where I_{Lr} represents the resonant tank inductor current, I_{Lm} denotes the transformer's magnetizing inductor current and the secondary-side MOSFET current (I_s). Notably, the primary-side current is the summation of the I_{Lm} and I_s referred to the primary. However, as I_{Lm} the flows solely in the primary side, it does not contribute to the power transferred from the primary side source to the secondary side load. During this mode, the fsw aligns with the fr. Upon the turn-off of switches Q11 and Q13, the resonant current diminishes to the magnetizing current's value, halting further power transfer to the secondary side. By appropriately delaying the turn-on time of switches Q12 and Q14, the circuit achieves HV side ZVS and enables a soft commutation of the synchronous rectifier on the LV side. The achievement of ZCS is clearly evident in the fourth waveform of Fig. 9. A notable advantage of the LLC converter topology is its potential for substantially reduced switching losses, primarily through HV-side ZVS. However, it's crucial to note that ZVS considerations determine the recommended design area to be limited to the right side of the resonant gain curves. ZVS cannot be attained in the capacitive region, and without it, switching losses escalate, nullifying the efficiency benefits of using an LLC converter. The third waveform shown in Fig. 9 corresponds to the voltage across the resonant tank capacitor, denoted as V_{Cr}. In fault-free or normal operation, it's evident that operation at series resonance yields a single operational point. To accommodate both input and output variations, the fsw must be adjusted away from resonance. FO LLC converters can be designed by ensuring either fsw ≥ fr or fsw ≤ fr, or by varying fsw on either side around fr. Optimal operation lies in the vicinity of the series resonant frequency, where the advantages of the LLC converter are maximized, serving as the design objective.

the end of the driving pulse width, power transfer ceases despite the continued presence of I_{Lm}. Operating below the series resonant frequency still allows for primary ZVS and ZCS. However, secondary-side MOSFETs operate in discontinuous current mode, necessitating higher circulating current in the resonant tank circuit and resulting in increased conduction losses. It's important to note that primary ZVS may be compromised if the switching frequency decreases too much, leading to elevated switching losses and associated challenges. The transition from FB to HB operation at the same output power effectively and introduces a nearly doubled peak-to-peak AC voltage with a DC offset in the V_{Cr}. Operating in this mode increased the current stress on the Lr and the voltage stress on the Cr. Therefore, careful consideration of the resonant tank's physical components, including the resonant inductor current rating and capacitor voltage rating, is essential when designing the proposed converter.

Figures 11(a), (b), and (c) display waveforms illustrating the output voltage, output current, and input voltage under fault-free and fault conditions, along with variable frequency modulation for three distinct input voltage levels: 220V, 400V, and 440V. This demonstration aims to showcase the achievement of FO mechanism across different input voltage levels. The setup incorporates a safety switch (Sf) for LV side reconfiguration within FO-LLC topologies. The safety switch utilizes a Bi-directional Power Switch (BPS) Configuration with Back-to-Back Connected N-MOSFETs in Common-Drain Configuration.

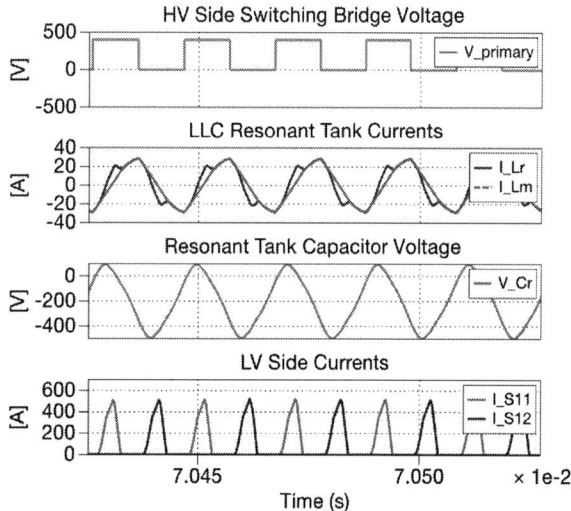

Fig. 10 Simulation waveforms (a) for fault-free, and (b) under fault operation.

Figure 10 illustrates the key waveforms under fault conditions. In the first waveform, the transition to HB operation results in unidirectional primary side voltage due to the HV side FB functioning as HB. Next, the resonant tank currents with ZVS capabilities, even under fault conditions, are depicted. The third waveform depicts the V_{Cr}, while the fourth waveform shows the I_S with ZCS capability. As the I_{Lr} decreases to the I_{Lm} value before

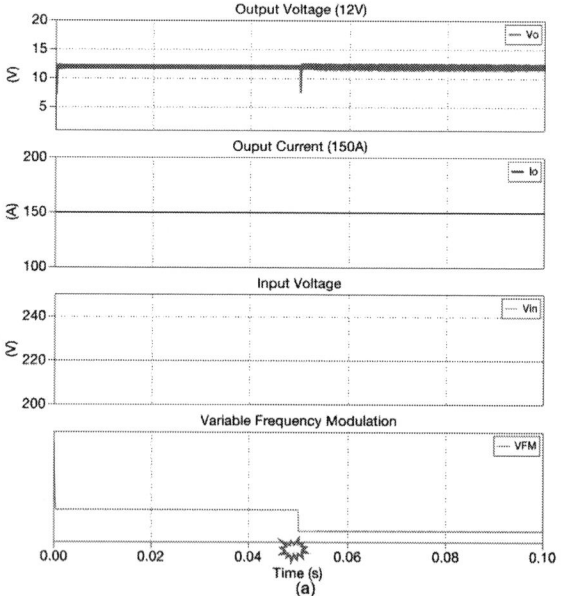

PCIM Europe 2024, 11– 13 June 2024, Nuremberg   DOI: 10.30420/566262405

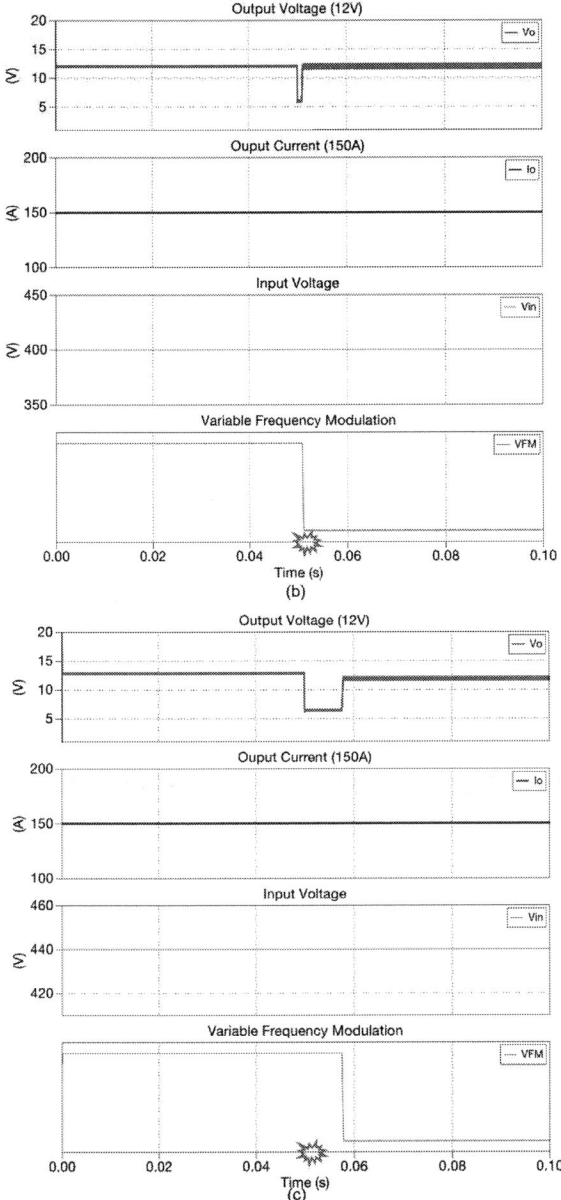

Fig. 11 Output waveforms at 220V, 400V and 440V.

At 50 ms (0.05 s), a simulated SC fault is induced in the LLC module under a constant current load, which is rectified after 5 ms. The initial waveforms of Figures 11(a), (b), and (c) depict the output voltage of the LLC converter module with input voltages of 220V, 400V, and 440V, respectively. In Figures 11(a) and (b), both before and after the fault occurrence, the output voltage remains steady at 12V. However, for input voltages of 220V and 400V, a voltage dip of 6.5V lasting approximately 0.45 ms and 0.8 ms, respectively, occurs at the moment of fault insertion. In Figure 11(c),

with an input voltage of 440V, normal operation yields an output voltage of 12.8V. However, at the time of fault insertion, a 7V voltage dip persists for 8 ms. Following reconfiguration, the output voltage stabilizes at 12V. After the fault is resolved, Figures 11(a), (b), and (c) show no spikes or peaks, attributed to the utilization of back-to-back MOSFETs. Additionally, a noticeable increase in voltage ripple is observed post-reconfiguration. Despite significant variations in the source voltage or other circuit impedances, the constant current load of 150A ensures consistent current delivery.

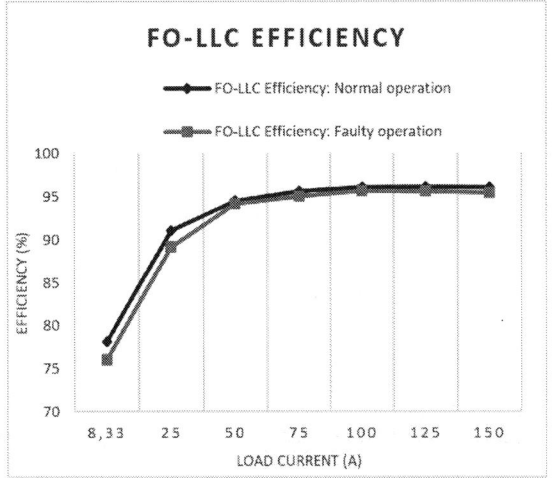

Fig. 12 FO-LLC Efficiency vs. load current.

Efficiency simulations, depicted in Fig. 12, was conducted for the FO-LLC under various load conditions using PLECS software. This analysis incorporated the thermal characteristics of automotive SiC HV side and LV side switches, specifically the ROHM-SCT3017ALHR for HV and VISHAY-SQJQ160E for LV, both operating at a case temperature of 85°C. Results revealed that efficiencies exceeding 90 % were consistently achieved for loads surpassing 300 W under both normal and faulty conditions. Despite encountering additional conduction and switching losses due to the presence of back-to-back connected MOSFETS under faulty conditions, the efficiencies remained at 96 % and higher, highlighting the robustness and resilience of the system.

5 Conclusion

This paper presents advanced fail-operational (FO) LLC topologies specifically designed for automotive drive (AD) applications, emphasizing two distinct reconfiguration approaches to maintain consistent output voltage and current. Through the design and simulation of a 1.8 kW 400 / 12 V LLC

2858

module using PLECS software, it is confirmed that the converter promptly restores its output voltage despite a fault. The proposed method ensures the preservation of soft-switching operation and high efficiency for a wide range of input voltage variations, even if a fault occurs in one of the HV side switches, significantly enhancing the reliability, availability, and safety of future AD systems while eliminating the need for external backup batteries. Future efforts will concentrate on crafting a compact high-current LLC converter utilizing the outlined fail-operational architecture and constructing an experimental prototype to validate the concept's feasibility in compliance with ASIL standards.

References

[1]. N. Lee, J. Y. Lee, Y. J. Cheon, S. K. Han, and G. W. Moon, "A highpower-density converter with a continuous input current waveform for satellite power applications," IEEE Trans. Ind. Electron., vol. 67, no. 2, pp. 1024–1035, Feb. 2020.

[2]. A. Barzkar and M. Ghassemi, "Electric power systems in more and all electric aircraft: A review," IEEE Access, vol. 8, pp. 169314–169332, Sep. 2020.

[3]. Kotb, R.; Chakraborty, S.; Tran, D.-D.; Abramushkina, E.; El Baghdadi, M.; Hegazy, O. Power Electronics Converters for Electric Vehicle Auxiliaries: State of the Art and Future Trends. Energies 2023, 16, 1753.

[4]. L. Ferreira Costa and M. Liserre, "Failure Analysis of the dc-dc Converter: A Comprehensive Survey of Faults and Solutions for Improving Reliability," in IEEE Power Electronics Magazine, vol. 5, no. 4, pp. 42-51, Dec. 2018, doi: 10.1109/MPEL.2018.2874345.

[5]. S. Yang, A. Bryant, P. Mawby, D. Xiang, L. Ran, and P. Tavner, "An industry-based survey of reliability in power electronic converters," IEEE Trans. Ind. Appl., vol. 47, no. 3, pp. 1441–1451, May/Jun. 2011.

[6]. E. Jamshidpour, P. Poure, E. Gholipour, and S. Saadate, "Single-switch DC–DC converter with fault-tolerant capability under open- and shortcircuit switch failures," IEEE Trans. Power Electron., vol. 30, no. 5, pp. 2703–2712, May 2015.

[7]. J. L. Soon, D. D. C. Lu, J. C. H. Peng, and W. Xiao, "Reconfigurable nonisolated DC–DC converter with fault-tolerant capability," IEEE Trans. Power Electron., vol. 35, no. 9, pp. 8934–8943, Sep. 2020.

[8]. J. L. Soon and D. D. C. Lu, "Design of fuse–MOSFET pair for fault tolerant DC/DC converters," IEEE Trans. Power Electron., vol. 31, no. 9, pp. 6069–6074, Sep. 2016.

[9]. H. Wang, X. Pei, Y. Wu, and Y. Kang, "A general fault-tolerant operation strategy under switch fault for modular series–parallel DC–DC converter," IEEE J. Emerg. Sel. Topics Power Electron., vol. 9, no. 1, pp. 872–884, Feb. 2021:

[10]. P. Tu, S. Yang, and P. Wang, "Reliability- and cost-based redundancy design for modular multilevel converter," IEEE Trans. Ind. Electron., vol. 66, no. 3, pp. 2333–2342, Mar. 2019.X. Pei, S. Nie, and Y. Kang, "Switch short-circuit fault diagnosis and remedial strategy for full-bridge DC–DC converters," IEEE Trans. Power Electron., vol. 30, no. 2, pp. 996–1004, Feb. 2015.

[11]. X. Pei, S. Nie, and Y. Kang, "Switch short-circuit fault diagnosis and remedial strategy for full-bridge DC–DC converters," IEEE Trans. Power Electron., vol. 30, no. 2, pp. 996–1004, Feb. 2015.

[12]. L. Costa, G. Buticchi, and M. Liserre, "A fault-tolerant series-resonant DC–DC converter," IEEE Trans. Power Electron., vol. 32, no. 2, pp. 900–905, Feb. 2017.

[13]. J. Huang, G. Chen, and H. Shi, "A cost-reliability trade-off fault-tolerant series-resonant converter combining redundancy and reconstruction," IEEE Trans. Power Electron., vol. 36, no. 10, pp. 11543–11554, Oct. 2021.

[14]. G. Chen, L. Chen, Y. Deng, K. Wang, and X. Qing, "Topology reconfigurable fault-tolerant LLC converter with high reliability and low cost for more electric aircraft," IEEE Trans. Power Electron., vol. 34, no. 3, pp. 2479–2493, Mar. 2019.

[15]. A. M. Prince and A. Ayad, "Analysis of Fault-Tolerant and Fully Redundant HV/LV DC-DC Converters for Battery Electric Vehicles," 2023 25th European Conference on Power Electronics and Applications (EPE'23 ECCE Europe), Aalborg, Denmark, 2023, pp. 1-9, doi: 10.23919/EPE23ECCEEurope58414.2023.10264479.

[16]. Infineon Technologies Asia Pacific, Design Guide for LLC Converter with ICE2HS01G (Application Note, V1.0, July 2011).

Thermal and Reliability Optimization of Clips in SiC MOSFET Power Modules

Zexiang Zheng[1], Jianwei Lv[1], Yiyang Yan[1], Jiaxin Liu[1], Yue Wu[2], Zhipeng He[2], Cai Chen[1], Yong Kang[1]

[1] State Key Laboratory of Advanced Electromagnetic Technology, Huazhong University of Science and Technology, China

[2] State Key Laboratory of HVDC, Electric Power Research Institute, CSG, China

Corresponding author: Cai Chen, caichen@hust.edu.cn
Speaker: Zexiang Zheng, zzx_huster@hust.edu.cn

Abstract

Clip-bonded power modules are extensively utilized due to their high current-carrying capacity and reliability. In the paper, an optimized thick copper-molybdenum clip was proposed to achieve low thermal resistance and high reliability. The impact of clip thickness and material on thermal performance and reliability is analyzed through simulations and experiments. The transient dual-interface method is employed to measure the thermal resistance of modules with clips of different thicknesses and materials. The thermal resistance of the modules with a 3 mm Cu clip and a 3 mm Cu15Mo85 clip is reduced by 13.4% and 10.4% respectively. Moreover, finite element simulations of temperature cycling are conducted to assess the reliability of the solders with different clips. The solder reliability with the 3 mm Cu15Mo85 clip is higher than that with the 3mm Cu clip and is close to that with the 0.1 mm Cu clip.

1 Introduction

Silicon carbide (SiC) power modules are widely used and are vital components in high-power applications. The bonding method between SiC chips and substrates is one of the main factors affecting the performance of power modules [1]. Aluminum (Al) wire bonding is currently the most widely used bonding method due to its low cost and high maturity. However, the additional parasitic inductance and on-resistance introduced by aluminum wires is a problem for SiC applications [2]. Moreover, the high operating temperatures of SiC chips can cause aluminum wires to break and fall off [3]. Therefore, Cu Clip bonding has gained widespread adoption, due to its high reliability and low on-resistance. Meanwhile, the clip enhances the heat dissipation capability of the module by providing additional channels on the chip surface [4]-[6].

There is some research on Clip bonding in power modules [4]-[8]. In [4] and [5], the electrical and thermal characterization of copper-clip-bonded IGBT modules was investigated. It shows that Clip can reduce the on-resistance, parasitic inductance, and thermal resistance of power modules compared to bonding wires. In [6], a novel Clip structure is proposed to improve the dynamic current sharing characteristics of parallel SiC chips. In [7], the lifetime of Cu Clips with different thicknesses under power cycling was researched. It shows that the lifetime of the power module will decrease as the Clip thickness increases. In [8], a highly reliable and low-cost Clip bonded module is proposed.

However, comprehensive research on the thermal and reliability of clip-bonded power modules is still lacking. The thermal conductivity of traditional Cu Clip is not ideal due to the limited thickness of the Clip. Moreover, the CTE mismatch between Cu and SiC leads to increased stress in the solder layer. Therefore, it is necessary to simultaneously optimize the thermal and reliability design of Clip bonded modules.

In this paper, a finite element simulation-based comprehensive optimization of the Clip's thermal and reliability was performed. Subsequently, a thick copper-molybdenum (CuMo) alloy clip is proposed, which can reduce the thermal resistance of the module while enhancing the reliability of the solder layer. Thermal resistance testing is performed to verify the optimization of thermal performance.

PCIM Europe 2024, 11– 13 June 2024, Nuremberg   DOI: 10.30420/566262406

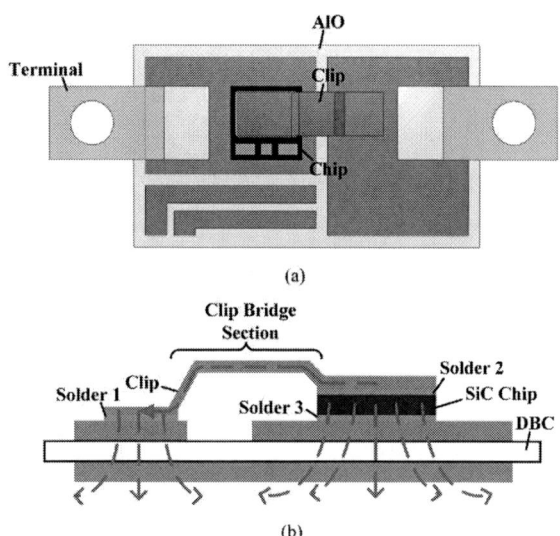

(a)

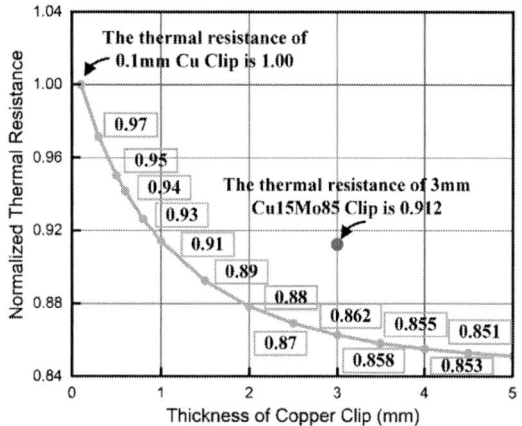

(b)

Fig. 1 (a) Baseline power module. (b) Cross-section of Clip bonding.

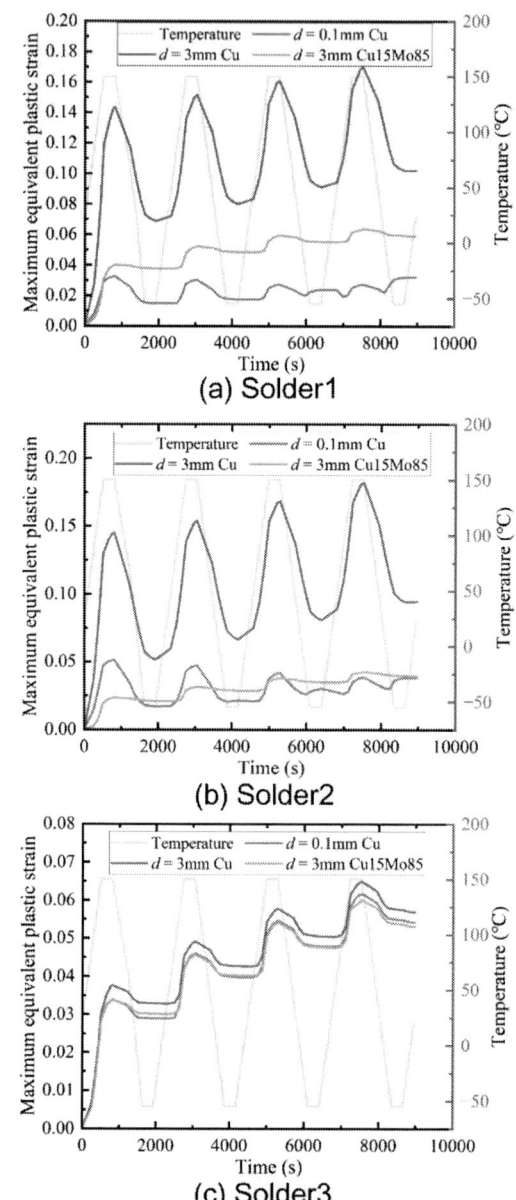

(a) Solder1

(b) Solder2

(c) Solder3

Fig. 3 Maximum equivalent plastic strain of solders of each clip-bonded module under TC.

Fig. 2 Normalized thermal resistance comparison.

2 Clip-Bonded Module Overview

In Fig. 1, a single-chip SiC power module is shown. The materials of DBC layers are Cu/Al₂O₃/Cu. The type of SiC MOSFETs is S4661 from ROHM with a small size of 5 mm ×5 mm ×0.15 mm. The rated voltage, current, and ON-state resistance of MOSFETs are 1200 V, 130 A, and 11 mΩ, respectively. The power source pad is connected to the direct bond copper (DBC) substrate using a clip. The power source pad can be soldered directly due to the gold plating on the surface. The solder used in the power module is Sn–3.0Ag– 0.5Cu (SAC305).

In the baseline module, the influences of the clip thickness and Molybdenum (Mo) content in copper-molybdenum (CuMo) alloy on the module's thermal resistance and solder reliability are explored.

3 Clip Optimization - Thermal Resistance and Reliability

3.1 Thermal Optimization

Clip bonding contributes to the reduction of thermal resistance. The calculation expression of thermal resistance (R_{th}) is as follows:

2861

PCIM Europe 2024, 11– 13 June 2024, Nuremberg DOI: 10.30420/566262406

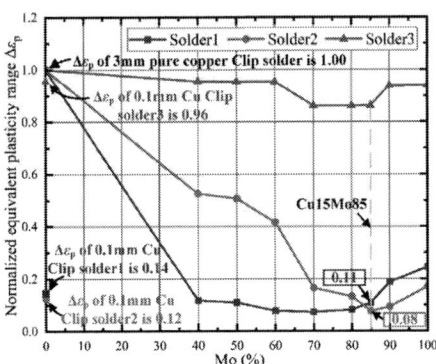

Fig. 4 Normalized equivalent plastic strain range comparison of different material clips.

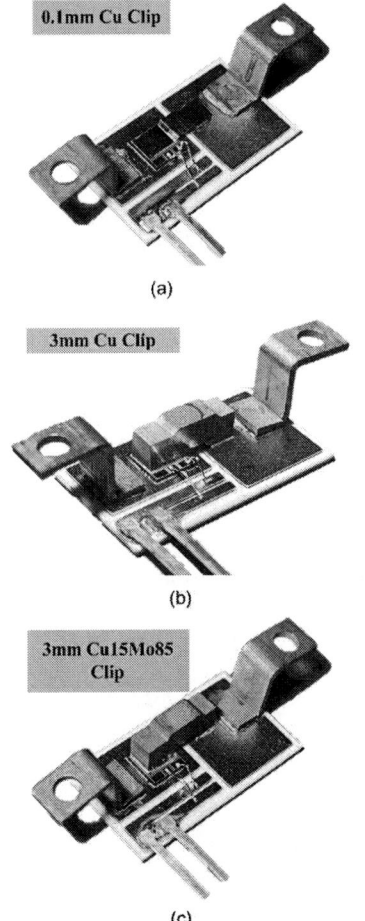

Fig. 5 Modules of different clips. (a) 0.1mm Cu clip. (b) 3mm Cu clip. (c) 3mm Cu15Mo85 clip.

$$R_{\mathrm{th}} = \frac{l}{\lambda A} \qquad (1)$$

where l is the length of the thermal conduction path, A is the area of thermal diffusion, and λ is

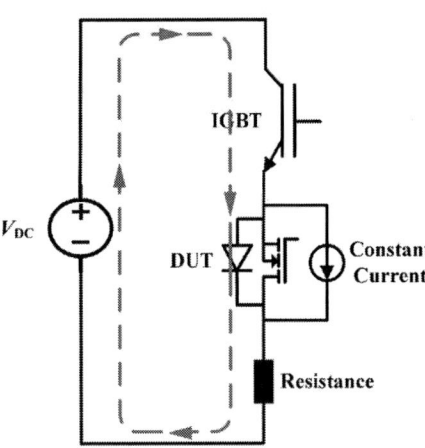

Fig. 6 Thermal Resistance Measurement Diagram

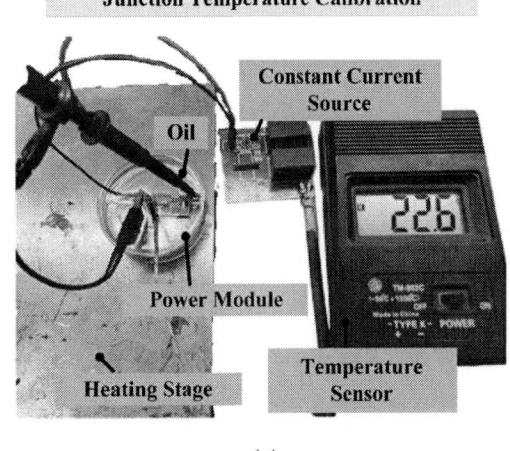

(a)

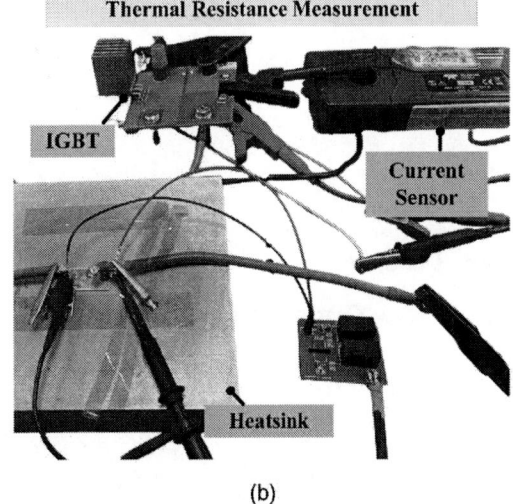

(b)

Fig. 7 (a) Temperature calibratio. (b) Thermal resistance test.

2862

PCIM Europe 2024, 11– 13 June 2024, Nuremberg   DOI: 10.30420/566262406

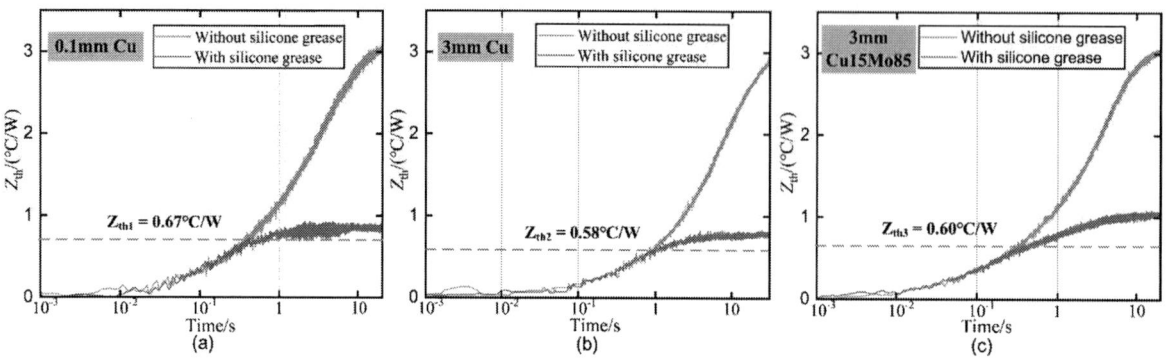

Fig. 8 Test results of thermal resistance. (a) 0.1mm Cu clip. (b) 3mm Cu clip. (c) 3mm Cu15Mo85 clip.

the thermal conductivity of the material. The thickness of the clip is the main factor affecting A. Therefore, the thickness plays a crucial role in determining the effectiveness of this top heat dissipation pathway. To evaluate the impact of clip thickness on thermal performance, a steady-state thermal finite element method (FEM) simulation is conducted using ANSYS Workbench. In the simulation, only the top source electrode with dimensions of 3 mm×4 mm is welded with the clip. The SiC MOSFET is the heat source and 60 W is applied. The equivalent coefficient of convective heat transfer was assumed to be 4000 W/m²·K.

When the thickness of the clip is 0.1 mm, the thermal resistance is normalized to 1. Fig. 2 illustrates the relationship between clip thickness and normalized thermal resistance. The cross-sectional area of the bridge section in the middle of the thin clip is smaller than the welding area on both sides, restricting the heat flow through the clip. As the clip thickness increases, the cross-section of the clip bridge also increases, resulting in a gradual decrease in thermal resistance. When the clip thickness reaches 3 mm, the thermal resistance is reduced by 13.8% compared to the 0.1 mm clip. However, further increases in thickness do not significantly improve thermal resistance. The weld area of the chips between the clips may limit this. Therefore, to maximize thermal resistance improvement, a clip thickness of 3 mm is chosen.

3.2 Reliability Optimization

Solder is the major failure-inducing factor for power modules. To assess solder reliability, a temperature cycling (TC) FEM simulation is performed. According to the Joint Electron Device Engineering Council (JEDEC) standard, -55 °C to 150 °C, was taken as the range of cycle temperature. The ramp rate was 15 °C /min, and the total duration was 8962 s. The Anand model is

employed to describe the stress and strain characteristics of the solder. The obtained range of equivalent plastic strain ($\Delta\varepsilon_p$) from the last temperature cycle is used to evaluate solder lifetime, according to the Coffin-Manson model [9]. The number of cycles (N_f) to failure is as follows [10]:

$$N_f = c_1 \Delta\varepsilon_p^{-c2} \qquad (2)$$

where c_1 and c_2 are constants related to material properties. It shows that $\Delta\varepsilon_p$ can qualitatively evaluate the lifetime of the solder layer. A smaller value of $\Delta\varepsilon_p$ corresponds to a longer solder lifetime. Although there are some assumptions on the lifetime model, all simulations are performed under the same conditions. Therefore, the results of different simulations can be compared to characterize the properties of the different structures.

Fig. 3 illustrates the $\Delta\varepsilon_p$ of three solders (solder1, solder2, and solder3, as shown in Fig.1(b)) under TC for two different Cu clips: a 0.1 mm Cu clip and a 3 mm Cu clip. It shows that the solder's lifetime of the 3 mm Cu clip is greatly reduced due to the increase in clip thickness. To enhance solder reliability, a proposed solution involves substituting the Cu clip with a CuMo clip. The $\Delta\varepsilon_p$ values of the three solders of pure copper clip are normalized to 1, and the normalized $\Delta\varepsilon_p$ values of the CuMo clip solders with varying Mo content are presented in Fig. 4. The CuMo clip demonstrates a considerable decrease in $\Delta\varepsilon_p$ for solder1 and solder2 compared to the Cu clip, while solder3 remains unaffected. Notably, Cu15Mo85 is identified as the optimal material for improving solder reliability based on $\Delta\varepsilon_p$ with different Mo contents. The green curve in Fig. 3 represents the equivalent plastic strain curve of the solder for the Cu15Mo85 clip, exhibiting reliability similar to that of the 0.1 mm Cu clip, as depicted in Fig. 4. Moreover, As shown in Fig. 2, the 3 mm

2863

Cu40Mo60 clip provides a notable enhancement in thermal resistance compared to the 0.1 mm Cu clip.

4 Experimental Results

In this section, the thermal resistances of the module bonded with 0.1 mm Cu clip, 3 mm Cu clip, and 3 mm Cu15Mo85 clip (as shown in Fig.5) were measured to verify the improvement of the thermal resistance by the transient dual-interface method [11]. The measurement principle is illustrated in Fig. 6. When the auxiliary IGBT is conducted, the device under test (DUT) is heated by a high current. A constant current source (10mA) is connected in parallel across the terminals of the DUT. When the IGBT is turned off, the junction temperature of the chip is obtained from the module's drain-source voltage. The junction temperature is calibrated in an oil bath, as shown in Fig. 7(a). The thermal impedance test platform is shown in Fig. 7(b). The load is resistive. The DC voltage is 200 V. Two different working conditions can be set: coating with thermal silicone grease and without silicone grease. The separation point of the curves of the two working conditions is the thermal resistance from the junction to the case (Z_{th}).

The thermal resistance results are shown in Fig. 8. Based on the separation points observed in the curves, the thermal resistances of the 0.1 mm Cu clip, 3 mm Cu clip, and 3 mm Cu15Mo85 clip are 0.67 °C/W, 0.58 °C/W, and 0.60 °C/W respectively. Compared with the module thermal resistance of the 0.1 mm Cu clip, the thermal resistance of the 3 mm Cu clip and 3 mm Cu15Mo85 clip is reduced by 13.4% and 10.4% respectively.

It indicates that the thicker Clip can effectively reduce the thermal resistance of the module. Although the CuMo Clip increases thermal resistance compared to pure copper Clips, it remains a favorable choice due to the reliability improvements. Morover, considering the poorer solderability of CuMo alloy, electroplating copper on the surface of the Clip could be considered for practical applications.

5 Conclusion

This paper presents the comprehensive optimization of the clip structure and materials to reduce module thermal resistance while improving the reliability of the solder layers. To effectively minimize thermal resistance, a 3 mm thick clip is utilized. Additionally, to alleviate the solder lifetime problem caused by the thick Cu clip, the optimized Cu15Mo85 material is used. Simulations and experiments show that the thermal resistance of the module bonded with 3 mm Cu clip and 3 mm Cu15Mo85 clip is 13.4% and 10.4% lower than that of 0.1 mm Cu clip. The solder layer life of the 3 mm Cu15Mo85 clip module is similar to that of the 0.1 mm Cu clip module.

References

[1] C. Chen, F. Luo, and Y. Kang, "A review of SiC power module packaging: Layout, material system, and integration," in *CPSS Transactions on Power Electronics and Applications*, vol. 2, no. 3, pp. 170-186, Sept. 2017.

[2] Y. Yan, C. Chen, Z. Wu, J. Guan, J. Lv, and Y. Kang, "A High Power Density Double-Side-End Double-Sided Bonding SiC Half-Bridge Power Module," in *IEEE Transactions on Transportation Electrification*, vol. 9, no. 2, pp. 3149-3163, June 2023.

[3] C. Durand, M. Klingler, D. Coutellier and H. Naceur, "Power Cycling Reliability of Power Module: A Survey," in *IEEE Transactions on Device and Materials Reliability*, vol. 16, no. 1, pp. 80-97, March 2016.

[4] W. -H. Chi, H. -C. Chen and H. -K. Liao, "High Reliability Wire-Less Power Module Structure," *2018 13th International Microsystems, Packaging, Assembly and Circuits Technology Conference (IMPACT)*, Taipei, Taiwan, 2018, pp. 71-74.

[5] Q. Zhu, A. Forsyth, R. Todd, and L. Mills, "Thermal characterisation of a copper-clip-bonded IGBT module with double-sided cooling," *2017 23rd International Workshop on Thermal Investigations of ICs and Systems (THERMINIC)*, Amsterdam, Netherlands, 2017, pp. 1-6.

[6] L. Wang, T. Zhang, and F. Yang, "Cu Clip-Bonding Method With Optimized Source Inductance for Current Balancing in Multichip SiC MOSFET Power Module," in *IEEE Transactions on Power Electronics*, vol. 37, no. 7, pp. 7952-7964, July 2022.

[7] R. Kato, Y. Ikeda, and T. Yamada, "The Effect of the Power Module Structure (Cupper Clip Thickness) on Reliability Under Power Cycling Test," *2023 IEEE CPMT Symposium Japan (ICSJ)*, Kyoto, Japan, 2023, pp. 184-187.

[8] X. Li, Y. Wang, M. Packwood, H. Neal, S. Malasani and M. Morshed, "A Standard Low Voltage Power Module Platform with High

Reliability and Low Cost," *PCIM Europe 2023;* Nuremberg, Germany, 2023.

[9] P. Ning, J. Liu, D. Wang, Y. Zhang, and Y. Li, "Assessing the Fatigue Life of SiC Power Modules in Different Package Structures," in *IEEE Access*, vol. 9, pp. 12074-12082, 2021.

[10] G. B. Thomas, J. Bressers, and D. Raynor, "Low cycle fatigue and life prediction methods," in *High Temperature Alloys for Gas Turbines*, Liège, Belgium, 1982, pp. 291_317.

[11] M. Pan, E. Deng, Y. Zhao, J. Liu, P. Liu, and Y. Huang, "Transient Thermal Impedance Measurement of HPD Power Modules," *2023 IEEE 6th International Electrical and Energy Conference (CIEEC)*, Hefei, China, 2023, pp. 2183-2187.

PCIM Europe 2024, 11– 13 June 2024, Nuremberg   DOI: 10.30420/566262407

Condition Monitoring of a GaN Full-Bridge by Means of Forward Voltage Measurements in Continuous Operation

Michael Vogt[1], Alexander Brunko[1], Gerrit Braun[2], Klaus Rigbers[2], Nando Kaminski[1]

[1] University of Bremen, Institute for Electrical Drives, Power Electronics, and Devices (IALB), Otto-Hahn-Allee NW1, 28359 Bremen, Germany
[2] SMA Solar Technology AG, Sonnenallee 1, 34266 Niestetal, Germany

Corresponding author: Michael Vogt, mvogt@uni-bremen.de
Speaker: Michael Vogt, mvogt@uni-bremen.de

Abstract

Due to their exceptional properties, wide bandgap (WBG) devices are steadily gaining ground in the field of power electronics. Various load profiles in automotive and photovoltaic applications expose WBG components to demanding load conditions, which promote the occurrence of trapping effects, thus leading to changes in the drain-source resistance $r_{DS,on}$. In fact, the change in $r_{DS,on}$ during operation at nominal power can be a crucial issue for the reliability of WBG components and can cause instability in the system. This study focuses on the monitoring of $r_{DS,on}$ of a gallium nitride (GaN) full-bridge in continuous 140 kHz switching operation under various operational cases.

1 Background: Trapping Effects in GaN

Wide bandgap (WBG) devices significantly reduce energy loss in the rapidly expanding field of power electronics, all while enhancing power density. Thus, weight-sensitive applications in the automotive industry and highly efficient systems such as photovoltaics benefit from the application of WBG technology.

The on-state resistance of a gallium nitride (GaN)-transistor plays a central role regarding the efficiency in power electronic applications. Therefore, it is important to understand the conditions and effects under which it increases. At a physical level, the variation of $r_{DS,on}$ is primarily caused by two mechanisms: Fundamentally, an increase in semiconductor temperature results in a reduction in electron mobility [1]. The second known effect is the so-called trapping effect, which will be discussed in more detail below.

In the device transitions from the on to the off-state (Fig. 1a), an effect occurs by which charges accumulate near the gate connection towards the drain side in the passivation layer. This effect is also referred to as surface trapping [2]. It is caused by material defects, which are primarily influenced

by historical stresses of the device and the production process itself. Wafer structures, epitaxial growth conditions, and other process steps impact the dynamic characteristics of the heterojunction semiconductor.

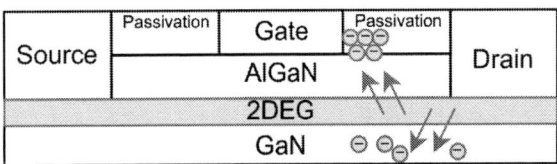

(a) Switching to off-state

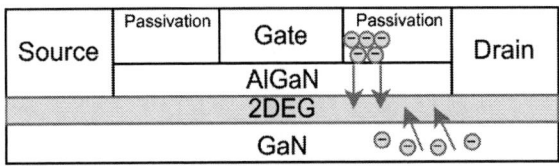

(b) Switching to on-state

Fig. 1: Trapping-effect [2]–[5].

Accumulation of the trapped charges can lead to depletion of the channel in the gate-drain region, resulting in a positive shift in the threshold voltage ($V_{GS,th}$), also known as virtual gate [5]. Another effect is that due to the trapped electrons, there is a weakening of the 2-dimensional electron gas

(2DEG), leading to a higher turn-on resistance in the transient turn-on phase. This is also referred to as the current-collapse phenomenon (Fig. 1b). Ultimately, this effect increases switching loss due to higher dynamic on-resistance and prolonged turn-on transition time, thus limiting the operating frequency range of the device [6].

To reduce the conductivity of the GaN buffer layer and increase the breakdown voltage, carbon (C) doping is a widely used technology solution. The acceptor traps created by C doping enable electrons to also settle in deeper layers of the buffer, contributing to an amplification of the current collapse and dynamic $r_{DS,on}$ during the off-state [7], [8]. This effect could lead to long-term degradation of the device [3].

2 Test Setup

An overview of the entire measurement setup is provided in the block diagram in Fig. 2. It should be noted that in this representation, blue arrows represent optically insulated signals, red arrows represent thermal quantities, and black arrows represent electrical signals. Fundamentally the setup can be divided into the full-bridge converter structure (yellow area) and the measurement board (blue area).

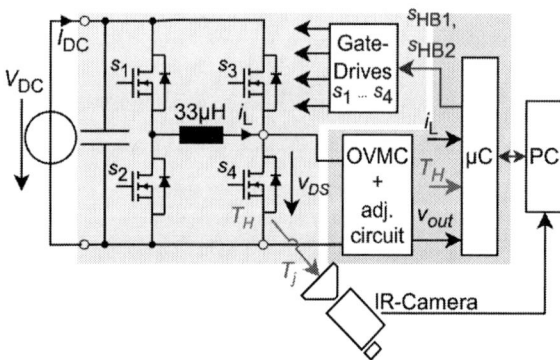

Fig. 2: Block diagram of the entire measurement setup. The full-bridge converter structure is within the yellow area. The blue area marks the developed measurement board.

The measurement board developed within the scope of this paper is used to control the full-bridge by providing pulse width modulation (PWM) signals at a carrier frequency of 140 kHz, which are then transmitted to the half-bridge drivers via optical fibres (see section 2.2). Furthermore, the measure-

ment board includes an on-state voltage measurement circuit (OVMC) (see section 2.1) to measure the drain-source voltage in the on-state ($v_{DS,on}$) of the low-side switch of one half-bridge. Additionally, a current transducer is used to measure the load current i_L, and a thermocouple voltage converter is used to measure the heat sink temperature T_H underneath the device under test (DUT) using a K-type sensor.

Apart from that, an infrared (IR)-camera is used to measure the junction temperature of the DUT ($T_j \approx T_{IR}$).

Additionally, a photo of the real test setup is shown in Fig. 3.

Fig. 3: Photo of the test setup. Similar to Fig. 2, the full-bridge converter structure is outlined in yellow, while the developed measurement board which sits on top of the converter is highlighted in blue. The red circles mark the location of the two half-bridges.

2.1 On-state Voltage Measurement Circuit

2.1.1 Main Circuit

In Fig. 4, the circuit configuration of the utilised on-state voltage measurement circuit (OVMC) is illustrated. This circuit design was not developed within the context of this paper; rather, it was adapted from [9]. Consequently, only a concise overview of its functionality is provided herein, with a more comprehensive description available in the mentioned paper.

2867

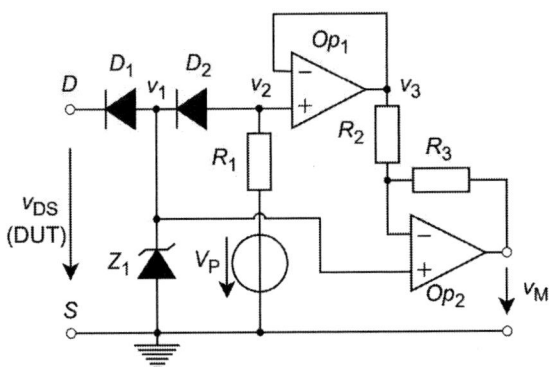

Fig. 4: On-state voltage measurement circuit (OVMC), as proposed in [9].

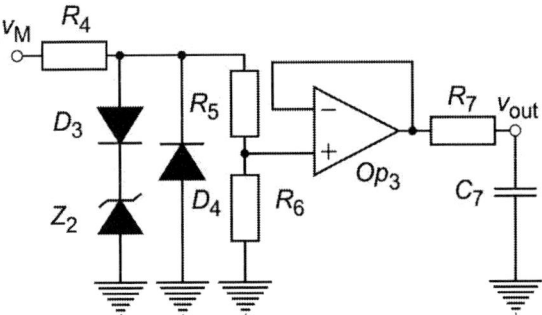

Fig. 5: Signal adjustment circuit.

During the off-state of the DUT a high potential difference v_{DS} may exist across the drain and source terminals. Diode D_1 effectively blocks this elevated voltage, ensuring node v_1 remains at a low potential. Conversely, when the DUT transitions into the conducting state, a current path is established between the drain and source, resulting in a small voltage drop. Upon fulfilling condition

$$v_{DS} < V_P - v_{D_1} - v_{D_2} \tag{1}$$

a current pathway is established from the voltage source V_P through resistor R_1, diode D_1, and diode D_2. As a consequence of the voltage drop across D_1, node v_1 attains a potential of $v_{DS} + V_{D_1}$. However, for the measurement of the voltage across the drain and source terminals (v_{DS}), it is necessary to eliminate the contribution of voltage v_{D_1}. To accomplish this, the identical diode D_2 is incorporated within the circuit. Given that both diodes derive their operating voltage from the common source V_P, the equality between V_{D_1} and V_{D_2} can be assumed. The functionalities attributed to Op_1 and Op_2 entail the subtraction of the diode voltage from potential v_1, therefore the output voltage of the circuit results in

$$v_M = v_{DS,on}. \tag{2}$$

2.1.2 Signal Adjustment Circuit

Since the analog digital converter (ADC) of the microcontroller (µC) requires an input signal between 0 V and 3.3 V, the measurement signal must be kept within this range (absolute maximum ratings between −0.3 V and 4 V [10]). For this purpose, the circuit from Fig. 5 is implemented. Diode D_3 and Zener diode Z_2 are connected to limit the upper voltage.

D_4 is a Schottky diode with a low forward voltage of 0.38 V and serves as the lower voltage limiter. The resistors $R_5 = R_6$ form an additional voltage divider, which halves the signal. Since the ADC must be supplied by a low-impedance current source to quickly charge the sample-hold capacitors, the signal is once again amplified by the voltage follower Op_3. R_7 and C_7 form an additional low-pass filter to provide the ADC with its sample and hold (SH) circuit with a stable voltage source for a short period. Due to the voltage divider, the output voltage v_{out} in the on-state corresponds to half of $v_{DS,on}$. Therefore, the following applies for the on-state:

$$v_{DS,on} = 2 \cdot v_{out} \tag{3}$$

2.2 Control of the PWM-Signals

The objective of the control is to generate a sinusoidal current i_L with a frequency of 50 Hz through the inductive load to simulate a quasi-real load.
The implementation of sinusoidal modulation is achieved by using two PWM functions, each driving a half bridge at a carrier frequency of 140 kHz. Although a full bridge consists of 4 controllable switches, only two control signals are required here, as there is one gate driver with only one input signal per half-bridge. This activates the low-side (LS) switch with a low signal at the input and the high-side (HS) switch with a high signal. The hardware dead time of the gate driver is set to 200 ns.
The duty cycle of the two PWM functions is varied sinusoidally over time as shown in Fig. 6. The crucial control parameter is the phase shift φ between the two PWM signals. This configurable parameter φ is a measure for the RMS value as well as the peak current $\hat{i}_L$ of the load current.

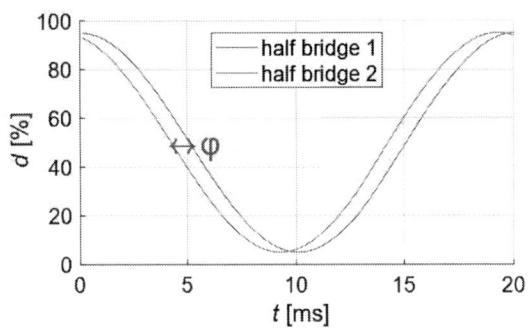

Fig. 6: Control of the duty cycle resulting in a sinusodial-like current waveform. The angle φ is the control variable.

3 Examined GaN Full-Bridge

The structure of the full-bridge under investigation consists of a newly developed hybrid circuit board structure [11]. On the one hand, there is a so-called insulated metal substrate (IMS) board, on which the GaN power semiconductor is soldered in a surface-mounted device (SMD) package [12], and on the other hand, there is a main board (made of usual FR4 material) on which the remaining components are installed, such as filter components, galvanically insulated gate driver, or other additional circuits. The connection between the IMS board and the main board is made via a ball grid array (BGA). A schematic cross-section of the structure is outlined in Fig. 7. The IMS board is attached to a heat sink using fastening screws and thermal paste. Additionally, a K-type thermocouple is attached into the heat sink to monitor its temperature during operation.

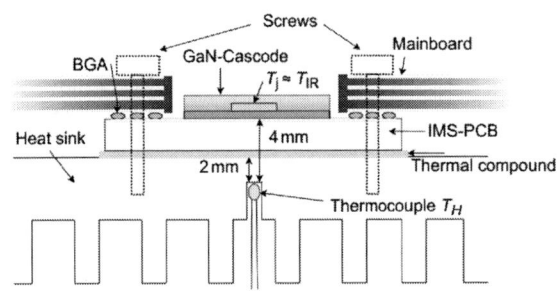

Fig. 7: Illustrated cross-section of a hybrid circuit board structure for power electronics [11].

As already indicated in Fig. 2, the temperature of the DUT is examined during operation using an IR camera. Since this method only measures the case temperature, a part of the case has been removed using a high-power laser, as shown in Fig. 8. By removing the housing, the chip temperature T_j can be measured more accurately with the IR-camera.

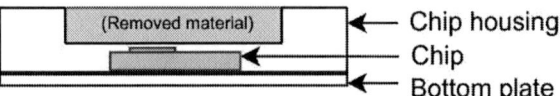

Fig. 8: Modified housing of the DUT for a more accurate measurement of T_j with an IR-camera.

4 Measurements

4.1 Exemplary Data Acquisition

The example in Fig. 9 illustrates the data collected during operation. Initially, the current of i_L in (a) is measured using a current clamp. Here, the current through the inductor is shown with a frequency of 50 Hz, highlighted by the timescale of ± 40 ms. When focusing on the peak current $\hat{i}_L$, the waveforms in (b) and (c) emerge. It is important to note the different time scales here, as the converter's switching events need to be examined in the microsecond range.

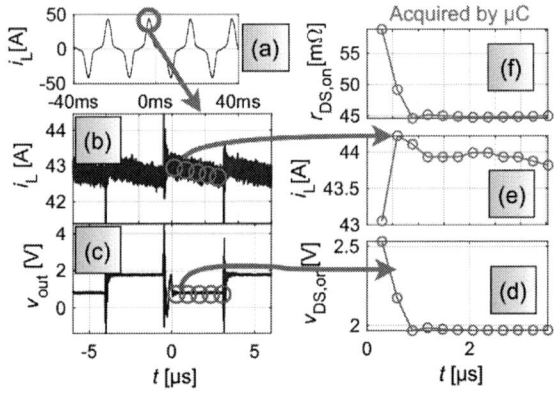

Fig. 9: Example measurement showing which data is collected during operation.

In (c) the output of the OVMC is displayed with the voltage v_{out} applied to the input of the ADC of the µC. The µC is configured to capture 12 measurement points in the 3.5 µs on-state interval. Both measurements of i_L and $v_{DS,on}$ are done in parallel, represented in (d) and (e). The on-resistance of the DUT can be determined using Eq. (4) as shown in graph (f).

$$r_{DS,on} = \frac{v_{DS,on}}{i_L} \tag{4}$$

4.2 Load Profiles

For continuous operation measurements, the full-bridge is operated with a DC-link voltage V_{DC} of 400 V. The source V_{DC} has to provide only a relatively small current I_{DC} to cover the power losses of the load. The test starts from load profile 1, progressively transitioning to load profile 2 and 3, as shown in table 1. The full-bridge is exposed to each profile for a duration of 168 h (24 h · 7).

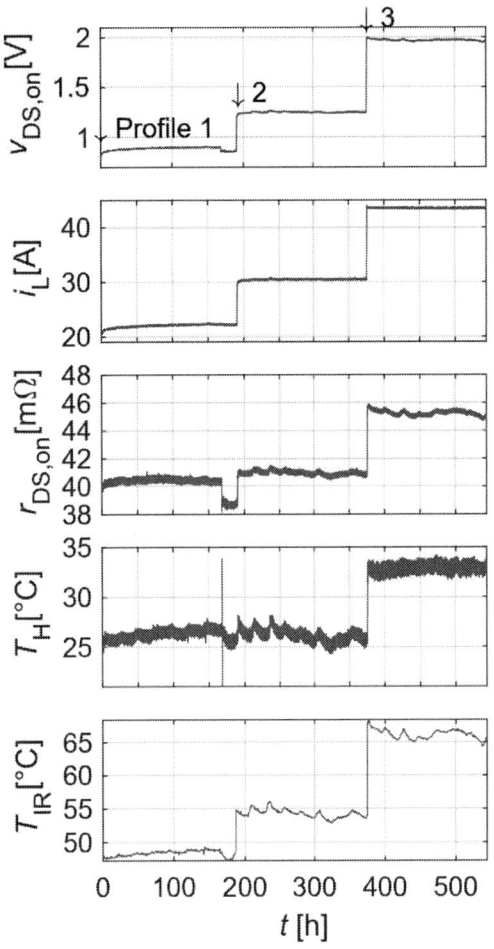

Fig. 10: Overview of entire measurement. Total duration 544 h.

Load-profile	1	2	3
φ [°]	7.8	8.3	9
V_{DC} [V]	400	400	400
I_{DC} [mA]	118	156	230
$I_{L,RMS}$ [A$_{RMS}$]	9.7	14.36	20.57
$\hat{\imath}_L$ [A]	22.71	30.94	43.5

Tab. 1: Load profiles that are tested in section 4.3, φ according to Fig. 6.

4.3 Overview

Fig. 10 illustrates the overall trend for $v_{DS,on}$, i_L, $r_{DS,on}$, T_H, and T_{IR} based on load profiles 1-3. The starting points of the measurement with respective load profiles are marked in the course of $v_{DS,on}$. For the on-state measurements, the 7th value of a 12-row measurement (cf. Fig. 9) is plotted over the operating time. Fundamentally, it can be observed that by increasing the load profile, a higher peak current i_L flows through the component. Proportionally, a measurable influence on the drain-source voltage $v_{DS,on}$ is evident. When considering the on-state resistance $r_{DS,on}$, it is noticeable that it increases when increasing the load. For instance, it amounts to approximately 39 mΩ for load profile 1, 41 mΩ for load profile 2, and around 45 mΩ for load profile 3. A proportional increase in the measured chip temperature T_{IR} using a IR camera is also recorded due to the increasing load. Although the measured trends may appear very constant at first glance, a closer examination within the timeframe of operation with load profile 3 is investigated further in the following section 4.4.

4.4 Detailed View of Load Profile 3

In this section the collected data from load profile 3 is further investigated as seen Fig. 11, as it represents the load profile with the highest load.

The junction temperature measured by the IR-camera T_{IR} exhibits noticeable fluctuations in a 24 h-interval. This is attributed to variations in room temperature, resulting in T_{IR} ranging between 64 °C and 68 °C. Additionally, it is observed that the heat sink temperature T_H in load profiles 1 and 2 (see Fig. 10) varies in proportion to T_{IR}. However, in load profile 3, the heat sink temperature remains very constant with an average temperature of approximately 33 °C.

The trend of $r_{DS,on}$ shows significant similarity to the observed temperature T_{IR}.

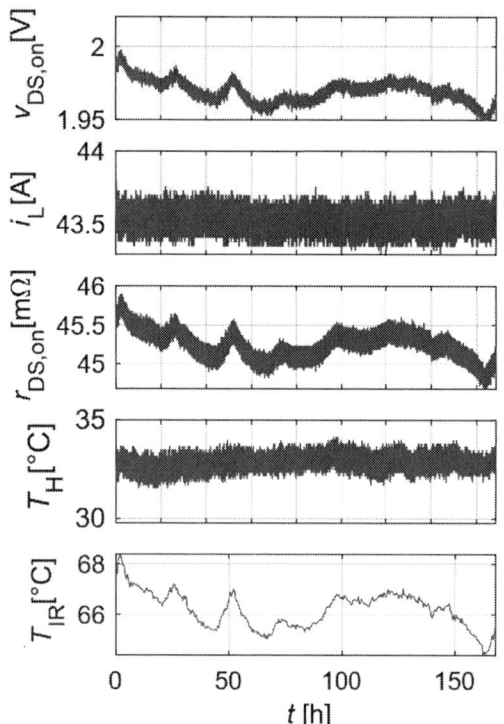

Fig. 11: Measurement of load profile 3. Total duration of 168 h.

This is expected considering the temperature dependency of $r_{DS,on}$ and the room temperature-dependent cooling performance. Nevertheless, the measurements do not reveal an increasing trend in on-resistance, thus providing no evidence of degradation based on these initial tests. Possible reasons for this observation could be an insufficient test time and load current, resulting in insufficient stress on the switches to induce degradation effects related to the trapping effect. This assumption is supported by the measured temperature T_{IR}, which reaches a maximum of only 68 °C, whereas power switches are typically designed to operate at much higher temperatures.

4.5 Detailed View Initial Phase

In the following section, the initial phase of the measurements with load profiles 1-3 in the interval from 0 h to 4 h is examined, as depicted in Fig. 12. During this initial phase of the measurement, some transient phenomena can be observed. For instance, it takes approximately 0.5 h for the heat sink temperature to reach a constant value, because the heat sink heats up slowly due to its large thermal mass, influenced by the heat dissipation from the two half-bridges.

Based on $r_{DS,on}$, an increase of approximately 1.3 % for each load profile is observed in the first hour, before they level off. This could be attributed to the described trapping effect, as after a certain runtime, all available electron traps are filled, leading to no further increase in $r_{DS,on}$ and eventually reaching saturation [4].

The decrease in current profile i_L at the beginning of load profile 3 can be well explained by the increasing $r_{DS,on}$ and thus the increase in conduction resistance. However, this explanation does not apply to load profiles 1 and 2, where, despite the increase in resistance, no decrease in current is observed.

5 Conclusion

Within the scope of this paper, a measurement and control board has been designed and successfully put into operation. The main function of this board is to control a full-bridge circuit consisting of GaN cascodes with a carrier frequency of 140 kHz to generate an adjustable sinusoidal current in an inductive load. Additionally, the board features a current sensor and a circuit that allows monitoring of the drain-source voltage in the on-state, enabling determination of the channel resistance $r_{DS,on}$ during switching operation.

The full-bridge circuit has been tested under various load profiles for 168 h each. Within the first hour of operation of each load profile, an increase in channel resistance of approximately 1.3 % is observed. The subsequent course of the measurement shows a clear temperature dependency of this parameter, however, no increasing trend is discernible.

It should be mentioned that the number of tested samples and the test duration are not sufficient to make a definitive statement about the long-term reliability of the system or the GaN components. The converter is intended to be operated in future investigations with higher load inductance, higher DC-link voltage (up to 500 V), higher load current, and thus higher temperature. Additionally, the variation of the switching frequency may be of interest, as it is a determining factor in the power semiconductor's power loss.

Nevertheless, this experimental setup demonstrates the possibility of capturing the parameter $r_{DS,on}$ during fast switching operation using regularly available hardware. This could be used for temperature monitoring, loss calculation, and condition monitoring of WBG power electronic components.

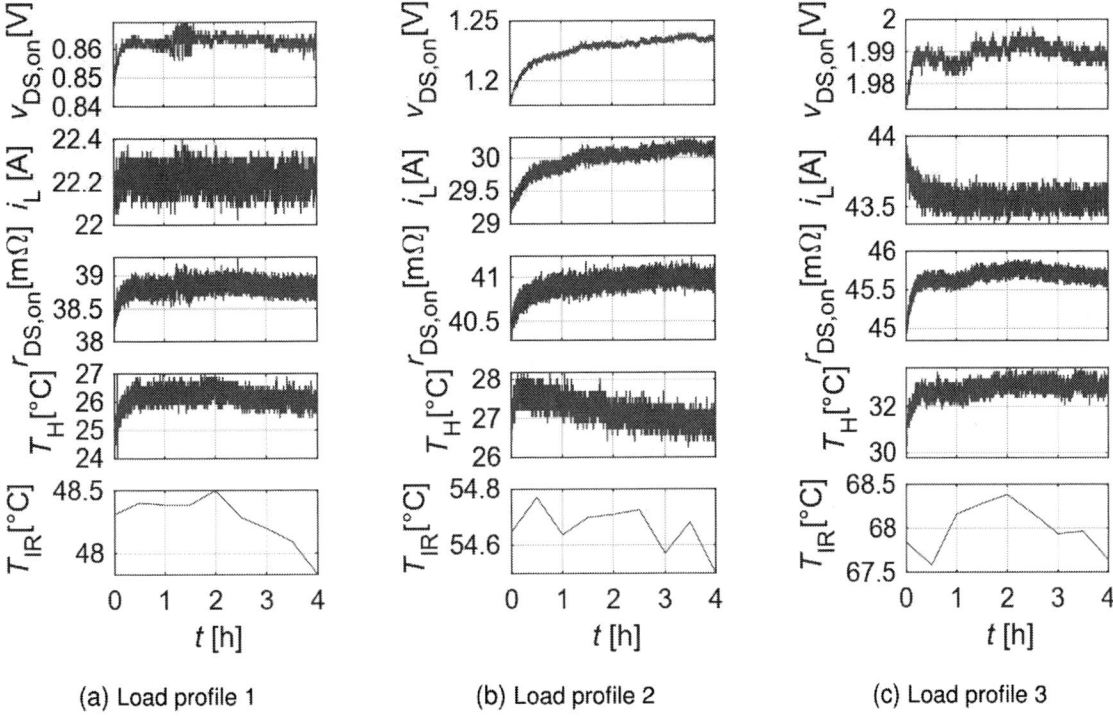

(a) Load profile 1   (b) Load profile 2   (c) Load profile 3

Fig. 12: Measurements in the initial phase 0 h to 4 h.

References

[1] R. Hou, Y. Shen, H. Zhao, H. Hu, J. Lu, and T. Long, "Power Loss Characterization and Modeling for GaN-Based Hard-Switching Half-Bridges Considering Dynamic on-State Resistance," *IEEE Transactions on Transportation Electrification*, vol. 6, no. 2, pp. 540–553, 2020. DOI: 10.1109/TTE.2020.2989036.

[2] H. Sayed, G. S. Kulothungan, and H. S. Krishnamoorthy, "Characterization of GaN HEMTs' Aging Precursors and Activation Energy Under a Wide Range of Thermal Cycling Tests," *IEEE Open Journal of the Industrial Electronics Society*, vol. 4, pp. 123–134, 2023.

[3] E. A. Jones, F. Wang, and B. Ozpineci, "Application-based Review of GaN HFETs," in *2014 IEEE Workshop on Wide Bandgap Power Devices and Applications*, 2014, pp. 24–29. DOI: 10.1109/WiPDA.2014.6964617.

[4] M. González-Sentís, P. Tounsi, A. Bensoussan, and A. Dufour, "Degradation Indicators of Power-GaN-HEMT under Switching Power-Cycling," *Microelectronics Reliability*, vol. 100-101, p. 113412, 2019, 30th European Symposium on Reliability of Electron Devices, Failure Physics and Analysis. DOI: https://doi.org/10.1016/j.microrel.2019.113412.

[5] L. Gill, S. DasGupta, J. C. Neely, R. J. Kaplar, and A. J. Michaels, "A Review of GaN HEMT Dynamic ON-Resistance and Dynamic Stress Effects on Field Distribution," *IEEE Transactions on Power Electronics*, vol. 39, no. 1, pp. 517–537, 2024. DOI: 10.1109/TPEL.2023.3318182.

[6] H. Huang, Y. C. Liang, G. S. Samudra, T.-F. Chang, and C.-F. Huang, "Effects of Gate Field Plates on the Surface State Related Current Collapse in AlGaN/GaN HEMTs," *IEEE Transactions on Power Electronics*, vol. 29, no. 5, pp. 2164–2173, 2014. DOI: 10.1109/TPEL.2013.2288644.

[7] N. Zagni, A. Chini, F. M. Puglisi, M. Meneghini, G. Meneghesso, *et al.*, ""Hole Redistribution" Model Explaining the Thermally Activated RON Stress/Recovery Transients in Carbon-Doped AlGaN/GaN Power MISHEMTs," *IEEE Transactions on Electron Devices*, vol. 68, no. 2, pp. 697–703, 2021. DOI: 10.1109/TED.2020.3045683.

[8] A. Bouchour, A. El Oualkadi, P. Dherbécourt, O. Latry, and A. Echeverri, "Investigation of the Aging of Power GaN HEMT under Operational Switching Conditions, Impact on the Power Converters Efficiency," *Microelectronics Reliability*, vol. 100-101, p. 113 403, 2019, 30th European Symposium on Reliability of Electron Devices, Failure Physics and Analysis. DOI: https://doi.org/10.1016/j.microrel.2019.113403.

[9] M. Guacci, D. Bortis, and J. W. Kolar, "On-state Voltage Measurement of Fast Switching Power Semiconductors," *CPSS Transactions on Power Electronics and Applications*, vol. 3, no. 2, pp. 163–176, 2018. DOI: 10.24295/CPSSTPEA.2018.00016.

[10] *TMS320F2837xD Dual-Core Microcontrollers Data Sheet*, https://www.ti.com/lit/pdf/sprs880o, visited on 23th January 2024, Texas Instruments Incorporated, Jan. 2021.

[11] G. Braun and D.-H. Moldenhauer, "Hybrid Circuit Board Structure for Power Electronics," in *2022 24th European Conference on Power Electronics and Applications (EPE'22 ECCE Europe)*, 2022.

[12] S. Kloetzer and J. Honea, "Modern High Voltage GaN FETs - Power Applications and Impact of Packaging Technology," in *Components of Power Electronics and their Applications 2023; ETG Symposium*, 2023, pp. 31–38.

PCIM Europe 2024, 11– 13 June 2024, Nuremberg   DOI: 10.30420/566262408

A Simple and Low Cost Overcurrent Protection System Based on Commercial Shunt for Wide-Bandgap Devices

Emanuele Martano[1] ⓘ, Yoann Pascal[2] ⓘ, Marco Liserre[2,3] ⓘ, Giovanni Busatto[1] ⓘ, Annunziata Sanseverino[1] ⓘ

[1] University of Cassino and Southern Lazio, Italy
[2] Fraunhofer Institute for Silicon Technology ISIT, Germany
[3] Kiel University, Germany

Corresponding author: Emanuele Martano, emanuele.martano@unicas.it
Speaker: Emanuele Martano, emanuele.martano@unicas.it

Abstract

Due to current measurement limitations in wide bandgap (WBG)-based systems, designing short circuit (SC) protection for gallium nitride (GaN) power devices is a challenging task. Currently, the most effective protection systems are based on alternative methods to current measurement. However, due to GaN's susceptibility to short circuit conditions, these methods require the design of complex circuits to meet the strict intervention time requirements. Protection methods based on direct current measurement could represent an easier solutions and help safeguard systems also from overcurrent problems. This is especially important in high-power applications where overcurrent issues can be particularly troublesome. This paper proposes a simple and fast overcurrent protection (OCP) system based on direct current measurement for 650V E-mode GaN HEMTs. The circuit is designed to quickly identify when the current exceeds the recognized abnormal limit and automatically shuts off the device under test (DUT).

1   Introduction

Compared to their Si and SiC counterparts, GaN HEMTs exhibit different short-circuit characteristics, resulting in an overall inferior SC withstand capability. Experimental results show that the SC withstand time is highly dependent on the DC bus voltage. In addition, when the VDC is greater than 300V, the withstand time is only a few hundred nanoseconds [1]. Therefore, because of this vulnerability to high voltage, fast SC protection is essential to protect GaN device power systems. Various solutions for SC protection circuits have been proposed in the literature, but the most established technology in this area is desaturation-based systems. These systems use the rise in drain-source voltage to detect the short circuit [2], [3]. Although this technique provides excellent response dynamics, the high switching speed of GaN HEMTs causes interference problems due to high dv/dt and high di/dt. This leads to complex and expensive circuit designs. An indirect approach for short circuit detection is proposed in [4]. This method is suitable for GaN HEMT applications due to its ultra-fast response time. It is a solution for half-bridge cell-based converters that detect the sudden voltage drops over the DC bus decoupling capacitors in the event of a short circuit. Although the detection times are incredibly fast, implementing this method is challenging due to the use of a high-pass filter for the voltage drop detection. The filter's parameters depend on the parasitic elements of the circuit, making its project closely related to the converter's design. Additionally, not all converters use half-bridge cells, which further complicates the implementation process.

While an abrupt surge in drain current is a discernible trait of a short-circuit, there are currently no safeguard measures for GaN devices that are dependent on direct current monitoring. To ensure high precision without compromising accuracy, a method based on direct current measurement should use a high-bandwidth current sensor and avoid introducing excessive stray inductance in the power loop. Several attempts have been made in this direction, including the solution presented in [5]. Their proposal is based on a coaxial shunt, which enables precise detection of short circuits or overcurrent conditions through simple control logic. However, the coaxial shunt introduces a non-negligible inductance into the circuit when working

2874

with short rise times, such as in the case of GaN HEMTs. Additionally, it is an expensive and bulky solution that is not well suited for integration into newly developed power converters.

This paper proposes a simple and cost-effective method for protecting power devices from overcurrent conditions by directly measuring the drain current. The current monitoring system uses SMD shunts and a passive network to compensate for their stray inductance. The method is validated by both simulation and experiment, with the latter demonstrating an intervention time of only 68 ns to shut down the protected device.

2 The RL Compensation Network

Even though the low cost, simplicity and integration of SMD shunt would make it the best choice for a short-circuit protection system, it should be noted that the stray inductance of the shunt severely limits its bandwidth, thus affecting current detection during fast switching transients. For this reason, sensors such as Rogowski coils and co-axial shunts are generally preferred over commercially available SMD shunts to ensure accurate current measurement in WBG converters. An approach to overcome this problem has been proposed in [6] using a simple passive network to compensate for the effect of the inductance associated with the shunt. As the origin of this study lies in proposing the use of this compensation network to detect overcurrent conditions, in Fig. 1 the diagram used in [6] is presented to illustrate how this network successfully addresses the issue. The voltage Vo, which is proportional to the voltage drop across the shunt, is obtained using an oscilloscope with an input impedance of 50Ω. Impedance matching between the circuit and oscilloscope is ensured by the 50Ω resistor in series. The measurement scheme is completed by including the shunt resistance Rs, its stray inductance Ls, as well as the resistor Rc and inductor Lc employed to balance the effects of Ls.

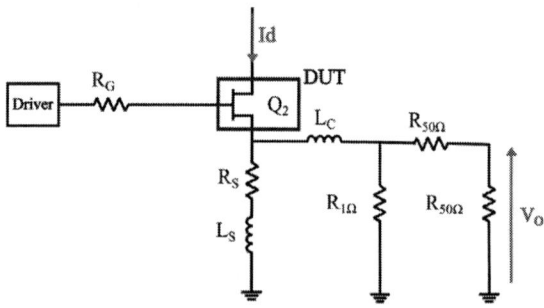

Fig. 1 Compensation network for stray inductance in SMD shunts

By equating the coefficients of the imaginary parts, The relationship between the measured voltage (Vo) and the drain current (Id) can be determined by approximating the equivalent resistance of $R_{1\Omega}$ // $2R_{50\Omega}$ to $R_{1\Omega}$ itself:

$$V_O(\omega) = \frac{1}{2} \frac{R_s + j\omega L_s}{R_s + R_{1\Omega} + j\omega(L_s + L_c)} R_{1\Omega} I_d(\omega) \quad (1)$$

Combining the resistances as common factor:

$$V_O(\omega) = \frac{1}{2} \frac{R_s(1 + \frac{j\omega L_s}{R_s})}{(R_s + R_{1\Omega})(1 + \frac{j\omega(L_s + L_c)}{R_s + R_{1\Omega}})} R_{1\Omega} I_d(\omega) \quad (2)$$

It is possible to find the value of L_c that satisfies the equality:

$$L_c = L_s \frac{R_{1\Omega}}{R_s} \quad (3)$$

Combining (3) and (2):

$$V_O = \frac{1}{2} \frac{R_s}{R_s + R_{1\Omega}} R_{1\Omega} I_d \quad (4)$$

The Eq.(4) is a linear relationship and no longer shows frequency dependence.

3 The Overcurrent Protection Circuit

Since the voltage across resistor $R_{1\Omega}$ is twice Vo (see Fig. 1), this voltage can be used to monitor the current Id, as shown in the following equation:

$$V_{R1\Omega} = \frac{R_s}{R_s + R_{1\Omega}} R_{1\Omega} I_d \quad (5)$$

By setting the limit value of the current Id, it is possible to obtain the corresponding voltage value according to Eq. 5. This voltage value can be used as a reference at the input of a comparator to determine when the current Id exceeds the set limit. This approach is utilized in the proposed overcurrent protection system, as shown in Fig.2.

To achieve overcurrent conditions, the OCP is designed to turn on the DUT in short circuit conditions. This is done by turning on Q1 before Q2 (DUT) and keeping it on during the conduction phase of Q2 [7]. The comparator generates a fault

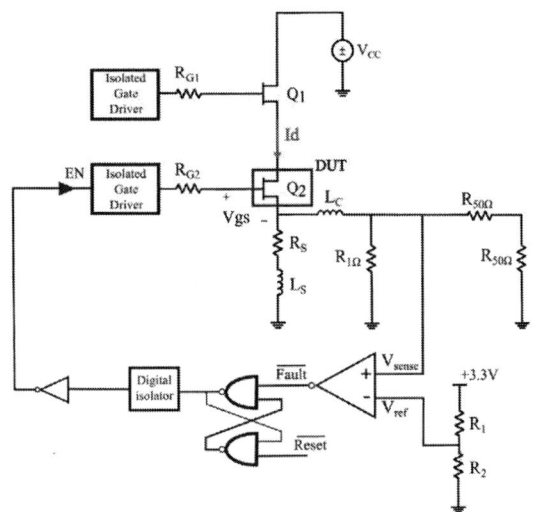

Fig. 2   OCP simplified electric scheme

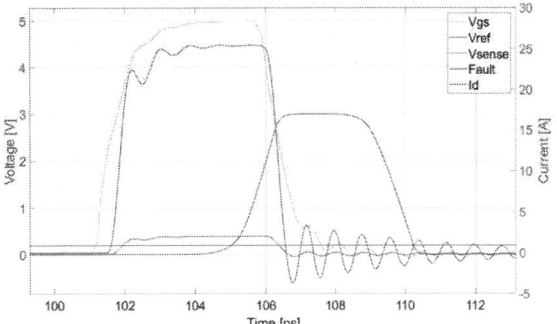

Fig. 3   Simulated OCP characteristics waveforms

signal and disables the driver of the protected device when the sense signal (V_{sense}) indicates that the current exceeds the normal operating limit. It is crucial to have a latch circuit on the output of the op-amp to prevent the DUT driver from being enabled until the system is reset. Otherwise, if the V_{sense} voltage drops below the V_{ref} threshold value, the device will reactivate and repeatedly turn the DUT on and off in short-circuit conditions. To ensure driver isolation, the digital signal on the output of the detection system is also isolated.

3.1 Simulation Results

The validity of the proposed method was first verified by reproducing the circuit shown in Fig. 2 in the LTspice environment.
The simulation was performed with two 8 A - 650V GaN HEMTs, using the model provided by the manufacturer. The gate resistors were both set to 10Ω. The Id current limit was 14A, and the V_{gs} and V_{CC} values were 5V and 50V, respectively. The other circuit parameters, including L_s, L_c and R_s, were chosen based on the experimental setup described in the following section. Figure 3 shows the simulated waveforms.
As can be seen, when V_{sense} exceeds the V_{ref} threshold, a fault signal is activated. The fault signal is in the high logic state approximately 4 ns after the 14 A limit is exceeded. This delay is caused by the operational amplifier. Activating the fault signal disables the DUT driver, so as soon as this signal begins to rise toward the logic high state, the gate voltage and drain current start returning to zero. It's important to note that due to the SR

latch at the comparator output, the driver will not be re-enabled when the fault signal returns to the off state once the current Id falls (and therefore V_{sense}) below the overcurrent threshold.
The simulations showed that the system's total intervention time was about 4 nanoseconds. However, delays caused by the LR latch, digital isolator, inverter, and gate driver were not accounted for in the simulation. Moreover, the driver was simulated using a voltage-controlled square-wave voltage generator with rise times of less than one nanosecond.
Based on the data provided by the manufacturers and the simulation results, the preview indicated an intervention time for the real circuit of no more than 97 ns.

3.2 Experimetal Validation

Based on the schematic shown in Fig. 2, a prototype was built and is shown in Fig. 4. The PCB includes a GaN half-bridge composed of two 650 V GaNSystem HEMTs. The DUT has a current rating of 8 A, while the upper-side device has a rating of 60 A. Two SMD pads (JP_3) on the bottom side allow the connection of a coaxial shunt (T&M SDN-414-10) in series with up to three SMD shunt resistors in parallel.
The inductance required for compensation was determined by measuring the stray inductance of a single SMD shunt using a vector network analyzer (VNA). This measurement was made by disabling the entire board and measuring directly at the ends of the shunt soldered to the PCB. Nevertheless, a tunable inductor was selected to allow for precise adjustment of the compensation network. This choice was made because measuring the leakage inductance via VNA is a complicated and imprecise process. Additionally, the stray inductance of the SMD shunt can be also frequency-dependent because of edge effects and parasitic

PCIM Europe 2024, 11– 13 June 2024, Nuremberg   DOI: 10.30420/566262408

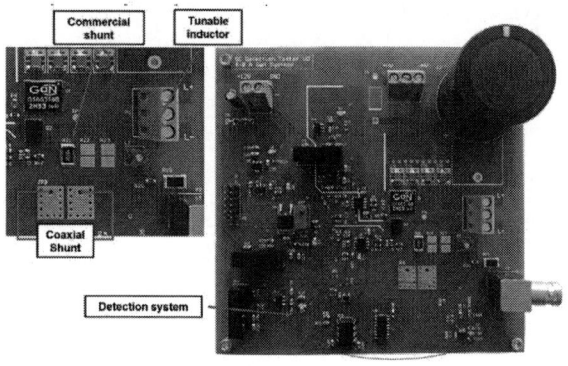

Fig. 4   Picture of the OCP validation prototype

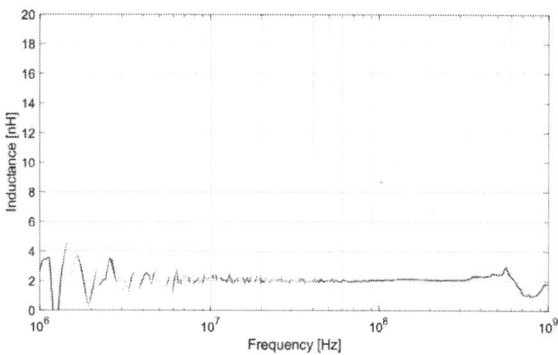

Fig. 5   SMD shunt inductance measurement with VNA

capacitances in the circuit. The results of the above measurement are shown in Fig. 5.

A double SMD jumper (JP_1 and JP_2) has been inserted to enable or disable the detection system. This involves comparing the output of the compensation network to that of the coaxial shunt, without any intervention from the detection system.

The oscilloscope used is a Lecroy HDO6104 and the logic is controlled by a Zedboard (Xilinx ZynqTM-7000 All Programmable SoC). Table 1 summarizes other characteristic parameters of the tests.

3.2.1 Current Measurement

Before testing the potential of the detection system, it is necessary to guarantee the accuracy of

Table 1. Experimental test parameters

V_{CC}	V_{gs}	T_{on}	$R_{G1,2}$	R_S	R_{COAX}
50 V	5V	400 ns	20 Ω	33 mΩ	10 mΩ

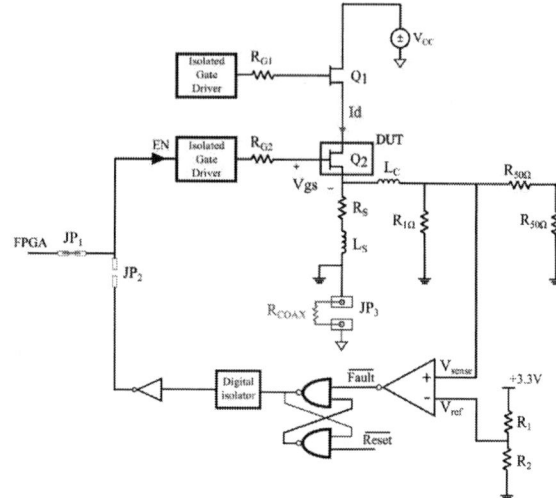

Fig. 6   OCP calibration configuration

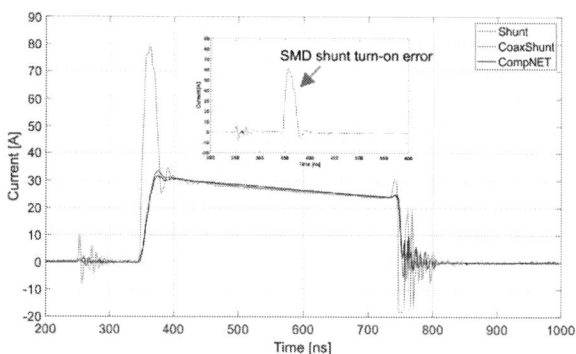

Fig. 7   Current measurement comparison obtained via SMD shunt, coaxial shunt and RL compensation network

the current measurement system made by the SMD shunt with compensation network. In this sense, the coaxial shunt (R_{COAX}) in series allowed for a comparison with the current measuring instrument that has the greatest bandwidth on the market. Figure 7 shows the current waveforms obtained during a short-circuit test with the circuit configuration of Fig. 6, i.e. with the detection system disconnected. The compensated network's measurement shows exceptional ability to track the high dynamics associated with switching on and off transients. This ability is not present when voltage is measured directly at the ends of the shunt.

A VNA measurement was then performed to determine the actual value of the compensation inductance required to achieve a current waveform comparable to that of the coaxial shunt. This

2877

was done because the inductance value to be inserted into the network could be affected not only by the estimate of the shunt inductance and the tolerance of the compensation inductor, but also by the parasitic inductance associated with the PCB traces. Therefore, the inserted value may differ significantly from the value determined by the project, especially if this value is in the order of tens of nH. As with the stray inductance of the shunt, the measurement for the tunable inductor was conducted with the circuit switched off and the probe placed at its ends.

Figure 8 illustrates that the compensation inductance value is approximately 60 nH up to 400 MHz. This value demonstrates both the accuracy of the SMD shunt inductance measurement conducted during the design phase and the minimal impact of PCB traces due to the optimized measurement layout. In fact, Equation 3 confirms that 60 nH is the expected value based on the approximately 2 nH stray inductance of the SMD shunt. Beyond 400 MHz, the compensation network can no longer guarantee its task due to the drastic change in its inductance value. As a result, the potential system bandwidth is limited to 400 MHz.

The compensation inductance was also measured by removing the tunable inductor from the circuit and using an impedance analyzer with a 120 MHz bandwidth. The result shown in Fig. 9 indicates a quite constant inductance value of 60 nH. This value, although obtained in a limited spectrum, is a counter-proof of the accuracy of VNA measurements.

3.2.2 Evaluation of the OCP

To test the potential of the overcurrent protection system, a short circuit test was carried out with the detection system activated. For this purpose, JP_1 was opened while JP_2 and JP_3 were both short-circuited, as shown in Fig. 10. The current threshold for system intervention was set at 14 A, and the remaining parameters were the same as those shown in Table 1. Figure 11 demonstrates that by applying a short-circuit pulse of 400 ns, the DUT

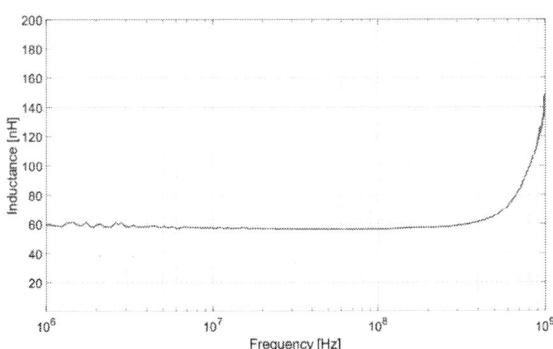

Fig. 8 Compensation inductance measurement with VNA

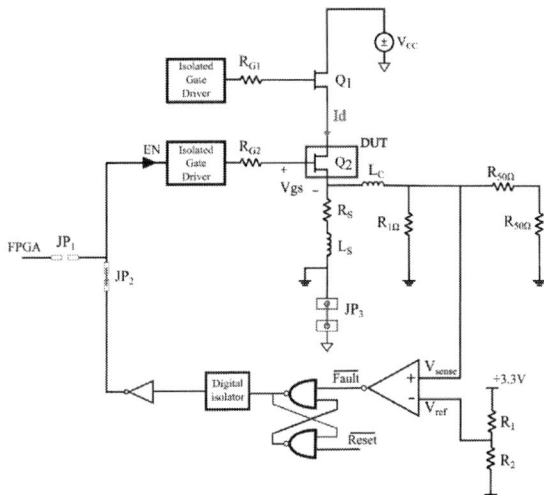

Fig. 10 OCP evaluation configuration

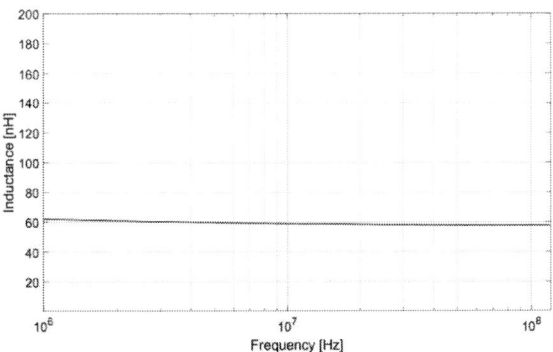

Fig. 9 Compensation inductance measurement with LCR

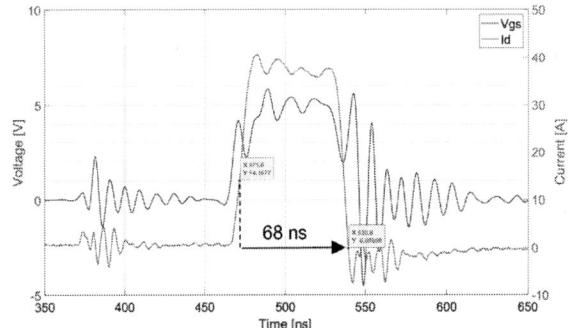

Fig. 11 DUT drain current and gate voltage during a short test of 400 ns activating the overcurrent protection

switches off after approximately 70 ns. This occurs because the output voltage of the compensation network has exceeded the system intervention threshold. In particular, since the drain current exceeds 14 A, the system takes 68 ns to completely shut down the device. Disabling the gate driver prevents the device from powering up according to the PWM signal, thus avoiding possible restarts under short conditions. To restore normal operation and re-enable the driver, it is necessary to press the SR latch reset button on the FPGA. As just demonstrated, the protection system's intervention time did not exceed the maximum value expected during the simulation phase. In accordance with [8], such a fast intervention time can contribute to preserve the functioning of even the least robust GaN power devices in the event of a short circuit.

4 Conclusions

This article presents an overcurrent protection system based on direct current measurement for wide bandgap devices. Current monitoring is achieved using a passive network consisting of an SMD shunt, inductor and SMD resistor. The small size and easy integration of these three elements make this method practical and suitable for the compact requirements of WBG converters, providing an easier to implement alternative to commonly used measurement methods such as the Rogowski coil and coaxial shunt. In addition, the cost can be significantly reduced compared to the coaxial shunt.

The method has been validated through both simulation and experiment. In particular, the overall response time showed during the experimental test was only 68 ns, a short enough time to avoid damage to GaN devices.

References

[1] H.Li et al., "Robustness of 650-V Enhancement-Mode GaN HEMTs Under Various Short-Circuit Conditions", IEEE Trans. Ind. Appl, vol. 55, no. 2, 2019.

[2] S. Kenzelmann et al., "R. Hou, J. Lu, Z. Quan, and Y. W. Li, "A simple desaturation-based protection circuit for GaN HEMT with ultrafast response", IEEE Trans. Power Electron., vol. 36, no. 6, pp. 6978–6987, Jun. 2021.

[3] J. Wu, W. Meng, F. Zhang, G. Dong, and J. Shu, "A Short-Circuit Protection Circuit With Strong Noise Immunity for GaN HEMTs", IEEE Trans. on Power Electron., vol. 36, no. 2, 2021.

[4] X. Lyu et al., "A Reliable Ultrafast Short-Circuit Protection Method for E-Mode GaN HEMT," in IEEE Transactions on Power Electronics, vol. 35, no. 9, pp. 8926-8933, Sept. 2020.

[5] W. Zhang, F. Wang, Z. Zhang and a. B. Holzinger, "Fast Wide-bandgap Device Overcurrent Protection with Direct Current Measurement," 2019 10th International Conference on Power Electronics and ECCE Asia (ICPE 2019 - ECCE Asia), Busan, Korea (South), 2019, pp. 1-6.

[6] C. Abbate et al., "An Accurate Switching Current Measurement Based on Resistive Shunt Applied to Short Circuit GaN HEMT Characterization", Appl. Sci., 2021, 11, 9138.

[7] Abbate, C.; Busatto, G.; Sanseverino, A.; Tedesco, D.; Velardi, F. Experimental study of the instabilities observed in 650 V enhancement mode GaN HEMT during short circuit. Microelectron. Reliab. 2017, 76–77, 314–320.

[8] M. Landel, C. Gautier, D. Labrousse, S. Levebvre, F. Zaki and Z. Khatir, "Dispersion of Electrical Characteristics and Short-Circuit Robustness of 600 V Emode GaN Transistors," PCIM Europe 2017; International Exhibition and Conference for Power Electronics, Intelligent Motion, Renewable Energy and Energy Management, Nuremberg, Germany, 2017, pp. 1-9.

PCIM Europe 2024, 11– 13 June 2024, Nuremberg　　　DOI: 10.30420/566262409

SVM-Based Fault-Tolerant Control for a Cascaded H-Bridge Multi-level Converter under Multiple Open-Circuit Switch Faults

Dong Xie[1] ⓘ, Hongjian Lin[2] ⓘ, Chunxu Lin[3] ⓘ, Thomas Basler[1] ⓘ, Xinglai Ge[3] ⓘ

[1] Chemnitz University of Technology, Chair of Power Electronics, Chemnitz, Germany
[2] City University of Hong Kong, Department of Electrical Engineering, Hong Kong, China
[3] Southwest Jiaotong University, School of Electrical Engineering, Chengdu, China

Corresponding author:　Dong Xie, dong.xie@etit.tu-chemnitz.de
Speaker:　　　　　　　　Dong Xie, dong.xie@etit.tu-chemnitz.de

Abstract

To improve the operation ability of the cascaded H-bridge multilevel converter (CHBMC) under multiple open-circuit (OC) faults, this paper has explored a space vector modulation (SVM)-based fault-tolerant control method. Based on the selectable vector range in the postfault conditions, the assigned output vector for each power cell is modified to reduce the current harmonics and to maintain the DC voltage balancing as much as possible with the smooth voltage level transitions. Moreover, the level number of total port voltage is not affected by the multiple OC faults, thereby balancing the voltage stresses on the switches of the different power cells. Simulation and experimental results have shown the applicability of the proposed method.

1　Introduction

Due to the modularity, control flexibility, and high tolerance, the cascaded H-bridge multilevel converter (CHBMC) is properly applied in medium- and high-voltage conversions such as solid-state transformers, medium voltage (MV) high-power converters, and STATCOMs [1]-[3]. The CHBMC contains a significant number of power switches that are particularly vulnerable, making them prone to open-circuit (OC) faults, caused by the gate driver failure or bond wire lift-off, or short-circuit (SC) faults caused by the high voltage breakdown or high junction temperature [4]. These types of faults pose significant challenges to the safe and efficient operation of the system. Due to the fact that OC faults do not lead to an immediate system breakdown, numerous studies have proposed software-based diagnostic methods to identify the OC fault location [5]-[7]. However, this is merely the foundation of system fault tolerance control. The key to maintain long-term reliable operation lies in how to adjust system control to address the effects of different OC faults.

Typical fault-tolerant techniques often involve bypassing the faulty power cells and maintaining system operation through additional compensation or circuits such as methods in [8] and [9]. These methods are complicated, and the remaining output capacity of faulty power cells is not considered. Regarding this, the space vector modulation (SVM) has been widely applied in both normal and fault-tolerant control of three-phase systems due to their flexible vector selection and maximized power output [10], [11]. Recently, it has also shown great potential for OC diagnosis and DC voltage balancing in the single-phase CHBMC [12]. Building upon the foundation laid in [12], this paper further explores the fault tolerance capabilities of the SVM under multiple OC fault conditions.

2　System Overview

The single-phase CHBMC with three H-bridge power cells is shown in Fig. 1. The power switches (T_{11} to T_{34}) and anti-paralleled diodes (D_{11} to D_{34}) create different current paths, resulting in different port voltages (u_{a1b1} to u_{a3b3}). In the grid side, L_N and R_N are the filter inductor and parasitic resistor, u_N and i_N are the grid voltage and current, while u_{a1b3} is the total port voltage. At the dc-side, C_i, R_i, and u_i represent the DC-link capacitor, the resistive load, and the DC voltage, respectively.

Figure 2 describes the block diagram of the main control and SVM process. E is established as the reference value for DC voltage, denoted as $U^*_{dc} = E$. The normalized value of u_{aibi} represents the voltage level for each power cell, defined as

$V_i = u_{aibi} / E$. The total voltage level of all power cells is defined as V. The transient current control (TCC) is used to obtain the reference modulated voltage V_{ref}, and the normalized value $V'_{ref} = V_{ref} / E$, which is further applied to the SVM for generating switch driving signals (s_{i1}, s_{i2}, s_{i3}, s_{i4}). V_{max} is the maximum value of the total voltage level in one vector zone and V_{min} is the minimum one. t_{max} and t_{min} are the dwelling times of V_{max} and V_{min} during one switching cycle T_d. A more detailed vector selection process in the pre-fault condition can be found in [12]. The specific vector modification is investigated as follows, considering the different OC faults in the single-phase CHBMC.

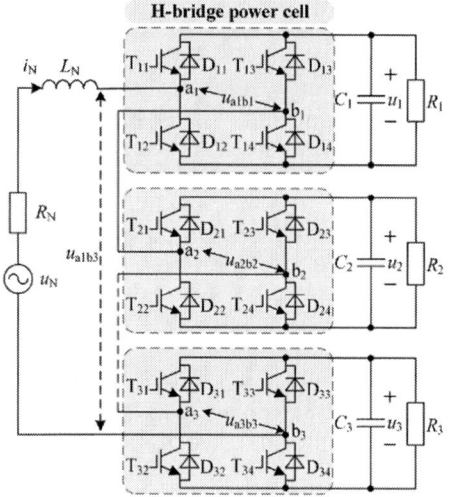

Fig. 1 Topology of the single-phase CHBMC.

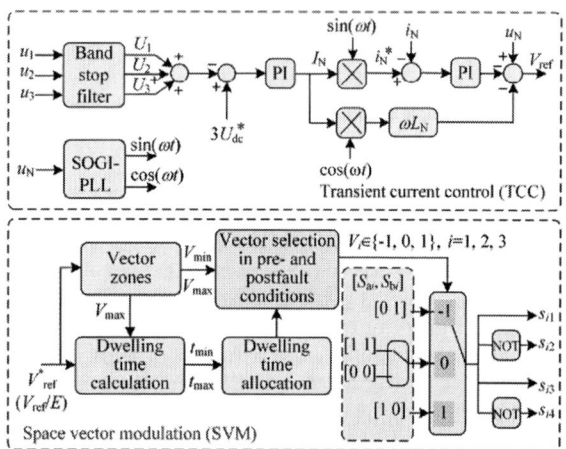

Fig. 2 Block diagram of the main control and SVM.

3 Assigned vector modification

In the assigned vector, the relationship between V_i and switching pairs $[S_{ai}, S_{bi}]$ is expressed as $V_i = (S_{ai}\text{-}S_{bi}) \in \{-1, 0, 1\}$. Then, $V \in \{-3, ..., 3\}$ for

the CHBMC with 3 power cells. Defining that k, $k+1$, and $k+m$ are the vector switching moments, under the pre-fault condition, V_i at each power cell should be constrained to

$$C_{in}: \quad V_{imin} \le V_i \le V_{imax}, i = 1, 2, 3 \quad (1)$$

where V_{imin} = max $\{V\text{-}(3\text{-}i)\text{-}V_1\text{-}V_2...\text{-}V_{i\text{-}1}, \; -1\}$ and V_{imax} = min $\{V\text{+}(3\text{-}i)\text{-}V_1\text{-}V_2...\text{-}V_{i\text{-}1}, \; 1\}$. The same constraint can be applied to other switching moments. When the OC faults occur, available voltage levels in the power cells with OC switches are affected and the constraint C_{in} should also be modified to avoid the output of absent voltage levels and ensure balanced DC voltages. The detailed description is as follows.

3.1 Multiple OC switch faults in one power cell

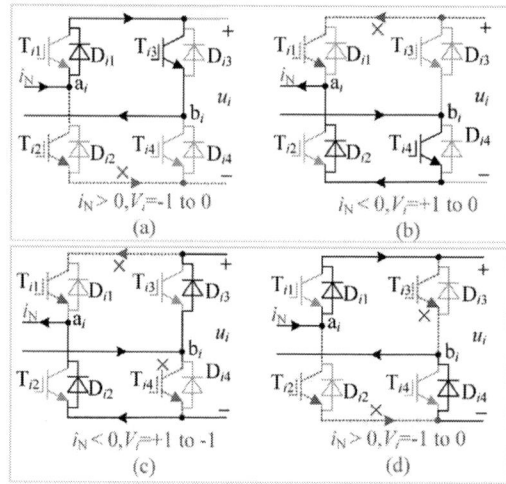

Fig. 3 Current paths under (a) T_{i2} OC fault, (b) T_{i1} OC fault, (c) T_{i1} and T_{i4} OC faults, and (d) T_{i2} and T_{i3} OC faults.

Case 1) (T_{i1} or T_{i4}) and (T_{i2} or T_{i3}) OC faults: the positive and negative current are affected. (T_{i2} or T_{i3}) OC faults result in the absence of $V_i = -1$ when $i_N > 0$, only outputting zero voltage level like Fig. 3(a). Conversely, $V_i = 1$ cannot be adopted when $i_N < 0$ due to (T_{i1} or T_{i4}) OC faults, as shown in Fig. 3(b). Thus, the constraint of V_i should be modified as follows to avoid the failed current paths.

$$\begin{cases} V_{imin1} \le V_i \le V_{imax}, \; i_N > 0 \\ V_{imin} \le V_i \le V_{imax1}, \; i_N < 0 \end{cases}, i = 1, 2, 3 \quad (2)$$

where V_{imin1} = max$\{V\text{-}(3\text{-}i)\text{-}V_1\text{-}V_2...\text{-}V_{i\text{-}1}, 0\}$ and V_{imax1} = min$\{V\text{+}(3\text{-}i)\text{-}V_1\text{-}V_2...\text{-}V_{i\text{-}1}, 0\}$.

Therefore, the output voltage level is limited to [0, 1] when $i_N > 0$. The other healthy power cells can sequentially adjust the assigned voltage levels based on (1) to maintain the total voltage level unchanged during $i_N > 0$. Similarly, when $i_N < 0$, all voltage levels can be adjusted by using (1) and (2) to finish vector modification. This process can be online achieved.

Case 2) (T_{i1} and T_{i4}) OC faults: only the negative current is distorted. This fault type causes the absence of $V_i = 1$ and 0 when $i_N < 0$, and only $V_i = -1$ is the normal current path in Fig. 3(c). The range constraint of V_i is adjusted as

$$\begin{cases} V_{imin} \leq V_i \leq V_{imax}, \ i_N > 0 \\ V_i = -1, \ i_N < 0 \end{cases}, i = 1, 2, 3 \qquad (3)$$

Case 3) (T_{i2} and T_{i3}) OC faults: only the positive current is distorted. $V_i = -1$ and 0 when $i_N > 0$ is absent and only $V_i = 1$ is the normal current path in Fig. 3 (d). The range constraint of V_i is adjusted as

$$\begin{cases} V_i = 1, \ i_N > 0 \\ V_{imin} \leq V_i \leq V_{imax}, \ i_N < 0 \end{cases}, i = 1, 2, 3 \qquad (4)$$

3.2 Multiple OC switch faults in different power cells

In the CHBMC, multiple OC switches are likely to occur in two different power cells. Different fault cases are discussed by using power cells i and j as examples.

Case 4) (T_{i1} or T_{i4}) and (T_{j2} or T_{j3}) OC faults: the whole current is distorted. $V_i = 1$ cannot be adopted when $i_N < 0$ due to (T_{i1} or T_{i4}) OC faults and $V_i = -1$ is absent when $i_N > 0$ due to (T_{j2} or T_{j3}) OC faults. The range constraints of V_i and V_j should be changed as

$$\begin{cases} V_{jmin1} \leq V_j \leq V_{jmax}, \ i_N > 0 \\ V_{imin} \leq V_i \leq V_{imax1}, \ i_N < 0 \end{cases}, i, j = 1, 2, 3, i \neq j \ (5)$$

where $V_{jmin1} = \max \{V-(3-j)-V_1-V_2...-V_{j-1}, \ 0\}$ and $V_{jmax} = \min \{V+(3-j)-V_1-V_2...-V_{j-1}, 1\}$. Like Case 1), the difference is that two faulty power cells should adjust the voltage level limitation rather than one faulty power cell. The total voltage level can also be maintained by further adjusting the voltage levels of healthy power cells based on constraint (1).

Case 5) (T_{i1} or T_{i4}) and (T_{j1} or T_{j4}) OC faults: the negative current is distorted. $V_i = 1$ *and* $V_j = 1$ cannot be adopted when $i_N < 0$ due to these two OC faults. The range constraints of V_i and V_j should be changed as

$$\begin{cases} V_{jmin} \leq V_j \leq V_{jmax1}, \ i_N < 0 \\ V_{imin} \leq V_i \leq V_{imax1}, \ i_N < 0 \end{cases}, i, j = 1, 2, 3, i \neq j \ (6)$$

where $V_{jmin} = \max \{V-(3-j)-V_1-V_2...-V_{j-1}, \ -1\}$ and $V_{jmax1} = \min \{V+(3-j)-V_1-V_2...-V_{j-1}, \ 0\}$. During the negative current period, the faulty power cells change the output voltage levels by using (6) while keeping the same limitation as (1) when $i_N > 0$.

Case 6) (T_{i2} or T_{i3}) and (T_{j2} or T_{j3}) OC faults: the positive current is affected. $V_i = -1$ and $V_j = -1$ cannot be adopted when $i_N > 0$ due to these two OC faults. The range constraints of V_i and V_j are changed as

$$\begin{cases} V_{jmin1} \leq V_j \leq V_{jmax}, \ i_N > 0 \\ V_{imin1} \leq V_i \leq V_{imax}, \ i_N > 0 \end{cases}, i, j = 1, 2, 3, i \neq j \ (7)$$

When $i_N > 0$, the voltage levels of failed power cells are still constrained by (1). Through adjusting the range constraints of voltage levels in the faulty power cells, the originally assigned vector is modified to ensure normal current paths and balanced DC voltages as much as possible. The cases of more than two OC switch faults can be similarly discussed to adjust the output voltage levels of each power cell as long as there are available redundant vectors to ensure total voltage levels.

4 Simulation and experimental results

4.1 Simulation results

The topology and algorithms in Fig. 1 and Fig. 2 are simulated in PLECS. The simulation parameters are listed in Table 1, which are the same as in the experimental setup. The OC switches are emulated by isolating the driving signals. In the following analysis, the faulty switches are considered to have been identified by existing diagnostic algorithms in [6] and [7], while the performance of the proposed method in terms of fault tolerance is mainly analyzed.

Table 1 Simulation and experimental parameters

Parameter	Simulation and experimental values
Grid voltage u_N (RMS)	110 V
Line inductor L_N	3 mH
Parasitic resistor R_N	0.1 Ω
Reference DC voltage U^*_{dc}	70 V
DC-link capacitor C_i	2.8 mF
Rated equivalent load R_l	20 Ω
H-bridge cell number	3
Grid frequency f_g	50 Hz
Switching cycle T_d	0.25 ms
Sampling/control cycle T_s	25 µs

Firstly, the two OC switches in one power cell are emulated, as described in Case 1). In Fig. 4, T_{11} and T_{13} OC faults cause a current distortion and the decrease of DC voltages at t_0, as well as affect the port voltages due to the absence of $V_1 = -1$ in $i_N > 0$ and $V_1 = 1$ in $i_N < 0$. After activating the vector modification in (2), the THD of grid current is significantly reduced from 26.79 % to 4.51 %, almost the same as in the pre-fault condition, and DC voltages also return to the reference value after 160 ms. In the fault state and tolerance state, the voltage level transition has no skip, and the total port voltage recovers to a approximately fault-free state, equalizing the stress time of the voltage on the remaining switches of the different power cells as much as possible.

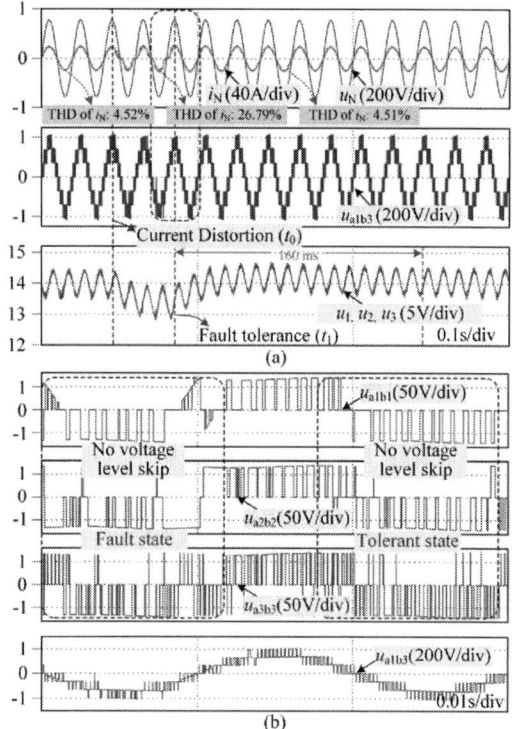

Fig. 4 Simulation results for T_{11} and T_{13} OC faults.

Furtherly, Case 2) is analyzed in power cell 2, where T_{21} and T_{24} OC faults occur, as shown in Fig. 5. The current harmonics are reduced but the fluctuation of DC voltages gets larger due to less available vectors after missing $V_2 = 1$ and 0 in $i_N < 0$. After $i_N > 0$, DC voltages are regulated to a balanced state. In any case, low harmonic grid currents and unaffected total voltages are obtained, reducing the impact of faults to a lower level.

Then, OC faults in two power cells are considered in Fig. 6, where T_{12} and T_{33} are set as OC switches. As discussed in Case 6), only positive grid current and port voltage are affected. After implementing the constraint (7) for faulty power cells 1 and 3 when $i_N < 0$, the THD of grid current is changed from 19.79 % to 4.53 %, almost the same as that in the pre-fault state, and the total port voltage u_{a1b3} is restored to the normal values. The DC voltages are also balanced despite of not performing the fault tolerance. However, the additional harmonics

caused by the fault are removed from the DC voltages after applying the proposed SVM-based fault-tolerant control.

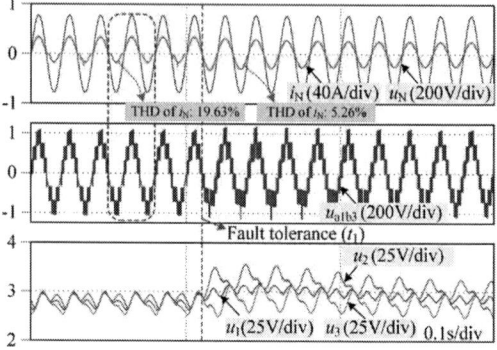

Fig. 5 Simulation results for T_{21} and T_{24} OC faults.

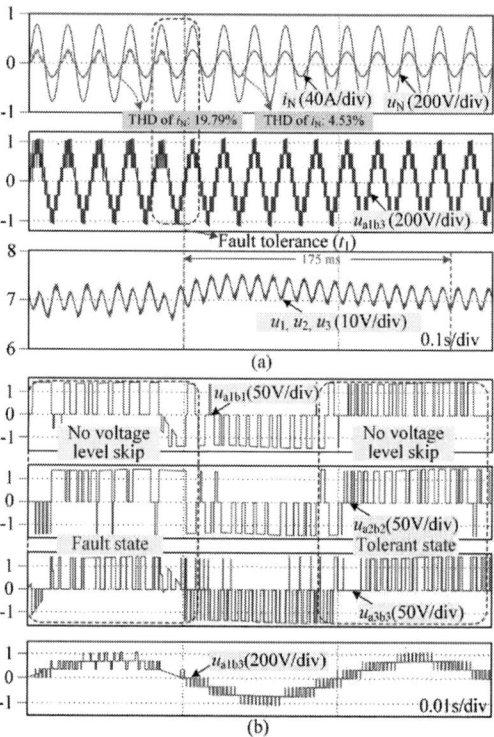

Fig. 6 Simulation results for T_{12} and T_{33} OC faults.

Moreover, the dynamic balancing performance of the DC voltage is investigated during the fault-tolerant period. As shown in Fig. 7, the load resistors of the different cells are successively adjusted as $(R_1 = R_2 = R_3 = 20\ \Omega)$, $(R_1 = R_2 = 20\ \Omega,\ R_3 = 40\ \Omega)$, $(R_1 = R_2 = 20\ \Omega,\ R_3 = \infty)$, $(R_1 = R_3 = 20\ \Omega,\ R_2 = 40\ \Omega)$, and $(R_1 = R_2 = 40\ \Omega,\ R_3 = 20\ \Omega)$. During different load transition processes, the fast dynamic response capability is well maintained to track the reference value. Therefore, the proposed method has a good DC voltage balancing capability to adapt to different load conditions, even for extreme unbalanced loads such as ($R_1 = R_2 = 20\ \Omega$, $R_3 = \infty$).

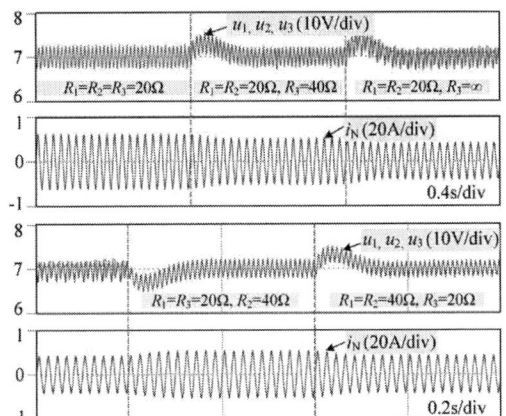

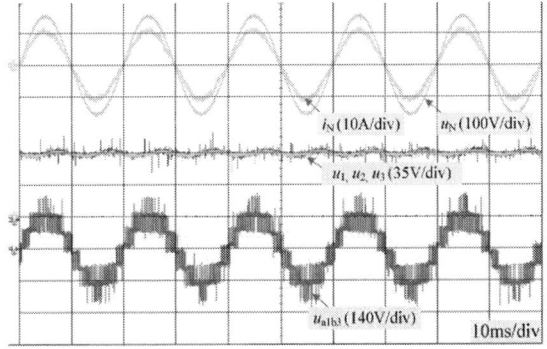

Fig. 7 DC voltages under the different load conditions after introducing the proposed method.

4.2 Experimental results

When applying the control and SVM in Fig. 2 to the single-phase CHBMC, Fig. 8 shows the experimental results. The grid-side unit power factor and balanced DC voltages are obtained to produce a total port voltage with seven levels. Such results demonstrate the good performance of the algorithm under normal conditions.

Fig. 8 Experimental results under the normal state.

For the multiple OC faults in one power cell, T_{11} and T_{13} OC faults are taken as an example to analyze the fault-tolerant performance. As seen in Fig. 9, after triggering these OC faults, the THD of grid current is increased from 6.61 % to 31.24 % and all DC voltages are slightly decreased. The port voltage u_{a1b1} and total port voltage u_{a1b3} have obvious voltage distortion due to the failed current paths when $V_1 = -1$ for $i_N > 0$ or $V_1 = 1$ for $i_N < 0$. After performing the proposed vector modification in Case 1), the grid current is recovered to the fault-free state with a low THD of 6.27 %. Meanwhile, all DC voltages are balanced and track to the reference value after 32 ms. Moreover, the distorted parts in the port voltage u_{a1b1} and total port voltage u_{a1b3} are restored by avoiding the output of absent voltage levels under the vector adjustment.

Afterward, considering multiple OC faults in two power cells, Fig.10 has illustrated experimental results for T_{21} and T_{33} OC faults, where the grid voltage is reduced to obtain the total port voltage with only five voltage levels, analyzing the effect of different voltage levels on the fault-tolerant function. Similarly, the distorted current appears during the fault period. u_{a2b2} has the abnormal output when $V_2 = 1$ for $i_N < 0$, while u_{a3b3} is affected when $V_3 = -1$ for $i_N > 0$. After implementing the vector modification based on (5) in Case 4), the grid current harmonics are comparable to the normal state, and the distortions in the port voltages no longer appear. All the operational performance returns to be similar to the normal condition.

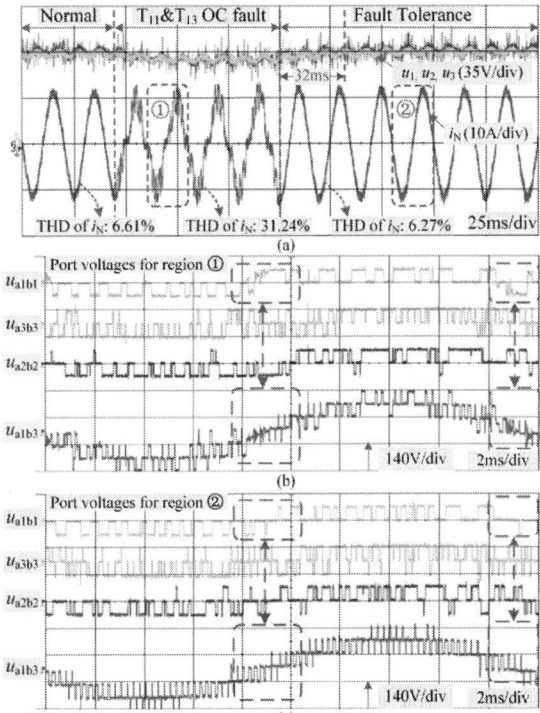

Fig. 9 Experimental results for T_{11} and T_{13} faults.

5 Conclusion

To improve the fault-tolerant control performance of the CHBMC under multiple OC fault conditions, this paper has proposed an SVM-based method to recover the operating performance. Utilizing the favorable conditions of redundant vectors in the SVM, assigned vector modifications under different multiple OC faults can effectively reduce the current distortion and maximize power output ability. However, the switching loss under different fault-tolerant states needs to be further analyzed.

The possibility of realizing a balanced power distribution of different cells also needs to be explored to improve the lifetime of remaining switches.

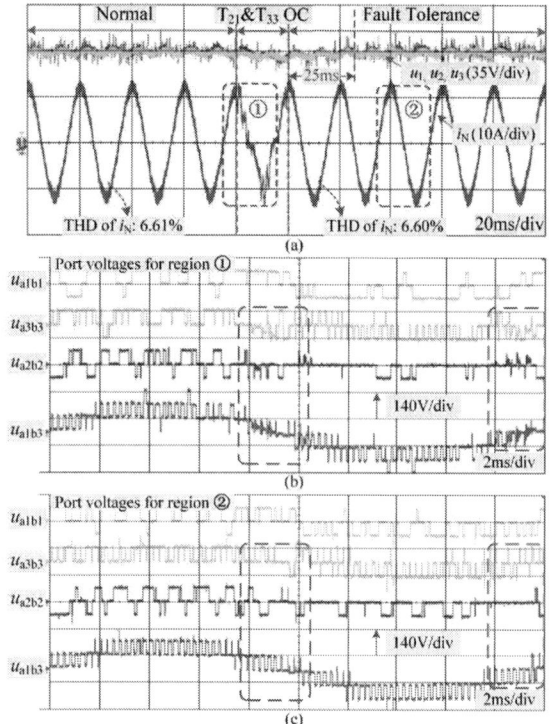

Fig. 10 Experimental results for T_{21} and T_{33} faults.

References

[1] D. Xie et al., "Diagnosis and resilient control for multiple sensor faults in cascaded H-bridge multilevel converters," *IEEE Trans. Power Electron.*, vol. 38, no. 9, pp. 11435-11450, Sept. 2023.

[2] I. P. Lobos and D. Dujic, "Estimating auxiliary power supply consumption of the modular multilevel converter submodule for the condition health monitoring," *PCIM Europe 2022; International Exhibition and Conference for Power Electronics, Intelligent Motion, Renewable Energy and Energy Management*, Nuremberg, Germany, 2022, pp. 1-9.

[3] L. Maharjan, T. Tajyuta, K. Maruyama and M. Tamate, "Discussions on fault-tolerant operation of modular multilevel converters using mechanical bypass switches," *2021 23rd European Conference on Power Electronics and Applications (EPE'21 ECCE Europe)*, Ghent, Belgium, 2021, pp. P.1-P.9.

[4] U. Choi, F. Blaabjerg, and K. Lee, "Study and handling methods of power IGBT module failures in power electronic converter systems," *IEEE Trans. Power Electron.*, vol. 30, no. 5, pp. 2517-2533, May 2015.

[5] N. Brahmendra Yadav Gorla, S. Kolluri, M. Chai and S. Kumar Panda, "A new model-based fault detection and localization scheme for cascaded H-bridge multilevel converter," *2020 IEEE Energy Conversion Congress and Exposition (ECCE)*, Detroit, MI, USA, 2020, pp. 4981-4987.

[6] D. Xie, C. Lin, H. Lin, and T. Basler, "SVPWM-Based dc voltage balancing and fault diagnosis method for a cascaded H-bridge multilevel converter," *2023 25th European Conference on Power Electronics and Applications (EPE'23 ECCE Europe)*, Aalborg, Denmark, 2023, pp. P.1-P.9.

[7] H. Wang, J. Wu, M. Ma, and Q. Chen, "A model-based power switch fault diagnosis strategy for cascaded H Bridge converter," *Microelectron. Reliab.*, vol. 150, 115095, Nov. 2023.

[8] A. Raki, Y. Neyshabouri, M. Aslanian, and H. Iman-Eini, "A fault-tolerant strategy for the safe operation of cascaded H-bridge multilevel inverter under faulty condition," *IEEE Trans. Power Electron.*, vol. 38, no. 6, pp. 7285–7295, Jun. 2023.

[9] N. Bisht and A. Das, "Maximizing the postfault output voltages of cascaded H-bridge STATCOM with reserved cell," *IEEE Trans. Ind. Appl.*, vol. 60, no. 1, pp. 584-595, Jan.-Feb. 2024.

[10] W. Wang, Y. Zhang, C. Xiao, Y. Li, C. Ma, and J. Liu, "A fast and generalized space-vector modulation scheme for cascaded H-bridge multilevel converters under faulty conditions," *2023 11th International Conference on Power Electronics and ECCE Asia (ICPE 2023 - ECCE Asia)*, Jeju Island, Korea, Republic of, 2023, pp. 1066-1071.

[11] H. Lin, M. Mehrabankhomartash, F. Yang, M. Saeedifard, J. Yang, and Z. Shu, "A flexible space vector modulation scheme for cascaded H-bridge multilevel inverters under failure conditions," *IEEE Trans. Ind. Electron.*, vol. 69, no. 12, pp. 11856–11867, Dec. 2022.

[12] D. Xie, C. Lin, H. Lin, W. Liu, Y. Du, and T. Basler, "OC switch fault diagnosis, pre- and postfault DC voltage balancing control for a CHBMC using SVM concept," *IEEE Trans. Power Electron.*, vol. 39, no. 1, pp. 677-692, Jan. 2024.

PCIM Europe 2024, 11– 13 June 2024, Nuremberg   DOI: 10.30420/566262410

Revolutionizing Mobility: The Second Life of Onboard Charging Systems in Commercial Vehicles

Viswanathan Ganesh [1,2], Ajay Krishna Voppu Muralikrishna [1,3], Björn Isaksson[3], Yujing Liu [1]

[1] Department of Electrical Engineering, Chalmers University of Technology, Gothenburg, Sweden
[2] Department of Architectural Engineering, Pennsylvania State University, University Park, PA, USA
[3] SiNIX GROUP AB, Gothenburg, Sweden

Corresponding author: Ajay Krishna Voppu Muralikrishna, ajay.krishna@sinix.se
Speaker: Ajay Krishna Voppu Muralikrishna, ajay.krishna@sinix.se

Abstract

This study investigates the potential for repurposing onboard charging electrical power systems (OCEPS) in commercial vehicles to enhance sustainability and economic efficiency within the electric vehicle (EV) industry. Beginning with market analyses to assess demand for such innovations, the research evaluates the readiness for transitioning to repurposed OCEPS, aiming to overcome existing design limitations and improve energy efficiency and reliability. A critical examination of current onboard charger technology provides opportunities for extending the lifecycle of these chargers beyond the vehicle's lifespan. This exploration supports the development of a sustainable business model that integrates circular economy principles, economic viability, and environmental stewardship by optimizing resource use and minimizing waste. The study proposes design enhancements to increase technical efficiency and introduces a business model that harmonizes economic and ecological objectives, thereby contributing to a more sustainable and circular economy in the electric mobility sector. The findings offer a systematic framework for repurposing onboard chargers, highlighting their potential in fostering technological advancement, economic growth, and environmental balance. This research presents a comprehensive approach to leveraging second life opportunities for onboard chargers, advocating for broader sustainability practices within the electric mobility ecosystem and addressing the challenges of electronic waste.

1 Introduction

The automotive industry is increasingly embracing electric vehicles (EVs) due to environmental and technological advancements. This shift necessitates the development of extensive charging networks to address range anxiety among users. As a result, EVs now have larger batteries, which, while extending their range, also increase their weight. The goal is to evaluate the lifespan of current OCEPS units and consider their repurposing as Off Board DC chargers. This involves analyzing their competitiveness in the market and formulating a cost-effective scaling strategy that is environmentally sustainable and economically beneficial, in line with circular economy principles. This approach also leverages the modules' existing certifications, minimizing waste and contributing to

e-waste reduction.

The main challenge with AC/DC converters lies in their non-linear loads, which introduce unwanted frequencies into the current, thus increasing Total Harmonic Distortion (THD) and degrading power quality. To mitigate the electronic noise caused by higher switching frequencies, a precisely designed filter is essential, yet issues with harmonic distortions and noise have arisen due to incorrect filter parameters. Additionally, the high power demand of semiconductor-equipped boxes leads to significant thermal losses, proportional to the square of the current, elevating the temperature of silicon devices and affecting their performance. Therefore, an efficient cooling system is necessary to keep component temperatures within safe limits.

Re-purposing these devices requires a thorough

understanding of each component's lifespan to ensure reliable operation. Components need to be easily replaceable and accessible, with the possibility of adding new elements to broaden the device's applications and enhance its redundancy, allowing it to adapt to various uses. The redesigned product should be compact, lightweight, and capable of scaling to higher capacities while complying with all regulatory standards. It must also offer competitive pricing and performance advantages over existing market offerings.

This article presents a comprehensive study on re-purposing onboard charging electrical power systems for commercial vehicles, with a focus on advancing sustainable transportation solutions. Through a systematic literature review, we examine the design, operational efficiency, and re-purposing viability of existing systems. Our research methodology encompasses a thorough electrical performance evaluation of the current onboard charging system, including simulations and practical testing to assess key performance metrics like voltage regulation and power efficiency. We also delve into theoretical calculations to understand the impact of circuit parameter variations on system efficiency and performance in the context of a Second Life application.

A significant portion of our work is dedicated to thermal analysis of Silicon Carbide (SiC) power modules, aiming to estimate their operational longevity and efficiency under various conditions. Addressing the challenges associated with re-purposing, such as technical constraints and regulatory compliance, is a critical aspect of our study. We advocate for the principles of the circular economy by identifying components with re-purposing potential, thus emphasizing the environmental benefits of reduced waste and enhanced resource efficiency. The integration of additional necessary components, supported by a detailed mechanical design facilitated by Computer-Aided Design (CAD) tools, is also discussed. Our market analysis underscores the economic feasibility and environmental implications of the proposed Second Life system, positioning it as a viable and competitive solution in the market.

2 Existing Product

The onboard charger (OBC) is essential in electric vehicles (EVs), enabling direct AC grid charging.

Its adoption in the automotive sector is driven by its convenience compared to the costly and bulky off-board charging methods [1, 3, 4, 12]. Unidirectional OBCs are preferred for their simpler hardware requirements and reduced impact on battery longevity [9]. However, the evolution of EV technology has showcased the vehicles' potential as mobile energy sources. Bidirectional OBCs, capable of supporting vehicle-to-grid (V2G) functions, allow for the return of excess energy to the grid during peak demands [5][7]. Furthermore, these OBCs enhance the utility of EVs by enabling vehicle-to-home (V2H), vehicle-to-load (V2L) power supply during grid outages, and vehicle-to-vehicle (V2V) energy sharing in emergencies [6, 8, 11]. The architecture of OBCs is generally categorized into Two Stage and Single Stage designs, with the former featuring separate AC/DC and DC/DC converters connected by a DC link, and the latter employing only an AC/DC converter, thus omitting the DC link [2].

Figure 1, depicts the high level electrical diagram of existing OCEPS. Similarly, Figure 2 provides an overall schematic diagram of the existing control that has voltage, current control and implements Phase Locked Loop (PLL). Table 1 provides a summary of critical components and operational frequency for switching components.

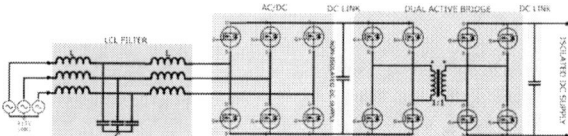

Fig. 1: Existing OCEPS - High level [10]

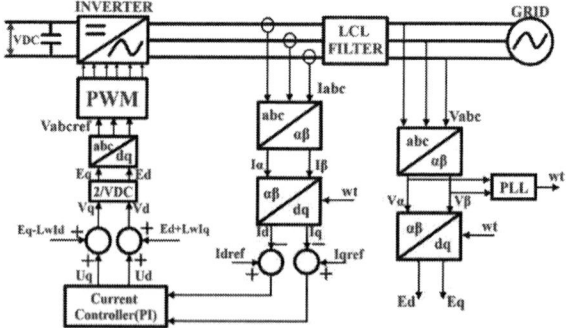

Fig. 2: Overall OCEPS Controls [10]

Parameter	Value	Unit
Switching Frequency	30	Khz
Resonant Frequency	3	Khz
Filter Inductor	59.35	μH
Filter Capacitor	44.12	μF
DC Link Capacitance	1150	μF

Tab. 1: Component Value of OCEPS [10]

2.1 Software implementation

The OCEPS was simulated in MATLAB SIMULINK based on the values of LCL filter and DC link as listed in Table 1. (Fig 3)

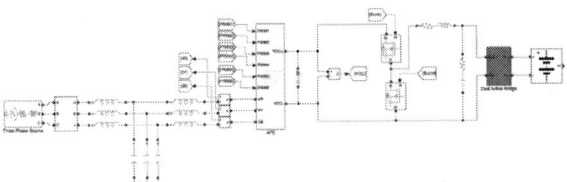

Fig. 3: OCEPS Model [10]

Figure 4 shows the software implementation of controls as depicted in Figure 2.

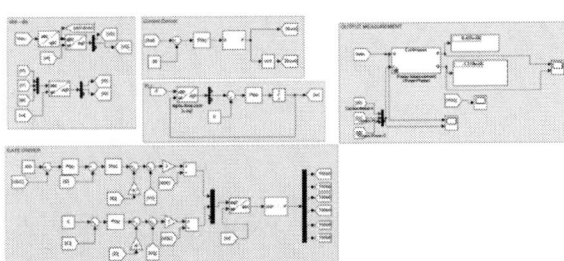

Fig. 4: OCEPS Control [10]

THD of existing OCEPS to be around 0.96%, in accordance with the IEEE standards. (Fig 5)

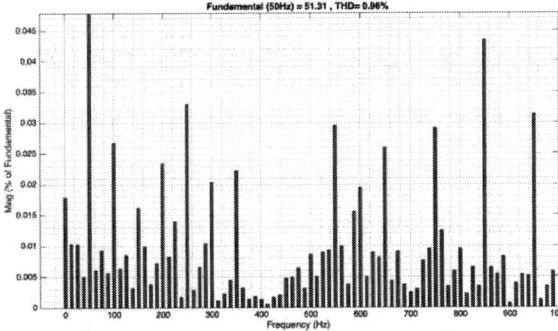

Fig. 5: Total Harmonic Distortion (THD) [10]

2.2 Lifetime calculation

The current power module underwent simulation and hardware performance analysis (Figure 6), revealing thermal problems and a reduced lifespan when evaluated according to Wohler's and Miner's rules. Wohler's rule utilizes the rainflow algorithm to estimate the fatigue experienced by a component throughout its entire lifespan. This involves transforming the varying stress from uni-axial loading into a constant stress amplitude. The rainflow algorithm gets its name from the pattern of rain rising and falling on a roof, mimicking the stress cycles a component undergoes.

Miner's rule, in contrast, is used to assess the cumulative damage, represented by D, as a fraction of the total finite life fatigue that a component can withstand before failing. If a component is broken, its D value exceeds 1; otherwise, it remains below 1.

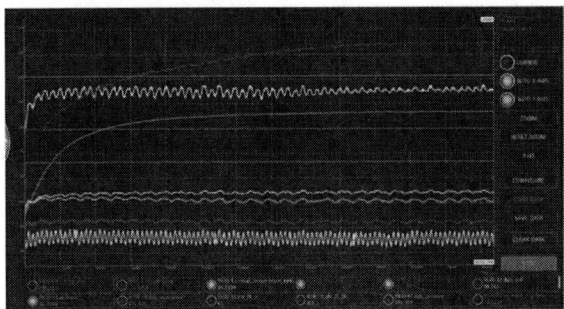

Fig. 6: Stress test of existing power module [10]

$$D = \sum_{i=1}^{k} \frac{n_i}{N_{fi}} = 1 \qquad (1)$$

Where N_{fi} represents the number of cycles until failure, and N_i indicates the number of cycles experienced under a specific stress load.

$$T_{life} = \frac{T}{D} \qquad (2)$$

Where **T** represents the time period of load.

By applying a streamlined approach to calculate lifespan, the Modified Coffin-Manson model can be employed, utilizing the curves shown in Figure 7 and following the formula presented in Equation 3.

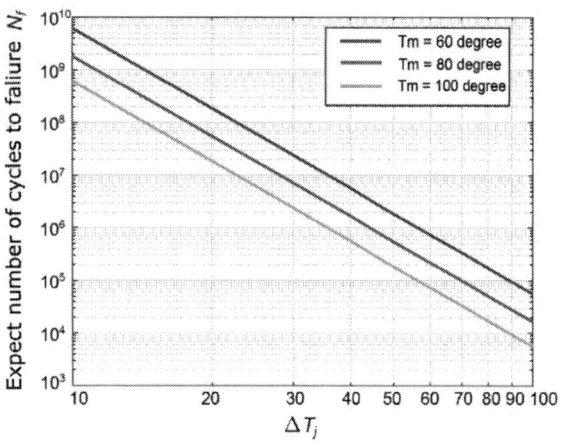

Fig. 7: Wohler Curve [10]

$$N_f = A\Delta T_j^{\alpha} e^{\left(\frac{E_a}{k_B T_m}\right)} \tag{3}$$

Where ΔT_j denotes the temperature difference across the junction, A and α are parameters of the model. E_a represents the activation energy, K_B is the Boltzmann constant, and T_m refers to the mean temperature measured in Kelvin.

	A	α	E_a	k_B
Value	302500	-5.039	9.89×10^{-20} J	$1.38 \times 10^{-23} JK^{-1}$

Tab. 2: Values of Modified Coffin Manson Lifetime Model

The simulation employs an acceleration factor where 1 day corresponds to 1 week, resulting in an estimated lifespan of approximately 1.66 years. Taking into account a reliability factor of 90%, the Mean Time To Failure (MTTF) is calculated as

$$MTTF = \frac{Lifetime}{log(2)} = \frac{1.66}{log(2)} = 5.51 \; years$$

3 Component study

An comprehensive study was performed to compare the existing power module on factors like cost, performance, peak temperature, losses and efficiency (Fig 8 - Fig 10).

Based on linear weighed method for cost, peak temperature, losses and efficiency, the tested module were provided normalized scores where 10 is the highest and 0 is the lowest score (Fig 11). This

method of ranking helps in selecting the overall best module for the application.

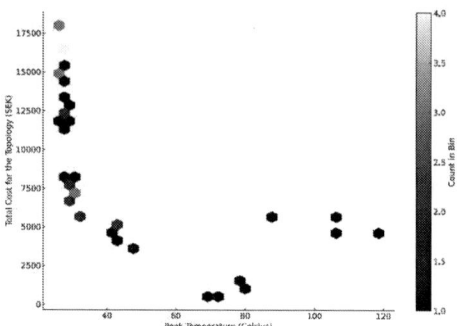

Fig. 8: Hexbin plot of peak temperature vs total cost for the topology

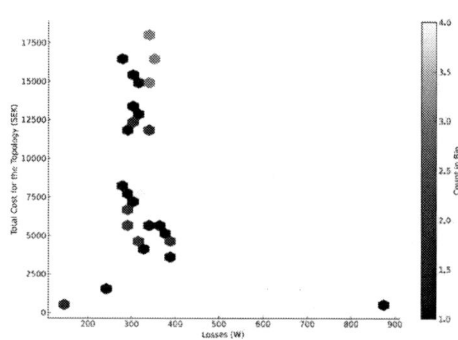

Fig. 9: Hexbin plot of total losses vs total cost for the topology

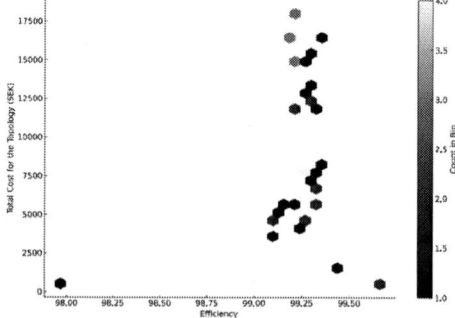

Fig. 10: Hexbin plot of efficiency vs total cost for the topology

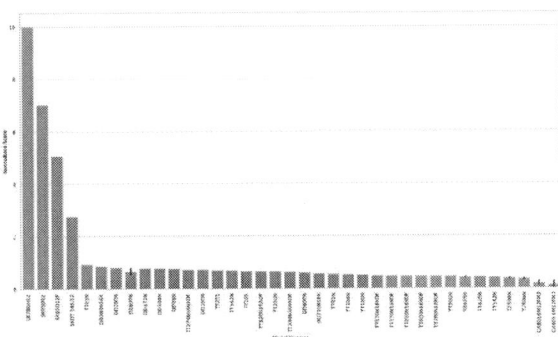

Fig. 11: Normalized score of all tested model considering equal weighted for cost, peak temperature, losses and efficiency

4 Second Life Application

Considering an EV truck with OCEPS, which has operated up to its MTTF, it's notable that 90% of these systems remain functional beyond this point. These operational OCEPS units can be transitioned into off-board DC chargers (Fig 12 - Fig 14), with the added advantage of bypassing the requirements for Noise, Vibration, and Harshness (NVH) testing.

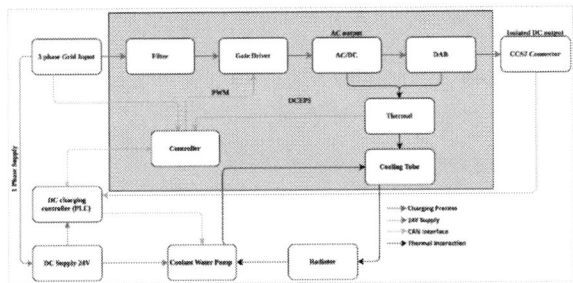

Fig. 12: High level diagram of repurposed OCEPS to DC charger [10]

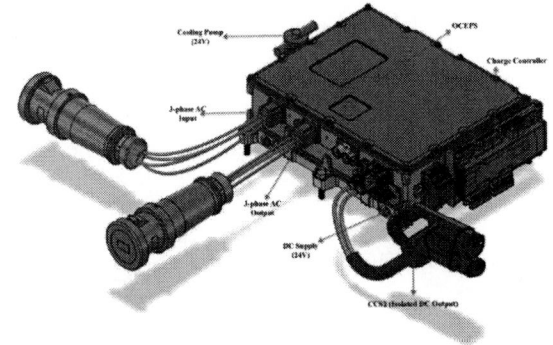

Fig. 13: Repurposed OCEPS [10]

Fig. 14: Scalable design of Repurposed DC [10]

5 Feasibility Analysis

The principle of reusability is fundamentally backed by the values of sustainability and the circular economy, enabling a company to scale and generate profit efficiently. Therefore, a clear understanding of cost split up of components is required to effectively perform feasibility analysis.

Parts	Value (%)
OCEPS	40.42
EXTERNAL BODY	2.25
PLC COMMUNICATION	44.91
AC-DC	1.20
PUMP	0.75
ASSEMBLY KIT	0.75
RADIATOR AND TUBING	1.95
ASSEMBLY	1.80
DISSASSEMBLY	2.69
TESTING	1.80
MISCELLANEOUS	1.50
TOTAL	100.00

Tab. 3: Value of Components (Repurposed OCEPS)

The table 3 outlines the component value distribution in repurposing Onboard Electric Power Systems (OCEPS), with the bulk of value coming from the OCEPS (40.42%) and PLC communication (44.91%). These key components significantly overshadow others like the external body, AC-DC converters, and various assembly parts, contributing to less than 15% combined.

	EKO-ENERGETYKA	DESIGNWORK	HELIOX	KEMPOWER 1/2	PHIHONG	Proposed DC
Input Specs (3Φ)	80A, 360-440V 43kVA	63A, 200-460V 44kVA	63A, 400V 43kVA	63A, 380-480V 43.6kVA	60A, 380-415V 30kVA	63A, 230-400V 43.6kVA
Output Power/Current	40kW 67A	42kW 120A	40kW 67A	40/20kW 60A	30kW 50A	43kW 59A
Voltage Range (V)	150-950	250-1000	100-1000	920-1000	150-950	500-850
Cooling Method	Air	Air	Air	Air	Air	Liquid
Dimensions (DxWxH)mm	600x860x1060	500x240x730	500x500x900	670x640x1220	589x490x740	750x500x170
Weight (kg) & Cost Ratio	160 1.32	52 1.41	120 1.63	120/127 1.96/2.02	80 1.04	52 1.00
Power Density	30.29 kW/Cost 73.13 kW/m^3 0.25 kW/kg	29.87 kW/Cost 479.45 kW/m^3 0.807 kW/kg	24.48 kW/Cost 177.77 kW/m^3 0.33 kW/kg	20 kW/Cost 76.46 kW/m^3 0.322 kW/kg	28.92 kW/Cost 138.35 kW/m^3 0.375 kW/kg	43 kW/Cost 722.68 kW/m^3 0.827 kW/kg
Safety & Compliance	All models comply with electrical safety, residual current, overvoltage, over temperature, and EMC standards					

Tab. 4: Market Analysis and Comparison

In a condensed market analysis comparing the proposed direct current (DC) system against competitors in similar power categories (Tab 4), the proposed DC system showcases significant advantages. It features competitive input specifications and leads in output power at 43kW, demonstrating its efficiency. Unique to this system is its liquid cooling method, contrasting with the air-cooled approaches of its rivals, indicating enhanced thermal management. Its compact dimensions and light weight underscore its installation ease and space-saving benefits. Notably, it offers the best cost efficiency and highest power density, both in terms of cost and physical size, affirming its superior value and performance. Safety features and compliance with electrical standards are fully met, ensuring its reliability and market readiness.

6 Conclusion

This study explores the potential of repurposing on-board charging electrical power systems (OCEPS) in commercial vehicles to foster sustainability and economic efficiency in the electric vehicle (EV) industry. It assesses market demand for such innovations and evaluates the feasibility of transitioning to repurposed OCEPS, aiming to address current technological limitations and enhance energy efficiency and reliability. By examining the lifecycle of onboard charger technology, the research identifies opportunities to extend its usability beyond a vehicle's lifespan, integrating circular economy principles into a sustainable business model that balances economic and ecological goals. The proposed design improvements and business strategy underscore the importance of optimizing resource use and minimizing waste. The findings offer a framework for the repurposing of onboard chargers, showcasing their role in advancing technology, economic growth, and environmental sustainability within the electric mobility sector, while also addressing electronic waste challenges.

Acknowledgment

The authors wish to express their gratitude to Volvo Group Trucks Technology (Volvo GTT), Sweden, for providing access to their testing facilities and allowing the publication of the test results. This work is a condensed version of a master's thesis submitted to both Volvo GTT and Chalmers University of Technology (For detailed version check out [10]). The support and guidance from these institutions have been invaluable. The authors also extend their thanks to all contributors to this research.

References

[1] Dylan C Erb, Omer C Onar, and Alireza Khaligh. "Bi-directional charging topologies for plug-in hybrid electric vehicles". In: *2010 Twenty-Fifth Annual IEEE Applied Power Electronics Conference and Exposition (APEC)*. IEEE. 2010, pp. 2066–2072. DOI: 10.1109/APEC.2010.5433520.

[2] Viswanathan Ganesh, VM Ajay Krishna, S Senthilmurugan, and S Hemavathi. "Modeling of Electric Vehicle DC Fast Charger". In: *Electric Transportation Systems in Smart Power Grids*. CRC Press, 2023, pp. 269–292.

[3] Byung-Kwon Lee, Jong-Pil Kim, Sam-Gyun Kim, and Jun-Young Lee. "An isolated/bidirectional PWM resonant converter for V2G (H) EV on-board charger". In: *IEEE Transactions on Vehicular Technology* 66.9 (2017), pp. 7741–7750. DOI: 10.1109/TVT.2017.2678532.

[4] Il-Oun Lee. "Hybrid PWM-resonant converter for electric vehicle on-board battery chargers". In: *IEEE Transactions on Power Electronics* 31.5 (2015), pp. 3639–3649. DOI: 10.1109/TPEL.2015.2456635.

[5] Zhengyang Liu, Bin Li, Fred C Lee, and Qiang Li. "Design of CRM AC/DC converter for very high-frequency high-density WBG-based 6.6 kW bidirectional on-board battery charger". In: *2016 IEEE Energy Conversion Congress and Exposition (ECCE)*. IEEE. 2016, pp. 1–8. DOI: 10.1109/ECCE.2016.7855024.

[6] Mahmoud Nassary, Mohamed Orabi, Maged Ghoneima, and Mohamed K El-Nemr. "Single-phase isolated bidirectional AC-DC battery charger for electric vehicle–review". In: *2019 International conference on innovative trends in computer engineering (ITCE)*. IEEE. 2019, pp. 581–586. DOI: 10.1109/ITCE.2019.8646528.

[7] Sepehr Semsar, Theodore Soong, and Peter W Lehn. "On-board single-phase integrated electric vehicle charger with V2G functionality". In: *IEEE Transactions on Power Electronics* 35.11 (2020), pp. 12072–12084. DOI: 10.1109/TPEL.2020.2982326.

[8] Seyedfoad Taghizadeh, M Jahangir Hossain, Noushin Poursafar, Junwei Lu, and Georgios Konstantinou. "A multifunctional single-phase EV on-board charger with a new V2V charging assistance capability". In: *IEEE Access* 8 (2020),

pp. 116812–116823. DOI: 10.1109/ACCESS.2020.3004931.

[9] Kotub Uddin, Matthieu Dubarry, and Mark B Glick. "The viability of vehicle-to-grid operations from a battery technology and policy perspective". In: *Energy policy* 113 (2018), pp. 342–347. DOI: 10.1016/j.enpol.2017.11.015.

[10] Ajay Krishna Voppu Muralikrishna and Viswanathan Ganesh. "Second Life for Commercial Vehicles Onboard Charging Electrical Power System". In: (2023). DOI: 20.500.12380/307560.

[11] Chen Wei, Jianwen Shao, Binod Agrawal, Dongfeng Zhu, and Haitao Xie. "New surface mount SiC MOSFETs enable high efficiency high power density bi-directional on-board charger with flexible DC-link voltage". In: *2019 IEEE Applied Power Electronics Conference and Exposition (APEC)*. IEEE. 2019, pp. 1904–1909. DOI: 10.1109/APEC.2019.8721866.

[12] Yang Xiao, Chunhua Liu, and Feng Yu. "An effective charging-torque elimination method for six-phase integrated on-board EV chargers". In: *IEEE Transactions on Power Electronics* 35.3 (2019), pp. 2776–2786. DOI: 10.1109/TPEL.2019.2924538.

PCIM Europe 2024, 11– 13 June 2024, Nuremberg   DOI: 10.30420/566262411

A Behavioral Transient Model for IGBT Device with Anti Parallel Freewheeling Diode

Shiwu Zhu[1], Chunlin Zhu[1], Qi Huang[3], Ken Zhang[3], Katsuaki Saito[2], Wei Gong[1], Huiling Zuo[3], Junli Xiang[3]

[1] Nexperia, UK

[2] Nexperia, Japan

[3] Nexperia, China

Corresponding author: SHIWU ZHU, William.zhu@nexperia.com
Speaker: SHIWU ZHU, William.zhu@nexperia.com

Abstract

This paper presents a comprehensive behavioral model for Nexperia's new Insulated Gate Bipolar Transistor (IGBT) products, which is specifically designed for SPICE simulation. The model is constructed upon the classical "Mosfet + PNP transistor" framework for the IGBT and incorporates subcircuits to capture characteristics related to parasitic capacitances, di/dt and dv/dt adjustment, tail current and diode reverse recovery, as well as junction temperature simulation. Calibrated using datasheet information, the model not only achieves high accuracy for device fundamental characteristics but also operates with a fast simulation speed, enabling accurate topology-level simulation analysis. The detailed model structure and calibration guidelines are introduced in the paper. The proposed model and methodology can be extended to any IGBT device.

1   Introduction

It has been widely recognized that SPICE-based simulation analysis plays a pivotal role in power electronics system design. Over the past decades, numerous models have been proposed in the literature [1]-[2] and can be classified into two main categories: device-physics models and behavioral models. Device-physics models [3]-[4] rely on solving one-dimensional ambipolar diffusion equations. These models can provide very good accuracy in transient behavior but require prohibitively high computational demands, posing challenges in terms of convergence and speed.

In contrast, behavioral models [5]-[10], based on curve-fitting techniques and calibrations against datasheet specification, achieve a balance between accuracy and computational demands. These models not only provide reasonable accuracy for fundamental device characteristics but also offer rapid simulation speeds. However, existing behavioral models for IGBT devices often lack clarity regarding calibration methods or are tailored to specific operating points.

Nexperia recently unveiled a series of Insulated Gate Bipolar Transistor (IGBT) products, as depicted in Fig. 1. These IGBT products feature a robust and cost-effective carrier-stored, trench gate advanced field stop construction. In this paper, we present a behavioral model specifically developed for the new IGBT products. This paper not only reveals the intricate model structure but also provides insights into calibration guidelines. Additionally, comparison results for fundamental device characteristics and switching waveforms drawing from simulations and experimental tests conducted on the new NGW40T65H3DFP IGBT device [11] are also provided.

Fig. 1   Nexperia 650V discrete IGBT products

	IGBT IC$_{nom}$ 100°C	Diode IF$_{nom}$ 100°C	TO-247-31	Application
IGBT M3 + full rated Diode	40 A	40 A	NGW40T65M3DFP	Drive
	50 A	50 A	NGW50T65M3DFP	Drive
	75A	75A	NGW75T65M3DFP	Drive
IGBT H3 + full rated Diode	40A	40 A	NGW40T65H3DFP	Solar / UPS/ Welding/Charger
	50A	50 A	NGW50T65H3DFP	Solar / UPS/ Welding/Charger
	75A	75 A	NGW75T65H3DFP	Solar / UPS/ Welding/Charger

2   Model description

2.1   Model interface

The proposed model, depicted in Fig. 2 comprises the following four distinct components.
- IGBT behavioral model

PCIM Europe 2024, 11– 13 June 2024, Nuremberg   DOI: 10.30420/566262411

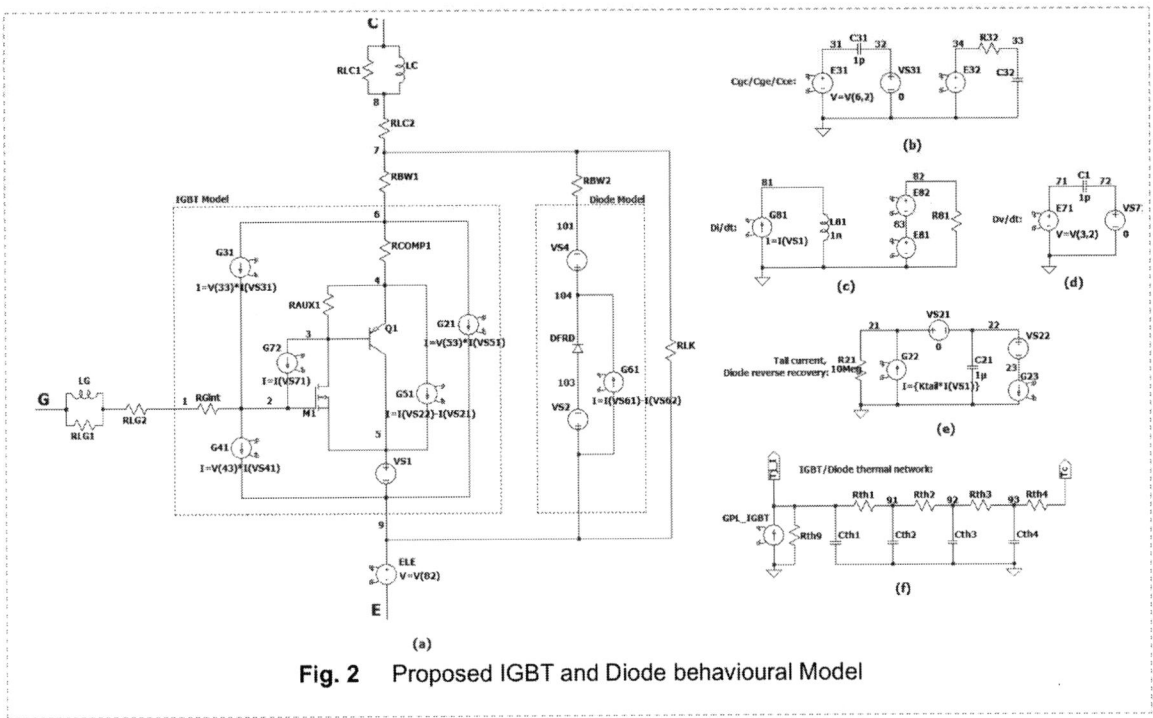

Fig. 2 Proposed IGBT and Diode behavioural Model

- Diode behavioral model
- Package impedance
- Thermal model

The model interface, illustrated in Fig. 3(b), employs conventional IGBT interfaces denoted by C (collector), G (gate), and E (emitter). Additionally, it incorporates temperature-related connections: Tj_I (junction temperature of the IGBT), Tj_D (junction temperature of the diode), and Tc (case temperature). These temperature connections function as voltage pins, with a 1 V potential difference corresponding to a temperature difference of 1 °C.

```
.SUBCKT NGW40T65H3DFP_L1 C G E Tj_I Tj_D Tc
+params: DeltaVGEth={0} DidtFactor={0} DvdtFactor={0}
```

(a)

(b)

Fig. 3 Model name syntax (a) and pin interfaces (b)

Furthermore, the model exposes three user-accessible parameters as shown in Fig. 3(a):

- DeltaVGEth: adjusts the gate emitter threshold voltage embedded in the model, facilitating device paralleling analysis. Its adjustable range is defined as ±0.5 V.

- DidtFactor and DvdtFactor: used to fine-tune the di/dt and dv/dt speeds at switching transients. These parameters offer an adjustable range of ±100% relative to the model-predefined di/dt and dv/dt speeds.

The proposed model was developed within the SPICE simulation environment LTspice. LTspice is a free software tool by Analog Devices and is widely employed in electronics and power electronics design and simulation analysis.

2.2 IGBT behavioral model

As depicted in Fig. 2(a), the proposed IGBT behavioral model is constructed upon the classical "Mosfet + PNP transistor" framework, utilizing LTspice's built-in device models to achieve efficient simulation speeds. Moreover, the model's fundamental structure provides flexibility for future enhancements, accommodating more detailed device physics characteristics, such as variable depletion capacitance and latch-up effects.

In Fig. 2(a), the Mosfet device (M1) adheres to a typical VDMOS model, while the PNP transistor (Q1) follows the classical Gummel-Poon model. Key modelling parameters for the Mosfet and PNP transistor are summarized in Table 1 and Table 2. These parameters collectively define essential

2894

IGBT static characteristics, including transconductance (g_m), collector-emitter saturation voltage (V_{CEsat}), and gate-emitter threshold voltage (V_{GEth}). Although some parameters exhibit coupling, most device characteristics of interest can be effectively defined using one or two parameters according to the author's calibration experience.

Table 1 IGBT Mosfet part key parameters

LTSpice parameter name	Parameter description	Units
Vto	Zero-bias threshold voltage	V
Ksubthres	Subthreshold conduction parameter	-
Kp	Transconductance parameter	A/V²

Table 2 IGBT PNP transistor key parameters

LTSpice parameter name	Parameter description	Units
Is	Transport saturation current	A
Ibe	Base-emitter saturation current	
Bf	Ideal maximum forward beta	-
Nf	Forward current emission coefficient	-
Ikf	Corner for forward beta high current roll-off	A
Eg	Energy gap for temperature effect on Is	eV
Xtb	Forward and reverse beta temperature exponent	-
Xti	Temperature exponent for Is	-
Tf	Ideal forward transit time	sec

During model calibration, Mosfet Kp and PNP transistor Bf determine the g_m value. Adjusting these parameters aligns the simulated g_m slope with the datasheet curve. Mosfet Vto enables shifting of the g_m curve along the gate-emitter voltage (V_{GE}) axis, as illustrated in Fig. 4(a). Adjusting Kp, Bf and Vto, zone ② of g_m curve can be aligned with the datasheet curve. Detailed adjustments to the Ksubthres value align both the g_m curve zone ① and the V_{GEth} with datasheet values. The temperature coefficient Xtb accounts for Bf's temperature dependence. For calibration at different temperatures, Vto and Ksubthres should also be designed as temperature-dependent variables.

The PNP transistor's saturation current (Is) significantly impacts leakage current at elevated junction temperatures and voltages. Temperature factors, Eg and Xti, further influence its behavior. Specifically, Ib defines the initiation point of the V_{CEsat} curve (referred to as zone ①, as depicted in

Fig. 4(b)), Ikf adjusts the slope of V_{CEsat} at high current levels (zone ② in Fig. 4(b)) and Nf permits lateral shifts of the curve along the V_{CE} axis.

In Fig. 2(a), resistor RCOMP1 compensates for simulated V_{CEsat} at high current levels, similarly as Ikf. Both Ikf and RCOMP1 exhibit temperature dependence, facilitating V_{CEsat} calibration across different temperatures. Furthermore, the PNP transistor parameter Tf represents the time constant associated with minority-carrier charge storage in the base region, significantly influencing turn-off energy (Eoff).

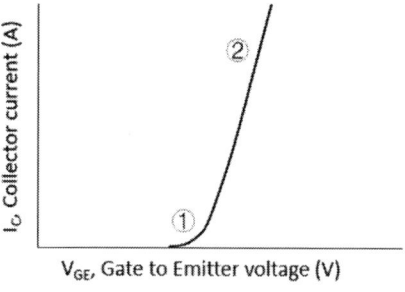

(a)

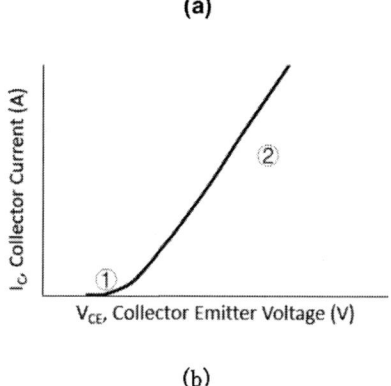

(b)

Fig. 4 Illustrations of g_m and V_{CEsat}

The IGBT's dynamic behavior is closely tied to parasitic capacitances. To simulate these effects, the proposed model employs voltage-dependent current sources G21, G31, and G41 which modelling the collector-emitter capacitance (C_{ce}), collector-gate capacitance (C_{gc}), and gate-emitter capacitance (C_{ge}), respectively. Taking C_{gc} as an example, it is generated using a lookup table method, as illustrated in Fig. 2(b). Voltage source E31 measures the dv/dt between nodes 6 and 2, converted to a charging current via the serially connected 1 pF capacitance C31. Voltage-dependent voltage source E32 stores capacitance values (e.g., C_{gc}) from datasheet curves. Resistor R32 and capacitor C32 form a filtering circuit, ensuring smooth capacitance transitions and simulation convergence. By multiplying the current through

VS31 with the voltage at node 33, G31 effectively models the equivalent effect of C_{gc}. The underlying relations are expressed by Eq. (1). Similarly, C_{ce} and C_{ge} are generated using G21 and G41, respectively.

$$
\begin{aligned}
G31 &= V(33) \cdot I(VS31) \\
&= C_{gc\_filtered} \cdot \left(1pF \cdot \frac{dV(6,2)}{dt}\right)
\end{aligned} \quad (1)
$$

The dv/dt adjustment function is implemented via current source G72, as depicted in Fig. 2(a). This function follows the same design principles as the above explained capacitor equivalent method, with its subcircuit illustrated in Fig. 2(d). Notably, G72 connects between M1 drain and source terminals, exerting no influence on the values of C_{ce}, C_{gc}, and C_{ge}. By adjusting the value of C1, we can fine-tune the switching transient dv/dt. Additionally, leveraging the "Limit" function inherent in SPICE allows us to control the positive or negative effectiveness of G72, thereby selectively adjusting the dv/dt value during turn-on and/or turn-off transients.

As mentioned in reference [9], the standard PNP transistor model inadequately captures the switching behavior of IGBTs due to high-level carrier injection and non-quasi-static effects. To address this limitation, the model employs an equivalent sub-circuit, as depicted in Fig. 2(e), specifically designed to simulate the IGBT tail current effect. In Fig. 2(a), current source G51 mirrors the equivalent tail current generated by the sub-circuit in Fig. 2(e), in which, C21 is a 1uF capacitor, and G23 is a discharge current source with the value of $V(22) \times 1\,\mu F$ / Taub. The parameter Taub represents the discharge time, calibrated to match the tail current duration.

Mathematically, the sub-circuit in Fig. 2(e) is described by Eq. (2). Specifically, $K_{tail} \times I(VS1)$ mirrors the IGBT emitter current, with K_{tail} calibrated based on the tail current peak value. To account for temperature-related effects, both Taub and K_{tail} should be temperature-dependent.

$$
\begin{aligned}
G51 &= I(VS22) - I(VS21) \\
&= V(22) \cdot \frac{1uF}{Taub} - K_{tail} \cdot I(VS1)
\end{aligned} \quad (2)
$$

2.3 Freewheeling Diode

The Freewheeling Diode behavioral model is constructed based on the standard Berkeley SPICE semiconductor diode which is integrated in LTspice. Additionally, an auxiliary circuit similar to the IGBT tail current subcircuit, is employed to replicate the reverse recovery charge effect (Fig. 2(a)

and Fig. 2(e)). Specifically, voltage controlled current source G61 mirrors the equivalent reverse recovery current generated by the sub-circuit in Fig. 2(e), following the principles outlined in Eq. (2).

It is noteworthy that the LTspice Diode model inherently supports reverse recovery modeling. However, the author observed that its inclusion significantly impacts simulation convergence. To mitigate this, the proposed model sets Diode parameters CJO (zero-bias junction capacitance) and Tt (transit time) both to zero, effectively deactivating its reverse recovery modeling.

Table 3 enumerates key parameters defining the Diode's static behavior. Is determines the leakage current, similar to the PNP transistor and marks the starting point of the forward saturation voltage curve (V_F, zone ①, Fig. 5). Eg and Xti represent temperature coefficients associated with Is. N shifts the V_F curve along the voltage axis. Ikf defines the slope of the V_F curve at high current levels (zone ②, Fig. 5). Tikf accounts for the temperature dependence of Ikf. Rs exhibits effects similar to Ikf and is also temperature-dependent.

Table 3 Diode model key parameters

LTspice parameter name	Parameter description	Units
Is	Saturation current	A
N	Emission coefficient	-
Ikf	High-injection knee current	A
Rs	Ohmic resistance	Ω
Eg	Activation energy	eV
Xti	Saturation current temperature exponent	-
Tikf	Linear Ikf temp coeff.	/°C

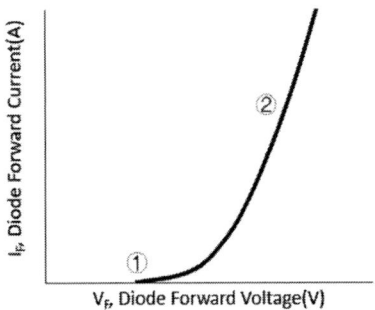

Fig. 5 Illustration of V_F

2.4 Package impedance

In IGBT devices, package impedance—comprising parasitic resistors and inductors—exerts a significant influence on both static and dynamic characteristics. To address this, we employed Ansys Q3D Extractor to extract equivalent resistance and inductance values for typical parasitic components. As depicted in Fig. 2(a), terminal inductors (LC, LG, and ELE) represent the equivalent inductance at terminals C, G and E. Terminal resistors (RLC2 and RLG2) account for terminal parasitic resistance. Bonding wire equivalent resistances (RBW1 and RBW2) model the resistance associated with bonding wires. Additionally, the model includes arbitrary parallel resistors (RLC1 and RLG1) to enhance modelling convergence. Furthermore, resistor RLK compensates for high-accuracy leakage current.

Among all parasitic components, the most influential parasitic component is the common stray inductance, particularly during turn-on and turn-off transients. This inductance arises from the coupling of the gate drive loop and the power emitter path. Essentially, it represents a parasitic inductance formed by wire bonds and copper tracks on the substrate. The common stray inductance negatively impacts gate drive voltage, thereby slowing down the switching speed in most power device packages. However, in multi-die paralleling designs, it can also exhibit positive feedback due to interference among paralleled dies. By leveraging this inductance value, the proposed model can modulate turn-on and turn-off transient speeds.

In the proposed model, the emitter inductance is represented by a voltage-controlled voltage source, denoted as ELE, as illustrated in Fig. 2(a). As depicted in Fig. 2(c), the positive transient voltage of ELE is determined by E81 while the negative transient voltage originates from E82. Specifically, E81's voltage is computed using Eq. (3), where K_{PLe} represents the absolute value of the equivalent common stray inductance during turn-on transient. Similarly, E82 employs a comparable approach, incorporating a distinct factor K_{NLe} for common stray inductance which becomes effective at turn-off transient. This approach effectively regulates di/dt at both turn-on and turn-off transients at the circuit level. By adjusting K_{PLe} and K_{NLe} through the model interface parameter DidtFactor, as mentioned in paragraph 2.1, the model user gains flexibility in controlling the di/dt value during switching transients.

$$
\begin{aligned}
E81 &= K_{PLe} \cdot V(81) \\
&= K_{PLe} \cdot 1nH \cdot \frac{dI(VS1)}{dt}
\end{aligned}
\tag{3}
$$

2.5 Thermal model

To achieve thermal and electrical interactive simulation, the proposed model incorporates two sub-circuits for simulating junction temperature for both IGBT and Diode. Figure. 2(f) illustrates a subcircuit example for real-time simulation of IGBT die junction temperature. It employs a voltage-controlled current source, GPL_IGBT, to compute the transient power loss of the IGBT die. It multiplies the transient voltage between the collector and emitter terminals by the transient current flowing through the emitter terminal. The resulting power loss value is then fed into a junction-to-case equivalent Cauer thermal network.

The choice of the Cauer thermal network over the Foster thermal network is deliberate. Unlike the Foster network, the Cauer network does not require the user to specify thermal capacitance values for external power dissipation components (such as thermal interface material and heatsinks) but still generates reasonable temperature fluctuations for the junction-to-case.

In Fig. 6, a typical application circuit for the thermal model was presented. Parameters Rhs and Chs represent the heatsink thermal impedance, while Rint denotes the thermal resistance of the thermal interface material. The voltage values of nets Tj_IGBT, Tj_Diode, and Tc provide the simulated junction temperature for IGBT and Diode dies, as well as the device's copper case temperature.

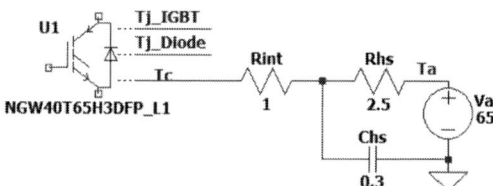

Fig. 6 An application example of the thermal model

3 Calibration Results

The proposed device model has been verified across a range of Nexperia's new IGBT products. Figures 7 to 13 present fundamental characteristics comparisons between simulations and experimental tests based on the IGBT product NGW40T65H3DFP [11].

Figures 7 and 8 provide a comparison of the measured and simulated V_{CEsat} for the IGBT and V_F for the Diode with respect to collector current (I_C). The calibration span for the IGBT collector current and Diode forward current extends up to 80A, which is twice the device's nominal current (I_{Cnom}). The discrepancy between the measured and simulated values of V_{CEsat} and V_F remains within ±3.0% across this specified current range.

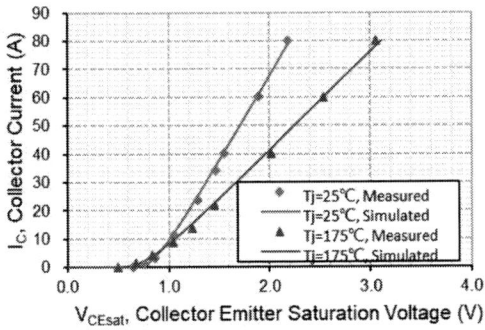

Fig. 7 I_C as a function of V_{CEsat} with Tj=25℃ and 175℃ at V_{GE}=15V

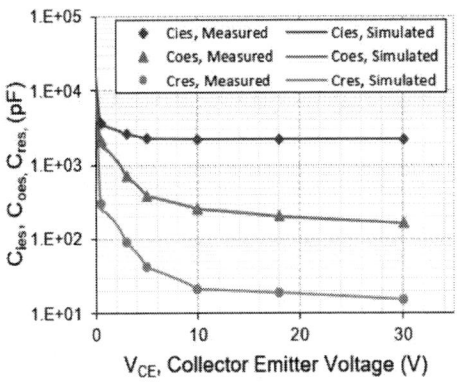

Fig. 9 C_{ies}, C_{oes}, C_{res} as a function of VCE

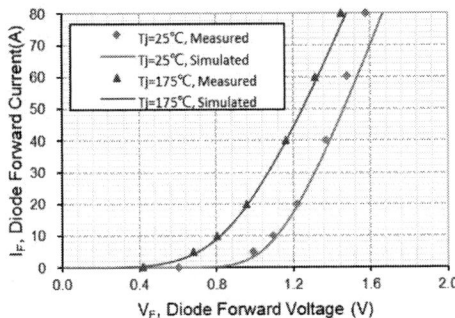

Fig. 8 I_F as a function of V_F with Tj=25℃ and 175℃

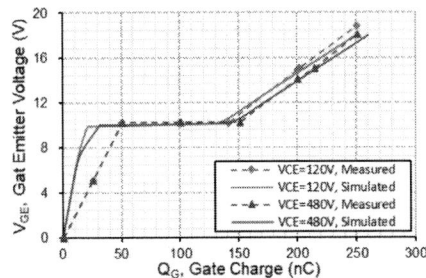

Fig. 10 V_{GE} as a function of Q_G with V_{CE} at 120V and 480V while with Tj=25℃

Fig. 9 illustrates the comparison results for input capacitance (C_{ies}), output capacitance (C_{oes}), and reverse transfer capacitance (C_{res}). These capacitances are derived from C_{gc}, C_{ge}, and C_{ce}, as expressed in Eq. 4. Notably, C_{gc}, C_{ge}, and C_{ce} are incorporated into the model using the previously mentioned lookup table method, promising a high accuracy of the simulated capacitances.

$$\begin{cases} C_{ies} = C_{gc} + C_{res} \\ C_{oes} = C_{ce} + C_{res} \\ C_{res} = C_{gc} \end{cases} \quad (4)$$

Fig. 10 depicts a gate charge (Q_G) comparison plot between simulated and measured values with respect to V_{GE}. The total gate charge difference remains below 2% for V_{GE} voltage up to 18V.

Figures 11 and 12 present the turn-on transient power loss (E_{on}) and turn-off transient power loss (E_{off}), along with their algebraic sum value (E_{total}). Fig. 11 illustrates power losses with respect to I_C, while Fig. 12 pertains to external gate resistance (R_g). These power losses are calibrated up to 2x I_{Cnom}, with the following accuracy targets: >95% within 1x I_{Cnom}, >80% between 1x and 2x I_{Cnom}, and >80% for R_g in the range of 0.5x to 2x R_{gnom} (where R_{gnom} represents the external gate resistance used in datasheet testing). Given that most applications operate within the specified I_C and R_g range, the proposed calibration approach is reasonable. Users can further enhance accuracy at their specific operating points by adjusting model parameters DidtFactor and DvdtFactor. Additionally, E_{on}, E_{off}, and E_{total}—relative to collector-emitter voltage (V_{CE}) and junction temperature (Tj)—are calibrated with similar accuracy but are not explicitly depicted in the paper.

Fig. 13 provides a typical waveform comparison between tested and simulated turn-on and turn-off transients at 400V/40A. Notably, both the V_{CE} and I_C waveforms align well with the experimental results. However, differences exist in the V_{GE} waveforms, particularly in the duration of the miller plateau. These discrepancies likely arise from variations in depletion capacitance between the simulation model and the physical device. For improved gate waveform accuracy, a detailed depletion capacitance model becomes essential.

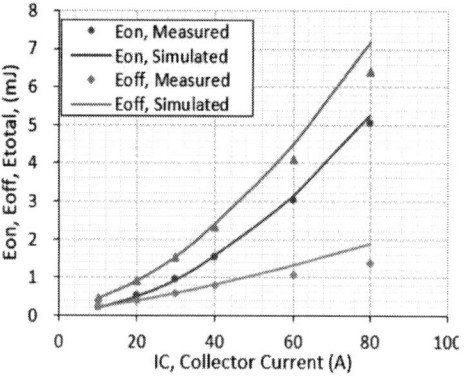

Fig. 11 E_{on}, E_{off}, E_{total} as a function of I_C at Tj=175℃

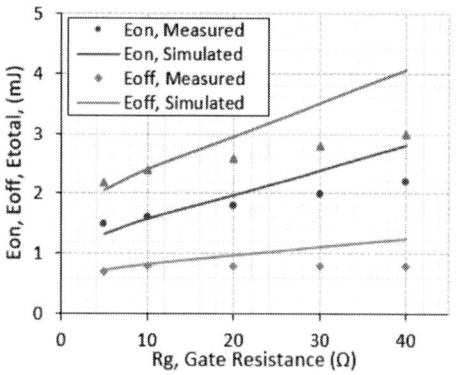

Fig. 12 E_{on}, E_{off}, E_{total} as a function of R_g at Tj=175℃

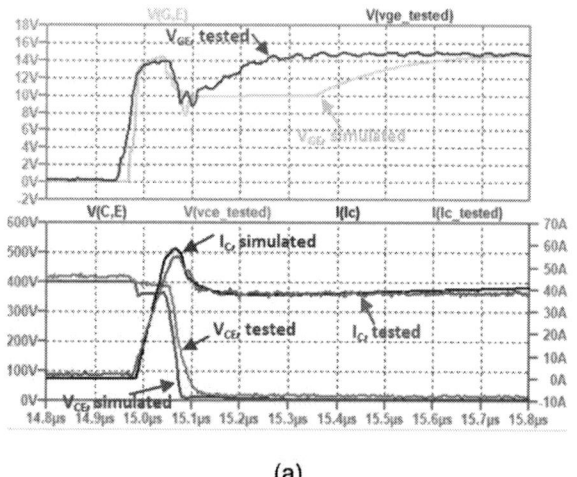

(a)

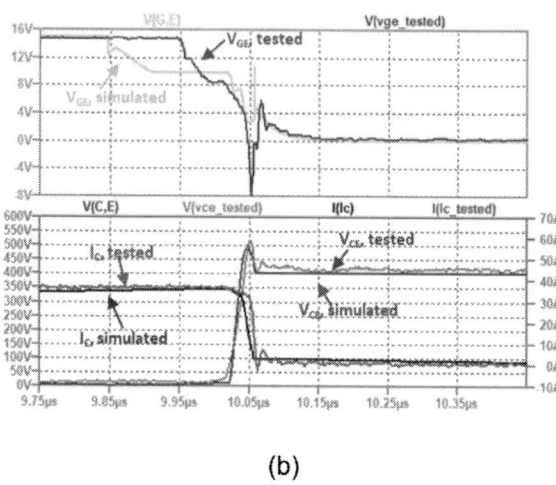

(b)

Fig. 13 Transient waveforms comparison between tested and simulated for turn-on (a) and turn-off (b)

4 Conclusion

This paper presents a comprehensive IGBT behavioral model which is constructed upon the classical "Mosfet + PNP transistor" framework and incorporates subcircuits to capture characteristics related to parasitic capacitances, di/dt and dv/dt adjustment, tail current and diode reverse recovery, as well as junction temperature simulation. The model is calibrated using datasheet information, and not only achieves high accuracy for device fundamental characteristics but also operates with a fast simulation speed, enabling accurate topology-level simulation analysis. The proposed model and methodology can be extended to any IGBT device.

References

[1] Kuang Sheng, B. W. Williams and S. J. Finney, "A review of IGBT models," in IEEE Transactions on Power Electronics, vol. 15, no. 6, pp. 1250-1266, Nov. 2000, doi: 10.1109/63.892840.

[2] Igic, Petar & Jankovic, Nebojsa. (2014). Review of advanced IGBT compact models dedicated to circuit simulation. Facta universitatis - series: Electronics and Energetics. 27. 13-23. 10.2298/FUEE1401013I.

[3] A. R. Hefner and D. M. Diebolt, "An experimentally verified IGBT model implemented in the Saber circuit simulator," in IEEE Transactions on Power Electronics, vol. 9, no. 5, pp. 532-542, Sept. 1994, doi: 10.1109/63.321038.

[4] R. Kraus, P. Turkes and J. Sigg, "Physics-based models of power semiconductor devices for the circuit simulator SPICE," PESC 98 Record. 29th Annual IEEE Power Electronics Specialists Conference (Cat. No.98CH36196), Fukuoka, Japan, 1998, pp. 1726-1731 vol.2, doi: 10.1109/PESC.1998.703414.

[5] F. . -F. Protiwa, O. Apeldoorn and N. Groos, "New IGBT model for PSPICE," 1993 Fifth European Conference on Power Electronics and Applications, Brighton, UK, 1993, pp. 226-231 vol.2.

[6] A. Maxim, D. Andreu and J. Boucher, "High accuracy SPICE behavioral macromodeling of insulated gate bipolar transistor (IGBT)," APEC '98 Thirteenth Annual Applied Power Electronics Conference and Exposition, Anaheim, CA, USA, 1998, pp. 749-755 vol.2, doi: 10.1109/APEC.1998.653982.

[7] K. Asparuhova and T. Grigorova, "IGBT Behavioral PSPICE Model," 2006 25th International Conference on Microelectronics, Belgrade, Serbia, 2006, pp. 203-206, doi: 10.1109/ICMEL.2006.1650931.

[8] I. Baraia, J. Galarza, J. A. Barrena and J. M. Canales, "An IGBT behavioural model based on curve fitting methods," 2008 IEEE Power Electronics Specialists Conference, Rhodes, Greece, 2008, pp. 1971-1977, doi: 10.1109/PESC.2008.4592232.

[9] T. Mizoguchi, Y. Sakiyama, N. Tsukamoto and W. Saito, "High Accurate IGBT/IEGT Compact Modeling for Prediction of Power Efficiency and EMI Noise," 2019 31st International Symposium on Power Semiconductor Devices and ICs (ISPSD), Shanghai, China, 2019, pp. 307-310, doi: 10.1109/ISPSD.2019.8757656.

[10] D. Ludwig, G. Alia, A. Biswas and M. Cotorogea, "Behavioral compact models of IGBTs and Si-diodes for data sheet simulations using a machine learning based calibration strategy," PCIM Europe digital days 2020; International Exhibition and Conference for Power Electronics, Intelligent Motion, Renewable Energy and Energy Management, Germany, 2020, pp. 1-8..

[11] Nexperia, Datasheet NGW40T65H3DFP, Rev. 1.0, 3 July 2023

PCIM Europe 2024, 11– 13 June 2024, Nuremerg   DOI: 10.30420/566262412

Parameter Extraction for an ANN-assisted IGBT Model in Transient Simulations

Huaiyuan Zhang[1], Abby Shih[2], Stefan Haensel[3], Zeeshan Umar[3], Steven Lee[1], Felix Zeyss[3]

[1] Keysight Technologies, USA
[2] Keysight Technologies, Germany
[3] Siemens, Germany

Corresponding author: Huaiyuan Zhang, huaiyuan.zhang@keysight.com
Speaker: Abby Shih, abby.shih@keysight.com

Abstract

The development of power IGBT devices require an accurate transient model over multiple currents and voltages. With the Artificial Neural Network (ANN) assisted diode model and the modified Angelov-based capacitance model, a new IGBT model describes both the overshoot of collector current and collector-emitter tail voltage at different bias conditions. Given a completed parameter extraction workflow, a comprehensive good fitting of the transient current and voltage waveforms is verified at both the turn-on and turn-off process for currents from 30 to 600 A and voltages from 100 to 820 V.

1 Introduction

The IGBT devices attracts lots of industrial interests in power modules and electrical vehicles. The accurate modeling of switching behavior of power devices can help to accelerate the development cycles, analyze the power loss and determine the electromagnetic compatibility [1–3]. In contrast, unlike only one or several bias conditions are dealt with in most literature, such as [4–11], the actual applications of power modules over a wide range of currents and biases challenge the stability and accuracy of an IGBT model. We recently proposed a new ANN-assisted reverse recovery characteristics of diode model for IGBT devices[12]. The collector current (i_C) of IGBT is well fitted up to 600 V at currents from 30 to 500 A. In this paper, with further developments and parameter extraction flow, the improved ANN-assisted IGBT model can be extended for both switching on and switching off characteristics at even higher voltages.

The IGBT devices paralleled with the freewheeling diodes in an 1200 V/260 A power module under the double pulse testing (DPT) setup are measured for switching characteristics. The input gate loop are converted by a gate driver at the turn-on and tun-off process, respectively. Fig. 1(a) shows the equivalent DPT circuit for IGBT transient simulations.

The packaging parasitic networks surrounding the IGBT devices are considered but not shown in the schematic. V_{DC} is the DC supply voltage. V_{pulse} is the pulse voltage source. $R_{G,ext}$ is the extrinsic gate resistance. L_{stray} is the lumped stray inductance. L_{load} and R_{load} are the load inductance and resistance, respectively. L_s is the lumped inductance at the source side. I_L is the load current. i_C, v_{GE} and v_{CE} are the measured collector current, gate-emitter voltage and collector-emitter voltage of the IGBT at lower side, respectively. Fig. 1(b) shows the internal capacitances in the IGBT model. C, G, E, G' and B' are the collector, gate, emitter, internal gate and base nodes, respectively. D' and S' are also shown for internal nodes of the MOS model. $R_{G,int}$ is the intrinsic gate resistance. C_{GD}, C_{GS}, C_{BE}, C_{Diode} are the gate-drain, gate-source, base-emitter and diode capacitances, respectively. The paper is organized as follows: Sec. 2 describes the switching waveforms in eight stages from a turn-on process to a turn-off process. Sec. 3 proposes a modified Angelov-based formulations of C_{GD} model for v_{CE} tail, the importance of PNP DC parameters for tail current and ANN assisted reverse recovery of diode model for current overshoots, which can fit the transient waveforms at different bias conditions. A direct parameter extraction flow is also provided for each parasitic elements in the proposed IGBT

2901

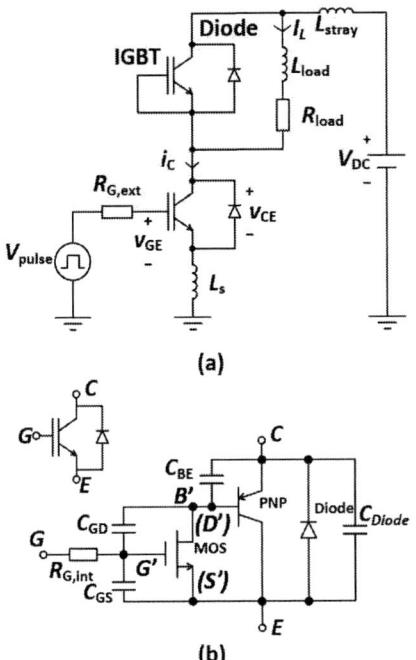

Fig. 1: A schematic of (a) the equivalent DPT circuit for IGBT transient simulation and (b) IGBT symbol and the internal capacitances in the IGBT model. The parasitic packaging network is considered but not shown in the schematic.

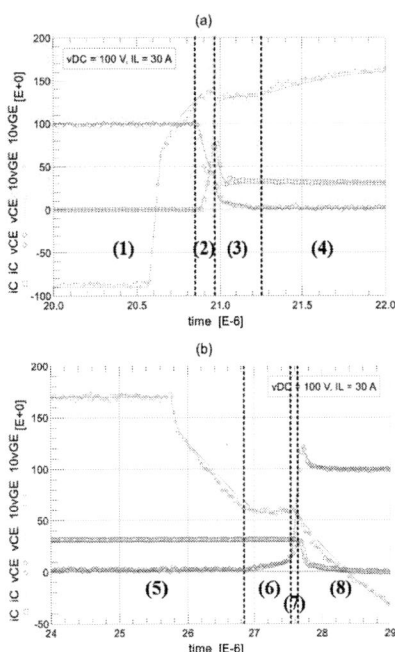

Fig. 2: The measured (symbols) and simulated (lines) v_{GE}, v_{CE} and i_C waveforms at (a) a turn-on and (b) a turn-off process for $V_{DC} = 100$ V and $I_L = 30$ A. The eight stages are shown.

model. Sec. 4 shows an overall good fitting of v_{GE}, v_{CE} and i_C waveforms at different bias conditions. Sec. 5 draws the conclusions.

2 Typical transient v_{GE}, v_{CE} and i_C waveforms at both the turn-on and turn-off process

The switching process can be generally separated in eight stages as follows: Fig. 2(a) shows transient IGBT waveforms in four stages of a turn-on process. Stage (1): v_{GE} increases from -9 V to the threshold voltage around 12 V, with the slope determined by the input capacitance and gate resistance. The input capacitance at this stage is assumed constant and linear capacitance because the Miller's capacitance C_{GD} is smaller compared to C_{GS} [13]. The kink in the simulated rising v_{GE} slope is due to the difference of $R_{G,ext}$ and $R_{G,int}$. Since v_{GE} is below the threshold voltage, i_C is zero and v_{CE} keeps constant.

Stage (2): v_{GE} becomes larger than the threshold voltage, resulting in an increase of i_C. The IGBT operates in the saturation region. The deduction of v_{CE}, $\triangle v_{CE}$, is calculated by

$$\triangle v_{CE} \approx V_{DC} - (L_{stray} + L_s) \cdot \frac{di_C}{dt}. \quad (1)$$

As i_L keeps constant during this short time, the diode current falls with the same i_C slope. The reverse recovery begins when the diode current falls below zero. Then i_C becomes larger than i_L and continues to reach its peak which also points to the local v_{GE} peak [13].

Stage (3): When i_C reaches the peak, the diode starts to block voltage and v_{CE} resumes to decrease in the saturation region. After the reverse recovery process is finished and i_C equals to i_L, IGBT enters the ohmic region and v_{CE} further decreases to its on-state value. The changing dropping slope of v_{CE} at this stage is due to the change of increasing C_{GC} [13]. Due to C_{GC}, v_{GE} is held constant at this stage.

Stage (4): After v_{CE} decreases to its on-state value, i_C keeps constant and is not controlled by v_{GE}. Therefore, v_{GE} continues to increase to its static steady value.

Fig. 2(b) shows transient IGBT waveforms in four stages of a turn-off process. Stage (5): v_{GE} decreases to the Miller plateau. IGBT operates in the ohmic region and i_C remains unchanged.

Stage (6): IGBT still works in the ohmic region, so v_{CE} increases with a small slope and both v_{GE} and i_C keep constant.

Stage (7): IGBT enters the saturation region, v_{CE} increases with a rapid slope due to a smaller C_{GC} and reaches V_{DC} at the end of this stage. The small deduction of v_{GE} and i_C is due to the bypass current flowing through C_{BE} and C_{Diode} [13].

Stage (8): After the blocking voltage of diode decreases to zero, i_C shows an initial rapid fall due to the removal of the MOSFET current. The i_C drop induces a voltage deduction through L_{stray} and L_s, hence the overshoot of v_{CE} is observed. A further slowly dropping i_C is due to remaining decaying excess carriers in the base. v_{GE} reduces to V_{pulse} with a slope determined by gate resistance and input capacitances.

Fig. 2(a) and (b) presents accurate transient v_{GE}, v_{CE} and i_C waveforms for V_{DC} = 100 V and I_L = 30 A. However, actual power application always requires an accurate transient model over multiple voltages and currents. Therefore several improved formulations in IGBT model are developed below.

3 The proposed models for IGBT over multiple biases

3.1 A modified Angelov-based C_{GD} formulations for v_{CE} tail

The effect of the IGBT gate-collector capacitance is mainly on v_{GE} slope in stage (4) and (8) and v_{CE} slope in stage (3), (6) and (7). The modified Angelov-based formulations [14] are proposed for C_{GD} of MOSFET in IGBT as

$$C_{GD}(V_{D'S'}, V_{G'S'}) = C_{gdpi} + C_{GD1}(V_{D'S'}, V_{G'S'}) + C_{GD2}(V_{D'S'}, V_{G'S'}), \quad (2)$$

where C_{gdpi} is the constant C_{GD}. $C_{GD1}(V_{D'S'}, V_{G'S'})$ is the conventional depletion capacitance, which is dependent on gate-source voltage $V_{G'S'}$ and drain-source voltage $V_{D'S'}$ of MOSFET and given as

$$C_{GD1}(V_{D'S'}, V_{G'S'}) = C_{gd0} \left(1 + \frac{V_{D'S'}}{V_j}\right)^{-M} \cdot (1 + \tanh \phi_2), \quad (3)$$

where C_{gd0} is the zero bias C_{GD}. V_j is the built-in potential. The bias dependent term $\tanh \phi_2$ is expressed by

$$\tanh \phi_2 = P_{40} + P_{41}V_{G'S'} - P_{111}V_{D'S'}, \quad (4)$$

where P_{40}, P_{41} and P_{111} are coefficient parameters. $C_{GD1}(V_{D'S'}, V_{G'S'})$ is extracted from the IGBT C-V characteristics, together with C_{BE}. $C_{GD2}(V_{D'S'}, V_{G'S'})$ is the modified Angelov-based capacitance and expressed by

$$C_{GD2}(V_{D'S'}, V_{G'S'}) = C_{gd02}(1 + \tanh \phi_1 - P_{111}) \cdot (1 + \tanh \phi_2), \quad (5)$$

where C_{gd02} is the model parameter. The bias dependent term $\tanh \phi_1$ is

$$\tanh \phi_1 = P_{302} - P_{312}V_{D'S'}, \quad (6)$$

where both P_{302} and P_{312} are coefficient parameters. $C_{GD2}(V_{D'S'}, V_{G'S'})$ is extracted from transient simulations.

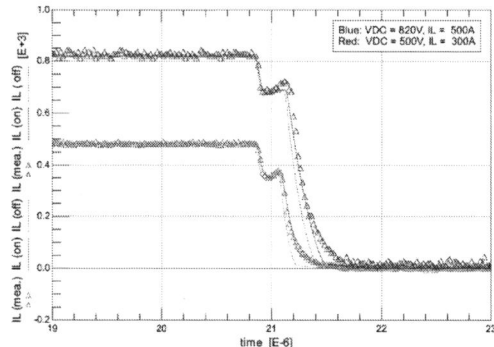

Fig. 3: The measured (symbols) and simulated (lines) v_{CE} versus time at a turn-on process for V_{DC} = 820 V, I_L = 500 A and V_{DC} = 500 V, I_L = 300 A. The solid lines and dashed lines correspond to the simulated v_{CE} with and without $C_{GD2}(V_{D'S'}, V_{G'S'})$ term in C_{GD}, respectively.

Fig. 3 shows the measured and simulated v_{CE} versus time at a turn-on process for V_{DC} = 820 V, I_L = 500 A and V_{DC} = 500 V, I_L = 300 A. Both C_{gdpi} and $C_{GD1}(V_{D'S'}, V_{G'S'})$ terms in C_{GD} have been determined in C-V characteristics. With the extra $C_{GD2}(V_{D'S'}, V_{G'S'})$ term extracted in transient v_{CE} waveforms, an improvement of second falling v_{CE} is observed at two different bias.

3.2 Importance of PNP DC parameters in i_C tail

In stage (8) of a turn-off process, the IGBT device is switched off and the i_C drops rapidly. The tail current is generated due to the recombination of excess electrons and holes in the base region of PNP.

Besides the carrier recombination lifetime of PNP, the DC PNP model parameters are also efficient in fitting both tail current and IGBT static steady transfer characteristics.

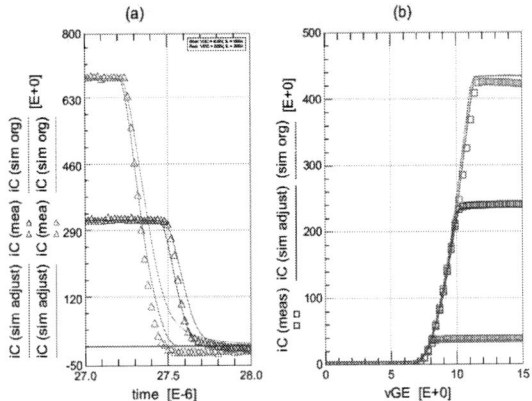

Fig. 4: (a) The measured (symbols) and simulated (lines) i_C versus time at a turn-off process for V_{DC} = 820 V, I_L = 500 A and V_{DC} = 500 V, I_L = 300 A. (b) The measured (symbols) and simulated (lines) I_C-V_{GE} at a static steady transfer characteristics. The solid lines and dashed lines correspond to the IGBT model with the adjusted and raw PNP parameters, respectively.

Fig. 4(a) shows the measured and simulated i_C versus time at a turn-off process for V_{DC} = 820 V, I_L = 500 A and V_{DC} = 500 V, I_L = 300 A. With the adjusted DC PNP parameters, an improvement on tail currents at both biases is achieved. Fig. 4(b) shows the measured and simulated I_C-V_{GE} at a static steady transfer characteristics. Although the clear difference is observed in transient tail currents between using the adjusted and using the raw PNP parameters, the difference is still hardly seen in DC I-V characteristics of the same bias condition. Therefore, the DC PNP parameters are quite useful in fitting the tail current, while keeping accuracy in DC I-V characteristics.

3.3 ANN assisted reverse recovery of a diode model in i_C overshoot

In stage (2) and (3) of a turn-on process, the overshoot of i_C is determined by the reverse recovery of the freewheeling diode model. For a single bias, the diode model in [15] with two model parameters, carrier lifetime τ and diffusion transit time T_M, can accurately fit i_C overshoot with the following

equations:

$$i_D(t) = \frac{q_E(t) - q_M(t)}{T'_M}, \qquad (7)$$

$$0 = \frac{dq_M(t)}{dt} + \frac{q_M(t)}{\tau} - \frac{q_E(t) - q_M(t)}{T'_M}, \qquad (8)$$

$$q_E(t) = I_S\tau\left[\exp\left(\frac{v_F(t)}{nV_T}\right) - 1\right], \qquad (9)$$

where $i_D(t)$ is the diode current changing with time, $q_E(t)$ is the injected charge, $q_M(t)$ is the charge in base region, I_S is the saturation current, n is the non-ideal coefficient and V_T is the thermal voltage. At high bias and current, the depletion width (W_{dep}) modulation effect on T'_M is also considered by [16]

$$T'_M = T_M(1 - W_{dep})^2, \qquad (10)$$

$$W_{dep} = \sqrt{\frac{\phi_{bi} - v_F}{V_{depl}(1 + |i_D|/I_{depl})}}, \qquad (11)$$

where ϕ_{bi} is the built-in potential. Both V_{depl} and I_{depl} are model parameters. The overshoot shape varies largely at different currents and voltages, which implies both τ and T_M are current and voltage dependent. Therefore, an ANN assisted diode model is previously proposed in [12] with the following procedure: Both τ and T_M are extracted from the i_C overshoot at each bias condition. Then I_L and V_{DC} dependent τ and T_M are modeled by ANN after a training process.

Fig. 5 shows the measured and simulated transient i_C waveforms at a turn-on process at I_L from 30 to 500 A and V_{DC} = 820 V. An overall good fitting of the overshoot of i_C is achieved with the ANN assisted diode model. If without the depletion width modulation effect, a large deviation is observed around the i_C peak, which validates the importance of depletion width effect in T_M especially at both high current and voltages.

3.4 A general parameter extraction procedure

Due to the similar mechanism between turn-on and turn-off process, parasitic elements affect several transient stages of both the turn-on and turn-off process [13]. Therefore, a direct parameter extraction procedure to distinguish the effect of each element and extract at each step is required.

Table. 1 shows a parameter extraction procedure of the proposed ANN assisted IGBT model in transient simulations, from the turn-off process to the

PCIM Europe 2024, 11– 13 June 2024, Nuremberg   DOI: 10.30420/566262412

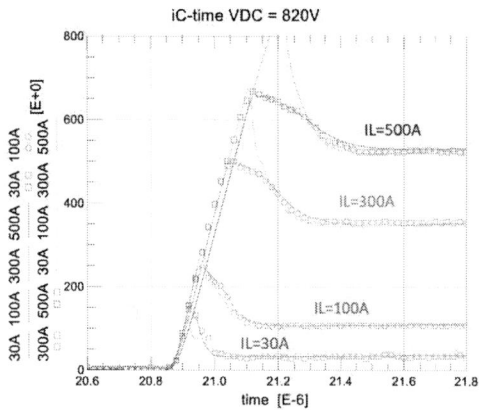

Fig. 5: The measured (symbols) and simulated (lines) transient i_C waveforms at a turn-on process at I_L from 30 to 500 A and V_{DC} = 820 V. The solid and dotted lines are the simulated i_C with and without depletion width modulation effect in the ANN assisted diode model, respectively.

Elements	Extraction procedure
L_{load}, R_{load}	i_C in stage (5) if turn-on or stage (4) if turn-on
$R_{G,ext}$, $R_{G,int}$	v_{GE} in stage (5) if turn-on or stage (1) if turn-off
C_{GD}	v_{CE} in stage (6) and stage (7)
PNP model	i_C in stage (8)
L_{stray}	v_{CE} in stage (2)
L_s	i_C in stage (2)
C_{GS}	i_C in stage (2)
Diode model	i_C in stage (2) and stage (3)

Tab. 1: The parameter extraction procedure of the proposed ANN assisted IGBT model in transient simulations.

turn-on process. The ANN assisted diode model is extracted lastly for the overshoot of i_C in stage (6) and (7). The majority of IGBT model parameters are still extracted from DC I-V and C-V characteristics. Both C_{BE} of the PNP model and C_{Diode} of the diode model in Fig. 1(b) are determined in C-V characteristics, hence not included in Table. 1. Similar to the extra $C_{GD2}(V_{D'S'}, V_{G'S'})$ term in C_{GD} in Sec.3.1, the modified Angelov-based formulations are also added in C_{GS} for i_C slope in stage (2). L_s is extracted from the i_C slope difference at different bias [17].

4 Results

In the Power Electronics Model Generator (PEMG) tool of Keysight Integrated Circuit Characterization and Analysis Program (ICCAP) [18], the majority of model parameters are extracted from DC I-V and C-V measurements and the package parasitics are extracted from the off-state S-parameters measurements. The proposed extraction procedure for transient simulations is followed as below: Due to the actual switching input resistance in a gate driver, $R_{G,ext}$ is extracted separately with 30 Ω in the turn-off process and 4 Ω in the turn-on process. Both L_{load} and R_{load} is extracted according to the initial current in the turn-off process or the final steady current in the turn-on process, respectively. Both C_{GD} and PNP model parameters are extracted in the turn-off process. $R_{G,int}$, L_{stray}, L_s and C_{GS} model

parameters are extracted in the turn-on process. For the diode model, the τ and T_M are extracted from the overshoots of i_C under each bias condition [15]. The dependence of both τ and T_M on I_L, V_{DC} is modeled by ANN after a training process [12].

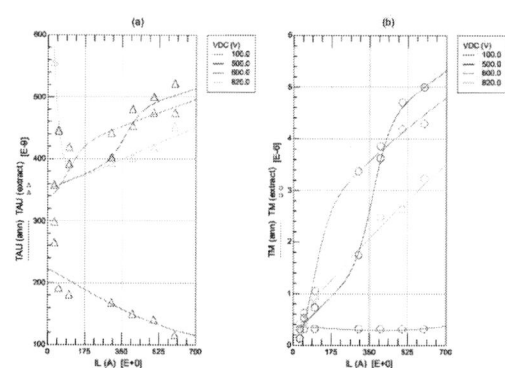

Fig. 6: The extracted (symbols) and ANN exported (lines) (a) τ and (b) T_M versus I_L from 20 to 700 A at V_{DC} of 100, 500, 600 and 820 V. T = 25°C.

Fig. 6 shows the extracted and ANN exported (a) τ and (b) T_M versus I_L from 20 to 700 A at V_{DC} of 100, 500, 600 and 820 V. $T = 25°C$. The ANN is comprised of 1 layer and 4 neurons per layer. Besides an overall good fitting is achieved by the

2905

trained ANN, a reasonable estimation of both τ and T_M in both the interpolation and extrapolation ranges of I_L at each V_{DC}.

Fig. 7(a)-(f) shows the measured and simulated transient v_{GE}, v_{CE} and i_C waveforms at both the turn-off and turn-on process under three typical biases. The other biases are not shown to save spaces. Due to the ANN assisted diode model, the overshoots of i_C at each bias are fitted. The discrepancies in the initial simulated v_{GE} is due to the simplicity of driving circuit at the gate side, compared to the actual gate driver. A comprehensive good fitting on multiple bias at both the turn-on and tun-off process validates the parameter extraction procedure and the accuracy of proposed IGBT model in transient simulations.

5 Conclusion

In this paper, typical transient v_{GE}, v_{CE} and i_C waveforms from the turn-off to turn-on process are analyzed at a single bias. However, the actual application of power module requires the transient IGBT model working in multiple bias conditions. A modified Angelov-based C_{GD} are developed for transient v_{CE} tail at different bias. whose similar formulations in C_{GS} are developed for transient i_C slope. Besides the carrier lifetime of PNP model in IGBT, an adjusted DC PNP model parameter leads to an improvement on i_C tail, while the DC I-V transfer characteristics keep unchanged in the same bias range. To fit i_C overshoot over multiple currents and voltages, the two model parameters, τ and T_M, in a simple diode model are extracted under each bias. After an ANN training, the current and voltage dependent τ and T_M are modeled. Then the overshoot of i_C at each bias is well captured by the ANN assisted reverse recovery of diode model in IGBT. A direct parameter extraction flow is also proposed from the turn-off and turn-on process. The completed IGBT model is validated by an overall good fitting of the transient v_{GE}, v_{CE} and i_C waveforms over currents from 30 to 600 A and voltages from 100 to 820 V.

6 Acknowledgment

The authors would like to thank R&D engineers from the Keysight Power Electronics (PE) squad for their helpful discussions.

References

[1] F. Blaabjerg, M. Liserre, and K. Ma, "Power electronics converters for wind turbine systems," in *IEEE Energy Convers. Congr. Expo.*, Mar.-Apr. 2011, pp. 281–290. DOI: 10.1109/ECCE.2011.6063781.

[2] S. Kouro, M. Malinowski, K. Gopakumar, J. Pou, L. G. Franquelo, *et al.*, "Recent advances and industrial applications of multilevel converters," *IEEE Trans. Ind. Electron.*, vol. 57, no. 8, pp. 2553–2580, Aug. 2010. DOI: 10.1109/TIE.2010.2049719.

[3] S. Debnath, J. Qin, B. Bahrani, M. Saeedifard, and P. Barbosa, "Operation, control, and applications of the modular multilevel converter: a review," *IEEE Trans. Power Electron.*, vol. 30, no. 1, pp. 37–53, Jan. 2015. DOI: 10.1109/TPEL.2014.2309937.

[4] A. T. Bryant, X. Kang, E. Santi, P. R. Palmer, and J. L. Hudgins, "Two-step parameter extraction procedure with formal optimization for physics-based circuit simulator IGBT and p-i-n diode models," *IEEE Trans. Power Electron.*, vol. 21, no. 2, pp. 295–309, Mar. 2006. DOI: 10.1109/TPEL.2005.869742.

[5] A. T. Bryant, L. Lu, E. Santi, J. L. Hudgins, and P. R. Palmer, "Modeling of IGBT resistive and inductive turn-on behavior," *IEEE Trans. Ind. Appl.*, vol. 44, no. 3, pp. 904–914, May-June 2008. DOI: 10.1109/TIA.2008.921384.

[6] A. R. Hefner, "An investigation of the drive circuit requirements for the power insulated gate bipolar transistor (IGBT)," *IEEE Trans. Power Electron.*, vol. 6, no. 2, pp. 208–219, Apr. 1991. DOI: 10.1109/63.76807.

[7] A. R. Hefner and D. M. Diebolt, "An experimentally verified IGBT model implemented in the Saber circuit simulator," *IEEE Trans. Power Electron.*, vol. 9, no. 5, pp. 532–542, Sep. 1994. DOI: 10.1109/63.321038.

[8] P. R. Palmer, E. Santi, J. L. Hudgins, X. Kang, J. C. Joyce, and P. Y. Eng, "Circuit simulator models for the diode and IGBT with full temperature dependent features," *IEEE Trans. Power Electron.*, vol. 18, no. 5, pp. 1220–1229, Sep. 2003. DOI: 10.1109/TPEL.2003.816194.

[9] F. Iannuzzo and G. Busatto, "Physical CAD model for high-voltage IGBTs based on lumped-charge approach," *IEEE Trans. Power Electron.*, vol. 19, no. 4, pp. 885–893, Jul. 2004. DOI: 10.1109/TPEL.2004.830085.

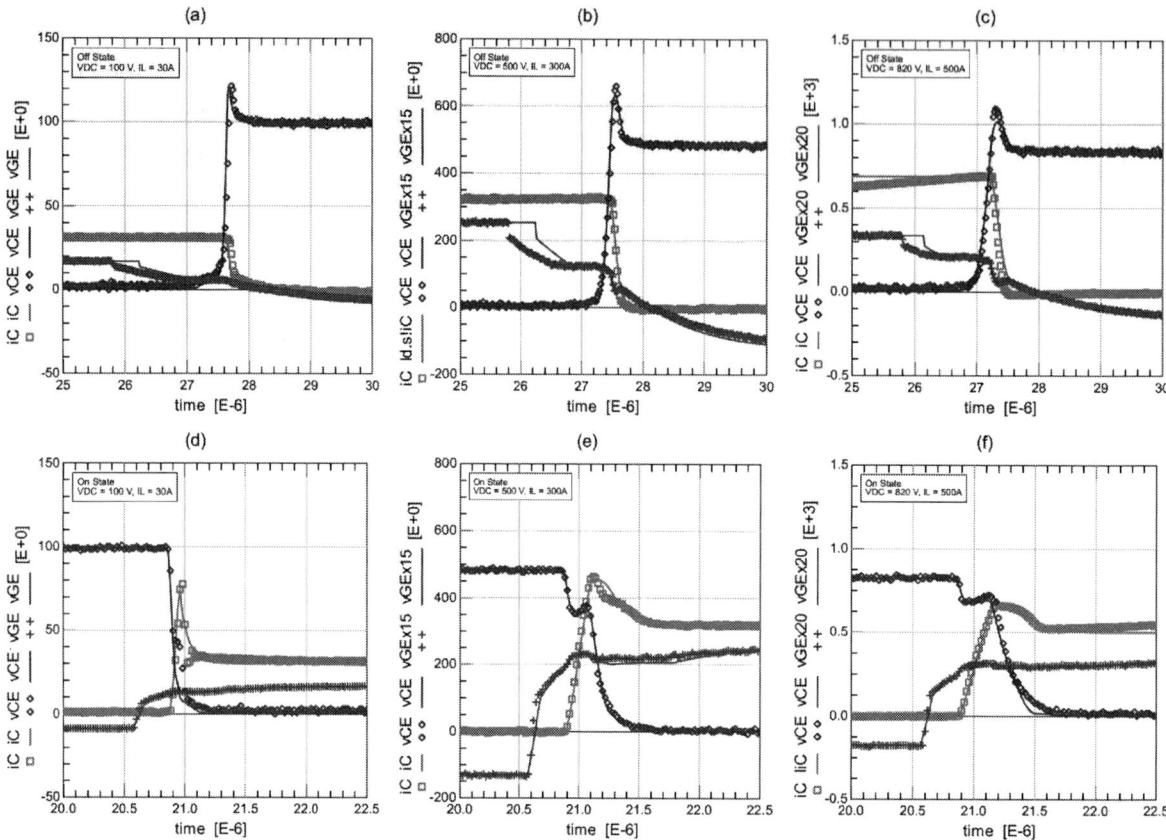

Fig. 7: The measured (symbols) and simulated (lines) transient v_{GE}, v_{CE} and i_C waveforms of the turn-off process at (a) $V_{DC} = 100$ V, $I_L = 30$ A, (b) $V_{DC} = 500$ V, $I_L = 300$ A, (c) $V_{DC} = 820$ V, $I_L = 500$ A and the turn-on process at (d) $V_{DC} = 100$ V, $I_L = 30$ A, (e) $V_{DC} = 500$ V, $I_L = 300$ A, (f) $V_{DC} = 820$ V, $I_L = 500$ A.

[10] M. Miyake, D. Navarro, U. Feldmann, H. J. Mattausch, T. Kojima, *et al.*, "HiSIM-IGBT: a compact Si-IGBT model for power electronic circuit design," *IEEE Trans. Electron. Devices*, vol. 60, no. 2, pp. 571–579, Feb. 2013. DOI: 10.1109/TED.2012.2226181.

[11] A. Tone, Y. Miyaoku, M. M.-Mattausch, H. Kikuchihara, U. Feldmann, *et al.*, "HiSIM_IGBT2: modeling of the dynamically varying balance between MOSFET and BJT contributions during switching operations," *IEEE Trans. Electron. Devices*, vol. 66, no. 8, pp. 3265–3272, Aug. 2019. DOI: 10.1109/TED.2019.2919799.

[12] H. Zhang, A. Shih, S. Haensel, Z. Umar, S. Lee, and F. Zeyss, "An ann assisted reverse recovery of diode model for switching on characteristics of igbt devices," in *Proc. Int. Exhibition Conf. Power Electron., Intell. Motion, Renewable Energy Manage. (PCIM)*, May 2023, pp. 1–9. DOI: 10.30420/566091012.

[13] J. Wang, H. Chung, and R. Li, "Characterization and experimental assessment of the effect of parasitic elements on the MOSFET switching performance," *IEEE Trans. Power Electron.*, vol. 28, no. 1, pp. 573–590, Jan. 2013. DOI: 10.1109/TPEL.2012.2195332.

[14] I. Angelov, L. Bengtsson, and M. Garcia, "Extensions of the Chalmers nonlinear HEMT and MESFET model," *IEEE Trans. Microw. Theory Techn.*, vol. 44, no. 10, pp. 1664–1674, Oct. 1996. DOI: 10.1109/22.538957.

[15] P. O. Lauritzen and C. L. Ma, "A simple diode model with reverse recovery," *IEEE Trans. Power Electron.*, vol. 6, no. 2, pp. 188–191, Apr. 1991. DOI: 10.1109/63.76804.

[16] C. L. Ma, P. O. Lauritzen, and J. Sigg, "Modeling of power diodes with the lumped-charge modeling technique," *IEEE Trans. Power Electron.*, vol. 12, no. 3, pp. 398–405, May 1997. DOI: 10.1109/63.575666.

[17] S. Musumeci, A. Raciti, A. Testa, A. Galluzzo, and M. Melito, "Switching-behavior improvement of insulated gate-controlled devices," *IEEE Trans. Power Electron.*, vol. 12, no. 4, pp. 645–653, Jul. 1997. DOI: 10.1109/63.602559.

[18] *PathWave Device Modeling (ICCAP) Modeling Handbook.* Keysight Technol., Santa Rosa, CA, USA, 2024.

Fabrication of 600V RC-IGBT using 300mm Wafer

Masaki Ueno[1], Shohta Oh[1], Takahiro Nakatani[1], Haruhiko Minamitake[1] Taiki Hoshi[1]
Hidenori Koketsu[1], Yusuke Miyata[1], Yuta Asai[1], Ai Sugamoto[1], Kenji Suzuki[1]
[1] Mitsubishi Electric Corporation, Japan

Corresponding author: Masaki Ueno , ueno.masaki@dn.mitsubishielectric.co.jp
Speaker: Masaki Ueno , ueno.masaki@dn.mitsubishielectric.co.jp

Abstract

We have successfully manufactured 600V RC-IGBT using 300mm wafers with advanced wafer process. This device has not only comparable or better characteristics, but also less variation in electrical characteristics than the 200mm. We are now ready to provide a stable supply of power device to meet the growing demand for a wide range of applications from home appliances to automotive. Furthermore, we will be able to provide even higher performance devices in the future by using the fine pattern technology of 300mm wafer process.

1 Introduction

1.1 Background

The Paris Agreement was adopted in 2015 to solve the global issue of climate change, and efforts are underway around the world to achieve carbon neutrality by 2050. Furthermore, power devices are highly expected as one of the key technologies to contribute to the improvement of energy efficiency. Among power devices, Reverse Conducting IGBT (RC-IGBT), which combines IGBT and diode on a single chip, has been attracting attention in recent years as a device that can greatly contribute to making power modules smaller, lighter, and more powerful, and its adoption is expanding in a wide range of fields, including white goods, industrial motor drives, and electric vehicles. Expansion of supply capacity is essential to meet growing demand. One solution is to increase production efficiency by using a large diameter wafer. This increases the number of chips that can be produced on a single wafer and improves production efficiency. Currently, power devices are mainly manufactured on 200 mm Si wafers, but mass production technology for 300 mm wafers is beginning to emerge[1]. In order to ensure a stable supply of power devices to the market, we also plan to start mass production of 300mm wafers with high production efficiency in 2024. We are pleased to report that we have achieved production of 600V RC-IGBT on 300mm wafers using our advanced wafer process.

1.2 Our approach

Through our successful history of improving IGBTs performances, we have been providing high performance IGBT, from planar gate structure to trench gate one, especially CSTBT™[2-4]. CSTBT™ consists of a triple layered structure in the mesa region between the trench gates, i.e., N+Emitter/ P-Base/ CS (Carrier Stored)-layer. CS-layer acts as a potential barrier that blocks holes from being collected by the emitter, As a result, the hole concentration in the drift region near emitter side increases and V_{CEsat} can be reduced. Furthermore, we have been developing RC-IGBT based on CSTBT™ structure in order to meet the demands of the user side. In terms of the productivity, we have been mass producing IGBTs from 125mm to 200mm so far. With the rapid growth in global demand for power devices expected in the future, we have also established a manufacturing technology using 300mm Si wafers.

In this paper, we report results for the 600 V class used for air conditioners and other equipment. 600V-class power devices are extremely thin at several tens of microns and prone to warping and cracking, making even 200mm wafers extremely difficult to handle. We have solved these problems through several test runs and are well on our way to starting mass production of the 600V class.

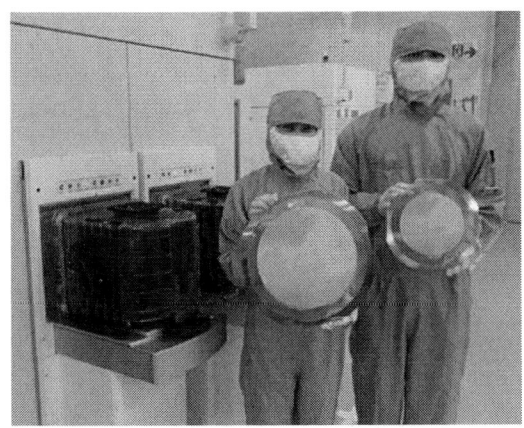

Fig. 1 Photographs of devices fabricated on 300mm and 200mm wafers.

https://www.mitsubishielec-

tric.com/news/2023/0829.html

Mitsubishi Electric announced in a public relations release that it completed the installation of a production line for 300mm Si wafers at its Fukuyama Plant on August 29, 2023. Figure 1 shows pictures of devices fabricated at 300mm and 200mm wafers as described in the press release page.

In addition, we actively introduced the automation of the manufacturing process in our 300 mm wafer production line to prevent humans from touching thin wafers as much as possible. This has prevented wafer cracking and improved the overall efficiency of the manufacturing process. And automation has dramatically improved productivity by reducing work errors through labor-saving effects.

Figure 2 shows a cross-sectional views of the 2nd generation (2nd Gen.) [2-4] RC-IGBT in mass production and the 3rd generation (3rd Gen.)[5] RC-IGBT developed on 300mm wafer. The 3rd Gen. RC-IGBT was adopted for devices fabricated on 300mm wafers to improve not only productivity, but also device performance. The diode structure is optimized, considering the tradeoff relationship between V_F-E_{rr}. The anode concentration is reduced compared with 2nd Gen. and P+contact pattern is shrunk to a size where the contact resistance between the emitter electrode and Si substrate does not increase.

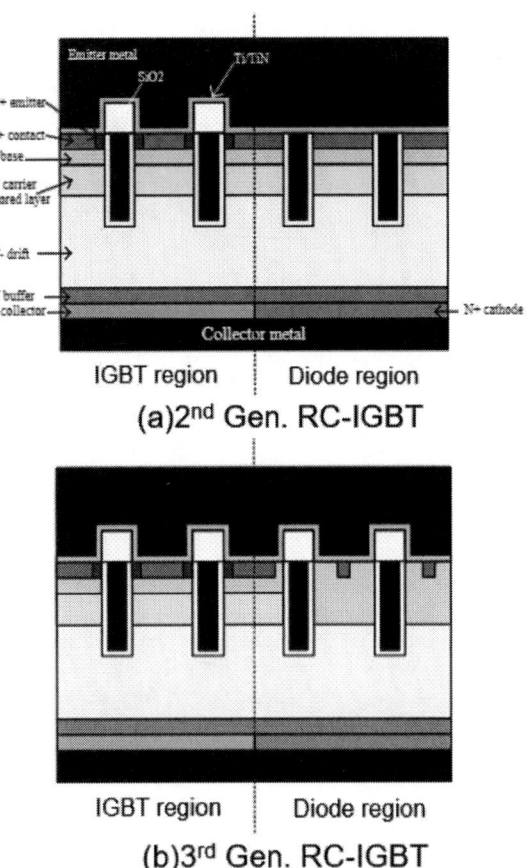

Fig. 2 Cross sectional views of 2nd Gen. RC-IGBT and 3rd Gen. RC-IGBT.

2 Static characteristics

The electrical characteristics of the 200 mm and 300 mm RC-IGBT with a rated voltage of 600 V and a rated current of 15 A are shown. Fig. 3 and 4 show the probability distributions of V_{CEsat} and V_F of a single wafer respectively. It shows that V_{CEsat} and V_F of 300mm are small compared with 200mm. This result suggests that the variation of characteristic values within the wafer plane was suppressed despite the larger wafer diameter.

This was made possible by new high performance equipment and new process technology at the new plant.

Figure 5 shows the I_C-V_{CE} of 300mm RC-IGBT characteristics. It has been confirmed that the device operates normally from 25℃ to 150℃ without snapback.

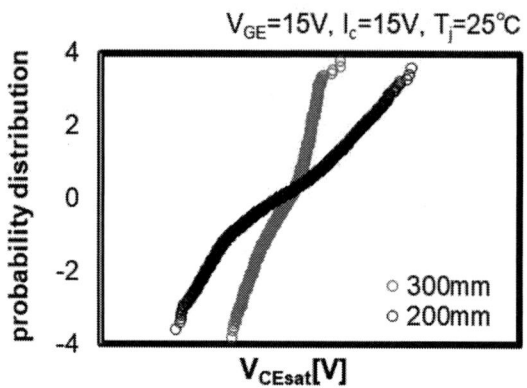

Fig. 3 V_{CEsat} probability distribution

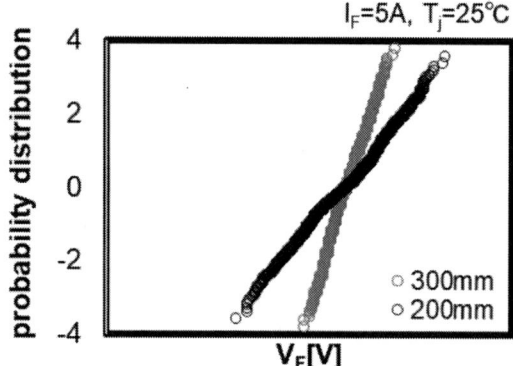

Fig. 4 V_F probability distribution

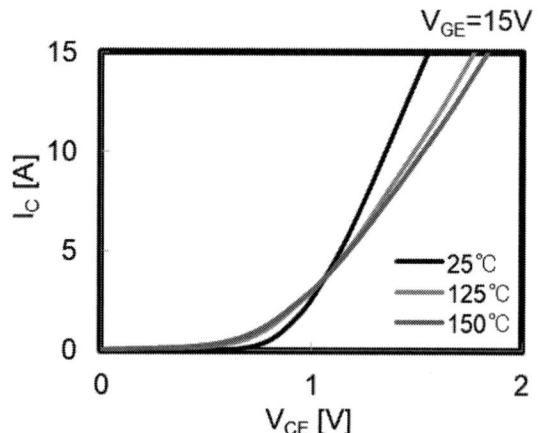

Fig. 5 I_c and V_{CE} curve of Temperature dependence

Figure 6 shows the I_{Csat} and $V_{GE(th)}$ correlation. Devices fabricated at 300mm and 200mm wafers have a sufficient I_{Csat} of 30A or more, which is twice the rated current of 15A. The relationship between I_{CEsat} and V_{GEth} is equivalent for devices fabricated at 300mm and 200mm wafers.

Fig. 6 I_{Csat} and $V_{GE(th)}$ correlation.

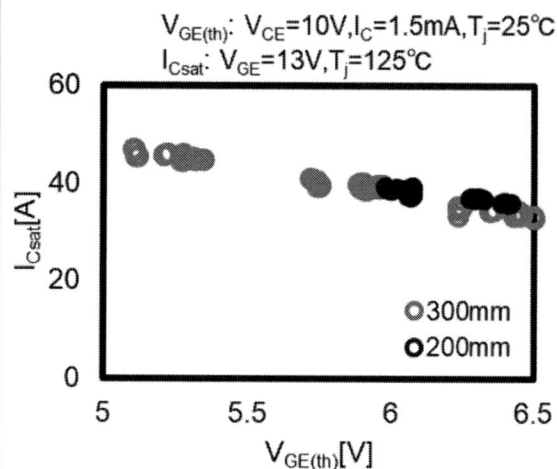

3 Dynamic characteristics

Figure 7 shows the turn-on waveforms of the devices fabricated at 300mm and 200mm wafers. V_{CE} fall-off of 300mm devices is faster than that of 200 mm wafers because 300mm devices have lower gate capacitance. As a result, energy loss (E_{on}) during turn-on is reduced.

Figure 8 shows the turn-off waveforms of the devices fabricated at 300mm and 200mm wafers. These waveforms are in perfect agreement.

Figure 9 shows the recovery waveforms of the devices fabricated at 300mm and 200mm wafers. Devices fabricated at 300 mm wafers show a slightly faster rise in V_{KA} than devices fabricated at 200 mm wafers.

Figure 10 shows the total switching loss of E_{on}, E_{off} and E_{rr} added together for 300mm and 200mm wafers. As a result, devices fabricated at 300 mm wafers were found to have 10% lower the total switching loss than devices fabricated at 200 mm wafers.

Figure 11 shows the reverse bias safety operation area(RBSOA) waveforms of a devices fabricated at 300mm and 200mm wafers. Even at 10 times the rated current, there is no breakdown and no oscillation has been observed.

Figure 12 shows the short-circuit safe operating area(SCSOA) waveforms of a devices fabricated at 300mm and 200mm wafers. The waveforms are roughly equivalent without oscillation, indicating that even 2.8 μsec does not cause breakdown.

Figure 13 shows the correlation between the short-circuit pulse time width (T_w) and $V_{GE(th)}$ of a devices fabricated at 300mm and 200mm wafers. The correlation between T_w and $V_{GE(th)}$ shows the same trend for devices fabricated at 300mm and 200mm wafers.

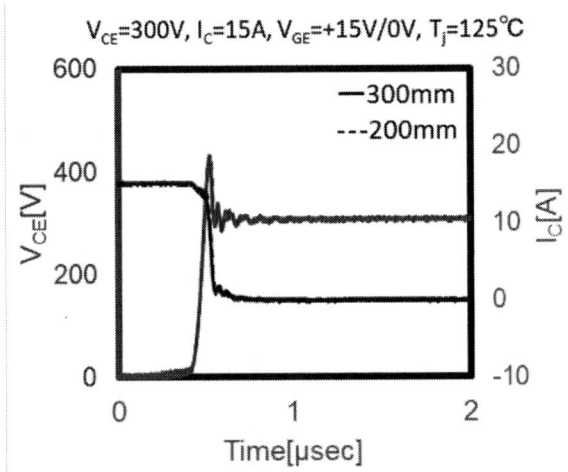

Fig. 7 Turn-on waveform

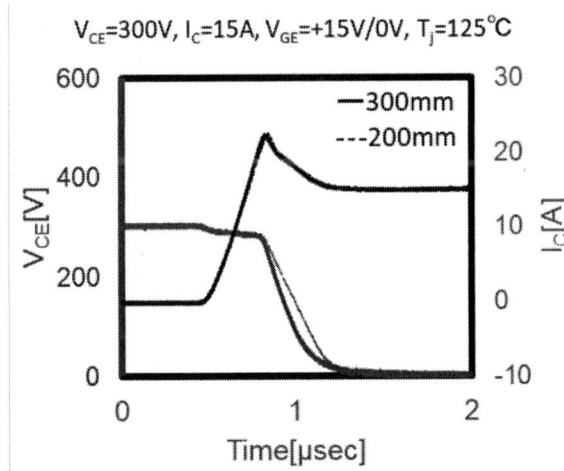

Fig. 8 Turn-off waveform.

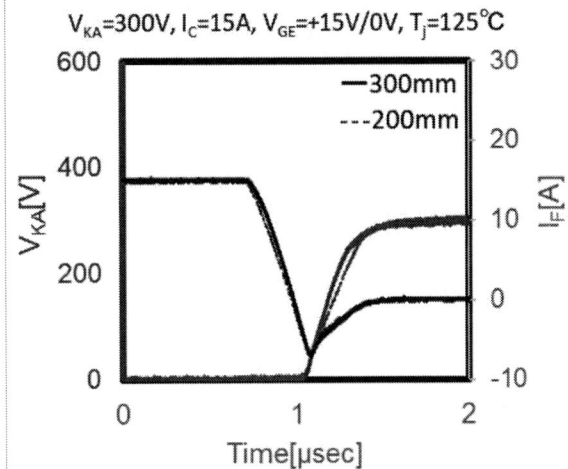

Fig. 9 Recovery waveform.

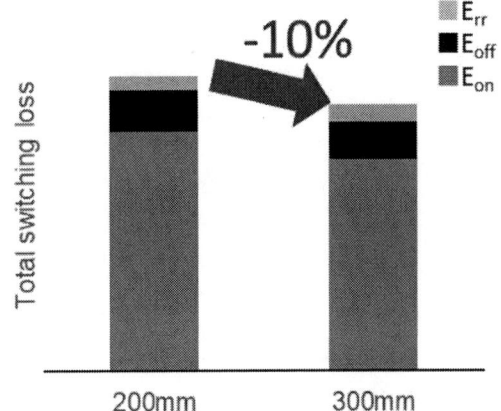

Fig. 10 Total switching loss characteristics.

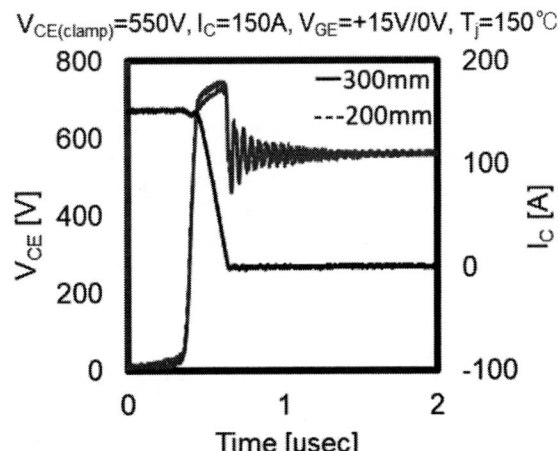

Fig. 11 RBSOA waveform.

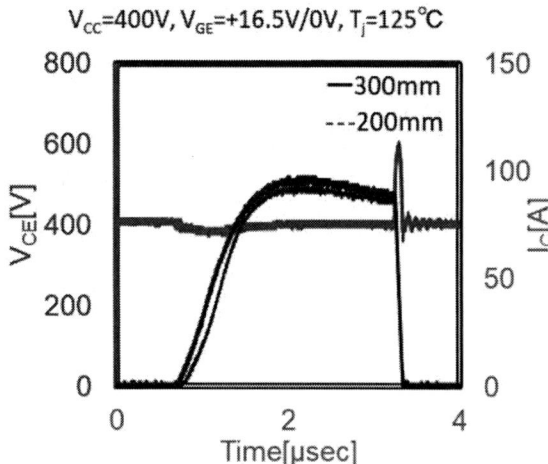

V_{CC}=400V, V_{GE}=+16.5V/0V, T_j=125°C

Fig. 12 SCSOA waveform.

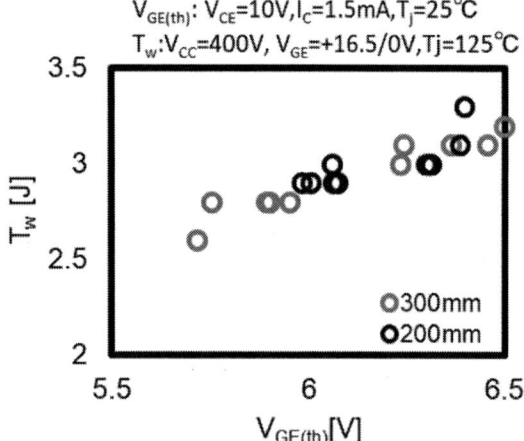

$V_{GE(th)}$: V_{CE}=10V, I_C=1.5mA, T_j=25°C
T_w: V_{CC}=400V, V_{GE}=+16.5/0V, Tj=125°C

Fig. 13 $V_{GE(th)}$ - T_w characteristic.

we intend to use this technology to expand our 300 mm lineup and develop high-performance equipment. From now on, we will also provide a stable supply of high quality, state of the art power devices to meet the ever-increasing demand for power devices.

References

[1] H. Schulze, H. Öfner, F. Niedernostheide, F. Lükermann and A.Schulz, "Fabrication of I GBTs using 300 mm magnetic Czochralski substrates", IET Power Electron2019, ISSN 1755-4535

[2] R. Kamibaba, K. Konishi, Y. Fukada, A. N arazaki and M. Tarutani,"Next Generation 6 50V CSTBT™ with improved SOA fabricate d by an Advanced Thin Wafer Technology". Proc. ISPSD'15, pp29-32(2015)

[3] T. Yoshida, T. Takahashi, K. Suzuki, and M. Tarutani, "The second-generation 600V RC-IGBT with Optimized FWD" Proc. ISPSD'16, pp159-162(2015).

[4] S. Soneda and A. Furukawa, "High avalan che capability specific Diode part structure of RC-IGBT based upon CSTBT™", Proc. I SPSD, pp.132-135, 2018.

[5] K.Sakaguchi, K.Konishi, K.Eguchi and S.Soneda, "Reduction of Junction Temperature with Local Lifetime Control and High Density Arranged Diode for 3rd Gen. 650V RC-IGBT" Proc. ISPSD'23, pp215-218(2022)

4   Conclusion & Outlook

Based on the above results, we have obtained a technology to produce devices in the 600V class even on 300mm wafers with advanced wafer process. In addition, we have succeeded in launching a process for 3rd Gen. RC-IGBT, which are expected to reduce the total switching loss from the 2nd Gen. RC-IGBT. And we have acquired a process line that can suppress the variation of characteristics within a wafer plane on a 300 mm wafer. Furthermore, we have already made 3rd Gen. RC-IGBT on wafers from several wafer vendors and have confirmed that the electrical characteristics and the gate reliability of those wafers are comparable. The issue of procuring the bare wafers for the products has also been resolved. Next step,

PCIM Europe 2024, 11– 13 June 2024, Nuremberg   DOI: 10.30420/566262414

Next Level of Power Module Solution for PV C&I String Inverter with 1200 V H7 Technology in Easy3B Package

Tilo Poller[1], Ma Si Chao[1], Heike Prinz-Ruether[1], Maximilian Kummetat[1], Christian R. Müller[1]

[1] Infineon Technologies AG, Germany

Corresponding author: Tilo Poller, tilo.poller@infineon.om
Speaker: Tilo Poller, tilo.poller@infineon.om

Abstract

Recently the new 1200 V TRENCHSTOP™ IGBT7 H7 and the 1200 V emitter-controlled 7 Rapid diode have been presented [1]. These are fast switching devices with low dynamic and conduction losses. To use the full capability of these technologies within a power module, an optimisation of the commutation loops regarding stray inductance for cos φ = -1 to 1 is necessary. A power module with a 3-level neutral-point clamped 2 (NPC2) topology within Easy3B package is developed with these latest chips for a 1000 VDC 150 kW PV string inverter in commercial and industrial (C&I) application. This paper shows how the module enables a low inductive commutation path with a proper PCB design for such high-power and current rating. Therefore, the normal operation as well as a target application failure mode are analysed. And it is described how the module is optimised for these modes. For both operation modes the stray inductance is reduced by 45% in comparison to a reference design, which is enabling the usage of such a high current module. Via an application simulation it is shown that the module efficiency increases by 0.15% in comparison to a common design. It shows also that the converter power output can increase from 125 kW up to 165 kW.

1 Commutation loops within NPC2 topology

The F3L500R12W3H7_H11 is designed for converters with an NPC2 topology. Fig 1 shows the schematic topology. For the IGBT T1 and T2 a parallelisation of three new 170 A 1200 V TRENCHSTOP™ IGBT7 H7 chips is used [1]. This IGBT technology is designed for fast switching operations. Antiparallel to these IGBTs are the diodes D1 and D2, which are three 100 A 1200 V emitter-controlled 7 Rapid diodes [1]. The IGBTs T3 and T4 are two 200 A 950 V TRENCHSTOP™ IGBT7 S7 IGBTs [2], which are also fast switching devices. Antiparallel to these devices are D3 and D4, which are 950 V emitter-controlled rapid diodes [2]. For this module the main target application is PV C&I string inverters. Here two main operation conditions have to be considered:

The normal operation mode happens for a cos φ = 0 to 1, which means an energy flow from the solar field to grid occurs. In this operation mode IGBT T1 is the active switch (see Fig. 1). The freewheeling path is provided over the diode D3 and the IGBT T4. Therefore, the IGBT T4 is turned-on. For the negative sine half wave, the

IGBT T2 is the active device, and the freewheeling path is over diode D4 and IGBT T3.

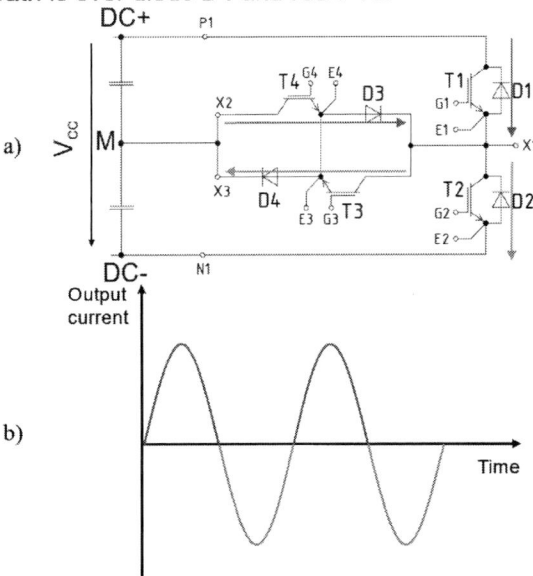

Fig. 1 Main commutation loops within a NPC2 topology cos φ = 0 to 1 (normal operation mode). a) Current flow within the commutation loops. b) Time shape of the output current

Besides this, the converter also must withstand several failure modes. As example a low voltage ride through (LVRT) might occur, which corresponds to a cos φ = -1 to 0. Fig. 2 shows the current flow for this case.

In the scenario either the IGBT T3 or T4 are the active switches. The freewheeling paths are provided by the diodes D1 or D2. Under this operation condition currents above the nominal module current have to be turned off.

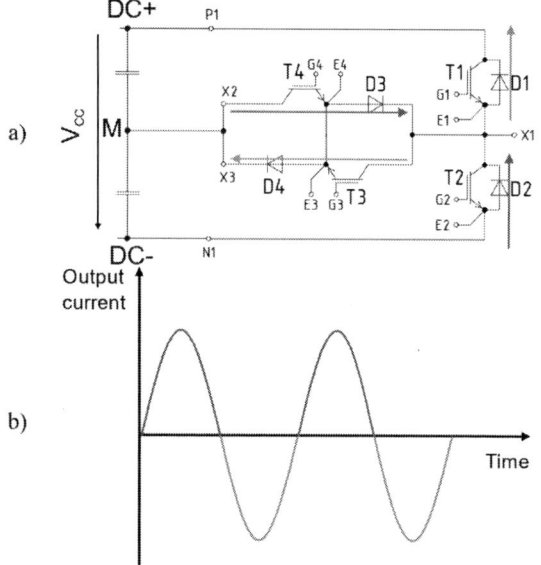

Fig. 2 Main commutation loops within a NPC2 topology for cos φ = -1 to 0 (failure mode). a) Current flow within active commutation loops. b) Time shape of the output current

2 Commutation loop optimisations

The F3L500R12W3H7_H11 uses the Easy3B package [3], which has a high flexibility to adjust the pin-out to the application requirements. To enable high nominal currents the package provides high-current pins [4] which have significantly higher current rating as the standard pins. Due to its flexibility, the power module allows a low inductive arrangement of the devices and pins within the package as described in [5].

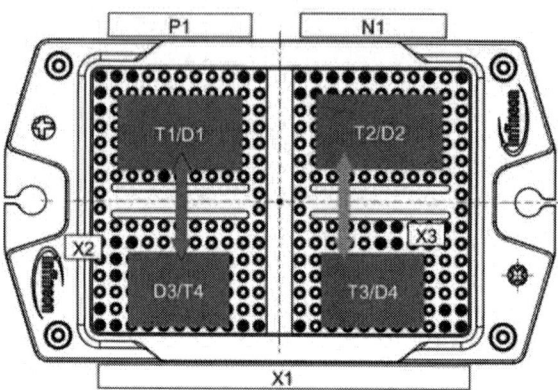

Fig. 3 Arrangement of the switches within the F3L500R12W3H7_H11 for the normal operation mode. Terminal and devices named according to Fig. 1. Arrows represent current flow.

To enable a low inductive path for the normal operation mode, the circuit is split accordingly in functional blocks and arranged on the substrates as depicted in Fig. 3. The green blocks show, how the devices are placed within the module and the arrows represent the current flow during the commutation according to Fig. 1. Due to the positioning the commutation for the normal operation mode occurs only inside the module.

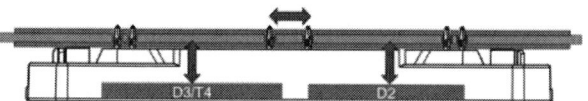

Fig. 4 Current flow for commutation over the PCB. Orange/Green/Orange: PCB with copper layer, purple arrows: current flow between the commutation partners

In former designs the commutation loop for the failure mode includes the PCB which is presented in Fig 4. The arrows represent the current flow over the PCB, for the commutation partners T4 and D2. The schematic also applies for the commutation between T3 and D1. Resulting in a high stray inductance for these commutation loops.

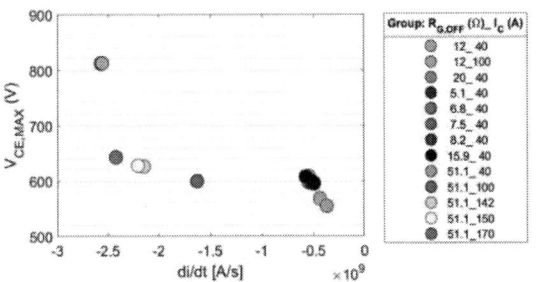

Fig. 5 Overvoltage at T3/T4 in dependency of di/dt. Grouped by $R_{G,OFF}$ and switched current

Fig. 5 shows the overvoltage in dependency of the di/dt during turn-off at T3/T4 for such a configuration. Within the graph the symbols are grouped to the external turn-off gate resistor $R_{G,OFF}$ and the switched current I_C. The rated nominal current for the F3L500R12W3H7_H11 is 500 A. The graph shows that at almost 35% of the rated current and a $R_{G,OFF}$ of 51.1 Ω the overvoltage exceeds the rated IGBTs blocking voltage of 650 V for the T3/T4. Further investigation showed that a $R_{G,OFF} > 100$ Ω is needed, to turn-off a current which is at twice nominal current. As a result, the switching speed of active device T3 and T4 has to be significantly decreased to omit an electrical overstress e.g. high overvoltage during switching. This results in high switching losses for the involved devices, which must be take into account for the thermal converter design as well as the lifetime calculation. In this example the determined total stray inductance, with regards to the measurement setup as well as the module, is 60 nH.

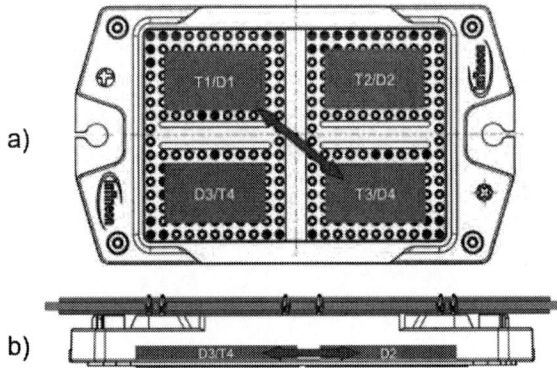

Fig. 6 Commutation within the Easy3B for failure mode; a) top view; b) side view including the PCB. Arrows shows the commutation path for cos φ = -1 to 0

In comparison, the F3L500R12W3H7_H11 provides a commutation path for the failure mode within the module (see Fig. 6.). In this case, a total stray inductance (module together with PCB) of 21 nH is determined in double-pulse test. For the normal operation mode the total stray inductance is 12 nH. In comparison to the reference design this is a reduction of 45%.

3 Switching behaviour module

The 1200 V TRENCHSTOP™ IGBT7 H7 and the 1200 V emitter controlled 7 Rapid diodes are optimized for fast switching operation. Although these new chip technologies offer low switching losses, they still offer quite low conduction losses [1]. The typical DC-link voltage from DC+ to DC- (see Fig. 1) is 1000 V. Due to the NPC2 topology and the capacitive voltage divider, the voltage between DC+ and the neutral point M is 500 V typically; as well as between the neutral point M and the DC-. Therefore, the switching losses of F3L500R12W3H7_H11 are determined at a DC- link voltage of 500 V. Fig. 7 and Fig. 8. show switching curves for the normal operation and the failure mode operated at implemented nominal current of the IGBTs, i.e., for 1200 V H7 at 500 A and for 950 V S7 at 400 A, and nominal gate resistances. For both operation modes a clean switching behaviour is observed. The switching event itself triggers the resonant circuit, which is formed by the device output capacitances and the measurement setup stray inductance. As result fast transients in form of oscillations are visible after the switching events. From the measurements the overvoltage and di/dt for T3/T4 are extracted in dependency of the switched current. The graph can be found in Fig. 9. The data points are grouped by the switched current. For these measurements a $R_{G,OFF}$ of 22 Ω is used, which is significant lower as for the reference design (> 100 Ω). In comparison to Fig. 4 the graph shows that a turn-off up to twice the nominal current (800 A) is possible without violating the overvoltage criterion. Furthermore, higher di/dt's are reached in comparison to the reference design from Fig. 4. At a current of 160 A the di/dt is 40% higher. At twice nominal current a di/dt of - 6 kA/µs is reached.

PCIM Europe 2024, 11– 13 June 2024, Nuremberg   DOI: 10.30420/566262414

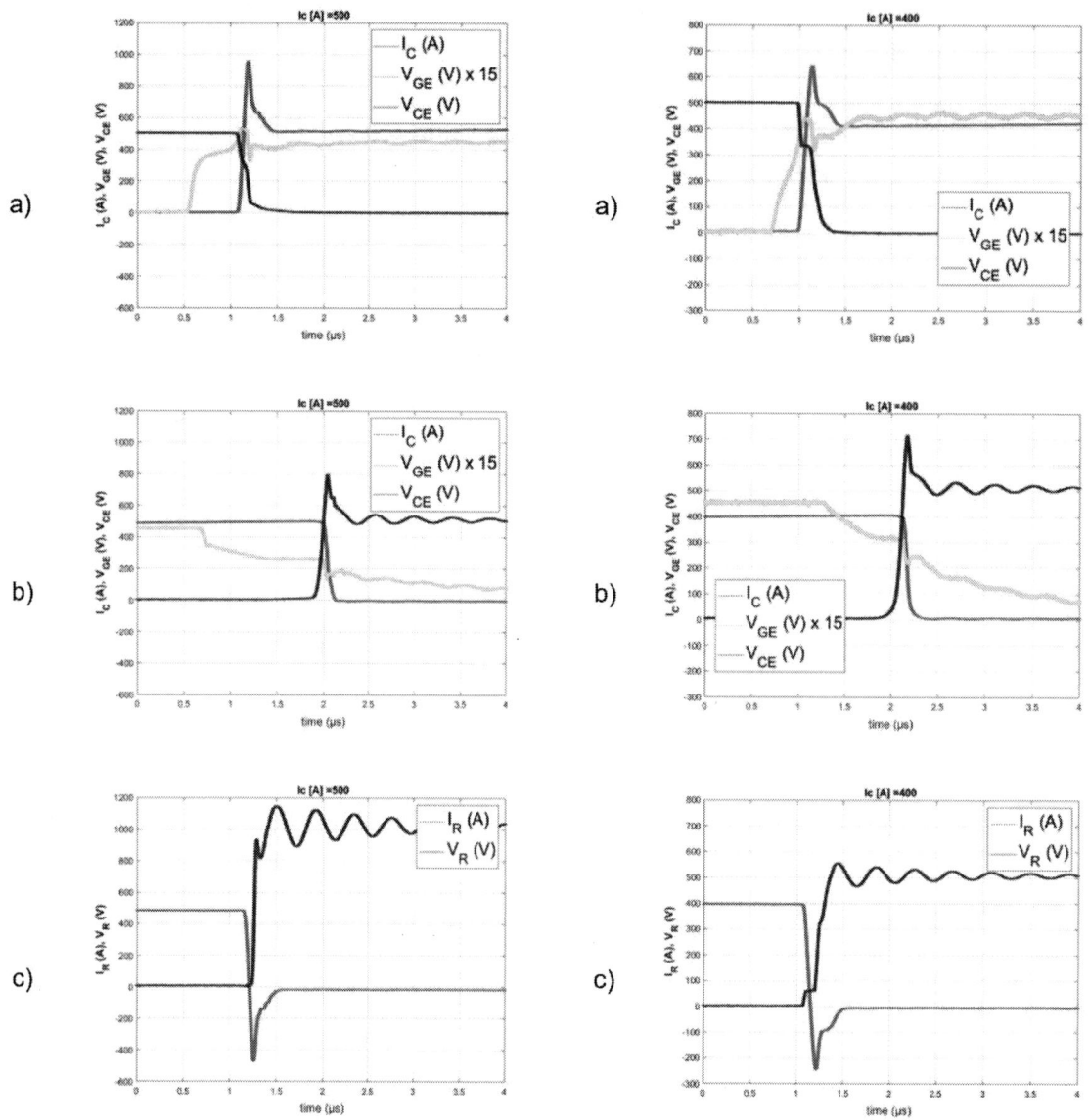

Fig. 7 Switching curves of the normal operation mode. a) Turn-On IGBT T1/T2 b) Turn-Off IGBT T1/T2 c) Diode recovery diode D3/D4

Fig. 8 Switching curves of the failure operation mode. a) Turn-On IGBT T3/T4 b) Turn-Off IGBT T3/T4 c) Diode recovery diode D1/D2

PCIM Europe 2024, 11– 13 June 2024, Nuremberg   DOI: 10.30420/566262414

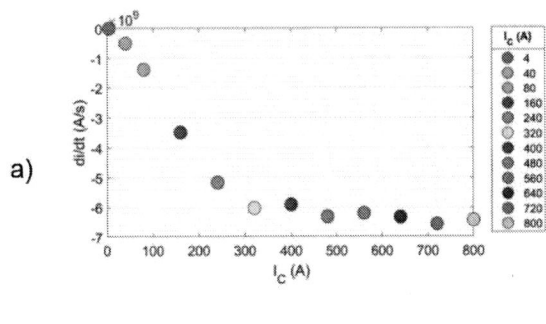

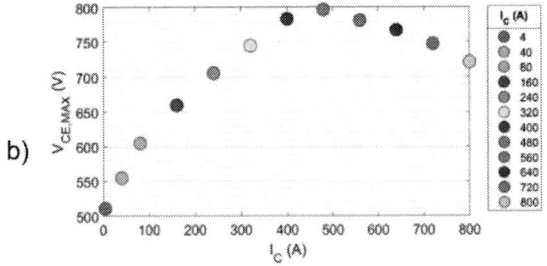

Fig. 9 a) di/dt and b) overvoltage at T3/T4 for the F3L500R12W3H7_H11 in dependency of the switched current at the recommended $R_{G,OFF,}$ = 22 Ω. Grouped by current

4 Application simulation

Earlier, the maximum power rating of PV string inverters was 125 kW for both commercial and industrial applications, supported by the power module F3L400R07W3S5_B59 with a Trenchstop™ 5 IGBT. In the following part, the performance increase of the combination 1200 V TRENCHSTOP™ IGBT7 H7 with emitter controlled 7 Rapid diode within the in new NPC2 power module F3L500R12W3H7_H11 is simulated. It is worth to mention that both modules provide different 3-level topologies, i.e., an NPC1 topology in F3L400R07W3S5_B59 and an NPC2 topology in F3L500R12W3H7_H11. As both modules are using the same Easy3B package and are operated under identical application conditions, this is a valid and fair comparison from system perspective. Anyhow, one should keep in mind that both power module differ in number of switches and device's blocking voltages significantly.
Based on the static and dynamic characteristics of both power modules, the power losses and temperature simulation are evaluated with PLECS simulation tool. The scenario of high DC link voltage and low DC link voltage are considered for the simulation over the whole output power range. The working conditions at low DC link voltage of V_{DC} = 580 V represents the main working mode of

$$\eta = \left(1 - \frac{3 \times P_{losses\_module}}{P_{output}}\right) \times 100\%$$

the PV string inverter in C&I application. To compare both systems the module efficiency η is calculated. The definition of this parameter is:

The module efficiency describes the relation between the total power losses within the modules (in this case one module for each phase) and the total power output of the converter. Under these working conditions and the same output power level the converter with the F3L500R12W3H7_H11 shows an increased overall efficiency in the range 0.1~0.15% compared to the reference design (see Fig. 10). This results in a higher energy generation of the PV inverter as benefit. A comparison on the same efficiency level shows that total output power can be increased in the range of a factor of 2 to 3.

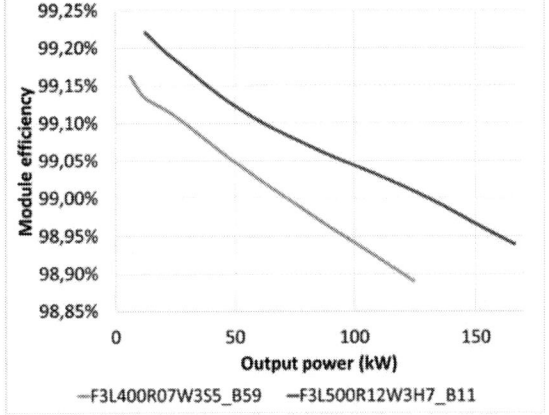

Fig. 10 Module efficiency over output power of power module F3L500R12W3H7_H11 and the reference power module F3L400R07W3S5_B59. Conditions: V_{DC} = 580 V, V_{AC} = 400 V, f_{sw} = 16 kHz, T_h = 90 °C, DPWM

2918

PCIM Europe 2024, 11– 13 June 2024, Nuremberg   DOI: 10.30420/566262414

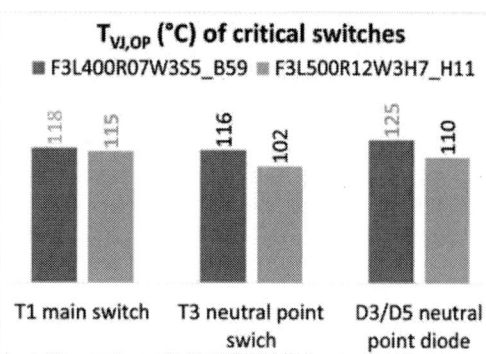

Fig. 11 $T_{VJ,OP}$ of the power module F3L500R12W3H7_H11 and the reference power module F3L400R07W3S5_B59 at P_{out} = 125 kW. Conditions: V_{DC} = 800 V, V_{AC} = 400 V, f_{sw} = 16 kHz, T_h = 90 °C, DPWM

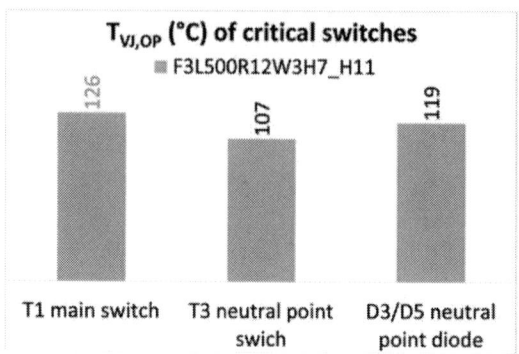

Fig. 12 $T_{VJ,OP}$ of the power module F3L500R12W3H7_H11 and the reference power module F3L400R07W3S5_B59 at P_{out} = 165 kW. Conditions: V_{DC} = 800 V, V_{AC} = 400 V, f_{sw} = 16 kHz, T_h = 90 °C, DPWM

Under the working conditions with high DC link voltage of 800 V, the highest thermal stress applies to the chips inside the reference module as well as for the F3L500R12W3H7_H11. For the target applications the typical $T_{vj,op}$ under normal operation conditions is set into the range of 125 to 130 °C to ensure a sufficient thermal safety margin. At 125 kW output power (see Fig. 11), the operation temperature $T_{vj,op}$ of the neutral-point diode in F3L400R07W3S5_B59 reaches 125 °C and limits the power rating of the inverter. As shown in section 2 the turn-off overvoltage of the 650 V devices, at this power rating, becomes another limiting factor for further power increase. Attributed to the optimized chip configuration in the F3L500R12W3H7_H11 with NPC2 topology, the maximum chip $T_{vj,op}$ in the module is 115 °C.

As the device junction temperature is below 125 °C, the F3L500R12W3H7_H11 allows to increase the converter output power. Therefore, the simulation model output power is increased until $T_{vj,op}$ = 125°C is reached. The simulation result shows that a power output of up to 165 kW is achieved. This is a boost of about 39%. Fig. 12 shows for this scenario the calculated $T_{vj,op}$. In the simulation T1 is the device with the highest $T_{vj,op}$ = 126 °C.

5   Conclusion

In the previous chapters the module optimisation for fast switching devices within a NPC2 topology is discussed. Due to analysis of the operation modes, typical for PV C&I string inverters, the main commutation loops are identified. Based on this, the commutation paths within the module are optimised and the stray inductance is reduced by 45% compared to the reference design. This reduction allows a usage of lower gate resistances to switch currents up to twice the nominal current. and leads to smaller switching losses for the discussed operation modes. In an application simulation, it is calculated that the chip $T_{vj,op}$ of the F3L500R12W3H7_H11 is 10 K cooler as a common reference design for a 125 kW converter, which is also leads directly to an increased module efficiency by 0.15%. By usage of the typical $T_{vj,op}$ range of 125 to 130 °C the output power of the converter can be increased by 39%.

References

[1] J. Cerezo and A. K. Sekar, "1200 V TRENCHSTOP™ IGBT7 H7 and Emitter-Controlled EC7 Rapid Diode Technologies Define an Enhanced Benchmark for Improved Energy-Efficient, Fast-Switching Inverter Applications," in *PCIM*, Nuremberg, 2023.

[2] C. R. Müller, A. Colvero Schittler and M. Si Chao, "950 V IGBT and Diode Technology Integrated in a Low Inductive ANPC Topology for 1500 VDC PV String Inverter," *Bodo's Power Systems,* pp. 34-37, 2020.

[3] C. R. Mueller, A. Colvero Schittler and J. Laven, "New 950-V IGBT and diode technology integrated in a low-inductive ANPC topology for solar applications," in *PCIM Europe*, Nuremberg, 2019.

[4] "AN2009-01 Assembly Instructions for the Easy Modules," Infineon Technologies AG, Munich, 2023.

[5] R. Bayerer and D. Domes, "Power Circuit design for clean switching," in *CIPS*, Nuremberg, 2010.

PCIM Europe 2024, 11– 13 June 2024, Nuremberg   DOI: 10.30420/566262415

Analysis of MOSFET Switching Losses in Resonant Converters Using Electrical and Thermal Measurements and Loss Trends with MOSFET Size Variation

Alfio Scuto[1], Giuseppe Sorrentino[1], Marco Ventimiglia[1], Gaetano Belverde[1].

[1] STMicroelectronics

Corresponding author: Alfio Scuto, alfio.scuto@st.com
Speaker: Alfio Scuto, alfio.scuto@st.com

Abstract

The present work aims to implement a thermal method to measure the MOSFET power losses in an LLC resonant converter using silicon superjunction (SJ) MOSFETs with different die sizes. The difficulty in carrying out this type of thermal measurement lies in measuring a temperature as close as possible to the junction temperature. In this work, the junction temperature was measured through the temperature measured on an additional die that was copacked with the MOSFET for this purpose.

On zero voltage switching (ZVS) converters, it is difficult to measure switching power losses only using oscilloscopes because there is no correlation between the value measured and the temperature.

This quantity is measured by subtracting the known contributions from the total thermal losses. To better quantify the on-state losses, an on-state voltage measurement circuit (OVMC) is implemented to ensure that there are no significant errors that would otherwise be obtained from a high voltage probe. This circuit allows measurement of the conduction losses, the diode losses, and the $R_{DS(on)}$, which can be used also as temperature-sensitive electrical parameters (TSEP) to estimate the junction temperature from electrical measurement.

1 Introduction

In resonant or zero voltage switching (ZVS) converters, the stored energy of the output capacitance C_{OSS} is not completely lost during switching, unlike what happens in hard-switching converters. While this energy cannot be measured through oscilloscopes or electrical simulations, it is very important to be able to quantify it to choose the correct device and optimize converter efficiency.

The amount of energy lost also depends on the shape of the V_{DS} voltage during charging and discharging of the capacitance, which leads to having a capacitive hysteresis whose quantification has been treated in several works through thermal [1] or electrical methods [2].

This phenomenon is not only present on silicon SJ MOSFETs, but also on silicon carbide (SiC) and gallium nitride (GaN) MOSFETs [3].

Superjunction MOSFET technology is still improving and offering cost-effective products that can meet high-power-density requirements. For this reason, we chose to focus our work on these products, and investigate the previously mentioned

losses in devices with different die sizes belonging to the MDmesh™ DM6 technological family.

2 Device prototypes

Precisely measuring the junction temperature is difficult with noninvasive methods. While an NTC was integrated into a power module in another study [4], this work involves measuring the temperature with an additional small MOSFET die (MD) copacked with the power MOSFET, soldered close to the main die to ensure highly similar temperatures at thermal steady states. The additional die was configured as a diode to enable its function as a thermal sensor; i.e., shorting the gate and source pads, and exploiting its linear relationship in temperature with fixed current. The reason for choosing a MOSFET as a temperature sensor is due to the ease of assembly, but any device can be used (NTC, PTC, diode, etc.). The assembled device with internal view and its circuital configuration is shown in Fig. 1.

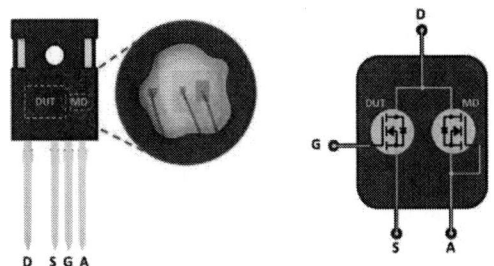

Fig. 1 Assembled device with its pinout and circuital configuration. Dice dimensions not in scale.

External access to the MD diode anode is provided via wire-bonding to the 4th lead of the TO247-4 package (pin A of Fig. 1). While commercial products in TO247-4 packages generally have a different pinout, using the 4th pin as a Kelvin source to optimize MOSFET switching [5], this device is configured differently for the purpose of the present work.

To study the MOSFET losses in resonant converters using different die device sizes, three different devices in 650 V MDmesh DM6 technology were assembled. Table 1 shows the three assembled devices, indicating the starting device (the device with which the die is shared), the typical $R_{DS(on)}$, and the die size as a percentage of the largest die.

	Starting device	Typ. $R_{DS(on)}$	Die size
Dev. A	STWA46N65DM6AG	55 mΩ	Ref.
Dev. B	STWA38N65DM6AG	68 mΩ	80%
Dev. C	STB30N65DM6AG	102 mΩ	57%

Table 1 DUT characteristics.

The future goal is to estimate the junction temperature exploiting only one of the temperature-sensitive electrical parameters (TSEP) [6] of the MOSFET, to use any type of device without the need for special assemblies. Using an OVMC to measure the $V_{DS(on)}$ for the precise losses measurement, the $R_{DS(on)}$ was chosen as TSEP for temperature estimation.

3 MOSFET thermal measurements

Before being able to take measurements in the application, some thermal characterizations are mandatory:

- MD forward voltage vs temperature
- DUT body-diode forward voltage vs temperature
- Junction to ambient thermal resistance
- DUT $R_{DS(on)}$ characterization vs temperature.

3.1 MD and DUT body diode thermal characterization

With a fixed 5 mA current, the MD forward voltage V_f is characterized as a function of the temperature. This measurement is performed using the Analysis Tech Phase 12 thermal analyzer with its "device calibration bath" while the device is immersed in a uniform temperature mineral oil bath and the MD V_f is measured over a range of bath temperatures (40 to 120°C). The relationship is linear, and its regression equation allows the use of the diode as a thermal sensor.

The same characterization is also made also for the body-diode of the DUT. This allows us to perform the measurement in the next section.

3.2 Junction to ambient thermal resistance measurement

The thermal resistance is measured by the Phase 12 thermal analyzer, starting from the following formula:

$$R_{th} = \frac{T_j - T_{amb}}{P_{diss}} \quad (1)$$

where T_j is the junction temperature, T_{amb} is the ambient temperature measured through a thermocouple and P_{diss} is the power dissipated in the device that is fixed by the thermal analyzer.

T_j is measured using the JEDEC 51-1 dynamic mode, where the MOSFET is polarized to dissipate some amount of power, and a negative voltage with fixed current is provided every 30 seconds to polarize the body-diode and calculate the junction temperature starting from the V_f measurement and the characterization curve obtained in the previous section.

This measurement was performed on all the DUTs without heatsinks and with two devices dissipating the same amount of power and placed in the same half-bridge board with the secondary-side diode fan enabled to emulate application board operation.

This test also allows us to verify the temperature measurement with the MD by comparing it with T_j measured using the JEDEC 51-1 dynamic mode. This allows the determination of how the measurement made with the MD follows the T_j and obtain a correction factor.

3.3 DUTs $R_{DS(on)}$ characterization

The present work is also exploited to validate the temperature estimation through the $R_{DS(on)}$ measurement to completely eliminate the MD diode in future measurements. The drain-source resistance $R_{DS(on)}$ is obtained by measuring drain-source voltage $V_{DS(on)}$ and drain current I_D.

To obtain a formula that converts $R_{DS(on)}$ into temperature, the drain-source resistance was characterized versus drain current I_D and junction temperature T_j using the Keysight B1506A curve tracer with an external heater with drain currents up to 20 A.

The temperature in this phase was measured through the MD diode using the previous characterization without the correction factor previously performed because the device is heated externally and the two dies are therefore assumed to be in thermal equilibrium. These $R_{DS(on)}$ measurements are used to obtain Tj as a function of $R_{DS(on)}$ and I_D, as performed in [7] [8]. The regression used is the following 2nd order polynomial function:

$$T_j\left(R_{DS(on)}, I_D\right) =$$
$$= a_1 + a_2 \cdot R_{DS(on)} + a_3 \cdot R_{DS(on)}^2 + \qquad (2)$$
$$+ a_4 \cdot I_D + a_5 \cdot R_{DS(on)} \cdot I_D$$

with $a_{1,N}$ the polynomial coefficient of the function, which depends on the DUT.

The result for the DEV. A is plotted in Fig. 3.
Using Eq. 2 with the measured value, the maximum error obtained is 1.5°C.

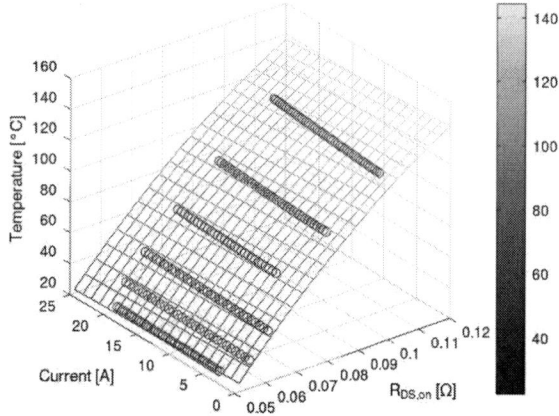

Fig. 2 Equation (2) regression in function of $R_{DS(on)}$ and I_D. The blue circles are the measured value.

4 Circuits explanation

4.1 LLC converter

The converter used in this work is a 1500 W LLC converter with 48 V output.

To better understand the waveform and the functioning, a small summary is required.

LLC converter is a high-power step-down converter. In this case, a half-bridge (HB) converter is used, and its simplified schematic is shown in Fig. 3. The two MOSFETs Q_1 (high-side HS) and Q_2 (low-side LS) switch at high frequency with the same ON-time T_{ON}. The operation exploits the resonance of the LLC circuits, consisting of L_R, L_M, and C_R. The output power changes with the switching frequency: as the frequency decreases, the power increases.

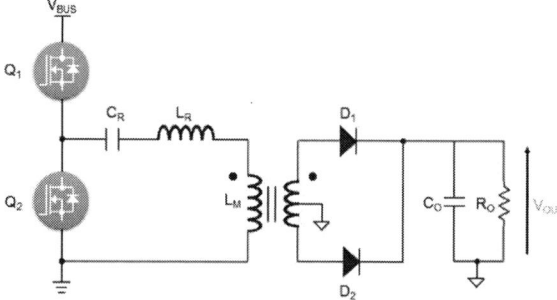

Fig. 3 LLC converter schematic.

Figure 4 shows the typical LLC converter switching period with V_{DS} and the three MOSFET operating areas. During dead-time, the body-diode conducts, which allows the achievement of zero voltage switching ZVS (blue area). The MOSFET then turns on and the current passes through its channel with a drain voltage that follows the current shape (yellow area). The MOSFET is than turned off (green area), and in this zone the complementary MOSFET follows the behavior just described. As will be seen, at low power, the current shape is triangular and as the frequency nears the peak of the resonance curve, the current shape becomes sinusoidal.

In LLC resonant converters, the power dissipated on the MOSFET can be split into the following components:

1. P_{diode}: power losses in body diode.
2. P_{cond}: conduction losses due to $R_{DS(on)}$.
3. P_{driver}: gate losses during commutation.
4. $P_{sw(off)}$: losses during switching-off.

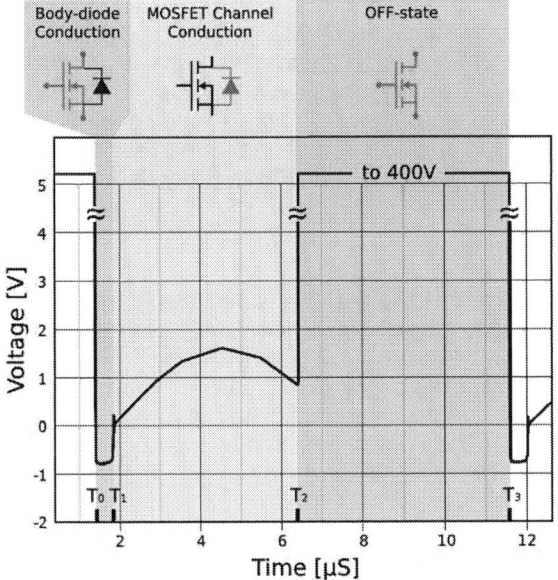

Fig. 4 Typical LLC converter switching period V_{DS} highlighted the different MOSFET operating areas.

$P_{SW(on)}$ losses are not present as, during switching-on, the conduction of the diode allows ZVS, so power dissipation power is zero.
The diode and conduction losses are calculated as follows:

$$P_{diode} = \frac{1}{T_{sw}} \int_{T_0}^{T_1} V_{DS} \cdot I_D \, dt \qquad (3)$$

$$P_{cond.} = \frac{1}{T_{sw}} \int_{T_1}^{T_2} V_{DS} \cdot I_D \, dt \qquad (4)$$

where T_{sw} is the switching period (T_0-T_3).
V_{DS} is measured using an OVMC because the use of a high-voltage probe leads to inaccurate measurement on such a small signal.
The gate loss is calculated as follows:

$$P_{driver} = Q_G \cdot V_{GS} \cdot f_{SW} \qquad (5)$$

With Q_G taken from the datasheet, V_{GS} is the peak-to-peak voltage of the gate signal, and f_{sw} is the switching frequency.
The switching loss $P_{SW,off}$ is dissipated when the MOSFET turns off, in T_2, due to the charge of output capacitance C_{OSS}. This cannot be measured with an oscilloscope, however, because it leads to an incorrect calculation in terms of temperature in the DUT.
To calculate this, a thermal approach is used, starting from temperature measurements. First, the total power dissipation is thermally calculated:

$$P_{tot} = \frac{T_j - T_{amb}}{R_{th}} \qquad (6)$$

using R_{th} measured in section 3.2.

The switching loss is thus calculated:

$$P_{sw,off} = P_{tot} - P_{diode} - P_{cond} - P_{driver} \qquad (7)$$

4.2 On-voltage measurement circuit

In application, the $R_{DS(on)}$ measurement is derived from the relationship between drain voltage and current measurements. The current measurement is performed with a Tektronix TCP0030A current probe. The on-state voltage is instead measured with an OVMC circuit that limits the V_{DS} voltage variation (Fig. 5). This increases the precision of the voltage measurement by setting a smaller vertical scale.

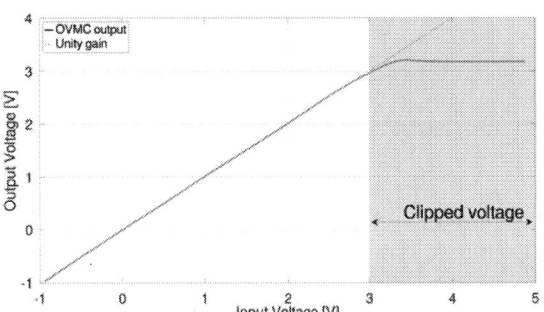

Fig. 5 OVMC output V_m vs input voltage.

As previously mentioned, the estimation of the junction temperature starting from $R_{DS(on)}$ is validated through the temperature measurement conducted on the MD.

In the HB LLC, the MD was not used as thermal sensor of the LS MOSFET because the drain voltage and hence the MD cathode swings from 0 V to the BUS voltage (400 V). The voltage transient at the middle point, associated with the stray elements, was considered a potential source of noise. For this reason, the measure of the T_j with the MD was only performed on the HS MOSFET, where the MD cathode voltage is fixed at V_{BUS}. The OVMC circuit is therefore designed to work on the HS MOSFET.

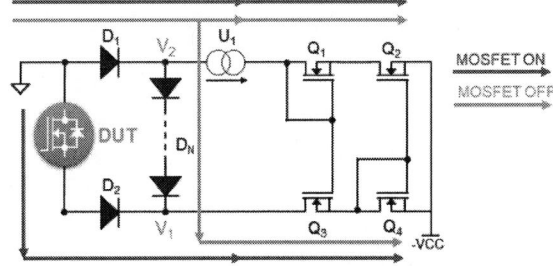

Fig. 6 Proposed OVMC schematic.

The basic OVMC schematic is shown in Figure 6. The starting circuit is proposed in [9] [10], but in this study, several changes were made. The entire schematic is mirrored to obtain the reference point on the drain rather than on the source, as in [11] [12]. Transistors Q_1, Q_2, Q_3, and Q_4 are MOSFETs instead of BJTs, and form a Wilson current mirror instead of a cascode one. Finally, the biasing resistor is replaced with a constant current driver.

The chosen D_1 and D_2 diodes have a maximum voltage of 600 V; to raise the maximum OVMC voltage, simply choose D_1 and D_2 diodes with higher voltages or insert more diodes in series to them.

Diode D_1 is always polarized by U1 at about 10 mA. When the DUT is ON, D_2 is polarized with the same current as D_1 thanks to the current mirror (purple line in Figure 6). Instead, when DUT is OFF, the series of diodes D_N are polarized with the same current as U_1 (light blue line in Figure 6). The number of diodes D_N defines the clamp voltage of the OVMC. In this work, there are six diodes.

After the schematic shown in Fig. 6, there is also a differential stage that provides the difference between V_2 and V_1, and so the OVMC output V_m is:

$$V_m = V_2 - V_1 = -V_{D1} - (-V_{DS,on} - V_{D2}) \quad (5)$$

$V_{D1} \approx V_{D2}$ because V_{D1} and V_{D2} are polarized by the same current; hence:

$$V_m = V_{DS,on} + V_{D2} - V_{D1} = V_{DS,on} \quad (6)$$

All this circuit is powered by an isolated DC-DC converter with dual 12 V output.

A significant problem when reading the OVMC output V_m is that this signal is referred to the drain voltage of the HS MOSFET, and this point is usually connected to the input voltage. To read this voltage, a low-voltage differential probe with a high common-mode voltage must be used. Usually, low-voltage differential probes have low common-mode voltage, unless an opto-isolated probe is used, as performed in [10]. To avoid this problem, the oscilloscope ground is connected to the drain reference to cancel the common-mode voltage of the low voltage differential probe, as shown in Fig. 9.

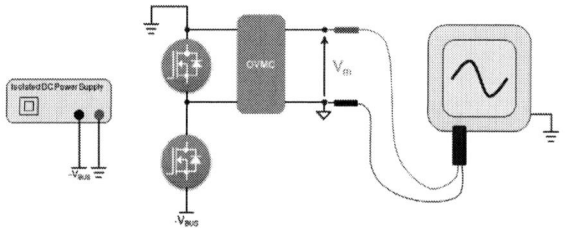

Fig. 7 Ground connection between half-bridge board and oscilloscope.

Equation (6) contains an error due to parasitic components. The OVMC measures the ON-state voltage across DUT leads; therefore, V_m voltage is $V_{DS(on)}$ plus the contribution of the series inductance due to the DUT wire bonding and leads. There is also a constant error due to the differential stage. Hence, the ON-state voltage is:

$$V_{DS(on)} = V_m - V_{offset} - L_{stray} \frac{dI_D}{dt} \quad (7)$$

V_{offset} and L_{stray} are calculated at different power levels to change the dI_D/dt, as performed in [13], to obtain the $V_{DS(on)}$ in phase with the I_D, and therefore with the same zero crossing. The waveforms without and with correction are shown in Fig. 8 and Fig. 9. The correction is made directly with the oscilloscope, which leads to a slight ripple in the resulting waveform due to the calculation of the derivative, as shown in Fig.9, but this is not an issue for loss measurements.

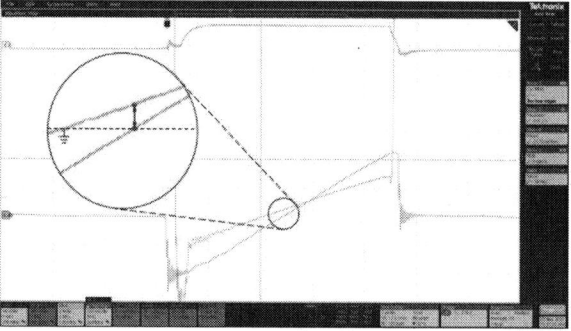

Fig. 8 LLC waveforms without stray inductance compensation. Gate voltage in blue, drain current in green, V_m output in yellow.

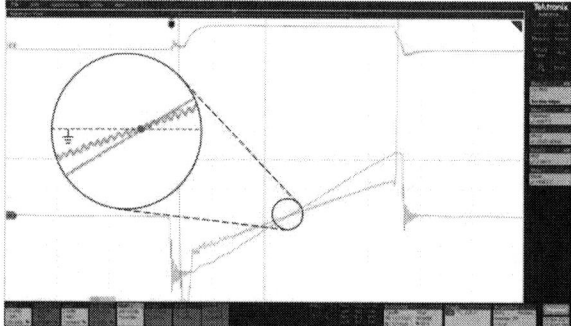

Fig. 9 LLC waveforms with stray inductance compensation.

5 Results

Figure 10 shows the application waveforms where it is possible to see the three different operating areas as shown previously in Fig. 4. One difference between the real and the simulated waveform is that the activation period of OVMC, hence the diode and conduction areas, is slightly smaller than the V_{DS} (red curve) due to the small rounded corner near the switching-on and switching-off.

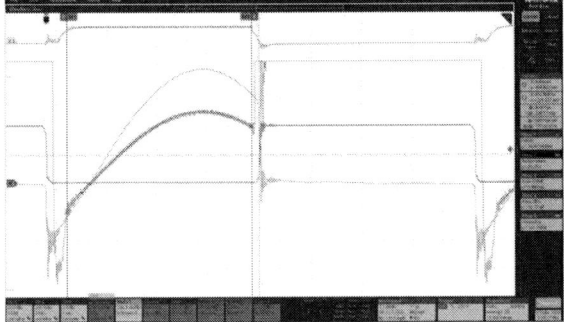

Fig. 10 One LLC period waveform with DEV. A at 800 W of output power. The gate voltage in cyan, the drain voltage in red, the drain current in green, and the compensated V_m in yellow.

The measurements have been made at different power levels until the temperature reaches approximately 100°C and the maximum output power between the DUTs is therefore different.

In the next sections, the different loss contributions, the $R_{DS(on)}$ measurements, and the junction temperature estimations are analyzed with the three DUTs of different die sizes.

5.1 Body-diode losses

The body-diode losses are measured as discussed in section 4.1. The results for the three DUTs are shown in Fig. 11.

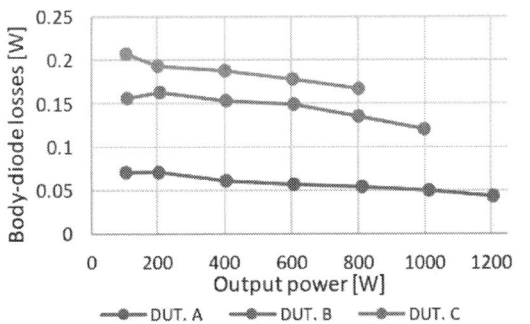

Fig. 11 Diode losses for the three DUTs in the LLC converter.

The decreasing trend as the die size increases is consistent with the data present in the datasheet. In fact, the V_{SD} of the diode reduces as the size of the die grows, even at the higher temperature. Moreover, in the device with smaller die size the lower C_{OSS} helps them switch faster; the rounding

of V_{DS} is also smaller and therefore the diode activates sooner causing longer conduction period of the diode for the same dead-time condition.

The decreasing trend as the output power increases is instead consistent with the decreasing V_{SD} as the temperature increases, and also due to the decreasing switching frequency f_{sw}.

5.2 Conduction losses

Unlike other losses, the conduction losses increase with the square of the current:

$$P_{cond} = I_D^2 \cdot R_{DS(on)} \tag{8}$$

Therefore, we expect a dissipated power that increases as the output power increases.

The results are shown in Fig. 12.

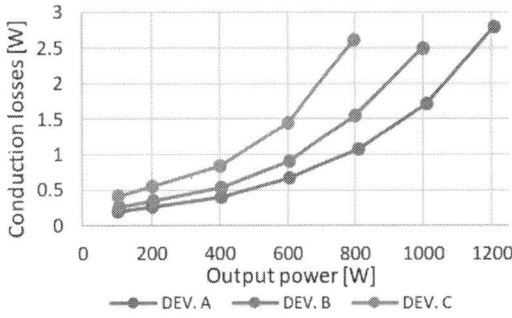

Fig. 12 Conduction losses for the three DUTs in the LLC converter.

The conduction losses of course heavily depend on the $R_{DS(on)}$ of the device, and therefore, the larger the die size, the greater the maximum output power for the same thermal resistance R_{th}.

5.3 Switching losses

As mentioned in section 4.1, the switching losses measured via oscilloscope introduce errors. Figure 13 shows the temperature measured in DEV. A during its functioning in application (blue curve) and the temperature calculated (orange curve) starting from the losses measured with the oscilloscope, except for the P_{driver} that is calculated using datasheet value (Eq. 5):

$$T_{calc} = R_{th} \cdot P_{diss}^* + T_{amb} \tag{9}$$

with

$$P_{diss}^* = P_{diode} + P_{cond} + P_{driver} + P_{sw(off)}^* \tag{10}$$

where $P_{sw(off)}^*$ is measured as shown in Fig. 14 between the cursors and calculating:

$$P_{sw(off)}^* = \frac{1}{T_{sw}} \int_A^B V_{DS} \cdot I_D \, dt \tag{11}$$

PCIM Europe 2024, 11– 13 June 2024, Nuremberg   DOI: 10.30420/566262415

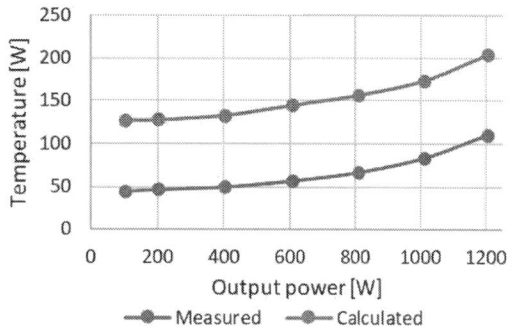

Fig. 13 Measured and calculated temperature vs output power.

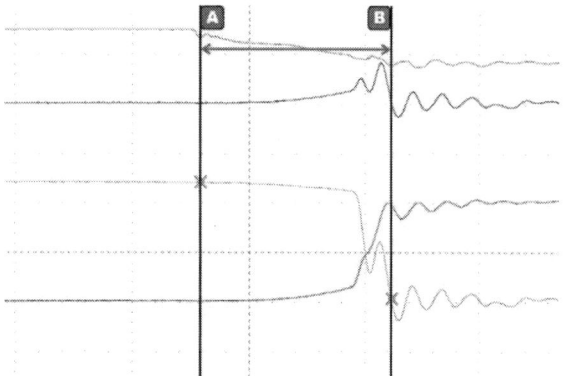

Fig. 14 Switching-off losses measured with oscilloscope: Gate voltage in blue, power in violet, drain current in green and drain voltage in red.

The switching losses with the thermal method seen in section 4.1 and using Eq. 6 and Eq. 7, are shown in Fig. 15 and Fig. 16. As can be seen, as the device die size increases, the switching power, and therefore the energy, increases.

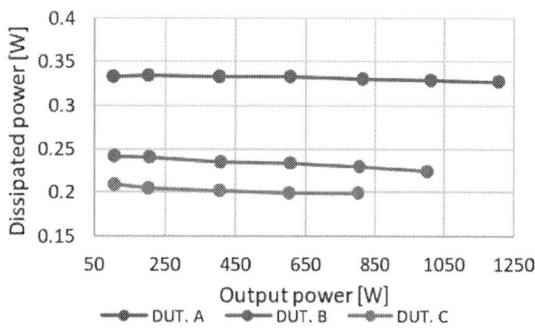

Fig. 15 Thermal switching-off losses vs output power.

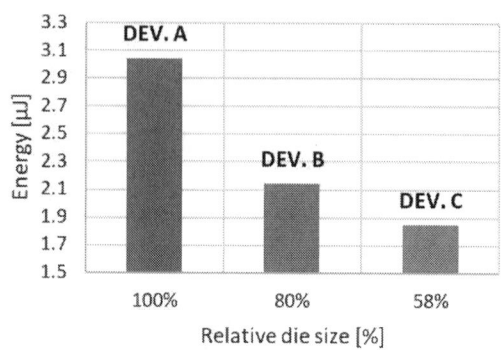

Fig. 16 Average thermal switching-off energy vs die size.

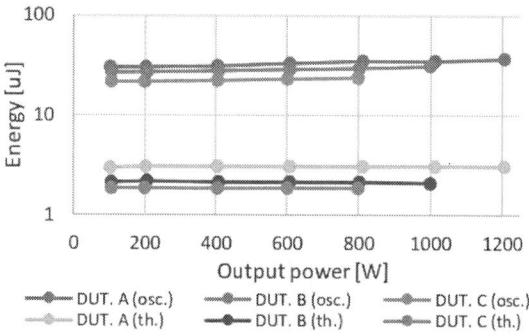

Fig. 17 Oscilloscope measurements and thermal switching-off energy vs output power.

Figure 17 shows the switching-off energy measured with an oscilloscope and then thermally measured. The two different values, for the same device, have the same increasing trend with die size. The thermal energy measurement with respect to the oscilloscope measurement is in the range of 7-10% for all the DUTs.

5.4 $R_{DS(on)}$ analysis

To verify the $R_{DS(on)}$ measured in application through the OVMC and the current probe, and therefore the loss measurements, the $R_{DS(on)}$ is compared to the datasheet values. In particular, the normalized $R_{DS(on)}$ is calculated and plotted with the MD temperature value varying and compared with the typical $R_{DS(on)}$ vs temperature curve taken from the relevant datasheet. The results are shown in Fig. 18-20.

This comparison reveals a maximum error of 3%, which confirms the veracity of the values measured in this way.

PCIM Europe 2024, 11– 13 June 2024, Nuremberg   DOI: 10.30420/566262415

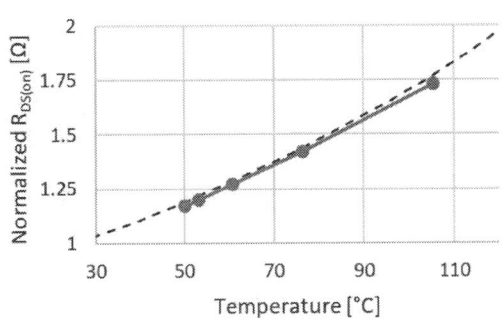

Fig. 18 Datasheet normalized $R_{DS(on)}$ (dashed line) vs measured $R_{DS(on)}$ in application (blue line) with DUT. A.

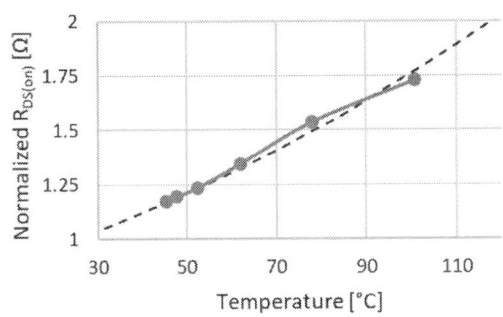

Fig. 19 Datasheet normalized $R_{DS(on)}$ (dashed line) vs measured $R_{DS(on)}$ in application (orange line) with DUT. B.

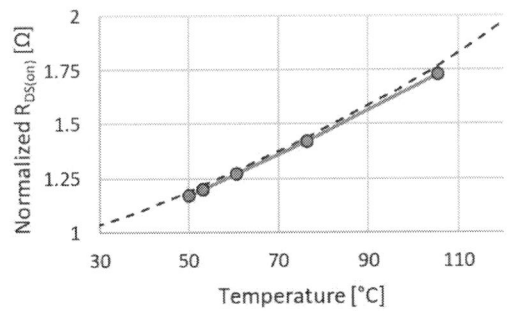

Fig. 20 Datasheet normalized $R_{DS(on)}$ (dashed line) vs measured $R_{DS(on)}$ in application (gray line) with DUT. C.

5.5 Temperature estimation

As previously mentioned, $R_{DS(on)}$ is also used as a TSEP to estimate temperature from electrical measurements. Eq. 2 is used to obtain the temperature value and it is implemented directly in the oscilloscope. Figure 21 shows LLC waveforms with Eq.2 in brown.

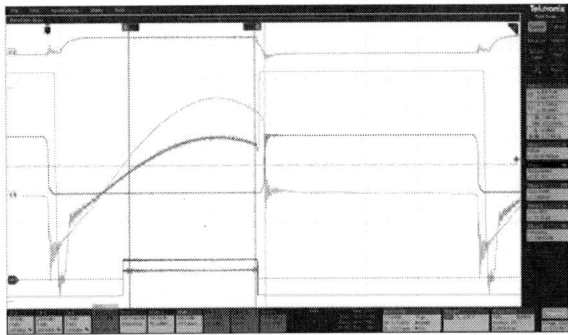

Fig. 21 LLC waveforms with temperature estimation equation. Gate voltage in blue, drain current in green, compensated V_m output in yellow, temperature estimation in brown and $R_{DS(on)}$ in violet.

Temperature estimation is taken as the mean between the two cursors. All the temperature values taken for the three DUTs are shown in Fig. 22, where proximity to the unitary line indicates an estimate closer to the measured value. The maximum error obtained from the measurement is approximately 1.5°C, close to the one obtained in section 3.3 during T_j model creation.

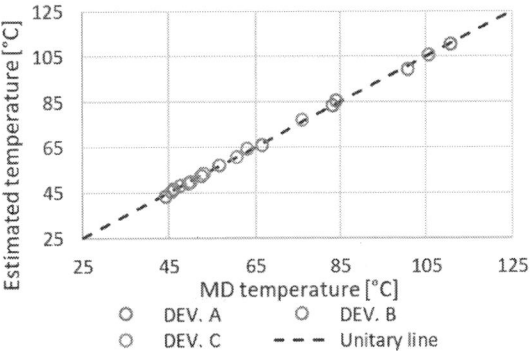

Fig. 22 Estimated temperature vs temperature read through MD during LLC converter functioning.

6 Conclusion

Using a thermal method, the performances of DM6 technology MOSFETs in different die sizes was investigated. The results demonstrate the capabilities of this MDmesh technology in LLC converters, with low losses during switching-off. The study also shows how the switching losses increase as the die size increases.

A small die, copacked with the power MOSFET, is used as a thermal sensor to measure temperatures close to the die. This allows accurate thermal analysis and calculation of the real power losses of the MOSFET.

2928

This can be crucial for semiconductor designers seeking to correlate device parameters with switching losses, ultimately allowing better solution designs for resonant converters.

The results of power loss measurements using this thermal method highlights the marked difference between switching energy measured with oscilloscopes and real dissipation in MOSFETs during LLC converter operation. The difference between the two measurements is around 10%, leading to major error when performing measurement only with an oscilloscope.

The use of OVMC allowed us to obtain more accurate body-diode and conduction loss measurements, as well as estimation of the junction temperature from electrical measurement. This noninvasive method allowed us to limit the error from temperature estimation to an accuracy that is comparable with the use of standard methods, such as thermal IR cameras, but strictly related to the die and not to the external part of the device.

The results obtained are in line with the datasheet values and the correlation with thermal measurements validate the method presented.

Future research will include benchmarks with devices of different families and will include silicon carbide or gallium nitride.

References

[1] J.B. Fedison M. Fornage, M.J. Harrison D.R. Zimmanck, "Coss Related Energy Loss in Power MOSFETs Used in Zero-Voltage-Switched Applications" *Applied Power Electronics Conference and Exposition (APEC)*, 2014.

[2] J.B. Fedison, M.J. Harrison, "COSS Hysteresis in Advanced Superjunction MOSFETs" *Applied Power Electronics Conference and Exposition (APEC)*, 2016.

[3] D. Bura, T. Plum, J. Baringhaus, R.W. De Doncker, "Hysteresis Losses in the Output Capacitance of Wide Bandgap and Superjunction Transistors" *EPE'18 ECCE Europe*, 2018.

[4] A. Scuto, G. Sorrentino, M. Ventimiglia, G. Belverde D. Nardo et al., "Assessment of MOSFET switching losses in an LLC converter by a calorimetric method" *IEEE Conference on Applied Power Electronics Conference and Exposition (APEC)*, 2023.

[5] S.A. Rizzo, N. Salerno, "Actual Reasons Involving Turn-Off Losses Improvement With Increasing Load and Gate Resistance in MOSFETs Enhanced With Kelvin Source" *IEEE Transactions on Industrial Electronics,* vol. 71, no. 1, pp. 369-379, 2024.

[6] S.A. Rizzo, G. Susinni, F. Iannuzzo, "Intrusiveness of Power Device Condition Monitoring Methods: Introducing Figures of Merit for Condition Monitoring" *IEEE Industrial Electronics Magazine,* vol. 16, no. 1, pp. 60 - 69, March 2022.

[7] F. Stella, G. Pellegrino, E. Armando, D. Daprà, "Online Junction Temperature Estimation of SiC Power MOSFETS Through On-State Voltage Mapping" *IEEE TRANSACTIONS ON INDUSTRY APPLICATIONS,* 2018.

[8] Q. Zhang, G. Lu, Y. Yang, P. Zhang, "A High-Frequency Online Junction Temperature Monitoring Method for SiC MOSFETs Based on ON-State Resistance With Aging Compensation" *IEEE TRANSACTIONS ON INDUSTRIAL ELECTRONICS,* 2023.

[9] R. Gelagaev, P. Jacqmaer, J. Driesen, "A Fast Voltage Clamp Circuit for the Accurate Measurement of the Dynamic ON-Resistance of Power Transistors" *IEEE TRANSACTIONS ON INDUSTRIAL ELECTRONICS,* 2015.

[10] M.C.J. Weiser, K.M. Barón, T. Fink, I. Kallfass, "A Fast ON-State Drain-to-Source Voltage Amplifier for the Dynamic Characterization of GaN Power Transistors" *Applied Power Electronics Conference and Exposition (APEC)*, 2023.

[11] C. Roy, N. Kim, J. Gafford, B. Parkhideh, "On-State Voltage Measurement of High-Side Power Transistors in Three-Phase Four-Leg Inverter for In-Situ Prognostics" *Energy Conversion Congress and Exposition (ECCE)*, 2021.

[12] C. Roy, N. Kim, D. Evans, A.P. Sirat, J. Gafford et al., "A Half-Bridge On-State Voltage Sensor for In-Situ Measurements" *IEEE Energy Conversion Congress and Exposition (ECCE)*, 2022.

[13] M. Guacci, D. Bortis, J.W. Kolar, "On-State Voltage Measurement of Fast Switching Power Semiconductor" *CPSS TRANSACTIONS ON POWER ELECTRONICS AND APPLICATIONS,* 2018.

PCIM Europe 2024, 11– 13 June 2024, Nuremberg   DOI: 10.30420/566262416

OptiMOS 6 135 V for High Power Motor Drive Applications

Kunal Jha [1], Tien Quang Tran [2], Kapil Kelkar[1], Stefan Tegen[3], Josef Mohammed[2]

[1] Infineon Technologies, USA
[2] Infineon Technologies, Austria
[3] Infineon Technologies, Germany

Speaker: Kunal Jha, Kunal.Jha@infineon.com

Abstract

Decarbonization has become a necessity to tackle climate change and electrification is a small step in achieving this goal. Infineon Technologies is developing a new Best-in-Class OptiMOS™ 6 135 V technology for high-power motor-drive applications such as LEVs, e-forklifts, power & gardening tools etc. which predominantly use 72 - 84 V batteries. It has been designed to replace existing 135 - 150 V MOSFETs being used for these applications by providing significant improvements in $R_{DS(ON)}$ and cost, helping improve the system efficiency. This paper demonstrates OptiMOS™ 6 135 V switching and thermal performance improvements in power & gardening tools over OptiMOS™ 5 150 V and Best-in-Class parts from competitors.

1 Introduction

Decarbonization initiatives are impacting every industry in various ways, and gasoline engine tools and vehicles are few of the prominent targets for electrification. There is a big push to electrify these using battery-powered Brushless DC (BLDC) motors, Permanent Magnet Synchronous Motors (PMSM), etc. which require an efficient controller to enable this trend. To accomplish this, Infineon Technologies has designed OptiMOS™ 6 135 V to address high power motor drives such as LEVs, e-forklifts, power & gardening tools, etc.
A typical 3-phase motor drive has three half-bridges connected to the motor as shown in Fig. 1. In power tools, these half-bridges consist of MOSFETs which are predominantly driven using block-commutation to drive the motor since it is easier and cheaper to implement than Field Oriented Control (FOC). The power carrying capabilities of these systems has been increasing by using higher voltage batteries and as a result, the popularity for 72 – 84 V batteries for Outdoor Power Equipment (OPE) is growing. To efficiently handle these higher powers and have a good thermal performance, paralleling of multiple MOSFETs is becoming more common. The peak power in these tools can be as high as 3 – 4 kW. Today, for 72 – 84 V batteries, the typical choice of MOSFETs is 150 V. These MOSFETs can be

overkill for power & gardening tools as the breakdown voltage of the MOSFET is high and these parts also reduce the efficiency by increasing the conduction losses (due to higher $R_{DS(ON)}$). In this paper, we have demonstrated the product level and application performance comparison between the Best-in-Class (BiC) 135 V and 150 V MOSFETs from competitors against the state-of-the-art 135 V technology from Infineon Technologies.

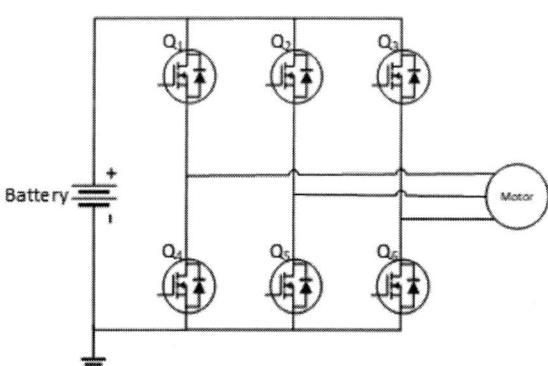

Fig. 1: 3-phase motor-drive schematic showing three half-bridges connected to the motor

2 Performance Comparison

2.1 Technology Overview

The motivation behind the development of 135 V technology was to provide the right balance between the voltage design margin and lower conduction losses by achieving BiC $R_{DS(ON)}$. This reduction in $R_{DS(ON)}$ also results in higher I_D ratings as shown in Fig. 2 and Fig. 3

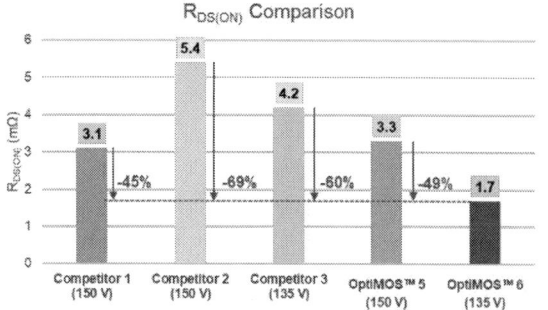

Fig. 2: $R_{DS(ON)}$ comparison of OptiMOS™ 6 135 V against competitors

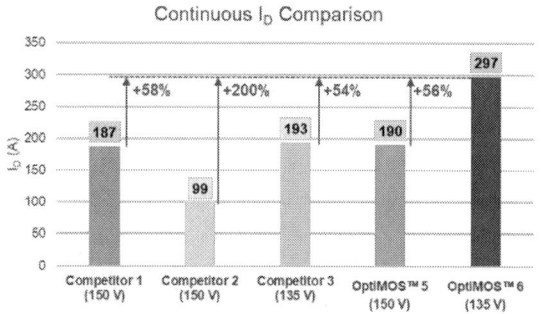

Fig. 3: I_D comparison of OptiMOS™ 6 135 V against competitors

OptiMOS™ 6 135 V with a typical $R_{DS(ON)}$ of 1.7 mΩ is 45-69% better than competitors' 150 V parts and 60% lower than competitor 135 V part. It also has approximately 50% lower $R_{DS(ON)}$ compared to OptiMOS™ 5 150 V. This improvement in $R_{DS(ON)}$ also results in up to a 200% increase in I_D rating of 297 A compared to competitors' 150 V parts, and 54% increase compared to competitor 135 V part. It also has a 56% higher I_D rating compared to OptiMOS™ 5 150 V.

In addition to these benefits, OptiMOS™ 6 135 V also has much tighter $V_{GS(th)}$ spreads as shown in Fig. 4. This is beneficial especially in paralleling multiple MOSFETs to scale up the power level of the system and have better dynamic current sharing between parallel MOSFETs.

OptiMOS™ 6 135 V has a $V_{GS(th)}$ spread of 1 V which is 50% lower than competitors' 150 V parts which have a 2 V spread, and 30% lower than competitor 135 V part. It also has 40% tighter $V_{GS(th)}$ spread than OptiMOS™ 5 150 V.

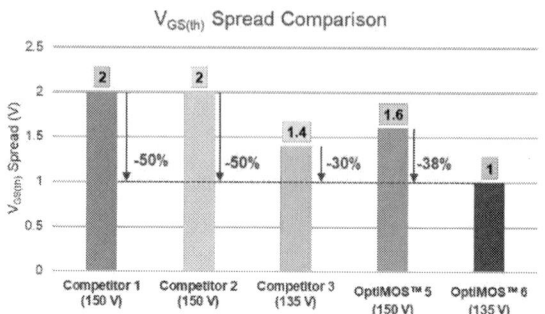

Fig. 4: $V_{GS(th)}$ spread comparison of OptiMOS™ 6 135 V against competitors

2.2 Diode Performance

Diode performance, especially softness, is an important part to ensure optimal switching performance. To evaluate the diode performance of OptiMOS™ 6 135 V against competitors, all the parts were tested in a double pulse test setup. The double pulse schematic is shown in Fig. 5 where the low-side MOSFET, Q_2, is switching and the body diode of the high-side MOSFET, Q_1, undergoes reverse recovery at the start of the second pulse.

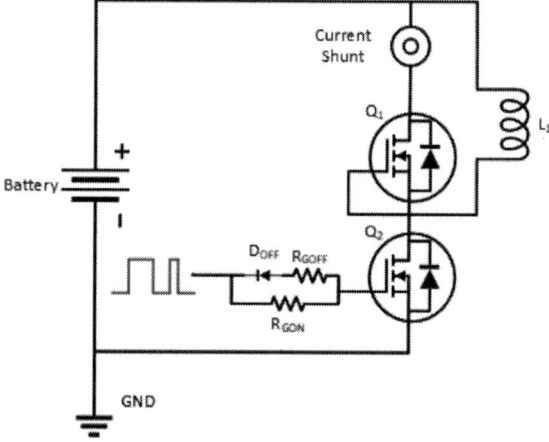

Fig. 5: Double Pulse circuit schematic showing the gate drive circuit and current shunt

All the parts were tested in this setup, shown in Fig. 6 at the test conditions mentioned in table 1.

Parameter	Notation	Value	Units
Input Voltage	V_{IN}	84	V
Forward Current	I_F	50	A
dI/dt	dI/dt	500 - 1000	A/μs
Inductor	L	10	μH

Tab. 1: Double Pulse test conditions

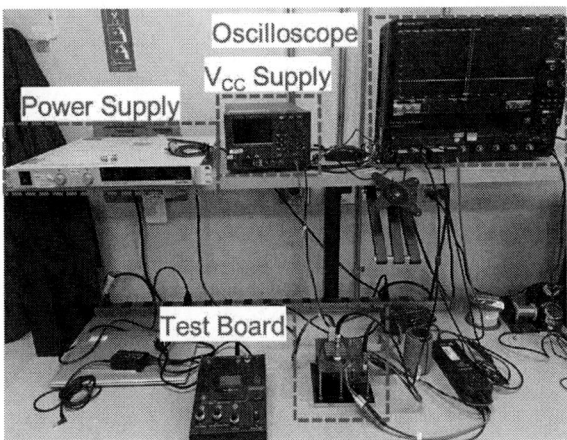

Fig. 6: Double Pulse setup to test diode performance

The dI/dt of the diode forward current was controlled by tuning the R_{GON} of Q_2 based on Eq. (1) to Eq. (3).

$$I_{GON} = \frac{V_{CC} - \frac{(V_{PL} + V_{GS(th)})}{2}}{R_{GD} + R_{GON} + R_{INT}} \quad (1)$$

where I_{GON} is the average gate current during drain current ramp up [1], V_{CC} is the gate supply voltage, V_{PL} is the Miller plateau voltage, R_{GD} is the gate driver's internal resistance, R_{GON} is the external gate resistance and R_{INT} is the MOSFET's internal resistance.

Therefore to increase I_{GON}, R_{GON} should be reduced and vice versa. I_{GON} can then be used for calculating the time taken for current to ramp up in Q_2, which is equal to the time taken for the forward current of the diode to decrease to zero using Eq. (2)

$$\Delta t = \frac{Q_{GS2}}{I_{GON}} \quad (2)$$

where Δt is the time taken for the current to ramp up in Q_2, which is same as diode forward current I_F to decrease to zero, Q_{GS2} is gate-source charge between $V_{GS(th)}$ and V_{PL} [2] in the gate-charge waveform as shown in Fig. 7.

$$\frac{dI_F}{dt} = \frac{\Delta I_F}{\Delta t} = \frac{I_F - 0}{\Delta t} = \frac{I_F \times I_{GON}}{Q_{GS2}} \quad (3)$$

where ΔI_F is the change in current in the diode going from I_F to zero. From Eq. (3), it is clear that to increase dI/dt, I_{GON} should be increased and vice versa which can be controlled by tuning R_{GON}.

In this test, diode softness, reverse recovery charge (Q_{RR}), peak reverse recovery current (I_{RR}), and V_{DS}

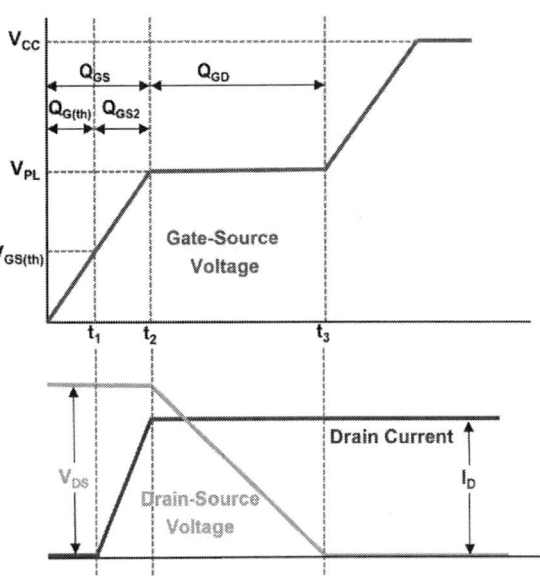

Fig. 7: Typical gate charge waveform showing switching times

peak across the body diode of Q_1 were measured for all parts and plotted as shown in Fig. 8 to Fig. 11 OptiMOS™ 6 135 V consistently shows the highest softness at all dI/dt conditions. It also has the lowest Q_{RR}, I_{RR} and V_{DS} peak. The lower V_{DS} peak can help switch from 150 V to 135 V without compromising the switching performance. This makes OptiMOS™ 6 135 V a better solution for motor drive applications even at higher switching speeds.

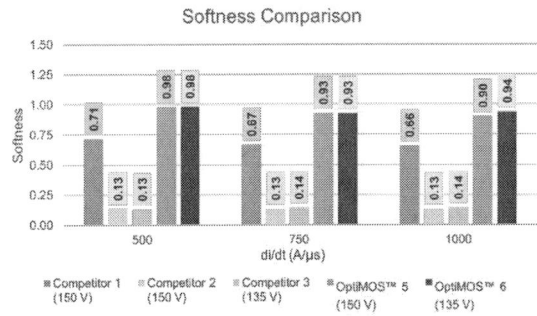

Fig. 8: Softness comparison of OptiMOS™ 6 135 V against competitors

Competitors 2 and 3 show very poor softness and as a result have the highest V_{DS} peaks. There is a risk of avalanching the MOSFETs at higher dI/dt conditions which can prematurely fail the system. These parts also show highest Q_{RR} and I_{RR} at all dI/dt conditions making them unsuitable for fast switching applications and thereby increasing switching losses in the system. Competitor 1 shows

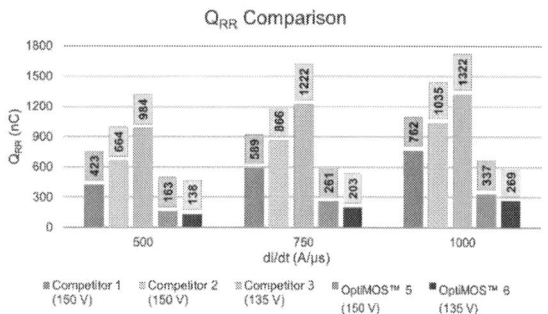

Fig. 9: Q_{RR} comparison of OptiMOS™ 6 135 V against competitors

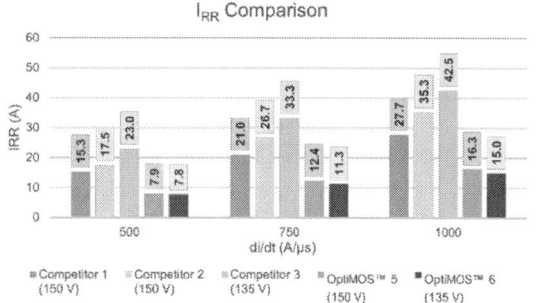

Fig. 10: I_{RR} comparison of OptiMOS™ 6 135 V against competitors

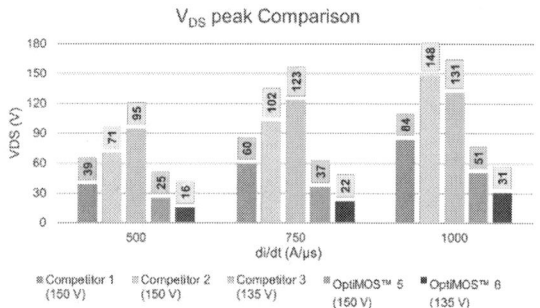

Fig. 11: V_{DS} peak comparison of OptiMOS™ 6 135 V against competitors

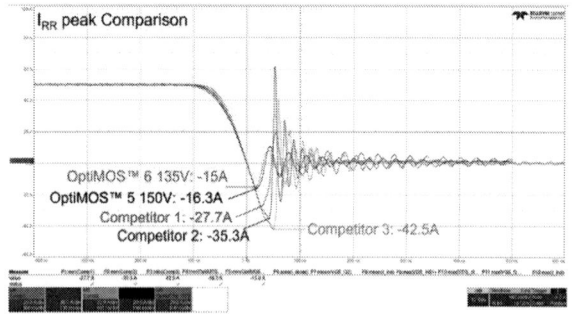

Fig. 12: I_{RR} waveform comparison at 1000 A/µs showing lowest I_{RR} peak and ringing for OptiMOS™ 6 135 V

lower softness and higher V_{DS} peak compared to both OptiMOS™ 5 150 V and OptiMOS™ 6 135 V. For competitors 1, 2 and 3, these drawbacks paired with the higher $R_{DS(ON)}$ make them less efficient for battery powered motor drive applications.

The waveform comparisons for all parts are shown in Fig. 12 and Fig. 13. In addition to higher V_{DS} peaks, all competitor parts also show higher ringing which can cause EMI issues and will require additional circuitry to resolve. Competitors 2 and 3 samples also show very high current spikes after diode reverse recovery since the diode softness is very low and can be harmful for the system.

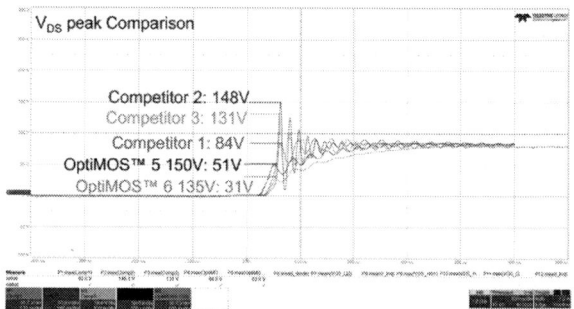

Fig. 13: V_{DS} peak waveform comparison at 1000 A/µs showing lowest V_{DS} peak and ringing for OptiMOS™ 6 135 V

2.3 Synchronous Buck Converter Test

All the parts were tested on a synchronous buck converter test setup to evaluate the switching and thermal performance. A synchronous buck converter can be considered as a single phase of a three-phase inverter. In this test, the high-side MOSFET, Q_1, has both switching and conduction losses while the low-side MOSFET, Q_2, has conduction and diode losses as it undergoes Zero Voltage Switching (ZVS). The schematic, test board and the setup are shown in Fig. 14 - Fig. 16 respectively.

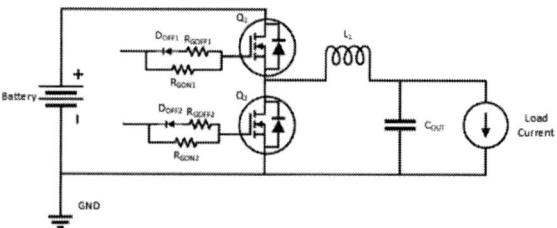

Fig. 14: Synchronous Buck Converter schematic

The test conditions are mentioned in table 2. The output current was increased until the parts reached case temperature (T_C) of 100 °C or it

2933

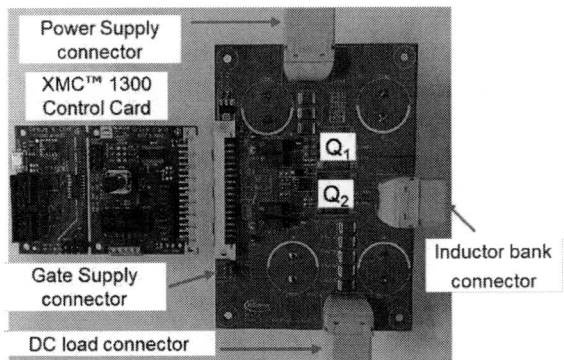

Fig. 15: Synchronous Buck Converter test board

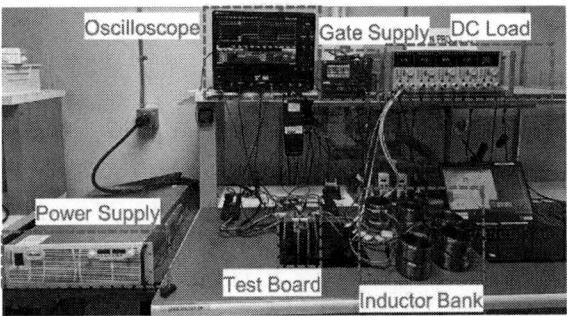

Fig. 16: Synchronous Buck Converter test setup to monitor switching and thermal performance

reached max load current of 40 A, whichever comes first. The parts were run at each current level for 10 minutes to ensure thermal equilibrium is reached before increasing the current level. Switching frequency of 10 kHz was selected as this is the most common frequency for block commutation in power & gardening tools. The switching performance was tuned using the external R_{GON} and R_{GOFF} to have max V_{DS} peak of approximately 108 V and ensure no induced turn-on occurs for Q_2 during diode reverse recovery.

Parameter	Notation	Value	Units
Input Voltage	V_{IN}	84	V
Switching Frequency	f_{sw}	10	kHz
Output Voltage	V_{OUT}	42	V
Load Current	I_{LOAD}	10-40	A
Dead Time	t_D	2	µs
Inductor	L	500	µH
Run Time		10	min

Tab. 2: Synchronous Buck Converter test conditions

The thermal performances (T_C vs. Load Current) for both Q_1 and Q_2 are shown in Fig. 17 and

Fig. 18. Both Q_1 and Q_2 of OptiMOS™ 6 135 V show the lowest temperatures over the complete load range. Competitor 1 and OptiMOS™ 5 150 V parts can carry 37.5 A load current when they reach T_C of 100 °C while OptiMOS™ 6 135 V shows 20 °C lower temperature on Q_1 and 25 °C lower temperature on Q_2 at the same current level. Low-side MOSFET, Q_2 shows a larger improvement since it experiences mainly conduction losses, which is reduced by significant improvement in $R_{DS(ON)}$.

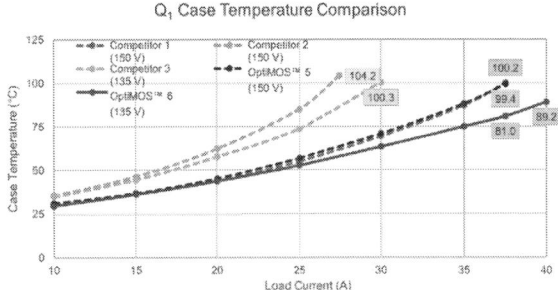

Fig. 17: High Side MOSFET case temperature comparison in Synchronous Buck Converter

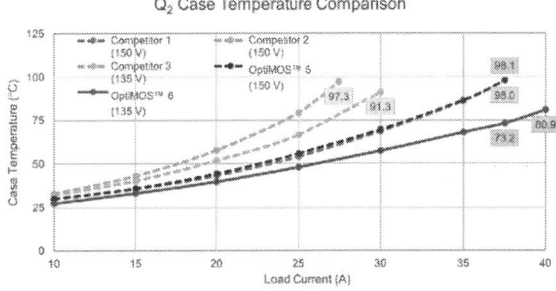

Fig. 18: Low Side MOSFET case temperature comparison in Synchronous Buck Converter

The power loss for all the parts was calculated using the switching times and waveforms [3] from test data, shown in Fig. 19.

Competitor 1 and OptiMOS™ 5 150 V show similar power losses which is reflected in the thermal performance as well. However, competitor 2 shows lower losses compared to competitor 3, even though it has higher T_C. This is because competitor 3 has lower system R_{TH} compared to competitor 2, which compensates for the higher power loss. OptiMOS™ 6 135 V has similar power loss as OptiMOS™ 5 150 V at low loads but once the load current increases and conduction losses dominate, OptiMOS™ 6 135 V shows reduction in total power

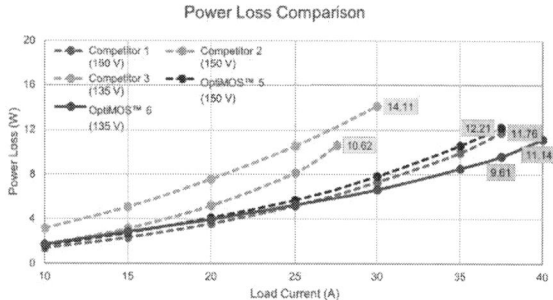

Fig. 19: Comparison of total MOSFET power loss in Synchronous Buck Converter

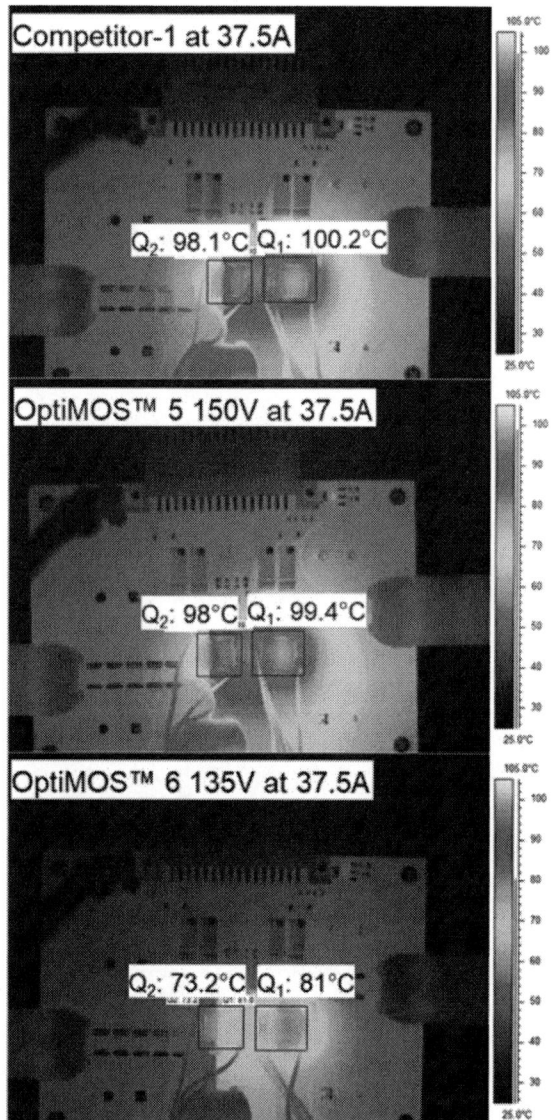

Fig. 20: Thermal Images at 37.5 A showing lower temperatures for OptiMOS™ 6 135 V

loss and as a result it can carry higher currents. This makes OptiMOS™ 6 135 V an attractive product for increasing the power capability of the system while also improving the efficiency. The thermal images for competitor 1, OptiMOS™ 5 150 V and OptiMOS™ 6 135 V in Fig. 20 show a 20 °C lower temperature on Q_1 and 25 °C lower temperature on Q_2 for OptiMOS™ 6 135 V

2.4 3-Phase Motor-Drive Test

A 3-phase motor-drive inverter board was designed and assembled as shown in Fig. 21 with 4 MOSFETs per phase (12 MOSFETs total). Each phase has two high-side MOSFETs and two low-side MOSFETs, with the phase connection between them to have as symmetric a layout as possible. This will ensure better dynamic current sharing between the parallel MOSFETs and minimize the stress on them. Power loss analysis for block-commutation for all the parts [4] at the test conditions mentioned in table 3 are shown in Fig. 22.

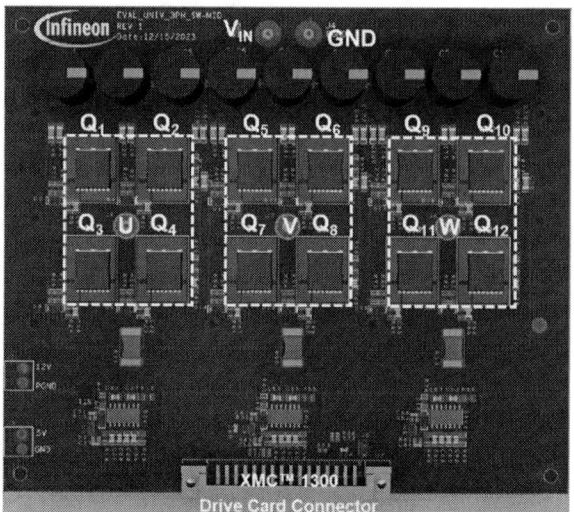

Fig. 21: 3-phase motor-drive inverter board with 12 MOSFETs

Parameter	Notation	Value	Units
Input Voltage	V_{IN}	84	V
Switching Frequency	f_{sw}	10	kHz
Rotor Speed	n	3000	rpm
Input Power	P_{IN}	3000	W
Run Time		10	min

Tab. 3: 3-phase motor-drive test conditions

OptiMOS™ 6 135 V parts have approximately 20 % improvement in power loss at 3000 W input power

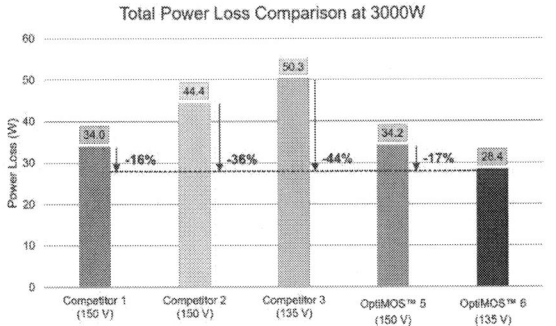

Fig. 22: Total power loss comparison at 3000W input power

compared to competitor 1 and OptiMOS™ 5 150 V. However, compared to competitors 2 and 3, OptiMOS™ 6 135 V has 36 - 44 % lower power loss. The main improvement area being conduction losses as shown in Fig. 23. OptiMOS™ 6 135 V shows up to 75 % improvement in conduction losses due to significantly better $R_{DS(ON)}$.

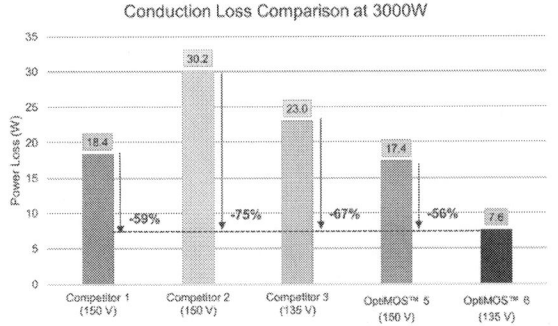

Fig. 23: Conduction loss comparison at 3000W input power

These lower power losses also improve the thermal performance of the system and allow for higher power levels. This makes OptiMOS™ 6 135 V a very attractive product for high-power motor-drive applications compared to existing solutions from Infineon and competitor 135 V and 150 V parts.

3 Conclusion

OptiMOS™ 6 135 V is a new Best-in-Class technology from Infineon Technologies, which shows signif-

icant improvements compared to existing products available in the market. With up to 70 % improvement in $R_{DS(ON)}$, resulting in up to 200 % increase in I_D rating, it can help increase the power capability of the system. It also has a tighter $V_{GS(th)}$ range, making it easier to parallel multiple MOSFETs and scale up the system.

The double pulse test showed the improvements in diode performance, especially softness, Q_{RR} and V_{DS} peak at low and high switching speeds, allowing for faster switching without the risk of avalanche. The softer diode also helps reduce the ringing on V_{DS} and can improve the EMI performance.

The synchronous buck converter test showed the improvement in power loss and thermal performance with up to 25 °C lower temperatures compared to Infineon and competitor 150 V parts and approximately 35 % higher power carrying capability compared to competitor 135 V part.

The power loss analysis on 3-phase motor-drive showed up to 75 % reduction in conduction losses which resulted in up to 44 % lower total power loss. This makes OptiMOS™ 6 135 V an attractive product for high-power motor-drive applications where 72 - 84 V batteries are used and can help achieve decarbonization goals by efficiently electrifying LEVs, e-forklifts, power & gardening tools, etc.

References

[1] "Mosfet fast switching: Motivation, implementation, and precautions," vol. 1, no. AN 2309 PL51 2309 120502, pp. 1–24, 2023, Infineon Technologies.

[2] "Use gate charge to design the gate drive circuit for power mosfets and igbts," vol. 1, no. AN 944, 2012, Infineon Technologies.

[3] "Calculation of power loss (synchronous)," no. AEK59-D1-0065-2, 2016, ROHM.

[4] H. Amirkhanian and S. Oknaian, "Power loss breakdown in bldc drives applications using matlab," in *PCIM Europe 2018; International Exhibition and Conference for Power Electronics, Intelligent Motion, Renewable Energy and Energy Management*, 2018, pp. 1–5.

PCIM Europe 2024, 11– 13 June 2024, Nuremberg   DOI: 10.30420/566262417

Auto Power-SOI: Shaping the Future of Battery Monitoring Technology

Alex Lim

Soitec, Singapore

Corresponding author: Alex Lim, alex.lim-jan-pang@soitec.com
Speaker: Alex Lim, alex.lim-jan-pang@soitec.com

Abstract

Cell-to-pack (CTP) is the prevailing method for EV battery assembly due to its compact size, cost-effectiveness, streamlined assembly with fewer connections, and enhanced reliability. The primary challenges in Battery Monitoring ICs (BMIC) for CTP lie in supporting more series-connected monitoring cells and achieving greater measurement precision. This presentation outlines how Auto Power-SOI technology facilitates the development of advanced BMICs that excel in precision, integrate high and low voltage components on a single die, and meet stringent functional safety demands, addressing the evolving needs of CTP technology in EV batteries.

1   Introduction

The automotive industry's shift towards electric vehicles (EVs) has accelerated the demand for high-performance, reliable, and safe battery systems. As EVs gain mainstream adoption, automakers and battery manufacturers are under pressure to develop advanced battery technologies that offer longer driving ranges, faster charging times, and enhanced safety features.

At the forefront of innovation are next-generation Battery Monitoring ICs (BMICs) supporting cell-to-pack (CTP) battery pack configurations. BMICs monitor and manage individual battery cells' performance, ensuring safe and efficient operation of EV battery systems.

CTP configurations integrate battery monitoring, battery management and other functions directly into the battery pack, eliminating the need for separate modules. This approach offers several advantages:

A. Smaller size: Compact battery pack designs crucial for space-constrained EVs.
B. Reduced bill of materials: Fewer components lead to cost savings and improved manufacturing efficiency.
C. Simplified assembly: Streamlined process with fewer interconnections and potential failure points.

D. Heightened reliability: Minimized interconnections reduce the risk of failures associated with module-to-module connections.

Automakers like BYD, SAIC and others are leading the way in adopting CTP configurations, which are vital for ensuring the functional safety and reliability of next-generation EV battery systems. This adoption has accelerated further, with CTP-equipped vehicles capturing a remarkable 48.6% market share from January to October 2023. Projections indicate that by the end of 2023, CTP configurations will surpass the 50% mark, becoming the predominant choice for new energy vehicle battery systems. [1]

2   Cell-To-Pack Battery Monitoring IC Trends and Challenges

2.1   CTP BMIC Trends

With the evolution of CTP technology, the trends of BMICs are summarized below.

2.1.1   Compact and Efficient Design

CTP technology demands BMICs to be capable of monitoring a higher number of cells in series and to provide smaller solution size. Traditional battery pack configurations often have a limited number of cells connected in series within a

2937

module, typically up to 18 cells. However, CTP architectures eliminate the modular approach, allowing for a higher number of cells to be connected in series within a single pack, often exceeding 100 cells or more.

This increase in the number of cells in series poses a significant challenge for BMICs, as they need to be designed to handle higher voltage levels and more complex monitoring requirements. As a result, each BMICs need to be capable of monitoring a higher number of cells in series (up to 30 cells). This next generation BMICs must be able to withstand these higher voltage levels while maintaining accurate cell monitoring and management capabilities.

BMIC designs are trending towards integration and miniaturization. Semiconductor manufacturers are developing highly integrated BMICs that combine multiple functionalities, such as cell voltage monitoring, temperature monitoring, cell balancing, and communication interfaces, into a single chip. This integration helps reduce the overall component count, board space, and wiring complexity, enabling more compact and streamlined battery pack designs. Additionally, miniaturization efforts are focused on reducing the package size and footprint of BMICs through advanced packaging technologies and die shrinks.

2.1.2 High Measurement Precision

As CTP configurations integrate a larger number of cells in series, precise and accurate cell monitoring over wide temperature ranges and harsh environments becomes crucial. Even minute measurement errors can accumulate, impacting battery performance, safety, and driving range. Consequently, BMICs are prioritizing achieving the highest measurement precision by minimizing error sources.

Advanced analog-to-digital converter (ADC) architectures, calibration techniques, and error compensation algorithms are employed to reduce measurement inaccuracies. High-resolution ADCs, up to 16-bit or higher, enable precise voltage monitoring, while on-chip temperature sensors and signal conditioning circuits compensate for thermal and environmental effects.

Error sources like offset, gain, and non-linearity errors are minimized through trimming, auto-

calibration, and digital compensation. Robust design practices, like isolating analog and digital domains, mitigate noise and interference.

Stringent accuracy specifications, such as ±1mV cell voltage measurement error over -40°C to +125°C, ensure optimal battery management and extended driving ranges.

2.1.3 Functional Safety

Functional safety (FuSa) is a top priority in the automotive industry, particularly for electric vehicles (EVs), where battery systems are critical components. Ensuring the safe operation of these systems is paramount to prevent hazards that could lead to injuries or fatalities. To address this, BMICs are integrating advanced safety features and redundancy measures to meet the strictest safety requirements, such as Automotive Safety Integrity Level D (ASIL D), the highest level defined by ISO 26262 for automotive functional safety.

ASIL D certification mandates rigorous safety measures, including hardware and software redundancy, fault detection and mitigation mechanisms, and stringent design processes. BMICs designed for ASIL D compliance incorporate redundant voltage measurement channels, error-correcting codes for communication interfaces, and built-in self-test (BIST) capabilities to detect and recover from faults.

2.1.4 Communication Protocol

BMICs are incorporating advanced communication protocols, such as Controller Area Network (CAN), Local Interconnect Network (LIN), and Serial Peripheral Interface (SPI), to enable seamless and efficient interaction with the vehicle's Battery Management System (BMS) and other critical control systems.

Additionally, advanced communication protocols in BMICs facilitate Over-the-Air (OTA) firmware updates, allowing manufacturers to remotely update the chip's software and introduce new features or performance enhancements without the need for physical intervention. This capability enhances the overall flexibility and longevity of the vehicle's battery management system, contributing to improved safety, efficiency, and user experience throughout the vehicle's lifespan.

2.1.5 Advanced Diagnostics

BMICs are incorporating advanced diagnostic capabilities to detect and report battery faults, anomalies, and degradation at an early stage. Some of these diagnostic features include voltage monitoring, temperature sensing, cell balancing, impedance tracking algorithms etc. By continuously analyzing battery data, BMICs can identify potential issues like cell imbalances, thermal runaway risks, or capacity fade. Upon detecting anomalies, BMICs can promptly report these conditions to the Battery Management System (BMS) and vehicle control units. This early detection enables timely maintenance interventions, preventing further damage and enhancing overall battery pack safety and longevity in electric vehicles.

2.2 CTP BMIC Challenges

As CTP technology evolves, the challenges faced by BMICs can be summarized as follows.

2.2.1 Increased Complexity

Monitoring an increasing number of series-connected cells in CTP significantly complicates the design of BMICs. Engineers face the challenge of balancing simplicity in BMIC architecture while ensuring robust functionality for accurate cell monitoring, balancing, fault detection, thermal management, and communication with vehicle systems. Achieving this balance is crucial for cost-effective and reliable battery management solutions in electric vehicles.

2.2.2 Cost Considerations

Although CTP technology has the potential to reduce the bill of materials (BOM), BMICs must strike a balance between cost-effectiveness and incorporating advanced features. This trade-off presents a significant challenge. CTP architectures provide a cost-reduction pathway by integrating critical functions. However, the inclusion of sophisticated capabilities within BMICs, such as comprehensive diagnostics, precision monitoring, and advanced balancing algorithms, inevitably increases complexity and associated expenses. Therefore, designers of BMICs must navigate a complex trade-off between reducing manufacturing costs and delivering durable and feature-rich battery management solutions that meet the strict requirements of electric vehicle applications.

2.2.3 Redundancy and Fail-Safe Systems

Designing BMICs that balance functional safety, redundancy, and cost is a challenging task. To meet strict safety standards, these circuits must include fault detection and mitigation mechanisms, as well as redundant architectures to prevent single-point failures. Concurrently, the exigency of curtailing manufacturing expenditures imposes constraints on the complexity and resource footprint of the BMIC. Reconciling these seemingly antithetical objectives demands a judicious optimization of the system's intricate tradespace, necessitating a delicate balance between safety assurances, reliability enhancements, and economic viability to forge cost-effective yet uncompromising battery management solutions.

3 Auto Power-SOI Technology: A Game-Changer

Auto Power-SOI is emerging as an important enabler in addressing the evolving trends and challenges associated with BMICs for the latest CTP battery configuration in EVs.

3.1 Auto Power-SOI for Compact and Efficient Design

3.1.1 Higher Monolithic Integration of High-Voltage and Low-Voltage Blocks on the Same Die

Auto Power-SOI technology facilitates the monolithic integration of high-voltage and low-voltage functional blocks on the same die, while enabling a reduced die area footprint. This is achieved through the deployment of Silicon-on-Insulator (SOI) substrates in conjunction with Deep Trench Isolation (DTI) techniques. The buried oxide layer of SOI wafers provides robust dielectric isolation between the high-voltage and low-voltage domains, mitigating latch up and minimizing parasitic coupling. Complementarily, DTI structures form effective vertical isolation barriers, allowing high-voltage and low-voltage devices to coexist within close proximity on the die. This integration approach yields compact BMIC implementations with reduced form factors, benefiting CTP battery configurations in EVs where board area constraints are critical. Figure 1 illustrates the comparison of required device-to-device isolation spacing based on currently available technologies in the market.

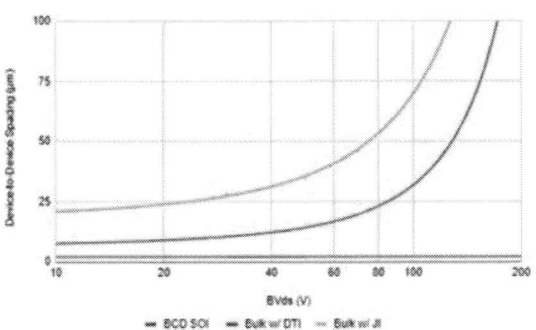

Fig. 1 Device-to-device isolation spacing based on BCD with Auto Power-SOI and DTI, BCD with bulk and DTI and BCD with bulk and junction isolation

3.1.2 High Integration of Main Functions and Redundancy on the Same Die

For high functional safety systems, redundancy blocks are required to prevent failures in main function blocks. Conventionally, two separate ICs were needed to ensure redundancy and meet stringent FuSa requirements like ASIL D. However, this approach increases system size, cost, and failure-in-time (FIT) rate due to higher bill-of-materials (BOM).

As shown in Fig. 2, Automotive Power-SOI enables monolithic integration of main and redundant blocks on a single die by leveraging deep trench isolation (DTI) and buried oxide (BOX) layers for full electrical isolation. Main and redundant blocks reside on separate silicon islands, meeting FuSa redundancy mandates.

This monolithic BMIC solution approach significantly reduces system size and cost compared to multi-chip approaches. Lower BOM count also optimizes system FIT rates, facilitating compliance with stringent automotive safety integrity levels like ASIL D.

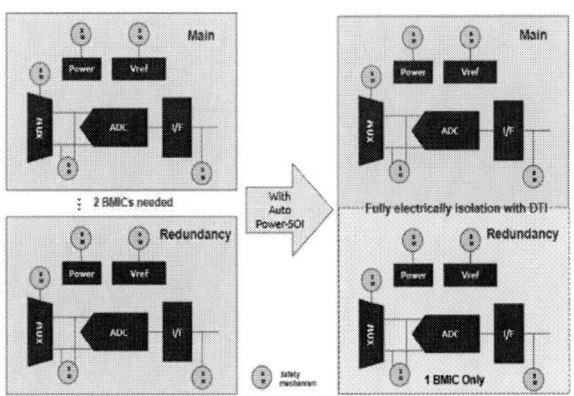

Fig. 2 Monolithic integration of main and redundancy blocks on the same die with DTI and BOX

3.2 Auto Power-SOI for Higher Measurement Precision

3.2.1 Low Leakage Current for High Measurement Accuracy

Auto Power-SOI structure, comprising a BOX layer, minimizes leakage currents at elevated temperatures (e.g. 125°C) compared to conventional bulk silicon substrates. The BOX acts as an insulator, suppressing substrate leakage paths that increase exponentially with temperature in bulk devices. Figure 3 shows the comparison of leakage currents for both Auto Power-SOI with DTI and bulk substrate with junction isolation [2].

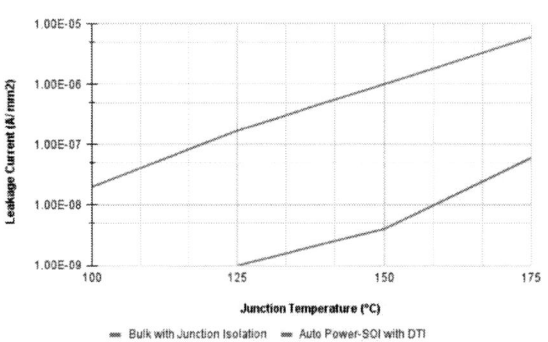

Fig. 3 Comparison of leakage currents for Auto Power-SOI with DTI and bulk substrate with junction isolation [2]

Auto Power-SOI's remarkable low leakage current, even at elevated junction temperatures exceeding 125°, simplifies the design of high-precision analog circuits. Auto Power-SOI's suppressed leakage currents at high temperatures, eliminating need for temperature compensation circuitry, facilitate design of precision reference circuits with reduced drifts over temperature (voltage, gain, etc.). This enhances measurement accuracy of analog-to-digital converters (ADCs) by minimizing temperature-induced errors, critical for reliable high-temperature automotive BMICs.

3.2.2 High Noise Immunity and Less Cross-talk for High Measurement Accuracy

Leveraging BOX and DTI in Auto Power-SOI, the technology mitigates noise coupling and crosstalk issues prevalent in highly integrated BMICs with multiple measurement channels. The intrinsic isolation provided by the BOX/DTI stack significantly reduces capacitive and substrate noise coupling between adjacent channels. This enhances measurement accuracy for ADCs by minimizing inter-channel interference as channel counts scale higher in next generation BMICs.

3.3 Auto Power-SOI for Higher FuSa

3.3.1 Low FIT with High Integration

Auto Power-SOI enables higher monolithic integration, reducing the chip count per system. Fewer components decrease the overall system failure-in-time (FIT) rate by minimizing interconnect and other failure points. This chip-level integration enhances system reliability, a critical requirement for achieving automotive functional safety targets. The monolithic integration facilitated by Auto Power-SOI's DTI & BOX isolations directly improve the mean time between failures (MTBF) for safety-critical applications such as BMICs.

3.3.2 High Robustness & Reliability

Auto Power-SOI's intrinsic robustness against latch-up, electrostatic discharge (ESD), electromagnetic compatibility (EMC) issues and electromagnetic interference (EMI) minimizes systematic faults stemming from the manufacturing process technology. The SOI structure with BOX and DTI layers mitigates these reliability concerns. This robust process reduces the burden of implementing design techniques to tolerate process-induced faults, simplifying the development of high functional safety (FuSa) compliant systems adhering to stringent automotive standards like ISO 26262.

4 Conclusion

In summary, Auto Power-SOI is a pivotal enabling technology for advanced BMIC designs targeting CTP battery architectures in EVs. The technology's monolithic integration capabilities, facilitated by DTI and BOX layers, allow more compact, power-efficient and reliable BMIC implementations.

Auto Power-SOI's inherent robustness against latch-up, ESD, EMC/EMI issues enhances systematic reliability. Reduced chip counts decrease overall system FIT rates, while integrated redundancy provisions assist in achieving stringent automotive functional safety requirements like ISO 26262 ASIL-D.

As CTP configurations proliferate and BMIC complexity increases, Automotive Power-SOI remains a key driver enabling next-generation EV battery management solutions to meet evolving performance, safety and efficiency demands.

References

[1] Research In China, Jan 2024, "Passenger Car CTP (Cell to Pack), CTC (Cell To Chassis) and CTB (Cell to Body) Integrated Battery Industry Report, 2024", http://researchinchina.com/Htmls/Report/2024/73951.html

[2] 藤井圭一, 2017, "SOI 技術による機能安全対応車載電池監視用 IC", https://holdings.panasonic/jp/corporate/technology/technology-journal/pdf/v6302/p0111.pdf

PCIM Europe 2024, 11– 13 June 2024, Nuremberg   DOI: 10.30420/566262418

Understanding the Impact of IEC 60747-17 on Capacitive and Magnetic Couplers

Shu-Ee Ong[1], Keith Coffey[2]

[1] Skyworks Solutions, Inc., United States

[2] Skyworks Solutions, Inc., United States

Corresponding author: Shu-Ee Ong, shuee.ong@skyworksinc.com
Speaker: Shu-Ee Ong, shuee.ong@skyworksinc.com

Abstract

Capacitive and magnetic couplers are devices that provide safety isolation for high-voltage power systems. They have been around for more than a decade, yet the first international component standard that regulates these devices, IEC 60747-17, was published only 4 years ago. Before IEC 60747-17, these CMOS-based isolator devices were certified to various component and system standards, including VDE 0884-11 and IEC 62368-1. This paper discusses these past standards and how IEC 60747-17, with working voltage lifetime estimation and bipolar surge testing, marks a significant evolution in safety standards and isolation requirements for these devices.

1 Introduction to Capacitive and Magnetic Couplers

1.1 Isolation Technology

Isolators, also known as "couplers", are used in automotive, telecom, industrial, and medical applications to safeguard people and equipment from high-voltage hazards. Their primary function is to provide galvanic isolation between two or more power domains, thereby enabling signal transmission without direct electrical conduction.

Isolators serve many purposes across various applications. They can mitigate ground loops by isolating the common ground connection between circuits, which in turn reduces electrical noise and electromagnetic interference. Isolators can also facilitate level shifting, enabling communication between circuits operating at significantly different supply voltages. However, the most prevalent application of isolators is to protect low-voltage circuits or human operators on one side of the isolation barrier from potentially hazardous voltages on the other side.

Isolators can incorporate various isolation technologies, such as optical, capacitive, or magnetic isolation. Optical isolators, which have been in use since the 1960s, use light to transfer signals between two sides. These devices are typically fabricated using gallium-arsenide process technologies [1].

On the other hand, capacitive and magnetic isolators, introduced in the 2000s, are CMOS-based devices that transfer digital signals across an isolation barrier [1]. Owing to their underlying technology, capacitive and magnetic isolators may also be called "digital isolators" or "CMOS isolators."

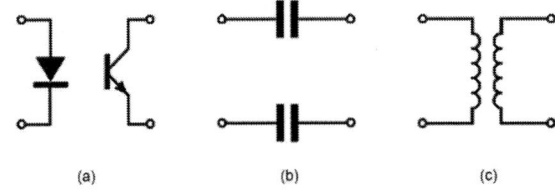

Fig. 1   Types of Isolation Technology: (a) Optical,

(b) Capacitive, (c) Magnetic

The durability of the isolator's insulation, or "isolation barrier", is crucial for ensuring safety against electric shock in many applications. Various factors, such as lightning strike surges, electrostatic discharge (ESD), overvoltage, overheating, and aging can cause degradation or breakdown of the isolator's insulation.

2 Isolation Standards

2.1 Existing Isolation Standards

Until the publication of IEC 60747-17 in September 2020, digital isolators were certified to various regional standards such as UL 1577 and VDE 0884-10, or adjacent international standards like IEC 60747-5-5. Depending on their application, these isolators were also certified to system (or equipment) standards such as IEC 62368-1, IEC 60601-1, and IEC 61010-1.

2.1.1 Component Standards

Component standards apply specifically to chip-level devices, while system standards pertain to the entire electronic system that houses the component of interest. In this context, component standards outline the requirements and testing procedures for the manufacturing and rating of digital isolators. System designers using component-certified devices in their system can more easily achieve system certification.

When digital isolators were first developed, they were certified to IEC 60747-5-5, an international component standard originally meant for optical isolators. However, due to the differences in technology, digital isolators were removed from that standard in 2010 and transitioned to system standard IEC 60950, which governs safety in IT equipment [1]. This shift marked the industry's progression towards advanced isolation technologies beyond traditional optocouplers.

In the United States, the component standard UL 1577 primarily applies to optical isolators but has been extended to include digital isolators. This standard remains active today and is referenced in most CMOS isolator datasheets. The German regional standard VDE 0884-10 was the first component standard that applied specifically to capacitive and magnetic isolators but was retired in 2020. It laid the groundwork for component standards VDE 0884-11 and IEC 60747-17, ensuring that digital isolators meet high-voltage requirements and end-equipment certifications, instilling confidence in their safe operation within various systems.

The inconsistency between isolation component standards across the world has led to confusing and costly certification processes for CMOS isolators over the last two decades. To simplify this process, IEC 60747-17 was developed as the first international component standard for these devices. It mandates lifetime estimation and bipolar surge testing, making it the most stringent component standard to date [2].

2.1.2 System Standards

System standards comprise a set of requirements and methods to assess isolation effectiveness in an end system, typically on a PCB. An isolator device can be certified to the safety portion of a system standard by undergoing proof testing, which involves component-level type tests that validate their safety capability. System designers who use a certified component save time and money by avoiding months of testing.

As previously mentioned, IEC 60950 is an international safety standard for IT equipment, and its guidelines on electrical insulation and creepage distances were expanded for digital isolators. In 2020, IEC 60950 became obsolete, and IEC 62368-1 took its place. This new standard outlines more stringent requirements for telecom and IT equipment and is adopted worldwide by end customers in a wide range of industries [3]. This standard provides broad international coverage for existing and future system designs, so virtually all digital isolators today are certified to IEC 62368-1.

IEC 60747-17 combines the most stringent requirements of these system standards, such as bipolar surge and creepage/clearance measurements. As a result, IEC 60747-17 should be readily accepted by these system standards, and isolation manufacturers will no longer have to certify their parts to specific system standards.

2.2 Introduction to IEC 60747-17

IEC 60747-17 was released in September 2020, and it became the first international component standard for digital isolators. It is largely based on VDE 0884-11, which was the first standard to define strict requirements regarding lifetime predictions of the isolation barrier [2]. This new standard also provides more details on the surge and partial discharge tests. This section will discuss several key features of IEC 60747-17, particularly requirements that differ from previous component and system standards.

Today, digital isolators that meet the IEC 60747-17 reinforced rating are considered among the highest-performing isolators available in the market.

2.2.1 Time-Dependent Dielectric Breakdown Test

A distinctive feature of IEC 60747-17 is its requirement for lifetime modeling, which is achieved through Time-Dependent Dielectric Breakdown (TDDB) testing. This test was first introduced in VDE 0884-11, and it is a rigorous requirement for digital isolator testing.

The TDDB test employs accelerated lifetime analysis to determine the isolator's maximum rated repetitive voltage, VIORM [ref IEC 60747-17]. The results offer manufacturers and system designers a comprehensive assessment of the isolator's barrier insulation performance over time, temperature, and applied voltage. This test is conducted in-house by the device manufacturer on 432 samples, as outlined under subgroup 6 of the type tests [4].

Data is gathered by measuring the "time-to-failure" at various voltages and temperatures. A failure is fined as an insulation resistance of less than 2 MΩ for basic insulation, or 4 MΩ for reinforced insulation, measured with 500 V_{DC} applied across the isolation barrier [2]. Samples from three different lots are tested at room temperature and at the maximum operating temperature. These sample groups are tested at a minimum of three different 50/60 Hz AC voltages, chosen such that the resulting Mean Time To Failure (MTTF) values for the different voltages span two orders of magnitude. The Weibull analysis, a proven method for determining reliability characteristics, is used to create the lifetime model. IEC 60747-17 references IEC 62539 for the Weibull analysis calculations.

For devices that use SiO_2 insulation, the exponential lifetime model should be used, as shown in Fig. 2. For devices that use thin-film polymer technology, a non-linear model should be applied, as shown in Fig. 3. These lifetime projection models display the probability curves that can be extrapolated to determine the maximum rated repetitive peak isolation voltage (V_{IORM}) of the isolator.

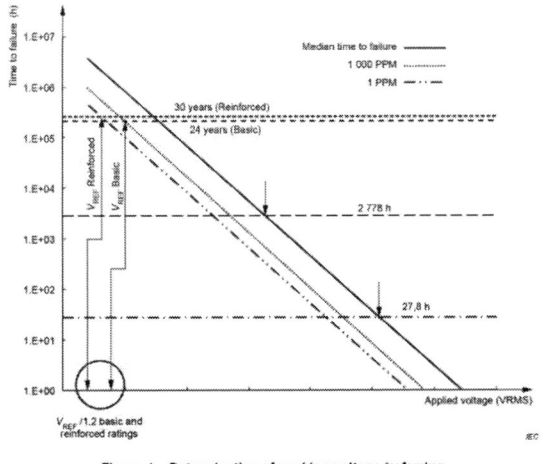

Figure 4 – Determination of working voltage (referring to method in 5.5.5.8 for exponential model)

Fig. 2 TDDB Exponential Model, taken from IEC 60747-17 Standard

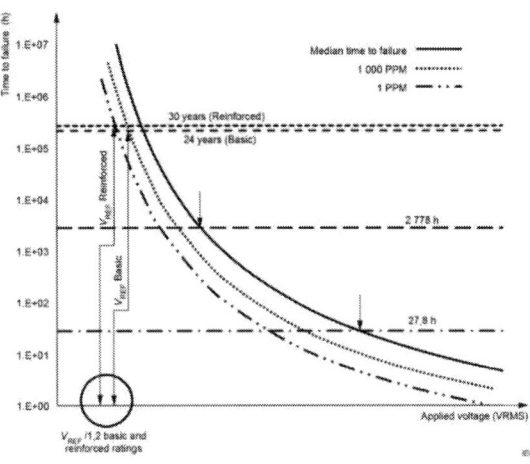

Figure 5 – Determination of working voltage (referring to method in 5.5.5.8 for non-linear model)

Fig. 3 TDDB Non-linear Model, taken from IEC 60747-17 Standard

2.2.2 Surge and Impulse Tests

Equipment connected to power mains, earth ground, or other cabling that runs outside of the building can be subject to large voltage surges due to lightning strikes. To ensure that isolators in safety-related applications continue to protect against electrical shock during and after such surge events, safety standards may necessitate components to undergo surge testing using waveforms designed to mimic those from lightning strikes.

Under IEC 60747-17 type testing subgroup 1, a surge test is performed using a standardized "1.2/50 us" waveform, as depicted in Fig. 4 [2]. The surge isolation voltage, V_{IOSM}, is defined as the highest instantaneous value that the device can withstand during that test. Surge test voltages are typically in the thousands of volts, so it is common to test the device in non-conductive oil to prevent arcing outside of the package.

To an end system designer, the performance of a digital isolation barrier in oil may not be directly applicable. To address this, IEC 60747-17 introduced the V_{IMP} definition and test requirements, which more directly map to the end system surge requirements [2]. The impulse voltage, V_{IMP}, is defined as the highest peak value of impulse voltage without flashover and solid insulation breakdown. It must be tested in air, which has a breakdown voltage of approximately 1 kV_{RMS}/mm between two conductors. As a result, clearance, the shortest distance through air between two conductors, typically provides the physical limitation for V_{IMP} [5]. A 4mm wide package can support a V_{IMP} of 4 x 1.414

= 5.65 kV$_{PEAK}$, even though its surge voltage could be much higher because the device is submerged in oil and would not arc in that scenario.

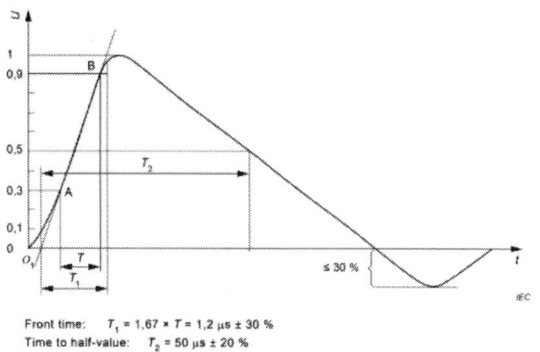

Front time: $T_1 = 1,67 \times T = 1,2\ \mu s \pm 30\ \%$
Time to half-value: $T_2 = 50\ \mu s \pm 20\ \%$

Figure 2 – 1,2/50 µs surge pulse according 61000-4-5:2014 allowed as equivalent impulse for isolation testing

Fig. 4 Surge Testing Pulse, taken from IEC

60747-17 Standard

For IEC 60747-17, the maximum surge isolation voltage V_{IOSM} for basic insulation is equal to 1.3 times the rated impulse voltage V_{IMP} [2]. The surge test voltage requirement for reinforced-rated isolators is the larger of V_{IMP} x 1.3 or 10 kV. The IEC 60747-17 safety standard requires that digital isolators maintain a minimum insulation resistance of $10^9\ \Omega$ after being subjected to 25 surge discharges in one polarity followed by 25 surge discharges in the opposite polarity. The maximum repetition rate is 12 discharges/minute, and samples must be discharged for 1 – 2 hours before the polarity change.

Bipolar surge testing differs from unipolar surge testing, which allows different samples to be tested for each polarity. By requiring both polarities to be surge tested on the same parts, the 10 kV reinforced requirement induces significant stress on the isolation barrier, enforcing the stringent requirements of this new standard.

2.2.3 Partial Discharge Test

The partial discharge test, also known as the apparent charge test, is another measure of the insulation integrity of an isolation barrier. Partial discharge refers to the localized electrical discharge that happens in the insulation between two sides of an isolator, and many standards specify a maximum threshold for the electrical discharge that these isolators must meet. This test also ensures that the devices' mold compound has no voids and is uniformly distributed throughout. In IEC 60747-17, a partial discharge test must be performed for both production and type tests [2].

The test is conducted by ramping the applied voltage up to a specified initial test voltage for partial discharge, $V_{pd(ini)}$, for a specified initial time, t_{ini} [2]. Then, the voltage is ramped down to the apparent charge measuring voltage, $V_{pd(m)}$, for a specified partial discharge stress time, t_{st}, during which the apparent charge is measured. The apparent charge must be less than a specified threshold value to pass the test. For IEC 60747-17, the voltage values and time intervals depend on what test – routine, sample, or type test – is being performed. The standard outlines these different "methods" and indicates which one to follow for each partial discharge test that is run. Figures 5 and 6 show the waveforms of two different methods.

In general, $V_{pd(ini)}$ is related to the maximum transient voltage V_{IOTM}, and $V_{pd(m)}$ is related to the maximum repetitive peak isolation voltage V_{IORM}. The apparent charge threshold is 5 pC.

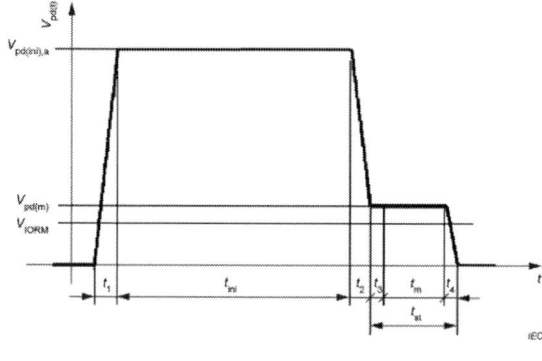

Time intervals for method a

Fig. 5 Partial Discharge Measurement Time Intervals, Method a, taken from IEC 60747-17 Standard

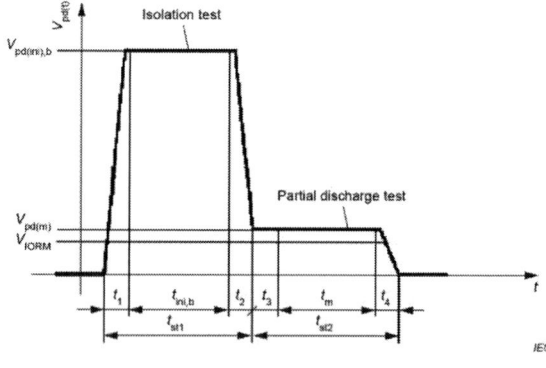

Time intervals for method b1

Fig. 6 Partial Discharge Measurement Time Intervals, Method b1, taken from IEC 60747-17 Standard

For IEC 60747-17, a new requirement for routine tests can be found in the isolation test requirements in Section 6.4 [2]. The partial discharge initial voltage, $V_{pd(ini)}$, used to be 1 x V_{IOTM}, but now it is 1.2 x V_{IOTM}. The duration of this initial voltage remains at 1 second. A change in the routine test has a significant impact, as every single device will have to go through this more stringent requirement during production. It is expected that manufacturers would have to lower their V_{IOTM} isolation rating for some isolation devices.

2.2.4 Clearance and Creepage Requirements

Clearance and creepage are critical parameters in digital isolators, as the insulation width sets limitations on the isolation voltage ratings. Clearance refers to the shortest distance in air between two conductors, in this case, the metal leads. On the other hand, creepage refers to the shortest distance along a surface between two conductors. For digital isolators, the package type determines the clearance and creepage values, which significantly influence the isolation ratings.

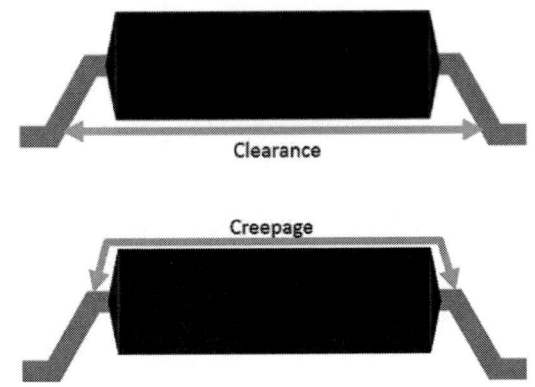

Fig. 7 Clearance and Creepage

Under type test subgroup 5 of IEC 60747-17, test houses are required to measure the creepage and clearance values of 10 samples, adhering to the IEC 60664-1 measurement technique [4]. This is a new requirement that was not present in previous digital isolation standards.

3 The Impact of IEC 60747-17

3.1 Advancements in Digital Isolation

IEC 60747-17 serves as a comprehensive standard for digital isolator devices. Manufacturers across the industry are progressively aligning their products with this standard. Due to the addition of stringent test requirements and modifications to existing tests, the transition has been gradual. For instance, the TDDB test necessitates over two months for data collection. Numerous components, previously certified as reinforced under other standards, fail to meet all the criteria for a reinforced rating under IEC 60747-17, particularly the 10 kV bipolar surge requirement. Consequently, these products will be reclassified as basic insulation under IEC 60747-17.

Following the ratification of IEC 60747-17, digital isolator manufacturers have adopted this standard to guide their chip design. Being the first standard specifically designed for digital isolators worldwide, it has instigated a global shift in engineering processes among isolator manufacturers to ensure compliance with the new requirements. This standard has catalyzed advancements in isolation technology, with contemporary digital isolators achieving 6 kV$_{RMS}$ isolation. This enhancement allows systems employing isolators to operate at higher voltages and power levels while maintaining safety. System designers worldwide benefit from a uniform safety framework.

3.2 IEC 60747-17 Adoption Timeline

Looking ahead, digital isolator manufacturers have begun certifying to the new standard, and this trend is expected to continue. Some manufacturers are likely to transition mature products, certified under various system and component standards, to IEC 60747-17 to simplify their customers' certification processes.

This certificate will be recognized by the system standards of all countries, eliminating the need for most system designers to directly certify to any system standards henceforth. IEC 60747-17 is currently coexisting with UL1577 (US component standard) and GB4943.1 (Chinese standard). However, these standards are expected to become less relevant and eventually phased out over the next few years, subject to the discretion of the respective countries. It is anticipated that IEC 60747-17 will emerge as the sole digital isolator standard, unifying the many standards that came before it.

References

[1] Skyworks. 2022. Isolator vs. Optocoupler Technology. Skyworks. https://www.sky-worksinc.com/-/media/Skyworks/SL/documents/public/white-papers/isolator-vs-opto-coupler-technology.pdf

[2] IEC. 2020. IEC 60747-17 International Standard, Edition 1.0: Semiconductor Devices – Part 17: Magnetic and capacitive coupler for basic and reinforced insulation

[3] IEC. 2018. IEC 62368-1 International Standard, Edition 3.0: Audio/video, information and communication technology equipment – Part 1: Safety requirements

[4] Skyworks. 2022. AN1167: Safety Considerations for Skyworks Series Capacitive Isolators. Skyworks. https://www.skyworksinc.com/-/media/SkyWorks/SL/documents/public/application-notes/AN1167.pdf

[5] IEC. 2020. IEC 60664-1 International Standard, Edition 3.0: Insulation coordination for equipment within low-voltage supply systems - Part 1: Principles, requirements and tests

PCIM Europe 2024, 11– 13 June 2024, Nuremberg   DOI:10.30420/566262419

Paris Law Applied to Wire Bonds Degradation Using Crack Growth Measurement

Merouane Ouhab[1] , Pierre-Yves Pichon[1]

[1] Mitsubishi Electric R&D Centre Europe (MERCE), France

Corresponding author: Merouane Ouhab, m.ouhab@fr.merce.mee.com
Speaker: Merouane Ouhab, m.ouhab@fr.merce.mee.com

Abstract

Power Modules (PMs) with a standard packaging technology reveal weaknesses after long operation time due to thermomechanical fatigue. Top electrical interconnections, or wire bonds, used in PMs are the most critical components prone to degradation leading to the device failure. In this paper, a crack growth lifetime model dedicated to wire bonds fatigue is constructed to estimate the PM lifetime under Power Cycling Tests (PCTs) conditions. The proposed model considers lifetime dependence on junction temperature extent (ΔTj) and heating time (ton) as main load parameters. The PCTs are conducted under high switching frequency and high DC bus voltage conditions. The tested PM integrates a single-wired chip instrumented with auxiliary sensors to estimate the crack growth rate in-situ during the PCTs. In parallel, a nonlinear fracture parameter is evaluated using an electro-thermo-mechanical Finite Element Model (FEM). Combination of crack length measurement and the selected crack-tip parameter allows to confirm the Paris law and finally to build the fracture-mechanics lifetime model.

1   Introduction

Lifetime modelling using Physics-of-Failure (PoF) has gained a large interest in the field of power electronics reliability, thanks to its low cost and reduced time of construction. This type of models came either to replace or to support statistical and empirical models to: reduce products qualification time, in-service lifetime estimation (digital twin) or Design-for-Reliability (DfR). Regarding wire bonds lift-off as a major cause of failure in standard Power Modules (PMs), several physics-based approaches were proposed in the literature. Generally, they can be classified under two main methods. In the first method [1-3], the Coffin-Manson equation (power law type) is scaled on fatigue tests data such as Power Cycling Tests (PCTs), Thermal Cycling Tests (TCT) or external mechanical tests. These models use a specific damage parameter (cyclic extent of stress, strain or energy), which is selected in agreement with the materials properties (elasticity, plasticity and/or creep) and the loading conditions (high/low temperature, short/long heating time). The damage parameter is evaluated either by Finite Element Analysis (FEA) or by simplified analytical solutions. Even if this type of approaches is easy for

implementation using an already computed parameters by the FEM, it requires several experimental data for scaling to extrapolate the results with confidence. Another drawback is related to the high sensitivity of the damage parameter regarding the evaluation zone, particularly its dependency to the meshing size. The second modelling method relies upon fracture mechanics, where two different models are mostly used in the literature. The first is based on the Cohesive Zone Method (CZM) where the crack length is calculated dynamically as a function of the subjected load cycles. This method requires specific input properties (example: cohesive strength and critical separation) to describe a gradual traction-separation at a given interface, these properties can be estimated using a sophisticated mechanical test setup (traction-separation test). Moreover, the CZM requires unachievable calculation performances to simulate a fatigue test profile. The second model is more direct and easier for implementation [4], where the Paris equation is used together with a preselected fracture parameter (stress intensity factor, energy release rate, creep J integral … etc.) to describe the crack propagation at the wire-metallization interface. Based on a

2948

literature review about these models one can highlight a lack in terms of accurate crack length measurement to emphasize or not the adequation of the Paris linear relationship between the preselected fracture parameter and the crack growth rate. This is due to the difficult accessibility to the wire bond interface and the limited available measurements to characterize separately the wire bond health-state specially under online conditions [5]. Within this context, we carry out in our study several PCTs on a single-wired PM under high switching frequency and high DC bus voltage. The PCTs target different ΔTj and ton values. The Devices Under Tests (DUTs) are electrically instrumented with auxiliary emitters to estimate wire bond health-state accurately. Cyclic creep J integral (ΔJ^*) is evaluated as a crack-tip parameter using electro-thermo-mechanical simulations. Correlation between measured crack growth rate and the fracture parameter ΔJ^* confirms Paris law adequation to describe wire bond crack propagation for both load parameters ΔTj and ton.

2 Instrumentation, characterization and PCTs results

As demonstrated in Fig. 1 (a), the tested IGBT PM, rated 1200V/50A, is connected to a dummy PM via an inductance according to a back-to-back topology. Herein, each IGBT has 5 thick aluminum wires (w1-w5) of 400um diameter, 6 wire-bonds (wb1-wb6) and two aluminum metallization pads, see Fig. 1 (b). The top IGBT (T1) represents the DUT in this study where only one conducting emitter wire (w3) is kept connected. The remaining 4 wires (w1, w2 w4 & w5) are all cut to control the current flow through the studied wire, i. e. w3. As the metallization degrades with the PCT, the on-state resistance of the DUT (RCE) is expected to include two contributions, the first is due to crack propagation in the wire bond and the second contribution is due to metallization degradation, more details are provided in the next section. Aiming to limit the metallization contribution, an auxiliary emitter voltage (Vwb4) is instrumented close to the wire of interest (w3) using an external probe that is positioned on wire-bond 4 (wb4) as illustrated in Fig. 1 (c). The DUT is black-painted to monitor its temperature using a thermal camera, in order to adjust the load current to achieve the desired ΔTj at a fixed ton/toff.

As shown in Table 1, four PCTs were carried out under Pulse Width Modulation (PWM) technique. The tests targeted three different values of ΔTj and two different values of ton in order to establish a lifetime model dependent on these two parameters.

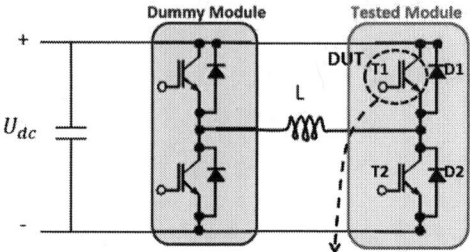

a) Power cycling circuit topology

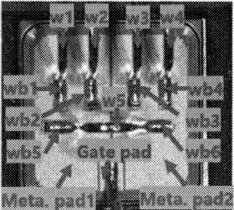

b) DUT before wires cutting

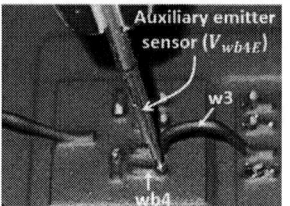

c) Instrumented & black-painted DUT

Fig. 1: PWM power cycling test conditions, (a) circuit topology, (b) DUT before wires cutting, (c) Instrumented and black-painted DUT after wires cutting

During the PCTs, IGBT T1 and diode D2 are the main switching devices in the tested module according to the switching pattern illustrated in Fig. 2 where both devices dissipate in-phase conduction and switching losses. To be mentioned, all the IGBT gates are controlled with a 0V/15V signal.

Target ΔTj	ton/toff	Th	I	fsw	D	Udc
(°C)	(s)	(°C)	(A)	(kHz)	(%)	(V)
50	3/3	20	15	30	50	600
70	3/3	20	20	30	50	600
90	3/3	20	24	30	50	600
70	12/12	20	19	30	50	600

Table 1 PCTs conditions

By the end of the cooling stage (ton+toff), a characterization current I_C=25A is injected through the DUT for a period of 20µs in order to measure V_{CE} and V_{wb4E} at a known temperature T=Tjmin without causing any overheating to the device, thus, to avoid any secondary effect due to die-solder degradation.

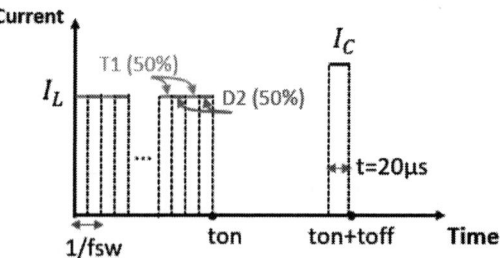

Fig. 2 Switching pattern during the PCTs

In the following, the test at $\Delta Tj=70°C$ and ton=3s is selected as a reference test, therefore, all the tests will be normalized to its number of cycles to failure (Nf). In Fig. 3 (a), measured temperature field is shown at t=3s (hot state) for the reference test using the thermal camera. As it can be noted, the temperature is not homogeneous over the die top surface. The hottest point is located close to the position of wire-bond 3 (wb3). In Fig. 3 (b) & (c), junction temperature and wb3 local temperature are evaluated, respectively, for all the PCTs using IR thermal data averaged over the surfaces depicted in Fig. 3 (a). As it can be seen, wb3 experiences temperature swings higher than the targeted junction temperature swing by at least 15°C.

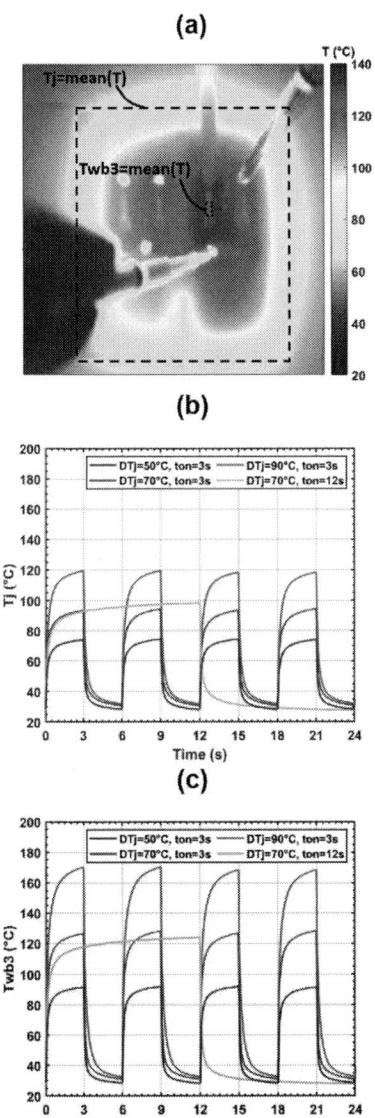

Fig. 3 PCTs temperature profiles, (a) IR thermal map at t=3s for reference test, (b) Junction temperature, (c) Wire-bond 3 local temperature

Fig. 4 shows a resistance increase measured between the auxiliary and the main emitters (ΔR_{wb4E}). In Fig. 4 (a) the curves are plotted to highlight the effect of ΔTj at constant ton=3s, and Fig 4 (b) for the effect of ton at constant $\Delta Tj=70°C$. The numbers of cycles are normalized to the number of cycles to failure (Nf) measured for the reference test. As expected, when ΔTj or ton rise, the crack at wb3 grows faster therefore ΔR_{wb4E} increases faster due to constriction effect at the interface between w3 and metallization pad 2. The cause of failure in the four PCTs is due to a short-circuit between the gate and the emitter due to a potential over-heating at wb3 position and possible local damage in the semiconductor internal structure. This was observed on IR thermal images of the DUTs at the End-of-Life (EoL) under zero load current and turned-on gate, where a hotspot appeared close to wb3 due to a leakage current coming from the gate supply voltage.

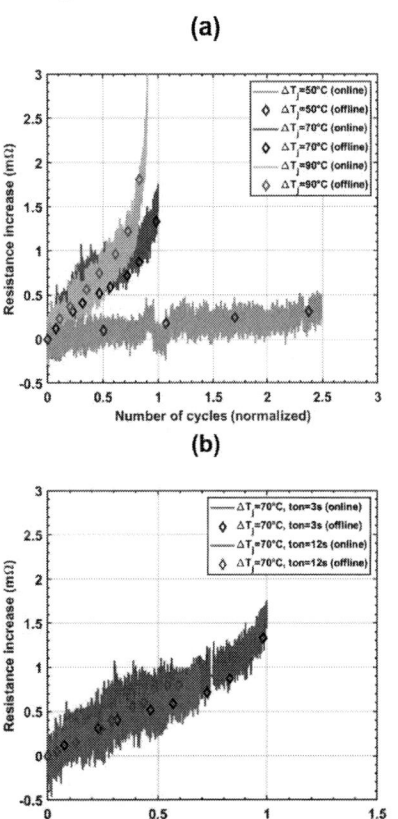

Fig. 4: Resistance increase measured between the auxiliary & the main emitters (ΔR_{wb4E}) during all PCTs, (a) ΔTj effect, (b) ton effect

ΔR_{wb4E} was measured each cycle (online) during the PCT using a conditioning circuit board and an acquisition system from National Instrument (NI). Moreover, each set number of cycles, the ΔR_{wb4E}

was measured offline using an external current source under thermal steady-state conditions and a constant load current I_C=20A. As it can be noted from Fig. 4 (a) & (b), online and offline measurements provide similar results.

3 Crack growth estimation

In this section, acquired measurements of ΔR_{wb4E} during PCTs are used to estimate the crack length in wb3. For more insight about this method, a qualitative analytical study is introduced in the following. As demonstrated in the schemes of Fig. 5, the constriction resistance is due to current lines constricted between two connected bodies due to reduced contact surface when the crack propagates. The two bodies in this work represent the wire and the metallization. The realistic configuration of the wire-metallization contact shown in Fig. 5 (b) can be simplified into the equivalent configuration illustrated in Fig. 5 (c) where a circular shape is assigned to the bond and the crack areas. If the contribution of metallization degradation is neglected, the resistance increase ΔR_{wb4E} due to a remaining bond length $a_{b\,eq}$ (radius of the bond area in Fig. 5 (c)) can be decomposed analytically into two terms as follow:

$$\Delta R_{wb4E} = \Delta R_{cons} + \Delta R_{meta} \tag{1}$$

Where:

$$\Delta R_{cons}\left(a_{b\,eq}\right) = \frac{\rho}{4\,a_{b\,eq}} \left(1 - \frac{a_{b\,eq}}{a_{b0\,eq}}\right)^{3/2} \tag{2}$$

And:

$$\Delta R_{meta}\left(a_{b\,eq}\right) = \frac{\rho}{2.\pi.t}\left(ln\left(\frac{a_{b0\,eq}}{a_{b\,eq}}\right)\right.$$
$$\left. + \frac{1}{4}\frac{a_{b\,eq}^2 - a_{b0\,eq}^2}{b^2}\right) \tag{3}$$

Here, ΔR_{cons} is the additional resistance due to constriction effect inside the wire material, ΔR_{meta} is the additional resistance due to additional metallization length flowed by the load current I_L, ρ the electrical resistivity of the wire and metallization aluminum material, t is the metallization thickness, b metallization length (reduced to an equivalent radius of the metallization top surface) and $a_{b0\,eq}$ is the initial equivalent radius of the bond area. In [5], detailed boundaries conditions and analytical development steps can be found.

From the previous equations, the resistance increase ΔR_{wb4E} has a forward relation with the

equivalent crack length that can be expressed simply by:

$$a_{c\,eq} = a_{b0\,eq} - a_{b\,eq} \tag{4}$$

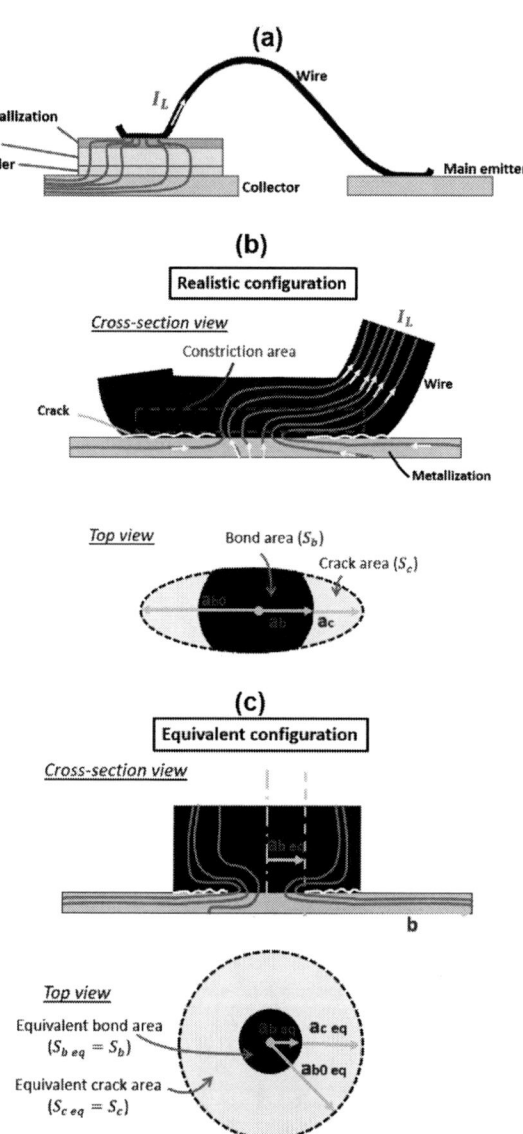

Fig. 5 Constriction effect, (a) Interconnecting wire in a real environment, (b) Constriction effect in the real wire-bond configuration, (c) Constriction effect in the equivalent configuration

Due to the relatively complex geometry of the wire bond, it was necessary to develop a 3D electrothermal Finite Element Model (FEM) of the studied DUT. The power module 3D geometry was built using a contactless optical profilometer, microsection and Scanning Electron Microscopy (SEM)

to identify different materials and layers thicknesses as listed in Table 1. The electrothermal properties of different materials are provided in Table 2.

Layer	Material	Thickness (µm)	Dimensions (mm×mm)	Mesh size (µm)
Wire	Aluminum	400	-	0.1-50
Metallization	Aluminum	5	6.5x6.4	1-1000
IGBT	Silicon	130	7x7	100-1000
Diode	Silicon	130	7x5	100-1000
IGBT solder	Sn96Ag35	70	7x7	100-1000
Diode solder	Sn96Ag35	70	7x5	100-1000
Substrate top layer	Copper	250	-	10-1000
Substrate ceramic	AlN	640	52x39	100-1000
Substrate bottom layer	Copper	250	49x36	10-1000
Substrate-solder	Sn96Ag35	300	49x36	10-1000
Baseplate	Copper	3000	120x60	300-3000
Thermal interface material (TIM1)	Graphite	50	120x60	300-3000
Adaptation plate	Copper	8000	150x140	500-5000
Thermal interface material (TIM2)	Thermal grease	50	150x140	500-5000
Heatsink	Copper	16000	500x140	1000-1000

Table 1 Power module dimensions, materials and meshing size

	Density (kg.m-3)	Electrical resistivity (Ω.m)	Thermal conductivity (W.m-1.K-1)	Specific heat (J.kg-1.K-1)
Al	2700	2.4×10^{-8} @20°C	237	900
Si	2330	12.8×10^{-3} @25°C	150	710
Cu	8930	17×10^{-9}	400	390
AlN	3300	-	170	720
SnAg	7360	12×10^{-8}	55	240
TIM1	-	-	7	-
Tim2	-	-	2	-

Table 2 Materials electrothermal properties

The DC load current injected during offline characterization I_C=20A was simulated at different crack length values. The crack in the FEM is represented by a 1µm height notch inserted at the interface between the wire and the metallization supposing a circular shape progression of the crack as illustrated in Fig. 6.

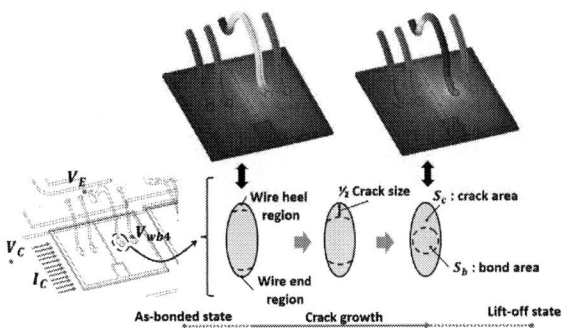

Fig. 6 Crack shape implementation in the FEM

The simulation results are shown in Fig. 7. We can note a slight increase of ΔR_{wb4E} as the equivalent crack length increases until 90% of the initial equivalent bond length ($a_{b0\ eq}$). Beyond this value, ΔR_{wb4E} increases very fast to reach very high values by 99% of $a_{b\ eq}$. Based on wire contact analysis made on several DUTs using optical microscopy, we could measure a mean initial bond length of $a_{b0\ eq}$=272µm and a mean pre-crack value of $a_{c\ eq}$=35µm. Therefore, it is expected that ΔR_{wb4E} starts to increase in Fig. 7 after the pre-crack value.

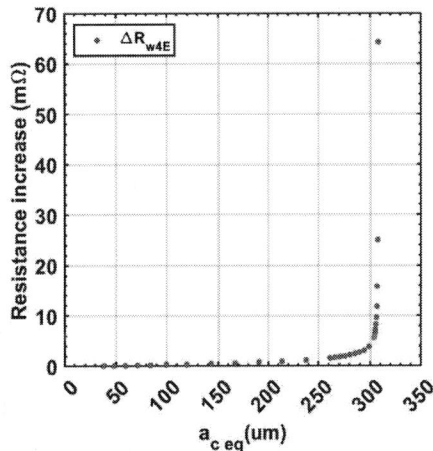

Fig. 7 Resistance increase between the auxiliary and the main emitters (ΔR_{wb4E}) as a function of the equivalent crack length

Combination of measured data of ΔR_{wb4E} during PCTs (shown in Fig. 4) and simulation data from Fig. 7, the equivalent crack length was estimated as shown in Fig. 8 (a) & (b) considering both dependencies of ΔTj and ton respectively. As it can be noted, for all the PCTs the crack length increases fast during the beginning of the lifetime, and then it increases with a remarkable deceleration during the mid of the lifetime. By the End-of-Life, the crack length accelerates until failure.

(a)

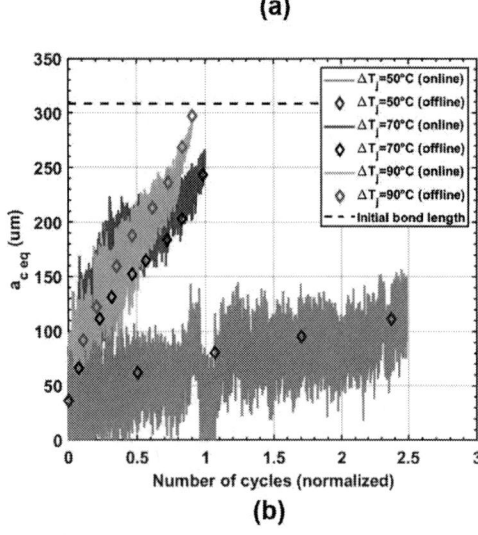

(b)

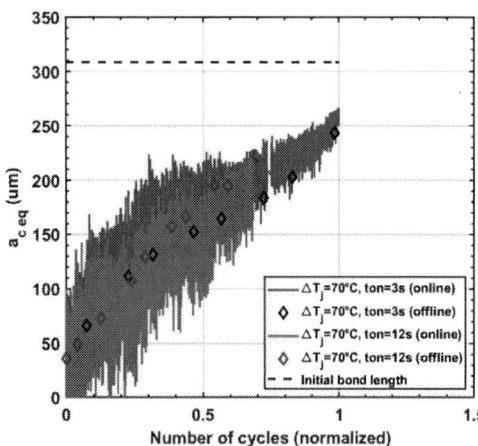

Fig. 8 Crack growth estimation during PCTs, (a) ΔT_j effect, (b) ton effect.

4 Lifetime modelling

The proposed lifetime model in this work is based on a time crack growth model that can be expressed as follow:

$$\frac{da}{dt} = \alpha [J^*]^\beta \tag{4}$$

Under PCT conditions, the cycle crack growth rate can be estimated by integration of equation (4) over the heating time period, this gives the following equation:

$$\frac{da}{dN} = \int_0^{t_{on}} \frac{da}{dt} dt \tag{5}$$

Considering both PCT parameters ΔT_j and ton, equation (2) becomes:

$$\frac{da}{dN} = \int_0^{t_{on}} \alpha \big[J^*(\Delta T_j, t_{on}) \big]^\beta dt$$
$$= \alpha \Delta J^*(\Delta T_j, t_{on}, \beta) \tag{6}$$

Where: ΔJ^* represents the cyclic creep J integral, a is the crack length at the wire/metallization interface, α and β are material parameters.

The fracture parameter J^* is evaluated using the following definition:

$$J^* = \int_\Gamma \left(W^* dx_2 - T_i * \left(\frac{\partial \dot{u}_\iota}{\partial x_1} \right) * ds \right) \tag{7}$$

Where:

$$W^* = \int_0^{\dot{\varepsilon}_{ij}} \sigma_{ij}\, d\dot{\varepsilon}_{ij} \tag{8}$$

And:

$$T_i = \sigma_{ij} n_j \tag{9}$$

Here, W^* represents the strain energy rate density associated with the stress point σ_{ij} and the strain rate $\dot{\varepsilon}_{ij}$. Γ is a curve surrounding the crack tip, Ti represents the traction vector. nj (with: j=1,2) represents the normal vector to the contour path Γ. u is the displacement vector, ds is an element of arc length along Γ.

Using a 3D Electro-Thermal (E-T) FEM of the studied PM, realistic temperature distributions were simulated under each PCT conditions for 3 cycles to reach a thermal steady-state. For sake of simplicity, the temperature field at the wire contact zone is mapped between the 3D E-T FEM and a 2D Thermo-Mechanical (T-M) FEM as illustrated in Fig. 9.

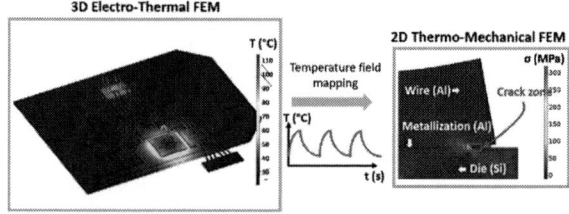

Fig. 9 Electro-Thermo-Mechanical simulation scheme

Based on the electro-thermo-mechanical FEM, ΔJ^* at the crack-tip is calculated considering different crack lengths. As illustrated in Fig. 10 (a) & (b), the curves of ΔJ^* are plotted to show the effect of

ΔTj and ton respectively. For different PCTs conditions, as the crack length increases ΔJ^* decreases. This trend is in opposition to what is commonly observed during mechanical fatigue test under controlled stress amplitude, where the fracture parameter extent (creep J integral, J integral or stress intensity factor) is an increasing function with the crack length inside the test specimen.

(a)

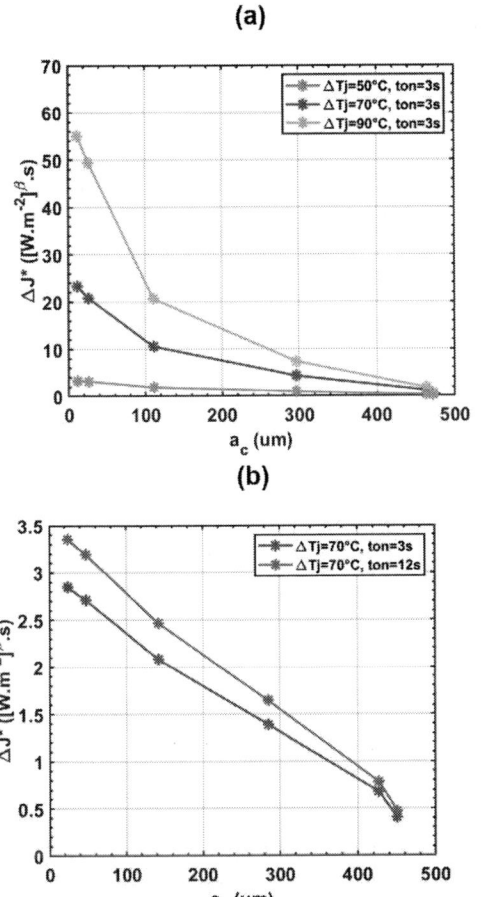

(b)

Fig. 10 Cyclic creep J integral as function of the crack length, (a) ΔTj effect, (b) ton effect

As shown in Fig. 11, correlation between measured crack growth rate and evaluated ΔJ^* allowed to check Paris law adequation and to establish the constants α and β based on equation (6).

The validation of the Paris law indicates that the fracture mechanics is a valid approach for modelling the crack propagation in the wire-bonds, and that the implementation selected in this paper (i.e. both experimental crack length assessment and electro-thermo-mechanical FEM) is relevant.

As a matter of interest, α and β depend on the type of the stressor parameter whether it is temperature or time. In this study we measure approximately:

- For ΔTj effect $\alpha=528.5$ & $\beta=2$
- For ton effect $\alpha=84.65$ & $\beta=0.82$

The values given for parameter α are normalized to the lifetime of the reference test. The constants calculated for ΔTj effect are expected to be the same for the minimum junction temperature as they represent temperature parameters. Further tests are needed to verify this speculation.

The values of β are applicable to different power module references with the same aluminium wire-bond material. These parameters can be used for lifetime modelling to reduce the required number of scaling power cycling test.

(a)

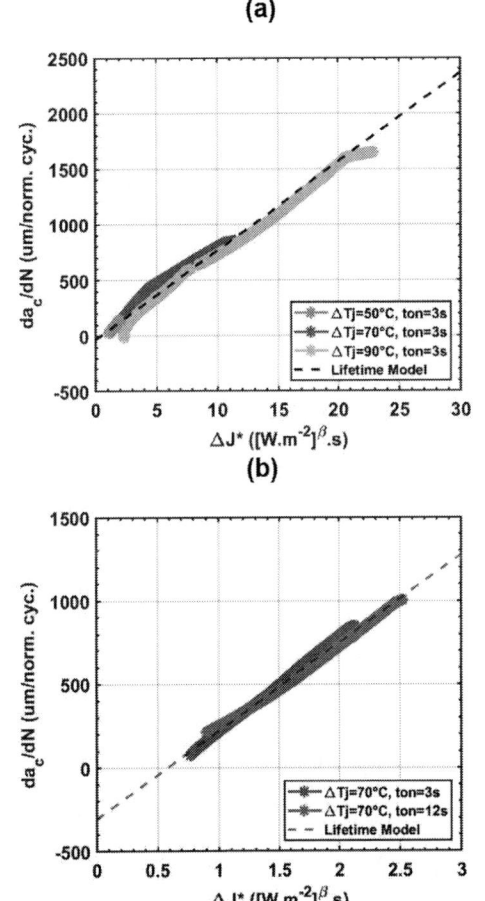

(b)

Fig. 11 Crack growth rate correlation with ΔJ^*, (a) ΔTj effect, (b) ton effect

As it is shown in Fig. 12, using the developed lifetime model, the crack growth for each PCT is reconstructed by integration of equation (6). Despite the following accumulated sources of error:

- Variabilities related to the DUTs (wire bond shape, electrothermal behaviour …etc.)
- Crack estimation accuracy and sensitivity
- Crack growth rate derivation
- Hypothesis on the circular shape of the crack
- Variabilities related to materials properties

The proposed model delivers reasonable results as a promising tool for furthermore investigations. In particular, it is capable of not only providing the number of cycles to failure Nf, but also the crack length which is responsible for current and temperature redistribution and evolution of damage-sensitive electrical parameters such as Vce or Vee.

(a)

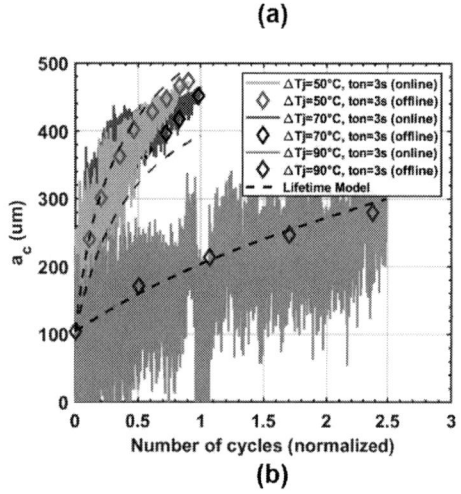

(b)

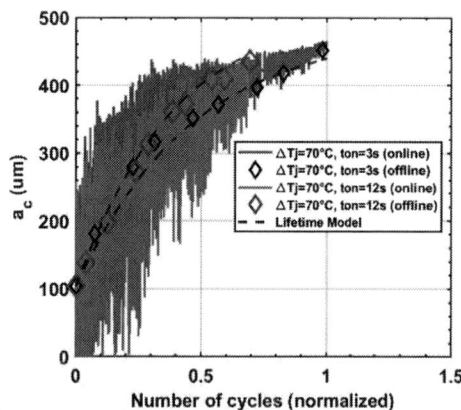

Fig. 12 Lifetime model estimations, (a) ΔTj effect, (b) ton effect

5 Conclusion

Single-wired DUTs were instrumented with an auxiliary emitter voltage sensor and then tested under conditions close to application. Resistance increase measured between the auxiliary and the main emitters allowed to better estimate the crack length of the wire bond non-destructively and at the PCT rate offering a high sampling level. Correlation between measured crack growth rate and ΔJ* allowed to find reasonably the Paris law parameters and validate the use of the selected fracture-mechanics implementation. Validation of estimated crack propagation and lifetime by the proposed model is to be extended to other load conditions. Finally, this law can be easily generalized to several parallel wires within a device. The proposed model finds application during prototyping, design stage or in-service lifetime estimation since it allows, with a very reduced number of power cycling tests, to evaluate and explain the lifetime in regions difficult to test.

6 References

[1] J. Bielen, J. . -J. Gommans and F. Theunis, "Prediction of high cycle fatigue in aluminum bond wires: A physics of failure approach combining experiments and multi-physics simulations," EuroSimE 2006 - 7th International Conference on Thermal, Mechanical and Multiphysics Simulation and Experiments in Micro-Electronics and Micro-Systems, Como, Italy, 2006, pp. 1-7, doi: 10.1109/ESIME.2006.1644022.

[2] N. Dornic et al., "Stress-Based Model for Lifetime Estimation of Bond Wire Contacts Using Power Cycling Tests and Finite-Element Modeling", in IEEE Journal of Emerging and Selected Topics in Power Electronics, vol. 7, no. 3, pp. 1659-1667, Sept. 2019, doi: 10.1109/JESTPE.2019.2918941.

[3] Rajaguru, Pushparajah & Lu, Hua & Bailey, Chris. (2014). Application of nonlinear fatigue damage models in power electronic module wirebond structure under various amplitude loadings. Advances in Manufacturing.

[4] M. Ouhab, N. Degrenne, Y. Ito and S. Izuo, "Physics-of-Failure Model to Explain the Heating-Time Effect on IGBT Power Modules Lifetime," PCIM Europe 2023, Nuremberg, Germany, 2023, pp. 1-6, doi: 10.30420/566091007.

[5] P. -Y. Pichon and M. Ouhab, "Wirebond state-of-health measurements using specific, non-destructive electrical and thermal methods," in IEEE Journal of Emerging and Selected Topics in Power Electronics, doi: 10.1109/JESTPE.2023.3314077.

PCIM Europe 2024, 11–13 June 2024, Nuremberg   DOI: 10.30420/566262420

Condition Monitoring Technique of Power Electronic Modules via Square-Wave Gate Signal Modulation

Isabel Austrup ©[1], Fabian Janoth ©[1], Rik W. De Doncker ©[1]

[1] Institute for Power Electronics and Electrical Drives (ISEA), RWTH Aachen University, Germany

Corresponding author: Isabel Austrup, isabel.austrup@isea.rwth-aachen.de
Speaker: Isabel Austrup, isabel.austrup@isea.rwth-aachen.de

Abstract

As demonstrated in previous work, an efficient way to detect and localize various degradation modes in power modules is to excite periodic losses via sinusoidal gate-source voltage v_{GS} modulation. This allows extracting the phase shift between v_{GS} and the resulting drain-source voltage v_{DS} as an indicator of the thermal path's state of health, as it changes with degradation at characteristic bandwidths. This paper demonstrates that the excitation does not have to be sinusoidal, but can also be a square wave, allowing an easier hardware implementation. It thus simplifies the integration of the diagnostic circuitry into the gate driver. Moreover, fewer frequencies need to be excited individually, since a square wave excites the odd harmonics, that can be evaluated for degradation detection as well. Thus, measuring time can be reduced.

1 Introduction

Power electronics are used in many applications such as renewable energy systems or the electrification of transport. Reliability is therefore a major concern [1]–[3]. Condition monitoring techniques to estimate the state of health and thus avoiding unexpected failures are an alternative to over-sizing power electronic converters [4]–[6].

Various approaches to detect aging of power electronic modules have been presented in the literature. Degradation in the chip solder leads to an increased thermal impedance between the chip and the heat sink. Therefore a larger proportion of the heat is dissipated via the silicone gel towards the ambient air instead of the cooling liquid. Monitoring the heat dissipated to both the cooling liquid and the ambient air therefore enables detecting chip solder degradation [7], [8].

For forced air cooled systems, [9] proposes to monitor the thermal network's natural frequency via sensors measuring the case temperature between the power module and the heat sink as well as the ambient temperature during the turn-off process. The cooling curve is then fitted to a thermal network to obtain the natural frequencies. Changes in these natural frequencies allow to detect and distinguish between degradation of the thermal interface material (TIM) and the cooling fan.

Detecting and localizing degradation in the thermal path of the power module itself as well as the cooling system can also be achieved via thermal impedance spectroscopy in the frequency domain. This method injects sinusoidal losses at multiple frequencies and measures the temperature response. Changes of amplitude and phase of the frequency dependent transfer function $\underline{Z}_{th}(j\omega)$ as relation between power excitation P_{loss} and temperature response T_j at characteristic bandwidths is an indicator of degradation within the thermal path [4], [10], [11].

All degradation diagnostic methods presented above need temperature sensors or the determination of the junction temperature T_j via temperature sensitive electric parameters (TSEPs), leading to additional costs and potentially complex extraction circuits [5], [12], [13].

A method that allows detecting and localizing aging of the thermal path without the need to determine junction or case temperature was described [14] and is shown schematically in Fig. 1. For this method, periodic losses are excited in the device by superimposing a sinusoidal small-signal voltage v_{mod} onto a constant gate-source voltage V_0 that keeps the device in conduction mode and allows

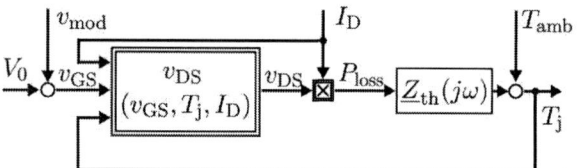

Fig. 1: Modulation of the gate voltage v_{GS} to detect degradation of the thermal path by measuring the phase shift between the excitation v_{GS} and the electrical response v_{DS}.

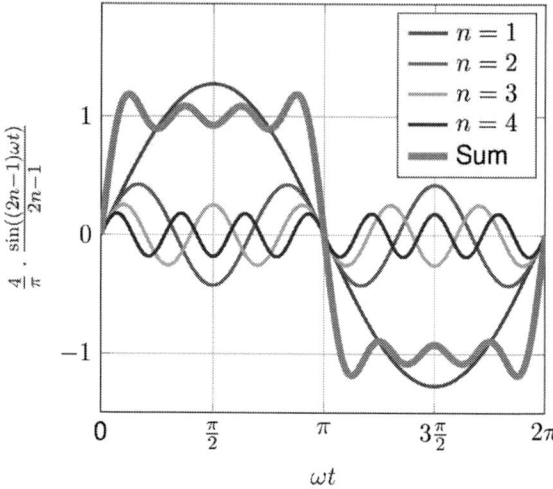

Fig. 2: Exemplary addition of four sinusoidal waves according to Eq. (1) leading to a square wave.

it to conduct a constant current I_{D}. Due to the v_{GS} dependence of the on-state resistance $R_{\mathrm{DS,on}}$ and thus v_{DS}, this modulation leads to periodic losses within the device. These losses in turn result in a periodic temperature response whose phase is degradation dependent, as described above. The degradation dependency is reflected in the drain-source voltage v_{DS}, as the on-state resistance $R_{\mathrm{DS,on}}$ is not only dependent on the gate-source voltage v_{GS}, but also temperature dependent [15], [16]. Therefore, instead of the phase shift $\angle \underline{Z}_{\mathrm{th}}$ between P_{loss} and T_{j}, the phase shift between v_{GS} and v_{DS} is evaluated. To ensure a constant temperature sensitivity of $R_{\mathrm{DS,on}}$ throughout the measurement, the ambient temperature T_{amb} should be kept constant. It could be shown that changes in the thermal path, such as degradation of the TIM or the cooling system, lead to changes in the phase shift between v_{GS} and v_{DS} at characteristic frequencies. These characteristic frequencies allow the localization of single, as well as superimposed degradation modes [14], [17]. The changes of the phase shift, which is determined using a Fast Fourier Transform (FFT), for different degradation modes are in the sub-degree range. However, those sub-degree changes can be detected easily by a simple measuring circuit described in [14]. Two main drawbacks of the proposed methods are the necessity of a sinusoidal voltage generation as well as the long measuring time, since the frequencies considered are in the millihertz range.

To overcome these drawbacks, this paper proposes using a rectangular instead of a sinusoidal gate-source voltage modulation v_{mod}. This makes the generation of a sinusoidal voltage obsolete and thus decreases the hardware effort for the implementation of the gate-signal modulation. Moreover, the odd-integer harmonics of a square-wave's fundamental frequency are excited as well.

Therefore, several frequencies can be excited at the same time, allowing a reduction in measuring time.

This paper first describes the detection of degradation using a rectangular gate-source voltage excitation. Afterwards, experimental results compare the rectangular with a sinusoidal gate excitation. Finally, the utilization of the harmonics is analyzed on the basis of the measurement results.

2 Degradation Detection Method

The following section describes how a rectangular gate voltage with a variable frequency can be used to detect degradation of the thermal path. According to Fourier analysis, a square wave $x(t)$ consists of an infinite sum of sine waves. Equation (1) shows how a square wave with a duty cycle of 50 % excites the odd integer multiples of the fundamental frequency:

$$x(t) = \frac{4}{\pi} \sum_{n=1}^{\infty} \frac{\sin((2n-1)\omega t)}{2n-1}. \tag{1}$$

The addition of four exemplary sine waves according to Eq. (1) is visualized in Fig. 2. It can be seen that the fundamental frequency ($n = 1$) is the most dominant. However, sine waves with higher frequencies, the odd-integer harmonics, are excited as well when applying a square wave onto the gate. Using FFT, the different frequency components of the square wave excitation signal -the gate-source

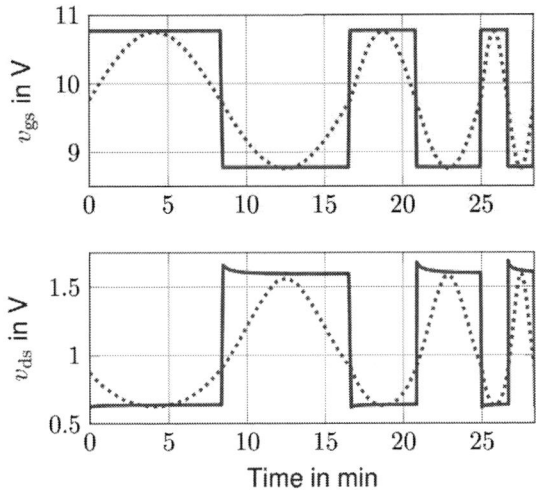

Fig. 3: Course of v_{GS} and v_{DS} with sinusoidal and rectangular excitation for three exemplary frequencies.

voltage v_{GS} - can be extracted. Analogue to the excitation signal, the frequency components of the electrical response - the drain-source voltage v_{DS} - can be extracted. The phase shift between the extracted wave forms of the excitation signal and the electrical response can be used to determine the state of health of the device under test (DUT) according to [14].

This enables shortening the measurement time. As several higher harmonic frequencies can be directly excited and evaluated with each fundamental frequency, fewer fundamental frequencies need to be excited overall during the excitation frequency sweeps. To achieve a high resolution of the measurement, the fundamental frequencies $f_{\text{fund},k}$ of the square wave can be chosen as follows to ensure that no double excitation of the frequencies by fundamental and harmonics occurs:

$$f_{\text{fund},k} = f_{\text{fund},0} \cdot 2^k, k \in \mathbf{N}, \tag{2}$$

with $f_{\text{fund},0}$ being the lowest considered frequency. In addition to a shortened measurement time, the hardware requirements can be reduced significantly, as only two fixed voltage levels are required. The hardware realization for a sinusoidal modulation presented in [14] requires a 12 bit digital-to-analogue converter to produce the sinusoidal voltage. When applying a rectangular voltage to the gate, it is sufficient to have a circuit that can switch between two voltages. Multilevel gate drivers have already been presented in literature [18]–[20].

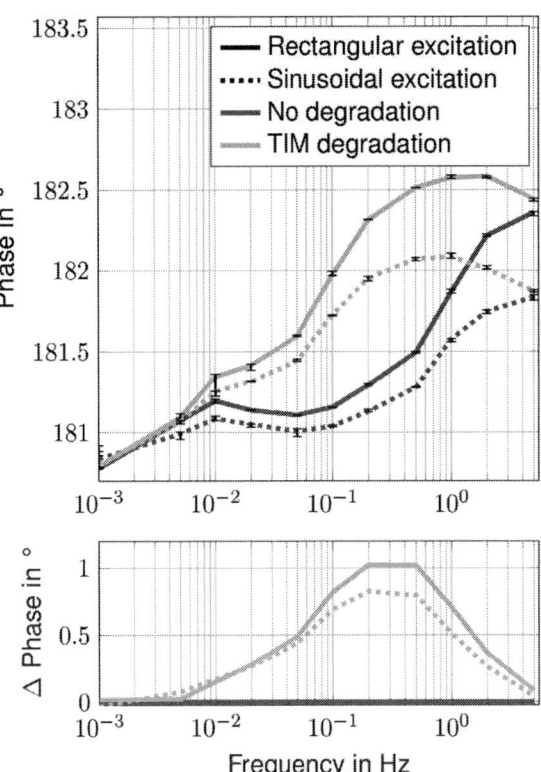

Fig. 4: Phase shift between v_{GS} and v_{DS} at different degradation modes with sinusoidal and rectangular excitation.

Thus, the excitation circuitry for the degradation diagnostics can be integrated into existing circuits with little additional effort.

3 Experimental Verification

In the following, the experimental verification of the method described in section 2 is shown. Therefore, the square-wave excitation is compared to a purely sinusoidal excitation as proposed in [14]. Moreover, the usage of higher order harmonics to increase the bandwidth of the measurement is exterminated.

3.1 Experimental Setup

For the experimental verification, the device under test (DUT), which is the low-side SiC power MOSFET of an Infineon module (FF45MR12W1M1_B11), is conducting a constant current of 5 A. The module is screwed onto a heat sink with a fan using a fixed torque of 1 N m. In order to evaluate the functionality of the proposed method and to compare its sensitivity to a sinusoidal modulation, the aging of the TIM is

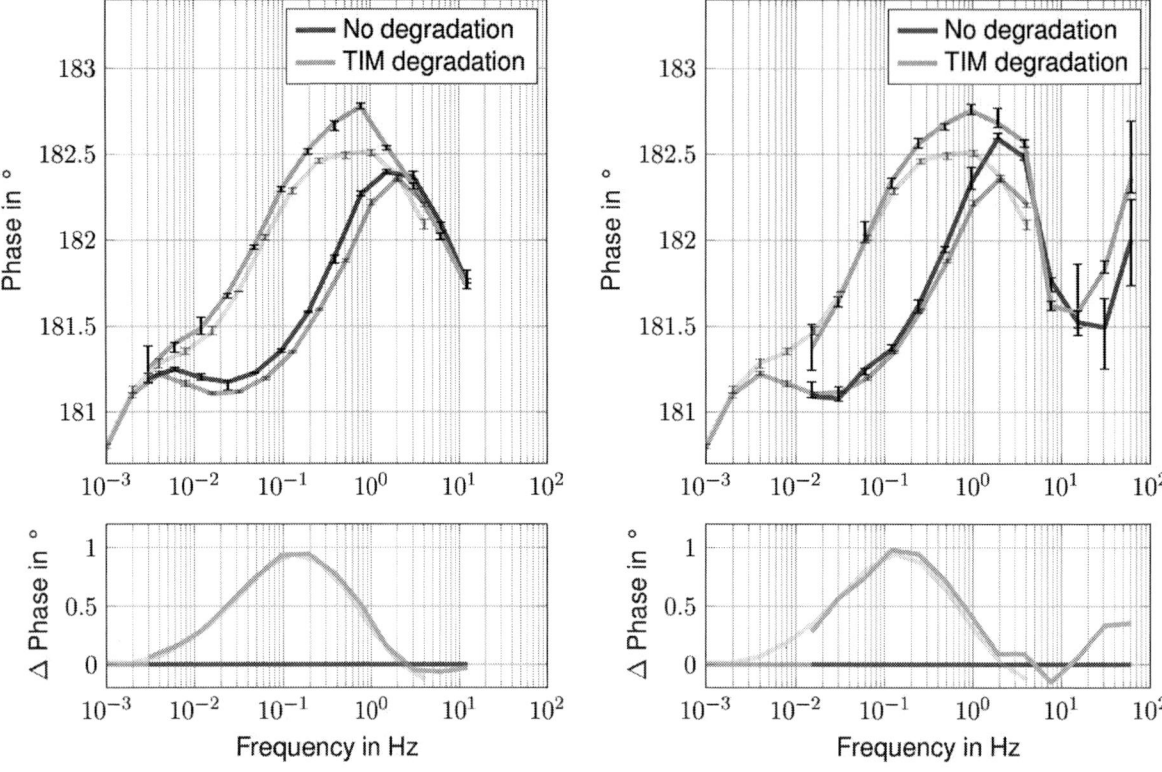

Fig. 5: Phase shift of the 3rd harmonic component between v_{GS} and v_{DS} at different degradation modes compared to the fundamental component (transparent lines) with rectangular excitation.

Fig. 6: Phase shift of the 15th harmonic component between v_{GS} and v_{DS} at different degradation mode compared to the fundamental components (transparent lines) with rectangular excitation.

emulated by the deliberate removal of the thermal grease. The module is therefore connected to the heat sink without any TIM.

For all measurements, the test bench presented in [14] has been used. The excitation voltage applied to the gate has an amplitude of 1 V and an offset of 10 V and can be either rectangular or sinusoidal. Its variable frequency ranges from 1 mHz to 5 Hz. The voltages were sampled 600 times per period, both for the sinusoidal and the rectangular excitation. In [17] it was shown that the sensitivity of the proposed method is influenced by the ambient temperature T_{amb}. Therefore, the measurements were conducted in a climate chamber that kept T_{amb} constant at 25 °C. To prove the reproducibility of the results, all measurements were repeated three times. The highest and lowest value of the measurements are indicated by error in the experimental results shown.

3.2 Experimental Results

The following section presents the experimental results. First, the square-wave excitation is compared to a sinusoidal excitation. Afterwards, the usage of harmonics to increase both the bandwidth and the resolution of the measurement is evaluated.

3.2.1 Comparison between Sine- and Square-Wave Excitation

Using a square wave instead of a sinusoidal modulation reduces the hardware requirements of the excitation circuitry. Their sensitivities are compared experimentally in the following. Figure 3 shows the course of v_{GS} and the resulting v_{DS} for three exemplary frequencies for both sinusoidal and rectangular excitation. The phase shift between the fundamental frequencies of the excitation signal and the electrical response are depicted in the upper plot of Fig. 4. The solid lines show the phase shift for the rectangular excitation while the dotted

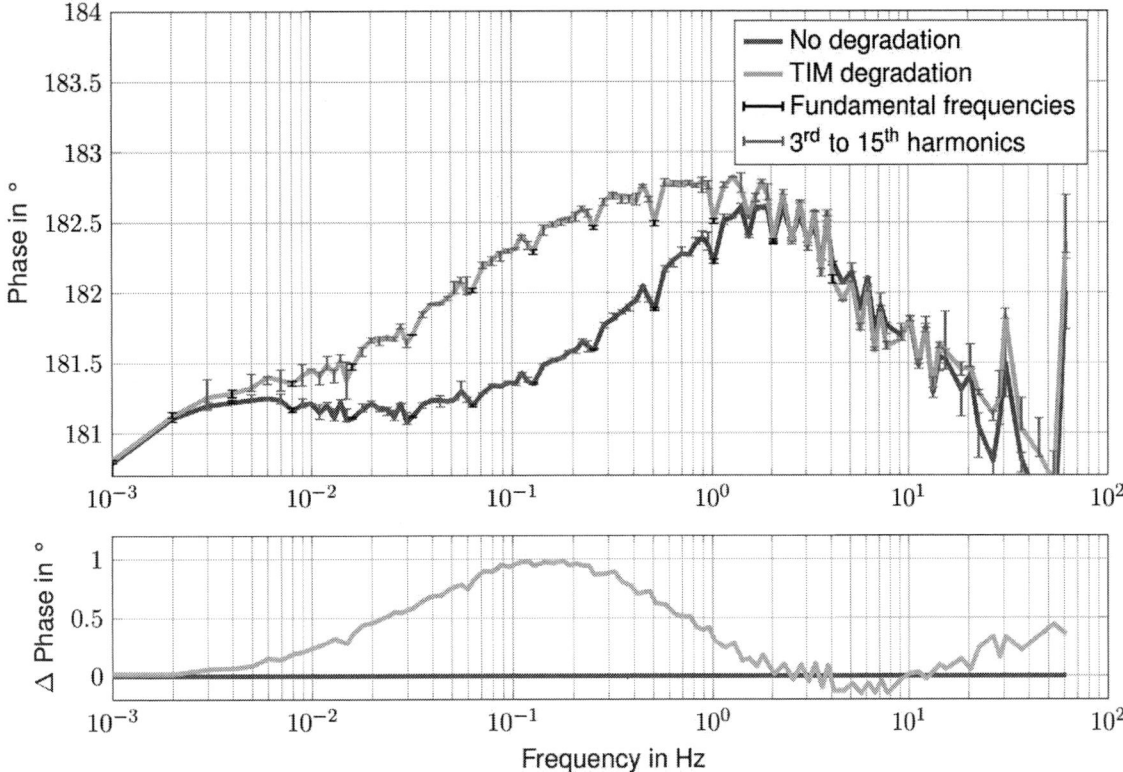

Fig. 7: Simultaneous evaluation of the 1st to 15th harmonic component of the excitation frequencies to increase the bandwidth and the frequency resolution of the measurement.

lines are the phase shift of the sinusoidal excitation for the non-degraded setup (blue) as well as for the thermal interface degradation (yellow). The lower plot of Fig. 4 shows the change in phase due to degradation. It can be seen, that for both excitation forms, the change in TIM results in a change in the phase shift with a peak at 200 mHz. This proves, that not only the sinusoidal, but also the rectangular excitation is suitable to detect degradation within the thermal path of a power module.

The difference in phase shift and sensitivity between the degraded and non-degraded condition is even larger for the rectangular excitation. Different amplitudes of the sinusoidal signal and the 1st harmonic of the rectangular signal, as well as different temperature profiles that result from the excitation could be the origin of this change in sensitivity. Future work should further investigate the causes.

3.2.2 Evaluation of Utilizing Higher Order Harmonics

In addition to the lower hardware requirements, the utilization of the harmonics enables to extend the bandwidth and the resolution of the measurement. According to Eq. (1), all odd-integer harmonic frequencies are excited, when using a square wave with a duty cycle of 0.5 at the gate. To demonstrate both the bandwidth and the resolution extension, the fundamental frequencies of the excitation signal were chosen according to Eq. (2) with a minimum frequency $f_{\text{fund},0}$ of 1 mHz and k reaching from 1 to 12, resulting in a maximum frequency of 4.096 Hz for the square-wave modulation. Figure 5 shows the phase shift exemplary for the 3rd harmonic. The measurement results of the 1st harmonic are displayed via the transparent lines. It can be seen, that the absolute value of the phase shift differs between the evaluation of the 1st and the 3rd harmonic. The change in phase due to degradation, however, shows the same sensitivity in the same frequency range. Thus, the higher order harmonics are suitable to be used for detecting degradation. Since the frequencies of the 3rd harmonic are three times higher than those of the fundamental frequencies, they can be used to extend the bandwidth. In this case the maximum frequency could be extended from 4.096 Hz to 12.288 Hz.

To increase the bandwidth even further, the phase shift between the 15th harmonic components is considered in Fig. 6. It can be seen that there is no loss of sensitivity for the detection of TIM degradation, which is measurable at frequencies between 2 mHz and 2 Hz. For the frequencies above 10 Hz, the deviation between the three measurements in the same aging condition increases, as indicated by the error bars. The extent to which this affects the detection of degradation closer to the chip, that is expected to be seen in this frequency range, needs to be investigated in future work.

As the frequencies of all harmonics differ, the frequency resolution of the measurement can be increased in addition to the bandwidth increase, when combining the measurement results from all harmonics. Figure 7 shows the measurement taking the 1st to 15th harmonics into account. The increased frequency resolution is indicated by the larger number of measurement points. The fundamental frequencies of the rectangular excitation voltage $f_{\text{fund,k}}$ are indicated by the black error bars, while the higher order harmonics of these fundamental frequencies are shown in gray. As already seen above, the phase shift between v_{GS} and v_{DS} differs for the different harmonics, resulting in the ripple that can be seen in Fig. 7. When evaluating the change in phase due to degradation, however, the ripple decreases. Therefore, using the harmonics of a square wave modulation enables to increase the resolution as well as the bandwidth. For applications, where the high frequency resolution is not required, the number of different fundamental frequencies $f_{\text{fund,k}}$ can be reduced to shorten the measurement time.

4 Conclusion

This work demonstrates that degradation within the thermal path of a power electronic module can be detected by applying a square-wave voltage with variable frequency onto the gate of a SiC MOSFET conducting a constant drain current and measuring the phase shift between the fundamental frequencies of v_{GS} and v_{DS}. A rectangular gate voltage

excites not only the fundamental frequency but also the odd-integer harmonics. Taking these harmonics into account allows an increased frequency resolution as well as an increased bandwidth, as fewer frequencies need to be excited separately. Moreover, the low hardware requirements are promising for future integration of the method into the gate driver.

References

[1] F. Blaabjerg, K. Ma, and D. Zhou, "Power electronics and reliability in renewable energy systems," in *2012 IEEE International Symposium on Industrial Electronics*, 2012, pp. 19–30. DOI: 10.1109/ISIE.2012.6237053.

[2] Y. Yang, A. Sangwongwanich, and F. Blaabjerg, "Design for reliability of power electronics for grid-connected photovoltaic systems," *CPSS Transactions on Power Electronics and Applications*, vol. 1, no. 1, pp. 92–103, 2016. DOI: 10.24295/CPSSTPEA.2016.00009.

[3] O. Alatise, A. Deb, E. Bashar, J. Ortiz Gonzalez, S. Jahdi, and W. Issa, "A review of power electronic devices for heavy goods vehicles electrification: Performance and reliability," *Energies*, vol. 16, no. 11, 2023. DOI: 10.3390/en16114380.

[4] C. H. van der Broeck, S. Kalker, T. A. Polom, R. D. Lorenz, and R. W. De Doncker, "In-situ thermal impedance spectroscopy of power electronic modules for localized degradation identification," in *PCIM Europe 2019; International Exhibition and Conference for Power Electronics, Intelligent Motion, Renewable Energy and Energy Management*, 2019, pp. 1–8.

[5] Y. Avenas, L. Dupont, N. Baker, H. Zara, and F. Barruel, "Condition monitoring: A decade of proposed techniques," *IEEE Industrial Electronics Magazine*, vol. 9, no. 4, pp. 22–36, 2015. DOI: 10.1109/MIE.2015.2481564.

[6] S. Yang, D. Xiang, A. Bryant, P. Mawby, L. Ran, and P. Tavner, "Condition monitoring for device reliability in power electronic converters: A review," *IEEE Transactions on Power Electronics*, vol. 25, no. 11, pp. 2734–2752, 2010. DOI: 10.1109/TPEL.2010.2049377.

[7] Z. Hu, B. Hu, L. Ran, *et al.*, "Monitoring power module solder degradation from heat dissipation in two opposite directions," *IEEE Transactions on Power Electronics*, vol. 37, no. 8, pp. 9754–9766, 2022. DOI: 10.1109/TPEL.2022.3157464.

[8] T. F. Baumann, K. Papastergiou, R. Murillo Garcia, and D. Peftitsis, "A condition monitoring method for solder layer degradation of liquid-cooled power semiconductors," in *PCIM Europe 2023; International Exhibition and Conference for Power Electronics, Intelligent Motion, Renewable Energy and Energy Management*, 2023, pp. 1–8. DOI: 10. 30420/566091215.

[9] J. Zhang, X. Du, W. Xiao, and S. Zheng, "Condition monitoring the health status of forced air cooling system using the natural frequency of thermal network," *IEEE Transactions on Power Electronics*, vol. 34, no. 11, pp. 10 408–10 413, 2019. DOI: 10.1109/TPEL.2019.2923801.

[10] T. A. Polom, C. H. van der Broeck, R. W. De Doncker, and R. D. Lorenz, "Exploiting distinct thermal response properties for power semiconductor module health monitoring," *IEEE Journal of Emerging and Selected Topics in Power Electronics*, vol. 9, no. 4, pp. 4865–4878, 2021. DOI: 10.1109/JESTPE.2020.3022775.

[11] C. H. van der Broeck, T. A. Polom, and R. W. De Doncker, "Diagnosing power module degradation with high-resolution, data-driven methods," in *2021 IEEE Energy Conversion Congress and Exposition (ECCE)*, 2021, pp. 3607–3614. DOI: 10.1109/ECCE47101.2021.9594921.

[12] S. Kalker, L. A. Ruppert, C. H. van der Broeck, *et al.*, "Reviewing thermal-monitoring techniques for smart power modules," *IEEE Journal of Emerging and Selected Topics in Power Electronics*, vol. 10, no. 2, pp. 1326–1341, 2022. DOI: 10. 1109/JESTPE.2021.3063305.

[13] J. O. Gonzalez and O. Alatise, "Challenges of junction temperature sensing in sic power mosfets," in *2019 10th International Conference on Power Electronics and ECCE Asia (ICPE 2019 - ECCE Asia)*, 2019, pp. 891–898. DOI: 10.23919/ ICPE2019-ECCEAsia42246.2019.8797281.

[14] I. Austrup, C. H. van der Broeck, T. B. Albert, S. Kalker, and R. W. de Doncker, "Diagnosing degradation in power modules using phase delay changes of electrical response," in *2022 IEEE 7th Southern Power Electronics Conference (SPEC)*, 2022, pp. 1–6. DOI: 10.1109/SPEC55080.2022. 10058410.

[15] K. Hong, X.-Y. Chen, Y. Chen, *et al.*, "Experimental investigations into temperature and current dependent on-state resistance behaviors of 1.2 kv sic mosfets," *IEEE Journal of the Electron Devices Society*, vol. 7, pp. 925–930, 2019. DOI: 10.1109/JEDS.2019.2937837.

[16] H. Sheng, Z. Chen, F. Wang, and A. Millner, "Investigation of 1.2 kv sic mosfet for high frequency high power applications," in *2010 Twenty-Fifth Annual IEEE Applied Power Electronics Conference and Exposition (APEC)*, 2010, pp. 1572–1577. DOI: 10.1109/APEC.2010.5433441.

[17] I. Austrup, T. B. Albert, C. H. van der Broeck, and R. W. De Doncker, "Identifying superimposed degradation effects in power electronic modules," in *PCIM Europe 2023; International Exhibition and Conference for Power Electronics, Intelligent Motion, Renewable Energy and Energy Management*, 2023, pp. 1–7. DOI: 10.30420/566091218.

[18] T. B. Albert, L. Radon, and R. W. De Doncker, "Enabling active thermal control via an adaptive multi-voltage gate driver," in *PCIM Europe 2024; International Exhibition and Conference for Power Electronics, Intelligent Motion, Renewable Energy and Energy Management*, 2024.

[19] S. Zhao, A. Dearien, Y. Wu, *et al.*, "Adaptive multilevel active gate drivers for sic power devices," *IEEE Transactions on Power Electronics*, vol. 35, no. 2, pp. 1882–1898, 2020. DOI: 10.1109/TPEL. 2019.2922112.

[20] L. Wang, B. Vermulst, J. Duarte, and H. Huisman, "Thermal stress reduction of power mosfet in electric drive application with dynamic gate driving strategy," in *2021 IEEE Applied Power Electronics Conference and Exposition (APEC)*, 2021, pp. 720–727. DOI: 10.1109/APEC42165.2021. 9487385.

PCIM Europe 2024, 11–13 June 2024, Nuremberg   DOI: 10.30420/566262421

Statistics-based Lifetime Simulation Methodology for Power Modules incorporating Degradation Models

Karthik Debbadi[1], Martin Votava[1], Yoann Pascal[1], Gopal Mondal[2], Sebastian Nielebock[2], Marco Liserre[1,3]

[1] Fraunhofer Institute for Silicon Technology, Germany
[2] Siemens AG, Germany
[3] Kiel University, Germany

Corresponding author: Karthik Debbadi, karthik.debbadi@isit.fraunhofer.de
Speaker: Karthik Debbadi, karthik.debbadi@isit.fraunhofer.de

Abstract

The standard approach for the lifetime estimation of power modules (PM) is based on lifetime models derived from accelerated lifetime test (ALT) data. In the mission profile-based simulation, the lifetime model is included, but it does not account for the impact of degradation on the PM model. Therefore, in this paper, a new simulation methodology is proposed for mission profile-based lifetime studies incorporating degradation models, and a statistical analysis is applied that leads to $\approx$20% better lifetime estimation for the chosen use case.

1   Introduction

As power modules (PMs) constitute one of the weakest links in the reliability of power electronic systems [1], their lifetime estimation from the design phase of power converters is of importance to ensure reliable operation meeting the requirements in terms of mean time to failure (MTTF), for a given mission profile. This is further emphasized when considering the extension of power electronics to application with low exploitation margins, and, to an even higher extent, to mission critical applications, where failures must be avoided.

On the other hand, thermal cycling and, to a lesser extent, high temperatures, have been identified as the main stressors for PMs, leading to the development and standardization of lifetime characterization procedures [2]. These are based on active and passive thermal cycling tests, the results of which are fitted lifetime models that link the characteristics of the applied thermal cycles and the number of dycles to failure.

The standard approach to mission-profile lifetime estimation consist in 1/partitioning the mission profile into a repeating motif, 2/performing a thermo-electrical simulation of this motif to estimate device temperatures, 3/extracting relevant characteristics

(amplitude, mean value, etc.) of the thermal cycles composing the motif, 4/estimating the degradation of each of those cycles using a lifetime model, and consequently, by linear accumulation, the degradation of the entire motif, and finally 5/estimating how many times the motif can be repeated before the PM fails.

Extensive research has aimed to improve this process, e.g., by considering the effect of device dispersion of characteristics through Monte Carlo simulations [3] or the time resolution of the motif [4]. However, this conventional approach neglects the impact of system degradation on its operation. In fact, aging impacts losses and cooling performance. Therefore, each motif has a different temperature profile and consequently different damage. This study presents a simulation methodology that addresses this limitation and is combined with a Monte Carlo simulation-based statistical analysis to demonstrate the impact of the approach compared with a more straightforward one.

More specifically, the approach includes a feedback loop in the standard simulation so that system aging is integrated into the simulation model to update the characteristics of the devices throughout their lifetime. However, to limit the computational effort and time, considering that the motif is significantly shorter than the degradation timescale a priori, the model is updated only a few times during the lifetime, and the thermoelectrical simulation must be run as many times as possible.

Section 2 provides a review of the degradation mechanisms considered and their modeling based on published work, and an overview of lifetime models. This is used as the basis for the development of our simulation environment, as described in Section 3. Section 4 analyses the simulation results, demonstrating the relevance of the approach. Section 5 concludes the paper.

2 Literature Review

Temperature, and more importantly thermal swings, have been identified as the main stressors for PMs [1]. Repeated cyclic thermal expansion and relaxation conditions lead to fatigue and wear-out of the interconnections inside the PMs. Therefore, dominant degradation mechanisms are studied using accelerated lifetime tests, preferably active power cycling tests or passive thermal cycling tests (easier to implement but less representative). These tests provide an empirical estimation of the number of cycles to failure (N_f) based on the defined failure threshold for specific relevant failure precursors. The dominant degradation mechanisms can be classified on the basis of the power semiconductor technology used into device-level and packaging-level. Device-level degradation is mainly relevant for SiC and GaN technologies, which are less mature than silicon, while packaging-level degradation, which consists of discrete and power module packaging, is relevant for all technologies. This study focuses on Si IGBT-based PMs. Because of the maturity of silicon die manufacturing processes, the analysis is therefore confined to power module packaging technology.

The main degradation mechanisms are described in Section 2.1, followed by a review of lifetime models in Section 2.2. The phenomenological modeling of typical degradation precursors on-state voltage drop V_{ce} and thermal resistance R_{th} are in Section 2.3.

2.1 Degradation Mechanisms inside a Power Module

2.1.1 Bondwire failure

Temperature-induced stress and the resulting crack propagation are the primary mechanisms leading to bond wire liftoff failures [5]. As thermal swings occur during operation, due to significant Coefficient of Thermal Expansion (CTE) mismatch between Silicon (Si) and Aluminium (Al) induces strain and stress at the interface between the bondwire and the chip. The cyclical nature of these thermal swings leads to stress-strain hysteresis energy loss contributing to crack initiation and propagation along the small grain boundaries as shown in [5]. Typical crack initiates at the sides of the bondwire and propagates towards the center along this grain boundary. Once it reaches the center, the bond wire lifts off [6], [7].

During ALTs, the bondwire liftoff can be electrically detected by measuring $V_{ce,sat}$ (here termed as V_{ce}) [8] and the failure criteria usually employs the threshold of +5% increment of V_{ce} [9]. Because other failure mechanisms also influence the measured Vce, offline microscopy techniques are employed to determine the exact cause [7].

2.1.2 Solder Fatigue

Another major failure mechanism for power modules is solder fatigue and cracking between the module substrate and the baseplate and/or the device chip and substrate [5]. This mechanism arises because the silicon die and copper substrate have different CTE, resulting in a shear stress in the solder layer and eventually cracks (void), which reduce the effective area for extinguishing the heat conduction out of the PM. This effective increase in thermal impedance due to void creation increases the average die temperature and lead to hot spots, which eventually accelerates the process of void creation and extension [7]. The detailed mechanism is described in [5].

In ALTs, the steady state thermal impedance equals the thermal resistance R_{th} between the junction to case for the PM as the temperature of the baseplate is maintained constant during the test. An increase of +20% in R_{th} in the cycling tests can be termed as the indicator for failure of power module due to solder fatigue.

2.2 Lifetime Models

Lifetime models provide the correlation between number of cycles to failure (N_f) for a given accelerated lifetime test condition to the real operating conditions.

For power modules, these models can be classified in general into empirical models and Physics-of-failure (PoF) lifetime models [10]. The PoF modeling approach for failures includes analysis and modeling of the actual failure mechanism with the stress and strain evolution over the thermal cycles

[11]. This approach provides better insights in the physical process leading to failure, however, would require detailed knowledge of the geometry, material compositions and bonding technologies which limits its use case as manufacturers would seldom share this information in datasheets. There are multiple PoF lifetime modeling approaches available which are well-detailed in [11]–[14].

On the other hand, the empirical models are statistical in nature with the abstraction of the physical processes for deformation mechanisms inside the power module due to thermal cycles. However, they facilitate there advantage with less complexity under defined ALT conditions close to the mission conditions. They describe the dependency of N_f on ALT conditions such as the ΔT_j, $T_{j,mean}$, cycle frequency, heating and cooling times, load current, inclusive of the power module's properties such as blocking voltage class and geometry of bondwire.

The published empirical models can be classified into variants of Coffin-Manson models [6], Norris-Landzberg models [15], and Bayerer's model [16]. The Norris-Landzberg model is suited to solder joints for printed circuit boards under thermal stress cycling conditions whereas the Bayerer's model is suited to understand the technology trends to improve the lifetime of power modules under power cycling tests. The focus is limited to Coffin-Manson models as discussed below as they fit better for the power cycling results used in this work.

According to the Coffin-Manson model formula in (1), the degradation of the power module assembly is due to the thermal swing (ΔT_j) and empirically relating the ΔT_j with number of cycles to failure N_f from the power cycling (PC) tests with the fitted parameters (α, n). The relationship also highlights for the higher amplitude of thermal swing, the numbers of cycles to failure are reduced with the negative sign for n.

$$N_f = \alpha \cdot (\Delta T_j)^{-n} \tag{1}$$

An improved Coffin-Manson law was proposed as the outcome of LESIT study [6] which also incorporates the Arrhenius law to include the effect of the mean junction temperature $T_{j,m}$. Here the activation energy E_A is a fitted parameter along with α, n based on the PC tests for the PM technology used. It is given by (2).

$$N_f = \alpha \left(\Delta T_j\right)^{-n} \cdot e^{\frac{E_A}{k_b T_{j,m}}} \tag{2}$$

were k_b being the Boltzmann constant.

However, the state-of-art, in terms of packaging technologies, has considerably evolved since the study was conducted in the 1990s and the fitted parameters provided by this study are not relevant anymore. The model's equation is nonetheless still applicable using manufacturer-provided PCT data for use in lifetime evaluations.

For low cyclic stress below the yield strength, caused by low-amplitude thermal cycles, the operation is in the elastic range and minimal influence on degradation is seen. However, when the stress increases above the yield strength, the operation also influences the plastic deformation which is irreversible, leading to degradation. The improved Coffin-Manson model in (2) does not account for this effect and will therefore induce inaccuracies in the lifetime estimation. A general Coffin-Manson model was proposed in 2012 accounting for the elastic strain model [17]. The elastic behavior can be accounted with a simplification such as a portion of thermal swing given by the introduction of ΔT_0 and the Coffin-Manson law can be modified further as General Coffin-Manson law as in (3).

$$N_f = \alpha \cdot (\Delta T_j - \Delta T_0)^{-n} \tag{3}$$

The General Improved Coffin-Manson model was proposed to include the influence of the mean cycle temperature $T_{j,m}$ given by (4) [18].

$$N_f = \alpha \cdot (\Delta T_j - \Delta T_0)^{-n} e^{\frac{E_A}{k_b T_{j,m}}} \tag{4}$$

2.3 Degradation Models

The primary indicators available in literature for bondwire lift-off is on-state voltage (V_{ce}) and for solder layer degradation due to cracks is thermal resistance (R_{th}) as described in 2.1.1, and 2.1.2 respectively, although they are somewhat coupled. These two parameters are modeled for introducing degradation-related effects into the simulation model.

2.3.1 V_{ce} degradation model

The failure criteria are usually set to a 5% increase in the V_{ce} from its initial value. However, after a

1% of increase due to the degradation in the bond-wire, already the process has reached the lift-off phase and 5% is a mere indication of the final failure. For this reason, the threshold can be set to 1.2% as per the fitted data extracted from the published literature [19]. From the ALT data in [19] at $\Delta T_j = 70\,°$, $T_{j,mean} = 90\,°C$, the evolution of normalized on-state voltage v_{ce} w.r.t the damage as percent (directly correlated to number of cycles to failure N_f) can be seen in top plot of Fig. 1. The initial value of V_{ce} before the start (V_{ce0}) of tests is considered for the normalization, and it changes for each device. Since in real mission profiles, there are different thermal cycles, it is assumed that normalized number of cycles to failure N_f for the PCT data as 100% of lifetime consumption based on the linear accumulation of the damage. The degradation model is given by (5) based on [19].

$$V_{ce} = V_{ce0} \cdot v_{ce}$$
$$v_{ce} = a_{v_{ce}} + b_{v_{ce}} \cdot D + \frac{c_{v_{ce}}}{1 - d_{v_{ce}} \cdot D} \tag{5}$$

where $D \in [0, 100]$ % is the accumulated damage, and $(a, b, c, d)_{v_{ce}}$ are a set of parameters characterizing the degradation path, i.e. its evolution with degradation. $a_{v_{ce}}$ and $c_{v_{ce}}$ are mutually dependent on the relationship $a_{v_{ce}} + c_{v_{ce}} = 1$ to ensure $v_{ce}(D = 0) = 1$. The initial value of on-state voltage is given by V_{ce0}.

2.3.2 R_{th} degradation model

The thermal resistance R_{th} is an indicator of the degradation in the solder layers over the lifetime operation. The available data in literature provides the number of cycles to failure N_f w.r.t absolute change in R_{th} [20]. A 20% increase in thermal resistance is usually accepted as failure criteria [9]. The degradation model provided in [20] considers the Coffin-Manson-Arrhenius based model for thermal resistance and linearizes for different phases of operation. In this work, the accelerated lifetime test data of thermal resistance from [20] is normalized to r_{th} which reduces the degrees of freedom and allows a second-order polynomial model as per (6). In Fig. 1, the bottom plot reveals the evolution of normalized thermal resistance r_{th} w.r.t total accumulated damage D. The spread range for the path shows the overall range in which the degradation path model can propagate for statistical simulations.

$$R_{th} = R_{th0} \cdot r_{th}$$
$$r_{th} = a_{r_{th}} \cdot D^2 + b_{r_{th}} \cdot D + 1 \tag{6}$$

where $D \in [0, 100]$ % is the accumulated damage, $(a, b)_{r_{th}}$ is a set of parameters characterizing its degradation path, i.e. its evolution with degradation and R_{th0} is the initial value of thermal resistance.

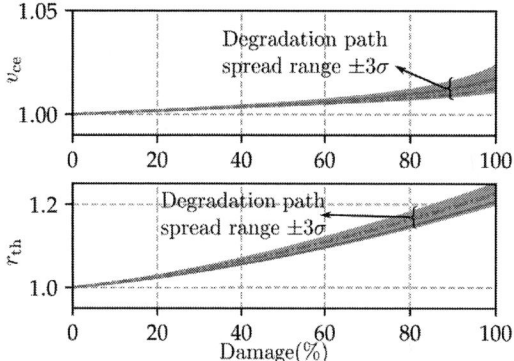

Fig. 1: Degradation evolution for selected few w.r.t Damage

3 Simulation Environment

A simulation of the PM under test for the given mission profile, and assuming that it repeats itself without change over the entire lifetime of the system would hide effects such as the acceleration of aging throughout lifetime due to the increase in V_{ce} and R_{th}. Instead, to take into consideration the effect of aging on the long-term behavior of the PM under test, a looped simulation is conducted. In such simulation, the thermo-electrical parameters (namely R_{th} and V_{ce}) of the PM are updated throughout its lifetime. However, given the slow degradation process and to reduce the computational burden, these parameters are only updated at a very low rate. This time during which the simulation are not updated is hereafter referred to as "iteration period". This iteration period is constituted of a single motive (or "unit mission profile") repeating itself N times, and continuously through lifetime although the simulation environment could consider more complex profiles, e.g., varying from one iteration period to the other.

A high-level flow diagram of the simulation methodolgy is shown in Fig. 2 which can be processed using the following steps.

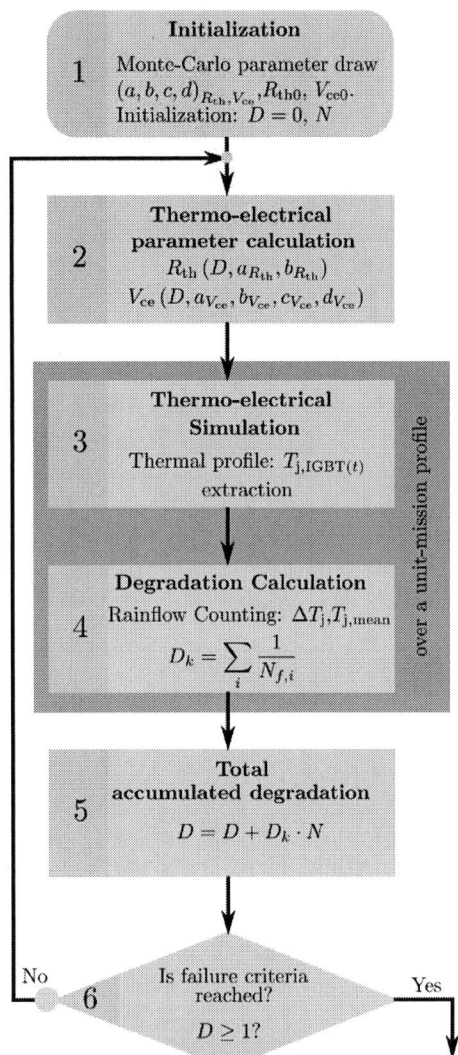

Fig. 2: Flow diagram of the simulation methodology

1. **Monte Carlo drawing:** The draws for the initialization of new power module degradation parameters for the lifetime simulation

2. **Thermo-electrical parameter calculation:** Process the power module parameters for given total damage D to calculate the parameters for unit-mission profile simulation execution.

3. **Thermo-electrical simulation:** Perform a switching model based power electronic application simulation to extract the thermal profile.

4. **Degradation calculation:** Calculate the damage D_k due to one unit-mission profile on the power module, using lifetime model.

5. **Total accumulated degradation:** Extend the previous calculated damage to total accumulated damage with N as acceleration factor.

6. **Looping decision criteria:** Based on the overall damage on the power module, decide on the next action for simulation.

A more detailed description of the process of the simulation methodology is given below.

3.1 Monte Carlo Analysis

Besides consideration of individual case of one PM, it is necessary to consider various variances which are close to reality for the simulation for the prediction of damage and overall lifetime of the PM. In this context, the relevant sources of variances in the real value of V_{ce}, R_{th}, degradation path taken by each specimen as per the degradation model. A normal distribution is used within bounds given by $\pm 3\sigma$ around the mean μ for all the parameters for the generation of the Monte Carlo inputs of each specific simulation.

The Tab. 1 provides the information of the used values for the Monte Carlo analysis.

Param	Mean (μ)	Standard deviation (σ)
$a_{v_{ce}}$	9.97e-01	2.24e-03
$b_{v_{ce}}$	6.70e-05	5.13e-06
$c_{v_{ce}}$	2.15e-03	2.14e-03
$d_{v_{ce}}$	8.32e-03	4.02e-05
V_{ce0}	1.75 V	7.67%
$a_{r_{th}}$	1.32e-05	2.20e-07
$b_{r_{th}}$	1.28e-03	8.53e-05
R_{th0}	0.22 $\frac{K}{W}$	3.33%

Tab. 1: Values for the normal distribution of degradation modelling parameters. $c_{v_{ce}} = 1 - a_{v_{ce}}$.

In Fig. 1, an illustrative range for the degradation path evolution range for normalized values v_{ce}, r_{th} are shown for overall damage range applied in the Monte Carlo simulations.

Apart from these parameters, the initial value of V_{ce} and R_{th} are based on Monte Carlo implementation as per the datasheet limits.

The central value of the normal distributions (with draws outside the $\pm 3\sigma$ being discarded) were chosen to match the experimental results from [19], [20], whilst the standard deviations were chosen to match datasheet specification or as reasonable expectations given data from the literature in the case of the degradation path.

As illustrated in Step 1 of Fig. 2, the Monte Carlo parameters are initialized during the start of each lifetime simulation along with reset of damage D. The acceleration factor N is predefined for all the draws.

5 and 6 contain too many degrees of freedom to ensure that one failure criteria will be reached for $D = 1$ and that none will be reached before that. To avoid this, an additional constraint is added. After drawing the $(a, b, c, d)_{v_{ce}, r_{th}}$ parameters following Gaussian distributions, the first failure criteria to be reach (either V_{ce} or R_{th}) is identified, and the associated D-value, hereafter named D_0 is extracted. Then, the parameters of the degradation path are modified as follows:

- if the failure criteria associated with R_{th} was reached first, the associated degradation model is modified as follows: $a_{r_{th}} := \frac{a_{r_{th}}}{D_0^2}, b_{r_{th}} := \frac{b_{r_{th}}}{D_0}$ whereas $(a, b, c, d)_{v_{ce}}$ are not changed.

- if the failure criteria associated with V_{ce} was reached first, the associated degradation model is modified as follows: $b_{v_{ce}} := \frac{b_{v_{ce}}}{D_0}, d_{v_{ce}} := \frac{d_{v_{ce}}}{D_0}$ whereas $(a, b)_{r_{th}}$ and $(a, c)_{v_{ce}}$ are not changed.

This ensures that the first failure criteria is reached when the total degradation reaches unity. Data given in Tab. 1 describe the parameter distributions before this additional normalization.

In Step 2 of Fig. 2, the degradation models for the V_{ce} and R_{th} as described in degradation modelling section respectively are implemented in the post-processing script used for the simulation.

3.2 Unit-mission profile simulation

Step 3 & 4 in Fig. 2 constitute the unit-mission profile simulation with thermoelectrical modelling of the power converter and calculation of the degradation based on the used lifetime model and counting algorithm. The use case for this simulation was a standard two-level 3-phase grid-tied inverter with Si-IGBT power module and a symmetrical load is assumed. The mission profile chosen was simplified for the purpose of the evaluating the influence of degradation. A standard thermal model provided by manufacturer of the power module was used for the simulation, in most cases a foster model (provides physical abstraction and simplifies implementation) is defined and provided in the datasheet. To keep the approach simple and standard, the thermal non-linearities due to internal and external couplings in the system for the power module have been ignored. An extension of the approach could be a more complex thermal model based on the real system constraints which improve primarily the $T_{j, IGBT}$ estimation accuracy for lifetime analysis.

In degradation calculation, the rainflow counting method is applied in comparison to other counting methods due to its wide acceptance. Here the rainflow counting calculates the number of cycles considering both the ΔT_j and $T_{j, mean}$ to use with general improved Coffin-Manson lifetime model for degradation calculation ($d_i = 1/N_{f,i}$) for each combination of $[\Delta T_j, T_{j, mean}]$. Accepting linear damage accumulation, Miner's rule is applied to calculate D_k, the total damage induced during the unit-mission profile, as per step 4. Independent of the lifetime model used, thermal cycles at the fundamental and switching frequencies of the power converter lead to negligible damage.

3.3 Total damage estimation in simulation

In the post-processing phase of each unit-mission profile simulation, step 5 in Fig. 2 allows for speeding up the simulation with a factor N. The underlying assumption for performing the step 5 is that non-linear damage can be split into smaller entities where the damage can be assumed linear and constant. This assumption is valid as long as the error induced in the degradation path is very low while selecting the value of N.

In step 6, the failure conditions set by the threshold of V_{ce} and R_{th} are checked, if one of these threshold is exceeded, it is assumed as the end of life for the power module. This can also be validated with D reaching the value of 1. If outcome of step 6 is no, the simulation is looped back to step 2 with the calculation next value of V_{ce} and R_{th}.

4 Results

This work focuses on methodology and analysis of the impact of the integration of the degradation model into life analysis, compared to a conventional approach, rather than on actual life estimation. Consequently, in this section, the results of the analysis are summarized with the normalization of the different parameters (e.g., lifetime) instead of absolute values.

When quantifying only the influence of degradation models on the lifetime simulations, it can be inferred that the influence of increase in R_{th} due to degradation plays a significant role in the non-linear evolution of damage with lifetime seen in Fig. 3 in comparison to V_{ce} increase. The degradation path taken by V_{ce} over the lifetime shows a comparatively low increase in amplitude (in the order of 1% vs. up to 20% for R_{th}), and consequently induces a low change in temperature profile. Due to the nature of the degradation paths of these parameters, there is a strong correlation through power losses and thermal model where both the degradation mechanisms interact, and it is challenging to decouple the influence of one mechanism on the other. Based on the above degradation modelling, a statistical analysis is performed to show the influence on the overall lifetime estimation under mission profile based methodology.

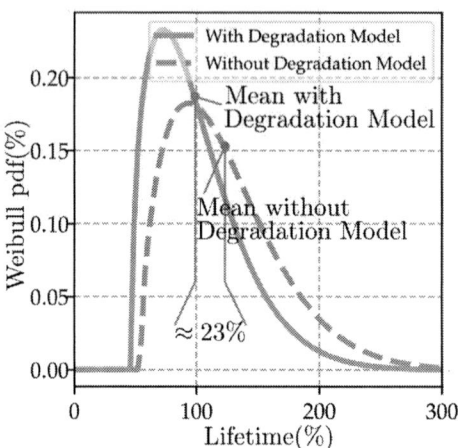

Fig. 4: Comparison of Weibull lifetime with and without degradation model

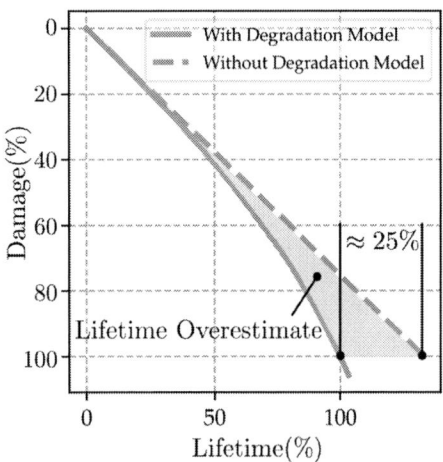

Fig. 3: Comparison of Damage with and without degradation model, for a representative simulated sample.

As highlighted earlier, the non-linear nature of the evolution of the damage over the lifetime with the incorporation of the degradation model into simulation workflow can be seen in Fig. 3. The standard mission profile-based lifetime estimation methodology might yield a higher lifetime, an example comparison in Fig. 3 shows 25% overestimate. High resolution temperature binning was used during rainflow counting to ensure that the discretization of $\Delta T_j, T_{j,mean}$ would introduce negligible error in the lifetime estimation.

The nature of the used approach for simulation considering the switching model of the power converter led to longer simulation times for each lifetime simulation under one Monte Carlo draw which limited the overall simulations to just above 100. This limitation of the simulation can be overcome by using an av-

erage model of the power converter which can lead to good gains in reduction of the simulation times to support around 10000 lifetime simulations as a representative for Monte Carlo simulations. This might lead to higher accuracy in Weibull lifetime estimation in comparison to the one seen in Fig. 4, wwhich shows that incorporating degradation modeling will lead to a lower estimate in comparison to the standard approach which overestimates by around 23% for this case.

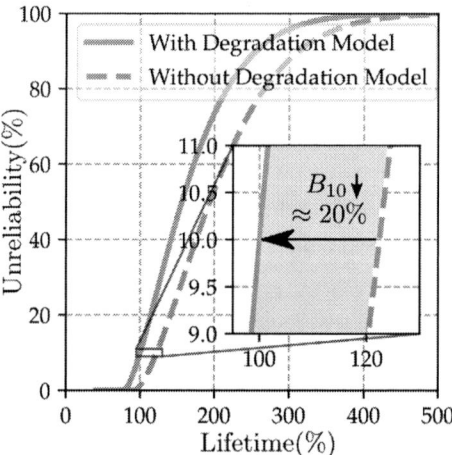

Fig. 5: Comparison of B_{10} lifetime from Unreliability function with and without degradation model

With the impressions from the available state-of-art [21], a standard approach followed for mission profile-based simulations to provide a B10 lifetime (i.e., the time after which 10% of devices have, statistically, reached a failure threshold) extracted from the unreliability function from the available data.

This function can be evaluated with the cumulative probability distribution evaluated from the Weibull lifetime data of Fig. 4. A reduction of 20% in the overall B_{10} lifetime is observed for chosen use case with the incorporation of degradation modeling into the standard mission profile based lifetime simulation studies.

Although the paper focuses on the impact of R_{th} and V_{ce} on the lifetime estimation, the approach holds high potential for generalization, e.g., considering additional parameters such as body diode voltage degradation, and devices or sub-systems, e.g., DC-link capacitors. On the other hand, other degradation parameters, such as the gate leakage current, which might be a good indicator of degradation, have a negligible impact on the temperature profile and therefore do not need to be included in the model update step of the iteration loop. Furthermore, for improved accuracy and simulation speed, adaptive N-values based on D_k, could be considered apart from the usage of average models.

5 Conclusion

Mission-profile-based lifetime estimations relying on thermoelectrical simulations, rainflow counting, and the combined use of a lifetime model and Miner's rule typically assume that the thermoelectrical model of the system under study is constant throughout its lifetime. However, system degradation leads to changes in various parameters such as the thermal resistance R_{th} and the on-state voltage drop V_{ce}, which in turn has an impact on the thermal profile and affects aging.

This paper describes a mission profile-oriented, simulation-based approach to estimate the lifetime considering the effect of aging on the thermoelectrical parameters. For the considered use case with publicly available data, the predicted lifetime with the incorporation of degradation is reduced by a minimum of 20% compared to the standard approach without degradation influence for all single PM, Weibull mean, and B_{10}.

Although the focus is on the impact of two specific parameters (namely, R_{th} and V_{ce}) on the lifetime estimation of power modules, the approach can be generalized to integrate other parameters such as diode forward characteristics, and DC-link capacitors.

References

[1] J. Falck, C. Felgemacher, A. Rojko, M. Liserre, and P. Zacharias, "Reliability of Power Electronic Systems: An Industry Perspective," *IEEE Industrial Electronics Magazine*, vol. 12, no. 2, pp. 24–35, Jun. 2018. DOI: 10.1109/MIE.2018.2825481.

[2] P. Jacob, M. Held, P. Scacco, and W. Wu, *Reliability Testing and Analysis of IGBT Power Semiconductor Modules*. May 1995, pp. 4/5. DOI: 10.1049/ic:19950531.

[3] Y. Zhang, "Mission Profile based System-Level Lifetime Prediction of Modular Multilevel Converters," Ph.D. dissertation, Aalborg University, Denmark, Aalborg, 2019.

[4] I. Vernica, H. Wang, and F. Blaabjerg, "Impact of Long-Term Mission Profile Sampling Rate on the Reliability Evaluation of Power Electronics in Photovoltaic Applications," in *2018 IEEE Energy Conversion Congress and Exposition (ECCE)*, Portland, OR: IEEE, Sep. 2018, pp. 4078–4085. DOI: 10.1109/ECCE.2018.8558092.

[5] S. Yang, D. Xiang, A. Bryant, P. Mawby, L. Ran, and P. Tavner, "Condition Monitoring for Device Reliability in Power Electronic Converters: A Review," *IEEE Transactions on Power Electronics*, vol. 25, no. 11, pp. 2734–2752, Nov. 2010. DOI: 10.1109/TPEL.2010.2049377.

[6] M. Held, P. Jacob, G. Nicoletti, and P. Scacco, "Fast Power Cycling Test for IGBT Modules in Traction Application,"

[7] M. Ciappa and W. Fichtner, "Lifetime prediction of IGBT modules for traction applications," in *2000 IEEE International Reliability Physics Symposium Proceedings. 38th Annual (Cat. No.00CH37059)*, San Jose, CA, USA: IEEE, 2000, pp. 210–216. DOI: 10.1109/RELPHY.2000.843917.

[8] J. Lehmann, M. Netzel, R. Herzer, and S. Pawel, "Method for electrical detection of bond wire lift-off for power semiconductors," in *ISPSD '03. 2003 IEEE 15th International Symposium on Power Semiconductor Devices and ICs, 2003. Proceedings.*, Cambridge, UK: IEEE, 2003, pp. 333–336. DOI: 10.1109/ISPSD.2003.1225295.

[9] G. Coquery and R. Lallemand, "Failure criteria for long term Accelerated Power Cycling Test linked to electrical turn off SOA on IGBT module. A 4000 hours test on 1200A–3300V module with AlSiC base plate," *Microelectronics Reliability*, Reliability of Electron Devices, Failure Physics and Analysis, vol. 40, no. 8, pp. 1665–1670, Aug. 2000. DOI: 10.1016/S0026-2714(00)00191-8.

[10] H. Wang, M. Liserre, F. Blaabjerg, P. De Place Rimmen, J. B. Jacobsen, *et al.*, "Transitioning to Physics-of-Failure as a Reliability Driver in Power Electronics," *IEEE Journal of Emerging and Selected Topics in Power Electronics*, vol. 2, no. 1, pp. 97–114, Mar. 2014. DOI: 10.1109/JESTPE.2013.2290282.

[11] I. F. Kovacevic-Badstuebner, J. W. Kolar, and U. Schilling, "Modelling for the lifetime prediction of power semiconductor modules," in *Reliability of Power Electronic Converter Systems*, H. S.-h. Chung, H. Wang, F. Blaabjerg, and M. Pecht, Eds., Institution of Engineering and Technology, Dec. 2015, pp. 103–140. DOI: 10.1049/PBPO080E_ch5.

[12] O. Schilling, M. Schäfer, K. Mainka, M. Thoben, and F. Sauerland, "Power cycling testing and FE modelling focussed on Al wire bond fatigue in high power IGBT modules," *Microelectronics Reliability*, vol. 52, no. 9-10, pp. 2347–2352, Sep. 2012. DOI: 10.1016/j.microrel.2012.06.095.

[13] P. Steinhorst, T. Poller, and J. Lutz, "Approach of a physically based lifetime model for solder layers in power modules," *Microelectronics Reliability*, vol. 53, no. 9-11, pp. 1199–1202, Sep. 2013. DOI: 10.1016/j.microrel.2013.07.094.

[14] S. Deplanque, W. Nuchter, B. Wunderle, R. Schacht, and B. Michel, "Lifetime Prediction of SnPb and SnAgCu Solder Joints of Chips on Copper Substrate Based on Crack Propagation FE-Analysis," in *7th. Int. Conf. on Thermal, Mechanical and Multiphysics Simulation and Experiments in Micro-Electronics and Micro-Systems*, Como, Italy: IEEE, 2006, pp. 1–8. DOI: 10.1109/ESIME.2006.1643976.

[15] A. Hanif, Y. Yu, D. DeVoto, and F. Khan, "A Comprehensive Review Toward the State-of-the-Art in Failure and Lifetime Predictions of Power Electronic Devices," *IEEE Transactions on Power Electronics*, vol. 34, no. 5, pp. 4729–4746, May 2019. DOI: 10.1109/TPEL.2018.2860587.

[16] D. R. Bayerer, T. Herrmann, D. J. Lutz, and M. Feller, "Model for Power Cycling lifetime of IGBT Modules – various factors influencing lifetime,"

[17] H. Wang, K. Ma, and F. Blaabjerg, "Design for reliability of power electronic systems," in *IECON 2012 - 38th Annual Conference on IEEE Industrial Electronics Society*, Montreal, QC, Canada: IEEE, Oct. 2012, pp. 33–44. DOI: 10.1109/IECON.2012.6388833.

[18] Y. Zhang, H. Wang, Z. Wang, Y. Yang, and F. Blaabjerg, "Impact of lifetime model selections on the reliability prediction of IGBT modules in modular multilevel converters," in *2017 IEEE Energy Conversion Congress and Exposition (ECCE)*, Cincinnati, OH: IEEE, Oct. 2017, pp. 4202–4207. DOI: 10.1109/ECCE.2017.8096728.

[19] N. Degrenne and S. Mollov, "Experimentally-Validated Models of On-State Voltage for Remaining Useful Life Estimation and Design for Reliability of Power Modules," 2018.

[20] V. Samavatian, H. Iman-Eini, and Y. Avenas, "Reliability Assessment of Multistate Degraded Systems: An Application to Power Electronic Systems," *IEEE Transactions on Power Electronics*, vol. 35, no. 4, pp. 4024–4032, Apr. 2020. DOI: 10.1109/TPEL.2019.2933063.

[21] I. Vernica, H. Wang, and F. Blaabjerg, "Design for reliability and robustness tool platform for power electronic systems — Study case on motor drive applications," in *2018 IEEE Applied Power Electronics Conference and Exposition (APEC)*, San Antonio, TX, USA: IEEE, Mar. 2018, pp. 1799–1806. DOI: 10.1109/APEC.2018.8341261.

PCIM Europe 2024, 11– 13 June 2024, Nuremberg   DOI: 10.30420/566262422

Power Cycling Results for Reliability Studies of SiC-Inverters

Robert Keilmann [1,2], Florian Lippold [2], Regine Mallwitz [1,2]

[1] Cluster of Excellence SE²A – Sustainable and Energy-Efficient Aviation,
Technische Universität Braunschweig, Germany
[2] Institute for Electrical Machines, Traction and Drives (IMAB),
Technische Universität Braunschweig, Germany

Corresponding author: Robert Keilmann, r.keilmann@tu-braunschweig.de
Speaker: Robert Keilmann, r.keilmann@tu-braunschweig.de

Abstract

Amidst the electrification of aviation, this paper discusses the reliability of silicon carbide (SiC) metal–oxide–semiconductor field-effect transistors (MOSFETs). With the help of the widely used power cycling test (PCT), the bond wire connections in particular are aged cyclically in this study. The AQG 324 used in the automotive industry serves as a reference for defining test conditions and threshold values. The paper illustrates step by step how the parameters of the underlying LESIT lifetime model can be found from the PCT data. The lifetime parameters found are classified and checked for plausibility with the help of existing literature.

1 Introduction

Numerous studies show the need to electrify aircraft in order to reduce the climate impact of air traffic. It is conceivable that after a period of hybridization, the complete electrification of the powertrain of short-range aircraft will become technically feasible [1]–[3]. While numerous studies are examining the issue of efficiency and power density [4]–[7], the question of the reliability of electric drive systems is becoming ever louder. Publications such as [8] call for up-to-date data on the lifetime of silicon carbide (SiC) semiconductors. Wide bandgap (WBG) semiconductors are expected to play a decisive role in electrified aviation. On the one hand, this is due to the higher efficiency and higher power density of power electronics with built-in WBGs compared to their silicon (Si) counterparts. On the other hand, SiC, for example, is reported to be more robust against cosmic radiation than Si [9], which will become a relevant issue as flight altitude and on-board power supply voltage increase.

This publication is structured as follows: Section 2 discusses the main points concerning the reliability of semiconductors, and SiC in particular, in the aerospace environment. The publication focuses on the cyclic aging of bond wire connections. To this end, important model assumptions on which the test is based are listed in section 3. Section 4 describes the PCT with the set parameters, discusses the results, and shows in detail how to fit the parameters of the underlying lifetime model. Finally, there is a discussion of the results as well as a summary and an outlook for future investigations.

2 Main Aspects of SiC Reliability in Aviation

Four main factors affect cosmic radiation exposure: altitude, latitude, solar activity, and solar proton events. Altitude reduces atmospheric shielding, increasing radiation; at flying altitudes, it can be up to 100 times greater than at sea level. Latitude influences exposure due to the Earth's magnetic field, with stronger shielding at the equator and weaker at the poles, resulting in double the radiation at the poles compared to the equator. Solar activity, cycling every 11 years, can intensify cosmic radiation during periods of low solar activity. Solar proton events, marked by large ejections of charged particles from the sun, can cause abrupt spikes in cosmic radiation levels [10].

Single Event Burnout (SEB) and Single Event Gate Rupture (SEGR) occur in power devices when in the off-state with high voltage across drain and source, due to charge deposits surpassing dielectric strength. Protection against SEB and SEGR relies on derating, as no effective shielding exists against cosmic rays. SEB leads to catastrophic failures, while SEGR results in gate rupture, both posing significant challenges in power device reliability. Cosmic radiation-induced failures, including SEB and SEGR, are unpredictable and depend on factors like altitude and operating voltage [11].

However, the long-term reliability assessment of power electronics is primarily based on the influence of degradation. More and more studies on degradation under the influence of humidity can be found in the literature. This is particularly relevant as the power modules may be exposed to condensation and high humidity levels in the areas of use during their service life. If hermetic sealing of the housing of the power electronics cannot be ensured and boiling out of the water cannot take place, this circumstance remains a further field for future investigations. The literature reports that humidity reduces the lifetime [12] and increases the leakage current [13].

Furthermore, SiC exhibits a characteristic degradation mechanism known as threshold voltage (V_{th}) shift at chip level which stems from the SiC-characteristic Gate Oxide Degradation [14]–[16]. The V_{th} shift in SiC power MOSFETs presents challenges, impacting die degradation and introducing new failure mechanisms during power cycling tests [17]. It is reported that this shift could affect die resistance, potentially leading to errors in online junction temperature (T_j) estimation methods, highlighting the need for innovative approaches such as independent T_j estimation using optical fibers [17]. Moreover, it induces V_{th} hysteresis, affecting switching characteristics and potentially increasing the risk of device failure, especially at higher switching speeds [18]. Nevertheless, current research suggests that the impact might be minimized with a positive voltage applied at the gate during turn-on and a negative voltage during turn-off, which are assumed to mitigate these effects [19].

Despite WBG semiconductors showcasing resilience at temperatures exceeding 500 °C, their vulnerability lies in interconnection and packaging technologies. Interconnection and packaging technology presents numerous challenges and potential failure modes, including mismatch of the coefficient of thermal expansion (CTE) between distinct layers of the power module, solder fatigue, contact degradation, polymer decomposition, diffusion phenomena, intermetallic formation, and creep. Failures often stem from thermo-mechanical stresses, with

bond wire fatigue and delamination as crucial failure modes. Bond wire fatigue results from cyclic loading during thermal cycling, leading to mechanical degradation in the form of bond wire lift-off or heel cracking, and potential open circuit failures. Delamination involves the separation of layers within the power module stack up, influenced by a combination of factors like CTE mismatch, mechanical stress, and solder degradation. [14], [15]

3 Definition of Model Boundaries and Model Assumptions

An exemplary mission profile for a short-range aircraft is shown in Fig. 1. The underlying model used in [20] to generate it is used to analyze novel airframe technologies, a novel type of electric machines, and high energy density batteries. The aborting of the landing maneuver and the approaching of an alternative airport must be taken into account in particular when carrying out investigations to estimate the range of the aircraft. This is the reason for the steep increase in required power at around minute 130 in Fig. 1.

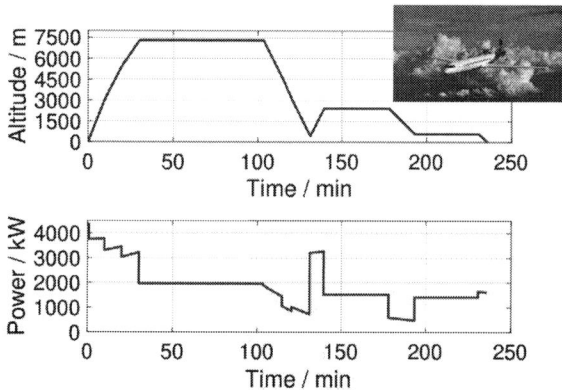

Fig. 1: Mission profile for a short range aircraft, adapted from [20].

Especially in the non-distributed propulsor approach pursued in [20], the individual power demands per inverter are very high. There are no all-SiC modules on the market that could be used for this application. For a distributed propulsor approach, where multiple propulsors are placed along the wing, the individual power requirement comes into a manageable power range. A scaled demonstrator for such a distributed propulsor approach is shown in [2].

To supply the aircraft with the necessary electrical power for propulsion, either a set of scalable

SiC modules [21] is required or the approach as in [2] is developed with the scaled demonstrator. Hence, the study presented here also follows a scaled approach. Therefore, commercially available SiC-multichip modules in a six-pack configuration with 1200 V SiC-chips are being used for the PCT. As a simplified model assumption, it is assumed that the components offer a similar joining and packaging technology as their high-power counterparts.

4 Power Cycling Test and Data Analysis

This section first documents the basic PCT parameters, the assumed lifetime model and the method for determining the virtual junction temperature. Subsection 4.2 presents the PCT results, which are then used for lifetime model parameter fitting in subsection 4.3. The obtained parameters are then discussed and classified in subsection 4.4 with the help of existing literature.

4.1 Testplan and Power Cycling Test Parameters

The test plan shown in Tab.1 was developed under consideration of the ECPE Guideline AQG 324 for the qualification of power modules for use in power electronics in motor vehicles. The guideline is an important tool, as there are still no standardized specifications for lifetime tests for power electronics in the aviation sector.

Tab. 1: Test Plan for the PCT

	$t_{on} = t_{off}$	$T_{j,max}$	$T_{j,min}$	ΔT_j	T_{jm}
M1	4 s	150 °C	50 °C	100 K	100 °C
M2	1 s	130 °C	70 °C	60 K	100 °C
M3	4 s	120 °C	20 °C	100 K	70 °C

The measurement series M1, M2, and M3 are planned in a way that the temperature swings lay in the high cycle fatigue area, characterized by elastic deformation. At the same time, M1 and M2 are designed around the coolant temperature assumed in [22] of around $60\,°C$. The maximum junction temperature $T_{j,max}$ is set in a way that it does not exceed the maximum junction temperature of the devices specified in the datasheet.

When assuming the the LESIT model [23] shown in (1), three distinct measurement series M1, M2, and M3 are necessary for parameter fitting. The LESIT model explicates the number of cycles until failure

N_f, encompassing a composite of a Coffin-Manson term, describing failure in fatigue arising from thermal strain at interfaces or joints, and an Arrhenius factor, illustrating an accelerated failure mechanism correlated with an elevated mean temperature.

$$N_f = A \cdot (\Delta T_j)^\alpha e^{\left(\frac{E_a}{k_B \cdot T_{jm}}\right)} \quad (1)$$

In (1), ΔT_j denotes the junction temperature swing:

$$\Delta T_j = T_{j,max} - T_{j,min} \quad (2)$$

T_{jm} represents the mean junction temperature defined as

$$T_{jm} = \frac{T_{j,max} + T_{j,min}}{2} \quad (3)$$

whereas k_B is the Boltzmann constant. The parameters A, α, and E_a are intrinsic to the model and are determined through the fitting of (1) to the experimental PCT data. E_a specifically signifies the activation energy characterizing the deformation process of the dominating failure mechanism.

There are several other lifetime models [24], a very prominent model is the CIPS08 [25] which employs device parameters that remain unknown without notice from the manufacturer. Despite its initial application to silicon devices with aluminum (Al) bond wires, research demonstrates that the LESIT model extends its versatility to encompass a broad spectrum of semiconductors, including gallium nitride (GaN) cascodes [26], making it a fitting choice for the focus of this study.

These preliminary considerations justify the choice of parameters in Tab. 1. M1 and M2 are chosen so that a variation of ΔT_j takes place, while the mean junction temperature T_{jm} is the same for M1 and M2. M3 represents a modified M1. Here, the same temperature swing $\Delta T_j = 100\,K$ is selected, but the mean junction temperature T_{jm} is lowered by $30\,K$ to determine the activation energy E_a. The switch-on and switch-off times t_{on} and t_{off} are selected in such a way that the bond wires in particular are cyclically loaded.

The junction temperature is monitored via the so-called VSD method [27], [28]. The underlying principle is that the voltage drop across the intrinsic body diode is measured at a fixed measuring current and compared against a calibration curve. The underlying mechanism is illustrated in Fig. 2 and described in more detail in [27], [28]. Increasing the measuring current I_{meas} leads to a reduction in measurement noise. However, selecting a current

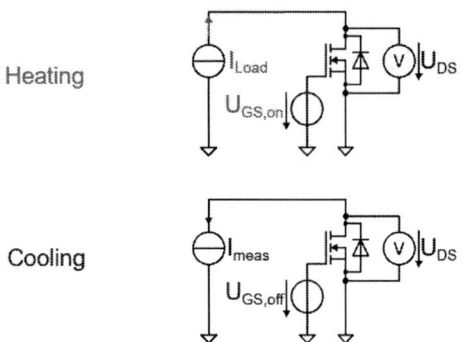

Fig. 2: Definition of electrical parameters during PCT.

that is too high is associated with non-negligible self-heating and therefore incorrect measurement. For the test series, an I_{meas} of $20\,\mathrm{mA}$ is selected after a preliminary measurement series with a very low noise level. I_{meas} lies between the chosen measuring currents in [27] and [28]. The load current I_{Load} is set so that the desired temperature swing can be set with the individual component parameters but remains constant over the entire measurement. The gate voltages $U_{\mathrm{GS,on}}$ and $U_{\mathrm{GS,off}}$ are set according to the datasheet recommendation as shown in Tab. 2.

Tab. 2: Device Parameter Setting for the PCT

Parameter	Value
I_{Load}	set individually
$U_{\mathrm{GS,on}}$	$15\,\mathrm{V}$
$U_{\mathrm{GS,off}}$	$-5\,\mathrm{V}$
I_{meas}	$20\,\mathrm{mA}$

Contrary to the specifications in AQG 324, the thermal resistance is not measured over the test period using an additional copper block with an integrated measuring point. In the present case, the cooling curve of the components determined every 1000 cycles using the VSD method is compared with the very homogeneously temperature-distributed cooling and heating plate temperature of the test bench.

Unlike prescribed in AQG 324, not every topological switch of a module is subjected to the same test plan. M1 and M2 are distributed to different topological switches of the same module. The switch positions are systematically swapped between the modules. M3 is run on modules that have not seen any other cyclic load.

4.2 Power Cycling Test Results

The end-of-life (EOL) criteria from AQG 324 are used for all measurement series. This means that a component is considered to have failed if the drain-source voltage U_{DS} has increased by $5\,\%$ compared to its original value or if the thermal resistance R_{th} has increased by $20\,\%$ compared to its original value.

As the U_{DS} measurements are subject to clearly recognizable cycle-to-cycle fluctuations, a median filter is applied to them. The R_{th} measurement, which takes place every 1000 cycles, would otherwise result in highly visible dips in the measurement series. The R_{th} measurement results are not filtered.

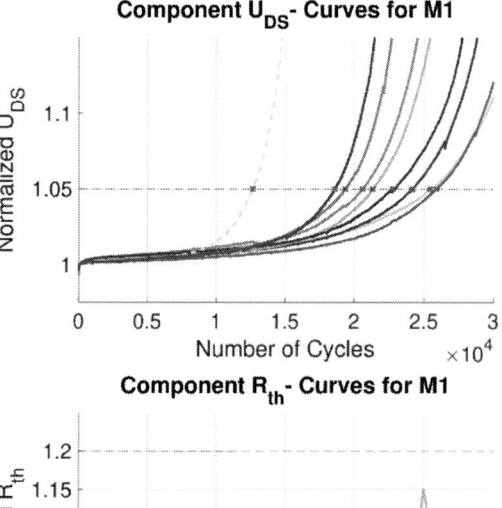

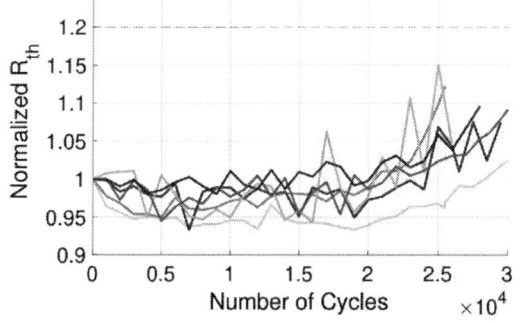

Fig. 3: Course of normalized U_{DS} and R_{th} for M1.

The PCT results for M1 are shown in Fig. 3. The EOL criteria are highlighted with a dashed gray line for the normalized U_{DS} and the normalized R_{th}. The R_{th} measurement data indicate that the devices fail due to bond wire fatigue as the R_{th} of no component reaches the threshold of 1.2 within the measurement. One early failure is recognizable and highlighted with a dashed line within the U_{DS} measurement in Fig. 3.

PCIM Europe 2024, 11– 13 June 2024, Nuremberg   DOI: 10.30420/566262422

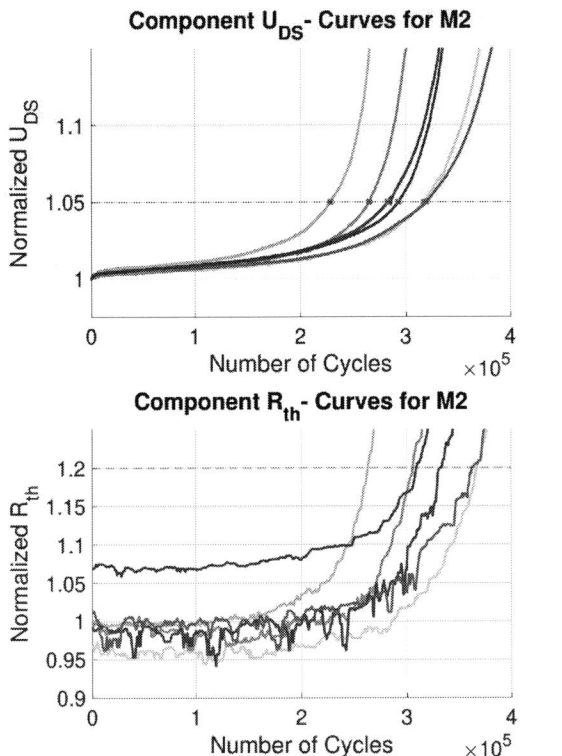

Fig. 4: Course of normalized U_{DS} and R_{th} for M2.

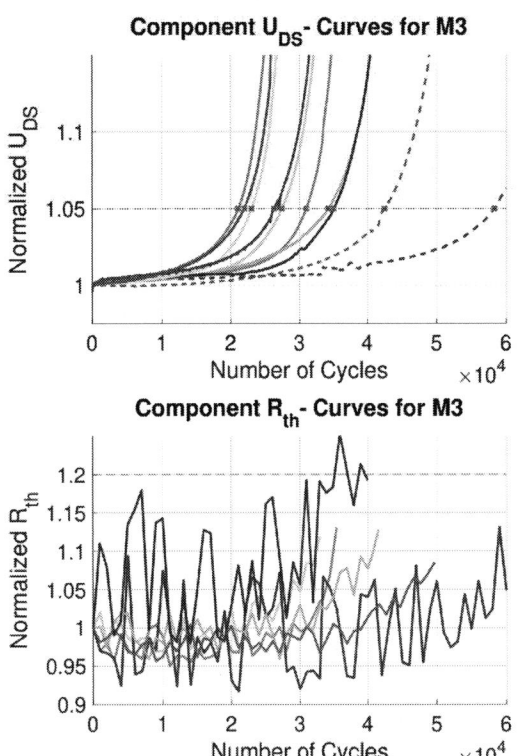

Fig. 5: Course of normalized U_{DS} and R_{th} for M3.

The measured values for M2 are shown in Fig. 4. No early failures or long runners can be identified. The number of cycles until failure is longer for this series of measurements with reduced temperature swing ΔT_j than for M1. As a result, the R_{th} measurements also appear somewhat smoother. In contrast to M1, the steep increase in the normalized R_{th} can be observed for all components for a high number of cycles. This is primarily due to the long measurement time; each of the components examined fails according to the 5 % U_{DS} EOL criterion. Based on the color coding in Fig. 4, it is clear that the increase in R_{th} is only significant for the respective component after the U_{DS} threshold has been exceeded. This observation indicates that all components also fail due to bond wire fatigue.

The results from measurement series M3, illustrated in Fig. 5, show two long runners - both illustrated with dashed lines in the U_{DS} curves. The last component to fail in M3 differs greatly from the other components in terms of load cycles. The second last failed component stands out from the remaining ones, as the time of failure is very late concerning the overall population. Another compo-

nent is not of interest because of its failure time, but mainly because of the measured R_{th}. This component seems to exceed the EOL criterion for U_{DS} and R_{th} at almost the same time. A closer analysis shows that EOL criterion of 5 % U_{DS} increase is met just slightly before the 20 % R_{th} increase. This means that the decisive failure mode cannot be identified beyond doubt. The high fluctuation of the R_{th} measurement must be emphasized here, which makes evaluation difficult. This high fluctuation is due to noise in the temperature measurement. A larger measuring current in the VSD method could help here. However, the test plan described in subsection 4.1 and the set parameters are kept constant for the entire measurement campaign so as not to jeopardize the comparability of the results.

4.3 Lifetime Model Parameter Fitting

Comparing M1 with M2 using the test plan outlined in Tab. 1 reveals that among the parameters considered in the LESIT model, only the temperature swing ΔT_j varies. Assuming that the activation energy remains constant under the same failure mechanism (as discussed in subsection 4.2), the

2976

factors A and the exponential term can be merged. Thus, taking the logarithm of (1) yields the expression:

$$\ln N_f = \ln A + \frac{E_a}{k_B \cdot T_{jm}} + \alpha \cdot \ln \Delta T_j \quad (4)$$

which becomes a linear equation in ln-plane. This gets more visible when substituting:

$$y = a + b \cdot x \quad (5)$$

Here, y represents $\ln N_f$, $\ln A + \frac{E_a}{k_B \cdot T_{jm}}$ is substituted into a, b equals α, and the variable x represents $\ln \Delta T_j$.

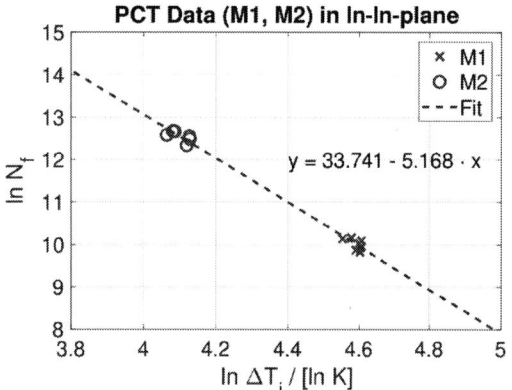

Fig. 6: Fitting for the exponent α.

Fitting the data points of the two measurement series M1 and M2 to the linear function (5), illustrated in Fig. 6, results in the required exponent α for the LESIT model from (1) via the slope of the dashed line and back-substitution. The early failure, marked in Fig. 3 with a dashed line, was not considered for this analysis.

The technology factor A and the activation energy E_a are still unknown. To determine E_a, a substitution is made in (1), in which A, ΔT_j and α are assumed to be constant. The constancy of ΔT_j is given between M1 and M3 in the test plan (see Tab. 1). For this,

$$K = A \cdot (\Delta T_j)^\alpha \quad (6)$$

is set so that

$$N_f = K \cdot e^{\left(\frac{E_a}{k_B \cdot T_{jm}}\right)} \quad (7)$$

results by inserting (6) into (1). Equation (7) can also be converted into a linear function. To do

so, (7) must be logarithmized so that the following expression results:

$$\ln N_f = \ln K + \frac{E_a}{k_B} \cdot \frac{1}{T_{jm}} \quad (8)$$

Equation (8) is intentionally put into the given form to make the substitution with the linear equation in (5) clearer. Here, y represents again $\ln N_f$, $\ln K$ is substituted into a, b equals $\frac{E_a}{k_B}$, and the variable x represents $\frac{1}{T_{jm}}$.

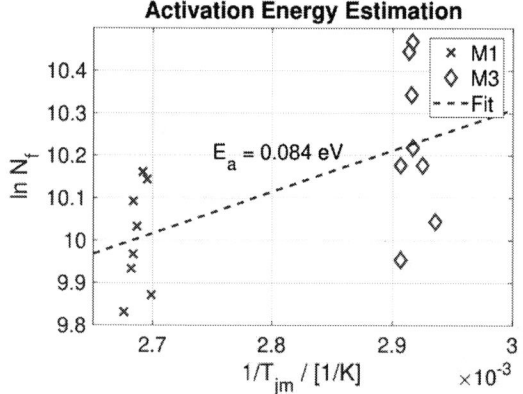

Fig. 7: Fitting for the activation energy E_a.

In Fig. 7, the slope of the dashed line corresponds to the quotient of the activation energy E_a and the Boltzmann constant k_B. If $k_B = 8.617 \cdot 10^{-5} \frac{eV}{K}$ is assumed, E_a can be estimated using the PCT data as shown in Fig. 7. The mean junction temperature T_{jm} is calculated using (3) after a median filter has been applied to the temperature data and – taking AQG 324 as a reference – the 100th sample of the respective measurement series has been used for the calculation. For the estimation of E_a the early failure in M1 and the two long runners in M3 were not considered.

The last parameter to be determined is the technology factor A. One way to determine this parameter is to use (4), (5), and Fig. 5. The fitting parameter a contains the required parameter as well as the just estimated activation energy, the Boltzmann constant, and the mean junction temperature. The relationship can be written as

$$a = \ln A + \frac{E_a}{k_B \cdot T_{jm}} \quad (9)$$

and by rearranging and applying the exponential function to both sides of the equation

$$A = e^{a - \frac{E_a}{k_B \cdot T_{jm}}} \quad (10)$$

can be obtained. In this example, the median of all mean junction temperatures from M1 considered in the evaluation is used as T_{jm}. This means that all the unknowns from (1) have now been found. These parameters are listed in Tab. 3.

Tab. 3: Parameters obtained from the PCT

Parameter	Value
A	$3.295 \cdot 10^{13}\,\mathrm{K}^{-\alpha}$
α	-5.168
E_a	$0.084\,\mathrm{eV}$

The function from (1) instrumented with the parameters found is transferred together with the data points from the PCT into a joint diagram in Fig. 8.

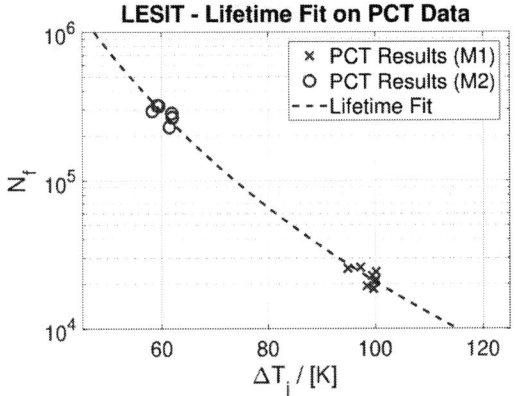

Fig. 8: Lifetime fit for the SiC-multichip modules.

Figure 8 shows high level of agreement between the experimental data and the descriptive model derived from it. The variables found are discussed and classified in the following subsection 4.4.

4.4 Discussion

The LESIT model used in this study is a simple model with fewer parameters than the CIPS08 model. What was not investigated in this study was the influence of the load current and the temperature swing duration. These parameters can influence the expected lifetime, but the additional exponents associated with these parameters are small compared to the α derived here [25]. The consideration of such parameters and the extension of the measurement campaign in the direction of smaller temperature swings massively increases the measurement times and the occupancy of the test bench infrastructure involved [29]. Normalization of N_f via the set PCT parameters is also

sometimes carried out in the literature [30], but was not used in this study.

Due to the multitude of lifetime models, which often vary in their formulation between the LESIT and CIPS08 models, classifying the technology factor A is not trivial. Simpler, and much more interesting, are the parameters α and E_a.

In [31], α is given in a range from -2.766 to -5.621. The exponent derived within the scope of this study can thus be plausibly validated.

In a tabulary organized form, [32] provides the range of values for E_a from 0.066 [33] to 0.080 [34]. All these values apply to Al-bonds in (high) power modules in insulated gate bipolar transistors (IGBTs). For molded packages, these values are higher [32]. This demonstrates that the value for E_a derived within the context of the study at hand aligns with the known values for bond wire fatigue found in the literature.

While the sample size of cyclically aged semiconductor switches is not very large, the failure distribution is already perceptible in Figs. 3, 4, and 5. Another significant contribution of the publication is the provision of data and model parameters for SiC semiconductors within modules. This facilitates more precise modeling, even at the system level, as model parameters derived solely from IGBT data are no longer necessary.

5 Conclusion and Outlook

This paper provided an overview of radiation-induced and thermo-electric failure mechanisms, which are particularly critical in the aviation environment. It also discussed why SiC, despite its characteristic additional degradation mechanisms, is considered a viable candidate for electrified aircraft applications. Even for short-range aircraft, propulsor powers are remarkably high. Current approaches utilize scaled models or demonstrators. Therefore, focusing on SiC-multichip modules in a six-pack configuration, given their better manageability, was proposed. For future reliability studies, it must be initially assumed that the joining and packaging technology mirrors that of larger power modules to be used in aviation.

Using AQG 324 as a reference, a PCT was conducted to derive LESIT model parameters, establish EOL criteria, and evaluate failure modes. The paper then detailed the derivation of these parameters from PCT data and compared them to literature values.

Future investigations aim to consider the influence of load current and cycle duration, as numerous long temperature swings occur in the mission profile of the anticipated short-range aircraft. Moreover, a larger sample size would enable a more precise failure distribution estimation, thus enhancing future reliability assessments of inverters and converters.

References

[1] L. Radomsky, R. Keilmann, D. Ferch, and R. Mallwitz, "Challenges and opportunities in power electronics design for all- and hybrid electric aircraft: A qualitative review and outlook," in *Deutscher Luft- und Raumfahrtkongress (DLRK) 2023*, Stuttgart, Germany, 2023.

[2] D. Fischer, R. Rohn, and R. Mallwitz, "Design approach of an integrated sic-inverter for an electric aircraft," in *2023 IEEE International Conference on Electrical Systems for Aircraft, Railway, Ship Propulsion and Road Vehicles & International Transportation Electrification Conference (ESARS-ITEC)*, 2023, pp. 1–6. DOI: 10.1109/ESARS-ITEC57127.2023.10114813.

[3] M. Meindl, J. Zettelmeier, C. Bentheimer, F. Hilpert, and M. März, "Aircraft drive inverter design at physical limits," in *2023 IEEE Workshop on Power Electronics for Aerospace Applications (PEASA)*, 2023, pp. 1–6. DOI: 10.1109/PEASA58318.2023.10235585.

[4] J. Ebersberger, M. Hagedorn, M. Lorenz, and A. Mertens, "Potentials and comparison of inverter topologies for future all-electric aircraft propulsion," *IEEE Journal of Emerging and Selected Topics in Power Electronics*, vol. 10, no. 5, pp. 5264–5279, 2022. DOI: 10.1109/JESTPE.2022.3164804.

[5] J. Ebersberger, R. J. Keuter, B. Ponick, and A. Mertens, "Power distribution and propulsion system for an all-electric regional aircraft," in *2023 IEEE International Conference on Electrical Systems for Aircraft, Railway, Ship Propulsion and Road Vehicles & International Transportation Electrification Conference (ESARS-ITEC)*, 2023, pp. 1–7. DOI: 10.1109/ESARS-ITEC57127.2023.10114868.

[6] J. Kugener, A. Pal, and S. Kazula, "Preliminary design tool for power converters in electric aircraft propulsion systems," in *Deutscher Luft- und Raumfahrtkongress (DLRK) 2023*, Stuttgart, Germany, 2023.

[7] L. Radomsky and R. Mallwitz, "Review, comprehensive analysis and derivation of analytical power loss calculation equations for two- to three-level midpoint clamped inverter topologies with hybrid switch configurations," *Energies*, vol. 16, no. 18, 2023. DOI: 10.3390/en16186710.

[8] Y. Cao, L. Fauth, R. J. Keuter, A. Sangwongwanich, M. Novak, *et al.*, "Inverter design for future electrified aircraft propulsion systems under consideration of wear-out failure and random failure," in *2023 25th European Conference on Power Electronics and Applications (EPE'23 ECCE Europe)*, 2023, pp. 1–10. DOI: 10.23919/EPE23ECCEEurope58414.2023.10264306.

[9] C. Felgemacher, S. V. Araújo, P. Zacharias, K. Nesemann, and A. Gruber, "Cosmic radiation ruggedness of si and sic power semiconductors," in *2016 28th International Symposium on Power Semiconductor Devices and ICs (ISPSD)*, 2016, pp. 51–54. DOI: 10.1109/ISPSD.2016.7520775.

[10] A. M. Ferrick, N. Bernstein, A. Aizer, and L. Chinitz, "Cosmic radiation induced software electrical resets in icds during air travel," *Heart Rhythm*, vol. 5, no. 8, pp. 1201–1203, 2008. DOI: 10.1016/j.hrthm.2008.04.018.

[11] P. Morey and M. Carpita, "On the cosmic ray influence on the electronics design of a high altitude electric aircraft," in *2022 24th European Conference on Power Electronics and Applications (EPE'22 ECCE Europe)*, 2022, pp. 1–8.

[12] Y. Wang, E. Deng, L. Wu, Y. Yan, Y. Zhao, and Y. Huang, "Influence of humidity on the power cycling lifetime of sic mosfets," *IEEE Transactions on Components, Packaging and Manufacturing Technology*, vol. 12, no. 11, pp. 1781–1790, 2022. DOI: 10.1109/TCPMT.2022.3223957.

[13] D.-P. Sadik, H.-P. Nee, F. Giezendanner, and P. Ranstad, "Humidity testing of sic power mosfets," in *2016 IEEE 8th International Power Electronics and Motion Control Conference (IPEMC-ECCE Asia)*, 2016, pp. 3131–3136. DOI: 10.1109/IPEMC.2016.7512796.

[14] A. Hassan, Y. Savaria, and M. Sawan, "Electronics and packaging intended for emerging harsh environment applications: A review," *IEEE Transactions on Very Large Scale Integration (VLSI) Systems*, vol. 26, no. 10, pp. 2085–2098, 2018. DOI: 10.1109/TVLSI.2018.2834499.

[15] Y. Wang, Y. Ding, and Y. Yin, "Reliability of wide band gap power electronic semiconductor and packaging: A review," *Energies*, vol. 15, no. 18, 2022. DOI: 10.3390/en15186670.

[16] J. Kim, S. Kwak, and S. Choi, "Impacts of sic-mosfet gate oxide degradation on three-phase voltage and current source inverters," *Machines*, vol. 10, no. 12, 2022. DOI: 10.3390/machines10121194.

[17] H. Luo, F. Iannuzzo, and M. Turnaturi, "Role of threshold voltage shift in highly accelerated power cycling tests for sic mosfet modules," *IEEE Journal of Emerging and Selected Topics in Power Electronics*, vol. 8, no. 2, pp. 1657–1667, 2020. DOI: 10.1109/JESTPE.2019.2894717.

[18] Y. Cai, H. Xu, P. Sun, J. Ke, E. Deng, et al., "Effect of threshold voltage hysteresis on switching characteristics of silicon carbide mosfets," *IEEE Transactions on Electron Devices*, vol. 68, no. 10, pp. 5014–5021, 2021. DOI: 10.1109/TED.2021.3101459.

[19] X. Zhong, H. Jiang, G. Qiu, L. Tang, H. Mao, et al., "Bias temperature instability of silicon carbide power mosfet under ac gate stresses," *IEEE Transactions on Power Electronics*, vol. 37, no. 2, pp. 1998–2008, 2022. DOI: 10.1109/TPEL.2021.3105272.

[20] S. Karpuk and A. Elham, "Influence of novel airframe technologies on the feasibility of fully-electric regional aviation," *Aerospace*, vol. 8, no. 6, p. 163, Jun. 2021. DOI: 10.3390/aerospace8060163.

[21] H. Schefer, W.-R. Canders, J. Hoffmann, R. Mallwitz, and M. Henke, "Cryogenically-cooled power electronics for long-distance aircraft," *IEEE Access*, vol. 10, pp. 133 279–133 308, 2022. DOI: 10.1109/ACCESS.2022.3228161.

[22] R. J. Keuter, F. Niebuhr, M. Nozinski, E. Krüger, S. Kabelac, and B. Ponick, "Design of a direct-liquid-cooled motor and operation strategy for the cooling system," *Energies*, vol. 16, no. 14, 2023. DOI: 10.3390/en16145319.

[23] M. Held, P. Jacob, G. Nicoletti, P. Scacco, and M. H. Poech, "Fast power cycling test of igbt modules in traction application," *Proceedings of Second International Conference on Power Electronics and Drive Systems*, vol. 1, pp. 425–430, 1997.

[24] M. Thoben and T. Reiter, "Guideline for lifetime calculation of power modules annex of ecpe guideline aqg 324," *ECPE GUIDELINE AQG 324 Rev. 03.1/2021*, 2021.

[25] R. Bayerer, T. Herrmann, T. Licht, J. Lutz, and M. Feller, "Model for power cycling lifetime of igbt modules - various factors influencing lifetime," in *5th International Conference on Integrated Power Electronics Systems*, 2008, pp. 1–6.

[26] F. Lippold and R. Mallwitz, "Lifetime model adjustments for gan cascodes as a base for inverter lifetime estimation," *Power Electronic Devices and Components*, vol. 5, p. 100 039, 2023. DOI: 10.1016/j.pedc.2023.100039.

[27] F. Hoffmann and N. Kaminski, "Evaluation of the vsd-method for temperature estimation during power cycling of sic-mosfets," *IET Power Electronics*, vol. 12, no. 15, pp. 3903–3909, 2019. DOI: 10.1049/iet-pel.2018.6369.

[28] Y. Zhang, Y. Zhang, Z. Xu, Z. Wang, H. Wong, et al., "A guideline for silicon carbide mosfet thermal characterization based on source-drain voltage," in *2023 IEEE Applied Power Electronics Conference and Exposition (APEC)*, 2023, pp. 378–385. DOI: 10.1109/APEC43580.2023.10131449.

[29] J. Lutz, C. Schwabe, G. Zeng, and L. Hein, "Validity of power cycling lifetime models for modules and extension to low temperature swings," in *2020 22nd European Conference on Power Electronics and Applications (EPE'20 ECCE Europe)*, 2020, pp. 1–9. DOI: 10.23919/EPE20ECCEEurope43536.2020.9215609.

[30] P. Heimler, N. Thönelt, J. Lutz, and T. Basler, "Impact of bond wire configuration on the power cycling capability of discrete sic-mosfet devices," in *2022 24th European Conference on Power Electronics and Applications (EPE'22 ECCE Europe)*, 2022, pp. 1–9.

[31] A. Otto, X. Liu, R. Eichhorn, T. Basler, J. Lutz, and S. Rzepka, "Study of power cycling tests superimposed with passive thermal cycles on igbt modules," in *PCIM Europe 2023; International Exhibition and Conference for Power Electronics, Intelligent Motion, Renewable Energy and Energy Management*, 2023, pp. 1–10. DOI: 10.30420/566091024.

[32] G. Zeng, L. Borucki, O. Wenzel, O. Schilling, and J. Lutz, "First results of development of a lifetime model for transfer molded discrete power devices," in *PCIM Europe 2018; International Exhibition and Conference for Power Electronics, Intelligent Motion, Renewable Energy and Energy Management*, 2018, pp. 1–8.

[33] U. Scheuermann and R. Schmidt, "Impact of load pulse duration on power cycling lifetime of al wire bonds," *Microelectronics Reliability*, vol. 53, no. 9, pp. 1687–1691, 2013, European Symposium on Reliability of Electron Devices, Failure Physics and Analysis. DOI: 10.1016/j.microrel.2013.06.019.

[34] O. Schilling, M. Schäfer, K. Mainka, M. Thoben, and F. Sauerland, "Power cycling testing and fe modelling focussed on al wire bond fatigue in high power igbt modules," *Microelectronics Reliability*, vol. 52, no. 9, pp. 2347–2352, 2012. DOI: 10.1016/j.microrel.2012.06.095.

GaN Cascode in High Speed Driven Air Compressors for Automotive Fuel Cells

Florian Lippold[1] 0000-0001-5214-7600, Regine Mallwitz[1] 0000-0002-8176-0380

[1] Institute for Electrical Machines, Traction and Drives, TU Braunschweig, Germany

Corresponding author: Florian Lippold, f.lippold@tu-braunschweig.de
Speaker: Florian Lippold, f.lippold@tu-braunschweig.de

Abstract

At first some advantages of GaN cascodes for automotive applications are described and Lifetime models introduced. An existing lifetime model based on the LESIT-model is updated with new measured data. Here Power Cycling test (PCT) results with low temperature swings with about 30 K on discrete GaN cascodes are presented. With curve fitting new parameters for the lifetime model are derived. With the new parameters the lifetime of an inverter for automotive auxiliary applications is again estimated for different missionprofiles and switch configurations. The estimated lifetimes are compared and discussed.

1 Introduction

For Fuel Cells an air compressor is required which provides the required oxygen for the reaction. To keep this compressor very compact the speed of the turbine should reach 100,000 rpm. The electrical power of the compressor should be 15 kW peak. A switching frequency of at least 50 kHz was aimed. For this high switching frequency Wide band gap semiconductors are in focus. For this power range discrete packages are possible. A comparison of different semiconductors was made in [1]. A 2-level three phase bridge topology was chosen for the inverter.

GaN cascodes are for automotive applications very interesting because of robustness aspects and a higher threshold voltage compared to standard GaN HEMTs. Also the TO-247 package is in automotive applications very common. Furthermore a normally off device is required. And a higher maximum junction temperature Tjmax about 175°C is possible compared to standard silicon chips with Tjmax about 150°C.

The GaN cascode, shown in **Figure 1**, consists of a normally-on HEMT and a silicon MOSFET. The MOSFET turns the cascode on and off. The Si-MOSFET is placed on top of the HEMT [2] and [3], depicted in **Fig. 2**.

For the lifetime estimation of the switches of the inverter different lifetime models can be used for example CIPS [4] or some adapted models like in [5].

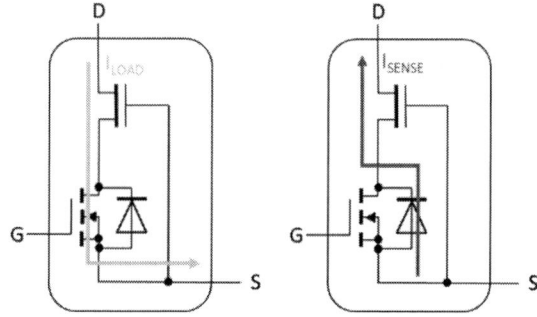

Fig. 1   Load and Sense current in the cascode [3].

Fig. 2   Inner structure of the GaN cascode.

The LESIT lifetime model [6] from the year 1997 is still very useful for lifetime estimations, when no detailed information of the inner structure of the semiconductor switch like bond wire diameter or number of bond wires are given. After loss calculations and counting the temperature swings evoked by the power losses of the mission profile the lifetime can be estimated. The Methodology is described in detail in [2]. Here GaN cascodes in discrete TO-247 package with Rdson = 35 mΩ, with first Power cycling test results in [7], are again tested with the Power cycling test but with lower temperature swings ΔTj. The temperature measurement of the junction temperature is done with the Vsd-Method. A sense current about 20 mA flows from source-pin to the drain-pin, see **Fig. 1**. The voltage drop and calibration curves are used for the indirect temperature measurement of the junction temperature Tj.

2 Test Parameter of the Power Cycling test

For Power cycling tests often high temperature swings ΔTj higher than 60 K are carried out [7]. This is done to accelerate the test. The stress for the inner structure of the devices under test (DUT) is very high. Cracks in the silicon chip of the GaN cascode are found in [2]. For lifetime calculations with lower temperature swings the lifetime curve is extrapolated. But this is not exact. For generating this data additional Power cycling tests with low temperature swings are carried out. The targeted test parameter of the tests M0 and the earlier tests M1 and M2 are shown in **Table 1**.

Measurement Series	M0	M1	M2
ton = toff	0,5 s	0,5 s	4 s
Iload	25 A	35,5 A	35,5 A
Tjmin	90°C	70°C	40°C
Tjmax	120°C	140°C	170°C
ΔTj	30 K	70 K	130 K
Tjm	105°C	105°C	105°C

Table 1 Test Parameter of the Power Cycling tests

The test parameter can not be chosen free because of their coupling. To get the same mean junction temperature T_{jm} for all measurement series the load current and the cooling of measurement series M0 were adjusted.

The mechanical Test set up is unchanged to earlier work and is depicted in Fig. 3, [7]. The structure was build according to the AQG 324 [8]. Four DUTs are tested at the same time. Under each DUT a copperplate with integrated thermocouple Type K is placed for measuring the temperature under the chip at the case Tc/s, as a reference temperature for the thermal impedance Rthj-c/s measurement, which is repeated in fixed intervals.

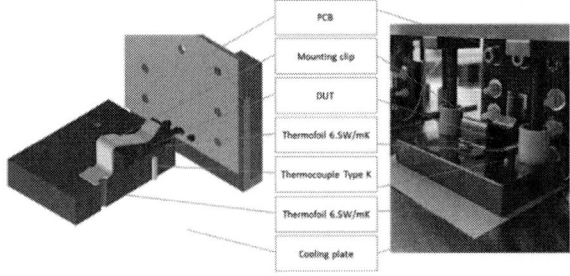

Fig. 3 Mechanical Test set up [7].

The Power cycling tests are carried out with a Mentor Graphics/Siemens Power Tester 2400 A and a liquid cooler Julabo Presto, which provides a stable tempering of the cooling plate.

3 Power Cycling Test Results and Lifetime model

3.1 Power cycling results

The stop criteria was set to a junction temperature Tjmax = 250°C. At this point the End of life (EoL) criteria with an increased drain-source voltage (DUTs 1, 2 and 4), **Fig. 4**, and an increased thermal impedance Rthj-c/s (DUTs 2 and 4), **Fig. 5**, is reached. The test was stopped after 10,137,326 cycles. DUT 3 (yellow curve) did not reach EoL at this time.

Although the test parameters for the PCT could be set relatively precisely, a large spread in the number of cycles from 6 to 10 million cycles can be recognised.

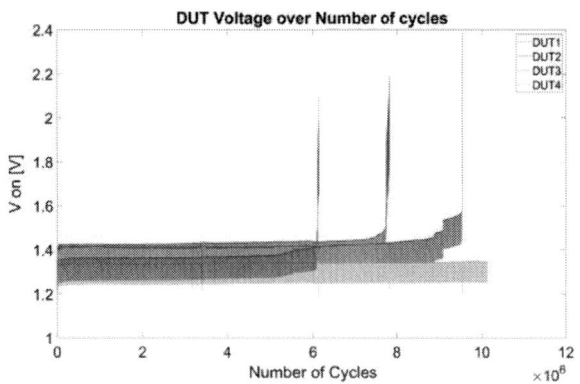

Fig. 4 DUT Voltage Von over number of cycles.

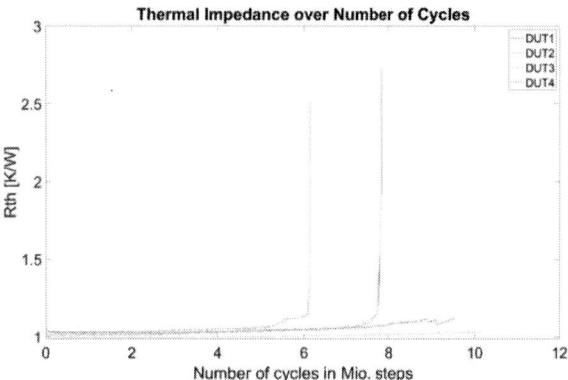

Fig. 5 Thermal Impedance Rthj-c/s over number of cycles.

3.2 Lifetime modelling and parameter update

The LESIT parameters, formula shown in equation (1), are modified to get a good agreement with the measured data. The exponent α_{new} and factor A_{new} are obtained by curve fitting. The activation energy Ea = 9.891·10^-20 J and the Boltzmann-constant kB = 1.38065·10^-23 J/K are unchanged.

$$N_f = A_{new} \cdot \Delta T_j^{\alpha_{new}} \cdot e^{\frac{E_a}{k_b \cdot T_{jm}}} \qquad (1)$$

DUT 3 was not used for the model, because it did not reach EoL.

The Number of cycles reached for M0, M1 and M2 are the orange triangles shown in **Figure 6**.

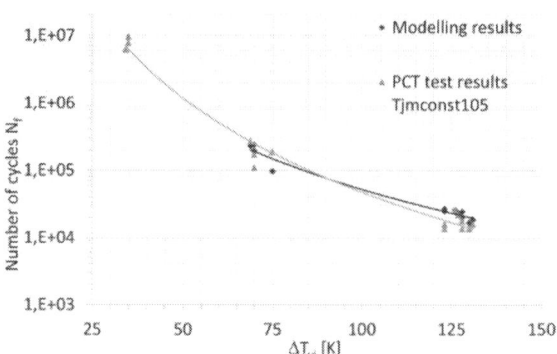

Fig. 6 Number of cycles of measurement series M0, M1 and M2 (orange) and calculated values (purple).

The purple squares in **Figure 6** represent the calculated number of cycles with the parameter fitted in [7]. It can be seen that for low temperature swings under 90 K the calculated number of cycles are underestimated and above 90 K the calculated number of cycles are overestimated.

With the additional data with temperature swings around 30 K the curves in **Figure 6** were fitted and the parameter A_{new} and α_{new} generated. All parameters are shown in **Table 2**:

Used measurement series for curve fitting	A_{new}	α_{new}
M1, M2	6018	-3,661
M0, M1, M2	1,000,000	-4,784

Table 2 Curve fitting parameters

4 Lifetime estimation and Discussion of earlier calculations

With the parameters in **Table 2** the repetition rates for a 2 Level inverter are calculated with Equation 2 and 3 [9, 10].

$$Q_i\left(\Delta T_j, T_{jmin}\right) = \frac{N_i\left(\Delta T_j, T_{jmin}\right)}{N_{fi}\left(\Delta T_j, T_{jmin}\right)} \quad (2)$$

$$N_r = \frac{1}{\sum_{i=1}^{n} Q_i} = \frac{1}{\left(\frac{N_1}{N_{f1}} + \cdots + \frac{N_n}{N_{fn}}\right)} \quad (3)$$

The used missionprofiles for a city ride is shown in **Figure 7** and a Highway ride in **Figure 8**.

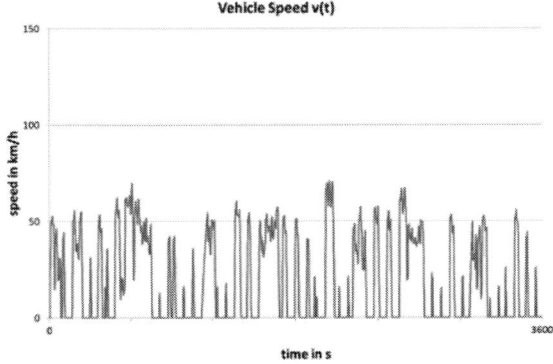

Fig. 7 Calculated repetition rates for a city ride

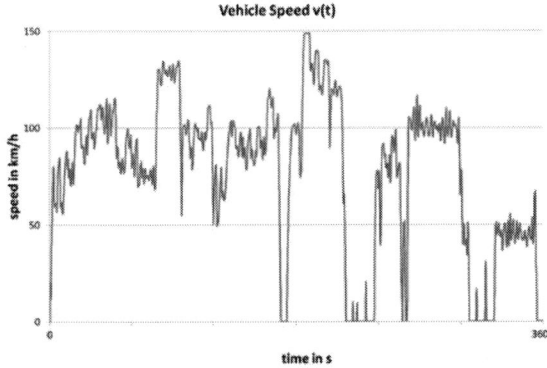

Fig. 8 Calculated repetition rates for a Highway ride

With this missionprofiles, the current per switch and their losses are calculated. With the losses and the thermal impedance from the cascode from junction to case Rthj-c/s = 0.8 K/W and from case to ambient Rthc-a = 1 K/W the chip temper-

ature is calculated. For counting the temperature swings, the rainflow algorithm is used [11].

The load current is divided in the calculation to see, whether a single chip (35 mΩ) or a parallel chip (2*35 mΩ) configuration is possible. The switching frequency is 50 kHz for this application. The results are shown in **Figure 9**.

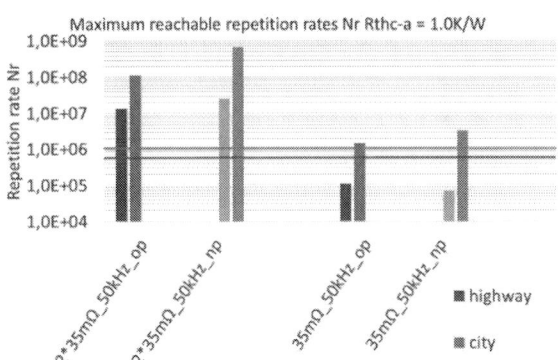

Fig. 9 Calculated repetition rates for a Highway (green) and a City (blue) missionprofile.

For low temperature swings from the mission profile the calculated repetition rates are increasing for city profile and highway profile with (2*35 mΩ) chip configuration. The reason is the lifetime curve extrapolated in **Figure 6** which underestimated the number to failure for low temperature swings under 90 K with the old lifetime model parameters.

But high temperature swings above 90 K are leading to lower repetition rates which can be seen at the single chip configuration (35 mΩ) and the highway profile. So, a single chip configuration is for the highway mission profile not sufficient and even gets worse with the updated lifetime model parameters.

5 Conclusion

GaN cascodes in discrete TO-247 package can be an interesting alternative for silicon switches for automotive applications with medium power requirements. The Power cycling tests carried out with low temperature swings show high numbers of cycles, which is often not correct modelled. Curve fitting of the additional data yield to new parameters for the LESIT-Model for GaN cascodes. With this the lifetime of a 2 Level inverter with a high switching frequency is estimated for two chip configurations and two different mission profiles. The parameters of the lifetime model show a great influence to the results of the lifetime estimation.

References

[1] N. Langmaack, F. Lippold, D. Hu, R. Mallwitz: Analysing Efficiency and Reliability of High Speed Drive Inverters Using Wide Band Gap Power Devices. MDPI Journal Machines, 2021,
https://doi.org/10.3390/machines9120350

[2] Florian Lippold, Regine Mallwitz, "Lifetime Model Adjustments for GaN Cascodes as a Base for Inverter Lifetime Estimation", Power Electronic Devices and Components, Volume 5, June 2023,
https://doi.org/10.1016/j.pedc.2023.100039

[3] Xu C., Yang, F., Pu, s., Akin, B.: Performance Degradation of GaN HEMTs under accelerated power cycling tests. CPSS Transactions on power electronics and applications Vol 3 No 4 December 2018, pp. 269-277.

[4] R. Bayerer, T. Herrmann, T. Licht, J. Lutz and M.Feller, "Model for power cycling lifetime of IGBTmodules - various factors influencing lifetime," in Proc. of the 5th CIPS, pp. 1-6, 2008.

[5] G. Zeng, L. Borucki, O. Wenzel, O. Schilling, J. Lutz: First results of development of a lifetime model for transfer molded discrete power devices. PCIM Power Conversion and Intelligent Motion, Nuremberg, 2018

[6] Held, M.; Jacob, P.; Nicoletti, G.; Scacco, P.; Poech, M.H.: Fast Power Cycling Test for IGBT Modules in Traction Application, Power Electronics and Drive Systems 1997, Conference Proceedings

[7] Florian Lippold, Philipp Hauenschild, Regine Mallwitz, "Comparison of Power Cycling Results of discrete GaN Cascodes for Automotive Power Electronics with high Temperature Swings", 2022 24th European Conference on Power Electronics and Applications (EPE'22 ECCE Europe)

[8] ECPE Guideline AQG 324: Qualification of power modules for use in power electronics converter in motor vehicles, Release no.: 02.1/2019, ECPE European Center of Power Electronics e.V., Nuremberg, p. 28

[9] M. Ciappa: Lifetime Modeling and Prediction of Power Devices, Proceedings of CIPS 2008, Nuremberg, Germany, 2008

[10] J. Lutz, H. Schlangenotto, U. Scheuermann, R. D. Doncker: Semiconductor Power Devices, 2nd ed., Springer International Publishing, Cham, Switzerland, 2018, p. 530

[11] K. Mainka, O. Schilling, M. Thoben: Lifetime calculation for power modules, application and theory of models and counting methods, Proceedings of ECPE 2011, Nuremberg, Germany, 2011

PCIM Europe 2024, 11– 13 June 2024, Nuremberg   DOI: 10.30420/566262424

Prognostic Analysis of IGBT Health: Real-Time On-State Voltage Prediction through Machine Learning

Tanya Thekemuriyil[1] ⓘ, Jaspera Dominique Rohner[1] ⓘ, Renato Amaral Minamisawa[1] ⓘ

[1] University of Applied Sciences and Arts Northwestern Switzerland, Switzerland

Corresponding author: Tanya Thekemuriyil, tanya.thekemuriyil@fhnw.ch
Speaker: Tanya Thekemuriyil, tanya.thekemuriyil@fhnw.ch

Abstract

This study demonstrates a machine learning method to estimate the on-state voltage of semiconductors in a converter prototype exposed to dynamically changing mission profiles of electric vehicles (EVs) and photovoltaic (PV) systems under different ambient temperatures. The real-time monitoring of on-state voltage is an indicative measure of the junction temperature for the predictive maintenance of power converters. The method offers novel insights into uncertainties and feature importance for the on-state voltage predictions using available parameters of a converter. The approach is industrial-ready and applicable to various converter systems regardless of their specifications, achieving a remarkable absolute prediction error of 0.75%.

1   Introduction

Power semiconductors, such as Silicon Insulated-Gate Bipolar Transistors (Si IGBTs) and Silicon Carbide Metal-Oxide-Semiconductor Field-Effect Transistors (SiC MOSFETs), are crucial for the reliable operation of power electronic converters in various applications like electric vehicles (EV) and renewable energy systems. These components operate at high frequency, voltage, and current, leading to continuous thermal-mechanical stress and eventual degradation of bonded contacts. Semiconductors are among the most vulnerable components to failure in power converters. Hence, predictive maintenance, achieved through monitoring chip junction temperature, is essential for estimating remaining lifetime and scheduling maintenance before system breakdown [1]-[4].

To monitor the junction temperature in IGBTs, thermal modelling, thermal sensors and current/voltage measurement methods are usually employed. These methods become feasible for monitoring constant load drives in indoor applications, where operating power and ambient temperature remain relatively constant unlike in EV and Photovoltaic (PV) systems [5]. EV and PV inverters commonly operate with variable loads and in diverse ambient temperatures [6]-[7]. Additionally, these applications frequently function in sub load regimes, such as driving EVs in cities and non-peak irradiation profiles in PV systems [8]. Dynamic operating conditions and harsh ambient conditions forces accumulation of stress and gradual degradation in power semiconductors that are inaccessible with such measurement methods and set point junction temperatures [9]-[11].

In this work, we report on the use of machine learning methods for predicting collector-emitter voltage (V_{ce}) in a real IGBT-based inverter exposed to EV drive cycle and PV solar profile under different ambient temperatures generated in a climate chamber. The available parameters of an operating converter prototype are recorded, processed, and trained to create models based on three different machine learning algorithms: Support Vector Machine (SVM), K-nearest neighbors (K-NN) and Decision Tree.

The goal is to develop a modern industrial solution: we try to reproduce the operating conditions as realistic as possible to predict variations of V_{ce} in different operating conditions and not the specific maximum junction temperature set point. V_{ce} is the temperature sensitive electrical parameter (TSEP), which is directly proportional and fast responsive to changes in the junction temperature [1][3][4]. Our method does not require calibration of the V_{ce} to the junction temperature because it aims to predict variations over low and high loads, and thus does not require a specific maximum temperature set point. This makes the method to

2986

be applicable to any converter irrespective of its unique properties. The methodology provides important insights for implementation of the machine learning based monitoring directly in converter applications, further mitigating additional hardware and high investment risks.

The main contributions of the work are first, to determine the uncertainties achieved in using each machine learning model, second, to enlist the most important feature parameters that yield best predictions for EV and PV application and third to assess the uncertainty to ensure reliable and robust monitoring systems.

The methodology is outlined as follows: (i) identifying relevant input attributes for the converter in PV and EV applications based on the parameter with the highest variability, (ii) gathering data from an in-house-built single-phase inverter exposed to solar irradiation and electric vehicle drive profiles in varying ambient temperatures simulated within a climate chamber, (iii) applying filtering to the collected data using both physical and statistical principles based on a statistical figure of merit devised within the study, (iv) training machine learning models using the pre-processed data to establish the relationship between the input attributes and the prediction target, and (v) comparing and assessing the errors associated with different methods to identify the most suitable method and input attributes for specific PV and EV applications.

2 Machine Learning Method of V_{ce} Calibration

The most aimed method for machine learning based condition monitoring of power electronic converter is to measure the most sensitive parameters and other entities with available sensors and accessible data, with minimal addition of hardware and cost. All power electronic converters have some sensitive parameters and other related entities that are already measured. PV inverters measure the amplitude and fundamental frequency of the output current for Phase Locked Loop (PLL) or Maximum Power Point Tracking (MPPT) [12]. Similarly, EV inverters have parameters that are already monitored to control the torque and speed of the motor [13]. Temperature sensors are also present in the converters particularly close to semiconductors as a safety measure to limit the temperature below tolerances [14].

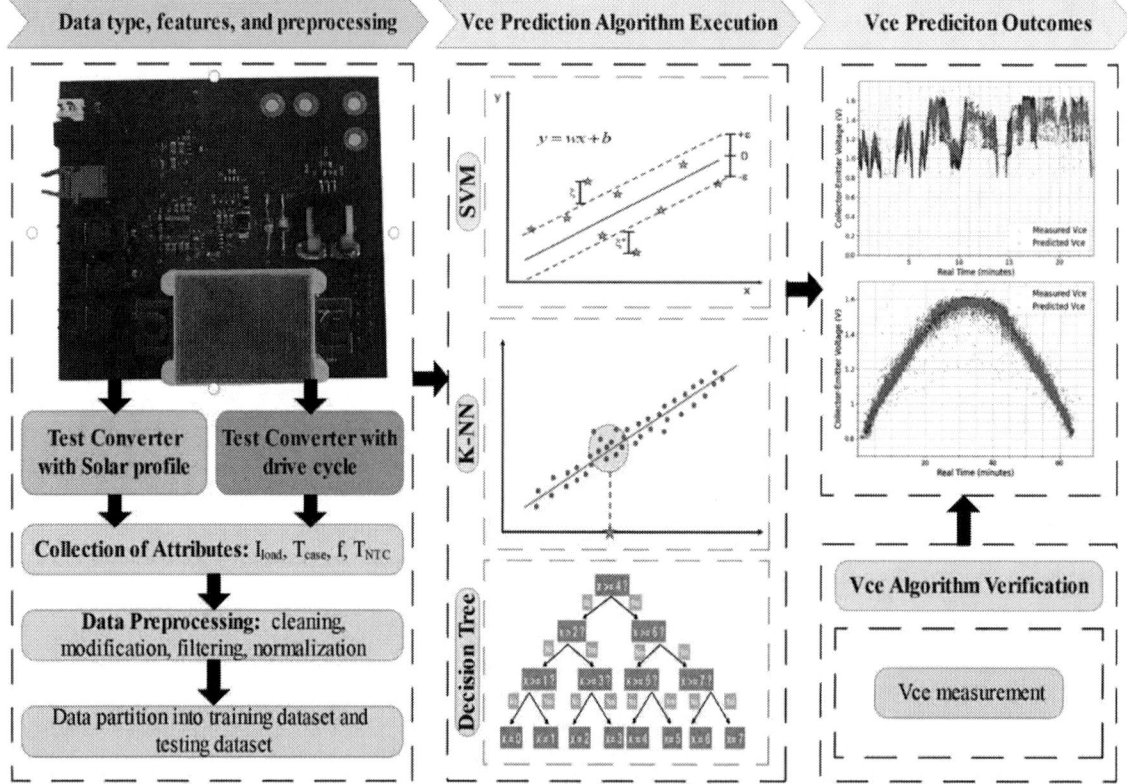

Fig. 1   Framework for prediction of on-state voltage.

Apart from being easily accessible, RMS value of output current (I_{load}), case temperature (T_{case}), fundamental output frequency (f) and IGBT module temperature based on negative-temperature-coefficient (NTC) thermistor integrated in module (T_{NTC}) are most susceptible parameters to variations under the dynamic operating profile and harsh ambient conditions of EV and PV inverters. Therefore, these parameters are considered as input attributes to train different machine learning models.

Supervised learning algorithms like Support Vector Machine (SVM), K-nearest neighbors (KNN) and Decision Tree are used in this work to train the dataset from a converter for an electric vehicle and photovoltaic application, accurately predicting the V_{ce} of an IGBT. SVM models can capture nonlinear relationships between attributes and outputs. Given the relatively small number of features and the likelihood of measurement outliers due to sensor dependency, SVM kernels are favored for achieving high prediction accuracy. On the other hand, the KNN algorithm is suitable for this problem due to its compatibility with a smaller feature set and its resilience to missing attributes that may occur during prolonged converter operation. Decision Trees offer high efficiency and can handle datasets with missing attributes, making them also an optimal choice for large datasets with fewer features and nonlinear relationships as in this study [15].

The selected parameters are considered as input attributes to train various models and the prediction is the on-state collector emitter voltage (V_{ce}) of the IGBT in the converter. We have built a monitoring circuit that allows precise measurement of the on-state voltage during converter operation to train and test the models with high precision and accuracy. The work presented here is based on an in-house-built single-phase inverter with IGBT modules. The framework of the proposed system is shown in Fig. 1.

3 Technology Implementation in Applications

The topology of the in-house inverter with the monitoring circuit is illustrated in Fig. 2. The single-phase inverter consists of the Si IGBT module FS35R12W1T4_B11 (1.2 kV and 35 A) and has a maximum input voltage of 60 V and output current of 35 A. The main PCB consists of DC-link capacitors and IGBT module in addition to the circuits monitoring the attributes and is shown in Fig.3.

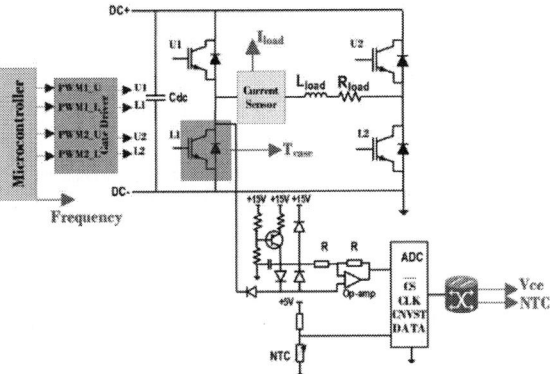

Fig. 2 Schematic of single-phase inverter with V_{ce} measurement circuit.

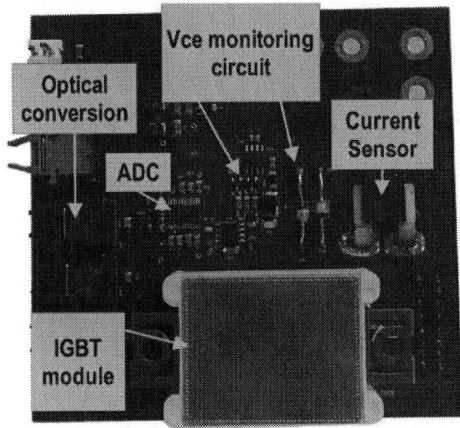

Fig. 3 PCB of single-phase inverter board with integrated V_{ce} measurement circuit.

The on-state voltage is measured using the circuit as defined in [16] and [17]. The temperature of the module is measured using the integrated NTC. The isolated K-type thermocouple sandwiched between the heatsink and IGBT module measures the case temperature of the semiconductor. A Hall effect-based current sensor ACS770LCB-050B-PFF-T in series to the R-L load measures the phase current. TMS320F28379D microcontroller receives all the measured signals and additionally generates Pulse Width Modulated (PWM) signals based on mission profile of PV and EV inverter. Therefore, the controller already contains information about the fundamental frequency and switching frequency of converter operation. SCiCoreDriver from Semicode amplifies the PWM pulses to drive the gates of the IGBT module. Fig. 4 (a) shows the experimental setup mounted over the cooling fins and connected to the external load. Two types of experiments are completed on the test setup: The first experiment is a simulation of a solar inverter, and the second dataset is a simulation of an electric car drive cycle. The converter prototype is operated in the closed climate chamber programmed based on a weather profile and is shown in Fig. 4 (b).

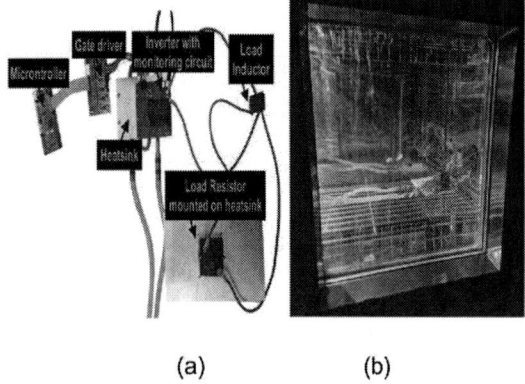

(a)   (b)

Fig. 4   (a) Experimental setup (b) Experimental setup in operation in the Climate Chamber.

In the solar experiment, the goal is to vary the temperature between 18°C and 27°C, and the input power of the converter to simulate a day of the inverter. The solar power output is assumed to scale inversely linearly with the solar zenith angle. For the solar inverter simulation, the sun zenith angle and temperature data are chosen to be that of a normal summer day in Western Europe. While the temperature and angle data stretch over 15 hours, the actual simulation is shortened to 1 hour span.

For the driving cycle simulation, the frequency and power output data are borrowed from the World-wide Harmonized Light Vehicles Test Cycle (WLTC). Note that the power output data is normalized to fit within the specifications of the setup here. The experiment is also conducted in the temperature chamber with the same temperature data as the solar inverter simulation. The driving cycle is repeated four times.

4   Data Extraction and Processing

Let $m \in \mathbb{N}$ be the total number of measurements made since the starting of the simulation. Vectors $V, I, T_{\text{NTC}}, T_{\text{amb}} \in \mathbb{R}^m$ contain m temporally ordered measurements of collector-emitter voltage, load current, NTC temperature and case temperature. The vectors $f, \tilde{I} \in \mathbb{R}^m$ contain the fundamental frequency and calculated RMS current. The RMS-current is calculated with the moving RMS (1) as,

$$\tilde{I}_i = \sqrt{\frac{1}{\omega} \sum_{k=i-\omega+1}^{i} I_i} \tag{1}$$

where $\omega \in \mathbb{N}$ is the RMS window size. A datapoint refers to a 6-tuple $T_i = (V_i, I_i, \tilde{I}_i, T_{\text{NTC},i}, T_{amb,i}, f_i)$. Before teaching the neural network, the data is filtered both according to physical and statistical properties. Following physical filtering is conducted.

1. Any data points with negative current (IGBT blocking) are removed from the dataset.

2. Any data points with a power output of less than 3 W are removed. The IGBT is not generating enough heat in this state for it to affect semiconductor lifetime.

3. Data points with impossibly high V_{ce} and a V_{ce} of under 0.8 V are removed.

The set of remaining data points is D_{phys} containing $n \in \mathbb{N}$ data points as 6-tuples. The statistical filtering algorithm is a very simple algorithm based on the fact that two measurements taken within a short time span of each other ("temporal neighborhood") will have similar I and Vce values. The algorithm calculates a weighted average of the distances between the measurement and its temporal neighborhood on a normalized Voltage-Current-plane. Let $r \in \mathbb{N}$, $G \in \mathbb{R}^{2r+1}$ be a discrete Gaussian kernel with $2r + 1$ elements and $\sigma \in \mathbb{R}$. G can be written as in (2),

$$G_x = exp^{-\frac{(x-r+1)^2}{2\sigma^2}} \tag{2}$$

$\|.\|_\infty$ is the maximum norm. The normalized vectors $\overline{V_{ce}} := \frac{V_{ce}}{\|V_{ce}\|_\infty}$ and $\bar{I} := \frac{I}{\|I\|_\infty}$ form the complex voltage-current plane $\overline{V_{ce}} + j\bar{I}$ with complex identity j. For the temporal neighborhood $N \in C^{2r+1}$ of radius r around the ith measurement $N_i = (V_{i-r} + jI_{i-r} ... V_i + jI_i ... V_{i+r} + jI_{i+r})$ and the measurement vector $M \in C^{2r+1}$, $M_i = (V_i + jI_i ... V_i + jI_i ... V_i + jI_i)$, we define vector $\Gamma \in \mathbb{R}^n$ as in (3),

$$\Gamma_i = G^T.|M_i - N_i| \tag{3}$$

Note that $|.|$ calculates the element-wise absolute value. Γ is the vector of the weighted distances between every measurement and its temporal neighborhood. Γ values can be considered figures of merit of the corresponding data points. Now, let $\rho \in \mathbb{N}$ be the amount of data points we want to statistically filter from D_{phys}. We define the final filtered 6-tuple set D_{stat} as according to (4).

$$D_{stat} = \left\{ T_x \in D_{phys} : x \in argmin_{Z \subset [1,n] \cap \mathbb{N}, |Z| = n - \rho} \sum_{i \in Z} \Gamma_i \right\}. \quad (4)$$

D_{stat} contains the $n - \rho$ data points with the lowest corresponding Γ values. Now the data is ready for machine learning. Window-size ω is set to 5. For both the solar and driving cycle simulations a ρ of $0.2n$ is experimentally deemed appropriate. As such, a fifth of elements in D_{phys} is removed. Temporal radius r is experimentally chosen as 15 for the solar inverter simulation and 20 for the driving cycle simulation. The Gaussian kernel's σ constant is chosen as $2r$.

5 Machine Learning based Model Implementation

The algorithms chosen for this project descend from the scikit-learn library for python [18]. The filtered data points are randomly grouped into training data and test data in 80:20 parts respectively. In the K-nearest neighbors (KNN) algorithm, each test data point calculates its Euclidean distance from all other data points, sorts them, selects the nearest neighbors, and estimates its output based on their weighted average, with the value of k being updated accordingly. The SVM model is initialized with regularization and kernel parameters, and then the model undergoes iterative training and testing while continuously updating the SVM parameters until optimization is achieved. In decision tree construction, the predictors are correlated with target output values, and a decision tree node is built at the minimal leaf size. The node is then either split or terminated, leading to the expansion of the tree, with the leaf size being updated accordingly.

Following parameters are determined to be optimal for the solar dataset:

1) K-NN: The number of neighbors is set to 5 as no significant improvement is observed for more than 5 neighbors.

2) Support Vector Machine: A linear kernel is chosen. All data is first scaled using a standard scaler.

The epsilon parameter is set to 0 and the regularization parameter is set to 1. The stopping tolerance is 10^{-6}.

3) Decision Tree: Since the impact on memory usage is minimal during training, no limits are set on the maximum leaf nodes, features, or tree depth.

For the drive cycle simulation, three drive cycles are used to train data and the fourth is used to test data. Following parameters are found to be optimal for the drive cycle dataset:

1) K-NN: The number of neighbors is set to 15 as it requires three times as many neighbors to plateau the mean squared error for the driving data due to the ambiguousness in data.

2) SVM: A linear kernel is chosen. All data is first scaled using a standard scaler. The epsilon parameter is set to 0 and the regularization parameter is set to 15. The regularization parameter deviation from the previous set stems from the fact that this dataset is slightly more dependent on features other than the RMS-current (I_{load}). The stopping tolerance of 10^{-7} is chosen.

3) Decision Tree: Since the impact on memory usage is minimal during training, no limits are set on the maximum leaf nodes, features, or tree depth.

Machine learning algorithms are applied to the filtered dataset using various combinations of attributes. In the case of a PV inverter, the fundamental frequency of the phase current remains constant and is therefore omitted as an attribute. Similarly, both temperature sensors are not considered as attributes simultaneously. The results of implementing different machine learning algorithms to predict V_{ce} for various combinations of input parameters (I_{load}, T_{NTC}, T_{case}, f) are summarized in Fig. 5. The percentage mean absolute error, representing the percentage of the mean absolute error relative to the mean V_{ce} values, is calculated and compared across different scenarios. While the RMS load current consistently emerges as a crucial predictor for on-state device voltage, the significance of other parameters varies depending on the application.

The most accurate result in the solar profile (0.75% of mean absolute error) is found when using the IGBT-internal temperature sensor as a second predictor for the K-nearest neighbor algorithm. Solar inverters operation is dependent on the solar irradiation profile and on the module temperature, T_{NTC}. However, since T_{NTC} is a parameter that is rarely available in converters, T_{case} as the second predictor is preferred.

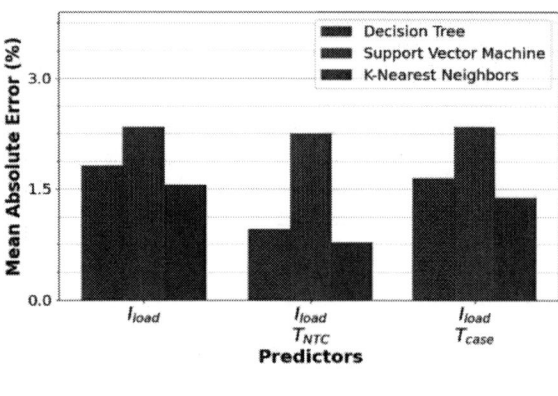

(a)

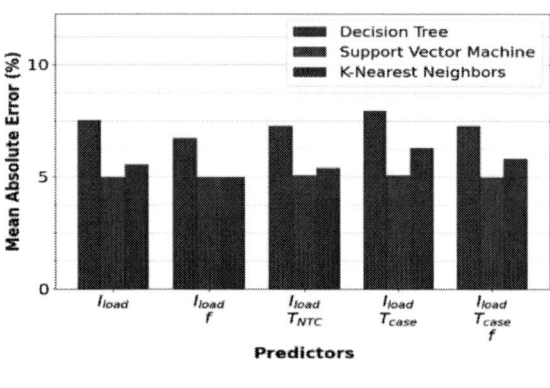

(b)

Fig. 5 (a) Mean absolute error for different combination of predictors for three selected algorithms for inverter in PV application (b) mean absolute error for different combination of predictors for three selected algorithms for inverter in EV application.

The least percentage mean absolute error of 5% is observed in the driving profile when the frequency of the output current is used as the second predictor to the load current using K-NN algorithm. The EV inverter is a variable frequency drive that constantly changes frequency to control speed and hence f becomes an important second predictor.

Figure 6 shows the measurements and the prediction results for the K-NN, SVM and Decision Tree algorithms on a time axis respectively using predictors I_{load} and T_{case} for the solar profile. It is evident that the predicted V_{ce} follows the solar load profile. The level of dispersion is different for each algorithm and indicates their effectiveness in predicting the output parameter accurately. Figure 7

is a set of three 2-dimensional histograms of dispersion plots for predictors I_{load} and T_{case} using each algorithm. The less visible a point is, the lower is the frequency of the prediction. The higher a point deviates from the dashed line, the lower is the accuracy of the prediction. The discrete steps as seen in the plots are due to the fact that the power supply has a limited current resolution.

Figure 8 shows the measurements and the prediction results for the K-NN, SVM and Decision Tree algorithms respectively on a time axis using predictors I_{load} and f for the driving profile. The frequency becomes an important predictor for any variable frequency drive in electric vehicles. Figure 9 represents three histograms of the dispersion plots for predictors I_{load} and f using algorithms K-NN, SVM and Decision Tree respectively. The histogram shows that the prediction density is the highest around the dashed line. The dataset contains a lack of V_{ce} measurements around 1.4 V and has no relation to the prediction accuracy.

The solar inverter dataset performs better than the driving cycle dataset across the board due to the lack of change in the modulated frequency. As such, the inductor's current "smearing" effect on the IV-plane is lower and more predictable in the solar inverter. The dynamic operation of the automotive inverter, characterized by rapid frequency transitions, leads to smearing of the measured load current. In the setup presented here, measurements are taken at a considerably lower sampling rate, making it challenging to capture the fast-changing operating conditions of the mission profile accurately. The predictive accuracy could possibly be improved by increasing the sampling frequency significantly to gain more insights and details for the regression to latch onto.

While single prediction values may not be accurate enough to hard-determine IGBT health, the histograms show that averaging enough single prediction values could allow for accurate V_{ce} prediction. From an engineering point of view, saving every single prediction value just to monitor IGBT failure would be overkill. Instead, the authors propose to average and statistically analyze the quality of large amounts of predictions to make one qualitative wear prediction.

The results also show that depending on application, a semiconductor-module-internal NTC should be a pertinent factor in choosing a device for high-reliability power electronics. It is also possible that

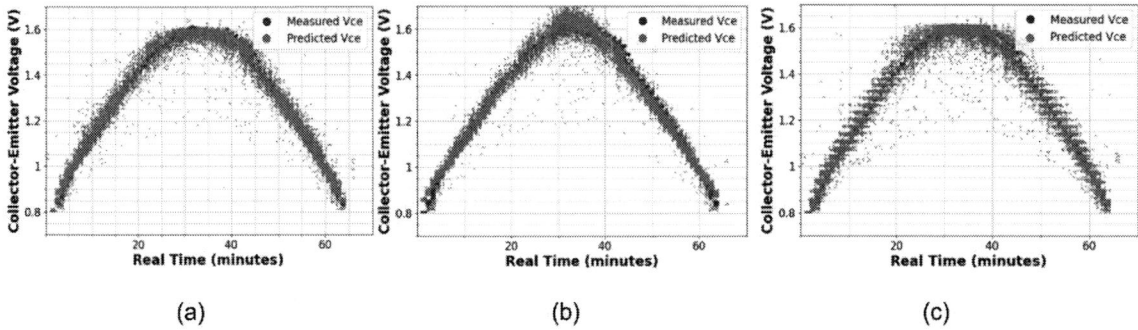

Fig. 6 Solar profile driven prediction results using I_{load} and T_{case} as predictors:(a) measured and predicted V_{ce} over time for K-NN (b) measured and predicted V_{ce} over time for SVM (c) measured and predicted V_{ce} over time for Decision Tree.

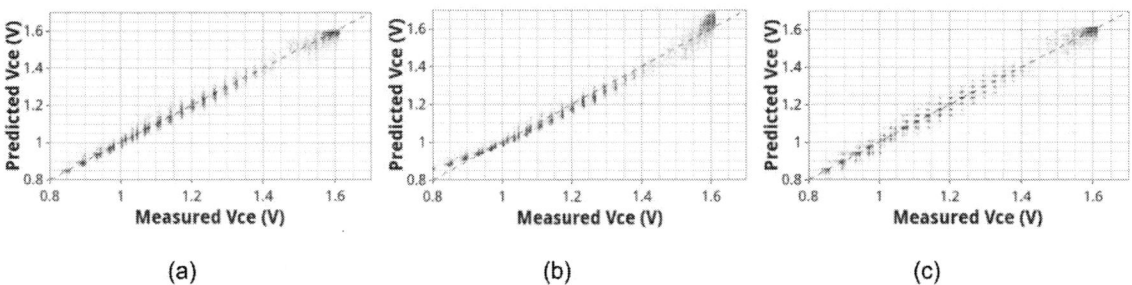

Fig. 7 2-dimensional histogram of the dispersion plot between measured and predicted on-state voltage for the solar profile using predictors I_{load} and T_{case} for (a) K-NN, (b) SVM and (c) Decision Tree.

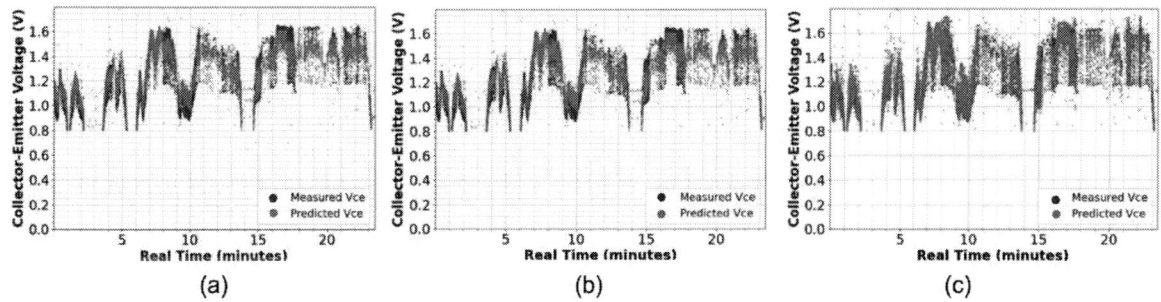

Fig. 8 Electric vehicle driving profile driven prediction results using I_{load} and f as predictors: (a) measured and predicted V_{ce} over time for K-NN (b) measured and predicted V_{ce} over time for SVM (c) measured and predicted V_{ce} over time for Decision Tree.

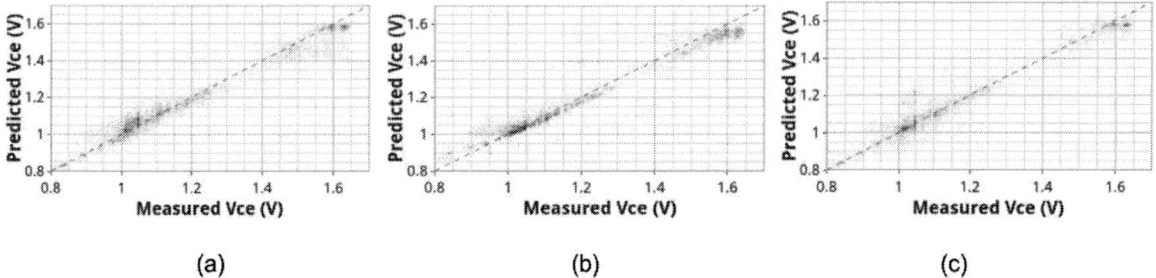

Fig. 9. 2-dimensional histogram of the dispersion plot between measured and predicted on-state voltages for the electric vehicle profile using predictors I_{load} and f for (a) K-NN, (b) SVM and (c) Decision Tree.

placing an external NTC near the switching device on the PCB could have a similar effect on prediction accuracy. From the implementation of the machine learning models in two application cases, it is obvious that the choice of input attributes depends on the application. In particular, the most varying parameters of the application field are to be considered to create the best model. The implementation of the machine learning algorithms into the controller of the converter is simple. The important parameters of the converter influencing the on-state voltage of the semiconductor are already available in the processor during converter operation. Trained machine learning models can accurately predict the online V_{ce} of the IGBT and can be used for temperature monitoring of the semiconductor in converters particularly operating in varying load conditions. The technique presented here clearly surpasses other existing methods because of its simplicity and lack of reliance on additional hardware and pre-calibrations. We hope our work will support engineers to deploy the methodology into their specific inverters, since the demonstration of the technology was successful, and its first insights presented here.

6 Conclusion

In this work we report on the investigation of the use of the machine learning algorithms to predict the on-state collector-emitter voltage of the IGBT module during normal converter operation of electric vehicles and photovoltaics. Since V_{ce} is an important temperature sensitive parameter of the semiconductor, on-state voltage has enough potential to monitor the junction temperature and wear-out status in converters operating over the range of its nominal power. Data collected from a conventional inverter together with the measurements from the online monitoring circuit can train and test the data driven models to predict any unseen V_{ce} data points. The single-phase inverter integrated with the sensors is tested in electric vehicle driving and solar irradiation profiles over a weather cycle to fit into different machine learning models. As most important conclusions, for the PV case, KNN method yields uncertainties below 1% if the load current and the NTC temperature are used as predictors. For the EV case, the uncertainty is around 5% when the load current and the frequency are used as predictors.

References

[1] M. H. M. Sathik, J. Pou, S. Prasanth, V. Muthu, R. Simanjorang and A. K. Gupta, "Comparison of IGBT junction temperature measurement and estimation methods-a review," 2017 Asian Conference on Energy, Power and Transportation Electrification (ACEPT), Singapore, 2017, pp. 1-8, doi: 10.1109/ACEPT.2017.8168600.

[2] W. Lai, Yunhai Wei, Minyou Chen, Hongjian Xia, Dan Luo et al., "In-Situ Calibration Method of Online Junction Temperature Estimation in IGBTs for Electric Vehicle Drives," in IEEE Transactions on Power Electronics, vol. 38, no. 1, pp. 1178-1189, Jan. 2023, doi: 10.1109/TPEL.2022.3204547.

[3] N. Fritz, T. Kamp, T. A. Polom, M. Friedel and R. W. D. Doncker, "Evaluating On-State Voltage and Junction Temperature Monitoring Concepts for Wide-Bandgap Semiconductor Devices," in IEEE Transactions on Industry Applications, vol. 58, no. 6, pp. 7550-7561, Nov.-Dec. 2022, doi:10.1109/TIA.2022.3191632.

[4] H. Hu, Z. Wang, Y. Zhou, D. Zhou, G. Xin and X. Shi, "Online Junction Temperature Monitoring for Discrete SiC MOSFET Based on On-state Voltage at High Temperature," 2022 IEEE Transportation Electrification Conference and Expo, Asia-Pacific (ITEC Asia-Pacific), Haining, China, 2022, pp. 1-4, doi: 10.1109/ITECAsia-Pacific56316.2022.9941977.

[5] Y. Yang, Q. Zhang and P. Zhang, "A Fast IGBT Junction Temperature Estimation Approach Based on ON-State Voltage Drop," in IEEE Transactions on Industry Applications, vol. 57, no. 1, pp. 685-693, Jan.-Feb. 2021, doi: 10.1109/TIA.2020.3030753

[6] Q. Chai, C. Zhang, Y. Xu and Z. Y. Dong, "PV Inverter Reliability-Constrained Volt/Var Control of Distribution Networks," in IEEE Transactions on Sustainable Energy, vol. 12, no. 3, pp. 1788-1800, July 2021, doi: 10.1109/TSTE.2021.3065451.

[7] Ranjith Kumar Gatla, Guorong Zhu, Jianghua Lu, Sainadh Singh Kshatri, Gireesh Kumar Devineni, "The impact of mission profile on system level reliability of cascaded H-bridge multilevel PV inverter",Microelectronics Reliability, Volume 138, 2022, 114639, ISSN 0026-2714,https://doi.org/10.1016/j.microrel.2022.114639.

[8] E. Knischourek and D. Gerling, "Analysis of electric vehicle driving cycles for inverter efficiency improvement at partial load," 2015 IEEE 11th International Conference on Power Electronics and Drive Systems, Sydney,

NSW, Australia, 2015, pp. 503-508, doi: 10.1109/PEDS.2015.7203539.

[9] A. Sangwongwanich, H. Wang and F. Blaabjerg, "Impact of Mission Profile Dynamics on Accuracy of Thermal Stress Modeling in PV Inverters," 2020 IEEE Energy Conversion Congress and Exposition (ECCE), Detroit, MI, USA, 2020, pp. 5269-5275, doi: 10.1109/ECCE44975.2020.9235750.

[10] Ghaderi D, Pourmahdavi M, Samavatian V, Mir O, Samavatian M. Combination of thermal cycling and vibration loading effects on the fatigue life of solder joints in a power module. Proceedings of the Institution of Mechanical Engineers, Part L: Journal of Materials: Design and Applications. 2019;233(9):1753-1763. doi:10.1177/1464420718780525

[11] Davood Ghaderi, Reliability analysis for TO-247 multilayered power module packaging under mechanical oscillation based on finite element method, Microelectronics Reliability,Volume 118, 2021, 114046, ISSN 0026-2714, https://doi.org/10.1016/j.microrel.2021.114 046.

[12] M. Fatima, A. S. Siddiqui and S. K. Sinha, "Design of Single Stage Inverter Control for Single-Phase Grid Connected Solar PV System," 2022 8th International Conference on Advanced Computing and Communication Systems (ICACCS), Coimbatore, India, 2022, pp. 223-227, doi: 10.1109/ICACCS54159. 2022.9785179.

[13] R. Gora, R. Biswas, R. K. Garg and U. Nangia, "Field Oriented Control of Permanent Magnet Synchronous Motor (PMSM) Driven Electric Vehicle And Its Performance Analysis," 2021 IEEE 4th International Conference on Computing, Power and Communication Technologies (GUCON), Kuala Lumpur, Malaysia, 2021, pp. 1-6, doi:10.1109/GUCON50781.2021.9573 814.

[14] C. Zhang, S. Srdic, S. Lukic, K. Sun, J. Wang and R. Burgos, "A SiC-Based Liquid-Cooled Electric Vehicle Traction Inverter Operating at High Ambient Temperature," in CPSS Transactions on Power Electronics and Applications, vol. 7, no. 2, pp. 160-175, June 2022, doi: 10.24295/CPSSTPEA.2022.00015.

[15] Malti Bansal, Apoorva Goyal, Apoorva Choudhary, A comparative analysis of K-Nearest Neighbor, Genetic, Support Vector Machine, Decision Tree, and Long Short Term Memory algorithms in machine learning, Decision Analytics Journal, Volume 3, 2022, 100071, ISSN 2772-6622, https://doi.org/ 10.1016/j.dajour.2022.100071.

[16] S. Bęczkowski, P. Ghimre, A. R. de Vega, S. Munk-Nielsen, B. Rannestad and P. Thøgersen, "Online Vce measurement method for wear-out monitoring of high power IGBT modules," 2013 15th European Conference on Power Electronics and Applications (EPE), Lille, France, 2013, pp. 1-7, doi: 10.1109/EPE.2013.6634390.

[17] U. -M. Choi, I. Trintis, F. Blaabjerg, S. Jørgensen and M. L. Svarre, "Advanced power cycling test for power module with on-line on-state VCE measurement," 2015 IEEE Applied Power Electronics Conference and Exposition (APEC), Charlotte, NC, USA, 2015, pp. 2919-2924, doi: 10.1109/APEC.2015.7104765.

[18] F. Pedregosa, G. Varoquaux, A. Gramfort, V. Michel, B. Thirion, O. Grisel, M. Blondel, P. Prettenhofer, R. Weiss, V. Dubourg, J. Vanderplas, A. Passos, D. Cournapeau, M. Brucher, M. Perrot and E. Duchesnay, "Scikit-learn: Machine Learning in Python", Journal of Machine Learning Research, Vol. 12, pp. 2825-2830, 2011.

PCIM Europe 2024, 11– 13 June 2024, Nuremberg   DOI: 10.30420/566262425

Robustness Analysis of Temperature-Sensitive Electrical Parameters of IGBTs

Laurids Schmitz [1], Tetsuya Kojima [2], Rik W. De Doncker [1]

[1] Institute for Power Electronics and Electrical Drives (ISEA), RWTH Aachen University, Germany
[2] Advanced Technology R&D Center, Mitsubishi Electric Corporation, Hyogo, Japan

Corresponding author: Laurids Schmitz, post@isea.rwth-aachen.de
Speaker: Laurids Schmitz, post@isea.rwth-aachen.de

Abstract

Temperature-sensitive electrical parameters (TSEPs) are a promising quantity for online high-bandwidth junction temperature measurements. Besides their sensitivity to temperature, these parameters are also susceptible to parasitic influences, such as dc-link voltage, load current, gate supply voltages, making their application challenging. Thus far, this has only been examined in a limited scope for individual parameters. Based on double-pulse measurements, this paper aims to evaluate and compare multiple TSEPs with regard to their robustness in the presence of such parasitic influences.

1 Introduction

Power semiconductors as part of power electronic systems are increasingly used in high-maintenance applications like railway traction drives or wind power generation [1], [2] and are in general one of the most critical components in converters [3]. Since most aging effects have thermal origins, this makes monitoring the junction temperature important as it enables the detection of possible degradation mechanisms and allows state-of-health (SoH) estimation of power modules [4].

Several methods allow the measurement or estimation of the junction temperature. Infrared thermography is only used in laboratory environment and module-integrated physical sensors can often only extract the average substrate temperature. In contrast, temperature-sensitive optical parameters (TSOPs) and TSEPs have shown promising characteristics. They offer high bandwidths and do not require any additional space in the module [4].

Thus far, the evaluation of TSEPs focused mostly on their sensitivity with respect to temperature and less on their robustness with regards to other parasitic influences. These, however, have a massive impact on the precision and feasibility of TSEP-based temperature estimation if they are neglected or not taken into account properly.

A comparative study on the temperature sensitivity of TSEPs and their feasibility has been shown in [5]. In [6], parasitic influences like gate supply voltage

and gate resistance are already examined, but only the turn-on delay time and the current rise time are considered in detail.

In this paper, a sensitivity analysis with regard to temperature T, dc-link voltage V_{dc}, load current I_{L}, turn-on and turn-off gate resistance $R_{\mathrm{g,on}}$, $R_{\mathrm{g,off}}$, and positive and negative gate supply voltage $V_{\mathrm{g,on}}$, $V_{\mathrm{g,off}}$, is conducted based on double-pulse measurements. A larger set of TSEPs, consisting of turn-off delay $t_{\mathrm{d,off}}$, turn-on delay $t_{\mathrm{d,on}}$, on-voltage $v_{\mathrm{ce,on}}$, rising and falling slopes of current, $\mathrm{d}i/\mathrm{d}t_{\mathrm{rise,1090}}$, $\mathrm{d}i/\mathrm{d}t_{\mathrm{fall,max}}$, and voltage, $\mathrm{d}v/\mathrm{d}t_{\mathrm{rise,max}}$, $\mathrm{d}v/\mathrm{d}t_{\mathrm{fall,max}}$, as well as the Miller-plateau voltage during the turn-off transition $v_{\mathrm{ge,m,h}}$ is analyzed.

2 Measurement Setup

The selected TSEPs are evaluated for Mitsubishi's insulated-gate bipolar transistor (IGBT) power module CM600DE-66X ($3.3\,\mathrm{kV}$, $600\,\mathrm{A}$) which is used in railway applications. The measurements are conducted with the double-pulse test bench introduced in [7]. The module temperature is controlled using a temperature control unit based on Peltier elements as introduced in [8] and is varied from $25\,^\circ\mathrm{C}$ to $120\,^\circ\mathrm{C}$. The dc-link voltage is varied from $100\,\mathrm{V}$ to $900\,\mathrm{V}$ in $200\,\mathrm{V}$ steps, the load current from $75\,\mathrm{A}$ to $275\,\mathrm{A}$ in $50\,\mathrm{A}$ steps.

To evaluate the influence of the external gate resistances, their values are varied by $\pm50\,\%$ of the nominal values specified in the power module data

2995

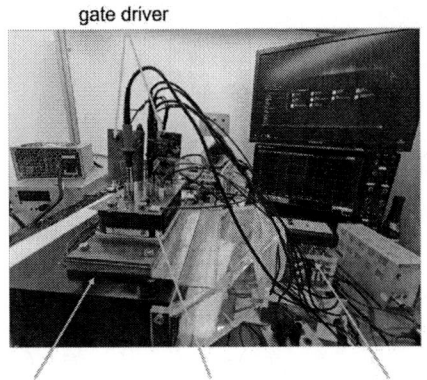

temperature control unit IGBT module double pulse module

Fig. 1: The measurement setup for the double-pulse tests.

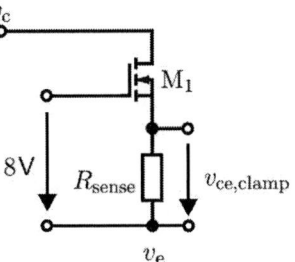

Fig. 2: Simplified schematic diagram of the clamping circuit used for $v_{\mathrm{ce,on}}$ measurements.

sheet ($2.2\,\Omega/50\,\Omega$). The gate supply voltages are varied by $\pm 2\,$V with respect to their nominal value ($-10\,$V/$15\,$V).

The gate resistances $R_{\mathrm{g,on}}$ and $R_{\mathrm{g,off}}$ are varied simultaneously, as they either influence the turn-on or the turn-off switching transition. Thus, for the sensitivity analysis of every switching-transition related TSEP, only the relevant gate-resistance is considered.

The setup for the double-pulse measurements is depicted in Fig. 1. The power module is mounted on the heat plate of the Peltier element-based temperature control unit. The printed circuit board (PCB) on top of the power module is an adapter board. It establishes the electrical connections between the power module and the double-pulse test bench. It also allows the connection of the high-side and low-side drivers to the power module and enables the measurement of all required voltages using passive probes. Voltages are measured with *PP021* probes by *LeCroy* and the current is measured with the Rogowski coil *CWT Mini* using the *WaveRunner 8208HD* oscilloscope with a sample rate of $10\,$GS/s. To be able to also assess the forward voltage $v_{\mathrm{ce,on}}$ within the same double-pulse test setup, a MOSFET based clamping circuit as depicted in Fig. 2 which has already been described in [9] is included on the adapter PCB. Using this clamping circuit, the voltage probe does not see the entire dc-link voltage in the off-state of the evaluated IGBT, as the voltage is limited by the gate voltage of the clamping MOSFET M_1. By setting the gate voltage of transistor M_1 to $8\,$V, all voltages above that voltage minus M_1's threshold voltage are clipped.

This allows for $v_{\mathrm{ce,on}}$ measurements with a much higher resolution compared to measurements without such a clamping circuit.

3 Measurement Results

In Fig. 3, the extraction of the evaluated TSEP candidates is visualized for an exemplary operating point for the turn-on and turn-off switching transition, respectively.

The turn-off delay $t_{\mathrm{d,off}}$ analyzed in this paper is defined as the duration from v_{ge} dropping by $10\,\%$ to i_{e} dropping by $10\,\%$ with respect to their previous steady state values. The resulting delay is highlighted in Fig. 3 (a).

For the voltage and current slopes, the maximum slopes during the switching transient, $\mathrm{d}i/\mathrm{d}t_{\mathrm{fall,max}}$ and $\mathrm{d}v/\mathrm{d}t_{\mathrm{rise,max}}$, are used for the evaluation. Both are highlighted in Fig. 3 (a). Due to the tail current of i_{e}, it was refrained from using other possible definitions, such as the average slope between $10\,\%$ to $90\,\%$ of the transition.

The turn-on delay $t_{\mathrm{d,on}}$ is defined opposite to the turn-off delay as the duration from v_{ge} rising by $10\,\%$ to i_{e} rising by $10\,\%$ with respect of their steady-state values after the switching transition, as illustrated for the exemplary waveform in Fig. 3 (b).

For the slopes during the turn-on event, for the voltage slope, the maximum slope within the part of the voltage drop after the current commutation, $\mathrm{d}v/\mathrm{d}t_{\mathrm{fall,max}}$, is used as illustrated in Fig. 3 (b). By doing so, the initial voltage drop, caused by the voltage induced over the parasitic inductances because of the rise of i_{e} is not represented by this parameter. For the current slope, the average slope between $10\,\%$ and $90\,\%$ of the steady state value of i_{e} after the switching transition, $\mathrm{d}i/\mathrm{d}t_{\mathrm{rise,1090}}$, is used. The reason is that for some operating points (as the one visualized in Fig. 3 (b)), the current overshoot is clipped, which could lead to distortions

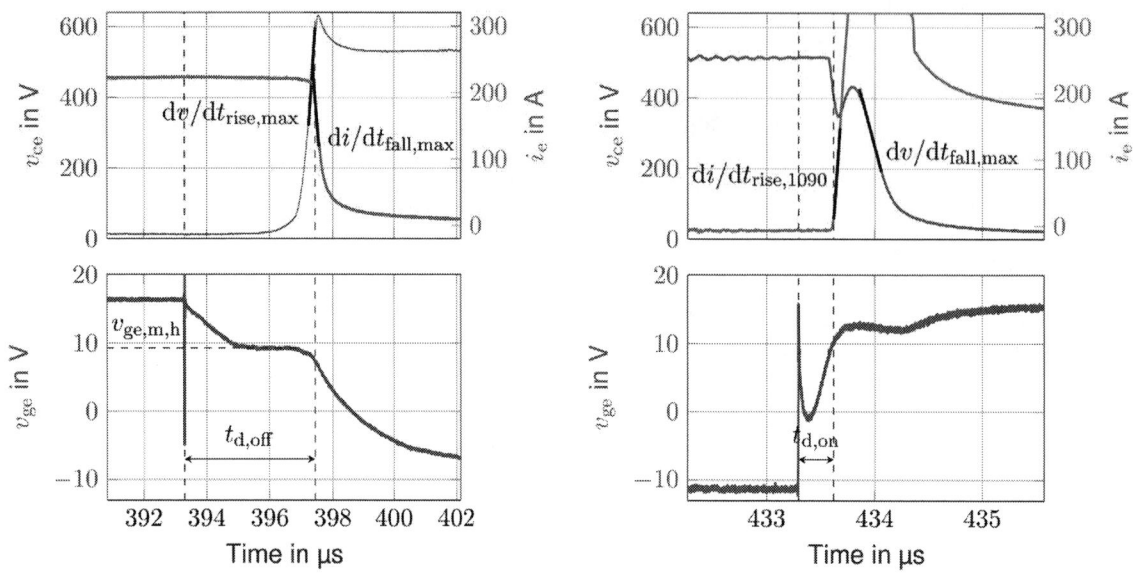

(a) Acquisition of the TSEPs $v_{g,m,h}$, $t_{d,off}$, $di/dt_{fall,max}$ and $dv/dt_{rise,max}$.

(b) Acquisition of the TSEPs $t_{d,on}$, $dv/dt_{fall,max}$ and $di/dt_{rise,1090}$.

Fig. 3: Acquisition of the selected TSEPs for the turn-off (a) and turn-on transition (b).

in the following analysis. Due to the high linearity of the current slope, no big differences are to be expected between the selected slope $di/dt_{rise,1090}$ and a potential maximum slope $di/dt_{rise,max}$ analysis for a measurement without clipping.

The on-voltage $v_{ce,on}$ is sampled from the clamped collector-emitter voltage $v_{ce,clamp}$ right before the gate-emitter voltage is starting to drop.

As described in section 2, double-pulse tests are conducted for a large set of operating points with different T, V_{dc}, I_L, $V_{g,off}$, $V_{g,on}$, $R_{g,off}$, $R_{g,on}$ with $R_{g,off}$ and $R_{g,on}$ being varied at the same time, as they either influence the turn-on or the turn-off transition. For each operating point, the presented TSEPs are extracted. The data is then used to derive their sensitivity with regard to the desired influence temperature T and the parasitic influences. This process is shown at the hands of the TSEP $t_{d,off}$ in the following.

A selection of the extracted values for the turn-off delay $t_{d,off}$ can be seen in Fig. 4. In all figures, $t_{d,off}$ is plotted over current I_L. Fig. 4 (a) shows the temperature dependency which is used for the temperature sensitivity calculation. The parasitic influences are parameterized in Fig. 4 (b-e). The extracted values from the measurements are marked with the bigger dots. To be able to interpolate $t_{d,off}$ at each current level which subsequently allows a sensitivity analysis with respect to I_L and the

other influences, the extracted values are fitted to functions $t_{d,off,fit}(I_L)$. The dotted lines in Fig. 4 represent these fitted functions $t_{d,off,fit}(I_L)$. To derive the sensitivity of $t_{d,off}$ with respect to I_L, the derivative of the fitted functions at an arbitrary operating point OP is calculated:

$$s_{t_{d,off},I_L}\bigg|_{OP} = \frac{dt_{d,off,fit}(I_L)}{dI_L}\bigg|_{OP} \quad (1)$$

To obtain the sensitivity of $t_{d,off}$ with respect to the other influences, $t_{d,off,fit}(I_L)$ is sampled at equidistant current levels and at each of these current levels a function is fitted with respect to any of the remaining influences. For instance, to determine the temperature sensitivity of $t_{d,off}$, Fig. 4 (f) is derived from Fig. 4 (a): The big dots represent the sampled values from Fig. 4 (a) whereas the dotted line represent the fitted functions with respect to temperature, $t_{d,off,fit}(T)$. The derivative of this function at an arbitrary operating point yields the sensitivity of $t_{d,off}$ with respect to T, $s_{t_{d,off},T}\big|_{OP}$.

A selection of the resulting sensitivities for $t_{d,off}$, $s_{t_{d,off}}$ at $T = 100°C$, $V_{dc} = 500V$, $V_{g,off} = -10V$, $V_{g,on} = 15V$, $R_{g,off} = 50\Omega$ and $R_{g,on} = 2.2\Omega$, are visualized in Fig. 5. Each dot represents the sensitivity $s_{t_{d,off}}$ at different current levels with respect to the different influences. In Fig. 5 (a) and (b), it can be seen that the absolute sensitivity $s_{t_{d,off}}$ with respect to T, V_{dc} and I_L decreases for higher

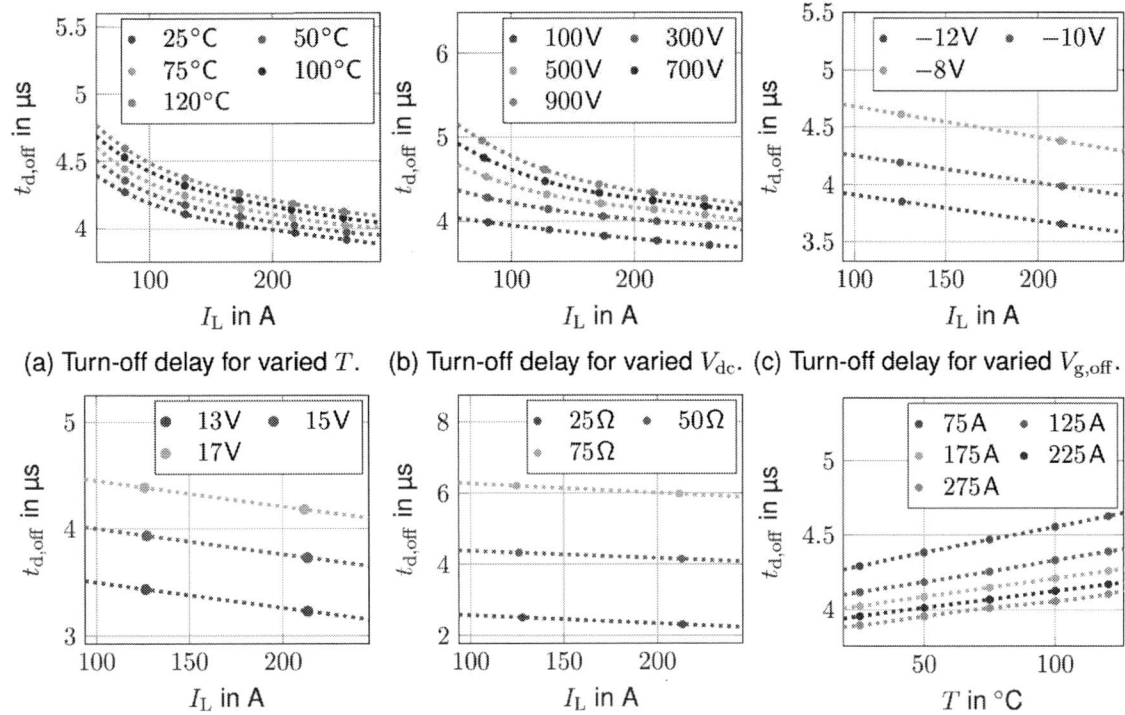

(a) Turn-off delay for varied T.   (b) Turn-off delay for varied V_{dc}.   (c) Turn-off delay for varied $V_{\mathrm{g,off}}$.

(d) Turn-off delay for varied $V_{\mathrm{g,on}}$.(e) Turn-off delay for varied $R_{\mathrm{g,off}}$.   (f) Turn-off delay for varied I_{L}.

Fig. 4: Measurement results for several influences on the turn-off delay of IGBTs for (if not varied) $T = 100\,°\mathrm{C}$, $V_{\mathrm{dc}} = 500\,\mathrm{V}$, $I_{\mathrm{L}} = 225\,\mathrm{A}$, $V_{\mathrm{g,off}} = -10\,\mathrm{V}$, $V_{\mathrm{g,on}} = 15\,\mathrm{V}$, $R_{\mathrm{g,off}} = 50\,\Omega$, $R_{\mathrm{g,on}} = 2.2\,\Omega$ (markers represent extracted values from the measurements for (a-e) and the interpolated values for (f), dotted lines represent the fitted functions).

current levels, while it decreases less for the other influences (Fig. 5 (c) and (d)).

Analogous to the described procedure to determine the sensitivities $s_{t_{\mathrm{d,off}}}$, similar steps are conducted to derive the sensitivities for the remaining evaluated TSEPs. With the fitted functions for each TSEP, their sensitivity can be determined for arbitrary operating points. The results for one specific operating point, $T = 100\,°\mathrm{C}$, $V_{\mathrm{dc}} = 500\,\mathrm{V}$, $I_{\mathrm{L}} = 225\,\mathrm{A}$, $V_{\mathrm{g,off}} = -10\,\mathrm{V}$, $V_{\mathrm{g,on}} = 15\,\mathrm{V}$, $R_{\mathrm{g,off}} = 50\,\Omega$, $R_{\mathrm{g,on}} = 2.2\,\Omega$ are summarized in Tab. 1.

The sensitivities of all turn-on related parameters with respect to the turn-off resistance $R_{\mathrm{g,off}}$, and vice-versa, are shown as "-" in the table.

As the TSEPs and influences and thus also the sensitivities have different units, two adjustments are made to allow for better comparability, and thus understanding of the values: First, the sensitivities are re-scaled to show the change in the TSEP in the presence of a $1\,\%$ deviation of the respective influence from its nominal operating point value. E.g., a deviation of V_{dc} of $5\,\mathrm{V}$ for the nominal value of

$500\,\mathrm{V}$. Although still an arbitrary choice, this allows for a much better comparison between the different influences. Second, for all parasitic influences, the resulting deviations in the TSEP values are translated into temperature estimation errors. Tab. 2 shows the resulting temperature estimation errors in the case of a $1\,\%$ deviation of the respective influence for each TSEP. This table allows a comparison of the robustness of all TSEPs with respect to all parasitic influences. Small values indicate a high robustness of the TSEP towards a certain influence, whereas high values signal a high susceptibility to that influence. For example, row 2 of Tab. 1 shows that the turn-off delay $t_{\mathrm{d,off}}$ is very susceptible to changes in the gate supply voltages $V_{\mathrm{g,on}}$ and $V_{\mathrm{g,off}}$ as well as the turn-off gate resistance $R_{\mathrm{g,off}}$ while showing less reaction to changes in V_{dc} and I_{L}.

To simplify the comparison of the robustness of the TSEPs with regard to the different influences, the values from Tab. 2 are visualized in polar plots. The estimation errors in K for the turn-off transition

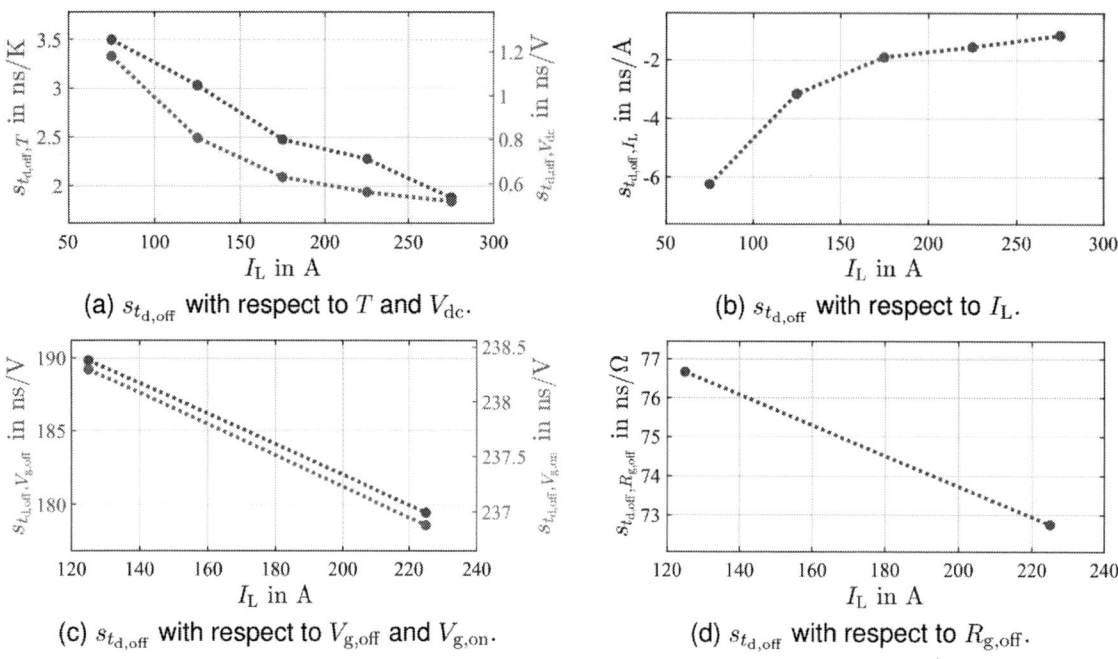

(a) $s_{t_{d,off}}$ with respect to T and V_{dc}.

(b) $s_{t_{d,off}}$ with respect to I_L.

(c) $s_{t_{d,off}}$ with respect to $V_{g,off}$ and $V_{g,on}$.

(d) $s_{t_{d,off}}$ with respect to $R_{g,off}$.

Fig. 5: Determined sensitivities $s_{t_{d,off}}$ over I_L at $T = 100°C$, $V_{dc} = 500\,V$, $V_{g,off} = -10\,V$, $V_{g,on} = 15\,V$, $R_{g,off} = 50\,\Omega$, $R_{g,on} = 2.2\,\Omega$.

Tab. 1: Sensitivities s in $\mathrm{unit\,TSEP}/\mathrm{unit\,influence}$ for the selected TSEPs and influences.

Sensitivity/Influence	T in K	V_{dc} in V	I_L in A	$V_{g,on}$ in V	$V_{g,off}$ in V	$R_{g,on}$ in Ω	$R_{g,off}$ in Ω
$s_{v_{ce,on}}$ in $\mathrm{mV}/$unit infl.	3.16	0.02	3.35	-24.80	-1.04	1.00	-0.32
$s_{t_{d,off}}$ in $\mathrm{ns}/$unit infl.	2.28	0.56	-1.55	236.87	179.44	-	72.75
$s_{di/dt_{fall,max}}$ in $\mathrm{A}/\mathrm{\mu s}/$unit infl.	1.93	-0.02	-2.30	-1.04	11.13	-	6.73
$s_{dv/dt_{rise,max}}$ in $\mathrm{V}/\mathrm{\mu s}/$unit infl.	-2.00	2.22	0.91	-5.37	-48.12	-	-16.85
$s_{v_{gs,m,h}}$ in $\mathrm{mV}/$unit infl.	-5.84	-0.40	8.49	46.69	0.94	-	14.77
$s_{t_{d,on}}$ in $\mathrm{ns}/$unit infl.	-0.04	-0.03	0.11	-23.57	-15.46	71.96	-
$s_{di/dt_{rise,1090}}$ in $\mathrm{A}/\mathrm{\mu s}/$unit infl.	-3.83	1.83	4.12	336.55	6.90	-948.64	-
$s_{dv/dt_{fall,max}}$ in $\mathrm{V}/\mathrm{\mu s}/$unit infl.	0.45	-2.03	1.29	-142.52	-18.32	203.23	-

related TSEPs are presented in Fig. 6 whereas the estimation errors for the turn-on transition related TSEP and the static TSEP $v_{ce,on}$ are shown in Fig. 7 and 8. Points closer to the middle of the polar plots indicate a better robustness of the TSEP with regard to the certain influence whereas points further to the outside indicate a high susceptibility to that influence and thus, a low robustness. Note the different scales: Due to their much higher susceptibility with regard to some influences, $t_{d,on}$ and $dv/dt_{fall,max}$ are presented separately in Fig. 8.

Fig. 6 shows that all turn-off transition related TSEPs show a significant susceptibility with regards to $R_{g,off}$ and $V_{g,off}$ as expected. The turn-off delay, additionally, is also susceptible to $V_{g,on}$ as it influ-

ences the duration till the Miller plateau is reached during the turn-off transition. $dv/dt_{rise,max}$ is also susceptible to V_{dc} whereas $v_{gs,m,h}$ is most susceptible with regard to the current level. The Miller plateau $v_{gs,m,h}$ and the current slope $di/dt_{fall,max}$ show the best overall robustness.

As expected, the turn-on transition related TSEPs all have a significant susceptibility with regard to $R_{g,on}$ and $V_{g,on}$. Similarly to the turn-off delay, the turn-on delay also depends on the second gate supply voltage, $V_{g,off}$ in this case, as well. $di/dt_{rise,1090}$ shows the best robustness of the turn-on related parameters.

Last, the on-voltage $v_{ce,on}$ shows a high robustness, while only being significantly impacted by the influ-

Tab. 2: Deviation in temperature estimation given a 1 % deviation of the influence with respect to its nominal value.

ΔT(TSEP)/Infl. Δ	V_{dc} 5 V	I_{L} 2.25 A	$V_{\mathrm{g,on}}$ 0.15 V	$V_{\mathrm{g,off}}$ 0.1 V	$R_{\mathrm{g,on}}$ 0.022 Ω	$R_{\mathrm{g,off}}$ 0.5 Ω
$\Delta T(v_{\mathrm{ce,on}})$ in K	0.04	2.39	-1.18	-0.03	0.01	-0.05
$\Delta T(t_{\mathrm{d,off}})$ in K	1.23	-1.53	15.61	7.88	-	15.98
$\Delta T(\mathrm{d}i/\mathrm{d}t_{\mathrm{fall,max}})$ in K	-0.05	-2.68	-0.08	0.58	-	1.74
$\Delta T(\mathrm{d}v/\mathrm{d}t_{\mathrm{rise,max}})$ in K	-5.54	-1.02	0.40	2.40	-	4.21
$\Delta T(v_{\mathrm{gs,m,h}})$ in K	0.35	-3.27	-1.20	-0.02	-	-1.26
$\Delta T(t_{\mathrm{d,on}})$ in K	3.82	-6.93	94.96	41.53	-42.51	-
$\Delta T(\mathrm{d}i/\mathrm{d}t_{\mathrm{rise,1090}})$ in K	-2.39	-2.42	-13.19	-0.18	5.45	-
$\Delta T(\mathrm{d}v/\mathrm{d}t_{\mathrm{fall,max}})$ in K	-22.58	6.45	-47.50	-4.07	9.94	-

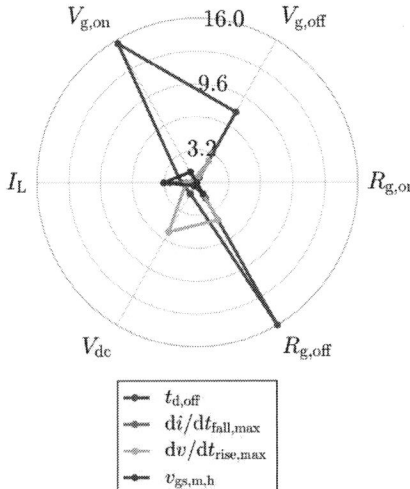

Fig. 6: Temperature estimation error in K for a 1 % influence deviation with respect to its nominal value.

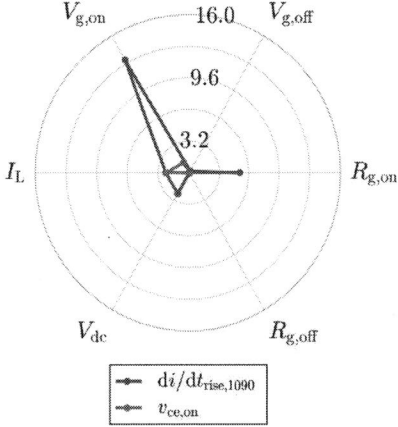

Fig. 7: Temperature estimation error in K for a 1 % influence deviation with respect to its nominal value.

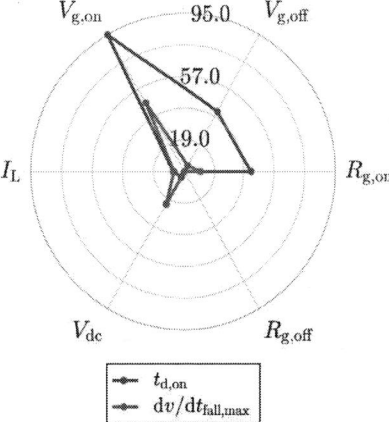

Fig. 8: Temperature estimation error in K for a 1 % influence deviation with respect to its nominal value.

ences I_{L} and $V_{\mathrm{g,on}}$.

4 Conclusion

The presented measurements of the selected TSEPs have shown that they are all susceptible to several other parasitic influences, which has not been studied in detail before. If the influences are not accounted for, this makes it difficult to use the TSEPs' existing and desired sensitivity with regards to temperature for the junction temperature estimation, as it degrades the accuracy.

This paper lays the foundation for a better understanding of such influences which allows to compensate them through calibration. To be able to compare the different TSEPs and their robustness with regard to the different parasitic influences with each other, the resulting temperature estimation error in K for a 1 % influence deviation with respect to its nominal value was introduced.

The Miller plateau $v_{\mathrm{gs,m,h}}$, the current slope $\mathrm{d}i/\mathrm{d}t_{\mathrm{fall,max}}$ and the forward voltage $v_{\mathrm{ce,on}}$ showed the highest robustness towards the evaluated influences.

The presented sensitivities can be used to determine precision requirements for the different influences during the design phase. Additionally, it allows to select suitable parameters for sensor fusion concepts in which the information of more than one parameter is combined to yield higher robustness. As the next step, different aging mechanisms can be added to the analysis. They can be either treated as parasitic influence for temperature estimation as in this paper or - if degradation detection is the goal - as desired influence.

References

[1] Z. Chen, J. M. Guerrero, and F. Blaabjerg, "A review of the state of the art of power electronics for wind turbines," *IEEE Transactions on Power Electronics*, vol. 24, no. 8, pp. 1859–1875, 2009. DOI: 10.1109/TPEL.2009.2017082.

[2] D. Ronanki, S. A. Singh, and S. S. Williamson, "Comprehensive topological overview of rolling stock architectures and recent trends in electric railway traction systems," *IEEE Transactions on Transportation Electrification*, vol. 3, no. 3, pp. 724–738, 2017. DOI: 10.1109/TTE.2017.2703583.

[3] S. Yang, A. Bryant, P. Mawby, D. Xiang, L. Ran, and P. Tavner, "An industry-based survey of reliability in power electronic converters," *IEEE Transactions on Industry Applications*, vol. 47, no. 3, pp. 1441–1451, 2011. DOI: 10.1109/TIA.2011.2124436.

[4] S. Kalker, L. A. Ruppert, C. H. van der Broeck, J. Kuprat, M. Andresen, *et al.*, "Reviewing thermal-monitoring techniques for smart power modules," *IEEE Journal of Emerging and Selected Topics in Power Electronics*, pp. 1–1, Mar. 2021. DOI: 10.1109/jestpe.2021.3063305.

[5] H. Yu, X. Jiang, J. Chen, Z. J. Shen, and J. Wang, "Comparative study of temperature sensitive electrical parameters for junction temperature monitoring in sic mosfet and si igbt," in *2020 IEEE 9th International Power Electronics and Motion Control Conference (IPEMC2020-ECCE Asia)*, 2020, pp. 905–909. DOI: 10.1109/IPEMC-ECCEAsia48364.2020.9367830.

[6] D. Herwig, T. Brockhage, and A. Mertens, "Impact of parasitics in power modules and gate drivers on TSEP-based temperature estimations," in *2020 23rd International Conference on Electrical Machines and Systems (ICEMS)*, 2020, pp. 804–809. DOI: 10.23919/ICEMS50442.2020.9291091.

[7] J. Gottschlich, M. Kaymak, M. Christoph, and R. W. De Doncker, "A flexible test bench for power semiconductor switching loss measurements," in *International Conference on Power Electronics and Drive Systems (PEDS)*, IEEE, Jun. 2015, pp. 442–448. DOI: 10.1109/PEDS.2015.7203495.

[8] G. Engelmann, M. Laumen, K. Oberdieck, and R. W. De Doncker, "Peltier module based temperature control system for power semiconductor characterization," in *International Power Electronics and Motion Control Conference (PEMC)*, IEEE, Sep. 2016, pp. 957–962. DOI: 10.1109/EPEPEMC.2016.7752123.

[9] N. Fritz, T. Kamp, T. A. Polom, M. Friedel, and R. W. De Doncker, "Evaluating on-state voltage and junction temperature monitoring concepts for wide-bandgap semiconductor devices," *IEEE Transactions on Industry Applications*, vol. 58, no. 6, pp. 7550–7561, 2022. DOI: 10.1109/tia.2022.3191632.

PCIM Europe 2024, 11– 13 June 2024, Nuremberg　　　DOI: 10.30420/566262426

Observation of Thermal-Resistance Increase of Degraded IGBT Modules by $V_{CE(sat)}$ Measurement in a Chopper Circuit

Kazunori Hasegawa[1] , Hisaki Ueda[1], Kanta Hara[1], Nobuyuki Shishido[2] , Satoshi Nakano[3], Wataru Saito[3] , Tamotsu Ninomiya[4]

[1] Kyushu Institute of Technology, Japan
[2] Kindai University, Japan
[3] Kyushu University, Japan
[3] NPERC-J, Japan

Corresponding author:　Kazunori Hasegawa, hasegawa@ele.kyutech.ac.jp
Speaker:　　　　　　　Kazunori Hasegawa, hasegawa@ele.kyutech.ac.jp

Abstract

This paper presents an observation of the increase in thermal resistance of IGBT modules degraded by power-cycling test. The observation is based on a $V_{CE(sat)}$ measurement setup and thermal analysis by a three-dimensional structure model of the module. The $V_{CE(sat)}$ measurement setup consists of a sensing circuit and a low-cost IoT platform "Leafony," which provides the junction temperature T_j profile of an IGBT chip inside the module. The three-dimensional structure model produces predicted T_j profile as a bench mark, in which the transient thermal impedance is evaluated by heat transfer analysis. The comparison between the measured and predicted profiles shows increases in thermal resistance in degraded IGBT modules, where two intentionally-degraded modules were created by a power tester. The demonstrated simple platform and T_j prediction are useful to monitor the health condition on the IGBT modules.

1　Introduction

Insulated-gate bipolar transistors (IGBTs) used in high-power inverters always face a significantly long operating time and thus have to be highly reliable, applications of which include flexible AC transmission system (FACTS), medium-voltage industrial motor-drives, electric and hybrid-electric vehicles (EV/HEVs), and so on.

A questionnaire survey of reliability in power electronic converters from industry showed that power semiconductor devices including IGBTs were chosen by 31% of fragile components responsible for converter failure [1]. To overcome this problem, much attention has been paid to failure analyses, simulation methods, and condition monitoring of IGBTs and other semiconductor devices [2]-[11].

Faire of IGBTs results not only from catastrophic ones but also from ageing degradation. Fig. 1 is a cross-section view inside an IGBT module that comprises a copper base plate, solder layers, an insulated substrate, IGBT/diode chips, and aluminum bonding wires. These materials have different coefficients of thermal expansion each other,

which yields mechanical stress when the IGBT/diode chips are conducting. This mechanical stress will result in bonding-wire lift-off, solder clack, base plate crack, and so on. The solder clack is a major degradation, which increases the thermal resistance between an IGBT chip inside the module and the ambient. Hence, the junction temperature is a sensitive indicator of the degradation. Many monitoring methods have been demonstrated in the previous works [2]-[8], which deal with junction temperature measurement, thermal cycle characteristics, and life time estimation. They, however, do not sufficiently investigate how to evaluate the degradation level with an appropriate benchmark.

The authors of this paper have developed a power-cycling degradation monitoring method of an IGBT module with $V_{CE(sat)}$ measurement to estimate the junction temperature [11]. The method also combines junction temperature prediction with a three-dimensional structure model.

This paper presents experimental verifications of increase in thermal resistance of degraded IGBT modules with the degradation monitoring method, where the three-dimensional structure model is

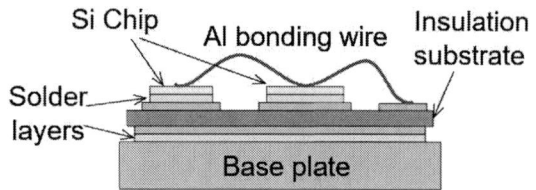

Fig. 1 Cross-section view inside an IGBT module

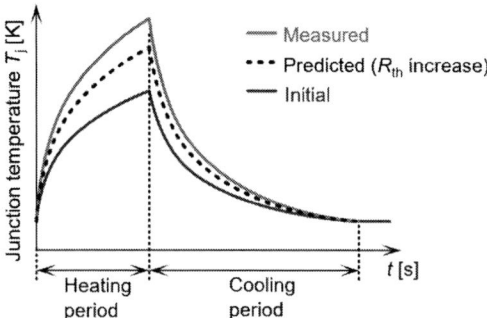

Fig. 2 Measured, predicted, and initial temperature profiles of the junction temperature T_j

used for predicting the junction temperature T_j as a benchmark. The degradation level is confirmed from comparison between the measured and predicted T_j profiles of one of IGBT chips in the module with a chopper circuit providing a constant current.

2 Junction Temperature Estimation with $V_{CE(sat)}$ Measurement

2.1 Temperature Profile of T_j

Fig. 2 shows predicted, measured, and initial temperature profiles of the junction temperature T_j, where the heating period shows that the device under test (DUT) is conducting, whereas the cooling period does that no current flows into the DUT, i.e., the DUT produces no power loss. The rise of the T_j is proportional to the thermal resistance R_{th} if the dissipated power and heating time of the IGBT chip are constant. The temperature profile is predicted by the three-dimensional structure model as discussed in the next section. One can obtain the predicted profile according to the thermal resistance that is the benchmark of the lifetime because the structure model combines the thermal resistance. Thus, the failure of the IGBT module is observed if the measured junction temperature becomes more than the predicted one.

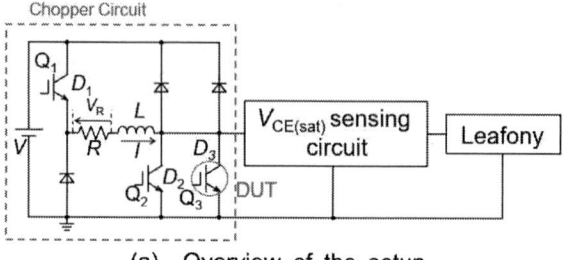

(a) Overview of the setup.

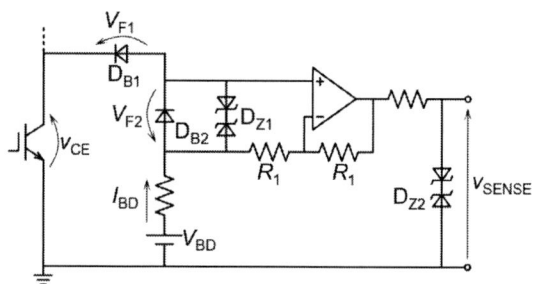

(b) Configuration of the $V_{CE(sat)}$ sensing circuit.

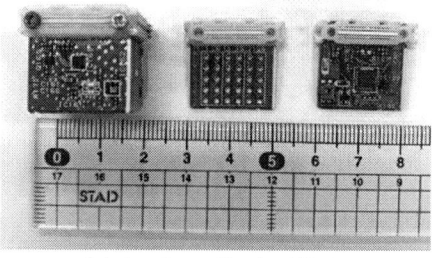

(c) Leafony Basic Kit 2.

Fig. 3 $V_{CE(sat)}$ measurement setup.

2.2 $V_{CE(SAT)}$ Measurement Setup

Fig. 3 shows the $V_{CE(sat)}$ measurement setup that consists of a chopper circuit, a $V_{CE(sat)}$ sensing circuit, and an IoT platform "Leafony [12]." The chopper circuit consisting of a 1200-V 75-A six-in-one IGBT module (Infineon FS75R12KT4B15) provides a constant current to the DUT. The reason why Q_2 is connected in parallel with the DUT is to make the load current I constant even though the DUT is in cooling period [11]. The constant load current also guarantees that the current flowing into the DUT constant.

Fig. 3(b) shows the sensing circuit that consists of an operational amplifier and blocking diodes D_{B1}. This circuit eliminates a high voltage in the off state of the DUT and produces the same voltage (v_{SENSE}) as $V_{CE(sat)}$ only in the on-state [2].

Fig. 3(c) is a photo of the Leafony Basic Kit 2, which is a tiny IoT platform having a 32-bit microcontroller (STM32), a 12-bit analog-to-digital (A/D) converter, a universal serial bus (USB), and Bluetooth interfaces [9]. The sensing circuit transfers $V_{CE(sat)}$ to the Leafony only when the DUT turns on, where the Leafony receives the synchronous signals from a field-programmable gate array (FPGA) that also produces turn-on/off signals for Q_1, Q_2, and Q_3.

2.3 Degraded Modules Used for Devices under Test

This paper introduces two intentionally-degraded modules that were produced by a power tester (Siemens POWERTESTER 1500A). The thermal resistances of the two modules are higher than those of the non-degraded one by +10% and +28%, respectively. The relation among the junction temperature T_j, collector current I_C, and $V_{CE(sat)}$ is necessary to obtain the thermal profile. Hence, this paper measured the relation with a semiconductor curve tracer (IWATSU CS-3200) before testing the two modules.

3 Junction Temperature Prediction

Fig. 4(a) shows the three-dimensional structure of finite element model that is constructed for the prediction, where the bonding wires, the case and the silicone gel are neglected due to their minimal contribution to heat transfer from the DUT to the ambient. Fig. 4(b) shows the cross-section of the multilayered structure, where the material properties of those materials are summarized in Table 1.

The junction temperature $T_j(t)$ is estimated by the transient thermal impedance $Z_{th\_jc}(t)$ and heat generation $P(t)$ as follows:

$$T_j(t) - T_c(t) = \int_0^t P(\tau) Z_{th\,jc}(t - \tau) d\tau \quad (1)$$

The transient thermal impedance of the DUT (Q3) was evaluated through heat transfer analysis using the three-dimensional structural model. The heat generation $P(t)$ is the sum of the conduction loss (the product of $V_{CE(sat)}$ and the load current) and the switching loss calibrated by using the experimental result in advance. The temperature rise is confined to the vicinity of the IGBT chip of interest, so the case temperature $Tc(t)$ can be considered to be constant and equal to the ambient temperature. For the two degraded module, the junction temperature was predicted under the assumption that the thermal impedance would be uniformly increased by 10% and 28%, respectively.

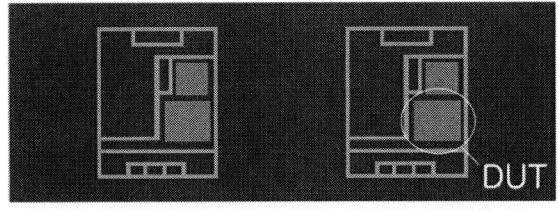

(a) The whole of three-dimensional structure

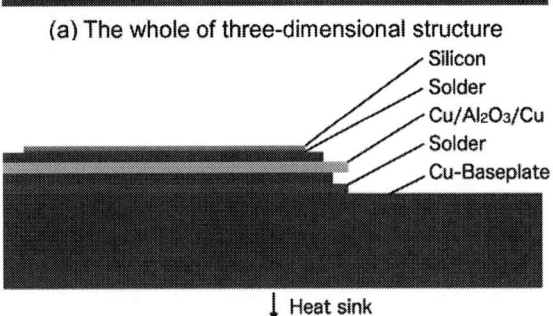

(b) Cross section of the multi-layered structure

Fig. 4 Three-dimensional structure model of the test module.

Table 1 Thickness and material properties.

	Thickness [mm]	Density [kg/m³]	Specific heat [J/kgK]	Thermal conductivity [W/mK]
Silicon	0.13	2340	729	139
Solder	0.08	7300	226	73
Cu	0.3	8960	386	397
Al₂O₃	0.35	3900	850	35
Cu	0.35	8960	386	397
Solder	0.3	7300	226	73
Cu baseplate	3.00	8960	386	397

4 Experimental Results

Fig. 5 shows temperature profiles of +10% degraded module with different load currents of 50 and 70 A and different switching frequencies of 20 and 30 kHz, along with initial and predicted ones. The initial profile indicates that with non-degraded module.

Fig. 6 shows those of +28% degraded module with different load currents and switching frequencies, along with predicted ones. The temperature rise increased according to the thermal resistances of the modules. These results demonstrate the experimental verification of the measurement setup for temperature profile measurements. The reason why the measured profiles included spikes was that the $V_{CE(sat)}$ sensing circuit is somewhat affected by noises from the chopper circuit.

The predicted profiles well agreed with the measured ones for the 10% increased modules. On the

PCIM Europe 2024, 11– 13 June 2024, Nuremberg   DOI: 10.30420/566262426

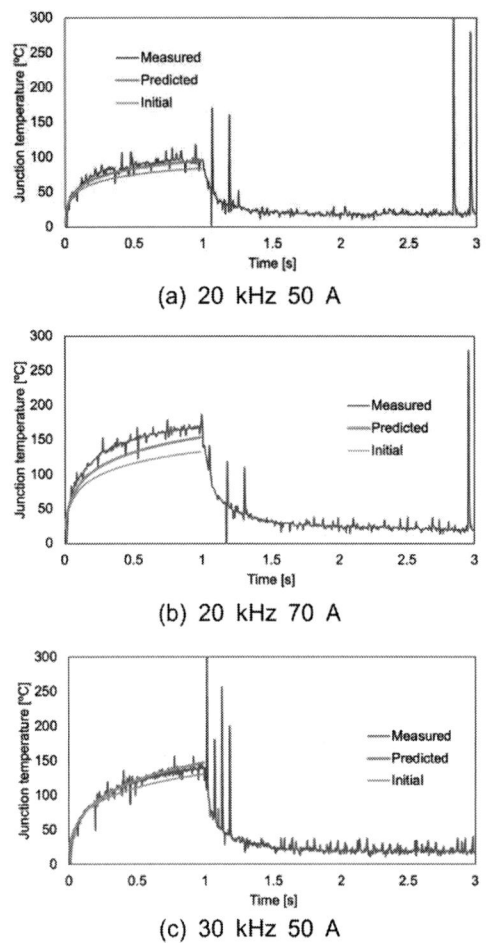

(a) 20 kHz 50 A

(b) 20 kHz 70 A

(c) 30 kHz 50 A

Fig. 5 Temperature profile with +10% degraded module.

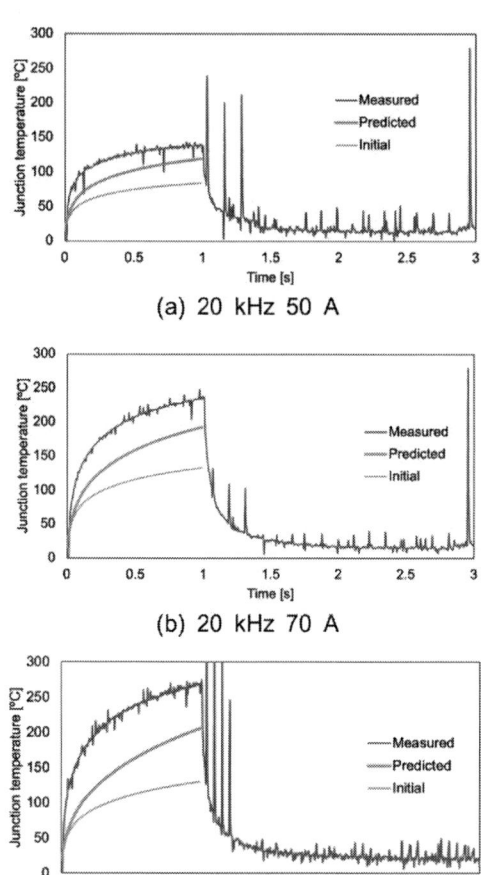

(a) 20 kHz 50 A

(b) 20 kHz 70 A

(c) 30 kHz 50 A

Fig. 6 Temperature profile with +28% degraded module.

other hand, for the 28% increased modules, the predicted profiles are totally underestimated. This result indicates that the temperature observation consisting of both the $V_{CE(sat)}$ monitoring and the T_j prediction can detect the critical deterioration of power modules by checking the gap between those temperatures.

5 Conclusion

This paper has provided observation of thermal-resistance increase of degraded IGBT modules with a junction temperature profile measurement setup that consists of a sensing circuit and the low-cost IoT platform "Leafony." The combination of the demonstrated setup and predicted temperature profiles using a three-dimensional structure model is useful for detecting increased thermal resistance in degraded IGBT modules. Two intentionally-degraded IGBT modules with different

thermal resistances were tested with a chopper circuit and a $V_{CE(sat)}$ measurement setup. Experimental results confirmed that the junction temperatures of the two modules increased according to the increase in thermal resistance as expected. Comparing the temperature profiles between the experiment and predictions provides information about the health condition of the IGBT modules.

Acknowledgment

This paper is based on results obtained from a feasibility study commissioned by the New-generation Power Electronics and System Research Consortium Japan (NPERC-J).

This paper was supported by the joint research project in research institute of applied mechanics (RIAM), Kyushu university.

3005

References

[1] S. Yang, A. Bryant, P. Mawby, D. Xiang, L. Ran, and P. Tavner, "An Industry-Based Survey of Reliability in Power Electronic Converters," IEEE Trans. Ind. Appl., vol. 47, no. 3, pp. 1441-1451, May/Jun. 2011.

[2] S. Bęczkowski, et. al., "Online Vce measurement method for wear-out monitoring of high power IGBT modules," 15th EPE, pp. 1-7, 2013.

[3] U.-M. Choi, F. Blaabjerg, S. Jørgensen, S. Munk-Nielsen, B. Rannestad, "Reliability Improvement of Power Converters by Means of Condition Monitoring of IGBT Modules," IEEE Trans. Power Electron. vol. 32, no. 10, pp. 7990–7997, Oct. 2017.

[4] H.-C. Yang, R. Simanjorang, and K. Y. See, "A Method of Junction Temperature Estimation for SiC Power MOSFETs via Turn-on Saturation Current Measurement," IEEJ J. Industry Applications, vol. 8, no. 2, pp. 306--313, 2019.

[5] F. Gonzalez-Hernando, et. al., Wear-Out Condition Monitoring of IGBT and MOSFET Power Modules in Inverter Operation, IEEE Trans. Ind. Appl. vol. 55, no. 6, pp. 6184-6192, Nov./Dec. 2019.

[6] A. Kundu, et. al., "Power Module Thermal Characterization Considering Aging Towards Online State–of– Health Monitoring," IEEE APEC, pp. 1128-1134, Mar. 2023.

[7] H. Yamasaki, K. Hata, and M. Takamiya, "Estimation of Both Junction Temperature and Load Current of IGBTs from Output Voltage of Gate Driver", IEEJ J. Industry Applications, vol. 12, no. 3, pp. 392-400, 2023.

[8] T. Thekemuriyil, J.D. Rohner, and R. A. Minamisawa, "Machine learning-based prediction of on-state voltage for real-time health monitoring of IGBT," Power Electronic Devices and Components, vol. 6, 100049, Oct. 2023.

[9] T. Ozawa, T. Yamamoto, H. Sugiura, Y. Kondo, "Circuit Simulation Method for Insulated-Gate Bipolar Transistor Short-Circuit Operation with High Accuracy," IEEJ J. Industry Applications, vol. 12, no. 4, pp. 835-841, 2023.

[10] H. Nakano, T. Muranaka, Y. Nabetani, "Minimizing Thermal Imbalance of RC-IGBT by Bonding Technology," IEEJ J. Industry Applications, vol. 11, no. 6, pp. 771-778, 2022.

[11] K. Hasegawa, K. Hara, N. Shishido, S. Nakano, W. Saito, T. Ninomiya, "Power-Cycling Degradation Monitoring of an IGBT module with $V_{CE(sat)}$ Measurement in Continuous Operation of a Chopper Circuit," Power Electronic Devices and Components, vol. 7, 100061, Apr. 2024.

[12] Leafony website [Online]. Available: https://docs.leafony.com/en/

PCIM Europe 2024, 11– 13 June 2024, Nuremberg   DOI: 10.30420/566262427

Modulation Technique for Reduced AC Content of the DC Link Current in Three-Phase Two-Level Inverters

Steffen Frei, Gerd Griepentrog

TU Darmstadt, Germany

Corresponding author: Steffen Frei, steffen.frei@lea.tu-darmstadt.de
Speaker: Steffen Frei, steffen.frei@lea.tu-darmstadt.de

Abstract

DC link capacitor lifetime plays a crucial role in the design of power inverters. One of the main contributors to capacitor aging is the operation at high temperatures which comes with high capacitor currents. While higher switching frequencies allow for a down-sizing of the DC link capacitance, the required current rating is unaffected. To counteract this, a modulation technique is presented that reduces the DC current ripple significantly. Furthermore, the effects on DC bus current ripple, DC bus voltage ripple, output voltage quality and switching losses are evaluated based on simulations.

1 Introduction

One of the lifetime-critical components in power inverters is the DC link capacitor, especially when operated at high temperatures due to high capacitor current loads. [1] While recent advances in wide band-gap power semiconductors allow for higher switching frequencies and thus lower DC voltage ripple or lower DC link capacitance, the DC current ripple resulting from the discrete switching states of the power semiconductors remains unchanged in amplitude. Thus, a down-sizing of DC link capacitance is hindered and the gap between design requirements for current carrying capability and DC voltage ripple is increased.

Changes in capacitor technologies may be used to mitigate effect, however specialized modulation techniques can be used to reduce the rms current in the capacitors as well. In this paper, such a technique is presented and its effects on DC voltage ripple, semiconductor switching losses and harmonic distortion are discussed. Additionally, an outlook on the usage of the technique in a dual three-phase system is given.

2 Mathematical Description of DC Link Capacitor Current

A mathematical description of the DC link capacitor current is needed to accurately evaluate modulator performance. The hardware topology is shown

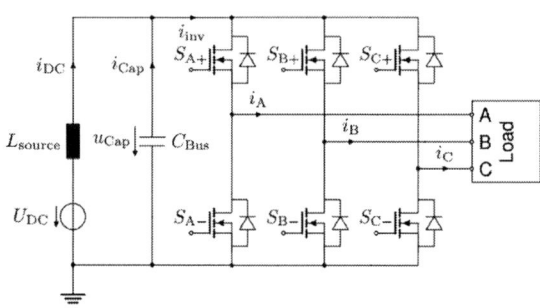

Fig. 1: Two-level inverter for three-phase loads

in Fig. 1. The two-level inverter consists of three half-bridges consisting of two semiconductor switches and two anti-parallel free-wheeling diodes each. They are labeled from A to C. The dead-time intervals are assumed to be negligibly short, the gate drive signals S_{X+} and S_{X-} of any half-bridge can therefore be derived from a single control signal S_X. When $S_X = 1$, the top switch is conducting, when $S_X = 0$, the bottom switch is conducting. Using Kirchhoff's first law, the inverter input current i_{inv} can be calculated as the sum of currents in the top semiconductors.

$$i_{inv} = S_A \cdot i_A + S_B \cdot i_B + S_C \cdot i_C \qquad (1)$$

From eq. 1, it can be concluded that whenever one of the control signals S_X changes state, a sudden change in i_{inv} occurs. Assuming that the power source has an infinite source inductance L_{source}, it is only capable of supplying the DC current i_{DC}. To

3007

account for the AC content in i_{inv}, a DC link capacitor C_{Bus} is added. As the DC link capacitor does not dissipate energy, i_{DC} must match the mean value of i_{inv}.

$$i_{\text{DC}} = \bar{S}_{\text{A}} \cdot i_{\text{A}} + \bar{S}_{\text{B}} \cdot i_{\text{B}} + \bar{S}_{\text{C}} \cdot i_{\text{C}} \quad (2)$$

In eq. 2, $\bar{S}_{\text{X}}$ is the mean value of the phase control signal over one modulation period. The phase currents i_{X} are considered to be approximately constant during the respective modulation period. The above description can be further simplified. Using the *Clarke transformation*, the three phase output voltages and currents can be transformed into the two-dimensional orthogonal voltage or current plane, commonly referred to as the *alpha-beta reference frame*.

$$\begin{bmatrix} \bar{u}_\alpha \\ \bar{u}_\beta \end{bmatrix} = \frac{2}{3} \cdot U_{\text{DC}} \cdot \begin{bmatrix} 1 & -\frac{1}{2} & -\frac{1}{2} \\ 0 & \frac{\sqrt{3}}{2} & -\frac{\sqrt{3}}{2} \end{bmatrix} \begin{bmatrix} \bar{S}_{\text{A}} \\ \bar{S}_{\text{B}} \\ \bar{S}_{\text{C}} \end{bmatrix} \quad (3)$$

$$\begin{bmatrix} i_\alpha \\ i_\beta \end{bmatrix} = \frac{2}{3} \cdot \begin{bmatrix} 1 & -\frac{1}{2} & -\frac{1}{2} \\ 0 & \frac{\sqrt{3}}{2} & -\frac{\sqrt{3}}{2} \end{bmatrix} \begin{bmatrix} i_{\text{A}} \\ i_{\text{B}} \\ i_{\text{C}} \end{bmatrix} \quad (4)$$

In eq. 3, the mean values of the gate signals were multiplied by U_{DC} in order to calculate the mean phase output voltages. $\bar{u}_\alpha$ and $\bar{u}_\beta$ correspond to the mean output voltages during one modulation step and should match the components of a given voltage reference vector $\vec{u}_{\text{ref}}$. Taking into account that the Clarke transform is amplitude invariant, the mean electrical power and thus the mean inverter input current can be calculated.

$$i_{\text{DC}} = \bar{i}_{\text{inv}} = \frac{\bar{P}_{\text{inv}}}{U_{\text{DC}}} = \frac{3}{2} \cdot \frac{\bar{u}_\alpha \cdot i_\alpha + \bar{u}_\beta \cdot i_\beta}{U_{\text{DC}}} \quad (5)$$

Finally, using Kirchhoff's first law, the instantaneous current in the DC link capacitor i_{C} can be calculated.

$$i_{\text{Cap}} = i_{\text{inv}} - i_{\text{DC}} \quad (6)$$

In other words, under the aforementioned assumptions, the DC link capacitor current is identical to the AC content in the inverter input current.

3 State of the Art Techniques

3.1 Classification of Techniques

In general, *pulse width modulation* (PWM) techniques can be classified in synchronous and asynchronous modulation. This investigation focuses on asynchronous modulation techniques only, which

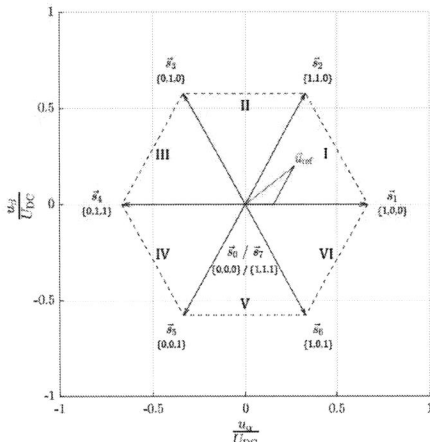

Fig. 2: Graphical representation of space vector pulse width modulation

modulate the output voltage at a fixed frequency and independently from the fundamental wave. In order to ensure properly suppressed output current harmonics, a sufficiently high modulation frequency factor ($m_{\text{f}} \geq 15$), defined as the ratio of modulation to fundamental frequency, must be used. If this condition is met, the instantaneous voltage and phase current values can be assumed as constant during a modulation step.

The most commonly used technique making use of time discrete reference values is the so-called *space vector pulse width modulation* (see Fig. 2). The eight possible switch state combinations can be expressed as voltage space vectors $\vec{s}_0$ to $\vec{s}_7$ using the Clarke transform. For a given $\vec{u}_{\text{ref}}$, the modulation task is simplified to finding a linear combination of space vectors to match the reference. As for most reference voltages there is an infinite number of solutions, the number of used voltage vectors needs to be constrained. Classically, *adjacent voltage vector pulse width modulation* (APWM) is used. This technique uses only the space vectors that are the closest to the reference voltage. In the example in Fig. 2, the selected vectors would be $\vec{s}_0$, $\vec{s}_1$, $\vec{s}_2$ and $\vec{s}_7$. The voltage vector hexagon is divided into six *sectors* and denoted by roman numerals in the plot. Using APWM has the advantage of reduced instantaneous voltage deviations, which result in lower output current ripple.

As the voltage vectors $\vec{s}_0$ and $\vec{s}_7$ represent a short-circuit of all three phases, they are referred to as *zero voltage vectors*, while all others are called *active voltage vectors*. $\vec{s}_0$ and $\vec{s}_7$ are redundant in the voltage plane. Therefore, the distribution

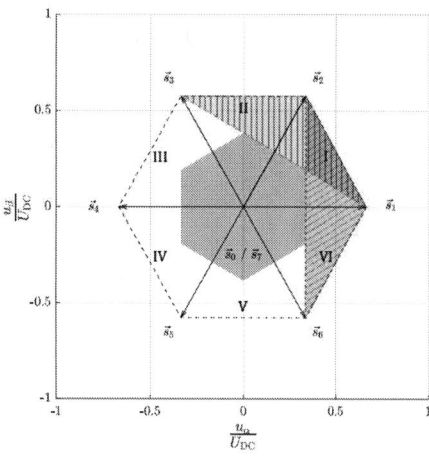

Fig. 3: DCPWM voltage vector triples for sector I

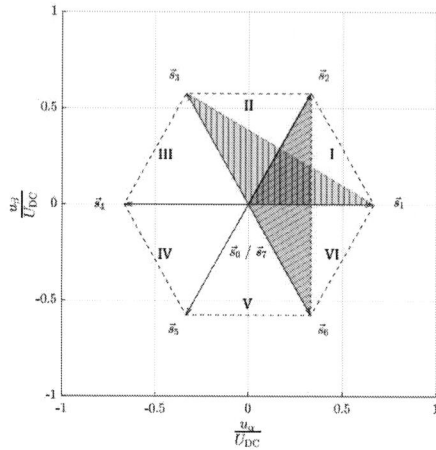

Fig. 4: Ext-DCPWM voltage vector triples for sector I

of zero voltage times between these two vectors is a degree of freedom. In *symmetric space vector pulse width modulation* (SVPWM), the duty cycle of the two voltage vectors is selected to be equal. Techniques that only use either $\vec{s}_0$ or $\vec{s}_7$ are non-continuous in phase and therefore called *discontinuous pulse width modulation* (DPWM) techniques. These techniques clamp one phase during the modulation cycle which reduces switching losses. Which half-bridge is clamped can e.g. be decided based on load angle or the phase currents. One example for such a technique is called *generalized discontinuous pulse width modulation* (GDPWM). One downside of the DPWM techniques is their increased output current harmonic distortion in some operating points. [2][3]

For a given output voltage reference vector, the modulation degree m_a is defined as the ratio between the output voltage amplitude and the maximum possible output voltage amplitude (eq. 7). The maximum modulation degree is thus $2/\sqrt{3}$.

$$m_a = \frac{2 \cdot |\vec{u}_{\text{ref}}|}{U_{\text{DC}}} \quad (7)$$

3.2 Double Carrier Techniques

In [4], a modulation technique specialized on reducing the input current ripple at high modulation degrees is presented. The technique is based on the assumption of high power factors. In such cases, a non-zero output current implies non-zero active power and thus non-zero mean inverter input current. Using the zero-voltage vectors, which produce a zero inverter input current, causes a high momentary deviation from the average input current. Avoiding these voltage vectors is therefore benefi-

cial. In order to achieve that, three adjacent active voltage vectors are used when possible. For sector I, the available voltage vector triples are ($\vec{s}_1$, $\vec{s}_2$, $\vec{s}_3$) and ($\vec{s}_1$, $\vec{s}_2$, $\vec{s}_6$), as displayed in Fig. 3. Where both spanned triangular regions overlap, the voltage vector triple clamping the phase with the higher current magnitude is selected to reduce the switching losses. All reference voltage vectors lying within the light gray hexagon are too short to reach any of the spanned triangular areas. In those cases, the technique defaults to GDPWM, thereby omitting the capacitor current reduction. One noteworthy aspect of this technique is that for the switching signal generation a second, inverted carrier signal needs to be used. The technique has therefore been referred to as *double carrier pulse width modulation* (DCPWM). [4][5]

In order to also decrease the input current ripple for low modulation degrees, a modulation scheme called *extended double carrier pulse width modulation* (Ext-DCPWM) is proposed in [5]. The technique replaces GDPWM used by DCPWM at low modulation degrees by using voltage vector triples consisting of two non-adjacent active voltage vectors and one fitting zero-voltage vector. For sector I, the respective triples would be ($\vec{s}_0$, $\vec{s}_1$, $\vec{s}_3$) and ($\vec{s}_2$, $\vec{s}_6$, $\vec{s}_7$), as shown in Fig. 4. The used voltage vector triple is selected in a way to maximize the time where the instantaneous input current direction matches that of the average input current. In contrast to DCPWM, this technique can not fully avoid using the zero-voltage vectors, however the zero-voltage vector duty cycle is reduced in comparison to SVPWM as the spanned triangular areas are closer to the origin. [5]

In order to enable efficient implementation on DSP systems, a simplified approach called *unified double carrier pulse width modulation* (Uni-DCPWM) using space vector math is proposed in [6]. As the switching state selection is done based on the minimization of switching losses, the DC link capacitor current ripple results differ slightly from those of Ext-DCPWM.

3.3 Performance Analysis

In order to evaluate the performance of Ext-DCPWM, the DC link capacitor rms current $i_{\mathrm{C,rms}}$ is calculated in dependance of the modulation degree m_{a} and the load angle φ. For SVPWM, this can be easily performed by using the analytic formula in [7]. As the DC link capacitor rms current scales linearly, the result is plotted in relation to the output rms current $i_{\mathrm{S,rms}}$ in Fig. 5a. The maximum current occurs at power factors close to positive or negative unity and medium modulation degrees. The maximum DC link rms current is $65\ \%$ of the output current rms value.

As the switching state selection policy for Ext-DCPWM is non-continuous, an analytic solution for the DC link capacitor current cannot easily be determined. Instead, a numeric calculation is performed over a fine grid of modulation degree m_{a}, load angle φ and reference voltage vector angle γ. Finally, the rms of all samples in the γ dimension is computed as an approximation for the capacitor rms current. The result is plotted in Fig. 5b. In Fig. 5c, the same result is given in relation to the capacitor rms current using SVPWM. It can be observed that the technique is effective at reducing the capacitor rms current in areas of high load factor magnitudes ($\varphi \approx 0$ rad or $\varphi \approx \pm\pi$ rad). In one of the regions featuring the highest rms current for SVPWM, the reduction is as high as $43.3\ \%$.

What can also be observed, however, is that in the regions of low power factor magnitudes ($\varphi \approx \pm\frac{\pi}{2}$ rad) the technique performs worse than SVPWM. The regions of improved performance are edged in red. With a maximum value of $110\ \%$ of the output rms current, the maximum capacitor rms current occuring anywhere in the operating range is $69.2\ \%$ higher than that of SVPWM. In applications using reactive power (e.g. permanent magnet synchronous machines in field weakening, reactive power compensation), this is not acceptable. In the plots (Fig. 5b and 5c), the constant values $m_{\mathrm{a}} = 2/3$ and $m_{\mathrm{a}} = 4/\sqrt{27}$ are marked as white dashed lines. For $m_{\mathrm{a}} \geq 4/\sqrt{27}$, three adjacent active

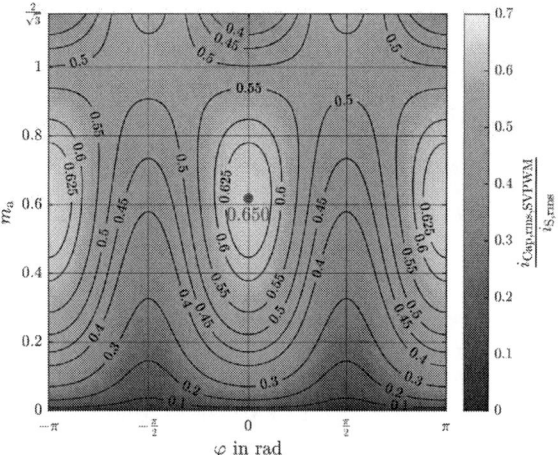

(a) SVPWM DC link capacitor rms current vs. output rms current

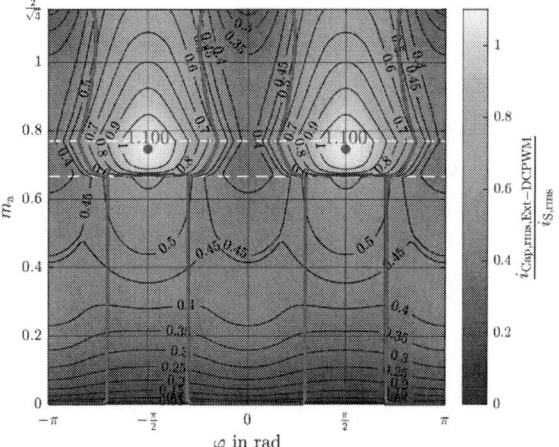

(b) Ext-DCPWM DC link capacitor rms current vs. output rms current

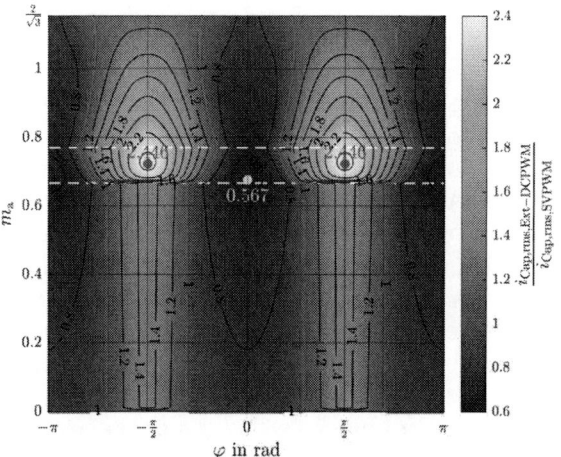

(c) Ext-DCPWM DC link capacitor rms current vs. SVPWM DC link capacitor rms current

Fig. 5: Comparison of DC link capacitor rms current of SVPWM and Ext-DCPWM

voltage vectors are used (see Fig. 3). For $m_a \le {}^2/_3$, two non-adjacent active voltage vectors and one zero-vector is used (see Fig. 4). In between, a combination of both is used.

4 Generalized approach

4.1 Modulator Description

The original publications describe the DCPWM and Ext-DCPWM as carrier-based solutions. However, the carrier-based approach results in a complex pattern generation. Therefore, the original approaches are generalized and described in space vector notation. The proposed modulation technique can thus be referenced to as *generalized double carrier pulse with modulation* (GDCPWM). GDCPWM is described representatively for sector I only. [4][5]

It shall be noted that a carrier-based approach with a similar modulator output called *multi-carrier generalized discontinuous pulse width modulation* (MC-GDPWM) is presented in [8]. The technique switches between the use of a single and dual carriers, hence the name.

4.1.1 Constraint Definition

If a voltage reference vector lies within sector I, it lies in the right half-space defined by the line crossing the tips of $\vec{s}_3$ and $\vec{s}_6$. To produce any voltage vector in that half-space, either $\vec{s}_1$, $\vec{s}_2$ or both need to be utilized. This is a first constraint for the selection of voltage vectors.

Secondly, the selected space vector combinations need to span a region that at least partially overlaps sector I in order to be used for modulation. For example, the voltage vector triple $(\vec{s}_2, \vec{s}_3, \vec{s}_4)$ fulfills the first condition, yet is not useful in sector I.

The third constraint is the number of switching events in a modulation cycle. In SVPWM, each half-bridge switches twice, resulting in a total of six switching events per modulation cycle. To keep the switching losses below the level of SVPWM, the number of switching cycles must be below or equal to four. As the modulation scheme is supposed to be symmetric, the number of switching events in a full modulation cycle is equal to twice that of a simple transition from one voltage vector to the other. Therefore, using two active voltage vectors opposite of each other would exceed four switching events. Therefore, $\vec{s}_0$ and $\vec{s}_7$, $\vec{s}_1$ and $\vec{s}_4$, $\vec{s}_2$ and $\vec{s}_5$ or $\vec{s}_3$ and $\vec{s}_6$ may never be used within the same switching cycle. It was empirically determined that allowing more switching events per modulation cycle does not result in a further reduction of capacitor current

ripple. This constraint does therefore not restrict the DC link capacitor rms current minimization.

4.1.2 Candidate Voltage Vector Combinations

With these constraints, six voltage vector combinations remain as candidates. Two of them are redundant, specifically the ones used for DPWM techniques, $(\vec{s}_0, \vec{s}_1, \vec{s}_2)$ and $(\vec{s}_1, \vec{s}_2, \vec{s}_7)$. The triple clamping the phase with the higher current magnitude is selected to further reduce switching losses. The remaining four voltage vector combinations, which happen to be the same ones used for Ext-DCPWM, are mutually exclusive in two pairs, as they each overlap only half of sector I. Thus, a list of three voltage vector triples is generated following a simple scheme (see Fig. 6):

- If $|i_1| \ge |i_3|$, A1 $(\vec{s}_1, \vec{s}_2, \vec{s}_7)$ is added as a candidate. Otherwise, A2 $(\vec{s}_0, \vec{s}_1, \vec{s}_2)$ is added.

- If $u_{\text{ref},\alpha} \ge \frac{1}{3}$, B1 $(\vec{s}_1, \vec{s}_2, \vec{s}_6)$ is added as a candidate. Otherwise, B2 $(\vec{s}_2, \vec{s}_6, \vec{s}_7)$ is added.

- If $u_{\text{ref},\beta} \ge \frac{2 - 3 \cdot u_{\text{ref},\alpha}}{3 \cdot \sqrt{3}}$, C1 $(\vec{s}_1, \vec{s}_2, \vec{s}_3)$ is added as a candidate. Otherwise, C2 $(\vec{s}_0, \vec{s}_1, \vec{s}_3)$ is added.

4.1.3 Capacitor Current Minimization

In the next step, the best suited combination has to be selected based on the associated rms capacitor current. In a first step, the squared capacitor current $i_{C,x}$ is calculated for all used voltage vectors $\vec{s}_x$ by solving and squaring eq. 6. The results for an example voltage reference vector $\vec{u}_{\text{ref}}$ and output current vector $\vec{i}_S$ is shown in Fig. 7.

In the next step, the duty cycle τ of each voltage vector in each voltage vector triple is calculated. A generalized formula for three generic voltage vectors $(\vec{s}_x, \vec{s}_y, \vec{s}_z)$ can be determined using linear algebra:

$$\vec{s}_x = \begin{bmatrix} u_{\alpha,x} \\ u_{\beta,x} \end{bmatrix} \quad \vec{s}_y = \begin{bmatrix} u_{\alpha,y} \\ u_{\beta,y} \end{bmatrix} \quad \vec{s}_z = \begin{bmatrix} u_{\alpha,z} \\ u_{\beta,z} \end{bmatrix} \quad (8)$$

$$\begin{bmatrix} \tau_x \\ \tau_y \\ \tau_z \end{bmatrix} = \begin{bmatrix} \vec{s}_y \times \vec{s}_z + \vec{u}_{\text{ref}} \times \vec{s}_y - \vec{u}_{\text{ref}} \times \vec{s}_z \\ \vec{s}_z \times \vec{s}_x + \vec{u}_{\text{ref}} \times \vec{s}_z - \vec{u}_{\text{ref}} \times \vec{s}_x \\ \vec{s}_x \times \vec{s}_y + \vec{u}_{\text{ref}} \times \vec{s}_x - \vec{u}_{\text{ref}} \times \vec{s}_y \end{bmatrix} \cdot \kappa \quad (9)$$

$$\kappa = \frac{1}{\vec{s}_x \times \vec{s}_y + \vec{s}_y \times \vec{s}_z + \vec{s}_z \times \vec{s}_x} \quad (10)$$

Next, the mean square capacitor current $\overline{i_C^2}$ for each voltage vector triple can be calculated using the following formula:

$$\overline{i_{\text{Cap}}^2} = \tau_x \cdot i_{C,x}^2 + \tau_y \cdot i_{C,y}^2 + \tau_z \cdot i_{C,z}^2 \quad (11)$$

PCIM Europe 2024, 11– 13 June 2024, Nuremberg   DOI: 10.30420/566262427

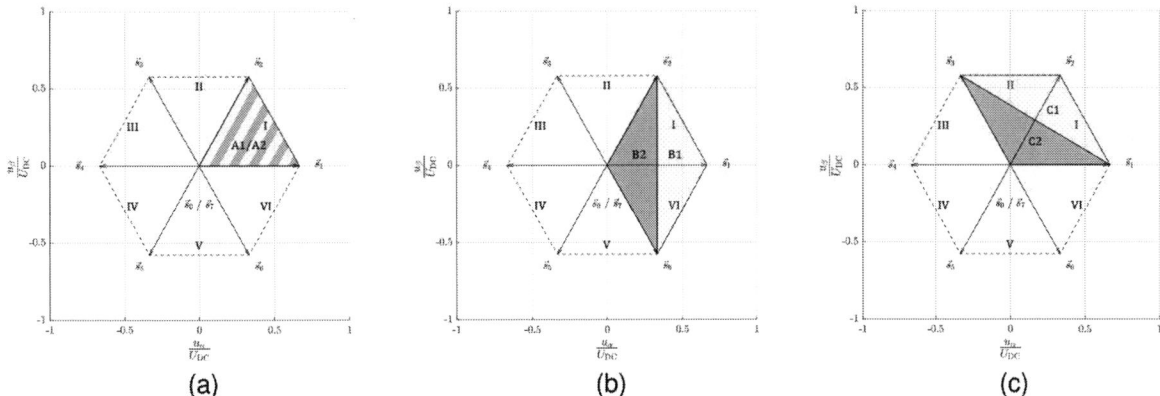

(a)   (b)   (c)

Fig. 6: Candidate voltage vector combinations for sector I

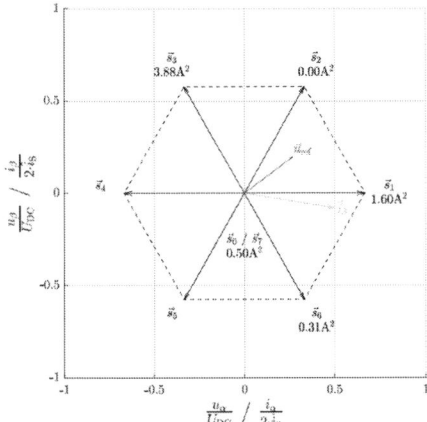

Fig. 7: Squared current deviation results

As the square root function is monotonically increasing, minimizing the mean squared current is equivalent to minimizing the rms current. Therefore, computationally expensive square root computations can be avoided. The voltage vector triple resulting in the lowest value of eq. 11 is selected for the modulation. The duty cycles can be reused from eq. 9.

4.1.4 Switching Pattern Generation

In order to minimize the switching losses, the order in which the determined voltage vectors are applied needs to be optimized. The number of switching events is minimal when each voltage vector transition involves switching only one phase. This is the case for transitions in between neighboring active voltage vectors, between $\vec{s}_0$ and active voltage vectors with odd subscript and between $\vec{s}_7$ and active voltage vectors with even subscript. Following the constraint that the modulation cycle shall be symmetric in respect to the center of the modula-

tion cycle, the following voltage vector orders are allowed in sector I:

A1: $(\vec{s}_1, \vec{s}_2, \vec{s}_7, \vec{s}_2, \vec{s}_1)$ or $(\vec{s}_7, \vec{s}_2, \vec{s}_1, \vec{s}_2, \vec{s}_7)$

A2: $(\vec{s}_0, \vec{s}_1, \vec{s}_2, \vec{s}_1, \vec{s}_0)$ or $(\vec{s}_2, \vec{s}_1, \vec{s}_0, \vec{s}_1, \vec{s}_2)$

B1: $(\vec{s}_6, \vec{s}_1, \vec{s}_2, \vec{s}_1, \vec{s}_6)$ or $(\vec{s}_2, \vec{s}_1, \vec{s}_6, \vec{s}_1, \vec{s}_2)$

B2: $(\vec{s}_2, \vec{s}_7, \vec{s}_6, \vec{s}_7, \vec{s}_2)$ or $(\vec{s}_6, \vec{s}_7, \vec{s}_2, \vec{s}_7, \vec{s}_6)$

C1: $(\vec{s}_1, \vec{s}_2, \vec{s}_3, \vec{s}_2, \vec{s}_1)$ or $(\vec{s}_3, \vec{s}_2, \vec{s}_1, \vec{s}_2, \vec{s}_3)$

C2: $(\vec{s}_1, \vec{s}_0, \vec{s}_3, \vec{s}_0, \vec{s}_1)$ or $(\vec{s}_3, \vec{s}_0, \vec{s}_1, \vec{s}_0, \vec{s}_3)$

For each combination of voltage vectors, two sequences are possible, which are identical yet shifted by half a modulation cycle. The one featuring the higher (i.e. more positive) capacitor current $i_{\text{Cap,x}}$ in the center of the modulation cycle is arbitrarily selected.

4.2 Performance Evaluation

The performance of the GDCPWM technique is evaluated analogous to the procedure in Section 3.3. In Fig. 8a, the DC link capacitor rms current is displayed in relation to the output rms current. The regions of maximum capacitor currents in both SVPWM ($m_a \approx 0.62$, $\varphi \approx n \cdot \pi, n \in [-1, 0, 1]$) and Ext-DCPWM ($m_a \approx 0.75$, $\varphi \approx \pm \frac{\pi}{2}$) are uncritical using GDCPWM. The maximum capacitor rms current now occurs at pure reactive power ($\varphi \approx \pm \frac{\pi}{2}$) and maximum modulation degree. Such operating points occur e.g. in field weakening operation of permanent magnet synchronous machines. However, these operating points are seldom combined with maximum rated output current and may therefore not be critical in the actual application. The global maximum DC link capacitor rms current is

3012

PCIM Europe 2024, 11– 13 June 2024, Nuremberg   DOI: 10.30420/566262427

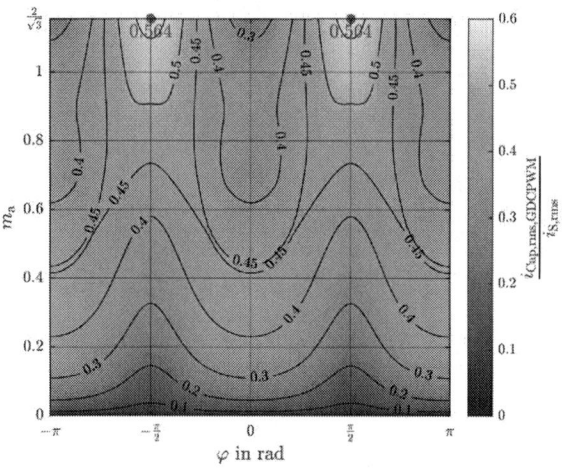

(a) GDCPWM DC link capacitor rms current vs. output rms current

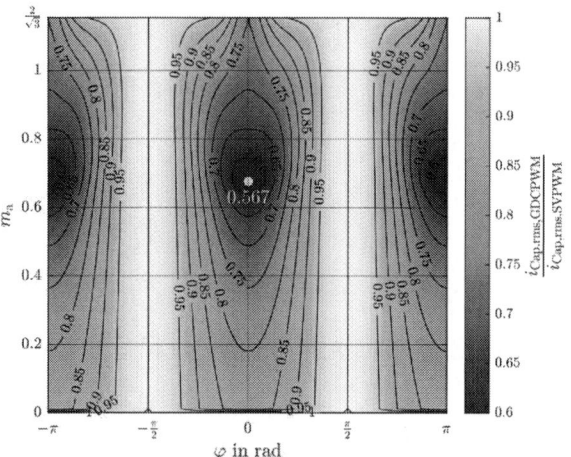

(b) GDCPWM DC link capacitor rms current vs. SVPWM DC link capacitor rms current

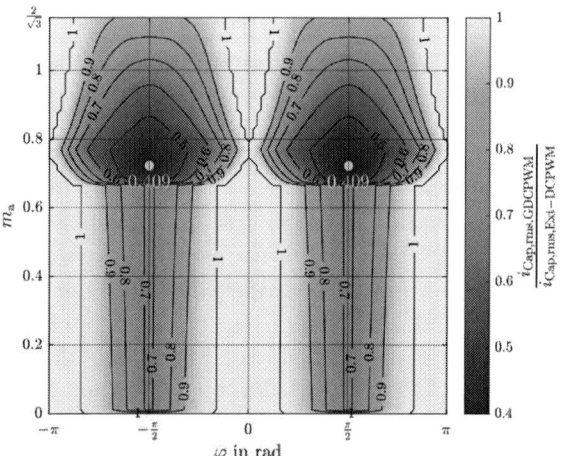

(c) GDCPWM DC link capacitor rms current vs. Ext-DCPWM DC link capacitor rms current

Fig. 8: Comparison of DC link capacitor rms current of GDCPWM to SVPWM and Ext-DCPWM

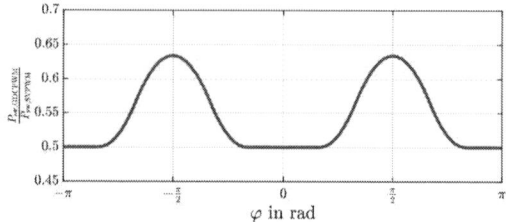

Fig. 9: Switching loss ratio between GDPWM and SVPWM. Contrary to Ext-DCPWM, the switching losses are independent from the modulation degree m_a

lowered to $56.4\,\%$ of the output rms current and thus lowered by $13.2\,\%$ in comparison to SVPWM. Therefore, different to Ext-DCPWM, GDCPWM globally decreases the AC content in the DC link current. At pure reactive power or zero modulation degree, the GDCPWM exactly matches the results for SVPWM (see Fig. 8b). In Fig. 8c, the GDCPWM capacitor rms current is plotted in relation to that of Ext-DCPWM. It can be observed that for high magnitude power factors, GDCPWM matches the performance, while for low magnitude power factors, GDCPWM performs better than Ext-DCPWM.

5 Secondary Effects

5.1 Switching Losses

As discussed in Section 4.1.2, the modulator reduces the switching losses in the power semiconductors by reducing the number of switching events from six to four in one modulation cycle. As in GDCPWM, the phase that carries the maximum current magnitude is clamped and the switching losses scale roughly linear with the switched current magnitude, the maximum switching loss reduction is even higher than one third. In order to obtain an estimate for the switching losses generated by GDCPWM ($P_{\mathrm{sw,GDCPWM}}$) in relation to those produced by SVPWM ($P_{\mathrm{sw,SVPWM}}$), a full electrical revolution with a suitably high switching frequency is simulated for both GDCPWM and SVPWM and the current magnitudes for each semiconductor and switching event are accumulated and divided. This step has been performed in analogy to [9], where it is described in more detail. In contrast to the results presented for Ext-DCPWM in [9], the switching loss ratio of GDCPWM displayed in Fig. 9 is independent from the modulation degree. GDCPWM achieves a switching loss reduction of $50\,\%$ at high magnitude power factors and $37\,\%$ at low power factors in comparison to SVPWM.

3013

5.2 Output Current Harmonic Distortion

One drawback of DPWM techniques is their increase in output current harmonic distortion. In analogy to [10], the rms value of the current harmonics in phase A ($i_{A,AC,rms}$) is determined instead of the total *total harmonic distortion* (THD), as the THD is dependent from the fundamental amplitude. As no closed-form solution was found, the value was determined numerically. The ratio of current harmonics rms value for GDCPWM and SVPWM is shown in Fig. 10a. As shown in [10], the result for SVPWM is dependent on m_a and independent from φ. GDCPWM, however, shows a strong dependence in both dimensions. GDCPWM performs worse than SVPWM in terms of harmonic content in all operating points. At $(m_a \approx 0.4, \varphi \approx n \cdot \pi, n \in [-1, 0, 1])$, the current harmonics rms value is more than tripled.

However, as the switching losses are significantly reduced for GDCPWM, the modulator frequency can be increased in comparison to SVPWM at no cost. Fig. 10b shows the adjusted current harmonics rms value when the modulator frequency is adjusted accordingly. Using this approach, the maximum increase in current harmonics rms value is reduced to $57.5\,\%$. GDCPWM performs significantly better than SVPWM at modulation degrees above 1. [10]

5.3 DC Link Voltage Ripple

The maximum permissible DC voltage ripple is an important design criterion for the inverter DC link capacitor. In [11], the following voltage ripple coefficient is introduced:

$$\Delta k_{pp} = \frac{C_{Bus} \cdot f_{Mod}}{\hat{I}_S} \cdot \Delta u_{Cap,pp} \qquad (12)$$

In eq. 12, C_{Bus} denotes the DC link capacitance, f_{Mod} represents the modulation or carrier frequency and $\hat{I}_S$ is the output current amplitude. In contrast to the maximum peak-to-peak DC voltage ripple $\Delta u_{Cap,pp}$, this coefficient is independent from design parameters and offers better comparability. The simulation was performed using infinitely high source and load inductance.

For SVPWM, the maximum DC link voltage ripple occurs at zero power factor and maximum modulation degree with a voltage ripple coefficient of $\Delta k_{pp} = 0.251$ (see Fig. 10c). Due to the lower mean switching frequency, the voltage ripple coefficient for GDCPWM is significantly increased.

However, when the modulation frequency is adjusted to keep the switching losses identical to those of SVPWM, the voltage ripple coefficient for GDCPWM becomes smaller than that of SVPWM in almost the entire operating range (see Fig. 10d). In fact, the maximum occuring voltage ripple coefficient for GDCPWM with adaptive modulation frequency is $\Delta k_{pp} = 0.169$ and thus $32.8\,\%$ smaller than that or fixed-frequency SVPWM.

5.4 Computational Burden

The algorithm described in Section 4.1 simplifies the voltage vector selection by reducing the number of possible voltage vector combinations to a minimum, thus avoiding a brute force optimization approach. All computations can be calculated with low additional effort. The resulting switching pattern adheres to the double carrier approach presented in [4] and can thus be generated using hardware timers. Therefore, it is assumed that the computational burden is small when a modern *digital signal processor* (DSP) is used.

5.5 Use for Dual Three Phase Inverters

As part of the algorithm described in Section 4.1, the inverter input current deviation from the mean input current is calculated for all relevant voltage vectors. In Section 4.1.4, this value is used to sort the voltage vectors in a way that the current deviation for the voltage vector used in the center of the modulation cycle is more positive than that at the beginning and the end of the modulation cycle. If a second inverter is used, this can be utilized to further decrease the DC link current ripple. By sorting the voltage vectors in the second inverter in the opposite way, i.e. such that the voltage vector resulting in the more negative inverter input current deviation is in the center of the modulation cycle, constructive interference is partially reduced. Fig. 11a shows the DC link capacitor current for dual standard GDCPWM, whereas Fig. 11b shows the result when the voltage vector order in inverter 2 is inverted. Modulation degree m_{a2} and load angle φ_2 of inverter as well as the ratio of inverter output currents $i_{S1,rms}$ and $i_{S2,rms}$ have been fixed to a random operating point. Even by visual comparison, a promising DC link capacitor current reduction is achieved. Because of the high dimensionality of the problem, however, this will be a topic for a standalone future investigation.

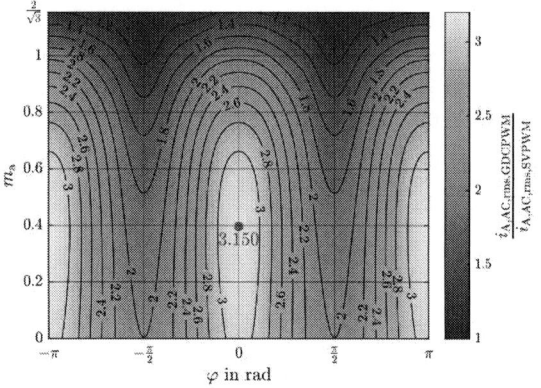

(a) Ratio of rms output current harmonics of GDCPWM and SVPWM using fixed modulation frequency

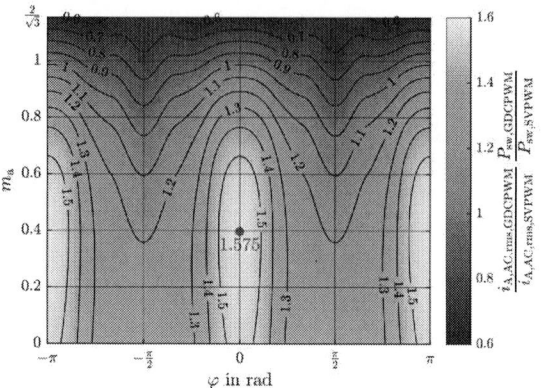

(b) Ratio of rms output current harmonics of GDCPWM and SVPWM using adaptive modulation frequency

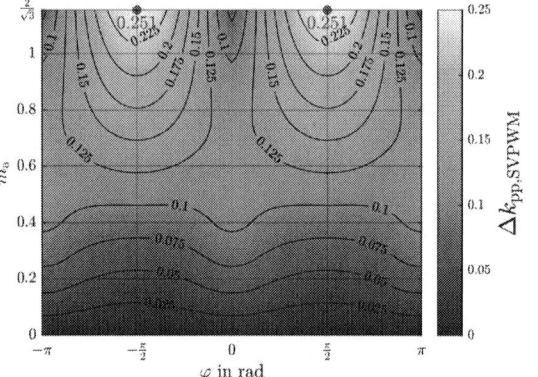

(c) DC voltage ripple coefficient for SVPWM using fixed modulation frequency

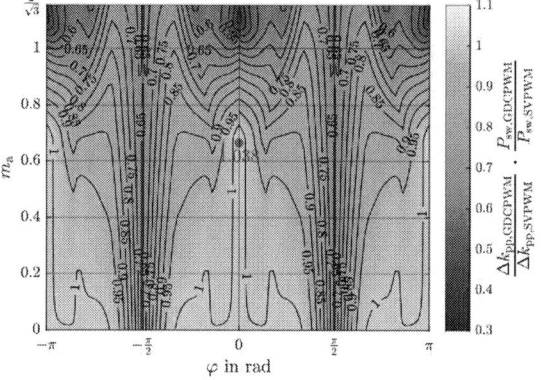

(d) Ratio of DC voltage ripple coefficients of GDCPWM and SVPWM using adaptive modulation frequency

Fig. 10: Comparison of output current harmonic content and DC voltage ripple of GDCPWM and SVPWM

6 Conclusion and Outlook

In this investigation, existing modulation techniques focused on reducing the DC link current ripple were presented and evaluated. While they perform well at high magnitude load factors, they perform poorly at low load factors. To the author's best knowledge, a novel generalized space-vector approach is proposed that allows the usage of zero-voltage vectors and performs well in all operating points. Used with adaptive modulation frequency, the DC link capacitor current and voltage ripple can be greatly reduced in comparison to SVPWM while generating only a small increase in output current harmonics. All results have been independently validated using a custom numeric model as well as power-electronic simulation, yet still need to be validated in hardware, which will be performed next. Additionally, an analysis of the frequency spectra of input and output current is necessary to understand the impact on filtering requirements and harmonic losses in passive components and loads. Further research into using the technique with dual three-phase inverter systems also seems promising.

7 Acknowledgments

This investigation was conducted as part of the *Double-E-Drive for Long Range* (DE4LoRa) project funded by the German Federal Ministry for Economic Affairs and Climate Action (reference number 19I20027I).

Supported by:

Federal Ministry
for Economic Affairs
and Climate Action

on the basis of a decision
by the German Bundestag

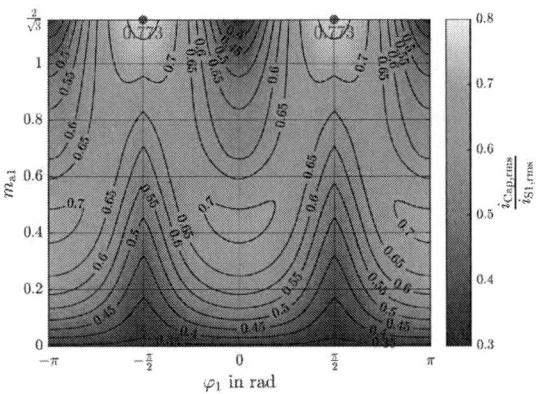 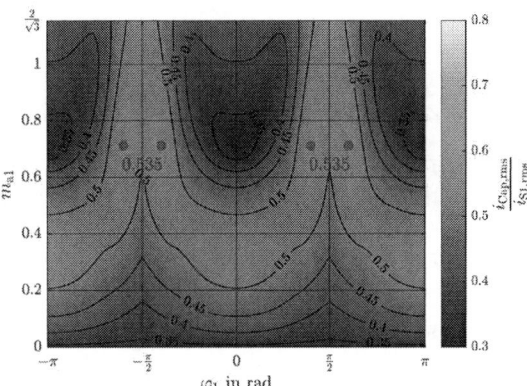

(a) Both inverters operating in standard GDCPWM   (b) Reversed voltage vector order in inverter 2

Fig. 11: Comparison of DC link rms current for dual inverters ($m_{a2} = 0.6$; $\varphi_2 = -\frac{\pi}{4}$; $\frac{i_{S2,rms}}{i_{S1,rms}} = 0.75$)

References

[1] H. Wang and F. Blaabjerg, "Reliability of capacitors for dc-link applications in power electronic converters—an overview," *IEEE Transactions on Industry Applications*, vol. 50, no. 5, pp. 3569–3578, 2014. DOI: 10.1109/TIA.2014.2308357.

[2] A. Hava, R. Kerkman, and T. Lipo, "A high-performance generalized discontinuous pwm algorithm," *IEEE Transactions on Industry Applications*, vol. 34, no. 5, pp. 1059–1071, 1998. DOI: 10.1109/28.720446.

[3] T. D. Nguyen, J. Hobraiche, N. Patin, G. Friedrich, and J.-P. Vilain, "A direct digital technique implementation of general discontinuous pulse width modulation strategy," *IEEE Transactions on Industrial Electronics*, vol. 58, no. 9, pp. 4445–4454, 2011. DOI: 10.1109/TIE.2010.2102311.

[4] J. Hobraiche, J.-P. Vilain, P. Macret, and N. Patin, "A new pwm strategy to reduce the inverter input current ripples," *IEEE Transactions on Power Electronics*, vol. 24, no. 1, pp. 172–180, 2009. DOI: 10.1109/TPEL.2008.2006357.

[5] T. D. Nguyen, N. Patin, and G. Friedrich, "Extended double carrier pwm strategy dedicated to rms current reduction in dc link capacitors of three-phase inverters," *IEEE Transactions on Power Electronics*, vol. 29, no. 1, pp. 396–406, 2014. DOI: 10.1109/TPEL.2013.2251005.

[6] T. D. Nguyen, N. Patin, and G. Friedrich, "Pwm strategy dedicated to the reduction of dc bus capacitor stress in embedded three phase inverter," in *2011 IEEE Vehicle Power and Propulsion Conference*, 2011, pp. 1–6. DOI: 10.1109/VPPC.2011.6043068.

[7] J. Kolar, T. Wolbank, and M. Schrodl, "Analytical calculation of the rms current stress on the dc link capacitor of voltage dc link pwm converter systems," in *1999. Ninth International Conference on Electrical Machines and Drives (Conf. Publ. No. 468)*, 1999, pp. 81–89. DOI: 10.1049/cp:19990995.

[8] J. Lee, M.-W. Kim, and J.-W. Park, "Carrier selection strategy of generalized discontinuous pwm method for current reduction in dc-link capacitors of vsi," *IEEE Transactions on Power Electronics*, vol. 37, no. 9, pp. 10428–10442, 2022. DOI: 10.1109/TPEL.2022.3167941.

[9] T. D. Nguyen, N. Patin, and G. Friedrich, "A pwm strategy dedicated to rms current reduction in dc link capacitor of an embedded three phase inverter," in *Proceedings of the 2011 14th European Conference on Power Electronics and Applications*, 2011, pp. 1–9.

[10] D. G. Holmes and T. A. Lipo, "Zero space vector placement modulation strategies," in *Pulse Width Modulation for Power Converters: Principles and Practice*. 2003, pp. 259–336. DOI: 10.1109/9780470546284.ch6.

[11] T. Chen, S. Li, and B. Fahimi, "Analysis of dc-link voltage ripple in voltage source inverters without electrolytic capacitor," in *IECON 2018 - 44th Annual Conference of the IEEE Industrial Electronics Society*, 2018, pp. 1041–1048. DOI: 10.1109/IECON.2018.8591719.

Common Mode Currents in Resonant Circuits Generated with a Delta-Sigma Modulated Voltage Source Inverter

Tobias Haas[1], Theo Zeißel[1]

[1] Technische Hochschule Würzburg Schweinfurt, Germany

Corresponding author: Tobias Haas, tobias.haas@thws.de
Speaker: Tobias Haas, tobias.haas@thws.de

E03 Pulse Width Modulation Methods
Preferred presentation form: Oral presentation

Abstract

According to parasitic effects, resonant circuits occur in drive systems. The inverter's switching behaviour determines the resonance's excitation. In comparison to conventional modulation techniques, the delta-sigma procedure generates a beneficial spectral voltage to excite low common mode currents.

1 Introduction

Voltage source inverters are prevalent in electric drive systems. Especially in industrial applications, the power electronic is commonly used to control electric drives. In compliance with the switching frequency, an inverter allows generating arbitrary voltages at the fundamental frequency. Besides the desired behaviour, common mode disturbances are generated through the switching operations. Hence, parasitic capacitances are polarised, e.g. between a conductor and shielding mesh or from an electric drive's windings to the stator housing. As a result, the parasitic capacitances and inductances generate series resonant circuits. Driven common mode currents depend on the voltage excitation, generated by an inverter and on the system's impedance.
Inverters with high DC link voltage and switching frequencies are able to generate strong disturbances. The resulting excitation of parasitic paths may violate regulations by disturbing or damaging other components in the grid. For example a drive's bearing can be damaged [1]. Even control units or measurement equipment might be influenced through electromagnetic interference.
Thereby a reduction of common mode currents is desirable. This goal can be achieved by choosing the correct modulation of an inverter. Common techniques are pulse width or space vector modulations. One advantage is the simple implementation and handling of these techniques. On the other hand, high peaks occur in the output voltage's spectrum [2]. In case that a series resonant circuit is excited through those peaks, high currents are driven. In comparison to these approaches, the delta-sigma modulation shows different spectral behaviour with vales instead of peaks [3]. Hence, the common mode current caused by a series resonant circuit can be reduced.
This paper shows an application with a two-level voltage source inverter feeding an inductive load. A series resonant circuit is created through additional foil capacitors in the system. As a consequence, the foil capacitors are excited instead of parasitic capacitances. This provides the advantage, that the system's behaviour is adjustable. The inverter is modulated with a pulse width modulation approach and the delta-sigma technique, to show the influence on common mode currents.

2 Delta-Sigma Modulation

In the last decades, a variety of researches investigated variants of the delta-sigma modulation in combination with power electronic systems [4], [5], [3], [6]. The technique can be classified as a pulse frequency modulation. Commonly used procedures work with a fixed pulse frequency and varying duty cycle (pulse width modulation).

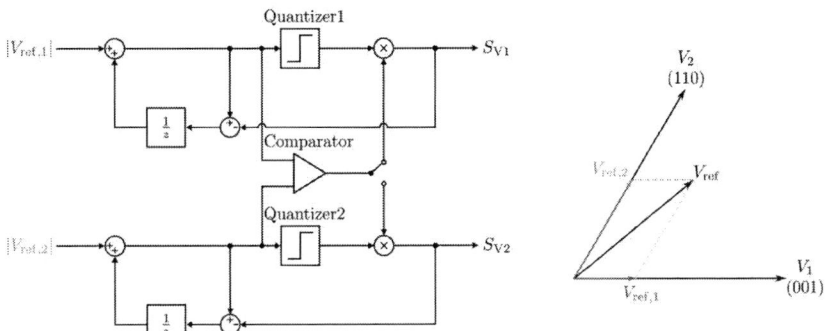

Fig. 1: Delta-sigma Modulation in combination with space vector modulation

In this paper a delta-sigma modulation scheme, first presented in [3] is used. Figure 1 shows the operating principle. On the right side, the split of a specific reference voltage vector V_{ref} in space dimensions is depicted. Instead of determining the switching duration over one pulse period with the split reference vectors $V_{ref,1}$ and $V_{ref,2}$, as it is the case in conventional space vector pulse width modulation, the vectors' magnitudes are used as input quantity for a delta-sigma modulation. Hence, the values of one reference vector is summed up until the threshold value of the quantizer is exceeded.

3 Test Setup

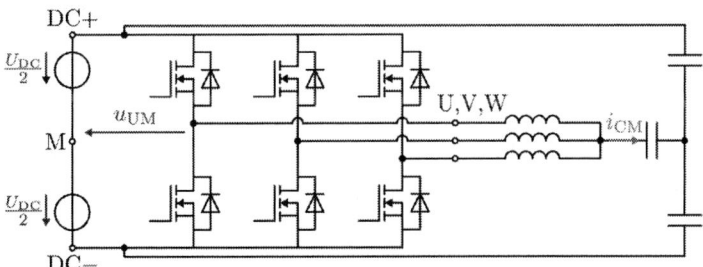

Fig. 2: Schematic

In the test setup, a two-level voltage source inverter is controlled with the shown delta-Sigma modulation. To compare the approach to a conventional procedure, a pulse width modulation with an injected third harmonic within the reference signal is used. The setup's common mode path is determined by the connected inductors and capacitors. Two paths, from U,V,W to $DC+$ and $DC-$ enable a common mode current. The components are designed to create a resonant frequency at 50 kHz.

Both modulation techniques generate different voltages u_{UM} and common mode currents i_{CM} (Fig. 3). The delta-sigma approach generates a vale at the space vector's switching frequency, which is set to 50 kHz. In comparison to that, the pulse width modulation shows a peak at the semiconductor's switching frequency (10 kHz) and its multiples. Hence, the excitation of common mode currents is lower with the delta-sigma procedure.

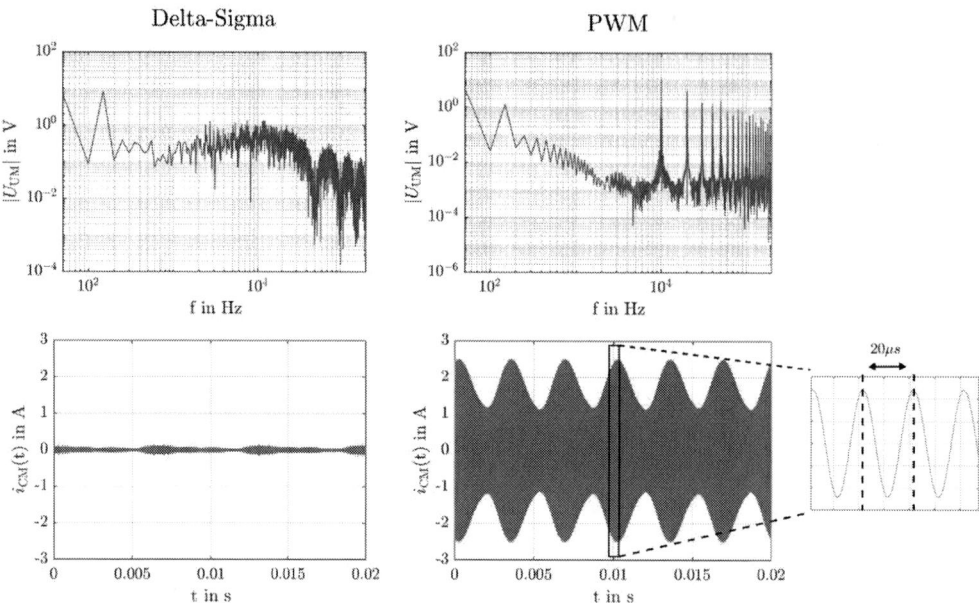

Fig. 3: Common Mode Currents generated from Delta-Sigma space vector modulation (on the left) and pulse width modulation (on the right)

4 Conclusion

The use case of a delta-sigma modulation is shown. In dependency of the common mode path's behaviour, different modulation techniques show various currents. Instead of changing the procedure, a pulse width modulation with another switching frequency can also improve the common mode disturbances. Anyway, this can excite another resonance in higher frequency regions.

The final paper will show different applications and use cases of both modulation techniques. Pros and cons are discussed and the interaction between switching behaviour and common mode resonances is emphasized. Simulative results will be validated by experimental setups.

References

[1] A. Muetze, "Bearing currents in inverter-fed ac-motors," en, Dissertation, Technische Universitaet Darmstadt, Darmstadt, 2003.

[2] S. Bernet, *Selbsgeführte Stromrichter am Gleichspannungszwischenkreis*, de. Springer, 2012. DOI: 10.1007/978-3-540-68861-7.

[3] A. Hirota, B. Saha, S.-P. Mun, and M. Nakaoka, "An advanced simple configuration delta-sigma modulation three-phase inverter implementing space voltage vector approach," IEEE Power Electronics Specialists Conference, 2007.

[4] A. Mertens, "Performance analysis of three-phase inverters controlled by synchronous delta-modulation systems," IEEE Transactions on Industry Applications, 1994.

[5] A. Mertens and V. Ganesan, "Three-phase sigma-delta modulation using zero-sequence components," IEEE European Conference on Power Electronics and Applications, 2005.

[6] B. Jacob and M. Baiju, "Space vector based pulse density modulation scheme for two level voltage source inverter," IEEE Conference on Industrial Electronics and Applications, 2011.

PCIM Europe 2024, 11– 13 June 2024, Nuremberg   DOI: 10.30420/566262429

Evaluation of New Modulation Scheme for 3L-ANPC Using Both Current Paths in Zero State

Felix Eichler[1], Matthias Meissner[1], Markus Meissner[1], Prof. Dr.-Ing. Steffen Bernet[1]
Andreas Giessmann[2]

[1] Dresden University of Technology, Germany
[2] SEMIKRON Danfoss Elektronik GmbH & Co. KG, Germany

Corresponding author: Felix Eichler, felix.eichler@tu-dresden.de
Speaker: Felix Eichler, felix.eichler@tu-dresden.de

Abstract

Several control methods are available for the Active Neutral Point Clamped (ANPC) converter, with the objective of achieving a more balanced distribution of power loss across all switches. This paper evaluates a control method which uses both current paths in zero state and intermediate pulses for removing remaining charge carriers due to zero voltage switching. Measurements of switching losses of an existing ANPC phase leg in the MW range where taken into account. Static parameters were taken from the datasheet. The new modulation scheme is compared to the PWM-1 method regarding thermal behavior of all switches in the ANPC converter using MATLAB. Low modulation depths, power factors close to zero and high currents were shown as advantageous due to reducing and distributing the losses of the inner switches.

1   Introduction

In [1] the 3L-ANPC-converter for medium voltage applications was introduced. This enabled power loss balancing between the switches within the converter. Nevertheless, the use of both zero states is not allowed in the classic modulation approach.

Using both current paths in zero state can help to distribute the power dissipation and thus reduce the temperature difference between the inner switches of the 3L-ANPC-converter [2,3,4,5]. At a low modulation index with a high percentage of zero states, the temperature of the inner semiconductor switches can be reduced. This modulation scheme is applicable for current limit mode or low voltage ride though (LVRT) [4].

In [3] both neutral paths (n-paths) are turned off at the same time. As a result, switching transients in small and long commutation loop occur. Using the long commutation loop leads to higher switching losses and high voltage stress for the active power device. This paper evaluates the modulation scheme (modified PWM-1-00) presented in [6] where intermediate pulses were placed to reduce overvoltage stress. This enables the converter to use both n-paths. Furthermore, by allowing two n-paths, long commutation loops can be avoided. A

novel modulation scheme for different operating points of a MW-range 3L-ANPC-converter is analyzed in this paper. This modulation method is described in detail in Chapter 2. In Chapter 3, semiconductor losses are identified and modeled based on measurements. The types of losses are discussed in Chapter 4. The simulation model is shown in Chapter 5, with the corresponding results presented in Chapter 6.

2   Advanced modulation scheme for usage of both current paths in zero state

The new modulation scheme is derived from PWM-1 [7,8]. This modulation uses a single n-path. Current commutates between T1 (see Fig. 3a for designators) and T5 for positive output voltages, while T2 is permanently turned on. The pulse pattern for PWM-1 can be seen in Fig 1a. For negative output voltages T4 and T6 are being switched, T3 is permanently switched on. Because they are conducting half period the conduction losses are highest in T2 and T3.

To reduce and distribute conduction losses during zero state the second n-path is switched on with

3020

T3 and T6 for positive or with T2 and T5 for negative output voltages. In [3] both n-paths in zero state turn off simultaneously. For an ANPC converter, simultaneous switching of both n-paths causes a long and a short commutation. The longer commutation has a higher stray inductance. This is particularly true for ANPC converters in the MW-range, where a phase-leg is built from multiple discrete half-bridge modules. For these MW-range converters switches T1/T5 and T4/T6 are well connected to the DC link, while T2/T3 are typically connected with a higher stray inductance. This can lead to high overvoltage during switching. To address this, larger gate resistances are used for switches T2 and T3. However, even in this case, simultaneous switching of both n-paths is problematic. The longer commutation path is turned off more slowly than short one. Hence, the full load current is switched to the longer n-path. Therefore, the subsequent switching from zero state to Vout+ must be very slow to limit overvoltage. This is resulting in very high switching losses, which occur after every use of the double n-path. Hence, the simultaneous switching of both n-paths cannot be practically implemented in these setups. Therefore, a dead time is added in [5,9]. Additionally, in this implementation the switches of the second n-path are not switched at the same time. The IGBT that was not conducting is switched on last and switched off first [5]. Figure 1b shows the associated pulse pattern.

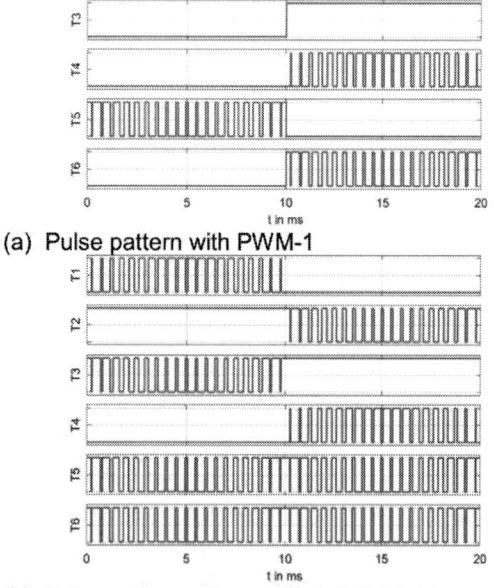

(a) Pulse pattern with PWM-1

(b) Pulse pattern with modified PWM-1-00

Fig. 1 Pulse pattern

By incorporating a dead time into the modulation scheme, the second path is turned on under zero current condition (ZCS) and turned off during zero voltage condition (ZVS) [5,9]. Due to stray inductances, an overvoltage and therefore turn-off losses occur when switching off the second n-path.

When changing the output voltage from Vout+ to Vout-, at least one long commutation takes place in PWM-1 modulation. By allowing the state of two conducting current paths, it is possible to split one long commutation into one short commutation and two commutations between the n-paths. Figure 2 shows the two variants.

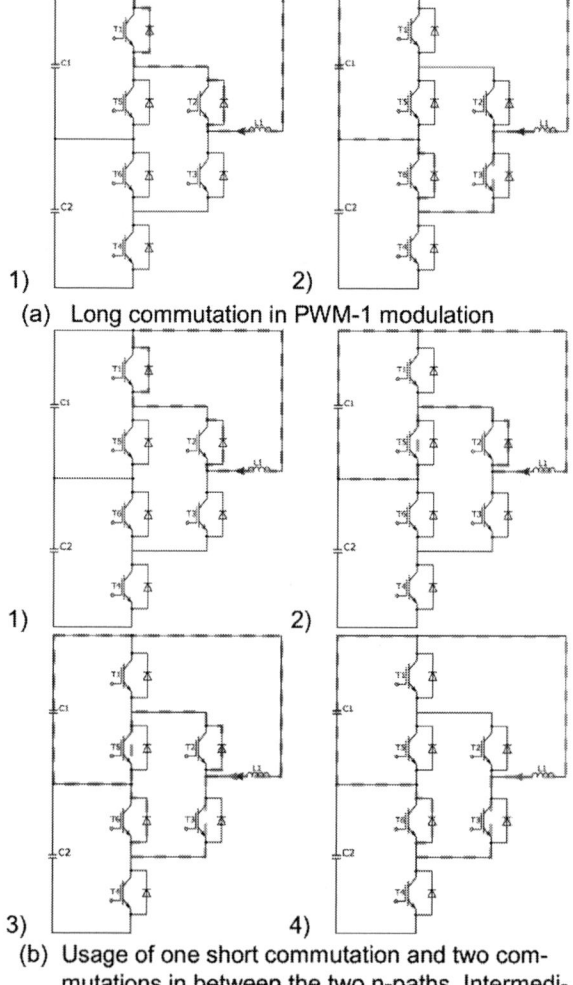

(a) Long commutation in PWM-1 modulation

(b) Usage of one short commutation and two commutations in between the two n-paths. Intermediate pulses are not shown.

Fig. 2 Comparison of the two variants to commutate from positive output voltage two zero voltage output with lower n-path conducting

3 Measurement and parameter extraction

Measurements were done with a 3L-ANPC phase leg in the MW-range with IGBT-Modules SKM1400GB12M7S2I4 [10] designed by Semikron Danfoss (see Fig. 3).

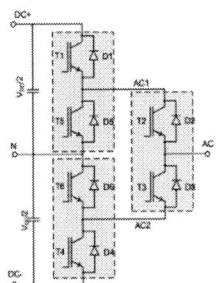

(a) schematic

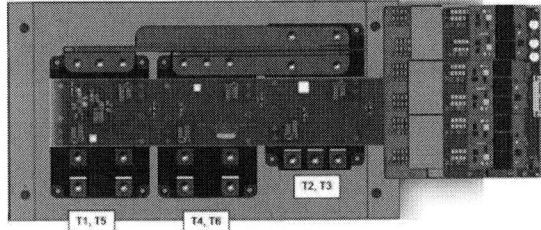

(b) stack design

(c) stack

Fig. 3 ANPC topology based on 3 standard half-bridge power modules

All emitter currents and collector-emitter voltages were measured at different DC voltages, output currents and junction temperatures. According to the extracted switching losses Eq. 1 and 2 [11] is parameterized using least nonlinear squares method. Equation 1 is applied to turn-on and turn-off energy losses.

$$E_{sw} = E_{ref} \left(\frac{V_{DC}}{V_{ref}}\right)^{k_u} \left(\frac{I_C(t)}{I_{ref}}\right)^{k_i} \left(1 + k_T(T_j - T_{ref})\right) \quad (1)$$

$$P_{cond} = \frac{1}{T_P} \int_0^{T_P} \Big(V_{CE0\,ref} + k_{T\,v}(T_j - T_{ref}) + \\ \big(r_{CE\,ref} + k_{T\,r}(T_j - T_{ref})\big) \cdot i_C(t) \Big) \cdot i_C(t)\,dt \quad (2)$$

Static parameters were taken from the datasheet [10].

The parameters are shown in Table 2. The ANPC has a symmetrical structure. Thus, the losses of the switches of the upper part (T1, T5, T2) and the lower part (T4, T6, T3) are assumed as equivalent.

Symbol	Description
T_P	– period (here: 20 ms)
E_{ref}	– reference energies for switching on and off
V_{DC}	– DC link voltage
$I_C(t_i)$, $I_C(t)$	– collector current (at moment of switching or as function of time)
V_{ref}, I_{ref}, T_{ref}	– reference values
T_j	– junction temperature
ki	– current coefficient
k_u	– voltage coefficient
k_T	– temperature coefficient
V_{CE0}, V_{F0}	– (collector-emitter-)threshold voltage
r_{CE}, r_F	– on-state/ forward slope resistance

Table 1: List of variables used in Eq. 1 and 2

Transistor/ Diode	k_u	k_i	k_T/ K^{-1}	E_{ref}/ mJ
		Turn-on		
T1/T4 (short)	1.21	1.63	4.0e-3	125
T5/T6 (short)	1.31	1.53	3.9e-3	140
T5/T6 (2N)	-	-	-	-
T2/T3 (long)	1.55	1.36	5.0e-3	174
T2/T3 (2N)	-	-	-	-
D1/D4 (short)	1.43	1.45	0.0e-3	10
D2/D3 (long)	-	-	-	-
D5/D6 (short)	0.74	1.14	1.2e-3	9
		Turn-off		
T1/T4 (short)	0.97	1.23	1.7e-3	164
T5/T6 (short)	1.10	1.25	1.5e-3	151
T5/T6 (2N)	0.00	2.00	0.4e-3	53
T2/T3 (long)	1.16	1.11	1e-6	834
T2/T3 (2N)	0.00	1.53	0.8e-3	45
D1/D4 (short)	0.91	0.91	1.2e-3	48
D2/D3 (long)	0.61	1.04	5.0e-3	28
D5/D6 (short)	1.00	1.11	8.3e-3	49

	Static	
	V_{CE0}, V_{F0}/ k_T	r_{CE}, r_F/ k_T
IGBT (25°C)	0,86 V	0,49 Ω
k_T IGBT	-0,088 mVK^{-1}	2,08 mΩK^{-1}
Diode (25°C)	1,39 V	0,58 Ω
k_T diode	-2,48 mVK^{-1}	2,0 mΩK^{-1}

Table 2 Dynamic and static parameters. Dynamic parameters were determined from double pulse tests, I_{ref} = 1000 A, V_{ref} = 800 V and T_{ref} = 25°C

Conduction losses in both n-paths are dependent on current distribution, influenced by stray inductance and forward voltage drop of the conducting devices. An observed current distribution is illustrated in Fig. 4. Current equalization can be approximated by an exponential function dependent on the temperature of all switches. As this modulation scheme targets even temperature distribution, this assumption is accurate enough. Nevertheless, the time constant for the alignment of the currents is taken into account.

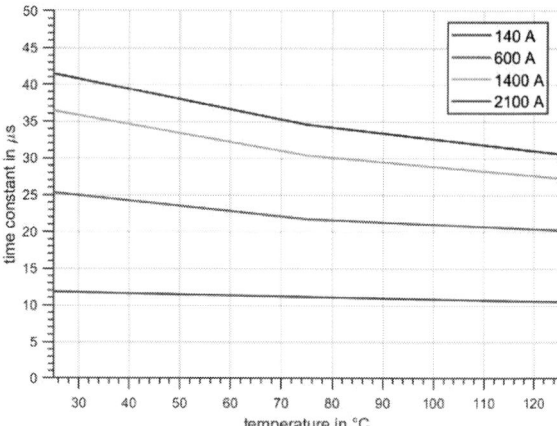

Fig. 5 time constant over temperature and current

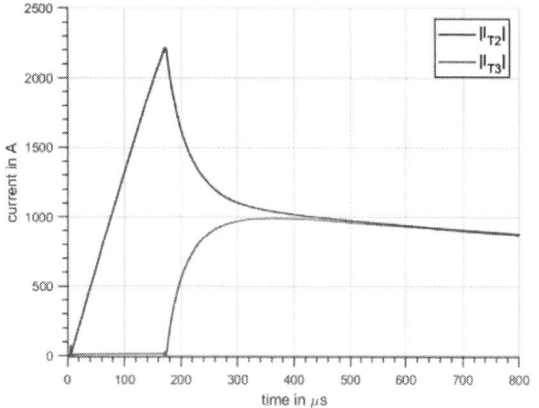

Fig. 4 Current distribution in both neutral paths during double pulse test.

4 Losses

The new modulation scheme aims to reduce the conduction losses during zero state. Also long commutation loops can be avoided, what is essential for 3L-ANPC-converters build up with three half-bridge modules. Nevertheless, this leads to altered switching losses compared to PWM-1. Additional, losses induced by the extra pulse for overvoltage reduction, must be taken into account.

4.1 Conduction losses

With PWM-1 modulation conduction losses are highest in T2 and T3. Due to the use of two n-paths, the losses can be reduced and distributed over the inner switches in zero voltage output state. The distribution of current over two paths is quite symmetric but needs some time to settle. Since the equivalent circuit consists of a current source and two RL-elements in parallel the current curves will be described with exponential functions in the model.
Figure 5 shows the time constants with respect to temperature at different currents. Time constants are ascending with the output current and descending with the temperature. The temperature dependency is small at low currents. This is due to different temperature coefficients in the forward characteristics of diode and IGBT. For low currents this coefficient tends to zero.

This current and temperature dependency was not implemented in the model because the effect is negligible small at low switching frequencies.
Using the Eq. 3 for conduction losses in the IGBT and diode respectively, a normalized saved power at evenly distributed current can defined by Eq. 4.

$$P_{\mathrm{cond}} = V_{\mathrm{CE0}} \cdot I_{\mathrm{C}} + r_{\mathrm{CE}} \cdot I_{\mathrm{C}}^2 \qquad (3)$$

$$\Delta P = \frac{1}{2\left(\dfrac{V_{\mathrm{CE0}}}{r_{\mathrm{CE}}I_{\mathrm{C}}} + 1\right)} \qquad (4)$$

Savings depend on the load current and the ratio of threshold voltage to slope resistance. In Fig. 6 curves for different ratios $V_{\mathrm{CE0}}/r_{\mathrm{CE}}$ over continuous current are shown. For semiconductors with dominant resistive part the modulation scheme is more beneficial. The ratios at 25°C for the used modules is 2400 A for the diodes and 1750 A for the IGBTs. At 150°C it is 1300 A for diodes 1000 A for IGBTs. The relative saved energy is higher at high junction temperatures.

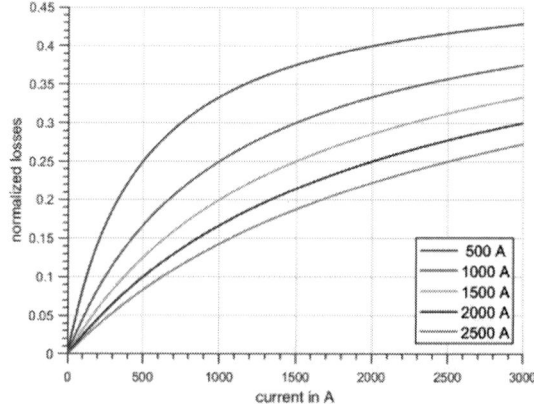

Fig. 6: Normalized saved conduction power loss over continuous current for different ratios $V_{\mathrm{CE0}}/r_{\mathrm{CE}}$.

4.2 Switching losses

Typically, PWM-1 modulation has a long and a short commutation. Long commutation has higher turn-off losses, but appears only twice per period. Higher turn-off losses result from the higher stray inductance. Thus the switch needs to be turned off slower. The new modulation scheme avoids the long commutation path. By using two current paths, long commutation can be split into three steps: short commutation and two commutations from one to the second current path. A new type of turn-off loss can be seen: switching off under ZVS where only stray inductance leads to an overvoltage and therefore losses. These losses are very low compared to the others. Losses in T5/T6 and T2/T3 differ slightly. Figure 7 shows all turn-off losses over current.

$$P_{sw} = \frac{1}{T_P} \sum_n E_{SW}(n) \qquad (5)$$

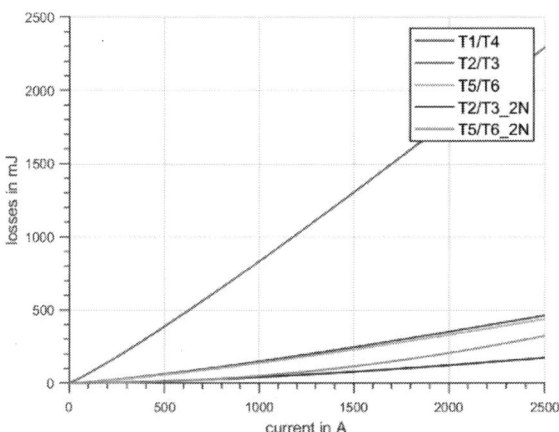

Fig. 7 Turn-off losses over transistor current for the transistors. T2/T3 is long commutation, T1/T4/T5/T6 short commutation and _2N is for turning off the second current path.

4.3 Losses due to extra pulses

Extra pulses were introduced to remove charge carriers in T2 and T3 after ZVS condition. The output capacitances of the participating switches are charged and small currents flow. Figure 8 shows the voltages and currents of the switches T3, T4 and T6. The losses can be positive or negative depending on charging or discharging the capacitance. In Fig. 8 the turn-off losses for T3 are 10 mJ when T4 turns on. Therefore, they are neglected. Other losses are measured as less than zero due to the discharged capacities and are therefore also neglected.

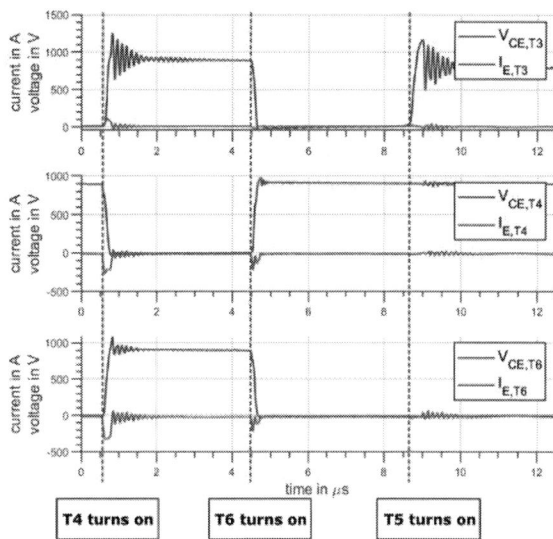

Fig. 8 Currents and voltages of T3, T4 and T6 at 2100 A load current

5 Simulation

Converter Losses are simulated using Matlab/ Simulink. The model structure is shown in Fig. 9. Gate signals are computed in the block *modulation* based on a given output voltage of the converter, switching frequency and DC link voltage. It can either output PWM-1 modulation or modified PWM-1-00 modulation. Phase currents are given as well.

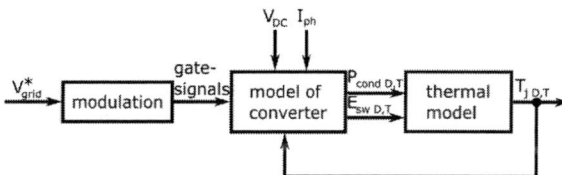

Fig. 9 Structure of simulation model

The block *model of converter* calculates the switching and conduction losses of the transistors and diodes. Based on the direction of the load current and the gate signals, the model detects whether a diode or transistor of a switch is conducting. Conduction losses are computed from Eq. 2 and parameters from table 2. As the turn-on and turn-off times are below the temporal resolution the switching losses are computed as energies. Whenever a transition of a gate signal is detected, switching energy losses will be calculated from half the DC link voltage, the switched current and the temperature according to Eq. 1 and parameters from table 2 for the IGBTs and diodes respectively. Figure 10 shows example curves of losses.

PCIM Europe 2024, 11– 13 June 2024, Nuremberg   DOI: 10.30420/566262429

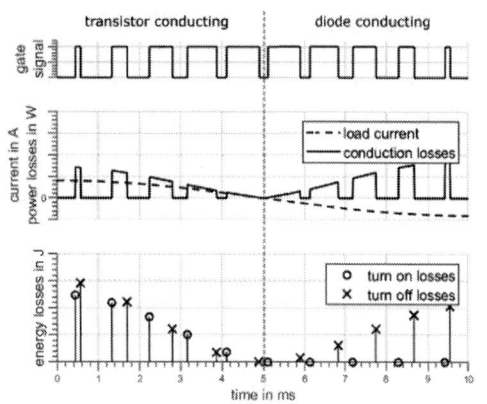

Fig. 10 Principle of *model of converter* demonstrated at T1 and D1. The block computes conduction power losses and switching energy losses based on the current, temperature and gate signal.

Conduction and switching losses are fed into the block *thermal model*. It consists of the Cauer model of the whole stack which includes the half-bridge modules and the heat sink. The modules are thermally coupled. The calculated junction

temperatures of all IGBTs and diodes are fed back to the model of the converter.

Operational points with different currents, power factors and modulation depths were simulated. To reduce simulation time, turn-on losses of the diodes were neglected since they are small compared to other losses and are equal between both investigated modulations.

6 Results

6.1 Static simulation

Figure 11 shows the simulated conduction losses with PWM-1 and modified PWM-1-00 modulation in steady state. Conduction losses are most evenly distributed over the inner switches in Fig. 11a where the modulation depth is low. In Fig. 11b conduction losses are evenly distributed with both modulation schemes but could be reduced using the second n-path. At regular operational points like Fig. 11c the advantage is very small. At high currents using two n-paths becomes more advantageous, which can be slightly seen in Fig. 11c. The outer switches are not affected.

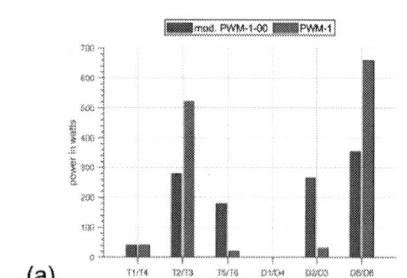

(a)
Power factor 0.9, I_{ph} = 900 A, 10% of nominal voltage

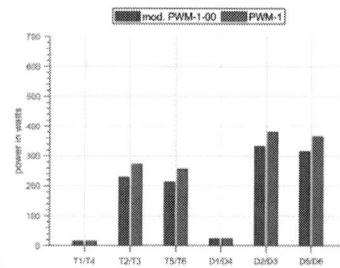

(b)
Power factor 0.0, I_{ph} = 900 A, 10% of nominal voltage

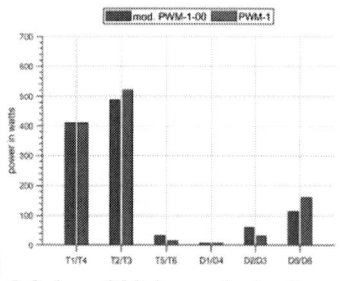

(c)
Power factor 0.9, I_{ph} = 900 A, nominal voltage

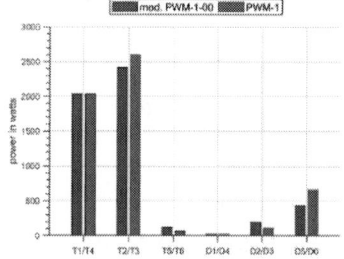

(d)
Power factor 0.9, I_{ph} = 2700 A, nominal voltage

Fig. 11 Simulated conduction losses at f_{SW} = 4 kHz, $V_{DC}/2$ = 750 V, V_{LLN} = 950 V

3025

PCIM Europe 2024, 11–13 June 2024, Nuremberg   DOI: 10.30420/566262429

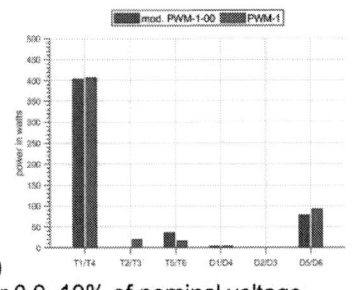

(a)
Power factor 0.9, 10% of nominal voltage

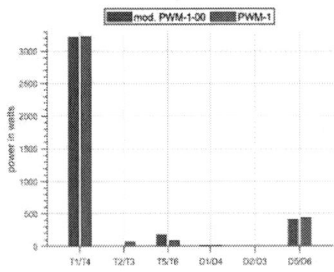

(b)
Power factor 0.0, I_{ph} = 900 A , 10% of nominal voltage

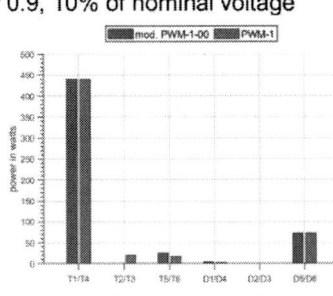

(c)
Power factor 0.9, I_{ph} = 900 A , nominal voltage

(d)
Power factor 0.9, I_{ph} = 2700 A, nominal voltage

Fig. 12 Simulated switching losses at f_{SW} = 4 kHz, $V_{DC}/2$ = 750 V, V_{LLN} = 950 V

Figure 12 shows the switching losses of all transistors and diodes. For T2/T3 the switching losses are reduced in every operational point as the long commutation loop can be avoided. Switching losses in T5/T6 are higher with new modulation scheme since they experience short commutation and commutation from one to the other n-path additionally. Although the same current is switched in D5/D6 and T1/T4 the losses are smaller when using two n-paths. Because they experience less

conduction losses, the junction temperatures are lower, therefore losses are also lower according to Eq. 2.

In Fig. 13 total losses are shown. In all operating point losses could be reduced. However, at low modulation depths, power factors close to zero and high currents the new modulation is most beneficial.

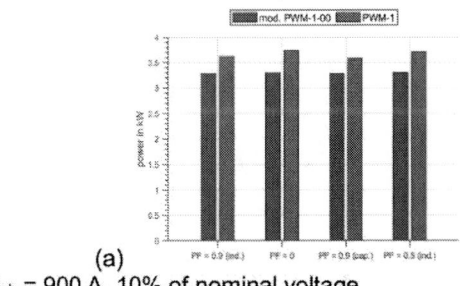

(a)
I_{ph} = 900 A, 10% of nominal voltage

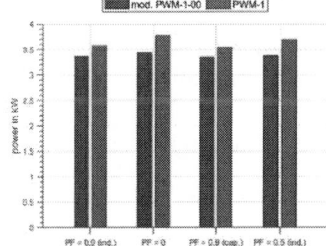

(b)
I_{ph} = 900 A, 50% of nominal voltage

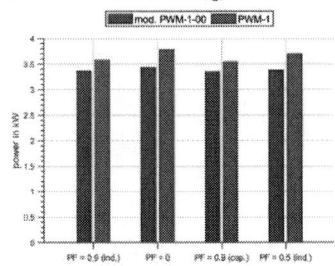

(c)
I_{ph} = 900 A , nominal voltage

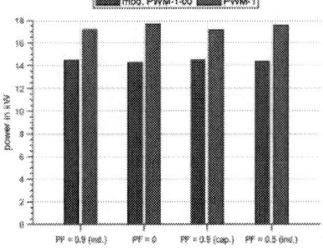

(d)
I_{ph} = 2700 A , 10% nominal voltage

Fig. 13 Simulated total losses at f_{SW} = 4 kHz, $V_{DC}/2$ = 750 V, V_{LLN} = 950 V

3026

6.2 Thermal simulation

Under normal conditions (high power factor, high modulation depth) the new modulation has comparable losses to the PWM-1. As shown in the previous chapter operating points with low voltage, high current and power factor close to zero are beneficial. Three overload scenarios are simulated, which mimic real-life situations for inverters used in grid support:

- Current limitation (CL) with low modulation depth and high power factor (Fig. 14a),
- Low voltage ride through (LVRT) with low modulation depth, power factor of zero and high current (Fig. 14b),
- Supply of spinning reserve (SR) with high modulation depth and high power factor (Fig. 14c).

All simulations start in thermal equilibrium at 55°C junction temperature. Overload event starts at 0 s by applying the given modulation depth, power factor and current. The maximum junction temperature according to the datasheet is 150°C for the transistors and 175°C for the diodes. Using modified PWM-1 max. junction temperature is reached later in LVRT and CL in overload condition compared to classic PWM-1. Figure 14a shows the temperature curves for the CL mode. Overload capacity is limited by the diodes D5/D6. With the new modulation the maximum junction temperature is reached at 3.5 s instead of 0.1 s. In case of LVRT shown in Fig. 14b the diodes D2/D3 reach their maximum junction temperature first. With the new modulation the time could be extended from 1.5 s to 3.0 s. Figure 14c shows the curves for supply of spinning reserve. The maximum junction temperature of T1/T4 is reached at the same time as with PWM-1 modulation. This is caused by the rare usage of the second n-path in this scenario.

In LVRT transistors T5/T6 showing the highest losses. One would expect that they have the highest junction temperatures. In fact, transistors T2/T3 have the highest junction temperature. This is because the half-bridge module with T1/T5 and T4/T6 respectively have the lowest losses. Therefore, the base plate temperature and the junction temperatures of T5/T6 are staying cooler then T2/T3.

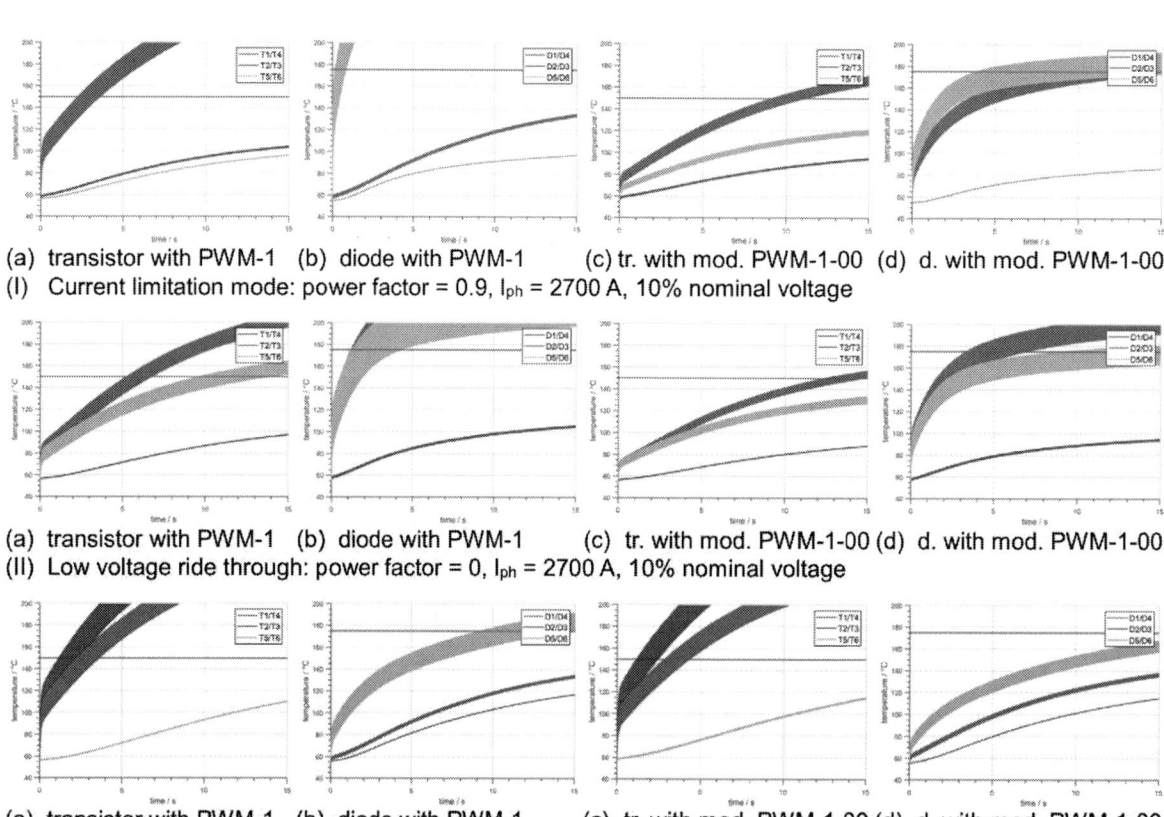

(a) transistor with PWM-1 (b) diode with PWM-1 (c) tr. with mod. PWM-1-00 (d) d. with mod. PWM-1-00
(I) Current limitation mode: power factor = 0.9, I_{ph} = 2700 A, 10% nominal voltage

(a) transistor with PWM-1 (b) diode with PWM-1 (c) tr. with mod. PWM-1-00 (d) d. with mod. PWM-1-00
(II) Low voltage ride through: power factor = 0, I_{ph} = 2700 A, 10% nominal voltage

(a) transistor with PWM-1 (b) diode with PWM-1 (c) tr. with mod. PWM-1-00 (d) d. with mod. PWM-1-00
(III) Supply of spinning reserve: power factor = 0.9, I_{ph} = 2700 A, nominal voltage

Fig. 14 Comparison of IGBT and diode junction temperature with standard and modified PWM-1 modulation. Red line marks the maximum junction temperature.

7 Conclusion

Switching losses were determined by double pulse testing short commutation, long commutation and commutation within the n-paths in an 3L-ANPC phase leg at different voltages, currents and temperatures. The influence of the intermediate pulses to reduce overvoltage when using both n-paths was investigated. For all measured losses analytical equations were given. Conduction and switching losses for an inverter at several operational points were simulated for PWM-1 and modified PWM-1-00 with intermediate pulses. Thermal dependency of the losses was considered. It could be shown that with the modified PWM-1-00 modulation at low modulation depths and power factors around zero the losses of the inner switches are distributed more evenly. Thus at regular operation usage of both n-paths shows no additional overall losses compared to PWM-1 modulation. Moreover, the new modulation scheme enables 3L-ANPC-converters build of three standard half-bridge modules to avoid long commutation with high turn-off losses and to use both n-paths.

However, at low modulation depths e.g. in current limit mode or power factors close to zero e.g. at low voltage ride through the new modulation can help to distribute conduction losses and therefore extend the current overload capability of the converter.

8 Reference

[1] T. Bruckner and S. Bernet, "The active NPC converter for medium-voltage applications," Fourtieth IAS Annual Meeting. Conference Record of the 2005 Industry Applications Conference, 2005., Hong Kong, China, 2005, pp. 84-91 Vol. 1

[2] G. Zhang, Y. Yang, F. Iannuzzo, K. Li, F. Blaabjerg, H. Xu, "Loss Distribution Analysis of Three-Level Active Neutral-Point-Clamped (3L-ANPC) Converter with Different PWM Strategies", IEEE, 2016

[3] Y. Jiao and F. C. Lee, "New Modulation Scheme for Three-Level Active Neutral-Point Clamped Converter With Loss and Stress Reduction" IEEE Transactions on Industr. Elect., vol. 62, no. 9, 2015

[4] A. Giessmann, C. Schmidt., M. Spang, "The Simultaneous Use of Two Alternative Neutral Point Current Paths in a 3-Level ANPC Topology – Benefits and Challenges in High Power Applications", PCIM Asia 2021

[5] Giessmann, A., Schmidt, C. (2022). Stromrichtereinrichtung mit einem Stromrichter und mit einer Steuereinrichtung. German Patent No. DE 102022119531A1. Munich, Germany. German Patent and Trade Mark Office

[6] F. Eichler, M. Meissner, M. Meissner, S. Bernet and A. Giessmann, "Voltage Stress Reduction for New 3L-ANPC Modulation Scheme Using Both Current Paths in Zero State," PCIM Europe 2023

[7] T. Brückner, S. Bernet, H. Güldner, „The Active NPC Converter and Its Loss-Balancing Control" in IEEE Trans. Ind. Electronics, Vol. 52, 2005, pp. 855-868

[8] D. Floricau, E. Floricau, G. Gateau, "Three Level Active NPC Converter PWM Strategies And Loss Distribution", IEEE, 2008

[9] H. Wang, X. Ma and H. Sun, "Active neutral-point-clamped (ANPC) three-level converter for high-power applications with optimized PWM strategy," PCIM Asia 2020

[10] Semikron, "IGBT M7 Modules" SKM1400GB12M7S2I4 datasheet, [Rev. 0.2 – 05.03.2021]

[11] Semikron, Application Manual Power Semiconductors (2nd edition), 2015-05-08

PCIM Europe 2024, 11– 13 June 2024, Nuremberg   DOI: 10.30420/566262430

An Innovative Synchronous Rectification Method for 11 kW CLLC Converters

Sanbao Shi[1] , Cheng Zhang[1] , Sidorov Vadim[2]

[1] Infineon Semiconductors (Shenzhen) Company Limited
[2] Infineon Technologies Austria AG

Corresponding author: Sanbao Shi, simon.shi@infineon.com
Speaker: Sanbao Shi, simon.shi@infineon.com

Abstract

Nowadays, CLLC topology was used widely in the bidirectional power electronic systems such as EV chargers, energy storages, on board chargers (OBC), and vehicle-to-grid (V2G) charger stations, due to a high efficiency, low electromagnetic interference (EMI), bi-directional capability, and a high-power density. However, a synchronous rectification control is not used widely because it can complicate a control system. As a result, the system performance is lower when it comes to efficiency and thermal behavior, especially when silicon carbide (SiC) MOSFETs are used in a converter. An innovative synchronous rectification method is introduced in this paper. In this method, the secondary-side synchronous rectification PWM signal is generated by the CLLC oscillation cavity current, the primary-side PWM signal and the system power. Results of a test on a 11 kW CLLC DC-DC converter show that this synchronous rectification method can increase the entire system efficiency from 97.1% to 98.1%. This rise can benefit the whole system and have a positive impact on the applications.

Keywords: CLLC, SI, SR, EV Charger, OBC

1   Introduction

A system composed of renewable energy equipment and energy storage units is effective in resolving the problem of connectivity and power quality in the grid [1]. A bidirectional converter plays an important role in renewable energy systems. Its bidirectional conversion characteristics enable it to act as a "bridge" for the entire system, as illustrated in Fig.1. Owing to its high efficiency and bidirectional power transmission, a symmetrical, bidirectional, full-bridge CLLC converter is usually used at the DC-DC stage of the bidirectional converter [2].

Compared to silicon (Si), SiC has a higher bandgap, higher breakdown field, and better thermal conductivity. These advantages of SiC MOSFETS help to improve performance of the converter.

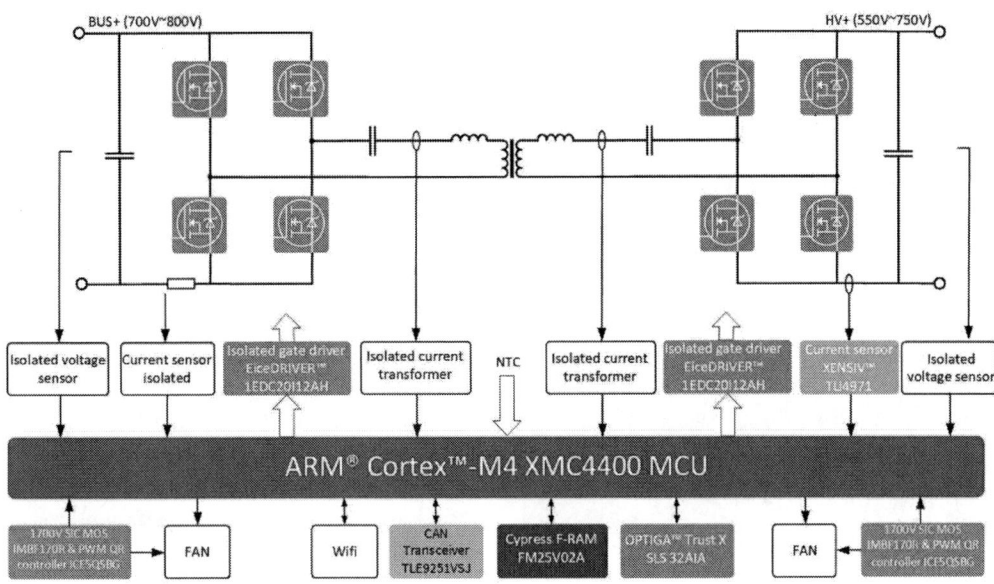

Fig. 1   Functional block diagram of a 11 kW CLLC DC-DC converter

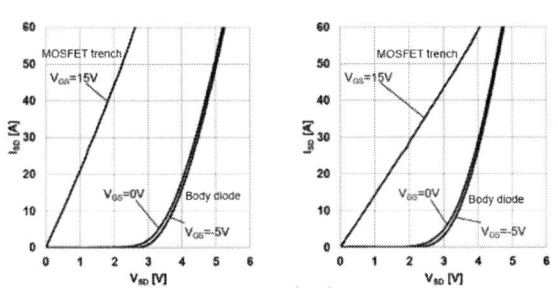

Fig. 2   Reverse current, I_{SD}, as a function of voltage at different gate voltages and temperature. T_j = 25°C (left), T_j = 175°C (right)

Unlike IGBTs, SiC MOSFETs can also conduct reverse current from source to drain [3] through the channel if a positive bias is applied to the gate. This mode of operation is called synchronous rectification (or third quadrant operation) and can be achieved with a positive voltage, typically, of +15 V at the gate. However, achieving synchronous rectification is a problem in the CLLC topology.

As shown in Fig.3 below, the current transformer CT1 is used to detect the current on the secondary bridge side. The output voltage of CT1 V_{ct} is rectified through diode bridge D1~D4, then the signal V_{rect} is compared with reference voltage V_{ref} in the comparator U1. After that output PWM signal $V_{comp.out}$ is generated. ADD gate U2 detect the

PWM signal, compare with MOSFET Q4 gate voltage $V_{gs.Q4}$, and then the MOSFET Q8 gate control signal $V_{gs.Q8}$ is generated.

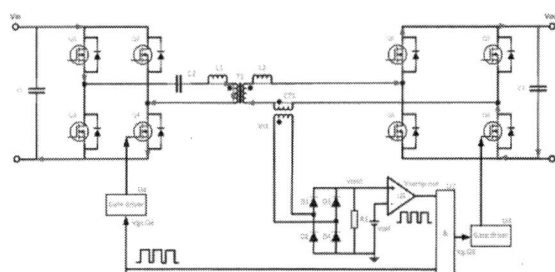

Fig. 3   Innovative synchronous rectification method

The microcontroller unit (MCU) disables the synchronous rectifier (SR) mode during low output load (burst mode) because the current waveform in CLLC are not clear during the discontinuous conduction mode (DCM). However, efficiency requirement during low load is extremely low.

A similar method for synchronous rectification has been described for the bidirectional series-resonant converter with fixed switching frequency [2]. The paper focuses on the synchronous rectification method for the CLLC converter with frequency control.

2   System Design

An 11 kW bidirectional CLLC reference board was designed, as shown in Fig. 4, using Infineon's CoolSiC™ MOSFET 1200 V discrete, IMZ120R030M1H [4], a wide body package gate driver IC,1EDC20I12AH [5], a 32-bit Arm® Cortex®-M4 industrial MCU, XMC4000 [6], and a high-precision coreless current sensor XENSIV™ TLI4971 [7] . The outer dimensions of the board, enclosed in the case, were 33.1 mm x 13.4 mm x 6 mm, which resulted in 4.1W/cm³ (5.5W/g) of power density.

Fig. 4   11 kW bidirectional CLLC converter

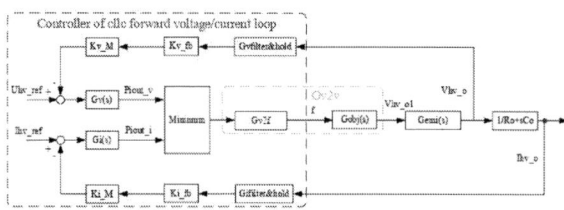

Fig. 5   Software control loop

The software control process is shown in Fig. 5. It includes the voltage control and current control loop. The MCU detects the output voltage and current and after comparing P_{iout_v} and P_{iout_i} generates the PWM frequency.

During low load, the control loop goes into the burst mode to improve efficiency. In forward mode, when PWM frequency is higher than 140 kHz, the PWM duty cycles start reducing.

3   Test Results

3.1   Test waveform

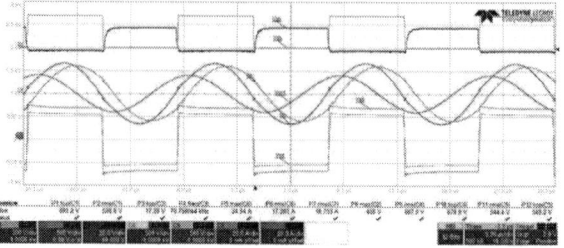

Fig. 6   Test waveform (BUS700V-HV550V/9.5 kW)

The functional test waveform is shown in Fig. 6. I_{pr} and I_{cr} are the primary and secondary side currents of the CLLC converter. The graph shows clean waveforms that can be used by the MCU to detect synchronous rectification signals.

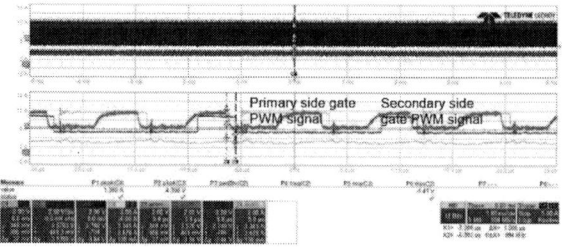

Fig. 7   Test waveform for synchronous rectification

Figure 7 shows the synchronous rectification PWM signal. The AND operation between the flip signal output of the comparator and the primary PWM signal is carried out inside the MCU. The MCU recognizes the high levels of comparator output and triggers an external interrupt. This combines with the current cycle of the PWM wavesending sequence in the interrupt service routine. The corresponding synchronous rectification drive is issued, but there is a small delay in the actual measurement due to software processing. The actual measurement current delay is about 1 μs, and software optimization is required to reduce this delay time.

3.2   Efficiency Test: Between BUS to HV in Forward Mode

According to the analyses above, the efficiency difference between BUS to HV (forward mode) is shown in Fig. 8. The system efficiency has increased about 1%.

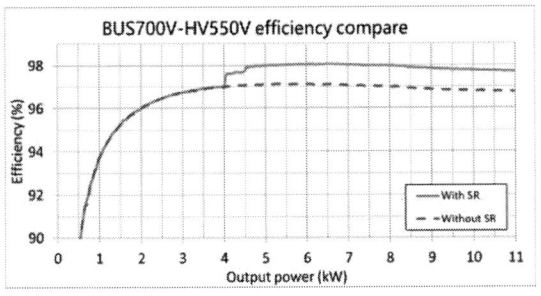

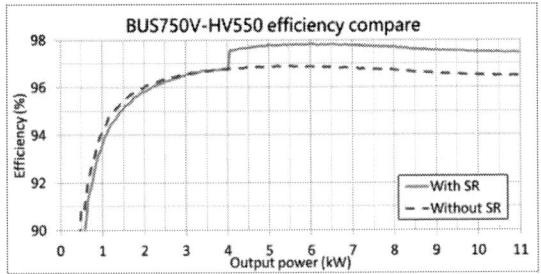

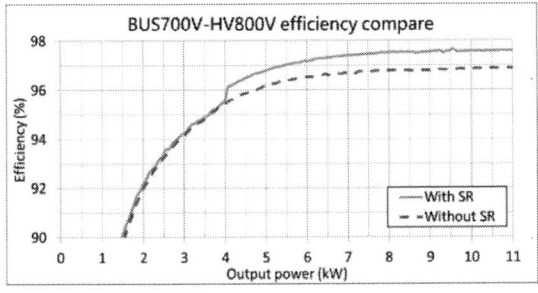

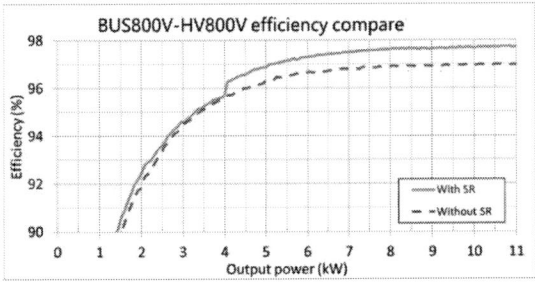

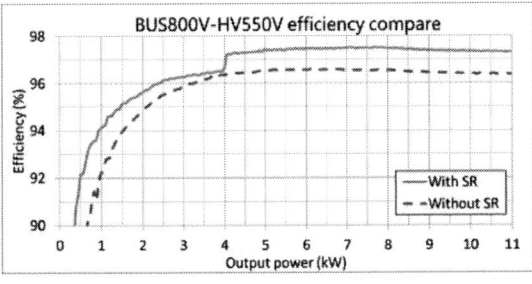

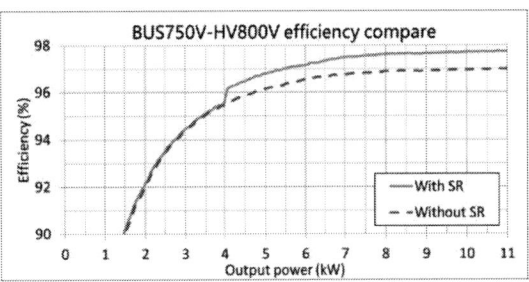

Fig. 8 Efficiency test results of the whole system during the forward mode

3.3 Efficiency Test: Between HV to BUS in Reverse Mode

During reverse mode also the efficiency increased by about 1%, as shown in Fig. 9.

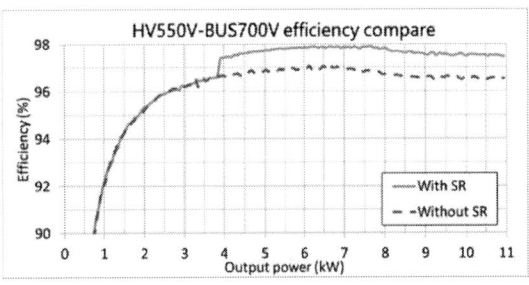

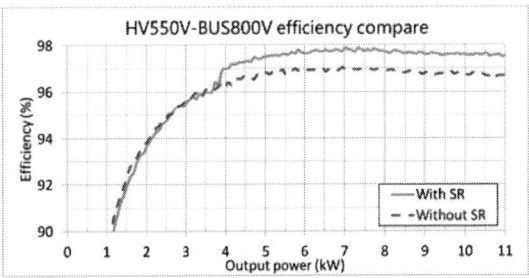

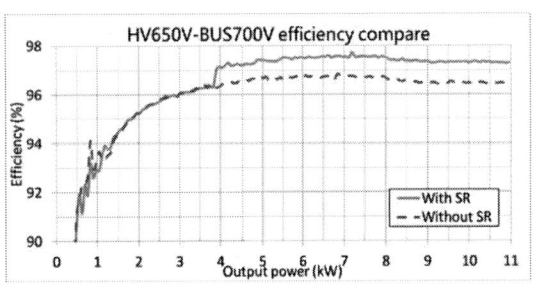

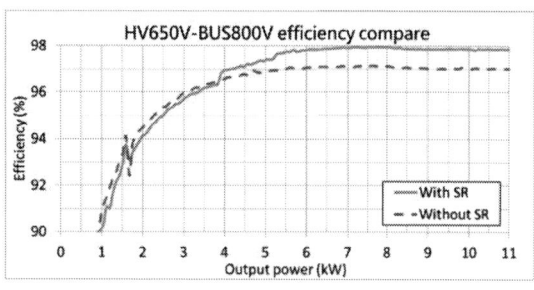

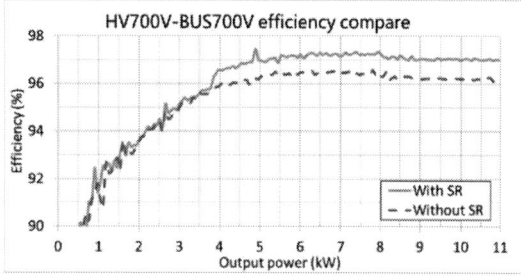

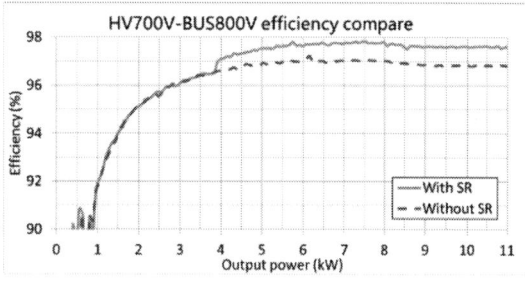

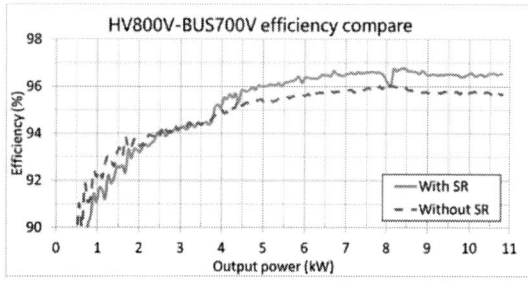

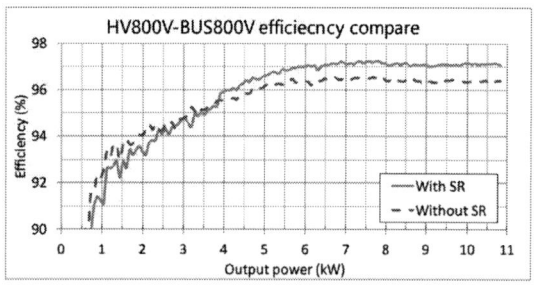

Fig. 9 Efficiency test results for the whole system during the reverse mode

4 Conclusion

An innovative synchronous rectification method for CLLC topology was introduced in this paper. The timing of the PWM control algorithm along with analog and digital solutions was also provided. The final functional test results have shown that this method can successfully increase the efficiency of the whole system by about 1%. This rise in efficiency can be very helpful in new energy applications, especially when SiC MOSFETs are used.

References

[1] Mladen Ivankovic, Jon Mark Hancock: Part I : LLC calculator, FHA analysis based on a vector algorithm, Infineon Technologies AN_201709_PL52_029, 2017

[2] Infineon Technology: 11 kW bi-directional CLLC DC-DC converter with 1200V and 1700V CoolSiC™ MOSFETs, UG_2020_31, 2020

[3] V. Sidorov, A. Chub and D. Vinnikov, "Bidirectional Isolated Hexamode DC–DC Converter," in IEEE Transactions on Power Electronics, vol. 37, no. 10, pp. 12264-12278, Oct. 2022.

[4] Infineon Technologies: CoolSiC™ 1200V SiC Trench MOSFET Silicon Carbide MOSFET, IMZ120R030M1H datasheet, 2018

[5] Infineon Technologies: EiceDRIVER™ 1EDC Compact Single channel IGBT gate driver IC in wide body package, 1EDCxxI12AH and 1EDCxxH12AH datasheet, 2017

[6] Infineon Technologies: XMC4400 Microcontroller Series for Industrial Applications, datasheet, 2018

[7] Infineon Technologies: TLI4971 high precision coreless current sensor for industrial applications in 8x8mm SMD package, TLI4971 datasheet, 2019

PCIM Europe 2024, 11– 13 June 2024, Nuremberg   DOI: 10.30420/566262431

Interleaved Asynchronous Delta-Sigma Modulation Concept for Dynamic Power Converters

Philipp Czerwenka[1], Jannik Maier[1], Eckhard Hennig[1], Gernot Schullerus[1], Ertugrul Sönmez[1]

[1] Electronics & Drives, Reutlingen University, Germany

Corresponding author: Philipp Czerwenka, philipp.czerwenka@reutlingen-university.de
Speaker: Philipp Czerwenka, philipp.czerwenka@reutlingen-university.de

Abstract

This paper proposes a novel synchronization concept for asynchronous delta-sigma modulators (ADSM), enabling its application for interleaved operation of parallel power converter structures. Two separate first-order ADSMs are coupled via a phase-locked loop, ensuring constant phase relationship between the modulators across their switching frequency ranges. Thus, the benefits of delta-sigma modulation such as spread spectrum qualities and lower average switching frequencies compared to pulse-width modulation can be used in dynamic interleaved power converters. In this paper, we derive system equations for the synchronization loop and confirm its functionality by simulations and measurements for a two-phase interleaved half-bridge.

1 Introduction

The application of pulse width modulation (PWM) for converting a value-continuous signal to a digital pulse pattern is well established in power converters, drive inverters and class-D amplifiers. PWM is simple to generate, requiring only a comparison with a triangle or sawtooth signal at the desired full-scale range. Recently, with the rise of fast-switching wide-bandgap semiconductors [1] such as silicon carbide and gallium nitride for high voltage devices, the drawbacks of PWM have become apparent. PWM generates concentrated spectral power at the switching frequency fundamental and its harmonics, which at higher switching frequencies of modern power electronics, leads to increased electromagnetic interference. Furthermore, as lower switching losses and faster switching speeds enable higher frequencies, the minimum producible pulse of such wide-bandgap devices does not suffice to produce the pulse lengths required at the extremes of duty cycles in the vicinity of 100 or zero percent.

These limitations have lead to emerging modulation types with spread-spectrum qualities. One of them is delta-sigma modulation, known from the field of analog-to-digital converters. A delta-sigma modulator (DSM) uses a control loop to convert an arbitrary signal into a bitstream with specified quantization steps. It therefore reduces its own quantization error leading to spread-spectrum qualities and noise-shaping properties for increased signal fidelity [2]. This modulation scheme has been already successfully applied in class-D amplifiers for audio applications [3].

The typically higher output current for power converters can be achieved by parallel connection of half bridges. In such cases, an interleaved operation is beneficial. Achieving interleaving in conjunction with delta-sigma modulation is challenging, as a fixed initial phase offset in the output signals of the modulators quickly deteriorates, because the modulators change their frequencies during operation. Only few concepts for combining the delta-sigma modulation technique with interleaving power electronics topologies exist. In [4] and [5], three-phase interleaved buck and boost converters with a single adaptive delta-sigma converter and constant phase shifts are presented. In [6], a quasi-interleaved non-standard buck converter topology driven by a delta-sigma modulator is presented. Such concepts require changing the power electronics or modulator structure with unclear spectral performance implications.

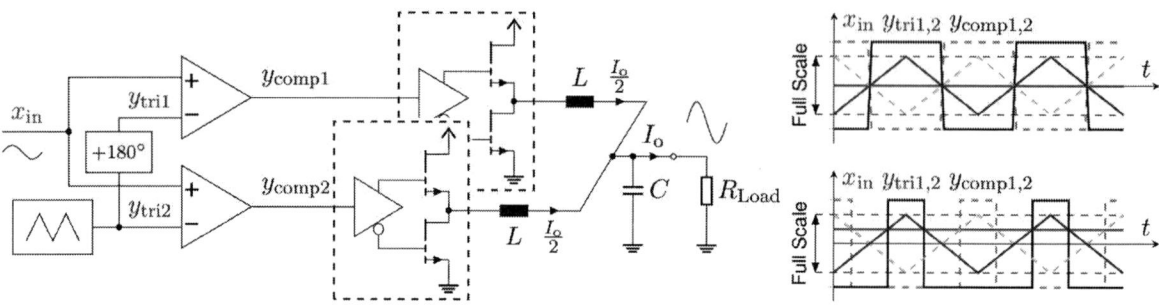

Fig. 1: Possible traditional system implementation of PWM-based two-stage interleaving power amplifier (left) with characteristic waveforms of the modulators (right)

This paper aims to provide a solution for the combined operation of delta-sigma modulation and interleaving parallelization. The concept expands on the true asynchronous delta-sigma modulator and uses only standard unmodified half-bridges as power stages. We propose a novel modulator system architecture consisting of two phase-synchronized asynchronous delta-sigma modulators (ADSM) for the application in dynamic interleaving power converters. The new concept has the following benefits:

1. Interleaving parallel connection delivers increased output currents and increases the effective switching frequency [7] [8], reducing ripple amplitudes.

2. Delta-sigma modulation reduces the average switching frequency compared to pulse-width based modulation methods at the same base frequency and for the same signal quality, reducing switching losses [9].

The paper is structured as follows: First we revise the theory for pulse-width and delta-sigma modulators in Section 2. In Section 3 we present the novel system concept. Transfer functions are derived and design formulae for the parameterization of such systems are given. Section 4 shows system level simulation results to verify the expected system behavior. Spectrum properties are discussed. In Section 5 we present an example of a working prototype using an interleaved ADSM with measurement results. Section 6 concludes this paper.

2 Modulator Theory

2.1 Pulse-Width Modulation

A analog PWM modulator consists of a reference triangle or sawtooth wave generator and a comparator. The input signal is compared to the reference by the comparator. If the input signal is above the reference signal, the output is logic high, otherwise low. PWM modulators are therefore simple to implement and require few components. The full-scale range of the modulator is determined by the reference signal amplitude. To the frequency spectrum, it is characteristic, that it exhibits significant harmonics at the multiples of the reference signal's frequency. In the passband, the PWM modulator features a constant noise floor with its magnitude dependent on the reference signal's frequency.

Achieving interleaving with PWM is also straightforward. The discussion is presented for two-stage interleaving for reasons of clarity without loss of generality. Figure 1 left shows an exemplary single-phase power amplifier system utilizing two-stage PWM modulated interleaving. The input signal is provided to two separate comparators with their reference signals phase-shifted by $\frac{360°}{n}$, where n is the number of parallel interleaving stages. No feedback is required. The constant phase relationship of the reference signals $y_{\text{tri}1,2}$ ensures a constant phase relationship of the modulated signals (see Fig. 1 right).

2.2 Delta-Sigma Modulation

Delta-sigma modulation is a family of signal processing methods, which are characterized by

- synchronous (running on a fixed-frequency base clock) or asynchronous (without external base clock) operation,

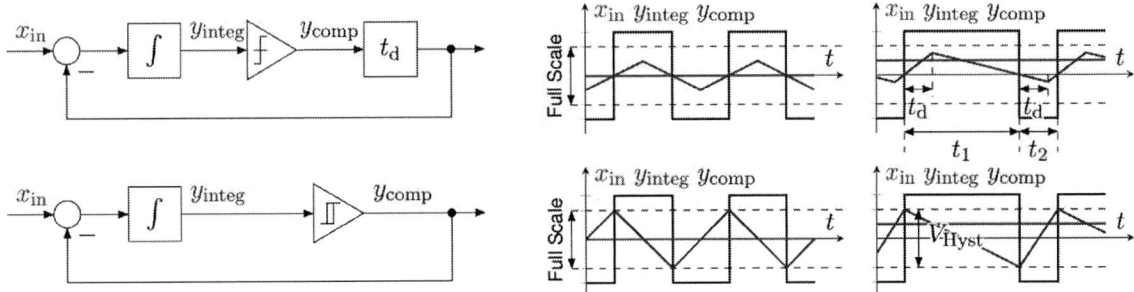

Fig. 2: Basic 1st order time-continuous asynchronous delta-sigma modulator structures of the delay-variant (top) and hysteresis-variant (bottom) with respective characteristic waveform (right)

- continuous-time or discrete-time integrator implementation,

- number of integration stages (order) and

- number of quantization steps.

This publication focuses on the asynchronous continuous-time first-order delta-sigma modulator (ADSM) using a two-level quantizer (e.g. comparator) to produce a binary output signal.

The main idea behind the delta-sigma modulator is to integrate the input-to-output error of a quantizer while controlling the integration loop delay. In the ADSM's frequency spectrum the switching frequency's harmonics are spread across a wide band, resulting in decreased single-frequency amplitudes. Additionally, a noise-shaping behavior is observed in the pass-band, where the noise floor magnitude decreases towards lower frequencies, resulting in increased signal fidelity.

Two system realizations of the ADSM have been described, yielding the same system behavior. The delay-based ADSM consists of an integrator, a comparator and a delay-line with delay t_d (see Fig. 2 top). Its characteristic time constant $T_c = 2t_d$ is independent of the integrator gain. The hysteresis-based ADSM replaces the comparator and the delay with a hysteresis comparator (see Fig. 2 bottom). Its timing is determined by the integration gain T_I and the hysteresis width V_{hyst}, such that $T_c = \frac{V_{hyst}}{T_I}$. While this makes the system behavior dependent on more parameters, a hysteresis comparator is often more simple to implement as a circuit compared to a precise transport delay. The full-scale range of both modulator variants is determined by the

comparator's output range and potential feedback gains. In the following we will focus on the delay-based variant because of its more relaxed timing requirements.

From the characteristic waveforms in Fig. 2, the following equations describing the ADSMs' behavior are derived geometrically. The widths of high and low pulses, t_1 and t_2 respectively, are

$$t_1 = \frac{T_c}{1 - x_{in}(t)} \quad t_2 = \frac{T_c}{1 + x_{in}(t)}, \quad (1)$$

where $x_{in}(t) \in (-1, 1)$ is the full-scale normalized input amplitude. Its information is linearly encoded in the output signal's duty cycle D with

$$D = \frac{t_1}{t_1 + t_2} = \frac{1 + x_{in}(t)}{2}. \quad (2)$$

As Eq. (2) illustrates, D is independent of the modulator's parameters given by T_c. Characteristically for the ADSM, the instantaneous output signal frequency f_{sw} changes based on the input signal's amplitude with

$$f_{sw} = \frac{1}{t_1 + t_2} = \frac{1 - x_{in}(t)^2}{2T_c}. \quad (3)$$

Equation (3) is the reason why two ADSMs as described above can not be extended to an interleaving system in the same straightforward way like the PWM modulator. Even under the assumption of an initial phase-shifted output, the changing instantaneous frequency prevents a constant phase relationship by passive methods. The following section therefore describes a concept of an active synchronization, which will overcome this limitation.

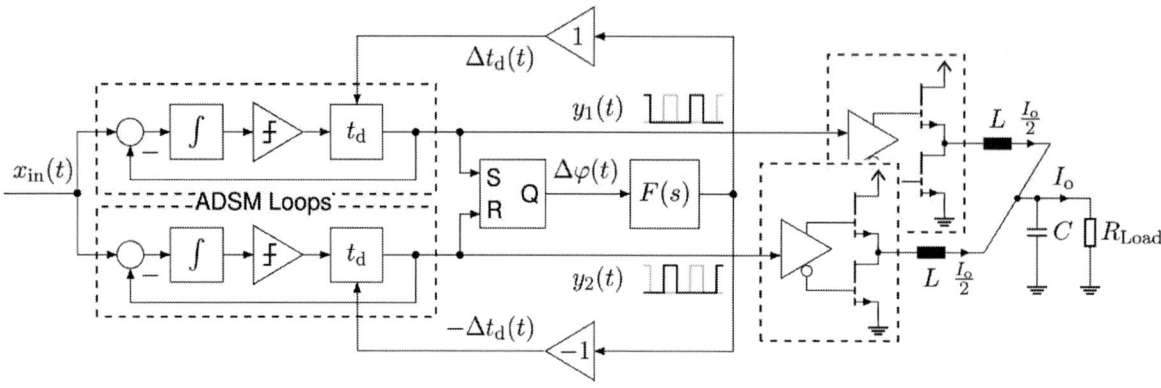

Fig. 3: Novel interleaved ADSM system concept with PLL-like synchronization control producing two $180°$ out-of-phase delta-sigma modulated bitstreams driving two half bridges with LC summing filter

3 Modulation System Concept

3.1 Block Diagram

Figure 3 shows the proposed concept of two phase-synchronized delay-based ADSMs. The signal paths are regular delay-based ADSMs receiving the same full-scale normalized input control signal $x_{in}(t)$. Due to different initial conditions and parameter variations, the phase difference $\Delta\varphi(t)$ of the output bitstreams $y_1(t)$, $y_2(t)$ is undetermined at startup. A synchronization control loop similar to a phase-locked loop consisting of an S-R latch and a continuous-time loop filter is added. The edge-triggered S-R latch is acting as a phase detector and produces a duty-cycle encoded offset proportional to the phase deviation $\Delta\varphi$ from $180°$. This offset is then converted to a delay time variation Δt_d by the loop filter $F(s)$, which in turn affects the delta-sigma delays ($t_d = t_{d0} \pm \Delta t_d$). Due to the inherent inverting relationship between delay time t_d and delta-sigma frequency f_{sw}, which will be shown in section 3.2, an increase in phase difference results in slowing down the faster delta-sigma path and speeding up the slower delta-sigma path and thus resulting in stable operation.

3.2 Transfer Function Derivation

To derive the signal and noise transfer functions for design of the synchronization loop, we use linearization assumptions inspired by phase-locked loop control theory [10]. We analyze the behavior of a delay-based ADSM path under the assumption of operation around a given operating point delay t_{d0}. We can then linearize Eq. (3), which yields

$$\Delta f_{sw} = -\frac{1 - x_{in}(t)^2}{4t_{d0}^2}\Delta t_d =: -K_f(x_{in})\Delta t_d \ . \quad (4)$$

Thus, for small deviations Δt_d in the neighborhood of the operating point t_{d0}, the ADSM behaves like a delay-controlled oscillator with gain $-K_f$. In the Laplace domain, a voltage-controlled oscillator can be represented as an integration of the input frequency, which yields a phase angle. For our model, we are interested in the variation of the frequency Δf_{sw} being integrated. The frequency variation due to delta-sigma operation, which is an error parameter for this model, is represented by the addition of φ_{sw}.

In the locked state, the S-R latch phase detector is linearized similarly to a quantizer as a difference of the ADSMs output signals, a gain K_{PD} and the addition of an error term e_{PD}. K_{PD} models differences in the amplitude between the phase detector input and output signals and can be assumed to be 1 for full-scale normalized signals. e_{PD} modles the discretization error because of the value-discontinuous operation of the latch.

The loop filter is required to obtain the control signal Δt_d that encodes the timing information of the two ADSM paths. Defining the S-R latch output to $+1$ if the most recent rising edge transition has occurred in the upper ADSM path and -1 for the lower path, the controlling nature of the loop will bring Δt_d towards zero, resulting in identical time differences between all edge transitions of the output signals. This in turn means that there is a constant instantaneous phase difference of $180°$ between the two ADSM paths, which is the desired value for two-stage interleaving. Theoretically, choosing a value other than zero will result in a

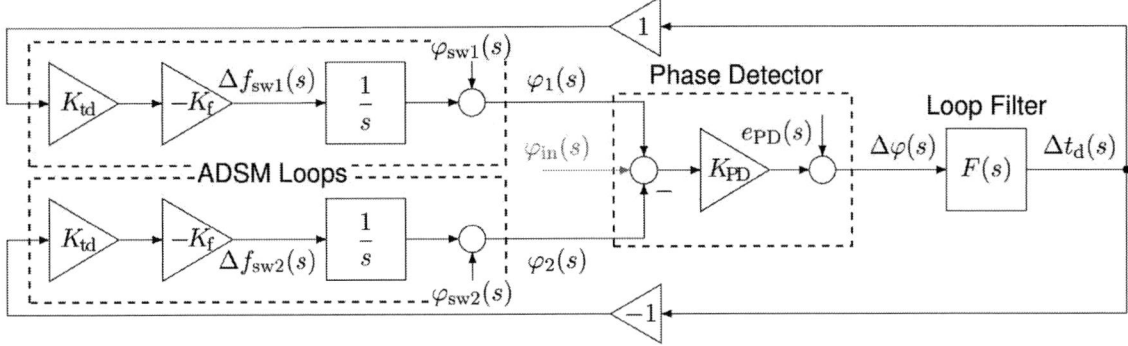

Fig. 4: Linearized model of phase-synchronized ADSM system from Fig. 3

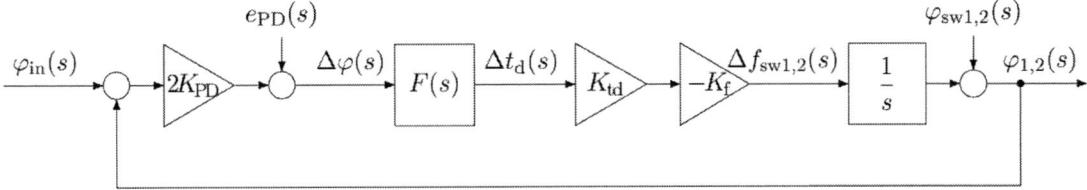

Fig. 5: Single ADSM loop model in the phase-synchronized system obtained by rearrangement from Fig. 4

phase relation $\neq 180°$. By design of the application of the S-R latch, an explicit input signal φ_{in} is not required for this loop.

The result from the previous discussion is the Laplace-space linearized equivalent control block diagram in Fig. 4 for a delay operating point t_{d0}. Using simplification techniques for block diagrams, we can further simplify and obtain the equivalent control block diagram for a single channel in Fig. 5. As the PLL-control loop was chosen to affect both ADSM channels, a factor of 2 is introduced here, providing an additional system gain. Theoretically, this is not necessary and the system could be built with the control loop affecting only a single channel, effectively creating a master-slave topology.

From Fig. 4 we can derive the synchronization loop transfer function for a desired signal and the attenuation of the undesired phase noise of the unsynchronized ADSM paths. The signal transfer function $G_s(s)$ in Eq. (5) is computed such that an arbitrary phase offset φ_{in} (zero in our case, see above) is the input and one of the modulator output signal's phase φ_1 is the output. The calculation for the other output φ_2 is analogous.

$$G_s(s) = \frac{\varphi_1}{\varphi_{in}} = \frac{2K_{PD}K_fK_{td}F(s)}{s + 2K_{PD}K_fK_{td}F(s)} \quad (5)$$

Parameter	Sym	Value
Max. switching frequency	f_{sw0}	$100\,\text{kHz}$
Max. input freqeuncy	f_{in}	$1\,\text{kHz}$
Max. norm. input amplitude	$x_{in,max}$	$-1\,\text{dBFS}$
Synch. loop damping factor	D	$\frac{1}{\sqrt{2}}$

Tab. 1: Example system requirements of this paper

The noise transfer function $G_n(s)$ in Eq. (6) measures the influence of the phase noise of a frequency change in the upper ADSM output φ_{sw1} on this modulator output signal's phase φ_1.

$$G_n(s) = \frac{\varphi_1}{\varphi_{sw1}} = \frac{s}{s + 2K_{PD}K_fK_{td}F(s)} \quad (6)$$

3.3 Example Design Parameterization

We present an example design parameterization which is used for the simulation and experiments in Section 4 and 5. Table 1 shows the system specification.

1. We calculate the delay operating point t_{d0} from the maximum switching frequency $f_{sw0} = \frac{1}{4t_{d0}}$, $t_{d0} = 2.5\,\mu s$. The lowest switching frequency at $x_{in,max}$ with Eq. (3) is $f_{sw,min} = 20.5\,\text{kHz}$.

2. For the loop, filter we choose a first-order low-pass filter with transfer function

$$F(s) = \frac{1}{1 + \tau_{LF}s}. \quad (7)$$

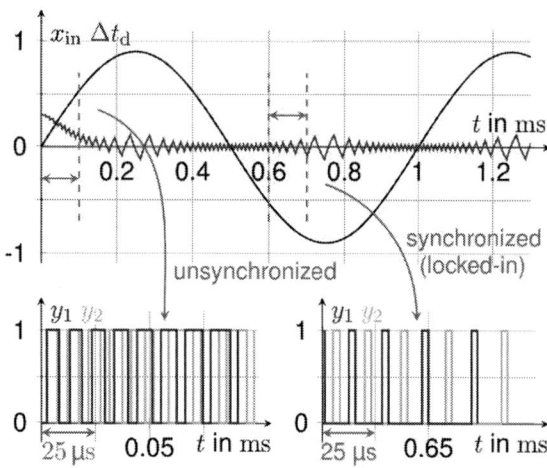

Fig. 6: Simulation of dual interleaved ADSM as per Fig. 3, control signal x_{in} and loop filter output Δt_d with ADSM output signals y_1, y_2 below

Its resonant frequency is given by

$$f_{r,LF} = \frac{1}{2\pi\tau_{LF}} \, . \tag{8}$$

3. Substituting Eq. (7) into Eq. (5), we see that the dynamics of the synchronization loop are of second order. By coefficient comparison with a general transfer function for second-order systems and rearrangement we can derive a damping factor D and the resonant frequency of the synchronization loop $f_{r,sync}$. We choose $f_{r,sync}$ between the maximum f_{in} and $f_{sw,min}$ and is here chosen to be $f_{r,sync} = 10\,\text{kHz}$.

4. The result is the parameter K_{td}, that determines the amount of shift delay for a full scale loop filter signal of the ADSM. We thus obtain the design formulas for defining the synchronization loop dynamics

$$K_{td} = \frac{\pi f_{r,sync}}{K_f(x_{in})K_{PD}D} \quad \tau_{LF} = \frac{1}{4Df_{r,sync}} \, . \tag{9}$$

5. From Eq. (4) we see that K_f is not constant but changes with the instantaneous input amplitude x_{in}. This means for a worst-case analysis we use $x_{in} = 0$ and $x_{in} = x_{in,max}$. For the given requirements K_f, is in the range from $8.23 \times 10^9\,\frac{1}{s^2}$ to $40 \times 10^9\,\frac{1}{s^2}$. K_{PD} can be assumed to be 1 for a consistent full-scale normalization. Thus, K_{td} is in the range between $1.1\,\mu s$ to $5.4\,\mu s$. For this example we select $K_{td} = 3\,\mu s$ and calculate $\tau_{LF} = 35\,\mu s$ ($f_{r,LF} = 4.55\,\text{kHz}$).

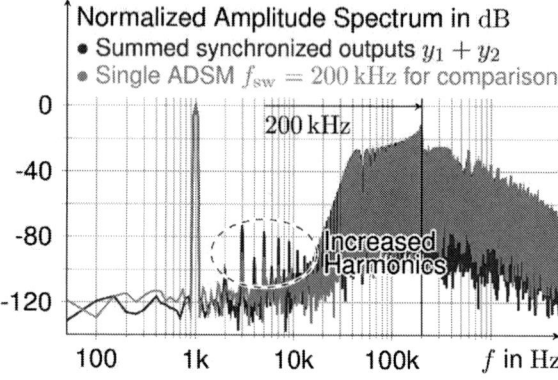

Fig. 7: Simulation of dual interleaved ADSM as per Fig. 3 showing spectrum of output signal sum compared to free running ADSM with interleaved frequency addition working

4 Simulation Results

For our simulations we synchronize two ADSMs with $100\,\text{kHz}$ maximum free-running frequencies f_{sw} and system settings as described in Section 3.3. Consider the loop filter output signal Δt_d shown in Fig. 6. After an initial synchronization period, this delay control signal approaches zero and keeps the ADSMs in constant phase to one another.

When compared to a separate modulator at twice the free running maximum frequency, the spectral content from the sum of the two $100\,\text{kHz}$ modulators for a sine-wave input shows that interleaving frequency addition is working at the cost of additional spectral distortion (Fig. 7).

5 Hardware Verification

Experimental validation of the synchronization concept was performed by a real-time digital hardware emulation of the modulators [11]. Figure 8 shows the utilized measurement setup. An FPGA-based emulation system models the behavior of the two ADSM paths and the synchronization loop. Furthermore it also generates complementary outputs with inserted dead-time for each modulator path. With these signals two identical gallium-nitride (GaN) based half-bridge modules are controlled. Each half-bridge module drives a second-order LC low-pass filter with an effective cutoff frequency of $5\,\text{kHz}$. The outputs of the filters are connected together and a $1\,\Omega$ load is applied at this node.

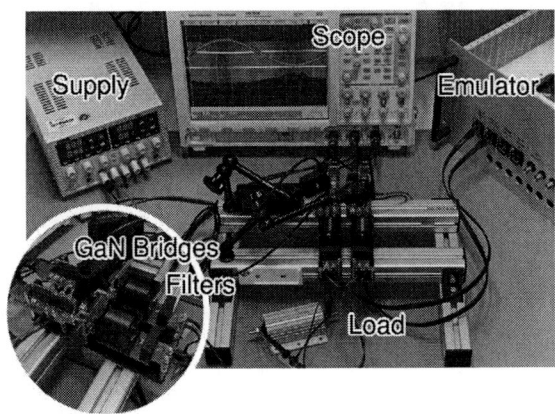

Fig. 8: Measurement and hardware setup for real-time modulator emulation for evaluation of the synchronization principle

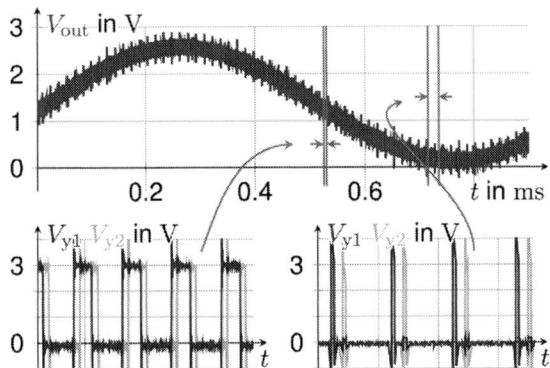

Fig. 9: Measurement results of the unsynchronized modulator emulation (see Fig. 8) for a sine-wave input

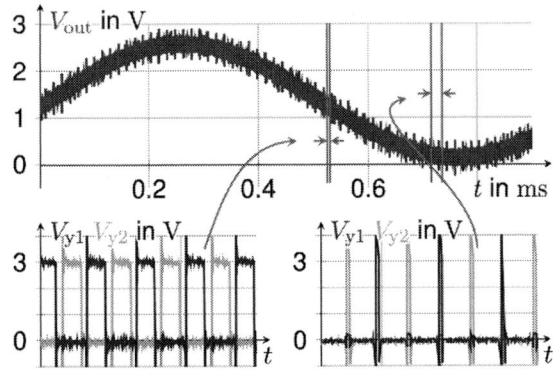

Fig. 10: Measurement results of the modulator emulation (see Fig. 8) for a sine-wave input with active synchronization

To reproduce the behavior of the continuous-time synchronized ADSM system, the emulation platform runs at an oversampling ratio > 1000 such that we assume quasi-analog behavior. Additionally, the modulator system settings correspond to the simulation from Section 4. As shown in the measurement results in Fig. 9 without the synchronization loop enabled, the modulators phase difference is undetermined (note the different time scales of the output signal sections). Any parameter drift in an analog system would lead to a chaotic behavior of this phase difference, but can not be reproduced by the digital system. Once synchronization is enabled in Fig. 10, the output signals of the modulators subsequently lock in to a constant phase relationship and phase drift due to switching frequency variation is compensated.

6 Conclusion

In this paper, a novel synchronization concept for using ADSMs in interleaved power electronic topologies was presented. It combines the ADSM advantages of spread spectrum modulation, lower average switching frequency and therefore reduced switching losses with the interleaved concept for high power dynamic switching converters. We have shown that the synchronization based on a structure similar to a phase-locked loop is able to maintain a constant phase relationship between two free-running ADSMs across their changing switching frequencies at the cost of increased spectral distortion.

Future work will focus on expanding the synchronization concept to more than two phases and transferring the concept to the hysteresis-based ADSM realization for a simplified hardware implementation.

Acknowledgment

This work was supported by the German Federal Ministry of Education and Research, FH-Kooperativ Project 13FH063KX1 "Clean Motor Supply".

References

[1] A. Lidow, M. de Rooij, J. Strydom, D. Reusch and J. Glaser, *GaN Transistors for Efficient Power Conversion*, 3rd edition, Hoboken, USA, John Wiley & Sons, Inc., 2020.

[2] A. Garcia-Tormo, E. Alarcon, A. Poveda and F. Guinjoan, "Low-OSR asynchronous Σ-Δ

modulation high-order buck converter for efficient wideband switching amplification," *IEEE Internat. Symposium on Circuits and Systems (ISCAS)*, Seattle, USA, 18-21 May 2008, pp. 2198-2201.

[3] K. Kang, J. Roh, Y. Choi, H. Roh, H. Nam and S. Lee, "Class-D Audio Amplifier Using 1-Bit Fourth-Order Delta-Sigma Modulation," in *IEEE Transactions on Circuits and Systems II: Express Briefs*, vol. 55, no. 8, pp. 728-732, Aug. 2008.

[4] F. S. Alargt, A. S. Ashur and A. H. Kharaz, "Novel adaptive delta modulation controller for Interleaved Boost DC-DC converters," *5th Internat. Conf. on Systems and Control (ICSC)*, Marrakesh, Morocco, 25-27 May 2016, pp. 20-25.

[5] F. S. Alargt, A. S. Ashur and A. H. Kharaz, "Adaptive delta modulation controller for interleaved buck DC-DC converter," *52nd Internat. Universities Power Engineering Conf. (UPEC)*, Heraklion, Greece, 28-31 Aug. 2017, pp. 1-6.

[6] S. Hadavi, M. J. Sanjari, A. H. Yatim and G. B. Gharehpetian, "A novel approach for interleaved buck step-down converter switching using the sigma-delta method," *IEEE Conf. on Energy Conversion (CENCON)*, Johor Bahru, Malaysia, 13-14 Oct. 2014, pp. 78-83.

[7] N. Pallo, S. Coday, J. Schaadt, P. Assem and R. C. N. Pilawa-Podgurski, "A 10-Level Flying Capacitor Multi-Level Dual-Interleaved Power Module for Scalable and Power-Dense Electric Drives," *IEEE Applied Power Electron. Conf. and Expo. (APEC)*, New Orleans, USA, 15-19 March 2020, pp. 893-898.

[8] P. S. Niklaus, J. W. Kolar and D. Bortis, "100 kHz Large-Signal Bandwidth GaN-Based 10 kVA Class-D Power Amplifier With 4.8 MHz Switching Frequency," in *IEEE Transactions on Power Electron.*, vol. 38, no. 2, pp. 2307-2326, 2023.

[9] A. Mertens, "Performance analysis of three-phase inverters controlled by synchronous delta-modulation systems," in *IEEE Transactions on Industry Applications*, vol. 30, no. 4, pp. 1016-1027, July-Aug. 1994.

[10] R. E. Best, *Phase-Locked Loops. Design, Simulation and Applications*, 6th edition, McGraw Hill, 2007.

[11] P. Czerwenka, T. Wolfer and E. Hennig, "Hardware-Plattform für die FPGA-Emulation von Analog/Mixed-Signal-Systemen," *64. Workshop der Multi Project Chip Gruppe Baden-Württemberg*, Esslingen, Germany, June 2022.

PCIM Europe 2024, 11– 13 June 2024, Nuremberg   DOI: 10.30420/566262432

High Resolution Mixed-Signal Pulse Width Modulator for High-Frequency DC-DC Converters

Tim McRae[1], Thomas Ebel[1], Kasper Paasch[1]

[1] University of Southern Denmark, Centre for Industrial Electronics, Denmark

Corresponding author: Tim McRae, mcrae@sdu.dk
Speaker: Tim McRae, mcrae@sdu.dk

E03 Pulse Width Modulation Methods
Preferred presentation form: Oral presentation

Abstract

This paper introduces a mixed-signal pulse-width modulator which uses a low clock-frequency digital pulse-width modulator and a binary-weighted resistive divider, ramp, and comparator to generate switching signals with sub-clock-cycle accuracy. A voltage reference is set prior to a desired switching event and a compared against a ramp, generating an asynchronous gate turn-off signal. The effective resolution of the PWM is increased shifting from a time-resolution required to a voltage resolution requirement. A discrete experimental prototype is built to verify the operation of the circuit. 20 ps time resolution is achieved on FPGA without the need for a PLL.

1 Synopsis

Both digital and analog controllers have been used in industry for quite some time [1]. Often analog controllers are described as being efficient in terms of energy consumption during steady-state operation [2]–[4], component count, and ease of implementation, and resolution, while also being somewhat limited in their complexity. Digital control methods on the other hand are seen as more flexible with the possibility for more complex control methods [3], but are more energy intensive and have limitations in terms of resolution or clock speed in fully integrated solutions. Mixed-signal control design has been a popular method of combining these two distinct controller types to gain the advantages of both while avoiding their drawbacks [5], [6].

One of the main challenges with digital control is resolution, for both the output voltage sensing and the digital pulse-width modulator (DPWM). Without proper consideration, mismatches in resolution between these two components can lead to limit-cycle oscillations in controlled variables [4]. One way to avoid this is to ensure the resolution of the DPWM is high enough such that the output voltage is able to fall within the zero error bin of the ADC [5], [7]. This can be achieved by using a high frequency internal clock for a counter-based DPWM, but this leads to higher energy consumption in steady-state and is limited by the maximum bandwidth of a PLL. Other methods to overcome this challenge include delay-line based PWMs, which are able to achieve very high resolution [8]. Some of these delay-line based PWMs also include an analog controlled supply voltage to reach even higher resolutions [9]. Access to a flexible, high-resolution DPWM should be available which can interface with a fast prototyping system, while also being viable for integration. This paper proposes a novel mixed-signal PWM which can be implemented both discretely and fully integrated with different structures. By combining both a counter-based DPWM with a variable reference ramp, a simple hardware implementation for a high resolution mixed-signal

3042

PCIM Europe 2024, 11– 13 June 2024, Nuremberg   DOI: 10.30420/566262432

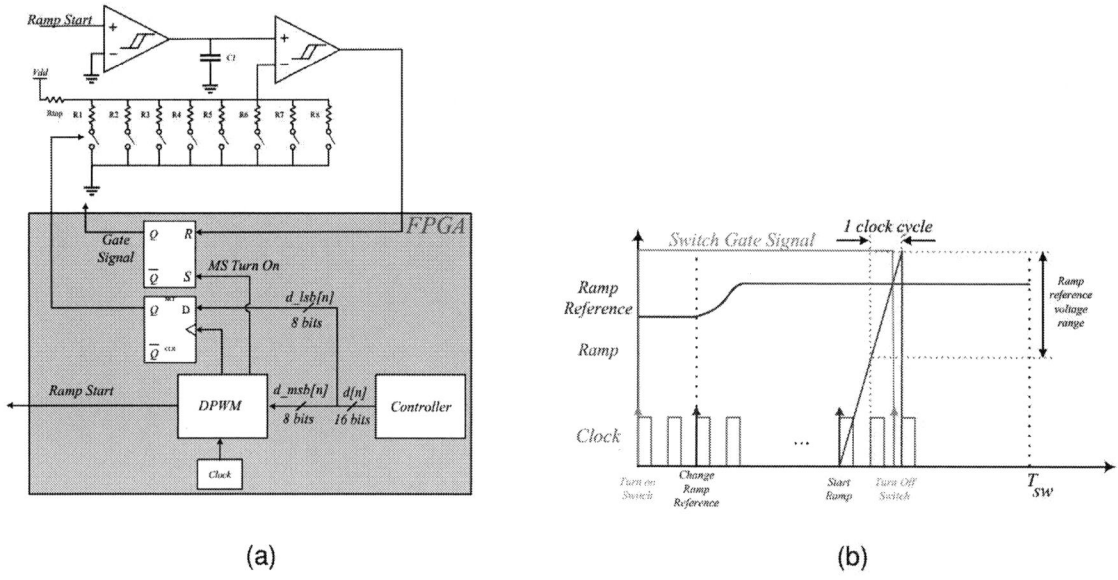

(a)   (b)

Fig. 1: a) System level diagram of the proposed pulse width modulator. b) Timing diagram to implement sub-clock-cycle resolution

pulse-width modulator (MSPWM) is achieved. Experimental results indicate an average time resolution of $29ps$, equivalent to a conventional counter-based DPWM with a clock frequency of $34GHz$.

2   Principle of Operation and Preliminary Results

A diagram of the proposed MSPWM can be seen in Fig. 1a. The MSPWM is comprised of a low frequency counter-based DPWM, a binary-weighted resistive divider, a ramp-generator, a comparator, and an SR-latch. This system is able to generate sub-clock cycle resolution within a selected clock cycle within the switching period.

The DPWM provides clocks to all sub-blocks, generating the timing sequence shown in Fig. 1b.The duty-cycle provided to the MSPWM is divided into two components, d_{MSB}, the most significant bits sent to the DPWM, and d_{LSB}, the least significant bits sent to the resistive divider. The switch gate signal starts by setting SR latch at the beginning of the switching cycle. The ramp reference is set to a new value, and the ramp begins by setting the comparator input high, the ramp enters the ramp reference voltage range at the beginning of the selected switching cycle. The ramp is compared to the ramp reference and generates a reset signal which forces the switch gate signal low. The addition of the analog portion of the PWM results in an effective time resolution of

$$\Delta t = \frac{1}{f_{clk}2^{n_{LSB}}} \tag{1}$$

where Δt is the minimum time resolution, f_{clk} is the digital clock frequency, and n_{LSB} is the number of least significant bits sent from the duty ratio $d[n]$ to the resistive divider. An experimental prototype was made to test the effective time resolution of the proposed PWM. Oscilloscopes with a high enough bandwidth or a fast enough sample rate were not available to verify or characterize the performance of the circuit. Instead, the resolution is tested by operating a phase-shifted full bridge DC-DC converter and measuring the change in steady-state output voltage, ΔV_{LSB}, across the full range of the analog portion of the MSPWM. Initial results can be seen in Fig. 2m demonstrating an average time resolution of $29ps$ across the 8-bit range of the LSBs of the duty ratio.

3043

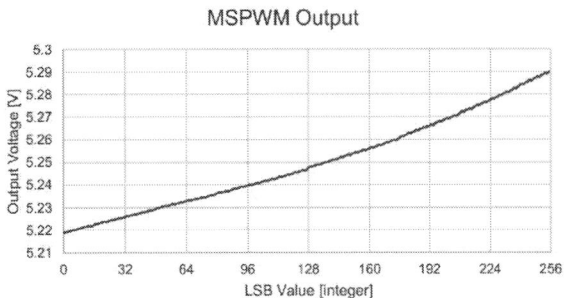

Fig. 2: Preliminary experimental results showing the change in the output voltage a phase-shifted full-bridge converter corresponding the change in LSBs of the duty ratio.

The final version of the paper will include an explanation and discussion of system operation along with details of the experimental setup and testing apparatus. Discussion of design criteria to enable reproduction, along with limitations and extension to accommodate systems with multiple switches will also be provided.

References

[1] F. G. R. Ramos, T. C. Pimenta, and L. H. Ferreira, "A mixed-signal pulse width modulator for portable smps applications," *Integration*, vol. 55, pp. 265–273, 2016. DOI: https://doi.org/10.1016/j.vlsi.2016.07.005.

[2] B. Patella, A. Prodic, A. Zirger, and D. Maksimovic, "High-frequency digital pwm controller ic for dc-dc converters," *IEEE Transactions on Power Electronics*, vol. 18, no. 1, pp. 438–446, 2003. DOI: 10.1109/TPEL.2002.807121.

[3] Z. Lukic, K. Wang, and A. Prodic, "High-frequency digital controller for dc-dc converters based on multi-bit /spl sigma/-/spl delta/ pulse-width modulation," in *Twentieth Annual IEEE Applied Power Electronics Conference and Exposition, 2005. APEC 2005.*, vol. 1, 2005, 35–40 Vol. 1. DOI: 10.1109/APEC.2005.1452883.

[4] M. D. Hagen and V. Yousefzadeh. "Applying digital technology to pwm control-loop designs." (2008), [Online]. Available: https://api.semanticscholar.org/CorpusID:17097730.

[5] A. Prodic and D. Maksimovic, "Mixed-signal simulation of digitally controlled switching converters," in *2002 IEEE Workshop on Computers in Power Electronics, 2002. Proceedings.*, 2002, pp. 100–105. DOI: 10.1109/CIPE.2002.1196722.

[6] O. Trescases, Z. Lukic, W. T. Ng, and A. Prodic, "A low-power mixed-signal current-mode dc-dc converter using a one-bit /spl delta//spl sigma/ dac," in *Twenty-First Annual IEEE Applied Power Electronics Conference and Exposition, 2006. APEC '06.*, 2006, 5 pp.-. DOI: 10.1109/APEC.2006.1620615.

[7] A. Syed, E. Ahmed, and D. Maksimovic, "Digital pwm controller with feed-forward compensation," in *Nineteenth Annual IEEE Applied Power Electronics Conference and Exposition, 2004. APEC '04.*, vol. 1, 2004, 60–66 Vol.1. DOI: 10.1109/APEC.2004.1295788.

[8] T. Carosa, R. Zane, and D. Maksimovic, "Implementation of a 16 phase digital modulator in a 0.35 /spl mu/m process," in *2006 IEEE Workshops on Computers in Power Electronics*, 2006, pp. 159–165. DOI: 10.1109/COMPEL.2006.305669.

[9] D. Costinett, M. Rodriguez, and D. Maksimovic, "Simple digital pulse width modulator under 100 ps resolution using general-purpose fpgas," *IEEE Transactions on Power Electronics*, vol. 28, no. 10, pp. 4466–4472, 2013. DOI: 10.1109/TPEL.2012.2233218.

PCIM Europe 2024, 11– 13 June 2024, Nuremberg   DOI: 10.30420/566262433

Implementation and Control of Optimized Pulse Patterns for Salient Permanent Magnet Synchronous Machines in Electric Vehicles

Maximilian Hepp[1], Kim Kaiser[1], Michael Saur[1], Wolfgang Wondrak[1], Mark-M. Bakran[2]

[1] Mercedes-Benz AG, Advanced Development Inverter, Germany
[2] University of Bayreuth, Department of Mechatronics, Germany

Corresponding author: Maximilian Hepp, maximilian.m.hepp@mercedes-benz.com
Speaker: Maximilian Hepp, maximilian.m.hepp@mercedes-benz.com

Abstract

Optimized pulse patterns (OPP) can be applied to improve the overall efficiency of the powertrain in electric vehicles. However, controlling OPPs to obtain a fast step response and to reject errors and disturbances is challenging. Therefore, this paper presents the implementation of OPPs in the control structure for electric drives in vehicles. Different methods are presented such as open-loop and closed loop-control of OPPs. Measurements are carried out to identify the modulation strategy with the highest efficiency considering both, the inverter and the electrical machine. These measurements are discussed in order to derive an optimal control strategy.

1 Introduction

Optimized pulse patterns (OPPs) are used to decrease the current harmonic distortion in 2-level inverter-fed AC drives while applying low switching frequencies. In contrast to the well known asynchronous space-vector modulation (SVM), OPPs are inherently a synchronous modulation scheme. SVM is usually applied to a symmetric up-down carrier modulator, with an update to the compare registers at the beginning of a carrier cycle. At the same time sampling of the phase currents is performed. The regular sampling in combination with SVM inherently obtains the fundamental component of the phase currents. The sampled currents can be used for fast current-control loops such as field-oriented control (FOC) using the fundamental wave for error tracking. For OPPs, in general, an inherent sampling of the fundamental is not possible, complicating the implementation in combination with current control or fast tracking response. Different methods are known to overcome this issue [1]–[3]. However, flux trajectory control methods, as in [4]–[6], are deemed more suitable due to their simplicity, dynamic responsiveness, and inherent capability to manage the overmodulation region effectively.

This article presents a suitable implementation of OPPs for electric vehicles using a 10 kHz control task based on flux trajectory control. It proposes a method to smoothly switch from SVM to OPP and presents a simplified concept for closed-loop OPP control for 2-level inverters.

By implementing the proposed methods, OPPs can be easily integrated into the electric powertrain. Their effects on the efficiency and overall performance of the electric powertrain are evaluated by measurements. Discussions are conducted based on these results to determine the optimal modulation strategy.

2 Introduction to OPPs

SVM is an asynchronous modulation scheme. In this context, the pulse number q is defined as the ratio $\frac{f_{sw}}{f_{el}}$, where f_{sw} is the switching frequency and f_{el} is the electrical frequency. Note that this ratio is a rational number. In contrast, the use of OPPs is inherently synchronous modulation, where the pulse number q is an integer. Various symmetry conditions can be imposed on OPPs, including quarter-wave symmetric waveforms or half-wave symmetric waveforms. In the following discussion, half-wave symmetric waveforms (HWS) are assumed because they inherently include quarter-wave symmetric waveforms, Fig. 1. Applying HWS $d = q - 1$ switching angles can be chosen.

The d switching angles α of the pulse patterns can

be determined by optimization algorithms. These algorithms aim to minimize the current harmonics or flux harmonics for a given number of pulses q, seeking an optimal switching configuration that results in the lowest possible harmonic distortion [7]–[9].

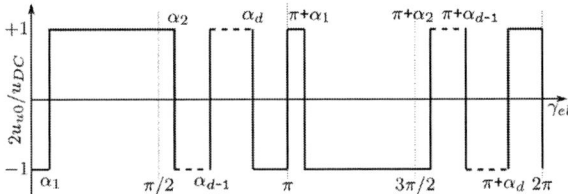

Fig. 1: Half-wave symmetric waveform.

3 Open-loop Control

3.1 Timer Implementation

A single timer module is used for both SVM and OPP. The timer period T_s (here $T_s = 100\mu s$) does not change and stays constant for both SVM and OPP. An up-counter is implemented and up to 16 values t_x ($t_1...t_{16}$) can be used in 16 counter compare registers (CR), see Fig. 2(a). For each compare event a switching state s_x ($s_1...s_{16}$) is assigned representing a voltage vector U_x (see. Fig. 3). The compare registers are updated at the beginning of each counter period and the phase currents are sampled. Using SVM seven values for the switching states and seven values for the CRs are calculated in each period. For OPPs it depends on the pulse pattern how many compare events happen during T_s.

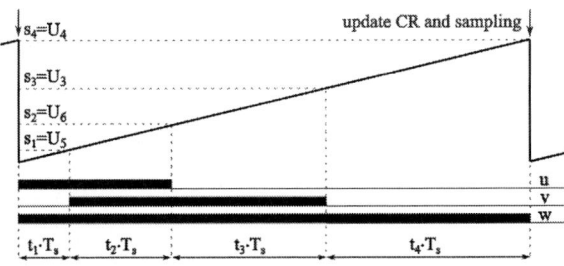

Fig. 2: Timer-module used for SVM/DBFC and OPP. In this example four compare events happen during one timer period T_s each assigned with a respective switching state s_x.

3.2 Synchronous Modulation

A three-phase pulse pattern s_{uvw} can be derived from the switching angles α obtained for a specific

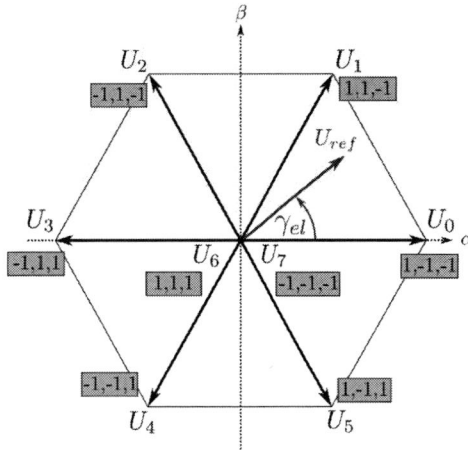

Fig. 3: Assignable voltage vectors U_0-U_7 of a two level inverter.

operating point, Fig. 4. This information is sufficient to apply the determined pulse pattern to the timer module, thus influencing the modulation of the inverter. However, the pulse pattern is stored using a flux trajectory. This approach is advantageous because it provides a seamless transition to open-loop operation in combination with deadbeat flux control (DBFC) and facilitates compatibility with closed-loop control.

In order to derive the normalized flux trajectory of a given pulse pattern s_{uvw}, Eq. (1) can be used:

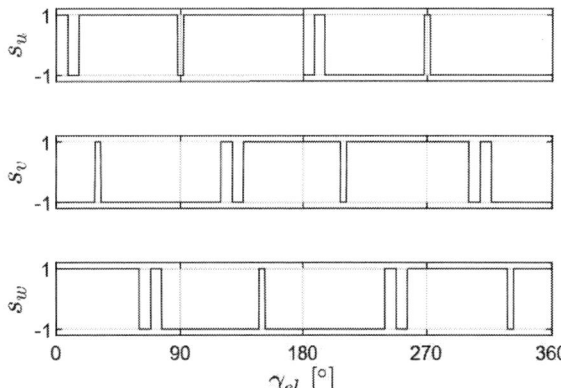

Fig. 4: Example of a half-wave symmetric three-phase pulse pattern $s_{uvw}(\gamma_{el})$ with $q = 5$, $m = 1.15$ and $\theta_u = 2.077$ for a two level inverter.

$$\psi_{\text{LUT}} = \psi_{\text{LUT},0} + \int_0^{2\pi} \mathbf{T_C}\, s_{uvw}(\gamma_{el})\, \mathrm{d}\gamma_{el} \qquad (1)$$

where $\psi_{\text{LUT},0}$ is chosen such that the resulting flux-trajectory is centred in the origin (0,0). $\mathbf{T_C}$ is the Clarke-Transformation:

$$\mathbf{T_C} = \frac{2}{3} \begin{bmatrix} 1 & -\frac{1}{2} & \frac{1}{2} \\ 0 & \frac{\sqrt{3}}{2} & -\frac{\sqrt{3}}{2} \end{bmatrix} \quad (2)$$

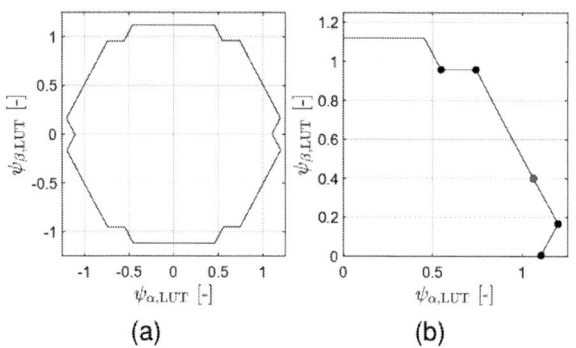

Fig. 5: (a) Flux-trajectory using Eq. (1) and the pulse pattern shown in Fig. 4. (b) First sector of the flux trajectory. The stored values are indicated with a dot. The red dot indicates a zero vector.

Subsequently, the corner points of the flux trajectory in the first sector can be stored in a look-up table (LUT) alongside the corresponding switching angles, Tab. 1.

$\psi_{\alpha,\mathrm{LUT}}$ [-]	1.104	1.197	1.062	1.062	0.740	0.547
$\psi_{\beta,\mathrm{LUT}}$ [-]	0.000	0.167	0.400	0.400	0.958	0.958
$\gamma_{el,\mathrm{LUT}}$ [°]	0.00	8.01	19.61	24.05	51.72	60.00

Tab. 1: Flux look-up table. Compared to Fig. 4 and γ_{el} the stored angle $\gamma_{el,\mathrm{LUT}}$ has an offset such that the flux values of the first sector are stored and $\gamma_{el,\mathrm{LUT}} \in [0°, 60°]$. In this example the offset is $-188.27°$.

From this information the switching instances for the first sector and all the other sectors can be derived online during the interrupt task of the microcontroller, Tab. 2.

s_x (U)	U_1	U_2	U_7	U_2	U_3
$\Delta\gamma_{el}$ [°]	8.01	11.60	4.44	27.67	8.28

Tab. 2: Switching Pattern for the first sector derived from Tab. 1.

Based on the electrical angle γ_{el}, the electrical speed ω_{el} and the timer period T_s from Tab. 2, together with $\gamma_{el,\mathrm{LUT}}$, the switching states s_x and the corresponding comparison times t_x for the upcoming timer period can be calculated.

With a mechanical speed of e.g. $n = 6000 \frac{1}{min}$ and a pole pair number of $z_p = 4$, the change of the electrical angle γ_{el} during one timer period $T_s = 100\mu s$ is $\Delta\gamma_{el} = 14.4°$. As an example, starting from the first sector, the resulting switching states s_x and the values t_x for the counter compare events can be calculated as in Tab. 3.

s_x (U)	U_1	U_2
$\Delta\gamma_{el}$ [°]	8.01	6.39
$t_x \cdot T_s$ [μs] at $n = 6000 \frac{1}{min}$	55.625	44.375

Tab. 3: s_x and t_x for the upcoming timer period.

The dynamic performance of the open-loop application of synchronous modulation has limitations. While modest changes or disturbances typically yield acceptable results characterized by minimal ringing and overshoot, significant torque increments or high-level disturbances result in degraded performance, Fig. 6.

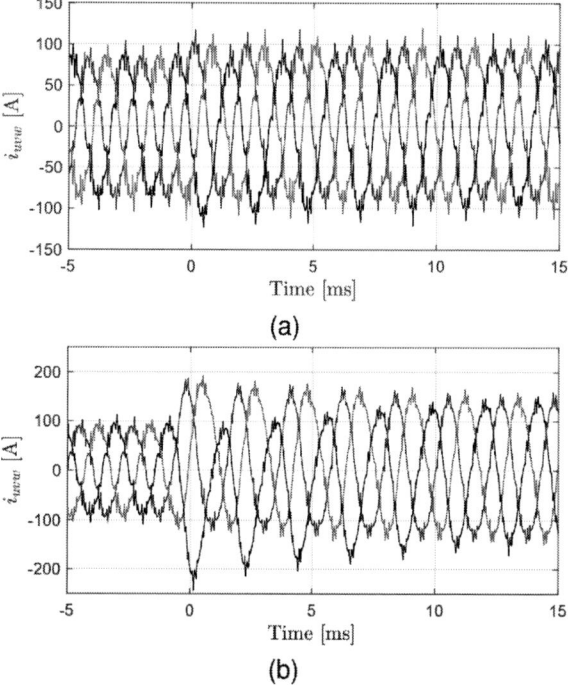

Fig. 6: Transient behavior of OPPs with $q = 5$ implemented in open-loop control at 7000 rpm, assessed through measurements conducted on a test bench. (a) torque step from 50 Nm to 60 Nm. (b) torque step from 50 Nm to 100 Nm.

3.3 Deadbeat Flux Control

DBFC, first introduced in [10] and further developed in subsequent works such as [11] and [12], provides the ability to continuously adjust the modulation index up to $m = \frac{4}{\pi} \approx 1.27$ (six-step operation). This control strategy is based on a flux trajectory tracking approach. In the over-modulation region I (OVM I, $1.15 < m \leq 1.21$), the reference flux trajectory is approximately circular. However, within the over-modulation region II (OVM II, $1.21 < m \leq 1.27$), the flux trajectory is characterized by a combination of circular and six-step hexagonal flux patterns.

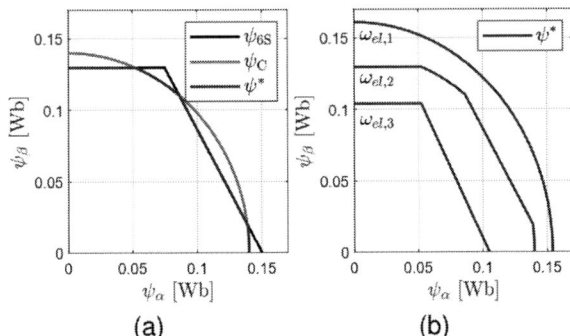

Fig. 7: (a) DBFC Flux-trajectory. (b) Different DBFC flux-trajectories for different electrical speeds ω_{el} ($\omega_{el,1} < \omega_{el,2} < \omega_{el,3}$).

The required voltage to be applied in the next timer period $k + 1$ can be calculated using the following expression:

$$\boldsymbol{u}_{\alpha\beta}(k+1) = \frac{\boldsymbol{\psi}^*_{\alpha\beta}(k+2) - \hat{\boldsymbol{\psi}}_{\alpha\beta}(k+1)}{T_s} + R_s \hat{\boldsymbol{i}}_{\alpha\beta}(k+1). \tag{3}$$

If the trajectory follows a circular path, the voltage is applied using SVM. Conversely, if the trajectory follows a hexagonal pattern, voltage application is performed using six-step modulation. In cases where the trajectory has characteristics of both circular and hexagonal patterns, a combination of both modulation techniques is used.

The estimated flux $\hat{\boldsymbol{\psi}}_{\alpha\beta}(k+1)$ and the current $\hat{\boldsymbol{i}}_{\alpha\beta}(k+1)$ is obtained by a flux and current observer similar to the one used and derived in [13], [14].

To calculate the reference flux $\boldsymbol{\psi}^*_{\alpha\beta}(k+2)$, three different scenarios can be distinguished:

1. $\psi_{\mathrm{C}} \leq \psi_{6\mathrm{S,min}}$:
The reference flux $\boldsymbol{\psi}^*_{\alpha\beta}(k+2)$ follows a circular

trajectory with a radius denoted as ψ_{C}:

$$\boldsymbol{\psi}^*_{\alpha\beta}(k+2) = \psi_{\mathrm{C}} \begin{bmatrix} \cos(\delta) \\ \sin(\delta) \end{bmatrix} \tag{4}$$

This is valid until the modulation index approaches $m \approx 1.21$, at which point the circular trajectory tangentially intersects the hexagon, with the inner radius $\psi_{6\mathrm{S,min}}$

$$\psi_{6\mathrm{S,min}} = \frac{\sqrt{3}}{2} \cdot \psi_{6\mathrm{S,max}}. \tag{5}$$

where

$$\psi_{6\mathrm{S,max}} = \frac{2\pi u_{\mathrm{DC}}}{9\omega_{el}}. \tag{6}$$

δ is the angle that accounts for the computational delay and the reference load angle θ^*_ψ

$$\delta = \gamma_{el} + 2T_s \omega_{el} + \theta^*_\psi. \tag{7}$$

Applying a Maximum-Torque-Per-Current (MTPC) strategy, the reference flux magnitude ψ_{MTPC} can be obtained by optimization methods using the parameters of the electrical machine. For high utilized permanent magnetic synchronous machines (PMSM) the magnetic parameters are of non-linear nature and depend on the currents [15]. The reference load angle θ^*_ψ to obtain the reference torque at the reference flux magnitude can be calculated using:

$$\theta^*_\psi = \mathrm{atan2}\left(\frac{\psi^*_q}{\psi^*_d}\right) \tag{8}$$

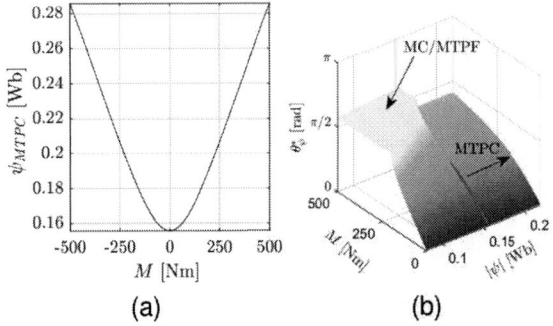

Fig. 8: (a) MTPC flux reference. (b) Reference load angle θ^*_ψ for motoring operation.

If the speed is too high, the induced voltage exceeds the available voltage and field-weakening is required. The maximum possible flux amplitude can be calculated using Eqs. (9)-(10) (including or excluding the ohmic voltage drop):

$$|\psi| = \frac{1}{\omega_{el}} \left(\left(m \frac{u_{DC}}{2} \right)^2 + R_s^2 |i_{dq}|^2 \right.$$

$$\left. -m \cdot R_s \cdot u_{DC} \cdot |i_{dq}| \cos\left(\theta_u - \theta_i\right) \right)^{\frac{1}{2}} \qquad (9)$$

$$|\psi|_{R_s=0} = \frac{1}{\omega_{el}} \cdot m \frac{u_{DC}}{2} \qquad (10)$$

The flux hexagon is also scaled and rotated due to the ohmic voltage drop. While the rotation can be neglected [11], [12] the scaling is accounted by a scaling factor k_s:

$$k_s = \frac{|\psi|}{|\psi|_{R_s=0}} \qquad (11)$$

The flux amplitude of the circular trajectory is calculated using

$$\psi_C = \frac{1}{k_s} \cdot G_s \cdot \min\left(\psi_{MTPC}, |\psi|\right) \qquad (12)$$

G_s is a scaling factor accounting for the non-linearities in the modulation index, when having a combined circle and six-step trajectory if $m > 1.21$, Fig. 9(a). Otherwise $G_s = 1$.

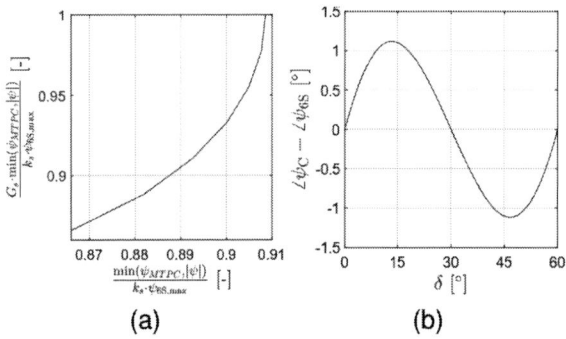

(a) (b)

Fig. 9: (a) Scaling factor G_s. (b) Flux angle difference of the circular trajectory and the hexagon trajectory.

2. $\psi_C \geq |\psi_{6S}(\delta)|$:
In this context, the reference flux $\boldsymbol{\psi}^*_{\alpha\beta}(k+2)$ for the first sector ($\delta \in [0, \pi/3]$) can be calculated by using Eqs. (13) and (14). The magnitude of the flux hexagon, which depends on δ is derived

from Eq. (15). Unlike previous work such as [11], [12], this approach includes the actual flux angle $\theta_{\psi_{6S}}(\delta)$, which is different from the circular flux angle equating to δ, Fig. 9(b). Ignoring the actual flux angle $\theta_{\psi_{6S}}(\delta)$ results in discontinuities. A transition to six-step modulation occurs when $\psi_C \geq \psi_{6S,\max}$. The calculations for all other sectors follow a similar procedure.

$$\psi_\alpha(\delta) = \psi_{6S,\max} - \frac{u_{DC}}{3} \cdot \frac{\delta}{\omega_{el}} \qquad (13)$$

$$\psi_\beta(\delta) = \frac{u_{DC}}{\sqrt{3}} \cdot \frac{\delta}{\omega_{el}} \qquad (14)$$

$$|\psi_{6S}(\delta)| = \sqrt{\psi_\alpha(\delta)^2 + \psi_\beta(\delta)^2} \qquad (15)$$

$$\theta_{\psi_{6S}}(\delta) = \mathrm{atan2}\left(\frac{\psi_\beta(\delta)}{\psi_\alpha(\delta)} \right) \qquad (16)$$

3. $\psi_C < |\psi_{6S}(\delta)|$:
Considering the flux angle $\theta_{\psi_{6S}}(\delta)$, the electrical angles at which the hexagon and the circle intersect can be calculated by setting $|\psi_{6S}(\delta)| = \psi_C$:

$$\delta_{int,el,1/2} = \frac{\pi}{6} \mp \frac{3\omega_{el}}{2u_{DC}} \cdot \sqrt{\psi_C^2 - \frac{\pi^2 u_{DC}^2}{27\omega_{el}^2}} \qquad (17)$$

The geometric intersection points of the flux, denoted as $\delta_{int,geo,1/2}$ can be determined by substituting Eq. (17) into Eq. (16). Subsequently, the necessary flux $\boldsymbol{\psi}^*_{\alpha\beta}(k+2)$ for the first sector can be approximated as follows:

$$\delta \in [0, \frac{\pi}{6}]: \quad \boldsymbol{\psi}^*_{\alpha\beta}(k+2) = \qquad (18)$$

$$\psi_C \begin{bmatrix} \cos\left(\delta \cdot \frac{\delta_{int,geo,1}}{\delta_{int,el,1}}\right) \\ \sin\left(\delta \cdot \frac{\delta_{int,geo,1}}{\delta_{int,el,1}}\right) \end{bmatrix}$$

$$\delta \in]\frac{\pi}{6}, \frac{\pi}{3}[: \quad \boldsymbol{\psi}^*_{\alpha\beta}(k+2) =$$

$$\psi_C \begin{bmatrix} \cos\left(\frac{\pi}{3} - \left(\frac{\pi}{3} - \delta\right) \cdot \frac{\frac{\pi}{3} - \delta_{int,geo,2}}{\frac{\pi}{3} - \delta_{int,el,2}}\right) \\ \sin\left(\frac{\pi}{3} - \left(\frac{\pi}{3} - \delta\right) \cdot \frac{\frac{\pi}{3} - \delta_{int,geo,2}}{\frac{\pi}{3} - \delta_{int,el,2}}\right) \end{bmatrix}$$

3.4 Smooth transition between DBFC and OPP

Because of their common foundation in flux-trajectory principles, both DBFC and open-loop applied OPPs as presented provide a seamless transition between the two approaches without requiring any changes or adjustments to the timer module.

This can be realized when the flux maintains identical amplitude and angle during the transition from DBFC to OPPs. However, as illustrated in Fig. 10, this condition is typically not satisfied in practice.

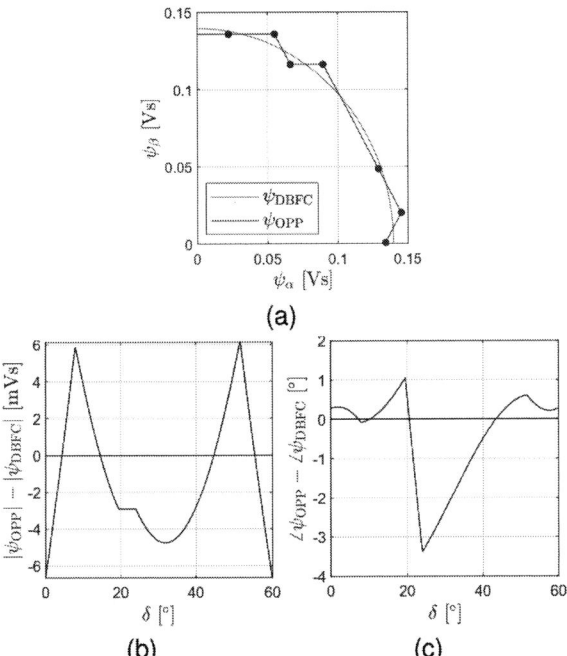

Fig. 10: (a) Flux trajectories of DBFC and OPP ($q = 5$) at $m = 1.15$. (b) Flux value difference of the OPP and DBFC trajectory. (c) Flux angle difference of the OPP and DBFC trajectory.

Therefore, it becomes necessary to adjust the reference flux $\psi_{\alpha\beta}^*(k + 2)$ using DBFC just prior to transitioning to open-loop synchronous modulation. The reference flux of the OPP trajectory can be retrieved from the LUT, Tab. 1, at the specific angle δ:

$$\psi_{\alpha\beta}^*(k + 2) = \frac{u_{\mathrm{DC}}}{2\omega_{el}} \cdot k_s \begin{bmatrix} \psi_{\alpha,\mathrm{LUT}}(\delta) \\ \psi_{\alpha,\mathrm{LUT}}(\delta) \end{bmatrix} \quad (19)$$

This value can be inserted into the equation (3) to obtain the reference voltages $u_{\alpha\beta}(k + 1)$.

However, there exists a restriction on the maximum voltage that $u_{\max}$ can be applied:

$$u_{\max} = \frac{2}{3} \frac{u_{\mathrm{DC}}}{\cos(\angle \boldsymbol{u}_{\alpha\beta}^*(k + 1)) + \frac{1}{\sqrt{3}} \sin(\angle \boldsymbol{u}_{\alpha\beta}^*(k + 1))} \quad (20)$$

$$|\boldsymbol{u}_{\alpha\beta}^*(k + 1)| \leq u_{\max} + u_{\mathrm{tolerance}} \quad (21)$$

If Eq. (21) is not satisfied, DBFC will continue to be employed for calculating the output voltages. However, if Eq. (21) holds true, the set-point voltages $\boldsymbol{u}_{\alpha\beta}^*(k + 1)$ computed using Eq. (19) will be applied to the inverter using SVM. In the next timer period $k + 2$, the modulation scheme changes to open-loop pulse pattern control. $\boldsymbol{u}_{\alpha\beta}^*(k + 2)$ and the following periods will then be calculated from the LUT as in Sec. 3.2. $u_{\mathrm{tolerance}}$ represents an application-specific parameter. If the tolerance is too small, the equation (21) may not be satisfied, depending on the pulse pattern used, or the transition may be significantly delayed. Conversely, a tolerance that is too large can lead to oscillatory behavior, as illustrated in Figure 6.

If the flux error

$$|\boldsymbol{\psi}_e| = |\boldsymbol{\psi}_{k+1}^* - \hat{\boldsymbol{\psi}}_{k+1}| \quad (22)$$

exceeds a certain threshold due to significant torque variations or significant disturbances, the modulation scheme can be switched back to DBFC until equation (21) is satisfied again. Similarly to $u_{\mathrm{tolerance}}$, $\boldsymbol{\psi}_e$ denotes an application-specific parameter.

The dynamic performance of the proposed control method can be seen in Fig. 11. Compared to Fig. 6 where open-loop control is applied no overshoots or major oscillations are present.

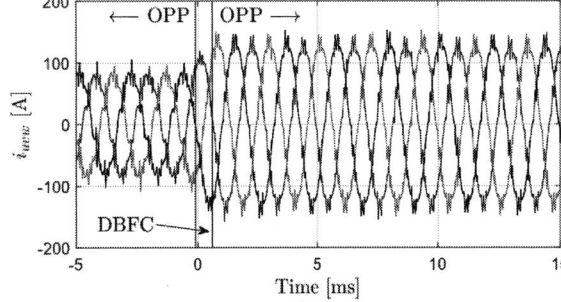

Fig. 11: A torque step from 50 Nm to 100 Nm is applied and measured while utilizing a combination of DBFC and open-loop pulse patterns.

4 Closed-loop Control

Flux error correction can also be achieved through closed-loop control methods when using OPPs. This has been addressed in [4]–[6]. However, these methods do not inherently preserve the space vector sequence of the pulse pattern. As a result, changes to the space vectors themselves may occur while adjusting the switching times. To keep the pulse pattern as close to the original as possible, the flux error and its correction are performed in the $\alpha\beta$-frame. This approach inherently preserves the correct order of the voltage space vectors and only modifies the switching times of these vectors.

In order to do so, the flux error is ψ_e is decomposed into the directions of the current occurring space-vector U_{now} and the next occurring space-vector U_{next} of the upcoming switching sequence obtained by Tab. 2. $\{U_{\text{now}}, U_{\text{next}}\}$ are active vectors $\in \{U_0, ..., U_5\}$ in contrast to the zero vectors $\{U_6, U_7\}$. For the first sector this is illustrated in Fig. 12.

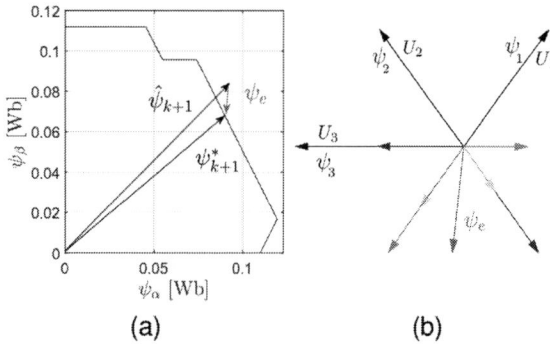

(a)   (b)

Fig. 12: (a) Flux error $\psi_e = \psi_{k+1}^* - \hat{\psi}_{k+1}$. (b) Two of three fundamental vectors (ψ_1, ψ_2, ψ_3) can be used to decompose the flux error ψ_e: ψ_1 and ψ_2 (·····), ψ_2 and ψ_3 (—) or ψ_1 and ψ_3 (—). The three flux vectors are based on the space-vectors U_1, U_2 and U_3 occurring in the first sector.

The decomposition can be done using one of Eqs. (23)-(25), depending on which directions are defined by $\{U_{\text{now}}, U_{\text{next}}\}$. k_s is a scaling factor accounting for the ohmic voltage drop, see Eq. (11).

$$\begin{bmatrix} \psi_1 \\ \psi_2 \end{bmatrix} = \frac{1}{k_s} \begin{bmatrix} 1 & \frac{1}{\sqrt{3}} \\ -1 & \frac{1}{\sqrt{3}} \end{bmatrix} \begin{bmatrix} \psi_{e,\alpha} \\ \psi_{e,\beta} \end{bmatrix} \tag{23}$$

$$\begin{bmatrix} \psi_1 \\ \psi_3 \end{bmatrix} = \frac{1}{k_s} \begin{bmatrix} 0 & \frac{2}{\sqrt{3}} \\ -1 & \frac{1}{\sqrt{3}} \end{bmatrix} \begin{bmatrix} \psi_{e,\alpha} \\ \psi_{e,\beta} \end{bmatrix} \tag{24}$$

$$\begin{bmatrix} \psi_2 \\ \psi_3 \end{bmatrix} = \frac{1}{k_s} \begin{bmatrix} 0 & \frac{2}{\sqrt{3}} \\ -1 & -\frac{1}{\sqrt{3}} \end{bmatrix} \begin{bmatrix} \psi_{e,\alpha} \\ \psi_{e,\beta} \end{bmatrix} \tag{25}$$

The pulse pattern for the upcoming timer period is taken from Tab. 2, and the switching times must be adjusted to keep the flux error at zero. Using Eq. (26) the necessary angle (or time) adjustment for the corresponding space vectors $\{U_{\text{now}}, U_{\text{next}}\}$ can be obtained.

$$\begin{bmatrix} \Delta\gamma_{el,e,\text{now}} \\ \Delta\gamma_{el,e,\text{next}} \end{bmatrix} = \frac{\omega_{el}}{u_{\text{DC}}} \begin{bmatrix} \psi_{\text{now}} \\ \psi_{\text{next}} \end{bmatrix} \tag{26}$$

Now the following algorithm can be applied to adjust the switching angles:

Premises: The cumulative sum of switching times during one timer period T_s must equal T_s. In terms of switching angles, it should be $T_s \cdot \omega_{el}$. Negative switching times or angles are not allowed.

Step 1: If there exists a combination of U_{now} with U_{next} in the upcoming timer period, proceed with Step 2. If there is a combination of U_{now} with a zero vector $\{U_6, U_7\}$ in the upcoming timer period, proceed with Step 5. Otherwise, no correction is feasible in the upcoming timer period.

Step 2: If $\text{sgn}(\Delta\gamma_{el,e,\text{now}}) \neq \text{sgn}(\Delta\gamma_{el,e,\text{next}})$ and both errors are non-zero the maximum error that can be reduced is $\Delta\gamma_{el,e} = \min(|\Delta\gamma_{el,e,\text{now}}|, |\Delta\gamma_{el,e,\text{next}}|)$. Otherwise proceed with step 5.

Step 3: Reduce the switching time of the space-vector(s) $\{U_{\text{now}}, U_{\text{next}}\}$ associated with the negative value from $\{\Delta\gamma_{el,e,\text{now}}, \Delta\gamma_{el,e,\text{next}}\}$ such that a minimum switching time $\Delta\gamma_{el,\text{min}}$ of the space vectors is still being kept. This ensures that the order of the space vectors remains unchanged. The negative value from $\{\Delta\gamma_{el,e,\text{now}}, \Delta\gamma_{el,e,\text{next}}\}$ is then increased by the same amount by which the switching time of the space vector(s) is reduced in total.

Step 4: Extend the switching time of the vector $\{U_{\text{now}}, U_{\text{next}}\}$ associated with the positive value from $\{\Delta\gamma_{el,\text{now}}, \Delta\gamma_{el,\text{next}}\}$ by the time that was reduced in Step 3. Subsequently, reduce the positive value from $\{\Delta\gamma_{el,e,\text{now}}, \Delta\gamma_{el,e,\text{next}}\}$ by the same amount of time.

Step 5: If zero vectors $\{U_6, U_7\}$ are present in the next timer period, they can be utilized to update the switching times.

(a) If both errors $\{\Delta\gamma_{el,e,\text{now}}, \Delta\gamma_{el,e,\text{next}}\}$ are negative, reduce the associated switching times of the space-vector(s) $\{U_{\text{now}}, U_{\text{next}}\}$ such that a minimum switching time $\Delta\gamma_{el,\text{min}}$ of the space vectors is maintained. Extend the switching time of one zero vector by the same amount that was reduced.

(b) If both errors $\{\Delta\gamma_{el,e,\text{now}}, \Delta\gamma_{el,e,\text{next}}\}$ are positive, reduce the switching times of the zero vectors by that error while ensuring that a minimum switching time $\Delta\gamma_{el,\text{min}}$ of the zero space vectors is maintained. Extend the total switching time of $\{U_{\text{now}}, U_{\text{next}}\}$ by the same amount that was reduced.

(c) If $\text{sgn}(\Delta\gamma_{el,e,\text{now}}) \neq \text{sgn}(\Delta\gamma_{el,e,\text{next}})$ applies, reduce the switching time of the space-vector(s) $\{U_{\text{now}}, U_{\text{next}}\}$ associated with the negative value out of $\{\Delta\gamma_{el,e,\text{now}}, \Delta\gamma_{el,e,\text{next}}\}$ such that a minimum switching time $\Delta\gamma_{el,\text{min}}$ of the space vectors is still maintained. Extend the switching time of one zero vector by the same amount that was reduced. Reduce the switching time of the zero space-vector(s) by the positive value from $\{\Delta\gamma_{el,e,\text{now}}, \Delta\gamma_{el,e,\text{next}}\}$ such that a minimum switching time $\Delta\gamma_{el,\text{min}}$ of the zero space-vector(s) is maintained. Extend the space vector(s) associated to the positive value from $\{\Delta\gamma_{el,e,\text{now}}, \Delta\gamma_{el,e,\text{next}}\}$ by the same amount that was reduced.

The example shown in Fig. 12 is used to illustrate the algorithm. A minimum switching time of $\Delta\gamma_{el,\text{min}} = 5\,\mu s$ is chosen using a timer period $T_s = 100\mu s$. In the first timer period $T_{s,1}$ the following pulse pattern is read from Tab. 2:

s_x (U)	U_2	U_3
$\Delta\gamma_{el}$ [rad]	0.2472	0.0167
$t_x \cdot T_s$ [μs] at $n = 6300\,\frac{1}{min}$	93.67	6.33

Tab. 4: Pulse pattern for the first timer period $T_{s,1}$.

Here, U_{now} is U_2 and U_{next} is U_3. Eq. (25) applies. The necessary angle adjustments in both directions can be calculated using Eq. (26):

$$\begin{bmatrix} \Delta\gamma_{el,e,2} \\ \Delta\gamma_{el,e,3} \end{bmatrix} = \begin{bmatrix} -0.1265 \\ 0.0728 \end{bmatrix} \text{rad}$$

Since both vectors U_{now} (U_2) and U_{next} (U_3) are present in the upcoming timer period and the sign of the errors is different, the switching times can be adjusted $\min(|\Delta\gamma_{el,e,2}|, |\Delta\gamma_{el,e,3}|) = 0.0728\,\text{rad}$. The switching time for U_2 (U_{now}) is decreased by that amount, ensuring that the minimum switching time $\Delta\gamma_{el,\text{min}}$ is preserved. Conversely, the switching time for U_3 (U_{next}) is increased by the same amount:

s_x (U)	U_2	U_3
$\Delta\gamma_{el}$ [rad]	0.1743	0.0896
$t_x \cdot T_s$ [μs] at $n = 6300\,\frac{1}{min}$	65.94	34.06

Tab. 5: Adjusted pulse pattern for the first timer period $T_{s,1}$.

For the second timer period $T_{s,2}$ the following pulse pattern is read from Tab. 2:

s_x (U)	U_3	U_2
$\Delta\gamma_{el}$ [rad]	0.1278	0.1361
$t_x \cdot T_s$ [μs] at $n = 6300\,\frac{1}{min}$	48.43	51.57

Tab. 6: Pulse pattern for the second timer period $T_{s,2}$.

U_{now} is U_3 and U_{next} is U_2. Eq. (25) applies. The necessary angle adjustments in both directions are:

$$\begin{bmatrix} \Delta\gamma_{el,e,3} \\ \Delta\gamma_{el,e,2} \end{bmatrix} = \begin{bmatrix} 0 \\ -0.0537 \end{bmatrix} \text{rad}$$

Since there are no zero vectors present and there is not a combination of positive and negative angle errors, no adjustment can be made in this timer period.

For the third timer period $T_{s,3}$ the following pulse pattern is read from Tab. 2:

s_x (U)	U_2	U_3	U_6
$\Delta\gamma_{el}$ [rad]	0.0036	0.2025	0.0578
$t_x \cdot T_s$ [μs] at $n = 6300\,\frac{1}{min}$	1.36	76.74	21.90

Tab. 7: Pulse pattern for the third timer period $T_{s,3}$.

U_{now} is U_2 and U_{next} is U_3. The necessary angle adjustments in both directions are:

$$\begin{bmatrix} \Delta\gamma_{el,e,2} \\ \Delta\gamma_{el,e,3} \end{bmatrix} = \begin{bmatrix} -0.0537 \\ 0 \end{bmatrix} \text{rad}$$

Since the switching time of U_2 (U_{now}) is already smaller than the minimum switching time $\Delta\gamma_{el,\text{min}}$, it cannot be further reduced, even though $\Delta\gamma_{el,e,2}$

is negative. Therefore, no adjustments are possible in this timer period.

For the fourth timer period $T_{s,4}$ the following pulse pattern is read from Tab. 2:

s_x (U)	U_6	U_3
$\Delta\gamma_{el}$ [rad]	0.0197	0.2442
$t_x \cdot T_s$ [μs] at $n = 6300\ \frac{1}{min}$	7.46	92.54

Tab. 8: Pulse pattern for the fourth timer period $T_{s,4}$.

U_{now} is U_3 and U_{next} is U_4. The necessary angle adjustments in both directions are:

$$\begin{bmatrix} \Delta\gamma_{el,e,3} \\ \Delta\gamma_{el,e,4} \end{bmatrix} = \begin{bmatrix} -0.0537 \\ 0.0537 \end{bmatrix} \text{rad}$$

The switching time of U_3 (U_{now}) is subsequently reduced by $\Delta\gamma_{el,e,3}$. At the same time, the switching time of the zero vector U_6 is increased by the same amount:

s_x (U)	U_6	U_3
$\Delta\gamma_{el}$ [rad]	0.0733	0.1906
$t_x \cdot T_s$ [μs] at $n = 6300\ \frac{1}{min}$	27.78	72.22

Tab. 9: Adjusted pulse pattern for the fourth timer period $T_{s,4}$.

For the fifth and the sixth timer period also no adjustments are possible. In the seventh timer period $T_{s,7}$ the residual flux error can be further reduced to zero. For $T_{s,7}$ the following pulse pattern is read from Tab. 2:

s_x (U)	U_4	U_7
$\Delta\gamma_{el}$ [rad]	0.1977	0.0661
$t_x \cdot T_s$ [μs] at $n = 6300\ \frac{1}{min}$	74.92	25.08

Tab. 10: Pulse pattern for the seventh timer period $T_{s,7}$.

U_{now} is U_4 and U_{next} is U_5.

$$\begin{bmatrix} \Delta\gamma_{el,e,4} \\ \Delta\gamma_{el,e,5} \end{bmatrix} = \begin{bmatrix} 0.0537 \\ 0 \end{bmatrix} \text{rad}$$

The switching time of U_4 (U_{now}) is increased by $\Delta\gamma_{el,e,4}$. At the same time, the switching time of the zero vector U_7 is decreased by the same amount: The procedure described in this example is depicted in Fig. 13. After seven timer periods T_s the estimated $\hat{\psi}_{\alpha\beta}$ flux trajectory converged towards the reference flux trajectory $\psi^*_{\alpha\beta}$.

s_x (U)	U_4	U_7
$\Delta\gamma_{el}$ [rad]	0.2507	0.0132
$t_x \cdot T_s$ [μs] at $n = 6300\ \frac{1}{min}$	95.00	5.00

Tab. 11: Adjusted pulse pattern for the seventh timer period $T_{s,7}$.

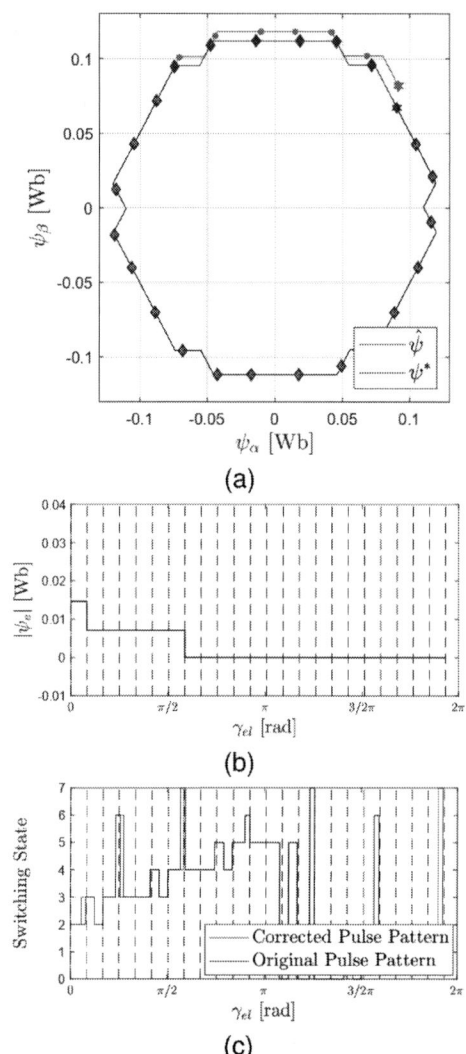

Fig. 13: (a) estimated $\hat{\psi}_{\alpha\beta}$ and reference $\psi^*_{\alpha\beta}$ flux trajectories. (b) Flux error ψ_e. (c) original and corrected pulse pattern. The dashed lines represented one timer interval T_s.

Although it exhibits greater overshoot than the combination of DBFC and open-loop control, the use of closed-loop control results in significantly fewer oscillations and faster reference tracking than using OPPs in open-loop control alone. The use of DBFC can be advantageous in scenarios where the flux error is high, while applying closed-loop pulse pattern control when only small adjustments need to be made.

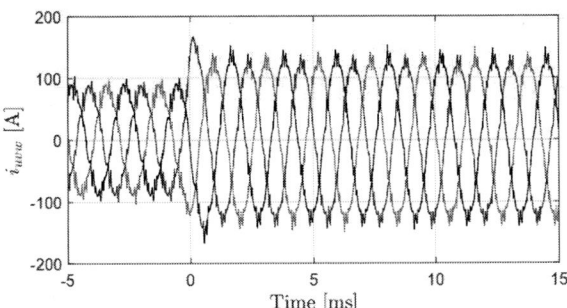

Fig. 14: A torque step from 50 Nm to 100 Nm is applied and measured while utilizing OPP closed-loop control.

5 Implementation of OPPs in Electric Drive Systems

5.1 Current Distortion Effects

For OPPs, the goal is to optimize the switching angles to minimize machine losses for a given number of pulses. However, this task is complex because machine losses include copper losses, iron losses, and, in the case of a PMSM, magnetic losses. Iron losses and magnetic losses are particularly difficult to model, especially from an analytical point of view. Thus, for most cases the current distortion is minimized. With synchronous modulation, a Fourier Series expansion can be utilized to calculate the current harmonics in electric machines depending on the switching angles α. Equation (27) is employed to calculate the Fourier coefficients a and b for the harmonic order ν.

$$\left. \begin{aligned} a_\nu(\boldsymbol{\alpha}) &= \frac{4}{\nu\pi} \sum_{i=1}^{d} (-1)^{i+1} \sin(\nu\alpha_i) \\ b_\nu(\boldsymbol{\alpha}) &= \frac{4}{\nu\pi} \left(1 + \sum_{i=1}^{d} (-1)^i \cos(\nu\alpha_i)\right) \end{aligned} \right\} \quad (27)$$

For isotropic machines the current harmonics can be calculated as:

$$i_{\mathrm{u},n,\mathrm{RMS,iso}} = \frac{u_{\mathrm{DC}}}{\sqrt{2}\cdot 2\,\omega_{\mathrm{el}}} \frac{1}{L} \underbrace{\sqrt{\sum_{\nu=6k\pm1}^{\infty} \frac{a_\nu^2 + b_\nu^2}{\nu^2}}}_{\mathrm{WTHD0}} \quad (28)$$

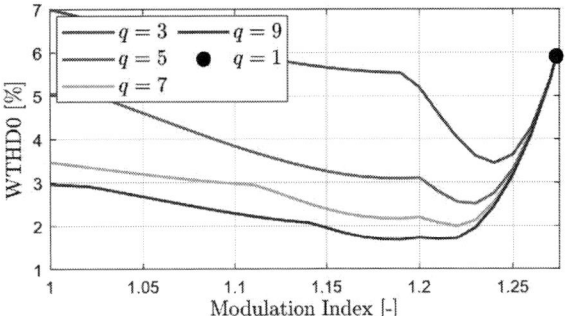

Fig. 15: WTHD0 for different pulse numbers q. The optimal modulation index m_{opt} for the lowest distortion varies for different pulse numbers: $m_{opt} = 1.22$ for $q = 9$, $m_{opt} = 1.22$ for $q = 7$, $m_{opt} = 1.23$ for $q = 5$, $m_{opt} = 1.24$ for $q = 3$.

Figure 15 illustrates the WTHD0 for different pulse numbers. For isotropic machines, the WTHD0 is directly related to the current harmonics, Eq. (28). For non-linear anisotropic PMSMs the current harmonics can be obtained using Eqs. (29)-(32), where L_{dd}, L_{qq} are the differential inductances and L_d, L_q are the absolute inductances [7].

$$G_n = \qquad\qquad (29)$$
$$\left[g_{\mathrm{qd},n}^2 + g_{\mathrm{qa},n}^2 + g_{\mathrm{da},n}^2 + g_{\mathrm{dd},n}^2 \right.$$
$$\left. + 2\left(g_{\mathrm{qd},n}\cdot g_{\mathrm{qa},n} + g_{\mathrm{da},n}\cdot g_{\mathrm{dd},n}\right)\right]\cdot \frac{1}{2}\left(a_{n-1}^2 + b_{n-1}^2\right)$$
$$+ \left[g_{\mathrm{qd},n}^2 + g_{\mathrm{qa},n}^2 + g_{\mathrm{da},n}^2 + g_{\mathrm{dd},n}^2 \right.$$
$$\left. - 2\left(g_{\mathrm{qd},n}\cdot g_{\mathrm{qa},n} + g_{\mathrm{da},n}\cdot g_{\mathrm{dd},n}\right)\right]\cdot \frac{1}{2}\left(a_{n+1}^2 + b_{n+1}^2\right)$$
$$+ G_{\mathrm{saliency}}$$

$$G_{\mathrm{saliency}} = \qquad\qquad (30)$$
$$\left(g_{\mathrm{qd},n}^2 - g_{\mathrm{qa},n}^2 + g_{\mathrm{da},n}^2 - g_{\mathrm{dd},n}^2\right)$$
$$\cdot \sqrt{a_{n+1}^2 + b_{n+1}^2}\cdot \sqrt{a_{n-1}^2 + b_{n-1}^2}$$
$$\cdot \cos\left(2\theta_{\mathrm{u}} + 2\cdot \mathrm{atan2}\,(b_1/a_1)\right.$$
$$\left. + \mathrm{atan2}\,(b_{n-1}/a_{n-1}) - \mathrm{atan2}\,(b_{n+1}/a_{n+1})\right)$$

$$\begin{bmatrix} g_{\mathrm{qd},n} & g_{\mathrm{qa},n} \\ g_{\mathrm{da},n} & g_{\mathrm{dd},n} \end{bmatrix} = \frac{1}{n^2 L_{\mathrm{dd}} L_{\mathrm{qq}} - L_{\mathrm{d}} L_{\mathrm{q}}} \begin{bmatrix} nL_{\mathrm{qq}} & L_{\mathrm{q}} \\ L_{\mathrm{d}} & nL_{\mathrm{dd}} \end{bmatrix} \quad (31)$$

$$i_{\mathrm{u},n,\mathrm{RMS}} = \frac{u_{\mathrm{DC}}}{\sqrt{2}\cdot 2\,\omega_{\mathrm{el}}} \underbrace{\sqrt{\sum_{n=6k}^{\infty} G_n}}_{\sigma} \quad (32)$$

Assuming a constant DC-link voltage u_{DC}, current harmonics for salient PMSMs depend on the following parameters aside optimal switching angles:

- The reduction of current harmonics is correlated with an increase in electrical speed ω_{el}, see Eqs. (28) and (32).

- For salient PMSMs, the reduction of current harmonics is correlated with an increase in torque or voltage phase angle θ_u up to π, see Eq. (30).

- The reduction of current harmonics correlates with a higher modulation index, up to an optimum depending on the pulse number q, see Fig. 15.

- As the pulse number q increases, there is a tendency for the current harmonics to decrease, see Fig. 15.

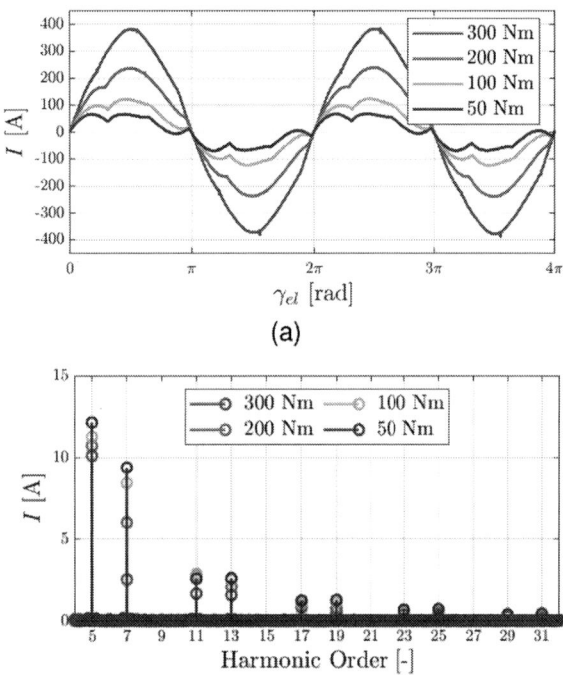

Fig. 16: (a) Six-step operation is used to obtain the phase current I for different torque levels. (b) FFT of the phase current I.

To gain a preliminary understanding of how current harmonics affect machine efficiency and whether it correlates with machine losses, the following experiment is performed: At a constant speed (here

7000 rpm), the voltage phase angle θ_u, or torque respectively, is incrementally increased using six-step modulation. Hence, the higher voltage phase angle inherently results in a reduction of current harmonics. Other factors affecting machine losses, such as the higher frequency of the fundamental current or flux at higher speeds, or higher d-current, are maintained constant. Figure 16 displays the resulting phase current and the corresponding Fast Fourier Transform (FFT).

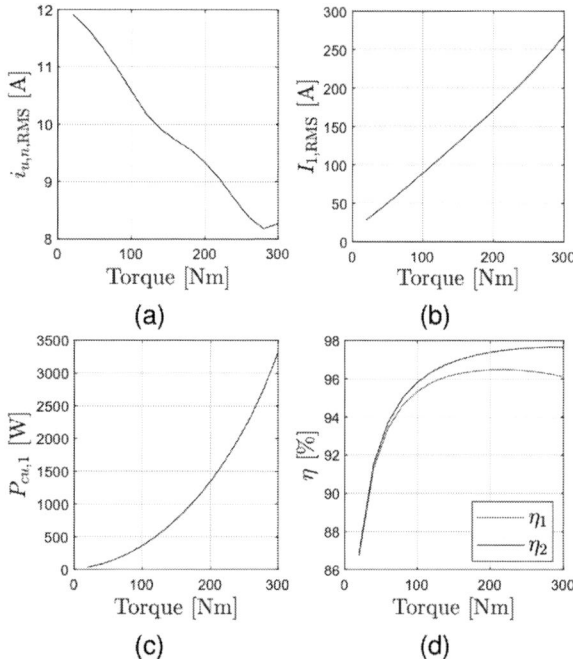

Fig. 17: (a) RMS value of the current harmonics. (b) Fundamental RMS current $I_{1,\mathrm{RMS}}$. (c) Fundamental copper losses $P_{\mathrm{cu},1} = 3 \cdot R \cdot I_{1,\mathrm{RMS}}^2$ ($R_{25} = 12\,\mathrm{m}\Omega$ at 25°C whereas $R = 15.3\,\mathrm{m}\Omega$ at 90°C). (d) Efficiency considering $P_{\mathrm{cu},1}$ (η_1) and neglecting $P_{\mathrm{cu},1}$ (η_2).

The experiment is further evaluated in Fig. 17. As expected, the RMS value of the current harmonics decreases with increasing torque (up to a voltage phase angle of π). Conversely, the RMS value of the fundamental current $I_{1,\mathrm{RMS}}$ increases approximately linearly, resulting in a quadratic increase in fundamental copper losses. The machine efficiency η_1, including the fundamental copper losses $P_{\mathrm{cu},1}$, increases with higher torque and decreases again when the torque exceeds about 200 Nm. If the efficiency η_2 is calculated without considering the fundamental copper losses, it can be seen that the efficiency increases due to the lower current harmonic content until the current harmonics start

Parameter	Value
pole pairs z_p	4
stator resistance R_s	12 mΩ at 25°C
nominal flux linkage ψ_p	153 mWb at 25°C
nominal absolute inductance L_d / L_q	387 / 748 μH
nominal differential inductance L_{dd} / L_{qq}	387 / 748 μH

Tab. 12: Parameters of the PMSM.

OP1 – 6		q [-]	f_{sw} [kHz]	m [-]	$I_{1,RMS}$ [A]	$i_{u,h,RMS}$ [A]	$P_{l,inv}$ [W]	$P_{l,machine}$ [W]
5600rpm 30Nm	SVM	26.8	10	1.15	25.90	4.99	163	2476
	OPP	9	3.36	1.15	27.08	5.13	47	2523
6300rpm 140Nm	SVM	23.8	10	1.15	118.81	4.37	457	3618
	OPP	9	3.78	1.22	110.67	3.72	201	3552
7000rpm 40Nm	SVM	21.4	10	1.15	55.11	5.08	248	3188
	OPP	9	4.20	1.22	44.71	4.01	110	2941
7000rpm 250Nm	SVM	21.4	10	1.15	232.98	3.47	1034	6916
	OPP	1	0.47	1.27	211.6	8.80	523	6926
8300rpm 60Nm	SVM	18.1	10	1.15	96.13	4.92	371	3917
	OPP	9	4.98	1.22	83.07	3.32	204	3726
9000rpm 170Nm	SVM	16.7	10	1.15	204.24	3.96	804	7413
	OPP	1	0.6	1.27	181.17	7.72	410	6711

Tab. 13: OPPs compared to SVM at six different operating points OP1-6.

to increase again. It can be concluded that optimization of current harmonics can indeed lead to a reduction in machine losses, making it a valid target for pulse pattern optimization.

5.2 Evaluation on different operating points

Figure 18 shows the WLTC driving cycle. The extra high segment of the driving cycle is particularly important as it represents driving on a motorway at speeds in excess of 80 km/h. The electric range is critical in such scenarios, as shorter distances are not typically driven on highways. In this example, 80 km/h corresponds to a machine speed of 5457 rpm, while the maximum speed is 131.3 km/h at 8956 rpm. In the following analysis, only the motor operation is considered. It is evident that the majority of the driving cycle occurs in the field-weakening region with $m > 1.15$.

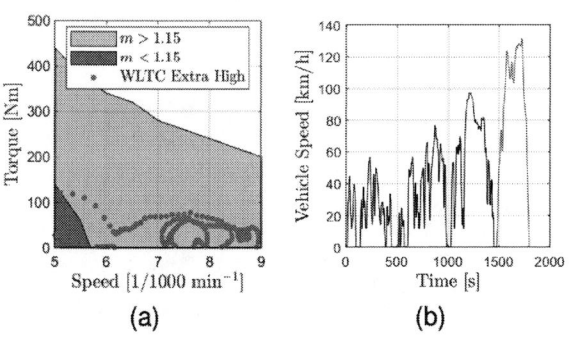

Fig. 18: (a) Example of a speed-torque map for an electric powertrain relevant to the WLTC Extra High. (b) WLTC Class 3. The Extra High part of the WLTC is marked in red.

For six different operating points (OP1-6) in the field weakening region OPPs are compared to regular SVM. The machine used in the following sections is presented in [16], Tab. 12.

For OPPs, different pulse numbers ($q \in \{1, 3, 5, 7, 9\}$) are utilized at their optimal modulation index m_{opt}, where they exhibit the lowest distortion,

see Fig. 15. Applying OPPs, it is inherently feasible to operate in the over-modulation region. Thus, it is possible to reduce the necessary field-weakening current and consequently decrease the fundamental RMS current. Table 13 presents the results of the OPP yielding the highest efficiency.

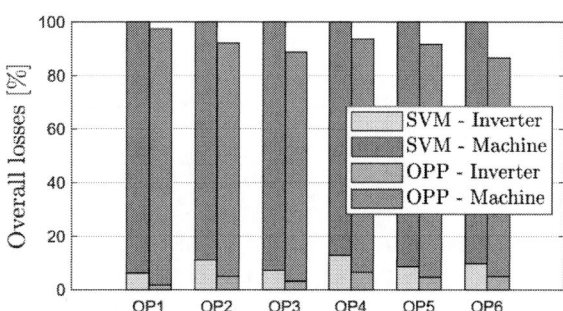

Fig. 19: Relative comparison of the overall efficiency between SVM and OPP for the six operating points OP1-6.

Figure 19 illustrates a direct comparison of the overall losses between SVM and OPPs. To directly see the benefits of OPPs, for OP3 and OP6 the waveforms are displayed in Figs. 20 and 21. Due to the higher modulation index, the amplitude of the currents decrease. At a much lower switching frequency, current distortion does not increase significantly for OP6, while current distortion is actually reduced for OP3. From the obtained results, it is possible to conclude that OPPs deliver superior performance in the field-weakening area compared to SVM with a modulation index of $m = 1.15$.

PCIM Europe 2024, 11– 13 June 2024, Nuremberg   DOI: 10.30420/566262433

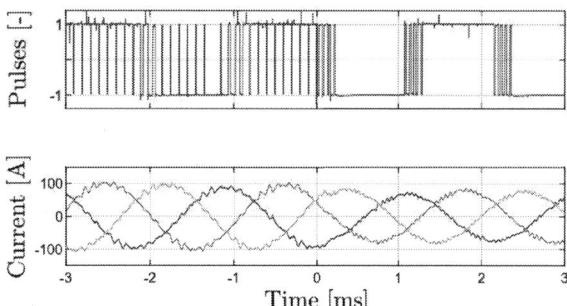

Fig. 20: Current waveform at 7000 rpm and 40 Nm (OP3) transitioning from SVM to OPP with $q = 9$.

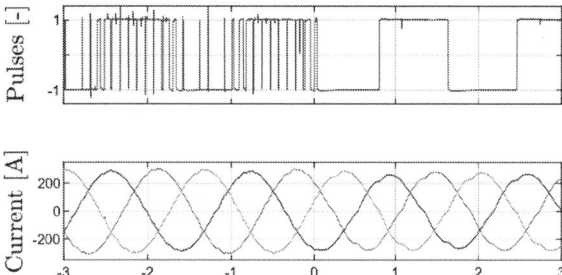

Fig. 21: Current waveform at 9000 rpm and 170 Nm (OP6) transitioning from SVM to OPP with $q = 1$.

5.3 Impact of OPPs on the overall electric drive train

Measurements are performed on a test bench across multiple operating points in the torque-speed map to evaluate the impact of OPPs on various aspects of the overall electrical drivetrain. This includes analysis of the DC link capacitor, air gap torque ripple and total losses. SVM is implemented with a modulation index of $m = 1.15$, while OPPs are applied with the respective optimal modulation index m_{opt} depending on the pulse number. Figure 22 shows the best results for the highest efficiency for all measured operating points (marked with a square ■). The colored area represents the practical application where the respective modulation scheme is implemented to minimize transitions between different pulse numbers.

As anticipated from prior discussions, six-step operation is suitable for higher torques and/or higher speeds. At low load, $q = 9$ gives the best results in this example.

The influence on the overall system is depicted in Fig. 23. Particularly noteworthy is the high calculated air-gap torque ripple when using six-step

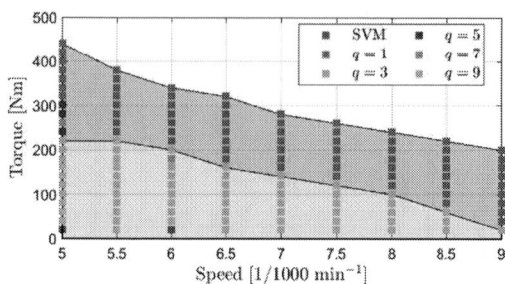

Fig. 22: Optimal outcomes achieved with various pulse numbers and SVM across the torque-speed map including six-step operation.

modulation. However, the air-gap torque is influenced by other factors such as cogging torque and spatial harmonics. Additional experiments must be conducted to evaluate how this influences the output torque of the machine and the NVH (Noise, Vibration, and Harshness) behavior of the drive train.

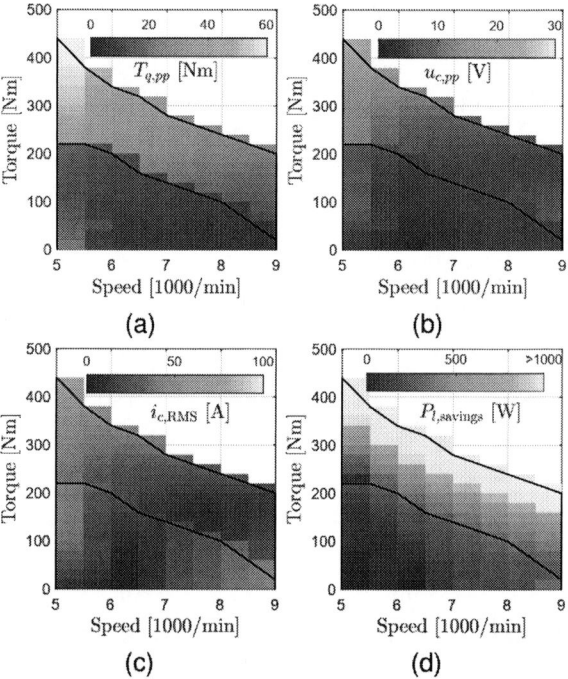

Fig. 23: (a-c) Numerical obtained torque air gap ripple $T_{q,pp}$, DC-link voltage ripple $u_{c,pp}$, and DC-link current stress $i_{c,\mathrm{RMS}}$ ($C_c = 460\mu F$). (d) Loss savings $P_{l,savings}$ compared to SVM with $m_{max} = 1.15$ obtained by experiments on a test-bench.

If it is not feasible to employ six-step modulation, Fig. 24 illustrates an alternative application of OPPs in the torque-speed map. In this scenario,

3057

$q = 3$ and $q = 5$ are employed in the high torque region to minimize overall losses.

Again, the influence on the overall system can be seen in Fig. 25. Compared to the modulation strategy of Fig. 22, $q = 5$ and $q = 3$ lead to higher current stress in the DC-link capacitor $i_{c,\mathrm{RMS}}$ and to a slightly higher voltage ripple $u_{c,pp}$ on the high voltage DC-bus. However, it is important to note that the current stress is not considered critical, as the maximum DC-link current occurs at $m = 0.6$ (assuming the power factor equals 1).

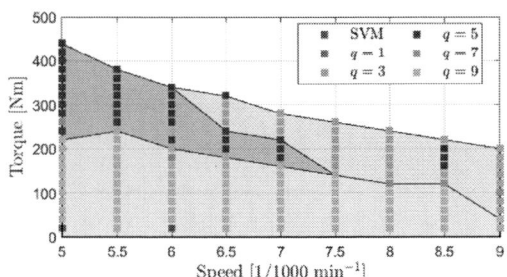

Fig. 24: Optimal outcomes achieved with various pulse numbers and SVM across the torque-speed map excluding six-step operation.

This experiment demonstrates the superiority of OPPs over SVM in the field weakening area, especially when analyzing the loss savings $P_{l,\mathrm{savings}}$ in Figs. 23 and 25. When using low switching modulation schemes such as OPPs, it is essential to consistently evaluate the impact on the overall system to ensure compliance with specified constraints.

6 Conclusion

In conclusion, three methods of controlling pulse patterns have been presented. It has been shown that the open-loop control principle, although functional, has poor dynamics and is prone to oscillations and overcurrents. When DBFC is used with high dynamic requirements, a transition between DBFC and open-loop control achieves excellent results. In addition, for small discrepancies between the desired and actual flux, a closed-loop control method has been introduced that strictly adheres to the sequence of space vectors and only modifies the switching times.

Experimental investigations have highlighted the factors that influence the current harmonics of the machine. Experiments have shown that optimizing pulse patterns to minimize current harmonics

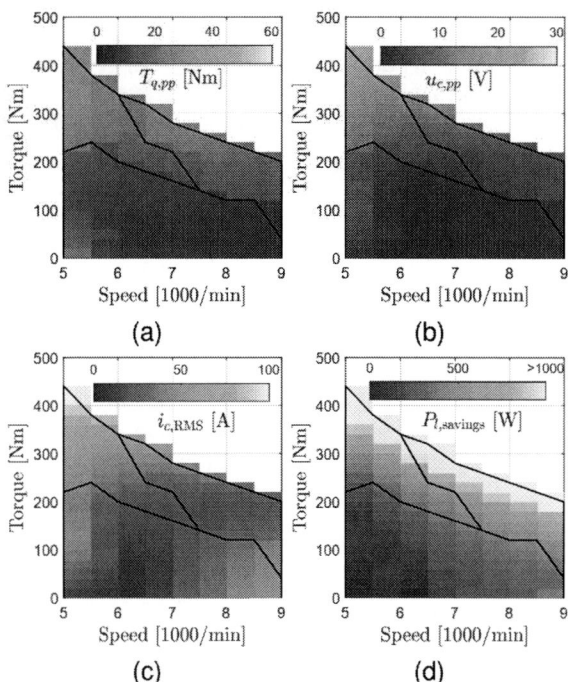

Fig. 25: (a-c) Numerical obtained torque air gap ripple $T_{q,pp}$, DC-link voltage ripple $u_{c,pp}$, and DC-link current stress $i_{c,\mathrm{RMS}}$ ($C_c = 460\mu F$). (d) Loss savings $P_{l,savings}$ compared to SVM with $m_{max} = 1.15$ obtained by experiments on a test-bench.

is a legitimate approach to improving machine efficiency. SVM has been compared with OPP in the relevant field weakening region, showing that OPP is the superior modulation method in this context. However, when using OPPs, it is important to consider the entire drive system, as OPP can result in operating limits being exceeded.

Subsequent works will concentrate on further simplifying the utilization of OPPs for electric vehicle traction drives.

References

[1] A. Fathy Abouzeid, J. M. Guerrero, A. Endemaño, I. Muniategui, D. Ortega, *et al.*, "Control strategies for induction motors in railway traction applications," *Energies*, vol. 13, no. 3, 2020. DOI: 10.3390/en13030700.

[2] K. Peter, F. Mink, and J. Böcker, "Model-based control structure for high-speed permanent magnet synchronous drives," in *2017 IEEE International Electric Machines and Drives Conference (IEMDC)*, 2017, pp. 1–8. DOI: 10.1109/IEMDC.2017.8002284.

[3] A. Birda, J. Reuss, and C. M. Hackl, "Simple fundamental current estimation and smooth transition between synchronous optimal pwm and asynchronous svm," *IEEE Transactions on Industrial Electronics*, vol. 67, no. 8, pp. 6354–6364, 2020. DOI: 10.1109/TIE.2019.2938490.

[4] N. Oikonomou and J. Holtz, "Stator flux trajectory tracking control for high-performance drives," in *Conference Record of the 2006 IEEE Industry Applications Conference Forty-First IAS Annual Meeting*, vol. 3, 2006, pp. 1268–1275. DOI: 10.1109/IAS.2006.256694.

[5] N. Oikonomou and J. Holtz, "Closed-loop control of medium-voltage drives operated with synchronous optimal pulsewidth modulation," *IEEE Transactions on Industry Applications*, vol. 44, no. 1, pp. 115–123, 2008. DOI: 10.1109/TIA.2007.912735.

[6] T. Geyer, N. Oikonomou, G. Papafotiou, and F. Kieferndorf, "Model predictive pulse pattern control," in *2011 IEEE Energy Conversion Congress and Exposition*, 2011, pp. 3306–3313. DOI: 10.1109/ECCE.2011.6064215.

[7] M. Hepp, M. Saur, W. Wondrak, and M.-M. Bakran, "Optimized pulse patterns for salient permanent magnet synchronous machines considering nonlinear magnetic effects," in *PCIM Europe 2023; International Exhibition and Conference for Power Electronics, Intelligent Motion, Renewable Energy and Energy Management*, 2023, pp. 1–10. DOI: 10.30420/566091246.

[8] A. Birda, J. Reuss, and C. M. Hackl, "Synchronous optimal pulsewidth modulation for synchronous machines with highly operating point dependent magnetic anisotropy," *IEEE Transactions on Industrial Electronics*, vol. 68, no. 5, pp. 3760–3769, 2021. DOI: 10.1109/TIE.2020.2984460.

[9] D. G. Holmes and T. A. Lipo, *Pulse width modulation for power converters: principles and practice.* John Wiley & Sons, 2003, vol. 18.

[10] M. S. Petit, B. Sarlioglu, R. D. Lorenz, B. S. Gagas, and C. W. Secrest, "Deadbeat flux vector control for dynamic six-step operation of synchronous machines," in *2019 IEEE Transportation Electrification Conference and Expo (ITEC)*, 2019, pp. 1–6. DOI: 10.1109/ITEC.2019.8790528.

[11] D. E. Gaona, H. El Khatib, T. Long, and M. Saur, "Overmodulation strategy for deadbeat-flux and torque control of ipmsm with flux trajectory control in the stationary reference frame," in *2020 IEEE Energy Conversion Congress and Exposition (ECCE)*, 2020, pp. 6087–6095. DOI: 10.1109/ECCE44975.2020.9236331.

[12] H. El Khatib, D. Gerling, and M. Saur, "Deadbeat flux vector control as a one single control law operating in the linear, overmodulation, and six-step regions with time-optimal torque control," *IEEE Open Journal of Industry Applications*, vol. 3, pp. 247–270, 2022. DOI: 10.1109/OJIA.2022.3222852.

[13] J. S. Lee, C.-H. Choi, J.-K. Seok, and R. D. Lorenz, "Deadbeat-direct torque and flux control of interior permanent magnet synchronous machines with discrete time stator current and stator flux linkage observer," *IEEE Transactions on Industry Applications*, vol. 47, no. 4, pp. 1749–1758, 2011. DOI: 10.1109/TIA.2011.2154293.

[14] J. S. Lee and R. D. Lorenz, "Robustness analysis of deadbeat-direct torque and flux control for ipmsm drives," *IEEE Transactions on Industrial Electronics*, vol. 63, no. 5, pp. 2775–2784, 2016. DOI: 10.1109/TIE.2016.2521353.

[15] S. L. Kellner and B. Piepenbreier, "General pmsm d,q-model using optimized interpolated absolute and differential inductance surfaces," in *2011 IEEE International Electric Machines and Drives Conference (IEMDC)*, 2011, pp. 212–217. DOI: 10.1109/IEMDC.2011.5994848.

[16] M. Hepp, J. Weigold, M. Ebli, A. Nisch, M. Saur, *et al.*, "Power electronics for automotive electric drivetrains – design and optimization," in *Components of Power Electronics and their Applications 2023; ETG Symposium*, 2023, pp. 155–164.

PCIM Europe 2024, 11– 13 June 2024, Nuremberg   DOI: 10.30420/566262434

A 3-Leg Interleaved TP PFC with a 90° Phase-Shifted Asymmetric Leg for Reduced Magnetics

Ali Tausif [1], Ahmet Faruk Bakan [1], and Serkan Dusmez [2]

[1] Yildiz Technical University, Istanbul, Turkey
[2] Huawei Technologies Duesseldorf GmbH, Nuremberg Research Center, Germany

Corresponding author: Serkan Dusmez, serkan.dusmez@huawei.com

Abstract

In this paper, a novel approach utilizing an asymmetric switching leg is introduced, in which the inductance of one leg is halved, and unconventional phase-shifts are applied between the legs. By canceling out the 2nd harmonic of the input current ripple present in the EMI spectrum while maintaining the 3rd harmonic at the same level as in the symmetrical 3-leg interleaved version, the differential-mode filter requirement remains the same while one of the inductor size is reduced. The proposed asymmetrically phase-shifted converter shows better performance in terms of volume of input inductors (17% reduction) and overall cost (11% reduction) compared to its conventional counterpart.

1 Introduction

To improve the power density of a GaN-based totem-pole (TP) power factor correction (PFC) converter, the switching frequency (f_{sw}) needs to be increased to minimize the size of the passive elements such as the input inductors and electromagnetic interference (EMI) filters [1,2]. However, simply increasing the f_{sw} isn't always possible due to increased losses in switching devices and increased core losses in magnetic elements. Apart from low efficiency, this may even hinder further reduction in the volume of the magnetics and necessitate additional volume for larger heat sinks.

Therefore, interleaving the converter is believed to be a solution to improve the DM noise, as it increases the effective switching frequency and reduces input ripple current, demanding a smaller DM filter. Additionally, interleaving divides the power into each interleaved leg, requiring smaller cores for boost inductors not only due to reduced copper losses but also due to low core-saturation requirements.

However, the notion that interleaving always reduces the size of the DM filter is not universally true and is highly dependent on the choice of f_{sw} and topology [3–5]. The peculiar relationship between the choice of f_{sw}, multi-leg/multi-level interleaving topologies, and the size of the converter is studied and presented in [6]. Provided all constraints related to the optimal switching frequency range are taken into account, interleaving may serve as one of the best options to improve the power density of the converter.

A 3-leg interleaved TP PFC converter emerges as a viable solution for high-power density designs. Operating at 135 kHz per leg with a 120° phase shift among the interleaving legs, it facilitates a noticeable reduction in magnetics size. Employing a symmetric 120° phase shift eliminates the first and second harmonics (135 kHz and 270 kHz, respectively), resulting in an effective input current ripple frequency of 405 kHz with reduced amplitude due to the cancellation effect. As clarified in [6], mitigating DM noise at 405 kHz allows for the use of a more compact DM filter.

In this study, a concept for further enhancing the power density of a 3-leg interleaved TP PFC is introduced. This enhancement is achieved by minimizing the size of the input inductor of one of the legs while maintaining the same DM noise at the 3rd harmonic as in the conventional symmetrical 120° phase-shifting case. The approach involves asymmetric switching legs, wherein one out of the three legs employs half the inductance compared to the other two, while the other two inductors maintain equal inductance. By using asymmetric legs and introducing a phase-shift other than 120°, it is possible to eliminate the desired harmonic from the

3060

PCIM Europe 2024, 11–13 June 2024, Nuremberg DOI: 10.30420/566262434

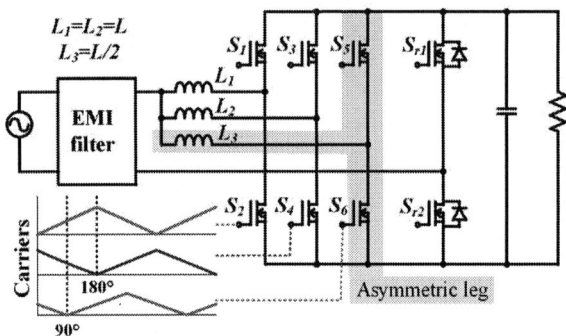

Fig. 1: Proposed 3-leg interleaved TP PFC.

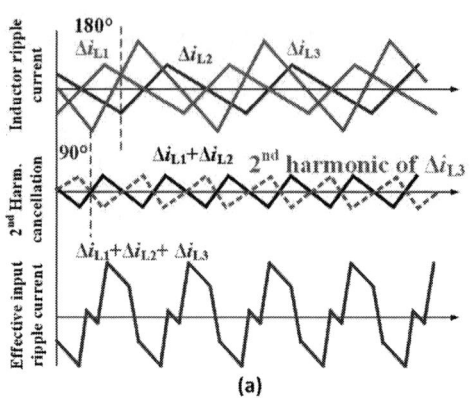

(a)

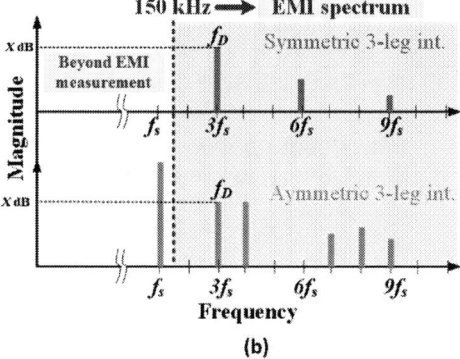

(b)

Fig. 2: Key waveforms of the proposed technique (a) input and inductors current, (b) FFT of the input current.

EMI spectrum measurement range (i.e., 150 kHz to 30 MHz). It must be noted that the EMI measurement range depends on the standards defined and categorized for each electronics equipment by the regulatory authority, and the categorization of standards is based on the applications. The standards' frequency range for the conducted emission measurements along with their application is provided in Tab. 1. This study targets TP PFC for industrial, automotive, and consumer electronics applications, whose conducted EMI test range is from 150 kHz to 30 MHz [7] and [8].

The proposed 3-leg interleaved TP PFC operates at 135 kHz per leg. In this setup, two symmetric legs with inductances equal to L are driven with a 180° phase-shift. Meanwhile, the third asymmetric leg, using an inductance of $L/2$, is driven with a 90° phase shift. This configuration aims to cancel out the 2nd harmonic, resulting in a 3rd harmonic current ripple equivalent to that in the conventional symmetrical phase-shifting case.

2 Proposed converter

2.1 Concept

In conventional 3-leg interleaved TP PFCs, a 120° phase-shift exists between high-frequency switching legs. In the proposed method, two symmetric legs exhibit a 180° phase shift, while the asymmetric leg is phase-shifted by 90° from the first leg's carrier (see Fig. 1). Symmetric legs are equipped with inductors of the same inductance, denoted as $L_1, L_2 = L$, while the asymmetric leg's inductance is half that of the other two, $L_3 = L/2$.

In the proposed converter, all legs operate at 135 kHz under Continuous Conduction Mode (CCM). The symmetric legs, akin to a 2-leg interleaved setup, generate an effective current with ripple at the 2nd harmonic of the f_{sw}. When combined with the third asymmetric leg, it cancels out the 2nd harmonic in the input current spectrum as shown in Fig. 2(a). Notably, the 1st harmonic in the current spectrum emerges at 135 kHz, unlike in the conventional method where it is canceled out. However, this first

Tab. 1: EMI standards for conducted emissions.

EMI std.	Conducted freq. range	Application
CISPR 11	150kHz-30MHz	Industrial
CISPR 22/24	150kHz-30MHz	Consumer's electronics
CISPR 25	150kHz-30MHz	Automotive
MIL-STD-461 CE101	30Hz-30MHz	Military
RCTA DO-160	50kHz-100MHz	Avionics

3061

harmonic is beyond the range of conducted EMI measurements, as depicted in Fig. 2(b). Therefore, canceling out the 2nd harmonic proves beneficial only when $75\,\text{kHz} < f_{\text{sw}} < 150\,\text{kHz}$. It must be noted that the proposed method can also be extended for multi-leg TP PFC topologies by carefully determining the phase shifts among legs as it may differ when the number of legs increases.

2.2 Simulation and experimental validations

The proposed concept has been successfully validated through simulations conducted with the following parameters: output power (P_o) of 3700 W, output voltage (V_o) of 400 V, grid voltage (V_g) of 220 V_{rms} at 50 Hz, and f_{sw} of 135 kHz. Simulations used inductor values of 80 µH for L_1 and L_2, and 40 µH for L_3. A conventional method with similar specifications, except for $L_1 = L_2 = L_3 = 80$ µH, was simulated for comparison. The simulation parameters are also shown in Tab. 2. Results, crucial for evaluating performance, are presented in Fig. 3 to 4.

According to the simulation results in Fig. 3(a), the input current in the proposed method, represented as a combination of i_{L1}, i_{L2}, and i_{L3}, exhibits higher ripple compared to the conventional method (see Fig. 4(a)). However, the first harmonic observed in the EMI measurement range remains at 405 kHz as shown in Fig. 3(b), with nearly the same magnitude as in the conventional case (see Fig. 4(b)), which demands the same DM filter size. Despite the high ripple in the proposed method's input current, caused by the 135 kHz switching frequency component, it does not occur within the EMI measurement range. Additionally, when a DM filter is utilized, its cutoff frequency at a very low value and finite insertion loss at f_{sw} result in significant attenuation of the 1st harmonic component, effectively avoiding any major total harmonic distortion (THD) issue.

The proof of concept has been similarly validated through an experimental prototype, shown in Fig. 5.

Tab. 2: Simulation parameters.

Output power (P_o)	3700 W
Output voltage (V_o)	400 V
Input voltage (V_g)	220 V_{rms}
Switching frequency (f_{sw})	135 KHz
Symmetric leg inductance(L_1, L_2)	80 µH
Asymmetric leg inductance(L_3)	40 µH

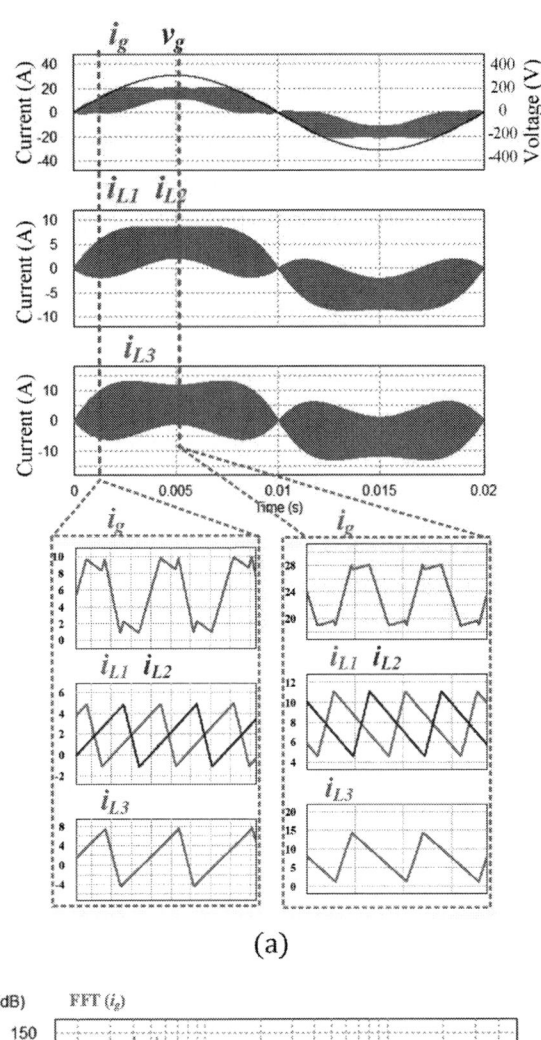

(a)

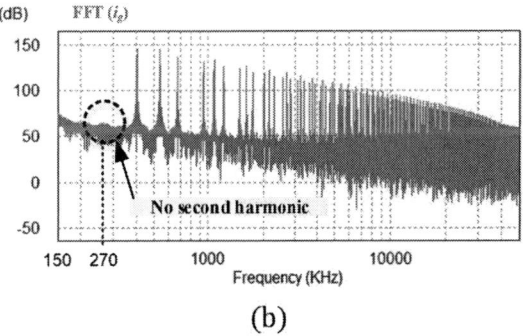

(b)

Fig. 3: Simulation results for the proposed method; (a) leg currents, (b) FFT of i_g.

The experimental results are specifically provided for its operation as a 3-leg DC-DC converter. The experimental waveforms, crucial for understanding its practical performance, are illustrated in Fig. 6. It is important to note that the current experimental prototype has not been optimized yet and is made

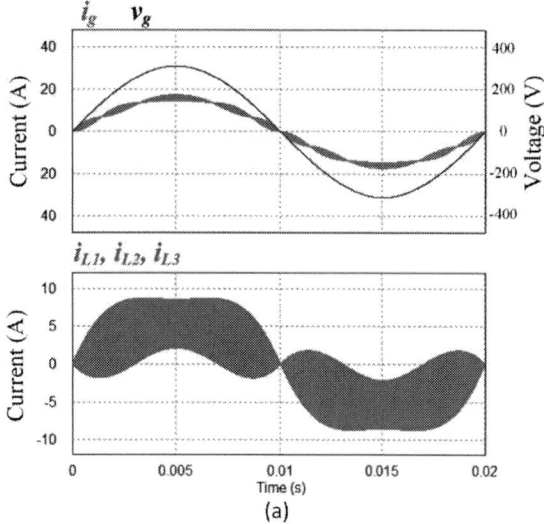

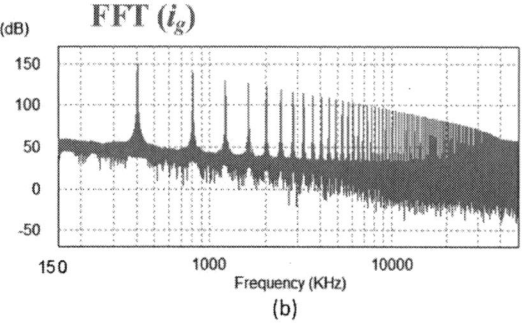

(a)

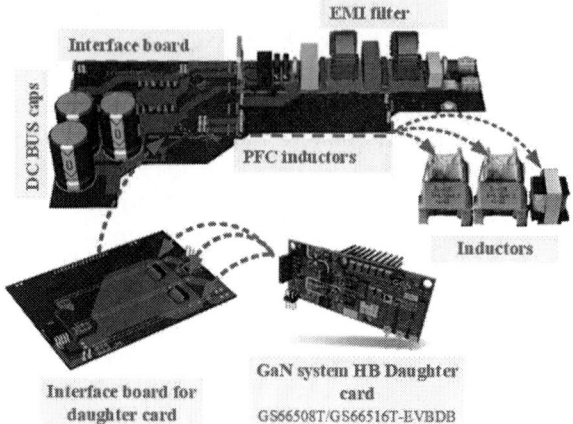

Fig. 5: Experimental prototype.

FFT (i_g)

(b)

Fig. 4: Simulation results for the conventional method; (a) leg currents, (b) FFT of i_g.

to run as a DC-DC converter.

Figure 6(a) shows the operation of the converter at duty cycle (d) of 0.6 along with the FFT of the input current waveform. It can clearly be seen that the 2nd harmonic is eliminated from the input current, occurring at 270 kHz, which is in the conducted EMI spectrum. Similarly, Figure 6(b) shows the results when $d = 0.25$. The FFT in this case also shows that the 2nd harmonic component of the input current is successfully canceled using the proposed method.

2.3 Analysis and comparison of volume, loss and cost

The design of the 3-leg TP PFC is achieved by estimating the DM noise at the designed frequency, which in the present case is 405 kHz (3rd harmonic of 135 kHz). Utilizing the framework presented in [6] for DM noise estimation and its filter volume determination, the DM noise level is estimated to be 149 dB, and the optimum volume of the filter

comes out to be 105 cm^3.

At $f_{sw} = 135$ kHz, the input inductance per leg is determined to be 80 µH. The choice of this value of inductance is based on the f_{sw} and the optimum DM filter size, which utilizes the best core from the available core database in terms of size. For 80 µH inductors, 2 E-cores (E341409) are stacked, and copper wire is wound with 20 turns and an air gap of 1.28 mm. The volume of the 80 µH inductor (without bobbin, i.e., core + winding) is about 26.5 cm^3. The designed inductor is shown in Fig. 7. The inductance for the asymmetric leg is halved, i.e., 40 µH, which is achieved by using a single stack of E341409 core, keeping all other parameters similar. The 40 µH inductor reduces to half the size of the 80 µH. The practically designed 40 µH inductor is also presented in Fig. 7.

The selection of FETs is based on the best performance at 135 kHz and RMS current distribution per leg. A 50mΩ GaN FET is shown to have superior performance for the proposed design specifications. The heat sink associated with symmetrical legs is UB3510B, occupying a size of 12.25 cm^3, while the heat sink associated with the asymmetric leg would be slightly larger due to excessive heat generated because of more loss. This additional loss is associated with the high RMS current flowing in the asymmetric leg as the ripple current is doubled due to halving the inductance. The volume estimation of the proposed converter and the conventional converter was conducted based on the volume of the EMI filter, PFC inductors, and heatsinks, as these components contribute significantly to the overall volume of PFCs. Since the EMI filter is the same in both converters, only the volume of PFC inductors

PCIM Europe 2024, 11– 13 June 2024, Nuremberg   DOI: 10.30420/566262434

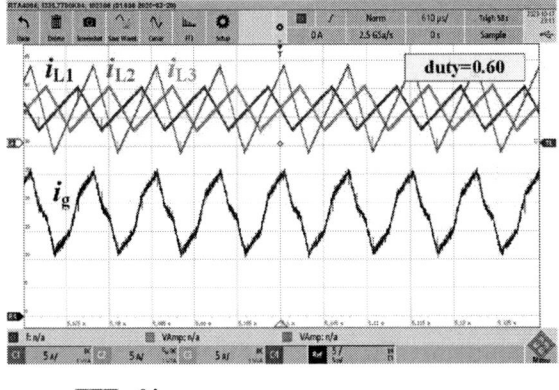

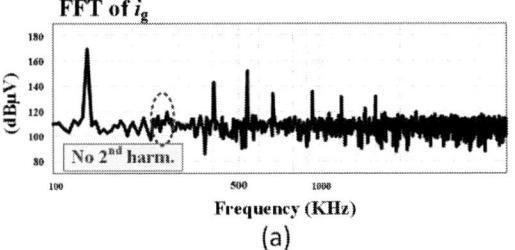

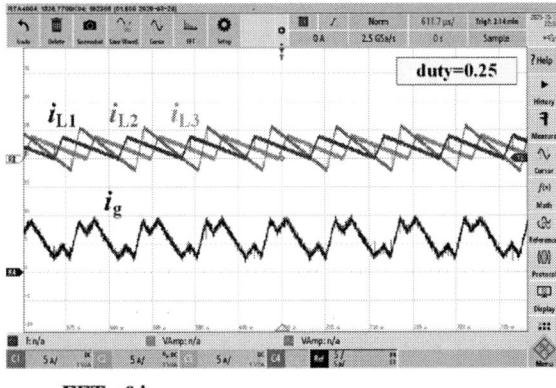

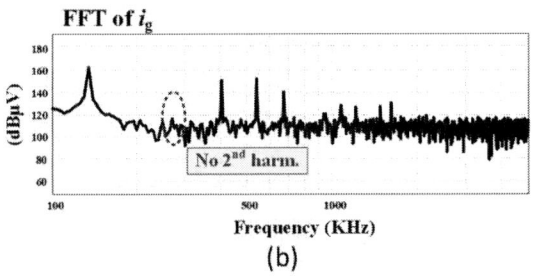

Fig. 6: Experimental waveforms of input and inductor currents with input current FFT for; (a) d=0.60 and (b) d=0.25.

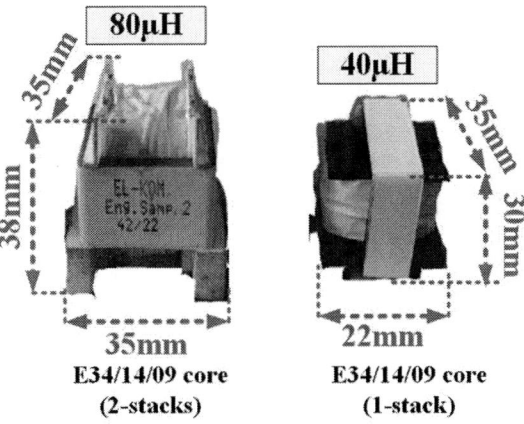

Fig. 7: Size comparison of 80μH and 40μH inductors.

resulting in a magnetics volume reduction totaling 17%, although there was a 3 cm³ increase in the volume of the heatsink due to higher conduction losses on the asymmetric leg. The overall improvement was approximately from 116 cm³ to 105 cm³, representing around a 10% reduction in the volume of the combined heatsinks and PFC inductor.

Regarding losses, the analysis considered PFC inductor losses and FET losses, estimated based on the models presented in references [9] and [10]. It was observed that losses increased from 44.73W to 51W as shown in Fig. 8(b). This increase can be attributed to the halving of inductance (L) in the case of the asymmetric leg, which doubled the current ripple due to the same switching frequency. Note that in the current design, for ease of implementation, 2-stacked cores used in the case of 80 μH were reduced to a single core to obtain 40 μH. Size comparison of the two is shown in Fig. 7. This reduction halved the reluctance while maintaining the air gap and N constant. The higher current ripple together with lower reluctance lead to increased ΔB, causing higher core losses. Additionally, the elevated ripple resulted in higher RMS current, potentially increasing copper losses and conduction losses in the FETs. Please note that, there is room for optimizing the 40 μH inductor using different core material and geometries.

The cost of a PFC converter primarily consists of the input EMI filter, PFC inductors, heatsinks, and FETs. In the current comparison, the FETs are identical in both converters, thus not factored into the analysis. The cost estimation is expressed in per unit. Fig. 8(c) illustrates that the proposed converter surpasses the conventional one with an

and heatsinks was considered for comparison. The proposed converter demonstrated an improvement in volume compared to the conventional 3-leg TP PFC, as depicted in Fig. 8(a). Specifically, the volume of inductors reduced from 79.6 cm³ to 66 cm³,

3064

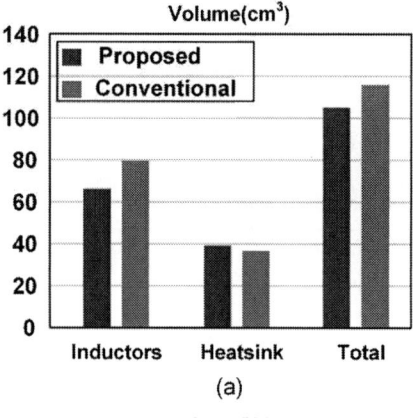

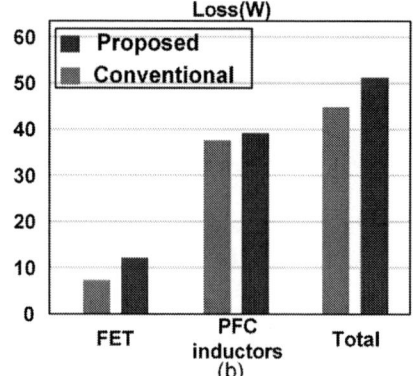

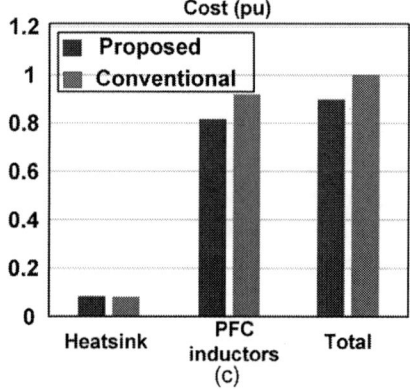

Fig. 8: Comparison of; (a) volume, (b) loss, and (c) cost.

11% improvement. The total per unit cost of the proposed converter is 0.89, indicating a more cost-effective solution.

3 Conclusion

In this paper, a novel method to improve the power density of 3-leg interleaved TP PFCs is presented, which utilizes the asymmetric leg with half the inductance and a $90°$ phase-shift. The proposed approach works on the principle of canceling out

the equivalent ripple generated by the two symmetric legs with the 2nd harmonic of the ripple generated by the asymmetric leg. As a result, the proposed converter is shown to have an improvement in terms of volume reduction of 17% of overall input inductors and also an overall cost reduction of around 11%.

The proposed method stands as an innovative technique to enhance the power density of hard-switched TP PFCs, albeit with a slight decrease in efficiency, which represents a tradeoff. The initial findings indicate that the proposed converter, even without optimization, demonstrates marginal improvements in two crucial performance metrics: volume and cost. However, with an optimized design, these parameters are expected to further enhance, rendering this approach a viable solution for power-dense PFCs.

References

[1] J. Mühlethaler, H. Uemura, and J. W. Kolar, "Optimal Design of EMI Filters for Single-Phase Boost PFC Circuits," in Proceedings of the 38th Annual Conference of the IEEE Industrial Electronics Society (IECON 2012), Montreal, Canada, October 25-28, 2012, pp. 2765-27701

[2] Q. Li, H. Y. Zhao and J. C. Song, "Design and loss analysis of the high frequency PFC converter," 2015 IEEE International Conference on Applied Superconductivity and Electromagnetic Devices (ASEMD), Shanghai, China, 2015, pp. 124-125, doi: 10.1109/ASEMD.2015.7453497.

[3] C. Wang, M. Xu, F. C. Lee and B. Lu, "EMI study for the interleaved multi-channel PFC", Proc. IEEE Power Electron. Specialists Conf., pp. 1336-1342, 2007.

[4] K. Raggl, T. Nussbaumer, G. Doerig, J. Biela and J. W. Kolar, "Comprehensive Design and Optimization of a High-Power-Density Single-Phase Boost PFC," in IEEE Transactions on Industrial Electronics, vol. 56, no. 7, pp. 2574-2587, July 2009, doi: 10.1109/TIE.2009.2020074.

[5] N. N. Esfetanaj, H. Wang, F. Blaabjerg and P. Davari, "Differential mode noise estimation and filter design for interleaved boost power

factor correction converters", Appl. Sci., vol. 11, no. 6, 2021.

[6] A. Tausif and S. Dusmez, "A Unified Differential Mode Noise Estimation Method and Filter Size Comparison in Single-Phase Multileg and Multilevel Totem-Pole PFC Converters,"IEEE Trans. on Power Electronics, vol. 38, no. 6, pp. 7197-7206, June 2023.

[7] Electromagnetic Compatibility (EMC) – Part 6 Generic Standards Section 3: Emission Standard for Residential Commercial and Light-Industrial Environments, 2006.

[8] "Vehicles boats and internal combustion engines - radio disturbance characteristics - limits and methods of measurement for the protection of on-board receivers", 2016.

[9] E. B. Bulut, M. O. Gulbahce, D. A. Kocabas and S. Dusmez, "Efficiency and Power Density Optimization of Three-Level TP PFC," PCIM Europe digital days 2021; Online, 2021, pp. 1-8.

[10] A. Lordoglu, M. O. Gulbahce, D. A. Kocabas and S. Dusmez, "A New Optimization Method for Gapped and Distributed Core Magnetics in LLC Converter," in IEEE Access, vol. 11, pp. 14061-14072, 2023, doi: 10.1109/ACCESS.2023.3242869

PCIM Europe 2024, 11– 13 June 2024, Nuremberg   DOI: 10.30420/566262435

Fault-Tolerant Operation Analysis of a Five-Phase Three-Level TNPC Inverter for Electric Aircraft Propulsion Systems

Chanuch Chaisakdanugull, Klaus F. Hoffmann

Helmut Schmidt University, Germany

Corresponding author: Chanuch Chaisakdanugull, chaisakc@hsu-hh.de
Speaker: Chanuch Chaisakdanugull, chaisakc@hsu-hh.de

Abstract

This paper presents an analysis of a five-phase three-level TNPC inverter focusing on its fault-tolerant capabilities and mitigation techniques for electric aircraft propulsion systems. The space vector pulse-width-modulation (SVPWM) schemes in both normal and post-fault operations accounting for short-circuit and open-circuit occurrences in individual power electronic switches in a single phase are discussed. In fault cases, the remaining switching vectors and their sequences are selected to produce the necessary voltage in the $\alpha\beta$-coordinate system. This selection aims to efficiently suppress undesired voltage components and complying with the rules focusing on minimizing switching losses. These methods are implemented and assessed through simulation, showing the ability and performance limitations of the inverter under various fault conditions.

1 Introduction

Multi-phase drives offer distinct advantages over standard three-phase systems, including reduced torque ripples, enhanced power density, and lower current ratings by distributing power among a higher number of phases [1], [2]. Additionally, their inherent phase redundancy provides superior fault-tolerant capabilities, making them highly promising for electric aircraft applications. To satisfy the demands for high power and torque density, a design and evaluation of a five-phase permanent magnet synchronous motor (PMSM) with a flux-barrier stator and fractional slot concentrated windings to enhance flux density and achieve higher torque have been conducted in [3]. Building upon this work and serving as an energy supply for such a motor, a three-level T-type Neutral-Point-Clamped inverter (TNPC) with additional disconnecting devices, as shown in Fig. 1, is analyzed.

The TNPC inverter has been known to provide outstanding efficiency in the medium switching frequency range, produce less voltage distortions, reduce harmonic losses of electric machines and generate lower common-mode voltage, compared to a two-level converter [4], [5]. These characteristics make it suitable for electric propulsion systems, in which high power and efficiency as well as

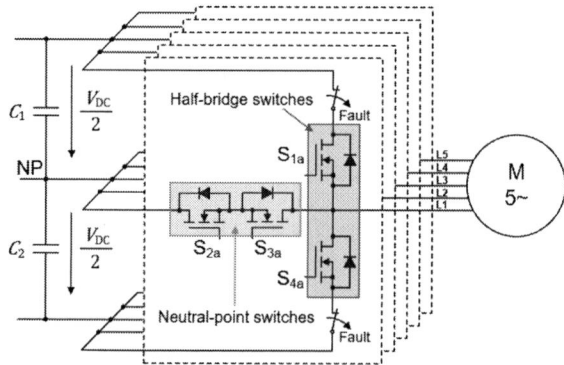

Fig. 1: Five-phase three-level TNPC inverter with disconnecting devices

strict regulations concerning the electromagnetic interference (EMI) are required. Moreover, thanks to its redundant switching states the topology can withstand certain types of switch failures within the inverter. Despite these advantages, it requires complicated control algorithms and modulation techniques caused by higher number of phases and levels, presenting a relative challenge compared to standard three-phase two-level counterparts.

A comprehensive analysis and implementation of space vector pulse-width-modulation

3067

(SVPWM) for a five-phase three-level inverter have been presented in [6], using vector space decomposition [7] to separate switching vectors into two orthogonal planes. By employing five vectors within a switching period, sinusoidal output voltages with minimized distortions can be achieved. Building on this foundation, [8] and [9] have proposed an improvement of this method to mitigate variations of the common mode voltage (CMV), resulting in 14 switching sequences in each sector. This improved modulation strategy was compared with other optimized carrier-based PWM techniques with signal injections, demonstrating no significant difference in voltage spectrum and current Total Harmonic Distortion (THD). Nevertheless, the thorough analysis of the space vector yields better understanding of its behavior under switch failures and its application for fault-tolerant operations. This paper extends the analysis of space vector diagrams to cover different fault scenarios. The pulse-width-modulation methods including vector selections and switching sequences are proposed and verified in the simulation.

2 Inverter Normal Operation

The output voltage of a five-phase three-level inverter can be projected onto two independent orthogonal sub-spaces, $\alpha\beta$ representing fundamental components and xy higher-order harmonics, as described by Eq. (2) and Eq. (3) in [6]. Given that each phase of the inverter can generate three different voltage levels, there are a total of $3^5 = 243$ possible voltage outputs, which can be reduced to 113 corresponding switching states forming a decagon shown in Fig. 2(a) [6]. In each switching period, five vectors are selected such that the desired voltage vector can be formed on the $\alpha\beta$-plane while simultaneously achieving zero in the xy-components. Examples of voltage vectors for the first sector on the xy-plane are depicted in Fig. 2(b). Through the selection of appropriate vector combinations and their dwell times, output voltages with minimal distortions can be attained. To ensure optimal operations, the following rules are applied in selecting vectors and arranging their sequences [6].

1. Constructing the desired vector in the $\alpha\beta$-plane should ensure that their combination in xy subspace leads to complete cancellation over the switching cycle.

2. The inverter is limited to making one transition

per switching event and adjustments in its level are allowed only in one step, either between -1 and 0, or 0 and 1, to reduce the switching losses.

3. Selection of redundant space vectors and adjusting their dwell time help balance the neutral point capacitor voltage ripple.

Applying the rules in conjunction with a modification outlined in [8] and [9] yields 14 switching patterns per sector. In each switching cycle, one of the patterns containing 5 vectors is applied based on the position of the reference vector on the $\alpha\beta$-plane. As an example, the switching sequence of a vector located in the sector 1A, in which the modulation index m defined in Eq. (1) is lower than 0.2628 and its angle ranges between 0 to 18°, is illustrated in Fig. 3. Within the switching period T_s, the inverter initiates with a redundant vector V_0, progresses from V_1 to V_4, returns to another state of V_0 and cycles back to the starting point for the next cycle. Moreover, only one level change in each transition is noticed. This algorithm achieves the maximum modulation index of 0.5256 marking a 5.12 % increase in DC-Link usage compared to the carrier-based PWM with a pure sinusoidal reference.

$$m = \frac{V_{\text{ref}}}{V_{\text{dc}}} \tag{1}$$

$$\vec{V}_{\alpha\beta} = \frac{2}{5}\left(v_{\text{L}1} + v_{\text{L}2} \cdot e^{j\frac{2\pi}{5}} + v_{\text{L}3} \cdot e^{j\frac{4\pi}{5}} + v_{\text{L}4} \cdot e^{-j\frac{4\pi}{5}} + v_{\text{L}5} \cdot e^{-j\frac{2\pi}{5}}\right) \tag{2}$$

$$\vec{V}_{\text{xy}} = \frac{2}{5}\left(v_{\text{L}1} + v_{\text{L}3} \cdot e^{j\frac{2\pi}{5}} + v_{\text{L}5} \cdot e^{j\frac{4\pi}{5}} + v_{\text{L}2} \cdot e^{-j\frac{4\pi}{5}} + v_{\text{L}4} \cdot e^{-j\frac{2\pi}{5}}\right) \tag{3}$$

3 Fault-Tolerant Operation

The vector diagram of a five-phase three-level inverter exhibits pairs of vectors with identical magnitudes and directions. This redundancy is beneficial to the operation in the event of faults or component failures. According to [10], faults in power semiconductors can be classified as open-circuit and short-circuit faults. In this paper, open-circuit in individual components as well as short-circuit faults in a single half-bridge switch,

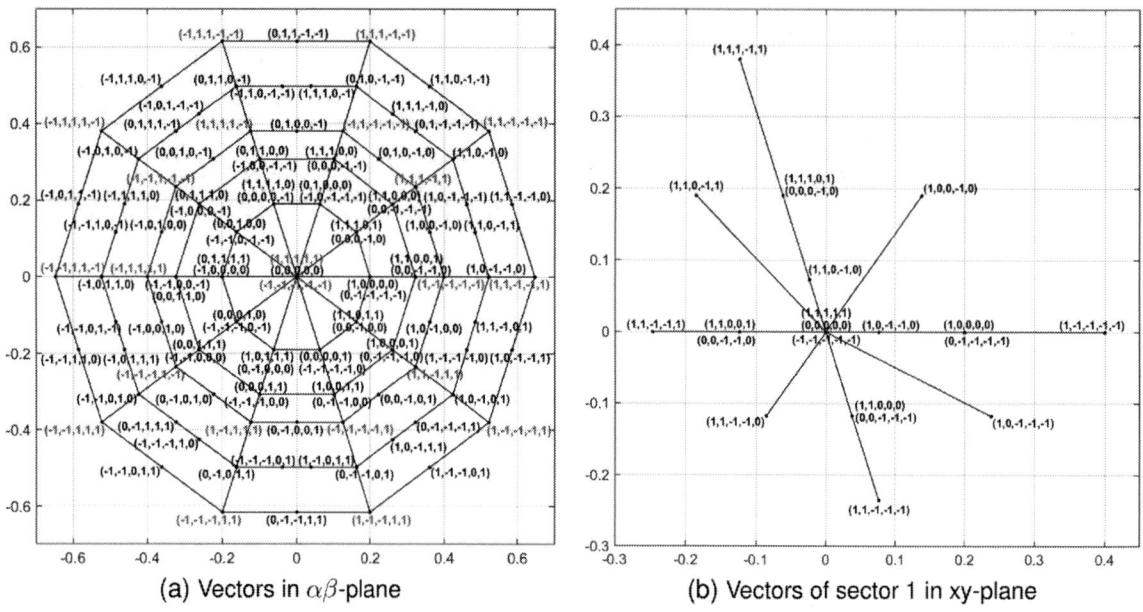

(a) Vectors in $\alpha\beta$-plane

(b) Vectors of sector 1 in xy-plane

Fig. 2: Switching vectors of a five-phase three-level inverter normalized with the DC-Link voltage

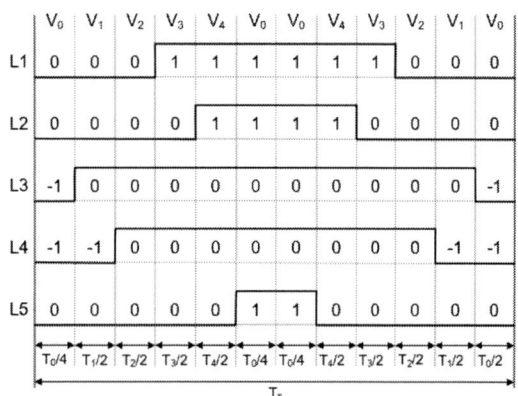

Fig. 3: Switching patterns of the sector 1A in normal operation

an inverter phase-leg and a single neutral-point switch are considered. Multiple fault types are not taken into account and only faults at phase L1 are assumed for simplicity. In order to cover all the mentioned situations, additional disconnecting devices are required in each inverter phase-leg, as depicted in Fig. 1. For this purpose, a method applying two fuses, two silicon controlled rectifiers (SCRs) and auxiliary capacitors to generate a transient short-circuit through the fuse as a disconnecting device has been proposed in [11]. Another potential solution involves the use of pyrofuses, such as the one described in the datasheet [12].

Both methods can disconnect the faulty switch from the circuit within a couple of milliseconds. However, one significant disadvantage are additional parasitic inductances in the commutation loop of the inverter. Therefore, great care has to be taken when switching with very high voltage slopes.

Assuming fast and effective fault detection and disconnection, the post-fault operation of the inverter can be divided into three different cases. The first case arises when one or both of the neutral-point switches experience an open-circuit fault. In such instances, the realization of switching state 0 becomes unattainable and the inverter can only operate with two-level states, highlighted in red in Fig. 2(a), switching the output between positive and negative polarities. This increases the current harmonics, EMI and also switching losses. Since the modulation method for this operating mode is similar to that of a two-level half-bridge inverter, it is neglected in this paper. Nevertheless, addressing the other two cases requires adjustments in vector selections and switching patterns, which are considered in the following sections.

3.1 Modulation Method for Phase L1 connected to NP

In the event of a short-circuit of a single half-bridge switch or an inverter phase-leg, the disconnection of faulty components from the DC bus has to be activated. Subsequently, post-fault operation can proceed by turning on both neutral-point switches to connect the phase to the neutral point NP. The corresponding vector diagram of this scenario can be seen in Fig. 5(a), comprising 10 sectors. Because each of the sector contains 5 vectors, achieving the desired output voltage while maintaining minimal distortions remains feasible. By applying consistent criteria for selecting switching orders as mentioned in section 2, the resulting switching sequences are summarized in table 1. It is crucial to note that the third selection rule cannot be applied in this scenario, which potentially results in larger variations in the DC-Link voltages.

An example of the switching sequence for sector 1 is shown in Fig. 4. During the switching period T_s, the inverter's state changes from the vector V_0 through V_1, V_2, V_3 and V_4 respectively, then reverses back from V_4 to V_0. The dwell time T_0 to T_4 can be calculated using the matrix equation in Eq. (4), where v_α and v_β represent the reference voltage components in $\alpha\beta$-plane normalized with the DC-Link voltage while v_x and v_y are set to zero. Since all five selected vectors in a switching period need to be active, their dwell times must always be greater than zero, thereby being constrained by the condition $T_0 > 0$ expressed in Eq. (5) for sector 1. As a result, the maximum modulation index is limited to $m_{\max} = 0.2628$ by which the maximum output voltage is reduced by half from the normal operation. This strategy can also be applied to half-bridge switch open-circuit and neutral-point switch short-circuit.

$$\frac{1}{T_s}\begin{bmatrix} T_0 \\ T_1 \\ T_2 \\ T_3 \\ T_4 \end{bmatrix} = \begin{bmatrix} V_{0,\alpha} & V_{1,\alpha} & V_{2,\alpha} & V_{3,\alpha} & V_{4,\alpha} \\ V_{0,\beta} & V_{1,\beta} & V_{2,\beta} & V_{3,\beta} & V_{4,\beta} \\ V_{0,x} & V_{1,x} & V_{2,x} & V_{3,x} & V_{4,x} \\ V_{0,y} & V_{1,y} & V_{2,y} & V_{3,y} & V_{4,y} \\ 1 & 1 & 1 & 1 & 1 \end{bmatrix}^{-1} \begin{bmatrix} v_{s,\alpha} \\ v_{s,\beta} \\ 0 \\ 0 \\ 1 \end{bmatrix}$$
(4)

$$m \le \frac{1}{3.6180\cos\theta_e + 1.1756\sin\theta_e}, \ 0 \le \theta_e < \pi/5$$
(5)

Sector	V_0	V_1	V_2	V_3	V_4
1	0,-1,-1,-1,-1	0,0,-1,-1,-1	0,0,-1,-1,0	0,0,0,-1,0	0,0,0,0,0
2	0,0,-1,-1,-1	0,0,0,-1,-1	0,0,0,-1,0	0,0,0,0,0	0,1,0,0,0
3	0,0,0,-1,-1	0,0,0,0,-1	0,0,0,0,0	0,1,0,0,0	0,1,1,0,0
4	0,0,0,0,-1	0,0,0,0,0	0,0,1,0,0	0,1,1,0,0	0,1,1,1,0
5	0,0,0,0,0	0,0,1,0,0	0,0,1,1,0	0,1,1,1,0	0,1,1,1,1
6	0,0,0,0,0	0,0,0,1,0	0,0,1,1,0	0,0,1,1,1	0,1,1,1,1
7	0,-1,0,0,0	0,0,0,0,0	0,0,0,1,0	0,0,0,1,1	0,0,1,1,1
8	0,-1,-1,0,0	0,-1,0,0,0	0,0,0,0,0	0,0,0,0,1	0,0,0,1,1
9	0,-1,-1,-1,0	0,-1,-1,0,0	0,0,-1,0,0	0,0,0,0,0	0,0,0,0,1
10	0,-1,-1,-1,-1	0,-1,-1,-1,0	0,0,-1,-1,0	0,0,-1,0,0	0,0,0,0,0

Tab. 1: Selected vectors in case L1 is connected to NP

3.2 Modulation Method for Phase L1 open

If phase L1 is open, the strategy outlined in [13] for shifting the axes of phases L2 and L5 can be employed. This leads to a reduced-order transformation represented by Eq. (6) [14], where the angle $\gamma = 2\pi/5$. Transforming the remaining 4 phases using the transformation matrix gives a two-dimensional $\alpha\beta$-plane and additional one-dimensional z-subspace illustrated in Fig. 5(b) and Fig. 6 respectively. The $\alpha\beta$-plane consists of 81 vectors and can be divided into 8 sectors, each containing 4 sub-sectors. On the contrary, z-subspace is one-dimensional but contains the same number of vectors. Unlike xy-components which are set to zero during normal operation, the voltage in z-component will be later utilized as a manipulated variable for current controllers and is thus non-zero in operation.

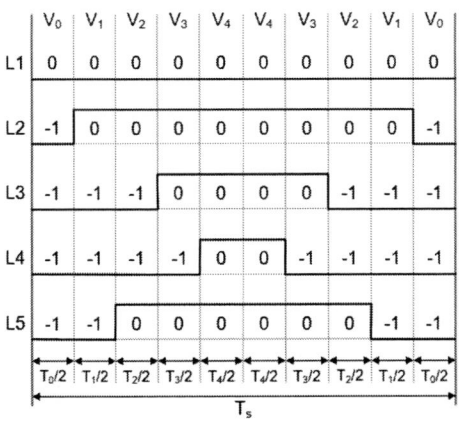

Fig. 4: Switching patterns of the sector 1 when phase L1 is connected to NP

$$\begin{bmatrix} v_\alpha \\ v_\beta \\ v_z \end{bmatrix} = \begin{bmatrix} \frac{\cos 0.5\gamma}{3.618} & \frac{\cos 2\gamma}{3.618} & \frac{\cos 3\gamma}{3.618} & \frac{\cos 4.5\gamma}{3.618} \\ \frac{\sin 0.5\gamma}{1.91} & \frac{\sin 2\gamma}{1.91} & \frac{\sin 3\gamma}{1.91} & \frac{\sin 4.5\gamma}{1.91} \\ \frac{\sin \gamma}{5} & \frac{\sin 4\gamma}{5} & \frac{\sin 6\gamma}{5} & \frac{\sin 9\gamma}{5} \end{bmatrix} \begin{bmatrix} v_{L2} \\ v_{L3} \\ v_{L4} \\ v_{L5} \end{bmatrix}$$
(6)

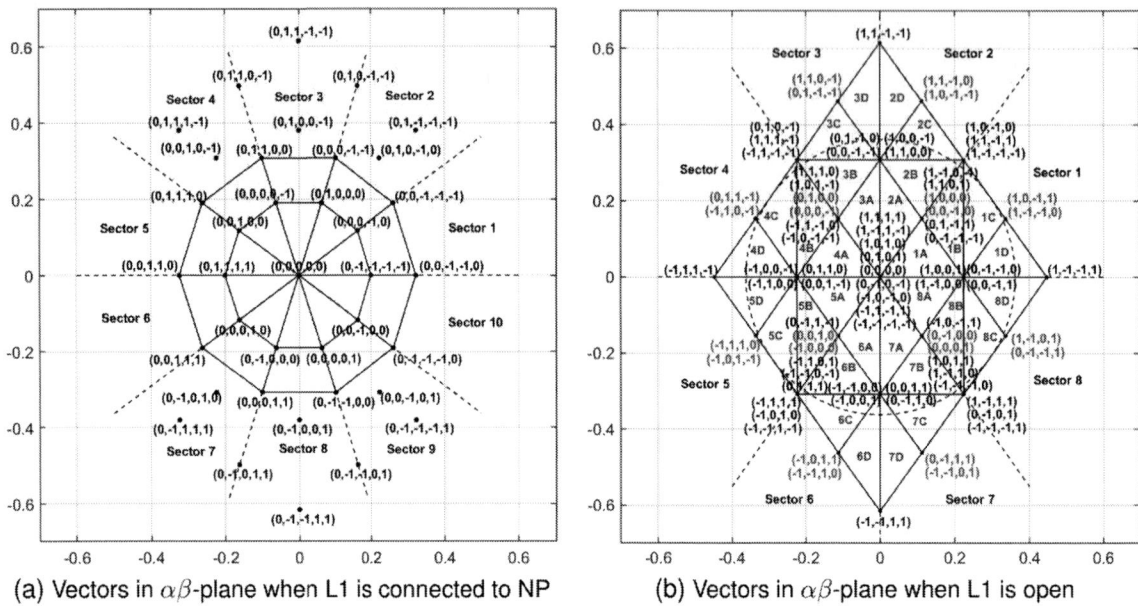

(a) Vectors in $\alpha\beta$-plane when L1 is connected to NP

(b) Vectors in $\alpha\beta$-plane when L1 is open

Fig. 5: Switching vectors in $\alpha\beta$-plane in post-fault operations normalized with the DC-Link voltage

Fig. 6: Z-component of the switching vectors normalized with DC-Link voltage when L1 is open

In each switching period, 5 vectors are selected: one single vector and two pairs of redundant vectors, thus forming a triangle within a sub-sector. All switching sequences are organized and can be summarized in table 2, starting with V_0, passing through V_1, V_2, V_3 and ending with V_0. It is noted that V_1 and V_3 highlighted in red in Fig. 5(b) are redundant vectors with the same magnitude and direction in $\alpha\beta$-plane but opposite directions in z-subspace. Therefore, manipulating their dwell time help generate voltage in z-subspace but has no influence in the $\alpha\beta$-plane. The dwell time can then be calculated as Eq. (7). Figure 7 depicts a switching pattern for the sector 1A. The maximum modulation index is defined as $m_{\max} = 0.3618$, corresponding to the maximum radius of the circle inscribed in the space vector diagram.

Sector	V_0	V_1	V_2	V_3	V_0
1A	0,-1,-1,0	0,0,-1,0	0,0,0,0	1,0,0,0	1,0,0,1
1B	0,-1,-1,0	0,0,-1,0	1,0,-1,0	1,0,0,0	1,0,0,1
1C	0,-1,-1,0	1,-1,-1,0	1,0,-1,0	1,0,-1,0	1,0,0,1
1D	0,-1,-1,0	1,-1,-1,0	1,-1,-1,1	1,0,-1,1	1,0,0,1
2A	0,0,-1,-1	0,0,-1,0	0,0,0,0	1,0,0,0	1,1,0,0
2B	0,0,-1,-1	0,0,-1,0	1,0,-1,0	1,0,0,0	1,1,0,0
2C	0,0,-1,-1	1,0,-1,-1	1,0,-1,0	1,1,-1,0	1,1,0,0
2D	0,0,-1,-1	1,0,-1,-1	1,1,-1,-1	1,1,-1,0	1,1,0,0
3A	0,0,-1,-1	0,0,0,-1	0,0,0,0	0,1,0,0	1,1,0,0
3B	0,0,-1,-1	0,0,0,-1	0,1,0,-1	0,1,0,0	1,1,0,0
3C	0,0,-1,-1	0,1,-1,-1	0,1,0,-1	1,1,0,-1	1,1,0,0
3D	0,0,-1,-1	0,1,-1,-1	1,1,-1,-1	1,1,0,-1	1,1,0,0
4A	-1,0,0,-1	0,0,0,-1	0,0,0,0	0,1,0,0	0,1,1,0
4B	-1,0,0,-1	0,0,0,-1	0,1,0,-1	0,1,0,0	0,1,1,0
4C	-1,0,0,-1	-1,1,0,-1	0,1,0,-1	0,1,1,0	0,1,1,0
4D	-1,0,0,-1	-1,1,0,-1	-1,1,1,-1	0,1,1,-1	0,1,1,0
5A	-1,0,0,-1	-1,0,0,0	0,0,0,0	0,0,1,0	0,1,1,0
5B	-1,0,0,-1	-1,0,0,0	-1,0,1,0	0,0,1,0	0,1,1,0
5C	-1,0,0,-1	-1,0,1,-1	-1,0,1,0	-1,1,1,0	0,1,1,0
5D	-1,0,0,-1	-1,0,1,-1	-1,1,1,-1	-1,1,1,0	0,1,1,0
6A	-1,-1,0,0	-1,0,0,0	0,0,0,0	0,0,1,0	0,0,1,1
6B	-1,-1,0,0	-1,0,0,0	-1,0,1,0	0,0,1,0	0,0,1,1
6C	-1,-1,0,0	-1,-1,1,0	-1,0,1,0	-1,0,1,1	0,0,1,1
6D	-1,-1,0,0	-1,-1,1,0	-1,-1,1,1	-1,0,1,1	0,0,1,1
7A	-1,-1,0,0	0,-1,0,0	0,0,0,0	0,0,0,1	0,0,1,1
7B	-1,-1,0,0	0,-1,0,0	0,-1,0,1	0,0,0,1	0,0,1,1
7C	-1,-1,0,0	-1,-1,0,1	0,-1,0,1	0,-1,1,1	0,0,1,1
7D	-1,-1,0,0	-1,-1,0,1	-1,-1,1,1	0,-1,1,1	0,0,1,1
8A	0,-1,-1,0	0,-1,0,0	0,0,0,0	0,0,0,1	1,0,0,1
8B	0,-1,-1,0	0,-1,0,0	0,-1,0,1	0,0,0,1	1,0,0,1
8C	0,-1,-1,0	0,-1,-1,1	0,-1,0,1	1,-1,0,1	1,0,0,1
8D	0,-1,-1,0	0,-1,-1,1	1,-1,-1,1	1,-1,0,1	1,0,0,1

Tab. 2: Selected vectors in case phase L1 is open

$$\frac{1}{T_s}\begin{bmatrix} T_0 \\ T_1 \\ T_2 \\ T_3 \end{bmatrix} = \begin{bmatrix} V_{0,\alpha} & V_{1,\alpha} & V_{2,\alpha} & V_{3,\alpha} \\ V_{0,\beta} & V_{1,\beta} & V_{2,\beta} & V_{3,\beta} \\ 0 & V_{1,z} & 0 & V_{3,z} \\ 1 & 1 & 1 & 1 \end{bmatrix}^{-1} \begin{bmatrix} v_{s,\alpha} \\ v_{s,\beta} \\ v_{s,z} \\ 1 \end{bmatrix} \quad (7)$$

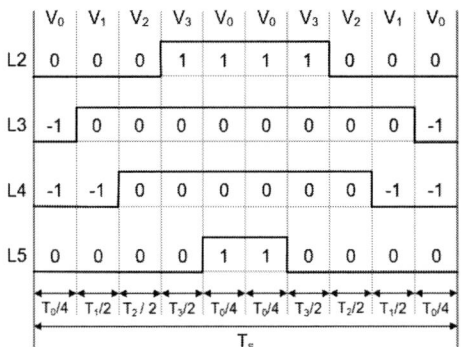

Fig. 7: Switching patterns of the sector 1A when phase L1 is open

4 Implementation and Validation

Fault-tolerant operations of a five-phase three-level TNPC inverter have been implemented and validated through simulation using MATLAB Simulink. The general control structure is depicted in Fig. 8. Five phase currents are measured and then transformed into $\alpha\beta$- plane using Eq. (8), followed by conversion to the dq rotating reference frame for current control via Park transformation using Eq. (9). PI current controllers are employed to maintain i_d at zero and regulate i_q as required for torque production. The voltage requests from the current controller are fed to the SVPWM-modulator to generate switching patterns for controlling the TNPC inverter. In the event of an open-circuit fault detection, the phase currents are transformed by reduced-order transformation as Eq. (10), and the current controller for z-component is activated to regulate i_z to zero. It is noted that both transformations yield the same result in $\alpha\beta$-plane if $i_\mathrm{L1} = 0$. The SVPWM modulator adjusts its modulation strategies according to each fault scenario.

$$\begin{bmatrix} i_\alpha \\ i_\beta \\ i_x \\ i_y \end{bmatrix} = \frac{2}{5}\begin{bmatrix} 1 & \cos\gamma & \cos 2\gamma & \cos 3\gamma & \cos 4\gamma \\ 0 & \sin\gamma & \sin 2\gamma & \sin 3\gamma & \sin 4\gamma \\ 1 & \cos 3\gamma & \cos\gamma & \cos 4\gamma & \cos 2\gamma \\ 0 & \sin 3\gamma & \sin\gamma & \sin 4\gamma & \sin 2\gamma \end{bmatrix} \begin{bmatrix} i_\mathrm{L1} \\ i_\mathrm{L2} \\ i_\mathrm{L3} \\ i_\mathrm{L4} \\ i_\mathrm{L5} \end{bmatrix} \quad (8)$$

$$\begin{bmatrix} i_\mathrm{d} \\ i_\mathrm{q} \\ i_\mathrm{xd} \\ i_\mathrm{yq} \end{bmatrix} = \begin{bmatrix} \cos\theta_e & \sin\theta_e & 0 & 0 \\ -\sin\theta_e & \cos\theta_e & 0 & 0 \\ 0 & 0 & \cos\theta_e & \sin\theta_e \\ 0 & 0 & -\sin\theta_e & \cos\theta_e \end{bmatrix} \begin{bmatrix} i_\alpha \\ i_\beta \\ i_x \\ i_y \end{bmatrix} \quad (9)$$

$$\begin{bmatrix} i_\alpha \\ i_\beta \\ i_z \end{bmatrix} = \begin{bmatrix} \frac{\cos 0.5\gamma}{3.618} & \frac{\cos 2\gamma}{3.618} & \frac{\cos 3\gamma}{3.618} & \frac{\cos 4.5\gamma}{3.618} \\ \frac{\sin 0.5\gamma}{1.91} & \frac{\sin 2\gamma}{1.91} & \frac{\sin 3\gamma}{1.91} & \frac{\sin 4.5\gamma}{1.91} \\ \frac{\sin\gamma}{5} & \frac{\sin 4\gamma}{5} & \frac{\sin 6\gamma}{5} & \frac{\sin 9\gamma}{5} \end{bmatrix} \begin{bmatrix} i_\mathrm{L2} \\ i_\mathrm{L3} \\ i_\mathrm{L4} \\ i_\mathrm{L5} \end{bmatrix} \quad (10)$$

4.1 PMSM and Inverter Model

The PMSM model implemented in the simulation is based on the voltage behind reactance (VSR) approach [15]. In this model, each phase of the motor is represented as an RL-load with a series voltage source representing the motor back-electromotive force (back-EMF). The model is simplified under the following assumptions:

- Mutual inductance between phases is neglected.

- The flux distribution of the permanent magnet is assumed to be sinusoidal.

- There is no magnetic saturation and the rotor structure is considered to be non-salient. The stator phase inductance is therefore constant.

The induced back-EMF depends on the rotor rotational speed ω_e, electrical angular position θ_e and the flux-linkage induced by the permanent magnet Ψ_m, as defined in Eq. (11). This simplified model can easily be implemented by using physical components, such as inductors, resistors and voltage sources. Furthermore, phase open-circuit can be directly incorporated into the model. The TNPC inverter is constructed using the MOSFET (ideal, switching) component, which is available in Simscape library of the Simulink software.

$$\begin{aligned} e_\mathrm{L1} &= -\omega_e \cdot \Psi_\mathrm{m} \cdot \sin(\theta_e) \\ e_\mathrm{L2} &= -\omega_e \cdot \Psi_\mathrm{m} \cdot \sin(\theta_e - 2\pi/5) \\ e_\mathrm{L3} &= -\omega_e \cdot \Psi_\mathrm{m} \cdot \sin(\theta_e - 4\pi/5) \\ e_\mathrm{L4} &= -\omega_e \cdot \Psi_\mathrm{m} \cdot \sin(\theta_e + 4\pi/5) \\ e_\mathrm{L5} &= -\omega_e \cdot \Psi_\mathrm{m} \cdot \sin(\theta_e + 2\pi/5) \end{aligned} \quad (11)$$

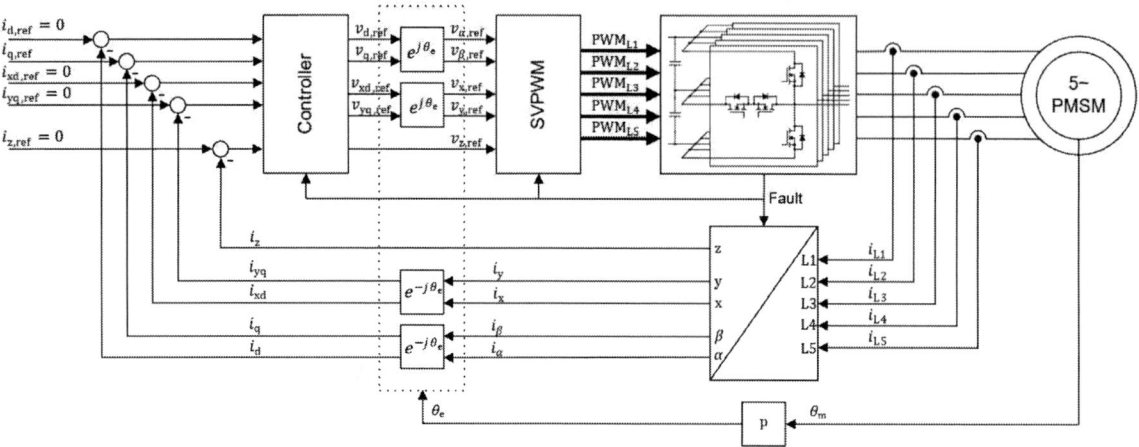

Fig. 8: Control structure of the inverter system

4.2 Simulation Results

The simulation is conducted using the parameters outlined in table 3. The motor speed is set to 1000 rpm with a load torque of 100 Nm. Figure 9 shows the phase currents and generated torque under normal operation. It is observed that the torque ripple is 2.39 Nm, and the phase currents are sinusoidal with the RMS value of 40.48 A.

Parameter	Value
Number of pole-pairs p	8
Permanent magnet flux linkage Ψ_{m}	0.08732 Vs
Stator phase resistance R_{s}	10 mΩ
Stator phase inductance L_{s}	0.243 mH
DC-Link capacitance C_1, C_2	2200 μF
DC-Link voltage V_{DC}	400 V
Switching frequency f_{s}	40 kHz

Tab. 3: Simulation parameters of the PMSM

The simulated results of phase L1 connected to the neutral point (NP) can be observed in Fig. 10. The simulation starts with the inverter operating in normal operation. At the time $t = 0.05$ s, the inverter phase-leg L1 is permanently connected to NP by the neutral point switches, transitioning the operating mode to the one mentioned in section 3.1. The phase currents remain sinusoidal, but their amplitudes fluctuate, as shown in Fig. 11, with the RMS current of phase L1 increased to 48.29 A. This fluctuation is due to the inverter losing its ability to balance the voltages in the

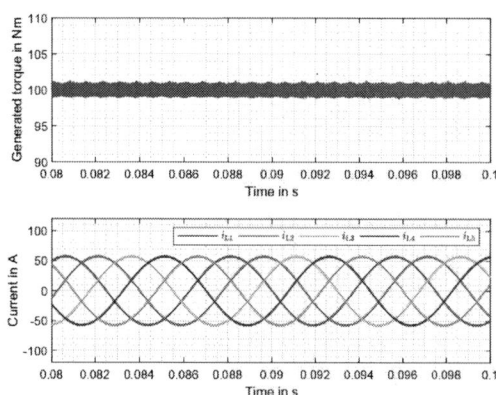

Fig. 9: Generated torque and phase currents in normal operation

DC-Link capacitors, resulting in a 39.45 V variation in the voltages of the upper and lower capacitors, as depicted in Fig. 14. Consequently, the torque ripple also increases to 4.87 Nm.

The open-phase fault is simulated by adding a variable resistor to phase L1. Its initial resistance is set to zero and is changed to 10 MΩ at the time $t = 0.03$ s. The fault-tolerant operation of the inverter is activated at $t = 0.045$ s. The transition from normal operation, through fault until post-fault operation is illustrated in Fig. 12. After the open-circuit occurs, the current in phase L1 drops to zero and the remaining phase currents become unstable, causing an increase in torque ripple. Once the modified modulation is activated, the torque and current return to normal. The generated torque and phase currents are clearly depicted in Fig. 13. The remaining phase currents have to be increased to achieve the same amount of torque. The simulation result shows the RMS current of phase L2 to be 55.97 A, which corresponds to a 38.26 % increase compared to the current in normal operation. This result is comparable to the fault-tolerant method implemented in [16] using hysteresis current controllers. Because i_z cannot be compensated completely, the torque ripple is increased to 5.02 Nm in this operating mode.

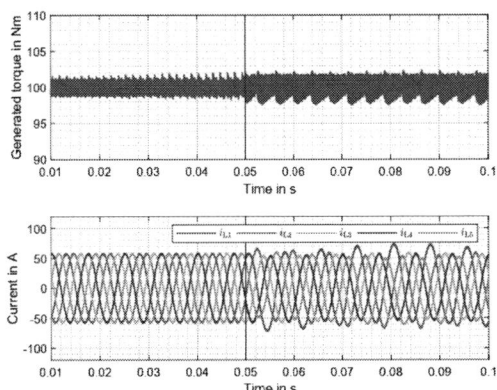

Fig. 10: Transition between normal operation and post-fault operation when L1 is connected to NP

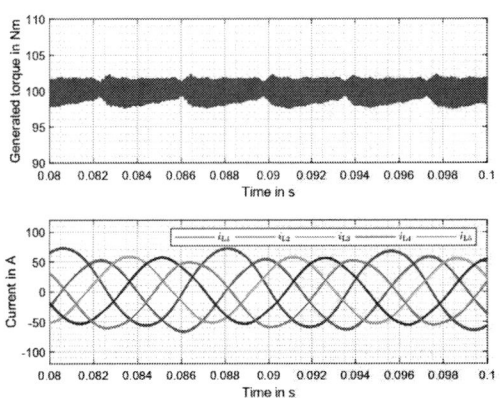

Fig. 11: Generated torque and phase currents when phase L1 is connected to NP

All the simulation results are summarized in table 4 for comparison among different inverter operations in terms of torque ripples, RMS currents, maximum modulation index and DC-Link voltage variations. It is evident that the operation with the phase connected to NP offers an advantage over the

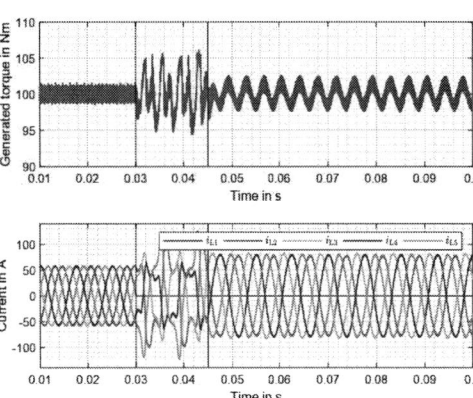

Fig. 12: Transition between normal operation and post-fault operation when L1 is open

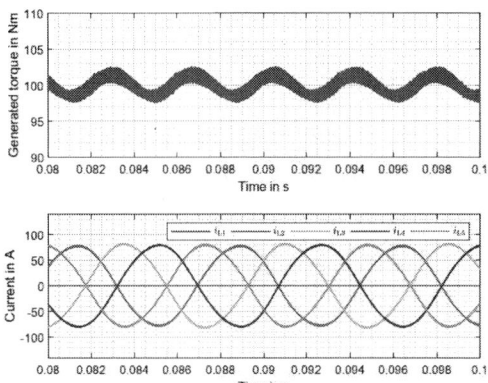

Fig. 13: Generated torque and phase currents when phase L1 is open

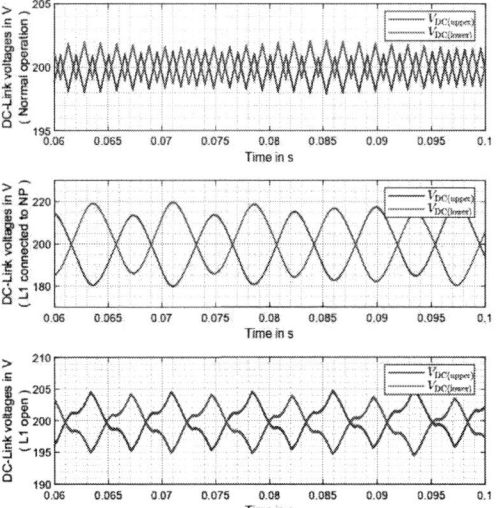

Fig. 14: Comparison between DC-Link voltages in different operations

operation with phase open in that all the currents remain available for torque production, with only a slight increase in RMS currents, in exchange for limitations in modulation index and greater variation in DC-Link voltages. On the contrary, the operation with phase open provides a higher modulation index, but the increase in the current may eventually be restricted by thermal constraints of the used power semiconductors.

Mode	T_{ripple}	$I_{\text{ph}}/I_{\text{N}}$	m_{max}	$\Delta V_{\text{DC}}/V_{\text{DC}}$
Normal	2.39 %	1	0.5256	0.91 %
L1 to NP	4.87 %	1.19	0.2628	9.86 %
L1 open	5.02 %	1.38	0.3618	2.45 %

Tab. 4: Summary of the simulation results

5 Conclusion

Fault-tolerant operations of the TNPC inverter have been tested and validated through simulation. The results demonstrate that despite its operational constraints, the TNPC with additional disconnecting devices exhibits robust fault tolerance for both short-circuit and open-circuit of power semiconductors in an inverter phase. If the TNPC phase is connected to the neutral point, it can sustain operation at full current rating but with reduced maximum voltage, which may impact the speed of the electric machine within the base speed range. The inverter itself also suffers from larger variation in the DC-Link voltages. Conversely, open-phase operation achieves higher maximum voltage and less neutral-point fluctuations. This operating mode, in the absence of one phase, carries more current to maintain the same torque level. Although single phase failures are securely handled, it is important to note that one single fault, such as short-circuit, may lead to additional faults. The inverter's ability to function under multiple fault scenarios will be addressed by experimental validation in the future activities of this research.

6 Acknowledgment

This research as part of the project ELAPSED is funded by dtec.bw – Digitalization and Technology Research Center of the Bundeswehr. dtec.bw is funded by the European Union – NextGenerationEU.

References

[1] E. Levi, "Multiphase electric machines for variable-speed applications," *IEEE Transactions on Industrial Electronics*, vol. 55, no. 5, pp. 1893–1909, 2008. DOI: 10.1109/TIE.2008.918488.

[2] A. Salem and M. Narimani, "A review on multiphase drives for automotive traction applications," *IEEE Transactions on Transportation Electrification*, vol. 5, no. 4, pp. 1329–1348, 2019. DOI: 10.1109/TTE.2019.2956355.

[3] D. Albán, G. Dajaku, and D. Gerling, "Evaluation of flux-barrier stator in five-phase pmsms for electric aircraft traction," in *2022 25th International Conference on Electrical Machines and Systems (ICEMS)*, 2022, pp. 1–6. DOI: 10.1109/ICEMS56177.2022.9983393.

[4] M. Schweizer, T. Friedli, and J. W. Kolar, "Comparative evaluation of advanced three-phase three-level inverter/converter topologies against two-level systems," *IEEE Transactions on Industrial Electronics*, vol. 60, no. 12, pp. 5515–5527, 2013. DOI: 10.1109/TIE.2012.2233698.

[5] R. Teichmann and S. Bernet, "Three-level topologies for low voltage power converters in drives, traction and utility applications," in *38th IAS Annual Meeting on Conference Record of the Industry Applications Conference, 2003.*, vol. 1, 2003, 160–167 vol.1. DOI: 10.1109/IAS.2003.1257498.

[6] L. Gao and J. E. Fletcher, "A space vector switching strategy for three-level five-phase inverter drives," *IEEE Transactions on Industrial Electronics*, vol. 57, no. 7, pp. 2332–2343, 2010. DOI: 10.1109/TIE.2009.2033087.

[7] Y. Zhao and T. Lipo, "Space vector pwm control of dual three-phase induction machine using vector space decomposition," *IEEE Transactions on Industry Applications*, vol. 31, no. 5, pp. 1100–1109, 1995. DOI: 10.1109/28.464525.

[8] O. Dordevic, M. Jones, and E. Levi, "A comparison of pwm techniques for three-level five-phase voltage source inverters," in *Proceedings of the 2011 14th European Conference on Power Electronics and Applications*, 2011, pp. 1–10.

[9] O. Dordevic, M. Jones, and E. Levi, "A comparison of carrier-based and space vector pwm techniques for three-level five-phase voltage source inverters," *IEEE Transactions on Industrial Informatics*, vol. 9, no. 2, pp. 609–619, 2013. DOI: 10.1109/TII.2012.2220553.

[10] R. Wu, F. Blaabjerg, H. Wang, M. Liserre, and F. Iannuzzo, "Catastrophic failure and fault-tolerant design of igbt power electronic converters - an overview," in *IECON 2013 - 39th Annual Conference of the IEEE Industrial Electronics Society*,

2013, pp. 507–513. DOI: 10.1109/IECON.2013.6699187.

[11] S. Bolognani, M. Zordan, and M. Zigliotto, "Experimental fault-tolerant control of a pmsm drive," *IEEE Transactions on Industrial Electronics*, vol. 47, no. 5, pp. 1134–1141, 2000. DOI: 10.1109/41.873223.

[12] Autoliv, *Battery disconnection switch – pss-1: Pyro-switch – automotive switch – high current device*, PSS-1 datasheet, Rev. 013, Feb. 2022.

[13] G. Liu, L. Qu, W. Zhao, Q. Chen, and Y. Xie, "Comparison of two svpwm control strategies of five-phase fault-tolerant permanent-magnet motor," *IEEE Transactions on Power Electronics*, vol. 31, no. 9, pp. 6621–6630, 2016. DOI: 10.1109/TPEL.2015.2499211.

[14] Q. Chen, L. Gu, Z. Lin, and G. Liu, "Extension of space-vector-signal-injection-based mtpa control into svpwm fault-tolerant operation for five-phase ipmsm," *IEEE Transactions on Industrial Electronics*, vol. 67, no. 9, pp. 7321–7333, 2020. DOI: 10.1109/TIE.2019.2944066.

[15] A. Bojoi, S. Ferrari, P. Pescetto, and G. Pellegrino, "Advanced circuital model for e-drive simulation, including harmonic effects and fault scenarios," in *PCIM Europe 2023; International Exhibition and Conference for Power Electronics, Intelligent Motion, Renewable Energy and Energy Management*, 2023, pp. 1–10. DOI: 10.30420/566091220.

[16] L. Parsa and H. Toliyat, "Fault-tolerant five-phase permanent magnet motor drives," in *Conference Record of the 2004 IEEE Industry Applications Conference, 2004. 39th IAS Annual Meeting.*, vol. 2, 2004, 1048–1054 vol.2. DOI: 10.1109/IAS.2004.1348542.

PCIM Europe 2024, 11– 13 June 2024, Nuremberg   DOI: 10.30420/566262436

CCM Totem-Pole PFC for Ultra-High Power Density USB-PD Chargers

David Meneses[1], Manuel Escudero[2] , Matteo-Alessandro Kutschak[2]
[1] Infineon Technologies Nordic AB Filial, Finland
[2] Infineon Technologies Austria AG, Austria

Corresponding author: David Meneses, david.menesesherrera@infineon.com
Speaker: Manuel Escudero, manuel.escuderorodriguez@infineon.com

Abstract

This paper presents a CCM totem-pole PFC to be used in USB-PD applications. The implementation of CCM instead of the traditional CrCM for this power range, simplifies the design and reduces the size of the EMI filter, enabling power density close to 90 W/in³ with a peak efficiency of 98.6%. On the contrary, the possibility of ZVS operation is considerably reduced compared to CrCM. This paper analyzes both aspects together with a robust implementation of bootstrapping for the GaN transistors, which considerably simplifies the driving and bias concept.

1   Introduction

The significant increase during the last years of the number of portable devices, such as mobile phones or laptops, drives the need for a universal power adapter for all of those devices which will not only improve the user experience, but ultimately also reduce electronic waste. These power adapters for portable devices are required to fulfill diverse standards covering stand-by power, power factor, harmonic distortion or electro-magnetic compatibility [1]. In addition, the increase in the voltage and power levels with standards like USB-PD [2], demand an increase in efficiency and power density.

For the 240 W output power (48 V / 5 A) defined by the USB-PD Extended Power Range (EPR), a multistage architecture with an active power factor corrector (PFC) and one or more DC-DC converters are typically used [3,4,5] to provide high performance input current, isolation and adapt to the wide input and output voltage range.

For the mid-power range, a classical boost PFC converter is typically operated in critical conduction mode (CrCM) or mixed mode [6], in which discontinuous conduction mode (DCM) is utilized to avoid excessive increase of switching frequency at light load. However, the variable switching frequency used in such an approach complicates the input filter design because of the need to comply with electromagnetic compatibility (EMC) standards.

A boost in performance of the classic boost PFC can be achieved by substituting the diode bridge with an active bridge [7]. Another alternative is the bridgeless totem-pole topology (Fig 1), which is simple and offers a reduced part count and full utilization of the PFC inductor and switches [8].

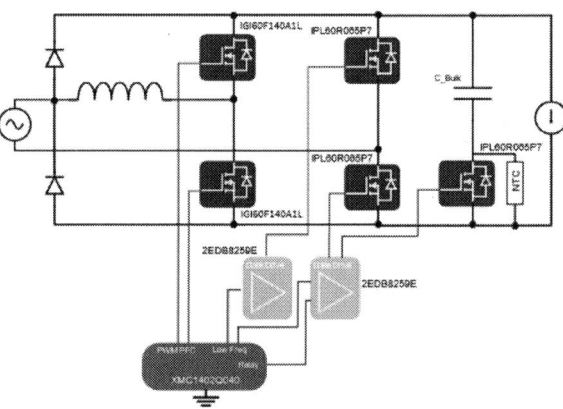

Fig. 1   Simplified schematic of the 240 W CCM Totem-pole design with Infineon components.

In contrast to the vastly used CrCM designs, this paper presents a bridgeless totem-pole PFC, with Gallium Nitride Gate Injection High Electron Mobility Transistors (GaN GIT HEMTs) integrated with its driver, operated in continuous conduction mode (CCM).

Considering cost and bias simplicity, the paper explores the implementation of bootstrapping for the GaN GIT HEMTs and its robustness not only in steady state but also in abnormal conditions. The possibility of ZVS operation at light load with the consequent efficiency increase is as well introduced. Before the experimental results, the paper presents a comparison between CCM, CrCM and

3077

triangular current mode (TCM) [9] in terms of EMI filter design and demonstrates that by using a fixed switching frequency (70 kHz), a reduction on the EMC filtering effort is possible. As a result, an outstanding power density of 87 W/in^3, with a peak efficiency of 98.6 % at 230 V and nominal output power is achieved.

2 PWM sequence for safe operation of CoolGaN™ IPS with bootstrapping

In the case of a first startup or a soft-start operation after the PFC is turned off due to abnormal conditions (voltage sag, line cycle drop out – LCDO, the PFC will always start at a negative AC cycle (in respect to the control ground). The reason behind this strategy is the bootstrap circuit of the low-frequency half bridge. If the converter waits for the proper AC polarity, the low-side switch of the return path can be switched on first and the bootstrap capacitor will have the required energy for the following AC half cycle. For the same reason, the low-frequency (LF) switches keep operation during an over-voltage protection (OVP) event (Fig 2).

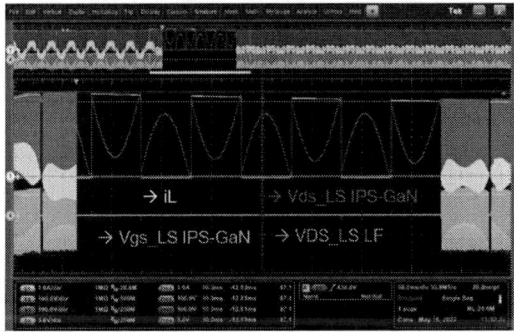

Fig. 2 Low-frequency pulses are applied during an over-voltage event (OVP).

On the operation of the high-frequency half-bridge, where a unipolar supply and bootstrap are being selected as supplying strategy for the high-side CoolGaN™ IPS, the microcontroller ensures that the high-frequency PWM always starts with the low-side switch. This enables the charging of the bootstrap capacitor to provide the following high-side switch pulse.

In addition, it must be noted that a change in the switching node on either half bridge (low-frequency - LF or high-frequency - HF) after the PWM was disabled affects the other switching node through the input capacitor and inductor network. Given the low voltage threshold of GaN devices and their Cgd-Cgs ratio, a mis-trigger of the device could occur and a shoot-through event could have

fatal consequences for the converter, especially if negative voltage driving is not used, as in the presented solution.

With these considerations, it is thus important to understand the proper sequence between the low- and high-frequency half bridges to avoid commutation issues. This sequence, which occurs after the PWM signals have been disabled at any condition, will be introduced in this section using the AC zero-crossing as an example, in which both HF and LF signals are momentarily disabled.

After a positive AC cycle in which the high-side (HS) switch of the LF half bridge was on, and the HS switch in the HF branch was acting as a boost switch, both switching nodes remain at a voltage close to the bulk voltage when the driving signals are disabled at AC zero crossing. In that case, starting the switching of the HF branch when the negative AC cycle starts would get the switching node to ground at a high dV/dt. As a consequence, the HS drain-source voltage moves from almost ground to the bulk voltage with the same dV/dt with a potential risk of short-circuit through the Cgd-Cgs feedback.

Therefore, in this case, the LF half bridge is commutated first. By switching on the LS MOSFET in the LF half bridge, the switching node of the HF branch is moved to ground from almost the bulk voltage. It can be observed that this change in the switching node induces a change in the LS gate voltage (red arrow on Fig 3), which is attenuated by the PFC inductor and the lower dv/dt motivated by high Rg,on in the LF half-bridge.

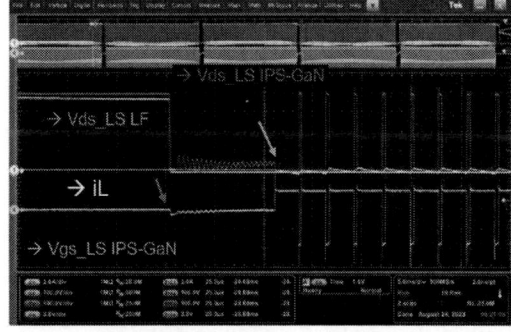

Fig. 3 Resume of operation with positive AC voltage: LF pulses are applied before HF PWM.

An opposite reaction is expected in the high side gate-source voltage, which is not a problem in this case but it could motivate a shoot-through if the LS HF gate source would be also enabled as explained above. After the change in the LF half bridge, the controller waits for a defined time in which the switching node oscillates around the AC voltage. Afterward, the LS switch in the totem-pole starts operation as a boost switch (pink arrow). At this point, the Cgd capacitance has been considerably reduced, and the amount of charge is also significantly lower due to the reduced change in the drain-source voltage, thus virtually eliminating the possible gate feedback.

When considering the opposite transition for the AC voltage, both of the switching nodes are close to zero after the negative AC half-cycle, because the LS MOSFET in the LF half-bridge was on and the LS of the HF half bridge was acting as a boost switch (long pulse for AC close to zero). Therefore, when the positive AC voltage cycle starts, the LS CoolGaN™ IPS can be switched on (red arrow on Fig 4) without triggering feedback in the HS device due to the very low voltage change. The switching of the HF branch forces the LF switching node to move from almost ground to the bulk voltage and the HS CoolMOS™ MOSFET can be triggered a defined number of switching cycles after this moment.

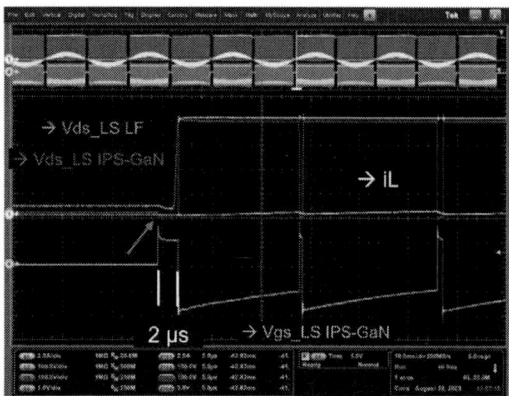

Fig. 4 Resume of operation with negative AC voltage: LF pulses are applied after the HF PWM.

Because during the positive AC cycle (from the controller ground perspective), the LS transistor of the HF branch operates as the boost diode, a very short pulse would be applied during the first cycles due to the low AC voltage. As already mentioned, the HS switch is driven using bootstrap capacitor and therefore, due to these very short pulses, the driver could potentially reach undervoltage lock out (UVLO). For that reason, the first pulse length is fixed to 2 µs in order to strengthen the supply voltage of the high-side driver.

In the case of abnormal conditions, such as OVP or LCDO, the same principle explained above is applied. The main difference is that the time without PWM and/or low-frequency pulses is significantly higher: from some hundreds of microseconds in the case of AC crossing to 20 ms in LCDO or even more in the case of OVP. In these cases, it is possible that the UVLO of the HS drivers is

triggered due to a voltage reduction on the bootstrap capacitor.

Figure 5 shows an example of discontinuous conduction mode (DCM) operation due to the lack of HS pulses for few switching cycles when the PWM resumes after OVP.

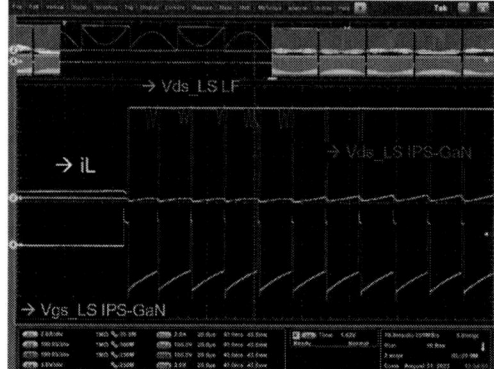

Fig. 5 DCM operation due to HS driver UVLO when the PWM resumes after OVP event.

3 Variable dead-time for natural ZVS in partial load condition

In a classic boost PFC operating in CCM, there are two commutations every switching period at the on and off moments of the boost switch. When the boost switch is turned off, the inductor current flows naturally toward the bulk capacitor through the boost diode. Therefore, zero voltage turn-on (ZVS) is possible if the diode is substituted by a switch as in a totem-pole PFC. However, when the transistor is turned on again, the current is hard-commutated from the diode to the switch.

A different modulation, the variable frequency triangular current mode (TCM) [9], is applied to a synchronous boost converter to obtain ZVS for both on and off transitions of the boost switch. With this modulation, ZVS is achieved by allowing the current to be reversed at the diode-to-switch transition. In CCM operation, this scenario is possible depending on the load conditions and the ripple current for a given AC input voltage.

For CCM totem-pole operation, the applied dead-time between the boost diode and boost switch is as short as possible in order to reduce diode conduction before the hard commutation occurs. Therefore, in those conditions in which the current reverts, this dead-time can be prolonged to enable resonance between the inductor and the Coss capacitors of the GaN HEMT implementing the totem-pole.

In the case of a PFC operated in CCM, the average inductor current depends on the AC voltage and load conditions. Therefore, the peak and valley inductor current depend also on those parameters as well as the switching frequency, output voltage, and inductance value. Then, it can be calculated at which power the valley current is below zero for a given angle of the AC voltage (1).

$$P_\alpha = \frac{V_{RMS}^2}{2 \cdot L} \cdot \left(1 - \frac{\sqrt{2} \cdot V_{RMS} \cdot \sin(\alpha)}{V_{DC}}\right) \cdot \frac{1}{f_{sw}} \quad (1)$$

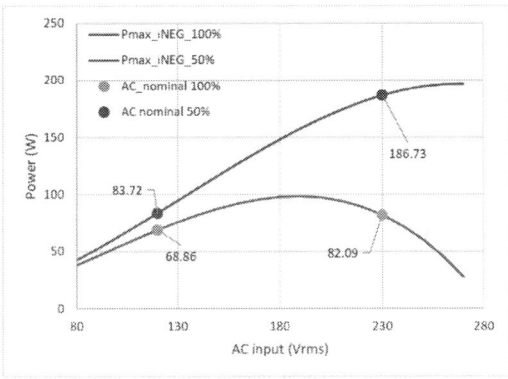

a)

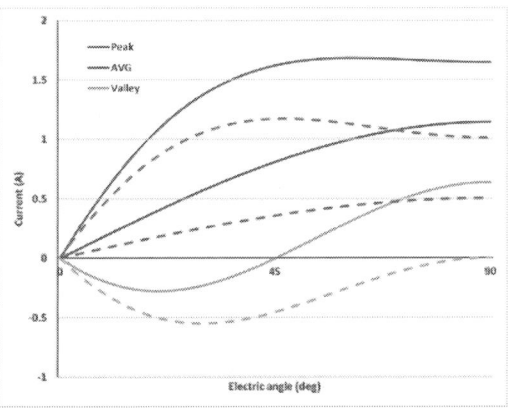

b)

Fig. 6 a) Output power to achieve reversed current and b) current envelopes for 230 V for 82 W and 186 W

The blue curve in Fig 6.a presents this calculation considering that the valley current stays reversed for the whole AC cycle (α = 90°), which would enable ZVS operation for the whole AC cycle. In that case, ZVS is possible for power levels under 69 W at 100 V AC and under 82 W at 230 V AC. When the output power is under those values, a longer dead-time should be applied to the diode-to-switch transition to enable resonance while a short dead-time is applied for higher power levels to minimize the diode conduction time. Note that full ZVS will be obtained when the energy in the inductor is higher than Qoss*Vbulk, otherwise partial ZVS is expected.

However, in high-line AC voltages, due to the lower current levels, it is easier to achieve the reversed

current condition and prevent the degradation in performance due to diode conduction. This opens the possibility to apply a longer dead-time until a power level in which the inductor valley current stays reversed for a percentage of the AC cycle. The red curve in Fig 6.a shows the power at which the inductor current reverts for half of the AC cycle (α = 45°).

Figure 6.b shows the inductor average current (red) as well as the peak (blue) and valley (green) envelopes for 230 V when the output power is 82 W (dashed lines), which corresponds to the condition in which the current changes polarity for the whole AC cycle, and for 186 W (solid lines), which corresponds to the scenario in which the current reverts during half of the AC cycle.

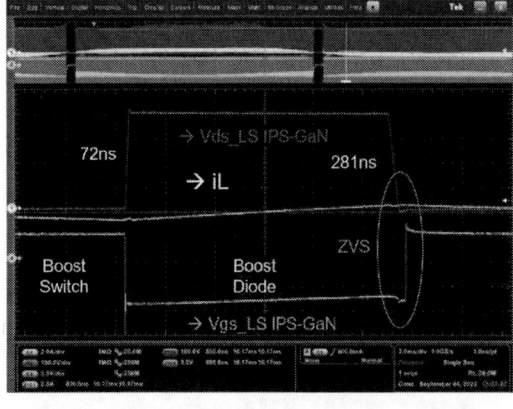

a)

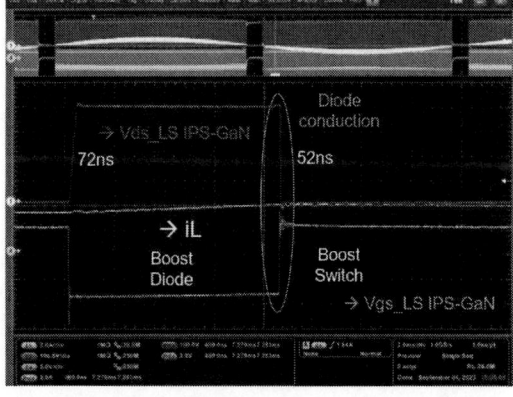

b)

Fig. 7 Switching waveforms at 140 W and a) 230 V and b) 100 V for a similar AC angle.

Switching waveforms are shown in Fig 7 to support this explanation. In this case, the inductor current and low-side GaN HEMT voltages are introduced for a similar angle in the AC with 140 W output power. At 230 V (Fig 7.a) it is clear that by applying a longer dead-time during the diode-to-switch transition (281 ns), it is possible to obtain ZVS operation due to the reversed inductor current. At 100 V (Fig 7.b) and same output power, however, because the current is not reversed, a short dead-time (52 ns) is used to minimize the diode conduction time of GaN HEMT.

The increase in efficiency at light load by applying this variable dead-time to enable ZVS operation when the current is reversed is shown in Fig 8. The measured efficiency when applying this strategy is shown in solid lines. The dashed lines show the measured efficiency when a short dead-time is applied as in the classical control approach.

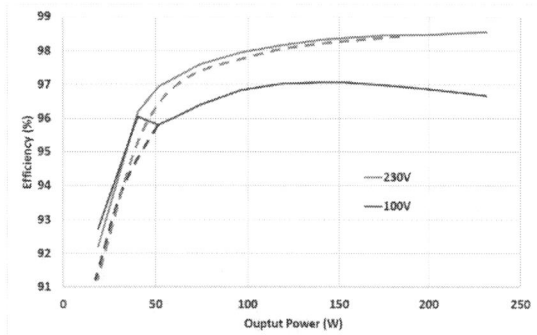

Fig. 8 Efficiency difference with (solid) and without (dashed) variable diode-to-switch deadtime.

4 EMI comparison between CCM and CrCM-TCM modulations

For the power range targeted by USB-PD chargers, typically critical conduction mode (CrCM) is used because no current sensing is required and therefore both the cost and complexity of the solution are reduced. Furthermore, the advantage in performance can be achieved by taking advantage of the soft-switching or valley-switching operation. However, the variable switching frequency in the system, together with an eventual change in mode of operation between critical and discontinuous (DCM) conduction mode depending on the AC voltage and load conditions, considerably increase the design effort as well as the size of the EMI filter. On the other hand, if CCM is chosen, despite the hard-switching operation, the fixed-frequency operation simplifies the EMI design.

This section compares a differential-mode EMI spectrum between both modes of operation. The EMI emissions have been modeled in MATLAB based on the inductor current ripple following the methodology described in [7] and including the input capacitor of the totem-pole PFC

4.1 EMI spectrum comparison

As a first step, inductor current ripple and switching frequency variation are computed for the three considered modulations: CCM, CrCM and TCM. TCM is included in the comparison because it could offer advantages in performance due to its capability of soft-switching operation in all the AC and load conditions. Four different RMS voltages are considered in the analysis (90 V, 115 V, 230 V and 265 V) at full load, since at the 240W nominal output power is where the inductor shows the highest ripple with lowest frequency range for the variable switching frequency modulations.

In the case of CCM, 70 kHz is considered as switching frequency as an optimal balance between performance and EMI. The PFC choke is implemented with a CH234060GT14 core and 105 turns (880 µH at zero amps) and its inductance variation with current has been considered to calculate the inductor current ripple.

In the case of CrCM or TCM, the current ripple is given by the converter operation. In the case of CrCM, because the inductor valley current is set to zero, the peak current must be two times the desired average current. This restriction, together with the inductance value set in the design, determines the switching frequency. The same applies for TCM, but the peak envelope of the inductor current must consider the reversed current used to achieve full ZVS operation.

The inductance for the CrCM and TCM is designed to provide a minimum switching frequency at full load (72 kHz) similar to the CCM switching frequency. Figure 9 shows the current ripple and switching frequency for CrCM at full-load operation with a fixed inductance of 120 µH. In the case of TCM, because the current ripple is higher due to the reversed current to obtain soft switching, the switching frequency is lower compared to CrCM. Therefore, the PFC inductance has been reduced to 60 µH, for -1 A as reversed current for ZVS operation.

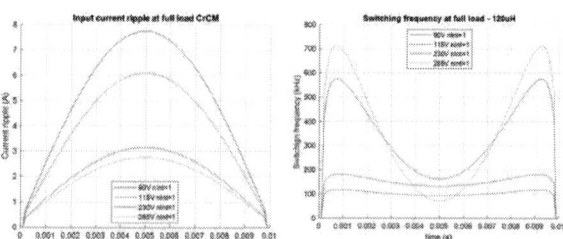

Fig. 9 Inductor current ripple and switching frequency for CrCM PFC with 120 µH inductance.

The three designs shown above are then compared from their EMI emissions, which have been calculated according to the methodology proposed in [7]. As example of the estimated EMI spectrum,

Figure 10 shows the EMI emission for 230 V. It must be noted that the input capacitor of the totem-pole PFC is included in the calculations as an attenuation of the inductor current ripple. Due to the higher ripple in CrCM and TCM designs, a higher capacitance is in place for those designs: 820 nF in respect to 470 nF for CCM.

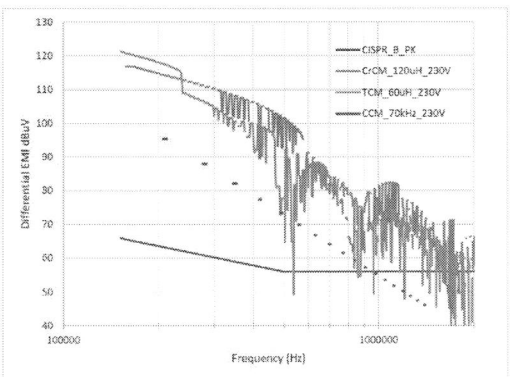

Fig. 10 Differential mode EMI estimation for CCM (red), CrCM (orange) and TCM (grey) at 230 V and 240 W.

Even though the CCM design implements the lowest input capacitance in front of the PFC choke, based on the estimated differential mode EMI, it is obvious that CCM presents the lowest emissions at a higher critical frequency. In this case, the first harmonic in the spectrum corresponds to the third harmonic of the switching frequency (210 kHz).

According to the estimations, for the worst case of the four AC voltage considered, the CCM design needs an attenuation of 38 dBµV at 210 kHz. In comparison, the CrCM design with 120 µH inductor would require 63 dBµV at 150 kHz, For the TCM design with 60 µH estimation is 61 dBµV at 150 kHz. Therefore, it is possible to reduce the EMI for CCM operation with respect to a higher switching frequency design in CrCM or TCM, as shown in the following section.

4.2 EMI filter design

Based on the previously presented estimation of required attenuation at the critical frequency obtained by modelling differential mode (DM) EMI emissions, a differential mode EMI filter has been designed for the three considered operating modes.

In the case of 70 kHz CCM operation, the asymmetric two-stage EMI filter design is shown in Fig 11. The three X capacitors are 470 nF each. On the PFC side, a 90 µH differential mode choke is implemented while the grid side of the filter implements a common-mode choke with a stray induct-

ance of 6 µH that provides differential mode filtering. A classical, and more conservative, approach is to design the filter with resonances at lower frequencies than the switching frequency. However, the asymmetric approach used in this design enables a size reduction with respect to the classic one and allows achieving the required attenuation (-38 dBµV – dashed horizontal blue line) at the critical frequency (210 kHz – dashed vertical orange line) and attenuates the switching frequency harmonic of the inductor current (70 kHz – vertical green line).

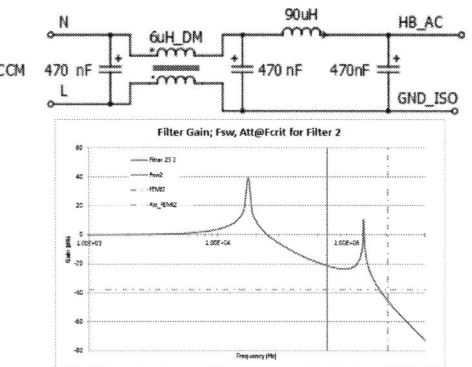

Fig. 11 DM EMI filter design for CCM operation.

For the CrCM and TCM designs, their differential-mode filter designs are presented respectively in Fig 12 and Fig 13. As already introduced, due to higher inductor ripple, the input capacitor for these modulations is 820 nF instead of 470 nF, as for CCM. Furthermore, the other two X capacitors are 560 nF instead of 470 nF. This increased total capacitance has not only an adverse effect in size but also jeopardizes the achievable power factor.

To have a fair and intuitive comparison, the same asymmetric approach with the same common-mode choke on the AC side and the same 90 µH differential-mode choke in the totem-pole side implemented in the CCM filter are kept for these two designs. With that consideration, an extra differential-mode choke is required for both CrCM (54 µH) and TCM (44 µH). It must be noted that a different design approach for CrCM and TCM could be followed. However, note that the filter resonances must be at a lower frequency than the minimum predicted switching frequency (green vertical line) to avoid unwanted interactions and oscillations in the input current.

It can be then concluded that despite the possibility of increasing the switching frequency with a lower inductance (possible reduction of size of the inductor) offered by the soft-switching operation of CrCM and TCM, the power density for those designs is still significantly limited by the EMI filter design.

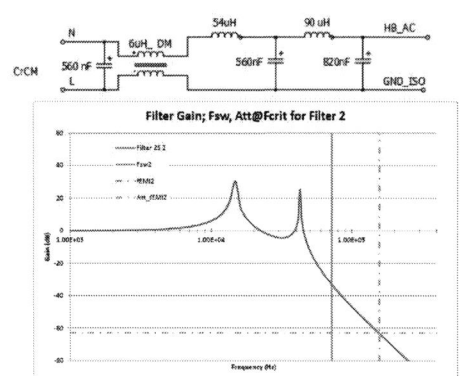

Fig. 12 DM EMI filter design for CrCM operation

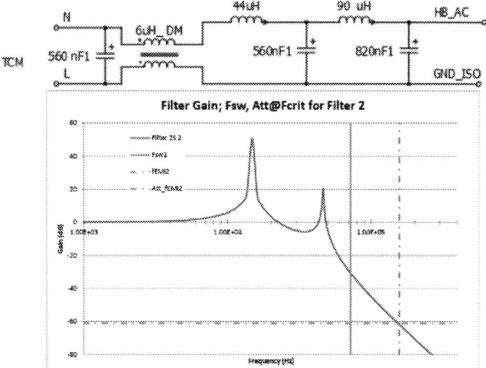

Fig. 13 DM EMI filter design for TCM operation

5 Experimental results

A 240 W CCM totem-pole converter, at 70 kHz, has been built. This reference design for USB-PD applications has a power density of 87 W/in³ (Fig 14). This universal input PFC has a 400 V DC link output and a hold-up time capability of 10 ms (120 µF). The PFC choke is implemented with a CH234060GT core and 105 turns (880 µH).

Fig. 14 Prototype of the 240 W CCM totem-pole PFC with CoolGaN™ IPS

The efficiency plots with 140 mΩ GaN GIT HEMT integrated with driver in the HF half-bridge, and 65 mΩ super-junction MOSFET for the LF half-bridge, for different AC voltages are shown in Fig 15. An outstanding efficiency of 98.6% is obtained

at 230 V input and full load. For the low-line, 97.5% efficiency is obtained at 120 V AC input. The abrupt change in efficiency at 120 V is due to the variable dead-time applied at light load operation, as presented in Section 3. Regarding input current performance, power factor is close to unity and THD under 4% is achievable at full load regardless the AC input voltage (Fig 16).

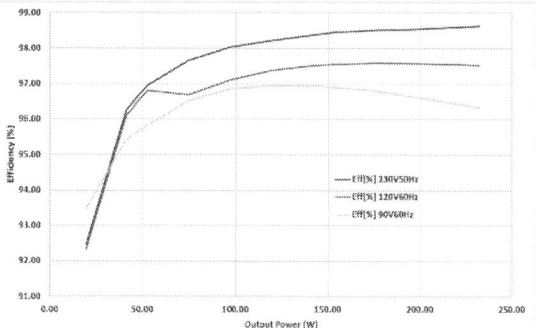

Fig. 15 Efficiency variation with load of the 240 W CCM totem-pole PFC.

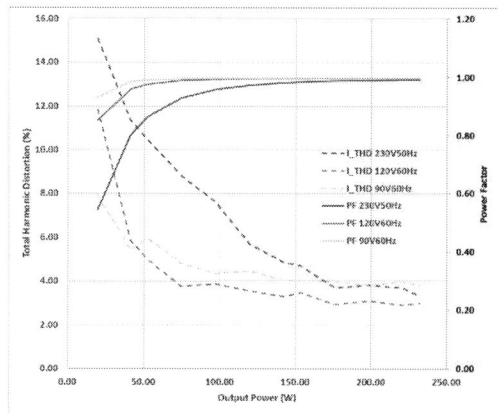

Fig. 16 THD and PF measurements for the 240 W CCM totem-pole PFC.

Despite hold-up time is not typically required for USB-PD or other battery charging applications, this 240 W CCM totem-pole has been tested for full load 10 ms and half-load 20 ms line cycle dropout (LCDO), and for different starting angles of the AC loss. The objective is to test the robustness of the selected driving scheme and sequence introduced in section 2. Fig 17 presents details of the 7th and 8th repetition for 10 times 10 ms LCDO test with 45 degree starting angle. As it can be concluded, together with the burst operation due to OVP presented in section 2, the operation of the totem-pole with GaN GIT HEMTS with unipolar and bootstrapping driving scheme is robust with the appropriate PWM sequence.

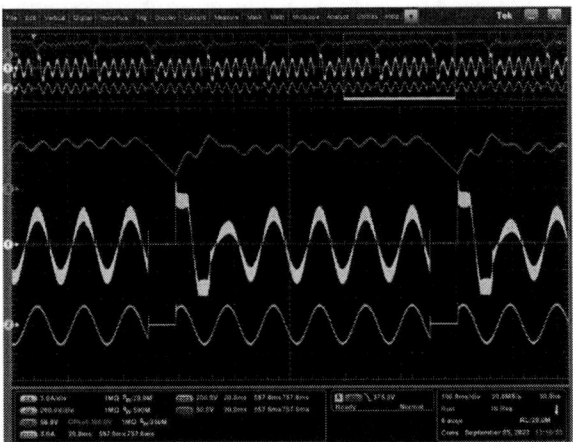

Fig. 17 Detail of the 7th and 8th 10 ms 45 degrees LCDO repetition for 100 V and 60 Hz at full load operation.

5.1.1 EMI measurement

The EMI filter resulted from the analysis presented in section 4 has been implemented in the reference design. Given the reduced input current, all the components to implement the filter in Fig 11 have been selected off-the-shelf. The EMI measurements conducted at 230 V and 240 W using resistive load, are presented in Fig 18. As it can be seen, the peak and average measurements are clearly under the CISPR 22 Class B limits.

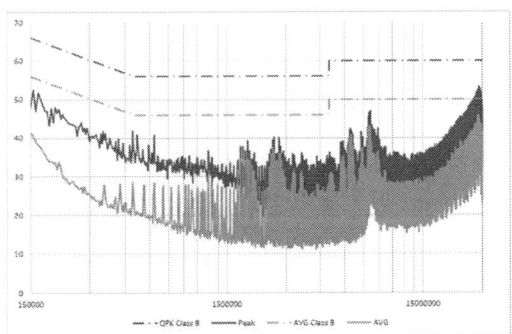

Fig. 18 Peak (blue) and average (orange) EMI measurement.

5.1.2 Thermal behaviour

The thermal behavior of the 240 W CCM totem-pole with integrated GaN GIT HEMTs has been tested with two different configurations. On one side, the PFC choke has been moved out from the main board to have access to the totem-pole HF switches (Fig 19). On the other side, the final form factor of the reference design has been recorded thermally (Fig 20). It must be noted that no external fan, neither heatsink were used in neither of the test to obtain the thermal captures.

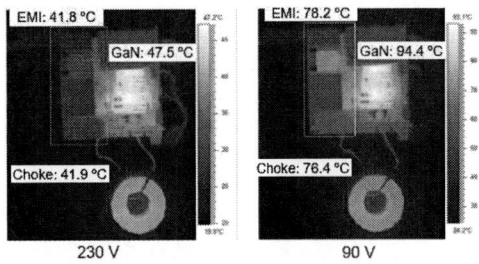

Fig. 19 Thermal measurements for 230V and 90V with the choke outside the main board.

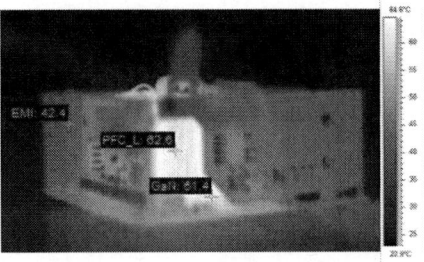

Fig. 20 Thermal measurement for 230 V with the ultra high power density form factor.

By comparing the thermal capture at 230 V when the choke is inside (Fig 20) and outside (Fig 19) the main board, it becomes evident that there is a thermal coupling between the choke and the GaN GIT HEMTs. Similarly, for low line operation, the GaN GIT HEMTs temperature considerably rises, even for open frame operation. Therefore, a proper cooling concept will be necessary for the final application.

5.1.3 Extended output power and efficiency with DCDC converter

If the proper cooling system is place when operating the proposed converter, it would be possible to operate the totem-pole with a maximum input current of 2.1 A, which is the measured input current at 120 V operation. In that case, it would be possible to get up to 480 W at 230 V operation, as shown in Fig 21. The presented results were obtained with the choke outside the main board as can be appreciated in the thermal capture for 230 V and 480 W.

The proposed CCM totem-pole PFC has been tested together with the hybrid-flyback DCDC converter presented in [5]. The presented efficiency measurements in Fig 22 were taken, for different AC voltages, at full load (240 W) and 48 V output voltage.

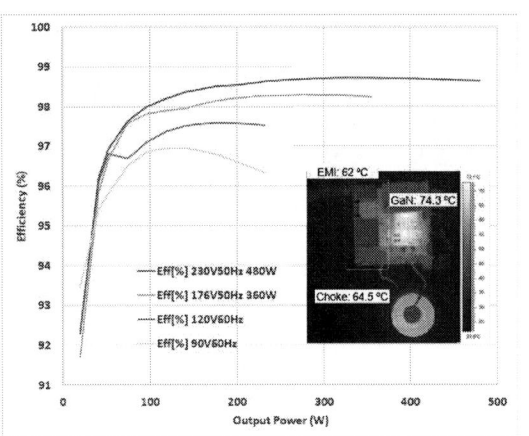

Fig. 21 Efficiency measurements for different AC voltages with 2.1 A maximum input current.

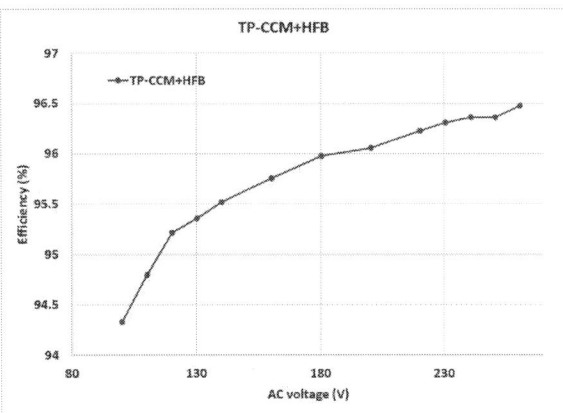

Fig. 22 Efficiency measurements of the proposed CCM totem-pole with the DCDC presented in [5] for different AC voltages and 48 V – 240 W output operation.

6 Conclusions

This paper presents a 240 W totem-pole PFC operated in CCM for USB-PD applications, which achieves a power density of 87 W/in^3 and a peak efficiency of 98.6%. Unipolar supply and bootstraping is selected for driving not only the line rectifiers but also the high-frequency half-brige, which is implemented with CoolGaN™ Integrated Power Stage. The paper introduces a sequence for robust operation of both half-bridges in steady-state and abnormal conditions. The outstanding power density of the presented converter is achieved by operating in CCM at fixed frequency, which enables an EMI filter size reduction compared to the more traditional CrCM operation, or the TCM modulation which enables full ZVS operation. Despite CCM operates with hard-switching at higher loads, the fact that a totem-pole is a rectified boost converter, enables ZVS operation for light load opera-

tion due to the reversed valley current of the inductor. The presented results include thermal captures which recommend the use of a cooling system for the final application. With that system in place, it could be possible to extend the output power for high-line operation, as long as the input current remains under 2.1 A. The totem-pole presented in this paper has been cascaded with a Hybrid-Flyback to show a possible high-performance USB-PD adapter.

References

[1] IEC Electromagnetic compatibility (EMC) - Part 3-2: Limits - Limits for harmonic current emissions (equipment input current <= 16 A per phase), International Electrotechnical Commission, IEC 61000-3-2:2014.

[2] www.usb.org. Universal Serial Bus Power De-livery Specification, revision 3.1, 26.May.2021.

[3] J. Weimer, D. Koch, M. Nitzsche, J. Haarer, J. Roth-Stielow and I. Kallfass, "Miniaturization and Thermal Design of a 170 W AC/DC Battery Charger Utilizing GaN Power Devices," in IEEE Open Journal of Power Electronics, vol. 3, pp. 13-25, 2022.

[4] M. J. Kasper, J. A. Anderson, G. Deboy, Y. Li, M. Haider and J. W. Kolar, "Next Generation GaN-based Architectures: From 240W USB-C Adapters to 11kW EV On-Board Chargers with Ultra-high Power Density and Wide Output Voltage Range," PCIM Europe 2022; Nuremberg, Germany, 2022, pp. 1-10.

[5] A. Medina-Garcia, M. Krueger, M. Schmid, J. Daimer and M. Schlenk, "Optimum Power Architecture for USB-PD EPR," PCIM Europe 2023; Nuremberg, Germany, 2023, pp. 1-7.

[6] X. Huang, Y. Lei, Y. Zhou, W. Du and J. Zhang, "An Ultra-High Efficiency High Power Density 140W PD3.1 AC-DC Adapter Using GaN Power ICs," 2023 APEC, Orlando, FL, USA, pp. 1252-1258,

[7] "Design of active bridge line rectification for SMPS", AN_1905_PL52_1905_101502, Infineon.

[8] Q. Huang and A. Q. Huang, "Review of GaN totem-pole bridgeless PFC," in CPSS Transactions on Power Electronics and Applications, vol. 2, no. 3, pp. 187-196, Sept. 2017.

[9] M. J. Kasper, J. A. Anderson, S. Weihe and G. Deboy, "Hybrid Fixed/Variable Frequency TCM Average Current Control Method Enabling ZVS MHz Operation of GaN HEMTs in PFC Stages," 2023 APEC, Orlando, FL, USA, pp. 1232-1237.

PCIM Europe 2024, 11– 13 June 2024, Nuremberg   DOI:10.30420/566262437

Comparison of Hybrid Si/SiC and SiC Two-Level and Three-Level Converters for Low-Voltage Low-Power Applications

Tim Augustin[1], Haofeng Bai[2], Muhammad Nawaz[1], Simon Round[3], Peter Steimer[4]

[1] Hitachi Energy Research, Västerås, Sweden
[2] Hitachi Energy, Västerås, Sweden
[3] Hitachi Energy, Turgi, Switzerland
[4] Hitachi Energy Research, Baden-Dättwil, Switzerland

Corresponding author: Tim Augustin, tim.augustin@hitachienergy.com
Speaker: Tim Augustin, tim.augustin@hitachienergy.com

Abstract

This study compares SiC two-level converters with hybrid Si/SiC and full-SiC three-level converter topologies for low-voltage and low-power applications in terms of electrical performance. Analytical loss models are derived for each topology and the loss models are iteratively solved with a thermal equivalent circuit. The results show that full-SiC three-level converters are slightly superior to hybrid Si/SiC equivalents in terms of output power capability and converter efficiency. Moreover, three-level converters do not offer major electrical performance improvements compared to two-level converters if SiC is used. The switching frequencies at which three-level converters are better than two-level converters are shifted to relatively high frequencies for SiC-MOSFETs compared to Si-IGBTs. For 400 V AC systems, two-level SiC converters are the most attractive option. For 690 V AC systems, three-level topologies can not be dismissed because 2300 V SiC chips are not available in the market yet.

1 Introduction

SiC MOSFETs receive significant attention by the power electronic community, mainly owing to their superior switching loss performance compared to Si-IGBTs. Previously, SiC MOSFETs were more of an academic subject, because of high costs. However, SiC chip cost decline consistently and news about manufacturers ramping up production or going for SiC converters become public frequently. Despite the high cost of SiC power modules, the total system cost of SiC converters are already competitive, because of higher efficiency and higher switching frequency. An alternative are hybrid Si/SiC converters using both technologies with an electrical performance similar to full-SiC converters with reduced semiconductor cost [1].

Traditionally, the question in converter design was not Si-IGBT or SiC MOSFET; it was which topology to use. Schweizer [2] compares Si-IGBT two-level (2L), three-level (3L) T-Type, and NPC topologies. Anthon [3] investigates the use of SiC-MOSFETs in T-Type converters. Häring [4] derives analytical loss models for various modulation schemes suitable for hybrid ANPC topologies. Zhang [5] compares hybrid ANPC topologies with two and four SiC-MOSFET positions. Feng [6] compares Si-IGBT, hybrid, and full-SiC ANPC topologies showing that full-SiC ANPC with a modulator paralleling the inner switches in the zero state has the highest efficiency.

Available research mainly focuses on either comparing chip technologies for a specific topology or various topologies for a specific chip technology. The purpose of this article is to systematically compare 2L and 3L topologies using SiC MOSFETs, including hybrid Si/SiC converters, for low-power applications in terms of electrical performance. A consistent set of semiconductors is chosen for comparable topologies and other system parts are kept constant for an overall fair comparison. The apparent power is approximately 100 kV A and both 400 V and 690 V AC systems are considered. The footprint of the power module is kept constant.

The article has the following structure. Section 2 describes the methodology applied. Section 3 explains the converter loss model. Section 4 includes the results. Section 5 discusses the findings. Section 6 concludes the report.

2 Methodology

2.1 Scope

The power electronic system does not only consist of the converter power module, but also, among others, of DC bus capacitors, busbars, AC filter, EMI filter, and cooling system. Losses occur in all system components and considering the relatively high efficiency of the power converter around 99 %, even small loss contributions can have a quite significant impact on the total system efficiency. To illustrate this, the losses of a 100 kW three-phase converter with 99 % efficiency are around 333 W per phase. A more complex converter might improve the efficiency to 99.2 % with 267 W losses per phase. A gate drive unit also exhibits losses, for instance its power supply around 10 W. A more complex converter could require four additional gate drive units per phase increasing the losses per phase by 40 W or 18 %. This simple calculation shows that only an evaluation of the power electronic system will give an accurate assessment. Being aware of the limitations, this article focuses exclusively on the converter topologies and terms like *losses* and *efficiency* only include the power semiconductors of the converter.

2.2 Topologies

This study analyzes converters implementing SiC devices for low-power applications. Hybrid Si/SiC means that each active switch position is either a Si-IGBT with anti-parallel diode or a SiC MOSFET operated with synchronous rectification. Diodes can be either Si or Schottky-SiC material, but this study focuses on the active switches. The study considers 3L T-Type, NPC, and ANPC converters as shown in Fig. 1 and excludes flying capacitor and other multilevel topologies. Each topology is denoted by its topology name and the switch position implemented with SiC MOSFETs. For instance, an ANPC, in which T5 and T6 are SiC MOSFETs and T1-T4 are Si-IGBTs, is referred to as ANPCT5T6. If all switches are SiC MOSFETs, the topology is referred to as full-SiC. The benchmark topologies are a full-SiC 2L converter (2LSiC) and a Si-IGBT 3L T-Type converter from another vendor.

The FlexyPak package from Hitachi Energy is well-suited for the above-mentioned topologies as it offers internal design flexibility. The direct pin connection of the FlexyPak yields a low stray inductance as required for fast-switching SiC-MOSFETs. The full-SiC 2L modules as shown in Fig. 2 have a block-

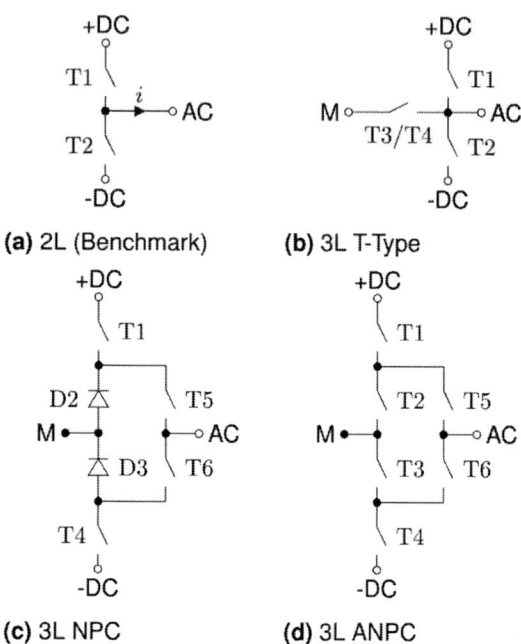

Fig. 1: Converter topologies considered in this study.

ing voltage of 1200 V and a current rating of 110 A and 220 A.

2.3 Harmonic Performance

The harmonic performance of 3L converters is superior to 2L converters at the same switching frequency. In practice, the 3L converters can be operated with a smaller output filter than 2L converters at a given switching frequency or their switching frequency can be reduced with the same output filter. A figure of merit for the harmonic performance is

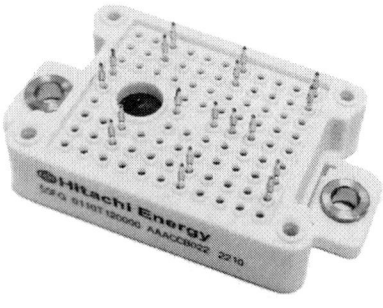

Fig. 2: Hitachi Energy's SiC half-bridge FlexyPak power module with a voltage rating of 1200 V and current rating of 110 A and 220 A.

Tab. 1: Specifications for the topology comparison.

Topology	2L	T-Type	NPC & ANPC
U_{ac}/V	400/690	400	690
U_{dc}/V	750/1500	750	1500
f/Hz		50	
f_{s}/kHz	36		15.3
U_{g}/V		15	
E_{FRV}	-	-	$0.1E_{\mathrm{on}}$
E_{rec}/mJ	1	1	0.5
R_{g}/Ω (SiC)		2	
R_{th}/K W^{-1}		0.055	
T_{amb}/°C		40	
T_{jmax}/°C		150	
ΔT_{m}/K		15	

the normalized weighted total harmonic distortion

$$\mathrm{WTHD0} = \frac{1}{U_1}\Bigg|_{M=1} \sqrt{\sum_{h=2}^{\infty} \frac{U_h^2}{h^2}} \qquad (1)$$

with the modulation index M, the harmonic order h, and the h-th order voltage harmonic U_h. To account for the total costs of the power electronics system, the same output filter is assumed for 2L and 3L converters. For an adequate comparison, the switching frequency of the 3L converters is reduced to yield the same WTHD0 as the 2L converter. Naturally-sampled PWM with triangular carrier was assumed and U_h is calculated according to the formulas included in [7].

2.4 Specifications

The specifications are summarized in Tab. 1 and are used in the model described in Section 3. Two pairs of AC system voltage U_{ac} and DC bus voltage U_{dc} are considered for grid-connected applications with grid frequency f. The same thermal resistance of the heatsink R_{th}, maximum junction temperature T_{jmax}, and safety margin ΔT_{m} are used effectively implying that the cooling requirements are kept constant for the comparison. Only for hybrid T-Type topologies, the gate resistance R_{g} is set according to the datasheet of the manufacturer and not as shown in Tab. 1 as they are the only topologies considered in which the current is actively commutated between Si-IGBTs and SiC-MOSFETs.

2.5 Modulation

For the modulation of the fast-switching devices, sinusoidal naturally-sampled PWM with triangular

carrier and without common-mode injection is used. The modulator is adapted for each topology to benefit the particular topology as much as possible, for instance that only SiC MOSFETs switch fast and that a parallel path is created to reduce the conduction losses, if possible. The devices operating at grid frequency are usually switched at the zeros of the reference.

3 Modelling

3.1 Iterative Solution

The losses in the power semiconductors are temperature-dependent. Thus, the loss model must be coupled with a thermal model. In this study, both analytical models are iteratively solved and the temperature-dependent electrical parameters are updated in every iteration.

3.2 Losses

Average loss models, as for instance explained in [8], are derived. The phase current is

$$i = \hat{I}\sin\left(\omega t + \phi\right) \qquad (2)$$

with the peak current $\hat{I}$, the electrical angular frequency $\omega = 2\pi f$, and the phase angle ϕ. The voltage drop across a switch is given by

$$u_{\mathrm{s}} = R_{\mathrm{on}}i + U_{\mathrm{s0}} \qquad (3)$$

with the on-resistance R_{on} and the threshold voltage U_{s0}, which is zero for SiC MOSFETs. Part of the switches operate at high switching frequency, whereas the other switches operate at grid frequency. The conduction losses of each switch position are calculated with

$$P_{\mathrm{c}} = \frac{1}{2\pi}\int_{\theta_1}^{\theta_2} D u_{\mathrm{s}} i \, \mathrm{d}\omega t \qquad (4)$$

with the duty cycle D expressed as continuious function according to the PWM pattern. The limits θ_1 and θ_2 of the integral must be formulated considering when the switch conducts and the dependency on the sign of ϕ. Fast-switching devices impose a commutation processes within slow-switching devices that actually do not switch during this time interval. Hence, D is often the same for fast- and slow-switching devices. To account for the temperature dependence, R_{on} and U_{s0} values from the datasheet are fitted to an exponential function and second-degree polynomial, respectively. R_{on} depends on the gate voltage U_{g}. For Si-IGBT, 15 V is an established value, but for SiC-MOSFETs, a wide range of 15 V to 20 V are stated by various manufacturers. R_{on} at a common gate voltage is given by

$$R_{\text{on}} = R_{\text{on}}^* R_{\text{on}}(U_{\text{g}}) \tag{5}$$

with the temperature-dependent R_{on}^* normalized by the reference temperature of the correction function $R_{\text{on}}(U_{\text{g}})$. The correction function is obtained by fitting the datasheet values for R_{on} for various gate voltages to either a first-degree or second-degree polynomial. The switching losses of the fast-switching devices are given by

$$P_{\text{s}} = \frac{f_{\text{s}}}{2\pi} \frac{U_{\text{c}}}{U_{\text{c,ref}}} \int_{\zeta_1}^{\zeta_2} \left(E_{\text{on}}(i) + E_{\text{off}}(i) \right) d\omega t \tag{6}$$

with the switching frequency f_{s}, the turn-on energy E_{on}, the turn-off energy E_{off}, the commutation voltage U_{c}, and the reference commutation voltage $U_{\text{c,ref}}$ at which the switching energies were measured. The limits ζ_1 and ζ_2 correspond to the period, in which the switch takes the switching loss. Note that U_{c} may not be the same as U_{dc}. Note also that SiC MOSFETs do not exhibit any switching loss for negative current because turning off the channel commutates the current internally to the body diode. To account for the temperature dependance of the switching energies, the corresponding datasheet values are fitted either to a first-degree or a second-degree polynomial. Similar to R_{on}, the switching energy of SiC-MOSFET must be adjusted for U_{g} and in addition for R_{g}:

$$E_{\text{on/off}} = E_{\text{on/off}}^* E_{\text{on/off}}(R_{\text{g}}) \frac{U_{\text{g,ref}}}{U_{\text{g}}}. \tag{7}$$

The function $E_{\text{on/off}}(R_{\text{g}})$ is obtained by fitting the datasheet values for the switching energies for various R_{g} values to a linear expression. The switching energy as function of U_{g} is not included in datasheets and the dependence is approximated with the factor $U_{\text{g,ref}}/U_{\text{g}}$ with the reference gate voltage $U_{\text{g,ref}}$ from the datasheet. The methodology for $E_{\text{on/off}}$ is also applied for the reverse recovery energy E_{rec} of the Si diode. Assuming that E_{on} and E_{off} depend linearly on i, which by definition is a sine function at fundamental frequency, (6) becomes

$$P_{\text{s}} = \text{sgn}\left(i(\zeta_1 + \phi)\right) \frac{1}{2\pi} f_{\text{s}} (-\cos(\zeta_2 + \phi) + \cos(\zeta_1 + \phi))$$
$$\times (E_{\text{on}} + E_{\text{off}}) \frac{\hat{I}}{\hat{I}_{\text{ref}}} \frac{U_{\text{c}}}{U_{\text{c,ref}}}. \tag{8}$$

with the reference current level $\hat{I}_{\text{ref}}$. If the switching energy depends quadratically on the current as approximately the case for the reverse recovery energy E_{rec} of Si diodes, the coefficients of the polynomial

$$E_{\text{rec}} = ai^2 + bi + c \tag{9}$$

can be determined by curve-fitting the corresponding data from the datasheet. The reverse recovery loss becomes

$$P_{\text{rec}} = \frac{f_{\text{s}}}{2\pi} \frac{U_{\text{c}}}{U_{\text{c,ref}}} \Big(-(\sin 2(\xi_2 + \phi) - \sin 2(\xi_1 + \phi))a\hat{I}^2/4$$
$$+ \text{sgn}\left(i(\xi_1 + \phi)\right)(-\cos(\xi_2 + \phi) + \cos(\xi_1 + \phi))b\hat{I}$$
$$+ (\xi_2 - \xi_1)(a\hat{I}^2/2 + c)\Big) \tag{10}$$

using ξ instead of ζ to differentiate the notation for the P_{rec} and P_{s}. The reverse recovery energy E_{rec} of SiC MOSFET in third quadrant operation are approximately independent of the current and (6) becomes

$$P_{\text{rec}} = \frac{f_{\text{s}}}{2\pi} (\xi_2 - \xi_1) E_{\text{rec}} \frac{U_{\text{c}}}{U_{\text{c,ref}}}. \tag{11}$$

The temperature dependence of E_{rec} of SiC MOSFETs is not considered because the data needed is not included in datasheets. This simplification does not create any noteworthy deviations because E_{rec} is relatively small compared to the other energy losses. The forward recovery losses of the IGBTs E_{FRV} as reported in [9] for ANPCT5T6 are approximated with the assumption included in Tab. 1. Other important aspects that affect the switching losses such as the commutation loop stray inductance and series resistance are not considered. Consequently, the results calculated in this study are an optimistic asymptotical case with the purpose of comparing the topologies. An example of how to solve the above equations is given in the Appendix. The loss models were validated against PLECS simulations for each topology.

3.3 Thermal

The thermal model is implemented with a lumped thermal circuit. The junction temperatures are given by the matrix

$$\mathbf{T}_{\text{j}} = (T_{\text{amb}} + R_{\text{th}} P_{\text{tot}}) \mathbf{I} + \mathbf{R}_{\text{thJH}} \circ \mathbf{P}_{\text{L}} \tag{12}$$

with the ambient temperature T_{amb}, the thermal resistance matrix from junctions to heatsink $\mathbf{R}_{\text{thJH}}$, the loss matrix $\mathbf{P}_{\text{L}}$, and the total losses

$$P_{\text{tot}} = \sum \mathbf{P}_{\text{L}}. \tag{13}$$

The first row of the matrices corresponds to the active switch positions and the second row to the diode positions. The reverse recovery losses of SiC MOSFETs are mapped into the active switch positions. The thermal resistance of empty positions are set to zero in $\mathbf{R}_{\text{thJH}}$ and the corresponding calculated loss values are discarded. The temperature mismatch of a topology is defined as

$$\Delta T = \max \mathbf{T}_{\text{j}} - \min \mathbf{T}_{\text{j}}. \tag{14}$$

4 Results

The active and reactive power capability is shown in Fig. 3a. For 400 V, 2L converters have the highest output power capability irrespective of phase angle because the conduction loss is lower than for 3L topologies that always include several switch positions in a conducting path. The reduced switching loss of 3L topologies is insufficient to compensate for the increased conduction loss at the switching frequency used for this analysis. The Si-IGBT T-Type has a relatively low output power capability and, to achieve the maximum datasheet currents, the cooling effort must be increased substantially compared to full-SiC or hybrid converters. For 690 V, the full-SiC ANPC and ANPCT2T3T5T6 display the highest output power capability since the modulation pattern parallels the SiC MOSFETs in the zero state, thus drastically reducing the conduction loss. The 2L SiC converter performs comparably well despite that the SiC chip technology is still inmature. In the future, we expect that the 2L converter will perform at 690 V at least as good as the full-SiC ANPC for the same reasons as explained for 400 V above. The other hybrid ANPC variants and the full-SiC NPC perform comparably.

For the 2L SiC converter, the output power capability is constant for all phase angles since the two switch position in the half-bridge always conduct equally long in the AC phase cycle. The opposite is true for 3L converters, where the current often shifts from the IGBT to its anti-parallel diode or vice-versa. The commutation path usually includes a diode exhibiting a reverse recovery loss and depending on the implementation of the power module, this can create great differences in losses for different phase angles. The loss parameters and thermal resistance of IGBTs, diodes, and SiC MOSFETs differ remarkably and the switching losses of Si components are changing by magnitudes with temperature. Consequently, the thermal margin shown in Fig. 3b varies substantially with phase angle for 3L converters. By proper power module design, the thermal mismatch can be limited as visible for the Si-IGBT T-Type module. The thermal mismatch limits the output power capability because increasing current in a switch position with thermal margin left would overheat another switch position. The application also affects the suitability of a topology or particular power module implementation since converters can work in completely different operating points with respect to phase angle.

The efficiency is shown Fig. 3c as a function of output power without reactive power. Positive output power corresponds to inverter operation and negative output power to rectifier operation. As expected, the Si-IGBT T-Type has the worst efficiency, especially for rectifier operation. For each voltage level, all hybrid 3L and full-SiC topologies display the same order of magnitude in efficiency with full-SiC topologies mostly performing slightly better. Increasing the converter voltage with constant output power increases the efficiency, because of the reduced converter current and conduction losses. This is excluding additional efficiency improvements in the busbar distribution system, cables etc. Since the output power capability of the power module increases correspondingly, it is generally recommended to adopt the higher system voltage.

The output power capability is shown in Fig. 4 as a function of the switching frequency (Fig. 4a) and of WTHD0 (Fig. 4b) without reactive power. With increasing switching frequency, the switching losses increase in proportion to the conduction losses and the increased total losses reduce the output power capability. Full-SiC ANPC and NPC and ANPCT1T2T3T4 perform more favorably at higher switching frequencies compared to the other topologies because the switching losses are distributed across more switch positions in these topologies. The results for ANPCT2T3T5T6 should be taken with a grain of salt at higher switching frequencies.

The efficiency is shown in Fig. 5 as a function of the switching frequency (Fig. 5a) and of WTHD0 (Fig. 5b) without reactive power. As the switching losses increase with switching frequency, the efficiency decreases accordingly. For the same reasons like the output power capability, full-SiC ANPC and NPC and ANPCT1T2T3T4 display superior efficiency at higher switching frequencies.

$U_{ac} = 400\,V,\ U_{dc} = 750\,V$ $U_{ac} = 690\,V,\ U_{dc} = 1500\,V$

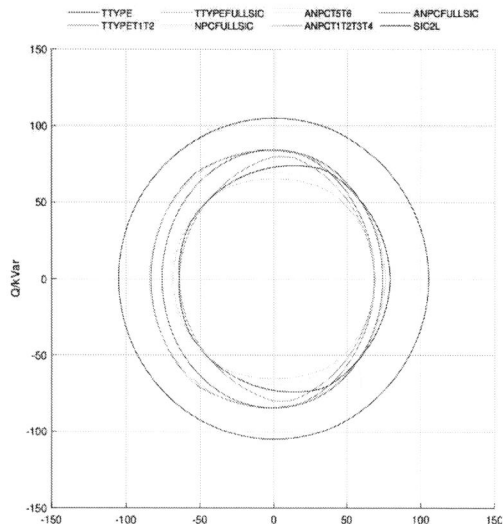 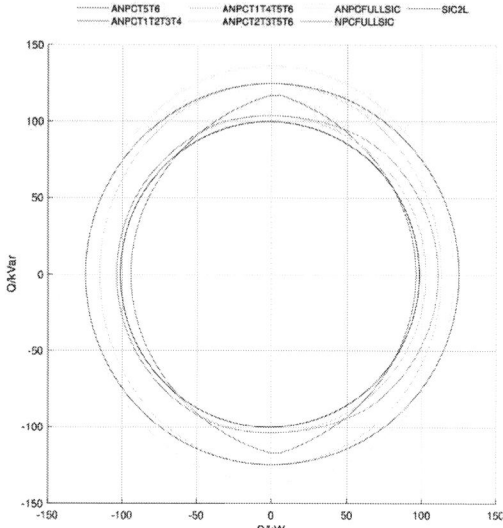

(a) Active and reactive power capability

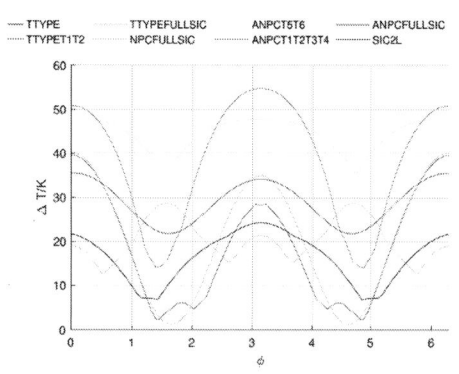 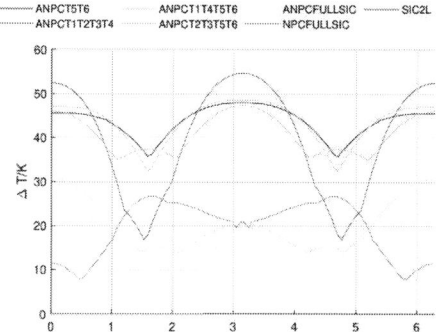

(b) Thermal mismatch as a function of phase angle

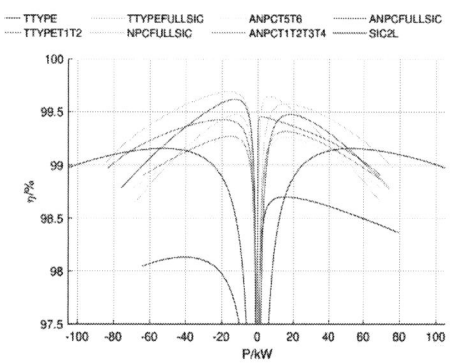 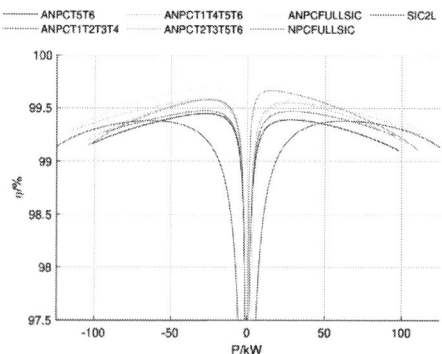

(c) Efficiency as a function of output power, no reactive power

Fig. 3: Power converter characteristics calculated for $U_{ac} = 400\,V$ and $U_{dc} = 750\,V$ (left) and $U_{ac} = 690\,V$ and $U_{dc} = 1500\,V$ (right). The switching frequency of the 2L topology is 36 kHz and of the 3L topologies 15.3 kHz for equivalent harmonic performance. The legend entries denote the respective topology and the switch positions realized with SiC MOSFETs, if any.

$U_{ac} = 400\,\text{V}, U_{dc} = 750\,\text{V}$

$U_{ac} = 690\,\text{V}, U_{dc} = 1500\,\text{V}$

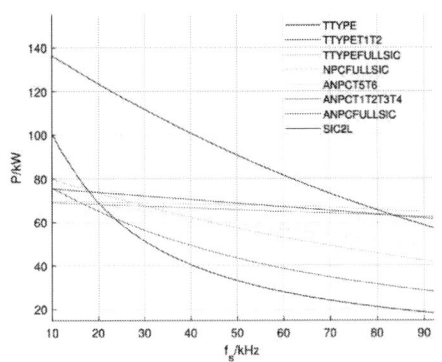

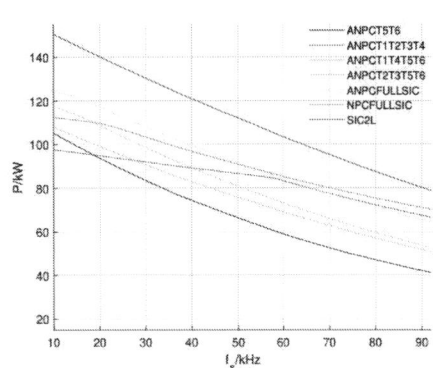

(a) As a function of switching frequency

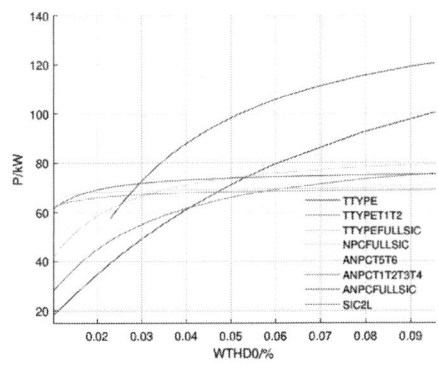

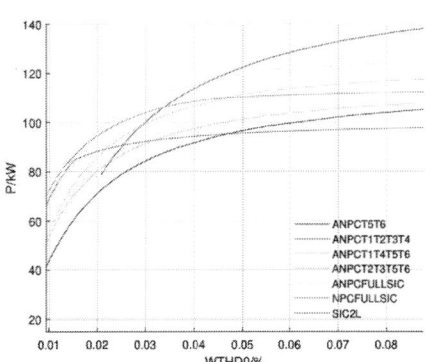

(b) As a function of WTHD0

Fig. 4: Output power capability switching frequency sweep without reactive power for $U_{ac} = 400\,\text{V}$ and $U_{dc} = 750\,\text{V}$ (left) and $U_{ac} = 690\,\text{V}$ and $U_{dc} = 1500\,\text{V}$ (right). The legend entries denote the respective topology and the switch positions realized with SiC MOSFETs, if any.

$U_{ac} = 400\,\text{V}, U_{dc} = 750\,\text{V}$ $U_{ac} = 690\,\text{V}, U_{dc} = 1500\,\text{V}$

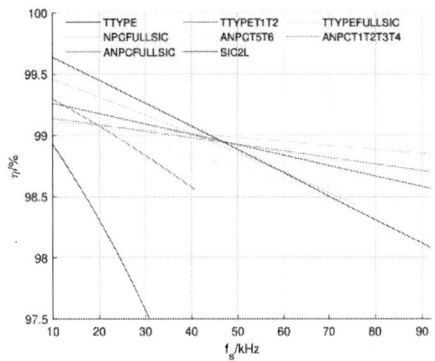 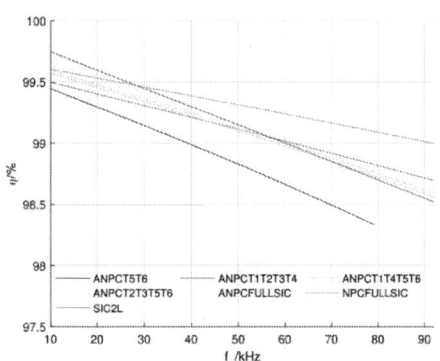

(a) As a function of switching frequency

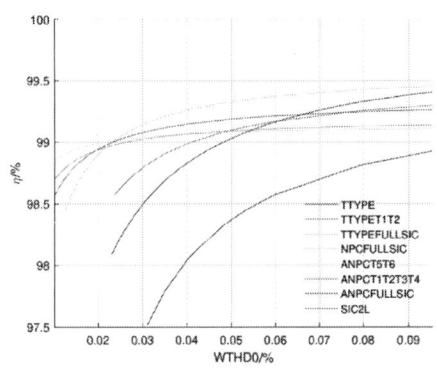 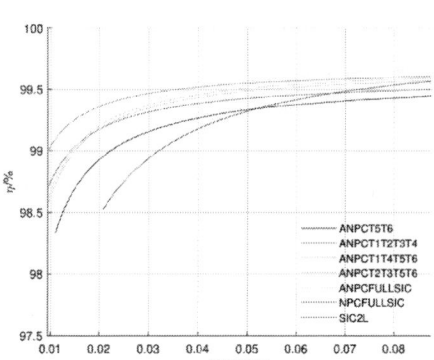

(b) As a function of WTHD0

Fig. 5: Efficiency switching frequency sweep at 47 kW and no reactive power for $U_{ac} = 400\,\text{V}$ and $U_{dc} = 750\,\text{V}$ (left) and $U_{ac} = 690\,\text{V}$ and $U_{dc} = 1500\,\text{V}$ (right). The legend entries denote the respective topology and the switch positions realized with SiC MOSFETs, if any.

5 Discussion

For a given current rating, both conduction and switching losses increase with higher voltage class devices if operated at the same voltage and current. 3L topologies have more components in the commutation path. Even if lower voltage class devices can be used in 3L topologies, the total conduction losses will be higher for a given voltage compared to 2L converters. Hence, the only benefits of the 3L topologies is that they facilitate converters with higher voltage and lower current rating. The lower current implies lower conduction losses at a given power and the better output harmonic performance allows operating with lower switching frequency and, consequently, lower switching losses. Both benefits are, however, irrelevant for low-power SiC converters up to 1500 V DC voltage. SiC-MOSFET devices are already available for 2 kV and 3.3 kV rating and if the demand becomes sufficient the market will provide 2.3 kV chips, thus facilitating 2L converters for 1500 V DC voltage. For typical switching frequencies, the switching losses are a smaller fraction of the total losses in SiC converters than for Si converters and decreasing the switching frequency in a 3L topology will not improve the converter efficiency substantially compared to 2L converters. Thus, the switching frequency of the 2L converter at which 3L converters are superior to 2L converters is increased to higher frequencies around 80 kHz for a 400 V AC system compared to Si converters. For a 690 V, the equivalent 2L switching frequency when 3L pays off is lower around 50 kHz. Irrespective of the AC system voltage, the full-SiC ANPC, ANPCT1T2T3T4, and full-SiC NPC are superior options at these higher switching frequencies. Considering the costs of the gate drives, the full-SiC NPC is likely the best option for high switching frequencies. Note that the whole analysis assumes a fixed power module size available such that the total semiconductor area is constant.

6 Conclusion

The results show that full-SiC 3L converters perform slightly better than hybrid Si/SiC 3L converters in terms of output power capability and efficiency. 2L-SiC converters actually perform similar to the best full-SiC 3L topology for the switching frequencies currently considered. The cost of the gate drives highly affects the attractiveness of a topology. Especially 2L-SiC converters are attractive, because of the low amount of gate drives required and their simplicity. We conclude that hybrid converters

may be interesting now from a cost point-of-view in certain cases, but when the SiC cost decrease in the future hybrid 3L topologies will be obsolete for most low-power applications. We conclude that 3L topologies will be obsolete for medium switching frequencies if SiC is used. Both conclusions are supported by the available power modules in the market. Despite the relatively high SiC cost, 2L SiC converters are already competitive, because of reduced total system cost. For 690 V, 3L converters, possibly hybrid, will still be used until 2300 V SiC chips will be commercially available in the market.

References

[1] B. Sahan, C. R. Mueller, A. Lenze, J. Czichon, and M. Slawinski, "Combining the benefits of SiC T-MOSFET and Si IGBT in a novel ANPC power module for highly compact 1500-V grid-tied inverters," in *PCIM Europe 2019*, 2019, pp. 1–6.

[2] M. Schweizer, T. Friedli, and J. W. Kolar, "Comparative Evaluation of Advanced Three-Phase Three-Level Inverter/Converter Topologies Against Two-Level Systems," *IEEE Trans. Ind. Electron.*, vol. 60, no. 12, pp. 5515–5527, 2013.

[3] A. Anthon, Z. Zhang, M. A. E. Andersen, D. G. Holmes, B. McGrath, and C. A. Teixeira, "The Benefits of SiC MOSFETs in a T-Type Inverter for Grid-Tie Applications," *IEEE Trans. Power Electron.*, vol. 32, no. 4, pp. 2808–2821, 2017.

[4] J. Häring, M. Gleissner, W. Wondrak, M. Hepp, and M.-M. Bakran, "Analytical Loss Calculation for ANPC Converters in Electric Drive Applications Using Different Modulation Strategies to Determine Efficiency and Overall Cost," in *PCIM Europe*, July 2020.

[5] L. Zhang, X. Lou, C. Li, F. Wu, Y. Gu, G. Chen, and D. Xu, "Evaluation of Different Si/SiC Hybrid Three-Level Active NPC Inverters for High Power Density," *IEEE Trans. Power Electron.*, vol. 35, no. 8, pp. 8224–8236, 2020.

[6] Z. Feng, X. Zhang, S. Yu, and J. Zhuang, "Comparative Study of 2SiC&4Si Hybrid Configuration Schemes in ANPC Inverter," *IEEE Access*, vol. 8, pp. 33 934–33 943, 2020.

[7] D. G. Holmes and T. A. Lipo, *Pulse Width Modulation for Power Converters : Principles and Practice*. Wiley-IEEE Press, 2003.

[8] F. Casanellas, "Losses in PWM inverters using IGBTs," *IEE Proc. - Electric Power Appl.*, vol. 141, pp. 235–239, September 1994.

[9] C. L. Kahraman, S. Lakshmeesha, S. Rosado, and T. Wijekoon, "Impact of forward recovery effects in different Si-IGBT technologies used in hybrid Si-IGBT, SiC-MOSFET based ANPC topology," in *IEEE Appl. Power Electron. Conf. and Expo. (APEC)*, 2022, pp. 1364–1370.

Tab. 2: Switching table for ANPCT1T2T3T4.

State	T1	T2	T3	T4	T5	T6
+	1	0	1	0	1	0
0(+)	0	1	0	0	1	0
0(-)	0	0	1	0	0	1
-	0	1	0	1	0	1

Appendix

In addition to the equations provided in the article, this Appendix provides hints to actually apply them. Some general comments:

- If a loss model for a comparable Si-IGBT topology is available and should be converted to SiC-MOSFET without external diode, U_{s0} of the anti-parallel diode must be set to zero. Adding the conduction losses of Si-IGBT and diode yields the losses of the SiC-MOSFET without external diode. This can also be done to verify any model derived by hand.

- Changes to the modulation scheme or switching sequences during the deadtime affect the loss calculation.

- The zero states for transitions from the positive and negative output state are often not identical.

- Equations taken from the literature should always be validated.

Here, an example is given for T1 in the ANPCT1T2T3T4. The first step is to setup a switching table for the modulator as given in Tab. 2. It is important to consider the deadtime because this can affect, for instance, the reverse recovery losses, however not in this case. The PWM pattern implies that T1 transitions between the + and 0(+) state with a sinusoidal duty ratio in the positive half cycle of the PWM reference

$$D = \begin{cases} M \sin \omega t & , \omega t \leq \pi \\ 0 & , \omega t > \pi \end{cases}. \quad (15)$$

In the next step, the limits θ, ζ, and ξ of the integrals in (4), (6), and (11) must be identified by understanding when the current commutates to the active switch or external diode of the switch position considered. Waveforms as shown in Fig. 6 are helpful to determine the limits. For SiC-MOSFETs without external diode, the conduction loss calculation is much simpler than for Si-IGBTs as the MOSFET channel always conducts when the switch position is turned on and conducts:

$$\theta_1 = 0 \qquad \theta_2 = \pi$$

The MOSFET only exhibits turn-on and turn-off losses for positive current and reverse recovery losses for negative current

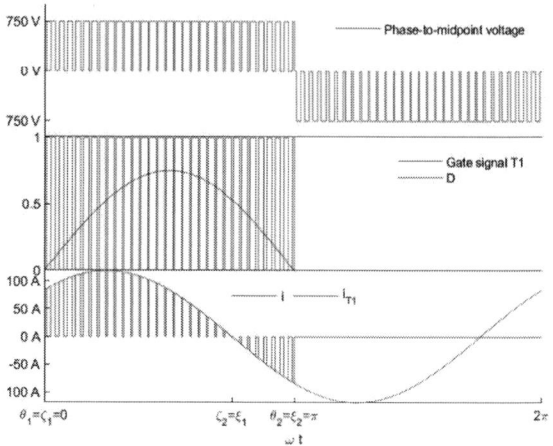

Fig. 6: Waveforms of T1 in ANPCT1T2T3T4 for a U_{dc}=1500 V system with $\phi = \pi/4$.

$$\zeta_1 = \begin{cases} 0 & , \phi \geq 0 \\ -\phi & , \phi < 0 \end{cases} \qquad \zeta_2 = \begin{cases} \pi - \phi & , \phi \geq 0 \\ \pi & , \phi < 0 \end{cases}$$

$$\xi_1 = \begin{cases} \pi - \phi & , \phi \geq 0 \\ 0 & , \phi < 0 \end{cases} \qquad \xi_2 = \begin{cases} \pi & , \phi \geq 0 \\ -\phi & , \phi < 0 \end{cases}$$

Solving (4), (6), and (11) yields

$$P_{\mathrm{c}} = \frac{M}{2\pi} R_{\mathrm{on}} \hat{I}^2 \left(1 + \frac{1}{3} \cos 2\phi\right), \quad (16)$$

$$P_{\mathrm{s}} = \frac{1}{2\pi} f_{\mathrm{s}} (1 + \cos \phi) (E_{\mathrm{on}} + E_{\mathrm{off}}) \frac{\hat{I}}{\hat{I}_{\mathrm{ref}}} \frac{U_{\mathrm{dc}}}{2U_{\mathrm{c,ref}}}, \quad (17)$$

and

$$P_{\mathrm{rec}} = \frac{|\phi|}{2\pi} f_{\mathrm{s}} E_{\mathrm{rec}} \frac{U_{\mathrm{dc}}}{2U_{\mathrm{c,ref}}}. \quad (18)$$

The integrals become more complex for IGBTs and other modulation schemes, for instance an ANPC using T2, T3, T5, and T6 for current sharing in the zero state. The electrical parameters are dependent on many factors like the gate voltage and more importantly temperature. If the thermal mismatch in the module is non-negligible, the equations should **never** be used without a coupled thermal module to draw any conclusions about a topology. The deviation of this average model from the average calculated in PLECS is extremely small and is caused by neglecting the deadtime. We compared the analytical model with a more detailed and complex transient calculation method using interpolated datasheet values and in average, the analytical model was more accurate while also being computationally more efficient.

PCIM Europe 2024, 11– 13 June 2024, Nuremberg   DOI: 10.30420/566262438

Analysis of Analogue Current and Flux Balancing for the Dual-Active-Bridge Converter

Christophe Basso

Future Electronics, France

Corresponding author: Christophe Basso, christophe.basso@futureelectronics.com

Abstract

The dual-active-bridge converter (DAB) is a popular structure found in the electric vehicle (EV) environment. Built around two full-bridges located on the primary and the secondary sides of the converter, the DAB lends itself well to controlling high-voltage battery charging up to several tens of kilowatts. Galvanic isolation with voltage scaling is obtained using a power transformer and an external inductor which ensures zero-voltage switching (ZVS) under certain conditions. Several modulation strategies exist but the single-phase approach, in voltage-mode control, represents a popular choice owing to its ease of implementation. In this mode, both full-bridges operate with 50% duty ratio, applying symmetrical voltages of equal duration across the transformer windings. Unfortunately, ohmic losses and timing errors affect the transformer magnetic operating point and can cause drift, leading to undesirable consequences: a dc-blocking capacitor is typically installed in the circuit to prevent flux runaway and transformer saturation. While this option ensures safe magnetic operation, the size and the cost of the dc-blocking capacitance can add significant cost to the bill of materials and affect long-term reliability. Considering the high ripple current circulating in this component, designers are often constrained to use many discrete parts connected in parallel. Equal current sharing across these discrete capacitors must be through appropriate PCB layout and reliable assembly. This paper describes a solution to remove the dc-blocking capacitance by monitoring the circulating currents and altering the bridges duty ratio in case of imbalance. Originally proposed as a digital implementation in [1], this document explores a fully analogue method.

1. Introduction

The dual-active-bridge converter appears in Figure 1. The structure biases the transformer by applying a bipolar waveform swinging between ± V_{HV}, effectively driving the transformer in quadrants I and III. The secondary side also hosts a full bridge for efficient synchronous rectification and it imposes secondary-side voltage toggling between ± V_{OUT}. Both bridges are operated in 50% duty ratio for the simplest control scheme and power flow is controlled by adjusting the overlap between primary and secondary waveforms. The series inductor L_r sets the maximum amount of processed power. This can be a standalone inductor or part of the transformer leakage inductance. When sufficient reactive current circulates, zero-voltage switching (ZVS) is possible. The typical operating waveforms are shown in Figure 2. Power switches used can typically be IGBTs or silicon

carbide (SiC) MOSFETs for improved efficiency and higher operating frequency.

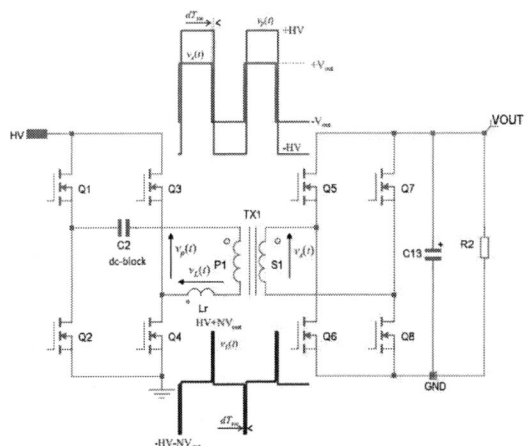

Figure 1: A DAB converter is made of two full-bridge structures driving a transformer.

A deadtime (DT) is inserted to prevent shoot-through currents during the commutations of the low- and high-side transistors. The control of such converter is realized by adjusting the phase delay between the two bridges whose duty ratio is fixed at 50%. This strategy, known as *single-phase-shift* (SPS) control represents a popular choice owing to its simplicity.

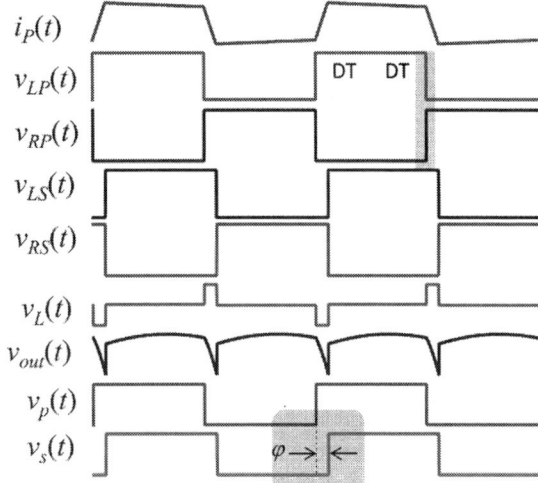

Figure 2: Power flow is controlled by adjusting the phase delay between the two bridges.

Other techniques exist, which combine phase shift and duty ratio modulation on both sides of the transformer. These different approaches are aimed at reducing the rms content of the circulating ac current in the high-voltage dc section and extending ZVS operations. Reference [2] offers a comprehensive analysis of these modulation schemes which, combined across different operating conditions, benefit efficiency. However, it comes at the cost of an increased complexity.

2. A Small-Signal Model

Before attempting to implement inner control loops for controlling the transformer magnetizing current, it is important to know the main *control-to-output* transfer function of the power stage H being driven. In other words, if a *stimulus* is applied at the control input, how does this excitation signal propagate through the power paths and produce a *response* at the output? Knowing this frequency-dependent expression identifies where poles and zeroes are located and how to neutralize their variability with adequate compensation. The easiest and most straightforward way to approximate the transfer

function of a switching converter, is to resort to the 1st-order approximation shown in Figure 3.

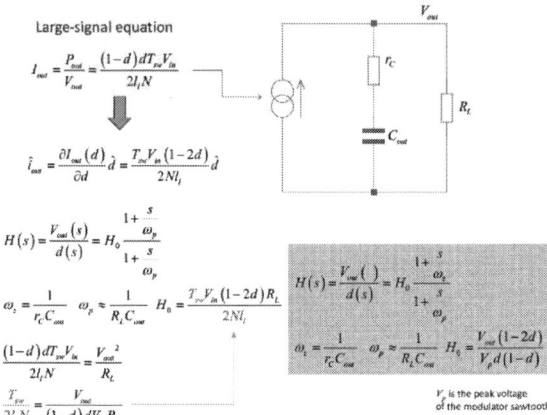

Figure 3: The transfer function of the DAB comes easily with this simple representation.

In this simplified model, a large-signal expression describes the current delivered by the converter. The variable *d* represents the control angle of the converter. Linearization is obtained using partial differentiation with respect to *d* and a simple equivalent model is constructed, showing how the output capacitor classically affects the response.

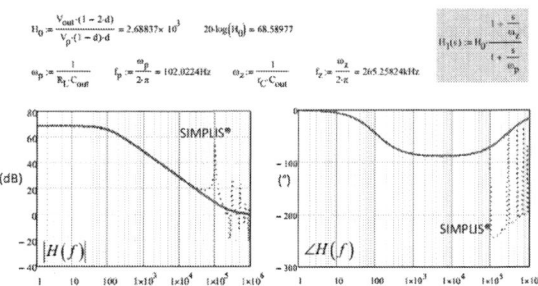

Figure 4: Responses from the Mathcad® sheet and SIMPLIS® are matching very well.

The transfer function is of 1st-order type and features a pole-zero pair. The dc gain H_0 depends on the output voltage and the selected control angle. This frequency-dependent expression has been plotted in Figure 4, together with the magnitude and phase obtained from a SIMPLIS® simulation. They are in excellent agreement.

3. The Dc-Blocking Capacitor

The transformer magnetic operating point is set by the magnetizing current, itself driven by the volt-seconds applied during the on- and off-times. When these volt-seconds are rigorously symmetrical, the magnetizing current is centered around zero and no magnetic drift is observable. Unfortunately, variations in $r_{DS(on)}$, propagation delays or a lack of granularity in the control timing [3] can generate a severe imbalance which, in a worst-case situation, provokes transformer saturation. If inserting a dc-blocking capacitor represents a simple solution, it is not a panacea considering the stress and cost of the solution.

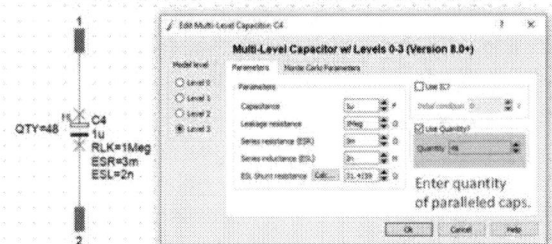

Figure 5: The dc-blocking capacitor requires careful consideration of parasitic contributors.

For example, calculations show that a typical rms current of 30 A can flow in a 500-V dc output 15-kW DAB converter supplied from a 600-V source. Such high current requires many parallel capacitors; in that case, 48 x 1-µF X7T parts in Figure 5. The PCB layout in this example must be well thought through to ensure adequate current sharing between the elements, especially considering parasitics of each capacitor.

4. The Origin of the Drift

The volt-seconds applied to the transformer depend on semiconductors' voltage drops which can change with temperature and from lots-to-lots variations. The lack of sufficient granularity in the modulator section can also cause severe mismatch and affect the magnetic equilibrium.

Finally, propagation delays distort the expected timings, changing the average voltage value resulting in dc drift. In this application, our interest lies with the magnetizing current i_m that we cannot directly measure. We thus need to *observe* it via a reconstruction mechanism, involving the primary- and secondary-side currents:

$$i_m(t) = i_p(t) - n \cdot i_s(t) \tag{1}$$

If we lump all ohmic elements into equivalent resistances in series with the primary and secondary sides of the converter [3], we can express the dc component of the magnetizing current as

$$I_{m,dc} = I_{p,dc} - n \cdot I_{s,dc} = \frac{V_{p,dc}}{R_{p,T}} - n\frac{V_{s,dc}}{R_{s,T}} \tag{2}$$

Using this equation, a simplified model of the converter can be built as illustrated in Figure 6. The high-voltage switching waveforms are split into ac and dc components. The ac part drives the core flux excursion whilst the dc part affects the minor loop location in the B-H curve.

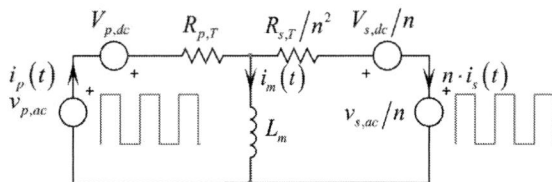

Figure 6: The dc current depends on the dc series voltage sources and limiting resistances.

From this representation, it becomes possible to express the dc magnetic drift as:

$$B_{dc} = \left(\frac{V_{p,dc}}{R_{p,T}} - n\frac{V_{s,dc}}{R_{s,T}} \right) \cdot \frac{N_p}{l_m} \cdot \mu_0 \mu_r \tag{3}$$

Wherein N_p represents the number of primary turns, l_m the mean magnetic path length and μ_r the material's relative permeability. In high-power designs, the ohmic elements are minimized for the highest possible efficiency. Therefore, any tiny dc shift in V_p and V_s will produce a potentially significant offset in the transformer leading to failure.

5. The Adopted Principle

The authors in [1] have implemented two control loops to reduce the magnetizing component to zero and also to keep the primary- and secondary-side currents to a minimum value. The control is digital and implies one-sided modulation for a simple and efficient compensation strategy. The proposed method departs purposely from this option to implement an analogue control in which the driving waveforms are modulated on both sides of the transformer as shown in Figure 7.

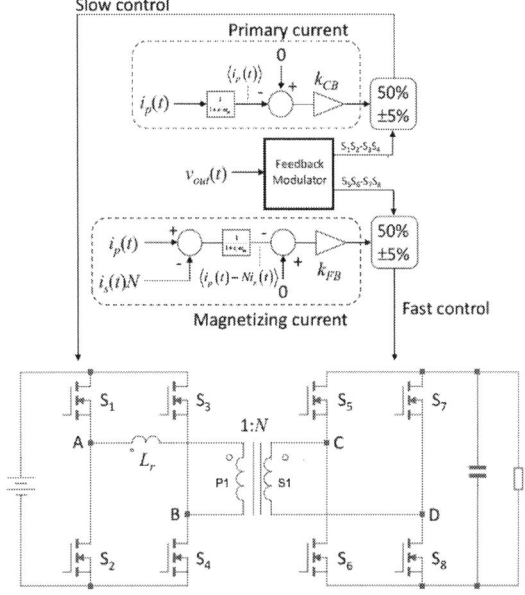

Figure 7: A duty ratio modulator is inserted after the main block for controlling the two switching sides.

For this purpose, a specific subcircuit is inserted in series with the two drive paths without disturbing the steady-state operation. The adopted principle is shown in Figure 8. When the clock arms flip-flop U_{17}, its output Q goes high as expected while its complemented output $\bar{Q}$ toggles low. Capacitor C_7 is now free to charge until the flip-flop reset occurs via U_{16}. The charging time is adjusted to be exactly half of the switching period. Therefore, if nothing alters the switching threshold set at 1 V by V_{10}, the duty ratio will be 50%. If a voltage is now applied at the modulation pin, the capacitor will charge longer or shorter, affecting the duty ratio by a fixed amount.

In the simulated design, a limit of a ±5% deviation was used, illustrated in Figure 8. The circuit design is used for controlling primary and secondary legs.

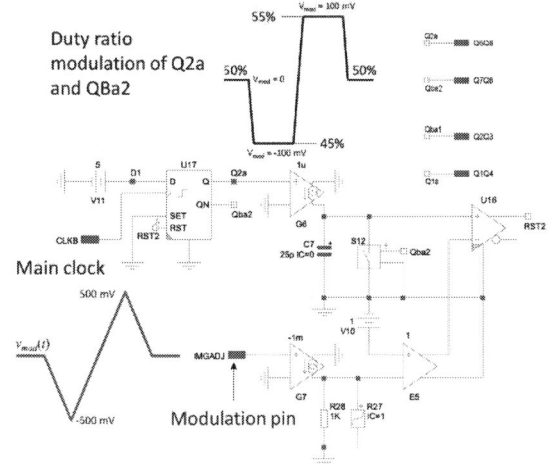

Figure 8: The modulator offers a ±5% modulation dynamic for the duty ratio.

The full system is made of two loops: one for monitoring the magnetizing current and one for the primary current. The magnetizing current is reconstructed by measuring the primary and secondary components, respectively noted $i_p(t)$ and $i_s(t)$. After subtracting the reflected secondary current from the primary current, a filter averages the resulting signal and compares it with a zero setpoint. The error voltage is amplified and drives the duty ratio modulator. The crossover frequency of this loop must be sufficiently high to quickly react in the presence of large transient conditions as it is important to avoid saturation and for the control loop to take immediate action. The second loop, which solely monitors the primary current, does not need to be fast, since a temporary dc shift in the primary or secondary currents is not harmful to the converter [4]. The response time can therefore be longer to avoid any interaction between the separate control loops.

6. Closing the Loops

Before simulating the entire structure, it is important to compensate the two control loops. As explained previously, the starting point for any compensation exercise is establishing the control-to-output transfer function. We can either measure it with SIMPLIS® or determine it symbolically with the equivalent linear circuit drawn in Figure 9.

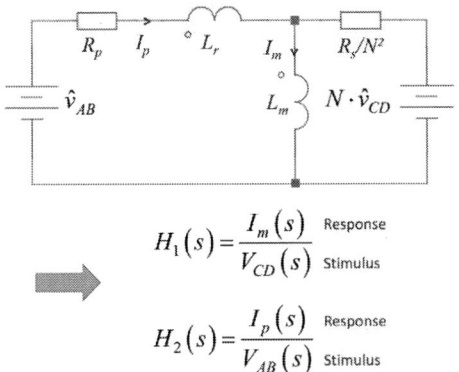

$$H_1(s) = \frac{I_m(s)}{V_{CD}(s)} \quad \text{Response / Stimulus}$$

$$H_2(s) = \frac{I_p(s)}{V_{AB}(s)} \quad \text{Response / Stimulus}$$

Figure 9: The equivalent circuit brings all secondary elements to the primary side.

The control-to-magnetizing-current transfer function is obtained by identifying v_{CD} as the stimulus – v_{AB} is constant – and i_m as the response. The resulting expression is a second-order polynomial:

$$H_1(s) = \frac{V_{out}}{1\,\text{V}} \frac{N}{R_s} \frac{1 + s\frac{L_r}{R_p}}{1 + s\left(\frac{L_m}{\frac{R_s}{N^2}\|R_p} + \frac{L_r}{R_p}\right) + s^2 \frac{L_m}{\frac{R_s}{N^2}\|R_p} \frac{L_r}{\frac{R_s}{N^2}+R_p}} \quad (4)$$

It is easy to confirm this expression with SIMetrix® as illustrated in Figure 10 and Figure 11.

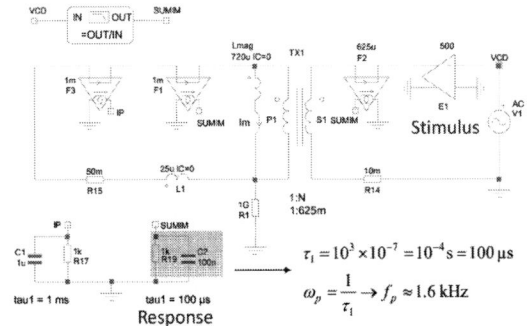

$$\tau_1 = 10^3 \times 10^{-7} = 10^{-4}\,\text{s} = 100\,\mu\text{s}$$

$$\omega_p = \frac{1}{\tau_1} \rightarrow f_p \approx 1.6\,\text{kHz}$$

Figure 10: The circuit is easily simulated with a SPICE engine.

The Bode plot shows a gain excess at 2 kHz with a phase lag of 135°. By providing a simple permanent attenuation of 35 dB, the loop will crossover at 2 kHz with a phase margin of 45°.

This is acceptable for this simulation example but design engineers will need to study the potential variability of the plot versus changes in parasitics with temperature and production lots.

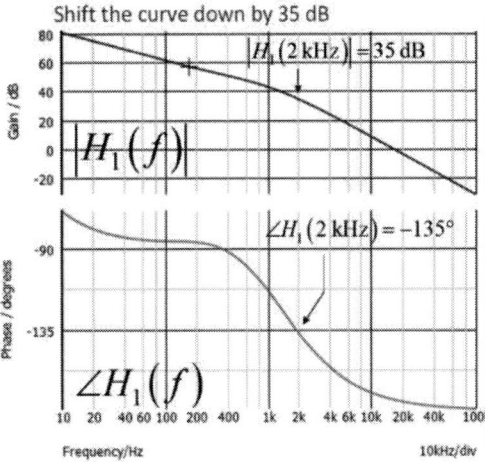

Figure 11: The ac response shows a large gain at 2 kHz and a phase lag of 135°.

The control-to-primary-current loop is determined through a simpler expression cascading two poles, if we consider the low-pass averaging filter featuring a 1-ms time constant:

$$H_2(s) = \frac{V_{in}}{1\,\text{V} \cdot \left(R_p + \frac{R_s}{N^2}\right)} \frac{1}{\left(1 + s\frac{L_r}{\frac{R_s}{N^2}+R_p}\right)(1 + s \cdot 1\,\text{ms})} \quad (5)$$

The simulation setup appears in Figure 12.

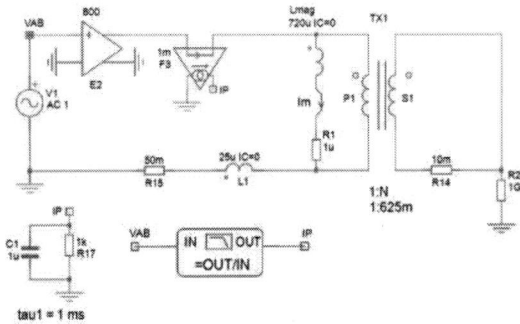

Figure 12: This second loop simulates easily also.

The magnetizing inductance disappears for this analysis as its state variable is zeroed by the first loop. Considering a zeroed i_m, then v_{CD} is also equal to 0 V on average. The ac response appears in Figure 13 and shows a moderate lag at 100 Hz where we will force crossover.

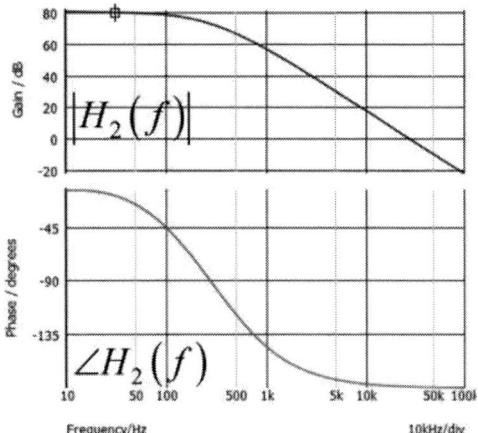

Figure 13: This second loop simulates easily also.

A simple constant attenuation is sufficient here as no phase boost is necessary. Please note the naturally-high dc gain, offering a small static deviation in closed-loop operations.

7. The Complete Picture

With the two inner loops closed, simulation of the entire converter can be considered. The DAB converter will be compensated using the automated macro given in [5]. The circuit appears in Figure 14 and Figure 15. Two current sensors are required for current reconstruction.

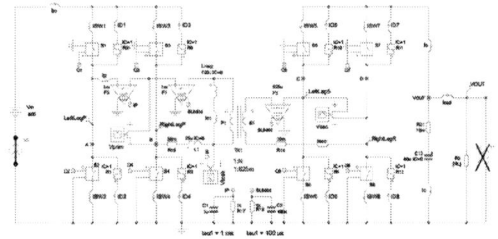

Figure 14: The DAB converter can be simulated with the two inner loops.

The converter delivers an output voltage of 500 V from an 800-V input source and produces 15 kW of power. The switching frequency is 100 kHz for this simulation. The switches are perfect models and fast SiC MOSFETs would be suitable for these experiments. For the sake of simplicity, a single phase-shift modulation is adopted which lends itself well to a simple simulation setup.

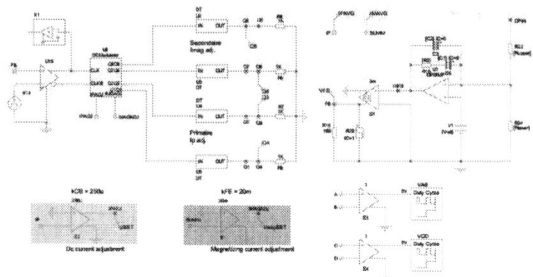

Figure 15: The modulators are supplemented by dead-time generators before driving the switches.

The first test consists of checking the variables when the inner loops are turned off. This is what is simulated in Figure 16.

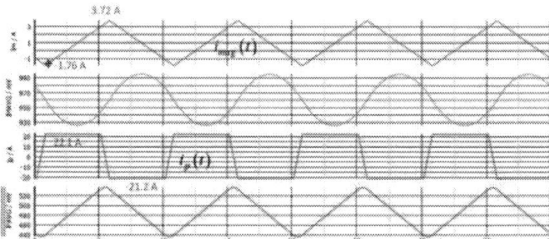

Figure 16: The magnetizing current shows a dc shift.

As can be observed, the magnetizing current is shifted by almost 1 A while the primary current shows a 500-mA dc offset. This amounts to -800 mA for the secondary-side current at this particular operating point.

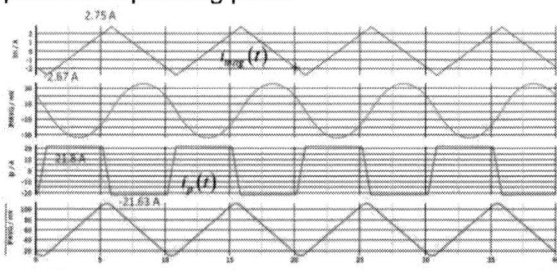

Figure 17: Once inner loops are active, the correcting actions cancel the dc drifts.

In Figure 17, the loops are activated and the magnetizing current average value drops to 1.1 mA while the primary- and secondary-side currents' dc shift are respectively 60.5 mA and 95 mA. This demonstrates a very good improvement compared to the original situation.

The next step is to check the transient response to a load change. This is what is shown in Figure 18 where the current is stepped from 15 A to 25 A with a slope of 1 A/μs. As observed from the top trace, the magnetizing current is kept perfectly to a value close to 0 A, in spite of the sudden output current deviation. The average primary current shows a 65-mA average value during the 25-A loading - which is extremely low.

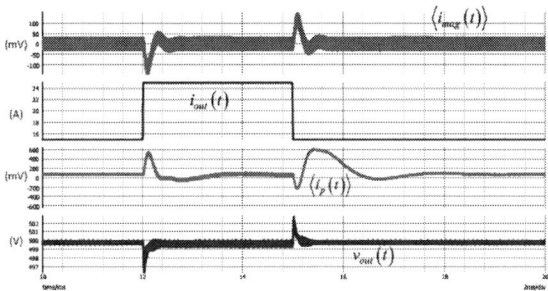

Figure 18: The fast loops ensure average currents in the transformer remain under control.

Conclusion

The inclusion of a dc-blocking capacitor in a voltage-mode-controlled dc-dc DAB converter can seriously hamper the cost and the size of the project as well as adversely affecting the transient response. Implementing additional loops controlling the averaged quantities of the magnetizing current as well as the primary- and secondary-side currents, represents an interesting alternative in digital or analogue implementation. This work details the theoretical approach and will be practically implemented as a next step.

References

1. M. Jovanović et al., _Novel Transformer-Flux-Balancing Control of Dual-Active-Bridge Bidirectional Converters_, APEC 2015, March 2015
2. M. Blanc et al., _Optimization of a Dc-Dc Dual Active Bridge Converter for Aircraft Application_, PCIM Conference, 16-18 May 2017
3. G. Ortiz et al., _Flux Balancing of Isolation Transformers and Application of "The Magnetic Ear" for Closed-Loop Volt–_ _Second Compensation_, IEEE Transactions, vol. 29, no. 8, Aug. 2014
4. J. Muhlethaler et al., _Core Losses Under the Dc Bias Condition Based on Steinmetz Parameters_, IEEE Transactions on Power Electronics, vol. 27, no. 2, 2012
5. C. Basso, Templates for SIMPLIS, http://powersimtof.com/Spice.htm

Design and Optimization of a Single-Stage Photovoltaic Microinverter with Integrated Magnetics

Jin Wen[1], Jiajia Guan[1], Chenhang Zeng[1], Yijie Huang[1], Yue Wu[2], Zhipeng He[2], Cai Chen[1], Yong Kang[1]

[1] Huazhong University of Science and Technology, China
[2] State Key Laboratory of HVDC, Electric Power Research Institute, CSG, China
Corresponding author: Jin Wen, wenjin@hust.edu.cn
Speaker: Jin Wen, wenjin@hust.edu.cn

Abstract

High-frequency link resonant inverter has the potential to achieve higher efficiencies due to its single-stage power conversion structure. An optimized design is carried out for a single-stage high-frequency link resonant inverter in this paper. First, the optimal phase shift angle between the primary and secondary sides is applied to reduce the RMS current of the resonant tank to reduce the conduction loss. And there is always a set of switches on the secondary that operate at low frequency, which can reduce switching losses. Then the transformer and resonant inductor in the converter are integrated, and the design is optimized based on loss and area. The final prototype is built to achieve a power density of 46 W/in³ and a full load efficiency of 95.3%.

1   Introduction

Research on renewable energy such as photo voltaics, wind power and energy storage has become a hot topic nowadays. Microinverters play a key role in solar photovoltaic systems. Their main function is to convert DC current from solar modules to AC power, helping users achieve energy independence, low-carbon, and environmentally friendly lifestyles. Single-stage high-frequency link inverters have received much attention in the field of micro photovoltaic inverters due to their features of fewer devices, electrically isolatable, and wider regulation range. This inverter structure was first specifically studied in [1], followed by various literatures on its different topologies and modulation strategies. In general, the modulation strategies are based on phase shifting, while the topologies are resonant type and dual active bridge [2]-[5]. In [6], an ac-grid-connected hybrid energy storage system is applied based on the deformation of this structure, and it is proposed that the resonant-type structure has lower current stress, higher hardware capacity, and better electromagnetic characteristic than the DAB-type. Therefore, this paper will optimize the design of resonant type high-frequency link inverter.

The main work of this paper will be divided into the following three parts: 1. optimization of the resonant current RMS of the converter to reduce the losses; 2. Integration and optimization of the magnetic components to increase the power density.

2   Basic Principles of the Topology

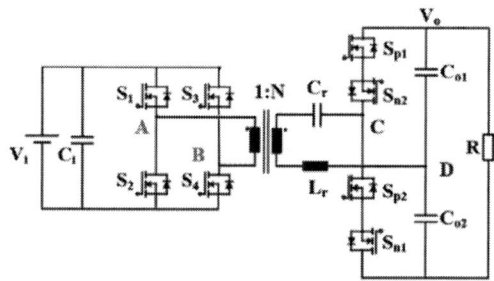

Fig. 1   Topology schematic

The topology structure of the resonant high-frequency link inverter is shown in Fig. 1. The primary side has a full-bridge structure, and the primary side energy is transferred to the secondary side through the transformer under the modulation of the primary side bridge. The secondary side includes a resonant tank and a cycloconverter. The cycloconverter consists of a capacitive half-bridge and a half-bridge including back-to-back switches.

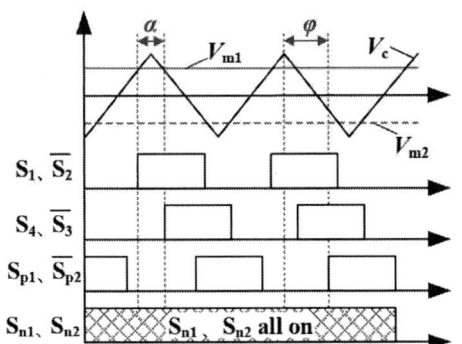

Fig. 2 Driver waveforms under positive half cycle of output voltage driver waveforms under positive half cycle of output voltage

The back-to-back switches can block voltage in both directions to generate AC voltage on the output side. The resonant elements are placed on the small current side of the secondary side, which reduces the loss on the resonant elements. The secondary side uses a separated capacitor bridge voltage doubler circuit, which has twice the boost ratio, reducing the number of transformer winding turns.

The corresponding driving waveforms are shown in Fig. 2. The primary switches always operate at a duty cycle of 50%, but there is a phase shift angle between the two bridges. Through the change of the phase shift angle α, the voltage applied to the primary side of the transformer can be changed, and the value of the output voltage can further be changed. The secondary side switches use positive group and negative group decoupling modulation. At the same time, only one group of switches operate at high frequency, while the other group of switches are all normally on. Shown in Fig. 2 is the secondary drive signal when the output voltage is a positive half cycle. During this time, S_{n1}、S_{n2} is usually on and S_{p1}、S_{p2} operates at high frequency. When the output voltage is negative, the opposite is true.

This modulation method provides a freewheeling path for the current at the commutation moment. The current can always freewheel through the normally conductive switches and the anti-parallel diode of the high-frequency switches during this period. The voltage peak during the commutation process of the secondary side cycle converter is greatly reduced. Moreover, the normally conductive switches provide a ZVS current flow path for the high-frequency switches, which can greatly reduce the switching loss under ZVS operation. The actual equivalent switching frequency is reduced, which greatly reduces switching losses and driving losses.

3 Optimization of the Resonant Current RMS

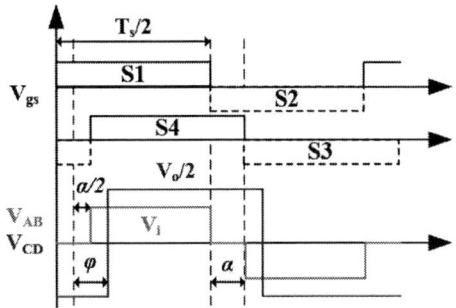

Fig. 3 Voltage waveforms of transformer port

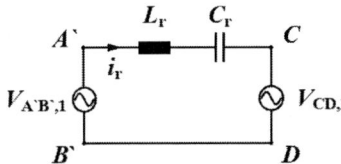

Fig. 4 Fundamental harmonic equivalent model

This topology uses a phase-shifting modulation strategy with two degrees of freedom: the primary in-phase shift angle and the primary and secondary out-phase shift angles. According to the driving waveforms, the voltage waveforms of the transformer port can be obtained, as shown in Fig. 3. Since the phase-shifting operation of the primary-side switches will generate a three-level voltage on the primary side of the transformer, the operating mode of the secondary-side switches generate a square wave voltage at the bridge arm port. To simplify the analysis, it can be approximately considered that the fundamental wave component of the two voltages is applied to the two ports of the resonant tank. The equivalent model shown in Fig. 4 can be obtained by equating the primary side of the transformer to the secondary side. When the secondary side is a resistive load, the gain of the circuit can be derived as shown in Eq. (1).

$$\frac{V_o}{V_{in}} = \frac{4N\cos(\frac{\alpha}{2})\sin\varphi}{\pi^2 Q(F_n - \frac{1}{F_n})}, \quad Q = \frac{Z_0}{R_L} = \frac{\sqrt{\frac{L_r}{C_r}}}{R_L}, \quad F_n = \frac{f_s}{f_r} \quad (1)$$

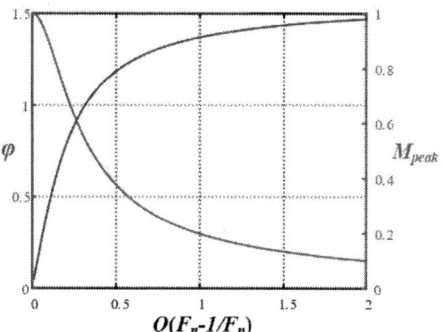

Fig. 5 Trend curves of φ and peak gain

At the same time, based on this model, the normalized RMS value of the resonant current can be derived, as shown in Eq. (2).

$$I_{r,pu} = \frac{I_{r,rms}}{V_o / R}$$
$$= \sqrt{\frac{2}{\pi^2 Q^2 (F_n - 1/F_n)^2} - \frac{2}{Q(F_n - 1/F_n)\tan\varphi} + \frac{\pi^2}{2\sin^2\varphi}} \quad (2)$$

The normalized current RMS is related to φ and not to α. Thus, a small current RMS can be realized by choosing a suitable φ. The optimum φ value exists under different loads and frequencies, but the peak gain requirement must be taken into account, and the trend curves of the two are shown in Fig. 5. The optimum φ that meets the peak gain requirement can be selected to minimize the RMS value of the resonant current, so that the losses can be reduced. The sinusoidal change of gain can be achieved by adjusting the α.

In fact, when the RMS of the resonant current is minimum, the corresponding CD port is purely resistive, that is, the secondary CD port voltage is in phase with the resonant current through modulation. Therefore, the minimum φ of the effective value of the resonant current can be calculated according to Eq. (3).

$$\varphi_{opt} = \arctan\left(\frac{1}{2}\pi^2 Q(f_n - \frac{1}{f_n})\right) \quad (3)$$

4 Integration and Optimization of the Magnetic Components

In this circuit topology, the resonant inductor and the transformer are connected in series. The two magnetic components can be integrated through the method shown in Fig. 6, eliminating the middle part of the winding, and the secondary winding of the transformer is reused with the inductor winding.

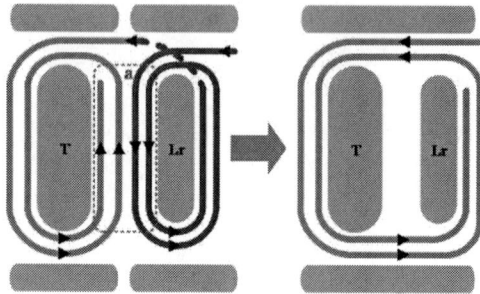

Fig. 6 Magnetic integration diagram

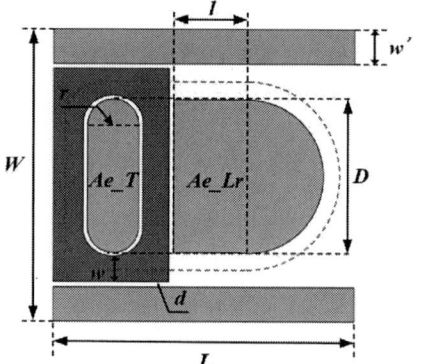

Fig. 7 Model of core size

Based on the idea of magnetic integration, the shape of the magnetic core is initially determined as shown in Fig. 7. According to the annotations in the figure, the magnetic core can be geometrically modeled as shown in Eq. (4). This formula can be used to calculate the core area for optimization.

$$\begin{cases} D = 2r + (AeT - \pi r^2)/2/r \\ l = (AeLr - \pi D^2/8)/D \\ L = 2r + l + D/2 + 4d + 3w \\ w' = AeT/2/(L - l - D/2 - 1.5w) \\ W = D + 4d + 2w + 2w' \end{cases} \quad (4)$$

The loss of magnetic includes core loss and winding loss. Core loss is calculated from loss volume density and volume. The loss volume density is evaluated through the modified Steinmetz equation as shown in Eq. (5). When calculating winding loss, the AC coefficient of the winding needs to be considered, which can be calculated by Dowell equation as shown in Eq. (6).

$$P_v = \frac{1}{T}\int_0^T k_i \left|\frac{dB(t)}{dt}\right|^\alpha (\Delta B)^{\beta-\alpha} dt \quad (5)$$

Where $k_i = \dfrac{K}{(2\pi)^{\alpha-1}\displaystyle\int_0^{2\pi}|\cos\theta|^\alpha 2^{\beta-\alpha} d\theta}$.

$$F_{\text{R}} = A \left(\frac{\sinh(2A) + \sin(2A)}{\cosh(2A) - \cos(2A)} + \frac{2}{3}(N_1^2 - 1)\frac{\sinh(A) - \sin(A)}{\cosh(A) + \cos(A)} \right) \qquad (6)$$

where $A = \dfrac{h}{\delta}\sqrt{\eta_r}$, h is the thickness of the winding, δ is the skin depth of the conductor, and N_1 and η_r are related to the structure of the winding.

The loss model of the integrated core can be established by combining this two. Based on the geometric model and loss model, the loss and projected area of the transformer can be comprehensively optimized, and the results are shown in Fig. 8. The optimization results show that a series of optimization points have been obtained, which can be selected according to actual needs. When favoring high efficiency, a larger magnetic core footprint will be required, and when favoring high power density, there will be a little loss in efficiency.

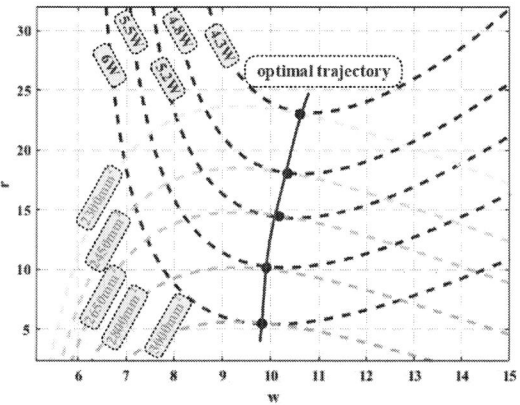

Fig. 8 Area and loss optimization curve of transformer

5 Simulation and experiment

The main parameters of the circuit are shown in Table 1. After determining the resonant frequency of the circuit, through the optimization of the effective value of the resonant current, the optimal phase shift angle between the primary and secondary sides and the quality factor Q of the resonant tank can be obtained, and then each resonance parameter can be calculated. In addition, the switching frequency is required to be always greater than the resonant frequency to achieve soft switching, and the

switching frequency is chosen as a compromise based on the switching loss.

The converter is simulated based on Matlab. The simulation results when the input is 30V and full load are shown in Fig. 9. The wavefo

Parameters	Value
Input voltage V_{in}	15~60V
Output voltage V_o	220V
Output power P_o	450W
Output frequency f_o	50Hz
Resonant inductor L_r	50uH
Resonant capacitor C_r	66nF
Resonant frequency f_r	88kHz
Switching frequency f_s	120kHz

Table 1 Circuit parameters

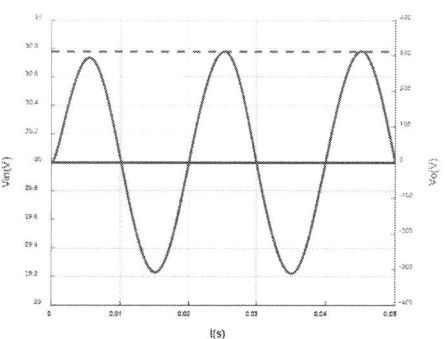

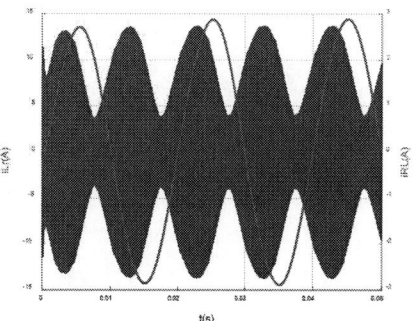

Fig. 9 Simulation results (full load, Vin=30V)

rms of the input and output voltage, output current and resonant current are given respectively. The output voltage and output current are both sinusoidal, and the resonant current has a sinusoidal envelope. Using Matlab to analyze the harmonics of the output voltage, the THD of the output voltage is only 1.25 %.

The overall structural design of the prototype is carried out for the topology as well as the magnetically integrated structure, as shown in Fig. 10, which adopts the structure of stacking the power board and the control board to make the converter more compact. The power board includes the switching devices on the primary and secondary sides and the corresponding decoupling capacitors and filter capacitors. The transformer with integrated resonant inductor and the filter inductor for the inverter output are also placed on the power board. The control board mainly includes the controller, auxiliary power supply, driver circuits and related sampling circuits, but the current sampling is placed on the power board. The separation of the power board and the control board can reduce electromagnetic interference and coupling between them. Thermal simulation of the entire converter was performed based on the designed structure and losses of corresponding parts. Fig. 11 shows the thermal simulation results of the converter with natural cooling and without heat sink. The high temperature of the primary side is because the primary side is the high-current side with high conduction losses, and a heat sink will be

Devices	Value
Primary devices	EPC2302
Secondary devices	GPI65030DFN
Gate drivers	1ED7275/NSi6602
Integrated transformer	DMR96 (2:14)
Filter inductor	NPA106060 (125 turns)
Controller	STM32F334C8T6
Isolated operational amplifier	NSi1312
Current Hall Sensor	NSM2012

Table 2   Device specifications of converter

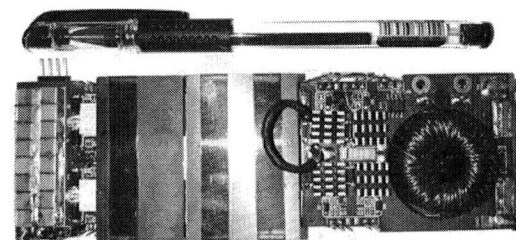

Fig. 12: Prototype pictures and dimensions

added in the subsequent experiments to improve its heat dissipation capability.

Table 2 gives the specific device specifications of the converter. Fig. 12 shows a 450W prototype and a pen. Its input voltage range is 15-60V and its output is 220V@50Hz with a power density of 46 W/in3. For the regulation range, it is worth noting that the converter can deliver full power when the input voltage is 25-35V. At other input voltages that are too low or too high, the power capability of the converter is going to be derated, but the voltage gain requirement can still be met. Fig. 13 shows the off-grid experimental waveform of the prototype, in which yellow is the resonant current, red is the secondary switch drive signal, blue is the output voltage, and green is the output current. Among them, (a) and (c) are the waveforms at light

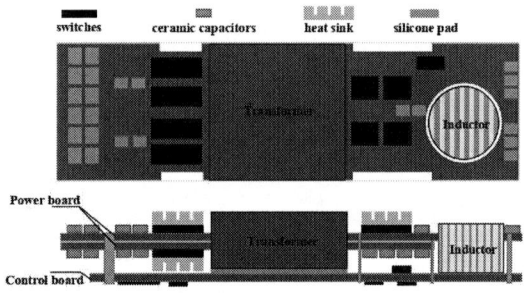

Fig. 10: Structural design diagram

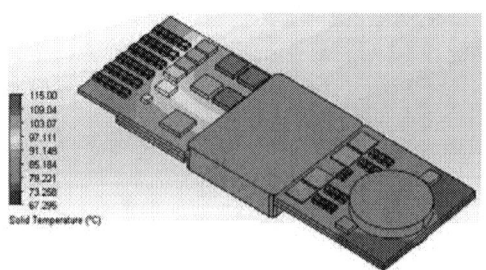

Fig. 11: Overall thermal simulation of converter

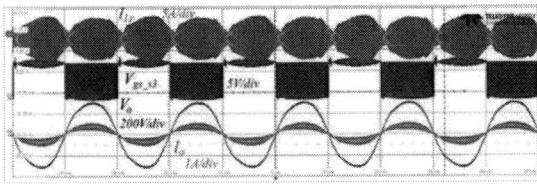

(a): V_in=15V，V_o=220Vac@60W

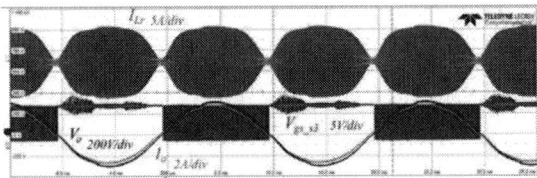

(b): V_in=30V，V_o=220Vac@450W

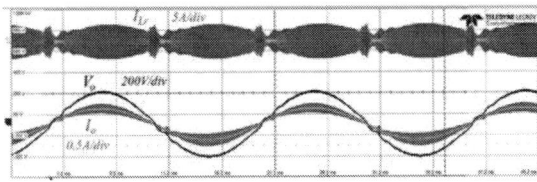

(c): V_in=60V，V_o=220Vac@60W

Fig. 13: Experimental results in off-grid mode

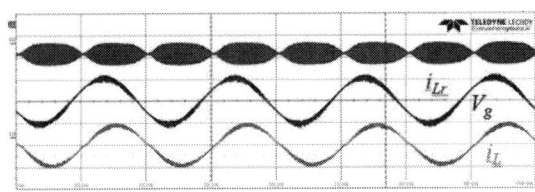

(a): PF=0.8

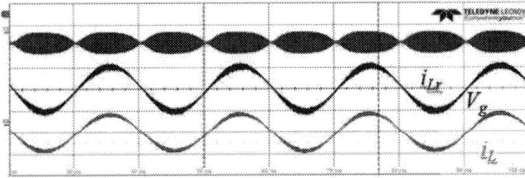

(b): PF=1

Fig. 14: Experimental results in on-grid mode

load when the input voltage is 15V and 60V respectively, while (b) is the waveform at full load when the input voltage is 30V. The prototype can achieve 220V AC output within the input range of 15V-60V DC, full load output from 25-35V, and THD<3% at 30V. When the prototype operates at full load with 30V input, the efficiency is measured to be 95.

Figure 14 shows the operating waveform of the converter in grid-connected mode. In the figure (a) is the waveform when pf=0.8, and (b) is the waveform when pf=1. The red waveform is the resonant circuit, the blue is the grid voltage, and the yellow is the grid-connected current. In grid-connected mode, the power factor can be adjusted while the current THD is <3%.

6 Conclusion

In this paper, the current RMS is optimized based on a single-stage resonant inverter, and the optimal design of magnetic integration is carried out, which reduces the loss and shrinks the size, and further improves the power density through the stacked structure. The optimized scheme is verified by simulation, and finally a prototype is built for experimental verification, which achieves a power density of 46 W/in3 and an efficiency of 95.3%.

References

[1] A. Trubitsyn, B. J. Pierquet, A. K. Hayman, G. E. Gamache, C. R. Sullivan and D. J. Perreault, "High-efficiency inverter for photovoltaic applications," 2010 IEEE Energy Conversion Congress and Exposition, Atlanta, GA, USA, 2010, pp. 2803-2810.

[2] H. Krishnaswami, "Photovoltaic microinverter using single-stage isolated high-frequency link series resonant topology," 2011 IEEE Energy Conversion Congress and Exposition, Phoenix, AZ, USA, 2011, pp. 495-500.

[3] D. R. Nayanasiri, D. M. Vilathgamuwa and D. L. Maskell, "Half-Wave Cycloconverter-Based Photovoltaic Microinverter Topology With Phase-Shift Power Modulation," in IEEE Transactions on Power Electronics, vol. 28, no. 6, pp. 2700-2710, June 2013.

[4] N. Kummari, S. Chakraborty and S. Chattopadhyay, "An Isolated High-Frequency Link Microinverter Operated with Secondary-Side Modulation for Efficiency Improvement," in IEEE Transactions on Power Electronics, vol. 33, no. 3, pp. 2187-2200, March 2018.

[5] S. Pal and S. Chattopadhyay, "Control of a Low-Voltage Microinverter with Secondary Phase-Shift Modulation for Grid-Connected Application," 2022 IEEE International Conference on Power Electronics, Drives and Energy Systems (PEDES), Jaipur, India, 2022, pp. 1-6.

[6] K. Wang, F. Wu, J. Su and G. Wang, "Three-Phase Single-Stage Three-Port High-Frequency Isolated DC–AC Converter," in IEEE Transactions on Power Electronics, vol. 38, no. 9, pp. 11113-11124, Sept. 2023.

PCIM Europe 2024, 11– 13 June 2024, Nuremberg   DOI: 10.30420/566262440

Experimental Investigation of Class Φ Inverter Under Various Load Conditions

Baptiste Daire [1], Christian Martin [1], Fabien Sixdenier [1], Charles Joubert [1], Loris Pace [1]

[1] Univ Lyon, Université Claude Bernard Lyon 1, INSA Lyon, Ecole Centrale de Lyon, CNRS, Ampere, France

Corresponding author: Baptiste Daire, baptiste.daire@univ-lyon1.fr
Speaker: Baptiste Daire, baptiste.daire@univ-lyon1.fr

Abstract

This document deals with class Φ inverter whose structure is at the origin class $\Phi2$ and EF_2 inverters. A tuning methodology is exposed and experimental demonstration of a 15 MHz - 25 W inverter is performed. The load resistor is then removed from its nominal value and the impact of this operation on the inverter's behavior is investigate. Experimental findings reveal that altering the load resistance induces a loss of Zero Voltage Switching (ZVS) or Zero derivative Voltage Switching (ZdVS), contingent upon whether the inverter is overloaded or underloaded. Furthermore, experimental results indicate the potential occurrence of reverse conduction in the switch, which could significantly impede the converter's overall performance. Notably, the measured efficiency demonstrates that superior performance is attainable by employing load resistance values higher than the nominal value, as opposed to lower values.

1 Introduction

In the field of power conversion, passive components (inductors and capacitors) account for most of the volume and mass of a converter. Operating at Very High Frequencies (VHF) leads to a drastic reduction of passive devices size and weight [1, 2]. One topology of inverter that can be used in order to implement VHF power converter is class Φ inverter [3–6] that operates in Zero Voltage Switching (ZVS) and Zero derivative Voltage Switching (ZdVS) and whose structure is at the origin of class $\Phi2$ [7] and class EF_2 [8, 9] inverters that found many applications in DC/DC power conversion, wireless power transfer or plasma generation. Indeed, its soft-switching operating conditions and simple driving requirement combined with the low voltage stress imposed to the switch compared to class E inverter [10–12] make this topology well-suited for high power capability operations in the VHF range (30-300 MHz). However, class Φ inverter is highly sensitive to load variations [13], posing a challenge to VHF power conversion community. While some modifications have been proposed to allow variable load operations by keeping ZVS only [14–16], there is no consensus to our knowledge on which soft-switching condition should be favored. Thus, this study aims to experimentally investigate the behavior of class Φ inverters under various load conditions. Initially, a design flow of a 15 MHz, 25 W nominal power class Φ inverter is proposed, followed by implementation and waveform analysis at the nominal operating point. Subsequently, the impact of deviating the load resistor from its nominal value while maintaining the same duty cycle is examined and discussed.

2 Generalities about Φ inverter

2.1 Operating principle

Class Φ inverter, depicted in figure 1, comprises two stages: the input and output stages. The input stage features a quarter-wavelength transmission line, assumed to be perfectly short-circuited for AC signals due to the input capacitor C_{in}. The output stage consists of an impedance $R + jL\omega$ and a series tuned filter L_sC_s, tuned to the switching frequency f_0 of the inverter. Additionally, a single source-grounded switch shunted with a capacitor C_p connects both stages. The quality factor of L_sC_s is presumed to be sufficiently high so that the output current $i_R(\omega t)$ flowing into the load can be expressed as:

$$i_R(\omega t) = I_R \cdot \sin(\omega t + \Phi) \qquad (1)$$

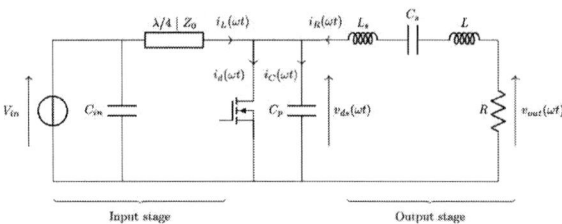

Fig. 1: Class Φ inverter

On the other hand, the quarter-wavelength input transmission line tends to impose two fundamental equations regarding the drain-to-source voltage $v_{ds}(\omega t)$ and the in-line current $i_L(\omega t)$ namely [17]:

$$
\begin{cases}
v_{ds}(\omega t) + v_{ds}(\omega t + \pi) = 2 \cdot V_{in} & (2) \\
i_L(\omega t) = i_L(\omega t + \pi) & (3)
\end{cases}
$$

As a consequence, and under the assumption that all elements are ideally tuned, theoretical waveforms in the inverter must adopt ones presented on figure 2.

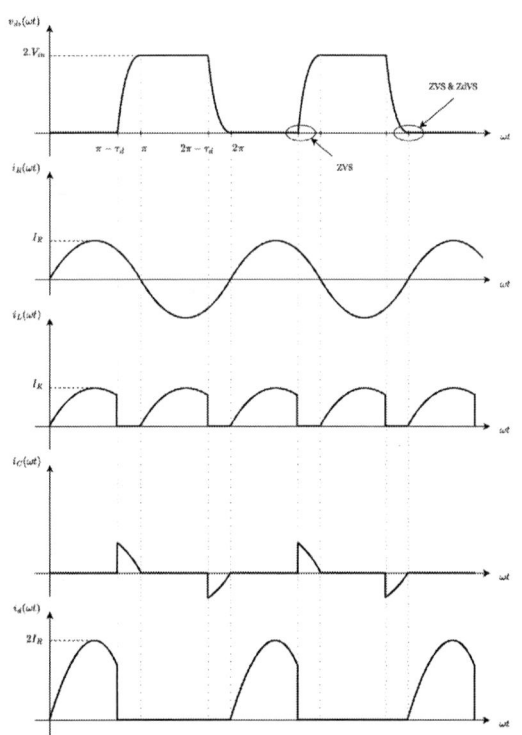

Fig. 2: Theoretical waveforms in a class Φ inverter

Note that the drain-to-source voltage of the switch reaches two times the input voltage and that the device operates in full soft-switching conditions:

– ZVS at the turn off.

– Both ZVS and ZdVS at the turn-on.

2.2 Design equations of class Φ inverter

According to [17], the design of a class Φ inverter is based on the impedance Z_{opt} that is measured between the drain and the source of the switch. Figure 3 provides a graphical explanation of the definition of Z_{opt}.

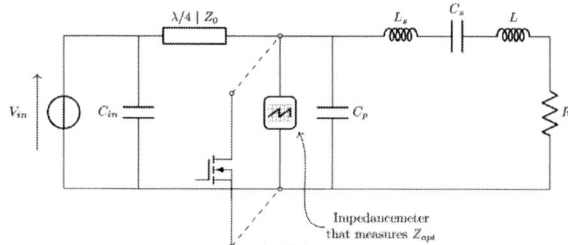

Fig. 3: Impedance Z_{opt} measured between the drain and the source of the switch

The optimal load impedance Z_{opt} that has to be presented to the transistor at the fundamental frequency f_0 and higher harmonics must respect conditions gathered in table 1.

(fundamental)	(even)	(odd)
f_0	$2nf_0$	$(2n+1)f_0$
$Z_{opt} \quad C_p \quad R$	Z_{opt}	$Z_{opt} \quad C_p$

Tab. 1: Optimal impedance Z_{opt} for fundamental, odd and even harmonics

Since the short-circuited quarter-wavelength transmission line presents a short-circuit for even harmonics and an open-circuit for odd harmonics, Z_{opt} is already satisfied for even harmonics. As a consequence, the design process of class Φ inverter consists of tuning the output stage so that a $R + jL\omega_0$ impedance is presented to the transistor at the switching frequency only. Thanks to the L_sC_s filter tuned to f_0, the output branch of the circuit theoretically appears as an open-circuit for higher harmonics to the transistor. Values of R and L are linked to the output power P_{out} delivered by the inverter, the value of C_p, the switching pulsation $\omega_0 = 2\pi f_0$ and the input DC voltage V_{in} by following equations [17]:

$$
\begin{cases}
R = \dfrac{2\,(1+\cos(\tau_d))^2}{\pi^2}\dfrac{V_{in}^2}{P_{out}} & (4) \\[2ex]
L = \dfrac{\tau_d - \frac{1}{2}\sin(2\tau_d)}{\sin^2(\tau_d)}\dfrac{R}{\omega_0} & (5) \\[2ex]
P_{out} = \dfrac{2}{\pi}\left(\dfrac{1+\cos(\tau_d)}{\sin(\tau_d)}\right)^2 \omega_0 C_p V_{in}^2 & (6)
\end{cases}
$$

Where the angle τ_d (in rad) corresponds to the duration that is required by the circuit to charge and discharge the shunt capacitor C_p as explained on figure 4 where the ideal drain-to-source waveform in a class Φ inverter is depicted.

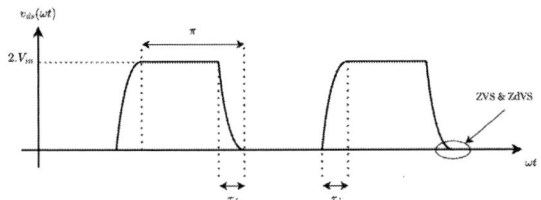

Fig. 4: Ideal drain-to-source voltage in class Φ inverter

This angle τ_d is required in order to obtain both ZVS and ZdVS switching conditions. Its value depends on the operating point of the inverter as well as the value of C_p. Indeed, equation (6) leads to the following expression of τ_d:

$$
\tau_d = 2 \cdot \mathrm{atan}\left(\sqrt{\frac{4 V_{in}^2 f_0 C_p}{P_{out}}}\right) \tag{7}
$$

Note that due to the theoretical drain-to-source waveform illustrated in Figure 4, the duty cycle D employed to drive the transistor must adhere to equation (8).

$$
D = \frac{1}{2} - \frac{\tau_d}{2\pi} \tag{8}
$$

3 Practical implementation of a 15 MHz - 25 W class Φ inverter

3.1 Design and implementation of the output stage

3.1.1 Switch selection

A 15 MHz switching frequency, 25 W output power and 25 V input voltage operating point has been chosen in order to design the converter. As the drain-to-source voltage of the transistor theoretically reaches $2 \cdot V_{in}$, a 100 V - 90 A transistor has been selected (GS61008P from GaN Systems) so that a theoretical safety margin of 2 is applied on the voltage constraint of the transistor.

When dealing with VHF power converters, drain-to-source capacitance of the switch is systematically absorbed by C_p capacitance. However, this capacitance is highly non-linear regarding the drain-to-source voltage that is applied on the switch, which hardened the design of such converters. However, a study was performed in [18], demonstrating that an equivalent time related and constant output $C_{oss(tr)}$ capacitance can be considered between the drain and the source of the switch. This capacitance being given by the following relation:

$$
C_{oss(tr)} = \frac{1}{V_{DS}}\int_0^{V_{DS}} C_{oss}(v)\,dv \tag{9}
$$

According to the datasheet of the device, a 385 pF time related output capacitance $C_{oss(tr)}$ can be expected between the drain and the source of the switch from 0 V to 50 V. For conveniance, C_p is only composed of the $C_{oss(tr)}$ output capacitance of the switch so that $C_p = 385$ pF can be used in previously exposed design equations.

3.1.2 Calculation of D, R and L

Therefore, equation (7) can be used to express the value of τ_d as:

$$
\tau_d = 1.3\ \mathrm{rad} \tag{10}
$$

Note that the ideal value of duty cycle D that theoretically provides soft switching conditions can be determined using equation (8) and is equal to 29.3%. Using this value of τ_d in equations (4) and (5) leads to an optimal load impedance Z_{opt} that should be presented to the transistor at the switching frequency f_0 of $R + jL\omega_0$ with:

$$
\begin{cases}
R = 8.14\ \Omega \\
L = 97\ \mathrm{nH}
\end{cases} \tag{11}
$$

3.1.3 Design of the $L_s C_s$ filter

We consider the following electrical circuit that represents the output stage of the inverter:

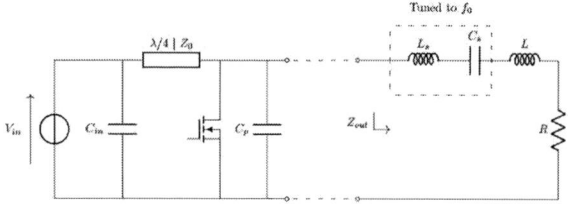

Fig. 5: Output stage of the inverter

Since the L_sC_s filter is tuned to f_0, the following relation can be written:

$$\omega_0^2 = \frac{1}{L_sC_s} \qquad (12)$$

In addition, the abslute value of the impedance Z_{out} at the switching frequency f_0 of the inverter can be expressed as:

$$|Z_{out}|_{\omega=\omega_0} = \sqrt{R^2 + (L\omega_0)^2} \qquad (13)$$

and the absolute value of Z_{out} for $f = 3f_0$ is equal to:

$$|Z_{out}|_{\omega=3\omega_0} = \sqrt{R^2 + \left((L+L_s)3\omega_0 - \frac{1}{3\omega_0 C_s}\right)^2} \qquad (14)$$

Because the goal of the L_sC_s filter is to block current harmonics at frequencies higher than f_0 i.e. to present a higher impedance at higher frequencies than the switching frequency of the inverter, we consider the following relation as a tuning criteria for designing the L_sC_s filter:

$$|Z_{out}|_{\omega=3\omega_0} = Q \cdot |Z_{out}|_{\omega=\omega_0} \qquad \text{with } Q > 1$$

Note that the choice of considering $3f_0$ instead of $2f_0$ as the reference frequency for designing the L_sC_s filter was motivated by the specific shape of the drain-to-source voltage that is applied at the input of the filter. Indeed, the theoretical quasi-square drain-to-source voltage is mainly composed of odd harmonics. Hence, chosing a $2f_0$ frequency in order to shape the filter would have been less relevant. Figure 6 presents a simplified depiction of the desired behavior concerning the absolute value of Z_{out}.

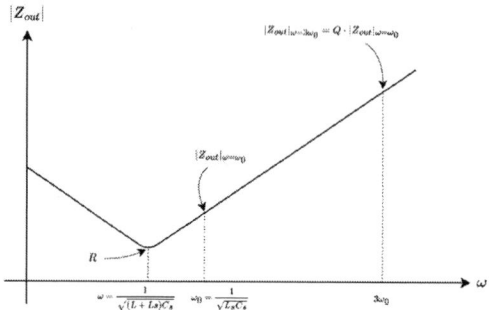

Fig. 6: Simplified desired behavior of $|Z_{out}|$

Using expressions obtained in equations (12), (13) and (14) leads the following values of L_s and C_s that have to be used in order to obtained an output impedance whose absolute value is Q times higher at 3 times the switching frequency than the one at the switching frequency f_0 of the inverter:

$$\begin{cases} L_s = \dfrac{3}{8\omega_0}\left(\sqrt{(Q^2-1)R^2 + (Q\cdot L\omega_0)^2} - 3L\omega_0\right) & (15) \\ C_s = \dfrac{1}{L_s\omega_0^2} & (16) \end{cases}$$

Chosing a high Q factor in above equation leads to better harmonic blocking. However, higher Q factors lead to bigger values of inductors whose implementation might be problematic at such frequencies. Hence, a $Q = 9$ factor is used here as it seems to be a good compromise, leading to the following theoretical values of L_s and C_s:

$$\begin{cases} L_s = 343 \text{ nH} \\ C_s = 328 \text{ pF} \end{cases}$$

To conclude this part dedicated to the analytical design of class Φ inverter, the following values of components and duty cycle can theoretically be used in order to satisfy a 15 MHz, 25 W, 25 V input voltage inverter:

Name	Theoretical value
R	8.14 Ω
L	97 nH
L_s	343 nH
C_s	328 pF
C_p	385 pF
D	29.3 %

Tab. 2: Theoretical values of components and duty cycle

Thus, table 3 lists all components that were used in the actual converter.

Name	Reference	Value
R	MP9100-7.50-1%	7.5 Ω
$L + L_s$	AWG16	438 nH @ 15 MHz
C_s	100B331JW600XC100	330 pF
C_{in}	CKG57NX7R2E105M500JH	2x1 µF
	CB037E0104KBA	2x0.1 µF
MOS	GS61008P	
Driver	UCC27611	

Tab. 3: Components used to realize the inverter

Caract. imped.	50 Ω (± 2 Ω)
Manufacturer	QAXIAL
Reference	RG303/U
Dielectric	PTFE
ε_r	2.04
$\tan(\delta)$	0.0003
A @ 100 MHz	12.6 dB/100m
Measured R_{DC}	78 mΩ/m

Tab. 4: Physical parameters of the coaxial cables used in order to realize the different input networks

Figure 7 depicts the absolute value of the measured impedance Z_{out}. Note that the behavior of the output stage is closed to the desired one at the switching frequency of the inverter (see tab. 2).

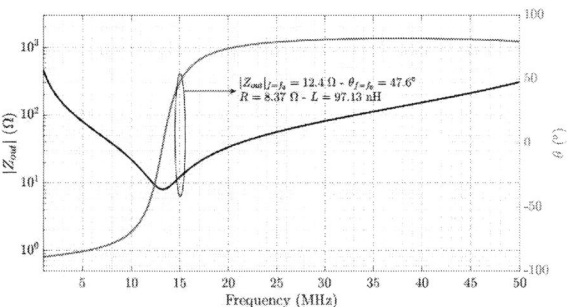

Fig. 7: Measured value of $|Z_{out}|$

One may notice that the behavior of the output stage at 3 times the switching frequency (i.e. 45 MHz) is quite far from the desired one. Indeed, absolute value of Z_{out} at 45 MHz is approximatively two times higher than the desired value of 110 Ω ($= Q \cdot |Z_{out}|_{\omega=\omega_0}$). This is mainly due to the fact that we approach the first resonance of the $L + L_s$ inductor. This phenomena has no impact on inverter's behavior, as shown later in the paper.

3.2 Practical implementation of the quarter-wavelength input transmission line

At the operating switching frequency, integration of the quarter-wavelength transmission line on the PCB is tough because of its length (approximatively 2.5 meters when using FR4). As a consequence, a PTFE (Poly Tetra Fluor Ethylene) coaxial cable was chosen to realize the input line. Table 4 lists physical caracteristics of the used cable.

When using a PTFE coaxial cable, the corresponding wavelength is about 14 meters so that a $\lambda/4$ length corresponds to a 3.5 meters coaxial cable. Figure 8 shows the measured impedance of the short-circuited cable as a function of frequency from 1 MHz to 150 MHz. As desired, the resonant frequency of the line is equal to 15 MHz.

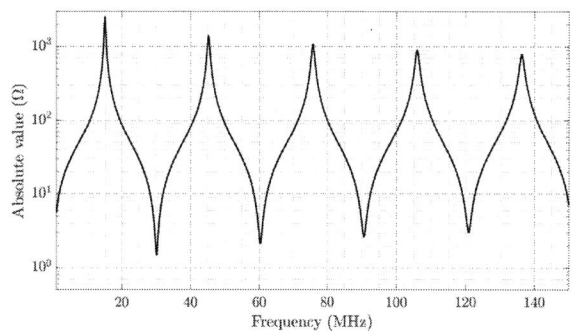

Fig. 8: Measured impedance of the short-circuited coaxial cable

Finally, pictures of the converter are presented on figure 9.

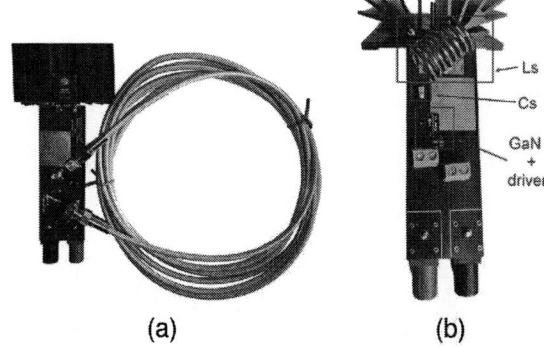

Fig. 9: Pictures of the converter : (a) Complete converter (b) Zoom on the PCB

4 Experimental results

All waveforms were obtained using a 1 GHz passive probe (TPP1000 from tektronix) with a MSO 5104 Mixed Signal Oscilloscope from Tektronix.

4.1 Nominal operating point

Figure 10 depicts drain to source and output voltages for nominal operating point. It is clear that both ZVS and ZdVS are achieved here. In addition, the maximum drain-to-source voltage reaches 2 x 25 V = 50 V as expected.

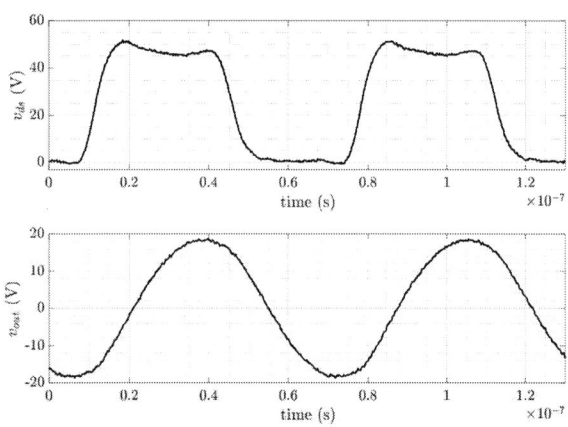

Fig. 10: Measured drain to source and output voltages at nominal operating point

In this specific case, the output power was measured to be equal to 23.3 W which is close to the desired value of 25 W (-6.8 %).

4.2 Variable load operation

4.2.1 Methodology

Load variation was implemented by deviating the load resistor from its nominal value, which is set at 7.5 Ω using an MP9100 resistor. Two cases are then defined:

- An overloaded case when the load resistance value is lower than its optimal one.

- An underloaded case when the load resistance value is higher than its optimal one.

All used resistors are from the MP9100 serie so that power rating, thermal drift, package inductance, and other caracteristics remained unchanged from a resistor to an other. During the experiment, circuit parameters especially duty cycle, remained the same.

4.2.2 Waveforms

Figure 11 shows the measured drain-to-source voltage when the inverter is being underloaded. In this specific case, circulating currents are not high enough so that ZVS is lost. However, it appears that ZdVS is preserved even if the duty cycle remains unchanged. Note that the symmetry of the drain-to-source voltage is due to equation (3).

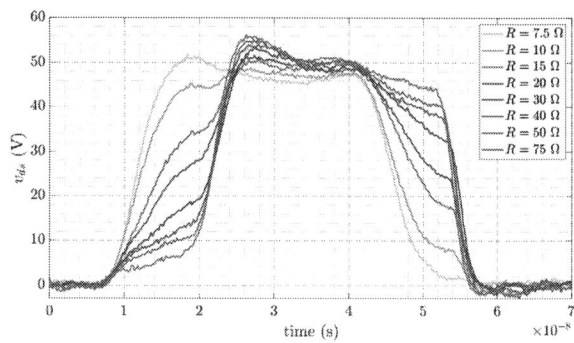

Fig. 11: Measured drain-to-source voltage in underloaded case

Figure 12 shows the measured drain-to-source voltage when the inverter is being overloaded. In this specific case, circulating currents are higher than required so that ZdVS is lost. Because the duty cycle remains unchanged, the switch enters in reverse conduction as shown on figure 12.

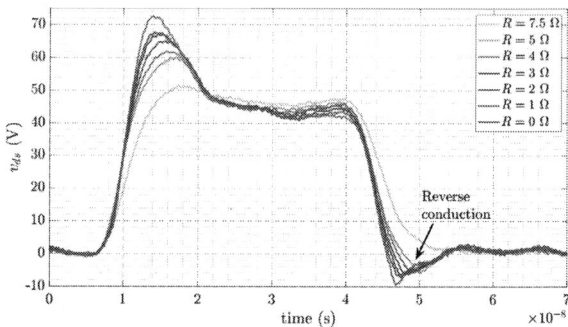

Fig. 12: Measured drain to source voltage in overloaded case

Because the drain-to-source voltage is clamped when the switch operates in reverse conduction, its value is close to zero: strictly speaking, ZVS is still achieved in this case. However reverse conduction is expected to increase switching losses which could be an issue at such frequencies.

4.2.3 Overall efficiency

Figure 13 shows measured efficiency of the inverter for various load conditions. It seems that the proposed class Φ inverter operates better when being underloaded rather than overloaded.

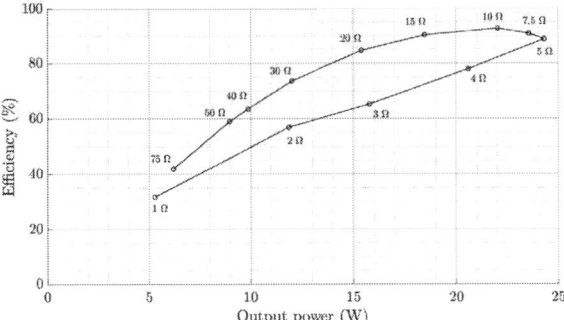

Fig. 13: Measured efficiency of the inverter

As explained previously, reverse conduction leads to higher losses in the switch, this may partly explain why overloaded class Φ inverter operates badly compared to underloaded inverter.

5 Discussion

Presented results highlight several phenomena :

- A change in the value of load resistor results in a lost ZVS or ZdVS depending on whether the inverter is overloaded or underloaded.

- Underloaded class Φ inverter seems to achieve ZdVS with no change in the duty cycle.

- In term of overall efficiency, the proposed class Φ inverter seems to operate better when being underloaded rather than overloaded

- In the specific case of an overloaded class Φ inverter, the switch enters in reverse conduction. Strictly speaking, this results in the maintain of ZVS but it is also believed that losses in the switch are greater because of reverse conduction.

This last point suggests that duty cycle management techniques that maintain ZVS such as those presented in [15, 19] and [20] could be beneficial in order to improve efficiency of overloaded class Φ inverter. In [20], an increase of 4.2 % in overall efficiency was measured when using a duty cycle management circuit in order to avoid reverse conduction in the switch. However, authors did not specify if additionnal consumption induced by the use of this additional circuit was taken into account in efficiency measurement. In any case, this increase may not be sufficient to compete with overall efficiencies that were measured when keeping ZdVS with no duty cycle management (see figure 13). In any

case, this underlines the need for more in-depth investigations. Because duty cycle adjustment can be performed in order to avoid reverse conduction in the switch , a new comparison between underloaded and overloaded class Φ inverter with duty cycle management could be performed. Hence, a more appropriate conclusion could be provide about wether class Φ inverter operates better when being overloaded or underloaded and about the need for additional circuit that performs duty cycle adjustment. In addition, several elements have not been taken into account in this work such as EMC issues or impact of load variation on the shape of the current flowing into the line, which could further affect overall performances of the inverter.

6 Conclusion

Class Φ inverter and its derivatives (Φ_2, EF_2) represent promising topologies for power transmission within the VHF range. Despite their utilization in VHF DC/DC power conversion and wireless power transfer applications, their performance under varied load conditions remains a significant concern necessitating thorough investigation. In this paper, a design flow of a 15 MHz - 25 W class Φ inverter is presented. Corresponding converter is implemented and its waveforms at the nominal operating point are presented. Then, load resistor is removed from its nominal value, impact of this operation on the behavior of the inverter is then investigated. Key observation are then exposed and discussed, enlighting that there is no evidence, in term of overall efficiency of the converter, that ZVS should be kept instead of ZdVS.

References

[1] Y. Wang, O. Lucia, Z. Zhang, S. Gao, Y. Guan, and D. Xu, "A review of high frequency power converters and related technologies," *IEEE Open Journal of the Industrial Electronics Society*, vol. 1, pp. 247–260, 2020. DOI: 10.1109/OJIES.2020.3023691.

[2] D. J. Perreault, J. Hu, J. M. Rivas, Y. Han, O. Leitermann, *et al.*, "Opportunities and challenges in very high frequency power conversion," in *2009 Twenty-Fourth Annual IEEE Applied Power Electronics Conference and Exposition*, 2009, pp. 1–14. DOI: 10.1109/APEC.2009.4802625.

[3] J. W. Phinney, D. J. Perreault, and J. H. Lang, "Radio-frequency inverters with transmission-line input networks," *IEEE Transactions on Power Electronics*, vol. 22, no. 4, pp. 1154–1161, 2007. DOI: 10.1109/TPEL.2007.900465.

[4] C. Liu, X. Li, Y. Zhao, T. Qi, X. Du, *et al.*, "Investigation of high-efficiency parallel-circuit class-ef power amplifiers with arbitrary duty cycles," *IEEE Transactions on Industrial Electronics*, vol. 68, no. 6, pp. 5000–5012, 2021. DOI: 10.1109/TIE.2020.2991924.

[5] C. Liu, H. Zhang, W. Chen, and F. M. Ghannouchi, "Novel design space of broadband high-efficiency parallel-circuit class-ef power amplifiers," *IEEE Transactions on Circuits and Systems I: Regular Papers*, vol. 69, no. 9, pp. 3465–3475, 2022. DOI: 10.1109/TCSI.2022.3177283.

[6] Z. Zhang, Z. Cheng, H. Ke, G. Liu, H. Sun, and S. Gao, "Design of broadband hybrid class EF power amplifier based on novel capacitance compensation structure," *Microwave and Optical Technology Letters*, vol. 62, no. 3, pp. 1069–1076, 2020. DOI: https://doi.org/10.1002/mop.32129.

[7] J. M. Rivas, Y. Han, O. Leitermann, A. Sagneri, and D. J. Perreault, "A high-frequency resonant inverter topology with low voltage stress," in *2007 IEEE Power Electronics Specialists Conference*, 2007, pp. 2705–2717. DOI: 10.1109/PESC.2007.4342446.

[8] Z. Kaczmarczyk, "High-efficiency class e, $hboxEF_2$, and $hboxE/F_3$ inverters," *IEEE Transactions on Industrial Electronics*, vol. 53, no. 5, pp. 1584–1593, 2006. DOI: 10.1109/TIE.2006.882011.

[9] S. Aldhaher, D. C. Yates, and P. D. Mitcheson, "Design and development of a class ef_2 inverter and rectifier for multimegahertz wireless power transfer systems," *IEEE Transactions on Power Electronics*, vol. 31, no. 12, pp. 8138–8150, 2016. DOI: 10.1109/TPEL.2016.2521060.

[10] G. D. Ewing, "High-efficiency radio-frequency power amplifiers," Ph.D. dissertation, Oregon State University, 1964.

[11] N. Sokal and A. Sokal, "Class e-a new class of high-efficiency tuned single-ended switching power amplifiers," *IEEE Journal of Solid-State Circuits*, vol. 10, no. 3, pp. 168–176, 1975. DOI: 10.1109/JSSC.1975.1050582.

[12] K. N. Surakitbovorn and J. M. Rivas-Davila, "On the optimization of a class-e power amplifier with gan hemts at megahertz operation," *IEEE Transactions on Power Electronics*, vol. 35, no. 4, pp. 4009–4023, 2020. DOI: 10.1109/TPEL.2019.2939549.

[13] Y. Yanagisawa, Y. Miura, H. Handa, T. Ueda, and T. Ise, "Characteristics of isolated dc–dc converter with class phi-2 inverter under various load conditions," *IEEE Transactions on Power Electronics*, vol. 34, no. 11, pp. 10887–10897, 2019. DOI: 10.1109/TPEL.2019.2898514.

[14] L. Roslaniec, A. S. Jurkov, A. A. Bastami, and D. J. Perreault, "Design of single-switch inverters for variable resistance/load modulation operation," *IEEE Transactions on Power Electronics*, vol. 30, no. 6, pp. 3200–3214, 2015. DOI: 10.1109/TPEL.2014.2331494.

[15] V. Massavie, G. Despesse, S. Carcouet, and X. Maynard, "Class φ2 zvs regulation applied to l-piezo inverter," in *2022 20th IEEE Interregional NEWCAS Conference (NEWCAS)*, 2022, pp. 490–494. DOI: 10.1109/NEWCAS52662.2022.9841975.

[16] Y. Jiang, J. Liang, H. Wang, Y. Liu, and M. Fu, "Load-impedance-insensitive design of high-efficiency class ef inverters," *IEEE Transactions on Power Electronics*, vol. 39, no. 2, pp. 1958–1962, 2024. DOI: 10.1109/TPEL.2023.3330515.

[17] A. Grebennikov, "High-efficiency class-fe tuned power amplifiers," *IEEE Transactions on Circuits and Systems I: Regular Papers*, vol. 55, no. 10, pp. 3284–3292, 2008. DOI: 10.1109/TCSI.2008.924123.

[18] M. Kasper, R. M. Burkart, G. Deboy, and J. W. Kolar, "Zvs of power mosfets revisited," *IEEE Transactions on Power Electronics*, vol. 31, no. 12, pp. 8063–8067, 2016. DOI: 10.1109/TPEL.2016.2574998.

[19] D. Zhang, D. Lyu, M. Zhang, M. Dong, R. Min, *et al.*, "Linear equivalent model for vhf class φ2 inverter based on spectrum quantification method to reduce gan reverse conduction loss," *IEEE Access*, vol. 9, pp. 61635–61645, 2021. DOI: 10.1109/ACCESS.2021.3074637.

[20] D. Zhang, D. Lyu, L. Li, Y. Wang, R. Min, and Q. Tong, "Optimized duty cycle control to reduce the reverse conduction loss of gan hemts in a very high frequency class φ2 inverter," in *2020 IEEE 9th International Power Electronics and Motion Control Conference (IPEMC2020-ECCE Asia)*, 2020, pp. 1069–1076. DOI: 10.1109/IPEMC-ECCEAsia48364.2020.9367939.

PCIM Europe 2024, 11– 13 June 2024, Nuremberg   DOI: 10.30420/566262441

Analysis, Modeling, Design, and Limitations of Current Injection based UPF Rectifier with Small DC-Link Capacitor

Ramkrishan Maheshwari[1], Ankur Srivastava[1], Prashant Surana[1], Thomas Ebel[1], Lasse Chr. Larsen[2], Egon Hansen[2]

[1] Centre of Industrial Electronics, University of Southern Denmark, Soenderborg, Denmark
[2] OJ Electronics, Denmark

Corresponding author: Ramkrishan Maheshwari, ramkrishan@sdu.dk
Speaker: Ramkrishan Maheshwari, ramkrishan@sdu.dk

Abstract

Unity power factor (UPF) rectifiers are typically used as AC-AD converters in various power electronics application. UPF rectifier based on current injection circuit is a low-cost solution. However, analysis for such circuits is not discussed in detail in literature. This paper analyses one of such circuit in details. Based on analysis, design guidelines for passive component selection are presented. Furthermore, the active device (MOSFET) stresses are also analysed in the paper which can further be used to select proper MOSFETs. In addition, the analysis also shows certain limitation of the topology under consideration. Such limitations provide details about the application where such UPF rectifiers can be used. The analysis presented in the paper is verified by a MATLAB Simulink model of the rectifier.

1   Introduction

A unity power factor (UPF) rectifiers are typically used as an ac-dc converter in applications where grid power quality requirements are strict [1]-[4]. To meet these requirements, a diode-bridge rectifier with a current injected circuit may be used [3]-[11]. The current injection (CI) circuits may use passive components [5]-[7] or active components to inject a controlled current [1]-[4], [8]-[11]. The passive components-based CI circuits are bulkier than the active components-based CI circuits. However, the active components-based CI circuits need to control a time-varying current (not dc). This requires high bandwidth controller which requires high sampling frequency. This translates to requirement of high switching frequency of the power semiconductor switches used in active components-based CI circuits. This requirement can be easily fulfilled by using SiC based power semiconductor devices [8], [11]. This makes the active components-based CI circuits more attractive in recent years.

Ref. [9] and [12] discuss the operation of the active components-based CI circuits. However, the inductor current and MOSFET current analysis are not present which are essential to design the circuit. These publication uses IGBT for the CI circuit. Ref [8] shows the operation of the circuit in practical implementation using MOSFETs. However, it does not present the analysis of the circuit. An accurate analysis for an active components-based CI circuit is presented in the paper. The analysis provides the details of the steady-state operation of the CI circuit. The analysis presents the current stress of the passive and active components used in the circuit. This stress analysis helps in selecting and optimizing the components in the circuits. Using the analysis, a design of the CI circuit is presented in the paper. Furthermore, the analysis also derives the expression of current to be injected to achieve the UPF operation for constant power load. This can also be considered as a limitation that the derived current expression achieves UPF operation for constant power load only.

The paper presents the steady-state analysis of the CI circuit in Section II. The analysis describes

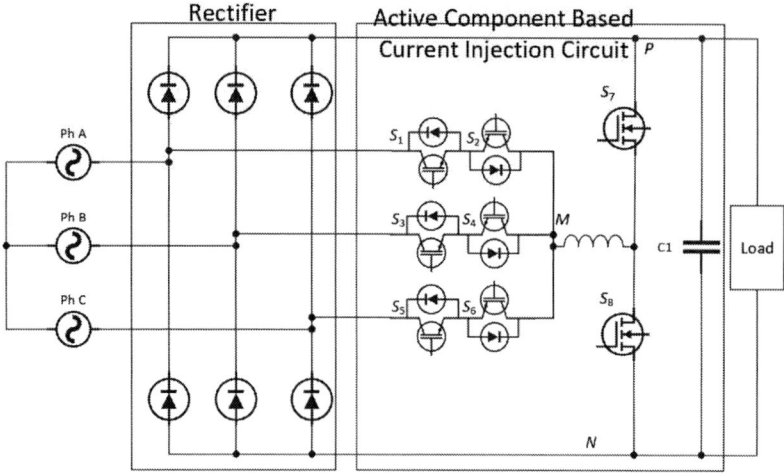

Fig. 1 Rectifier with active component-based CI injection circuit.

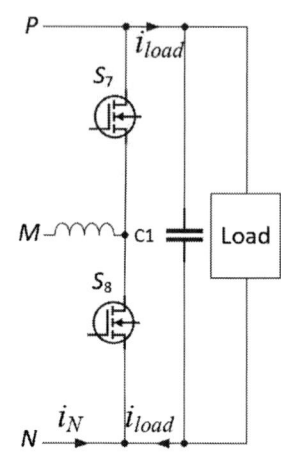

Fig. 2 Equivalent circuit for CI injection circuit

the load conditions for UPF operation. Using the analysis, the mathematical equations for current stress for MOSFETs and inductor are presented in Section III. Simulation results are presented in Section IV.

2 Steady-State Analysis

The rectifier circuit with active component-based CI circuit is shown in Fig. 1. The figure clearly marks rectifier circuit and the active component-based CI circuit. The CI circuit consists of six low switching frequency switches (S_1-S_6). These switches are turned on in a way that the phase with middle voltage amplitude among three phases is connected to the inductor terminal marked with M. The other end of the inductor is connected to the midpoint of the half bridge circuit formed by S_7 and S_8. The top and bottom terminal of the half-bridge circuit have the potential of phase with maximum and minimum voltage values, respectively. Therefore, an equivalent circuit can be drawn as shown in Fig. 2.

Switches S_7 and S_8 have high switching frequency of the order of 100 kHz, and only one of them is turned on at a given instant. If S_8 is turned on, during dT_s, the voltage across inductor is given by

$$v_L = v_{MN} = v_{mid} - v_{min} \qquad (1)$$

If S_7 is turned on, during $(1-d)T_s$, the voltage across inductor is given by

$$v_L = v_{PM} = v_{max} - v_{mid} \qquad (2)$$

where v_{max}, v_{mid}, and v_{min} are the maximum, middle, and minimum voltage values among three-phase voltages.

Due to high switching frequency operation, the net change in the inductor current in a switching period can be neglected. With this assumption, from (1) and (2)

$$\frac{d}{1-d} = \frac{v_{max}-v_{mid}}{v_{mid}-v_{min}} \quad \rightarrow \quad d = \frac{v_{max}-v_{mid}}{v_{max}-v_{min}}$$

$$(3)$$

For $-\pi/6 < \omega t < \pi/6$, if the six-pulse operation of diode bridge rectifier is considered,

$$v_{PN} = v_{max} - v_{min} = \sqrt{3} V_m \cos(\omega t)$$

$$(4)$$

and

$$v_{PM} = v_{max} - v_{mid} = \sqrt{3} V_m \sin\left(\omega t + \frac{5\pi}{6}\right)$$

$$(5)$$

where V_m is the peak of a phase voltage, and ω is the supply frequency in rad/s. Using (3)-(5),

$$d = \frac{\sin\left(\omega t + \frac{5\pi}{6}\right)}{\cos(\omega t)} \qquad (6)$$

After finding the function of duty cycle d, it is required to find the current in the inductor. For this, KCL will be applied at node N which yields

$$i_N + i_{load} + di_L = 0 \qquad (7)$$

where i_N, i_{load}, and i_L denote the current at terminal N, the load current, and the inductor current, respectively. To solve for i_L, the load current and the output current from rectifier are needed. The load current can be obtained by considering the constant power load (P_{load}) at the rectifier output and using (4),

$$i_{load} = \frac{P_{load}}{\sqrt{3}V_m \cos(\omega t)} \qquad (8)$$

An important assumption for (6) is that the current drawn by the capacitor is neglected at 300 Hz. Furthermore, if the current at terminal N is given by

$$i_N = I_m \sin\left(\omega t - \frac{2\pi}{3}\right) \qquad (9)$$

where I_m is the peak value of the current. Using (6)-(9) and $P_{load} = \frac{3\,V_m I_m}{2}$

$$i_L = I_m \sin(\omega t) \qquad (10)$$

A detailed step-by-step derivation of (10) is given in Appendix. To obtain the expression for i_L, the following assumption is made.

1. The load to the rectifier is constant power load.
2. The current flowing through the phase with the minimum voltage value is in-phase with its voltage.
3. Efficiency of the rectifier circuit is 1.

Using the abovementioned assumptions, it is shown that the inductor current is given by (10). This shows that current flowing through the phase with middle voltage value is in-phase with its voltage. This verifies the second assumption. Furthermore, the derivation is valid for constant power load. If the load is changed, the inductor current will not be equal to (10) and voltage and current of phase with middle value will not be in-phase, and the UPF operation will not be achieved.

Eq. (10) shows the current for $-\pi/6 < \omega t < \pi/6$. For the whole cycle, the inductor current can be plotted as shown in Fig. 3.

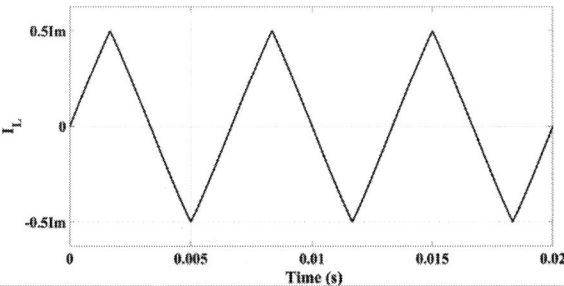

Fig. 1   Required inductor current.

3   Component Design Guidelines and Stress Analysis

3.1   Inductor

Using the analysis of the previous section, various equations to select the inductors and MOSFETs can be derived. To select the inductance value, the current ripple in a switching period is required which is given by

$$\Delta I_L = \frac{(v_{\text{mid}} - v_{\text{min}})}{L} d T_s \qquad (11)$$

where L denotes the inductance of inductor of CI circuit. Using (3)

$$\Delta I_L = (v_{\text{max}} - v_{\text{min}})(1 - d)d\frac{T_s}{L} \qquad (12)$$

This shows that the current ripple in the inductor is a function of time, and the maximum value of the current ripple is achieved for $d = 0.5$ which occurs at $\omega t = 0$. At this instant, the voltage $(v_{\text{max}} - v_{\text{min}})$ also has its maximum value $\sqrt{3}V_m$. Therefore, the maximum value of the current ripple is given by

$$\Delta I_{\text{max}} = \frac{\sqrt{3}V_m}{4}\frac{T_s}{L} \qquad (13)$$

This closed form equation can be used to find the inductance for a given maximum current ripple. As an example, for $\Delta I_{\text{max}} < 5\,\text{A}$, $V_m = 230\sqrt{2}$, and $T_s = 10\,\mu\text{s}$, L should be more than 281 µH. In this paper, an inductor with 300 µH inductance is chosen. Other important parameters to design the inductor are peak and rms current which can be obtained from Fig. 3, neglecting the current ripple in a switching period, as given by

$$I_{\text{pk},L} = \frac{I_m}{2} \qquad (14)$$

$$I_{\text{rms},L} = \sqrt{\frac{3}{\pi}\int_{-\frac{\pi}{6}}^{\frac{\pi}{6}} I_m^2 \sin^2(\omega t)\,d\omega t} = \frac{I_m}{\sqrt{2}}\sqrt{\left(1 - \frac{3\sqrt{3}}{2\pi}\right)} =$$

$$0.29 I_m \qquad (15)$$

3.2   MOSFETs (S7 and S8)

The MOSFETs of the half bridge switches at high frequency. The maximum blocking voltage seen by the MOSFETs is equal to $\sqrt{3}V_m$, and the rms current rating of the MOSFETs can be obtained by the following method in two steps. The first step is to find the rms current through MOSFET (S_8) in a switching period, neglecting the current ripple in a switching period, is given by

$$I_{rms,Tsw} = I_L\sqrt{d} = I_m \sin(\omega t)\sqrt{\frac{\sin\left(\omega t + \frac{5\pi}{6}\right)}{\cos(\omega t)}}$$

$$(16)$$

for $-\pi/6 < \omega t < \pi/6$. Using (16), the rms current through the MOSFET can be obtained by using

$$I_{rms,S_M} = \sqrt{\frac{3}{\pi} \int_{-\frac{\pi}{6}}^{\frac{\pi}{6}} I_{rms,Tsw}^2 \, d\omega t} = \frac{I_m}{2}\sqrt{\left(1 - \frac{3\sqrt{3}}{2\pi}\right)} =$$

$$0.208 I_m \qquad (17)$$

3.3 IGBTs (S_1 - S_6) [12]

Switches S_1 - S_6 have switching frequency of 50 Hz. The devices used to realize S_1 - S_6 are IGBTs and diodes. These device selections are based on the blocking voltage and rms and average current flowing through the devices. The maximum blocking voltage seen by the switches is equal to $\sqrt{3}V_m$. The expressions for the rms and average currents are given by

$$I_{rms,S_{IGBT}} = \sqrt{\frac{I_{rms,L}^2}{6}} = I_m\sqrt{\left(\frac{1}{12} - \frac{\sqrt{3}}{8\pi}\right)} \qquad \text{and}$$

$$(18)$$

$$I_{av,S_{IGBT}} = \frac{1}{\pi}\int_0^{\frac{\pi}{6}} I_m \sin(\omega t)\, d\omega t = \frac{I_m}{\pi}\left(1 - \frac{\sqrt{3}}{2}\right)$$

$$(19)$$

, respectively.

3.4 DC-link capacitor

From (8), the load current is a function of rectified output voltage from the diode bridge rectifier. The voltage has 300 Hz harmonic component, and the load current will also have 300 Hz harmonic component. This is required for the derivation of the inductor current given by (10). Therefore, the dc-link capacitor should not filter 300 Hz harmonic component of the voltage, and its capacitance will be decided by other requirements such as providing low impedance path for current component at the switching frequency. This means small dc-link capacitor can be used. In this paper, 2 µF capacitance is used. A film capacitor is suitable for such application.

4 Simulation Results

The simulation of CI based UPF rectifier has been performed using MATLAB Simulink tool. The system parameters are shown in Table I. The simulation has been performed with taking constant power load condition (10 KW). Input phase currents are show in Fig. 4(a), and the total harmonic

distortion (THD) is 0.62%. The FFT spectrum of phase A current, upto 2 kHz, is also shown in Fig. 4(b). The inductor current reference signal with actual currents, IGBTs and MOSFETs currents are shown in Fig. 5. The current stresses of different devices are summarized in Table II

Parameter	Value
Input grid voltage ($V_{g,L-L}$)	400 V
Grid frequency	50 Hz
Carrier switching frequency (f_s)	100 kHz
Inductor (L)	350 µH
Constant power load	10 kW
DC-link capacitor (C_{dc})	2 µF

Table 1 System parameters

Parameter		RMS	AVG
Inductor Current (A)	Calculated	5.97	
	Simulated	6.1	
MOSFET Current (A)	Calculated	4.28	
	Simulated	4.9	
IGBT Current (A)	Calculated	2.47	0.87
	Simulated	2.52	0.9

Table 2 Current stress of devices

5 Conclusion

The paper presents a detailed analysis of a rectifier circuit with active component-based CI circuit. The analysis helps selecting the value of the inductance for the CI circuit. Furthermore, it provides the expression of the current to be injected by the CI circuit for UPF operation. However, this operation has a limitation of the constant power load at the dc output. Based on this condition and the current expression in the inductor of CI circuit, the stress analysis for the switches is performed and analytic expressions for rms and average current through the switches are presented which can help optimizing the design. Finally, the analysis is verified using simulation results.

Acknowledgment

The authors gratefully acknowledge Erhvervsstyrelsen - the Danish Business Authority, EU's Regionalfond, and Energy Cluster, Denmark for supporting the project.

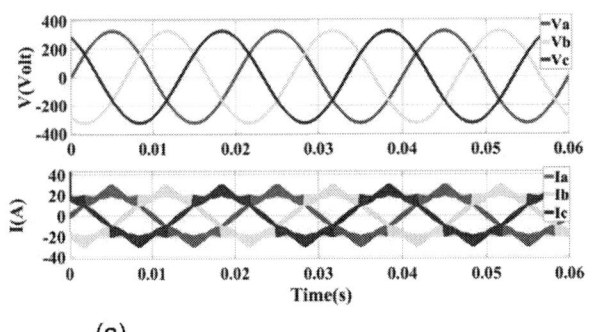
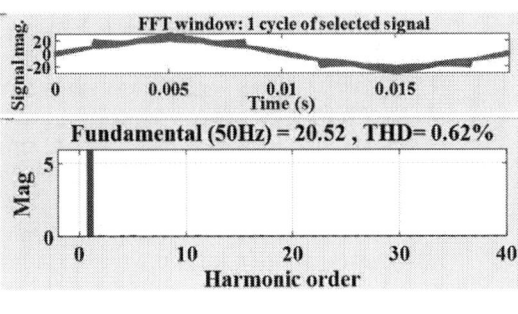

(a)   (b)

Figure 2. Simulation results for constant power load (a) input line currents (b) Harmonic spectrum of phase a current.

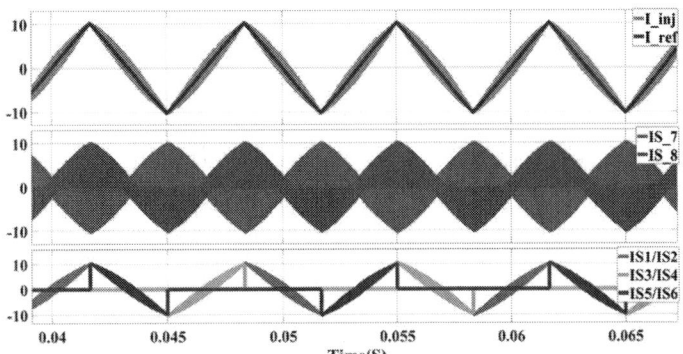

Figure 3. Key current waveforms.

Reference

[1] P. Saravana Prakash, R. Kalpana, and B. Singh, "Third harmonic current injection based front-end ac-dc converter for power quality improvement in DC distribution systems," in *Proc. PESGRE 2020*, Cochin, India, 2020, pp. 1-6.

[2] M. C. Ancuti, S. Musuroi, C. Sorandaru, M. Svoboda, V. M. Voian, and V. N. Olarescu, "Comparative analysis of vienna interleaved three-phase power factor correction rectifier over other ultra-efficient topologies," in *Proc. Int. Conf. ENERGY Environ. (CIEM)*, Timisoara, Romania, Oct. 2019, pp. 484–489.

[3] P. S. Prakash, R. Kalpana, B. Singh, and G. Bhuvaneswari, "Power quality improvement in utility interactive based AC–DC converter using harmonic current injection technique," *IEEE Trans. Ind. Appl.*, vol. 54, no. 5, pp. 5355–5366, Sep./Oct. 2018.

[4] J. I. Itoh and I. Ashida, "A novel three-phase PFC rectifier using a harmonic current injection method," *IEEE Trans. Power Electron.*, vol. 23, no. 2, pp. 715–722, Mar. 2008.

[5] H. Y. Kanaan and K. Al-Haddad, "Three-phase current-injection rectifiers: Competitive topologies for power factor correction," *IEEE Trans. Ind. Electron. Mag.*, vol. 6, no. 3, pp. 24–40, Sep. 2012.

[6] H. Y. Kanaan, K. Aramouni, J. Naoufal, and K. Al-Haddad, "Experimental implementation of a passive current-injection high power factor three-phase rectifier," in *Proc. 21st IEEE Int. Symp. Industrial Electronics (ISIE'12)*, Hangzhou, China, May 28–31, 2012, pp. 334–339.

[7] P. Pejovic and Z. Janda, "An improved current injection network for three-phase high-power factor rectifiers that apply the third harmonic current injection," *IEEE Trans. Ind. Electron.*, vol. 47, no. 2, pp. 497–499, Apr. 2000.

[8] D. Chilachava, E. Temesi, M. Vazsonyi, and G. Ipach, "Highly efficient PFC topology using constant power control enabling higher power density and cost savings in passives," in *Proc. PCIM Europe 2022*, Nuremberg, Germany, 2022, pp. 1-6.

[9] H. Yoo and S.-K. Sul, "A new circuit design and control to reduce input harmonic current for a three-phase ac machine drive system having a very small dc-link capacitor,"

[10] P. Pejovic, "Three-phase high power factor rectifier based on the third harmonic current injection with passive resistance emulation," in *Proc. Power Electron. Spec. Conf.*, Jun. 18–23, 2000, vol. 3, pp. 1342–1347.

[11] M. Makoschitz, M. Hartmann, and H. Ertl, "Hardware implementation and characteri-zation of SiC-based hybrid three-phase rec-tifier employing third harmonic injection," in *Proc. IEEE Appl. Power Electron. Conf. Expo.*, 2016, pp. 1–8.

[12] T. Soeiro, T. Friedli, and J. W. Kolar, "Three-phase high power factor mains inter-face concepts for electric vehicle battery charging systems," in *Proc. 27th Annu. IEEE Appl. Power Electron. Conf. Expo.*, 2012, pp. 2603–2610.

APPENDIX

Current reference calculation

$$i_L d + \frac{P_{out}}{\sqrt{3}V_m \cos(\omega t)} = -I_m \sin\left(\omega t - \frac{2\pi}{3}\right)$$

$$\Rightarrow i_L \frac{\sin\left(\omega t + \frac{5\pi}{6}\right)}{\cos(\omega t)} + \frac{P_{out}}{\sqrt{3}V_m \cos(\omega t)} = -I_m \sin\left(\omega t - \frac{2\pi}{3}\right)$$

$$\Rightarrow i_L \sin\left(\omega t + \frac{5\pi}{6}\right) = -I_m \sin\left(\omega t - \frac{2\pi}{3}\right)\cos(\omega t) - \frac{P_{out}}{\sqrt{3}V_m}$$

If efficiency is 1, and unity power factor is achieved.

$$P_{out} = \frac{3\,V_m I_m}{2}$$

$$i_L \sin\left(\omega t + \frac{5\pi}{6}\right) = -I_m \sin\left(\omega t - \frac{2\pi}{3}\right)\cos(\omega t) - \frac{\sqrt{3}}{2}I_m$$

$$i_L \sin\left(\omega t + \frac{5\pi}{6}\right) = -I_m \left(\sin(\omega t)\cos\left(\frac{2\pi}{3}\right) - \cos(\omega t)\sin\left(\frac{2\pi}{3}\right)\right)\cos(\omega t) - \frac{\sqrt{3}}{2}I_m$$

$$i_L \sin\left(\omega t + \frac{5\pi}{6}\right) = -I_m \left(\sin(\omega t)\left(-\frac{1}{2}\right) - \cos(\omega t)\frac{\sqrt{3}}{2}\right)\cos(\omega t) - \frac{\sqrt{3}}{2}I_m$$

$$i_L \sin\left(\omega t + \frac{5\pi}{6}\right) = \frac{1}{2}I_m \sin(\omega t)\cos(\omega t) + \frac{\sqrt{3}}{2}I_m \cos^2(\omega t) - \frac{\sqrt{3}}{2}I_m$$

$$i_L \sin\left(\omega t + \frac{5\pi}{6}\right) = \frac{1}{2}I_m \sin(\omega t)\cos(\omega t) - \frac{\sqrt{3}}{2}I_m \sin^2(\omega t)$$

$$i_L \sin\left(\omega t + \frac{5\pi}{6}\right) = I_m \sin(\omega t)\left(\frac{1}{2}\cos(\omega t) - \frac{\sqrt{3}}{2}\sin(\omega t)\right) = I_m \sin(\omega t)\sin\left(\omega t + \frac{5\pi}{6}\right)$$

$$i_L = I_m \sin(\omega t)$$

MOSFET RMS current calculation

$$I_{rms}^2 = \frac{3}{\pi}\int_{-\frac{\pi}{6}}^{\frac{\pi}{6}}\left(I_m^2 \sin^2(\omega t)\frac{\sin\left(\omega t + \frac{5\pi}{6}\right)}{\cos(\omega t)}\right)d\omega t$$

$$I_{rms}^2 = \frac{3I_m^2}{\pi} \int_{-\frac{\pi}{6}}^{\frac{\pi}{6}} (1 - \cos^2(\omega t)) \frac{\sin \omega t \cos \frac{5\pi}{6} + \cos \omega t \sin \frac{5\pi}{6}}{\cos(\omega t)} d\omega t$$

$$I_{rms}^2 = \frac{3I_m^2}{\pi} \int_{-\frac{\pi}{6}}^{\frac{\pi}{6}} (1 - \cos^2(\omega t)) \left(-\frac{\sqrt{3}}{2} \tan \omega t + \frac{1}{2} \right) d\omega t$$

$$I_{rms}^2 = \frac{3I_m^2}{\pi} \int_{-\frac{\pi}{6}}^{\frac{\pi}{6}} \left(-\frac{\sqrt{3}}{2} \tan \omega t + \frac{\sqrt{3}}{2} \sin \omega t \cos \omega t + \frac{1}{2} \sin^2 \omega t \right) d\omega t$$

$$I_{rms}^2 = \frac{3I_m^2}{2\pi} \int_{-\frac{\pi}{6}}^{\frac{\pi}{6}} (\sin^2 \omega t) d\omega t = \frac{3I_m^2}{2\pi} \left(\frac{\pi}{6} - \frac{\sqrt{3}}{4} \right)$$

PCIM Europe 2024, 11– 13 June 2024, Nuremberg   DOI: 10.30420/566262442

High-Efficient Isolated AC-DC Converter with Circulating Current Reduction for AC Adapters

Hiroki Watanabe[1], Jun-ichi Itoh[1], Naoto Izumoto[2], Kazunori Kidera[2], Kenji Okada[2]

[1] Nagaoka University of Technology, Japan
[2] Panasonic Co. Ltd, Japan

Corresponding author: Hiroki Watanabe, hwatanabe@vos.nagaokaut.ac.jp
Speaker: Hiroki Watanabe, hwatanabe@vos.nagaokaut.ac.jp

Abstract

This paper proposes a circulating current reduction of an isolated AC-DC converter based on a dual-active half-bridge converter. This converter has the capability of power factor correction, DC voltage control, and Zero Voltage Switching (ZVS) to reduce the switching loss. This circuit has the variable duty control in the secondary full-bridge rectifier to achieve the power factor correction. In this control method, the design of the modulation index is important to determine the performance of the power factor correction and conversion efficiency. This paper introduces the design method of the modulation index focusing on the circulating current reduction. As an experimental result, the conversion loss was reduced by 40.1%. Finally, the maximum efficiency is improved to 95.3% from 92.1%.

1   Introduction

Recently, the high-power density and high-efficiency power adapter is one of the technical trends owing to the wide spread of USB power delivery for home appliance applications [1]-[5].

Power adapters commonly consist of the AC-DC rectifier and isolated DC-DC converter for safety requirements and DC voltage regulation. The flyback converter is widely accepted because of its simple configuration and low cost [6]-[7]. In this case, the quasi-resonant or active clamp flyback converter candidates for the circuit topology of isolated DC-DC converter because these topologies provide Zero Voltage Switching (ZVS) to reduce the switching loss [8]-[11]. However, the flyback converter has a large conduction loss due to the triangular magnetizing current.

A Dual Active Bridge (DAB) converter has also been considered for high-power density isolated DC-DC converters [12]. The DAB converter delivers the input power to the load with low switching loss through ZVS capability. Furthermore, the current RMS decreases when the trapezoidal inductor current is obtained. A single-stage bridgeless half-bridge AC-DC converter was considered [13]. This converter combines the totem-pole PFC converter and isolated half-bridge converter. This converter provides a high-power factor, simple configuration, and easy control. On the other hand, high voltage rating devices are necessary due to the

voltage doubler configuration. Moreover, the modulation index influences the input current quality and the circulating current, which causes high conduction loss. There is a tradeoff between the input current distortion and the circulating current depending on the modulation index.

This paper proposes an isolated AC-DC converter based on the dual active half-bridge converter. The modulation index of the proposed circuit is optimized to obtain low input current harmonics and low circulating current. Experimental results demonstrate steady-state operation and efficiency.

2   Circuit Topology and Control

2.1   Circuit Configuration

Fig. 1 shows the circuit configuration of the isolated AC-DC converter, which consists of the diode-bridge rectifier and the dual active half-bridge converter. This converter has the capability of power factor correction without the PFC converter, galvanic isolation by the high-frequency transformer, and DC voltage control. The capacitors of C_1 and C_2 were chosen for small capacitance to make the full-wave rectification waveform of the grid voltage for power factor correction. The inductor current I_{trans} is automatically adjusted to $|\sin\omega t|$, which means the average current of I_{trans} is equal to grid current i_{ac}. Therefore, this converter achieves a high-power factor without feedback

3125

control. The phase shift control between the primary and secondary bridge circuit controls the transmission power. Finally, ZVS is achieved on all switching devices to reduce the switching losses.

2.2 Control and Modulation

Fig.2 shows the control block diagram of the isolated AC-DC converter. The half-bridge inverter generates the high-frequency square waveform voltage for the high-frequency transformer. The duty cycle of S_1 and S_2 is fixed to 0.5 because the symmetrical square waveform voltage is necessary to avoid the DC flex bias. The half-bridge inverter modulation applies the phase-shifted control to adjust the transmission power. Note that the PI controller determines the phase-shifted angle when the DC voltage control is implemented.

The variable duty signal gives the duty cycle of the secondary-side rectifier. The duty signal is expressed as

$$m = M_{\max} \left| \sin \omega t \right| \tag{1}$$

where d_{sec} is the duty signal of the secondary-side rectifier, $M_{\max}$ is the maximum modulation index and ω is the angular frequency of the single-phase grid. This modulation provides a three-level voltage on the secondary-side transformer, including a zero-voltage period. The input current of i_{ac} becomes a sinusoidal waveform because the isolated AC-DC converter behaves as the linear load owing to the secondary-side modulation.

Fig.3 shows the high-frequency transformer voltage and inductor current in the dual-active half-bridge converter. The two-transformer voltage and the phase shift angle between two transformer voltages give the inductor current. The inductor current has Mode 1 and 2 depending on the relationship between the phase shift angle δ, modulation index m, and the zero-voltage period α, as shown in Fig. 3 (a) and (b).

The inductor current RMS in Mode 1 reduces compared with Mode 2 because State III in Mode 2 increases the inductor current due to the primary-side transformer voltage. Mode 2 appears at a low modulation index and small phase shift angle condition. In this paper, the design criteria of the modulation index is discussed to reduce the circulating current based on Mode 1.

3 Design of Modulation Index

This paper introduces the reduction of the inductor current RMS by designing the maximum modulation index $M_{\max}$. Fig.4 shows the inductor current when the maximum modulation is changed. The

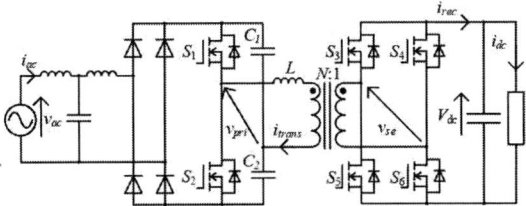

Fig.1. Circuit configuration of AC-DC converter.

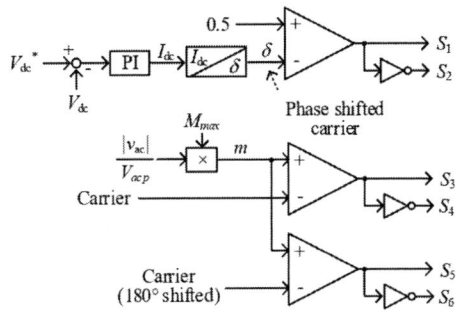

Fig.2. Control block diagram of AC-DC converter.

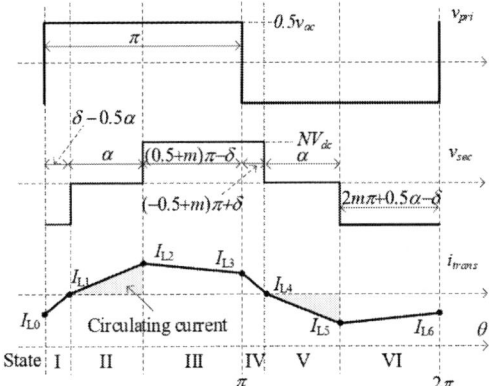

(a) Mode 1 ($\delta > 0.5\alpha$)

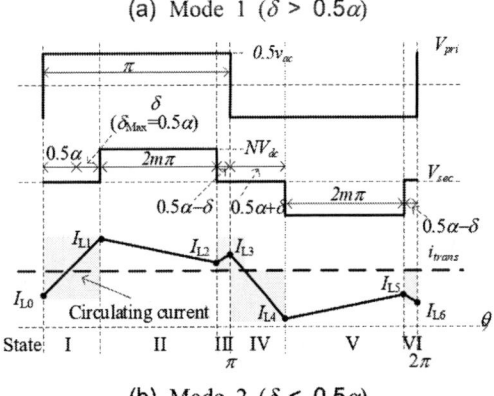

(b) Mode 2 ($\delta < 0.5\alpha$)

Fig.3. Transformer voltage and inductor current waveforms with different phase shift angle.

current I_{L4} determines the quality of the sinusoidal input current. The input current harmonics are minimal when I_{L4} is greater than or equal to zero. On the other hand, the negative current decreases the average current of i_{ac}. As a result, the input current distorts near the peak grid voltage.

This paper introduces the design method of the modulation index to obtain the sinusoidal input current. Moreover, the circulating current reduction is discussed, focusing on the modulation index and phase shift angle. Firstly, the inductor current in each state is discussed. Each current is expressed as

$$I_{L1} = I_{L0} + \frac{\left|0.5V_{acp}\sin\theta\right| + NV_{dc}}{L}\left(\delta - 0.5\alpha\right) \quad (2)$$

$$I_{L2} = I_{L1} + \frac{\left|0.5V_{acp}\sin\theta\right|}{L}\alpha \quad (3)$$

$$I_{L3} = I_{L2} + \frac{\left|0.5V_{acp}\sin\theta\right| - NV_{dc}}{L}\left\{(0.5+m)\pi - \delta\right\} \quad (4)$$

$$I_{L4} = I_{L3} + \frac{-\left|0.5V_{acp}\sin\theta\right| - NV_{dc}}{L}\left\{(-0.5+m)\pi + \delta\right\} \quad (5)$$

$$I_{L5} = I_{L4} + \frac{-\left|0.5V_{acp}\sin\theta\right|}{L}\alpha \quad (6)$$

$$I_{L6} = I_{L5} + \frac{-\left|0.5V_{acp}\sin\theta\right| + NV_{dc}}{L}\left(2m\pi + 0.5\alpha - \delta\right) \quad (7)$$

where V_{acp} is the peak grid voltage, δ is the phase shift angle, and L is the inductance. α is the zero-voltage period, and it is given by

$$\alpha = (1-2m)\pi \quad (8).$$

The initial current $-I_{L0}$ is the same as I_{L3}. Therefore, it is calculated by (2), (3), and (4). I_{L0} is expressed as

$$-I_{L0} = I_{L3} = \frac{1}{2L}\left\{\begin{array}{l}\left(2\delta - 0.5\alpha - (0.5+m)\pi\right)NV_{dc} + \\ \left\{(0.5+m)\pi + 0.5\alpha\right\}\left|0.5V_{acp}\sin\theta\right|\end{array}\right\} \quad (9).$$

I_{L4} should become greater than or equal to zero to obtain the sinusoidal input current. In this case, the circulating current decreases when I_{L4} is zero because the zero-voltage period is minimized. The maximum modulation index is given by Eq. (5) and (9), and it is expressed as

$$M_{max} = \frac{2}{\left(V_{acp} + 6NV_{dc}\right)\pi}\left\{\begin{array}{l}\left(2\delta - 0.5\pi\right)NV_{dc} + 0.25\pi V_{acp} \\ +\left(0.5\pi - \delta\right)\left(V_{acp} + 2NV_{dc}\right)\end{array}\right\} \quad (10).$$

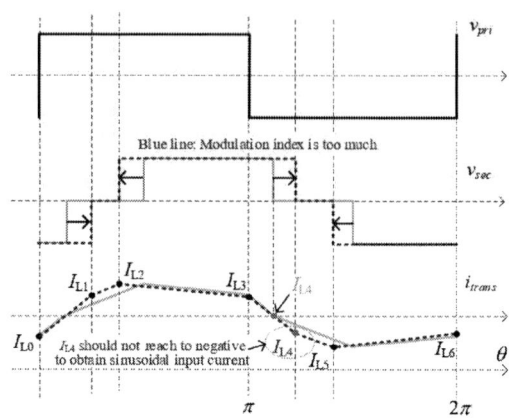

Fig.4. Inductor current waveforms when modulation index changed.

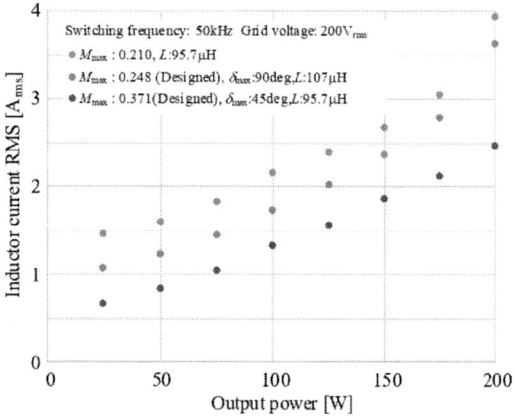

Fig.5. Inductor current RMS comparison when maximum modulation index changed.

Note that the term $\sin\theta$ becomes one because M_{max} is obtained during the peak grid voltage condition. Moreover, the zero-voltage period is approximately zero for delivering M_{max}. This is because the zero-voltage period depends on the modulation index.

Fig. 5 shows the inductor current RMS comparison when the maximum modulation index changed. The orange and blue plots use M_{max}, which is calculated by Eq. (10). According to Fig.5, the inductor current RMS of the blue plot decreased by 38% compared to the gray plot without the proposed design. It is also confirmed that the inductor current RMS decreases when M_{max} is high owing to a small zero-voltage period from the result of the orange and blue plots.

4 Experimental result

This chapter demonstrates some experimental results to confirm the validity of the circulating current reduction by the proposed modulation index design. Table I shows the experimental condition. The prototype circuit with a rated power of 200 W was tested. The switching frequency was set to 50 kHz in this experiment. The output capacitor of 9000 μF is used to obtain the constant DC voltage without the low-frequency component of twice the grid frequency.

Fig. 6 shows the experimental result of input and output waveforms. According to Fig. 6, the isolated AC-DC converter achieved high power factor correction without a PFC circuit. The DC voltage and current have a small ripple due to the single-phase power ripple. This ripple is reduced when output capacitance increases more than this condition.

Fig. 7 shows the experimental result of transformer voltage and inductor current waveforms when the modulation index is set to the designed value of 0.369. According to Fig. 7, The inductor current reaches approximately zero without the zero-voltage period.

Fig. 8 shows the efficiency curve when input voltage RMS and modulation index change. Compared to the same input voltage condition, the efficiency was improved in wide output power conditions with both voltage conditions. The conversion loss was reduced by 40.1% when the output power was 100 W and the input voltage of 230 V_{rms}. This is because the proposed modulation index design reduced the circulating current.

Fig. 9 shows the comparison of the inductor RMS current when the modulation index changed. According to Fig. 9, The inductor RMS current with the proposed modulation index design kept a small value compared with the low modulation index condition. The inductor RMS current was reduced by 26% when the output power was 150 W.

Fig. 10 shows the grid current harmonics evaluation result. According to Fig.10, harmonic components are less than the IEC 61000-3-2 class A standard.

5 Conclusion

This paper proposed a circulating current reduction of an isolated AC-DC converter based on a dual-active half-bridge converter. The design method of the modulation index focusing on the circulating current reduction was introduced. As the experimental results, it was confirmed that the maximum efficiency is improved to 95.3%. The conversion loss was reduced by 40.1% when the

Table I. Experimental condition.

Symbol	Quantity	Value
V_{ac}	Grid voltage	230 V_{rms}
P_{out}	Output power	200 W
f_{sw}	Switching frequency	50 kHz
L	Inductor	80 μH
δ	Phase shift angle	45 deg
$N_1{:}N_2$	Number of turn	63:9
L_m	Magnetizing inductance	8.4 mH
L_{leak}	Leakage inductance	48 μH
k	Coupling Coefficient	0.99
C_1, C_2	Arm capacitor	0.47 μF
C_{out}	Output capacitor	9000 μF

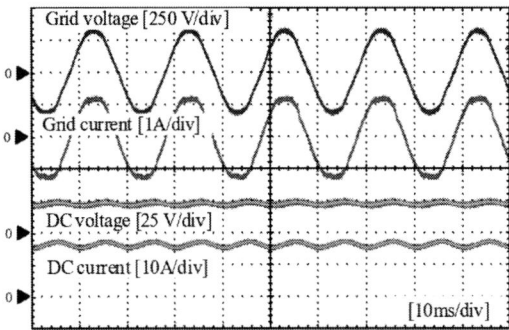

Fig.6. Input and output waveforms.

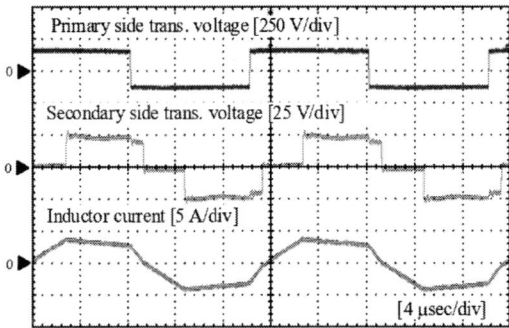

Fig.7. Transformer voltage and inductor current waveforms.

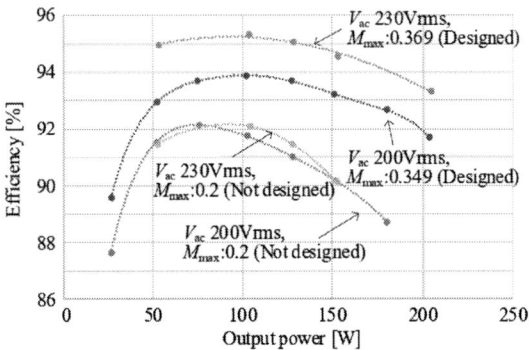

Fig.8. Efficiency comparison.

output power was 100 W with an input voltage of 230 V_{rms}. Finally, the inductor RMS current was reduced by 26% when the output power was 150 W.

References

[1] Y. Chen, Y. F. Liu, "Power Adapter with Line Voltage Control for USB Power Delivery", in Proc.2019 IEEE Applied Power Electron. Conf. Expo., May 2019, pp.2054-2060.

[2] Y.K. Lo, S.C. Yen, C.Y. Lin, "A High-Efficiency AC-to-DC Adaptor With a Low Standby Power Consumption", IEEE Trans. Ind. Electron., Vol.55, No.2, pp.963-965, Jan.2008.

[3] B.H. Lee, K.B. Park, C.E. Kim, G.W. Moon, "No-Load Power Reduction Technique for AC/DC Adapters", IEEE Trans. Power Electron., Vol.27, No.8, pp.3685-3694 Jan.2012.

[4] A. M. Garcia, M. Krueger, M. Schmid, J. Daimer, M. Schlenk, "Hybrid-flyback and GaN enable ultra-high power density 240W USB-PD EPR adaptor", in Proc.2023 IEEE Applied Power Electron. Conf. Expo., May 2023, pp.1259-1264.

[5] C. E. Kim, J. Baek, J. B. Lee, "Three-Switch LLC Resonant Converter for High-Efficiency Adapter With Universal Input Voltage", IEEE Trans. Power Electron., Vol.36, No.1, pp.630-638, Jun.2020.

[6] Z. Ma, Y. Lai, Q. Huang, Y. Yang, S. Wang, "Investigation and Mitigation of Radiated EMI due to Near-field Coupling in a High-density Active-clamp Flyback Power Adapter", in Proc. 2023 IEEE Energy Convers. Congr. Expo., Dec.2023, pp.2953-2958.

[7] J. Yao, Y. Li, S. Wang, X. Huang, X. Lyu, "Modeling and Reduction of Radiated EMI in a GaN IC-Based Active Clamp Flyback Adapter", IEEE Trans. Power Electron., Vol.36, No.5, pp.5440-5449, Oct.2020.

[8] S. Xu, W. Shen, Q. Qian, J. Zhu, W. Sun, H. Li, "An efficiency optimization method for a high frequency quasi-ZVS controlled resonant flyback converter", in Proc.2019 IEEE Applied Power Electron. Conf. Expo., May 2019, pp.2957-2961.

[9] C. Wang, S. Xu, L. Yu, Q. Qian, S. Lu, W. Sun, H. Li, "New digital control method for improving dynamic performance of a quasi-resonant flyback converter", in Proc.2019 IEEE Applied Power Electron. Conf. Expo., May 2019, pp.1788-1793.

[10] G. Mauromicale, A. Raciti, S. A. Rizzo, G. Susinni, F. Fusillo, A. Palermo, F. Scrimizzi, R.

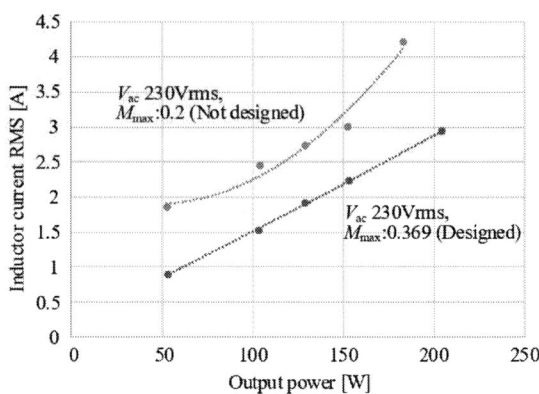

Fig.9. Inductor current RMS comparison when modulation index changed.

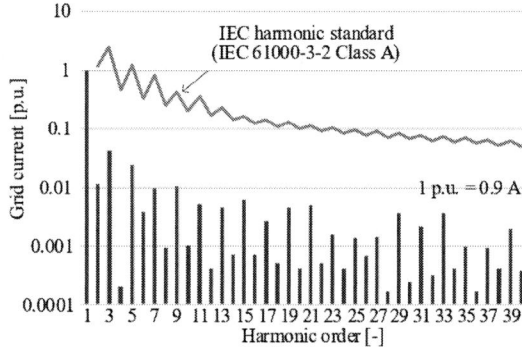

Fig.10. Grid current harmonics evaluation.

Scollo, "Si and GaN Devices in Quasi Resonant Flyback converters for Wall Charger Applications", in Proc. 2019 IEEE Energy Convers. Congr. Expo., Nov.2019, pp.3253-3258.

[11] S. Dey, M. B. Ray, H. Soni, R. Ghosh, M. Shah, "Comparison between Quasi-Resonant and Active Clamp Flyback topologies for GaN-based 65W Wall Charger Application", in Proc.2021 IEEE Applied Power Electron. Conf. Expo., July 2021, pp.1809-1814.

[12] G. Oggier, G. O. García and A. R. Oliva, "Modulation strategy to operate the dual active bridge DC-DC converter under soft switching in the whole operating range", IEEE Trans. Power Electron., Vol.26, No.4, pp.1228-1236, Sep.2010.

[13] A. M. Naradhipa, S. Kang, B. Kim, S. Choi, "A New Single-Stage Bridgeless Boost Half-Bridge AC/DC Converter with Semi-Active-Rectifier", in Proc.2019 IEEE Applied Power Electron. Conf. Expo., May 2019, pp.757-762.

PCIM Europe 2024, 11– 13 June 2024, Nuremberg   DOI: 10.30420/566262443

A Phase-Locked Loop (PLL) based Strategy for Accurate Blanking Times in Bridgeless Totem-Pole PFCs

Sandu Tigira[3], F. Javier Díaz[1] , Alberto Pigazo[1] , Francisco J. Azcondo[1] , Paula Lamo[2] , Christian Brañas[1] , Rosario Casanueva[1]

[1] Universidad de Cantabria, Spain

[2] Universidad Internacional de La Rioja, Spain

[3] Indra Sistemas S.A., Spain

Corresponding author: Sandu Tigira, stigira@indra.es
Speaker: Sandu Tigira, stigira@indra.es

Abstract

Bridgeless totem-pole Power Factor Correction (PFC) circuits experience current distortion around the ac voltage zero-crossings due to flaws in the synchronization with the grid and the reverse recovery phenomenon in the power devices switching at grid frequency, combined with the fast current change trough the active power device switching at high frequency. Blanking times and soft-start strategies are used to attenuate these effects, respectively. The incorporation of blanking intervals within the gate signals for the line frequency switching power devices is achieved by comparing of the measured grid voltage and a predetermined threshold level. This method facilitates the turn-off of the power device during each semi-cycle of the ac voltage before the occurrence of the zero-crossing event. However, grid voltage disturbances and a static threshold level result in suboptimal blanking times under changing operation conditions. A Phase-Locked Loop (PLL) is used not only to estimate the electrical grid angle and generating the instantaneous reference current but also for rejecting noise around the zero crossings. However, integrating a PLL within the multiloop controller implemented in a microcontroller (μC) is challenging since the computational burden associated with PLLs is relatively high compared to zero-crossing detector (ZCD) based controllers. This manuscript focuses on this issue and provides insights into PLL implementations on the TI C2000 μC.

1 Introduction

Bridgeless totem-pole PFC circuits achieve higher efficiency than the diode bridge + dc to dc converter PFC counterparts. However, the natural detection of the ac voltage zero-crossings is missed, and, accordingly, the controller must select the active switching power device and the on and off-state of the corresponding low-frequency device depending on the ac voltage polarity. Ensuring a smooth change of the current polarity requires taking into account various factors at the zero crossing of the grid voltage: the reverse recovery of the Si power MOSFET, which are the devices operating at the line frequency, the C_{oss} voltage across the power devices switching at high frequency, the duty cycle discontinuity of these devices and the nearly negligible voltage at the grid side. These elements converge to build the line current [1], trying to minimize its distortion. From the controller perspective, some of these issues can be addressed by including a blanking time for the gate signals of

the power devices switching at low-frequency, just before the zero crossing, so that the body diode acts as a rectifier device and generating a soft-starting sequence after each zero-crossing for the wide bandgap (SiC or GaN), i.e. high-frequency power devices, as shown in Fig. 1, during a certain number of switching periods, N [2]-[4].

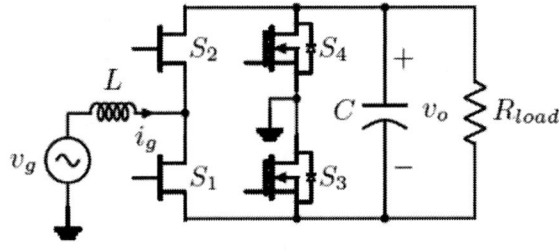

Fig. 1   Bridgeless totem-pole PFC.

Blanking times are usually generated by comparing the ac voltage measurement, v_g, with a previously defined threshold level, V_{th} (Fig. 2). By properly selecting V_{th}, depending on the power devices used and the converter operation conditions, the effect of the reverse recovery of the body diodes in the Si MOSFET line-frequency devices can be effectively mitigated. However, grid voltage disturbances, deteriorating both the amplitude and frequency, may result in larger or shorter blanking times, provided that V_{th} is constant. As the blanking time is longer, it increases the harmonic distortion of the line current, e.g. the 3rd harmonic. At the same time, a very short one cannot effectively reduce the line current distortion due to non-ideal current switching transients.

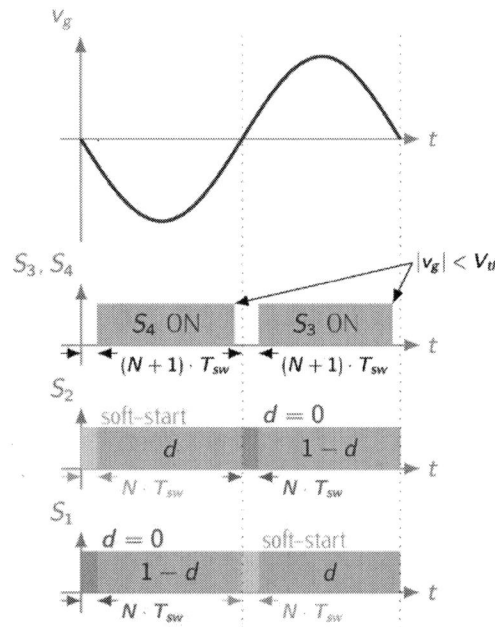

Fig. 2 Gate signals, including blanking times and soft-start.

Phase-locked loops (PLL) have been proposed in technical literature not only for assisting the generation of the reference signal [5] for the inner control loop but also to mitigate this synchronization with the grid issues around the zero voltage crossings [6], [7]. However, integrating a PLL within the digital controller of bridgeless TP PFCs is challenging due to the difficult compatibility of the computational burden associated with the PLL and the high switching frequency of WBD used.

Different approaches for integrating a digital PLL to control the blanking times of Si MOSFETs in a

bridgeless TP PFC are proposed. The digital controller runs on a microcontroller (μC) of the TI C2000 family.

2 Original Digital Controller

The digital controller in which the proposed modifications have been included corresponds to a 2.5 kW bridgeless TP PFC TDTTP-2500P100 by Transphorm. The digital controller is implemented in a TI F28335 μC. The original digital controller consists of two control loops, as depicted in Fig. 3. The outer control loop regulates the output voltage, v_o, by evaluating the power, P_o, required to keep v_o constant. The reference current, i^{ref}, for the inner control loop is obtained by dividing P_o by the rms value of the grid voltage, V_g, and multiplying the result with $|v_g|$. The V_g values are refreshed at each zero crossing of v_g, which is detected with a dedicated zero-crossing detector (ZCD) circuitry. The polarity signal provided by the ZCD is also used to obtain $|v_g|$ and $|i_g|$ out of v_g and i_g, respectively. The inner control loop includes a feedforward action to approximate the duty cycle required by the active GaN MOSFET, which is assisted by high pass filter action to correct deviations from i^{ref}. Blanking times for the Si devices and soft-start sequence for the GaN devices are imposed during the time resulting from the comparison of $|v_g|$ with a predefined threshold voltage, V_{th}.

The time diagram of the original controller is depicted in Fig. 4. Execution times are measured using the serial port of the F28335, populated through the UART connector in the TDTTP-2500P100 evaluation board. The pulse width modulator sets a 100 kHz switching frequency. It generates a high-priority interruption, int_1, which launches the analogue-to-digital conversion (T_{ADC} = 438 ns) and the inner control loop (T_{inner} = 338 ns). The outer control (T_{outer} = 325 ns) is executed at 25 kHz once every four sampling periods. The $|v_g|$ values are integrated, i.e. accumulated, with a low-priority interrupt, int_2, executed at 25 kHz. At the polarity signal events, the code launched by this interruption provides the accumulator value as V_g, and it is filtered out with a first-order low-pass filter. The execution time associated with this procedure around zero crossings, i.e. T_{rms} = 4.6 μs, occupies a large portion of the switching period, T_{sw} = 10 μs, making it difficult for the execution times of other blocks in the controller to fit in T_{sw}. The original code uses *float* types for mathematical operations in the controller and acceleration of the code execution through the Floating Point Unit (FPU) in μCs of the C2000 family is disabled. Fig.

PCIM Europe 2024, 11– 13 June 2024, Nuremberg   DOI: 10.30420/566262443

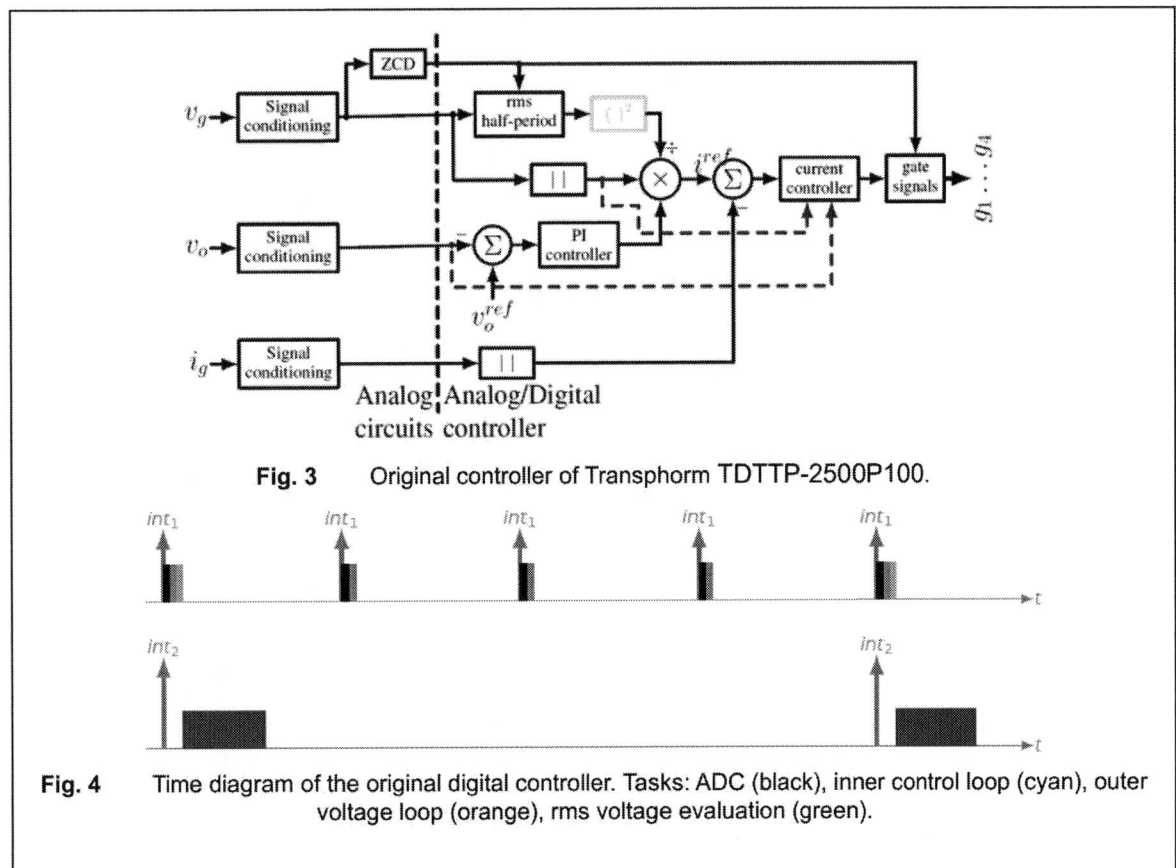

Fig. 3   Original controller of Transphorm TDTTP-2500P100.

Fig. 4   Time diagram of the original digital controller. Tasks: ADC (black), inner control loop (cyan), outer voltage loop (orange), rms voltage evaluation (green).

5 shows the obtained ac and dc voltage/current waveforms due to the original controller and the resulting blanking times.

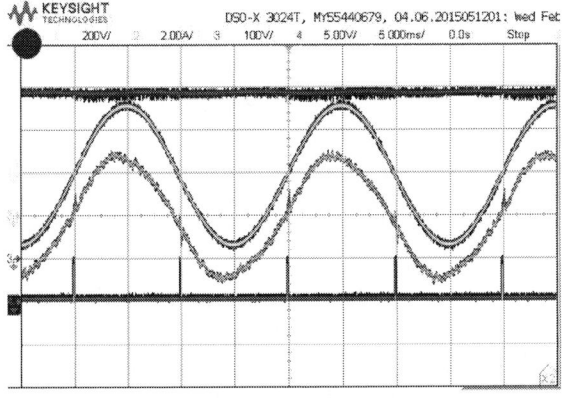

Fig. 5   Line current waveform due to the original controller. v_g: yellow, 200 V/div, i_g: green, 2 A/div, v_o: blue, 100 V/div, blanking time signal: magenta, 5 V/div, time scale: 5 ms/div.

3   Integrating a Phase-Locked Loop (PLL)

A second-order generalized-integrator (SOGI) phase-locked loop (PLL) has been used to integrate in the digital controller in Fig. 3. As a first approach, the raw C code of the PLL provided by an automated code generation tool has been used.

The obtained time diagram is shown in Fig. 6. The whole PLL is implemented in a floating point, and no hardware acceleration is used. The execution time associated with the PLL is 10.38 µs, which exceeds the switching period. The worst scenario is shown in Fig. 6, where the PLL code is executed after the outer voltage loop. As a result, the execution time of the routine exceeds the available execution time, and the subsequent execution of this routine is displaced by 1.48 µs. Consequently, the acquisition instant of v_g and i_g is also displaced, which deteriorates the synchronization of the ADC with the PWM. Moreover, the code associated with the low-priority interruption int_2 is displaced by 11.16 µs, deteriorating the calculus of the equivalent conductance at the ac side.

3132

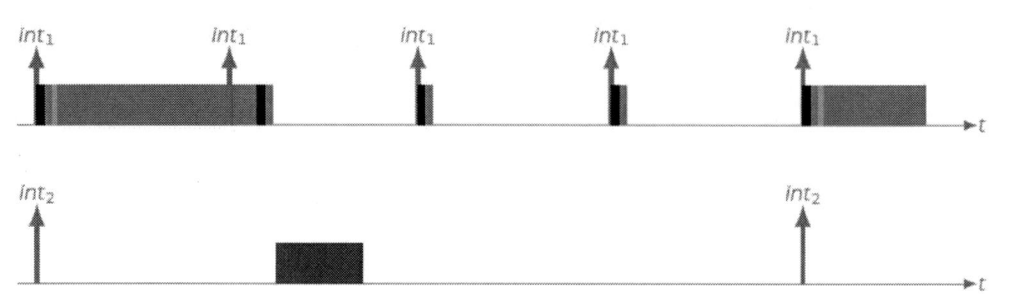

Fig. 6 Time diagram of the original digital controller with the PLL (1st approach). Tasks: ADC (black), inner control loop (cyan), outer voltage loop (orange), rms voltage evaluation (green), SOGI PLL (red).

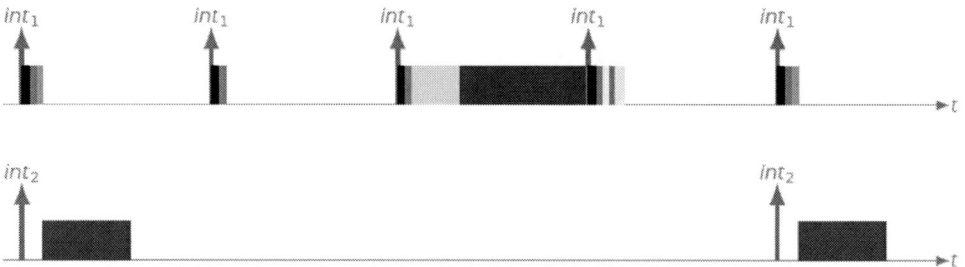

Fig. 7 Time diagram of the original digital controller with the PLL divided in sections (2nd approach). Tasks: ADC (black), inner control loop (cyan), outer voltage loop (orange), rms voltage evaluation (dark green), PLL phase detection (light green), PLL Normalization (blue), PLL PI (very light green), PLL oscillator (dark gray) and PLL SOGI cell (gray).

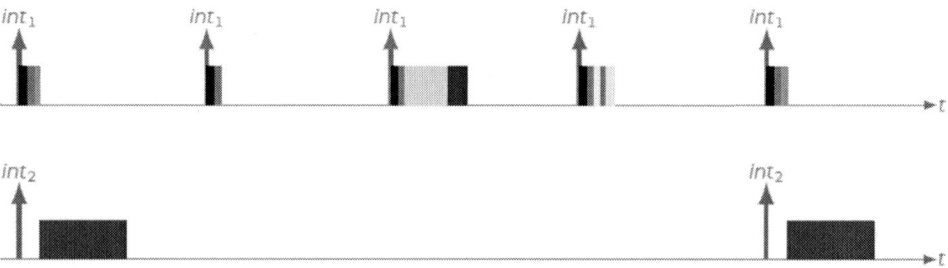

Fig. 8 Time diagram of the original digital controller with PLL (3rd approach). Tasks: ADC (black), inner control loop (cyan), outer voltage loop (orange), rms voltage evaluation (dark green), PLL phase detection (light green), PLL Normalization (blue), PLL PI (very light green), PLL oscillator (dark gray) and PLL SOGI cell (gray).

To alleviate these effects, the PLL code is divided into sections and distributed between successive samples of v_g, as shown in the timing diagram depicted in Fig. 7. Large code sections of the PLL are displaced two samples after the execution of the outer control loop. These sections are the Park transformation and the normalization, with $T_{dq} = 2.52$ µs and $T_{norm} = 6.72$ µs execution times, respectively. The successive sampling interval is used for evaluating the proportional-integral action ($T_{PI} = 258$ ns), the oscillator ($T_{osc} = 318$ ns) to obtain the grid angle estimation, and the SOGI cell ($T_{SOGI} = 472$ ns) within the PLL.

This second approach improves the controller execution sequence, but again, the ADC action after the normalization block of the PLL is displaced slightly.

For an accurate modification of the digital controller, with minor changes in both the original code and PLL one provided by the automated code generation tool, the microcontroller FPU has to be used, and minor cost-effective modifications, such as replacing normalization in $\alpha\beta$ reference frame,

i.e. $v_x, norm = \dfrac{v_x}{\sqrt{v_a + v_{beta}}}$,by dq one, i.e.

$v_q, norm = \dfrac{v_q}{v_d}$, have to be carried out.

The obtained time diagram is shown in Fig. 8. Using the FastRTS library of C2000 microcontrollers, the code for the Park transformation is slightly reduced from $T_{dq} = 2.52$ µs to 2.28 µs. The greatest reduction is due to the changes in the normalization block and the use of FPU acceleration. The execution time is reduced from $T_{norm} = 6.72$ µs to 973 ns. As a result, the PLL integrated in this way does not deteriorate the execution times of the original code and allows novel functionalities, such as pure sinusoidal line current or blanking times, to be effectively implemented.

4   Experimental results

The three approximations for integration of the PLL within the digital controller have been evaluated experimentally. The evaluation board is a 2.5 kW totem-pole TDPTTP2500P100 by Transphorm, controlled with a Texas Instruments F28335 controller. GaN devices are used for the high-frequency switching branch, and the switching frequency is set at 100 kHz. The PFC circuit is supplied with a 4.5 kW power source AMX 345 XTR from Pacific. The circuit performance is evaluated using a PPA5530 power analyzer from Newtons4th. The PLL programmed within the controller is a SOGI PLL, using a type-II SOGI cell, with a settling time equal to 0.5 s [8]. The threshold voltage is programmed to 7.5 V without PLL, and the blanking time is set to $T_{blk} = 108$ µs when the PLL is used. The tests are carried out with a pure sinusoidal grid voltage at 50 Hz, and the output power is 460 W.
Figure 9 shows the obtained results generating the blanking times with $|v_g| < V_{th}$. The measured line current and THD_I are 2.018 A and 5.403 %, respectively. The measured blanking time is $\mu = 109.3$ µs and $\sigma = 5.9$ µs.

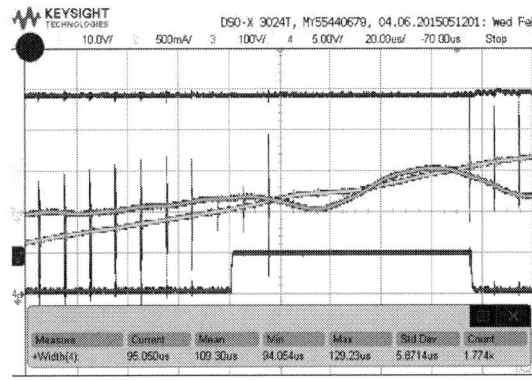

Fig. 9   Blanking times by comparing $|v_g| < V_{th}$ @ 50 Hz. v_g: yellow, 10 V/div, i_g: green, 500 mA/div, v_o: blue, 100 V/div, blanking time signal: magenta, 5 V/div, time scale: 20 µs/div.

Under the same conditions, the PLL-based approach with the first approximation performs similarly (Fig. 10). The measured line current and THD_I are 2.019 A and 5.358 %, respectively. The measured blanking time is $\mu = 105.4$ µs and $\sigma = 24.4$ µs. The effect shown in Fig. 6 deteriorates the consistency of this approach, resulting in a greater variability of the blanking times.

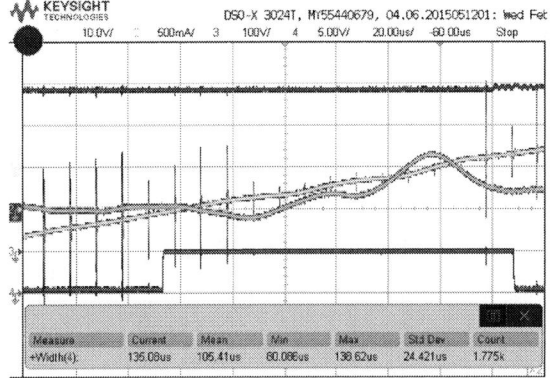

Fig. 10   Blanking time by the PLL (1st approach) @ 50 Hz. v_g: yellow, 10 V/div, i_g: green, 500 mA/div, v_o: blue, 100 V/div, blanking time signal: magenta, 5 V/div, time scale: 20 µs/div.

The consistency in applying the required blanking times increases (Fig. 11) using the second implementation (Fig. 7). The measured line current and THD_I are 2.019 A and 5.376 %, respectively. The measured blanking time is $\mu = 103.1$ µs and $\sigma = 22.0$ µs.

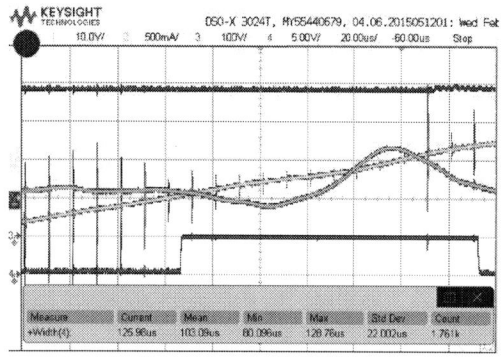

Fig. 11 Blanking time by the PLL (2nd approach) @ 50 Hz. v_g: yellow, 10 V/div, i_g: green, 500 mA/div, v_o: blue, 100 V/div, blanking time signal: magenta, 5 V/div, time scale: 20 μs/div.

The best performance is due to the third approach (Fig. 8). The measurements in Fig. 12 show that the blanking time is μ = 108.7 μs and σ = 3.6 μs. The measured line current and THD_I are 2.018 A and 5.317 %, respectively.

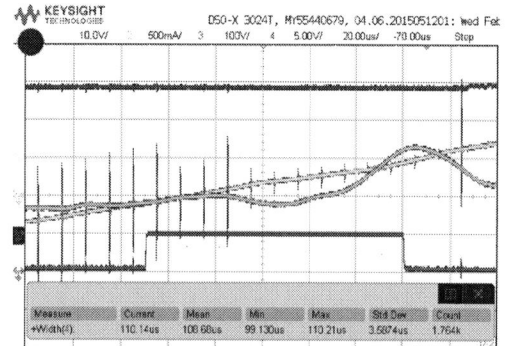

Fig. 12 Blanking time by the PLL (3rd approach) @ 50 Hz. v_g: yellow, 10 V/div, i_g: green, 500 mA/div, v_o: blue, 100 V/div, blanking time signal: magenta, 5 V/div, time scale: 20 μs/div.

5 Conclusion

PLLs are gaining interest due to their noise rejection capability in single phase bridgeless PFC circuits. This work has shown that including a software PLL within the digital controller is challenging when using low-cost microcontrollers, where parallelization capabilities are limited by the hardware used. Three possible PLL implementations have been compared experimentally. The results show that using dedicated hardware for floating point operations and properly allocating of the PLL functional blocks within the code allows the PLL to be integrated, reducing the variability of the blanking times and improving the THD_I.

Acknowledgment

This work was supported by the EU FEDER and the Spanish Ministry of Science and Innovation under the research project PID2021-128941OB-I00 (Efficient Energy Transformation in Industrial Environments) and by the Regional Government of Cantabria, Spain, and EU FEDER under the project 2023-TCN-008 (Ultra-Efficient Technologies for AI-based Anomaly Detection).

References

[1] B. Sun, "How to reduce current spikes at AC zero-crossing for totem-pole PFC," 2015, [Online]. Available: https://www.ti.com/lit/an/slyt650/slyt650.pdf?ts=1697002040345

[2] Z. Ye, D. Zhu, H. Yang, "Totem-Pole PFC Reliability and Performance Improvement with Advanced Controls," *IEEE Power Electron. Mag.*, vol. 9, no. 3, pp. 37–44, Sep. 2022, doi: 10.1109/MPEL.2022.3194285.

[3] J. W.-T. Fan, R. S.-C. Yeung, H. S.-H. Chung, "Optimized Hybrid PWM Scheme for Mitigating Zero-Crossing Distortion in Totem-Pole Bridgeless PFC," *IEEE Trans. Power Electron.*, vol. 34, no. 1, pp. 928–942, Jan. 2019, doi: 10.1109/TPEL.2018.2819422.

[4] B. Sun, "Control Challenges in a Totem Pole Bridgeless PFC," presented at the APEC 2023, Orlando, FL, Mar. 19, 2023.

[5] J. Sun, L. Zhu, R. Qin, D. J. Costinett, L. M. Tolbert, "Single-Phase GaN-Based T-Type Totem-Pole Rectifier With Full-Range ZVS Control and Reactive Power Regulation," *IEEE Trans. Power Electron.*, vol. 38, no. 2, pp. 2191–2201, Feb. 2023, doi: 10.1109/TPEL.2022.3215969.

[6] G.-Y. Lee, H.-C. Park, M.-W. Ji, R.-Y. Kim, "Digitalized Control Algorithm of Bridgeless Totem-Pole PFC with a Simple Control Structure Based on the Phase Angle," *Electronics*, vol. 12, no. 21, p. 4449, Oct. 2023, doi: 10.3390/electronics12214449.

[7] J. Sun *et al.*, "Mitigation of Current Distortion for GaN-Based CRM Totem-Pole PFC Rectifier With ZVS Control," *IEEE Open J. Power Electron.*, vol. 2, pp. 290–303, 2021, doi: 10.1109/OJPEL.2021.3071213.

[8] P. Rodriguez, A. Luna, M. Ciobotaru, F. Teodorescu, F. Blaabjerg, "Advanced Grid Synchronization System for Power Converters under Unbalanced and Distorted Operating Conditions," in *IECON 2006*, Nov. 2006, pp. 5173–5178. doi: 10.1109/IECON.2006.347807.

PCIM Europe 2024, 11– 13 June 2024, Nuremberg

DOI: 10.30420/566262444

Circulating Currents in Coupled Multi-Terminal Hybrid AC-DC Grids

Fabian Herzog [1], Benedict Mortimer [1], Rik W. De Doncker [1]

[1] Institute for Power Generation and Storage Systems, RWTH Aachen, Germany

Corresponding author: Fabian Herzog, post_pgs@eonerc.rwth-aachen.de
Speaker: Fabian Herzog, post_pgs@eonerc.rwth-aachen.de

Abstract

Coupling multi-terminal low voltage direct current (LVDC) and low voltage alternating current (LVAC) grids without isolation transformers is of high interest to reduce costs, material and losses. However, this introduces circulating currents between the LVAC grids. In this paper, the circulating currents are analyzed and reduction techniques are proposed. First, a simplified single phase system model is presented. Thereafter, the implementation of reduction methods derived from the model are introduced. Lastly, measurements from a test bench are presented where a significant reduction of circulating currents is achieved using the proposed methods.

1 Introduction

The coupling of LVDC and LVAC grids is gaining relevance with different use cases. The main goal is to reduce the loading of LVAC distribution grids and minimize the extension of LVAC grids. Approaches of integrating multi-terminal LVDC grids into existing LVAC grids use either 50 Hz transformers or medium-frequency transformers (MFTs) [1]. The classical setup with 50 Hz transformers needs a lot of resources, is costly and needs comparably large space. Novel solutions utilizing MFTs in solid-state transformers using rectifiers and a three-phase dual-active bridge converter reduce the needed resources and required space but are more complex [2]. Other approaches use modular multilevel converters to directly couple two grid sections of the same medium voltage AC grid through a medium voltage DC link [3].

In this paper, a transformerless approach using standard rectifiers is proposed to couple two LVAC grids with a LVDC grid. This system reduces the needed resources, the losses in the distribution grid and the power conversion and the required space compared to solutions utilizing transformers. However, this introduces circulating currents between the coupled LVAC grids through the grounded star point of the distribution grid transformers. The resulting system behaves comparable to parallel connected converters for which the reduction of circulating currents has been investigated in different

papers [4]–[9]. However, the reduction techniques mostly rely on a central controller or fast communication between distributed controllers.

In this paper, methods to reduce circulating currents with locally separated converters without any communication will be investigated. A common-mode voltage (CMV) model will be shown which is used to find mitigation concepts for the circulating currents. Afterwards, two of these concepts will be presented and implemented on a test bench replicating the proposed system. The first method synchronizes the converters without additional communication using a global navigation satellite system (GNSS) signal. The second method reduces the CMV by removing third-harmonic injection (THI) from the control. Finally, the two reduction methods will be implemented simultaneously and compared to the reduction of the individual methods.

2 Proposed System

The proposed system uses three-level neutral point clamped (3L-NPC) converters to supply the DC grid as active frontends. The system's voltage level is 700 V so that 650 V devices for three-level converters and 1200 V devices for two-level converters can be used. A schematic single line diagram of the system is shown in Fig. 1, where the star points of the transformers are connected through ground. The midpoints of the 3L-NPC converters are connected using a cable with the same diameter as the other cables. There is no additional grounding present in the LVDC grid.

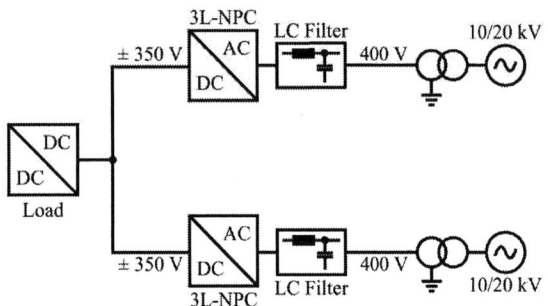

Fig. 1: Schematic single line diagram of the LVDC grid with two LVAC terminals.

The converter switches at a frequency of $f_{\text{sw}} = 8\,\text{kHz}$. The inductance of the LC filters is $L_{\text{f}} = 400\,\mu\text{H}$ and the capacitance is $C_{\text{f}} = 15\,\mu\text{F}$ with a series damping resistance of $R_{\text{d}} = 1\,\Omega$. The DC-link capacitance is $C_{\text{DC}} = 4\,\text{mF}$ for the two series capacitors of the 3L-NPC converters.

2.1 Control of 3L-NPC

The 3L-NPC converters are controlled using a model-based state feedback controller in the synchronous reference frame (SRF) aligned to the grid voltage using an SRF phase-locked loop. A cascaded control with an inner AC current control and an outer DC voltage control with PI regulators and command feedforward is implemented. For stable operation of the parallel converters a voltage droop of $\pm 7.5\,\%$ referred to the nominal power of the converters is added. Furthermore, sine-triangle pulse-width modulation (SPWM) with THI for an increase of the linear range of the output voltage by 15 % is used [10].

Additionally, a DC-link balancing algorithm is used, as shown in Fig. 2. The positive and negative DC-link voltages $U_{\text{DC,P}}$ and $U_{\text{DC,N}}$ of the 3L-NPC are subtracted and fed into a PI regulator. The output voltage U_{b} is either added or subtracted from the three-phase modulation signal of the cascaded controller, depending on the power flow direction.

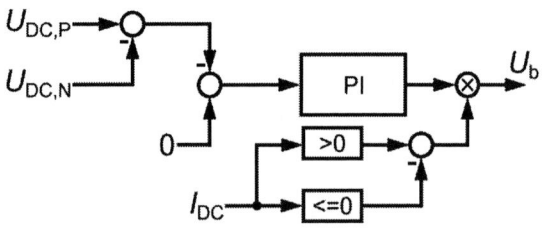

Fig. 2: DC-link voltage balancer.

3 Circulating Current Reduction

3.1 Common-Mode Model

A system model will be presented to find methods to reduce circulating currents in the proposed system. Analyzing the circuit yields that the circulating current is mostly caused by the converters' CMVs which can be calculated using Eq. (1). The phase voltages U_1, U_2 and U_3 are measured against the DC midpoint.

$$U_{\text{CM}} = \frac{U_1 + U_2 + U_3}{3} \tag{1}$$

However, for a precise calculation of the circulating current, the voltage between the DC midpoint and ground has to be added to U_{CM}, which is neglected here for simplification. Therefore, a simplified common-mode model as shown in Fig. 3 is introduced. The grounding of the transformer's star point is assumed to be not ideal, so a grounding resistance R_{g} is added. The total system inductance L_{tot} consists of the sum of the filter inductances, the line inductances and the transformer stray inductances divided by three due to the parallel connection. Additionally, the filter and transformer series resistances and the line resistances can be added to the grounding resistance R_{g}. The capacitors and resistors in the parallel path of the LC filters are neglected.

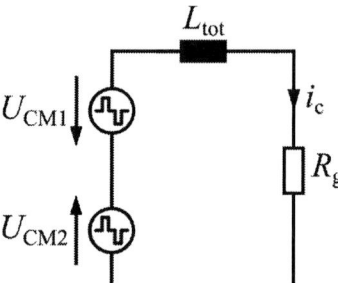

Fig. 3: Simplified common-mode model.

Using this model, a formula for the circulating current i_{c} can be derived. The difference of the CMVs U_{CM1} and U_{CM2} is switched across the inductance L_{tot} and R_{g} resulting in a differential equation. The solution of the differential equation is

$$i_{\text{c}} = (I_{\text{c,0}} - I_{\text{c,}\infty}) \cdot \text{e}^{-\frac{t}{\tau}} + I_{\text{c,}\infty}. \tag{2}$$

$I_{\text{c,0}}$ is the current at the switching time where the difference between the CMVs changes and τ is the time constant of the series connection of the passive elements. The current $I_{c,\infty}$ is the steady state current of the given combination which can be calculated using

$$I_{c,\infty} = \frac{U_{CM1} - U_{CM2}}{R_g}. \qquad (3)$$

From Eq. (2) and Eq. (3) it can be seen that the circulating current reaches its maximum for a maximum positive or negative U_{CM1} and the corresponding opposite for U_{CM2}. Furthermore, the circulating current vanishes for equal CMVs at the two converters.

3.2 Synchronization of the Converters

The common-mode model shows that equal CMV patterns result in a reduction of circulating currents. This can be achieved by synchronized switching of the two converters. However, the converters are locally separated which complicates direct communication between the converters to synchronize switching actions. Therefore, a Global Navigation Satellite System (GNSS) receiver is used which generates a globally synchronized pulse per second (PPS). The PPS signal is used to reset the carrier of the modulator of the converter controllers to achieve carrier synchronization without additional communication.

However, the synchronization using the PPS signal of GNSS receivers may have some limitations. The duty cycles of the two converters may have slight differences due to asymmetries in the two systems with deviations in filter or transformer inductances or measurement circuits. Additionally, the PPS signals are not perfectly synchronized with RMS differences between the signals in the range of tens of nanoseconds. Lastly, the oscillators generating the clock signals for the modulators may have small deviations leading to differences between two synchronization pulses.

3.3 Removal of Third Harmonic Injection

Further reduction techniques are needed if the synchronization of the switching actions of the converters is not ideal. A simple method to reduce the CMV generated by the converters is to remove THI from the converter control. Fig. 4 (a) shows a simulation of the CMV of the first converter with THI and Fig. 4 (b) shows the CMV without THI. It can be seen that the high CMV states last longer with THI than without. This is also reflected in the RMS value of the CMV which is 129.0 V in

Fig. 4 (a) and 108.8 V in Fig. 4 (b). Additionally, the frequency component at the converter's switching frequency, which is the highest component in the CMV, is 11.3 % lower without THI than with.

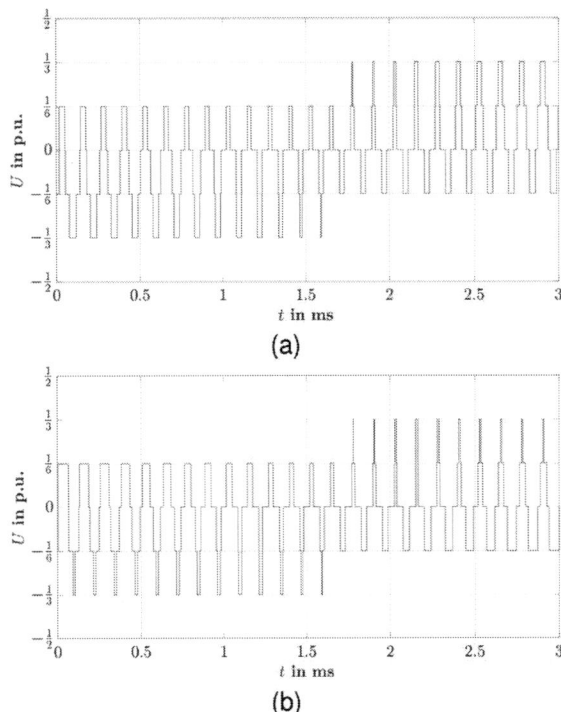

Fig. 4: Simulated CMV of the converter (a): with THI; (b): without THI.

However, removing THI reduces the output voltage's linear range which may negatively influence the converter's behavior at high loads and large load steps. Similarly, for converters controlled with space-vector pulse-width modulation, either the algorithm can be modified, or the modulator can be switched to SPWM without THI to reduce the CMV. Alternatively, more complex control modifications exist to further reduce the CMV [11], [12].

4 Test Bench Measurements

The test bench consists of two isolation transformers in Dyn5 configuration, where both star points are connected through a resistor, and two 3L-NPC converters with LC filters. While the 3L-NPC and the filters use the same components with the same rating, the transformers are rated differently and thus have different stray and main inductances and resistances. The 3L-NPC converters are controlled according to section 2.1 unless otherwise noted.

The general structure of the test bench is similar to Fig. 1, but instead of using medium-voltage to low-voltage transformers, the primary side is also low-voltage. The first measurements are conducted under no load conditions and using a grounding resistance of 25 Ω for each transformer. The test bench is shown in Fig. 5.

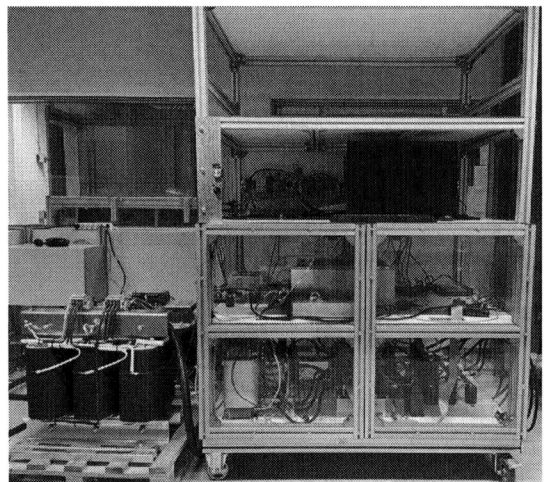

Fig. 5: Test bench for measuring circulating currents.

4.1 Reference Measurement

The reference measurement is conducted without any reduction of circulating current. Fig. 6 shows the switching pattern of one phase of the two converters. The converters are not switching simultaneously, indicating that the carriers have an offset against each other.

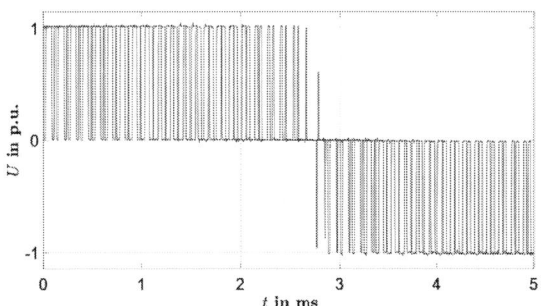

Fig. 6: Asynchronous switching pattern of one phase of the two converters.

As can be seen, the switching actions of the two converters are highly asynchronous. This leads to a circulating current which is shown in Fig. 7. The RMS and the peak value of the current is 3.73 A and 7.28 A, respectively, even though a large grounding

resistance is present. For lower grounding resistances the current would further increase according to Eq. (2) and Eq. (3). The largest frequency component of the circulating current is at the switching frequency of the converter which is 8 kHz in this case. Additional low frequency components can be observed with a smaller amplitude compared to the 8 kHz component. This current shape corresponds to the model shown in Fig. 3.

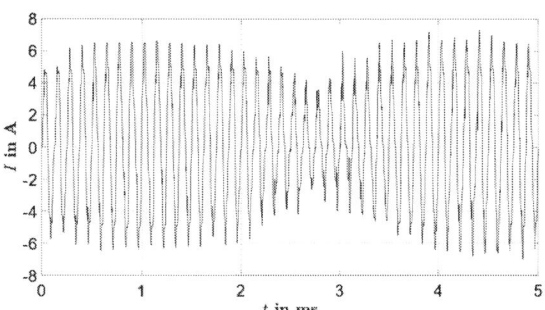

Fig. 7: Reference measurement of circulating current.

4.2 Reducing Circulating Currents

In the following, the proposed circulating current reduction methods are applied individually to investigate their impact. Thereafter, both reduction methods are combined and compared to the individual methods.

4.2.1 Synchronized Switching

A commercial GNSS module is connected to each control platform of the converters. The carriers of the controllers are reset with a frequency of 1 Hz using the PPS signal of the GNSS modules. The resulting switching pattern is shown in Fig. 8(a) with the corresponding circulating current in Fig. 8(b). Comparing the switching pattern in Fig. 8(a) with the reference in Fig. 6, the switching pattern is almost perfectly synchronized. Nevertheless, small deviations between the two converters are present which lead to a circulating current, as can be seen in Fig. 8(b). Compared to the reference case, a reduction of the RMS value of the circulating current by 90.3 % to a value of 0.36 A can be reached. Similarly, the peak value is reduced by 68.5 % to a value of 2.29 A.

4.2.2 Removal of Third Harmonic Injection

In the following, the GNSS module is removed. Instead, in the control of both converters THI is omitted so that they use regular SPWM. The resulting circulating current is shown in Fig. 9.

In this case, a reduction of the RMS value of the cir-

PCIM Europe 2024, 11– 13 June 2024, Nuremberg   DOI: 10.30420/566262444

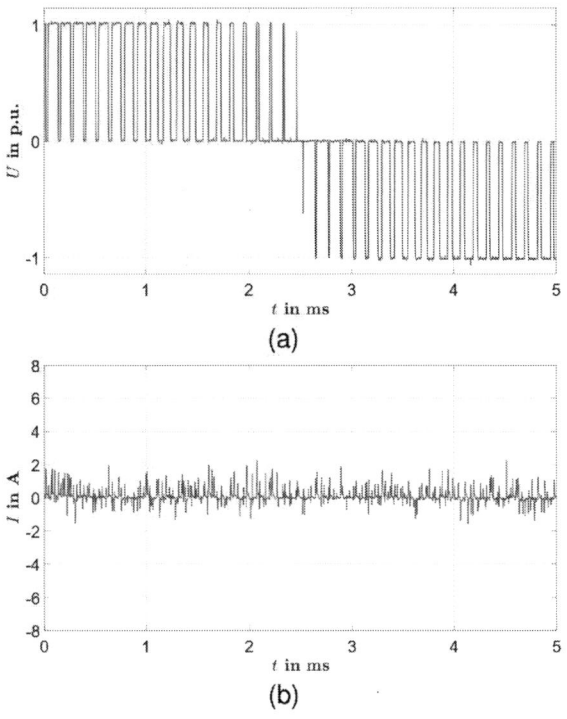

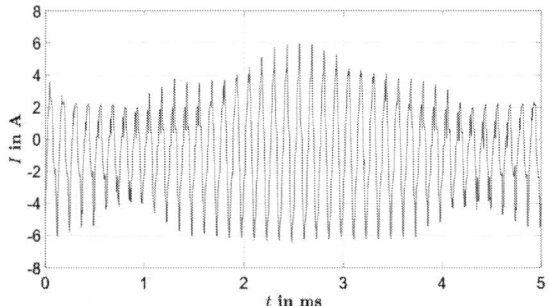

Fig. 8: Measurements with synchronized converters via GNSS (a): Switching pattern; (b): Circulating current.

culating current by 23.6 % to a value of 2.85 A can be reached. Similarly, the peak value is reduced by 12.1 % to a value of 6.4 A. However, the removal of THI has a much lower impact on the circulating current compared to the carrier synchronization.

Fig. 9: Circulating current with THI removed.

4.2.3 Combined Reduction Methods

Finally, both reduction methods will be used simultaneously to reduce the circulating current. Fig. 10 shows the measured circulating current with synchronized converters and THI removed. Generally, the current is similar to Fig. 8(b), where THI is not removed. This is also visible when comparing the two RMS and maximum values: with synchroniza-

tion and THI removed the RMS value and maximum value is 0.37 A and 1.91 A, respectively. This corresponds to a reduction of the RMS and maximum value of the circulating current by 90.1 % and 73.8 %, respectively.

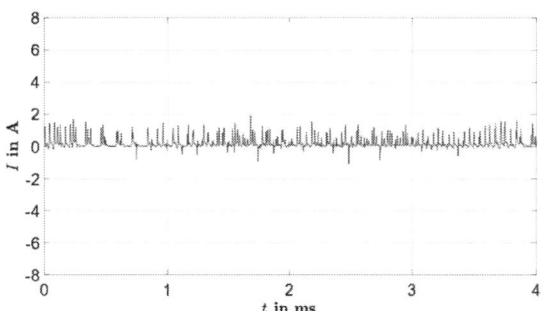

Fig. 10: Circulating current with synchronized converters and THI removed.

5 Conclusion

The coupling of LVAC distribution grids with LVDC grids without galvanic isolation is presented. First, a common-mode model for the proposed system is shown and formulas for the circulating current between the two distribution grids are derived. Thereafter, the synchronization of PWM carriers with a GNSS module and the removal of THI in the SPWM algorithm are presented to reduce the circulating current. These two methods are implemented on a test bench representing the proposed system with an increased grounding resistance of the star points of the transformers of 25 Ω. The synchronization of the carriers shows a significant reduction of the RMS value of the circulating currents by 90.3 % while the removal of THI results in a reduction of 23.6 % compared to a reference measurement without any reduction techniques implemented. Finally, the two methods are used simultaneously. This leads to similar reductions of the circulating current when only the carriers are synchronized. In this case, a reduction of 90.1 % was achieved.

For future investigations, the grounding resistance can be reduced to a more realistic value. Different reduction techniques for the CMV can be applied and further reduction methods can be implemented and combined with the ones proposed in this paper. Finally, differences in the AC voltage phase and amplitude, which can occur by different transformer connections or different loading of the transformers, can be included and investigated.

3140

Acknowledgment

The authors would like to thank the Federal Ministry for Economic Affairs and Climate Action of Germany for the financial support granted. They also thank their project partners for the equipment, insight as well as expertise they have provided, which contributed to this joint project IDEAL (FKZ: 01MV21008A).

References

[1] B. Mortimer, C. Olk, G. Roy, W. Tarnate, R. D. Doncker, et al., "Fast-charging technologies, topologies and standards 2.0," E.ON Energy Research Center Series, vol. 11, no. 1, 2019.

[2] C. Joglekar, B. Mortimer, F. Ponci, A. Monti, and R. D. Doncker, "Sst-based grid reinforcement for electromobility integration in distribution grids," Energies, vol. 15, no. 9, p. 3202, 2022. DOI: https://doi.org/10.3390/en15093202.

[3] A. Shekhar, Restructuring medium voltage distribution grids: parallel AC-DC reconfigurable links. 2020. DOI: https://doi.org/10.4233/uuid:a20ccf52-0b32-4f9c-924a-79b87b22505e.

[4] T.-P. Chen, "Zero-sequence circulating current reduction method for parallel hepwm inverters between ac bus and dc bus," IEEE Transactions on Industrial Electronics, vol. 59, no. 1, pp. 290–300, 2012. DOI: 10.1109/TIE.2011.2106102.

[5] S.-W. Kang, S.-Y. Choi, J.-H. Im, R.-Y. Kim, and S.-I. Kim, "Control strategy for suppression of circulating current using high-frequency voltage compensation in asynchronous carriers for modular and scalable inverter systems," IET Power Electronics, vol. 12, no. 14, pp. 3668–3674, 2019. DOI: https://doi.org/10.1049/iet-pel.2019.0461. eprint: https://ietresearch.onlinelibrary.wiley.com/doi/pdf/10.1049/iet-pel.2019.0461.

[6] Z. Ye, D. Boroyevich, J.-Y. Choi, and F. Lee, "Control of circulating current in two parallel three-phase boost rectifiers," IEEE Transactions on Power Electronics, vol. 17, no. 5, pp. 609–615, 2002. DOI: 10.1109/TPEL.2002.802170.

[7] S. Bella, A. Chouder, A. Djerioui, A. Houari, M. Machmoum, et al., "Circulating currents control for parallel grid-connected three-phase inverters," in 2018 International Conference on Electrical Sciences and Technologies in Maghreb (CISTEM), 2018, pp. 1–5. DOI: 10.1109/CISTEM.2018.8613377.

[8] S. Cho, Y. Jang, S. Jeon, and K.-B. Lee, "A reliable suppression method of high frequency circulating current in parallel grid connected inverters," in 2019 IEEE Energy Conversion Congress and Exposition (ECCE), 2019, pp. 1026–1031. DOI: 10.1109/ECCE.2019.8912758.

[9] H.-S. Jung and S.-K. Sul, "A design of circulating current controller for paralleled inverter with non-isolated dc-link," in 2017 IEEE 3rd International Future Energy Electronics Conference and ECCE Asia (IFEEC 2017 - ECCE Asia), 2017, pp. 1913–1919. DOI: 10.1109/IFEEC.2017.7992342.

[10] J. A. Houldsworth and D. A. Grant, "The use of harmonic distortion to increase the output voltage of a three-phase pwm inverter," IEEE Transactions on Industry Applications, vol. IA-20, no. 5, pp. 1224–1228, 1984. DOI: 10.1109/TIA.1984.4504587.

[11] Z. Ma, H. Niu, X. Wu, X. Zhang, and G. Lin, "An improved overmodulation strategy for a three-level npc inverter considering neutral-point voltage balance and common-mode voltage suppression," Sustainability, vol. 14, no. 19, 2022. DOI: 10.3390/su141912558.

[12] J. W. Kimball and M. Zawodniok, "Reducing common-mode voltage in three-phase sine-triangle pwm with interleaved carriers," in 2010 Twenty-Fifth Annual IEEE Applied Power Electronics Conference and Exposition (APEC), 2010, pp. 1508–1513. DOI: 10.1109/APEC.2010.5433431.

PCIM Europe 2024, 11– 13 June 2024, Nuremberg   DOI: 10.30420/566262445

Comparison of 4500 V State-of-the-Art XHP3 IGBT and Conventional IHV IGBT for 3300 V, 3-Level ANPC Medium-Voltage Drives

Xin Ma[1], Jens Czichon[2], Martin Knecht[2], Lifeng Chen[3], Marc Buschkuehle[2]

[1] Infineon Integrated Circuit (Beijing) Co., Ltd, China

[2] Infineon Technologies AG, Warstein, Germany

[3] Infineon Technologies Center of Competence (Shanghai) Co. Ltd, China

Corresponding author: Xin Ma, xin.ma@infineon.com
Speaker: Martin Knecht, Knecht.Martin@infineon.com

Abstract

The 3-level active neutral point clamped (3L-ANPC) topology is a commonly used inverter topology for medium-voltage drive (MVD) inverters. This topology has been extensively researched. It offers a flexible commutation loop and can reduce the uneven distribution of power losses among IGBTs. This paper presents the flexible 4500 V, high-power platform (XHP3) with insulated-gate bipolar transistor (IGBT) devices as an alternative to the conventional industrial high-voltage (IHV) IGBT module for constructing the scalable and modular 3L topology for the 3300 V 3L-ANPC MVD. We provide an overview of the 4500 V XHP3 and IHV IGBT products and carry out a quantitative comparison of the devices in terms of conduction losses, switching losses, thermal resistance and junction temperature based on the datasheet of the 450 A XHP3 and 800 A IHV IGBT. Furthermore, we present the recommended layout of the XHP3 IGBT for the 3L-ANPC power stack and analyze its advantages over the IHV IGBT solution. A thermal simulation study is conducted for these devices under the typical operating conditions of a 3300 V MVD system. The results demonstrate that with XHP3, the system benefits from a much higher power density and lower power dissipation.

1   Introduction

3300 V MVD inverters, based on the 4500 V IHV IGBT and 3L topology, are widely used in high-power motor drives for wind power, coal mining and many other industrial applications. The MVD inverters drive various types of loads such as rolling mills, fans, pumps, conveyors and so on. The 4500 V IHV IGBT is a single-switch device available in two different packages to accommodate the different current ratings. The base plate size of one package is 140 x 130 mm, and the other package 140 x 190 mm, both having the same height of 48 mm. Compared to the IHV IGBT, the newly developed XHP3 IGBT has a height of 40 mm and contains two switches. The general product comparison between XHP3 and IHV is summarized in Fig. 1, which includes the chip technology, the maximum current rating of each product, the maximum operating

temperature, and current density based on the base plate area of the IGBT.

As shown in Table 1 and Fig. 2(a), the XHP3 IGBT is designed in a half-bridge configuration with an isolation voltage of up to 10.4 kV, which is the same as the 4500 V IHV IGBT. Owing to the stacked design of the positive and negative busbars inside the module, the stray inductance of the module is minimized, typically 25 nH, which is significantly lower than that of the half-bridge configuration with IHV IGBT. This allows a reduced commutation inductance for both 2-level and 3-level converters.

XHP3 has an optimal arrangement of main and auxiliary terminals. The main power terminal layout is simplified, with the DC terminals on one side of the module, and the AC terminals on the other. This prevents the connection busbars from

3142

crossing over the gate driver board as in the case of the IHV module, thereby improving electromagnetic compatibility. The simplified access to the power terminals also

makes the design of the laminated DC busbar and connections between the IGBT modules much simpler.

Vces = 4500V	Conventional IHV IGBT		State of the art XHP3 IGBT
Chip technology	Trench/field stop IGBT3 and EC3		Trench/field stop IGBT4 and EC4
Rated current[A]	1200	800	450
$T_{vjop,max}$[°C]	125	125	150
Topology	Single switch	Single switch	Half bridge
Package outline			
Module stray inductance[nH]	18	20	25
Current density[A/cm²]	4.5	4.4	6.4
Isolation voltage[kV]	10.4	10.4	10.4

Table 1   Product comparison between 4500 V IHV

[2][3] and XHP3 [1] IGBT modules

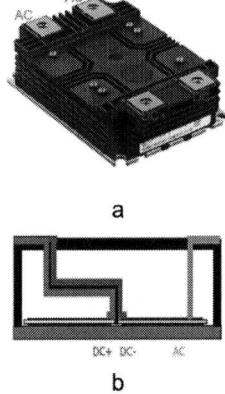

a

b

Fig. 1   a: XHP3 IGBT package; b: internal busbar

layout of XHP3

The 4500 V 450 A XHP3 IGBT FF450R45T3E4_B5(referred to as M1 in this paper) adopts the latest IGBT4 and EC4 diode chip technology. The maximum allowed operating temperature is 150°C, which is 25°C higher than that of the IHV IGBT. This helps to increase the output current of the device. The current density of M1 is 6.4 A per square centimeter, 45% higher than that of the IHV IGBT. As a result, the M1 shows great promise for achieving higher system power densities.

Figure 2(a) illustrates the simplified three-phase 3L-ANPC topology used in the 3300 V MVD inverter. Typically, six FZ800R45KL3_B5(referred to as M2 in this paper) are required to construct one phase leg, usually called a power stack. The power stack includes the connection busbars, IGBT gate drivers, heat sink and DC-link capacitors. In contrast, M1 contains two switches and only three M1 are required to make one phase leg. This simplifies the connection of busbars between the IGBTs as well as the mechanical design of power stack. For MVD inverters with varying power ratings, using the M1 in parallel is a simple and scalable solution. To achieve a higher power rating than M2, it is possible to use two M1 in parallel, as shown in Fig. 2(c). Similarly, using three M1 in parallel can result in a higher power rating compared to FZ1200R45KL3_B5. Therefore, using M1 is a better option for adjusting to the different power levels of the MVD inverter than using IHV IGBTs with different current ratings.

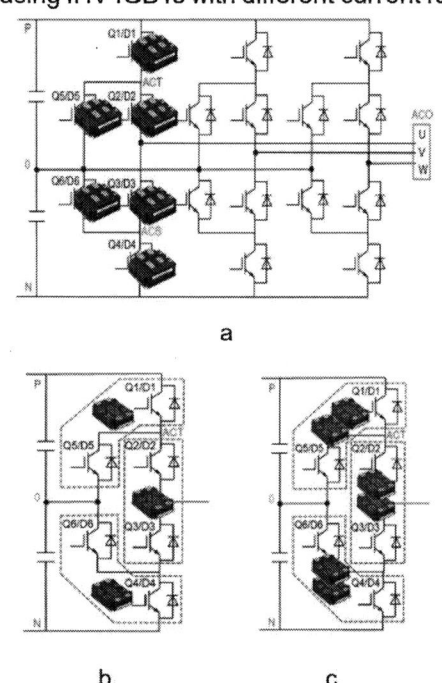

a

b   c

Fig. 2   Simplified 3L-ANPC topology, a: three-phase

leg with FZ800R45KL3_B5; b: one-phase leg with

FF450R45T3E4_B5; c: one-phase leg with two

FF450R45T3E4_B5 in parallel

2   Static and dynamic characteristics of FF450R45T3E4_B5 and FZ800R45KL3_B5

This section compares the static and dynamic characteristics of M1 and M2, as well as their thermal resistance, based on the specifications. To ensure a fair comparison, two M1 modules are used for comparison with one M2, where their equivalent rated currents only differ by 100 A.

2.1 Static characteristics

The M2 module has a current rating of 800 A, whereas two M1 modules have a combined rated current of 900 A. In reference to the collector-to-emitter voltage V_{ce} of M2 and two M1, at a current of 800 A and a junction temperature of 25℃, it was observed that the V_{ce} of M2 is 2.24 V while the V_{ce} of the two M1 is 2.50 V. The latter is 0.26 V lower than the former, a reduction of 10.4% as shown in Fig. 3(a). At a junction temperature of 125℃, the V_{ce} of M2 is 3.10 V, while the V_{ce} of the two M1 is 2.65 V. The latter is 0.45 V lower than the former, a decrease of 14.5% as shown in Fig. 3(b). For the forward voltage drop V_f of the diode, two M1 are similar to M2 at a junction temperature of 25℃, and slightly lower than M2 at a junction temperature of 125℃ over the entire current range, as shown in Fig 3(b).

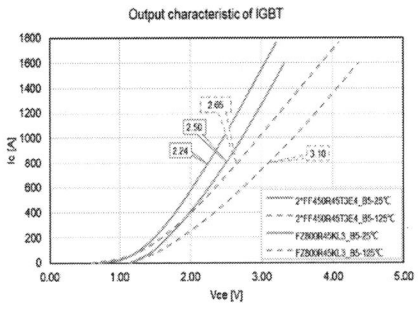

a

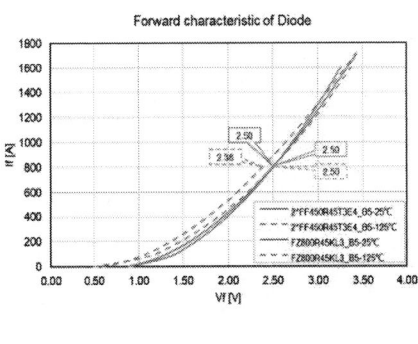

b

Fig. 3 Static characteristics comparison between one FZ800R45KL3_B5 and two FF450R45T3E4_B5,

a: output characteristics of IGBT; b: forward characteristic of diode

2.2 2.2 Dynamic characteristics

This section mainly covers the turn-on (E_{on}) and turn-off (E_{off}) energies of M2 and M1. To facilitate direct comparison, E_{on} and E_{off} in the specifications are divided by their respective rated currents. Specifically, they are divided by 800 A and 450 A for M2 and M1, as shown in Fig 4. Figure 4 displays E_{on} and E_{off} as a function of the turn-on resistor R_{gon} and turn-off resistor R_{goff} for a typical DC-link voltage of 2800 V and a junction temperature of 125℃. As M2 and M1 have different R_{gon} and R_{goff} ranges, their respective R_{gon} and R_{goff} values are normalized to their default values as specified in the specification. In Fig. 4(a), the E_{on} of M1 is significantly lower than that of M2. At default R_{gon} and 2 times default R_{gon}, E_{on} increases with R_{gon} to a much lesser extent than M2, being about 33% and 47% lower than M2. Figure 4(b) displays the comparison results of E_{off}, which is approximately 26% and 30% higher for M1 than for M2 at the default R_{goff}, and twice the default R_{goff}. It is evident from Fig. 4 that M1 has lower switching losses than M2 at its respective rated current and the same normalized R_{gon} and R_{goff}.

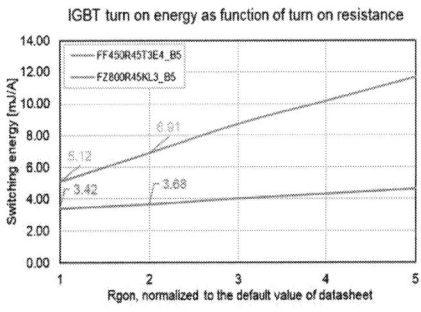

a

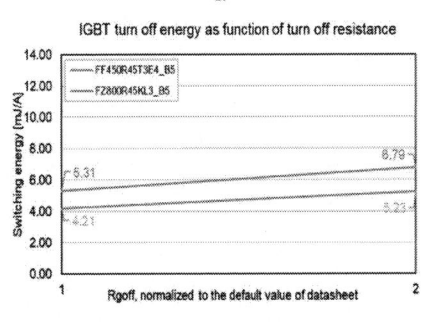

b

Fig. 4 Dynamic characteristics comparison, a: IGBT turn-on energy E_{on} as a function of R_{gon} at 800 A I_c for FZ800R45KL3_B5 and 900 A I_c for two FF450R45T3E4_B5, 2800 V_{dc} and T_{vj} 125℃; b: IGBT turn-off energy E_{off} as a function of R_{goff} at 800 A I_c for FZ800R45KL3_B5 and 900 A I_c for two FF450R45T3E4_B5, 2800 V_{dc} and T_{vj} 125℃

The thermal resistance of the IGBT module and heat sink indicates the system's ability to dissipate heat. Figure 5 displays the thermal resistance of the junction-to-case (R_{thjc}) and thermal resistance of case-to-heat sink (R_{thch}) for one M2 and two M1, along with their corresponding water-cooled heat sinks. Compared to M2, the two M1 have a 17.6% higher R_{thjc}, a similar R_{thch} and 15.8% lower R_{thha}. Overall, considering the junction-to-ambient thermal resistance as the sum of the three thermal resistances, the thermal resistance of the solution for two M1 is 0.0426 K/W as shown in Fig. 5, which is approximately 0.01 K/W lower than that of M2. The junction-to-ambient thermal resistance of the diode chips of two M1 is 23% lower than that of M2. This is due to the R_{thjc} and R_{thch} of the two M1 chips being 12% and 53% lower, respectively.

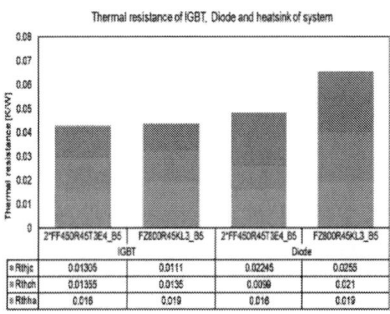

Fig. 5 Thermal resistance of FZ800R45KL3_B5, two FF450R45T3E4_B5 and heat sink

3 Comparison of the layout of 3L-ANPC power stack based on XHP3 and IHV IGBT

In the 3L-ANPC circuit, the major commutation paths [4] are mainly divided into short and long paths, as shown in Fig. 6. For example, during the positive half-cycle of the output-phase voltage and positive-phase current, the commutation path is short between the positive 'P' and the neutral point '0' of the DC bus, as illustrated in Fig. 6(a). In his scenario, Q2 remains active while the current is switched between Q1 and D5 depending on their respective states. This means that only two devices are involved in the commutation process. Similarly, in the long commutation path shown in Fig. 6(b), four devices are involved in the commutation, i.e. Q1, Q2, D3 and Q6. As a result, the stray inductance of the shorter commutation path is significantly lower than that of the longer commutation path.

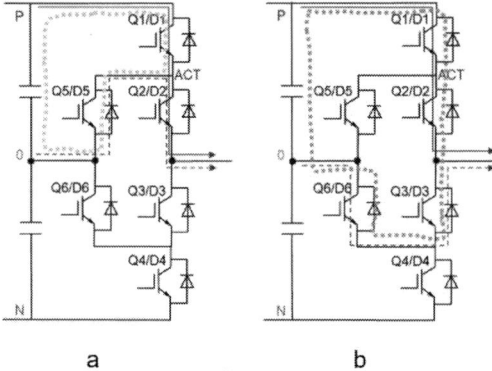

Fig. 6 3L-ANPC phase-leg commutation paths, a: short path between "P" and "0"; b: long path between "P" and "0". Both are during the positive half-cycle of the output-phase voltage and positive-phase current.

In addition to the type of commutation circuit, the type and layout of the IGBT greatly affects the stray inductance of the commutation circuit. As described in the previous section, it needs six M2 to construct the single-phase power stack. To minimize the stray inductance of the short commutation path, the layout of six M2 is shown in Fig. 7(a). For the upper phase leg, Q1D1 is close to Q5D5, and Q5D5 is close to Q2D2. The lower half of the phase leg has a symmetrical layout that mirrors the upper phase leg six pieces of copper busbars are required to make a laminated busbar to connect all the IGBTs and further reduce the stray inductance of the copper busbar. The dotted, colored rectangles represent different busbars. Clearly, the power stack with M2 will require a large area of laminated busbars. The entire dimensions of the busbars are similar to the base plate area of six M1.

Figure 7(b) shows the layout of a single-phase power stack with three M1. The upper left module consists of Q1D1 and Q5D5, the upper right module consists of Q4D4 and Q6D6, and the lower module consists of Q2D2 and Q3D3, with an overall left-right symmetrical structure. As Q1 and D5 are in the same IGBT, the stray inductance of the short commutation path is greatly reduced, even if only from the IGBT perspective. As only half of the upper modules are involved in

3145

commutation and M1 does not have stray inductance between its DC power terminals and AC power terminals, it is necessary to conduct tests in an actual system to determine the equivalent stray inductance for the long commutation path. For the parallel solution of two single-phase legs using M1, as shown in Fig. 7(c), the stray inductance of the IGBT module in both the short and long commutation paths will be halved, as compared to Fig. 7(b). Furthermore, the required connection busbar for M1 has been significantly reduced.

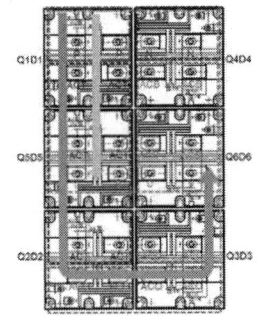

a

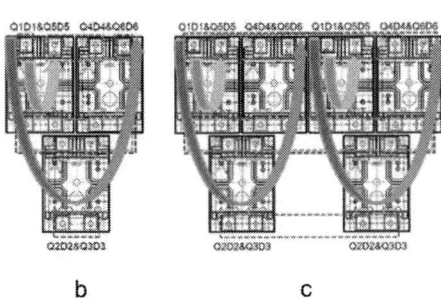

b   c

Fig. 7 IGBT layout of single phase 3L-ANPC power stack, a: six FZ800R45KL3_B5; b: three FF450R45T3E4_B5; c: six FF450R45T3E4_B5

In motor-drive applications, the switching power dissipation of IGBT Q1 and Q4 is a higher proportion of the total power dissipation at a positive power factor. According to Table 2 and Fig. 7, the M1 solution has lower stray inductance in short commutation paths compared to the M2 solution. Considering the maximum allowed power dissipation is 1500 kW and 1600 kW of the diode for M1 and M2 respectively, the allowed maximum power dissipation per ampere of the diode of M1 is 67% higher than that of the M2. Therefore, the M1 power stack allows for a faster diode recovery (and IGBT turn-on) speed compared to the M2 power stack, as the IGBT turn-on speed is limited by the diode's allowed maximum power dissipation during commutation [5]. This makes it feasible to

use smaller R_{gon} and R_{goff} for Q1 and Q4 to increase the switching speed, thereby reducing the switching power loss.

IGBT	IGBT quantity/single phase leg	Stray inductance of IGBT in short commutation path [nH]
FZ800R45KL3_B5	6	40
FF450R45T3E4_B5	3	25
FF450R45T3E4_B5	6	12.5

Table 2 Stray inductance of IGBTs in the short commutation path

The M1 solution not only offers the benefits mentioned above, but also reduces the necessary heat sink area, especially for the IGBT layout in Fig. 7(b). Assuming 10 mm distance between adjacent IGBT modules, Table 3 displays the required minimum dimensions and areas of the heat sink for the three layouts of IGBT shown in Fig. 7. The area of the heat sink in a power stack with three M1 is 61% smaller than that of a power stack with six M2. Although the heat sink area of the solution using six M1 is 5% larger than that of the solution using six M2, the former has a significantly higher power density, as explained in the following section.

IGBT layout	The minimum dimensions of heat sink, W*L[mm]	The minimum area of heat sink [mm²]
6* FZ800R45KL3_B5	270*440	118800
3* FF450R45T3E4_B5	210*290	45990
6* FF450R45T3E4_B5	430*290	124700

Table 3 The dimensions and area of heat sink

4 Thermal simulation study of FF450R45T3E4_B5 and FZ800R45KL3_B5 based on the typical operation condition of 3300 V MVD

The use of simulation tools to study and compare the performance of power devices in a system is an efficient evaluation method. Following the typical operating parameters of the 3300 V MVD in Table 4, this section presents a detailed simulation study of two M1 in parallel solution, and M2 using the power electronic system simulation software Plecs and a datasheet-based device thermal model. For the M1 solution, it is assumed that each module provides 50% of the system current. Due to the large influence of the gate driver resistance on the switching loss of the IGBTs, the simulation uses the resistance matched by the third-party driver supplier [6] for M2 to be closer to the actual application. M1 uses the same resistance correction factor as M2, e.g. the correction factor of R_{gon} is 2.5, as the R_{goff} of the gate driver of third party is very close to that of the datasheet; the correction factor of R_{goff} is set to 1.

Parameters	Value
V_{dclink} [V]	5000
V_{out} [Vrms]	3300
Power factor	0.85
F_{out} [Hz]	50
F_{sw} [Hz]	1000
Gate driver resistor	R_{gon}: 2.5*default of datasheet R_{goff}: 1* default of datasheet
R_{thha} per switch [K/W]	FF450R45T3E4_B5: 0.032 FZ800R45KL3_B5: 0.019
T_a [°C]	40
Overload condition	120%, 1 min/5 min 200%, 10 s/5 min

Table 4   Operation condition of 3300 V MVD

IGBT Q1 has the highest junction temperature in the MVD driving motor mode. Figure 8 shows the simulation curves of output current and junction temperature of IGBT Q1 of two types of IGBT modules. The maximum operating junction temperature of M2 is 125℃, and the corresponding steady-state output current is 373 Arms. The maximum operating junction temperature of M1 is 150℃, and the corresponding steady state output current of two M1 in parallel is 510 Arms, 36.7% higher than that of M2. The junction temperature, case temperature and heat sink temperature of Q1 using M1 are lower than those of M2 at the same output current. Taking the output current of 300 Arms as an example, these three temperatures are lower than those of M2 by 7.6℃, 9.7℃ and 7.2℃ respectively, as shown in Fig. 9(b). In Fig. 9(a), the power loss of a single-phase leg of the MVD is compared. The total power loss using M2 is 2220

W, whereas using M1 it is 2063 W, resulting in a reduction of 7.1% compared to M2.

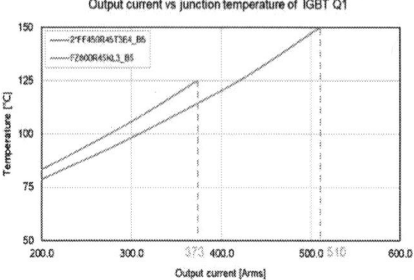

Fig. 8   Output current of IGBT module and junction temperature of IGBT Q1 under the typical operation conditions of MVD

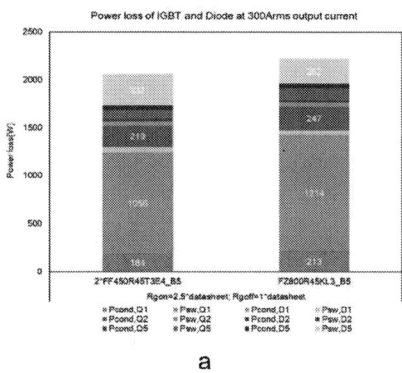

a

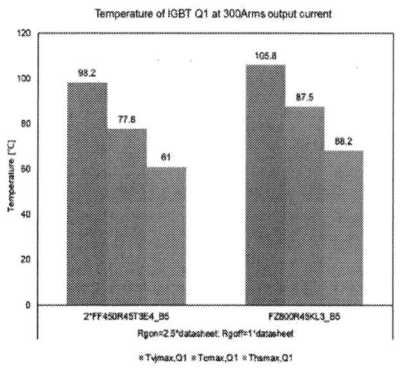

b

Fig. 9   Simulation result of 3300 V MVD at the output current of 300 Arms, a-power loss of the single-phase leg; b-junction, case and heat sink temperatures of IGBT Q1

In addition to evaluating the module's thermal performance under steady-state conditions, it is also important to assess its performance under overload conditions. For example, we can consider a steady-state current of 300 Arms and a 120% overload current of 360 Arms, which runs for

1 minute during the 5-minute load cycle. Figure 10(a) illustrates that the IGBT Q1 using M2 has a maximum junction temperature of 117.1°C, while the Q1 with M1 has a maximum junction temperature of 107.5°C, as shown in Fig. 10(b). The junction temperature margin for M2 is 7.9°C, while for M1 it is 42.5°C, compared to the maximum junction temperature allowed for each device.

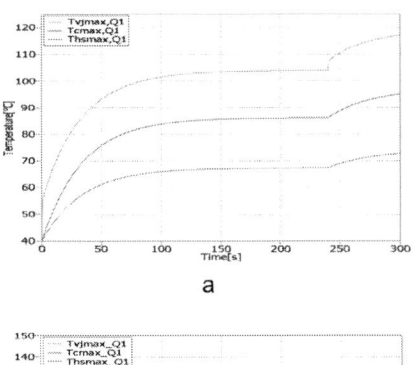

a

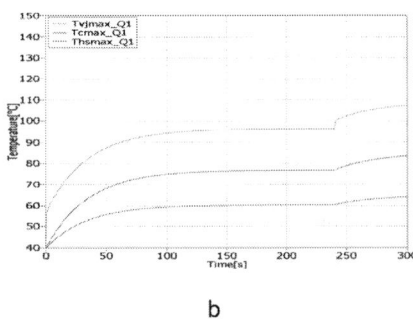

b

Fig. 10 Temperature simulation results for IGBT Q1 during 120% overload for 1 minute of a 5-minute load cycle, a- FZ800R45KL3_B5; b- FF450R45T3E4_B5

Short-term, high-current overloads are common in certain applications, including conveyor belt drives, resulting in a dramatic increase in the junction temperature of the IGBT Q1. The M2 experienced a junction temperature increase of 20°C when subjected to a 200% 10-second overload at 200 Arms steady-state current and 400 Arms overload, as shown in Fig. 11. It is important to note that this is much higher than the increase observed in the 120% overload condition. Similarly, Fig. 12 displays the simulation results for the output currents and junction temperatures of M2 and two M1 in parallel under varying steady-state output currents, but at the same 200% overload condition. The M2 has a steady-state output current of 261 Arms, while two M1 in parallel have a steady-state output current of 354 Arms, which is 35.6% higher than the former. This result is consistent with the data presented in Fig. 8.

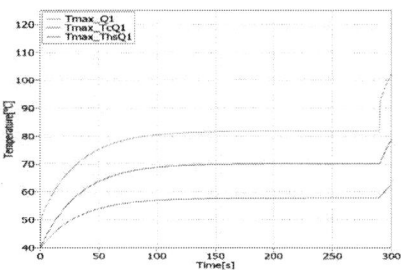

Fig. 11 Temperature simulation results for Q1 using FZ800R45KL3_B5 during 200% overload (400 Arms) for 10 seconds of a 5-minute load cycle

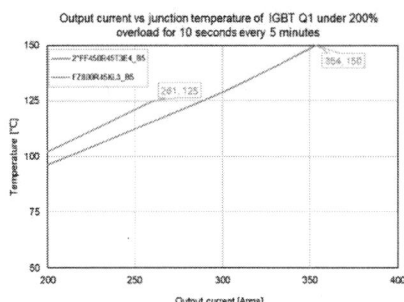

Fig. 12 Junction temperature simulation results for IGBT Q1 during 200% overload for 10 seconds of a 5-minute load cycle

The simulation study in this section demonstrates that using two parallel M1 results in up to approximately 36% increase in power density compared to M2. Additionally, the losses are slightly lower, allowing for a larger increase in the power density of the system. By analogy, three M1 have an output capacity of approximately 68% of six M2, resulting in an extremely high power density MVD.

5 Conclusion

This paper presents the FF450R45T3E4_B5, a new product from Infineon's latest-generation high-voltage 4500 V IGBT platform XHP3. A comparison between the FF450R45T3E4_B5 and the previous generation of high-voltage IHV IGBTs, the FZ800R45KL3_B5, shows the advantages of the former at both the product and system application levels. The FF450R45T3E4_B5 is highly scalable, aiding in the platform design of the MVD system, and has a much higher power density that can be achieved with fewer modules, thereby increasing the system's cost performance and competitiveness. Based on the Plecs simulation software and typical application conditions of 3300 V MVD, the simulation results demonstrate that the FF450R45T3E4_B5 can significantly improve the power density of the inverter. It should be noted that the simulation results presented in this paper are based on ideal working conditions. Therefore, the loss and output capability of the IGBT module in the actual system should be evaluated accordingly.

References

[1] Infineon Technologies
    data sheet, FF450R45T3E4_B5
[2] Infineon Technologies
    data sheet, FZ800R45KL3_B5
[3] Infineon Technologies
    data sheet, FZ1200R45KL3_B5
[4] Heng Wang, Xin Ma, Huibo Sun, Active neutral-point-clamped (ANPC) three-level converter for high-power applications with optimized PWM strategy, PCIM Asia, 2020.
[5] Sven S. Buchholz, Matthias Wissen, Thomas Schutze, Electrical performance of a low inductive 3.3kV half bridge IGBT module, PCIM Europe, 2015.
[6] Power Integrations
    data sheet SCALETM-2 1SP0335x2x1(C)-M2

PCIM Europe 2024, 11– 13 June 2024, Nuremberg   DOI: 10.30420/566262446

Generalized Switching Sequence for Voltage Balancing in a Flying Capacitor DC-DC Converter with Quasi-2-Level Modulation

José Andrés Aguilar Croston[1] , Piotr Dworakowski[1] , Besar Asllani[1]

[1] SuperGrid Institute, France

Corresponding author: José Andrés Aguilar Croston, jose-andres.aguilar@supergrid-institute.com
Speaker: José Andrés Aguilar Croston, jose-andres.aguilar@supergrid-institute.com

Abstract

A generalized voltage balancing algorithm is proposed for a multilevel flying capacitor (FC) bridge used in a dual active bridge (DAB) medium voltage DC-DC converter with quasi-2-level modulation. The operation of the flying capacitor converter (FCC) in a DAB is explained and the voltage imbalance problem is illustrated. The results of the proposed algorithm are presented in the form of a switching sequence that allows to balance the voltage of the flying capacitors in open loop during steady-state operation. The algorithm is generalized to be used with an n-level FCC in a half bridge or a full bridge configuration. The proposed method produces a balanced switching sequence based on fixed criteria such as symmetric use of converter and minimization of losses. Simulation results are presented for a 13-level FCC in a half bridge and full bridge configuration.

1   Introduction

In the realm of emerging DC networks, high power DC-DC converters will play a crucial role in efficiently transforming power. However, the maximum voltage rating of available semiconductor devices poses a major challenge in medium and high voltage applications, as it limits the achievable voltage of power electronic converters. When the voltage level exceeds the maximum rating of a single semiconductor device, one solution is to connect multiple devices in series to share the voltage across a valve but this can be challenging [1], [2]. Moreover, voltage sharing is not uniform among the devices due to manufacturing tolerances, common mode currents and gate signal desynchronization, leading to imbalanced voltage distribution and potential failure of the most stressed device. Therefore, alternative topologies, such as multilevel converters, can be used to overcome the limitations of the conventional two-level converter topology. For instance, in HVDC applications, multilevel voltage source topologies have completely superseded 2-level or 3-level topologies that used series connection of IGBTs.

In this article the topology considered for the DC-DC converter is the dual active bridge (DAB) which main properties are: galvanic isolation, bidirectional power-flow, high-power-density and possible soft switching [3], [4].

In two-level converters, rapid transitions between voltage levels can raise another issue, when used in an isolated DC-DC converter. A high steep rate of voltage change, denoted as dV/dt, can have harmful effects on the insulation of transformers [5], [6]. A solution to this problem would be a topology that allows the control of the dV/dt depending on the transformer limitations. A solution like this would allow to scale the two-level dual active bridge to higher DC voltage levels. A good prospect to address this challenge was proposed in [7] with an operation mode called quasi-2-level (Q2L).

Since its introduction the quasi-2-level modulation (Q2LM) has been subject to different studies [8]. A particular topology that seems promising for this modulation is the flying capacitor converter (FCC), firstly introduced in [9] to the knowledge of the authors. As a multilevel topology, it can be repurposed to function with Q2LM. The flying capacitor converter was subject of studies to evaluate the suitability of this topology into medium voltage (MV) applications [10], [11]. The quasi-2-level modulation employed in this topology allows the use of low voltage transistors, enabling converters at high voltage rating without exceeding the voltage ratings of individual semiconductors. Q2LM also allows the use of a smaller capacitors than the ones needed by any other more traditional modulation. This is thanks to the short voltage steps during voltage transitions. To ensure the operation of the FCC, meticulous consideration must be given to the voltage balancing of the flying capacitors. An inherent advantage of the FCC lies in its capability for open-loop voltage balance.

Studies have been done for the FCC operating under phase shifted pulse width modulation (PS-PWM) which is different from the application for DC-DC converters with Q2LM. However the principles shown in [12], [13] can be adapted to the specific waveforms and conditions seen in Q2LM. In [12], a full-bridge 3-level topology was shown and a Karnaugh-Veitch method was used to find the switching sequence that achieved the voltage balancing. In [13], a generalized and scalable PWM strategy for a FCC was proposed to ensure open-loop voltage balancing. The strategy was tested in simulation in a 3-phase 7-level FCC. A review regarding the natural voltage balance in a FCC was done in [14], showing applications in half-bridge, full-bridge, 1-phase and 3-phase operations. Nonetheless all these studies were done for a PWM, where the switching frequency is higher than the fundamental frequency.

In [15], a 5-level half-bridge flying capacitor bridge was studied in a DC-DC converter. A voltage balancing scheme was presented for the Q2L operation. However, a generalization was not done, and the full-bridge application was not considered.

The voltage balancing of the FCC has been studied for PWM and Q2LM operation. Equivalent studies have been done and generalized for the Modular Multilevel Converter (MMC) topology. In [16] an alternating modulation balance algorithm was proposed and studied for an MMC in a DAB. Further studies were done in [17] where various voltage-balancing algorithms in an MMC were compared.

This article proposes a generalized method to find a balanced switching sequence for open-loop voltage balancing of an *n*-level FCC operating with a Q2LM in an isolated DC-DC converter. The proposed method finds a switching sequence that provides natural balancing of the charge of flying capacitors. The generalization of the method allows the user to apply the balanced switching sequence to any number of flying capacitors in a half-bridge or full bridge Flying Capacitor Converter (FCC).

The rest of the paper is structured as follows: Section 2 presents the case study, in Section 3 the problem of voltage imbalance in a flying capacitor bridge operating in a DC-DC converter with Q2LM is shown, this is followed by the description of the method to create the balanced switching sequence in Section 4, simulation results are presented in Section 5 to validate the methods, and Section 6 presents the conclusions.

2 Case study

For this study a DC-DC isolated bidirectional converter for a MVDC-LVDC connection is considered. The converter schematic is shown in Fig. 1. The parameters used are shown in Table 1.

Table 1 Case study parameters

Nominal power (P_n)	250 kW
Medium voltage (V_{MV})	20 kV
Low voltage (V_{LV})	350 V
Switching frequency (F_{SW})	10 kHz

The topology is a Dual Active Bridge (DAB). The MV bridge is the main subject of study. For the MV bridge a multilevel topology is considered. The adopted topology is the flying capacitor converter (FCC). The FCC generalized topology is shown in Fig. 2. The multilevel topology is used with a quasi-2-level modulation.

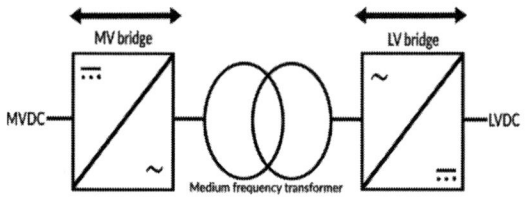

Fig. 1 Simplified single-line diagram of the isolated DC-DC converter.

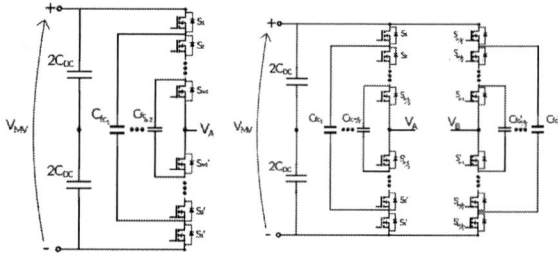

Fig. 2 Generalized *n*-level FCC topology in a half-bridge (left) and a full bridge (right). One valve is considered as the switches from S_1 to S_{n-1} in a half bridge and from S_1 to $S_{(n-1)/2}$ in a full bridge. There are two valves in a half bridge and four in a full bridge FCC.

3 Problem statement

The problem dealt with is the charge equilibrium of the flying capacitors. This is specific to the FCC in the medium voltage bridge. For simplicity, the FCC can be considered alone as an inverter.

The problem can be illustrated with an example. For instance, a 4-level half-bridge FCC. The schematic is shown in Fig. 3. To obtain the quasi-2-level modulation, 2 intermediate voltage levels are used.

The intermediate voltage levels are obtained with the help of the two flying capacitors C_{fc1} and C_{fc2}. Looking at the example, it can be deduced that several switching sequences are possible to produce the desired voltage waveform.

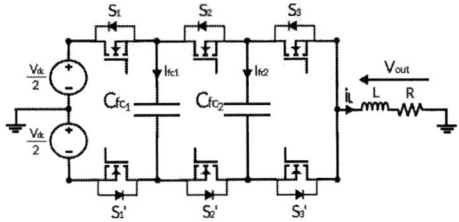

Fig. 3 4-level half-bridge FCC.

In this paper, a switching sequence is defined as a sequence of switching states. A switching state is the compilation of the state of all switches in the inverter at a given instant. The switching sequence will determine the output voltage and which capacitors are used. The goal is to use the switching sequence periodically, hence the switches states in a switching sequence must compose a whole number of voltage periods.

In a quasi-2-level modulation the voltage waveform includes a high voltage level, a low voltage level, and short intermediate voltage levels. The short intermediate levels are found in the transitions between the high and low levels. These intermediate levels have a length of t_{delay} and are also referred to as steps. In Fig. 4 there is a representation of the waveforms in a 4-level topology. It is shown that one period is composed of six switching states and that in this application the output current changes sign at some point at the extreme voltage levels.

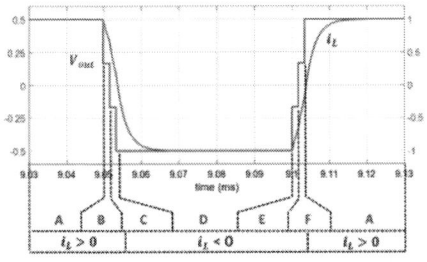

Fig. 4 Typical 1-phase 4-level Q2LM voltage and current waveforms.

For this example, it is possible to choose an arbitrary switching sequence as shown in Table 2. In a FCC a switching state can be represented with half of the switches because the high side switch and the low side switch are complementary. The charges at the flying capacitors are observed in Fig. 5 for the chosen switching sequence. It is

shown that the flying capacitors discharge at every intermediate voltage step. When looking closely at Table 2 it becomes clear the reason for the continuous discharge. Due to the inversion of the inductance current (i_L) the flying capacitors discharge at every intermediate level (B, C, E and F).

Table 2 Arbitrary switching sequence

State	S1	S2	S3	V_{out} (p.u.)	I_{fc1} (p.u.)	I_{fc2} (p.u.)
A	1	1	1	1/2	0	0
B	0	1	1	1/6	-1	0
C	0	0	1	-1/6	0	-1
D	0	0	0	-1/2	0	0
E	1	0	0	-1/6	1	0
F	1	1	0	1/6	0	1

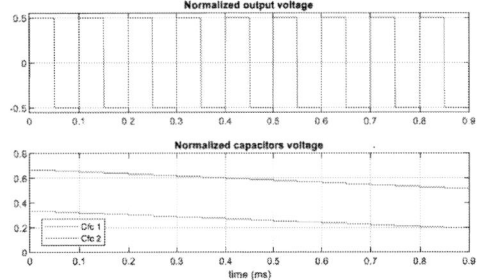

Fig. 5 Output voltage (top) and flying capacitors voltages (bottom) for an arbitrary switching sequence

3.1 Switching states and sequences

The method proposed in this article allows to find a switching sequence that balances the charge of the flying capacitors over a number of periods as it is shown in Fig. 6.

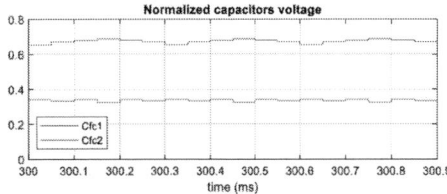

Fig. 6 Flying capacitors voltages for a balanced switching sequence

To construct a switching sequence, it is necessary to consider all the possible states as it is shown in Table 3 for the 4-level example. For this method it is supposed that there are forbidden states where one or more capacitors are short circuited. For this reason, it suffices to consider the top valve only. Since the other half will be the complement. The proposed method provides a sequence of state numbers based on the table of possible states.

Table 3 All switching states in the 4-level FCC with their state number

State number	S1	S2	S3	V_{out}	I_{fc1}	I_{fc2}
1	0	0	0	- 1/2	0	0
2	0	0	1	- 1/6	0	-1
3	0	1	0	- 1/6	-1	1
4	0	1	1	1/6	-1	0
5	1	0	0	- 1/6	1	0
6	1	0	1	1/6	1	-1
7	1	1	0	1/6	0	1
8	1	1	1	1/2	0	0

The total number of states can be expressed as a function of the number of levels (n) of the topology.

$$Number\ of\ states = 2^{n-1} \qquad (1)$$

By looking at the trends in the table of possible states (Table 3) it is possible to determine the number of possible sequences (PS) for 1 period as a function of the number of levels (n) of the topology.

$$PS = \prod_{k=0}^{n-1} \frac{(n-1)!}{k!\,(n-1-k)!} = \prod_{k=0}^{n-1} \binom{n-1}{k} \qquad (2)$$

In some cases, a balanced sequence is composed of more than one period. This means that the number of possible sequences will increase with the number of periods used to balance. However, many of the possible sequences will not result in a balanced sequence. The number of balanced sequences (BS) can be delimited by the following expression.

$$1 < BS < (PS)^m \qquad (3)$$

Where m is the minimum number of periods needed to balance the flying capacitor voltages.

It was found that m can be defined as:

$$m = \begin{cases} (n-1)/2, & \text{FB or odd } n \text{ in HB} \quad (4) \\ n-1, & \text{if } n \text{ is even in HB} \quad (5) \end{cases}$$

From the expression of BS it becomes clear that more criteria are needed to choose one specific balancing sequence. Among the possible criteria there are:

- Minimization of switching losses
- Balancing losses among switches
- Voltage ripple at each flying capacitor

It was observed that under different balancing sequences some semiconductor devices are under more stress than others. By adapting the switching sequence, it is possible to balance the losses among equivalent devices in a full bridge topology. To obtain losses balance among switching components in the same half-bridge the voltage ripple at each flying capacitor must be minimized.

In previous paragraphs it was shown that there are many possible balancing sequences. Some of them satisfy the previous criteria with many switches changing their state at every step. This may represent elevated switching losses. The proposed method only allows for one switch to change state at each voltage level change.

4 Proposed method

The principle is based on finding a balancing sequence that meets the selected criteria. This balancing sequence performs multiple consecutive charges for each flying capacitor and compensates for these charges with a large discharge. Adhering to this principle, at the end of the sequence, the charge of the flying capacitors remains constant. The total number of charges will depend on the number of levels the FCC has. However, all the charges are always compensated by a large discharge. It is worth noting that this method is effective under ideal conditions and in a steady state.

Thanks to the Boolean arrangement used to make the table of possible states (Table 3) it is possible to find the expressions to get the sequence of state numbers that result in a balanced switching sequence that respect the selected criteria.

4.1 Half bridge sequence

The plan behind the proposed sequence is to charge the capacitors at different switching periods for a t_{delay} amount of time and then discharge completely during a whole transition between high and low voltage levels. Repeating this principle for every flying capacitor is possible and is an approach that can be generalized for any number of levels (n) used in a flying capacitor converter.

For the half bridge sequence, two cases are considered:

I. If $(n-1)^2$ is divisible by $2 \cdot (n-1)$, otherwise said if n is an odd number.
II. If $(n-1)^2$ isn't divisible by $2 \cdot (n-1)$, otherwise said, if n is an even number.

4.1.1 Case I

First, a series must be defined. This series will allow the use of the Boolean features of the table with all the possible states. This series is called **base** and is composed by smaller series called b_i.

All the $\mathbf{b}_i$ series have the same length and the same absolute values but alternate sign. Each $\mathbf{b}_i$ series is a shifted version of the previous one with an opposite sign as it is shown in (7). This means that all the elements of the previous series were shifted one spot to the left, the first element is put at the end of the series and then they are all multiplied by -1.

$$\mathbf{b}_0 = [2^0\ 2^1\ 2^2\ ...\ 2^{n-3}\ 2^{n-2}] \qquad (6)$$

$$\mathbf{b}_i = (-1)^i \cdot [2^i\ 2^{i+1}\ ...\ 2^{n-2}\ 2^0\ ...\ 2^{i-1}]$$
$$i \in \mathbb{N}^* < (n-1) \qquad (7)$$

$$\mathbf{base} = [\mathbf{b}_0\ \mathbf{b}_1\ ...\ \mathbf{b}_i\ \mathbf{b}_{i+1}\ ...\mathbf{b}_{n-2}] \qquad (8)$$

With the series $\mathbf{base}$ it is possible to find a balanced sequence with a length of $(n-1)^2$ elements.

$$\mathbf{Seq_{hb}}(0) = 1 \qquad (9)$$

$$\mathbf{Seq_{hb}}(j) = \mathbf{Seq_{hb}}(j-1) + \mathbf{base}(j)$$
$$j \in \mathbb{N}^* \le (n-1)^2 \qquad (10)$$

The balanced sequence used is $\mathbf{Seq_{hb}}$ where j goes from 1 to $(n-1)^2$.

4.1.2 Case II

In this case, by repeating the previous method, it is possible to get a balanced sequence that represents a whole number of periods plus half a period. The first half of the sequence ($\mathbf{FHS_{hb}}$) can be expressed in the same way that the sequence in the Case I with a similar approach to obtain the series '*base*'.

$$\mathbf{FHS_{hb}}(0) = 1 \qquad (11)$$

$$\mathbf{FHS_{hb}}(j) = \mathbf{FHS_{hb}}(j-1) + \mathbf{base}(j)$$
$$j \in \mathbb{N}^* \le (n-1)^2 \qquad (12)$$

The balanced half sequence used is $\mathbf{FHS_{hb}}$ with j from 1 to $(n-1)^2$.

The simplest solution to complete this sequence is to mirror the balanced first half. This is done by grabbing the first half and rearranging its components in the reverse order. Therefore, the second half of the sequence ($\mathbf{SHS_{hb}}$) contains the same elements as the first half of the sequence but in reverse order to match the right voltage levels. The second half of the sequence is shorter by two terms compared to $\mathrm{FHS_{hb}}$. This is because the extreme values are already counted in the first half.

$$\mathbf{SHS_{hb}}(k) = \mathbf{FHS_{hb}}((n-1)^2 - k)$$
$$k \in \mathbb{N}^* \le (n-1)^2 - 1 \qquad (13)$$

Then the whole sequence will have a length of $2 \cdot (n-1)^2$.

Table 4 Balanced switching sequence values for a half bridge when there is an even number of levels

	First half	Second half
Length	$(n-1)^2 + 1$	$(n-1)^2 - 1$
$\mathbf{Seq_{hb}}$	$\mathbf{FHS_{hb}}$	$\mathbf{SHS_{hb}}$

The balanced sequence used is $\mathbf{Seq_{hb}}$ shown in Table 4. The sequence is composed of two parts, going from 0 to $(n-1)^2$ first and then from $(n-1)^2 + 1$ to $2 \cdot (n-1)^2$.

4.2 Full bridge sequence

For the full bridge sequence, two cases are also considered:

  I. If $(n-1)^2/2$ is divisible by $2 \cdot (n-1)$.
  II. If $(n-1)^2/2$ isn't divisible by $2 \cdot (n-1)$.

4.2.1 Case I

The full bridge topology can be treated as a specific case of the previous topology. It is considered that in a full bridge topology the number of levels (n) is an odd number. The proposed method proceeds in the same way and needs a series that allows the use of Boolean features in the table with all the possibilities.

In a similar way as the half bridge Case I sequence here the series $\mathbf{b}_i$ and $\mathbf{b}_i$ **prime** are constructed as alternating sign left-shifted versions of $\mathbf{b}_0$ and $\mathbf{b}_0$ **prime** respectively. In the full bridge sequence, there are two parts to each $\mathbf{B}_i$ series as shown in (19). This is due to the full bridge topology and each half of the $\mathbf{B}_i$ series corresponds to one half bridge. The compilation of all these series will enable the construction of the '*base*' series.

$$\mathbf{b}_0 = (-1) \cdot \left[2^0\ 2^1\ 2^2\ ...\ 2^{\frac{n-1}{2}-1}\right] \qquad (14)$$

$$\mathbf{b}_0' = \left[2^{\frac{n-1}{2}}\ 2^{\frac{n-1}{2}+1}\ ...\ 2^{n-2}\right] \qquad (15)$$

$$\mathbf{B}_0 = [\mathbf{b}_0\ \mathbf{b}_0'] \qquad (16)$$

$$\mathbf{b}_i = (-1)^i \cdot \left[2^i\ 2^{i+1}\ ...\ 2^{\frac{n-1}{2}-1}\ 2^0\ ...\ 2^{i-1}\right]$$
$$i \in \mathbb{N}^* < \frac{n-1}{2} \qquad (17)$$

$$\mathbf{b}_i' = (-1)^{i-1} \cdot$$
$$\left[2^{\frac{n-1}{2}+i}\ 2^{\frac{n-1}{2}+i+1}\ ...\ 2^{n-2}\ 2^{\frac{n-1}{2}}\ ...\ 2^{\frac{n-1}{2}+i-1}\right] \qquad (18)$$

$$i \in \mathbb{N}^* < \frac{n-1}{2}$$

$$\mathbf{B}_i = [\mathbf{b}_i \ \mathbf{b}_i'] \qquad (19)$$

$$\mathbf{base} = \left[\mathbf{B}_0 \ \mathbf{B}_1 \ ... \ \mathbf{B}_i \ \mathbf{B}_{i+1} \ ... \mathbf{B}_{\frac{n-1}{2}} \right] \qquad (20)$$

With the series **base** it is possible to find a balanced sequence with a length of $(n-1)^2/2$ elements.

$$\mathbf{Seq_{fb}}(0) = 2^{\frac{n-1}{2}} \qquad (21)$$

$$\mathbf{Seq_{fb}}(j) = \mathbf{Seq_{fb}}(j-1) + \mathbf{base}(j)$$

$$j \in \mathbb{N}^* \le \frac{(n-1)^2}{2} \qquad (22)$$

For this topology it is necessary to mirror the sequence even if the balanced sequence is a whole number of periods. This is done to balance the use of each half-bridge but not necessary to get a balanced switching sequence.

The balanced sequence used is **Seq_fb** and its mirrored counterpart. The sequence has a length of $(n-1)^2$ elements.

4.2.2 Case II

Similar to the half bridge Case II, in this case the number of necessary states to balance the voltage levels of the capacitors do not represent a whole number of periods. However, by repeating the mirroring method it is possible to get a balanced half sequence. This sequence can be expressed as (**FHS_fb**):

$$\mathbf{FHS_{fb}}(0) = 2^{\frac{n-1}{2}} \qquad (23)$$

$$\mathbf{FHS_{fb}}(j) = \mathbf{FHS_{fb}}(j-1) + \mathbf{base}(j)$$

$$j \in \mathbb{N}^* \le \frac{(n-1)^2}{2} \qquad (24)$$

Using the mirroring trick the second half of the sequence (**SHS_fb**) contains the same elements as the first half of the sequence but in reverse order to match the right voltage levels and keep the balancing voltages. The second half of the sequence is shorter by two terms compared to **FHS_fb**. This is because the extreme values are already counted in the first half.

$$\mathbf{SHS_{fb}}(k) = \mathbf{FHS_{fb}}\left(\frac{(n-1)^2}{2} - k \right) \qquad (25)$$

$$k \in \mathbb{N}^* \le \frac{(n-1)^2}{2} - 1$$

Then the whole sequence will have a length of $(n-1)^2$.

Table 5 Balanced switching sequence values for a full bridge in the case II.

	First half	Second half
Length	$\dfrac{(n-1)^2}{2} + 1$	$\dfrac{(n-1)^2}{2} - 1$
Seq_fb	**FHS_fb**	**SHS_fb**

The balanced sequence used is **Seq_fb** shown in Table 5. The sequence is composed of two parts, going from 0 to $(n-1)^2/2$ first and then from $(n-1)^2/2 + 1$ to $(n-1)^2$.

It must be pointed out that the proposed method provides sequences of indexes of states in the table with all possibilities. For these sequences to meet the established criteria, the table must be assembled in the binary manner indicated in the preceding sections. For example, how Table 3 was assembled. To name the switches in half bridge or full bridge, one must proceed as shown in Fig. 7 and Fig. 12. It is worth noting that the Boolean algebra properties of the converter remain the same for any notation chosen. However, the proposed method is specific to this notation, adaptations must take place if the notation is changed.

5 Simulation results

The simulations for this study were done on MATLAB/Simulink software environment. To validate the proposed method, it was tested for 13-levels in full-bridge and half bridge FCC topologies. The implementation of the switching sequence is done by adapting the control method in [12].

The objective of the simulations is to validate that the criteria proposed is respected. The balanced switching sequence was calculated and implemented before starting the simulations.

5.1 Full bridge (FB)

The full-bridge model is composed of an FCC with two half bridges like shown in Fig. 7, and each half-bridge is composed of 6 pairs of switching components. The switching components considered are 3.3 kV SiC MOSFETs. The RL load can be chosen arbitrarily.

Table 6 Specifications for the full bridge simulation

Voltage rating (Vdc)	10 kV
Flying capacitors capacitance (C_{fci})	450 nF
Number of capacitors	10
Number of levels (n)	13
Mirroring	YES

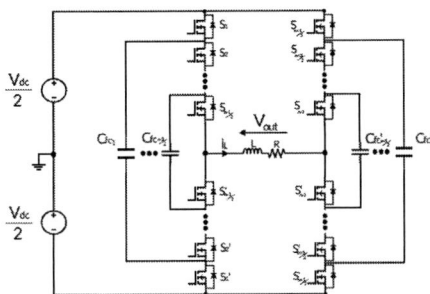

Fig. 7 Generalized full bridge FCC diagram

The first criterion is verified by looking at the voltage waveforms and the voltage intermediate levels in the quasi-2-level modulation. These waveforms are shown in Fig. 8. In Fig. 9 it is shown that all the flying capacitors have a balanced voltage and a similar voltage ripple. Furthermore, the second criterion is confirmed in Fig. 10 where the voltage of two equivalent flying capacitors at each half bridge are compared.

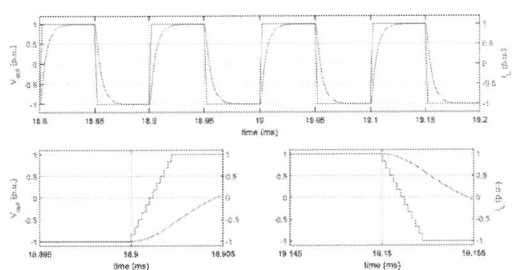

Fig. 8 Voltage (blue) and current (orange) waveforms for a 13-level full-bridge FCC

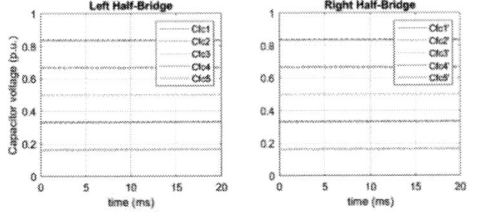

Fig. 9 Multiple switching periods of 10 flying capacitors' voltage operating in a 13-level full-bridge FCC.

The difference between the voltages shown in Fig. 10 is due to different initial starting voltage for each flying capacitor. The voltage ripple is the same for

the two equivalent capacitors at each half-bridge. It should be noted that the mirroring of the voltage sequence is evident by looking at the staircase form going up and down. The mirrored operation assures the similar use of each set of switching components at each half bridge.

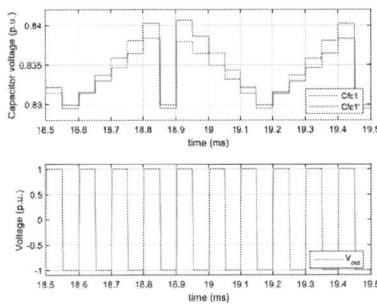

Fig. 10 Voltage oscillation (top) in two capacitors in a 13-level full-bridge FCC. Voltage output (bottom) of the FCC.

The third criterion is validated by looking at one whole switching sequence. At every transition it was found that only one pair of switches changed state. This operation minimizes switching losses. In Fig. 11 the states of every switch are followed during the complete switching sequence. It is shown that only one pair of switches change state from one switching state to the next one.

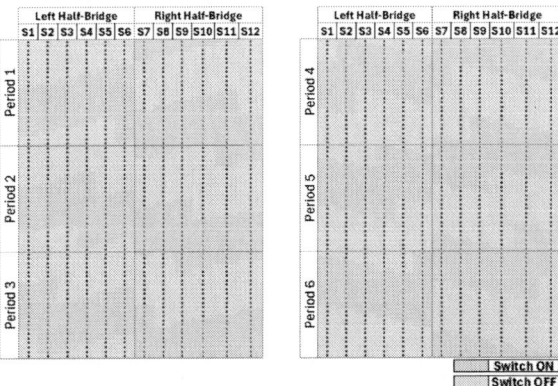

Fig. 11 Switch states in a complete balanced switching sequence including mirroring. Blue is used to indicate a switch in ON state and orange for OFF state. The switching sequence has a length of 6 periods.

5.2 Half bridge (HB)

The half-bridge model is composed of a FCC with 12 pairs of switching components like the one shown in Fig. 12. This model is the one considered and used for the case study proposed with a MV level of 20 kV. The RL load can be chosen arbitrarily.

Table 7 Specifications for the half bridge simulation

Voltage rating (Vdc)	20 kV
Flying capacitors capacitance (C_{fci})	90 nF
Number of capacitors	11
Number of levels (n)	13
Mirroring	NO

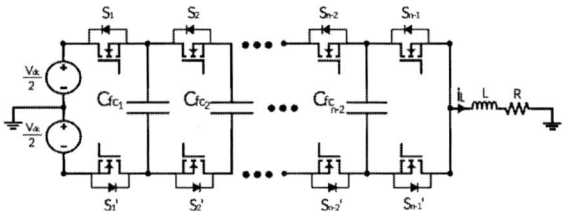

Fig. 12 Generalized half-bridge FCC diagram.

In this topology, the first criterion is verified by looking at the voltage waveforms and the voltage intermediate levels in the quasi-2-level modulation (Fig. 13). In Fig. 14 it is shown that all the flying capacitors have a balanced voltage.

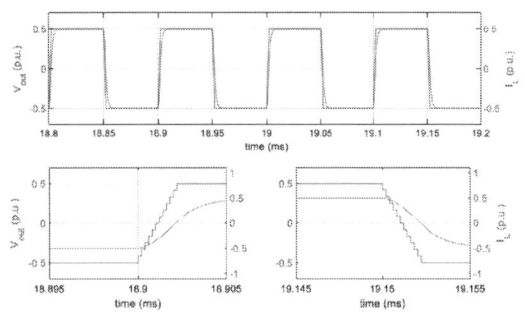

Fig. 13 Voltage (blue) and current (orange) waveforms for a 13-level half-bridge FCC.

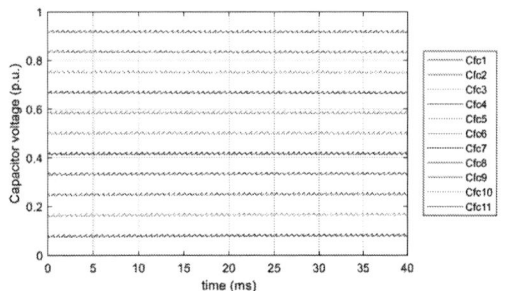

Fig. 14 Multiple switching periods of 11 flying capacitors' voltage operating in a 13-level half-bridge FCC.

In this case the voltage of one capacitor follows a staircase shape only going up because there is no mirroring of the sequence. This is shown in Fig. 15.

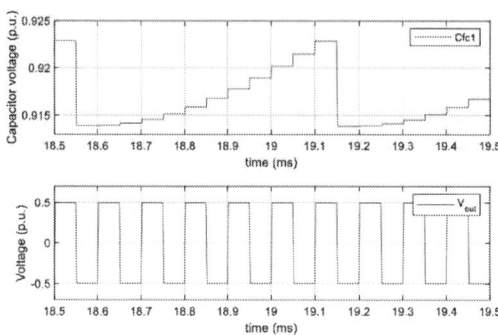

Fig. 15 Voltage oscillation (top) in one capacitor in a 13-level half-bridge FCC. Voltage output (bottom) of the FCC.

The third criterion is validated by looking at one whole switching sequence. Switching losses were minimized in the same way as before. At each step of one transition there is only one pair of switches that changes state. In Fig. 16 the states of every switch are followed during the complete switching sequence. It is shown that only one pair of switches change state from one switching state to the next one.

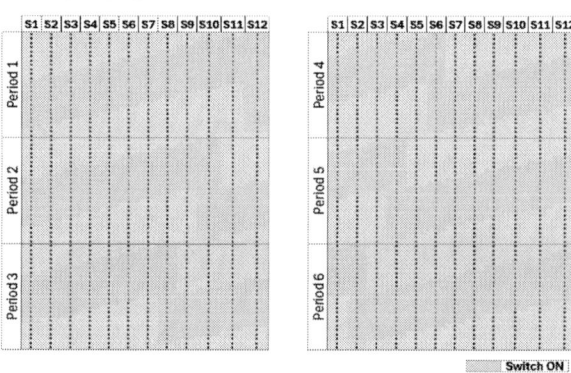

Fig. 16 Switch states in a complete balanced switching sequence. Blue is used to indicate a switch in ON state and orange for OFF state. The switching sequence has a length of 6 periods.

5.3 Analysis

For the proposed case study there are some considerations to take into account in the design. Considering a half bridge topology, it is better to have an odd number of levels. This will lead to shorter balanced switching sequences. Shorter balanced switching sequences may result in smaller capacitances needed for a constant voltage ripple.

Another important comparison can be made between the half bridge and full bridge topologies. Considering a constant voltage rating, a full-bridge topology will need more switching components and flying capacitors than a half bridge topology. However, the full-bridge topology would have twice as many voltage levels. This could be interesting for the regulation of the dV/dt with the quasi-2-level modulation.

Finally, an interesting result found while validating the third criterion during simulations is that switching losses could be ten times higher when an arbitrary balanced switching sequence is chosen without using this method or an equivalent one.

6   Conclusion

The Flying Capacitor Converter has new promising applications applied to MV DC-DC isolated converters. As a multilevel topology it can be used with a quasi-2-level modulation to regulate the dV/dt. One of the main challenges when using flying capacitors is the voltage balancing. This paper proposed a method to find a switching sequence offline. The proposed method was generalized for any number of voltage levels. The obtained switching sequence assures natural voltage balancing of flying capacitors in a full-bridge or half-bridge arrangement. The main limitation of the proposed method is that it considers a steady-state operation with ideal components. Real life components will introduce asymmetries to operation that must be compensated with closed-loop control. However, the proposed method would reduce the asymmetries to a minimum.

The proposed method produces a balanced switching sequence based on fixed criteria such as symmetric use of converter and minimization of losses. The proposed method was validated for the chosen criteria with 13-levels half bridge and full bridge simulations. It would be possible to find equivalent methods depending on the case study and criteria.

This study proposes a generalized switching sequence but opens the door to many possibilities related to the switching sequence in this type of converter. For a real-life application, it will be important to study the impact of the switching sequence on different criteria such as voltage ripple, flying capacitor size and individual stress on switching components.

Acknowledgements

This work was supported by a grant overseen by the French National Research Agency (ANR).

References

[1] C. Mathieu De Vienne, P. Lefranc, B. Asllani, P. O. Jeannin, and B. Lefebvre, "Experimental Investigation of a 10 kV-70A Switch with Six SiC-MOSFETs in a Series-Connection Configuration," *Mater. Sci. Forum*, vol. 1062, pp. 472–476, May 2022, doi: 10.4028/p-cg93gc.

[2] C. Mathieu de Vienne, C. Buttay, and M. Guillet, "Towards a Common Mode Free Packaging Solution for High Voltage Series Connected SiC MOSFET Switches," DE: VDE VERLAG GMBH, Jun. 2023. Accessed: Mar. 18, 2024. [Online]. Available: https://doi.org/10.30420/566091110

[3] R. W. A. A. De Doncker, D. M. Divan, and M. H. Kheraluwala, "A three-phase soft-switched high-power-density DC/DC converter for high-power applications," *IEEE Trans. Ind. Appl.*, vol. 27, no. 1, pp. 63–73, Feb. 1991, doi: 10.1109/28.67533.

[4] N. H. Baars, E. A. Lomonova, C. G. E. Wijnands, and J. Everts, *Three-phase dual active bridge converters: a multi-level approach for wide voltage-range isolated dc-dc conversion in high-power applications*. Eindhoven: Technische Universiteit Eindhoven, 2017.

[5] S. Anand, "Study and characterization of insulating material using partial discharge measurements for power electronics application in HVDC and MVDC."

[6] A. Cremasco, D. Rothmund, M. Curti, and E. A. Lomonova, "Voltage Distribution in the Windings of Medium-Frequency Transformers Operated With Wide Bandgap Devices," *IEEE J. Emerg. Sel. Top. Power Electron.*, vol. 10, no. 4, pp. 3587–3602, Aug. 2022, doi: 10.1109/JESTPE.2021.3064702.

[7] I. A. Gowaid, G. P. Adam, S. Ahmed, D. Holliday, and B. W. Williams, "Analysis and Design of a Modular Multilevel Converter With Trapezoidal Modulation for Medium and High Voltage DC-DC Transformers," *IEEE Trans. Power Electron.*, vol. 30, no. 10, pp. 5439–5457, Oct. 2015, doi: 10.1109/TPEL.2014.2377719.

[8] S. Milovanovic and D. Dujic, "Comprehensive Analysis and Design of a Quasi Two-Level Converter Leg," *CPSS Trans. Power Electron. Appl.*, vol. 4, no. 3, pp. 181–196, Sep. 2019, doi: 10.24295/CPSSTPEA.2019.00018.

[9] T. A. Meynard and H. Foch, "Multi-level conversion: high voltage choppers and voltage-source inverters," in *PESC `92 Record. 23rd Annual IEEE Power Electronics Specialists Conference*, Toledo, Spain: IEEE, 1992, pp. 397–403. doi: 10.1109/PESC.1992.254717.

[10] A. Tcai, T. Wijekoon, and M. Liserre, "Evaluation of Flying Capacitor Quasi 2-level Modulation for MV Applications," 2021.

[11] S. Mersche, D. Bernet, and M. Hiller, "Quasi-Two-Level Flying-Capacitor-Converter for Medium Voltage Grid Applications," in *2019 IEEE Energy Conversion Congress and Exposition (ECCE)*, Baltimore, MD, USA: IEEE, Sep. 2019, pp. 3666–3673. doi: 10.1109/ECCE.2019.8913201.

[12] A. Şchiop, "Capacitor Voltage Balancing Control for Flying Capacitor Multilevel Inverter," *J. Electr. Electron. Eng.*, vol. 7, pp. 35–38, Oct. 2014.

[13] G. Kampitsis, E. I. Batzelis, A. Kolokasis, E. Matioli, and B. C. Pal, "A Generalized Phase-Shift PWM Extension for Improved Natural and Active Balancing of Flying Capacitor Multilevel Inverters," *IEEE Open J. Power Electron.*, vol. 3, pp. 621–634, 2022, doi: 10.1109/OJPEL.2022.3209540.

[14] R. Stala, "Natural Capacitor Voltage Balance In Multilevel Flying Capacitor Converters. A Review Of Research Achievements," *POWER Electron. DRIVES ISSN 2451-0262*, 2016, doi: 10.5277/PED160201.

[15] P. Czyz, P. Papamanolis, T. Guillod, F. Krismer, and J. W. Kolar, "New 40kV / 300kVA Quasi-2-Level Operated 5-Level Flying Capacitor SiC 'Super-Switch' IPM," in *2019 10th International Conference on Power Electronics and ECCE Asia (ICPE 2019 - ECCE Asia)*, Busan, Korea (South): IEEE, May 2019, pp. 813–820. doi: 10.23919/ICPE2019-ECCEAsia42246.2019.8796998.

[16] B. Zhao, Q. Song, J. Li, Y. Wang, and W. Liu, "Modular Multilevel High-Frequency-Link DC Transformer Based on Dual Active Phase-Shift Principle for Medium-Voltage DC Power Distribution Application," *IEEE Trans. Power Electron.*, vol. 32, no. 3, pp. 1779–1791, Mar. 2017, doi: 10.1109/TPEL.2016.2558660.

[17] L. Zhang, J. Qin, Y. Zou, Q. Duan, and W. Sheng, "Analysis of Capacitor Charging Characteristics and Low-Frequency Ripple Mitigation by Two New Voltage-Balancing Strategies for MMC-Based Solid-State Transformers," *IEEE Trans. Power Electron.*, vol. 36, no. 1, pp. 1004–1017, Jan. 2021, doi: 10.1109/TPEL.2020.3000717.

PCIM Europe 2024, 11– 13 June 2024, Nuremberg   DOI: 10.30420/566262447

Optimization-based Sizing of a Modular Multilevel Converter based on 650 V GaN Modules for New LVDC/MVDC Grids

Gregoire Le Goff[1], Ilias Chorfi[1,2,3], Thierry Sutto[2], Corinne Alonso[1]

[1] LAAS-CNRS, Université de Toulouse, CNRS, UPS. France

[2] STMicroelectronics, Automotive & Discrete Group (ADG),France

[3] CEA Tech., France
Corresponding author: Corinne Alonso, alonsoc@laas.fr
Speaker: Gregoire Le Goff, glegoff@laas.fr

Abstract

This paper introduces a new co-design approach based on optimization for the modular multilevel converters (MMCs) using wide-bandgap (WBG) switches. This method takes into account several control strategies from the beginning and through the entire procedure. Gallium-Nitride (GaN) mature switches are attractive solutions to increase the compactness and efficiency in embedded applications. It could be a great opportunity for new grid-connected converters, thanks to their improved switching characteristics and lower on-resistance. Nevertheless, their 650 V rating requires the use of multilevel topologies to reach new 1.5 kV DC grids for eco-districts or, more generally, for MVDC grids. The proposed design method is validated using a high-fidelity simulation involving GaN half-bridge submodules (SM-HB).

1   Introduction

Wide-bandgap (WBG) components have shown disruptive performance, particularly in terms of losses and therefore efficiency, ability to increase switching frequency, and power density [1]. The work presented here is part of a wider research project aimed at developing, assessing, and revealing the potential of WBG components in multilevel converters. The first step in this direction was the development of a 3-level active neutral point clamped converter (ANPC) prototype based on GaN half-bridge modules [2]. The latter having shown interesting performances, work is now done to push further by dealing with the 5-level MMC based on the converter topology from [3].

1.1   Novelty of the proposed work

The main novelty lies within the development of a sizing procedure for the MMC involving a fast and simple optimization algorithm where the user can specify a control strategy. An MMC optimization co-design concept that both integrates control and sizing constraints is thus proposed. In the paper, the control strategy selection focuses on the circulating current for which the control strategy is chosen from the very beginning of the sizing procedure. The procedure is detailed for the two well-known extreme cases: circulating current suppression (CCS) [4]

and circulating current injection (CCI) [5] just right. In both cases, the steady-state power balance of the converter is verified. The optimization-based sizing procedure is quite versatile and suits a wide range of purposes such as DC-grid to AC-grid interconnection but also DC-grid to AC-load or even DC-grid to AC-drive with variable frequency.

Compared to the recent work from [6]–[8] the proposed work implements a new step to the sizing process. It involves an optimization that takes into account a control strategy. About the three following steps of the procedure, the paper extends the approach from [8] by considering a wider range of possible behaviors and interfaces with the DC and AC grid, it also uses a more accurate model for the arm voltages with the ability to account for impedances. The optimization approach by [6] goes up to the structural sizing of the MMC components but, as for [7], several choices of circulating current control strategies are not available. Unlike the proposed work, the global optimization is performed in a heuristic fashion where no formal optimization algorithm appears to be used. Work from [7] introduces an effective approach for the minimization of the capacitors by focusing on common-mode voltage and circulating current injection. However, here the proposed work embraces the sizing of the MMC from a more general point of view where a trade-off between different control strategies is enabled.

1.2 Outline

First, section 2 introduces the model for the sizing procedure. The latter is then described in section 3 as well as the formulation of the associated optimization problem. The method is applied afterward to the case of an MVDC-grid to LVAC-grid connection that meets the European grid standards. Section 4 shows the results obtained from high-fidelity simulation models that make it possible to validate the generic sizing approach. The concept of the prototype currently under production is also shown. More specifically, the sizing example is that of a typical PV-to-grid power injection, for instance, between a $1.5\ kV$ DC grid and the $230\ V$ root-mean-square (RMS) line-neutral, $50\ Hz$ AC grid.

2 Operational sizing-oriented model of the MMC

2.1 Voltage and current models

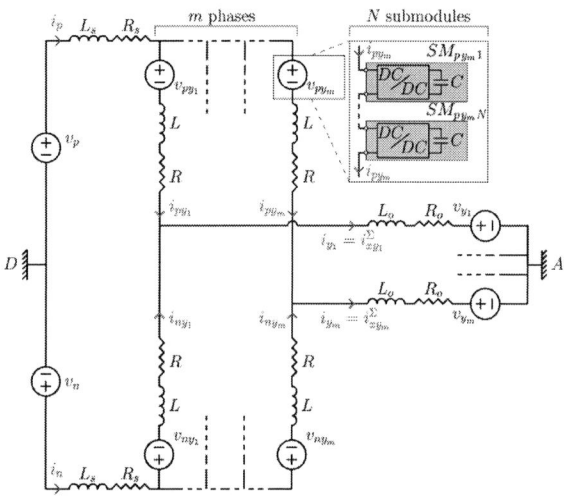

Fig. 1: MMC electrical diagram.

The approach is straightforward, based on models from [9], [10]. For electrical quantities such as arm currents and voltages, Kirchhoff's current and voltage laws (KCL and KVL, resp.) are applied to both upper and lower arms, which leads to:

$$\begin{cases} v_{py} = v_p - (v_{AD} + v_y) - \left(Z_o + \frac{Z}{2}\right) i_{xy}^{\Sigma} \\ \qquad - \frac{Z}{2} i_{xy}^{\Delta} - \frac{Z_s}{2} \sum_{y=y_1}^{y_m} \left(i_{xy}^{\Sigma} + i_{xy}^{\Delta}\right) \\ v_{ny} = v_n - (v_{AD} + v_y) - \left(Z_o + \frac{Z}{2}\right) i_{xy}^{\Sigma} \\ \qquad + \frac{Z}{2} i_{xy}^{\Delta} - \frac{Z_s}{2} \sum_{y=y_1}^{y_m} \left(i_{xy}^{\Sigma} - i_{xy}^{\Delta}\right) \end{cases} \quad (1)$$

Where y is the leg number, x represents the arm's DC pole connection, positive ($x = p$) or negative

($x = n$) arm, $i_{xy}^{\Sigma} := i_{py} + i_{ny} = i_y$, $i_{xy}^{\Delta} := i_{py} - i_{ny}$, $Z := R + L\ d/dt$, $Z_o := R_o + L_o\ d/dt$, and $Z_s := R_s + L_s\ d/dt$. v_{AD} is the voltage between neutral points. Also, arm current sums are determined by KCL: $i_p = \sum_{y=y_1}^{y_m} i_{py}$, $i_n = \sum_{y=y_1}^{y_m} i_{ny}$.

From (1), several steps can be applied, as in [9], to extract the dynamic behavior of the circulating current $i_{c_y} = (i_{xy}^{\Delta} - \sum_{y=y_1}^{y_m} i_{xy}^{\Delta}/m)/2$:

$$\left(R + L\frac{d}{dt}\right) i_{c_y} = \frac{1}{m} v_{c_y} \quad (2)$$

with m the total number of phases and v_{c_y} some voltage made of a linear combination of multiple v_{py} and v_{ny} voltages. The key point to note here is that v_{c_y} is the voltage that the MMC generates to control i_{c_y}.

2.2 MMC inner power and energy models

Then, the arm powers and energies are readily deduced from the product between the arm currents and voltages. Thus, the total power flowing through each leg is $p_y := p_{py} + p_{ny} := v_{py}i_{py} + v_{ny}i_{ny}$. Benefiting from the arm voltages from (1), p_y reads:

$$\dot{E}_y = p_y = \left[\frac{V_{DC}}{2}i_{xy}^{\Delta} - \frac{Z}{2}i_{xy}^{\Delta^2} - \frac{Z_s}{2}i_{xy}^{\Delta}\sum_{y=y_1}^{y_m} i_{xy}^{\Delta}\right] - \left[(v_{AD} + v_y)i_{xy}^{\Sigma} + \left(\frac{Z}{2} + Z_o\right)i_{xy}^{\Sigma^2} + \frac{Z_s}{2}i_{xy}^{\Sigma}\sum_{y=y_1}^{y_m} i_{xy}^{\Sigma}\right] \quad (3)$$

With p_{py} the power flowing through the positive arm, it's possible to deduce the energy ripples in this arm at the AC-grid fundamental period T_{AC} timescale: $\Delta E_{py}(t) = \int_0^{T_{AC}} p_{py}(t)dt$. The peak-to-peak energy ripple in the arm over one fundamental AC period is then determined by $\Delta E_{py} = \max_t (\Delta E_{py}(t)) - \min_t (\Delta E_{py}(t))$. Analogously, ΔE_{ny} is obtained. The maximum peak-to-peak energy ripple seen by the leg #y is then expressed as:

$$\Delta E_{xy} = \max (\Delta E_{py}, \Delta E_{ny}) \quad (4)$$

With δV_C^{Σ} the amplitude of the voltage ripple in the stack made up of N capacitors in series, associated with the energy ripple, it is possible to establish the link between these two quantities by defining the equivalent capacitive energy of all N capacitors in series. Let $\langle E_{xy} \rangle$ be the average energy contained in any arm of the converter:

$$\langle E_{xy} \rangle := N\frac{C}{2}(v_C^{nom})^2 = \frac{C}{2N}\langle V_C^{\Sigma} \rangle^2 \quad (5)$$

with $\langle V_C^{\Sigma} \rangle = Nv_C^{nom}$ the average value of the total voltage across the submodule stack and C the capacitor of any SM. The power fluctuation through each arm has a direct impact on energy variation, and therefore on the voltage V_C^Σ of the capacitor stack. By noting δV_C^Σ the amplitude of the voltage excursion in the stack, it is possible to express $\hat{E}_{xy}$ the energy contained in the arm when the stack voltage is at its maximum, and $\check{E}_{xy}$, the energy when the stack voltage reaches its minimum. These energies are then determined by analogy with (5):

$$\begin{aligned} \hat{E}_{xy} &= \tfrac{C}{2N}(\langle V_C^\Sigma \rangle + \delta V_C^\Sigma)^2 \\ \check{E}_{xy} &= \tfrac{C}{2N}(\langle V_C^\Sigma \rangle - \delta V_C^\Sigma)^2 \end{aligned} \quad (6)$$

Let ΔE_{xy} be the peak-to-peak energy variation within the arm under consideration:

$$\Delta E_{xy} = \hat{E}_{xy} - \check{E}_{xy} = 2\frac{C}{N}\langle V_C^\Sigma \rangle \delta V_C^\Sigma = 2\frac{C}{N}\langle V_C^\Sigma \rangle^2 \delta \bar{V}_C^\Sigma \quad (7)$$

with $\delta \bar{V}_C^\Sigma = \delta V_C^\Sigma / \langle V_C^\Sigma \rangle$, the relative amplitude of the voltage ripple in the stack compared with nominal operation.

2.3 MMC outer, AC-side power model

At the scale of the converter as a whole, the power delivered to the AC grid averaged over the fundamental AC period is expressed $\hat{P}_{AC} = \frac{m}{2}\hat{I}_{AC}\hat{V}_{AC}\cos\varphi$ in the case where the AC-side grid voltage amplitude $\hat{V}_{AC}$ is non-zero. Otherwise, the active power delivered to the AC grid is that delivered to the load R_o: $\hat{P}_{AC} = \frac{m}{2}R_o\hat{I}_{AC}^2$. Both equations can be summarized by:

$$\hat{I}_{AC} := \left[\sqrt{\frac{2\hat{P}_{AC}}{mR_o}} \cdot \left(\hat{V}_{AC} = 0\right) + \frac{2\hat{P}_{AC}}{m\hat{V}_{AC}\cos\varphi} \cdot \left(\hat{V}_{AC} \neq 0\right)\right] \quad (8)$$

3 Optimization-based sizing of the MMC

This section comprises the main contribution of the paper, in which the optimization-based co-design sizing procedure, presented in Fig. 2, is detailed. The procedure consists of five steps, each of them are explained in the following sub-sections.

3.1 Power conversion optimization

The first step is where optimization takes place to find the maximum power that can be converted by the MMC while taking into account a circulating current control strategy (CCCS), the DC and AC grid specifications at the interfaces, and minimizing losses in the converter arms.

The ultimate aim of the optimization is to determine the i_{xy}^Δ currents and the $\hat{P}_{AC}$ power to be converted

to the AC grid. As the current i_{xy}^Σ is imposed by the power to be delivered to the AC grid, the latter is not considered a decision variable. To seek for the minimum conduction losses in the arms, the cost function is defined according to:

$$\forall t \in [0; T_{AC}], \quad J := \sum_{y=y_1}^{y_m} i_{xy}^\Delta(t)^2 \quad (9)$$

The quantities that will limit the power and current are the power balance for one and the admissible current limit in each arm for the other. Limiting the currents in the arms is translated by the constraint:

$$\forall y \in \{y_1, \ldots, y_m\}, \begin{cases} i_{xy}^\Sigma(t) + i_{xy}^\Delta(t) \leq +2\,\hat{i}_{xy}^{lim} \\ i_{xy}^\Sigma(t) + i_{xy}^\Delta(t) \geq -2\,\hat{i}_{xy}^{lim} \end{cases} \quad (10)$$

For the power balance, the constraint will depend on the choice of CCCS. In the case of CCI, it is deduced from equation (3) where $p_y = 0$ shall be ensured:

$$\begin{aligned} &\forall y \in \{y_1, \ldots, y_m\}, \\ &\tfrac{V_{DC}}{2}i_{xy}^\Delta(t) - \tfrac{R}{2}i_{xy}^\Delta(t)^2 - \tfrac{R_s}{2}i_{xy}^\Delta(t)\sum_{y=y_1}^{y_m} i_{xy}^\Delta(t) \\ &= (v_{AD} + v_y(t))i_{xy}^\Sigma(t) + \left(\tfrac{R}{2} + R_o\right) i_{xy}^\Sigma(t)^2 \end{aligned} \quad (11)$$

Note that, compared to (3), the inductive part of impedances is neglected with respect to the resistive part, and the common mode current in the AC grid is assumed null (which translates to $\sum_{y=y_1}^{y_m} i_{xy}^\Sigma = 0$). This could be improved with further developments. In the CCS case, the constraints (11) are replaced by the global power balance at the scale of the converter, leading to:

$$\begin{aligned} &\sum_{y=y_1}^{y_m} \left(\tfrac{V_{DC}}{2}i_{xy}^\Delta(t) - \tfrac{R}{2}i_{xy}^\Delta(t)^2 - \tfrac{R_s}{2}i_{xy}^\Delta(t)\sum_{y=y_1}^{y_m} i_{xy}^\Delta(t)\right) \\ &= \sum_{y=y_1}^{y_m} \left((v_{AD}+v_y(t))i_{xy}^\Sigma(t) + \left(\tfrac{R}{2}+R_o\right) i_{xy}^\Sigma(t)^2\right) \end{aligned} \quad (12)$$

Those constraints involve quantities such as i_{xy}^Σ and v_y which are not part of the decision variables, just like V_{DC}, R, R_o, R_s, $\hat{i}_{xy}^{lim}$, f_{AC}, $\hat{V}_{AC}$, m, and φ. They will then be directly defined:

$$\forall y \in \{y_1, \ldots, y_m\}, \begin{cases} v_y(t) := \hat{V}_{AC}\sin(2\pi f_{AC}t - \varphi_y) \\ i_{xy}^\Sigma(t) := \hat{I}_{AC}\sin(2\pi f_{AC}t - \varphi_y - \varphi) \end{cases} \quad (13)$$

where $\hat{I}_{AC}$ is defined according to (8) and $\varphi_{y_k} := \frac{2\pi}{m}(k-1)$. The cost function, together with the constraints and parameter definitions described above, are then combined into a single optimization problem, (14) embodies this optimization sizing problem (OSP). In order to have a generic formalism,

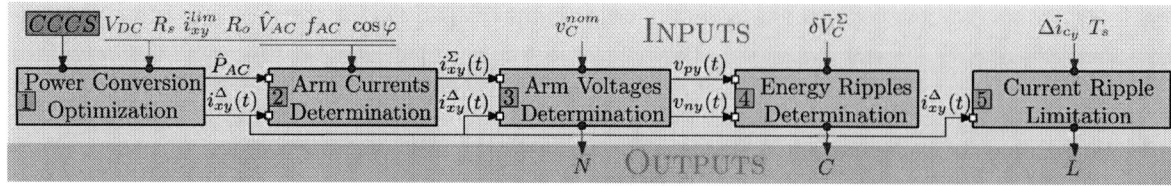

Fig. 2: MMC optimization-based sizing procedure (from left to right).

Tab. 1: Optimization sizing problem (OSP) developed for the first step of the sizing procedure, see Figure 2

$$
\left\{
\begin{array}{l}
\min_{\hat{P}_{AC}; \forall y \in \{y_1,\ldots,y_m\},\ \forall t \in [0;T_{AC}]\ i_{xy}^{\Delta}(t)}\ J = \sum_{y=y_1}^{y_m} i_{xy}^{\Delta}(t)^2 \\[4pt]
\text{subject to the constraints:} \\[4pt]
\forall y \in \{y_1,\ldots,y_m\},\ \forall t \in [0;T_{AC}],\ \sigma_{CC}\left(\frac{V_{DC}}{2}i_{xy}^{\Delta}(t) - \frac{R}{2}i_{xy}^{\Delta}(t)^2 - \frac{R_s}{2}i_{xy}^{\Delta}(t)\sum_{y=y_1}^{y_m} i_{xy}^{\Delta}(t)\right) \\[4pt]
\qquad\qquad = \sigma_{CC}\left((v_{AD}+v_y(t))i_{xy}^{\Sigma}(t) + \left(\frac{R}{2}+R_o\right)i_{xy}^{\Sigma}(t)^2\right) \\[4pt]
\forall t \in [0;T_{AC}],\ (1-\sigma_{CC})\sum_{y=y_1}^{y_m}\left(\frac{V_{DC}}{2}i_{xy}^{\Delta}(t) - \frac{R}{2}i_{xy}^{\Delta}(t)^2 - \frac{R_s}{2}i_{xy}^{\Delta}(t)\sum_{y=y_1}^{y_m} i_{xy}^{\Delta}(t)\right) \\[4pt]
\qquad\qquad = (1-\sigma_{CC})\sum_{y=y_1}^{y_m}\left((v_{AD}+v_y(t))i_{xy}^{\Sigma}(t) + \left(\frac{R}{2}+R_o\right)i_{xy}^{\Sigma}(t)^2\right) \\[4pt]
\forall y \in \{y_1,\ldots,y_m\},\ \forall t \in [0;T_{AC}],\ i_{xy}^{\Sigma}(t)+i_{xy}^{\Delta}(t) \leq +2\,\hat{i}_{xy}^{lim} \\[4pt]
\forall y \in \{y_1,\ldots,y_m\},\ \forall t \in [0;T_{AC}],\ i_{xy}^{\Sigma}(t)+i_{xy}^{\Delta}(t) \geq -2\,\hat{i}_{xy}^{lim} \\[4pt]
\forall y \in \{y_1,\ldots,y_{m-1}\},\ \forall t \in [0;T_{AC}],\ (1-\sigma_{CC})\left(i_{xy}^{\Delta}(t) - \frac{1}{m}\sum_{y=y_1}^{y_m} i_{xy}^{\Delta}(t)\right) = 0 \\[4pt]
\text{with the definitions:} \\[4pt]
\forall y \in \{y_1,\ldots,y_m\},\ \forall t \in [0;T_{AC}],\ i_{xy}^{\Sigma}(t) := \left[\sqrt{\frac{2\hat{P}_{AC}}{mR_o}}\cdot\left(\hat{V}_{AC}==0\right) + \frac{2\hat{P}_{AC}}{m\hat{V}_{AC}\cos\varphi}\right]\sin(2\pi f_{AC}t - \varphi_y - \varphi) \\[4pt]
\forall y \in \{y_1,\ldots,y_m\},\ \forall t \in [0;T_{AC}],\ v_y(t) := \hat{V}_{AC}\sin(2\pi f_{AC}t - \varphi_y) \\[4pt]
\text{with the boundaries:} \\[4pt]
\forall y \in \{y_1,\ldots,y_m\},\ \forall t \in [0;T_{AC}],\ -\hat{i}_{xy}^{lim} \leq i_{xy}^{\Delta}(t) \leq +\hat{i}_{xy}^{lim} \\[4pt]
0 \leq \hat{P}_{AC} \leq 1.25\,V_{DC}\,\hat{i}_{xy}^{lim}
\end{array}
\right.
$$

$$(14)$$

let σ_{CC} be 1 in the (CCI) case and 0 in the (CCS) case. Note that for both the CCI and CCS strategies, $\{\hat{P}_{AC}, i_{xy_1}^{\Delta},\ldots,i_{xy_m}^{\Delta}\}$ represent a total of $m+1$ decision variables for m equality constraints.

An important concept to keep in mind is that this optimization takes into account the maximum current $\hat{i}_{xy}^{lim}$ constraint that the selected WBG semiconductors can withstand, to achieve sizing as close as possible to actual realization capabilities. The optimization is then solved on a time scale of one T_{AC} period using *fmincon* available in Matlab. For that purpose, the time interval is discretized in 60 steps of about 333 μs each. The boundaries for the decision variables are imposed by the current limits. As outputs of the optimization, $\hat{P}_{AC}$ and the waveform of $i_{xy}^{\Delta}(t)$ are determined. The average value of $i_{xy}^{\Delta}(t)$ as well as its alternating amplitude, its frequency $2f_{AC}$ and its phase can be extracted for further processing.

3.2 Arm currents determination

From the first step, $i_{xy}^{\Sigma}(t)$ is trivially deduced from (13) and (8) and $i_{xy}^{\Delta}(t)$ is directly an output from the first step already. As the amplitude, frequency, phase, and continuous component of $i_{xy}^{\Sigma}(t)$ and $i_{xy}^{\Delta}(t)$ are known, their derivatives, $di_{xy}^{\Sigma}(t)/dt$ and $di_{xy}^{\Delta}(t)/dt$ can be derived analytically.

3.3 Arm voltages determination

Knowning $i_{xy}^{\Sigma}(t)$, $i_{xy}^{\Delta}(t)$, $di_{xy}^{\Sigma}(t)/dt$ and $di_{xy}^{\Delta}(t)/dt$, the arm voltages $v_{py}(t)$ and $v_{ny}(t)$ are directly determined by applying (1).

With v_C^{nom} the nominal capacitor voltage, $\hat{v}_{py} = \max_t(v_{py}(t))$, $\hat{v}_{ny} = \max_t(v_{ny}(t))$, $\check{v}_{py} = \min_t(v_{py}(t))$, $\check{v}_{ny} = \min_t(v_{ny}(t))$ and $V_C^{\Sigma} = \max(\hat{v}_{py},\hat{v}_{ny}) = \hat{v}_{xy}$ the total voltage required across the submodule stacks, the total number of submodules is determined as follows:

$$N = \left\lfloor \frac{V_C^{\Sigma}}{v_C^{nom}} \right\rfloor + 1 \tag{15}$$

This yields the nominal voltage of a complete stack: $V_C^{\Sigma\ nom} = N v_C^{nom} > V_C^{\Sigma}$. This voltage $V_C^{\Sigma\ nom}$ will also be considered as the average value, at the T_{AC} timescale, that the stacks will have to reach, then noted $\langle V_C^{\Sigma} \rangle$ in the following. Note that (15) gives the first sizing parameter of the MMC that is desired to be found.

3.4 Energy ripples determination

The previous steps yielded $i_{py}(t)$, $i_{ny}(t)$, $v_{py}(t)$ and $v_{ny}(t)$. The powers $p_{py}(t) = v_{py}(t)i_{py}(t)$ and $p_{ny}(t) = v_{ny}(t)i_{ny}(t)$ are then trivially deduced. From the energy ripple model (4) involving $p_{py}(t)$ and $p_{ny}(t)$, ΔE_{xy} is determined. According to (7), the SM capacity will have to be greater than a certain limit in order to guarantee total stack voltage ripples of less than $\delta \bar{V}_C^{\Sigma}$:

$$C \geq \frac{N \Delta E_{xy}}{2 \langle V_C^{\Sigma} \rangle^2 \delta \bar{V}_C^{\Sigma}} \tag{16}$$

3.5 Current ripple limitation

When sizing L, focus is on the circulating current since other currents see an electrical circuit with additional inductances resulting in smaller ripples. Sizing L with regard to limiting current ripples is therefore a worst-case sizing approach. The model of circulating currents in the worst-case dynamic (no resistance considered), is derived based on (2):

$$L \frac{di_{c_y}}{dt} = \frac{1}{m} v_{c_y} \implies \frac{\Delta i_{c_y}}{\Delta t} = \frac{v_{c_y}}{mL} \tag{17}$$

where the right-hand-side formula is the discretized version of the model at the Δt timescale. Δi_{c_y} is the circulating current deviation observed during Δt when v_{c_y} is applied.

Now assume that the voltage control is fully operational under all conditions, in a worst-case scenario the control would ask the MMC to apply the right voltage corresponding to the desired circulating current setpoint with a delay of one sampling period T_s. The current control being of very good quality it is considered that the maximum voltage error the control can generate is one arm voltage step: v_C^{nom}. Thus, to limit the Δi_{c_y} deviation, the armature inductance shall respect the condition:

$$L \geq \frac{T_s v_C^{nom}}{m \Delta i_{c_y}} = \frac{T_s v_C^{nom}}{m \Delta \bar{i}_{c_y} \cdot \max_t i_{xy}^{\Delta}(t)} \tag{18}$$

To limit the circulating current to a proportion of the maximum delta current i_{xy}^{Δ}, Δi_{c_y} is defined according to $\Delta i_{c_y} := \Delta \bar{i}_{c_y} \cdot \max_t i_{xy}^{\Delta}(t)$.

4 Validation of the sizing procedure

The proposed procedure is applied to the case of a typical PV-to-grid power supply, between a $1.5\ kV$ DC grid and the $400\ V$ RMS line-to-line, $50\ Hz$ AC grid. After validation in this single-phase case, it will be possible to extend it to three-phase.

4.1 Application of the optimization-based sizing procedure

The top part of Table 2 provides the input parameters of the sizing procedure. The parameters required by the first step of the procedure are fed into the corresponding algorithms crafted based on Section 3.1 and Section 3.2. According to the optimization, the maximum power that can be converted by the MMC while complying with the arm current limits and the available DC and AC grid voltages is $\hat{P}_{AC} = 1.53\ kW$ ($S_{AC}^{nom} = 1.64\ kVA$ at $\cos \varphi = 0.93$ power factor). As an outcome of both steps, Figure 3 shows the sigma and delta currents.

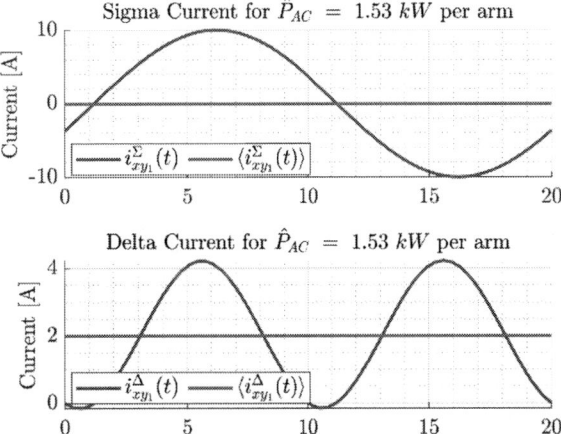

Fig. 3: Sizing procedure application - Steps 1 and 2.

From those results, the arm voltages can be deduced by applying the approach from Section 3.3. Figure 4 presents those voltages along with the corresponding arm currents. Thus, It is observed that the maximum voltage required in an arm is $V_C^{\Sigma} = \hat{v}_{xy} \simeq 1082\ V$. With capacitors having a nominal voltage of $v_C^{nom} = 400\ V$, only three submodules are necessary to reach V_C^{Σ}. However, for fault tolerance reasons, an additional hot-reserve

SM is considered. The total number of SMs per arm is then $N = 4$.

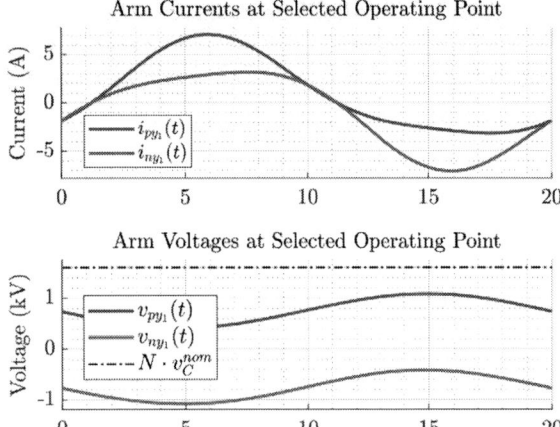

Fig. 4: Sizing procedure application - Step 3.

With the knowledge of $v_{py}(t)$, $v_{ny}(t)$, $i_{py}(t)$ and $i_{ny}(t)$, the powers $p_{py}(t)$ and $p_{ny}(t)$ flowing through the arms are deduced. Section 3.4 then explains how to derive the energy ripple for each arm. Their waveform over the AC-grid fundamental period is shown in Figure 5 and the maximum peak-to-peak energy ripple is found to be $\Delta E_{xy} \simeq 21\ J$. With an average stack voltage of $\langle V_C^\Sigma \rangle = N v_C^{nom} = 1.6\ kV$ and a desired stack voltage ripple of less than $\delta \bar{V}_C^\Sigma = 5\ \%$, the SM capacitor should be at least $329\ \mu F$, thus $C = 330\ \mu F$ is chosen.

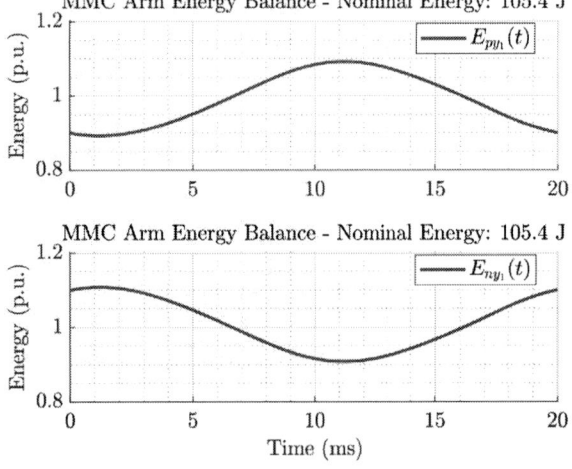

Fig. 5: Sizing procedure application - Step 4.

Being in a single phase case, $m = 1$. In order to limit the circulating current ripple Δi_{c_y} to less

than $\Delta \bar{i}_{c_y} = 15\%$ of the maximum delta current, the inductance formula derived in Section 3.5 is applied. From Figure 3, it is found that $\max_t i_{xy}^\Delta(t) \simeq 4.35\ A$. With a control sampling frequency of $f_s = 100\ kHz$, the arm inductance is chosen to be $L = 6.1\ mH$. The bottom part of Table 2 provides all the parameters resulting from the sizing procedure.

4.2 Simulation validation before prototyping

The sized MMC is simulated using a high-fidelity PLECS model of the converter and semiconductor losses, in order to validate the sizing before producing the prototype. The detailed losses model of the GaN switches is obtained through meticulous experimental electrical and thermal data measurement, enabling high-accuracy simulation of losses using PLECS. In the simulation, the system tracks an AC power setpoint that sweeps the operating range up to nominal power.

In order to carry out the simulation, not only must a CCCS be implemented, but also a complete MMC control architecture. The architecture implemented here is described in Figure 6. The higher-level control loops (energy, power, and current control) are based on that from [10]. As mentioned in [10], the current control can be improved for better reference tracking performances by implementing the transparent control architecture from [11]. This enhancement is applied here. In this case, the selected CCCS is the CCI, it is provided to the MMC by the energy control block. For lower-level control loops such as arm voltage control and capacitor balancing, the traditional nearest level control (NLC) and balancing capacitor algorithm (BCA) approaches are implemented here, respectively. The BCA is also referred to as submodule sorting techniques. First introduced by [3], the fifth chapter from [12] presents both the concepts from initial NLC and BCA techniques, that are used here, in further depth.

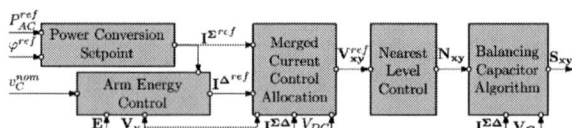

Fig. 6: MMC control architecture implemented

Figure 7 and Fig. 8 show the simulation results for the converter. The simulated losses in the MMC are comprised of switching losses $P_{MMC}^{loss-sw}$ and conduction losses $P_{MMC}^{loss-cond}$ in the semiconductor components, together with conduction losses in the

Tab. 2: Parameters of the MMC

Description	Value
Sizing procedure inputs - single-phase case	
DC bus voltage	$V_{DC} = 1.5\ kV$
DC bus impedance	$R_s = 0\ \Omega,\ L_s = 0\ H$
Number of AC phases	$m = 1$
AC voltage amplitude	$\hat{V}_{AC} = 230\sqrt{2} \simeq 327\ V$
AC grid frequency and period	$f_{AC} = 50\ Hz,\ T_{AC} = 20\ ms$
AC-side power factor	$\cos\varphi = 0.93$
AC-side impedance	$R_o = 0\ \Omega,\ L_o = 2.1\ mH$
Arm current limit	$\hat{i}_{xy}^{lim} = 5\sqrt{2} \simeq 7.07\ A$
Capacitor nominal voltage	$v_C^{nom} = 400\ V$
Stack voltage ripple limit ratio	$\delta\bar{V}_C^{\Sigma} = 5\ \%$
Circulating current ripple limit	$\Delta\bar{i}_{c_y} = 15\%$
Control sampling frequency and period	$f_s = 100\ kHz,\ T_s = 10\ \mu s$
Sizing procedure outputs - single-phase case	
MMC rated power	$S_{AC}^{nom} = 1.64\ kVA$
Number of SMs per arm	$N = 4$
SM capacitor	$C = 330\ \mu F$
Arm impedance	$L = 6.1\ mH$
Parameters estimated afterwards for simulation	
Arm resistance	$R \simeq 30\ m\Omega$

inductance of the converter arms $P_{MMC}^{loss-arm}$. To calculate the latter, the resistive component of the arm inductance was estimated analytically from the dimensions of the inductance and the materials of which it is made. It is found that $R \simeq 30\ m\Omega$. The losses associated with the WBG semiconductors are calculated using the accurate GaN semiconductor model previously mentioned. Overall, the total losses in the MMC are expressed as follows:

$$P_{MMC}^{loss-tot} = P_{MMC}^{loss-sw} + P_{MMC}^{loss-cond} + P_{MMC}^{loss-arm} \quad (19)$$

Details of the evolution of these different losses as a function of the power operating point reached in the simulation are shown in Figure 7. Note that for readability, losses are displayed after being averaged at the T_{AC} timescale. This averaging, which acts as a variable-window sliding filter at the beginning and end of the graph, is also at the origin of the irrelevant sine-shaped curves at these instants. Total converter losses are shown in black on the top plot with a maximum of $9.3\ W$ for a power delivered to the AC grid of $1.53\ kW$, i.e. a 99.39% efficiency. The evolution of the efficiency versus the operating point is shown in Fig. 8. It shows that the converter is over 98.50% efficient throughout the operating range from 10% to 100% of the rated power with an average of 99.41% on that range (for the full range

from 0% to 100% the average drops to 98.77%).

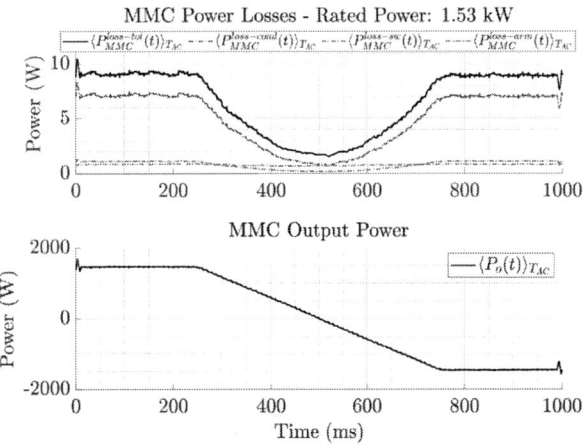

Fig. 7: High-fidelity simulated conduction, switching and arm power losses.

4.3 Prototyping of the converter

As the prototype is currently under production, the 3D computer-aided design (CAD) rendering is shown in Fig. 9. This prototype will enable the performances to be verified. Also, experimental results will be compared with that of the simulation with the aim to validate the optimization-based co-design procedure.

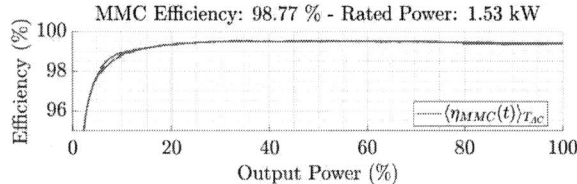

Fig. 8: High-fidelity simulated efficiency versus output power.

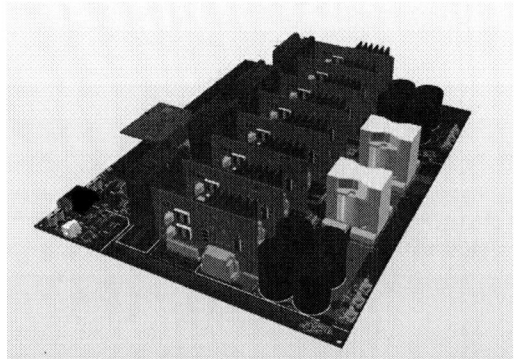

Fig. 9: 3D CAD rendering of the MMC prototype (40 x 30 x 7 cm).

5 Conclusion

A new MMC sizing method based on optimization has been presented. First, a suitable model is readily obtained from the electrical diagram of the converter, then the sizing procedure is detailed, with a particular focus on the optimization problem formulated. The approach was evaluated for a connection between a DC and an AC grid. The performances of the sized converter were validated through a simulation involving high-fidelity models of the GaN semiconductors. The high efficiency obtained is largely due to the WBG components. Future work includes analyzing the prototype's performance and comparing it with the simulations.

Thanks to the sizing procedure proposed here, it is now possible to design an MMC 1) in a way that is optimized to minimize losses due to circulating currents, while guaranteeing compliance with the converter's energy balance; 2) by choosing the desired control strategy for circulating currents, in all circumstances the converter's energy balance is guaranteed; 3) to size an MMC based on a current constraint not to be exceeded within the arms, submodules, and semiconductors. Such capabilities open up the possibility of designing MMCs for a wide range of operating conditions.

The development of the sizing procedure carried out here is a first step towards sizing and control co-design methods for MMCs. This study provides the groundwork for inclusion of other control strategies to be included as early as in the sizing stage. This will further enhance the alignment of the produced MMCs with their operational objectives while minimizing resource consumption, which will be the focus of future work.

References

[1] J. Millán, P. Godignon, X. Perpiñà, A. Pérez-Tomás, and J. Rebollo, "A Survey of Wide Bandgap Power Semiconductor Devices," *IEEE Transactions on Power Electronics*, vol. 29, no. 5, pp. 2155–2163, May 2014. DOI: 10.1109/TPEL.2013.2268900.

[2] I. Chorfi, C. Alonso, R. Monthéard, and T. Sutto, "A GaN-Based Three-Level Dual Active Half Bridge Converter With Active Cancellation of the Steady-State DC Offset Current," English, in *IECON 2022 – 48th Annual Conference of the IEEE Industrial Electronics Society*, Brussels, Belgium: IEEE, Oct. 2022. DOI: 10.1109/IECON49645.2022.9968579.

[3] A. Lesnicar and R. Marquardt, "An innovative modular multilevel converter topology suitable for a wide power range," en, in *2003 IEEE Bologna Power Tech Conference Proceedings*, vol. 3, Jun. 2003. DOI: 10.1109/PTC.2003.1304403.

[4] Q. Tu, Z. Xu, and L. Xu, "Reduced Switching-Frequency Modulation and Circulating Current Suppression for Modular Multilevel Converters," *IEEE Transactions on Power Delivery*, vol. 26, pp. 2009–2017, Aug. 2011. DOI: 10.1109/TPWRD.2011.2115258.

[5] H. Fehr and A. Gensior, "Model-Based Circulating Current References for MMC Cell Voltage Ripple Reduction and Loss-Equivalent Arm Current Assessment," in *2019 21st European Conference on Power Electronics and Applications (EPE '19 ECCE Europe)*, Sep. 2019, P.1–P.9. DOI: 10.23919/EPE.2019.8914767.

[6] A. Hillers and J. Biela, "Optimal design of the modular multilevel converter for an energy storage system based on split batteries," in *2013 15th European Conference on Power Electronics and Applications (EPE)*, Sep. 2013, pp. 1–11. DOI: 10.1109/EPE.2013.6634660.

[7] Z. Ke, J. Pan, M. A. Sabbagh, R. Na, J. Zhang, *et al.*, "Capacitor Voltage Ripple Estimation and Optimal Sizing of Modular Multi-Level Converters for Variable-Speed Drives," *IEEE Transactions on Power Electronics*, vol. 35, no. 11, pp. 12 544–12 554, Nov. 2020. DOI: 10.1109/TPEL.2020.2988403.

[8] F. Gruson, "Contribution au développement de convertisseurs modulaires et multiniveaux pour les réseaux électriques," Français, L2EP, Tech. Rep., 2022, p. 187.

[9] G. Le Goff, "Scalable Control Allocation Methods for the Modular Multilevel Converter: From Modelling to Real Time Implementation," en, PhD Thesis, Institut National Polytechnique de Toulouse (INPT), Nov. 2022.

[10] G. Le Goff, M. Bodson, and M. Fadel, "Control allocation for optimal and resilient operation of the MMC," in *2024 IEEE International Conference on Industrial Technology (ICIT)*, Bristol, UK, Mar. 2024.

[11] G. Le Goff, M. Bodson, and M. Fadel, "Model Reference Control of Constrained Overactuated Systems with Integral Compensation," in *2022 IEEE 61st Conference on Decision and Control (CDC)*, Dec. 2022, pp. 4507–4512. DOI: 10.1109/CDC51059.2022.9992648.

[12] K. Sharifabadi, L. Harnefors, H.-P. Nee, S. Norrga, and R. Teodorescu, *Design, Control, and Application of Modular Multilevel Converters for HVDC Transmission Systems*, en. Wiley-IEEE Press, Oct. 2016.

PCIM Europe 2024, 11– 13 June 2024, Nuremberg   DOI: 10.30420/566262448

A Novel Three-Phase Low-Switch-Count AC-DC Grid Converter Topology with Galvanic Isolation

Liska Steenbock [1], Jan Boris Loesenbeck[1]

[1] Hochschule Bielefeld - University of Applied Sciences and Arts, Germany

Corresponding author: Liska Steenbock, liska.zoehner-sell@hsbi.de
Speaker: Liska Steenbock, liska.zoehner-sell@hsbi.de

Abstract

This paper presents a new family of three-phase converters with galvanic isolation and a sinusoidal grid current. The proposed topologies allow to assemble a converter for cost-sensitive applications with only three switches or an efficiency-optimized converter which results in a higher switch count. The primary side as well as the secondary side can be assembled from partial topologies based on the specific needs. One of the possible circuits which only requires two switches on the primary side and one switch on the secondary side is presented in this paper and the behavior of the circuit is explained. The functionality was verified using simulations.

1 Introduction

Three-phase converters are used in a wide range of applications. Cost, voltage output type, efficiency and mains feedback play a role in the selection of the converter type. As the demand for electrical energy is increasing as well as the need for energy storage due to the volatile nature of renewable energies, new topologies are needed to satisfy each need [1].

If the aim is to save cost, a possibility is to reduce the number of switchable semiconductors. Various methods have been presented in [2]–[6]. In most methods the output voltage is not galvanically isolated.

Another option is to improve efficiency. In this case converters with a high number of switchable semiconductors are used. An exemplary AC-DC converter can be seen in Fig. 1. Based on [2], the novel grid converter topologies offer the possibility to assemble a converter from different blocks, which meet the corresponding requirements. The output voltage is galvanically isolated in every configuration and the grid current is always sinusoidal. One of these configurations, the P2S1, is presented in this paper and can be seen in Fig. 2. As the P2S1 uses only two high-voltage semiconductor switches and only one low-voltage semiconductor switch on the secondary side it is the most cost-efficient. The

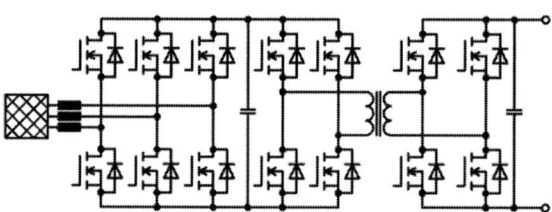

Fig. 1: Commonly used circuit for AC-DC converters.

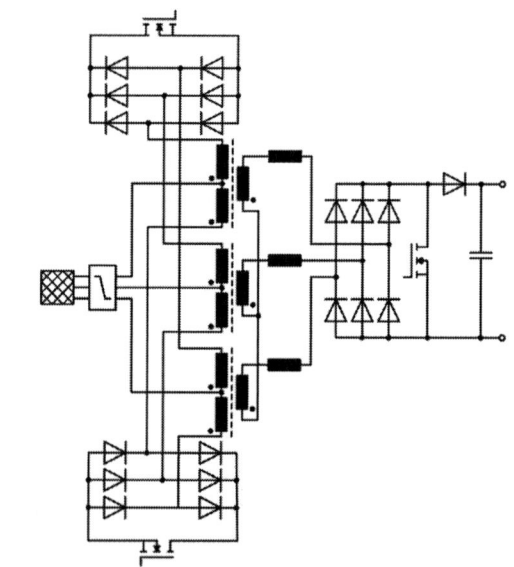

Fig. 2: The three-phase rectifier topology consisting of the P2 pulse inverter followed by an S1 rectifier.

3169

possibility to generate a pulse-frequent signal was verified in [7]. The aim of this paper is to verify a possible rectification using a secondary circuit with only one semiconductor switch via simulation.

2 Topologies

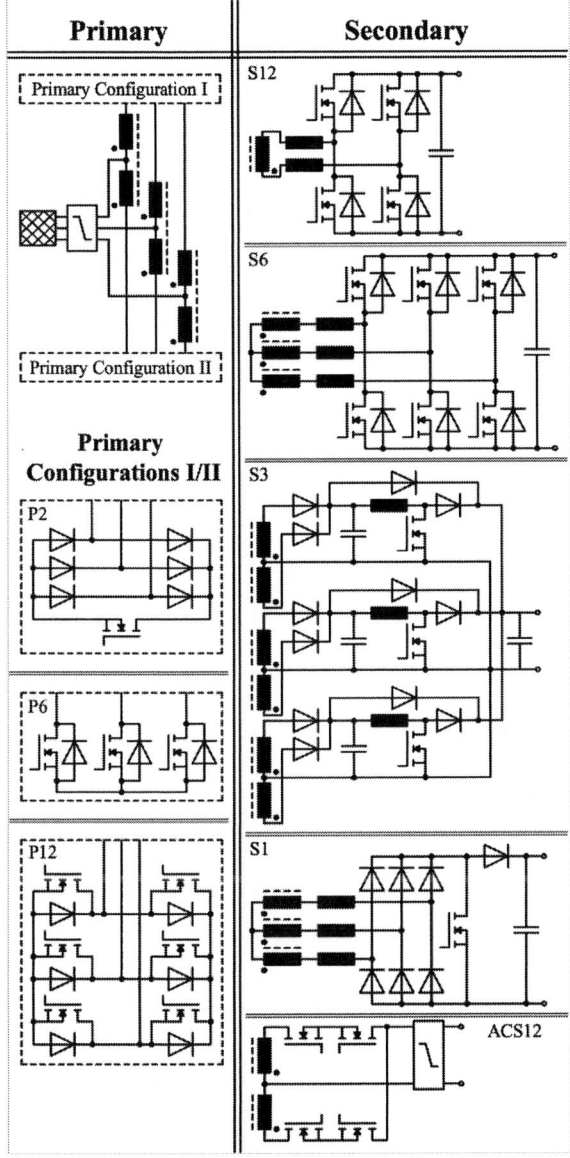

Fig. 3: Possible converter topologies primary side (P2, P6 and P12) as well as the secondary side (S12, S6, S1 and ACS12).

The converter can be divided into three parts: the primary windings of the three transformers and their low pass filtered connection to the grid, the blocks of the primary side and the blocks of the secondary side.

2.1 Primary side

Fig. 3 shows the possible circuit designs of the primary side on the left side. The transformer windings are wound in opposite directions and fed from a three-phase network. The other sides of the primary windings are connected to the blocks of primary configurations (P2, P6 or P12). There are always two of those blocks, one connected to the first windings, the other to the second windings. The transistors on the primary side need to be made to withstand high voltages as they need to switch more than double the peak voltage.

- **P2**: The circuit is composed of a diode bridge connected to a high-voltage transistor, requiring two switchable semiconductors to complete the circuit. For cost-sensitive applications, P2 can be used, as the primary circuit would only use two switchable semiconductor valves.

- **P6**: Each switch configuration consists of three transistors with one body diode each. P6 can be used for slightly higher efficiency because the six switches reduce conduction losses. The transistors can also be connected in delta.

- **P12**: Six transistors, each with a body diode, make up the switch configuration for the P12. The circuit is similar to the P6, but the load per switch is lower. Since the circuits are similar, so is their efficiency, but the P12 is considered to be even more efficient.

For this paper, the P2 configuration was chosen. Since it is the configuration with the least number of transistors, it is also the most cost-effective. The pulse frequency of the transistors, referred to as f_P, is significantly higher than the grid frequency f_G. Consequently, the input voltages can be considered as constant within one pulse period T_P. The two transistors operate alternately, with a pause between phases in which neither transistor conducts. This sequence results in a rectangular pulse pattern characterized by alternating polarity and a sinusoidal envelope. The curves shown in Fig. 4 and were verified in [7] via simulation and a prototype. Figure 4 shows the grid voltages and the envelope of the pulse frequent voltages measured phase to phase (a) as well as a few primary side pulses in more detail (b).

2.2 Secondary side

There are five possible circuits for the secondary side. The label is composed of an 'S' for the sec-

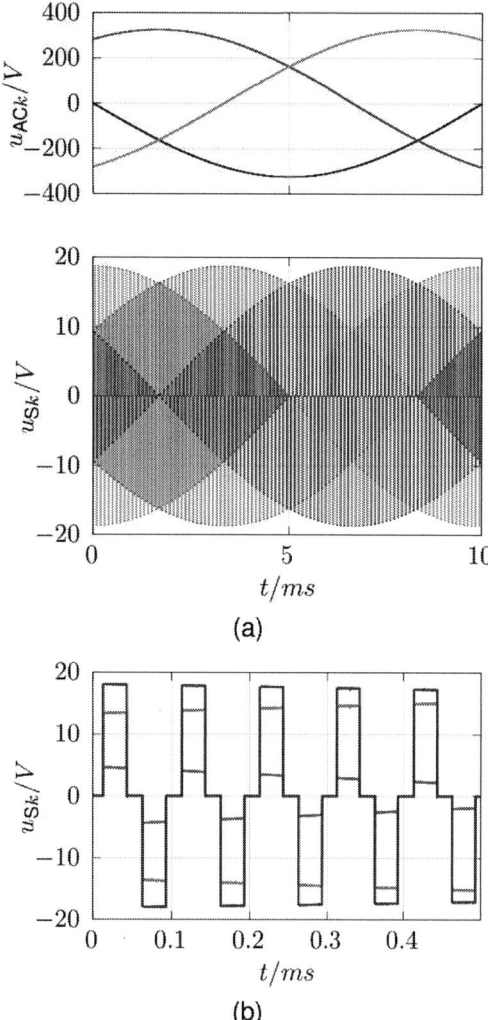

Fig. 4: The grid voltage u_{ACk} (top) and the three-phase phase to phase pulse-frequent voltage on the secondary side u_{Sk} with $k \in \{1, 2, 3\}$ (a). Primary-side pulses are shown in (b).

ondary side, followed by an index symbolizing the number of transistors required. One of the circuits does not produce a DC output voltage, but an AC voltage, which is indicated by the prefix 'AC'. The secondary side is magnetically coupled to the primary side via the windings of the push-pull flow converters. The switches on the secondary side don't need to be rated for high voltages because the pulse inverter produces a low output voltage. There are five possible circuits for the secondary side:

- **S12**: Figure 3 shows the circuit connected to one of the secondary windings. Three such

circuits are required for the entire converter, resulting in three DC voltages. The circuit consists of four transistors, each with an antiparallel diode, two inductors that are magnetized by the pulse-frequency currents generated on the primary side, and a smoothing capacitor. The output of the S12 is a DC voltage. By using GaN transistors for the twelve switches, very high efficiency can be achieved.

- **S6**: The output voltage of S6 is similar to S12, but only a single phase voltage is produced. The circuit uses six transistors with body diodes, three inductors, one for each phase, and a common smoothing capacitor. All three secondary windings are required to produce a single DC output voltage.

- **S3**: The S3 uses only three transistors to generate a single-phase DC output voltage. The circuit consists of three boost converters feeding a common smoothing capacitor. The circuit requires two secondary windings for each transformer. The efficiency is slightly lower than S6 and S12 because diodes have higher conduction losses than transistors. The efficiency could possibly be improved by replacing the diodes with synchronous rectification using GaN transistors.

- **S1**: The S1 generates a single-phase DC output voltage with a single switch. The circuit consists of a diode bridge to which a transistor is connected, a flyback diode, and a smoothing capacitor. Since only one transistor is required, it is the most cost-effective option, but the efficiency is expected to be slightly lower than the other topologies. The harmonic content of the input current depends on the ratio of the secondary output voltage to the DC output voltage. [8]

- **ACS12**: The ACS12 consists of two bipolar switches and a low pass filter. The circuit requires two secondary windings for each transformer. The bipolar switches can switch the output pulses so that an AC voltage can be generated on the output side. Compared to 50Hz transformers, a much smaller size can be realized.

The most cost-effective option, P2, was chosen for the primary side, and consequently, the secondary

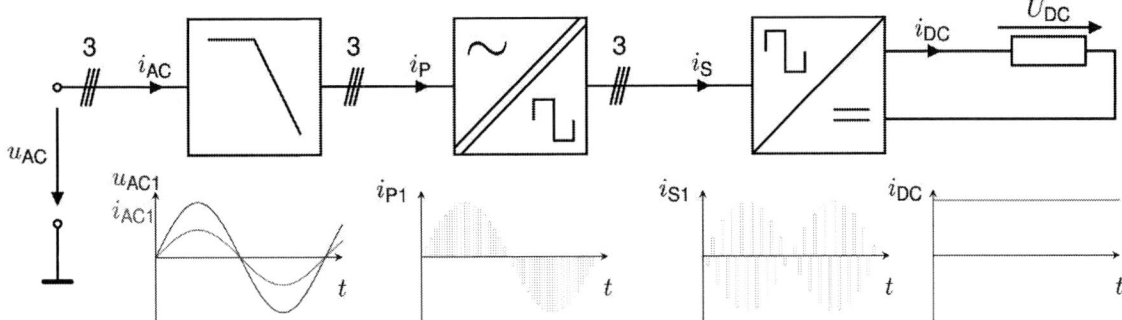

Fig. 5: Block diagram of the converter, consisting of the grid voltages and currents, a low pass filter, the pulse inverter and the subsequent rectifier. Below are the resulting waveforms (Grid voltage u_{AC1}, grid current i_{AC1}, current before the filter i_{P1}, secondary side current i_{S1} and rectified current i_{DC})

side was also fitted with the most cost-efficient circuit S1.

3 Functionality of P2S1

The P2S1 will be analyzed in more detail. The full circuit is shown in Fig. 2. The circuit uses only three switches to rectify an AC signal, resulting in a galvanically isolated DC output voltage. The block diagram of the circuit is shown in Fig. 5.

To derive the pulse frequent signals, simplifying assumptions are made. These include the consideration of an ideal network and ideal passive components. The diodes and transistors are treated as ideal switches, meaning they have no forward voltage when conducting and switched on. It is also assumed that they don't carry any current in the blocking, switched off state. A delay-free transition between the two states is also assumed.

3.1 Primary switching states

The frequency of the primary side switches f_P is much higher than the grid frequency f_G. There are three possible switching states for the primary side:

- **Switching state a_P - T_1 conducting, T_2 blocking:** A three-phase voltage system is formed across one of the primary windings of each of the three transformers. The conducting transistor T_1 and the upstream diode bridge induce a symmetrical three-phase voltage system on the secondary side. The voltages are transformed according to the transformation ratio of the transformers.

- **Switching state b_P - T_1 blocking, T_2 blocking:** The system is currentless and therefore no energy is transferred to the secondary side.

Since both diode bridges have an open switch, there can be no mesh circulation.

- **Switching state c_P - T_1 blocking, T_2 conducting:** The principle of operation for this switching state is analogous to switching state a_P. However, the voltages induced on the secondary side change polarity.

To prevent the switches from closing simultaneously, each state in which a transistor is conducting is followed by state b_P in which both switches are open. Thus, the switching sequence follows the pattern: a_P, b_P, c_P, b_P, a_P, b_P, c_P, and so on.

3.2 Secondary switching states

The switch on the secondary side switches only during the primary switching states a_P or c_P. The secondary side pulse frequency f_S must be faster than the primary side frequency f_P. To control the inverter, it is important to work in discontinuous current mode. It is also important that the secondary side switch is turned off before switching state b_P occurs.

For the secondary side there are four switching states during a pulse period.

- **Switching state a_S: Magnetization of the inductors:** Before turning on the secondary-side switch, the inductors are without current. Once the switch is turned on, the voltages across the inductors create a three-phase symmetrical voltage system that corresponds to the input voltages at the secondary windings. The circuit's current is carried by three diodes, with different diodes being used depending on the angles of the mains voltages. The sum of the positive currents is equal to the current flowing through the switch. The boost diode is in a blocking state, and there is no voltage across the diodes that are not involved in carrying the current. The switching current commutes to the boost diode during the turn-off process. Since the turn-off is such a small part of the pulse duration, it is neglected.

- **Switching state b_S: Demagnetization of three inductors:** The polarity and magnitude of the inductor voltages have changed, causing the induction currents to flow into the capacitor. The previously involved diodes continue to conduct the current, and the current through the boost diode still corresponds to the sum of the positive inductor currents. The output voltage is now present across the transistor and the diodes not involved in the current flow.

- **Switching state c_S: Demagentization of two inductors:** The third switching state occurs when one of the three inductors no longer carries any current. As a result, one of the diodes stops conducting current. The boost diode's current corresponds to the positive current of the remaining inductance currents.

- **Switching state d_S: All inductors are demagnetized:** The fourth switching state occurs when all inductors are demagnetized. The switching state is trivial, i.e. the inductors are currentless for the duration of the state. When the transistor is switched on again, a new pulse period begins.

The resulting pulse-frequent currents through the inductors can be seen in Fig. 6. The current starts during state d_S and ends in state d_S.

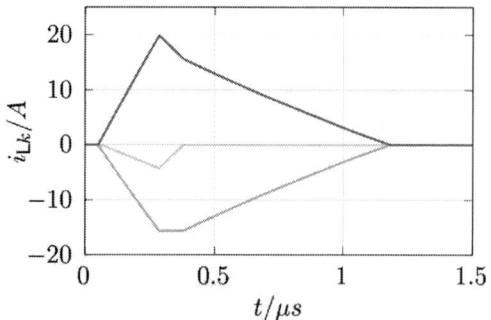

Fig. 6: Currents through the inductors i_{Lk} with $k \in \{1, 2, 3\}$.

4 Simulation

To verify the functionality the circuit was simulated using PLECS 4.8.1 from Plexim GmbH. The simulation concentrates exclusively on the electrical side of the converter. The simulation parameters are listed in table 1.

The transistors on the primary side are switched by a PWM controller that runs at a constant duty cycle D_P of 0.3 and a frequency f_P of 10 kHz. Each transistor conducts for 30 % of the pulse period, resulting in an overall primary side duty cycle of 0.6.

On the secondary side, a transistor is switched by a PWM controller with a duty cycle D_S of 0.07 and a frequency f_S of 300 kHz. The transistor conducts only during a primary side pulse. It is important to maintain discontinuous mode for the secondary side currents through the inductors. The duty cycle D_S of 0.07 was selected because it guarantees that the current flowing through the inductors is discontinuous.

The inductors are simulated in series with a resistor to represent a real inductance. Non-ideal components are used for the transistors. The grid current is smoothed by a low pass filter.

The primary AC voltage U_P of 230 V was converted into a rectified secondary voltage U_{DC} of 20 V. With a high transformation ratio of the transformers, the proportion of harmonics in the mains currents is reduced.

Parameter	Symbol	Value
Grid Voltage	U_P	230 V
Grid Frequency	f_G	50 Hz
Primary Frequency	f_P	10 kHz
Primary Duty Cycle	D_P	0.3
Transformer ratio	a	1:30
Secondary Frequency	f_S	300 kHz
Secondary Duty Cycle	D_S	0.07
Inductor	L	0.1 µH
Capacitor	C	500 µF
Load	R	15 Ω

Tab. 1: Simulation Parameters

The resulting currents for the primary side and the secondary side are displayed in Fig. 7. The resulting relative harmonic contents of the grid current can be seen in table 2. The grid currents and their harmonics indicate that the inverter allows for a sinusoidal mains current.

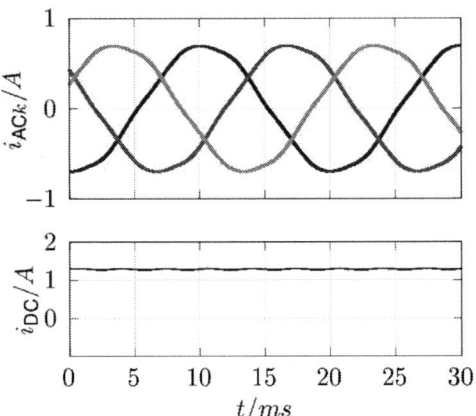

Fig. 7: Grid currents i_{ACk} with $k \in \{1, 2, 3\}$ (top) and DC output current i_{DC}(bottom).

Harmonic Order	Harmonics [%]		
	i_{AC1}	i_{AC2}	i_{AC3}
3	0.001	0.001	0.001
5	2.683	2.681	2.682
7	0.398	0.399	0.398
9	0.002	0.001	0.001
11	0.210	0.211	0.211
13	0.059	0.059	0.059
15	0.000	0.001	0.001
17	0.082	0.082	0.081

Tab. 2: Relative harmonic contents of the grid current i_{ACk} with $k \in \{1, 2, 3\}$

5 Conclusion

Due to the energy transition and the expansion of electrical grids, as well as the increased need for large-scale storage systems like electrolysers and batteries, there is a growing demand for cost-effective grid converters.

The presented three-phase grid converter topologies allow to assemble a circuit from different blocks. A converter with a low switch count can be chosen for cost-sensitive applications. Another topology can be assembled if high efficiency is the goal. The output voltage is galvanically isolated in every configuration and the grid current is always sinusoidal. The topology with the least amount of switches was chosen and further analyzed. The inverter's functionality was demonstrated through a simulation. The inverter can convert a three-phase alternating voltage signal into a rectified signal and maintain a sinusoidal grid current curve.

In summary, the P2S1 is suitable for generating a galvanically isolated rectified output signal at a low cost. To better showcase the functionality, it would be valuable to provide a prototype and consider its efficiency.

References

[1] W.-P. Schill, "Systemintegration erneuerbarer energien: Die rolle von speichern für die energiewende," *Vierteljahrshefte zur Wirtschaftsforschung*, vol. 82, pp. 61–88, Sep. 2013. DOI: 10.3790/vjh.82.3.61.

[2] J. B. Loesenbeck, "Dual switch current converter," WO 2012/013329 A2, 2012.

[3] M. Leibl, J. Huber, D. Menzi, and J. Kolar, "Synthesis of low-switch-count power converter topologies," May 2023, pp. 241–250. DOI: 10.23919/ICPE2023-ECCEAsia54778.2023.10213762.

[4] J. Dias and T. Lazzarin, "Single-phase hybrid boost ac–dc converters with switched-capacitor cells and reduced switch count," *IEEE Transactions on Industrial Electronics*, vol. PP, pp. 1–1, Jul. 2020. DOI: 10.1109/TIE.2020.3005060.

[5] G. Bhuvaneswari, B. Singh, and S. Madishetti, "Three-phase, two-switch pfc rectifier fed three-level vsi based foc of induction motor drive," Dec. 2012, pp. 1–6. DOI: 10.1109/PowerI.2012.6479577.

[6] P. Barbosa, F. Canales, and F. Lee, "Analysis and evaluation of the two-switch three-level boost rectifier," Feb. 2001, 1659–1664 vol. 3. DOI: 10.1109/PESC.2001.954357.

[7] L. Steenbock, A. Kirsch, and J. B. Loesenbeck, "Design, simulation, and construction of a three-phase grid converter with two switchable semiconductor valves," in *IEEE Power and Energy Student Summit 2023 (PESS)*, Bielefeld, Germany, Nov. 2023.

[8] J. Loesenbeck, "Untersuchung von netzpulsstromrichtern mit einem abschaltbaren halbleiterventil," VDI-Verlag, Düsseldorf, VDI-Fortschrittsbericht 392, 2010.

PCIM Europe 2024, 11– 13 June 2024, Nuremberg   DOI: 10.30420/566262449

Single-Stage LED Driver Based on Coupled Inductor Power Factor Correction and LLC Converter

Alireza Ramezan Ghanbari [1], Sayed Reza Afzali Arani [1], Heinz Seyringer [1], Dietmar Klien [2], Lukas Saccavini [2], Norbert Linder [2]

[1] V-Research, Austria
[2] Tridonic, Austria

Corresponding author and speaker: Alireza Ramezan Ghanbari, alireza.ghanbari@v-research.at

Abstract

In this paper, a single-stage LED driver using coupled inductors for power factor correction (PFC) is presented. The coupled inductors are utilized to charge and discharge the input inductor in each switching cycle, aiming to achieve an ohmic mains behavior result in a high power factor. The conditions of the ohmic mains behavior are analyzed. The coupled inductors are integrated into a half-bridge resonant LLC converter. The driver is designed so that the bus voltage is limited to the reflected secondary voltage, achieving a high power factor while keeping the bus voltage at an applicable level. Additionally, soft switching is achieved during PFC operation. The accuracy of the analytically determined PFC operation is investigated and approved by the experimental results.

1   Introduction

Along with the growing use of LEDs in several lighting systems due to their high efficiency and long lifetime, the requirements and performance of the LED driver, which is an interface between LEDs and power line, are becoming increasingly important [1].

To meet standards such as IEC 61000-3-2, power factor correction (PFC) converters are used in the LED driver industry [2]. Earlier methods for improving the power factor (PF) suggested applying an AC–DC input current shaper converter and a DC–DC converter for regulating the output. A bulk capacitor is applied between the two stages to buffer the input power. This method requires two independent power stages and controllers, so the topology and control become complicated, especially for low power applications, resulting in higher size and price [3]-[4].

Integrating two stages into a single-stage by sharing components, especially switches, leads to a single-stage LED structure that has fewer components and less complexity [5]-[18]. In recent years, this structure has attracted considerable attention. Many studies have focused on integrating the flyback converter with Boost, Buck, Sepic, and resonant converters due to the fewer circuit elements and simplicity of the circuit [6]-[9]. Although single-stage LED drivers using a flyback converter featuring only one switch, the stresses on the switch are relatively high, which limits this type of LED driver for low power applications.

Other studies integrate half-bridge LLC resonant converter with a Boost converter [10]-[18]. The LLC switches inherently provide soft switching which enhances efficiency. This topology achieves low total harmonic distortion (THD) and electromagnetic Interference (EMI) with good efficiency, but the bus voltage is usually twice of the peak of the input voltage, which is impractical for industrial applications.

In [13] to reduce the bus voltage, a hybrid pulse frequency modulation – asymmetric pulse width modulation (PFM-APWM) strategy is proposed. However, the method has complexity in control and results in asymmetric LLC resonant and output diodes currents, impacting efficiency. By using PFM-APWM hybrid control in interleaved Boost stage and full bridge LLC the asymmetric problem can be resolved. Nevertheless, this topology is not well-suited for low power applications [14]. Some studies proposed using interleaved Boost [15]-[16], Buck-Boost [17] and two Boost [18] converters with LLC converters. However, these topologies require large bulk capacitor or have EMI and THD problems. Coupled inductor as a current shaper has been utilized in various studies. In [19], achieving high PF is due to the DCM operation of the coupled inductors inside the dual buck-boost converter sub-circuit,

3175

which is integrated with the LLC resonant converter. However, it suffers from the drawback of high voltage across the bulk capacitor. Utilizing coupled inductors on the AC side of PFC converters may enhance PF. However, challenges like zero crossing and THD persist, and achieving ohmic mains behavior remains unclaimed by existing solutions [20]-[21].

To achieve low THD and EMI, as well as minimize both the bulk capacitor value and the bus voltage while incorporating dimming and maximizing the efficiency, this paper proposes a single-stage LED driver which integrates the coupled inductor PFC with the half-bridge resonant LLC converter.

2 The proposed topology

The proposed single-stage LED driver illustrated in Fig. 1 integrates a half-bridge LLC resonant converter with a coupled inductor to provide power factor correction. The driver is consisting of the coupled inductors L_{C1}, L_{C2} and L_{C3}, input buffer inductor L_f, input rectifier diodes D_1-D_4, two power switches S_1 and S_2 two bulk capacitors C_{b1} and C_{b2}, a resonant network consisting of C_r and L_r, a transformer T_r, output rectifier and a capacitor C_O along with LEDs. The proposed driver eliminates one power switch, one diode and PFC controller in comparison to the conventional two-stage LED driver. The important waveforms are shown in Fig. 2. As can be seen, inductor L_{C1} is periodically charged and discharged by the voltage across C_{b1} and C_{b2} which is $V_{bus}/2$ when switches S_1 and S_2 are turned on respectively. When S_1 is turned on, the voltage across L_{C1} becomes $-V_{bus}/2$. Due to the coupling the voltages across L_{C2} and L_{C3} become $-n_{C2}V_{bus}/2$ and $-n_{C3}V_{bus}/2$, respectively. In the condition where the sum of the input voltage and L_{C2} and L_{C3} voltages exceeds V_{bus}, a surge of input current is supplied to C_{b1} and C_{b2}. When S_2 is turned on, the voltage across L_{C1} becomes $V_{bus}/2$ and the voltages across L_{C2} and L_{C3} are $n_{C2}V_{bus}/2$ and $n_{C3}V_{bus}/2$, respectively. This condition results in L_f discharge. It will be demonstrated that the proposed topology can achieve PFC under specific conditions. The voltage across bulk capacitors is minimally higher than the input voltage and will be clamped to the reflected output voltage under light load condition.

2.1 Operation Principal

In this section, the theoretical operation of the proposed topology is described. The operation is analyzed during a switching period. Following

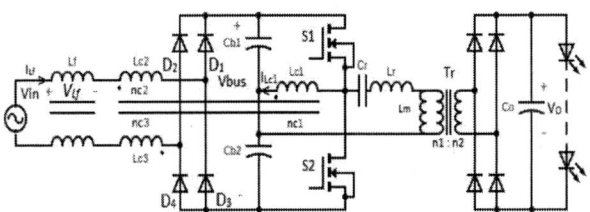

Fig. 1 Proposed single-stage LED driver topology

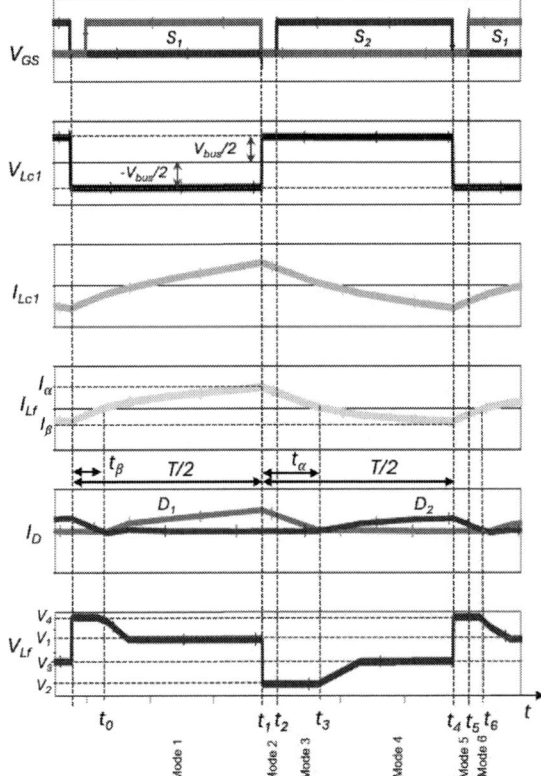

Fig. 2 Important waveforms

assumptions are considered to simplify the converter analysis:

- The input voltage is sinusoidal voltage and considered constant in a switching cycle since the line frequency is significantly lower than the switching frequency.
- The converter operates at steady-state condition.
- The voltage across C_{b1} and C_{b2} is constant throughout an input line cycle.
- The capacitors C_{b1} and C_{b2} have the same value and voltage, considered $V_{bus}/2$.

- The inductors L_{C2} and L_{C3} have the same value and the turn ratios n_{C2} and n_{C3} are equal and considered as n_C.

Mode 1 [$t_0 - t_1$]: At t_0, the switch S_1 is on and the voltage $-V_{bus}/2$ is applied to L_{C1} and to the resonant circuit. In this mode $-n_{C2}V_{bus}/2$ and $-n_{C3}V_{bus}/2$ is applied to L_{C2} and L_{C3} respectively and D_1 and D_4 are conducting. In this mode the voltage across, V_{Lf} is:

$$V_1 = V_{in} + (n_C - 1)V_{bus} \tag{1}$$

So, the L_f current, I_{Lf}, increases almost linearly. At this mode, a resonance occurs via C_{b1}, S_1, L_r, C_r and n_1.

Mode 2 [$t_1 - t_2$]: At t_1 the switch S_1 is turned off and the freewheeling inductor current of converter flows through the body diode of the S_2 and turns it on. The voltage across L_{C1} changes to $V_{bus}/2$ due to S_2 body diode being turned on. In this mode $n_{C2}V_{bus}/2$ and $n_{C3}V_{bus}/2$ is applied to L_{C2} and L_{C3} respectively. I_{Lf} decreases almost linearly while D_1 and D_4 are conducting. Therefore, V_{Lf} is:

$$V_2 = V_{in} - (n_C + 1)V_{bus} \tag{2}$$

Mode 3 [$t_2 - t_3$]: At t_2, S_2 is turned on under zero-voltage switching (ZVS) condition. The voltage $V_{bus}/2$ is applied to L_{C1} and to the resonant circuit. In this mode, another resonance occurs via C_{b2}, S_2, L_r, C_r and n_1 in the reverse direction. The voltage across L_f in this mode is the same as mode 2 and I_{Lf} decreases almost linearly. This mode ends when I_{Lf} reaches zero at t_3.

Mode 4 [$t_3 - t_4$]: At the beginning of this mode, I_{Lf} reaches zero and continues to decrease in the reverse direction. Consequently, D_2 and D_3 start to conduct. The voltage across L_{C1} is still $V_{bus}/2$ and V_{Lf} is:

$$V_3 = V_{in} - (n_C - 1)V_{bus} \tag{3}$$

Mode 5 [$t_4 - t_5$]: At t_4 the switch S_2 is turned off and the voltage across L_{C1} changes to $-V_{bus}/2$ due to S_1 body diode turned on. In this mode $-n_{C2}V_{bus}/2$ and $-n_{C3}V_{bus}/2$ is applied to L_{C2} and L_{C3} respectively and D_2 and D_3 are conducting. In this mode I_{Lf} increases almost linearly and V_{Lf} is:

$$V_4 = V_{in} + (n_C + 1)V_{bus} \tag{4}$$

Mode 6 [$t_5 - t_6$]: At t_5, S_1 is turned on under ZVS condition. The voltage $V_{bus}/2$ is applied to L_{C1} and to the resonant circuit. The voltage across L_f in this mode is the same as mode 5 and I_{Lf} decreases almost linearly. This mode ends when L_f reaches zero at t_6.

2.2 Analysis

In the following, an analysis of the proposed single-stage converter is outlined, along with a discussion of the design procedure for its elements. As shown in Fig. 2, the values of I_α and I_β, representing I_{Lf} positive and negative peaks, are calculated based on V_{Lf} voltage states as:

$$I_\alpha = \frac{V_2}{L_f}t_\alpha = \frac{V_1}{L_f}\left(\frac{T}{2} - t_\beta\right) \tag{5}$$

$$I_\beta = \frac{V_4}{L_f}t_\beta = \frac{V_3}{L_f}\left(\frac{T}{2} - t_\alpha\right) \tag{6}$$

From (5) and (6), t_α and t_β can be calculated as follows.

$$t_\alpha = \frac{V_1}{V_2}\left(\frac{T}{2} - t_\beta\right) \tag{7}$$

$$t_\beta = \frac{V_3}{V_4}\left(\frac{T}{2} - t_\alpha\right) \tag{8}$$

Substituting (8) in (7), the following equation is obtained.

$$t_\alpha = \omega\frac{T}{2}\left(\frac{1-\gamma}{1-\omega\gamma}\right) \tag{9}$$

$$t_\beta = \gamma\frac{T}{2}\left(\frac{1-\omega}{1-\omega\gamma}\right) \tag{10}$$

where $\omega = V_1/V_2$ and $\gamma = V_3/V_4$.

Also, by substituting (9) and (10) in (5) and (6), I_α and I_β can be calculated.

$$I_\alpha = \frac{\omega V_2 T}{2L_f}\left(\frac{1-\gamma}{1-\omega\gamma}\right) \tag{11}$$

$$I_\beta = \frac{\gamma V_4 T}{2L_f}\left(\frac{1-\omega}{1-\omega\gamma}\right) \tag{12}$$

The average input current can be calculated from I_α and I_β averages.

$$I_{Lf,avg} = I_{\alpha,avg} - I_{\beta,avg} = \frac{I_\alpha}{2T}\left(\frac{T}{2} - t_\beta + t_\alpha\right) - \frac{I_\beta}{2T}\left(\frac{T}{2} - t_\alpha + t_\beta\right) \tag{13}$$

Substituting (9)-(12) in (13) the following equation is obtained for $I_{Lf,avg}$:

$$I_{Lf,avg} = \frac{n_C T}{4L_f}V_{in} \tag{14}$$

The consideration of the I_α and I_β signs is essential in the calculation of (13). As can be seen from (14), the input impedance of the proposed converter is ohmic, allowing the input current to follow the input voltage without distortion. The PFC operation is dependent on the presence of I_β which has a negative current. Otherwise, the input current may experience distortion.

2.3 Design Consideration

The aim is to design an LED driver with PFC capability, regulation, and output dimming. As mentioned before, the proposed driver is the integration of the coupled inductor and the half-bridge resonant LLC converter. The coupled inductors L_{c1}, L_{c2} and L_{c3} are designed to be operated under the condition which provides PFC. The design equation can be expressed as follows. From (1) and (3) and considering V_{in} as a sinusoidal voltage source, the second term of (1) and (3) must be higher than zero [22]. This implies that the n_c must be higher than one.

To limit the current ripple in L_{C1} (ΔI_{Lc1}) which contributes to the conduction and core losses. The value of L_{C1} is calculated as:

$$L_{C1} = \frac{V_{bus} D T_{S,max}}{2\Delta I_{Lc1}} \quad (15)$$

where $T_{S,max}$ is maximum switching period which is related to the minimum switching frequency.

The half-bridge LLC resonant converter is designed to operate across the resonance frequency, providing a wide bus voltage range and dimming capability. According to fundamental harmonic analysis [23], the primary equivalent loads, R_{ac} is:

$$R_{ac} = \frac{8n^2 R_0}{\pi^2} \quad (16)$$

Where R_0 is the LED array equivalent resistance and n is the transformer turn ratio.

The resonant capacitor C_r can be calculated as:

$$C_r = \frac{1}{2\pi f_{rh} R_{ac} Q_r} \quad (17)$$

Where f_{rh} is the higher resonant frequency and Q_r is the quality factor of the resonant network.

Also, the resonant and magnetizing inductors are calculated as follows:

$$L_r = \frac{1}{4\pi^2 f_{rh}^2 C_r} \quad (18)$$

$$L_m = \frac{1}{4\pi^2 f_{rl}^2 C_r} - L_r \quad (19)$$

Where L_m is the magnetizing inductance of the transformer and f_{rl} is the lower resonant frequency. To balance efficiency and regulation m, defined as $(L_m + L_r)/L_r$, is set to 6. The switching frequency range is 55 kHz to 140 kHz.

The bulk capacitors are equal and calculated as:

$$C_{b1,2} = \frac{P_O}{\pi f_{line} V_{bus} \Delta V_{bus}} \quad (20)$$

3 Experimental Results

To validate the theoretical analysis, a universal input, 60 W prototype is manufactured. Table 1 shows the specification and main components of the proposed topology. The coupled inductor currents I_{LC1} and the buffer inductor current I_{Lf1} waveforms as well as the gate-source and drain-source voltages of the S_2 are shown in Fig. 3. As shown, L_{c1} and L_f are continuously charged and discharged to provide the PFC operation. The ZVS is achieved, enhancing overall efficiency. Fig. 4 shows the experimental results of input voltage and current under the input voltage of 230 V and full-load condition. The pure sinusoidal waveform of input current is achieved. The measured PF is 0.994 and THD is approximately 2%. The comparison in Fig. 5 illustrates the input current harmonics of the proposed LED driver under full load and 230V input voltage against the IEC 61000-3-2 Class-C standard. Each harmonic current is significantly lower than the standard values, aligning with the requirements of IEC 61000-3-2. The bus voltage of the proposed LED driver versus output power variation is shown in Fig. 6. As can be seen, this voltage is lower than 440 V under output power variation. The efficency of the manufactured prosed LED driver is shown in Fig. 7.

Component	Designators	Value or part
Input voltage	V_{in}	230 V_{rms}
Output voltage	V_O	48 V
Output current	I_O	1250 mA
Diodes	D_1 - D_4	ES3J
Power Switches	S_1 and S_2	STD10N60M2
Coupled Inductor	L_{c1}, L_{c2} and L_{c3}	700 μH, 1400 μH
Bulk Capacitor	C_{b1}, C_{b2}	22 μF/250V
Resonant network	L_m, L_r, C_r, R_{ac}, Q_r	2100 μH, 400 μH, 3.3 nF, 780Ω, 0.4
Resonant frequencies	f_{r1}, f_{r2}	55 kHz, 140 kHz

Table 1 Proposed Driver Specification

4 Conclusion

A single-stage LED driver with inherent PFC which integrates a half bridge LLC resonant converter with coupled inductors is proposed. It has been shown that the input impedance of the proposed driver is ohmic, enabling the input current to track the input voltage without distortion. The proposed driver bus voltage is not high and is limited to the reflected secondary voltage. The experimental results from the manufactured driver illustrate the attainment of high PF, low THD and ZVS.

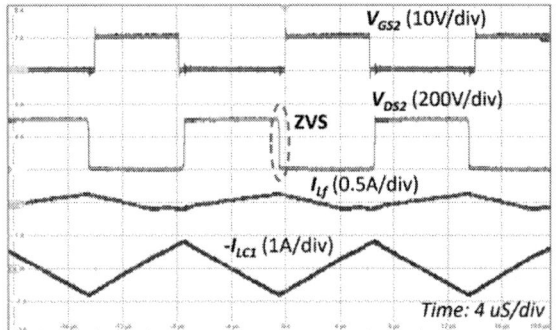

Fig. 3 Switching waveforms

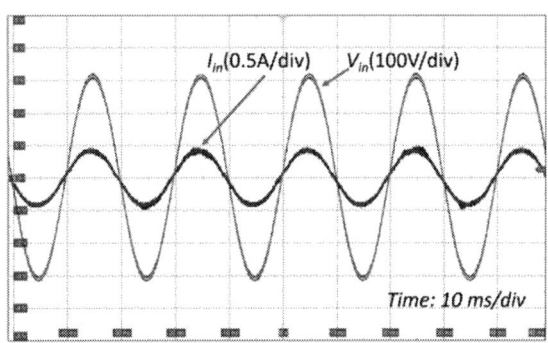

Fig. 4 Input Voltage and Current Waveforms

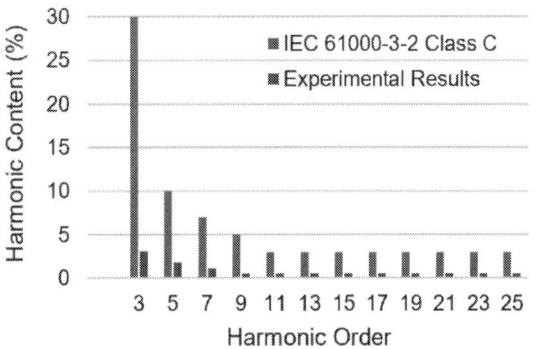

Fig. 5 Input current harmonics at 230V and full load

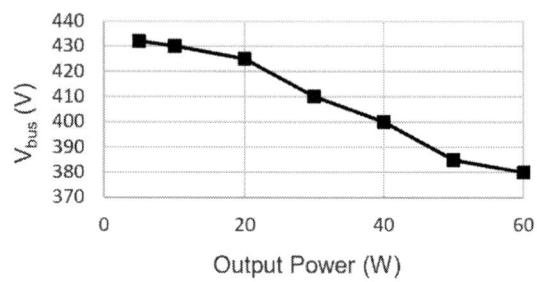

Fig. 6 Bus voltage VS output power

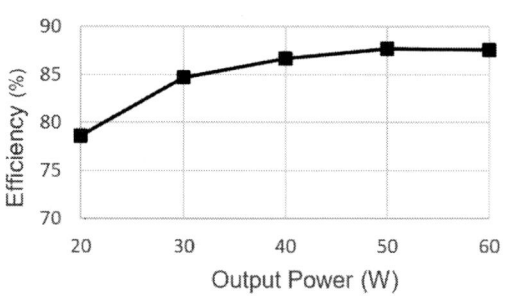

Fig. 7 The measured efficiency VS output power

References

[1] Y. Wang, J. M. Alonso and X. Ruan, "A Review of LED Drivers and Related Technologies," in IEEE Transactions on Industrial Electronics, vol. 64, no. 7, pp. 5754-5765, July 2017

[2] Electromagnetic Compatibility (EMC), Part 3–2: Limits–Limit s for Harmonic Current Emissions (Equipment Input Current ≤ 16A per Phase), International Electrotechnical Commission IEC 61000-3-2 Standard, 2020.

[3] L. Roggia, F. Beltrame, J. E. Baggio, and J. Renes Pinheiro, "Digital current controllers applied to the boost power factor correction converter with load variation," IET Power Electron., vol. 5, no. 5, pp. 532–541, May 2012.

[4] M. He, F. Zhang, J. Xu, P. Yang, and T. Yan, "High-efficiency two-switch tri-state buck-boost power factor correction converter with fast dynamic response and low-inductor current ripple," IET Power Electron., vol. 6, no. 8, pp. 1544–1554, Sep. 2013.

[5] A. Ramezan Ghanbari, E. Adib, and H. Farzanehfard, "Single-stage single-switch power

factor correction converter based on discontinuous capacitor voltage mode buck and flyback converters," IET Power Electron., vol. 6, no. 1, pp. 146–152, Jan. 2013.

[6] A. Abasian, H. Farzanehfard, and S. A. Hashemi, "A single-stage single switch soft-switching (S6) boost-flyback PFC converter," IEEE Trans. Power Electron., vol. 34, no. 10, pp. 9806–9813, Oct. 2019.

[7] B. Poorali, E. Adib, and H. Farzanehfard, "A single-stage single-switch soft-switching power-factor-correction LED driver," IEEE Trans. Power Electron., vol. 32, no. 10, pp. 7932–7940, Oct. 2017.

[8] Y. Wang, F. Li, Y. Qiu, S. Gao, Y. Guan, and D. Xu, "A single-stage LED driver based on flyback and modified class-E resonant converters with low-voltage stress," IEEE Trans. Ind. Electron., vol. 66, no. 11, pp. 8463–8473, Nov. 2019.

[9] D. Gacio, J. M. Alonso, A. J. Calleja, J. García, and M. Rico-Secades, "A universal-input single-stage high-power-factor power supply for HBLEDs based on integrated buck–flyback converter," IEEE Trans. Ind. Electron., vol. 58, no. 2, pp. 589–599, Feb. 2011.

[10] J.-I. Baek, J.-K. Kim, J.-B. Lee, H.-S. Youn, and G.-W. Moon, "A boost PFC stage utilized as half-bridge converter for high-efficiency DC–DC stage in power supply unit," IEEE Trans. Power Electron., vol. 32, no. 10, pp. 7449–7457, Oct. 2017.

[11] G. Zhang et al., "Control design and performance analysis of a double-switched LLC resonant rectifier for unity power factor and soft switching," IEEE Access, vol. 8, p. 44 511-44 521, 2020.

[12] C. -A. Cheng, H. -L. Cheng and T. -Y. Chung, "A Novel Single-Stage High-Power-Factor LED Street-Lighting Driver with Coupled Inductors," in IEEE Transactions on Industry Applications, vol. 50, no. 5, pp. 3037-3045, Sept.-Oct. 2014.

[13] H. Ma, G. Chen, J. H. Yi, Q.W. Meng, L. Zhang, and J. P. Xu, "A single stage PFM-APWM hybrid modulated soft-switched converter with low bus voltage for high-power LED lighting applications," IEEE Trans. Ind. Electron., vol. 64, no. 7, pp. 5777–5788, Jul. 2017.

[14] J. Yi, H. Ma, X. Li, S. Lu, and J. Xu, "A novel hybrid PFM/IAPWM control strategy and optimal design for single-stage interleaved boost-LLC AC–DC converter with quasi-constant bus voltage," IEEE Trans. Ind. Elec-

tron., vol. 68, no. 9, pp. 8116–8127, Sep. 2021.

[15] Y. Wang, Y. Guan, K. Ren, W. Wang, and D. Xu, "A single-stage LED driver based on BCM boost circuit and LLC converter for street lighting system," in IEEE Trans. Ind. Electron., vol. 62, no. 9, pp. 5446–5457, Sep. 2015.

[16] Y. Wang, Y. Guan, X. Zhang, and D. Xu, "Single-stage LED driver with low bus voltage," Electron. Lett., vol. 49, no. 7, pp. 455–457, Mar. 2013.

[17] Y. Wang, Y. Guan, J. Huang, W. Wang, and D. Xu, "A single-stage LED driver based on interleaved buck-boost circuit and LLC resonant converter," IEEE J. Emerg. Sel. Topics Power Electron., vol. 3, no. 3, pp. 732–741, Sep. 2015.

[18] D. Yu, X. Xie and H. Dong, "A Novel Quasi-Single-Stage Boost-LLC AC/DC Converter With Integrated Boost Cells for Achieving Low Bus Voltage for LED Driver," in IEEE Journal of Emerging and Selected Topics in Power Electronics, vol. 10, no. 4, pp. 4413-4424, Aug. 2022.

[19] C. -A. Cheng, H. -L. Cheng and T. -Y. Chung, "A Novel Single-Stage High-Power-Factor LED Street-Lighting Driver With Coupled Inductors," in IEEE Transactions on Industry Applications, vol. 50, no. 5, pp. 3037-3045, Sept.-Oct. 2014.

[20] H. Khalilian, H. Farzanehfard, E. Adib and M. Esteki, "Analysis of a New Single-Stage Soft-Switching Power-Factor-Correction LED Driver With Low DC-Bus Voltage," in IEEE Transactions on Industrial Electronics, vol. 65, no. 5, pp. 3858-3865, May 2018.

[21] S. Meleettil Pisharam and V. Agarwal, "Novel High-Efficiency High Voltage Gain Topologies for AC–DC Conversion With Power Factor Correction for Elevator Systems," in IEEE Transactions on Industry Applications, vol. 54, no. 6, pp. 6234-6246, Nov.-Dec. 2018.

[22] C. K. Tse, "Zero-order switching networks and their applications to power factor correction in switching converters," in IEEE Transactions on Circuits and Systems I: Fundamental Theory and Applications, vol. 44, no. 8, pp. 667-675, Aug. 1997.

[23] M. K. Kazimierczuk and D. Czarkowsk, Resonant Power Converters, 2nd ed. New York, NY, USA: Wiley, 2011.

PCIM Europe 2024, 11– 13 June 2024, Nuremberg   DOI: 10.30420/566262450

An Inverse Coupled DC-DC Boost Inductor with 2-kV SiC MOSFET Module for 1500V Solar Inverter MPPT

Yusi Liu[1], Andrew Yang[2], Ondrej Picha[3]

[1] onsemi, USA
[2] onsemi, Republic of Korea
[3] onsemi, Czech Republic

Corresponding author: Yusi Liu, Yusi.Liu@onsemi.com

Abstract

The demand for 1500V string solar inverters has increased rapidly in recent years. The dcdc boost converter is critical in the 1500V system since it enables the function of Maximum Power Point Tracking (MPPT). Thanks to a new 2kV SiC MOSFET technology, this paper offers a single-end two-level boost topology. As MPPT currents nowadays increase up to 80 A per string, an inverse coupled dcdc boost inductor with two windings is designed, prototyped, and validated. Compared with a traditional solution of two individual non-coupled inductors, it reduces total inductor weight and cost while keeping a similar ripple current.

1   Introduction

Compared with 1000V voltage systems, 1500V solar inverters became a mainstream solution at utility-scale solar power plants because they bring lower current (lower copper loss), higher output power, and lower levelized cost of electricity (LCOE). MPPT enables boosting the PV panel input voltage (full power MPPT input range is 860V~1300 V) to higher inverter dc bus voltage, which connects to 800V utility grid ac voltage through dc-ac inverters. Three-level boost topologies (symmetrical boost, flying capacitor boost, as shown in Fig. 1 (a) and (b)) are dominant in today's 1500V market. To reduce switching loss, boost diodes are usually SiC diode which has lower switching loss Err than Si diode. Either IGBT (better cost) or SiC MOSFET (better efficiency) are adopted by different system designs, while the device voltage ratings could be 1000V or 1200V.

Compared with the three-level topologies of Fig. 1 (a) and (b), which require complicated hardware and software, a single-end two-level topology is preferred, as shown in Fig. 1 (c), due to its simplicity. For blocking the maximum 1500 V dc bus voltage, a 2 kV SiC MOSFET device was developed. A 4-channel dc-dc boost power integrated module (PIM) is prototyped as shown in Fig. 2 (a) and (b), and its schematic circuit diagram is shown in Fig. 2 (c).

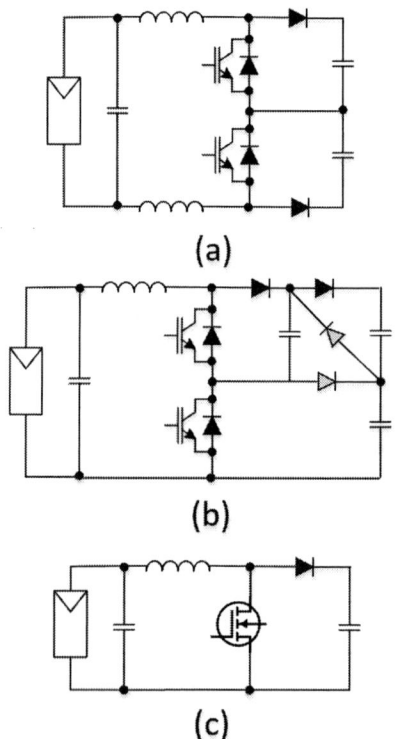

Fig. 1 1500V solar inverter dc-dc boost topologies and prototype.

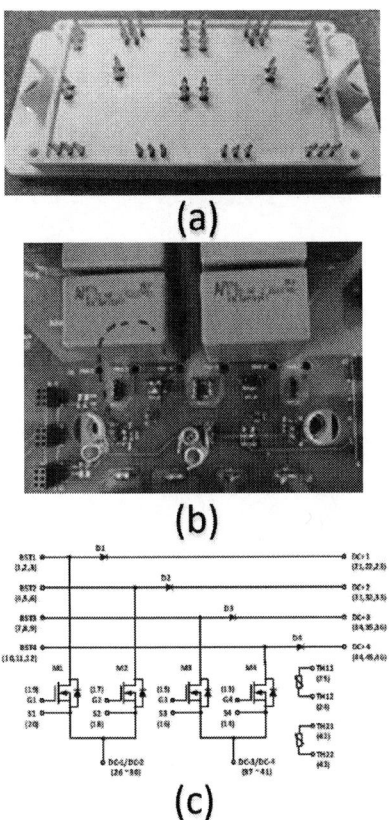

Fig. 2 Boost module prototype.

In each boost channel, R_{dson} of 2kV SiC MOSFET is 20 mΩ and SiC diode rated current is 50A. It is designed for 40A input current (half MPPT current of 80A). A single dc-dc inductor without coupling, which used POCO's NPS-125 powder core, was designed and massy used in the existing production as shown in Fig. 3. Most solar string inverters selected powder core for filter inductor due to lower loss, lower cost and high efficiency at light lower compared to silicone still magnetic core. Interleaving PWM is beneficial for reducing input and out current ripples and it is applied here.

Followed by PIM, inductors are the second most expensive components in a solar inverter system. It also takes considerable weight and space. Magnetic integration helps to reduce weight, and one option is to build two inductors into one inverse coupled inductor [1]. Its benefits were validated in three-level dcdc converter, by using ferrite core [3], and integrating powder core and Nano-crystalline core [4]. However, ferrite core has fixed inductance value and lower saturation permeability. In [4], complicated core and windings design brings high cost and manufacture challenges. Using a single powder core with variable inductance

is preferred at renewable energy applications [5]. In this paper, only powder core magnetic material with two windings is adopted, which makes it low-cost and easy at manufacture and assembly.

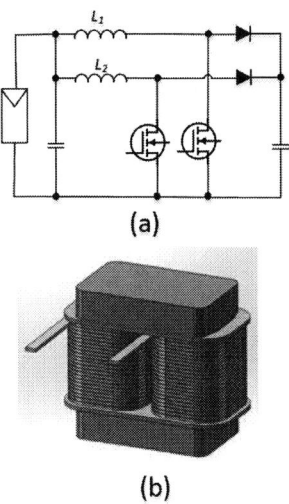

Fig. 3 Two inductors without coupling.

2 Coupled Inductor Design

Circuit diagram of two inverse coupled inductors L1 and L2 is shown in Fig. 4 (a). The physical design of coupled inductor is shown in Fig. 4 (b). Parameters of inductors are illustrated in Table 1.

Voltage of two inversed coupled inductors are shown in

$$\begin{cases} v_1 = L_1 \cdot \dfrac{di_1}{dt} - M \cdot \dfrac{di_2}{dt} \\ v_2 = L_2 \cdot \dfrac{di_2}{dt} - M \cdot \dfrac{di_1}{dt} \end{cases} \qquad (\ 1\ )$$

v_1 and v_2 are the voltages applied to the two corresponding inductors. Self- ($L_1=L_2=L$) and mutual- (M) inductance values from prototype measurement are illustrated in Fig. 5.

For a dcdc boost converter operates in continuous current mode (CCM), the output voltage can be expressed by input voltage and duty cycle as Vo = Vin/(1-D). In the 1500V solar string inverter applications, the minimum voltage of input PV panel during full power is around 900V. Therefore, the duty cycle D = (Vo-Vin)/Vo is always smaller than 0.5.

In case of duty cycle D ≤ 0.5, non coupled inductor input current ripple Δi_{in} and inductor current ripple Δi_L are well derived as follows [1]

$$i_{in} = \frac{V_{in}(2D - 1)}{L \cdot f_{sw}} \qquad (2)$$

$$\Delta i_L = \frac{V_{in}D}{L \cdot f_{sw}} \qquad (3)$$

For inverse coupled inductor, input current ripple $\Delta i'_{in}$ and inductor current ripple $\Delta i'_L$ are shown below

$$\Delta i'_{in} = \frac{V_{in}D(1 - 2D)}{(1 - D) \cdot (L - M) \cdot f_{sw}} \qquad (4)$$

$$\Delta i_L{'} = \frac{V_{in}D[(L+M)(1-2D)+(L-M)]}{2 \cdot (1-D) \cdot (L-M) \cdot (L+M) f_{sw}} \qquad (5)$$

	One single non-coupled inductor	Two non-coupled inductors	One coupled inductor
Powder core	POCO NPS-125		NPV-75
Initial self-inductance	360 uH @ 0 A		280 uH @ 0 A
Operating self-inductance	255 uH @ 40A		239 uH @ 40A
Winding turns	21 T	21+21 = 42 T	33+33 T
Inductor size	98x45x48 mm		80x44x93 mm
Weight of copper winding	0.3 kg	0.3 x 2 = 0.6 kg	0.5 kg
Weight of magnetic core	0.685 kg	0.685 x 2= 1.37 kg	0.848 kg
Total weight	0.985 kg	0.985 x 2= 1.97 kg	1.35 kg

Table 1 Comparison of non-coupled and inverse coupled inductors.

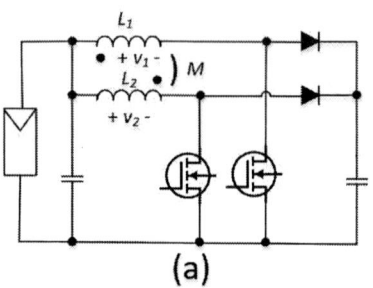

(a)

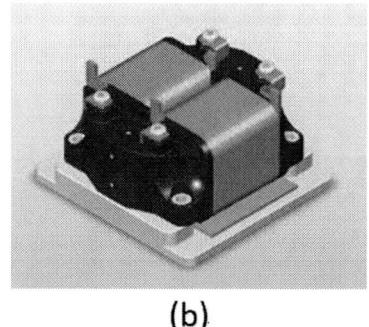

(b)

Fig. 4 inverse coupled inductor.

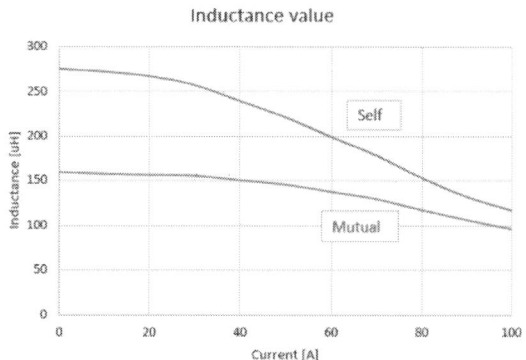

Fig. 5 Measured inductor values.

High inductor current ripple induces high core loss, winding copper loss and SiC MOSFET turn off loss. In this design, current ripple of the coupled inductor is designed to be similar to the non-coupled inductor. Fig. 6 shows a comparison of current ripple between non-coupled inductor (a) and coupled inductor (b). DCDC boost circuit parameters are shown in Table 2. Reduced inductor value as current increases, which is shown in Fig. 5, is applied at a lookup table function from PLECS simulation tool. For calculation based on (3) and (5), inductor values at rms value 40A are used. In this case, coupled inductor has slightly higher current ripple. Divided (3) by (5), different duty cycles have some impact on ripple current ratio, which is shown in Fig. 7.

PV input dc voltage	840 V
Output dc voltage	1200 V
Switching frequency	32 kHz
Duty cycle	0.3
Calculated and simulation current ripple of coupled inductor	37.3 A 38.14 A
Calculated and simulation current ripple of non coupled inductor	34.2 A 32.48 A

Table 2 Circuit simulation parameters.

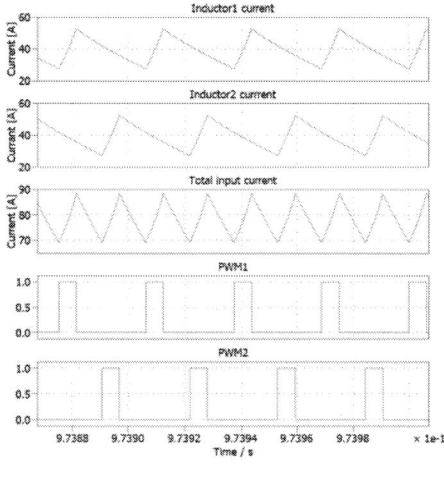

(a)

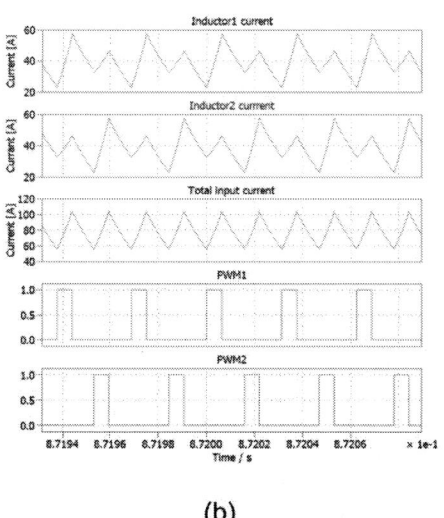

(b)

Fig. 6 Simulation current waveforms of (a) non-coupled inductor and (b) coupled inductor.

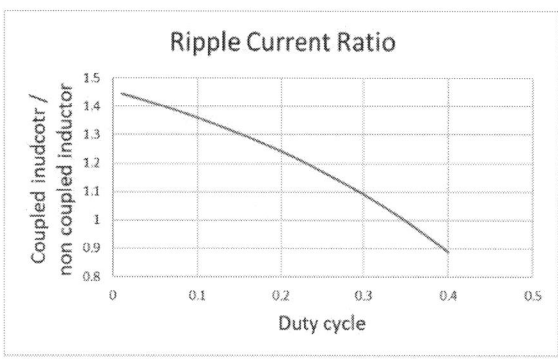

Fig. 7 Ripple current comparison of coupled inductor and non-coupled inductor.

3 Coupled Inductor Prototype

The inverse coupled inductor achieves similar current ripple while it shrinks the total weight of two single inductors from 1.97 kg to a single coupled inductor 1.35 kg as shown in Table 1. A prototype is manufactured by Magroots Technology Corporation (part number MIM251050-01M), as shown in Fig. 8. The oval-shaped magnetic core is made by two C-shaped cores and two I-shaped cores. There are four mounting holes for easy assembling on heatsink for better heat dissipation. 2kV SiC MOSFET switching dv/dt could achieve higher than 60 V/ns, so it is critical to design a winding with low parasitic capacitance. Flat copper wire is used to avoid overlapping high different voltage at adjacent two windings.

Fig. 8 Coupled inductor prototype.

Double pulse test (DPT) is widely used in dynamic switching evaluation [6]. It is used here for verifying dynamic self and mutual inductor at high voltage environment that is close to 1500V solar boost applications. Ten pulses are generated to calculate the inductor value through di/dt at different bias dc currents. The test schematic circuit is shown in Fig. 9.

Low side SiC MOSFET gate voltages Vgs (Ch1 and Ch2) and drain to source voltages Vds (Ch3 and Ch4) were measured by Tek passive voltage probe. Inductor currents of L1 and L2 were measured by PEM Rogowski coils (Ch5 and Ch6). Phase shift PWM signals of 180° were applied to two SiC MOSFET switches M1 and M2. Gate resistor value has a significant impact on switching speed. In this test, Rg = 6 Ohm is selected.

Turn on and off waveforms at 1200V dc bus voltage and 80A total inductor current is shown at Fig. 10 (a). During the period of M1 turn on while M2 turn off, coupled inductor L1 was energized at di/dt of 9.28 A/ns, and inductor L2 was de-energized at di/dt of 4.34A. The inductor sees doubled fsw which is 64 kHz here. At two times rated current of 160A, inductor values dropped as shown in Fig. 10 (b), which caused di/dt increase. The di/dt values from DPT validated the design of inductor values at Fig. 5.

SiC MOSFET turn off dv/dt is as high as 68 V/ns and turn on dv/dt is as high as 75 V/ns, which induces inductor current ringing at frequency of 60 Mhz. This makes EMI design more challenging than slow three-level IGBT solutions. One mitigation is increasing Rg to slowdown the dv/dt but it comes with a penalty of higher switching loss Eon and Eoff. Another mitigation is adding a damping resistor across the winding but it also reduces the system efficiency.

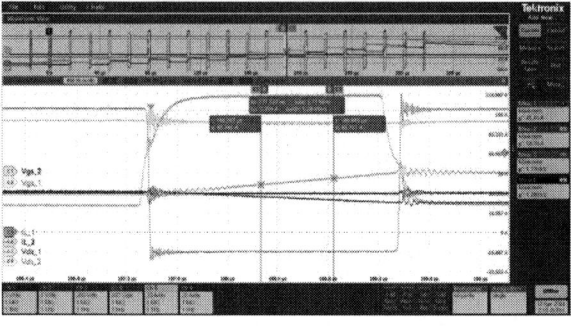

(a)

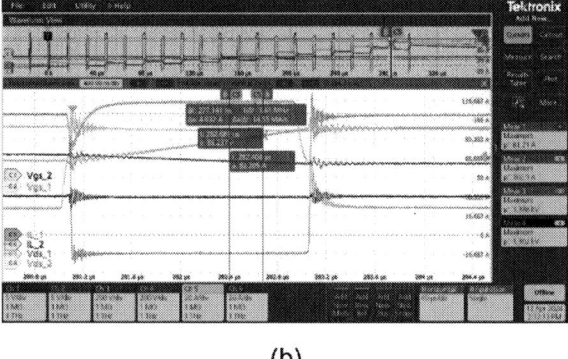

(b)

Fig. 10 DPT scope waveforms.

4 Summary and Conclusion

An inverse coupled inductor was designed for the latest 1500V solar inverter request of dcdc boost MPPT 80A input current. Compared with the traditional solution of two non-coupled inductors, it reduced the cost and total inductor weight from 1.97 kg to 1.35 kg (31% reduction). It also simplifies the assembling process and improves power density. A prototype was built, and experimental results validated its self and mutual inductor values.

References

[1] A. B. Mirza, A. I. Emon, S. S. Vala and F. Luo, "An E-Core Based Integrated Coupled Inductor for Interleaved Boost Converter," in IEEE Transactions on Industry Applications, vol. 59, no. 4, pp. 4199-4214, July-Aug. 2023

[2] Y. Liu, "A Full SiC MOSFET DCDC Boost Power Module Using 2 kV SiC MOSFET for 1500V String Solar Inverter Applications," 2024 IEEE Applied Power Electronics

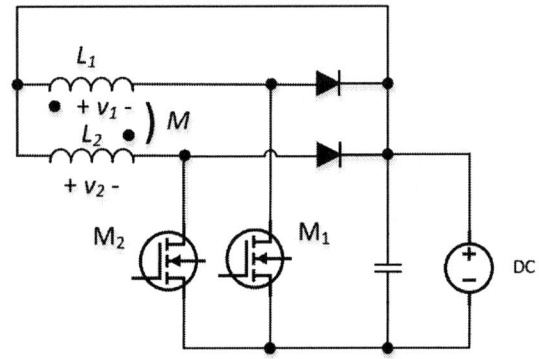

Fig. 9 DPT circuit for testing coupled inductor.

Conference and Exposition (APEC), Long Beach, CA, USA, 2024

[3] S. Lu, M. Mu, Y. Jiao, F. C. Lee and Z. Zhao, "Coupled Inductors in Interleaved Multiphase Three-Level DC–DC Converter for High-Power Applications," in IEEE Transactions on Power Electronics, vol. 31, no. 1, pp. 120-134, Jan. 2016, doi: 10.1109/TPEL.2015.2398572.

[4] R. Qin and F. C. Lee, "Modeling and design for integrated coupled inductors in interleaved three-level DC/DC converters," 2017 IEEE Energy Conversion Congress and Exposition (ECCE), Cincinnati, OH, USA, 2017, pp. 498-503, doi: 10.1109/ECCE.2017.8095824.

[5] Y. Liu, H. A. Mantooth, J. Carlos Balda and C. Farnell, "A Variable Inductor Based LCL Filter for Large-Scale Microgrid Application," in IEEE Transactions on Power Electronics, vol. 33, no. 9, pp. 7338-7348, Sept. 2018, doi: 10.1109/TPEL.2017.2764483.

[6] Z. Zhang, B. Guo, F. F. Wang, E. A. Jones, L. M. Tolbert and B. J. Blalock, "Methodology for Wide Band-Gap Device Dynamic Characterization," in IEEE Transactions on Power Electronics, vol. 32, no. 12, pp. 9307-9318, Dec. 2017.

PCIM Europe 2024, 11– 13 June 2024, Nuremberg DOI: 10.30420/566262451

Environmental Impact of Modular Power Electronics Systems Considering Diagnostic-Driven Unit Replacement.

Briac Baudais[1,2], Hamid Ben Ahmed[1], Gurvan Jodin[1], Nicolas Degrenne[2], Stéphane Lefebvre[3]

[1] SATIE

[2] Mitsubishi Electric R&D Centre Europe

[3] SATIE

Corresponding author:	Briac Baudais, briac.baudais@ens-rennes.fr
Speaker:	Briac Baudais, briac.baudais@ens-rennes.fr

Abstract

The combination of modularity and diagnosability is a natural candidate for mitigating environmental impacts (EI) throughout the lifecycle. This article proposes a method for quantifying Life Cycle Assessment (LCA), tailored to a power electronics product composed of power devices and gate drivers. It considers the effects of both modularity and diagnosability on the repair with replacement units (RU).

The results highlight that modularity and diagnosability are not relevant when several RU are subjected to failure almost concurrently (typically wearout failures of similar components), and need to be replaced during the intented lifespan. However, their significance is notable for faults qualified as "random" or "early", as these faults do not necessarily manifest throughout several RU. In certain situation involving this type of faults, system modularity and diagnostic accuracy can prove beneficial when implemented together, leading to a reduction in total EI over the service life, especially when considering mineral and metal ressource depletion.

1 Introduction

The goal of considering modularity and diagnostic techniques is to allow a more precise system reparation in case of failure. Indeed, modularity allows for the replacement of only the faulty parts, and diagnostics identify which modules to replace. Since the parts that remain operational after a failure are retained, there is no need to remanufacture the entire system, thus limiting the impacts due to manufacturing.

Modularity, considered a pillar of the circular economy [1], is increasingly favored to enhance the sustainability, repairability, and recyclability of products. The question then is to quantify the environmental gain from the addition of modularity and diagnosability. The evolution of EI over time for a PE system can be represented as in Figure 1. Initially, there are EI from the product's manufacturing. Then, the slope represents the EI during usage, i.e., losses during operation. Furthermore, in the event of a fault, EI increases, corresponding to the replacement of the faulty part. Diagnostics and architecture

(modularity) influence this jump. The more precise the diagnostics, the more the replacement is limited to the component affected by the fault, provided there is modularity allowing for it. An integrated architecture does not allow for separating the faulty component from the rest.

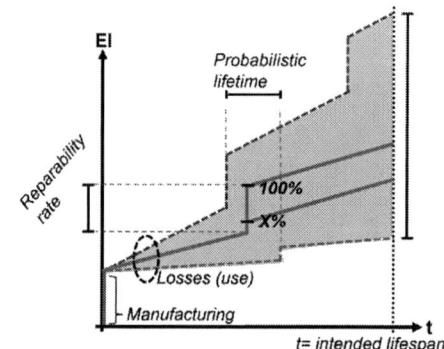

Fig. 1: Staircase curve.

This paper first presents the consideration of modularity and diagnosability in PE, then a generic method for quantifying EI. Finally, it proposes a case study, using the method, to highlight the relevance of these two characteristics in determining EI over the lifecycle.

2 Modularity and diagnostic in power electronics

2.1 Modularity

The development of PE systems has been marked by complex evolutions aimed at optimizing various parameters such as cost, efficiency, and robustness. This optimization leads to different product structures. Currently, in EV applications, integrated power modules, comprising all the power semiconductor switches of the inverter predominate [2], offering high efficiency and optimal power conversion. However, this integration can result in a total system loss if a part fails, due to the inability to separate the faulty part from the rest, which can impact EI over the lifecycle.

Modularity involves reducing integration (bringing together, in the same volume and on the same substrate, all the constituent elements of the converter [3]). High integration makes repair more difficult in case of failure. The power module solution is highly integrated, as all 6 chips are located on the same substrate, the DCB. There are more modular solutions achieved by playing on integration, such as using discrete components, which only contain a single chip. Figure 2 shows different types of integration at the conversion cell level for electric vehicle inverters.

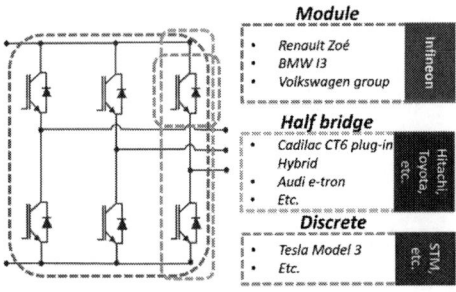

Fig. 2: Different design choices for inverter integration of electric vehicles [2].

Other modular structures exist, achieved by playing on design [4], which either allow the separation of functionalities into blocks or a matrix view of the converter.

2.2 Diagnostic

Diagnosability is of crucial importance, coupled with modularity to ensure specific fault repairs. Before exploring the various diagnostic techniques available, let us examine a common protection measure implemented in converters, as it can guide the diagnostic process.

Desaturation protection (desat) is frequently integrated to detect short circuits [5], a critical fault for power components. This mechanism reacts quickly by cutting off the control of a component when a desaturation threshold voltage is reached. This protection provides useful information for diagnosis in case of triggering during a fault.

Regarding diagnosis, attention must be paid to techniques for monitoring the lifespan of active components in PE. Various methodologies have emerged over the years [6]. Among them, monitoring temperature-sensitive electrical parameters (TSEP) proves to be interesting, and among TSEP, the gate resistance (R_{gin}) is particularly relevant [7]. Estimating the junction temperature from R_{gin} involves a calibration phase to determine the function linking the read TSEP and temperature, followed by the application of direct current (DC) to the system's gate and measuring the voltage between the gate and the emitter to estimate the temperature using the calibration function. The equivalent circuit is generally modeled by an RC circuit, as illustrated in Figure 3. A related method, called "Tjbalancing,"

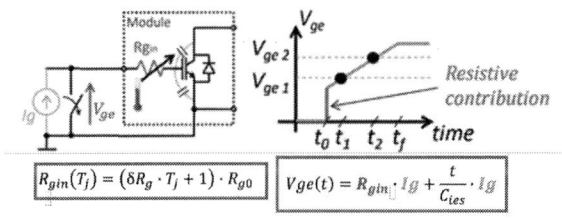

Fig. 3: Equivalent circuit with TSEP Rgin (left) and signal shape during Ig current injection (right) [7].

stands out for its use in systems with multiple chips in parallel [8]. This technique provides real-time estimation of the junction temperature for each chip using the TSEP R_{gin}. It then thermally balances the chips by individually modifying the commands of the components, thus mitigating temperature disparities between them.

Monitoring temperature variations resulting from potential failures can enable precise localization of issues and contribute to the proactive maintenance of power electronic systems.

3 Methodology

This section presents the general methodology for determining the EI of a PE product, taking into account replacement due to system faults, the use of diagnostic technology for more or less precise fault detection, and modularity. Figure 4 illustrates the methodology for creating the staircase curve figure 1. For manufacturing, an inventory of material flows and processes is conducted. For usage, the EI concern the energy lost during operation, and a typical operating cycle (e.g. WLTP for automotive application) with a loss model is used [9]. The EI related to manufacturing and usage are not further detailed in the document. The replacement modeling includes fault generation and diagnostic processes that allow for defining replacement scenarios.

The replacement modeling implies the concept of the Replacement Unit (RU), representing the smallest replaceable component in the system. It is crucial to clearly define this RU, as this definition guides the methodology. For example, in the case of a three-phase inverter built with discrete components such as TO247, the RU could be the discrete component itself. However, the same inverter built using an integrated power module would have the entire module as the RU.

3.1 Diagnostic : Replacement matrix

The objective of this section is to create replacement scenarios, i.e., for each fault, what replacement choice the diagnostic allows. To achieve this, we generate a matrix called the "Replacement Matrix (RM)." This matrix is an input to the algorithm shown in Figure 4.

To develop this matrix, it is necessary to model the observation of the diagnostic for each type of fault. Depending on the observation, the diagnostic can deduce the location of the failure, and the deduced part is then replaced, thus forming a replacement scenario. It is essential to understand the consequences of faults, as well as the perspective of the diagnostic. A fault tree can be constructed to link the fault, the effects, the observations of the effects, and the replacement scenario. If the same observation is made for failures of several/all RU, then the several/all RU need replacement. If an observed effect is linked to a specific RU or group of RU, then only the RU or the group is replaced. The replacement matrix $[RM]$ specific to a diagnostic technique (l) can then be constructed as shown in Figure 5. RU_i represents replacement unit i ranging from 1

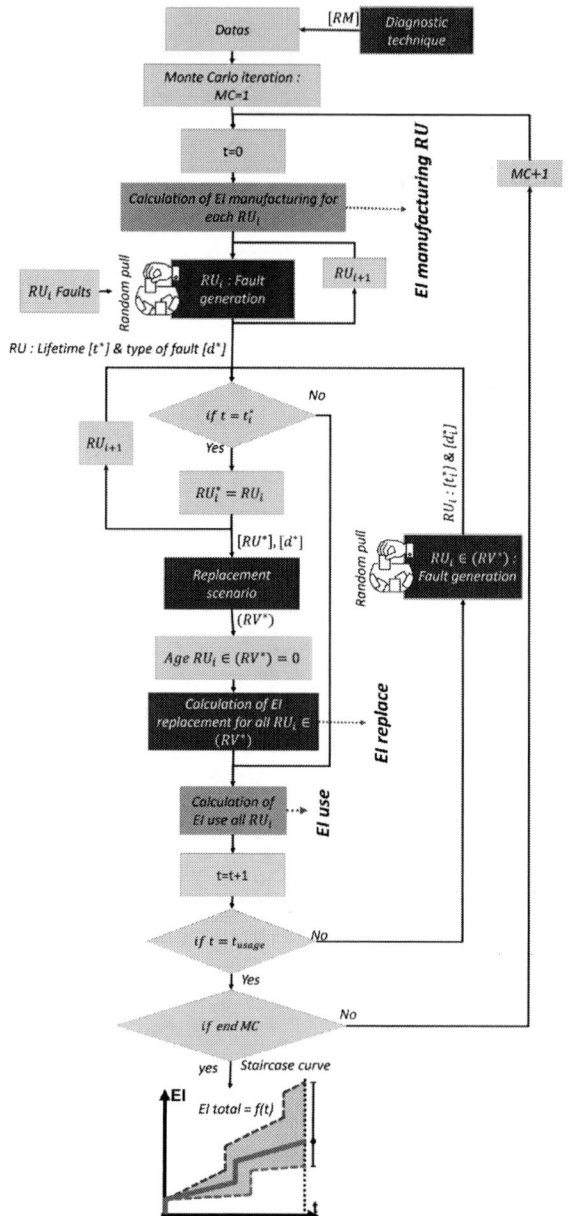

Fig. 4: Product life modelling with replacement and diagnostic.

to m, and $d_{k,i}$ represents fault k of replacement unit i, with a total number of faults of $n * m$ where n is the number of listed faults. The columns of the matrix are denoted by i and the rows by $j = k, i$. The replacement matrix is filled with 0 and 1, where 0 means no change is made, and 1 indicates a replacement. For example, in the matrix 5, if fault $d11$ of type 1 occurs on $RU1$, only $RU1$ is replaced, indicating high diagnostic accuracy for this specific fault. However, if fault $d21$ of type 2 occurs on $RU1$,

$$[RM]_{(l)} = \begin{bmatrix} \overbrace{}^{i} \\ \begin{array}{cccc} RU_1 & RU_2 & & RU_m \\ 1 & 0 & \cdots & 0 \\ 1 & 1 & \ddots & 0 \\ \vdots & \ddots & \ddots & \vdots \\ 1 & 1 & \ddots & 1 \\ 0 & 1 & \ddots & 0 \\ \vdots & \ddots & \ddots & \vdots \\ \cdots & \cdots & \cdots & \cdots \end{array} \end{bmatrix} \begin{array}{l} d_{1,1} \\ d_{2,1} \\ \\ d_{n,1} \\ d_{1,2} \\ \\ d_{n,m} \end{array} \quad k,i=j$$

Fig. 5: Replacement matrix generated from the fault tree, specific to a diagnostic type (l).

the diagnostic is less precise, and both $RU1$ and $RU2$ are replaced.

Each diagnostic technique has its own replacement matrix; they do not all have the same precision, some may better target the fault and result in fewer replacements for repair.

3.2 Fault Generation

First, an exhaustive list of all potential fault types for each RU needs to be established, along with determining the probability of the fault occurring over time. For example, a fault for a power chip could be a short circuit. The vector d of size $n*m$ with n as the total number of faults and m as the number of RU can be created as follows:

$$(\mathbf{d}) = \begin{pmatrix} d_{k,i} \\ \vdots \\ d_{n,m} \end{pmatrix} \tag{1}$$

With k as the fault number and i as the RU number. Mathematically, faults can be modeled based on statistical laws. The Weibull function is the one selected (with two parameters σ and β), equation 2, as it allows for reproducing the "bathtub curve". This curve characterizes the different failure phases during the product's lifetime (LT) of a PE product [10], encompassing the "infant mortality" phase $\beta < 1$ (related to insufficiently controlled design or manufacturing issues), the "useful life" period $\beta = 1$ (where failures occur independently of the component's age), and the "aging" stage $\beta > 1$ (resulting from wearout). The distribution function of each fault k is written in the form:

$$f_k(t) = 1 - e^{-(\frac{t}{\sigma})^{\beta}} \tag{2}$$

The algorithm for determining the fault type is presented in Figure 7, it consists of two key steps: fault generation and fault type selection.

Firstly, it is necessary to find the overall distribution function of the RU. It is necessary to associate faults within a RU, several types of associations exist [11]. Here, the series association is selected, meaning all subsystems must function for the overall system to be operational. In other words, there should be no faults, represented by the distribution function of an RU equation 3.

$$F_i(t) = 1 - \prod_{k=1}^{n} (1 - f_{k,i}(t)) \tag{3}$$

With $f_{k,i}$ being the distribution function of fault k of RU i and n the total number of faults for the RU. A random number compared to the RU's distribution function determines the fault time (t_i^*). A second random number is used to select the fault type $(d_{k,i}^* = d_j^*)$, based on the probabilities associated with each type at the time of the fault. This algorithm allows for each RU, both to determine the time at which a fault occurs (t_i^*) and also, what type of fault occurs (d_i^*). This enables the generation of two vectors t^* and d^* of size m representing respectively the occurrence times and the fault types for each RU.

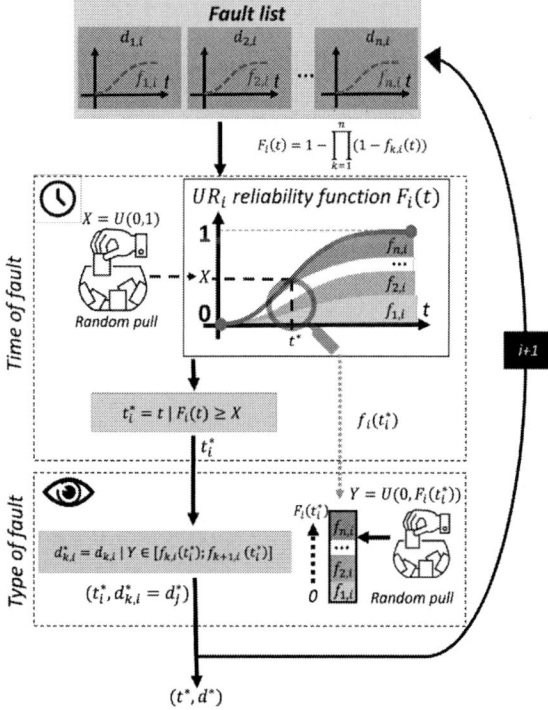

Fig. 6: Algorithm for generating a replacement unit fault.

3.3 Replace

The previous sections describe, on one hand, the creation of the replacement matrix based on faults and diagnostics for the input data, and on the other hand, the fault generation.

To calculate the EI related to replacement, it remains to link the two. For this purpose, depending on the fault type at a given time t, we will generate a replacement vector formed by a logical "OR" operation of all faults generated from the replacement matrix at time t^*, to avoid counting the replacement of a RU twice. This is expressed mathematically as follows:

$$RM^*(j,:) = \begin{cases} RM(j,:) & \text{if } d^*(i^*) = d(j) \\ 0 & \text{otherwise} \end{cases} \quad (4)$$

Where i^* represents the RU_i in fault, d^* the faults at t^* and d the vector of all type of faults.

$$(RV^*) = \bigvee_{j=1}^{m*n} RM^*(j,:) \quad (5)$$

Where RV^* represents the replacement vector when a fault arrives at t^*, $\bigvee$ the logical OR operator, and $RM^*(j,:)$ is the j-th line of RM^*.

Let's take an example: at time $t^* = t$, $RU1$ undergoes $d21$, and $RU2$ undergoes fault $d12$. The corresponding lines of the replacement matrix are as follows:

$$[RM^*] = \begin{matrix} & \scriptstyle RU_1 & \scriptstyle RU_2 & & \scriptstyle RU_m & \\ \begin{bmatrix} 0 & 0 & \cdots & 0 \\ 1 & 1 & \ddots & 0 \\ \vdots & \ddots & \ddots & \vdots \\ 0 & 0 & \ddots & 0 \\ 0 & 1 & \ddots & 0 \\ \vdots & \ddots & \ddots & \vdots \\ 0 & 0 & \cdots & 0 \end{bmatrix} & \begin{matrix} \scriptstyle d_{1,1} \\ \scriptstyle d_{2,1} \\ \\ \scriptstyle d_{n,1} \\ \scriptstyle d_{1,2} \\ \\ \scriptstyle d_{n,m} \end{matrix} \end{matrix}$$

Fig. 7: Example with $RU1$ undergoing $d21$ and $RU2$ undergoing $d12$ at time t.

The diagnosis indicates the need to replace $RU1$ and $RU2$. Using a logical OR, we obtain the following replacement vector:

$$(RV^*) = (1\ 1\ \dots\ 0) \quad (6)$$

Then, simply perform the matrix calculation of the replacement vector with the matrix of EI for the manufacturing of each component:

$$EI_{\text{replace}}(t) = (RV^*) \cdot (EI_{mf}) \quad (7)$$

$$(EI_{mf}) = \begin{pmatrix} EI_{mf\ \text{RU1}} \\ EI_{mf\ \text{RU2}} \\ \dots \\ EI_{mf\ \text{RUm}} \end{pmatrix} \quad (8)$$

With EI_{mf} representing the environmental impacts during manufacturing.

In conclusion, the developed methodology provides an approach to evaluate the EI related to the replacement of a PE system subdivided into RU. Through fault generation, determination of fault type, modeling of replacement scenarios via a diagnostic-dependent replacement matrix, and finally the calculation of EI associated with these replacements.

4 Case studies

To illustrate the method and quantify the environmental benefits of implementing modularity and diagnosability, a case study is developed here.

4.1 Systems studied: modularity and diagnostics

The analysis compare two topologies comprisingan IGBT conversion cell and its gate control board (driver). One is the "integrated" topology, illustrated in Figure 8, and the other is the "discrete" topology, represented in Figure 9 (a and b), which represents our modular cases.

The **integrated** topology is characterized by 2 RU, an IGBT power module consisting of six IGBT chips and six diodes, along with a control board. The power module is the FS820R08A6P2B (750V, 820A) with its driver, as presented in reference [9]. For the discrete topology, 2 proposals are studied:

Discrete (1): The simplest simplification involves virtualy splitting the integrated topology into 6 parts, where the total EI for manufacturing remains identical to those of the integrated one. However, there are now 12 replacement units (RU), including 6 components and 6 drivers, sharing these impacts.

Discrete (2): Represents a more realistic modeling of a modular system compatible with current rating of discrete component. It consists of 48 RU, including 24 discrete components and 24 drivers. The chips are paralleled four by four, each with its driver. The IGBTs are packaged in TO247, similar to the inverter of the Tesla Model S. The IGBT chip and the diode are in the same package, with a rating of 205A and a nominal voltage of 750V. The EI for the manufacturing of the IGBTs are deduced from the specific LCA of the TO247, which is not developed in the article. As for the drivers, given the lack of examples of discrete drivers, an assumption is made regarding their form, representing them in the form of cells arranged on the same printed circuit board, maintained at a sufficient distance to allow access without damaging neighboring cells in case of repair of the faulty component. This assumption is inspired by the proposal of standard conversion cells presented in reference [12], allowing individual repairs of the cells. The total EI for manufacturing of the drivers are considered equivalent to those of the integrated driver. Each driver RU evenly shares the total EI for manufacturing.

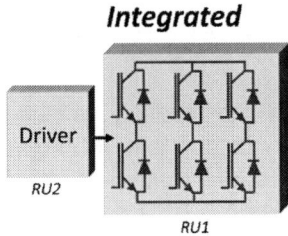

Fig. 8: Integrated topology.

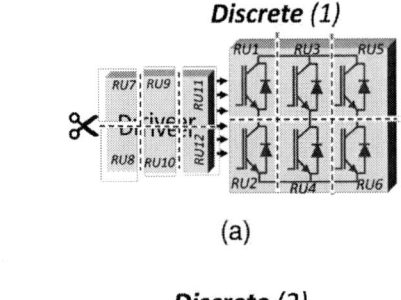

(a)

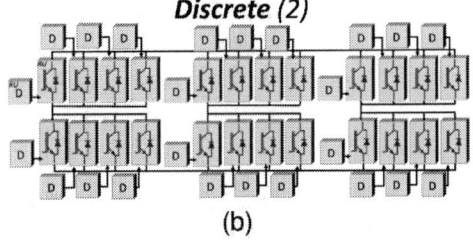

(b)

Fig. 9: Discrete topologies (1) and (2) of the case study.

4.2 Default type and probability

Based on a literature review [13], [14], a list of failures has been established for IGBT chips and drivers, with information regarding the stages of infancy, useful life, and end of life, all presented in Table 1.

Current literature presents limited availability of data regarding the parameters σ and β of the Weibull functions used to model the faults.

The three functions for infancy (Early-E), useful life (Random - R), and aging (Wearout - W) are modeled separately, and the values of the three functions are shown in Figure 10.

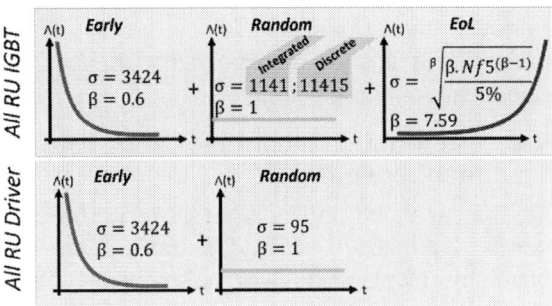

Fig. 10: Selected default probability functions.

The parameters for infancy-related faults are obtained from a reference at Semikron [15]. The values correspond to the entire PE system, with $\beta = 0.6$ and $\sigma = 3424$ years. In our case, these represent the values of the overall reliability function of the (IGBT+driver) for infancy-related faults.

The faults related to the useful life period all have $\beta = 1$. For the useful life faults of the integrated power module, the parameters used are derived from references [16]–[18]. σ is determined from the FIT (1 FIT corresponds to 1 failure per 10^9 hours). The total FIT of the power module (=100, so $\sigma = 1141$ years) appears to be 10 times higher than that of a discrete component, which is $\sigma = 11415$ years. This is logical, as the discrete IGBT RU consists of only one chip, hence less chance of experiencing failures.

For driver faults, the total number of FITs was set at 1200 [19], which corresponds to 95 years. The FIT of the integrated driver or the discrete driver is considered to be the same.

The parameters for end-of-life faults only concern the IGBT RU, with the modeling of system degradation, both for the IGBT chip and the diode. The values are determined from the chip lifetimes [20]. Firstly, input data are defined, including product usage (operating cycle) and component physics (electrical parameters, thermal parameters, and its lifetime model). Then, the input data are implemented into a loss model. The losses are used to calculate the junction temperature rise using a thermal model. A counting algorithm determines the number of temperature cycles undergone by the component, which are then compared to the lifetime model. This provides the maximum cycle number until failure denoted as $Nf5$, with the 5 indicating the failure rate to create the empirical model, i.e., 5% of the failed sample.

Using the functions for each RU, equation 3 allows retrieving the values of individual faults. Table 1 presents the different values obtained for each fault depending on the topology.

4.3 Diagnostic technique and replacement scenario

For the case study, both diagnostic techniques presented are utilized, namely the basic **desat** technique and **Tjbalancing** (which also implements desat technique).

Based on the selected techniques and as explained in the method, it is essential to model the diagnostic observation for each fault, which is not detailed in the article. Table 1 presents the replacement scenarios according to the type of fault and the diagnostic technique used. They are equivalent to the replacement matrix in 5.

4.4 Results

Figure 11 illustrates the evolution of the Mineral and Metal Resource Depletion (MRD) environmental impact over time for integrated and discrete topologies, taking into account fault generation and diagnosis (according to the probabilistic method presented earlier). MRD is presented because it is the impact most affected by manufacturing and replacement [9]. Additionally, the curves represent the median values of the 1000 Monte-Carlo simulations.

We can observe that diagnosis and modularity have little impact on the curves between integrated and discrete, as the entire system is replaced at 30 years. This is due to the type of fault, primarily related to wearout. Thus, the curves corresponding to integrated (desat and Tjbalancing) and discrete (1) (desat and Tjbalancing) have the same EI at 30 years, as both systems are entirely replaced once. The discrete solution (2) is the most polluting (9% higher than integrated), but this is solely explained by the manufacturing EI, which is more costly (26%).

Adding diagnostic functions is not always a solution to reduce EI. For this to be the case, the entire system should not be systematically replaced. Even if

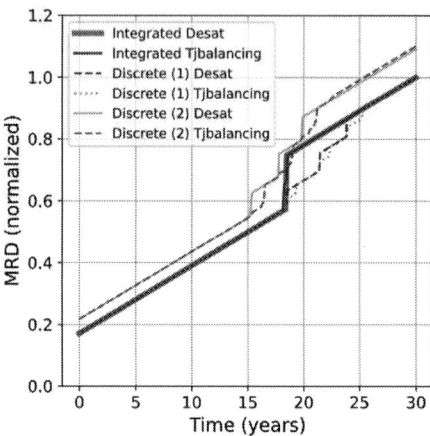

Fig. 11: Impact of diagnostics and modularity on MRD.

diagnosis allows for locating the fault and replacing only the faulty component, it brings no benefit in terms of EI if other components are also likely to fail shortly after.

This is why diagnosis is particularly relevant for faults related to the youth or useful life of the system, while faults related to wearout do not present significant interest, as the failure of one chip will likely be followed by failures of other chips in the system.

The data used for modeling faults were probably not representative of the actual probability of occurrence of faults, mainly due to a lack of in-depth studies in this area. Therefore, a definitive conclusion on the real interest of diagnosis and modularity cannot be formulated at this stage.

Having no precise vision of whether a fault will occur or not, we will now focus on the impacts of diagnosis and modularity depending on the type of fault.

According to the faults studied and the RU, based on Table 1, there are 4 levels of Fault Location Identification (FLI):

- Without FLI (e): The fault cannot be localized, resulting in the replacement of the entire system.

- Low FLI (d): Fault is localized at the level of one power arm with the associated drivers for that arm.

- Moderate FLI (c): Fault is localized at the level of one power switch with its driver.

- Strong FLI precisely localized at the level of

Tab. 1: Type and probability of fault for each RU in the case study, with replacement scenario.

RUi	Fault	Life period	σ (years) Integrated	Discret (1)	Discret (2)	β	Desat	Tjbalancing
IGBT	DC bus disconnection	Early	50 060	991 750	9 996 220	0.6	d	c
	Load disconnection	Early	50 060	991 750	9 996 220	0.6	d	c
	Bad thermal interface	Early	50 060	991 750	9 996 220	0.6	d	b
	Die SC IGBT	Random	4 564	45 660	45 660	1	d	d
	Die OC IGBT	Random	4 564	45 660	45 660	1	d	b
	Die SC diode	Random	4 564	45 660	45 660	1	d	d
	Die OC diode	Random	4 564	45 660	45 660	1	d	d
	IGBT degradation of the wire-bonds/ Degradation of the die-attach	End of life	22.8	28.9	28.9	7.59	d	b
	Diode degradation of the wire-bonds/ Degradation of the die-attach	End of life	20.4	25.8	25.8	7.59	d	d
Driver	Control board disconnection	Early	50 060	991 750	9 996 220	0.6	e	c
	Grid disconnection	Early	50 060	991 750	9 996 220	0.6	d	a
	Fault in high state (commands IGBT to close)	Random	190	570	2280	1	d	a
	Fault in Low state(commands IGBT to open)	Random	190	570	2280	1	e	c

Legend : Replace specific Driver (Strong FLI - a), Replace specific IGBT (Strong FLI - b), Replace specific IGBT and its Driver (Moderate FLI - c), Replace all IGBTs and Drivers leg (Low FLI - d), Replace all (Without FLI - e)

the power component alone or at the level of the driver alone.

The impact of these FLI levels on replacement is directly related to modularity. For example, in the case of moderate-level FLI, which corresponds to the following information: when a fault occurs, it may involve both the power component and its driver. In this scenario, for discrete topology (1), the specific IGBT and its driver are replaced, while for integrated topology, the entire system is changed because the RU do not allow for separation, even if information about the fault location is obtained. Furthermore, the Tjbalancing diagnostic technique allows for precise RU selection for certain faults, unlike the desat technique, which stops at the power arm level, thus offering less fine resolution.

Knowing this, the cost of replacement for different FLI levels and different topologies is determined, as shown in Figure 12. The EI at replacement of the integrated case without FLI is taken as a reference. Comparing the EI of the fault without FLI simply shows the difference in the system's manufacturing cost, as everything is replaced here. With the desat technique, when a fault with low FLI occurs, modularity reduces the EI at replacement compared to the integrated case by 67% (discrete(1)), 36% (discrete(2)). With the implementation of the Tjbalancing technique, the positive impact of modularity further increases. For faults with moderate FLI, there is a reduction of 83% (discrete(1)), 91% (discrete(2)). For faults with strong FLI, a reduction of 90% (discrete(1)), 92% (discrete(2)). In contrast, the integrated model achieves only a 37% reduction

at replacement for faults with strong FLI. So, having an accurate FLI benefits more to the most modular system, namely the discrete (2). Further studies

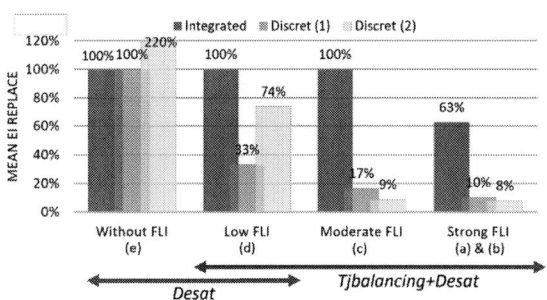

Fig. 12: EI at replacement between different modularities for different FLI, standardized EI compared to replacement of integrated without FLI.

are needed to determine the actual occurrence of faults and thus assess their diagnostic level, in order to conclude on the real utility of diagnostics and modularity. If, in reality, the most common faults are related to wearout, then diagnostics do not present a significant advantage. However, if the most frequent faults are those that can be located accurately and occur when the other parts of the system have high value (i.e. reliability), then it is worth to replace the single failed RU and modularity and diagnosability become relevant characteristics. To illustrate this point, Figure 13 highlights the impact of diagnostics and modularity on MRD over time. Three faults are simulated, occurring at 10, 15, and 25 years for each topology. Each fault corresponds to a certain FLI level; for example, the strong FLI level is represented by the failure of the thermal interface, leading to the replacement of the specific IGBT RU.

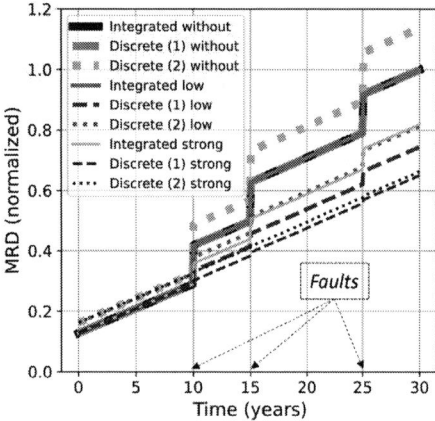

Fig. 13: Impact of diagnostics and modularity on MRD, in the proposed 3-defect scenario.

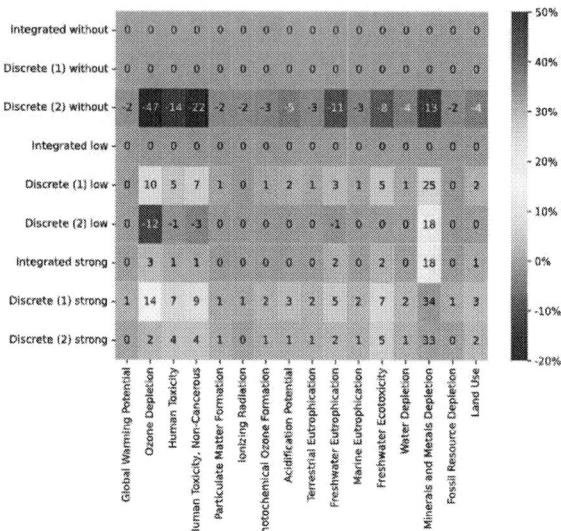

Fig. 14: Percentage gain (%) for each EI at 30 years of the different topologies with the different FLI levels compared with the integrated case with no FLI information.

As previously mentioned, without FLI, the difference is related to the manufacturing cost; modularity does not influence, with integrated and discrete (1) being at the same level, while discrete (2), with a higher manufacturing cost, increases MRD by 13% at 30 years. For a low diagnostic level, the total MRD reduction compared to the integrated case without diagnostics at 30 years is 25% and 18% respectively for discrete solutions (1) and (2). With a high diagnostic level, the total MRD reduction is 34% and 33%, but also 18% for the integrated solution with high diagnostics. Thus, two aspects emerge: the impact of diagnostics and that of modularity.

Figure 14 shows the same results at 30 years for all EI with environmental gain for different topologies with different FLI levels. The results are normalized relative to the integrated scenario without FLI information. When faults lead to strong FLI, all discrete topologies show a reduction in EI. However, for most EI, this reduction is low ($< 5\%$), mainly due to the significant contribution of the usage phase compared to manufacturing [9]. Also, discrete topology (2) leads to a significant increase in EI in the absence of FLI during a fault, with an increase of 47% for OD, 14% for HT, etc.

5 Conclusion

We have discussed some aspects of modularity and diagnosability in PE. Our approach involved a methodology for modeling EI integrating modularity and diagnosability. This method provides a modular view through RU, the smallest entities that can be replaced. It requires a thorough study of defects specific to each type of RU. The methodology implements a fault generation procedure and confronts them with diagnostic techniques to propose a replacement scenario for quantifying the EI.

An inverter composed only of power components and drivers served as a case study to implement this methodology, with one non-modular (integrated) topology and two modular topologies (discrete). The analysis of the results highlighted the influence of topology choice, as well as the presence or absence of diagnostics, on environmental indicators such as Mineral and Metal Resource Depletion (MRD).

Modularity and diagnostics have been found to be factors enabling EI reduction by targeting specific faulty components, although this reduction is not always guaranteed under all conditions. It is essential to note that the actual effectiveness of diagnostics and modularity depends on the frequency and type of faults observed, the easier to localize the faults, the more modularity will enable EI reduction. Also, we observed that, for wear-out faults, diagnostics may not bring significant improvement compared to an approach without diagnostics, as in all cases, the entire system is replaced. In an industrial context, this is relevant because converters are sized not to have wearout failures over the product's lifetime thanks to adequate cooling sizing for product usage.

The current case-study concentrates on the power semi-conductor devices and gate drivers only, which represent around 1/3rd of the MRD impact of a complete inverter [9]. Inverter-level modularity, to provide dismontability from other key components such as the casing and capacitors, may present additional advantages than can be adressed with the proposed metodology. Current literature presents limited availability of data regarding the σ and β parameters of Weibull functions used to model faults. Therefore, the different data used come from various sources and may not always precisely match the studied fault not the target application. Future research should thus focus on more comprehensive and application-specific data collection to assess the actual occurrence of faults and perform the life-cycle assessment of modularity and diagnosticability.

References

[1] M. Sonego, M. E. S. Echeveste, and H. Galvan Debarba, "The role of modularity in sustainable design: A systematic review," *Journal of Cleaner Production*, vol. 176, pp. 196–209, Mar. 2018. DOI: 10.1016/j.jclepro.2017.12.106.

[2] Y. Yang, L. Dorn-Gomba, R. Rodriguez, C. Mak, and A. Emadi, "Automotive Power Module Packaging: Current Status and Future Trends," *IEEE Access*, vol. 8, pp. 160 126–160 144, 2020. DOI: 10.1109/ACCESS.2020.3019775.

[3] C. Buttay, "Le Packaging en électronique de puissance," fr, 2015.

[4] T. T. Romano, T. Alix, Y. Lembeye, N. Perry, and J.-C. Crebier, "Towards circular power electronics in the perspective of modularity," *Procedia CIRP*, CIRP, vol. 116, pp. 588–593, Jan. 2023. DOI: 10.1016/j.procir.2023.02.099.

[5] S. Lefebvre and B. Multon, *MOSFET et IGBT : Circuits de commande*, fr, 2003.

[6] L. Foube, "Power Devices Health Condition Monitoring: A Review of Recent Papers," en, *PHM Society European Conference*, vol. 6, no. 1, pp. 15–15, Jun. 2021, Number: 1. DOI: 10.36001/phme.2021.v6i1.2808.

[7] J. Brandelero, J. Ewanchuk, and S. Mollov, "Online Virtual Junction Temperature Measurement via DC Gate Current Injection," in *CIPS 2018*, Mar. 2018, pp. 1–7.

[8] J. Brandelero, J. Ewanchuk, N. Degrenne, and S. Mollov, "Lifetime extension through Tj equalisation by use of intelligent gate driver with multi-chip power module," *Microelectronics Reliability*, vol. 88-90, pp. 428–432, Sep. 2018. DOI: 10.1016/j.microrel.2018.07.034.

[9] B. Baudais, H. Ben Ahmed, G. Jodin, N. Degrenne, and S. Lefebvre, "Life Cycle Assessment of a 150 kW Electronic Power Inverter," en, *Energies*, vol. 16, no. 5, p. 2192, Jan. 2023, Number: 5 Publisher: Multidisciplinary Digital Publishing Institute. DOI: 10.3390/en16052192.

[10] P. Blanquart and J.-c. Ronclin, *Fiabilité*, fr, 1989.

[11] T. T. L. Pham, "Contribution à l'étude de nouveaux convertisseurs sécurisés à tolérance de panne pour systèmes critiques à haute performance. Application à un PFC Double- Boost 5 Niveaux," fr, Ph.D. dissertation, INPT, Nov. 2011. DOI: 10/document.

[12] M. Rio, K. Khannoussi, J.-C. Crebier, and Y. Lembeye, "Addressing Circularity to Product Designers: Application to a Multi-Cell Power Electronics Converter," in *CIRP 2020*, ELSEVIER, May 2020. DOI: 10.1016/j.procir.2020.02.158.

[13] F. Richardeau and A. Gaillard, *Modes de défauts principaux et principes de sécurisation de l'onduleur de tension*, fr, 2017.

[14] A. Abuelnaga, M. Narimani, and A. Bahman, "A Review on IGBT Module Failure modes and Lifetime Testing," *IEEE Access*, vol. PP, pp. 1–1, Jan. 2021. DOI: 10.1109/ACCESS.2021.3049738.

[15] U. Scheuermann, *How to Define the Adequate Reliability Requirement for a Power Electronic System?* 2012.

[16] Infineon, *FIT-Rate Report FS820R08A6P2x*, 2022.

[17] I. Corporation, *RELIABILITY REPORT 2016 Power Semiconductor Devices*, 2016.

[18] Infineon, *FIT Failure Rate of AUIRGP4066D1-E*, 2022.

[19] J. Wylie, M. C. Merlin, and T. C. Green, "Analysis of the effects from constant random and wear-out failures of sub-modules within a modular multi-level converter with varying maintenance periods," in *EPE'17 ECCE Europe*, Sep. 2017, P.1–P.10. DOI: 10.23919/EPE17ECCEEurope.2017.8099246.

[20] N. Degrenne and S. Mollov, "Real-life vs. standard driving cycles and implications on EV power electronic reliability," in *IECON 2016*, Oct. 2016, pp. 2177–2182. DOI: 10.1109/IECON.2016.7793633.

Switching Performance Comparison of 3.3 kV SiC MOSFET and Si IGBT Power Modules for Railway Traction Systems

Zhuxuan Ma[1], Ahmed Ismail[1], Eric Allee[1], Ahmad Al-Hmoud[1], Feng Guo[1], Houqing Wang[1], Yue Zhao[1]

[1] University of Arkansas, USA

Corresponding author: Yue Zhao, yuezhao@uark.edu
Speaker: Yue Zhao, yuezhao@uark.edu

Abstract

This study provides an in-depth performance comparison between the latest 3.3 kV SiC MOSFET and Si IGBT power modules, particularly focusing on their switching characteristics. Encased in a standard 100 mm × 140 mm package, the SiC MOSFET module demonstrates superior switching speed, reduced losses, and outstanding high-temperature performance. Through rigorous experimental analysis, we offer substantial evidence advocating for SiC MOSFETs' superior suitability over Si IGBTs in railway traction systems. Both active and reactive power tests on a three-phase traction inverter substantiate SiC MOSFETs' effectiveness and efficiency at the system level.

1 Introduction

Silicon carbide (SiC) MOSFETs, renowned for their high-voltage and high-temperature resilience, have gained widespread acceptance in medium-voltage (MV) applications [1]-[10]. These devices' ability to handle high conduction and facilitate robust operations make them particularly advantageous for the railway industry, traditionally dominated by 1.5 kV and 3 kV DC buses. SiC MOSFETs offer a compelling alternative to Si IGBTs due to their lower switching losses, enhanced switching speed, and superior thermal conductivity [6]-[12]. Despite existing comparative analyses between SiC MOSFETs and Si IGBTs [13]-[20], the continual advancement in SiC technology and its specific application to railway systems necessitate a fresh examination. This study contrasts the latest SiC MOSFET [21] and Si IGBT modules [22], assessing their dynamic and static characteristics across various temperatures and employing a three-phase traction converter for a comprehensive system-level evaluation.

2 Characteristics Comparison

The packages of the latest 3.3 kV SiC MOSFET power module and a 3.3 kV Si IGBT module are depicted in Fig. 1. The dynamic characteristics are extracted for comparison purpose, through double-pulse test (DPT) at 1.8 kV under 25 °C and 150 °C with load current sweeping from 100 A to 800 A, representing the normal and harshest operating condition for the railway system, shown in Fig. 2.

The slew rate of dv/dt and di/dt for both power modules, shown in Fig. 3, represents the switching speed of these devices. The dv/dt and di/dt of the MOSFET are higher than IGBT under nearly all current levels with both low and high temperature conditions. What is more, with the current increases, all the slew rates of the MOSFET increases dramatically, except for the dv/dt_off under 150 °C condition. Meanwhile, the slew rates of the IGBT don't change much corresponding to different current under both temperature. Higher dv/dt of the MOSFET, illustrated in Fig. 3(a) and (c), results in a higher current spike for the current waveform during the turn-on transient, shown in Fig. 2. The current spike, however, also appears to the current waveform of the IGBT, which is caused by the large reverse recovery. As a result, the different dv/dt rates of IGBT and MOSFET don't result in

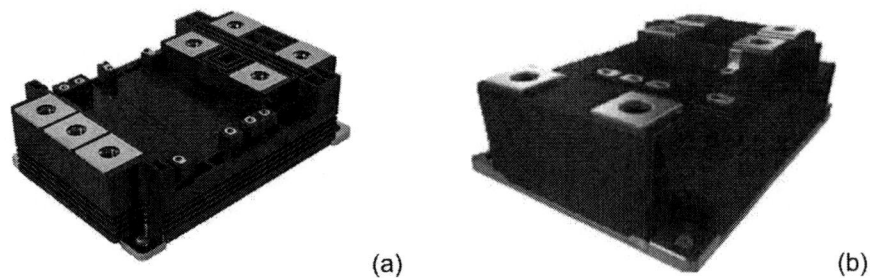

Fig. 1: The (a) 3.3 kV SiC MOSFET module and (b) 3.3kV Si IGBT module investigated in this work

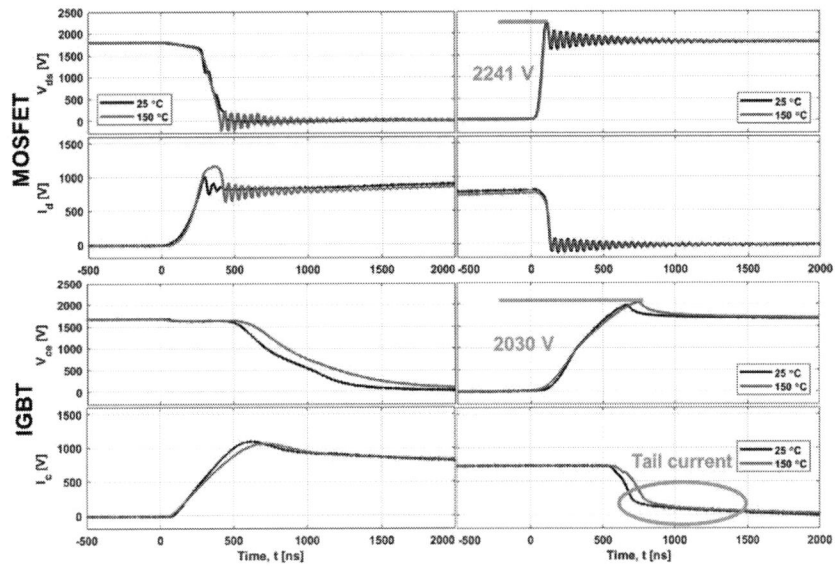

Fig. 2 DPT waveforms for MOSFET (top) and IGBT (bottom).

much different current spikes for these two devices. In comparison, the high di/dt rate of MOSFET, shown in Fig. 3(b) and (d), plays an important role on the high voltage overshoot of the SiC device during turn-off transient. As illustrated in Fig. 2, the voltage overshoot of the MOSFET under 1.8 kV, 800 A condition is around 2250 V at both temperatures, while the voltage overshoot of the IGBT device under the same conditions is lower than 2050 V. Though the voltage overshoot of MOSFET is much larger than that of the IGBT, the switching behavior of the MOSFET still doesn't influence the reliability of the traction inverter built by this type of device, since the voltage margin can become even larger with the 1.5 kV DC bus for the railway system.

The SiC MOSFET shows significantly lower switching losses than those of the IGBT at both temperatures under all current conditions as illustrated in Fig. 4. Also, the switching loss of the

IGBT under high temperature increases more significantly, compared to that of MOSFET, especially in heavy current condition. The tail current of the IGBT, shown in its turn-off waveform in Fig. 8, is more than 1 μs for both temperature, and this contributes a lot to its large turn off losses. The low switching speed and large reverse recovery current also make a huge difference on the high switching losses of the IGBT device.

As illustrated in Fig. 5, the on resistance of the MOSFET and IGBT corresponding to different current under high temperature and low temperature are measured using curve tracer. The on resistance of the SiC device is lower than that of the Si device in low current condition, while the situation becomes the opposite in the high current condition, under both temperature cases. As a result, the conduction loss of the converter using SiC device has lower conduction loss under

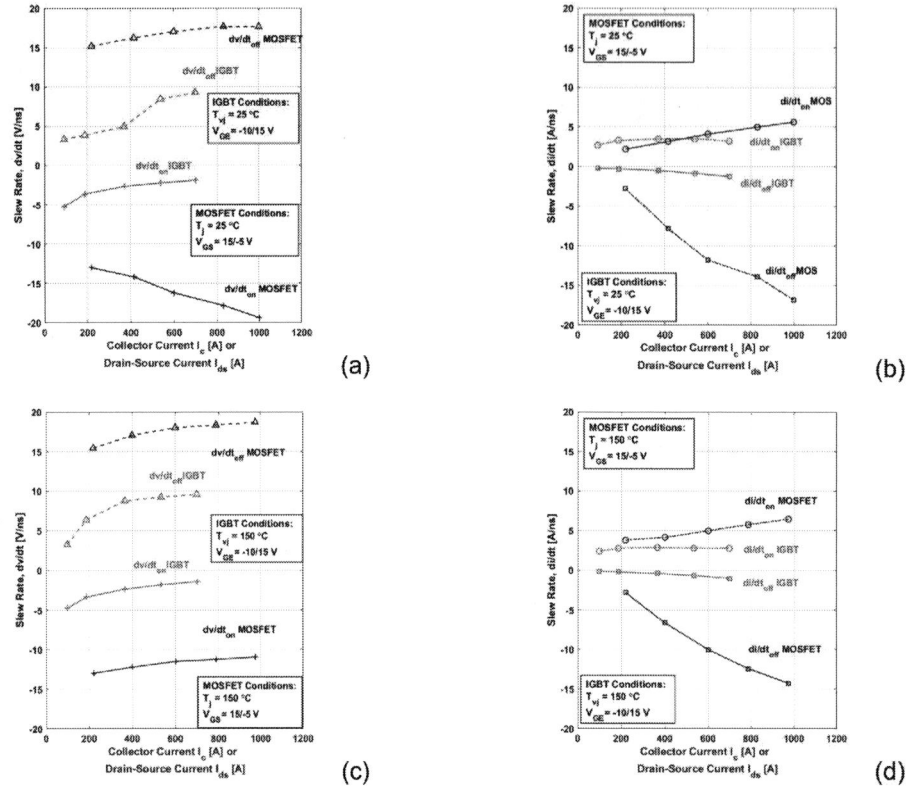

Fig. 3 Switching slew rates (dv/dt and di/dt) comparison at 25 °C (a), (b) and at 150 °C (c), (d).

light load condition than the one using Si device, while the conduction loss of the SiC converter has higher conduction loss than the Si one under heavy load.

3 Continuous Operating Test

The continuous operating experiment is conducted with the converter using both SiC MOSFET and Si IGBT. The circuit diagram of the three-phase two-level converter is shown in Fig. 6(a), and the fabricated three phase converter and the testing setup are shown in Fig.6(b), and (c).

To further compare the performance differences between the two devices, the waveforms of the converters based on these two modules are shown in Fig. 7. Because the Si-based converter cannot reach 177 kW with switching frequency being 4 kHz or higher, only the waveforms of 2 kHz operating frequency are illustrated. Though the waveforms of the SiC-based converter do not have much difference when it operates at the same frequency, the THD of the current

waveforms show a dramatic improvement with the 8 kHz condition. This can decrease the volume of the passive component and the loss of the whole system largely, crucial for railway application. This high-frequency high-power working ability of the SiC-based converter is enabled by the low switching loss of the switches. The high junction temperature caused by the switching loss of IGBT, on the other hand, prevents the further stretch of its high current conduction ability in a Si-based converter under high switching frequency operation.

Additionally, as the switching loss becomes the dominated part with high operating frequency, the converter based on SiC device has higher efficiency than the Si-based converter, shown in Fig. 8. This is because the lower switching losses of the MOSFET lead to a lower total loss of the converter, shown in Fig.4. However, due to the lower conduction loss of IGBT device, with high conduction current, the efficiencies of these two types of converter are still comparable with 1 kHz switching frequency at heavy load. As a result, the Si IGBT device can be a good economical

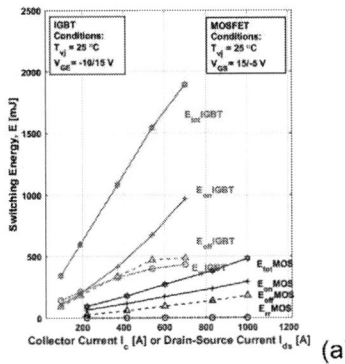 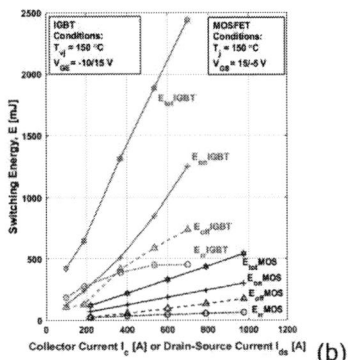

Fig. 4 Switching loss comparison at (a) 25 °C and (b) 150 °C.

Parameters	Values
Maximum active power	175 kW
DC voltage	1000 V
Switching frequency	1~8 kHz
Fundamental frequency	60 Hz
Line-line voltage	480 V_{rms}
AC filter	85 µH (IGBT)
AC filter	340 µH (MOSFET)
Load configuration	Δ
Load	3.95 Ω
Deadtime	3 µs (IGBT)
Deadtime	1.2 µs (MOSFET)
R_{th-ca}	8 mK/W
Coolant temperature	25 °C

Table 1 Operation parameters for the active power test using IGBT and MOSFET

alternative for the SiC MOSFET when the operating frequency is no more than 1 kHz.

In order to validate the operating ability of the SiC-based converter under rated DC bus voltage in real application, the reactive power tests are conducted with the experiment condition listed in Table 2. The test waveform of the converter under 8 kHz switching frequency and 1.2 µs operating conditions are illustrated in Fig. 9. The test results validate the operating ability of the converter built by MOSFET for railway application.

4 Conclusion

The comprehensive evaluation of 3.3 kV SiC MOSFET and Si IGBT modules in this study reveals the SiC MOSFET's distinct advantages for railway traction systems. The MOSFET's superior switching characteristics, lower losses, and robust performance under extreme conditions highlight its potential to revolutionize power conversion efficiency in railway applications. Furthermore, continuous operation tests confirm the MOSFET's superior performance, particularly at high switching frequencies, underscoring its viability as a superior alternative to traditional Si IGBTs. This analysis not only demonstrates the technological superiority of SiC MOSFETs but also establishes a compelling case for their broader adoption in advanced railway systems.

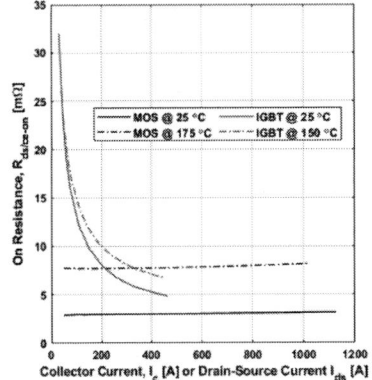

Fig. 5 Conduction loss comparison at different temperature

Parameters	Values
Reactive power	85 kVA
DC voltage	1800 V
Switching frequency	1~8 kHz
Fundamental frequency	60 Hz
Line-line voltage	1200 V_{rms}
Load configuration	Y
Load	10 mH and 127.7 µF in series
Deadtime	1.2 µs
$R_{th\text{-}ca}$	8 mK/W
Coolant temperature	25 °C

Table 2 Operation parameters for the reactive power test using MOSFET

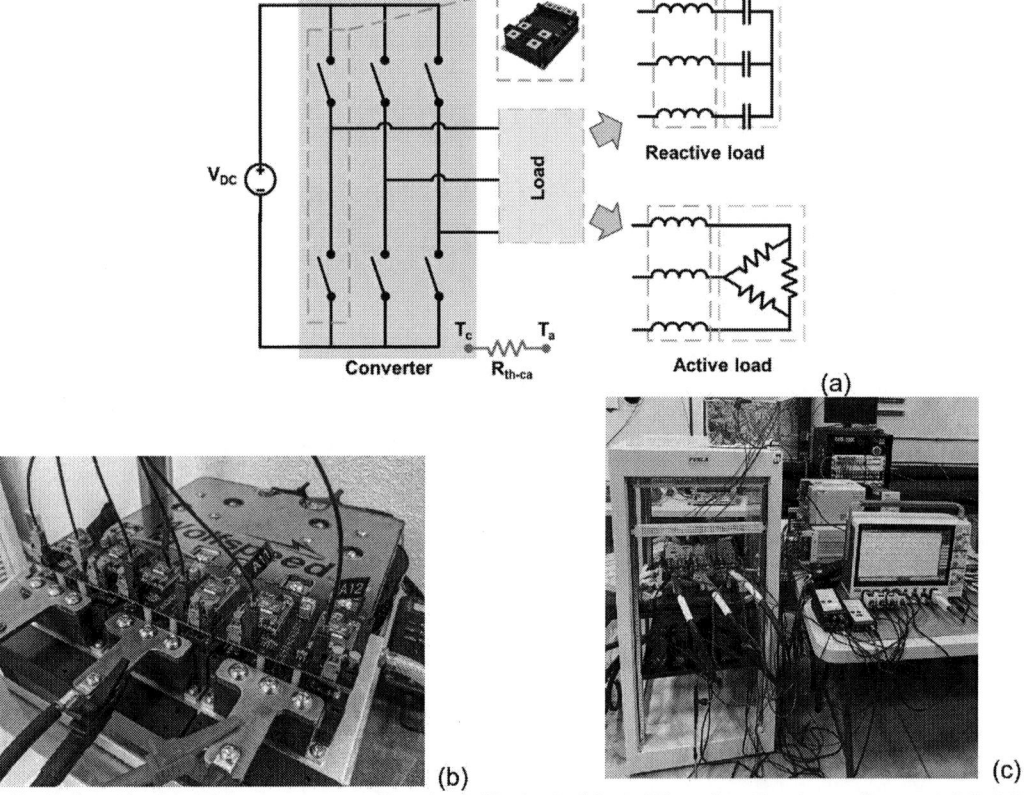

Fig. 6 Setup for the high power operating test, illustrated in (a) the circuit schematic, and (b), (c) the photo of the three-phase inverter prototype, and the experimental equipment.

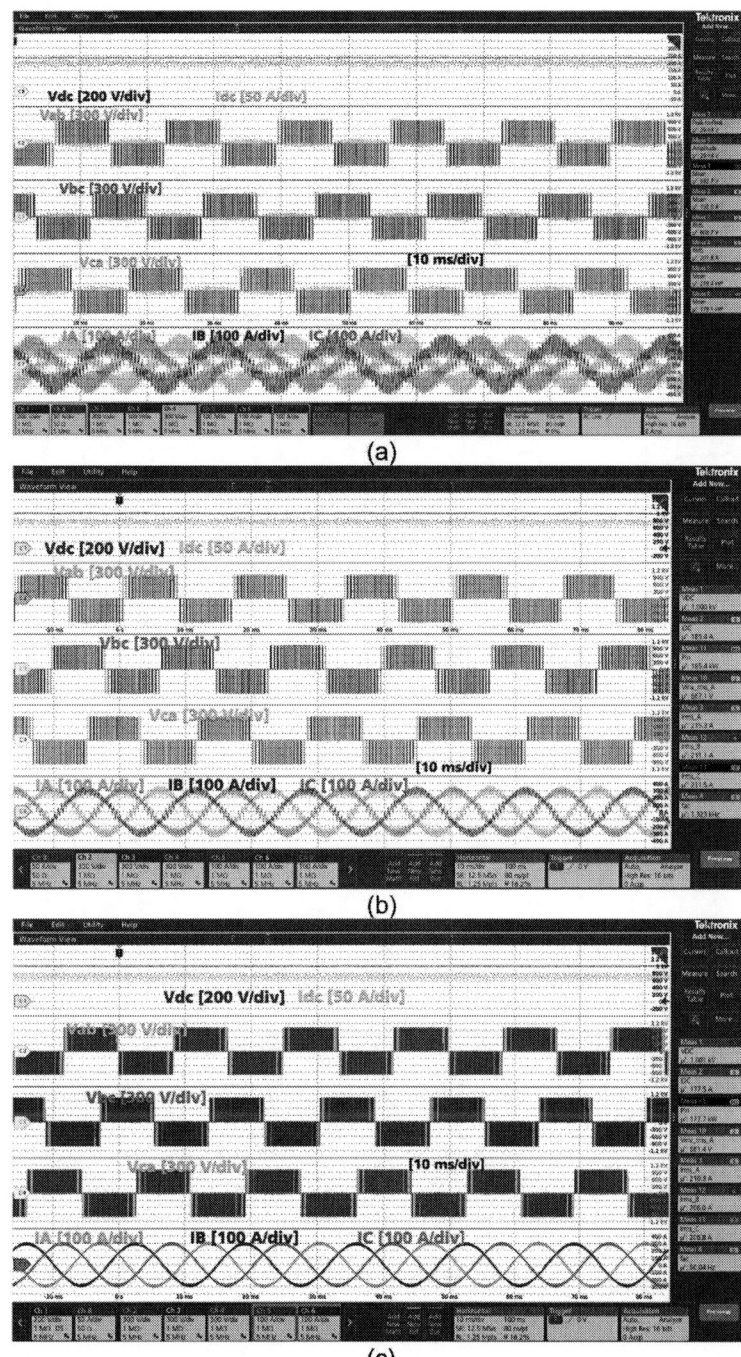

Fig. 7 Continuous test under 177 kW condition, (a) using IGBT module with 85 µH filter, f_{sw} = 2 kHz and $t_{deadtime}$ = 3 us, (b) using MOSFET module with 340 µH filter, f_{sw} = 2 kHz and $t_{deadtime}$ = 1.2 us, (c) using MOSFET module with 340 µH filter, f_{sw} = 8 kHz and $t_{deadtime}$ = 1.2 us.

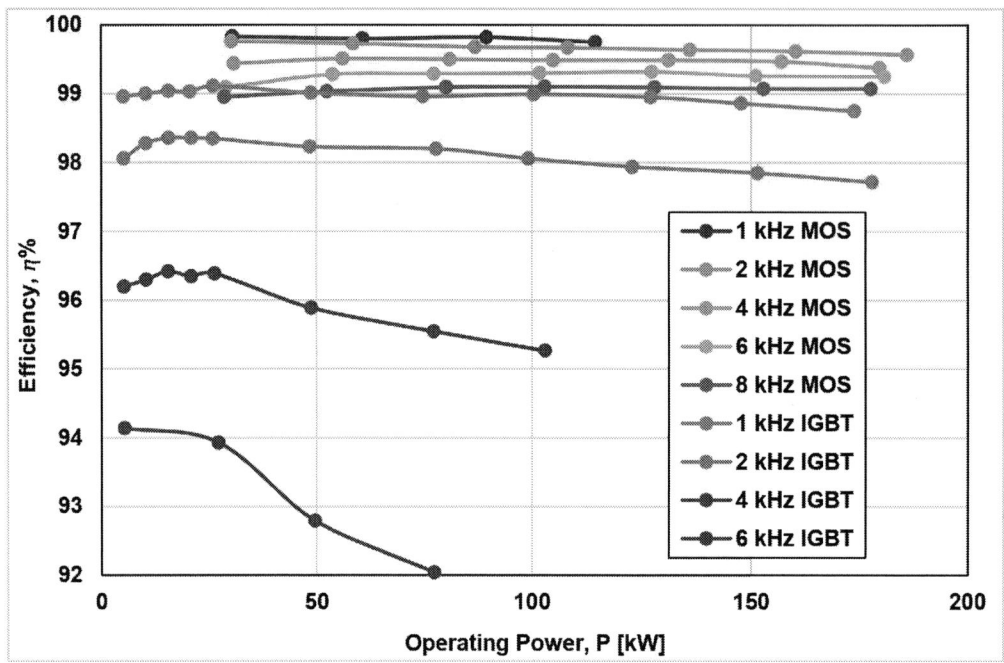

Fig. 8 Efficiency comparison between converters using MOSFET and IGBT under different switching frequency for the active power test.

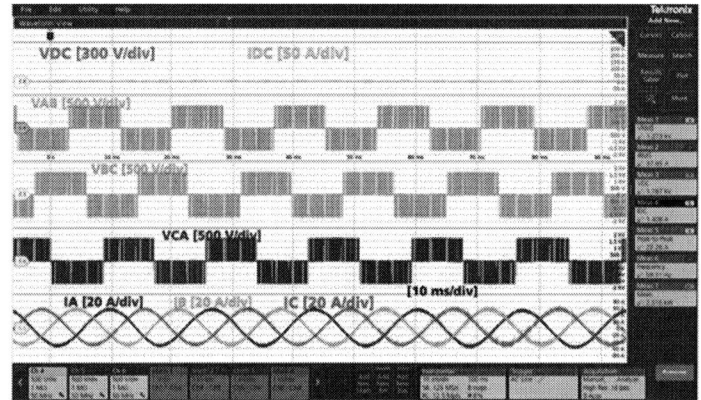

Fig. 9 Continuous test under 85 kVA condition, using MOSFET module when f_{sw} = 8 kHz and $t_{deadtime}$ = 1.2 us.

References

[1] H. Cao, L. Du, F. Guo, Z. Ma and Y. Zhao, "A Triple Active Bridge (TAB) Based Solid-State Transformer (SST) for DC Fast Charging Systems: Architecture and Control Strategy," 2023 IEEE Energy Conversion Congress and Exposition (ECCE), Nashville, TN, USA, 2023, pp. 855-860, doi: 10.1109/ECCE53617.2023.10362803.

[2] A. Rahouma et al., "A Hybrid Si/SiC MV Power Conditioning System for Ultra-Fast Electric Vehicle Charging Stations," 2023 IEEE Energy Conversion Congress and Exposition (ECCE), Nashville, TN, USA, 2023, pp. 2020-2025, doi: 10.1109/ECCE53617.2023.10362780.

[3] A. H. Ismail et al., "High-Density High-Power Converter using 3.3-kV All-Silicon Carbide Modules," 2023 IEEE Energy Conversion Congress and Exposition (ECCE), Nashville,

TN, USA, 2023, pp. 1831-1835, doi: 10.1109/ECCE53617.2023.10362446.

[4] F. Diao et al., "A Medium-Voltage Multilevel Hybrid Converter Using 3.3kV Silicon Carbide MOSFETs and Silicon IGBT Modules," 2023 IEEE Applied Power Electronics Conference and Exposition (APEC), Orlando, FL, USA, 2023, pp. 848-853, doi: 10.1109/APEC43580.2023.10131462.

[5] A. H. Ismail, H. Cao, A. Al-Hmoud, Z. Ma, X. Du and Y. Zhao, "A 3.3 kV Silicon Carbide MOSFET Based Building Block for Medium-Voltage Ultra-Fast DC Chargers," 2023 IEEE Applied Power Electronics Conference and Exposition (APEC), Orlando, FL, USA, 2023, pp. 1693-1700, doi: 10.1109/APEC43580.2023.10131643.

[6] A. Ismail, etc. "Characterization and System Benefits of Using 3.3 kV All-SiC MOSFET Modules in MV Power Converter Applications," 2023 PCIM Europe, Nuremberg.

[7] N. Soltau, E. Wiesner, E. Stumpf, S. Idaka and K. Hatori, "Electric-Energy Savings using 3.3 kV Full-SiC Power-Modules in Traction Applications," 2020 Fifteenth International Conference on Ecological Vehicles and Renewable Energies (EVER), Monte-Carlo, Monaco, 2020, pp. 1-5, doi: 10.1109/EVER48776.2020.9242996.

[8] T. Morikawa et al., "Enhancement of Switching Performance and Output Power Density in 3.3 kV Full SiC Power Module," PCIM Europe digital days 2021; International Exhibition and Conference for Power Electronics, Intelligent Motion, Renewable Energy and Energy Management, Online, 2021, pp. 1-5.

[9] X. Li et al., "High-Voltage Hybrid IGBT Power Modules for Miniaturization of Rolling Stock Traction Inverters," in IEEE Transactions on Industrial Electronics, vol. 69, no. 2, pp. 1266-1275, Feb. 2022, doi: 10.1109/TIE.2021.3059544.

[10] A. Rujas, V. M. Lopez, I. Villar, T. Nieva and I. Larzabal, "SiC-hybrid based railway inverter for metro application with 3.3kV low inductance power modules," 2019 IEEE Energy Conversion Congress and Exposition (ECCE), Baltimore, MD, USA, 2019, pp. 1992-1997, doi: 10.1109/ECCE.2019.8912684.

[11] H. Tajima, S. Ahmed, S. Mabuchi and Y. Abe, "SiC Hybrid Module based VVVF Inverter for Electric Railway," 2020 23rd International Conference on Electrical Machines and Systems (ICEMS), Hamamatsu, Japan, 2020, pp. 2106-2109, doi: 10.23919/ICEMS50442.2020.9290864.

[12] T. Sugahara and T. Ishida, "Traction Inverter Systems with SiC Power Modules for Railway Vehicles." Accessed: Apr. 09, 2024. [Online]. Available: https://www.advance.mitsubishielectric.com/advance/pdf/2021/173_TR3.pdf

[13] C. Ionita, etc. "Comparative assessment of 3.3kV/400A SiC MOSFET and Si IGBT power modules," 2017 IEEE Energy Conversion Congress and Exposition, Cincinnati

[14] T. Sakaguchi, M. Aketa, T. Nakamura, M. Nakanishi and M. Rahimo, "Characterization of 3.3 kV and 6.5 kV SiC MOSFETs," PCIM Europe 2017; International Exhibition and Conference for Power Electronics, Intelligent Motion, Renewable Energy and Energy Management, Nuremberg, Germany, 2017, pp. 1-5.

[15] A. Marzoughi, J. Wang, R. Burgos and D. Boroyevich, "Characterization and Evaluation of the State-of-the-Art 3.3-kV 400-A SiC MOSFETs," in IEEE Transactions on Industrial Electronics, vol. 64, no. 10, pp. 8247-8257, Oct. 2017, doi: 10.1109/TIE.2017.2694380.

[16] A. Marzoughi, R. Burgos and D. Boroyevich, "Characterization and Performance Evaluation of the State-of-the-Art 3.3 kV 30 A Full-SiC MOSFETs," in IEEE Transactions on Industry Applications, vol. 55, no. 1, pp. 575-583, Jan.-Feb. 2019, doi: 10.1109/TIA.2018.2865128.

[17] M. Hruska, P. Bhatnagar and M. Sleven, "Benefits of Using the New 1700V and 3300V High Power Modules for Traction Applications," PCIM Europe digital days 2021; International Exhibition and Conference for Power Electronics, Intelligent Motion, Renewable Energy and Energy Management, Online, 2021, pp. 1-6.

[18] Z. Ni, S. Zheng, M. S. Chinthavali and D. Cao, "Investigation of Dynamic Temperature-Sensitive Electrical Parameters for Medium-Voltage SiC and Si Devices," in IEEE Journal of Emerging and Selected Topics in Power

Electronics, vol. 9, no. 5, pp. 6408-6423, Oct. 2021, doi: 10.1109/JESTPE.2021.3054018.

[19] A. Marzoughi, R. Burgos and D. Boroyevich, "Investigating Impact of Emerging Medium-Voltage SiC MOSFETs on Medium-Voltage High-Power Industrial Motor Drives," in IEEE Journal of Emerging and Selected Topics in Power Electronics, vol. 7, no. 2, pp. 1371-1387, June 2019, doi: 10.1109/JESTPE.2018.2844376.

[20] Z. Ni, S. Zheng, M. S. Chinthavali and D. Cao, "Investigation of Dynamic Temperature-Sensitive Electrical Parameters for Medium-Voltage Low-Current Silicon Carbide and Silicon Devices," 2020 IEEE Energy Conversion Congress and Exposition (ECCE), Detroit, MI, USA, 2020, pp. 3376-3382, doi: 10.1109/ECCE44975.2020.9236121.

[21] Wolfspeed Inc., "3300 V LM Silicon Carbide Half-Bridge Power Modules," Accessed on: June 19, 2023. [Online]. Available: https://www.wolfspeed.com/3300v-lm-silicon-carbide-half-bridge-power-modules/

[22] Semikron Danfoss, "SKM450GB33F IGBT SEMITRANS 20," Accessed on: June 19, 2023. [Online]. Available: https://www.semikron-danfoss.com/products/product-classes/igbt-modules/detail/skm450gb33f-89100010.html

Comparison of Three-Level Inverter Topologies for MVDC Reversible Railway Substations

Luc Bimmel[1] , Philippe Ladoux[1], Erick Matheus Da Silveira Brito[1,2]

[1] University of Toulouse, Laboratory of Plasma and Energy Conversion – LAPLACE, France
[2] Universidade Federal de Viçosa, Institute of Exact and Technological Sciences - Brazil

Corresponding author: Luc Bimmel, luc.bimmel@laplace.univ-tlse.fr
Speaker: Luc Bimmel, luc.bimmel@laplace.univ-tlse.fr

Abstract

Rail is one of the most energy-efficient modes of transport and the lowest emitter of greenhouse gases. In the years to come, rail will play a key role in accelerating the transition to more sustainable mobility. In 2018, a new Medium Voltage DC (MVDC) electrification system, which also offers perspectives for the integration of renewable energy sources, was proposed. To achieve reversible power flow between the MVDC power system and the public grid, reversible substations are required. Considering the characteristics of HiPak IGBT power modules, this paper presents a comparative study between two three-level inverter topologies that can be used for this application.

1   Introduction

The last Intergovernmental Panel on Climate Change (IPCC) report published in March 2023 shows that greenhouse gas emissions are continuing to rise [1]. Expansion of rail networks is the best way to reduce flying over short and medium distances (rail emissions per passenger kilometer average around one-fifth of those of air travel). These last years, with the view of increasing railroad traffic while improving the power system efficiency, research works have shown the relevance of raising the nominal voltage of DC electrified railways to levels of 6 kV to 9 kV [2]-[3]. This new electrification system will include connection points for renewable energy sources and energy storage systems [4]. Even if the energy will mainly be consumed by trains, it is planned to offer voltage and frequency services to the upstream utility grid. Thus, at some points of the lines, active rectifiers will be used instead of conventional diode-rectifiers. In medium voltage DC, Three-Level Voltage Source Converters (3L- VSC) are able to control several megawatts of power, thus they appear to be a relevant solution for this application. In this paper, the topology of the reversible substation relies on a two-secondary transformer, according to Fig.1. Considering the characteristics of HiPak IGBT power modules, the paper presents a comparative study between 3L-Active Neutral Point Clamped (ANPC) and Flying Capacitor (FC) topologies and highlights the PQ capability of the new railway substation.

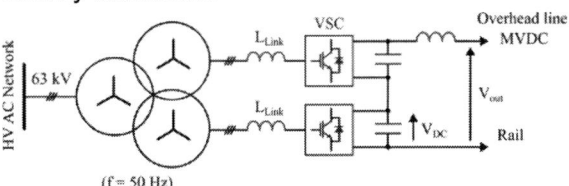

Fig. 1   Reversible railway substation based on two Voltage Source Converters.

2   Power sizing of a MVDC reversible railway substation based on three-level topologies

Nowadays, 3L topologies such as the Active Neutral Point Clamped (ANPC) and Flying Capacitor (FC) are commonly used in medium-voltage drive applications. However, the use of these topologies in MVDC railway substations has not yet been considered. In the study presented in this paper, IGBT power modules with a typical foot print of 140 mm x 190 mm in single switch configuration are considered [5].

2.1 Multilevel inverter topologies

By connecting the inverters in series, two different voltage levels were considered, in concordance with Tab. 1.

IGBT module voltage rating (V)	IGBT module current rating (A)	Inverter operating voltage, V_{DC} (V)	Substation output voltage, V_{out} (V)	Contact-Line Nominal Voltage (V)
3 300	1 800	3 500	7 000	6 000
4 500	1 500	5 000	10 000	9 000

Table 1 Three-Level topology – Correspondence between power module voltage rating and contact-line nominal voltage

The topology of the 3L-FC and 3L-ANPC are shown in Fig. 2.

Fig. 2 3-L Converter topologies: (a) FC; (b) ANPC.

with half the DC input voltage, $V_{DC}/2$. The ANPC requires 18 power modules whereas the FC only has 12. Figure 4 shows an example of output voltage and current waveforms.

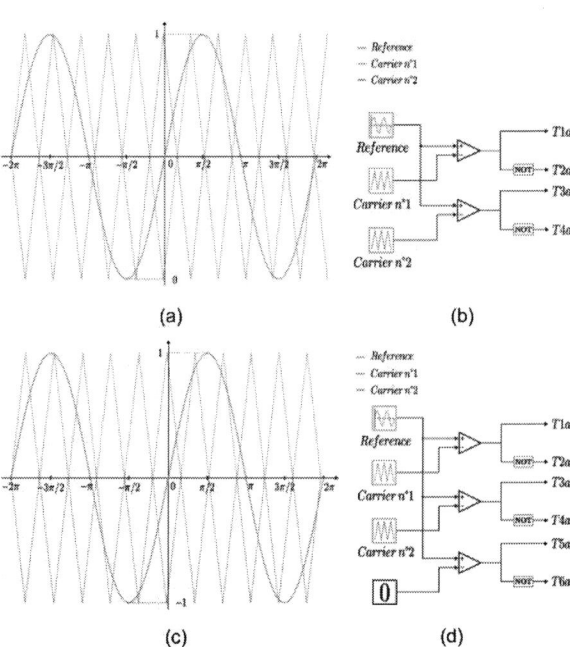

Fig. 3 PWM Control of converters: (a) FC modulation scheme; (b) FC block diagram; (c) ANPC modulation scheme; (d) ANPC block diagram.

Different control strategies are adapted to each topology. ANPC uses the Outer Switch Mode Modulation (OSMM) [6]. Switches $T1i, T2i, T3i$ and $T4i$ operate at high frequency, while switches $T5i$ and $T6i$ ($i = a, b, c$) work at low frequency. FC converter uses a phase shift PWM, according to Fig. 3. In both topologies, the semiconductors operate

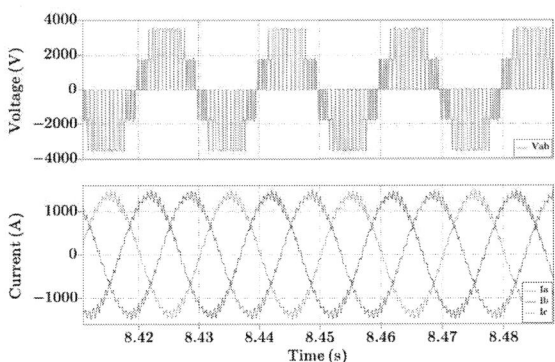

Fig. 4 FC inverter waveforms with 3 300 V / 1 80 0 A power modules, $P = 3\ MW$, $V_{DC} = 3\ 500\ V$, $f_{sw} = 750\ Hz$: (a) Line to Line voltage; (b) Output c urrent.

2.2 Nominal operating point

For the nominal power sizing an operating point inverter mode with a ratio Q/P of 35% ($\varphi = 19°$) was considered. A computational algorithm has been implemented to determine, the transformer secondary voltage (V), the link inductor (L_{link}), and can be calculated by Eq. (1):

$$L_{link} = \frac{V_{DC}}{12. \Delta I_{max}. f_a} \qquad (1)$$

where ΔI_{max} is the ripple of the peak value of fundamental component at the nominal point, f_a is the apparent switching frequency ($f_a = f_{sw}$ for ANPC topology and $f_a = 2f_{sw}$ for FC topology). The choice of switching frequency is a starting hypothesis. A parametric study on this frequency is carried out in section 4.

results given by the computational algorithm are presented in Tab. 2. It should be noted that the FC topology takes advantages of the interleaving of switching patterns and then requires a lower switching frequency than ANPC topology. Consequently, for the same thermal limits, the FC topology offers higher power levels.

3 Operating limits in the Active-Reactive Power plane

Once the nominal operating point is determined, the operating limits in the PQ plane were studied. Different limits were highlighted: The maximum VSC output voltage, the maximal junction temperature of the semiconductors (IGBTs or Diodes) and the performance of water cooled heatsink (4 kW max. per power module).

3.1 Loss calculations

The maximum junction temperature of semiconductors depends directly on losses. The computational algorithm calculates the various losses in each of the converters using Eq. (2) and Eq. (3):

$$P_{cond} = R_0 I_{rms}^2 + V_0 I_{avg} \qquad (2)$$

where R_0 is the slope resistance and V_0 is the threshold voltage of the IGBT.

$$P_{on/off/rec} = \frac{f_{sw}}{T} \frac{V_{CE}}{V_{CC}} \int_{t_1}^{t_2} E_{on/off/rec}\, dt \qquad (3)$$

where V_{CE} is the blocking voltage across the semiconductor, and V_{CC} is the blocking voltage associated with the switching energy curves $E_{on/off/rec}$ given in the datasheet.

Substation output voltage, V_{out} (V)	Topology	Nominal Operation in Inverter Mode (Q/P = 35%)	Maximum Active Power in Rectifier Mode (MW) (Q = 0)	Switching Frequency (Hz)
7 000 V	ANPC	4.32 MW / 1.52 MVAR	5.4	1500
	FC	6.28 MW / 2.16 MVAR	8.16	750
10 000 V	ANPC	5.08 MW / 1.76 MVAR	6.3	1000
	FC	6.56 MW / 2.22 MVAR	8.53	500

Table 2 Power sizing of a MVDC reversible railway substation based on three-level topologies (ANPC and FC).

For the calculations, the analytical expressions of conduction and switching losses and the thermal model of the power modules were considered. The maximum junction temperature has been fixed to 125°C with a heatsink temperature at 80°C. The

3.2 Loss distribution

A detailed distribution of semiconductor losses in the VSC topologies is shown in Fig. 5. Unlike FC topology [7], the distribution of losses is not homogeneous in the ANPC [8] and the power limitation is therefore imposed by switch $T1i$.

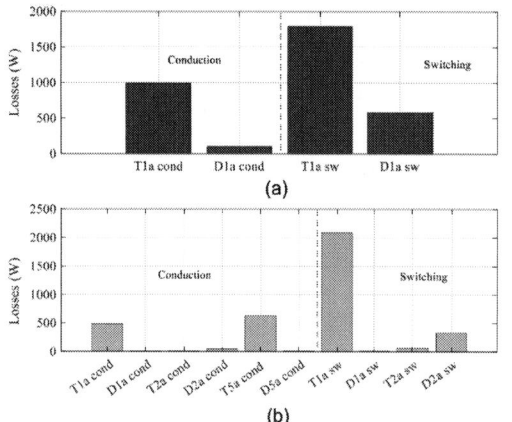

(a)

(b)

Fig. 5 Distribution of losses with 3 300 V / 1 800 A power modules ($\varphi = 19°$): (a) FC Converter; (b) ANPC Converter.

The thermal limits and the maximum output voltage of the VSCs are used to define the Active-Reactive power plane of the inverter. Different operating points can be identified. When the converter only supplies active power to the public grid, the operating point is noted P_{max}. The symmetric operation, corresponding to the rectifier mode (active power supplied to the railway line), is noted P_{min}. Furthermore, the nominal operating point of the VSC is considered with a ratio Q/P of 35%, and noted P_{nom}/Q_{nom}, in Fig. 6.

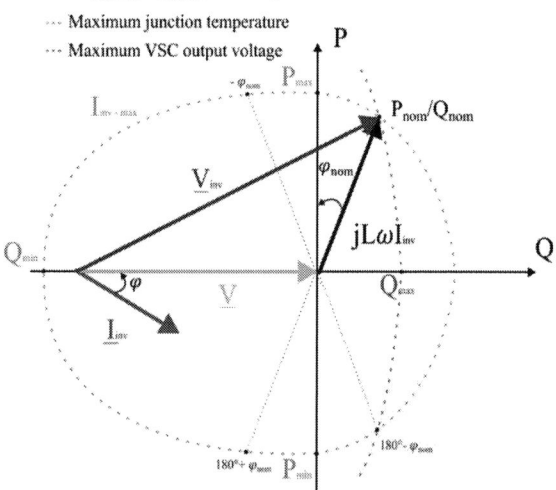

Fig. 6 Operating limits in the Active-Reactive Power plane.

Another limit has to be considered in the computational algorithm; it is linked to the cooling method chosen for the power modules. The selected liquid cooled cold plate can dissipate a maximum power of 4000 W per module. Figure 7 presents the distribution of semiconductor losses in the rectifier operating mode. In the case of the FC topology, the diodes show higher conduction losses than in the inverter operation. Consequently, at this point, the IGBTs and the diodes generate a level of losses which reaches the operating limit of the cooling system. Regarding the ANPC topology, the distribution of losses is different. Indeed, $T2i$ shows a high losses level and the maximum junction temperature is reached before the dissipation limit of the cold plate. In all cases, the FC topology achieves higher operating limits than the ANPC topology, as shown in Fig. 8.

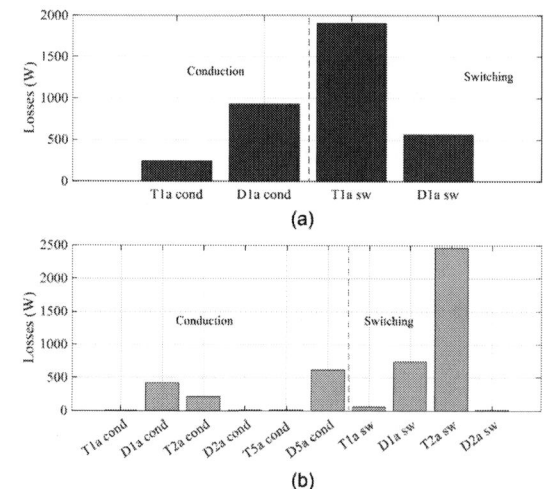

(a)

(b)

Fig. 7 Distribution of losses with 3 300 V / 1 800 A power modules ($\varphi = 180°$): (a) FC Converter; (b) ANPC Converter.

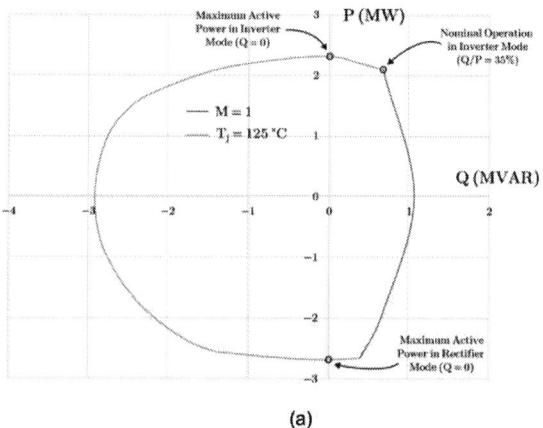

(a)

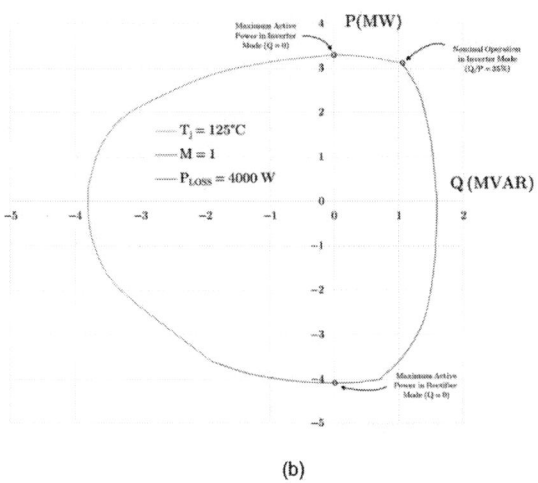

(b)

Fig. 8 Operating limits in the Active-Reactive Power plane for one VSC, with 3.3 kV / 1.8 kA power modules: (a) ANPC topology 1500 Hz; (b) FC topology 750 Hz.

4 Choice of switching frequency

The choice of the switching frequency is a key factor in sizing the converter. A lower frequency reduces losses in the semiconductors, while increasing the size of passive components.

The increase in switching frequency is an advantage when sizing passive elements of the converter. A comparative study of the different filtering elements as a function of the switching frequency is presented in Tab. 3. The comparison parameters correspond to the value of the passive components, and the energy stored in the capacitors (E_C) and in the inductors (E_L), calculated using Eq (4) and Eq (5):

$$E_C = \frac{1}{2} C V_C^2 \qquad (4)$$

where V_C is the voltage across the capacitor

$$E_L = \frac{1}{2} L_{link} \hat{I}_{abc}^2 \qquad (5)$$

where $\hat{I}_{abc}$ is the peak current in each phase.

Following the first approach, where the switching frequency was a priori fixed, a study was carried out to evaluate the impact of the switching frequency on the VSC sizing. As presented in Fig. 9, a maximum value of the nominal power can be achieved by varying the switching frequency.

V_{out} (V)	7 000 V			
Topology	ANPC		FC	
f_{sw} (Hz)	830	1500	500	750
L_{Link} (mH)	1.9 (× 6)	1.6 (× 6)	1.4 (× 6)	1.2 (× 6)
C_{bus} (mF)	16.15 (× 4)	8.81 (× 4)	7,2 (× 2)	2.24 (× 2)
C_k (mF)	X	X	15 (× 6)	7.55 (× 6)
E_L (kJ)	5.48	1.84	4.86	2.64
E_C (kJ)	396	214	226	96
P_{nom} (MW)	**5.5**	**4.22**	**6.82**	**6.28**

Table 3 7 kV DC substation based on two 3L-VSC in series (3 300 V / 1 800 A IGBT power modules).

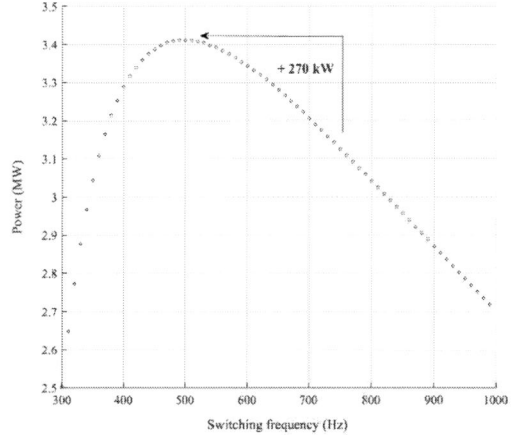

Fig. 9 Maximum power at nominal operating point as a function of switching frequency, for one FC, with 3 300 V / 1 800 A power modules.

Figure 10 shows that a gain on the operating area can be observed in all the Active-Reactive power plane. Nevertheless, as shown in Tab. 3, lowering the switching frequency to 500 Hz penalizes the sizing of the passive components. Finally, for the FC topology, it was considered that a switching frequency at 750 Hz offers a good compromise regarding the power rating and the size of the passive elements.

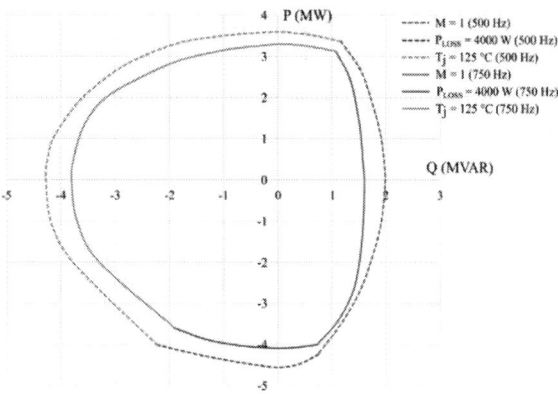

Fig. 10 Operating limits in the Active-Reactive Power plane for one FC, with 3.3 kV / 1.8 kA power modules: 500 Hz / 750 Hz.

This part focused on a 7 kV DC substation. However, the case of a DC electrification system at 10 kV was also considered, the calculation results are shown in Tab 4.

Substation output voltage, V_{out} (V)	10 000 V
Topology	FC
f_{sw} (Hz)	500
L_{Link} (mH)	3 (×6)
C_{bus} (mF)	2.04 (×2)
C_k (mF)	8.2 (×6)
Nominal Operation in Inverter Mode (Q/P = 35%)	6.56 MW 2.22 MVAR
Maximum Active Power in Rectifier Mode (Q=0)	8.53 MW

Table 4 10 kV DC substation based on two FC converters in series (4500 V/1500 A IGBT power modules).

5 Conclusion

In this paper, a comparison of three-level inverter topologies to achieve an MVDC reversible railway substations was presented. All the results performed by the computational algorithm were validated by time domain simulations in PLECS software. Thermal models of HiPak IGBT power modules were implemented. A study of the distribution of losses in converters showed that FC appeared to be less restrictive than the ANPC. Moreover, it was shown that the FC stores less energy in the passive elements. For both considered contact-line voltage level (7 kV and 10 kV), the nominal power of the reversible substation is around 6.6 MVA in inverter mode with a power factor of 0.94 and 8 MVA in rectifier mode at unity power factor. Based on the results of this study, it is appropriate to consider the design of the busbar of the inverter arms which must integrate the flying capacitors. Furthermore, the behavior under short-circuit on the contact line side must also be seriously studied in order to propose a solution avoiding the destruction of the power module diodes by overcurrent.

6 Acknowledgement

This work was funded by the French government as part of the France 2030 program (RACCOR-D project).

References

[1] H.-O. Pörtner et D. C. Roberts, « Climate Change 2022: Impacts, Adaptation and Vulnerability ».

[2] A. Verdicchio, P. Ladoux, H. Caron, et C. Courtois, « New Medium-Voltage DC Railway Electrification System », *IEEE Trans. Transp. Electrific.*, vol. 4, n° 2, p. 591-604, juin 2018, doi: 10.1109/TTE.2018.2826780.

[3] A. Verdicchio, P. Ladoux, H. Caron, et S. Sanchez, « Future DC Railway Electrification System - Go for 9 kV », in *2018 IEEE International Conference on Electrical Systems for Aircraft, Railway, Ship Propulsion and Road Vehicles & International Transportation Electrification Conference (ESARS-ITEC)*, Nottingham: IEEE, nov. 2018, p. 1-5. doi: 10.1109/ESARS-ITEC.2018.8607304.

[4] FUNDRES project. [Online]. Available on: https://fundres-project.eu/

[5] « Insulated Gate Bipolar Transistor (IGBT) and diode modules with SPT, SPT+, SPT++ and TSPT+ chips | Hitachi Energy ». [Online]. Available on: https://www.hita-

chienergy.com/products-and-solutions/semi-conductors/insulated-gate-bipolar-transistor-igbt-and-diode-modules

[6] Y. Jiao, « High Power High Frequency 3-level Neutral Point Clamped Power Conversion System », p. 302.

[7] P. Ladoux, P. Marino, G. Raimondo, et N. Serbia, « Comparison of high voltage modular AC/DC converters », in *International Symposium on Power Electronics Power Electronics, Electrical Drives, Automation and Motion*, Sorrento, Italy: IEEE, juin 2012, p. 843-848. doi: 10.1109/SPEEDAM.2012.6264569.

[8] G. Zhang, Y. Yang, F. Iannuzzo, K. Li, F. Blaabjerg, et H. Xu, « Loss distribution analysis of three-level active neutral-point-clamped (3L-ANPC) converter with different PWM strategies », in *2016 IEEE 2nd Annual Southern Power Electronics Conference (SPEC)*, Auckland, New Zealand: IEEE, déc. 2016, p. 1-6. doi: 10.1109/SPEC.2016.7846157.

PCIM Europe 2024, 11– 13 June 2024, Nuremberg   DOI: 10.30420/566262454

Control of Bidirectional Power Flow in Railway Catenary Overhead Lines

P.J. van Duijsen[1], D.C. Zuidervliet[1]

[1] THUAS, DC-Lab, Delft, The Netherlands

Corresponding author: Peter van Duijsen, p.j.vanduijsen@hhs.nl
Speaker: Peter van Duijsen, p.j.vanduijsen@hhs.nl

Abstract

An Interlink converter is presented, capable of transferring power in metro and tramway catenary overhead lines. The purpose of the converter is, to transport energy which is fed into the overhead lines from a solar farm, to the inner city. The reason for this is shortage of power in the public utility grid in the inner city. Also solar power, harvested along the tramway track can be fed into the overhead lines. By actively transporting electric power through the overhead lines, the voltage level remains at a nominal level at sections, where power is fed into the overhead lines. Also from the public utility grid in the inner city, there will be a lower power demand.

1   Introduction

Traditional railway grids are supplied from the public grid or a single power generator station, such as the hydro-power stations in, for example, Austria and Switzerland. The railway grid is mainly used to feed the locomotives. Therefore, the catenary overhead lines are used to transport the electric power from the power supply grid connection towards the locomotive. The power flow is purely passive based on supply and demand. With the introduction of solar power along the railway tracks, there is the possibility to directly feed the solar power into the overhead lines [1]. Also the electric brake energy from the locomotives is fed into the overhead lines [2]. To better utilize this electric energy, the control of the power flow in the overhead lines is introduced. In this study, only DC traction systems as in use for underground and tramway are considered. Since these overhead lines can be regarded as a DC grid, the same type of converters and control methods can be applied here. In this paper an Interlink converter is presented, that can be used to control a bidirectional power flow between two sections. Given the fact that the overhead lines are only fully utilized to the maximum power when the locomotives are accelerating or electrically braking, allows room to use the overhead lines for other applications to be powered from the overhead lines. Typ-

ically the overhead lines are only utilized to 20% of their maximum rating during nominal operating times, leaving 80% for transport of electric energy during the intervals when no locomotives are accelerating.

2   Power Congestion

To avoid power congestion in the inner city, the overhead lines from the public transport can be used to transport electric energy. Two sections, having nearly equal voltage levels are connected via power electronics converter as shown in Fig. 1

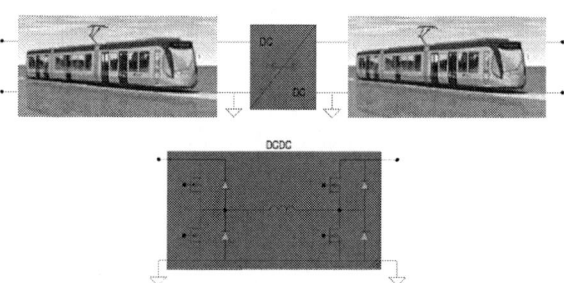

Fig. 1: Interlink converter connecting two tramway overhead line sections.

The converter can control the bidirectional power flow, by controlling the current in the inductor between the two half bridges. When different catenary overhead lines are not connected electrically, they cannot exchange electric energy. A reason why different tramway lines are not connected, can be the feeding from different locations and preventing overload from a central feeding station. However, when there are power congestion problems in the area, where the tramway has its feeder stations, extra feeder stations might help. Transport of electric energy through the overhead catenary lines is simply dictated by Ohms-law. The current in the overhead catenary line, only flows in the direction where the voltage is lowest. This could lead to power congestion problems at the feeder station, where the voltage on the catenary overhead line is the highest, see the red encircled section II in Fig 2.

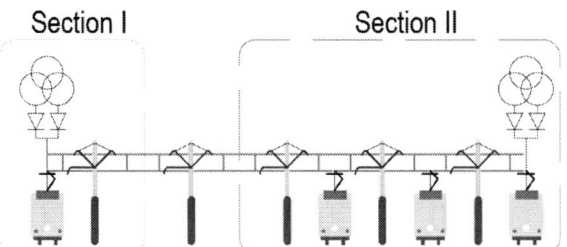

Fig. 2: Two tramway overhead line sections directly couple can lead to overload of the feeder station in Section II.

In Fig. 2, the feeder station in section II has to feed three trams, while the feeder station in section I only has to supply a single tram. This might lead to overload in feeder station II. The reason for this overload is, that the three trams in section II are close to the feeder station in section II, and thus have a lower resistance towards the feeder station in section II compared to the resistance of the catenary overhead line towards the feeder station in section I. To regulate the current flow in the catenary overhead lines, the Interlink can be placed between two sections.

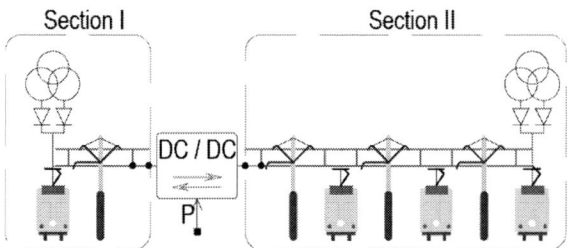

Fig. 3: Interlink reduced overload in feeder station in section II

The Interlink can actively regulate a power flow from the feeder station in section I to the feeder station in Section II, see Fig. 3. The Interlink is actively

regulating a current flow, from section I towards section II in Fig. 3. In this way, the current required to supply the four trams, can be evenly distributed between the two feeder stations in section I and section II.

In case there are power congestion problems in the electricity grid at one of the feeder stations, the Interlink can actively assist in regulating the power flow to be lower from this feeder station.

3 Single inductor DAHB

Typically DC grids are connected via a DAB [3]–[6]. Galvanic isolation, voltage scaling and simple control via Phase Shift Modulation [7] are the main features. Also three-phase transformers can be applied, for constant power flow, highest power density and smaller filters [8]. Why are we using the Dual Active Half Bridge [DAHB] structure with only a single inductor? There is no need for galvanic isolation in the returning line and the voltages on both sides are nearly equal. The rails are somehow in contact with the earth and therefore it is difficult to separate them electrically. Therefore an Interlink, with a single inductor can be applied, see Fig. 4.

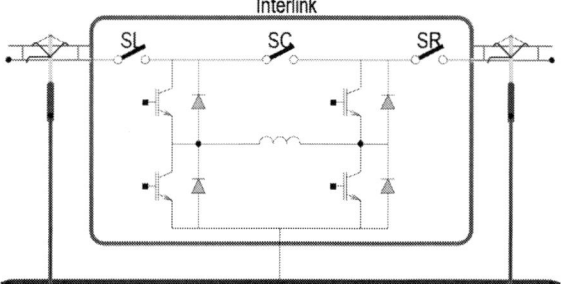

Fig. 4: Interlink with mechanical separators SL and SR, for galvanic isolation in case of a fault.

In Fig. 4 the interlink with mechanical contactors, is displayed the Interlink itself is used to regulate the current through the inductor L. The mechanical separators are used to galvanic isolate both sections of the overhead lines. The operation is following; first the connectors S_L and S_R closed, next the Interlink starts regulating the current to the desired current level. In the event of closing down the interlink, first the current is regulated down to 0. At the moment when the current through contactors S_L and S_R equals zero, both contactors can be opened. When both contactors SL and Sr. are opened both sections are electrically isolated. When two sections have to be connected and have

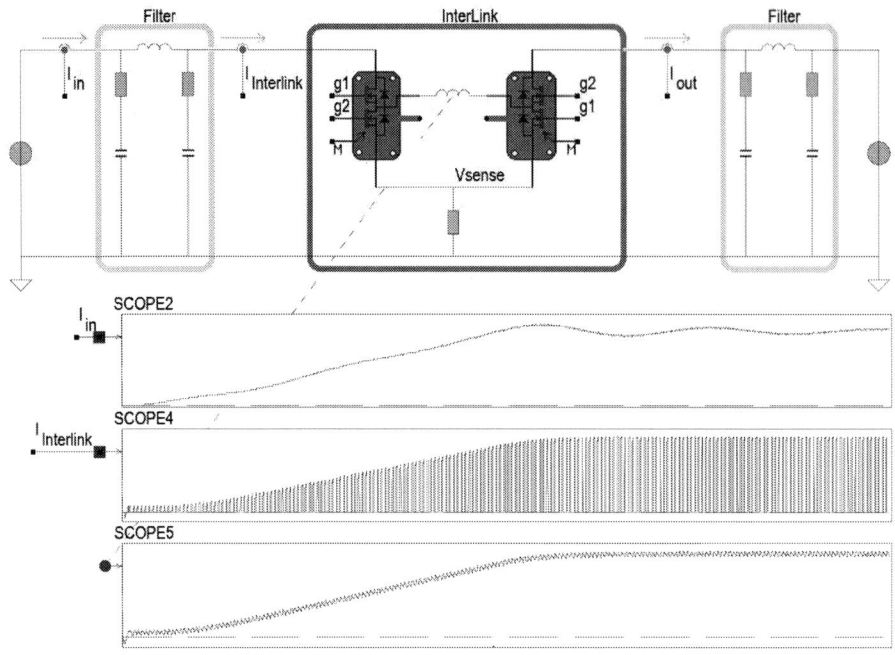

Fig. 5: Simulation in Caspoc [9] of a controlled inrush current in the Interlink, regulated as Four-switch Buck-Boost converter.

nearly equal voltage level, both sections could be connected directly. To do this first the current is regulated in such a way that the voltages of the left and right section are equal. When both section voltages are equal the contactor S_C can be closed. The interlink is in principle turned off, meaning the semiconductor switches are turned off and there is no inductor current and hence there is no loss in the Interlink.

3.1 Buck boost operation

The Interlink is controlled in the same way as a Dual Active Bridge [DAB] using Phase Shift Modulation [3], [10], [11]. It can also be operated as a Four-Switch Buck-Boost converter [12]–[14]. In that case a nearly DC current through the inductance can be regulated. This can be used if inrush currents have to be regulated, as can be seen in Fig 5. This control method is also used if a constant DC current has to be regulated.

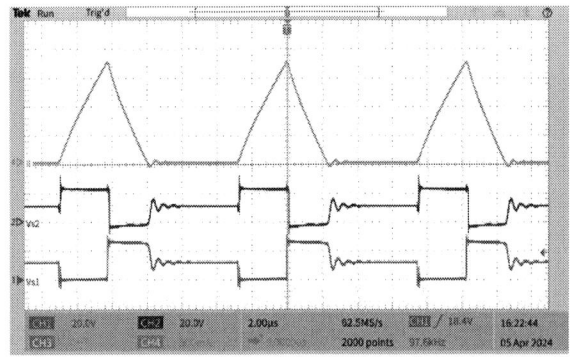

Fig. 6: Controlled discontinuous current flow from the right to the left half bridge in the Interlink. From top to bottom: Scaled Inductor current[blue], scaled voltage of the left[red] and right[blue] bridge.

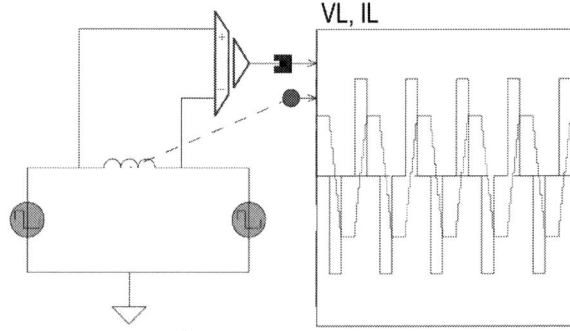

Fig. 7: Simplified operation principle of the DAB, simulated in Caspoc [9].

The interlink as displayed in Fig. 5, is working like a four switch buck boost converter. The main ad-

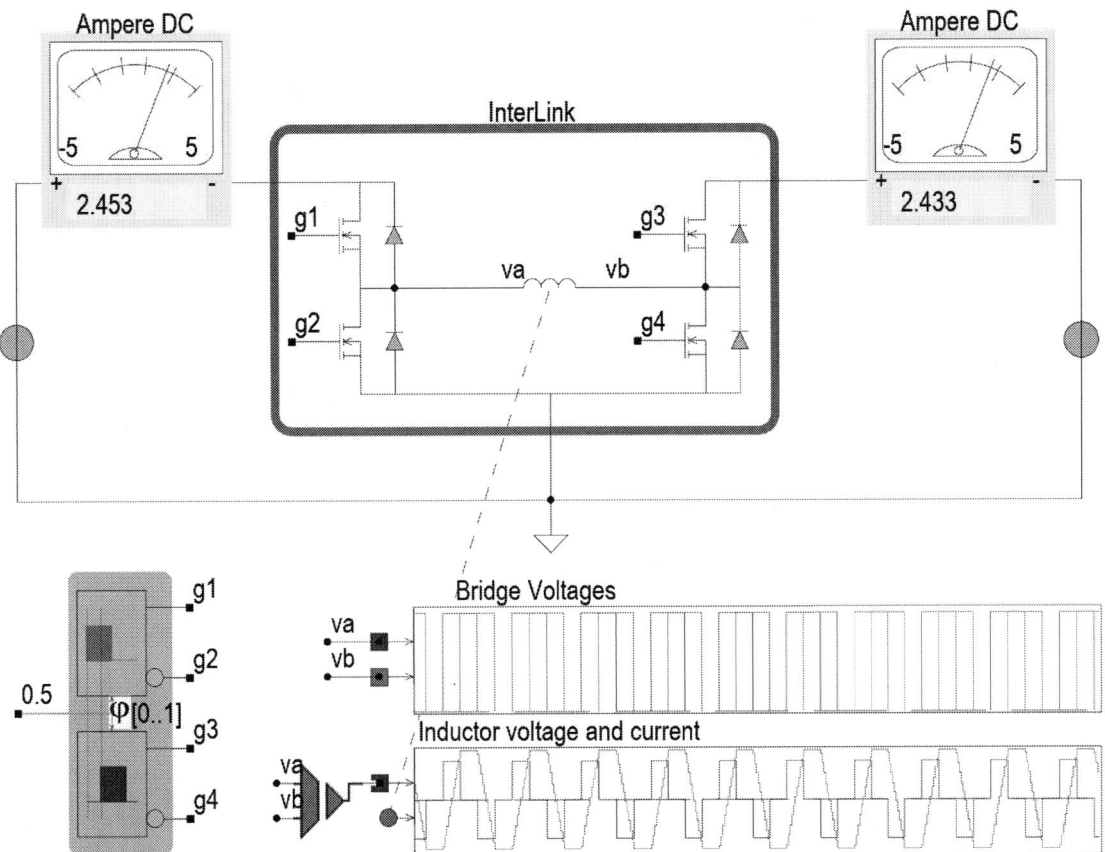

Fig. 8: Phase Shift Modulation of the DAHB, simulated in Caspoc [9]. An average current of $2.4A$ flows from left to the right, due to the phase shift of $90°$ between the left and right bridge voltages v_a and v_b. The bottom scope shows the indcutor voltage and current.

vantage that we have here, is that we can control a nearly constant current through the inductor. This allows us to regulate an inrush current, or in the case of maximum current flow, to regulate this current in a fast and simple way.

The disadvantage of this buck boost converter, is the fact that the inductor saturates due to the high DC bias current. Also the maximum current through the inductor, is twice the input or output current. In contrary to the Phase Shift Modulation, the inductor has to be fairly large. However, switching between buck boost operation and Phase Shift Modulation is only done by the control itself, the circuit itself is equal in both cases.

The simulation in Fig. 5 shows the inrush current when the interlink converter is started in buck boost operation mode. The filters on both sides of the converter take care that the input and output currents are continuous. A small current sense resistor is placed between the low-side Mosfets and

the ground connection, in order to measure the currents through the four switch buck boost converter. This measured voltage V_{Sense} is used as feedback to current-controller. Scope2 shows the input current of the interlink.

Scope4 shows the current as regulated by the left half bridge from the interlink. From Scope2 and Scope4, it is clear that the input current is filtered by the LC filter at the input. Scope5 shows the current through the inductor in the interlink. In contrary to Phase Shift Modulation, this current has a large DC bias.

The measurement of the bridge voltage and current during buck-boost operation is shown in Fig. 6. The current level is low, and therefore the converter operates in discontinuous mode. Because of this discontinuous operation mode, the bridge voltages are floating around $1/2V_{DC}$, during the time when $I_L = 0$.

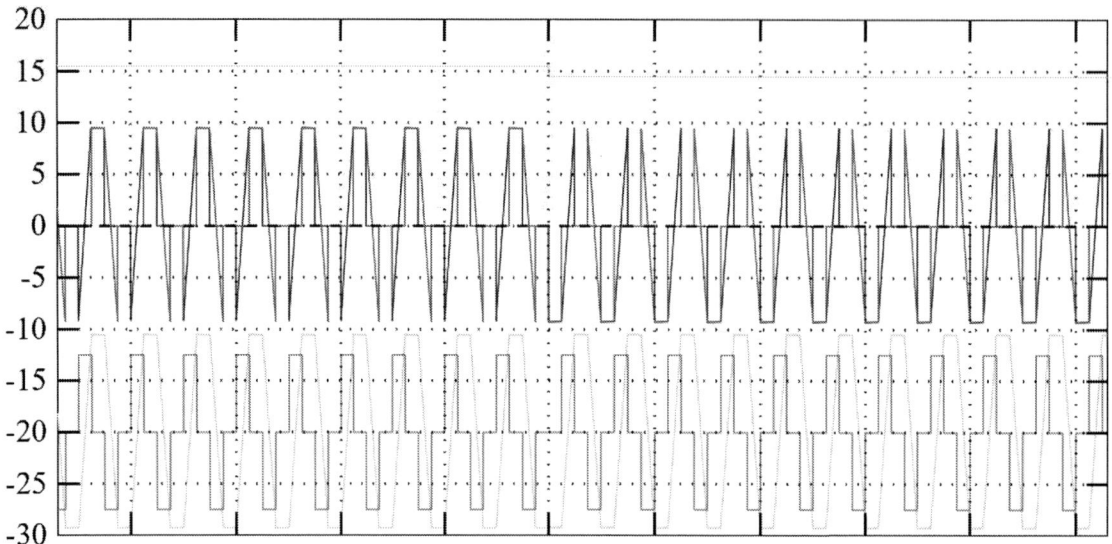

Fig. 9: Simulation in Caspoc [9], of the reversal of power flow in a DAHB. From top to bottom[5v/DIV, 5A/DIV, $20\mu s/DIV$]: Control signal[Lightblue], Input[Blue] and output current[Red], Inductor current[Green] and voltage[Purple].

3.2 Phase Shift Modulation

Phase Shift Modulation in the DAHB, is the same as in the DAB with coupled inductors. The control is implemented in the same way as phase shift control in full bridges [15]. The basic principle operation is shown in Fig 7 [16]. Here 2 control sources model the half bridges, on the left and the right side from the interlink. The duty cycle and amplitude of both square wave sources is equal but the phase shift between them is in this simulation equal to $90°$. The resulting current through the inductor is created depending on the voltage difference between the two sources. The voltage across the inductor and the current through the inductor are both displayed in the scope in figure 6. When the voltage difference is positive the current rises, when the voltage difference is negative the current decreases. During the time period when the two voltage levels are equal, the current level also remains equal. By modulating the phase shift between the two square waves the power transferred to the inductor can be controlled. A positive phase shift means a power flow from the left to the right. A negative phase shift means a power flow from the right to the left.

In figure 7 the Phase Shift Modulation of the dual active half bridge is simulated in Caspoc [9]. The two half bridges are modulated with equal duty cycle but with the phase shift of 90 degrees. The voltage of each half bridge is displayed in the upper scope, and he knew bottom scope the voltage across the inductor and current through the inductor are displayed. The input and output current are displayed by the analog meters which have a delay and therefore show the average current. In the scope "Bridge Voltages", the phase shift between the two voltages is feasible the difference of these two voltages, is displayed in the bottom scope, along with the inductor current.

Phase Shift Modulation in the DAHB, is the same as in the DAB with coupled inductors. The basic principle operation is shown in Fig. 7. Here 2 control sources model the half bridges, on the left and the right side from the interlink. The duty cycle and amplitude of both square wave sources is equal but the phase shift between them is in this simulation equal to $90°$. The resulting current through the inductor is created depending on the voltage difference between the two sources. The voltage across the inductor and the current through the inductor are both displayed in the scope in Fig. 7. When the voltage difference is positive the current rises, when the voltage difference is negative the current decreases. During the time period when the two voltage levels are equal, the current level also remains equal. By modulating the phase shift between the two square waves the power transferred to the inductor can be controlled. A positive phase shift means a power flow from the left to the right. A negative phase shift means a power flow from the

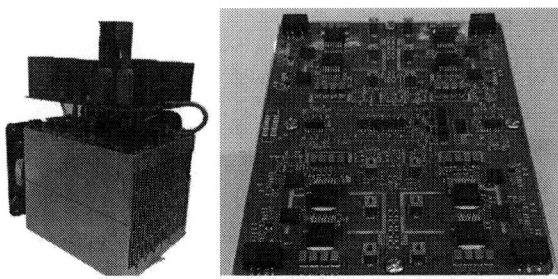

Fig. 10: PCB of the Interlink, $V_{max} = 1500volt$, $P_{max} = 5kW$. Left:Mounted pcb on heatsink, Right:top view of the PCB with the mosfets and gate drivers

right to the left.

In Fig. 8 the Phase Shift Modulation of the dual active half bridge is simulated in Caspoc [9]. The two half bridges are modulated with equal duty cycle but with the phase shift of $90°$. The voltage of each half bridge is displayed in the upper scope, and in the bottom scope, the voltage across the inductor and current through the inductor are displayed. The input and output current, are displayed by the analog meters, which have a time delay and therefore show the average current. In the scope "Bridge Voltages", the phase shift between the two voltages is visible as the difference of the two voltages v_a and v_b, is displayed in the bottom scope, along with the inductor current.

4 Interlink construction

The pcb of the Interlink is shown in Fig. 10 [17]. There are three layers of PCB in this design. The top layers holds the Film Capacitors for both side DC bus connections, the lower PCB contains the power Mosfets and are pressed against the heatsink, using the middle PCB and green press

Fig. 11: Side-view of the stacked PCB with Film capacitors on top and green blocks, to press the semiconductors against the heatsink(not shown in this photo).

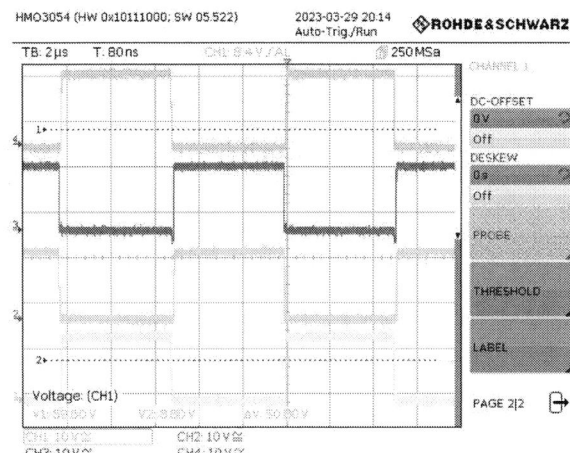

Fig. 12: Measurement of the voltage waveforms in the legs during switching. Measurement result during $5kW$ power transfer, from top to bottom[$2\mu s/DIV$, $10v/DIV$]: gate voltage and Mosfet Voltage for the left and right half bridges.

blocks, as can be seen in Fig. 11. Important is the thermal management. The cooling is probably oversized in Fig 10, as with this the prototype, the maximum allowable power was tested. The control electronics and microcontroller board are not included in Fig. 10.

5 Experimental set-up

The operation of the Interlink is verified using an experimental setup. In Fig. 11 the setup from Fig. 10 is mounted on a heatsink and a C2000 microcontroller [18], is employed to create the gating signals.

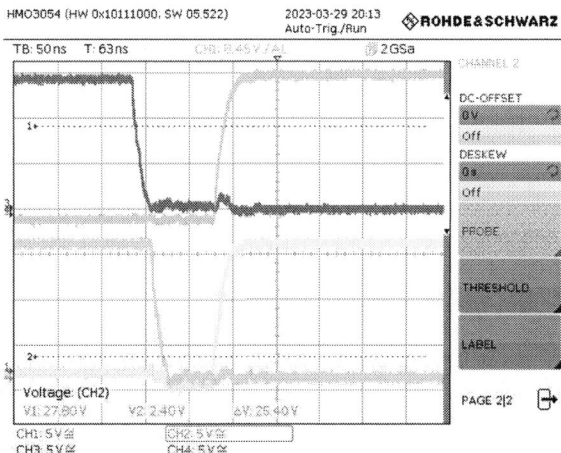

Fig. 13: Detailed view, showing the blanking time between the High-side and Low-side Mosfet.

The results from the experimental set-up are displayed in Fig. 12. Here the gating signals and the voltage at the switching node between the high-side and low-side Mosfet are displayed. A duty cycle of 50% is applied to achieve the maximum current through the inductor. The inductor itself is not displayed in Fig. 11, but was connected via external wiring. Leakage inductance from the connecting wires is not of concern, as it simply adds to the total inductance value of the inductor inside the Interlink. In Fig. 13, the voltage across the switching node between the high-side and low-side Mosfet, is shown in detail. The blanking time of around $60ns - 70ns$, is clearly visible in Fig. 13. Applying Phase Shift Modulation enables soft-switching during turn-on, without auxiliary circuits [19]. This reduces the switching losses, compared to the buck-boost modulation principle.

Conclusion

To control the power from one to another section in a public transportation overhead line, an Interlink converter was designed. The voltage levels in both sections are Because there is no galvanic isolation on the neutral line, a Dual Active Halve Bridge with a single inductor is applied. Both buck-boost operation and Phase Shift Modulation can be used to control the bidirectional power flow. Phase Shift Modulation is preferred to reduce losses in the converter. The converters was designed for $5kW$ and can operate in parallel. Measurement results confirm the operation of the converter. Using this type of converter, power from solar farms outside the city, can be transported into the inner city. Power congestion problems per section can be prevented, by controlling the power flow between ections.

References

[1] P. J. van Duijsen and D. C. Zuidervliet, "Distribution of renewable energy in light-rail traction grids," in *2022 International Symposium on Electronics and Telecommunications (ISETC)*, 2022, pp. 1–6. DOI: 10.1109/ISETC56213.2022.10010236.

[2] J. Sitar, V. Racek, and P. Bauer, "Dc/dc converter with active filter supplied from trolley net," in *2007 IEEE Power Electronics Specialists Conference*, 2007, pp. 383–389. DOI: 10.1109/PESC.2007.4342017.

[3] R. De Doncker, D. Divan, and M. Kheraluwala, "A three-phase soft-switched high power density dc/dc converter for high power applications," in *Conference Record of the 1988 IEEE Industry Applications Society Annual Meeting*, vol. 1, 1988, pp. 796–805. DOI: 10.1109/IAS.1988.25153.

[4] M. Kheraluwala, R. Gasgoigne, D. Divan, and E. Bauman, "Performance characterization of a high power dual active bridge dc/dc converter," in *Conference Record of the 1990 IEEE Industry Applications Society Annual Meeting*, vol. 2, 1990, pp. 1267–1273. DOI: 10.1109/IAS.1990.152347.

[5] M. Kheraluwala, R. De Doncker, and D. D.M., "Analysis, design and experimental evaluation of a high-power high-frequency dc/dc converter," in *EPE Conf. Records*, 1991, pp. 1568–1573. DOI: 10.1109/IAS.1993.299007.

[6] K. Vangen, T. Melaa, and A. Adnanes, "Soft-switched high-frequency, high power dc/ac converter with igbt," in *PESC '92 Record. 23rd Annual IEEE Power Electronics Specialists Conference*, vol. 1, 1992, pp. 26–33. DOI: 10.1109/PESC.1992.254687.

[7] M. Kheraluwala and R. De Doncker, "Single phase unity power factor control for dual active bridge converter," in *Conference Record of the 1993 IEEE Industry Applications Conference Twenty-Eighth IAS Annual Meeting*, vol. 2, 1993, pp. 909–916. DOI: 10.1109/IAS.1993.299007.

[8] R. De Doncker, D. Divan, and M. Kheraluwala, "A three-phase soft-switched high-power-density dc/dc converter for high-power applications," *IEEE Transactions on Industry Applications*, vol. 27, no. 1, pp. 63–73, 1991. DOI: 10.1109/28.67533.

[9] Simulation-Research, *Caspoc Simulation and Animation for Power Electronics and Electric Drives — caspoc.com*, https://www.caspoc.com/, [Accessed 08-04-2024].

[10] M. Kim, M. Rosekeit, S.-K. Sul, and R. W. A. A. De Doncker, "A dual-phase-shift control strategy for dual-active-bridge dc-dc converter in wide voltage range," in *8th International Conference on Power Electronics - ECCE Asia*, 2011, pp. 364–371. DOI: 10.1109/ICPE.2011.5944548.

[11] Ö. Ekin, G. Arena, S. Waczowicz, V. Hagenmeyer, and G. De Carne, "Comparison of four-switch buck-boost and dual active bridge converter for dc microgrid applications," in *2022 IEEE 13th International Symposium on Power Electronics for Distributed Generation Systems (PEDG)*, 2022. DOI: 10.1109/PEDG54999.2022.9923074.

[12] G. G. Oggier, G. O. García, and A. R. Oliva, "Switching control strategy to minimize dual active bridge converter losses," *IEEE Transactions on Power Electronics*, vol. 24, no. 7, pp. 1826–1838, 2009. DOI: 10.1109/TPEL.2009.2020902.

[13] X. Ren, Z. Tang, X. Ruan, J. Wei, and G. Hua, "Four switch buck-boost converter for telecom dc-dc power supply applications," in *2008 Twenty-Third Annual IEEE Applied Power Electronics Conference and Exposition*, 2008, pp. 1527–1530. DOI: 10.1109/APEC.2008.4522927.

[14] E. Gallo, F. Cvejić, G. Spiazzi, D. Biadene, and T. Caldognetto, "Average and small-signal model of the four-switch buck-boost converter under both duty-cycle and phase-shift modulation," in *2023 IEEE Applied Power Electronics Conference and Exposition (APEC)*, 2023, pp. 1299–1306. DOI: 10.1109/APEC43580.2023.10131591.

[15] R. Redl, L. Balogh, and D. Edwards, "Optimum zvs full-bridge dc/dc converter with pwm phase-shift control: Analysis, design considerations, and experimental results," in *Proceedings of 1994 IEEE Applied Power Electronics Conference and Exposition - ASPEC'94*, 1994, 159–165 vol.1. DOI: 10.1109/APEC.1994.316405.

[16] S. Koning, D. Zuidervliet, and P. van Duijsen, "Educational laboratory demonstrator for teaching dual active bridge control principles," in *2022 22nd International Symposium on Electrical Apparatus and Technologies (SIELA)*, 2022, pp. 1–4. DOI: 10.1109/SIELA54794.2022.9845754.

[17] *DC-Lab.org — dc-lab.org*, http://dc-lab.org, [Accessed 08-04-2024].

[18] *C2000 real-time microcontrollers — TI.com — ti.com*, https://www.ti.com/c2000, [Accessed 08-04-2024].

[19] R. Steigerwald, R. De Doncker, and H. Kheraluwala, "A comparison of high-power dc-dc soft-switched converter topologies," *IEEE Transactions on Industry Applications*, vol. 32, no. 5, pp. 1139–1145, 1996. DOI: 10.1109/28.536876.

PCIM Europe 2024, 11– 13 June 2024, Nuremberg   DOI: 10.30420/566262455

A Rail Traction Converter Platform Based on Power Module Implementations with 450 A, 600 A and 800 A 3.3 kV IGBT Modules

Ekrem R. Gunes[1,2], Abdulkerim Ugur[1], Osman S. Senturk[3]

[1] Rail Transport Technologies Institute, TÜBİTAK, Türkiye

[2] Istanbul Technical University, Türkiye

[3] OSSEN Software and Energy Inc. Co., Türkiye

Corresponding author: Ekrem R. Gunes, ekrem.gunes@tubitak.gov.tr
Speaker: Ekrem R. Gunes, ekrem.gunes@tubitak.gov.tr

Abstract

Compared to the single IGBT modules in the conventional package (140x190-130-73 mm footprint), the half-bridge IGBT modules in the new package with 140x100 mm footprint have advantages in terms of power scalability through low-, mid-, and high-power rail traction converter segments as well as DC bus bar design, paralleling, and compatibility with SiC technology for future. This study focuses on an 800 kW Electric Multiple Unit (EMU) (mid-power) and a 1250 kW locomotive (high-power) traction converter applications with two-level circuit topology employing 450 A, 600 A and 800 A 3.3 kV half-bridge IGBT modules in the 140x100 mm package. Using these IGBT modules with 2 and 4 IGBT modules in parallel, a power module platform with the common components is proposed. To compare the converter realization options based on this platform, simulation and experimental studies are conducted and their results are evaluated in terms of IGBT module junction temperatures, current sharing performances and commutation inductance values. This paper shows that the power modules with 2- and 4-paralleled 450 A, 600 A and 800 A IGBT modules with proper junction temperature, commutation inductance and current sharing can be accommodated as a platform solution in the rectifier and inverter circuits of the studied mid- and high-power applications.

1   Introduction

Rail vehicles can be roughly divided into three segments in terms of traction converter power requirement: low-, mid- and high-power segments. The low-power segment consists of light rail vehicles with <400 kW power rating of traction converter. The mid-power segment with >400 kW and <1000 kW power rating comprises metros and low-speed EMUs (commuter trains) in general. Having >1000 kW power capacity per traction converter stage, locomotives and high-speed electrical multiple units (EMUs) are in the high-power segment [1]. It should be noted that this categorization may be influenced by power stage configuration such as feeding from multiple transformer windings and driving multiple motors in parallel (See Fig. 1).

For these converters, usage of highly reliable, high-power switching devices (IGBTs, diodes) is essential. Hence, single IGBT modules with the conventional package (140x190-130-73 mm footprint) are often used in railway traction systems [2]-[4] due to their power capability and cost

despite that fairly new (140x100 mm footprint) IGBT half-bridge modules are getting attractive due to its scalability by paralleling, ever increasing current capability, easier busbar design with less commutation inductance and compatibility with SiC semiconductors [5], [6]. For the low- and mid-power segments, the IGBT half-bridge modules in the new package are more competitive against the single IGBT modules because of better current rating match with reasonable paralleled IGBT module count in addition to its abovementioned advantages. Moreover, the increasing IGBT module current rating in the new package from 450 A to 800 A makes its application to the high-power segment possible. Hence, these three power segments can be covered by the new IGBT half-bridge package as a single converter platform solution.

This study focuses on a typical 800 kW EMU (mid-power) and a 1250 kW locomotive (high-power) traction converter stages with two-level converter circuit topology employing 450 A, 600 A and 800 A 3.3 kV half-bridge IGBT modules in the 140x100 mm package. In this paper, first, the traction converter circuit options and the power module (PM) platform with these IGBT modules are introduced for the 800 kW and 1250 kW applications. Secondly, these circuits with the PM options are simulated electrically and thermally for each configuration to estimate semiconductor junction temperatures as well as their current capability at the limit temperatures. Next, the double-pulse test (DPT) results for the PM options are demonstrated to calculate commutation inductances and IGBT module current sharing. Finally, the continuous operation test (COT) results for the PM options are provided to show IGBT module current sharing.

This paper shows that the PMs with 2- and 4-paralleled 450 A, 600 A and 800 A IGBT modules with proper junction temperature, commutation inductance and current sharing can be accommodated in the rectifier and inverter circuits of the studied mid-power (800 kW) and high-power (1250 kW) applications as a PM platform solution. For the mid-power segment, 450 A and 600 A IGBT modules fit best while 600 A IGBT modules can provide longer lifetime. For the high-power segment, 800 A IGBT modules fit better as an alternative to the IGBT modules in the conventional package.

2 System Description

The circuit parameters of a typical EMU and locomotive traction converters are given in Table 1. The circuit topology is based on two-level voltage source converter (2L-VSC). In this topology; transformer-side voltage (1-ph 950 V and 1050 V), motor-side voltage (3-ph 1250 V) and the DC bus voltage (1800 V) are appropriate with the commonly available 3.3 kV IGBT voltage rating. As shown in Fig. 1, there is a three-phase full-bridge (TPFB) converter operating as inverter at motor side, which generally drives a single locomotive motor or two EMU motors in parallel [7], [8]. There are two alternatives for the rectifier circuit: single or double single-phase full-bridge (SPFB) converters operates as rectifier at transformer side. Between these two alternatives, single rectifier (SR) circuit has simplicity and cost advantages due to less control requirement while the double rectifier (DR) circuit has better power quality by phase-shifted pulse-width modulation, higher modularity by using the same PM in the rectifier and inverter circuits, and more redundancy by having two independent rectifiers.

The rectifier and inverter circuits are realized by power modules (PMs) composed of IGBT modules, gate drivers, busbars, capacitors, discharge resistors, current sensors, temperature sensors and cooling plates. Due to the cooling plate and

converter mechanical constraints, each power module can accommodate 4 IGBT modules with 100x140 mm footprint in this study (See Fig. 2). In order to match the rectifier and inverter circuit current ratings, the PMs can utilize 2 or 4 IGBT modules in parallel, which is represented in the power module circuit schematics in Fig. 3. A prototype of the PM is shown in Fig. 4. On the same PM platform, the various PM options are realized with only differentiations of IGBT module current rating and paralleling AC busbar. The rest of the PM components such as the capacitor bank, DC busbar, cooling plate and gate driver boards are suitable for each power module option. Hence, the traction converter options can be served by this PM platform, of which main components are listed in Table 2.

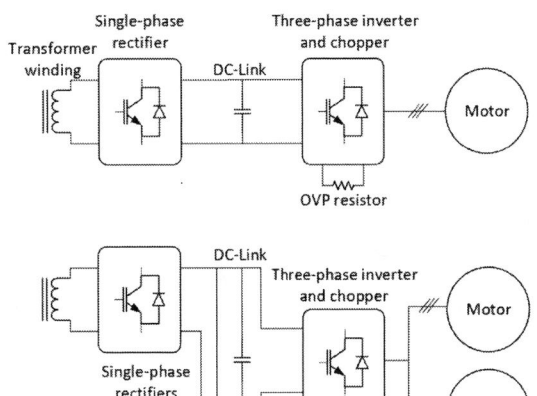

Fig. 1 Simplified traction converter schematics with (above) single rectifier (SR) and (below) double rectifiers (DR)

Specification	EMU	Locomotive
Nominal output power	800 kW 0.85 PF	1250 kW 0.85 PF
Nominal output voltage	1250 Vrms 0-200 Hz	1250 Vrms 0-200 Hz
Nominal output current	~450 Arms	~700 Arms
DC-link voltage	1800 VDC	1800 VDC
Nominal input voltage	950 Vrms 50 Hz	1050 Vrms 50 Hz
Input power factor	~1	~1
Nominal input current for single rectifier	~850 Arms	~1200 Arms
Nominal input current for double rectifier	~425 Arms	~600 Arms

Table 1 Traction converter circuit design parameters

Fig. 2 PM cooling plate with 4 IGBT modules (1-4)

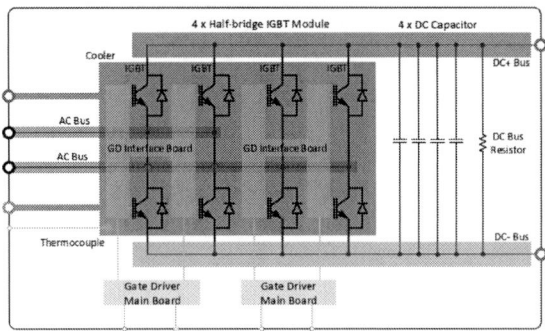

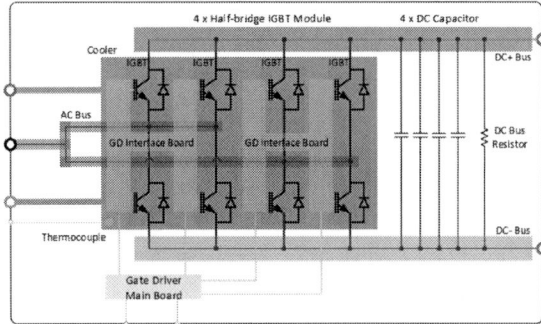

Fig. 3 Schematics of PMs with (top) 2 IGBT modules in parallel and (bottom) 4 IGBT modules in parallel

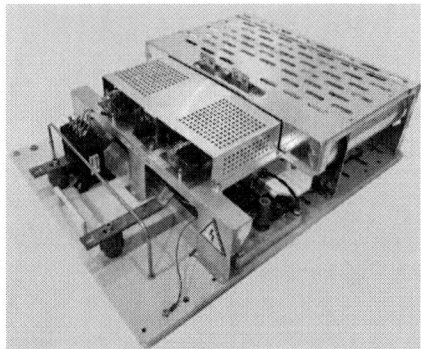

Fig. 4 Power module prototype

Table 3 lists the PM options which can be realized on the PM platform. The identification of a PM option is based on the current rating of the utilized IGBT module (450 A, 600 A, or 800 A) and the number of parallel IGBT modules. For example, the PM with 4-parallel 600 A IGBT modules is identified as PM604. For typical traction converter operating conditions, it is assumed

that PM output current is approximately half the rated current of the utilized IGBT modules in parallel. Please note that this approximation will be verified by the simulation results in Section 3.

In Table 4, the traction converter configurations for the 800 kW and 1250 kW circuits with the single-rectifier (SR) and double-rectifier (DR) cases are listed along with the suitable PM options regarding the match between the output current ratings given in Table 1 and Table 3.

Component	#	Producer - Part Number	Property
IGBT module	4	Hitachi MBM450FS33F, MBM600FS33G2, MBM800GS33G2	3.3kV 450 A, 600 A, 800 A [9]-[11]
GD interface board	2	Amantys/Poweronics XC000139	2x IGBT modules [12]
GD main board	1 or 2	Amantys/Poweronics AP03LA8	3x interface boards driving [12]
Capacitor	4	Electronicon E67.R31704W40	705uF, 25nH 1800 V [13]
Current transducer	1	PETERCEM HRS1000-T-014	1000 Arms
Cooling plate	1	Custom	550x175mm, 15-20 lt/min

Table 2 Power module main components

Power Module Identification	Parallel IGBT Module #	IGBT Module Rating [A]	PM Curr. Rating [A]	PM Output Curr. [Arms]
PM452	2	450	900	~450
PM454	4	450	1800	~900
PM602	2	600	1200	~600
PM604	4	600	2400	~1200
PM802	2	800	1600	~800
PM804	4	800	3200	~1600

Table 3 Power module options

	Rectifier	Inverter + Chopper
800 kW Double R.	2x PM452 2x PM602	2x PM452
800 kW Single Rectifier	2x PM454 2x PM604 1x PM802	2x PM602 2x PM802
1250 kW Double Rectifier	4x PM454 2x PM602 2x PM802	4x PM454 4x PM604 2x PM802
1250 kW Single R.	2x PM604 2x PM804	

Table 4 Traction converter configurations

3 Simulation

With the rectifier and inverter roles, the listed PMs in Table 3 are modelled and simulated electrically and thermally for the traction circuits at the conditions given in Table 5 by PLECS. In the electrical circuit modeling, ideal voltage sources and inductors at both input and output sides are used for the sake of simplification. In the power loss and thermal modeling of the IGBT modules, the datasheet parameters are utilized. In the cooler and module base plate thermal modeling, the simplified thermal models extracted by using CFD analyses' results are employed. In these models, IGBT chips and diode chips are aggregated over the PM cooling plate including the thermal coupling in between.

Parameter	800 kW TC	1250 kW TC
Output power	800 kW, 0.9 PF	1250 kW, 0.9 PF
Output voltage	1200 Vrms, 65 Hz	
Output current	425 Arms	670 Arms
DC bus voltage	1800 VDC	
Input voltage	1000 Vrms, 50 Hz	
Input current (DR/SR)	420/840 Arms, 1 PF	630/1260 Arms, 1 PF
Input inductance	2 mH	1.33 mH
Switching freq.	800 Hz	
Coolant	Water-glycol mix, 50%-50%	
Inlet coolant temp	55°C	
Coolant flow rate	15 lt/min	
Cooling plate and IGBT module base plate R_{th} values (aggr.)	9.17 K/kW (for IGBT chips), 10.79 K/kW (for diode chips), 4.96 K/kW (for coupling in between) 1.14 K/kW (for coolant)	

Table 5 Simulation parameters

For the traction converter applications under investigation, the simulation results of IGBT and diode junction temperatures and PM power losses are tabulated in Table 6 for the configurations given in Table 4.

For the 800kW converter, PM452/4 are suitable with sufficient temperature margin (40 K, 47 K) to the 150°C junction temperature limit. With higher temperature margin (52K, 54K) and less power loss, PM602/4 can be considered more reliable but somewhat underutilized. Having the 175°C limit, PM802 is significantly underutilized. Hence, PM452/4 as rectifier and PM452 as inverter seem the most fitting solution for the 800 kW converter. For the 1250 kW converter, PM454 has high temperature margin (54 K, 58 K). PM602 as rectifier has sufficient temperature margin (29 K) whereas PM604 as inverter has larger margin (62 K), which indicates underutilization. PM802/4 as rectifier and PM802 as inverter have signifi-

cant temperature margins (57 K, 46 K). Hence, PM802/4 as rectifier and PM802 as inverter seem the most fitting solution for the 1250kW converter. In order to find the PM output current limits in rectification and inversion modes, the simulations are conducted to reach the defined junction temperature limits of 125°C for 150°C-limited 450 A and 600 A IGBT modules and 150°C for 175°C-limited 800 A IGBT modules. Table 7 shows the obtained results of junction temperatures for PMs with 2 IGBT modules in parallel, PM power losses and output current limits along with the initial output current approximation. It is apparent that the rectifier operation gives more room (+15-31%) above the approximated output current than the inverter operation (+2.5-15%). Moreover, the 450 A IGBT module has larger output current margin than the others. Since the current sharing deviation between the parallel IGBT modules is <5% in practice, 2-in-parallel IGBT module case is also representative for the 4-in-parallel IGBT module case with almost the same thermal performance for the identical IGBT module loading.

	800 kW Traction Conv. – Rectifier			800 kW Traction Conv. – Inverter		
	T_{TJ}	T_{DJ}	P_{PM}	T_{TJ}	T_{DJ}	P_{PM}
	[°C]	[°C]	[W]	[°C]	[°C]	[W]
PM45x	95	103	3920	110	95	4370
PM60x	88	96	3490	98	87	3630
PM802	89	93	3570	99	86	3810
	1250 kW Traction Conv. – Rectifier			1250 kW Traction Conv. – Inverter		
	T_{TJ}	T_{DJ}	P_{PM}	T_{TJ}	T_{DJ}	P_{PM}
	[°C]	[°C]	[W]	[°C]	[°C]	[W]
PM454	86	92	3020	96	86	3340
PM60x	109	121	5710	88	81	2830
PM80x	109	118	5730	129	105	6310

Table 6 Simulation results at the operating points – IGBT and diode junction temperatures and PM losses

	Rectifier				Apprx.
	T_{TJ}	T_{DJ}	P_{PM}	$I_{out-lim}$	I_{out}
	[°C]	[°C]	[W]	[Arms]	[Arms]
PM452	112	125	5620	590, +31%	450
PM602	112	125	6050	710, +18%	600
PM802	135	150	8500	920, +15%	800
	Inverter				Apprx.
	T_{TJ}	T_{DJ}	P_{PM}	$I_{out-lim}$	I_{out}
	[°C]	[°C]	[W]	[Arms]	[Arms]
PM452	125	104	5530	520, +15%	450
PM602	125	104	5850	650, +8%	600
PM802	150	118	7980	820, +2.5%	800

Table 7 Simulation results at the junction temperature limits – IGBT and diode junction temperatures, PM power losses and the output current limits

4 Experiment

The test setup shown in Fig. 5 is used for both double-pulse tests (DPTs) and continuous operation tests (COTs), which includes single or two PMs (to form a full-bridge circuit), a DC power supply, single or two load inductors (to match PM output current), a cooling unit, a gate signal generator and a control unit. The test setup parameters are given in Table 8.

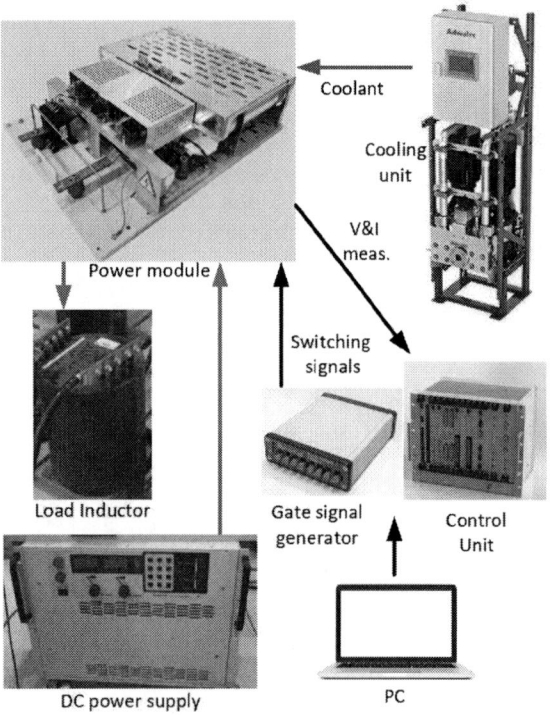

Fig. 5 Test setup for DPT and COT

	450 A IGBT Module	**600 A IGBT Module**	**800 A IGBT Module**
DPT Current	2x900 A 4x900 A	2x1200 A 4x1200 A	2x1600 A 4x1600 A
COT Current	2x225 Arms 4x225 Arms	2x300 Arms 4x300 Arms	2x400 Arms 4x400 Arms
DC bus vol.	1800 VDC		
Load inductance	250 µH (800 Arms) 2x 250 µH (2 in parallel)		
Cool. temp.	20-25 °C (DPT), 40-45 °C (COT)		

Table 8 Test setup parameters

For PM45x, PM60x and PM80x, the DPT results of IGBT turn-on, IGBT turn-off and diode turn-off waveforms are given in Fig. 6. These waveforms show the proper switching operation with <5% current sharing deviation and <500 V overvoltage. Only, PM604's 2nd and 4th IGBT's have slight current ringing against each other during turn on.

For calculating current sharing deviation, the current measurements at the middle of the second pulse are taken and used in Eq. (1) for IGBT and Eq. (2) for diode. The results are listed in Table 9. It is obvious that more paralleling causes more deviation; however, the deviations meet the target of <5%. It should be noted that the AC output cabling has certain influence on the deviation [3].

$$\Delta I_{C,max} = max(|I_{C,1-4} - I_{C,mean}|/I_{C,mean}) \quad (1)$$

$$\Delta I_{F,max} = max(|I_{F,1-4} - I_{F,mean}|/I_{F,mean}) \quad (2)$$

%	**Power Module**					
	452	**454**	**602**	**604**	**802**	**804**
DPT-IGBT	1.9	2.7	3.4	4.5	0.5	3.1
DPT-Diode	0.9	1.4	0.3	3.9	1.7	-
COT-IGBT	0.9	1.5	1.4	3.3	2.0	-
COT-Diode	0.5	3.0	0.3	2.9	0.5	-
Target	<5					

Table 9 Experimental results – Current sharing deviation from DPTs and COTs for each PM

Using the IGBT turn-on waveforms, the commutation inductance per IGBT module (L_S) is calculated by means of Eq. (3), where V_{drop} is the voltage drop between auxiliary collector and emitter terminals and $\Delta I/\Delta T$ is the collector current rate of change during IGBT turn-on.

$$L_s = V_{drop} * \Delta T/\Delta I \quad (3)$$

For each IGBT module position in PM, the calculated inductance values are given in Table 10. It is seen that the L_S values are less than or almost equal to the IGBT datasheet test value of 40 nH [9]-[11]. As expected, PMxx4 has around 25-35% more inductance than PMxx2 due to more confined use of the DC bus bar and the capacitors. On the other hand, the slight inductance difference between PM45x and PM60x/PM80x is considered to be related with IGBT module internally.

nH	**Power Module**					
Position	**452**	**454**	**602**	**604**	**802**	**804**
Module 1	28.8	36.0	30.5	39.6	32.0	38.9
Module 2	28.6	36.0	30.2	39.8	33.1	39.2
Module 3	-	36.4	-	40.5	-	40.6
Module 4	-	38.0	-	40.4	-	40.1
Datasheet	40					

Table 10 Experimental results – Commutation inductance for each IGBT module position (from left to right)

The COT results are given in Fig. 7 as the waveforms of the DC bus voltage, the load current and the IGBT-diode pair current for each IGBT module position. For PM454 and PM604, it is visible that there is a ringing current among the IGBT modules mainly due to the simple paralleling bar between Module 1-2 and Module 3-4.

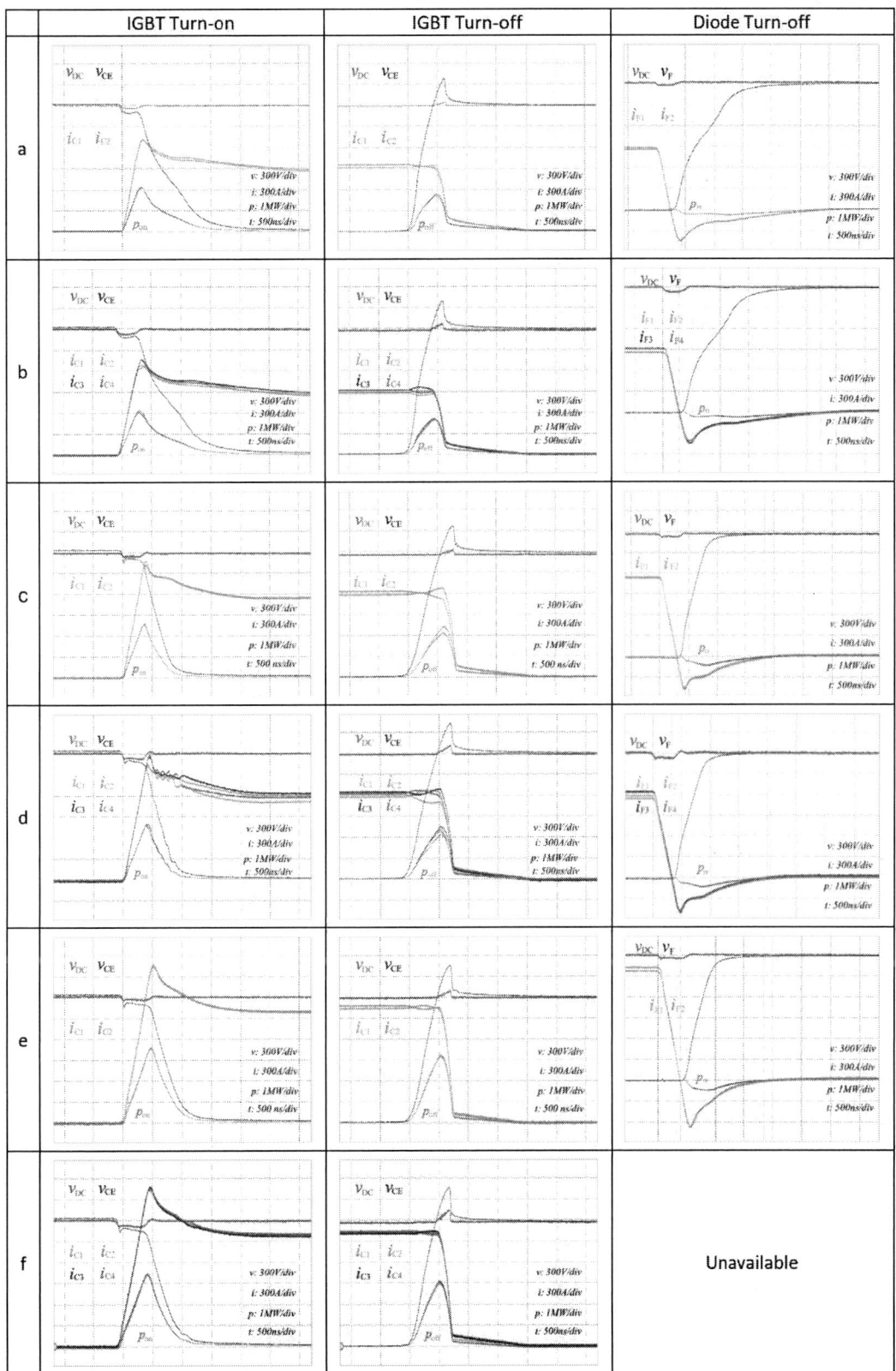

Fig. 6 DPT results: IGBT turn-on, IGBT turn-off and diode turn-off waveforms of v_{DC}, v_{CE}, i_{Cx}, v_F, i_{Fx} and p_x for (a) PM452, (b) PM454, (c) PM602, (d) PM604, (e) PM802, (f) PM804

PCIM Europe 2024, 11– 13 June 2024, Nuremberg   DOI: 10.30420/566262455

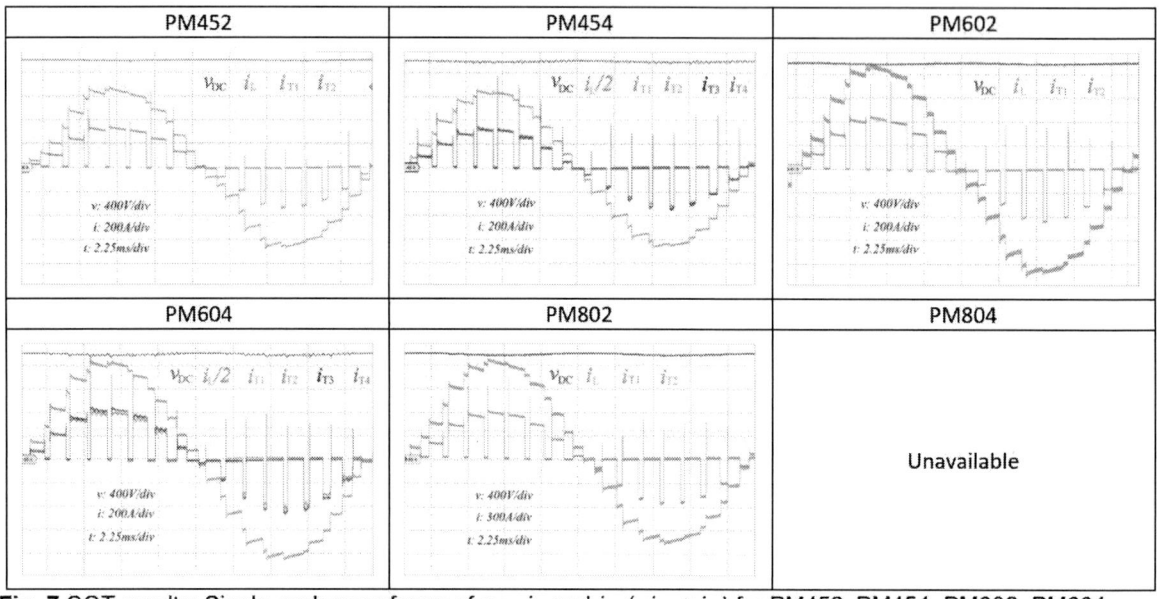

Fig. 7 COT results: Single-cycle waveforms of v_{DC}, i_L, and i_{Tx} (=i_{Cx} + i_{Fx}) for PM452, PM454, PM602, PM604, PM802 (PM804's test results could not be obtained until the submission date due to a test setup failure)

From the COT results, the current sharing deviation is calculated for the IGBT and diode current PWM pulses with the largest magnitude (around the sinusoid's peaks) similarly to the calculation for the DPT results. The current measurements at the middle of these pulses are taken and used in Eq. (1) for IGBT and Eq. (2) for diode. For the case with PM454 and PM604, where the ringing occurs, the measurement points are slightly shifted to avoid the influence of the ringing on current sharing. The results are listed in Table 9 showing that the deviations are within the target of <5%.

5   Conclusion

For high- and mid-power rail traction converters, the IGBT modules in the 140x100 mm package at various current ratings are suitable due to their power scalability by paralleling. In this study, a PM platform with 450 A, 600 A and 800 A half-bridge 3.3 kV IGBT modules is proposed to serve the traction circuits under investigation. By means of the simulation and experimental results, it has been shown that this PM platform is applicable in terms of semiconductor junction temperatures, current sharing deviation and commutation inductance. In particular, the PM with 450 A IGBT modules and the PM with 800 A IGBT modules are the most suitable for the 800 kW converter and the 1250 kW converter, respectively.

Acknowledgement

The authors gratefully thank to TÜBİTAK-RUTE and Power Electronics & Control Group Members for their support in this work as well as Neil Markham and Chris White from Hitachi EU for their support in IGBT module thermal modeling.

References

[1] Mohamed, Ahmed A., et al., "Transportation Electrification: Breakthroughs in Electrified Vehicles, Aircraft, Rolling Stock, and Watercraft." John Wiley & Sons, 2022.

[2] Soltau, N., et al., "High-Voltage IGBT Modules for High-Power High-Reliability Applications.", Bodo's Power Magazine, Sept. 2021.

[3] Weigel, Jan, et al., "Paralleling of High Power Dual Modules: Standard Building Block Design for Evaluation of Module Related Current Mismatch", ECCE Europe, 2018.

[4] T. Kushima, et al., "1800A/3.3kV IGBT Module using Advanced Trench HiGT Structure and Module Design Opt", PCIM Europe 2014.

[5] Benefits of Using the 1700V and 3300V High Power Modules for Traction Applications Praneet Bhatnagar & Chris White, Hitachi Europe Bodo's Power Magazine, Feb. 2022.

[6] D. Kawase, et al., "High voltage module with low internal inductance for next chip generation - next High Power Density Dual (nHPD2)", *PCIM Europe 2015*.

[7] M. M. Bakran et al., "Power Electronics Technologies for Locomotives", *Power Conversion Conference* - Nagoya, Japan, 2007.

[8] D. Ronanki, S. A. Singh and S. S. Williamson, "Comprehensive Topological Overview

3227

of Rolling Stock Architectures and Recent Trends in Electric Railway Traction Systems", IEEE Trans. on Transp. Electrification, vol. 3, no. 3, pp. 724-738, Sept. 2017.

[9] Silicon N-channel IGBT 3300V F version, MBM450FS33F Datasheet, Hitachi, Spec. No. IGBT-SP-14035 R6.

[10] Silicon N-channel IGBT 3300V G2 version, MBM600FS33G2 Datasheet, Hitachi, Spec. No. IGBT-SP-20010 R2.

[11] Silicon N-channel IGBT 3300V G2 version, MBM800GS33G2 Datasheet, Hitachi, Spec. No. IGBT-SP-21023 R3.

[12] Poweronics Amantys AP03LA8, Dual Channel Gate Drive and Module Interface Card, Datasheet

[13] Electronicon, GA85 Mesis DC Capacitors with Protection Devices, E67 Catalog, December 2022

PCIM Europe 2024, 11– 13 June 2024, Nuremberg   DOI: 10.30420/566262456

Comparison of Selected Megawatt-Level Traction Converter Power Module Implementations in terms of Commutation Inductance and Practicality

Abdulkerim Ugur[1], Ekrem R. Gunes[1,2], Osman S. Senturk[3]

[1] Rail Transport Technologies Institute, TÜBİTAK, Türkiye

[2] Istanbul Technical University, Türkiye

[3] OSSEN Software and Energy Inc. Co., Türkiye

Corresponding author: Abdulkerim Ugur, abdulkerim.ugur@tubitak.gov.tr
Speaker: Abdulkerim Ugur, abdulkerim.ugur@tubitak.gov.tr

Abstract

Rail vehicles utilize traction converter in high-, mid-, and low-power segments. The high power segment converters with above 1 MW are mainly utilized in locomotives and high speed EMUs. For these converters; semiconductors, capacitors and busbars are critical regarding power capability, electrical performance, reliability as well as practicality (ease of production, maintenance and testing). This study focuses on a 1250 kW locomotive traction converter with two-level circuit employing 3.3 kV IGBT modules in conventional package. In this paper, the converter circuit realization options with various DC-bus structure, power module busbars and capacitor types are introduced and experimentally evaluated. This paper shows that the converter realization with distributed DC-bus, two-layer power module busbar and cylindrical capacitors brings improvements in commutation inductance and practicality.

1   Introduction

Rail vehicles can be roughly divided into three segments in terms of traction converter power requirement: High-power, mid-power and low-power segments. The low-power segment consists of metros and light rail vehicles with less than 400 kW power rating of traction converter. The mid-power segment with above 400 kW power rating comprises low-speed EMUs (commuter trains). Having roughly above 1000 kW power capacity per traction converter stage, locomotives and high-speed electrical multiple units (EMUs) are in the high-power segment [1]. It should be noted that this categorization may be influenced by power stage configuration such as driving parallel motors and feeding common DC buses via isolated transformer windings.

For the high-power segment, traction converter is designed to handle power conversion at megawatt level. Therefore, usage of highly reliable, high power switching devices (IGBTs, diodes) and passive elements (capacitors, busbars) is essential for such applications. In this regard, single IGBT modules with 140x190 mm footprint are used conventionally in railway traction systems [2], [3] due to its power capability and cost despite that 140x100 mm IGBT phase-leg modules are getting

attractive due to its scalability by paralleling, ever increasing current capability, easier busbar design with less commutation inductance and compatibility with SiC semiconductors [4].

In traction converter design, DC-Link capacitors are selected carefully since capacitor self-inductance directly affects the voltage overshoot on IGBT modules during switching. Traditionally, box-shape (prismatic) large MKP type capacitors are used in high power class [5]. These capacitors have the ability of self-healing, which restores capacitor dielectric layer after a dielectric fault. Moreover, pressure switches may be provided to prevent rupture. On the other hand, having the same self-healing ability, cylindrical capacitors are advantageous in terms of paralleling, scaling, busbar design and enabling lower commutation inductance. Moreover, a capacitor failure can be isolated from other capacitors and rest of the power circuit if a safety mechanism is available [6].

Bridging IGBT modules and DC-link capacitors, DC busbars are critical since they influence commutation inductance strongly; hence, should be designed accordingly [3]. At power module level and converter level, various options with difference in ease of production, maintenance, and testing can be considered.

This study focuses on a 1250 kW locomotive traction converter stage with two-level circuit topology employing single IGBT modules in 3.3 kV conventional package (140x190 mm) at 1.8 kV DC bus. In this paper, first, the traction converter circuit topology and its realization options with various DC bus structures, power module busbars and capacitor types are introduced. Next, test methodology is discussed. Then, the experimental results are shown to compare the converter realization options.

This paper shows that the converter realization with distributed DC bus, simplified two-layer power module busbar and cylindrical capacitors brings improvements in terms of commutation inductance and ease of production, maintenance and testing.

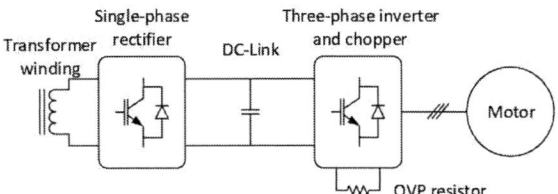

Fig. 1 Simplified schematic of a locomotive traction converter

2 System Description

The circuit parameters of the locomotive traction converter are given in Table 1. As shown in Fig. 1, the converter is composed of a single-phase rectifier, a DC-link, a three-phase inverter, a chopper and a overvoltage protection (OVP) resistor. The converter circuit topology is two-level voltage source converter (2L-VSC). Within this topology; transformer-side single-phase 1050 V_{rms}, motor-side 3-phase 1250 V_{rms} and DC-Link 1800 VDC are appropriate with the commonly available 3.3 kV IGBT voltage rating. The rectifier, inverter and chopper circuits are realized as power modules including IGBT modules, gate drivers, busbars, capacitors (for distributed DC bus structure only), discharge resistors (for distributed DC bus structure only), temperature sensors and cooling plates. Considering converter cabinet mechanical constraints and cooling performance, each converter circuit is composed of 2 rectifier and 2 inverter/chopper power modules with 4 IGBT modules per power module. Fig. 2 shows simplified representations of the inverter/chopper and rectifier power modules.

In power module design, simplicity and least amount of differentiation between rectifier and inverter power modules are targeted in order to ease converter cabinet design, production, maintenance and testing activities. In this study, the following converter realization options are studied:

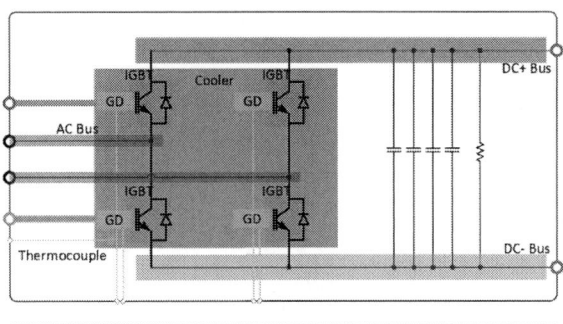

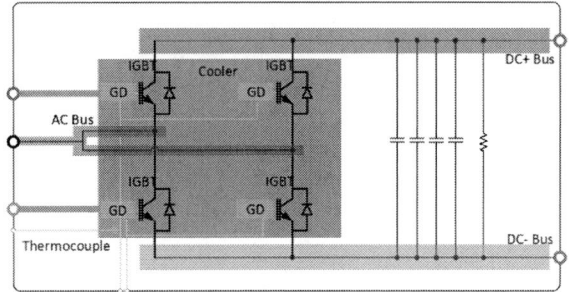

Fig. 2 Inverter/chopper power module (top), Rectifier power module (bottom)

- DC bus structure (Fig. 3): In the central DC bus structure, all capacitors and power modules are connected to each other via a laminated DC busbar backbone. In the distributed DC bus structure, each power module has its own DC busbar (without a laminated backbone) and capacitor(s) for low inductive commutation. Hence, each power module can be tested individually and, in case of failure, repaired individually.

- Power module busbar (Fig. 4): In the three-layer design; DC+, DC- and AC laminar busbars are stacked together to cancel the magnetic fields generated by the currents in both directions. In the two-layer design as proposed in [3]; DC+ and AC busbars are simplified by removing unused DC+ and AC busbar parts, which brings advantage with simpler structure, less material usage, and lower inductance.

Specifications	Value
Nominal output power	1250 kW, 0.85 PF
Nominal output voltage	1250 Vrms, 0-200 Hz
Nominal output current	~700 Arms
Nominal DC-link voltage	1800 VDC
Nominal input voltage	1050 Vrms, 50Hz, ~1 PF
Nominal input current	~1200 Arms

Table 1 Traction converter circuit parameters

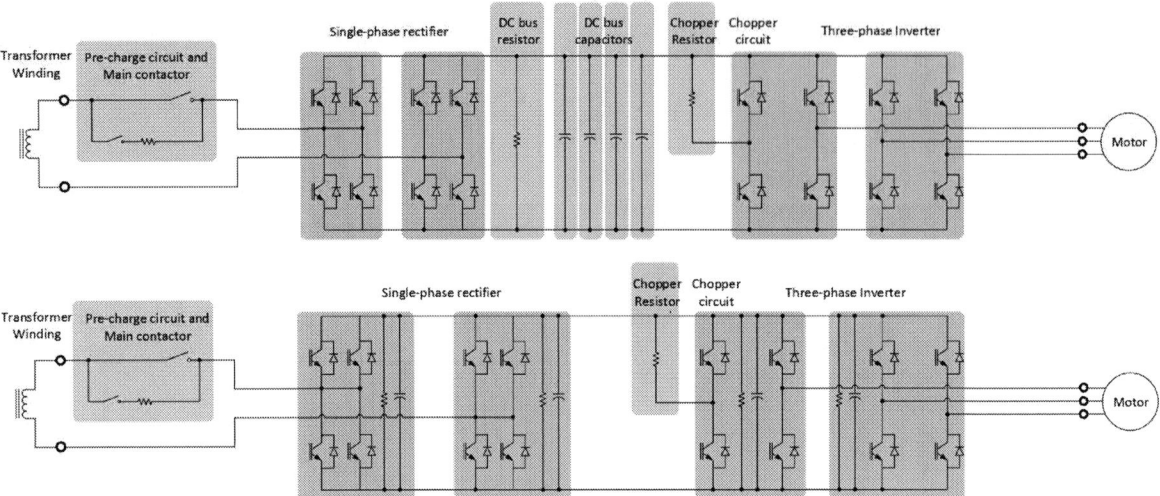

Fig. 3 DC bus structure: central (top) and distributed (bottom)

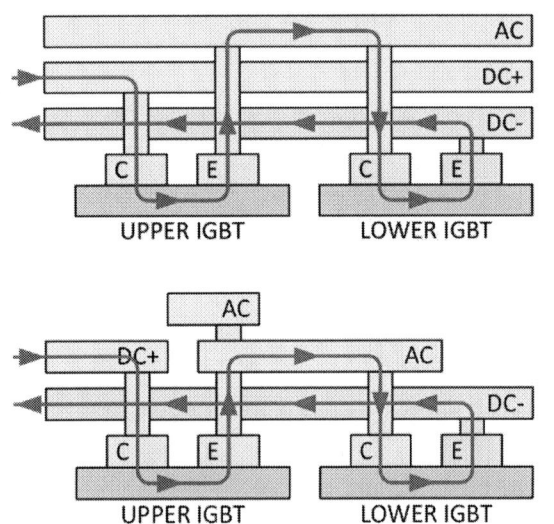

Fig. 4 Representations of busbars with commutation path: three-layer (top) and two-layer (bottom) busbars

- Capacitor type (Fig. 5): Prismatic capacitors have higher self-inductance compared to cylindrical capacitors due to internal cell layout and number of terminals. Moreover, cylindrical capacitors can be connected to a DC busbar in a more distributed manner, which is favorable in terms of stray inductance. Regarding their sizes, bulkier prismatic capacitors make production and maintenance more difficult comparatively. In case of a capacitor failure, least amount of cylindrical capacitors needs to be replaced thanks to the protection mechanism provided at each capacitor.

Fig. 5 Prismatic capacitor (top): Terminals and pressure switch. Cylindrical (bottom) capacitor: Terminals and overpressure protection device

3 Study

The cases given in Table 2 and shown in Fig. 6 are studied experimentally for rectifier and inverter power modules in terms of commutation inductance, overvoltage, and current sharing by means of double-pulse tests.

Case	DC bus	Cap. type	PM Busbar
1	Central	Prismatic	3-layer
2	Distributed	Prismatic	3-layer
3	Distributed	Cylindrical	3-layer
4	Distributed	Cylindrical	2-layer

Table 2 Study cases

The description of the study cases are as follows.

In Case 1, there are two rectifier and two inverter power modules connected (via fast connectors) to a common low-inductive laminar DC busbar backbone together with top and bottom prismatic capacitor pairs (See Fig. 6 (a)). Having a bulky backbone, bulky prismatic capacitors and necessity of a full converter setup for testing, this converter realization option is disadvantageous in terms of practical aspects such as production, maintenance and testing.

In Case 2, the busbar structure is simplified such that each power module has its own prismatic capacitor with <50 nH internal inductance and a simple DC busbar backbone shared by the power modules may be utilized, which improves the structure's practicality compared to Case 1. Hence, commutations are restricted to be within the power module via the low-inductive laminar three-layer busbar; however, the other 3 capacitors' contributions on commutation inductance reduction are lost.

In Case 3, the prismatic capacitor is replaced with 4 cylindrical capacitors with the same total capacitance and <25 nH internal inductance each. Hence, the commutation inductance is improved compared to Case 2 in addition to the improvement in production and maintenance due to less bulky capacitors.

In Case 4, the power module DC busbar is simplified to have 2 layers instead of 3 layers without sacrificing but even improving commutation inductance slightly. Moreover, 2-layer design brings the advantages in DC busbar production since it requires less material amounts and process complexity.

4 Experimental Results

The experimental setup is shown in Fig. 7. The test setup parameters and the test equipment list are given in Table 3. The switching characteristics of the power modules are evaluated by means of double pulse tests. To estimate IGBT commutation inductance, IGBT module V_{CE} terminal voltage and collector current are measured [7]. The commutation inductance value is estimated via Eq. (1) using the voltage drop and collector current change during IGBT turn-on. It should be noted that the calculated commutation inductance does not include the internal IGBT module inductance, which is given as 12 nH in its datasheet [7] since V_{CE} is measured at the main power terminals.

$$L_s = \frac{V_{drop}}{\Delta I}\Delta T \tag{1}$$

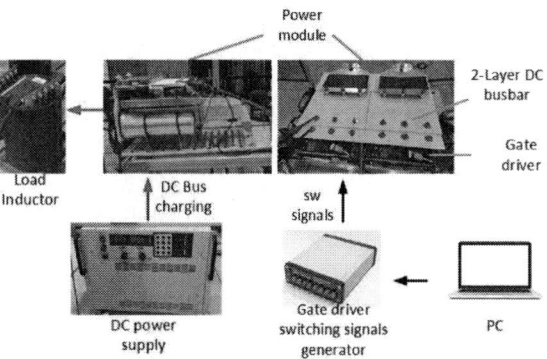

Fig. 7 Test setup

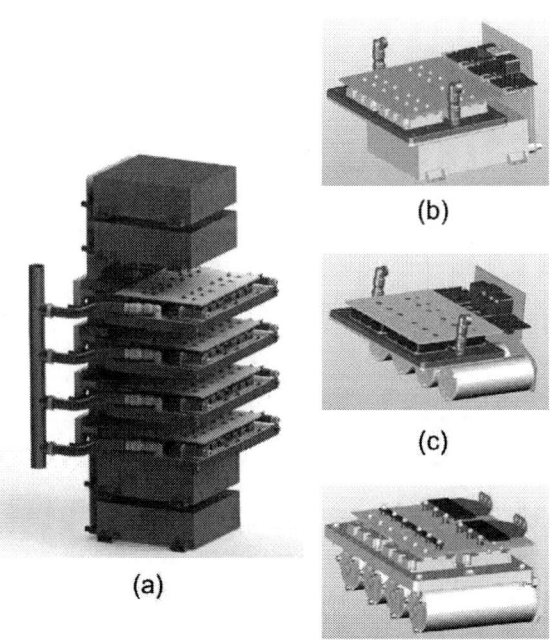

Fig. 6 (a) Case 1, (b) Case 2, (c) Case 3, (d) Case 4

	Case 1	
Parameter	**Inverter**	**Rectifier**
Load inductance	250 µH	125 µH
DC bus voltage	~1800 VDC	~1750 VDC
Switching current	~3400 A	~6000 A
Prismatic capacitor	4x Electronicon E59.G50-255010, 2500 µF	
	Case 2-3-4	
Load inductance	250 µH	125 µH
DC bus voltage	~1700 VDC	~1600 VDC
Switching current	~3000 A	~6000 A
Prismatic capacitor	1x Electronicon E59.G50-255010, 2500 µF	
Cylindrical capacitor	4x Electronicon E67.R31-704W40, 705 µF	

IGBT module	Hitachi MBN1500E33E2
IGBT gate driver	Power Integrations 1SP0635, $R_{g(on)}$: 2.8 Ω, $R_{g(off)}$: 4.1 Ω
Oscilloscope	Tektronix MSO4104C
Voltage probes	Tektronix THDP0100
Current probes	PEM CWT30R
Power supply	Magnapower TSD4000
Fiber optic gate signal generator	Amantys Power Insight Adapter

Table 3 Test setup parameters and test equipment list

IGBT turn-on and turn-off waveforms for each case are given in Fig. 8 and Fig. 9. The experimental results are summarized in Table 4. Benchmarking with Case 1, the commutation inductance is worsened in Case 2 by ~50% due to using single prismatic capacitor instead of 4. In Case 3, there is improvement in commutation inductance by ~10 % by using cylindrical capacitors instead of prismatic. In Case 4, the commutation inductance is further improved by using two-layer DC busbar instead of three-layer by 36% and 56% for inverter and rectifier modules, respectively. On the other hand, the influence of the commutation inductance variations from Case 1 to Case 4 is less pronounced on the turn-off overshoot voltage since the lower commutation inductance increases the turn-off di/dt, which somewhat compensates for the overvoltage value.

Additionally, in this table, the study cases are compared regarding practicality being ease in production, maintenance and testing. In Case 1, production, maintenance and testing are relatively difficult due to the backbone busbar and distributed capacitors. In Case 2; production, maintenance and testing become easier but the commutation inductance is increased and voltage turn-off overshoot is higher. In Case 3, by replacing the prismatic capacitors with cylindrical ones, the commutation inductance is reduced without sacrificing practicality. In Case 4, the two-layer busbar structure makes production easier compared to the three-layer structure. Overall, having more modular structure, smaller capacitors and simple DC bus bar, Case 4 is considered to serve as the most practical solution compared to the other three cases.

5 Conclusion

For high-power rolling stock traction converters such as locomotives, 140x190mm IGBT modules are still attractive due their power capability and cost. In this paper, the converter structures with distributed and central DC busbars, 3-layer and 2-layer power module busbars, and cylindrical and prismatic capacitors, are examined. It has been shown that the distributed DC busbar structure with two-layer power module busbar and cylindrical capacitors not only provides very good electrical performance in terms of commutation inductance but serves as the most practical solution in terms of production, maintenance and testability.

Acknowledgement

The authors gratefully thank to TÜBİTAK-RUTE and Power Electronics and Control Group Members for their support in this work.

	Case 1	Case 2	Case 3	Case 4
Commutation inductance Inverter	44nH (100%)	63.6 nH (+45%)	39.7 nH (-10%)	28.3 nH (-36%)
Commutation inductance Rectifier	81 nH (100%)	122.6 nH (>100 nH*, +51%)	63.6 nH (-11.5%)	35.6 nH (-56%)
Turn-off voltage overshoot Inverter	525 V 100%	605 V +15%	500 V -5%	470 V -10%
Turn-off voltage overshoot Rectifier	570 V 100%	800 V +40%	625 V +10%	555 V -3%
Current sharing (Rectifier only)	<%2	<%2	<%2	<%2
Production	Difficult	Moderate	Moderate	Easy
Maintenance	Difficult	Moderate	Easy	Easy
Testing	Difficult	Easy	Easy	Easy
Practicality	Low	Moderate	Moderate	High

* Commutation inductance value given in the semiconductor datasheet [7].

Table 4 Comparison of the study cases

PCIM Europe 2024, 11– 13 June 2024, Nuremberg   DOI: 10.30420/566262456

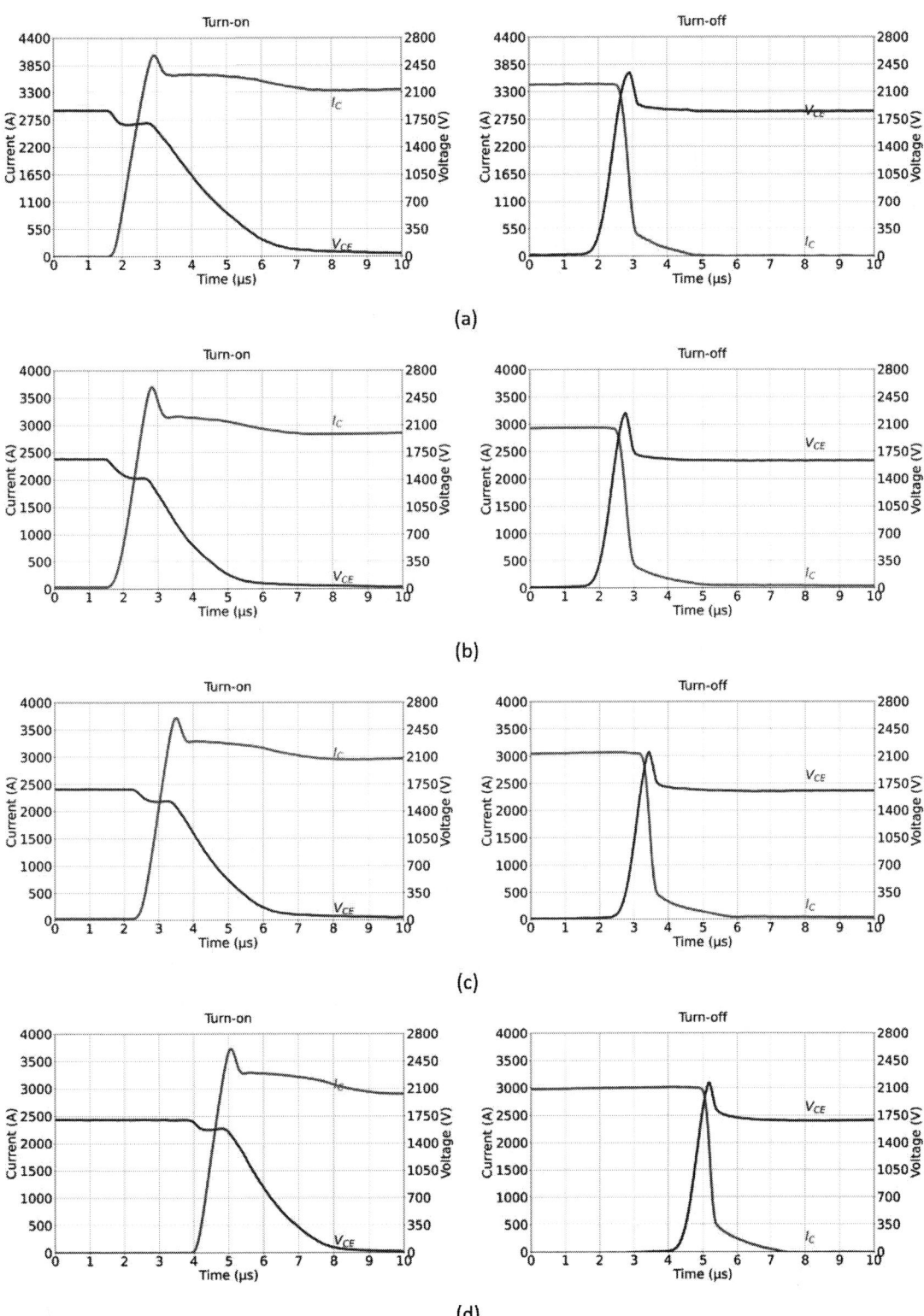

Fig. 8   Inverter IGBT turn-on and turn-off switching waveforms: (a) Case 1, (b) Case 2, (c) Case 3, (d) Case 4

PCIM Europe 2024, 11– 13 June 2024, Nuremberg   DOI: 10.30420/566262456

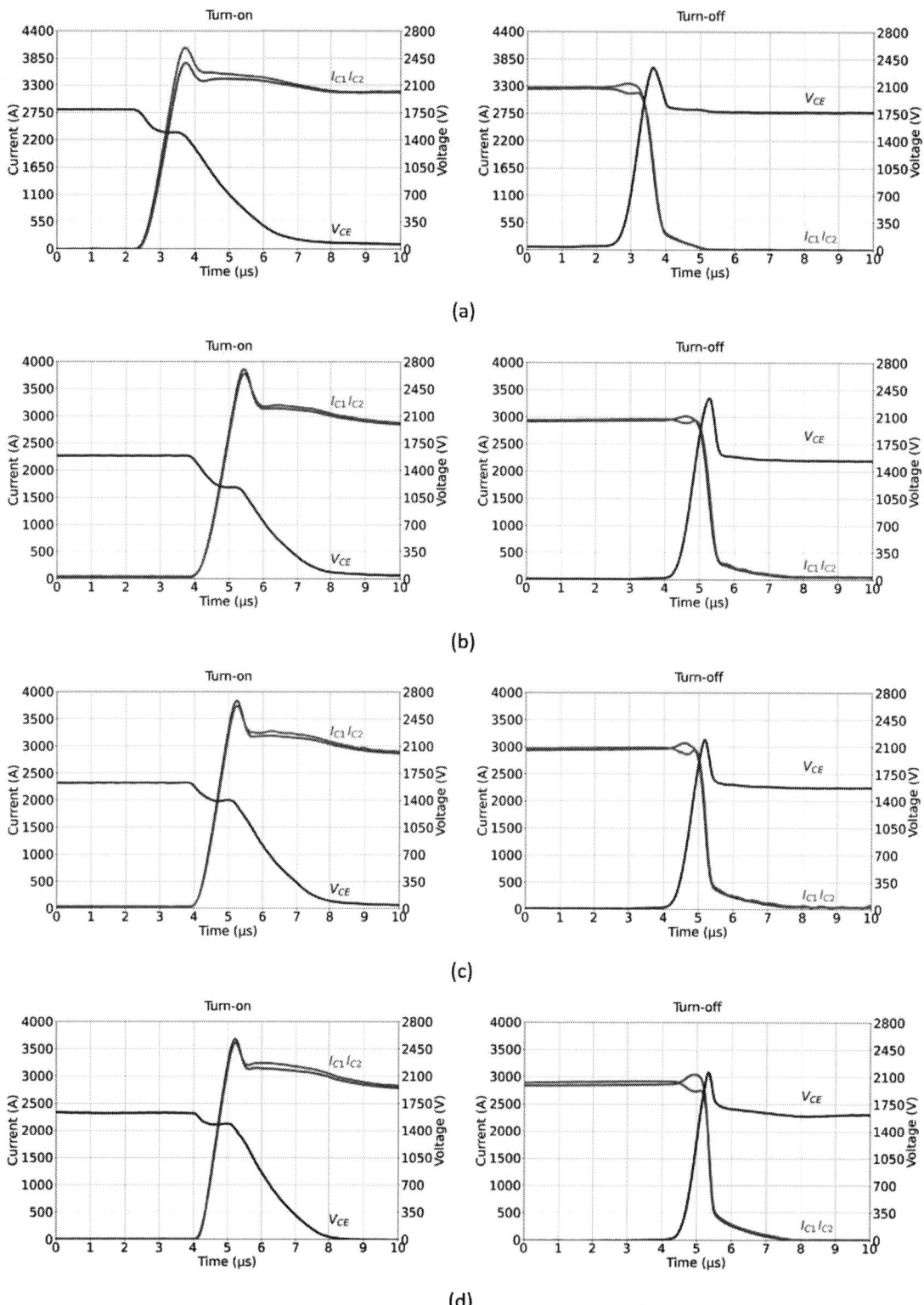

Fig. 9   Rectifier IGBT turn-on and turn-off switching waveforms: (a) Case 1, (b) Case 2, (c) Case 3, (d) Case 4

References

[1] Mohamed, Ahmed A., et al., "Transportation Electrification: Breakthroughs in Electrified Vehicles, Aircraft, Rolling Stock, and Watercraft." John Wiley & Sons, 2022.

[2] Soltau, N., Stumpf, E., Sakai, J., & Uemura, H. "High-Voltage IGBT Modules for High-Power High-Reliability Applications.", Bodo's Power Magazine, Sept. 2021.

[3] Weigel, Jan, et al. "Paralleling of High Power Dual Modules: Standard Building Block Design For Evaluation Of Module Related Current Mismatch", ECCE Europe, 2018.

[4] Tsuda, Ryo, et al. "LV100 High Voltage Dual Package in Paralleling Operation.", PCIM Europe, 2018.

[5] AC and DC Capacitors with Large Capacitances, E57/E59/SR17 Catalog, Electronicon, December 2022

[6] GA85 Mesis DC Capacitors with Protection Devices, E67 Catalog, Electronicon, December 2022

[7] Silicon N-channel IGBT 3300V E2 version, MBN15000E33E2 Datasheet, Hitachi, Spec.No.IGBT-SP-08002 R9

PCIM Europe 2024, 11– 13 June 2024, Nuremberg   DOI: 10.30420/566262457

Pitfalls and their Avoidability in the Double-Pulse Test

Nikolas Förster[1], Daniel Urbaneck[1], Benedikt Kohlhepp[2], Daniel Kübrich[2], Oliver Wallscheid[3],
Frank Schafmeister[1]

[1] Paderborn University, Germany
[2] Friedrich-Alexander-Universität Erlangen-Nürnberg, Germany
[3] University of Siegen, Germany

Corresponding author: Nikolas Förster, foerster@lea.upb.de

Abstract

Double-pulse tests are a common method to characterize the switching behavior of power semiconductors. However, there are many pitfalls when performing them. Especially the higher transients typically occuring with wide-bandgap semiconductors make the double-pulse test much more challenging as compared to setups with normal silicon semiconductors. This paper gives an overview of error sources and their troubleshooting. Using before and after comparisons, it is explained how faulty current sensors can be identified and the parasitic capacitance of the load-current-forming inductor can be decoupled from the switching process for clean switching. Furthermore it is shown, how a suitable voltage probe and oscilloscope can be selected for the given application. The theory of switching behavior will also be discussed in order to provide the basis for the experiments.

1 Introduction

Determining switching losses [1] by using a double-pulse test is the standard method today, in addition to the calorimetric measurement method [2, 3] (measuring the full losses and subtract the well-known conduction losses). There are various tutorials [4] for performing double-pulse tests, but special care must be taken to ensure that the test is performed correctly. There are pitfalls in many areas that are often not directly addressed in full depth. Especially when using modern wide bandgap semiconductors like SiC and GaN, which can achieve high $\mathrm{d}v/\mathrm{d}t$, the characterization becomes more challenging. This paper provides an overview of some failures based on before and after comparisons.

Fig. 1 shows the clamped inductive switching-test setup, also known as double-pulse test. The circuit is made to measure turn-on and turn-off losses of the device under test (DUT) M_1. In principle, the semiconductors here can be MOSFETs, SiC-MOSFETs or GaN semiconductors. SiC-MOSFETs will be used as the basic principle here. The body-diode of M_2 acts as commutation partner but the

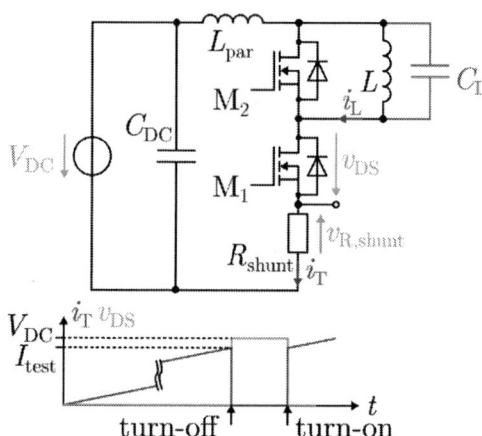

Fig. 1: Standard double-pulse test circuit.

SiC-MOSFET M_2 is kept in turn-off. It is also possible to measure the reverse-recovery losses of M_2 in this setup. The measured current i_T (by $v_{R,\mathrm{shunt}}$) and the measured voltage v_{DS} are referenced as shown (scope ground).

Fig. 2 shows a self-built double-pulse test bench for DC-link voltages V_{DC} up to $800\,\mathrm{V}$ and transistor currents i_T up to $120\,\mathrm{A}$. The test stand can be used universally, since a separate board (Fig. 3) can be plugged in for each type of semiconduc-

Fig. 2: Complete setup of double-pulse set-up.

Fig. 3: PCB for current measurement verification (SMD-shunt vs. coaxial shunt). SiC-MOSFETs of Wolfspeed (C3M0120065J, $1200\,\mathrm{V}$, $65\,\mathrm{m\Omega}$) are exemplary used here.

tor. This separate board contains all device specific things like the power semiconductors themselves, snubber capacitors, drivers, current and voltage measurement. The main board contains all general circuitry, like the DC-link capacitors, control hardware to generate the switch pattern and a communication interface to the computer. Thus, power semiconductors can be changed in a short period of time, what is the major purpose for this approach. A computer controls the voltage source to set the test voltage. By setting a target current I_{test}, the current i_T ramps up to this target current ($i_T = I_{\text{test}}$) and turns-off and -on the SiC-MOSFET. The data is read from the scope and is postprocessed on the computer. At the end of the measurement, data can be stored and evaluated by [5] or [6]. Safety is ensured by touch protection and automatic DC link discharge.

2 Switching behavior investigation

Before measurements are carried out, a theoretical analysis of the switching transitions is perfomed, like in [7, 8].

Given is a SiC-MOSFET with the following driver circuit (Fig. 4), which acts as M_1 in Fig. 1. The driver turns on the SiC-MOSFET by a voltage V_D and the gate resistor R_G. For simplification, the shown gate resistor R_G is the sum of the gate resistor and the internal driver resistance. The SiC-MOSFET's internal gate resistance is assumed to be zero. The SiC-MOSFET M_1 includes the voltage dependent

capacitances C_{GS}, C_{GD} and C_{DS}. These capacitances are summarized in the datasheets as

$$C_{\text{iss}} = C_{\text{GS}} + C_{\text{GD}} \tag{1}$$
$$C_{\text{oss}} = C_{\text{GD}} + C_{\text{DS}} \tag{2}$$
$$C_{\text{rss}} = C_{\text{GD}}. \tag{3}$$

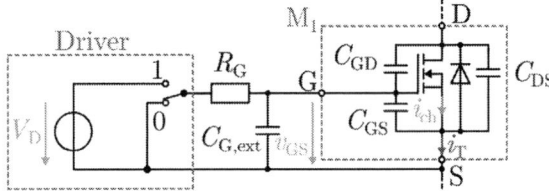

Fig. 4: SiC-MOSFET and it's parasitic capacitances.

The current i_T at the time instant of switching is I_{test} (c.f. Fig. 1). L_{par} represents the commutation loop inductance, which is recommended to be as small as possible. Note that the impact of L_{par} is neglected in the following, except for two small qualitative discussion parts in section 2.1 and 2.2.

2.1 Turn-on behavior

The turn-on process is divided into intervals I to IV. At the beginning of interval I, the driver switches from position 0 to position 1. The capacitance C_{iss} and the optional capacitance $C_{\text{GS,ext}}$ (parallel to C_{GS}) are charged via the gate resistor R_G. Due to the constant voltage V_{DS}, the voltage-dependent capacitances remain constant in this interval (c.f. Fig. 6a). The charging process therefore corresponds to an RC circuit. Interval I ends when the threshold voltage V_{th} is reached (c.f. Fig. 6b). From now on, the SiC-MOSFET begins to conduct

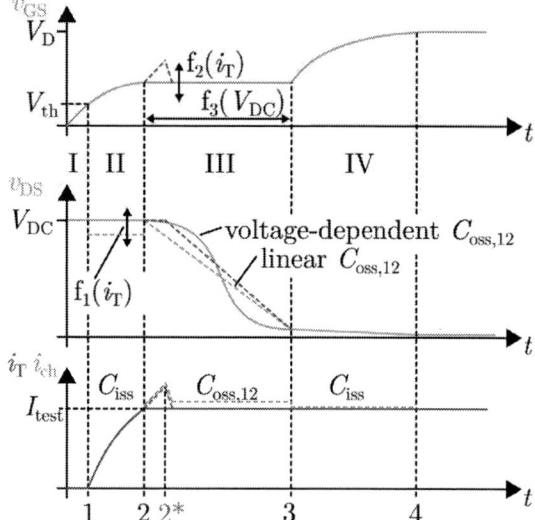

Fig. 5: SiC-MOSFET detailed turn-on behavior.

in interval II. As the gate voltage increases, the current through the SiC-MOSFET also increases and the current through the body diode of M_2 decreases accordingly. The transfer characteristic and the driver itself influences the i_T rise rate. Here, L_{par} plays the significant role, if is is not zero. Due to the current i_T increases in the SiC-MOSFET, a voltage now drops across L_{par}, so that $v_{DS} = V_{DC} - v_{L,par}$. The voltage $v_{L,par}$ is a function f_1 of i_T: $v_{L,par} = L_{par} \cdot \mathrm{d}i_T/\mathrm{d}t$. Interval II ends, once the current through the body-diode of M_2 is zero. Then, the body diode of M_2 turns off and the voltage v_{DS} is no longer clamped. The height of the following Miller plateau is therefore dependent on the current to be switched (f_2) and given by the transfer characteristic in Fig. 6b.

Due to the reverse-recovery effect of the body diode of M_2, a short-term short-circuit current is generated. By linking the channel current with the output characteristic, this couples to the gate voltage (blue marking in Fig. 5 and 6c). Note that high-voltage Si-MOSFETs, like superjunction types, in hard-switching applications (as investigated here) should only be paired with an explicit diode, as their intrinsic body diodes have a very poor reverse-recovery behavior.

In intervall III, the gate driver discharges C_{rss} while C_{DS} is discharged through the conducting channel itself [9]. Note, that due to small-signal parallel connection of M_1's and M_2's C_{oss}, both C_{oss}' must be transshipped. The parallel capacitance of M_1's

and M_2's C_{oss} is labeled as $C_{oss,12}$. Due to the non-linear behavior of the capacitances (c.f. Fig. 6a), the discharging is very slow at the beginning, fast in the middle and slow at the end of Interval III. Note, that the channel current $i_{ch,on}$ is higher than the external measured current i_T, as the discharging current of C_{oss} additionaly flows through the channel [10]. So, $i_{ch} = I_{test} + i_{Coss}$. This results in a higher v_{GS}, according to Fig. 6b. The transfer curve is typcially measured at slow-transition DC-currents, so this is a static current and represents the channel current i_{ch}. The lengh of interval III is mostly definded by V_{DC} and the capability of the driver and R_G to discharge C_{rss} (f_3). Once the discharging process has completed (point 3 in Fig. 6c is reached), the Miller plateau ends and the charging of C_{iss} resumes in interval IV. The charging process of C_{iss} is much slower than in interval I and II, as the C_{iss} has increased significantly due to the non-linearity (c.f. Fig. 6a). During interval IV the channel resistance is reduced to its static value $R_{DS,on}$. As can be seen from the given ex-

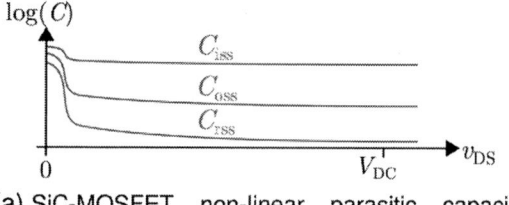

(a) SiC-MOSFET non-linear parasitic capacitances.

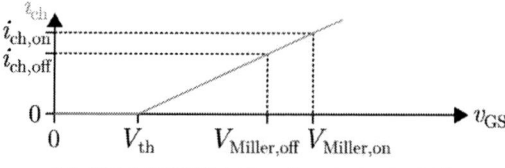

(b) SiC-MOSFET transfer characteristic.

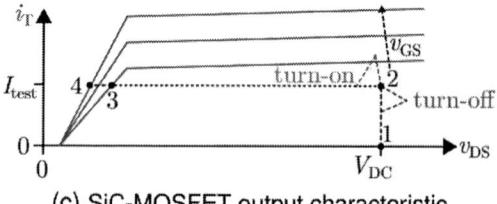

(c) SiC-MOSFET output characteristic.

Fig. 6: SiC-MOSFET important data sheet curves to explain the switching behavior.

planation, the switching speed (at given V_{DC}) to set $\mathrm{d}v_{DS}/\mathrm{d}t$ is controlled by the external gate resistance R_G. An external gate capacitance $C_{G,ext}$

parallel to C_{GS} allows for setting a certain $\mathrm{d}i_T/\mathrm{d}t$ during the switching process. So, $\mathrm{d}i_T/\mathrm{d}t$ is set by the external capacitance together with the R_G, and the $\mathrm{d}v_{DS}/\mathrm{d}t$ is set by R_G only.

The channel current i_{ch} during the switching event differs from the external measured transistor current i_T (definition in Fig. 4). Due to the discharing current of C_{oss}, the channel current i_{ch} is much higher then i_T. This is represented by the orange curve in Fig. 5. External capacitances $C_{DS,ext}$ help to linearize the switching process, but increases switching losses. This approach only makes sense in resonant converters, as there are highly reduced swithing losses. The parasitic capacitance C_L of the load inductor L increases the channel current i_{ch} especially during interval III, due to the same reason as C_{oss} increases the channel current.

Note that the interval limits shift slightly from point 2 to point 2* , what is shown in blue in Fig. 5 due to the reverse recovery effects. The voltage v_{DS} can start falling after the body diode of M_2 stops conducting, what is from the reverse-recovery current peak on.

[11] shows very detailed the current flow in the commutation part, as well as calculating the energy stored into the parasitic capacitances after a certain time of the commutation process.

2.2 Turn-off behavior

Fig. 7 shows the turn-off behavior of the devices. Due to consistency to Fig. 6c, the time step numbers and interval numbers are reversed. While discharging C_{iss} by the gate driver, in interval IV $R_{DS,on}$ increases accordingly to Fig. 6c. Due to the given switching current splits into $i_T = I_{test} = i_{ch} + i_{Coss}$, i_{ch} is lower than I_{test} and the gate voltage $v_{GE} = V_{Miller,off}$ stays nearly constant in interval III. Note, that the channel current i_{ch} during turn-off is lower than the current i_T measured from outside the device. This leads to a lower Miller-plateau voltage $v_{GS,ch}$ curve (orange), than the v_{GS} curve when using i_T (green). Due to the clamped voltage, there is no more discharge of the gate capacitance C_{GS}. The discharging process goes on through the Miller-capacitance C_{rss}. Once v_{DS} reaches V_{DC}, interval III ends. In case of non-zero L_{par}, an overshoot (purple) is produced by this unclamped stray inductances, as L_{par} resonates with $C_{oss,12}$. In interval II, the current commutates from M_1 to the body-diode of M_2. Interval II ends, as the current is zero due

to the $v_{GS} < V_{th}$. Interval I finalizes the discharging of the gate.

Note that the interval limits shift slightly from what is shown in Fig. 7 due to the additional effects (voltage overshoot).

[12] and [13] show variants to fastly bring the gate voltage v_{GS} below the threshold voltage V_{th}, to achieve nearly lossles turn-off. The basic idea is, fastly turing-off the SiC-MOSFET, so that due to the capacitor dividers of C_{DS} and C_{GD} there is a gate voltage that is $v_{GS} < V_{th}$. The remaining current fully charges the output capacitance C_{oss}, while the channel current i_{ch} is zero.

This approach makes the use of SiC-MOSFETs in resonant, zero-voltage switching (ZVS) converters very attractive, as the turn-off losses can be reduced to almost zero by the fast turn-off, and the turn-on losses can be made zero by selecting a resonant topology working in ZVS-region.

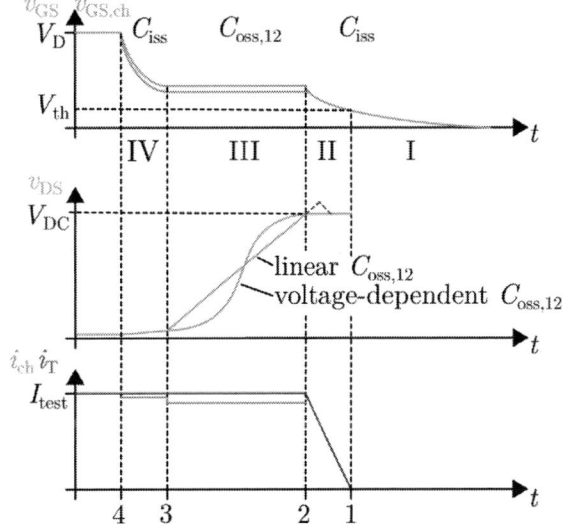

Fig. 7: SiC-MOSFET detailled turn-off behavior.

3 Common pitfalls in the double-pulse test

This chapter shows practical examples of how to perform the double-pulse tests and discusses common mistakes that can be overcome when properly assessed.

3.1 Selection of measurement equipment

Choosing the right measuring equipment for current and voltage measurements is not straightforward.

3240

Due to the high dv/dt and di/dt, switching transient measurement sets high standards on the measuring equipment due to the high voltage and current and a large bandwidth. Thus, a careful selection of measurement equipment (scope in combination with the probes) is essential. Fortunately, there is already quite a lot of literature available [14–16], where various voltage and current probes have been tested in practice. Particularly worth mentioning is [14], which contains a detailed overview of modern voltage probe models. [15–18] adress choosing or developing customized current sensors.

3.1.1 Voltage probe selection

Due to the high bandwidth requirement, passive probes are the only choice to acquire the switch node voltage, as differential probes having a voltage rating usually do not feature sufficient bandwidth. Tab. 1 summarizes the evaluated passive probes with their relevant properties. It is worth noting that the "PML 791" model is a low impedance probe ($5\,k\Omega$), which has a voltage rating of only 30 V RMS. Thus, it is only useful for gate-source voltage v_{GS} measurement or switch node voltage v_{DS} measurement at low voltage devices. In addition to the

	PML 711A	PML 721	PML 791
Bandwith (−3 dB)	500 MHz	500 MHz	1.5 GHz
Input impedance	10 MΩ 9.5 pF	20 MΩ 5.6 pF	5 kΩ 2.2 pF
Divider	10 : 1	20 : 1	100 : 1
Input coupling scope	1 MΩ	1 MΩ	50 Ω
Fall time	363 ps	346 ps	151 ps
$|dv/dt|_{max}$	55 V/ns	58 V/ns	132 V/ns

Tab. 1: Voltage probe characteristics (datasheet parameters and measurement results).

bandwidth of the probes and scope themselves, their connection to the circuit is relevant as well to achieve a high bandwidth of the system. Since connecting the probe's ground with an alligator clip does not meet the bandwidth requirements, the connection should made using an adapter intended for PCB mounting. This brings the coaxial structure of the cable or probe directly to the circuit board [19].

The high-impedance probes "PML 711A" and "PML 721" (Tab. 1) require an adjustment of the probe-internal trimming capacitor to the input capaci-

tance of the scope in order to achieve frequency-independent signal transmission (within the bandwidth). This is usually done using the calibration source on the scope's front panel (usually 1 V, 1 kHz, square wave signal). The probes should be trimmed to reach rectangular shape of the displayed waveform. However, compared to the calibration signal, the switch node voltage is in the range of several hundreds of volts with very high dv_{DS}/dt. In order to evaluate the bandwidth of the

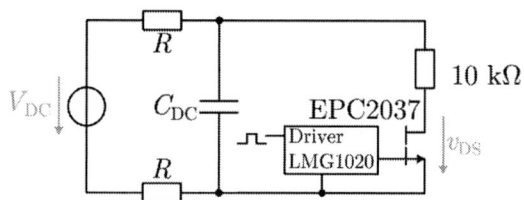

Fig. 8: Schematic of the proposed probe calibrator.

probe in combination with the scope, a signal featuring significant amplitude and high bandwidth is required. For this purpose, a special probe calibrator is designed and built. As a low voltage GaN-HEMT in combination with a high speed gate driver intended for Lidar applications reaches fall times in ps-range, those components built the basis for calibrator design [20]. Thus, a GaN-transistor with appropriate gate driver can be used to evaluate the bandwidth of the measurement equipment. Fig. 8 shows the schematic of the designed probe calibrator. The circuit resembles resistive switching of the GaN-transistor. As a result, it provides a fast falling edge in 100 ps-range. The rising edge of the signal v_{DS} is slow due to the charging of the output capacitance of the transistor by the $10\,k\Omega$ resistor. The relevant components used for the practical implementation can be found in the schematic.

Fig. 9 shows the PCB of the probe calibrator with the switching cell in the upper right corner. When layouting the PCB of the probe calibrator, care should be taken to ensure low capacitance of the switching node. The parasitic inductance is not that relevant as in power electronic converters, since the switched current is in the small range of several mA ($10\,k\Omega$ resistor as load). Furthermore, the probe's adapter should be placed very close to the switching cell (golden through-hole pads in Fig. 9). Additionally, the PCB includes linear regulators for the gate-driver and a pulse generator IC. Testing the probes from Tab. 1 results in the waveforms (falling edge) displayed in Fig. 10 using a voltage

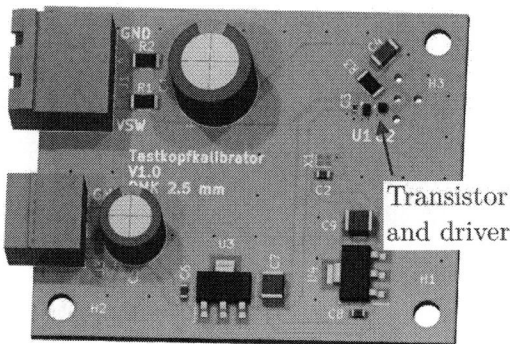

Fig. 9: PCB of the probe calibrator with connector (golden through-hole pads).

V_{DC} of 25 V with the proposed calibrator. The waveforms show a fast falling edge and some ringing afterwards. The oscillation frequency is 833 MHz and 1.12 GHz for the probe models "PML 711A" and "PML 721". The low impedance probe ("PML 791") shows ringing with 1.57 GHz. Post processing of the signals delivers the fall time and dv_{DS}/dt acquired by the probes and scope summarized in Tab. 1.

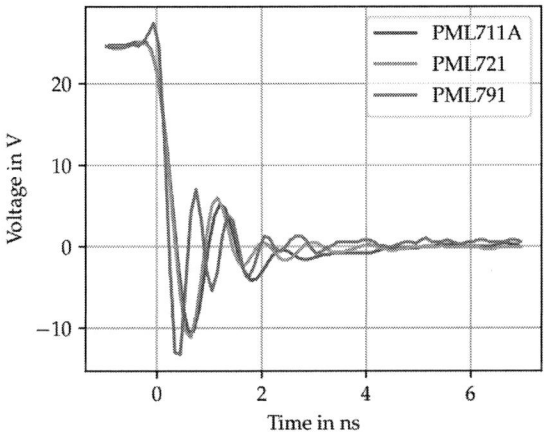

Fig. 10: Measured switch node voltage v_{DS} using the probes from Tab. 1.

3.1.2 Coaxial shunt investigation

For switching-loss measurement, the current measurement is critical. In [17, 18, 21] it was already shown, that the current measurement can be performed very accurately with an optimized PCB design and employing cheap SMD resistors with which the commutation loop is only minimally affected. To verify the current measurement, a comparison of the SMD shunt with a commercially available product should be performed. The coaxial shunt should be cut open so that the commutation

loop is only minimally affected by the series connection of the SMD resistor and the coaxial shunt. As the image in [21] shows, the coaxial shunt T&M SDN-414-05 can be cut open without concern.

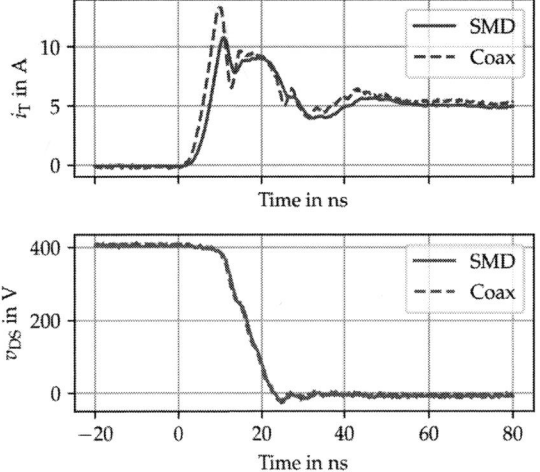

Fig. 11: Double-pulse test, same operating point measured my SMD shunt and coaxial shunt.

In order to verify the current measurement by means of the self-made SMD shunt, an own calibration board was developed with the SMD shunt, as well as the purchased coaxial shunt (see Fig. 3). Both sensors are in place when carrying out the measurements, to prevent that the change of the sensors impact the actual current waveform. After initial operation, it was found that the two current measurements differed significantly, see Fig. 11. By recording the measurements (SMD shunt vs. coaxial shunt) one after the other, the current waveforms were synchronized based on the voltage measurement v_{DS}. This makes significant differences in the subsequent power dissipation calculation for the SiC-MOSFET. For the coaxial shunts, it was figured out that different shunts (c.f. Fig. 12) of the same manufacturer and type lead to different current waveform results. The shunts were purchased in different years and shortened if necessary.

To check the quality of the shunts, they are to be measured with a vector network analyzer (VNA). The measurement objects are several times the identical shunt from different batches (purchased in different years). The impedance behavior is shown in Fig. 13. The clearly different frequency responses of the same product are striking. The 3 dB cut-off frequency of the shunt from 2023 is 1 GHz, while the other variants come along with

PCIM Europe 2024, 11– 13 June 2024, Nuremberg   DOI: 10.30420/566262457

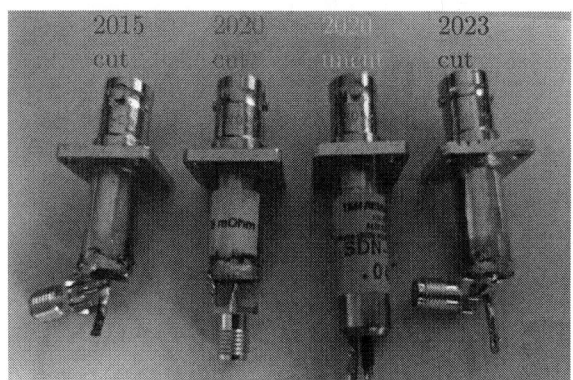

Fig. 12: Same coaxial shunts from different batches.

cut-off frequencies of 522 MHz (2015 cut), 62 MHz (2020 cut) and 232 MHz (2020 uncut). According to the data sheet, these should have a 3 dB cut-off frequency of 2 GHz. As a recommendation, measuring equipment with high frequency requirements should be characterized carefully before using it in double pulse test set ups. Since the end-of-line test usually only covers the DC resistance, but not the frequency behavior, poorly produced components are may not sort out during production.

The SMD shunt on the PCB in Fig. 3 has a cut-off frequency of 1 GHz in the measurement, which is slightly better than the simulated results (700 MHz) in [17]. As the SMD-shunts performance is based on minimizing current loops, one important part is the distance of the layers in the PCB layout. However, the spacing of the layers here (layout in Fig. 3) is 140 μm instead of 160 μm [17] and may lead to better results.

3.2 Reduce parasitic effects on switching behavior

In order to keep the influence of the parasitic capacitance C_L of the external load inductor L on the switching behavior as low as possible, the usually low-capacitance inductor is wound with a single-layer winding. Due to the mechanical setup, there is still a small winding capacitance (including terminal wires) C_L. It is also important not to twist the terminal wires from and to the choke in order to keep the parasitic capacitance low (keep them at distance). To decouple the remaining parasitic capacitance

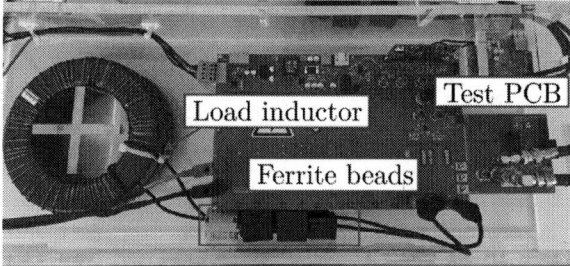

Fig. 14: Decoupling the parasitic capacitance C_L of the load inductor by use of ferrite beads.

C_L from the inductor and it's connection as much as possible, it is advisable to place ferrite beads around the connection wires (see Fig. 14). Note, that the ferrite beads are connected on each wire separately to attenuate the differential mode (DM) effect. There are four different ferrites with different frequeny characteristics, so the overall damping should be in a wide frequency range. In total, 8 ferrite beads, 4 per wire are used. Fig. 15 shows the comparative difference in the measurement results.

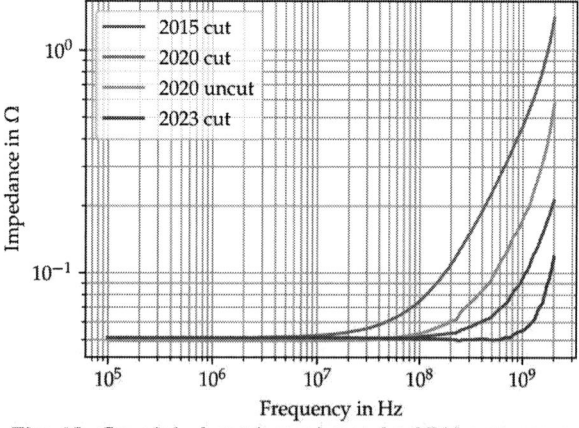

Fig. 13: Coaxial shunt impedance by VNA measurement of different batches.

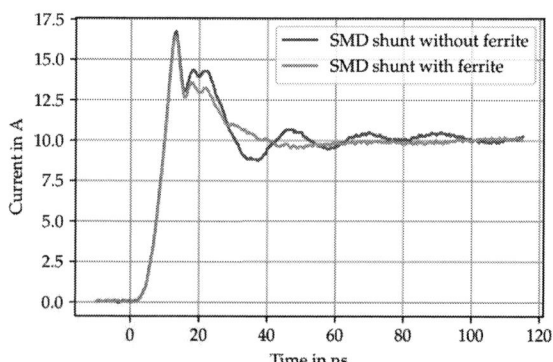

Fig. 15: Turn-on switching measurement (current) that shows the decoupling effect of the parasitic capacitance of the load inductor.

3243

3.3 Discussing useful concepts as terminating resistors for current measurments

To avoid reflections of the measured signal on the cable, it is advisable to install a serial terminating resistor in the signal path (c.f. Fig. 16a). In general, it would also be possible to use the termination in the oscilloscope. However, the risk of R_{shunt} breaking during a switching test and the load current flowing through the oscilloscope and destroying it would be very high. It is therefore better to choose solution (a), or to use an external $50\,\Omega$ terminating resistor in parallel with the $1\,\text{M}\Omega$ resistor inside the oscilloscope (c.f. Fig. 16b). In this way, only the external resistor would be damaged. The frequency behavior of the external terminating resistor should be checked beforehand.

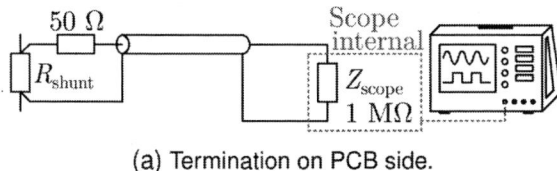

(a) Termination on PCB side.

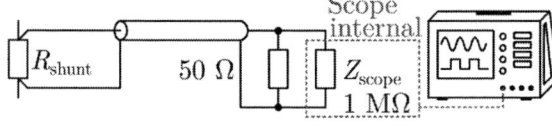

(b) Termination on scope side.

Fig. 16: Useful termination for self-made SMD resistor shunts.

3.4 Probe delay time compensation

Current and voltage measurements usually have different delay times. These should be calibrated (deskew function in scopes or postprocessing in computer), also with regard to the cable length. Even very short delay times in the range of nanoseconds have a considerable influence on the final results in the switching energy. The example in Fig. 17 using SiC-MOSFETs shows the uncompensated measurement $i_{\text{T,0ns}}$. According to section 2.1, current di_{T}/dt needs to fit during the first v_{DS} voltage drop. However, the time calibration should be carried out with a defined reference signal. In the given case, the voltage probe has a delay time of $2.2\,\text{ns}$. The compensated measurement is labeled as $i_{\text{T,2.2ns}}$. The difference in the switching energies for this measurment is approx. $25\,\%$ ($59\,\mu\text{J}$ for the uncalibrated measurement and $49\,\mu\text{J}$ for the calibrated measurement). The faster the switching, the greater the error due to uncalibrated measurement times in general.

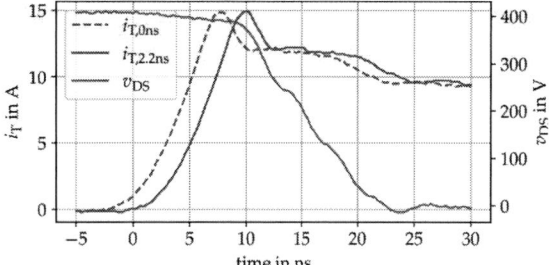

Fig. 17: Delay time correction between voltage v_{DS} and current i_{T} measurement with no delay correction ($E_{\text{on}} = 59\,\mu\text{J}$) and 2.2 ns time correction ($E_{\text{on}} = 49\,\mu\text{J}$). SiC-MOSFET C3M0120065J.

3.5 Scope signal path compensation

A very simple procedure, but often underestimated, is the signal path compensation (SPC), a feature of the oscilloscope itself. The signal path compensation corrects inaccuracies in the DC behavior caused by temperature changes and long-term drift [22]. In practice, the lack of compensation is noticeable if no current can physically flow through the transistor, but the measurement shows a non-zero offset (c.f. Fig. 18a).

During calibration, the complete signal path per channel is compensated. The process is time-consuming. It is usually recommended if the ambient temperature changes by more than 5 °C, or weekly if very high vertical accuracy is used. This is particularly the case with shunt measurements and is therefore recommended. Fig. 18 shows the

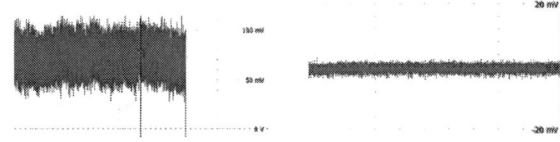

(a) DC ground offset before calibration.

(b) DC ground offset after calibration.

Fig. 18: Oscilloscope signal path compensation (SPC) before and after the compensation.

offset of the shunt channel ($0\,\text{V}$ applied) before and after performing the SPC. Note, that an offset of $50\,\text{mV}...100\,\text{mV}$ (c.f. Fig. 18a) used with a $83\,\text{m}\Omega$ shunt can lead to errors in range $0.6\,\text{A}...1.2\,\text{A}$.

3.6 Common-mode capacitance of the voltage source

It is also important to use a voltage source with a low common-mode (CM) capacitance, as this can have a significant influence on the measured switching transients. Fig. 19 shows a typical common-mode equivalent circuit of the double-pulse setup. All grey paths are parasitic elements. Especially $C_{\mathrm{CM,DC+}}$ and $C_{\mathrm{CM,DC-}}$ mainly consist of parasitic capacitances and/or internal Y-capacitors of the DC source. These capacitances can oscillate with L_{par} and therefore cause extra noise. A common

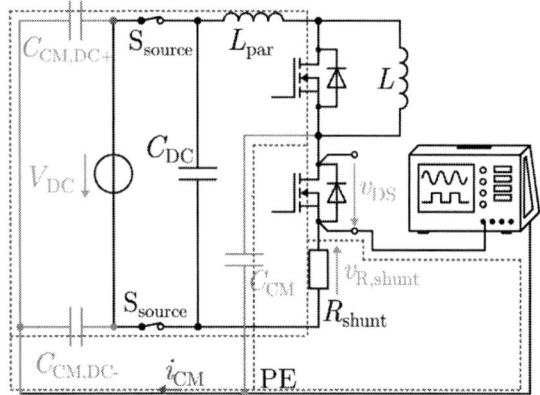

Fig. 19: Common-mode (CM) paths in the double-pulse setup.

practice, to minimize $C_{\mathrm{CM,DC+}}$ and $C_{\mathrm{CM,DC-}}$ is to decouple the voltage source after charging C_{DC}, before performing the test itself. This can be done by introducing mechanical relays S_{source} into the circuit. Another possibility is to insert a CM-choke into both DC-supply lines, or a protective earth choke into the PE-line. Another way to measure the common-mode current is to pass both DC-supply lines of the source through a high-frequency current probe and observe during switching.

The following measurement (Fig. 20) shows the test with the voltage source connected and disconnected. The source "FuG DC Power Supply MCA 1500-1500" used in this case appears to be very suitable, as the switching process does not change when the source is disconnected during the switching process.

3.7 Conclusion

This paper discusses the switching behavior of actual power semiconductors in their theory and

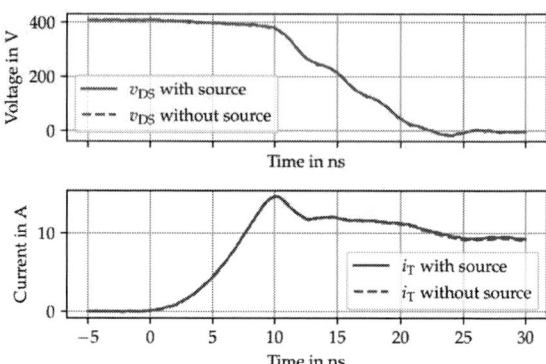

Fig. 20: Current measurement with and without voltage source connection.

points out common mistakes in performing double pulse tests. A rare but definitely possible issue is a malfunctioning current sensors, which can only be found by comparison with other measuring equipment. In order to avoid parasitic capacitances of the load inductor in affecting the switching behavior, this inductor should be decoupled from the switches by ferrite cores in the connecting wires. Furthermore, probes in combination with the employed oscilloscope should be tested for suitability considering the calibration signals close to the actual measurement signals as facilitated by a proposed calibrator circuit.

Acknowledgement

The authors would like to thank Jan Böcker for the discussions about switching events.

References

[1] R. Erickson and D. Maksimovic. *Fundamentals of Power Electronics*. Springer US, 2007.

[2] B. Kohlhepp, D. Kuebrich, R. Schwanninger, and T. Duerbaum. "Switching Loss Estimation of GaN-HEMTs by Thermal Measurement Procedure". In: *PCIM Europe; International Exhibition and Conference for Power Electronics, Intelligent Motion, Renewable Energy and Energy Management*. 2021.

[3] J. Weimer and I. Kallfass. "Soft-Switching Losses in GaN and SiC Power Transistors Based on New Calorimetric Measurements". In: *31st International Symposium on Power Semiconductor Devices and ICs (ISPSD)*. 2019.

[4] JEDEC Committee. *JC-70: Test Method for continuous-switching evaluation of Gallium Nitride Power Conversion Devices*. Tech. rep. 2021.

[5] N. Förster, P. Rehlaender, O. Wallscheid, F. Schafmeister, and J. Böcker. "An Open-Source Transistor Database and Toolbox as a Unified Software Engineering Tool for Managing and Evaluating Power Transistors". In: *IEEE Applied Power Electronics Conference and Exposition (APEC)*. 2022.

[6] C. Nöding. *ComCell*. URL: https : / / www . uni - kassel . de / eecs / fachgebiete / leistungselektronik / forschung - und - entwicklung/softwareentwicklung/comcell.

[7] D. Christen and J. Biela. "Analytical Switching Loss Modeling Based on Datasheet Parameters for mosfets in a Half-Bridge". In: *IEEE Transactions on Power Electronics* (2019).

[8] T. Dürbaum. "Switched Mode Power Supplies (Lecture, Friedrich-Alexander-Universität Erlangen-Nürnberg)". 2014.

[9] P. S. N. Perera. "Hard and Soft Switching Losses in Power Converters: Role of Transistor Output Capacitance". PhD thesis. EPFL, Switzerland, 2022.

[10] U. Jadli, F. Mohd-Yasin, H. A. Moghadam, P. Pande, J. R. Nicholls, and S. Dimitrijev. "Measurement of Power Dissipation Due to Parasitic Capacitances of Power MOSFETs". In: *IEEE Access* (2020).

[11] M. Kasper, R. M. Burkart, G. Deboy, and J. W. Kolar. "ZVS of Power MOSFETs Revisited". In: *IEEE Transactions on Power Electronics* (2016).

[12] D. Kübrich, T. Dürbaum, and A. Bucher. "Investigation of Turn-Off Begaviour under the Assumption of Linear Capacitances". In: *PCIM Europe; International Exhibition and Conference for Power Electronics, Intelligent Motion, Renewable Energy and Energy Management*. 2006.

[13] B. Nguyen, X. Zhang, A. Ferencz, T. Takken, R. Senger, and P. Coteus. "Analytic model for power MOSFET turn-off switching loss under the effect of significant current diversion at fast switching events". In: *IEEE Applied Power Electronics Conference and Exposition (APEC)*. 2018.

[14] S. Sprunck, M. Koch, C. Lottis, and M. Jung. "Suitability of Voltage Sensors for the Measurement of Switching Voltage Waveforms in Power Semiconductors". In: *IEEE Open Journal of Power Electronics* (2022).

[15] S. Sprunck, C. Lottis, F. Schnabel, and M. Jung. "Suitability of Current Sensors for the Measurement of Switching Currents in Power Semiconductors". In: *IEEE Open Journal of Power Electronics* (2021).

[16] S. Sprunck, M. Muench, and P. Zacharias. "Transient Current Sensors for Wide Band Gap Semiconductor Switching Loss Measurements". In: *PCIM Europe; International Exhibition and Conference for Power Electronics, Intelligent Motion, Renewable Energy and Energy Management*.

[17] J. Böcker, S. Schoos, and S. Dieckerhoff. "Experimental Comparison and 3D FEM Based Optimization of Current Measurement Methods for GaN Switching Characterization". In: *20th European Conference on Power Electronics and Applications (EPE'18 ECCE Europe)*. 2018.

[18] J. Böcker. "Analyse und Optimierung von AlGaN/GaN-HEMTs in der leistungselektronischen Anwendung". PhD thesis. TU Berlin, 2020.

[19] S. Biswas, D. Reusch, M. de Rooij, and T. Neville. "Evaluation of measurement techniques for high-speed GaN transistors". In: *2017 IEEE 5th Workshop on Wide Bandgap Power Devices and Applications (WiPDA)*. 2017.

[20] S. Sandler. "Faster-Switching GaN : Presenting a number of interesting measurement challenges". In: *IEEE Power Electronics Magazine* (2015).

[21] S. Sprunck. "Charakterisierung der Schaltverluste diskreter Wide Band Gap Leistungshalbleiter und Entwärmung kompakter Bauteile". PhD thesis. Kassel University, 2021.

[22] Tektronix. *4/5/6 Series MSO Help (User Guide)*. 2023.

PCIM Europe 2024, 11– 13 June 2024, Nuremberg   DOI: 10.30420/566262458

Modeling and Simulation of Fluxgate based Current Sensor

Yunus Çay [1,2], Erhan Demirok [1,3], Ozan Keysan [1,2]

[1] Center for Solar Energy Research and Applications, Turkey
[2] Middle East Technical University, Turkey
[3] İzmir Katip Çelebi University, Turkey

Corresponding author: Yunus Çay, yunus.cay@odtugunam.org
Speaker: Yunus Çay, yunus.cay@odtugunam.org

Abstract

Fluxgate-based current sensing is a prevalent method employed in residual-current devices (RCDs). However, the design and analysis of fluxgate transducers present challenges due to non-linear operational characteristics. Although approximations and assumptions have been utilized to address the non-linearities, the modeling and simulation of fluxgate transducers are not adequately elaborated in the literature. This paper proposes new modeling approaches and simulation-based analysis of fluxgate transducers. Modeling and simulation capability of the fluxgate current sensor may not provide designers with succinct analytical equations but valuable insights into the operation of non-linear circuits.

1 Introduction

The penetration of more power-electronic devices into daily lives, particularly solar inverters, and EV chargers, results in safety precautions to protect a person from electric shock; buildings, and properties from electrical fire hazards. The most prominent protection method is the installation of an RCD (residual current device). RCD provides protection against residual current passing through a person or a device. When the threshold is exceeded, RCD disconnects the device from the grid or source.

Sensitivity, fast response, and reliability are the features that RCDs commonly desire. The selection and installation of RCD, however, should also take into account application-specific requirements determined by standards. For instance, there are 2 trip thresholds for solar inverters as required by the DIN VDE 0126-1-1 due to the parasitic capacitance of PV panels. While a low threshold protects against rapid changes, a higher threshold is used for slowly rising leakage currents [1]. On the other hand, implementing RCDs in EV chargers may raise compatibility issues with diverse charging standards [2]. Unless the residual current limits and fault conditions are detected properly, false tripping of solar inverters or EV chargers may cause harm to the device or affect the grid and load stability. Also, false tripping of the device causes a waste of money and time due to less grid connection time for solar inverters and longer charging time in EV chargers. The requirements of reliability, sensitivity, and fast response of RCD, thus, orient the designers into fluxgate-based current sense to detect residual current.

In this paper, although the focus is on the low-frequency self-oscillating fluxgate current sensor [3], the resultant points can be utilized in other types of fluxgate current sensors. The operation principle of the fluxgate current sensor will be examined in Section 2. In Section 3, an example design of a fluxgate current sensor will be delivered. In Section 4, the magnetic core modeling approaches will be elaborated. Subsequently, the closed-loop operation details of the fluxgate current sensor will be analyzed in Section 5. Finally, Section 6 presents the experimental results for the open-loop fluxgate current sensor.

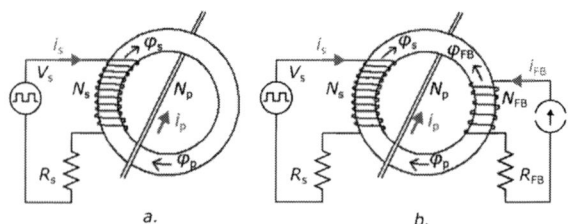

Fig. 1: Diagram of fluxgate based current sensor a. open-loop b. closed-loop

2 Working Principle

Isolated current measurement based on the flux-gate principle necessitates several key components: a magnetic core, a primary winding through which the measurement current flows, an excitation winding, and an excitation circuit for basic operation. Figure 1 a. shows the essential circuit representation for the open-loop fluxgate current sensor.

The measured current, i_p, flows through the single-turn primary winding. The excitation winding carrying excitation current, i_s, is excited with a bipolar square-wave voltage source. The excitation current increases and declines to a predetermined value that ensures core saturation in each half cycle, which is accomplished with a multi-vibrator. Hence, the excitation frequency is not fixed, as stated in the paper [4]. The excitation voltage alternates between positive and negative voltage controlled by a multi-vibrator circuit. This design is commonly called a self-oscillating fluxgate current sensor in the literature [5].

In the context of self-oscillating fluxgate current sensors, it is essential to grasp the concept of core saturation. Core saturation occurs when the core's permeability approaches that of a vacuum. This saturation state is induced periodically by the excitation circuit. Initially, when the primary winding carries zero current, the required magnetomotive force (MMF) to saturate the core in the positive and negative cycles is equal. However, as the primary current deviates from zero, an equilibrium shift occurs in the required MMF for saturation between the positive and negative cycles.

Figure 2 depicts the approximated BH curve of the T60006-L2020-W409 core and the corresponding excitation current waveform under various DC primary current values. In the graph, the curly braces labeled as H_+ and H_- denote the requisite H fields to saturate the core in the positive and negative directions, respectively. The red markers positioned at the far-right and far-left points on the BH curve indicate the saturation values for both B and H fields. Meanwhile, the red marker in the middle signifies the H field generated by the primary current. Figure 2 a. b. and c. represent the excitation current waveforms when the primary current value is zero, positive, and negative, respectively.

Non-zero primary current alters the saturation value for the excitation current, i_s. The asymmetry of the i_s waveform is closely linked to both the sign and magnitude of the primary current. This asymme-

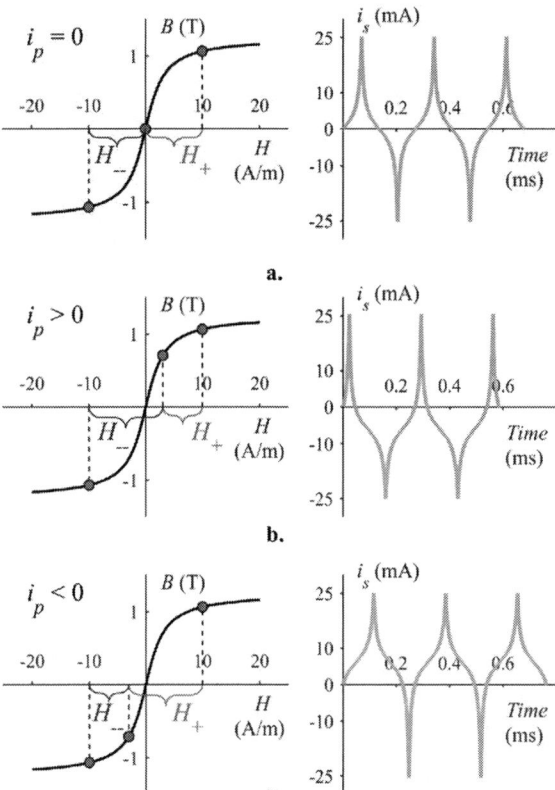

Fig. 2: Operating states of open-loop fluxgate current sensor with respect to primary current a. $i_p = 0$, b. $i_p > 0$, c. $i_p < 0$

try suggests that the average value of i_s and its sign can serve as a means to measure the primary current. Specifically, for the winding configuration depicted in Fig. 1 a., positive primary current leads to a negative average value of the excitation current, and vice versa.

Previously elaborated on the working principle of open-loop self-oscillating current sensors, it's worth noting that closed-loop fluxgate-based current sensors also exist in the literature [4], [6]. The closed-loop design shares the same fundamental topology as the open-loop fluxgate current sensor but incorporates feedback winding as seen in Fig. 1 b. This feedback winding carries a current that generates an MMF opposing that of the primary current, effectively creating a zero-flux state [4]. The effect of feedback current on excitation current can be seen in Fig. 3. The closed-loop operation of the sensor preserves the symmetry of i_s although the primary current is nonzero.

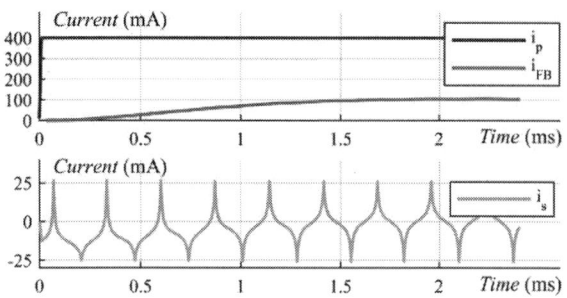

Fig. 3: Effect of the primary and the feedback current on the excitation current

Magnetic core	T60006-L2020-W409		
Excitation circuit		**Feedback circuit**	
N_s	30	N_{fb}	4
R_s	50 Ω	R_{fb}	20 Ω
V_{oh}	12 V	τ - LPF	0.36 m
V_{th}	1.25 V	T_i - Integ.	0.80 m
I_{th}	25 mA	k - Gain	4
Operational parameters			
f_s	3.67 kHz	I_{sat}	10 mA

Tab. 1: Selected values and resultant operational parameters of the example fluxgate current sensor

3 Example Excitation Circuit and Core Selection

Section 3 introduces an example simulation model for a fluxgate current sensor, providing the foundation for the subsequent sections dedicated to magnetic modeling and closed-loop operation analysis. It's essential to emphasize that the primary objective of this paper is not to present a detailed sensor design but to provide a simulation model that can facilitate the design process. As a result, certain sensor parameters, such as the number of turns in the excitation winding and the resistance of the excitation circuit, are not specifically tailored to meet the requirements of a particular design.

Figure 4 shows the circuit diagram of the fluxgate current sensor. The values of parameters shown in Fig. 4 and the selected core are given in Table 1.

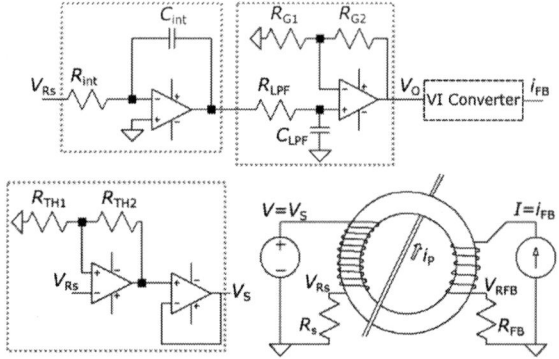

Fig. 4: Example circuit diagram for the fluxgate current sensor

4 Modeling Magnetization Characteristics

The operation of fluxgate-based current sensors relies on non-linear equations, often requiring approximations in the design process, such as frequency approximations for self-oscillating sensors [6], [7]. Although the analytical non-linear model [5] is suitable for investigating the open-loop configuration of fluxgate current sensors, it is insufficient when analyzing closed-loop operated fluxgate current sensors. Therefore, this paper presents modeling and simulation methods which may not yield succinct analytical equations, but they offer valuable insights into the operation of non-linear circuits.

This section focuses on modeling the magnetic core. Initially, two distinct approaches for representing the magnetic core within SPICE-based simulation software are examined: an arbitrary inductor model and an equivalent magnetic circuit model. Subsequently, modeling studies are detailed on BH curve approximations.

4.1 SPICE Models

The first approach to modeling a magnetic core is the arbitrary inductor model. BH characteristic of the core is determined by inductor flux with respect to the inductor current [8]. Equation (1a) presents the SPICE expression for the arbitrary inductor model, employing an arctan function to capture the BH curve's behavior. Notably, in the arbitrary inductor model, inductor current is denoted uniquely by the special keyword 'x.' The a_1 and a_2 are the resultant parameters of curve fitting of arctan BH approximation given in Eq. (1b). Furthermore, I_p represents the scaled primary current, i_p, as defined in Eq. (1c).

$$\lambda(x) = N_s A_e \left[\frac{a_1}{a_2} \arctan \left(a_2 \frac{N_s}{l_e} (x + I_p) \right) \right] \quad (1a)$$

$$B(H) = \frac{a_1}{a_2} \arctan(a_2 H) \quad (1b)$$

$$i_p = N_s I_p \quad (1c)$$

The SPICE schematic of the excitation circuit with the arbitrary inductor model is depicted in Fig. 5. The I_p value is parameterized to simulate changes in the primary current. It's important to note that the feedback current is not included in the arbitrary inductor modeling approach. Consequently, this approach is specifically employed for simulating open-loop fluxgate current sensors.

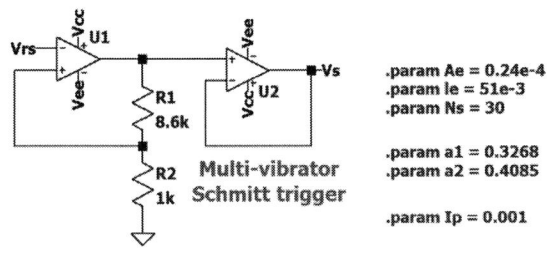

.param Ae = 0.24e-4
.param le = 51e-3
.param Ns = 30

.param a1 = 0.3268
.param a2 = 0.4085

.param Ip = 0.001

Flux = Ns*Ae*((a1/a2)*atan((a2*Ns/le)*(x+{Ip})))

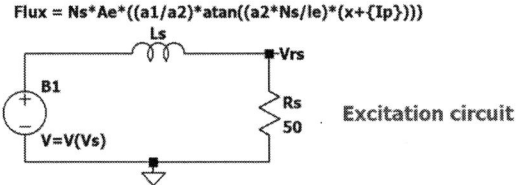

Fig. 5: SPICE schematic for arbitrary inductor model

The second approach for modeling the magnetic core is based on a magnetic circuit, a method elaborated in the studies presented within this paper [9]. In these studies, Alonso, Martínez, Perdigão, Cosetin and Prado have developed SPICE-based behavioural models that enable the construction of equivalent circuits for magnetic circuits. This approach considers reluctances that depend on the magnetic flux level. The SPICE schematic of the modeled fluxgate sensor with an equivalent magnetic circuit model can be seen in Fig. 6. Within the schematic, 'U3' and 'U4' represent SPICE models emulating the excitation winding and primary winding, respectively, while 'U5' corresponds to the variable reluctance SPICE model.

Variable reluctance model requires parameters for core size and total permeability as a function of magnetic flux density, B. This relationship is expressed in Eq. (2), where the total permeability is derived based on an approximation function of the BH curve.

$$R_m(B) = \frac{l_m}{\mu_t(B)A_m} \qquad (2)$$

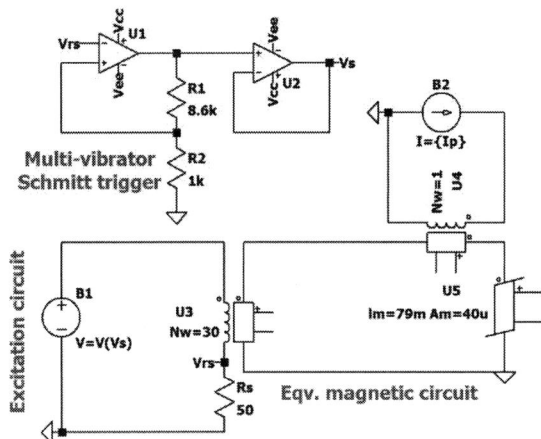

Fig. 6: SPICE schematic for equivalent magnetic circuit model

4.2 BH Modeling

Modeling the magnetization characteristics of the core is a prominent study for the simulation of fluxgate current sensors. Both the arbitrary inductor model and the equivalent magnetic circuit model require a function that expresses the magnetization in which the hysteresis effect is not taken into account. Anhysteretic magnetization (AM), also called ideal magnetization curve, is approximated [10]. Although ignorance of the hysteresis effect is done for simplicity [9], the AM remains a suitable approximation for high permeable cores, especially nano-crystalline cores, which often exhibit very narrow and practically linear BH loops [11].

In the literature, various methods are available for modeling the magnetization characteristics of materials. These models are typically classified into two categories: phenomenological and physical models, distinguished by whether they primarily describe the shape of the magnetization characteristic or delve into the underlying physics of the phenomenon [12]. The primary objective is to derive expressions for the BH characteristic. Therefore, both phenomenological and physical models could be utilized and curve-fitted for ease of use.

Functions	a_1	a_2	a_3	gof (R^2)
Langevin	1.242	1.361	-	0.9979
Arctan	0.327	0.409	-	0.9982
Tanh	1.049	0.264	0.004	0.9966
Brauers	1.33e-8	14.91	5.2	0.9763

Tab. 2: Resultant curve-fit parameters of various BH curve approximation functions

Table 2 presents the selected anhysteretic magnetization (AM) curve approximation functions with fit results based on the Eq. (3). The goodness of fit can be viewed numerically based on the R^2 value in Table 2 and visually with digitized BH data of the core and fitted curves in Fig. 8. Also, the differential permeability for each approximation is given in Fig. 8. While the Arctan and the Langevin approximations result in similar differential permeability, the Tanh approximation generates a similar shape to the former approximations but with a low peak value of differential permeability. On the other hand, the Brauers method preserves the linearity of BH characteristics up to the saturation point. Although all methods differ from each other with respect to differential permeability, the linearity of each approximation method for the open-loop sensor is almost the same as seen in Fig. 7. The resultant curve of non-linear analytical formulation [5] also coincides with other approximation methods as expected. The sensor output voltage, the mean voltage of V_{Rs} for open-loop sensor, curves coincide up to the non-linear region of the sensor with a small discrepancy. The sensitivity curves of approximation methods, however, differ from each other in the non-linear region.

Langevin : $B(H) = a_1 \left(\coth\left(\dfrac{H}{a_2} \right) - \dfrac{a_2}{H} \right)$ (3a)

Arctan : $B(H) = \dfrac{a_1}{a_2} \arctan\left(a_2 H \right)$ (3b)

Tanh : $B(H) = a_1 \tanh\left(a_2 H \right) + a_3 H$ (3c)

Brauers : $H(B) = \left(a_1 e^{a_2 B^2} + a_3 \right) B$ (3d)

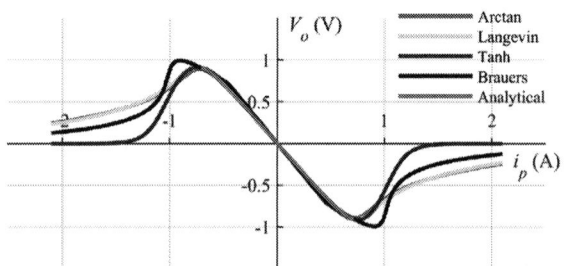

Fig. 7: Comparison of sensor sensitivities based on different approximation methods

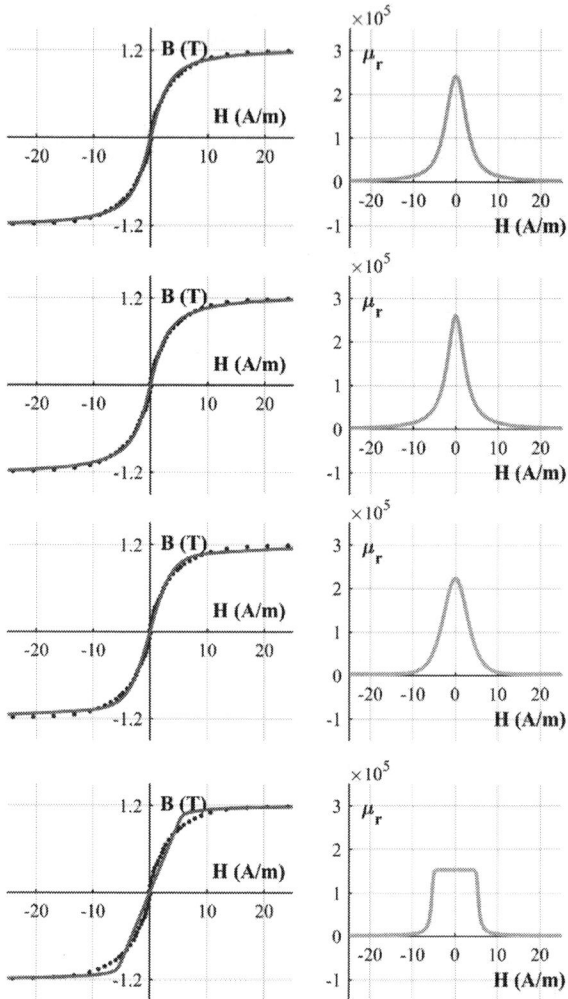

Fig. 8: Resultant approximation curves with respect to the digitized BH data and differential permeability a. Langevin, b. Arctan, c. Tanh, d. Brauers

5 Closed-loop Fluxgate Simulation

Fluxgate current sensors have been enhanced by the need to address the challenges of poor accuracy and linearity associated with open-loop configurations [4]. To enhance their accuracy and linearity, closed-loop configurations have become increasingly prevalent, often employing compensation coils called feedback windings. While the literature has extensively explored the enhancements achieved through closed-loop configurations [4], [6], [13], there remains a notable gap in research regarding comprehensive simulation approaches. This section aims to bridge this gap by analyzing the principles, operation, and linearity characteristics of the closed-loop fluxgate current sensors.

5.1 Feedback Winding Model and Circuit

Figure 9 shows the circuit diagram for the closed-loop operation of the fluxgate current sensor based on the equivalent magnetic circuit. The feedback winding is emulated with the component 'U6'. The turn number and feedback resistor are represented by the parameters N_w and R_{fb}, respectively. The sensitivity of the closed-loop current sensor is correlated with feedback winding turn number and feedback resistance as shown in Eq. (4). The demodulation circuit takes the voltage signal of the resistor, V_{Rs}, in the excitation circuit. Then, the processed voltage signal is supplied to feedback winding with a VI converter.

$$V_o = \frac{R_{fb}}{N_{fb}} i_p \quad (4)$$

$$V_o \approx \frac{N_s}{R_s} i_p \quad (5)$$

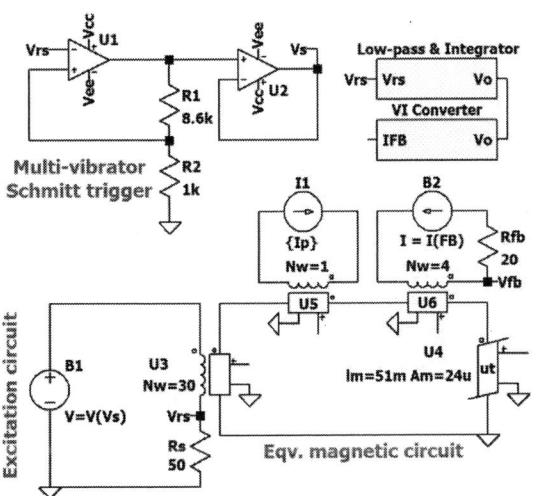

Fig. 9: SPICE schematic of closed-loop fluxgate current sensor

Figure 10 illustrates the open-loop and the closed-loop control block diagram of the fluxgate current sensor. The open-loop block diagram is established based on the assumptions regarding the sensor's sensitivity in Eq. (5). The demodulation module in open-loop configuration is represented by the low-pass filter (LPF). Since the demodulation in an open-loop configuration does not affect the operation of the excitation circuit, any kind of filter can be implemented. The closed-loop configuration, on the other hand, introduces a feedback loop and a demodulation circuit, which includes an integrator and a low-pass filter. This specific demodulation

circuit topology for closed-loop configuration is designed in the paper [6]. The product of the feedback current and the number of turns in the feedback winding are subtracted from the primary current. Subsequently, the demodulation circuit filters and integrates the voltage V_{Rs} to generate a current supplied to the feedback winding.

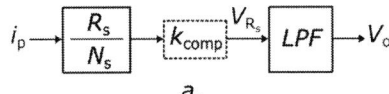

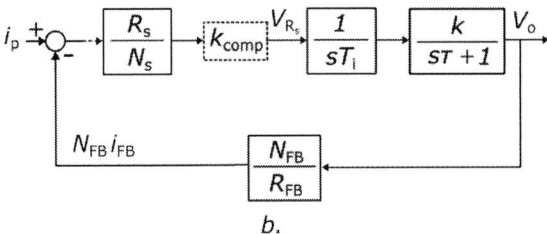

Fig. 10: Fluxgate current sensor block diagrams a. open-loop and b. closed-loop

The frequency response of the closed-loop configuration of the example fluxgate current sensor according to the block diagram in Fig. 10 b. is given in Fig. 11. The frequency response of the sensor is characterized with and without including compensation. The compensation block behaves like a gain block with a k_{comp} parameter. This parameter compensates for the assumption of the voltage V_{Rs} in Eq. (5). The compensation parameters for the selected BH approximation methods are given in Table 3. Although the detailed examination of compensation could result in a better understanding of the operation, the compensation parameter does not huge contribution to the design of closed-loop operation. Also, the compensation parameter closes to 1 with an increase in the difference between I_{sat} and I_{th}; and the sharp saturation characteristic.

Function	Sensitivity Simulation	Assumed Sensitivity	k_{comp}
Arctan	1.46 V/A	1.67 V/A	0.874
Langevin	1.47 V/A	1.67 V/A	0.880
Tanh	1.56 V/A	1.67 V/A	0.934
Brauers	1.58 V/A	1.67 V/A	0.946

Tab. 3: Compensation constant value with respect to BH approximation methods

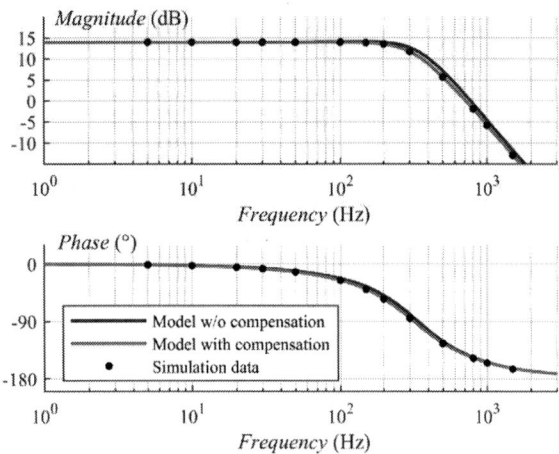

Fig. 11: Closed-loop bode diagram of the sensor

Region	Condition	I_{sat+} (mA)	I_{sat-} (mA)	i_p (mA)
Linear	$I_{th} > I_{sat+}$ $-I_{th} < I_{sat-}$	3	-17	210
Partial saturation	$I_{sat-} < I_{th} < I_{sat+}$ $I_{sat-} < -I_{th} < I_{sat+}$	-10	-30	600
Complete saturation	$I_{th} < I_{sat-}$ $-I_{th} > I_{sat+}$	-26	-46	1080

Tab. 4: Conditions for the linearity regions

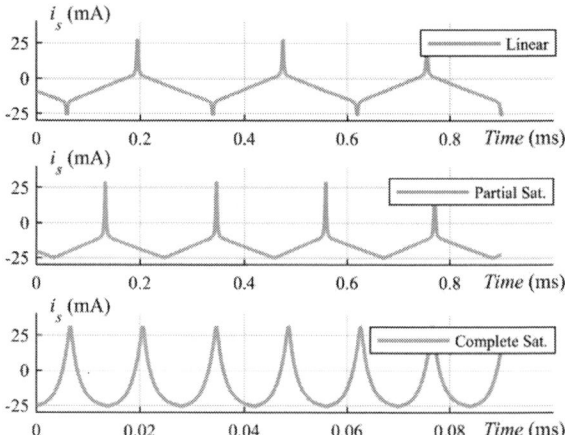

Fig. 13: Excitation current waveforms in a. linear b. partial saturation and c. complete saturation regions

5.2 Linearity of Fluxgate Current Sensor

The sensitivity of the open-loop fluxgate current sensor exhibits linear characteristics within a specific range of primary current values. However, when the primary current exceeds a certain threshold, deviations in linearity become evident. In the analysis, as detailed in [14], these deviations are categorized into three distinct regions based on the magnitudes of saturation current and threshold current. These classifications are referred to as the 'linear region,' where the magnetic core saturates in both positive and negative cycles; the 'partial saturation region,' where the magnetic core saturates only in one of the positive or negative cycles; and the 'complete saturation region,' where the magnetic core remains saturated throughout the operation. These regions are marked on the sensitivity curve of the example sensor based on Brauers B-H curve approximation in Fig. 12.

Table 4 shows the conditions for the linearity regions. One of the conditions for partial and complete saturation regions is sufficient for the respective linearity region. Also, i_p values for the waveforms in Fig. 13 are given in Table 4. The I_{sat+} and I_{sat-} changes with i_p as seen in Fig. 2.

The linearity comparison of open-loop and closed-loop fluxgate current sensors can be seen in Fig. 14. The partial and complete saturation regions are eliminated with the feedback current.

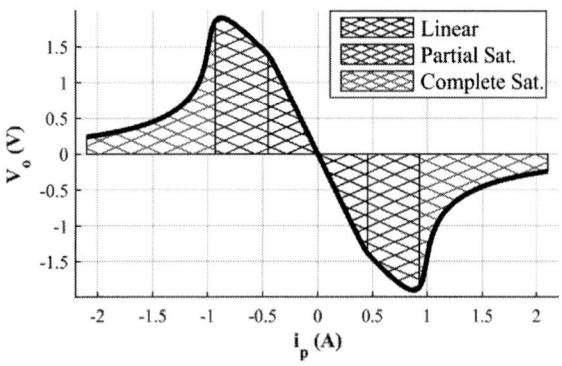

Fig. 12: Sensitivity regions of fluxgate current sensor with respect to primary current

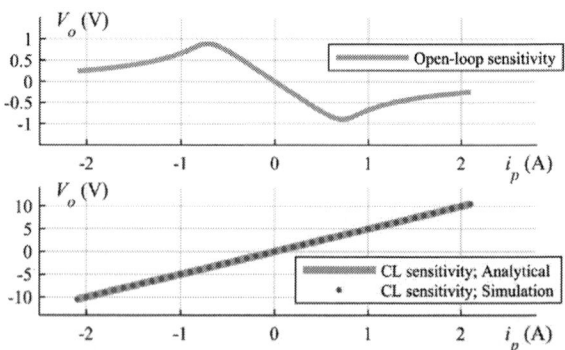

Fig. 14: Comparison of open-loop and closed-loop sensitivity of the example fluxgate current sensor

6 Experimental Results

The open-loop configured fluxgate current sensor is tested with the parameters in Table 1, except that excitation voltage, V_{oh}, and threshold, I_{th}, are determined as 10 V and 40 mA, respectively. The experimental waveforms of Vs and Is can be seen in Fig. 15. The effect of hysteresis could be detected from the experimental waveform of i_s compared to simulation results. The i_s steeply increases or decreases a value at which the core is not completely saturated before the excitation voltage alternates. The measured hysteresis of the magnetic core is shown in Fig. 16.

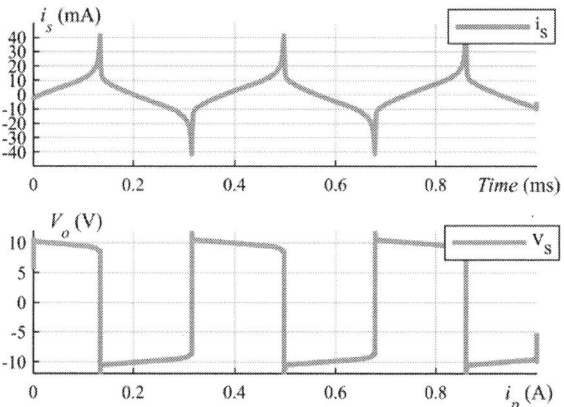

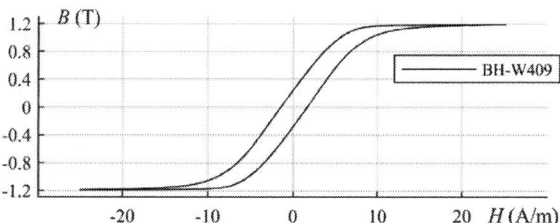

Fig. 15: Experimental excitation current, i_s, and excitation voltage, V_s, waveforms for $i_p = 0$

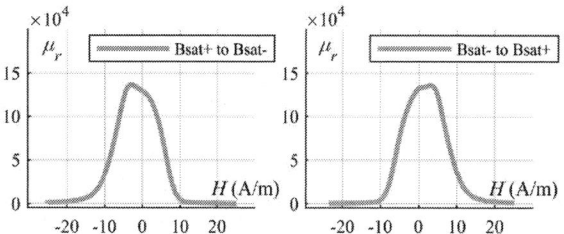

Fig. 16: BH characteristic of the core, T60006-L2020-W409

The differential permeability of the selected magnetic core could be seen in Fig. 17. The differential permeability has a dependence on the initial magnetic state of the core due to the hysteresis effect. Hence, the differential permeability has different values when the core is saturated from positive to negative and vice versa. The peak value and the shape of the measured differential permeability resemble the resultant BH characteristic of the Brauers approximation.

Fig. 17: Differential permeability of the core, T60006-L2020-W409

The final destination of experimental results is the comparison of sensor sensitivity between simulation results and measured sensitivity. The waveforms of simulation-based sensitivity and measured sensitivity are shown in Fig. 18. The simulations have been revised with the parameters used in the experiments. The experimental result and the simulation result based on the Brauers approximation have a good correlation, as expected from the similar differential permeability curves.

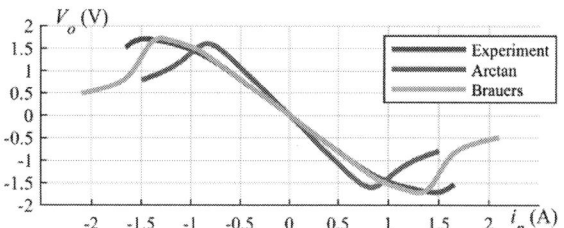

Fig. 18: Sensitivity comparison of the sensor based on experimental and simulation results

7 Conclusion

The simulation and modeling of the fluxgate-based current sensor are examined. The experimental results have confirmed the studies presented in the paper. Hence, valuable insights to understand the operation of fluxgate current sense could be gained with simulation and modeling studies. Besides, the simulation models could be significant resources to facilitate the design process and verify RCD in particular applications based on predefined fault cases. Overall, it's crucial to recognize that modeling and simulation cannot replace analytical models entirely. Nevertheless, simulation and modeling are essential for handling non-linear circuits effectively.

References

[1] SolarEdge. "Rcd selection for solaredge inverters." (2021), [Online]. Available: https : / /

knowledge-center.solaredge.com/sites/kc/files/application_note_ground_fault_rcd.pdf.

[2] A. N.Shete, V. Shukla, and S. V.Devane, "Residual current device for electric vehicle," *JETIR - Journal of Emerging Technologies and Innovative Research*, vol. 10, k454–k458, 5 2023.

[3] S. Ziegler, R. C. Woodward, H. H.-C. Iu, and L. J. Borle, "Current sensing techniques: A review," *IEEE Sensors Journal*, vol. 9, no. 4, pp. 354–376, 2009. DOI: 10.1109/JSEN.2009.2013914.

[4] X. Yang, Y. Li, W. Zheng, W. Guo, Y. Wang, and R. Yan, "Design and realization of a novel compact fluxgate current sensor," *IEEE Transactions on Magnetics*, vol. 51, no. 3, pp. 1–4, 2015. DOI: 10.1109/TMAG.2014.2358671.

[5] M. M. Ponjavic and R. M. Duric, "Nonlinear modeling of the self-oscillating fluxgate current sensor," *IEEE Sensors Journal*, vol. 7, no. 11, pp. 1546–1553, 2007. DOI: 10.1109/JSEN.2007.908234.

[6] T. R. Oliveira, "Design of a low-cost residual current sensor for lvdc power distribution application," in *2018 13th IEEE International Conference on Industry Applications (INDUSCON)*, 2018, pp. 1313–1319. DOI: 10.1109/INDUSCON.2018.8627238.

[7] I. Filanovsky and V. Piskarev, "Sensing and measurement of dc current using a transformer and rl-multivibrator," *IEEE Transactions on Circuits and Systems*, vol. 38, no. 11, pp. 1366–1370, 1991. DOI: 10.1109/31.99166.

[8] LTwiki, *The arbitrary inductor model*, 7 April, 2024, 2024.

[9] J. M. Alonso, G. Martínez, M. Perdigão, M. R. Cosetin, and R. N. do Prado, "A systematic approach to modeling complex magnetic devices using spice: Application to variable inductors," *IEEE Transactions on Power Electronics*, vol. 31, no. 11, pp. 7735–7746, 2016. DOI: 10.1109/TPEL.2016.2571845.

[10] M. Nowicki, "Anhysteretic magnetization measurement methods for soft magnetic materials," *Materials*, vol. 11, no. 10, 2018. DOI: 10.3390/ma11102021.

[11] K. Draxler and R. Styblíková, "Use of nanocrystalline materials for current transformer construction," *Journal of Magnetism and Magnetic Materials*, vol. 157-158, pp. 447–448, 1996, European Magnetic Materials and Applications Conference. DOI: https://doi.org/10.1016/0304-8853(95)01055-6.

[12] G. Mörée and M. Leijon, "Review of hysteresis models for magnetic materials," *Energies*, vol. 16, no. 9, 2023. DOI: 10.3390/en16093908.

[13] G. Velasco-Quesada, M. Román-Lumbreras, A. Conesa-Roca, and F. Jeréz, "Design of a low-consumption fluxgate transducer for high-current measurement applications," *IEEE Sensors Journal*, vol. 11, no. 2, pp. 280–287, 2011. DOI: 10.1109/JSEN.2010.2054831.

[14] P. Pejovic, "A simple circuit for direct current measurement using a transformer," *IEEE Transactions on Circuits and Systems I: Fundamental Theory and Applications*, vol. 45, no. 8, pp. 830–837, 1998. DOI: 10.1109/81.704822.

PCIM Europe 2024, 11– 13 June 2024, Nuremberg   DOI: 10.30420/566262459

Sigma-Delta based Current Acquisition with Reduced Settling Time

Joschka Randerath[1] , Jens Onno Krah[1]
[1] TH Köln, Cologne, Germany

Corresponding author: Joschka Randerath, Joschka.Randerath@th-koeln.de
Speaker: Joschka Randerath

Abstract

An approach of sigma-delta ($\Sigma\Delta$) based current acquisition with reduced settling time of the digital decimation filter and/or an increase of the effective number of bits (ENOB) is presented. Additionally, short circuits are also faster detectable. This is achieved by digitally compensating for the known parasitic inductance of the current sense shunt. The required digital signal processing can be performed by an FPGA or processor-based. This proposed method is compared to the commonly used sinc³ filter approach.

1   Introduction

The use of $\Sigma\Delta$-based current sensing technology offers many advantages for automation drives [1]. These dedicated analog-digital converters (ADCs) generate a digital bit stream that can be easily isolated for galvanic separation between the power circuit and the control circuit. Suitable 2nd order $\Sigma\Delta$ modulators are now available from many semiconductor manufacturers. Decimation filters generate the samples with the desired resolution or bandwidth from the high-frequency bit stream [2]. Two main technologies are used for such current sensing.

- Several semiconductor manufacturers such as Analog Devices, Broadcom or Texas Instruments offer integrated circuits with 2nd order $\Sigma\Delta$ modulators and galvanic isolation that use additional shunt resistors for current sensing.
- LEM offers open-loop Hall current transducers with integrated 2nd order $\Sigma\Delta$ modulators.

Semiconductor manufacturers' data sheets assume that the shunt resistor is ideal, i.e., without any inductance. However, real resistors always have parasitic inductance [3, 4]. The typical method to reduce the unwanted behavior of parasitic inductance is analog frequency compensation. The parasitic inductance causes the current measurement to become frequency-dependent, as described by the transfer function $H_S(s)$.

$$U_{\text{shunt}}(s) = (R_{\text{shunt}} + s\,L_{\text{shunt}})\,I_{\text{shunt}}(s) \qquad (1)$$

$$U_{\text{shunt}}(s) = R_{\text{shunt}}(s)\,I_{\text{shunt}}(s)\,H_S(s) \qquad (2)$$

$$H_S(s) = 1 + s\,T_{\text{shunt}} \quad \text{with} \quad T_{\text{shunt}} = \frac{L_{\text{shunt}}}{R_{\text{shunt}}} \qquad (3)$$

The frequencies above the corner frequency $f_C = 1/(2\pi\,T_{\text{shunt}})$ are amplified, which limits the maximum slew rate of the measured current [3]. If the parasitic inductance is known, the frequency response can be compensated with an analog first order RC low pass for pole-zero cancelation ($T_{\text{RC}} = T_{\text{shunt}}$). The transfer function $H_{\text{RC}}(s)$ for compensation is:

$$H_{\text{RC}}(s) = \frac{1}{1+s\,T_{\text{RC}}} \quad \text{with} \quad T_{\text{RC}} = R_C\,C_C \qquad (4)$$

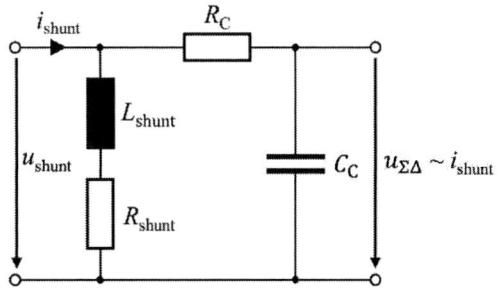

Fig. 1   *Analog low-pass filter (RC) for a dynamic compensation of the parasitic inductance L_{shunt}*

The corresponding schematic is shown in **Fig. 1**. The real shunt is described as a series connection of a resistor R_{shunt} and a parasitic inductance

3256

L_{shunt}. The impedance of the compensation network should be significantly higher than that of the selected shunt.

The voltage at the capacitor $u_{\Sigma\Delta}$ is proportional to the current i_{shunt}, with a constant gain of R_{shunt}. This voltage is used as the input signal for the $\Sigma\Delta$ modulator. The parasitic inductances of the resistor R_C and the capacitor C_C are usually very small and are neglected in this paper.

In [3], a current measurement method, using a low-inductance SMD shunt with a frequency compensation network is presented. In [5], such an RC-based compensation circuit with a time constant of $T_{\text{RC}} = 3.3\ \mu s$ (150 Ω, 22 nF) is used in a GaN-based inverter with up to 100 kHz switching frequency for feedback control of the current loop. In this configuration, the current sense shunt corner frequency of $f_C = 50$ kHz is dynamically compensated.

2 DSP-based frequency compensation

An alternative to an analog frequency compensation circuit is to implement digital signal processing (DSP) algorithms after digitization by the ADC [6, 7]. For the ADC the $\Sigma\Delta$ technology is advantageous because it converts the analog signal into a high-bandwidth digital bit stream that can be further processed digitally. A challenge of the $\Sigma\Delta$ technology is to account for the injected high-frequency quantization noise.

When implementing DSP algorithms, the Nyquist sampling theorem must be taken into account since steep-edge filters with their high phase shifts are critical in control engineering. The signal sampling time T_S should be significantly smaller than the time constant T_{RC} of the digitally implemented filter. In principle, it would be possible to filter the high-frequency bit stream, but this would result in a very complex DSP implementation.

A good compromise is to provide a first sinc³ filter with a decimation rate of $M_1 = 16$. This results in a current signal with a sampling time of $T_S = 0.8\ \mu s$ and a data word width of 13 bits. This signal is suitable for compensation with a 1st order infinite impulse response (IIR) filter as described in equation (5). The digital IIR low-pass filter compensates the parasitic inductance of the shunt in the same way as the analog compensation transfer function $H_{\text{RC}}(s)$.

$$G_{\text{IIR}}(z) = \frac{b_0}{1 - a_1 z^{-1}} \tag{5}$$

A second sinc³ filter with a decimation rate of $M_2 = 4$ can then be used to achieve a total decimation ratio of $M = M_1 \cdot M_2 = 64$. The sampling time of the result does not increase if the second sinc³ filter is implemented as an equivalent finite impulse response (FIR) filter, as shown in **Fig. 2**.

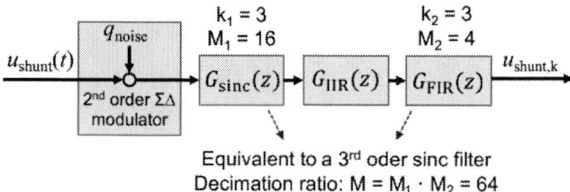

Fig. 2 *Filter configuration of the DSP-based frequency compensation*

Since the low-pass filter to compensate for the parasitic inductance also attenuates the quantization noise of the $\Sigma\Delta$ modulation, the effective number of bits (ENOB) of the measured current increases. This method is possible for inverters with significant smoothing inductances because high frequencies are amplified by the inductance L_{shunt}. The specified voltage measurement range of the $\Sigma\Delta$ modulator limits the maximum current slew rate:

$$u_{\text{shunt}}(t) = R_{\text{shunt}}\ i_{\text{shunt}}(t)$$
$$+ L_{\text{shunt}}\ \frac{\text{d}}{\text{dt}} i_{\text{shunt}}(t) \tag{6}$$

3 Shunts with an increased inductance and DSP-based frequency compensation

With shunts, the focus is usually on keeping the parasitic inductance as low as possible. However, with $\Sigma\Delta$ modulation a known parasitic inductance of current sense shunts can be used in combination with digital filtering to increase the ENOB and/or to reduce the settling time of the current sense. Therefore, the two approaches described above are combined to use an analog pre-emphasis filter with a digital de-emphasis filter. The schematic of the analog filter is identical to the schematic in **Fig. 1** but an increased inductance L_{shunt}

reduces the corner frequency of the high-pass filter.

$$\omega_{HP} = \frac{1}{T_{shunt}} = \frac{R_{shunt}}{L_{shunt}} \qquad (7)$$

This increased inductance can either be designed into the shunt itself or a ferrite can be placed around the shunt. In addition, the cutoff frequency of the RC low-pass filter ω_{LP} is increased so that it is greater than ω_{HP}.

$$\omega_{HP} < \omega_{LP} = \frac{1}{T_{RC}} = \frac{1}{R_C C_C} \qquad (8)$$

The total transfer function of the pre-emphasis filter $G_{PE}(s)$ is described in the following equation:

$$G_{PE}(s) = K_P \frac{1 + \dfrac{s}{\omega_{HP}}}{1 + \dfrac{s}{\omega_{LP}}} \qquad (9)$$

$$G_{PE}(s) = R_{shunt} \frac{1 + s \dfrac{L_{shunt}}{R_{shunt}}}{1 + s R_C C_C} \qquad (10)$$

The gain response of the analog pre-emphasis filter $G_{PE}(s)$ is shown by the blue plot in **Fig. 4**.

The analog voltage $u_{\Sigma\Delta}$ (see **Fig. 1**) is digitized by a 2nd order $\Sigma\Delta$ modulator, that injects high-frequency quantization noise. A sinc^K decimation filter converts the bit stream into data words. A digital IIR de-emphasis filter is used to attenuate the high frequencies and the quantization noise, as described in [7]. The cutoff frequencies of the de-emphasis filter are chosen so that the poles and zeros of the analog pre-emphasis and the digital de-emphasis cancel each other out. This amplitude response of the digital IIR de-emphasis is shown by the green plot in **Fig. 4**. The complete shunt current acquisition system, consisting of the series connection of the analog pre-emphasis filters, the ADC ($\Sigma\Delta$ modulator and sinc^K filter), and the digital de-emphasis filter, is shown in the following **Fig. 3**.

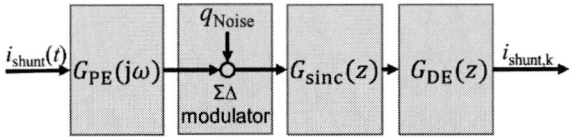

Fig. 3 *Shunt based current acquisition system*

The sinc^K transfer function can be neglected if the frequencies are lower than the cutoff frequency

f_{sinc} of the sinc^K filter. This is calculated using the following equation:

$$f_{sinc} = \frac{f_{\Sigma\Delta}}{M} \qquad (11)$$

The overall transfer function $G(s)$ results from the series connection of the pre-emphasis, sinc^K and de-emphasis transfer functions, as shown in red in **Fig. 4**.

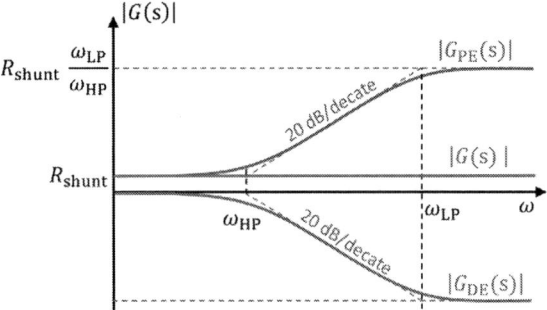

Fig. 4 *Gain response of the analog filter (blue), digital filter (green) and total current acquisition (red)*

In applications with large smoothing inductances, the corner frequency of the current measurement shunt can be reduced to a few kHz by a controlled increase of the shunt inductance. In applications with high switching frequencies and significant current slew rates, the voltage at the inductance is quite high and therefore the corner frequency must be higher than 10 kHz. The quantization noise is further attenuated as the corner frequency ω_{HP} decreases in comparison to the cutoff frequency ω_{LP}. This increases the ENOB of the converted current signal. The is an advantage over the digital compensation method discussed in Chapter 2 because it can be applied to inverters with small smoothing inductance as well. The RC low-pass filter attenuates high frequencies, which reduces the measuring voltage $u_{\Sigma\Delta}$ for high current slew rates. The slew rate of the measured current increased compared to the digital compensation from Chapter 2. The relationship between ENOB and the settling time is shown in **Fig. 5**. Several sinc^3 filter configurations with different decimation rates are shown by the green plot. The blue plot is also with sinc^3 filters, but with an increased shunt inductance and digital compensation. Either the ENOB is improved, and/or the settling time is reduced.

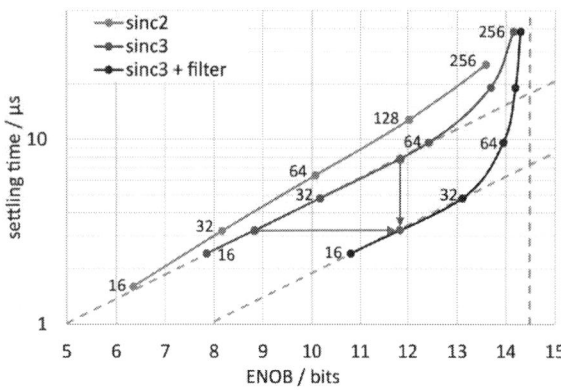

Fig. 5 *Settling time of the ADC as a function of the desired ENOB and the sinc^K filter configuration*

4 Dependency of the filter configuration on the resolution

To estimate the ENOB of the presented approach, a test setup was built. For this purpose, the circuit displayed in **Fig. 6** is used to apply a test voltage to the $\Sigma\Delta$ modulator. The resistances are $R_1 = 20\,\text{k}\Omega$, $R_2 = 10\,\Omega$, $R_3 = 1.2\,\text{k}\Omega$ and the capacitance is $C = 1.6\,\text{nF}$. A DC voltage source provides the DC voltage $u_{DC} \approx 36\,\text{V}$ to increase the voltage $u_{\Sigma\Delta}$ by an offset. An AC voltage source provides a $4\,\text{V}_{pp}$ sine with a frequency of $0.5\,\text{Hz}$ to prevent the $\Sigma\Delta$ modulator from reaching a steady state.

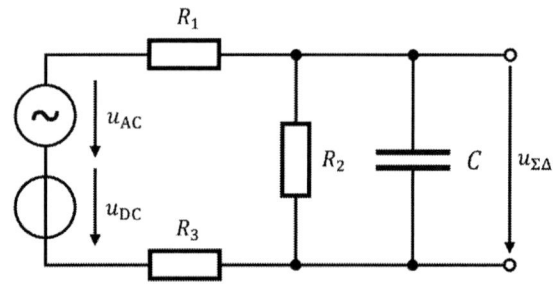

Fig. 6 *Circuit diagram for estimating the ENOB*

The voltage $u_{\Sigma\Delta}$ is converted to a 1-bit stream with a frequency of $20\,\text{MHz}$ by the 2nd order $\Sigma\Delta$ modulator AD7403 from Analog Devices. Its input voltage range is $\pm 0.25\,\text{V}$ [8]. A comparison is made between two different ADC filter systems. In the first filter system, the bit stream is converted into a 13-bit data word using a sinc^3 filter ($K = 3$) with $M = 16$. The sampling time is $T_S = 0.8\,\mu\text{s}$. The signal is then filtered by the 1st order low-pass filter with the transfer function $G_{IIR}(z)$. This filter is the low-pass filter from Chapter 3, which compensates for the parasitic inductance and suppresses the

high-frequency quantization noise. This system is shown in **Fig. 7**.

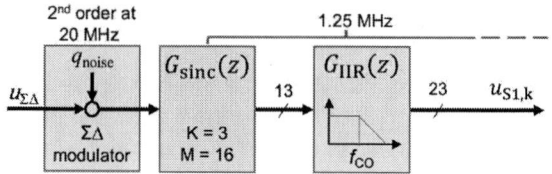

Fig. 7 *First filter system setup with only a sinc³ filter*

In the second filter system, the bit stream is converted into a 10-bit data word with a sinc^3 filter ($K_1 = 3$) with $M_1 = 8$. This results in a sampling time of $T_S = 0.4\,\mu\text{s}$. The signal is then also filtered with the transfer function $G_C(z)$. After the low-pass filter, a sinc^3 filter ($K_2 = 3$) and $M_2 = 2$ is implemented as an FIR filter. The sampling time T_S remains constant by $T_S = 0.4\,\mu\text{s}$. This system is shown in **Fig. 8**.

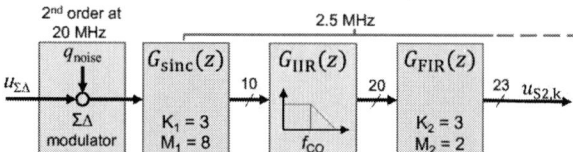

Fig. 8 *Second filter system with a sinc³ and a FIR filter*

To estimate the ENOB, the variances of the signals u_{S1} and u_{S2} are determined. For this purpose, the moving average of these signals is determined using a low-pass filter with a cutoff frequency of $f_{CO} = 200\,\text{Hz}$. The moving average is subtracted from the signal itself. The difference is then squared and averaged to determine the variance $\overline{s_x}^2$. The system for calculating the variance $\overline{s_x}^2$ is shown in **Fig. 9**.

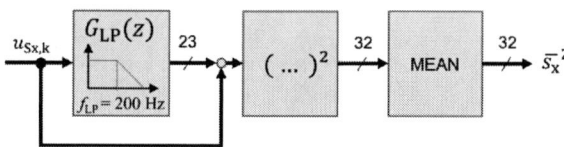

Fig. 9 *Filter system to calculate the variance*

The ENOB can be estimated using the variance and the following equation, where n_{mean} is the data width of the mean filter. In this configuration, $n_{mean} = 23$ bits.

$$n_{res} = n_{mean} - \frac{1}{2}\log_2\left(\overline{s_x}^2\right) \qquad (12)$$

Figure 10 shows the estimated ENOB of the first filter system (orange) and the second filter system (blue) as a function of the configured cutoff frequency f_{CO} of the low-pass filter $G_{IIR}(z)$. The dashed gray line indicates the ENOB when no low-pass filter is used. It can be seen that a lower cutoff frequency of $G_{IIR}(z)$ results in a higher ENOB. This is because the quantization noise is attenuated more effectively when the cutoff frequency is reduced.

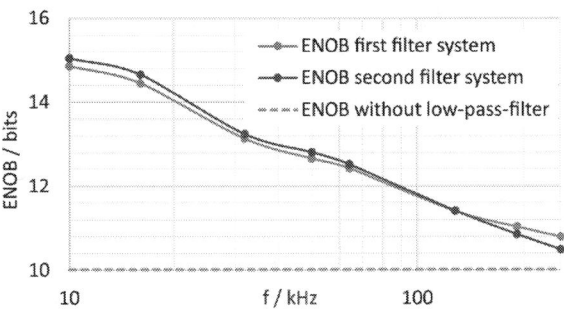

Fig. 10 *ENOB of the first filter system and second filter system as a function of the cutoff frequency*

It can also be seen that there is not much difference between the ENOBs of the first filter system and the second filter system. The second system is slightly better than the first system at lower cutoff frequencies and vice versa at higher cutoff frequencies. The advantage of the second system is that the combination of sinc³ and FIR filters is equivalent to the resolution of system 1 (13-bit resolution), but the sampling time is only half ($T_{S2} = 0.4\,\mu s$ instead of $T_{S1} = 0.8\,\mu s$).

5 Filter system configuration

In order to design the inductance of the pre-emphasis filter L_{shunt}, an inverter with a synchronous motor is considered as an example. The motor is shown as a simplified series connection consisting of the winding inductance L_{motor}, winding resistance R_{motor} and the motor speed-dependent back emf voltage u_{BEMF}, **Fig. 11**. The compensation resistor R_C and the compensation capacitor C_C have neglectable effects on the current flow, therefore, they are not shown in the figure.

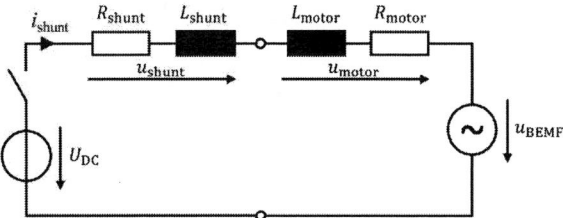

Fig. 11 *Single phase equivalent circuit of the shunt-based current acquisition in a synchronous motor application*

At high frequencies, such as with a voltage step, the resistors are negligible and the voltage is mainly at the inductances. The voltage at the shunt inductance should not exceed the maximum measuring range of the $\Sigma\Delta$ modulator. However, the possible voltage steps should utilize the measuring range as far as possible. If the back emf voltage u_{BEMF} is neglected and because the DC link voltage is significantly higher than the measurement voltage, the maximum inductance L_{shunt}, can be estimated using the following inequation.

$$L_{shunt} < \frac{U_{MR}}{U_{DC}} L_{motor} \qquad (13)$$

For example, for a typical motor application, with a DC link voltage of $U_{DC} = 600\,V$, a winding inductance L_{motor} of $10\,mH$ and a voltage measurement range of $U_{MR} = 0.25\,V$, the maximum inductance is approximately $4\,\mu H$ [8].

At low frequencies, the inductance is neglectable. The shunt is sized according to the maximum current to be measured and the voltage range of the ADC, with enough margin for the voltage due to the inductance.

Once the shunt inductance and the shunt resistance are known, the cutoff frequency for the low-pass filter can be determined. The analog pre-emphasis filter behaves like a proportional-differential (PD) element. **Fig. 12** shows the step response of the voltage $u_{\Sigma\Delta}$ to a current step.

The higher the low-pass cutoff frequency ω_{LP} is compared to the high-pass corner frequency ω_{HP}, the more the quantization noise is attenuated. However, this also increases the maximum voltage at the $\Sigma\Delta$ modulator, which must be considered. A good configuration is when the low-pass cutoff frequency is 2 to 10 times greater than the high-pass corner frequency.

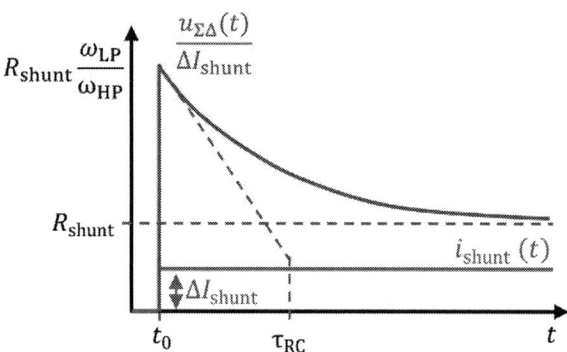

Fig. 12 *Measuring voltage $u_{\Sigma\Delta}$ as the response to a current step*

The voltage at the $\Sigma\Delta$ modulator is the shunt voltage minus the voltage at the compensation resistor. The voltage across the compensation resistor is the product of the modulator input current and the compensation resistor. Therefore input current and the compensation resistance should be as low as possible so that the $\Sigma\Delta$ voltage $u_{\Sigma\Delta}$ is as close as possible to the shunt voltage ushunt. The input current varies depending on the manufacturer and type. For example, the input current of the Broadcom ACPL-796J is 0.5 µA at an input voltage of 0 V [9]. The compensation resistor can be determined from the maximum tolerable distortion of the current value p_{error}, the measuring current $\hat{I}_{\text{M}}$, and the measuring voltage range of the $\Sigma\Delta$ modulator U_{MR} using the following equation.

$$R_{\text{C}} < p_{\text{error}} \frac{U_{\text{MR}}}{\hat{I}_{\text{M}}} \qquad (14)$$

The compensation capacitance is selected so that the desired low-pass cutoff frequency is achieved.

6 Experimental results

In an example setup, a square-wave current is measured via a shunt and a 2nd order $\Sigma\Delta$ modulator. The analog pre-emphasis filter has an inductance of approximately $L_{\text{shunt}} \approx 400\,\text{nH}$ with a shunt resistance of $R_{\text{shunt}} = 5\,\text{m}\Omega$. The compensation capacitor has a capacitance of $C_{\text{C}} = 430\,\text{nF}$ and the compensation resistor has a value of $R_{\text{C}} = 47\,\Omega$. The resulting cutoff frequencies are $f_{\text{HP}} \approx 2\,\text{kHz}$ and $f_{\text{LP}} \approx 8\,\text{kHz}$. This pre-emphasis filter configuration results in a 12 dB attenuation of the high-frequency quantization noise.

The DE10-Lite development board from terasIC, which is based on an Intel/Altera MAX10 FPGA with 50 000 Logic Cells [10], is used for the DSP and the algorithms are implemented in VHDL. The pre-emphasis filter and the 2nd order $\Sigma\Delta$ modulator AD7403 from Analog Devices with an effective measuring range of $\pm 0.25\,\text{V}$ are used for the current acquisition [8]. The 2nd order $\Sigma\Delta$ modulator and the pre-emphasis filter are realized on a separate PCB. The structure of the test system is shown in **Fig. 13**.

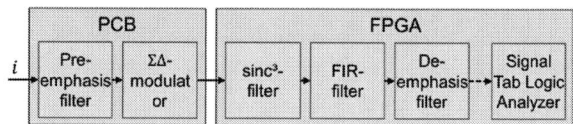

Fig. 13 *Test setup with current measurement and filter system*

In the FPGA, the 20 MHz bit stream is converted into a 16-bit digital word using a sinc³ filter (K = 3) with M = 8 and a sinc³ equivalent FIR filter with K = 3 and M = 8. The resulting sampling frequency is $f_{\text{S}} = 2.5\,\text{MHz}$ ($T_{\text{S}} = 0.4\,\mu\text{s}$).

Digital filters require FPGA resources in the form of logic cells and sometimes DSP blocks for multiplications. Here, most of the FPGA resources such as on-chip memory and logic cells are used by the Quartus Signal Tab Logic Analyzer. The following table shows the number of logic cells and DSP blocks used for the filter units.

DSP unit	Logic Cells	DSP blocks
sinc³ filter (K = 3, M = 8)	95	0
FIR filter (K = 3, M = 8)	230	0
deemphasis filter	412	0

Table 1 *Used Logic Cells and DSP blocks*

The current is generated by a laboratory function generator in parallel with the shunt and a series resistor. In this setup, a square-wave current with an amplitude of 1.5 A peak-to-peak is set at a frequency of 2 kHz. The current is measured with the TCP312 current probe in conjunction with the TCPA300 amplifier from Tektronix in combination with the Tektronix TBS 1072-EDU digital oscilloscope. The current signal processed within the FPGA after the de-emphasis filter is recorded with the Signal Tab Logic Analyzer from Intel/Altera. **Figure 14** shows the signal plots of the current. It can be seen that the current measured in the

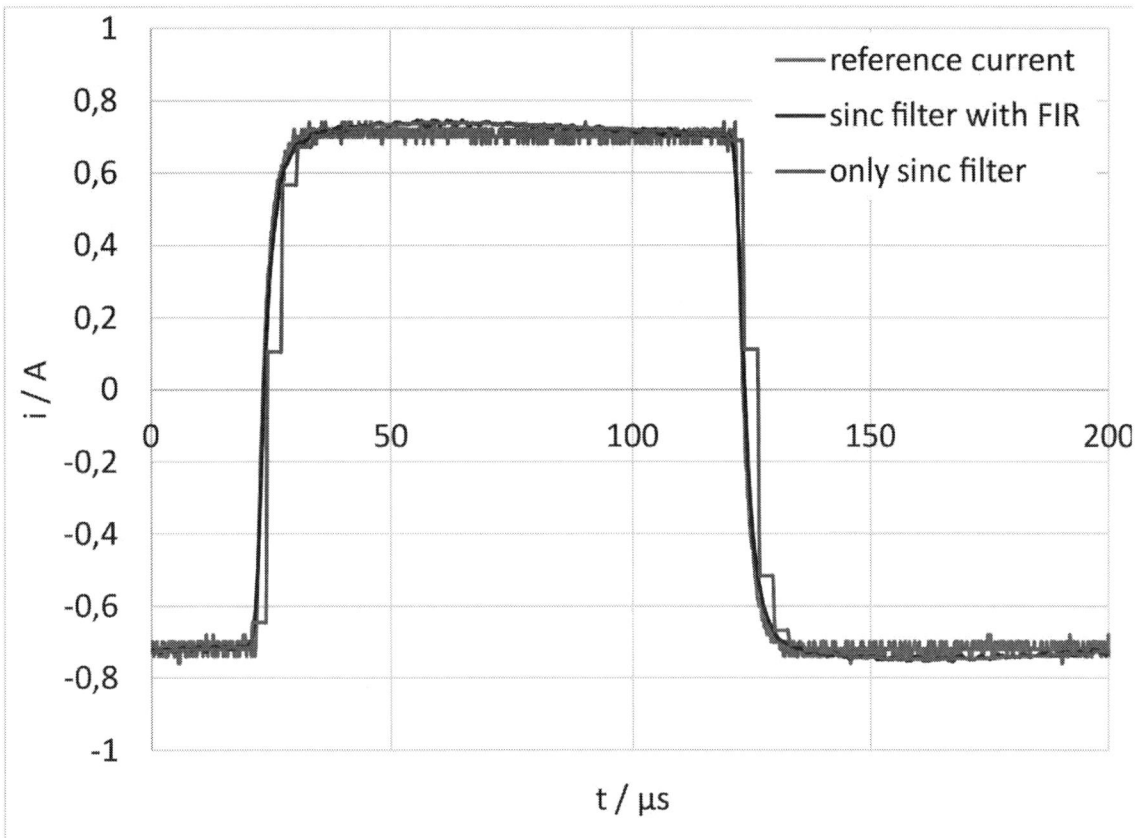

Fig. 14 *Measured current plot with reference measurement, ADC with FIR filter and ADC with sinc³ filter*

FPGA (blue) approximates the current through the shunt (orange). This means that the poles and zeros of the pre-emphasis filter are well compensated by the de-emphasis filter. Also shown in green is what the measured current would look like if a sinc³ filter with $M = 64$ were used instead of the combination of sinc³ and an FIR. Due to the sinc^K downsampling, the sampling time is increased to $T_S = 3.2$ µs.

7   Conclusion

Real shunts are not ideal because they come with parasitic inductances. The voltage drop across the inductance affects the current measurement. Since instead of the current the voltage parallel to the shunt is measured, the current deviation induces a voltage, which results in a systematic measurement error. Instead of an analog RC low-pass compensation, two digital filter systems are presented, which compensate for the influence of the parasitic inductance. The first option uses a digital IIR low-pass filter. The low-pass filter also attenuates the quantization noise, which increases

the ENOB. It is shown that the ENOB can be increased by around 2.8 bits. Since the analog measurement voltage is not attenuated, the current slew rate is limited, so this method only fits for applications with enough current smoothing inductance. The second filter configuration limits the measurement voltage using an analog pre-emphasis filter (a band-pass filter that allows low frequencies to pass). A digital de-emphasis filter compensates the poles and zeros of the analog filter. In addition, the high-frequency quantization noise is suppressed, which also leads to an increase in the ENOB and/or a reduction in the settling time. The filter configuration was successfully tested by a test setup with a square-wave current with an amplitude of 1.5 A peak-to-peak at a frequency of 2 kHz.

References

[1] A. Mertens and D. Eckardt, "Voltage and current sensing in power electronic converters using sigma-delta A/D conversion," in IEEE Transactions on Industry Applications, vol. 34, no. 5, pp. 1139-1146, Sept.-Oct. 1998

[2] M. Oljaca, T. Hendrick, "Combining the ADS1202 with an FPGA Digital Filter for Current Measurement in Motor Control Applications", Texas Instruments application note SBAA094.

[3] M. Meissner, J. Schmitz, F. Weiss and S. Bernet, "Current measurement of GaN power devices using a frequency compensated SMD shunt," PCIM Europe 2019; International Exhibition and Conference for Power Electronics, Intelligent Motion, Renewable Energy and Energy Management, Nuremberg, Germany, 2019.

[4] T. Wickramasinghe et al., "A Study on Shunt Resistor-based Current Measurements for Fast Switching GaN Devices," IECON 2019 - 45th Annual Conference of the IEEE Industrial Electronics Society, Lisbon, Portugal, 2019, pp. 1573-1578.

[5] V. Wagner, L. Geiger and C. P. Dick, "Stress Testbench for Two-Phase Interleaved Coupled Inductors in Three-Level Inverters using Cascaded Control," PCIM Europe 2023; International Exhibition and Conference for Power Electronics, Intelligent Motion, Renewable Energy and Energy Management, Nuremberg, Germany, 2023.

[6] A. Rath, C. Klarenbach, O. D. Djouosseu and J. O. Krah, "Fast Current Measurement based on Enhanced $\Sigma\Delta$ Technology" PCIM Europe 2012; International Exhibition and Conference for Power Conversion and Intelligent Motion, Nuremberg, Germany, 2012.

[7] J. Randerath, J. O. Krah and J. Holtz, "Inverter-Integrated Acquisition of the Current Through the DC-Link Based on the Measured DC-Link Voltage," PCIM Europe 2023; International Exhibition and Conference for Power Electronics, Intelligent Motion, Renewable Energy and Energy Management, Nuremberg, Germany, 2023.

[8] Analog Devices, Data sheet AD7403: https://www.analog.com/media/en/technical-documentation/data-sheets/AD7403.pdf

[9] Broadcom, Data sheet ACPL-796J: https://www.broadcom.com/products/opto-couplers/industrial-plastic/isolation-amplifiers-modulators/sigma-delta-modulators/acpl-796j-000e

[10] Terasic, DE10-Lite User Manual – Terasicc: https://www.terasic.com.tw/cgi-bin/page/archive.pl?Language=English&CategoryNo=234&No=1021&PartNo=4

PCIM Europe 2024, 11– 13 June 2024, Nuremberg   DOI: 10.30420/566262460

Characterisation of Wide-Bandgap Semiconductors in Double Pulse Testing Using Optically Isolated Probes

Lennart Hoffmann [1], Jens Friebe [1]

[1] University of Kassel, Department of Power Electronics, Germany

Corresponding author: Lennart Hoffmann, hoffmann@uni-kassel.de
Speaker: Lennart Hoffmann, hoffmann@uni-kassel.de

Abstract

As the utilization of wide-bandgab devices in power electronic systems continues to improve power density and efficiency, the knowledge of switching behaviour is crucial for optimized design of the driver circuit as well as module integrated parameter design. To address effects like inter-chip oscillations or parasitic turn-on events (PTO), measurements on the low-side (LS) switch are no longer sufficient. Due to fast switching events, measurements on high-side (HS) potential suffer from high common-mode (cm) stress, which requires devices with high common-mode rejection ratio (CMRR). In order to perform high-precision differential-mode (dm) measurements with high bandwith and low noise distortion for instance on HS-switch gate, high voltage optically isolated probes provide promising features for semiconductor characterisation. This paper provides a detailed application-specific selection of suitable measurement devices, based on datasheet evaluation and hardware verification considering electrical and thermal behaviour of the measurement devices, in order to assemble an optimized double pulse test setup for wide-bandgap devices.

1 Selection of suitable measurement equipment

Playing an essential role within the overall transition towards renewable energies, power electronic converters are facing various challenges for an increase in efficiency, power density and costs, whereas the use of wide bandgab semiconductors has been focus of research in the last decades. Driving the power systems towards higher switching frequencies, a significant reduction in volume and weight of passive components can be achieved [1]. Additionally, faster switching transitions allow a decrease in switching losses for hard-switching operations [1], [2], whereas the stress on the magnetic components increases as the inductor losses have to be taken special care of [3]. As wide bandgap semiconductors enable fast switching transitions, measurement devices with high bandwidth are required, to capture e.g. the gate-source (gs) voltages [4]. To accurately measure dynamic switching characteristics of the semiconductor device, high bandwidth of the overall measurement system becomes essential, especially for higher switching

frequencies and shorter switching times [5]. Some requirements on the measurement setup consisting of the oscilloscope and the measurement device can be derived by the amount of energy that the measurement signal carries up to a certain frequency as well as expected frequencies when parasitic LC-circuits are excited due to semiconductors switching with high transition rates. The first criteria can be expressed with the help of the maximum signal frequency f_{max} (see Eq.1), as it marks the frequency where the amplitude of the signal drops by $-6\,\text{dB}$ with respect to a $-20\,\text{dB}$ per decade rolloff as it applies to gaussian pulses [6]. As modern digital oscilloscopes come with a flat frequency response in contrast to a well known gaussian frequency response of analog oscilloscopes, signals below the $-3\,\text{dB}$ bandwith are measured more accurately, as the flat frequency response shows a lower attenuation of the measurement signals in that range. This implies, that a lower bandwidth is required to obtain the same measurement error than for a gaussian frequency response [5].

$$f_{\text{max}} = \frac{0.5}{T_{\text{r}}} \tag{1}$$

PCIM Europe 2024, 11– 13 June 2024, Nuremberg   DOI: 10.30420/566262460

Tab. 1: Datasheet evaluation of different suitable measurement devices.

Parameter	Unit	IsoVu TIVP1 [7]	FireFly [8]	DI10-ISO [9]	Bumblebee[10]
Bandwidth	GHz	1	1.3	1	0.4
Rise time	ps	450	280	435	1000
Propagation delay	ns	15	18.3	n.d.	12
CMRR @ DC	dB	160	180	150	80

$$f_r = \frac{1}{2\pi \sqrt{L_{\text{loop}} C_{\text{oss}}}} \qquad (2)$$

Different multipliers can be applied to the maximum signal frequency f_{max}, as described in [5], to estimate the required measurement bandwith with regard to a corresponding rise time measurement error. Assuming an error of 3 %, using an oscilloscope with flat frequency response, a multiplier of $1.4 f_{\text{max}}$ has to be considered [5]. Secondly, for many power electronic systems, the highest expected frequency within a switching transition occurs due to the excitation of a parasitic LC-circuit comprising of the power loop inductance L_{loop} and the equivalent output capacitance C_{oss} of the semiconductor (for a more precise estimation, the parasitic capacitance of the measurement probe C_{probe} has to be added) (see Eq. 2). The loop inductance can easily be determined by the voltage drop of the drain source (ds)-voltage during current commutation, whereas C_{oss} is defined in the semiconduc-

tors datasheet. Considering typical values for the loop inductance and output capacitance for silicon carbide (SiC) and gallium nitride (GaN) semiconductors, the required bandwith determined based on the maximal ringing frequency f_r is expected to be approximately 100 MHz and 500 MHz, respectively [11]. Often, separate measurements must be performed with a floating reference potential, as for HS-switch measurements or in multilevel converters [12]. In order to capture small signal differential signals in the presence of large cm-voltages, high CMRR is necessary for high-precision measurements [12]. Optically isolated devices provide such required CMRR over large areas of their bandwidth. Even the *Bumblebee* probe by *PMK*, being an active differential probe without optical isolation can be suitable for most measurements in terms of bandwidth, however suffers from low CMRR. Market leading devices promises bandwidth above 1 GHz and CMRR up to 180 dB, as listed in Tab. 1. As indicated in the Table, all devises are suitable for characterisation of SiC and GaN semiconductors in terms of bandwidth. However, their performance on floating reference potential with high cm-voltages and ambient conditions with temperature variance has to be investigated independently. For a precise analysis of the switching behaviour and semiconductor characteristics, different measurement settings and two optically isolated probes (*IsoVu TIVP* and *FireFly*) are investigated, to assemble an optimized measurement setup for each probe. In the first section, data sheet information are provided and two optically isolated probes are selected for an extended analysis based on their promised bandwitdh and CMRR. Furthermore, double pulse tests with different measurement setups as well as thermal measurements are performed for two optically isolated probes. Finally, Section 2 concludes the collected data and provides an outlook on further improvements of the measurement setup for each of the investigated probe.

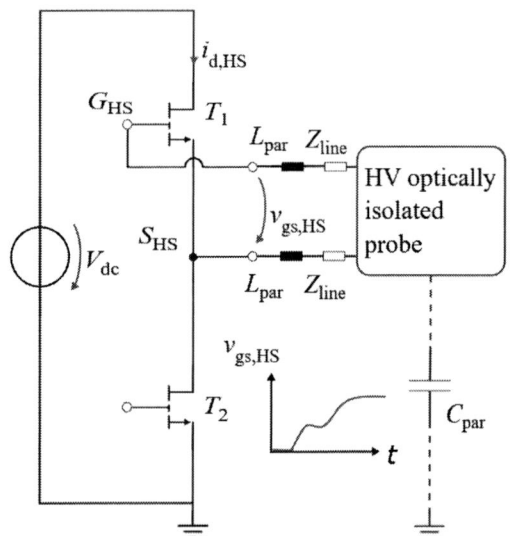

Fig. 1: Electrical equivalent circuit of with parasitic elements L_{par} and C_{par} showing a gs-voltage measurement using an optically isolated probe.

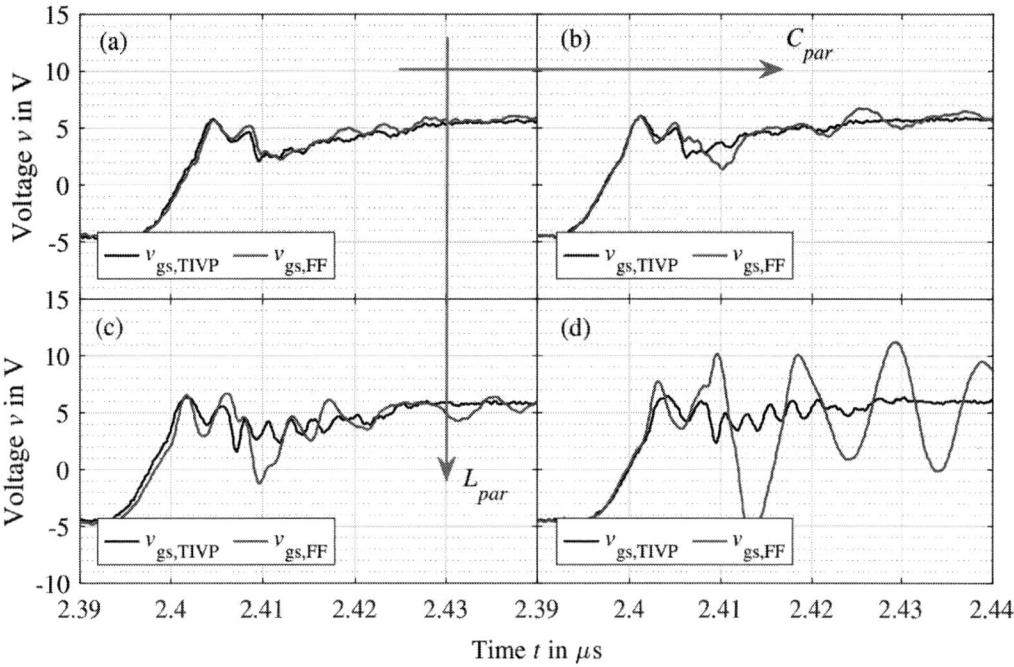

Fig. 2: Overview of gs-voltage $v_{gs,HS}$ measurements with a *IsoVu TIVP* and a *FireFly* probe on HS-switch at 400 V dc-link voltage V_{dc} and 10 A drain current i_d. The measurements are conducted with a connection to a chip-near reference point using (a) an MMCX-Adapter and sensor head placement with 7 cm distance to PE-potential, (b) an MMCX-Adapter and sensor head placement on grounded surface, (c) a Y-lead (5 cm length) and sensor head placement with 7 cm distance to PE-potential and (d) a Y-lead (5 cm length) and sensor head placement on grounded surface. All tests are performed with the evaluation board *GS66508T-EVBDB2* by *GaN Systems* [13].

1.1 Electrical Behaviour

Optically isolated differential probes provide promising features in terms of high bandwidth and high CMRR compared to conventional differential probes, as listed in Table 1. Two optically isolated probes are considered for different measurement setups, as their performance under different ambient conditions is investigated in the following section. Therefore, double pulse tests are performed, measuring the gs-voltage $v_{gs,HS}$ of the HS-switch as shown in Fig. 2. All tests are performed with the evaluation board *GS66508T-EVBDB2*, using GaN HEMTS with an $R_{ds,on}$ of $50\,m\Omega$ with a gate resistance for turn-on and turn-off of $R_{on} = 10\,\Omega$ and $R_{off} = 2\,\Omega$ respectively [13]. This setting enables switching transition rates up to $40\,V\,ns^{-1}$. This setting was selected, as it places the highest demands on the measurement regarding transition times and ringing frequency, thus requiring highest bandwidth and CMRR. The insights gained can be transferred for any other measurement setting using SiC semiconductors. For optimal measurement results re-

garding CMRR, [7] recommends to maximize the distance of the sensor head to any conductive surface to decrease the parasitic capacitance C_{par} (see Fig. 1). However, creating such a setup is not always possible due to limited access to the terminals of interest. Furthermore, the recommended use of MMCX adapters is not always feasible. Figure 2 provides an overview over the specific probe behavior for different parasitic capacitances C_{par} and inductances L_{par} which are induced in the cm-path. Two different connection options were tested to vary the input inductance L_{par} of the probes. Furthermore, different distances of the sensor head to Protective Earth (PE) potential are considered, varying C_{par}. Considering the setups displayed, Fig. 2a shows the best behaviour in terms of parasitic oscillations for both probes. However, increasing L_{par} leads to higher oscillations measured with a *FireFly* (FF) probe (see Fig. 2c). An increase in C_{par} reinforces this effect (see Fig. 2d). In contrast to this, the gs-voltage $v_{gs,HS}$ measured with an *IsoVu TIVP* probe is invariant in regard to these

PCIM Europe 2024, 11– 13 June 2024, Nuremberg   DOI: 10.30420/566262460

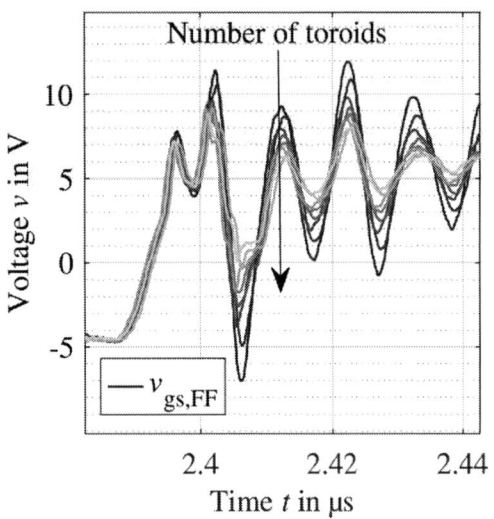

Fig. 3: Depiction of gs-voltage $v_{gs,HS}$ measurements with a *FireFly* probe, considering a different number of toroid cores on the sensor tip for increased line attenuation.

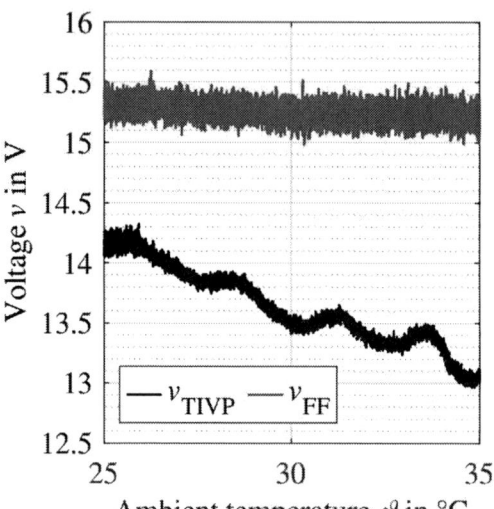

Fig. 4: Thermal behaviour of the investigated optically isolated probes. The ambient temperature is measured inside a thermally insulated box.

parasitic effects. It can be concluded, that measurements with the *FireFly* probe requires a higher awareness of these parasitic effects while designing a measurement setup. However, Fig. 3 displays the effect of toroid cores placed on the probe tip to attenuate cm-currents through the sensor tip, as recommended in [7], successfully limiting the cm-disturbance of the measured signal. It has to be marked, that manganese-zinc (MgZn)-ferrite cores were used to suppress the cm-induced oscillations. However, in this frequency range, nickel-zinc (NiZn)-ferrite cores will provide higher attenuation.

1.2 Thermal Behaviour

To imitate realistic conditions of a measurement chamber for semiconductor characterization, a thermally insulated box was used to perform dc-voltage measurements at different ambient temperatures. Figure 4 displays the conducted dc-measurements as the ambient temperature ϑ_{amb} is increased by $10\,\mathrm{K}$, utilizing heating resistors to control the temperature rise. A non negligible dc-drift of approximately 1 V can be observed for a *IsoVu TIVP* probe. It should be noted, that no self-calibration was performed during the test, although a self calibration is recommended after a temperature rise of 5 K [7]. However, this does not seem to be feasible in most kind of permanent test settings. Taking note, that no calibration is required for a tempera-

ture rise lower 5 V, a dc-drift of 0.5 V can already be observed. In contrast to that, the dc-drift of a *FireFly* probe occurs to be approximately ten times lower. Furthermore, the *FireFly* probe does not require any calibration after warm up. Additional measurements are performed without any external heat source, investigating the dc-drift of the probes caused by their selfheating. Figure 5 illustrates the conducted measurement for the *TIVP* probe,

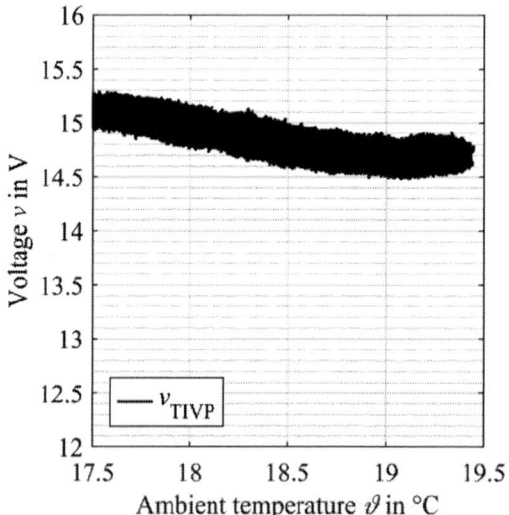

Fig. 5: Thermal behaviour of the *TIVP* probe caused by selfheating.

3267

showing a dc-drift of approximately 0.5 V after a temperature rise of 2 K. Note, that all temperature measurements are performed on the inside edge of the thermally insulated box, representing the average temperature inside the box. No Graph is shown for the *FireFly* probe, as no significant temperature rise was observed during a reasonable time period, thus creating no measurable dc-drift. It therefore seems necessary for the *TIVP* probe to ensure stable temperature control within an high-voltage (HV) measuring chamber, for example by supplying a controlled air flow. A promising approach is the construction of a multi-chamber system in which the measuring tip with DUT and the sensor head are located in two different chambers and can therefore be tempered separately. This setting also offers further advantages with regard to cm interference.

2 Conclusion and Outlook

A detailed comparison of two optically isolated probes was performed, investigating the effect of different parasitic elements within the measurement setup on the measurement results. Therefore, different connection options as well as distances of the sensor head to any conductive surface were considered. The evaluations show a susceptibility to parasitic influences of the setup for a *Fire-Fly* probe, while the measured curves of a *IsoVu TIVP* stay unaffected by the increase of parasitic inductance L_{par} and capacitance C_{par}. However, an effective suppression method for cm-disturbances was identified using several toroid cores for higher line attenuation, which can still be increased using NiZn-ferrite cores. With regard to thermally conditioned dc-drift, a *FireFly* probe shows sufficient temperature compensation without self calibration. Therefore, under consideration of the presented limitations of each probe, both probes are suitable for the measurement task. For both probes, an optimized measurement setup can be developed, taking limitations with regard to thermally induced dc-drift into account by using a multi-chamber measurement setup or controlled cooling of the sensor head of the *TIVP* probe or mitigating cm-voltage induced interference by minimizing parasitic elements within the measurement setup of a *FireFly* probe and considering suitable interference suppression measures as presented in this paper.

References

[1] T. Brinker, P. Mand, and J. Friebe, "Impact of gan-hemt combinations with different die-size on the efficiency of a single-phase photovoltaic differential buck inverter," in *2022 IEEE Energy Conversion Congress and Exposition (ECCE)*, 2022, pp. 1–8.

[2] J. Azurza Anderson, G. Zulauf, J. W. Kolar, and G. Deboy, "New figure-of-merit combining semiconductor and multi-level converter properties," *IEEE Open Journal of Power Electronics*, vol. 1, pp. 322–338, 2020.

[3] R. S. Yang, A. J. Hanson, B. A. Reese, C. R. Sullivan, and D. J. Perreault, "A low-loss inductor structure and design guidelines for high-frequency applications," *IEEE Transactions on Power Electronics*, vol. 34, no. 10, pp. 9993–10 005, 2019.

[4] S. Biswas, D. Reusch, M. de Rooij, and T. Neville, "Evaluation of measurement techniques for high-speed gan transistors," in *2017 IEEE 5th Workshop on Wide Bandgap Power Devices and Applications (WiPDA)*, 2017, pp. 105–110.

[5] *Understanding oscilloscope frequency response and its effect on rise-time accuracy*, 5988-8008EN, Keysight Technologies, Dec. 2017.

[6] H. Johnson and M. Graham, *High-speed Digital Design: A Handbook of Black Magic* (Prentice Hall Modern Semiconductor Design). Prentice Hall, 1993.

[7] *Tivp series isovu measurement system user manual*, 071369207, Rev. B, Tektronix Inc., Aug. 2023.

[8] *Firefly datasheet*, D-FF-1500, Rev. 05A, PMK GmbH, May 2023.

[9] *Dl-iso high voltage optically isolated probe*, Teledyne LeCroy Inc, Nov. 2022.

[10] *Bumblebee series datasheet*, D80-BUMBLEBEE-501, Rev. 01, PMK GmbH, Jan. 2023.

[11] B. Holzinger, M. Zimmermann, R. Takeda, and M. Hawes, "Understanding bandwidth requirements when measuring switching characteristics in power electronic applications," *Bodo's Power Systems Magazine*, 2021.

[12] P. S. Niklaus, R. Bonetti, C. Stäger, J. W. Kolar, and D. Bortis, "High-bandwidth isolated voltage measurements with very high common mode rejection ratio for wbg power converters," *IEEE Open Journal of Power Electronics*, vol. 3, pp. 651–664, 2022.

[13] *650v gan e-hemt evaluation board technical manual*, GS66508T/GS66516T-EVBDB2-UG, Rev. 200525, GaN Systems Inc., 2020.

PCIM Europe 2024, 11– 13 June 2024, Nuremberg   DOI: 10.30420/566262461

Non-Invasive Battery condition testing using electrical signals and Oscilloscopes

Srikrishna N.H, Madhusudan Acharya, Ramesh PE

[1] Tektronix, India

[2] Tektronix, India

[3] Tektronix, India

Corresponding author: Srikrishna N.H, krishna.n.h.sri@tektronix.com
Speaker: Srikrishna N.H, krishna.n.h.sri@tektronix.com

Abstract

Electrochemical Impedance Spectroscopy (EIS) is a powerful method for analyzing the electrochemical behavior of rechargeable batteries. By measuring the impedance response of a battery to an applied AC signal, EIS can provide valuable insights into the battery's charge and discharge characteristics, as well as its internal resistance and capacitance. To perform EIS measurements, an oscilloscope and signal generator are typically used to apply and measure the AC signal, while a specialized software is employed to analyze the resulting impedance data.

EIS is better method to other techniques commonly used in battery research, such as Open Circuit Voltage (OCV) and Accelerated Internal Resistance (AICR) measurements.

OCV is a simple method that measures the voltage of a battery when it is not supplying any current. However, OCV provide limited information about the battery's internal behavior and cannot reveal changes in its electrochemical properties over time. AICR measurements can be damaging to the battery and may not accurately represent its real-world behavior.

Designers want to observe voltage and current waveforms coming out of the battery, in a continuous mode. This paper talks about EIS using Oscilloscopes and Generator and post processing techniques.

1 Introduction

Electrochemical Impedance Spectroscopy (EIS) is a powerful technique for understanding the power delivery capability in rechargeable battery systems. This is a non-invasive method since measuring EIS spectra does not require dismantling of the battery cell, which is very advantageous in preventing sensitive samples from being contaminated with moisture and oxygen. Hence this non-destructive technique provides a considerable amount of information in a relatively short time span while preserving the integrity of the battery [1].

Also, industry looks at this method as powerful but simple technique to analyze the electrical properties of materials, including rechargeable batteries.

It provides insights into the electrochemistry of the battery and allows us to characterize its parasitic circuit elements.

EIS is commonly employed in battery research and development, in-line cell manufacturing, and off-line quality control [4].

To conduct this test, a small AC signal is applied over a wide frequency range and the response is measured. Measurement methodologies involving a span of frequencies are applied to individual cells, but they aren't applicable for larger sizes and larger operating power modules. In the automotive battery management system (BMS) space, the test is carried out on multi-cell batteries of varying capacities and mismatched cells must be carefully tracked. Cell mismatch can occur due to battery overcharge, discharge shorts, or simply because of aging.

The Randle's cell is one of the simplest cell models. Reasonable assumptions were made for the β coefficients, metal density and equivalent weight. This method is difficult for actual battery pack since there are many approximations, this is good for understanding the battery concepts [5].

For most cells, it is the supporting electrolyte (being an ionic conductor), that ends up dominating the magnitude of this resistance. Electronic conduction of the electrodes and external interconnection resistances are usually much smaller contributors to the total battery internal resistance.

3269

The passage of current through this real resistance will generate heat, which is a very important condition to consider. Many failure mechanisms of the cell show themselves through elevated measurements of the internal resistance. The characteristics of the internal resistance are generally very useful in determining the overall health of an electrochemical cell.

Fig. 1 shows the equivalent circuit model and Nyquist plot for a parallel R-C circuit where R is the Resistance, C is the Capacitance and Z is the impedance, also known as Warburg impedance (W) [2].

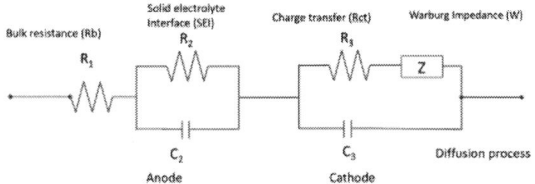

Fig. 1 Lithium-Ion battery equivalent circuit

Bulk resistance (R1) of the cell represents between electrolyte, separator, and electrodes.

R2 is and C2 represents resistance and capacitance of the interfacial layer.

R3 and C3 is charge transfer resistance and double layer capacitance.

W highlights diffusional effects of rechargeable battery material. [3].

The Randall's cell model assumes uniform behavior throughout the battery. This assumption may not hold true in real-world batteries, where the behavior can vary significantly across different regions of the battery.

Hence, Randle's cell is a basic starting point, EIS offers a deeper understanding of battery behaviour [6].

Researchers and engineers prefer non-invasive method due to its ability to reveal intricate electrochemical processes and provide valuable insights for battery design and optimization.

2 Understanding of EIS and Nyquist Plots

EIS can be used to test and debug rechargeable batteries in several ways. This can be used to detect changes in the battery's electrochemical properties over time, such as changes in its internal resistance or capacitance. By comparing the impedance response of a battery at different states of charge or under different operating conditions, EIS can provide valuable insights into the battery's behavior.

Also, EIS can be used to identify potential failure modes in rechargeable batteries. For example, EIS can detect the formation of a solid electrolyte interface (SEI) layer on the battery's electrodes, which can reduce the battery's performance over time. By detecting these failure modes early, EIS can help to extend the life of the battery and improve its overall performance [16].

EIS is a powerful method for testing and debugging rechargeable batteries, providing valuable insights into the battery's electrochemical behavior, and identifying potential failure modes [13]. EIS is becoming an increasingly popular tool in the field of battery research and development, as well as in the design of battery management systems for electric vehicles and other applications.

The Nyquist plot is a common way to represent EIS data [7]. It plots the negative imaginary component of impedance (vertical axis) against the real component of impedance (horizontal axis). Each data point on the Nyquist plot corresponds to a specific frequency at which impedance was measured.

A Nyquist plot is a graphical representation of the impedance response of a battery obtained through Electrochemical Impedance Spectroscopy (EIS) [14].

Fig. 2 shows an actual Nyquist plot captured from a Li-Ion battery module.

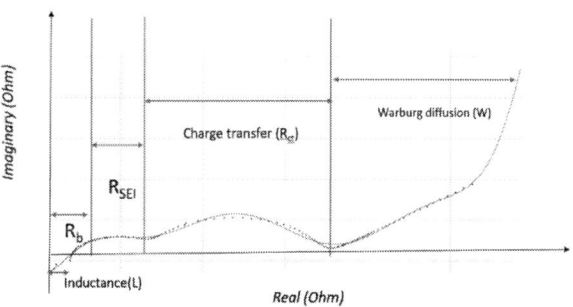

Fig. 2 Nyquist plot showing different states.

The Nyquist plot can be used to identify different states in the battery, including Rb, RSEI, RCT, and Warburg states. Rb state represents the bulk resistance of the electrolyte and is located on the real axis intercept of the Nyquist plot. RSEI state represents the resistance of the Solid Electrolyte Interface (SEI) layer on the electrode surface and is located on the semicircle of the Nyquist plot.

RCT state represents the charge transfer resistance at the electrode/electrolyte interface and is also located on the semicircle of the Nyquist plot. Warburg state represents the diffusion of ions in the electrolyte and is located on the sloping part of the Nyquist plot.[8]

By analyzing the Nyquist plot, researchers can obtain information about the different states in the battery and understand the underlying electrochemical reactions. This information can be used to optimize battery performance, identify potential failure modes, and develop new battery technologies.

3 Interpretation of Nyquist plot at different states

State of Health (SoH) diagnosis represents a key measure to guarantee safety and lifetime optimized operation of Lithium ion batteries (LIBs) in automotive applications. [10]

By analyzing the Nyquist plot, researchers can extract information about diffusion processes, electrode behavior, and more.

Understanding the root causes of battery degradation will help the BMS to prevent potential failures such as internal short-circuit due to overcharge or even thermal runaway.

Fig. 3 shows two different conditions of the Battery, first one is good and second is badly degraded.

There is a significant difference in the reaction resistance between a good and a degraded Lithium-ion battery.

The degradation of charge transfer resistance is particularly noticeable in the Nyquist plot as reaction resistance is longer than good state.

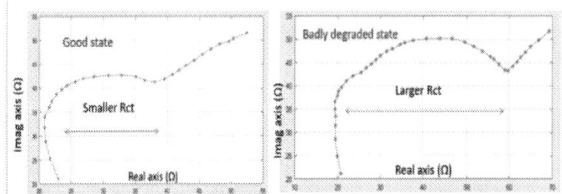

Fig. 3 Nyquist plots for battery degradation, (a) Healthy battery, (b) Degraded battery

Battery degradation analysis using Nyquist plots involves analyzing changes in the impedance response of a battery over time to identify degradation modes and assess the health of the battery.

As a battery degrades, its electrochemical properties change, resulting in changes in its impedance response. By analyzing the Nyquist plot obtained from EIS measurements, researchers can identify changes in the different states of the battery, such as the RSEI, RCT, and Warburg states [9,12].

For example, an increase in the RSEI state on the Nyquist plot may indicate the formation of an insulating layer on the electrode surface, which can limit the battery's performance. Similarly, an increase in the RCT state may indicate problems with the charge transfer at the electrode/electrolyte interface, which can also impact battery performance.

By monitoring changes in the Nyquist plot over time, researchers can identify trends in the battery's degradation and assess its remaining useful life. This information can be used to develop strategies to extend the battery's life, optimize its performance, or plan for its replacement.

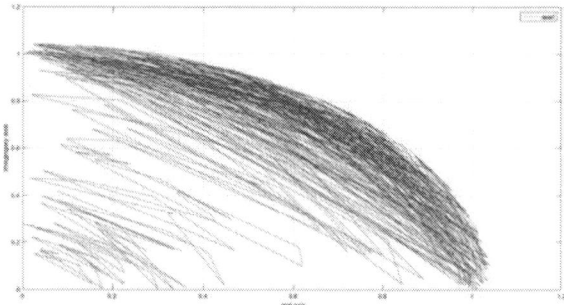

Fig. 4 Nyquist plot for a fully charged battery, does not have a true diffusion.

A Nyquist plot for a fully charged state battery condition obtained from Electrochemical Impedance Spectroscopy (EIS) measurements can provide valuable information about the battery's electrochemical behavior and performance.

In a fully charged state, the battery has a different impedance response compared to its other charge states. Specifically, the RSEI and RCT states on the Nyquist plot are typically lower in a fully charged battery compared to a partially charged or discharged battery.

Fig. 4 shows the Nyquist plot for a fully charged battery condition typically exhibits a smaller semi-circle and a shorter straight line compared to other charge states as observed in **Fig. 1**.

4 Test setup.

4.1 Setup using Oscilloscopes and Generator

The EIS setup consists of the measurement hardware and a fixture to connect to the battery under test (BUT). It has Generator (AFG31000), MSO 5/6 series scope, SMU: 2230. The analysis software runs on an Oscilloscope as shown in **Fig. 5**.

The first step is to connect the battery pack being tested to the oscilloscope and the AFG1000 (Arbitrary Function Generator from Tektronix). The AFG is used to generate the AC signal that will be applied to the battery, while the oscilloscope is used to measure the voltage and current waveforms.

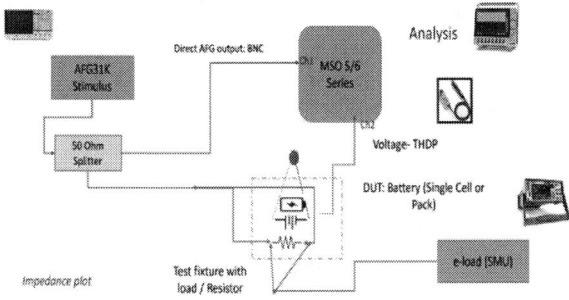

Fig. 5 Schematic test setup.

The stimulus from AFG will be sweep across range of frequencies, with start and stop frequency being configurable and number of PPD points. To improve the testing further, designers can specify amplitude profile at certain frequency ranges as show in **Fig. 6**.

The impedance of the battery can be calculated using the ratio of the voltage and current responses.

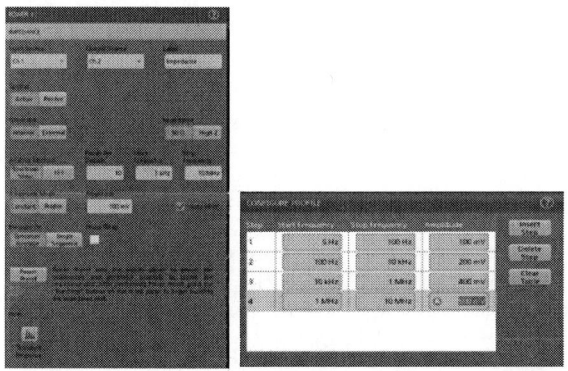

Fig. 6 Impedance Configuration amplitude profile

The impedance data obtained from the EIS measurement can be plotted on a Nyquist plot, which is a graphical representation of the impedance response. The Nyquist plot can be used to analyze the different states in the battery, such as the RSEI, RCT, and Warburg states. This is explained in the algorithm section.

4.2 Algorithm

4.3 Phase computation from Bode plot:

Control Loop Response computes phase difference between Input to Output source at each frequency within the swept band. Phase is the time shift between the input and output signals.

$$
\begin{aligned}
&\text{Phase @ swept frequency} \\
&= \text{FFT(Phase (Battery output Voltage)} \\
&- \text{Phase (Battery output Current))}
\end{aligned} \tag{1}
$$

Fig. 7 shows actual test setup where frequency is swept from 1Hz to 40MHz from AFG31000 generator using active splitter and DC block.

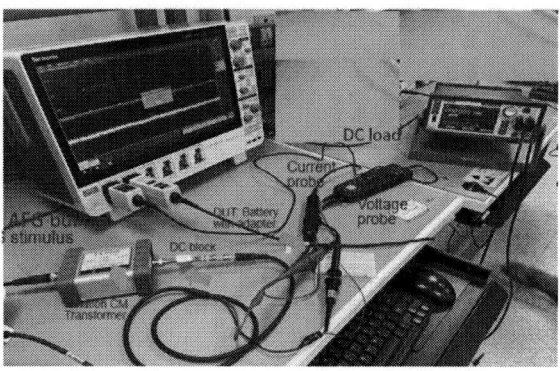

Fig. 7 Test setup showing different T&M instruments.

Impedance (Z) is computed as source gain magnitude ratio as

$$
(Z = G / (1 - G)) \tag{2}
$$

Where G is gain magnitude, defined as

$$
\begin{aligned}
&(G \text{ in dB} \\
&= \text{LOG10((Battery Output Voltage)} \\
&- (\text{AFG Direct (Input)Voltage)))}
\end{aligned} \tag{3}
$$

4.4 Computation of Nyquist plot:

Electrochemical impedance spectra (EIS) are frequently plotted in the Nyquist diagram for a wide range of frequencies from mHz to kHz. Typical spectra often resemble a composition of sometimes compressed or overlapping semi-circle regions.

Semicircles are usually attributed to certain processes using their characteristic frequency. The transition from one dominant process to a different characteristic process may be visible in the spectra via changes of the angle and amplitude.

$$(Zreal = Z * \cos(\emptyset)) \tag{4}$$

Equation (1) relates to resistance as it is real axis.

$$(ZImag = Z * \sin(\emptyset)) \tag{5}$$

Equation (5) relates to C(capacitance) and L(inductance)values in the circuit.

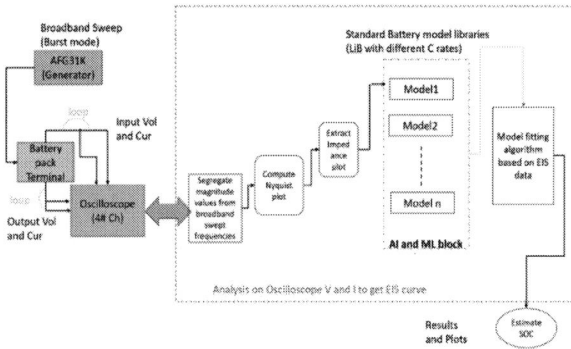

Fig. 8 Flowchart with modelling example

Following are the chemical parameters extracted from the Nyquist plot:

Different parameters are computed based on the semi-circle regions and taking the mean value on the real axis gives various resistance values.

Also, these parameters highlight importance of chemical properties of the battery.

- Rb: The total resistance of the electrolyte, separator, and electrodes. This value is significantly related to the SOH.

- RSEI: The first semicircle in the Nyquist plot, is associated with the generation of the interfacial layer deposited on the electrode. This value can be used to analyze the formation of the SEI layer, which is the result of the decomposition of the electrolyte.

- Rct: The second semicircle in the Nyquist plot, is related to the kinetics of the electrochemical reaction, which is changed by the surface coating, phase transition, band gap structure or particle size.

- Warburg impedance: Final straight line of the Nyquist plot, which is related to the diffusion of lithium ions.

- State of Charge (SOC) can be computed as

$$\left(SOC = \int \text{Batttery Current} \big/ \text{Rated Capacity}\right) \tag{6}$$

- The various above mentioned resistance parameters can be computed algorithmically by identifying the regions form the Nyquist plot as in **Fig. 2**. Computes the mean value in that specific region of interest.

where I is Battery Current from 0 to charging time.

State of Health (SOH): Ratio of maximum releasable capacity to the nominal rated capacity. Specified in percentage.

As explained in **Fig. 8** user can observe time domain waveforms acquired on Oscilloscope, basically is what is getting in and out of a Battery pack. These waveforms will be processed to get Nyquist plots. Display of dynamic output of time domain waveforms is a key part of the analysis which is one of the advantages of using Oscilloscope as testing instrument.

We compare and fit the extracted measured Battery impedance with the standard battery impedance curves [10]. This will help users to know dynamic quality of the battery in faster and accurate way.

To fit with appropriate battery type (Ex: LiB, LFP et al), we plan to use AI/ML method model and specific parameters are trained with machine-learning to measure battery SoC/SoH [15].

Designers will get to see the comparison plot of measured and standard impedance curves.

The study results from different experiments show that SoH of batteries can be predicted with high accuracy, and good reliability using machine learning approach to train some of these models. [17].

4.5 Overlapped Nyquist Plot:

Fig. 9 shows the overlapped Nyquist plot at different temperatures. The shape of the Nyquist defines the equivalent circuit, we can directly interpret the chemical composition and action by converting the shape to equivalent circuits.

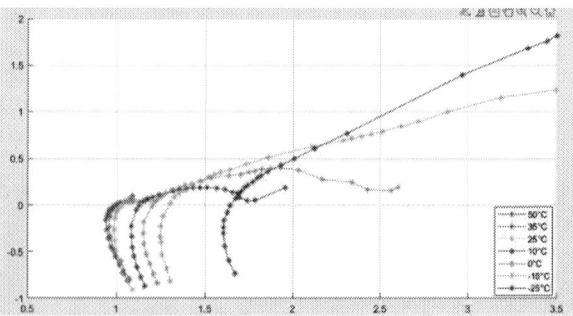

Fig. 9 Overlapped Nyquist MATLAB plot at different temperatures.

5   Results:

Fig. 10 shows results of 6V and 12V Lithium-Ion battery pack.

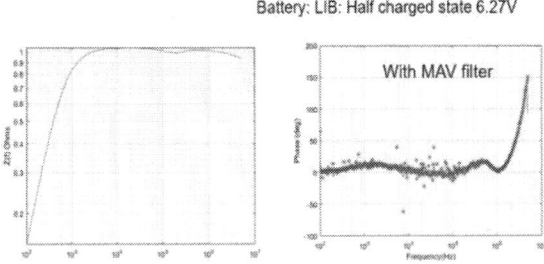

Fig. 10 Impedance and Bode Phase plot.

Fig. 11 and **Fig. 12** shows actual swept frequency from 10Hz to 10MHz phase and gain curves and impedance plots. This phase values will be used to compute Nyquist plot as explained in equation 4 and 5.

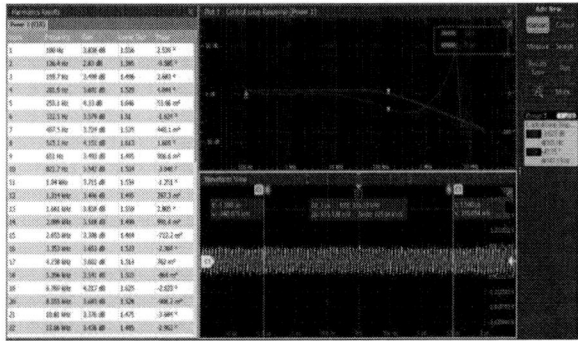

Fig. 11 Bode Phase plot with frequency values swept on Oscilloscope.

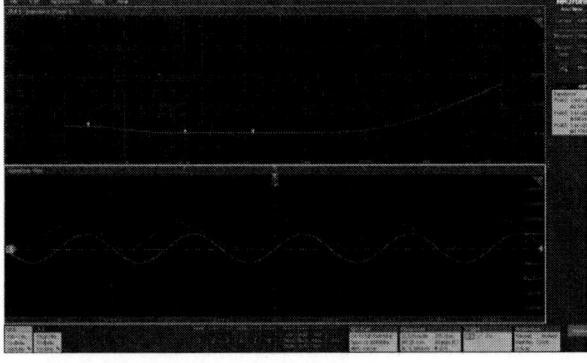

Fig. 12 Impedance plot with frequency values swept on Oscilloscope.

A Nyquist plot represents the complex transfer function of a system in a complex plane. The x-axis corresponds to the real part of complex Impedance value, while the y-axis represents the imaginary part. This is shown in **Fig. 13** with separating various chemical parameters of the rechargeable battery with actual results.

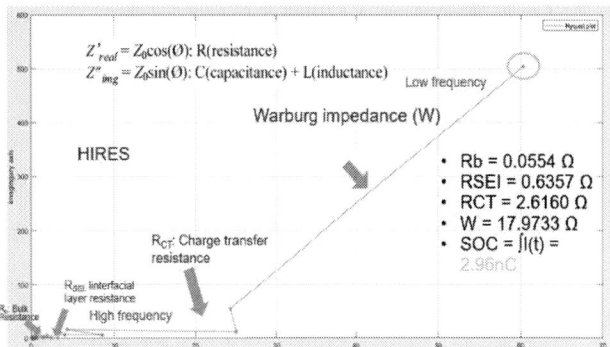

Fig. 13 Nyquist (Impedance and Bode Phase) plot.

5. Conclusion

With growing Electric Vehicle adoption in the market, there is a need for in-expensive testing methods and instruments to measure and observe static and in-vehicle battery state of health conditions.

Numerous electrochemical analysis methods can be used to investigate the internal processes of a battery. However, in the case of the EIS analysis, it is possible to obtain important values from each component in a lithium-ion cell using nondestructive method that does not damage the batteries.

By creating an accurate circuit model and implementing a consistent cell system for Electrochemical Impedance Spectroscopy (EIS) analysis, vital information about every component present in the lithium-ion battery can be obtained [11].

As a future work we would like to test the performance of the rechargeable battery using proposed low-cost solution (which contains general purpose Oscilloscope and built-in generator features) with dedicated EIS instruments.

References

[1] E. Unamuno, L. Gorrotxategi, I. Aizpuru, U. Iraola, I. Fernandez and I. Gil, "Li-ion battery modeling optimization based on Electrical Impedance Spectroscopy measurements," 2014 International Symposium on Power Electronics, Electrical Drives, Automation and Motion, Ischia, Italy, 2014,

pp. 154-160, doi: 10.1109/SPEEDAM.2014.6872015.

[2] M. Gaberscek, J. Moskon, B. Erjavec, R. Dominko, J. Jamnik, Electrochem. Solid-State Lett., 2008, 11(10), A170-A174.

[3] Modelling and Applications of Electrochemical Impedance Spectroscopy (EIS) for Lithium-ion Batteries.

[4] Textbook: Electrochemical Methods: Fundamentals and Applications, 3rd Edition, Allen J. Bard, Larry R. Faulkner, Henry S. White

[5] Electrochemistry experts 'Measuring battery internal resistance is easy!' from Tektronix

[6] "Impedance spectroscopy characterization of lithium batteries with different ages in second life application," 2020 IEEE International Conference on Environment and Electrical Engineering and 2020 IEEE Industrial and Commercial Power Systems Europe (EEEIC / I&CPS Europe), Madrid, Spain, 2020, pp. 1-6, doi: 10.1109/EEEIC/ICPSEurope49358.2020.9160616, https://ieeexplore.ieee.org/document/9160616

[7] Impedance Measurement Technique for Vehicle Battery Health Assessment Using On-Board CAN Signals, Sayandeep Dutta; Jayabrata Maity; Saumya R. Mohapatra; Munmun Khanra

[8] S. M. Lambert et al., "Rapid Nondestructive-Testing Technique for In-Line Quality Control of Li-Ion Batteries," in IEEE Transactions on Industrial Electronics, vol. 64, no. 5, pp. 4017-4026, May 2017, doi: 10.1109/TIE.2016.2643601.

[9] C. Pastor-Fernández, W. Dhammika Widanage, J. Marco, M. -Á. Gama-Valdez and G. H. Chouchelamane, "Identification and quantification of ageing mechanisms in Lithium-ion batteries using the EIS technique," 2016 IEEE Transportation Electrification Conference and Expo (ITEC), Dearborn, MI, 2016, pp. 1-6, doi: 10.1109/ITEC.2016.7520198.

[10] P. Weicker, A systems approach to Lithium-Ion mattery management. Artech House, 2014.

[11] Modeling and Applications of Electrochemical Impedance Spectroscopy (EIS) for Lithium-ion Batteries Woosung Choi1 , Heon-Cheol Shin2 , Ji Man Kim3 , Jae-Young Choi4,5, and Won-Sub Yoon1.

[12] Weiheng Li, Qiu-An Huang, Changping Yang, Jian Chen, Zhepeng Tang, Fangzhou Zhang, Aijun Li, Lei Zhang, Jiujun Zhang, "A fast measurement of Warburg-like impedance spectra with Morlet wavelet transform for electrochemical energy devices", Electrochimica Acta, vol.322, pp.134760, 2019.

[13] Alireza Rastegarpanah, Mohamed Ahmeid, Naresh Marturi, Pierrot S Attidekou, Muhammad Musbahu, Rohit Ner, Simon Lambert, Rustam Stolkin, "Towards robotizing the processes of testing lithium-ion batteries", Proceedings of the Institution of Mechanical Engineers, Part I: Journal of Systems and Control Engineering, vol.235, no.8, pp.1309,2021.

[14] E. Barsoukov, J.R. Macdonald, Impedance Spectroscopy, Theory, Experiment, and Applications, second ed., John Wiley Sons, New Jersey, 2005.

[15] J. Sihvo, D. -I. Stroe, T. Messo and T. Roinila, "Fast Approach for Battery Impedance Identification Using Pseudo-Random Sequence Signals," in IEEE Transactions on Power Electronics, vol. 35, no. 3, pp. 2548-2557, March 2020, doi: 10.1109/TPEL.2019.2924286.

[16] The understanding of solid electrolyte interface (SEI) formation and mechanism as the effect of flouro-o-phenylenedimaleimaide (F-MI) additive on lithium-ion batter, Nur Laila Hamidah, Fu Ming Wang, Gunawan Nugroho.

[17] M. A. Tonima et al., "Electrochemical Impedance Spectroscopy (EIS) and Machine Learning based Battery State of Health (SoH) Estimation," 2023 IEEE International Conference on Prognostics and Health Management (ICPHM), Montreal, QC, Canada, 2023, pp. 212-223, doi: 10.1109/ICPHM57936.2023.10194065.

PCIM Europe 2024, 11– 13 June 2024, Nuremberg   DOI: 10.30420/566262462

Instrumentation Requirements for Fast 130+ V/ns Switching of 1700 V, 35 mΩ SiC MOSFETs

Matthew Appleby[1] , Yushi Wang[1] , Qilei Wang[1] , Jiaqi Yan[1] , Harry C. P. Dymond[1] ,
Saeed Jahdi[1] , Bernard H. Stark[1]

[1]University of Bristol, United Kingdom

Corresponding author: Matthew Appleby, matt.appleby.2019@bristol.ac.uk

Abstract

This paper demonstrates the benefits, downsides, and instrumentation requirements of switching 1.7 kV, 35 mΩ SiC MOSFETs at 130+ V/ns, beyond the speed used by the device manufacturer for datasheet characterisation. Experimental results are obtained in a 1200 V, 50 A bridge leg, and comparisons are made between passive voltage probes, optically isolated differential probes, shunt current measurement, Rogowski coils, and Infinity Sensors. At 130 V/ns, a 24% improvement over the datasheet characterised switching loss is found, however the limitations of Rogowski coils and passive probes become significant. The methods demonstrated should permit design engineers to explore switching speed and efficiency limitations in their applications.

1   Introduction

Wide bandgap switches can reduce switching losses compared to traditional silicon switches due in part to their short switching duration. However, switching speeds must sometimes be limited in order to avoid reliability concerns through unwanted waveform features, including: excessive drain-source voltage overshoot leading to breakdown, power loop current overshoots and ringing [1] [2] [3] leading to EMI concerns [4], gate pick up leading to cross-talk [5], gate voltage overshoot leading to gate breakdown concerns [6], and excessive *dv/dt* potentially leading to latchup of the parasitic BJT [7]. By switching SiC MOSFETs faster, with gate resistors that are smaller than those used by the manufacturer for datasheet characterisation, significant efficiency and power density improvements could be achieved, but only if unwanted switching waveform features can be addressed, e.g. by improved layout, assist circuits, or advanced gate driving. This requires suitable instrumentation to observe unwanted, high-bandwidth features in switching waveforms. This paper demonstrates experimental switching speeds that are beyond those used for the datasheet characterisation of the 1700 V, 35 mΩ TO-247-4 SiC MSC035SMA170B4 MOSFET (Fig. 1), and shows where probes excel or run into limitations. The paper's contributions are:
- In Section 2, an analysis of the measurement requirements for fast SiC power converters, to include the key signals that need to be measured, bandwidth requirements, and voltage/current deskewing accuracy required to maintain power loss measurement accuracy.
- Different probing options, suitable for measurement in high-speed SiC circuits, are discussed and summarised in Section 3, including a helpful tip that can allow a Rogowski coil to be placed to measure source current without overlapping the switch node trace.
- In Section 4 a 1200 V, 50 A bridge-leg is demonstrated, switching at up to 3× datasheet characterisation speeds, reaching voltage slew rates of 200+ V/ns. Simultaneous measurements are demonstrated with different current and voltage probes, to allow comparison of different probe types and technologies. High-fidelity gate current measurements are shown using magnetic-field-based and shunt-based measurements.
- Limitations on switching speed are discussed in Sections 4.2 – 4.3, in terms of gate driver capability, device reliability and EMI.

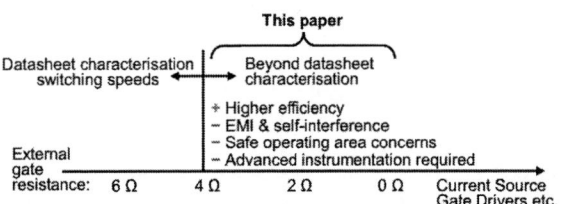

Fig. 1   Scope of this paper: switching beyond datasheet speeds.

2 Measurement considerations for fast transients of SiC MOSFETs

2.1 Target Measurements

Fig. 2 shows a simplified schematic of a half-bridge switching circuit, with an inductive load and the low-side device as the active switch. Fig. 3 demonstrates some of the unwanted features that can be observed in the switching waveforms of such half-bridge legs, especially when they are switched at high speed. Accurately observing these unwanted features is an important prerequisite for diagnosing whether a converter will have reliability issues, and a first step in resolving them.

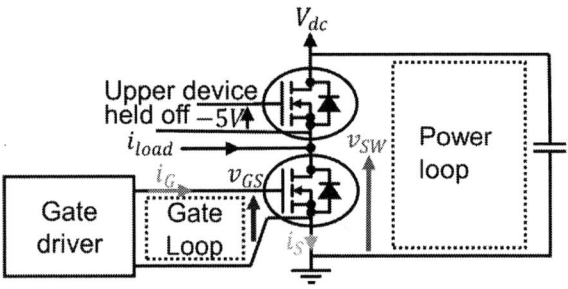

Fig. 2 Simplified circuit diagram of a half bridge switching under an inductive load.

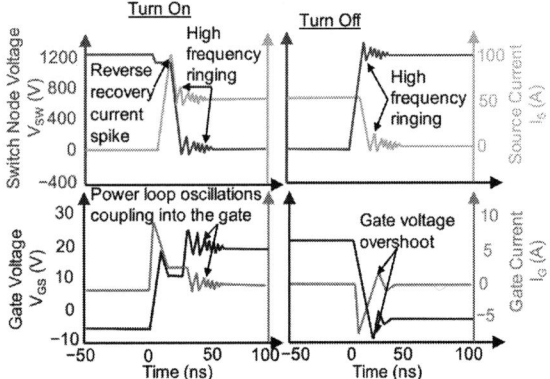

Fig. 3 Concept diagram showing waveform features at high switching speed. Left: low-side device turn-on. Right: low-side device turn-off.

A significant contributor to the current overshoot that occurs in the turn-on process of the active device, is the reverse recovery of the intrinsic body diode of the opposing MOSFET (in this case, the high-side MOSFET) [8]. The overshoot and subsequent oscillation can cause EMI concerns [1]. The peak source current may breach the maximum transient current capability of the MOSFET which

combined with large dv/dt can lead to BJT latch-up concerns [7].

High frequency ringing appears under fast switching transients due to the power devices' output capacitance resonating with the power loop inductance. For low-side switching the high side device's output capacitance resonates at turn-on, and the low-side capacitance at turn off. Increased switching speed increases the oscillation magnitude and duration.

When gate resistance is reduced, the gate loop can become underdamped causing transient gate voltage overshoot and subsequent gate loop ringing [6], observable with gate current and gate voltage instrumentation. Overshoot and oscillations from the power loop can be coupled into the gate loop via the reverse transfer capacitance, which may cause abnormal turn-on behaviour, and ultimately damage to the device. Additionally, in 3 pin packages, source current di/dt couples into the gate via the inductance of the common source pin. In packages with a Kelvin source, there will be some di/dt coupling, due to the gate loop encircling some area which will have some mutual coupling with the power loop. Peak gate current limitations were proposed in [6].

In order to observe these behaviours, to aid the design engineer in the optimisation process, the following waveforms should be measured: v_{SW}, i_S, v_{GS}, and i_G. Further, switching loss can be calculated from measured v_{SW} and i_S.

2.2 Measurement requirements

In this section some of the key requirements for accurate switching measurements are evaluated. Satisfying these requirements enables accurate power device characterisation which can mitigate converter failure.

When evaluating measurement apparatus, it must be taken in the wider context of the measurement system [9], including the signal, the measurement connection, the probe, the recording device - typically an oscilloscope, and the external environment that the measurement system is operating in. How a probe is connected to a circuit critically impacts its performance [10], therefore the probes and connection method will be evaluated together, where appropriate. As switching speed increases, the measurement environment becomes harsher, with larger di/dt and dv/dt, which can disrupt measurement. Faster switching edges require more bandwidth to observe, can induce common mode voltages that disrupt measurement, require higher magnetic-field and electric-field immunity and an increasingly accurate deskew for accurate loss characterisation.

2.2.1 Measurement Bandwidth

As switching speed increases, higher bandwidth probes and scopes are required to observe the resulting waveforms.

Firstly, considering the risetime of the signal of interest, the required measurement bandwidth is given by:

$$f_{3db} = 3 \times 0.35/T_{rise}, \qquad (1)$$

where f_{3db} is the required measurement bandwidth and T_{rise} is the 10%-90% rise time [10].

For a 1200 V dc-link, a measurement bandwidth of 110 MHz is required to observe a transient slewing at 100 V/ns and 220 MHz for 200 V/ns.

Secondly, the frequency of any post-edge ringing should be considered. For example, the power devices under investigation here have an output capacitance of 150 pF at 1200 V, which when coupled with the power loop inductance of approximately 20 nH can produce ringing at 92 MHz.

Thirdly, in respect of EMI performance, high frequency interference sources can reside in small waveform features [11], which will be filtered out of the measurement if low bandwidth instrumentation is used.

2.2.2 Common-mode rejection for gate-source measurements

When measuring the gate-source voltage, the common-mode performance of the probe used can be critical. Commonly, this requirement is considered in relation to the measurement of a high-side device, where the reference node (the source of the high-side MOSFET) is slewing rapidly relative to the negative side of the DC link. However, common-mode rejection can also be important for low-side devices. MOSFETs with kelvin connections can induce transient common-mode voltages between the kelvin and power ground, due to the source inductance of the MOSFET and the large di/dt under switching [12]. For example, 5 nH of source inductance can induce 50 V under a fast 10 A/ns switching event. To attenuate this common mode signal to 0.5 V error, a common-mode rejection ratio (CMRR) of 40 dB will be required [12] at up to 100 MHz as shown in equ. 2 [9].

$$CMRR = \left|\frac{A_{dm}}{A_{cm}}\right| = \left|\frac{V_{cm}}{V_{err}}\right| \qquad (2)$$

Where A_{dm} is the differential mode gain and is assumed to be 1, A_{cm} is the common mode gain, V_{cm} is the common mode signal, and V_{err} is the acceptable error in the measurement. As switching speed is increased, more common-mode rejection is required. High side gate measurements have a much stricter common mode immunity requirement due to the high side source voltage rapidly slewing during switching.

2.2.3 Limitation of gate voltage measurement

There are inherent limits on the accuracy of gate voltage measurements taken at the package terminals, due to the packaging inductance between the device terminals and the die, which in this case is 9 nH according to the manufacturer's SPICE model [13]. Gate di/dt has been measured in later sections to be 0.37 A/ns, which can induce 3.33 V across the packaging inductance, causing a misreading at the measurement terminals [14]. At low external gate resistance, the internal gate resistance of 0.85 Ω dominates, which complicates evaluation of the state of charge of the internal capacitance of the device. Gate current measurements can provide a useful alternative as the measurement is not impacted by these parasitics.

2.2.4 Magnetic-field and electric-field immunity

Voltage measurements require high magnetic-field immunity from the source current during switching, due to the ability of large di/dt events to radiate magnetic fields, in turn inducing a voltage in measurement loops. For a power loop with peak oscillation di/dt of 10 A/ns, mutual loop inductance must be less than 100 pH to ensure the interference induced in the measurement loop is less than 1 V. Mutual coupling is highly dependent on specific orientation and geometry. Similarly, the electric-field immunity of current probes should be considered, to minimise errors induced by high dv/dt in the vicinity of the probe. Immunity can be increased through specific probe design, shielding, and careful probe placement and orientation.

2.2.5 Probe features to reduce invasiveness of measurements

Measurement techniques should always strive to minimise the effects that they can have on the behaviour of the circuit under test.

Voltage probes should have low input capacitance to reduce current flowing into the front-end of the oscilloscope and to reduce loading on the circuit, whilst current probes should have as low insertion inductance as possible.

2.2.6 Considerations to maximise accuracy of switching-loss calculations derived from electrical signal measurements

Fast switching is primarily focused on reducing switching loss. Switching loss can be calculated from measured device voltage and current. Accurate characterisation requires the propagation delay of the current and voltage signals to be matched – i.e. the measured waveforms must be deskewed before their product is calculated to give the instantaneous power waveform. Increasingly accurate deskewing is required as switching speed increases [15], [16]. In Fig. 4, probe skew has been added to a 1200 V, 50 A low-side turn-on switching event as shown in Fig. 2. The increasing slope for different slew rates demonstrates how the deskew requirement becomes stricter as switching speed increases. For a ±5% switching loss accuracy, with the datasheet-employed 4 Ω gate resistance, probes must be deskewed to ±750 ps whereas at 0 Ω, ±250 ps deskew is required. In this work, all the waveforms are deskewed to within ±160 ps, giving <±5% error.

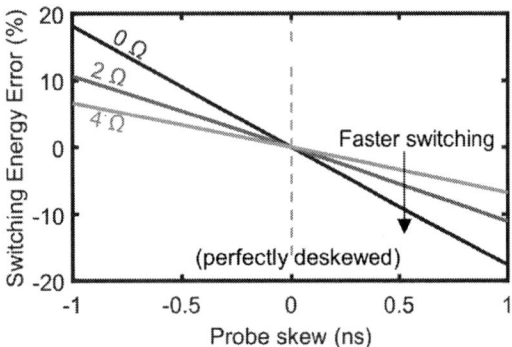

Fig. 4 Impact of deskewing current and voltage waveforms on derived switching energy, using experimental example data of a turn-on switching transient at 1200 V & 50 A at various switching speeds: 4 Ω: 70 V/ns & 7 A/ns; 2 Ω: 90 V/ns & 9 A/ns; 0 Ω: 130 V/ns & 12 A/ns.

3 Hardware Implementation

A double-pulse rig was used to test MSC035SMA170B4 1700 V 35 mΩ TO-247-4 SiC MOSFETs [17] at 1200 V and a load current of 50 A. The devices were driven by an Infineon Eicedriver 1ED3124MU12HXUMA1 [18] with +20 V and −5 V rails. The gate driver internal resistance is 0.45 Ω at turn on and 0.35 Ω at turn off, whilst the power device has an internal gate resistance of 0.85 Ω.

To reduce the power loop inductance the pins of the TO-247-4 package have been bent so that the package can be soldered onto the surface, eliminating through holes and thus allowing the return current to flow on an uninterrupted continuous ground plane on a close inner layer of the PCB. The power loop inductance was estimated from the ringing frequency to be 22.9 nH.

3.1 Comparison of measurement options

In this paper probing is reviewed using single-ended passive voltage probes, optically isolated voltage probes, commercial Rogowski coils and Infinity Sensors, which are types of a miniature magnetic field current sensor that is galvanically isolated, with high bandwidth, low cost, and low insertion inductance [19].

To demonstrate the trade-off of different measurement technologies, duplicate measurements of current and voltage on gate and power terminals are compared on a single capture of an 8 channel 2 GHz Tektronix MSO58B oscilloscope for two switching speeds. Fig. 5 details which probes were used in which locations of the circuit; the specific connections are shown in the photograph of Fig. 6. A summary of the probes and their specifications is provided in Table 1.

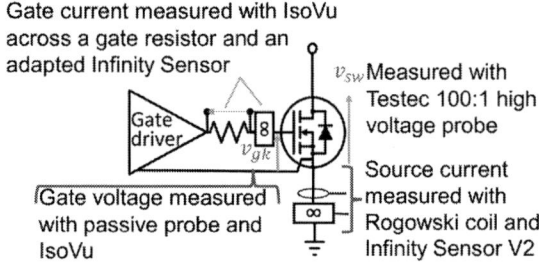

Fig. 5 Measurement diagram.

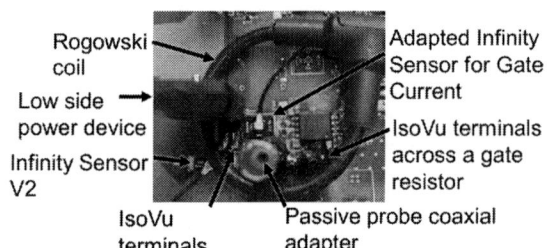

Fig. 6 Duplicate measurement for current and voltage on gate and power terminals.

3.1.1 v_{sw} measurement

A high-voltage passive probe is used to measure the low side switch node voltage. The probe has been used in a high-voltage coaxial probe adaptor from PMK to reduce the measurement loop inductance and improve its immunity to di/dt [10].

3.1.2 i_s measurement

The low-side source current is measured with a Rogowski coil and an Infinity Sensor.

To reduce the degree of dv/dt pick-up of the Rogowski coil, careful placement is required. In this circuit, enclosing only the source pin of the device would inevitably force the coil to be in close

Signal	Probe	Specification
v_{SW}	Testec HV250 passive probe	300 MHz 100:1 attenuation 2.5 kV max
i_S	Infinity Sensor V2	1 MHz to 1 GHz 100 mV/(A/ns) sensitivity 600 A/ns max
	Rogowski coil CWTMini 50HF 06 PEM	75 Hz – 50 MHz 50 mV/A sensitivity 8 A/ns max
v_{GS}	Tektronix IsoVu TIVP1L	1 GHz 50 V max tip 100 MHz CMRR: 92 dB
	PMK Firefly (Section 4.5)	>1.5 GHz 50 V max tip 100 MHz CMRR: 75 dB
	Keysight N2890A Passive Probe	500 MHz 10:1 attenuation 300 V max
i_G	Adapted Infinity Sensor	1 MHz – 300 MHz 1 V/(A/ns) sensitivity
	Infinity Gate Sensor (Section 4.5)	1 MHz – 500 MHz 0.67 V/(A/ns) sensitivity
	IsoVu TIVP1L across a gate resistor	1 GHz 50 V max 100 MHz CMRR: 92 dB

Table 1 Probes used in the study.

proximity to the switch node, which would increase the amount of *dv/dt*-induced noise in the measurement. If instead, the source, gate and Kelvin source pins are enclosed, the coil can be oriented away from the circuit switch node. As both the gate and Kelvin source pins are enclosed, the gate current cancels out and only the power-circuit source current is measured.

The Infinity Sensor is placed between the source and the power ground [19]; this arrangement minimises *dv/dt* seen by the sensor.

3.1.3 v_{GS} measurement

The gate voltage is measured with a passive probe in a coaxial adapter from PMK and using short enamel pair of jumper wires [10]. A second measurement is made via an IsoVu connected to the circuit using 2.54 mm pin-headers soldered into the gate and Kelvin terminals of the TO-247-4 package.

3.1.4 i_G measurement

The gate current is measured with an IsoVu across an SMD gate resistor, via a twisted pair of thin enamel wire [10], to 2.54 mm pin headers. The *di/dt* immunity is less strict for the gate current interconnects due to the gate resistor being 15 mm

away from the noise source. The gate current is also measured with an adapted Infinity Sensor design, based on the V2 design but with 10× the gain and 300 MHz bandwidth (measured with a Rhode & Schwarz ZVL Vector Network Analyser).

4 High Speed Switching Results

4.1 Comparison of measurement waveforms

Fig. 7 shows experimental waveforms of the turn-on transients in double pulse tests at 1200 V and 50 A. Two switching speeds are compared to show the measurement discrepancies of using different probes.

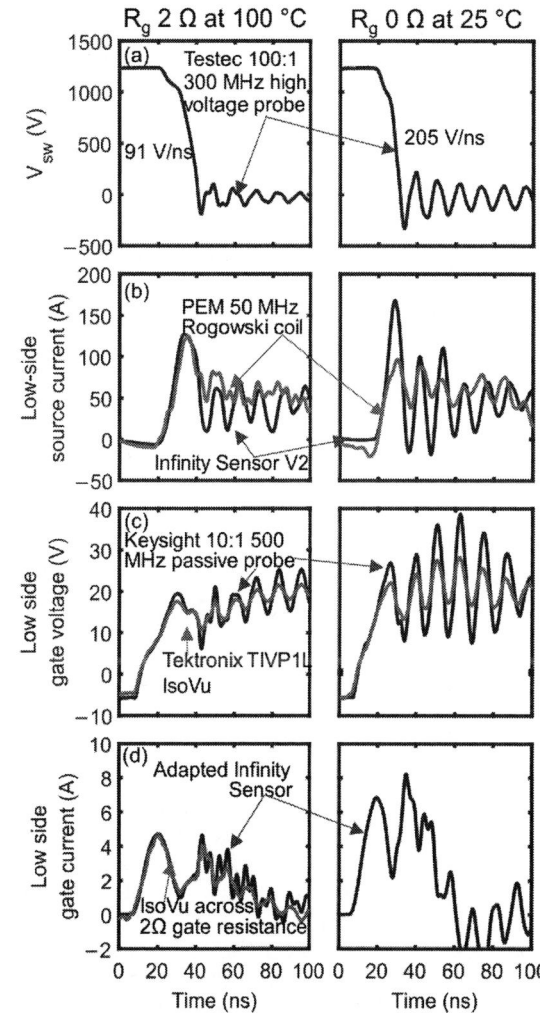

Fig. 7 Comparison of measurement technologies in circuit switching at 1200 V, 50 A, with case temperature at 100 °C and gate resistor of 2 Ω, to represent fast switching at a realistic converter temperature (left), and at 25 °C with gate resistor of 0 Ω, to achieve the maximum observed speed (right).

In Fig. 7(a), the switch node voltage is measured with a 300 MHz Testec HV250 100:1 passive probe. By reducing the external gate resistance and device temperature, the switching speed is greatly increased. The *dv/dt*, calculated from the 60% to 40% values of the swich node voltage, is 91 V/ns at 2 Ω, 100 °C, and 205 V/ns at 0 Ω, 25 °C. The ringing frequency is approximately 86 MHz for both speeds. Meanwhile, larger overshoot and oscillations are measured with the increased switching speed.

In Fig. 7(b), source current measurements are compared between a PEM CWTMini 50 MHz Rogowski coil and a Bristol 1 GHz Infinity Sensor V2 [19]. Significant discrepancies can be observed between these two current measurement instruments in both the current overshoot and ringing. At 2 Ω, 100 °C, there is a good match at the first peak of current, but the ringing captured by the Rogowski coil is attenuated. At 0 Ω, 25 °C, the discrepancy is more significant, where both peaks and ringing are mismatched. Such discrepancies may be caused by the limited bandwidth of the Rogowski coil used (50 MHz). In addition, some of these problems are a function of size, causing the Rogowski coil to overlap the switch node when enclosing the pin of the TO-247 device, leading to *dv/dt* pickup.

In Fig. 7(c), the low-side gate voltage to Kelvin source is compared for a simultaneous measurement with a Gen 2 IsoVu and a 500 MHz 10:1 N2890A passive voltage probe. Approximately 3× more ringing is observed with the passive probe than the IsoVu. This is likely due to the IsoVu probe's higher common-mode rejection of the voltage oscillations between Kelvin source and power ground and the improved magnetic-field immunity. In Fig. 7(d), the gate current at 2 Ω is measured with an adapted Infinity Sensor V2 and an IsoVu connected across the 2 Ω SMD gate resistor. The gate current profiles of two measurements have a good match, proving the accuracy of using a modified Infinity Sensor to measure the gate current. More oscillations exist on the Infinity Sensor measurement result than the IsoVu voltage probe due to the IsoVu having high CMRR. At 0 Ω, 25 °C, the external gate resistance is completely removed, and the measurement by IsoVu is not viable anymore. As can be seen from the figure, the gate current profiles are similar, and the peak of gate current increases from 5 A to 7 A when reducing the external resistance from 2 Ω to 0 Ω.

Considering the measurement accuracy, the following probes are used in later tests: a passive voltage probe for switch node voltage, an In-finity Sensor V2 for source current, an IsoVu for gate voltage, and the adapted Infinity Sensor for the gate current.

4.2 Impact of gate resistance with case temperature of 100°C

4.2.1 Turn-on

Double-pulse switching waveforms, for low side turn-on are shown in Fig. 8, with gate resistance being swept from of 4 Ω to 0 Ω, where 4 Ω is the value used by the manufacturer for datasheet characterisation. The power devices are held at 100°C, to emulate continuous operation. The results demonstrate how high bandwidth probing can identify onset of unwanted features in switching waveforms.

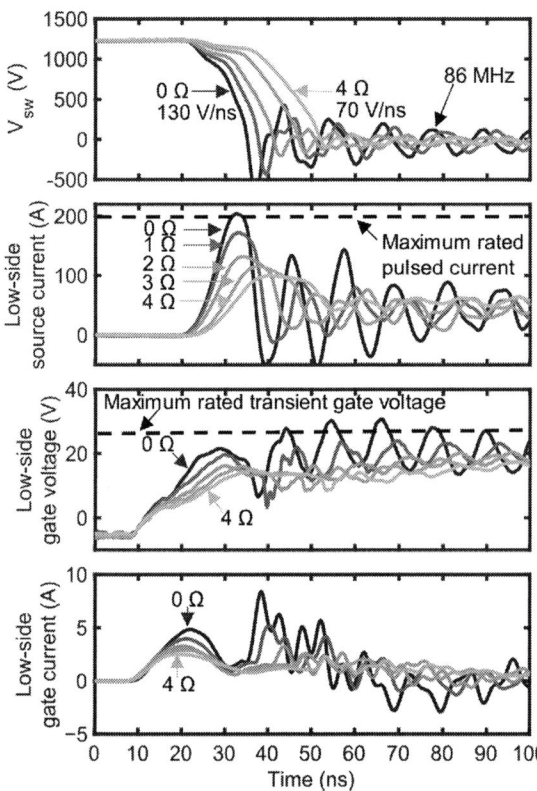

Fig. 8 Measured turn-on transients with external gate resistance of 4 Ω down to 0 Ω, and 100 °C case temperature.

The measured peak reverse recovery current is seen to increase from 90 A to 204 A; the maximum rated pulsed current limit in the datasheet is 200 A. The *dv/dt* increases from 68 V/ns at 4 Ω external gate resistance to 130 V/ns at 0 Ω, whilst the *di/dt* increases from 7 A/ns to 12 A/ns. The switch-node voltage undershoot becomes increasingly large with switching speed, raising concerns of exceeding the voltage rating of the high side device. The transient gate voltage increases from 20 V at 4 Ω to 30.7 V at 0 Ω; the transient limit of the gate voltage is 23 V, raising reliability concerns [6]. As gate resistance is minimised, the internal gate resistance becomes the dominant limitation in increasing switching speed at turn on, when using a step gate driver [20]. At room temperature (Fig. 7), significantly faster switching speed was observed (205 V/ns).

4.2.2 Turn-off

Double-pulse switching waveforms for low side turn-off are shown in Fig. 9, with gate resistance being swept from 4 Ω to 0 Ω.

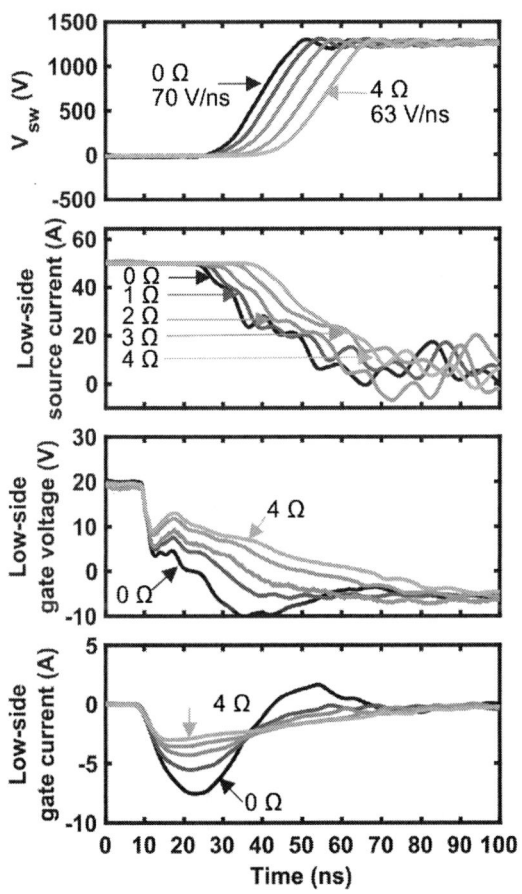

Fig. 9 Measured turn-off transients with external gate resistance of 4 Ω down to 0 Ω, and 100 °C case temperature.

Voltage slew rate increases from 63 V/ns at 4 Ω to 70 V/ns at 0 Ω, an increase in *dv/dt* of 14%. The relative lack of increase in switching speed, in comparison to turn on, is because the slew rate

here is determined mainly by the device output capacitance and the load current [21].

As gate resistance is reduced below 1 Ω, the gate becomes visibly underdamped leading to gate voltage undershoot raising gate oxide degradation concerns. The results suggest that operating the device with reduced gate resistance at turn off is not advisable due to a relative lack of improvement in switching energy with added reliability concerns due to the increased current oscillation and underdamping of the gate.

4.3 Spectral Analysis

The frequency spectrum of the switch node voltage was calculated by taking an FFT of the switching edges at full load current, turn off + turn on, in a single capture. This data is representative of a converter switching at 500 kHz with 50 % duty cycle.

Spectra are shown at different gate resistance in Fig. 10, demonstrating how the increased switching speed increases energy in the radiated emissions band 30 MHz to 1 GHz, as per EN 61000-6-3:2007 [22]. Two discernible ringing frequencies are apparent in the switch node voltage: 86 MHz and 236 MHz. These high-frequency oscillations raise EMI concerns as they may radiate efficiently, with ¼ wavelengths of 870 mm and 320 mm respectively.

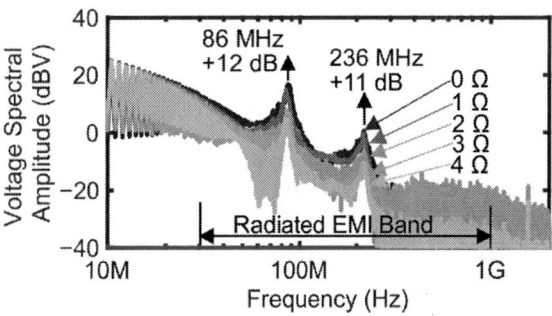

Fig. 10 FFT of switch node voltage shown in Fig. 8 and Fig. 9. The radiated emissions band 30 MHz to 1 GHz EN 61000-6-3:2007 is highlighted.

It can be observed from the switch node voltage waveforms that the slope increases with decreased gate resistance. As observed in the voltage spectrum Fig. 10, increased switching speed increases the −40 dB/dec corner point of the spectral components (f), as expected from the relationship shown in equ. 3.

$$f = 1/(\pi t_r) \tag{3}$$

Where t_r is the rise time, increasing the total spectral content with increasing switching speed [23]; the bandwidth limit of the voltage probe and noise floor of the scope limit the accuracy of the high frequency measurements above 300 MHz.

Fig. 11 demonstrates the impact of gate resistance on turn-on switching energy; a 24% improvement was observed over datasheet values, when the device was driven with zero external gate resistance. It has been shown here that a conventional voltage source gate driver can deliver much higher performance than datasheet characterisation for a 1700 V 35 mΩ device. However, as such improvements come at the cost of reliability and EMI concerns, care must be taken in PCB layout and filter design to mitigate these issues.

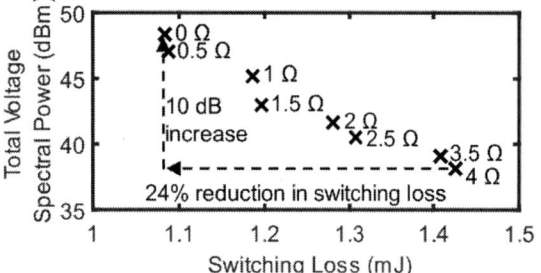

Fig. 11 Turn-on switching loss vs switch node voltage spectral power in a radiated emission zone from 30 MHz to 100 MHz *[24]*.

4.4 Demonstration of fast switching in a different power circuit

A new power board has been designed to demonstrate the measurement methods validated in Section 4.1. The new board utilises gate drivers on daughter boards connected through 8 × 2.54 mm pin headers. MSC035SMA170B4 SiC MOSFETs are used as in the preceding sections, in a half bridge topology with the low-side switch being driven in a double pulse test. The devices switch 1200 V and 50 A at a 25 °C case temperature, with 4.5 Ω or 0.5 Ω external gate resistances.

The measurement setup is shown in Fig. 12. The switch node voltage is measured with a Testec HV250 passive probe and source current is measured with an Infinity Sensor V2.

Gate voltage is measured with a PMK Firefly, a scope-agnostic optically-isolated voltage probe with >1.5 GHz bandwidth. This board uses MMCX connectors to connect the optically-isolated voltage probes into the circuit, to give the best measurement performance [25].

Gate current is measured with a 0.5 Ω current shunt using an IsoVu in an MMCX connector and using an Infinity Gate Sensor [26]. The sensor is an improved gate *di/dt* sensor compared to the sensor used in subsections 4.1-4.3. It is a verti-

cally-mounted PCB sensor with an improved sensing coil geometry to provide high external *di/dt* immunity [26].

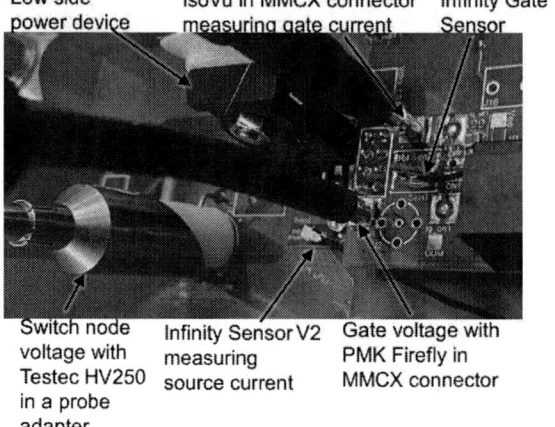

Fig. 12 Demonstration of probing in a different power board.

The measured performance is shown in Fig. 13. With 4.5 Ω gate resistor, v_{sw} slew rate is 41 V/ns and i_s slew rate is 5.1 A/ns, whilst with 0.5 Ω gate resistor, the slew rates increase to 130 V/ns and 11.2 A/ns. The Infinity Gate Sensor measurement is well correlated to the shunt+IsoVu measurement, delivering similar measurement fidelity for vastly reduced cost.

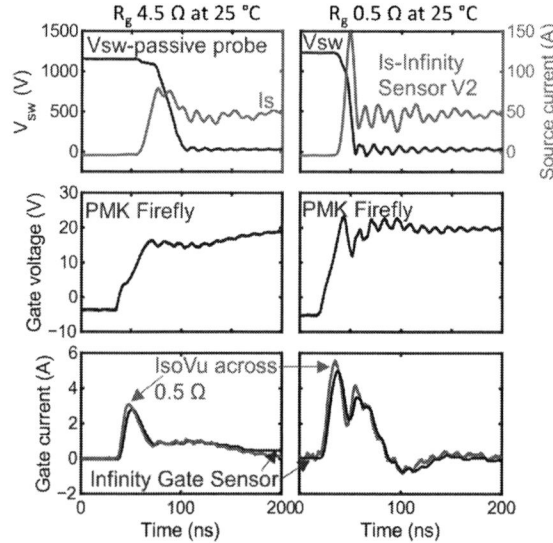

Fig. 13 Demonstration of switching waveforms at 1200 V, 50 A and 25 °C. Left: with 4.5 Ω gate resistance; Right: with 0.5 Ω gate resistance.

5 Conclusions

This paper has analysed measurement requirements for accurate characterisation of fast-switching SiC circuits, including bandwidth, radiated field immunity, common-mode rejection, and deskew accuracy. Various techniques have been demonstrated on a 1200 V, 50 A SiC MOSFET bridge leg switching at 130+ V/ns.

It has been shown that for fast-switching SiC devices with Kelvin source pins, a probe with high common-mode rejection is required in order to resolve the gate to source voltage with high fidelity, as the measurement delivered by a standard passive probe will be corrupted by the large common-mode voltage induced between power and Kelvin source pins. Further, it has been shown that in packages with relatively large gate-loop inductance, there can be a large discrepancy between the voltage measured at the terminals, and the actual gate-source voltage seen by the device at the die level. As such, it can be helpful to supplement the gate voltage measurement with a gate current measurement, so that, for example, the gate charge displacement may be accurately determined. Modified Infinity Sensors offer a low-cost means to make this measurement.

It has also been shown that commercially-available Rogowski coils may lack the bandwidth necessary to fully capture the high-frequency features of switching waveforms, requiring the use of alternatives such as the Infinity Sensor V2.

Unwanted switching features, such as overshoot and ringing, and their impact on SiC bridge legs have been analysed. Increasing switching speed beyond that used by the manufacturer for datasheet characterisation, by reducing the gate resistance down to 0 Ω, has been shown to reduce turn-on loss by over 24%. The internal gate resistance was found to be the dominant restricting factor in further increasing switching speed. However, such fast switching increases measurement demands and increases unwanted switching features, which would require mitigation through advanced techniques such as active gate driving.

Infinity Sensors are available from infinitysensor.com.

6 Acknowledgement

This work is supported by Siemens Gamesa Renewable Energy.

7 References

[1] S. Walder, X. Yuan, and N. Oswald, "EMI Generation Characteristics of SiC and Si Diodes: Influence of reverse recovery characterisitics", in *7th IET International Conference on Power Electronics, Machines and Drives (PEMD)*, Manchester, 2014.

[2] S. Jahdi, O. Alatise, L. Ran, and P. Mawby, "Accurate Analytical Modeling for Switching Energy of PiN Diodes Reverse Recovery", *IEEE Transactions on Industrial Electronics,* vol. 62, no. 3, pp. 1461 - 1470, 2015.

[3] M. Rahimo and N. Shammas, "Freewheeling Diode Reverse-Recovery Failure Modes in IGBT Applications", *IEEE Transactions on Industry Applications,* vol. 37, no. 2, pp. 661-670, 2001.

[4] B. Zhang and S. Wang, "A Survey of EMI Research in Power Electronic Systems with Wldebandgap Semiconductor Devices," *IEEE Journal of Emerging and Selected Topics in Power Electronics,* vol. 8, no. 1, pp. 629 - 643, 2020.

[5] S. Jahdi, O. Alatise, J. A. O. Gonzalez, R. Bonyadi, L. Ran, and P. Mawby, "Temperature and Switching Rate Dependence of Crosstalk in Si-IGBT and SiC Power Modules", *IEEE Transactions on Industrial Electronics ,* vol. 63, no. 2, pp. 849 - 863, 2015.

[6] O. Kreutzer, B. Eckardt, and M. Maerz, "Optimum gate driver design to reach SiC-MOSFET's full potential – speeding up to 200 kV/us", in *IEEE Workshop on Wide Bandgap Power Devices and Applications*, Virginia, 2015.

[7] N. Mohan, T. M. Undwerland and W. P. Robbins, *Power Electronics Converter Application and Design,* 3rd ed, ISBN: 978-0-471-22693-2 Wiley, 1995.

[8] B. J. Baligia, *Fundamentals of Power Semiconductor Devices*, 2nd ed, ISBN: 978-0-387-47314-7. Springer, 2008.

[9] P. S. Niklaus, R. Bonetti, C. Stäger, J. W. Kolar, and D. Bortis, "High-Bandwidth Isolated Voltage Measurements With Very High Common Mode Rejection Ratio for WBG Power Converters," *IEEE Open Journal of Power Electronics,* vol. 3, pp. 651-664, 2022.

[10] H. C. P. Dymond, Y. Wang, S. Jahdi, and B. H. Stark, "Probing Techniques for GaN Power Electronics: How to Obtain 400+

MHz Voltage and Current Measurement Bandwidths without Compromising PCB Layout", in *PCIM*, Nuremberg, 2022.

[11] N. F. Oswald, B. H. Stark, and N. McNeil, "IGBT Gate Voltage Profiling as a means of Realising an Improved Trade-Off Between EMI Generation and Turn-On Switching Losses", in *6th IET International Conference on Power Electronics, Machines and Drives (PEMD)*, Bristol, 2012.

[12] M. Zimmerman, B. Holzinger, R. Takeda, and T. Arai, "Understanding Probing Requirements When Measuring Dynamic Power Module Parameters", Bodo's Power Systems Magazine, 2022.

[13] MicroSemi, "MSCxxxSMA17020190212.lib," 6 November 2020. [Online]. Available: microsemi.com https://www.microsemi.com/document-portal/cat_view/56661-internal-documents/5674-spice-datasheets. [Accessed 20 March 2024].

[14] M. Hochberg, M. Sack, and G. Mueller, "Analyzing a Gate-Boosting Circuit for Fast Switching", in *IEEE International Power Modulator and High Voltage Conference (IPMHVC)*, San Francisco, 2016.

[15] D. Rothmund, D. Bortis, and J. W. Kolar, "Accurate Transient Calorimetric Measurement of Soft-Switching Losses of 10kV SiC MOSFETs", in *IEEE 7th International Symposium on Power Electronics for Distributed Generation Systems (PEDG)*, Vancouver, 2016 .

[16] Z. Zhang, B. Guo, F. F. Wang, E. A. Jones, L. M. Tolbert, and B. J. Blalock, "Methodology for Wide Band-gap Device Dynamic Characterization", *IEEE Transactions on Power Electronics,* vol. 32, no. 12, pp. 9307 - 9318, 2017.

[17] Microsemi, "MSC035SMA170B4 Silicon Carbide N-Channel Power MOSFET," March 2020. microchip.com https://ww1.microchip.com/downloads/en/DeviceDoc/Microsemi_MSC035SMA170B4_SiC_MOSFET_Datasheet_A.PDF. [Accessed 20 March 2024].

[18] Infineon, "EiceDRIVER™ 1ED31xxMU12H (1ED-X3 Compact)", 7 April 2020, infineon.com https://www.infineon.com/dgdl/Infineon-AN2019-20_1ED-X3Compact-ApplicationNotes-v01_00-

EN.pdf?fileId=5546d4627506bb320175211dff426207. [Accessed 20 March 2024].

[19] Infinity Sensor, "V2 Infinty Sensor datasheet", 11 November 2022. [Online]. infinitysensor.com https://www.infinitysensor.com/resources. [Accessed 20 March 2024].

[20] H. Gui, Z. Zhang, R. Chen, J. Niu, L. M. Tolbert, F. Wang, D. Costinett, B. J. Blalock, and B. B. Choi, "Gate Drive Technology Evaluation and Development to Maximize Switching Speed of SiC Discrete Devices and Power Modules in Hard Switching Applications", *IEEE Journal of Emerging and Selected Topics in Power Electronics,* vol. 8, no. 4, pp. 4160-4172, 2020.

[21] P. Anthony, N. McNeill, and D. Holliday, "High-Speed Resonant Gate Driver With Controlled Peak Gate Voltage for Silicon Carbide MOSFETs", in *IEEE Energy Conversion Congress and Exposition (ECCE)*, Raleigh, 2012.

[22] IEC, "EN 61000-6-3:2007 Electromagnetic compatibility (EMC) - Generic standards. Emission standard for residential, commercial and light-industrial environments", IEC, 2007.

[23] Bart Schröder, "How the heck do I measure a gate drive slewing at 70kV/us?", in *PCIM Europe; International Exhibition and Conference for Power Electronics, Intelligent Motion, Renewable Energy and Energy Management*, Nuremberg, 2017.

[24] N. Oswald, P. Anthony, N. McNeill, and B. H. Stark, "An Experimental Investigation of the Tradeoff between Switching Losses and EMI Generation With Hard-Switched All-Si, Si-SiC, and All-SiC Device Combinations", *IEEE Transactions on Power Electronics,,* vol. 29 , no. 5, pp. 2393-2407, May , 2014.

[25] Tektronix, "IsoVu Isolated Probes – Connectivity Options", September 2019 tek.com https://www.tek.com/en/documents/product-selector-guide/isovu(r)-isolated-probes-connectivity-options. [Accessed 20 March 2024].

[26] Y. Wang, Q. Wang, M. Appleby, J. Yan, H. C. P. Dymond, S. Jahdi, and B. H. Stark, "'Infinity Gate Sensor': a Differential Magnetic Field Sensor for Measuring Gate Current of SiC Power Transistors," in *PCIM* , Nuremberg, 2024.

PCIM Europe 2024, 11– 13 June 2024, Nuremberg   DOI: 10.30420/566262463

Conceptualization and Experimental Assessment of Design Aspects for 3-Level ANPC Inverters

Lukas Radomsky [1], Matthias Klintz[1], Regine Mallwitz [1]

[1] Institute for Electrical Machines, Traction and Drives, Technische Universität Braunschweig, Braunschweig, Germany

Corresponding author: Lukas Radomsky, l.radomsky@tu-braunschweig.de
Speaker: Lukas Radomsky, l.radomsky@tu-braunschweig.de

Abstract

The active neutral-point-clamped (ANPC) inverter topology offers many degrees of freedom, which must be considered in its design and control. The ANPC's sophisticated structure allows for redundant switching states, enabling several distinct modulation schemes. The selection of semiconductor components and the applied cooling may help distribute losses across the topological switches and offer a way to equalize junction temperatures. The design space is highly diverse, and very different approaches and designs can lead to similar performance. In addition, these degrees of freedom come with specific requirements, which must be considered in the hardware and especially the commutation loop design. This work, therefore, summarizes the design aspects for the 3-level ANPC topology. The design of a highly flexible 3-level ANPC experimental platform is described, which provides the basis for future experimental evaluations of the highlighted possibilities. As part of the commissioning of this platform, switching characteristics for the available commutations within the different modulation schemes are investigated experimentally in this work.

1   Introduction

Power electronic inverters are confronted with ever-increasing performance requirements as they enter new and challenging application areas, such as electric aircraft [1]. While some approaches to meeting these requirements focus on optimizing isolated components, optimizing the power electronic circuit topology holds significant improvement potential. In this context, multi-level topologies, particularly the ANPC topology, are intensively being researched [2]–[4] due to several reasons. In this topology, blocking voltage requirements for all switch positions are reduced, enabling the use of lower-voltage switches and offering an additional safety margin concerning environmental influences such as cosmic radiation-induced failures, which are relevant for aerospace applications. Furthermore, given its additional output voltage level, dv/dt can be reduced. This eases stress on insulation, helps reduce the size of filters, and shifts losses away from the already heavily stressed electrical machines. Efficiency is paramount in electrified aircraft applications due to the limited energy stor-age. Power switches generally are the primary loss contributors within the inverter. The amount of losses is influenced by the semiconductors applied and the operating strategy. Temperature balancing and hotspot minimization are crucial topics, especially in the ANPC topology, where the loss distribution among the semiconductors is non-uniform [5]. For the ANPC, multiple modulation strategies exist, which can be used to reduce or redistribute the switching and conduction losses among the switches or even to increase the output switching frequency [6]. Since different modulation strategies utilize different zero current paths, commutation loop design must be considered carefully [7].

In this work, first, Section 2 introduces the 3-level ANPC topology, its basic operational principle, and applicable modulation schemes. Subsequently, the challenge of power loss distribution and hot spot minimization is discussed in Section 3. Distinct approaches that may be used to homogenize loss distribution and equalize the junction temperatures are discussed. Section 4 deliberates the design of the experimental platform, which can be used to assess the spanned design space experimentally. Finally, an investigation of the commutation behavior, as well as a characterization of switching characteristics, is performed in Section 5. An overview of the state of the art in literature is provided in the context of the respective sections.

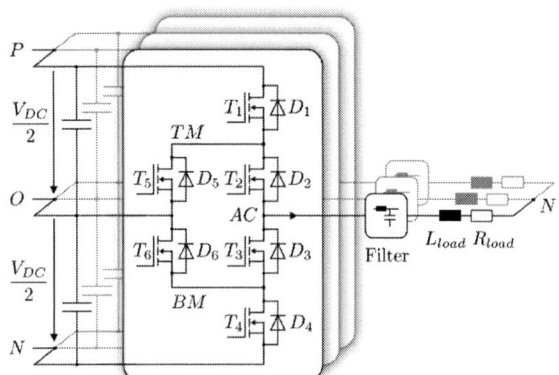

Fig. 1: Schematic phase leg of a 3-level ANPC inverter in an exemplary 3-phase configuration.

Tab. 1: 3-level ANPC phase leg switching states and naming according to original source [8].

State	T_1	T_2	T_3	T_4	T_5	T_6
P	1	1	0	0	0	1
$OU2$	0	1	0	0	1	0
$OU1$	0	1	0	1	1	0
OF	0	1	1	0	1	1
$OL1$	1	0	1	0	0	1
$OL2$	0	0	1	0	0	1
N	0	0	1	1	1	0

2 Modulation Schemes and Commutation Paths

The six active switches in a 3-level ANPC phase leg enable seven basic switching states. The switching states are defined in Table 1. Further usable switching states exist, and combining these led to the development of a large variety of modulation schemes. Depending on the modulation scheme, the direction of current flow, and the voltage conditions, different topological switches are subjected to conduction as well as switching loss. Besides the power loss distribution, which is influenced by the applied switching states, the voltage sharing across the switches is also impacted. For example, topological switch T_6 in the P-state is turned on to achieve equal sharing of voltage across the two

lower switches T_3 and T_4, which is also achieved by turning on T_4 during zero state $OU2$. The different switching states can be combined in numerous ways to create various modulation schemes. Two conventional modulation techniques exist: same-side clamping (SSCM) and opposite-side clamping modulation (OSCM). Additional commonly used techniques include full-path clamping modulation (FPCM) and double-frequency modulation (DFM), which are useful for switches that can be paralleled easily or offer superior switching properties. The main difference between the different modulation schemes is the utilization of different zero current paths, which have certain advantages and drawbacks, as discussed in [5].

3 Loss Distribution and Hot Spot Minimization

Midpoint-clamped inverter topologies, such as the ANPC topology, require additional considerations that are not required with the standard 2-level voltage source inverter topology. These include balancing the DC-link capacitor voltages and the distribution of power losses across the semiconductor devices, which, for example, is not an issue for Flying Capacitor (FC) topologies, which come with challenges and issues of their own. Balancing losses is especially relevant for the ANPC converter, where the additional clamping switches can actively redistribute these losses across the switches. The initial proposition of the active NPC converter in [8] was made directly in this context; however, it was limited by the semiconductor technology of the time.

To homogenize loss distribution among the topological switches, a number of approaches are at hand, a selection of which are discussed in this work and are summarized in Fig. 2. Since increased (or decreased) power losses translate to increased (or decreased) operating temperatures for a given cooling system, balancing the distribution of power losses as equally as possible among several power devices is immensely beneficial for efficient and reliable operation. Three approaches that may be used to homogenize the losses and equalize the junction temperatures are discussed in the following.

3.1 Active Loss Balancing

As stated in Section 2, the available switching states can be combined in numerous ways, resulting in different modulation schemes. Some of these modulation schemes, to a certain degree,

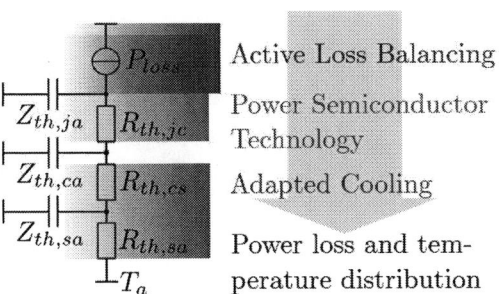

Fig. 2: Design vectors for a more uniform power loss and temperature distribution in the ANPC topology.

already distribute losses by themselves. FPCM utilizes zero state OF and clamps the DC-midpoint to the phase's AC output using two parallel current paths, and DFM also improves the unequal loss distribution [6], [9]. While in most cases, one distinct modulation scheme is used for one given converter design, it is also possible to intelligently arrange the available switching states [10] or switch between different modulation schemes in running operation to relieve heavily stressed switches and, thus, balance losses. This is exemplified by the simulation results in Fig. 3. These approaches require active temperature feedback, which can be implemented directly by sensing the device's temperatures or employing a thermal model. Publication [8] measures the phase currents, calculates switching and conduction losses, and, through a thermal converter model, junction temperatures. Different possibilities exist to directly estimate the semiconductor component temperature, at best, its junction temperature. Their feasibility strongly depends on the present setup and the chosen semiconductor package. In today's power modules, an NTC thermistor is often present. However, its primary purpose is over-temperature protection, and [11] found no general model for correlating the NTC reading to a semiconductor die's temperature. Another option is the use of dedicated external sensing elements. The setup used in [12] is built up from discrete semiconductor devices, and discrete thermocouples are used, which are placed through the mounting hole of the TO-247-packaged devices. Furthermore, sensorless temperature estimation using thermo-sensitive electrical parameters (TSEPs) is worth exploring. These parameters could, for example, be the threshold voltage or turn-off delay time of the power device. Precise measurement is possible since these parameters are usually directly dependent on the semiconductor junction temperature. A TSEP's characteristic, however, strongly depends on the semiconductor type applied and the aging of the semiconductor component, which makes further research necessary for this use case.

The inverter set up for this work incorporates a temperature measurement circuit, which uses external PT1000 temperature sensors that can be placed freely near the power semiconductor device. Additionally, as discussed in Section 4, the present mounting of the switches allows thermal infrared measurements on the housings to be conducted easily.

3.2 Power Semiconductor Technology

The selection of the semiconductor components to be used in the employed topology is a crucial design decision. The power losses of the semiconductor switches and the distribution of the overall losses over the different switch positions can be obtained computationally fast by using the set of analytical equations intensively discussed in [5]. In the ANPC topology, hybrid switch configurations become feasible solutions and could be used to relieve certain stressed topological switches. In SSCM, the outer switch positions T_1 and T_4, and the clamping switches T_5 and T_6, operate with switching frequency, whereas the inner switches T_2 and T_3 are turned on or off for half the fundamental period. In contrast, in OSCM, only the inner switches T_2 and T_3 operate at high frequency, while the four remaining ones switch at fundamental frequency. Both schemes allow for the possibility of using higher switching-performance components for the fast switching positions (T_1, T_4, T_5, T_6 in SSCM and T_2 and T_3 in OSCM) while using components with poor switching, but excellent conduction characteristics for the other switch positions. The switch configuration can be hybridized by varying the chip area for a given semiconductor technology or by mixing different semiconductor technologies. Within one semiconductor technology, the slow-switching positions could be equipped with larger chip-area switches, penalizing switching behavior by increasing the devices' parasitic capacitance while improving current-carrying capability and conduction characteristics. The results of this chip area optimization would be highly modulation scheme dependent as highlighted in [13]. Equally, cost-efficient Si IGBTs with superior conduction but poor switching characteristics could be employed

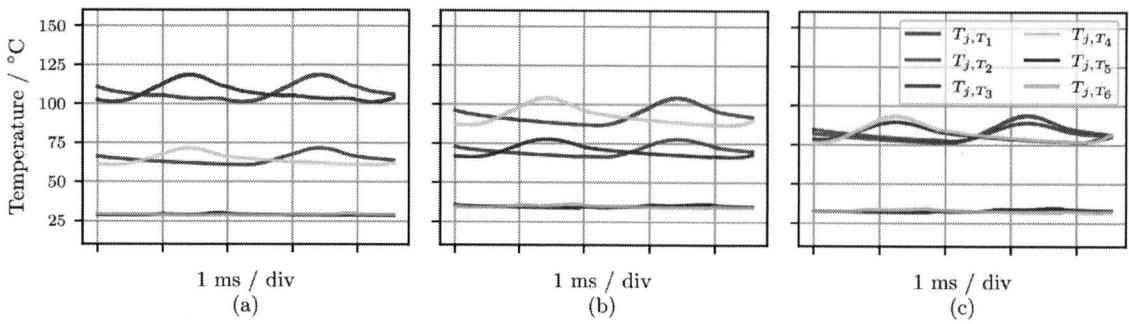

Fig. 3: Exemplary junction temperature waveforms in a 3-level ANPC phase leg with and without active loss balancing simulated in PLECS. All-SiC configuration. $f_{fund} = 250\,\mathrm{Hz}$. One heat sink per pair of switches with equal average losses. (a) OSCM. (b) SSCM. (c) Active loss balancing using modulation schemes OSCM and SSCM.

while keeping SiC MOSFETs or even GaN eHEMTs on the switch positions operating at high frequency. [14] describes an inverter optimized for OSCM using Si IGBTs and GaN-FETs. Similary [15] uses GaN HEMTs for positions T_2 and T_3. In this work, the challenge of the limited current carrying capability of state-of-the-art GaN switches is overcome by interleaving respective switches. [16] sets up an experimental hybrid Si/SiC ANPC inverter and [17] investigates the potential of using hybrid Si/SiC switches for switch positions T_2 and T_3.

The experimental platform described in Section 4 is designed to explore different switch types and configurations. This is aided by its design, where the semiconductor devices are mounted in an easily accessible and interchangeable way. This way, drop-in replacements and exchange of power semiconductor types to achieve hybrid topologies are possible. However, using the same platform, as done in this work might not fully exploit the potential for optimization due to the fixation on a single standardized package footprint and layout. This holds especially true for the use of GaN semiconductors, where parasitics are attenuated due to increasing switching frequencies and layout configuration becomes more and more critical, as highlighted for example in [18].

3.3 Adapted Cooling

Since increased losses translate to increased operating temperatures for a given cooling system, balancing the distribution of power losses among a larger number of power devices is beneficial for efficient and reliable operation. Besides, adapted cooling may offer a way to equalize junction temperatures for the differently stressed switch positions in

the 3-level ANPC inverter topology. Weight could be saved for cooling less stressed topological switches, while high-performance heat sinks may be used to cool highly stressed positions. The performance of the cooling system depends on various factors, including the type of material used, the structure of the heat sink, the arrangement of power loss sources, and the mounting conditions. Several different approaches are available to fabricate more sophisticated heat sinks. Advanced manufacturing techniques, such as additive manufacturing, enable the production of individualized heat sink structures using innovative or mixed materials. These structures may offer better thermal performance, lower weight, and compact design. Furthermore, advanced optimization algorithms could be utilized to refine the cooling system, enabling, e.g., mission profile-based heat sink design for junction temperature shaving or the integration of thermal capacities to mitigate thermally critical flight phases.

3.4 Combined Approaches

As outlined, various approaches to homogenizing power losses exist. They are often applied individually, but it is also possible to combine measures. It's crucial to acknowledge that the various measures differ in their compatibility with one another and that the potential of all balancing approaches is highly dependent on the operating point. Modulation schemes that may yield optimal results for an all-Si IGBT ANPC inverter might not necessarily be the most advantageous choice for configurations employing wide-bandgap (WBG) devices, such as SiC MOSFETs, or for hybrid setups. For example, the two conventional modulation schemes utilize either the upper or lower current path. FPCM,

PCIM Europe 2024, 11– 13 June 2024, Nuremberg   DOI: 10.30420/566262463

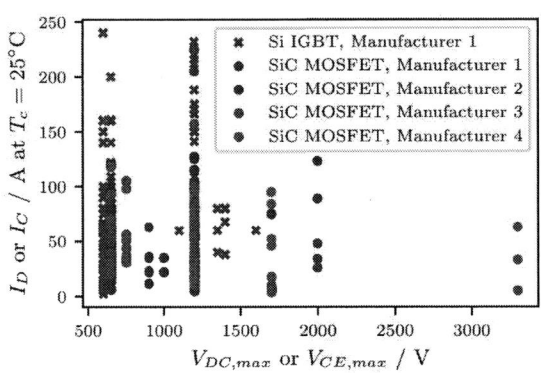

Fig. 4: Semiconductor landscape for different manufacturers' discrete SiC MOSFET and Si IGBT devices.

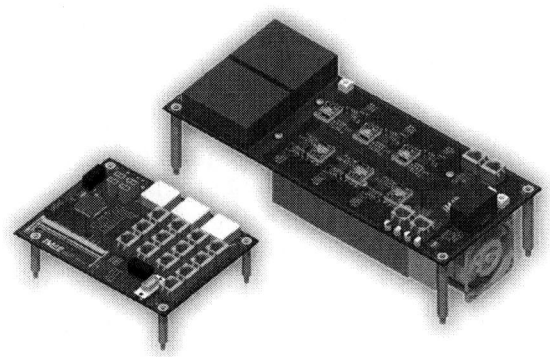

Fig. 5: Renders of realized modular experimental inverter platform. Control PCB to the left and 3-level ANPC phase leg PCB with attached heat sink to the right.

however, utilizes zero state OF to clamp the DC-midpoint to the phases' AC output. This zero state, in combination with a hybrid semiconductor technology configuration, may lead to the parallelization of, for example, Si IGBTs, which becomes challenging due to the component's temperature coefficient. The approach from [17] would offer the possibility to adapt the hardware for different modulation schemes and [19] specifically investigates junction temperature balancing algorithms for hybrid device NPC and ANPC structures. Furthermore, the arrangement of the power devices with respect to each other, the commutation loop design, and the resulting parasitics in the commutation paths are challenging since the hardware or power module is generally designed for one modulation scheme. Commutation loop design considerations are described in Section 4.3.

4 Experimental Inverter Platform

To explore the highlighted degrees of design freedom and to experimentally assess the unique behavior of the ANPC inverter topology, an experimental verfication platform is set up. Before the general setup of the platform is described with particular respect to the commutation loop design in Sections 4.2 and 4.3, first, the scaling of the platform is derived in the following Section 4.1.

4.1 Scaling Considerations

For a given application, environmental influences play a crucial role in evaluating the suitability of an inverter topology and the applied semiconductor switches. For electric aviation, high DC voltages are preferred for transmitting high power. The rated

and breakdown voltage of the power semiconductor result from the voltage requirements imposed by the inverter topology in which the semiconductors are employed. The rising concentration of cosmic radiation at high altitudes and increased risk of cosmic radiation-induced failures make an additional voltage safety margin necessary for semiconductor switches in aerospace applications [1]. This is a challenge, especially for WBG devices, where the development of breakdown voltages is still not as advanced as that of their silicon counterparts. Therefore, the 3-level ANPC inverter topology, which stresses the individual switches with half the DC link voltage only, is a viable option for increasing HVDC voltage without negatively influencing reliability. When comparing the required repetitive blocking voltage rating, considering a safety factor of 1.5...2.0 to account for cosmic ray rating, to today's available SiC power semiconductor voltage classes, it becomes apparent, that the required peak repetitive voltage rating for the semiconductor switches quickly outgrows the available switches on the market, which generally only reach up to a maximum of $3.3\,\text{kV}$. High DC-link voltages of up to the discussed $3\,\text{kV}$ [3] are, using today's available SiC devices, only realizable with 3- or more-level topologies. The experimental platform is scaled down regarding inverter output power and voltage level while still keeping a high DC-link voltage of $800\,\text{V}$, which is a voltage level under intense discussion for future mobility applications. Even though the inverter will exhibit different thermal characteristics compared to scaled-up systems, it is well suited to investigate the spanned design space. The findings and the considerations behind the platform's

3290

design and the results obtained with it can also be transferred to systems with higher voltages if this is desired later. If, to not restrict the hybridization possibilities, parallelization of semiconductor switches is to be avoided, the system's power limits can be directly related to the current-carrying capabilities of the individual semiconductor components and the selected voltage level. Different semiconductor configurations become feasible solutions for the 3-level ANPC topology. The decision was made to use discrete switches in a standard package to allow the greatest possible freedom in selecting semiconductor components for the experimental platform. Components in the TO-247-4 housing feature outstanding availability with a large number of manufacturers, switch types, and materials. The discrete switches available on the market (as of Q3 2023) were analyzed, and the landscape is shown in Fig. 4. it becomes apparent, that the widest selection is available for switches with a blocking voltage of $1200\,\text{V}$. Discrete SiC semiconductors are readily available up to a forward current of around $100...150\,\text{A}$ (at $T_j = 25\,°\text{C}$). However, this value is reduced by about half for the design-determining worst-case temperature of $T_j = 150\,°\text{C}$. In a three-phase system with a DC link voltage of $800\,\text{V}$ and a $V_{rms,line-line}$ of $490\,\text{V}$ ($m = 1$), an approximate nominal phase current of $50\,\text{A}$ is present for an output power varied around a nominal value of $40\,\text{kW}$.

4.2 Experimental Platform Design

A modular approach was chosen for the experimental platform, where the individual phase legs are assigned to individual PCBs. As pointed out, discrete switches packaged in standard TO-247-4 housings were used to maximize design flexibility without paralleling. The switches are mounted vertically on the bottom side of the PCB. In the first place, all switches of the phase leg are attached to a single heat sink, as shown in Fig. 5. This way, they stay easily accessible; for instance, Rogowski coils can be inserted, and thermal infrared measurements on the housings can be conducted. However, the switches could also be reconfigured into a 'TabUp' configuration, bending them to mount a cold plate from the bottom to the PCB, offering further design freedom in terms of cooling. The arrangement of the semiconductor switches and the PCB design strongly influence the phase leg's commutation loop. Therefore, the following Section 4.3 elaborates on this.

4.3 Commutation Loop Design Considerations

Optimizing the commutation loop for the 3-level ANPC inverter with its increased number of switching states is of paramount importance. Sources of inductance are the commutation loop stray inductance, the semiconductor device self-inductance, and the DC link capacitor self-inductance. While the latter two components can only be influenced by component selection, e.g., by selecting a low-inductance surface-mount package for the power semiconductors, the commutation loop stray inductance can be actively influenced by designing the interconnection of the switches or, in this case, the PCB layout. Previous Section 4.2 already gave first insights into the semiconductor selection, as well as the arrangement of the semiconductors. A schematic of the arrangement of the discrete semiconductor switches, the PCB design, and thus also the commutation loop design is provided in Fig. 6c. Various arrangements of the switches are conceivable. This work chooses an arrangement in which switches with symmetrical losses (T_1 and T_4, T_2 and T_3, T_5 and T_6) are positioned across from each other. The switches may be placed onto a single heat sink, however, this setup explicitly allows the switch positions with symmetrical losses to be installed on separate heat sinks. Individual heat sinks result in thermal isolation between the topological switches. This results in an uneven loss distribution directly causing an uneven distribution of temperatures between the semiconductors. Therefore, the maximum losses of a single device are only defined by the thermal characteristics of the employed power device and its associated heat sink. Concerning the commutation loop design, layers with opposing current flow directions are placed in overlapping, adjacent layers. If the current flows in opposing directions and both fields are highly interspersed, a significant reduction in stray inductance can be achieved by destructive interference.

Depending on the application, however, other setups and general arrangements of the semiconductors with respect to each other are also conceivable. Figure 6 shows the composition of the commutation loop for setting up a 3-level ANPC phase leg from half-bridge modules, from a dedicated single module, and from discrete devices, as done in this work. In Fig. 6a, insights into a 3-level ANPC power module [20] are given. A hybrid switch configuration applying different semiconductor materials is used.

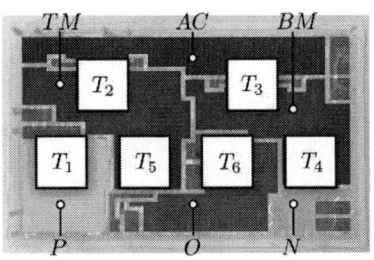

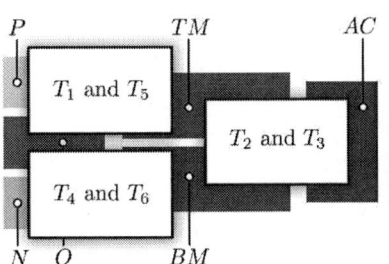

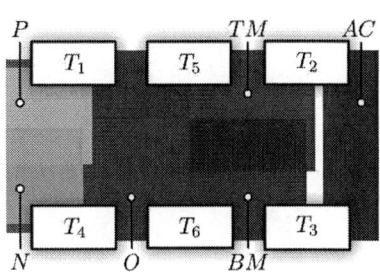

(a) 3-level hybrid ANPC power module [20]. The power module's structure is clarified with colors.

(b) Common arrangement of half-bridge modules for the build-up of a 3-level ANPC phase leg (e.g. [21]).

(c) 3-level ANPC phase leg build-up from discrete devices as done in this work.

Fig. 6: Schematic of the arrangement of the semiconductor switches and insights into the commutation loop design of a 3-level ANPC phase leg using different types of semiconductor packaging options. The depicted polygons are exemplary only and visualize the general interconnection of the devices. For the definition of the potentials P, O, N and TM, AC, BM see Fig. 1.

This power module utilizes SiC MOSFETs as inner switches T_2, T_3 while keeping Si IGBTs on the outer and midpoint switch positions T_1, T_4 and T_5, T_6. Due to the diverse modulation schemes, however, various optimized solutions in terms of module layout and power semiconductor types are feasible. As long as the availability of industrial multi-level converter and especially ANPC modules is limited [22], phase legs have to be constructed from smaller building blocks, such as discrete power semiconductor devices or half-bridge modules, to be able to explore the offered design possibilities fully. Research into innovative module concepts is essential for utilizing the high switching speeds offered by WBG technology, and multi-level power modules are deemed a necessary component for future high-power electric aircraft propulsion inverter concepts. Furthermore, constructing a phase leg from 2-level half-bridge modules (Fig. 6b) or split modules is possible. This approach, however, can lead to very long commutations across multiple power modules, which, especially for OSCM, may have a negative effect.

To reduce the commutation loop and thus minimize the loop inductance, a decoupling capacitor can be placed across T_2 and T_3. This decoupling capacitor stabilizes the intermediate DC-link voltage of the higher-frequency (HF) inner switches in OSCM and provides a low-impedance path for high-frequency noises. This is especially important for fast switching (hybrid) configurations, such as in [14], [15] and high power applications, where power modules are used which have to be connected by lengthy busbars [16]. The decoupling capacitor should be placed near the fast-switching switches. Still, caution is advised since this capacitor can cause additional resonances during HF switching when interacting with the layout stray inductance, leading to potential ringing and current spikes.

5 Switching Characterization

Compared to the switching characteristics in standard half-bridge configurations, the switching behavior within the 3-level ANPC inverter is much more complex due to the differing zero states and distinct commutation types. The possible commutations from one switching state to another exhibit unique switching behavior.

For the first commissioning of the phase leg and initial tests, an all-SiC configuration utilizing C3M0021120K 1200 V 100 A SiC Power MOSFETs from Wolfspeed was chosen. The double pulse tests (DPTs) were performed for one phase leg with the load inductor connected between node AC and the neutral point O (c.f. Fig. 1). A large inductor of $L_L = 187\,\mu\text{H}$ (c.f. Fig. 7) was used to keep the load current as constant as possible during the freewheeling interval. Since it is aimed at comparative measurements only, and additional stray inductance would be introduced by using current transducers for the measurement of i_D, a Rogowski coil was used to measure the drain current. While it is possible to use single-ended voltage probes to reduce loop inductance, differential probes are required for non-ground reference voltages. Three commutations associated with the three modulation schemes, OSCM, SSCM, and FPCM, were

Fig. 7: Setup used for the conduction of double pulse tests.

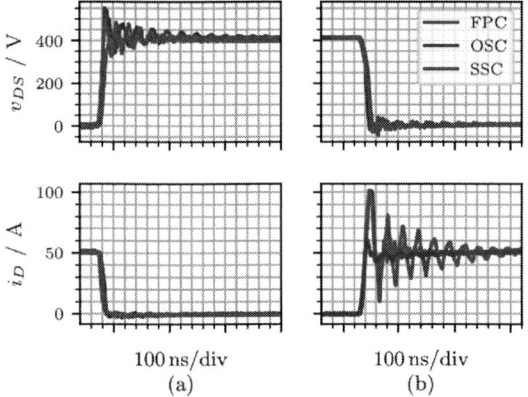

Fig. 8: Turn-off (a) and turn-on (b) commutation waveforms for OSCM, SSCM, and FPCM. $V_{DC}/2 = 400\,\mathrm{V}$, $R_{g,ext} = 12\,\Omega$. Waveforms for T_2 for OSCM and T_1 for SSCM and FPCM.

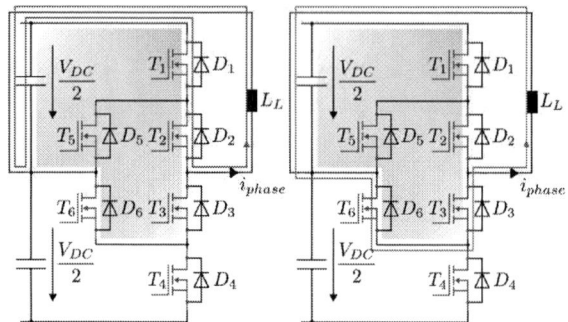

Fig. 9: Commutation loop for OSCM: State P to $OL1$.

investigated as shown in Fig. 9, 10 and 11. All commutations were investigated for a switching transition in the first quadrant. Turned-on and turned-off switches are colored green and red, respectively.

The commutation waveforms of drain-source voltage v_{DS} and the drain current i_D for the hard switching switch in each of the three investigated commutations are plotted in Fig. 8. Multiple DPTs were carried out for different i_D, whereby in Fig. 8 the waveforms for $V_{DC}/2 = 400\,\mathrm{V}$ and $i_{phase} = 50\,\mathrm{A}$ are shown. All measurements were carried out with an external gate resistance of $R_{g,ext} = 12\,\Omega$.

For the opposite-side commutation (OSC), the transition from switching state P to state $OL1$ was investigated. The associated commutation loop is highlighted in Fig. 9. To build up current in the inductor, T_1 and T_2 are turned on, representing the P-state. The current path is marked in pink color. For the commutation to the zero-state $OL1$, T_2 has to be turned off, suffering switching loss. The current

commutates from T_2 to D_3, where T_3 is the synchronous switch to D_3. D_3 suffers reverse-recovery losses. In contrast to the same-side commutation (SSC) discussed in the following, the current commutates to its non-neighboring path on the opposite side of the phase leg. The new current path is highlighted in blue, and the commutation loop is illustrated by the yellow-colored area in Fig. 9. This commutation features higher peak voltage during turn-off, as can be seen in Fig. 8. This can be explained by the higher commutation loop inductance, which includes the intrinsic parasitic inductance of the upper DC-link capacitor and the devices on switch positions 1, 2, 3, and 6. Furthermore, the parasitics of the interconnecting PCB are also part of this loop.

The SSC from P to $OU2$ is displayed in Fig. 10. Here, the current commutates from T_1 to the upper current path's D_5, whereas T_5 is the synchronous switch. T_1 and D_5 suffer switching and reverse recovery loss, respectively. The switch T_3 remains conducting to provide the freewheeling current path. The size of the commutation loop is reduced, as can be seen in Fig. 10. The switching loop inductance now only contains the DC-link capacitor's equivalent series inductance and self-inductance of the devices T_1(or D_1) and T_5(or D_5) in addition to the one of the PCB polygon pour. This results in a less severe voltage overshoot at the turn-off event.

It can be observed that the full-path commutation (FPC) displayed in Fig. 11 and the SSC feature similarities. This concerns the lower voltage overshoots during turn-off, as well as the more pronounced current oscillations during turn-on (see Fig. 8). The i_D overshoots depend on the type of commutation applied, where the less pronounced current ringing in OSC in contrast to SSC and especially FPC can

PCIM Europe 2024, 11– 13 June 2024, Nuremberg   DOI: 10.30420/566262463

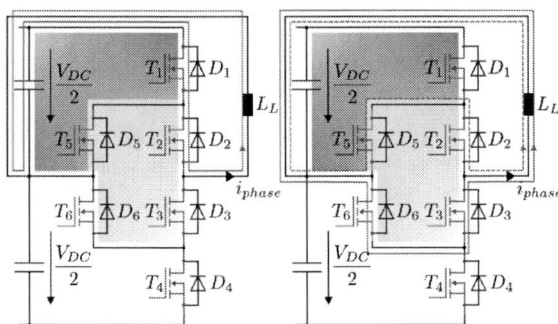

Fig. 10: Commutation loop for SSCM: State P to $OU2$.

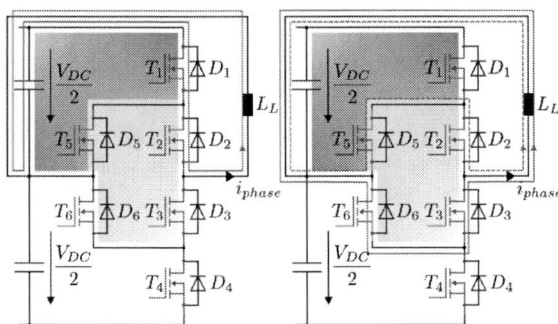

Fig. 11: Commutation loop for FPCM: State P to OF.

be explained by the differing capacitive charges during the switching events.

This similarity persists when analyzing the switching energies. Figure 12 shows the measured turn-off and turn-on switching energies for the different modulation schemes. To determine the switching loss energies, the instantaneous power, which is dissipated in the power semiconductor resulting from the multiplication of u_{DS} and i_D, is integrated over the duration of the switching interval. When comparing SSC and FPC to OSC, the OSC's higher turn-off and lower turn-on losses may be explained by the higher peak voltage during the turn-off event and the larger inductance of the commutation loop.

6   Conclusion

This paper provides a comprehensive overview of the design aspects for a more uniform power loss and temperature distribution within the 3-level ANPC inverter topology. The use of active loss balancing approaches, hybrid semiconductor configurations, adapted cooling, and combinations of these measures promise improved performance. The design of a highly flexible 3-level ANPC ex-

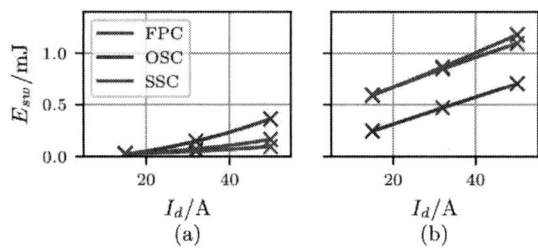

Fig. 12: Measured turn-off (a) and turn-on (b) switching energies for OSCM, SSCM, and FPCM. $V_{DC}/2 = 400\,\text{V}$, $R_{g,ext} = 12\,\Omega$.

perimental platform is described, and its scaling is reasoned. In this context, particular respect is paid to the commutation loop design, the arrangement of the switches, and the different types of semiconductor packaging options available to construct a 3-level ANPC phase leg. This setup will be used in future work to explore the highlighted possibilities and assess design aspects experimentally. Using this platform, switching characteristics for the multiple possible commutations for the different modulation schemes were investigated experimentally, highlighting the dependence of switching characteristics on the overall setup and its parasitics.

Funding   This work was supported by the Federal Ministry for Economic Affairs and Climate Action on the basis of a decision by the German Bundestag.

References

[1] L. Radomsky, R. Keilmann, D. Ferch, and R. Mall-witz, "Challenges and Opportunities in Power Electronics Design for All- And Hybrid Electric Aircraft: A Qualitative Review and Outlook," in *Deutscher Luft- und Raumfahrtkongress 2023*, DGLR e.V., Sep. 2023.

[2] D. Zhang, J. He, and D. Pan, "A Megawatt-Scale Medium-Voltage High-Efficiency High Power Density "SiC+Si" Hybrid Three-Level ANPC Inverter for Aircraft Hybrid-Electric Propulsion Systems," *IEEE Transactions on Industry Applications*, vol. 55, no. 6, pp. 5971–5980, 2019. DOI: 10. 1109/TIA.2019.2933513.

[3] H. Schefer, W.-R. Canders, J. Hoffmann, R. Mall-witz, and M. Henke, "Cryogenically-Cooled Power Electronics for Long-Distance Aircraft," *IEEE Access*, vol. 10, pp. 133 279–133 308, 2022. DOI: 10.1109/ACCESS.2022.3228161.

[4] N. F. Wahler, L. Radomsky, L. V. Hanisch, J. Göing, P. Meyer, *et al.*, "An Integrated Framework for Energy Network Modeling in Hybrid-Electric Aircraft Conceptual Design," in *AIAA AVIATION 2022 Forum*. DOI: 10.2514/6.2022-3741. eprint: https://arc.aiaa.org/doi/pdf/10.2514/6.2022-3741.

[5] L. Radomsky and R. Mallwitz, "Review, Comprehensive Analysis and Derivation of Analytical Power Loss Calculation Equations for Two- to Three-Level Midpoint Clamped Inverter Topologies with Hybrid Switch Configurations," *Energies*, vol. 16, no. 18, 2023. DOI: 10.3390/en16186710.

[6] D. Floricau, G. Gateau, A. Leredde, and R. Teodorescu, "The efficiency of three-level Active NPC converter for different PWM strategies," in *2009 13th European Conference on Power Electronics and Applications*, 2009, pp. 1–9.

[7] J. Häring, M. Hepp, W. Wondrak, and M.-M. Bakran, "Switching Behaviour of a SiC-MOSFET 3-Level ANPC Inverter with Different Modulation Schemes," in *PCIM Europe 2023*, 2023.

[8] T. Brückner, S. Bernet, and H. Guldner, "The Active NPC Converter and Its Loss-Balancing Control," *IEEE Transactions on Industrial Electronics*, vol. 52, no. 3, pp. 855–868, 2005. DOI: 10.1109/TIE.2005.847586.

[9] Y. Deng, J. Li, K. H. Shin, T. Viitanen, M. Saeedifard, and R. G. Harley, "Improved Modulation Scheme for Loss Balancing of Three-Level Active NPC Converters," *IEEE Transactions on Power Electronics*, vol. 32, no. 4, pp. 2521–2532, 2017. DOI: 10.1109/TPEL.2016.2573823.

[10] X. Jing, J. He, and N. A. O. Demerdash, "Loss balancing SVPWM for active NPC converters," in *2014 IEEE Applied Power Electronics Conference and Exposition - APEC 2014*, 2014, pp. 281–288. DOI: 10.1109/APEC.2014.6803322.

[11] M. Schulz and M. Xin, "Correlating NTC-Reading and Chip-Temperature in Power Electronic Modules," in *PCIM Europe 2015*, 2015.

[12] A. Poorfakhraei, M. Narimani, and A. Emadi, "Improving Power Density of a Three-Level ANPC Structure Using the Electro-Thermal Model of Inverter and a Modified SPWM Technique," *IEEE Open Journal of Power Electronics*, vol. 3, pp. 741–754, 2022. DOI: 10.1109/OJPEL.2022.3216214.

[13] J. Häring, M. Gleissner, W. Wondrak, and M.-M. Bakran, "Efficiency and Cost Comparison of B6 and Hybrid ANPC Converters for Traction Drives," in *2020 22nd European Conference on Power Electronics and Applications (EPE'20 ECCE Europe)*, 2020, pp. 1–10. DOI: 10.23919/EPE20ECCEEurope43536.2020.9215694.

[14] M. Najjar, A. Kouchaki, J. Nielsen, R. Dan Lazar, and M. Nymand, "Design Procedure and Efficiency Analysis of a 99.3% Efficient 10 kW Three-Phase Three-Level Hybrid GaN/Si Active Neutral Point Clamped Converter," *IEEE Transactions on Power Electronics*, vol. 37, no. 6, pp. 6698–6710, 2022. DOI: 10.1109/TPEL.2021.3131955.

[15] L. Fauth, C. Beckemeier, and J. Friebe, "A Hybrid Active Neutral Point Clamped Converter consisting of Si IGBTs and GaN HEMTs for Auxiliary Systems of Electric Aircraft," in *2021 IEEE Energy Conversion Congress and Exposition (ECCE)*, 2021, pp. 1917–1923. DOI: 10.1109/ECCE47101.2021.9595512.

[16] P. H. Trieu To and H.-G. Eckel, "Experimental 500kW Hybrid Si/SiC ANPC Inverter," in *PCIM Europe 2023*, 2023, pp. 1–10. DOI: 10.30420/566091086.

[17] D. Woldegiorgis, Y. Wu, Y. Wei, and H. A. Mantooth, "A High Efficiency and Low Cost ANPC Inverter Using Hybrid Si/SiC Switches," *IEEE Open Journal of Industry Applications*, vol. 2, pp. 154–167, 2021. DOI: 10.1109/OJIA.2021.3091549.

[18] E. Gurpinar, A. Castellazzi, F. Iannuzzo, Y. Yang, and F. Blaabjerg, "Ultra-low inductance design for a GaN HEMT based 3L-ANPC inverter," in *2016 IEEE Energy Conversion Congress and Exposition (ECCE)*, 2016, pp. 1–8. DOI: 10.1109/ECCE.2016.7855540.

[19] M. Novak, V. Ferreira, M. Andresen, T. Dragicevic, F. Blaabjerg, and M. Liserre, "FS-MPC Based Thermal Stress Balancing and Reliability Analysis for NPC Converters," *IEEE Open Journal of Power Electronics*, vol. 2, pp. 124–137, 2021. DOI: 10.1109/OJPEL.2021.3057577.

[20] Infineon Technologies AG Munich, *F3L11MR12W2M1 B74 EasyPACK™ Module*, Revision 0.20, https://www.infineon.com/dgdl/Infineon-F3L11MR12W2M1_B74-DataSheet-v00_20-DE.pdf?fileId=8ac78c8c80f4d32901814c8e7865287b [Accessed: 2023-09-01], 2022.

[21] J. Thoma, B. Volzer, D. Kranzer, D. Derix, M. Geiss, and A. Hensel, "Highly Compact 250 kVA Inverter Stack with 3.3 kV SiC MOSFETs," in *2021 23rd European Conference on Power Electronics and Applications (EPE'21 ECCE Europe)*, 2021, pp. 1–9. DOI: 10.23919/EPE21ECCEEurope50061.2021.9570608.

[22] I. Harbi, J. Rodriguez, A. Poorfakhraei, H. Vahedi, M. Guse, *et al.*, "Common DC-Link Multilevel Converters: Topologies, Control and Industrial Applications," *IEEE Open Journal of Power Electronics*, vol. 4, pp. 512–538, 2023. DOI: 10.1109/OJPEL.2023.3291662.

PCIM Europe 2024, 11– 13 June 2024, Nuremberg   DOI: 10.30420/566262464

Design of a High Power Density Inverter and FOC Implementation for UAVs

Matthias Neuner [1], Maurizio Incurvati[1], Davide Bagnara[1], Matthias Moroder[2], Moritz Moroder[2],

[1] MCI Management Center Innsbruck Internationale Hochschule GmbH, Austria
[2] FlyingBasket SRL, Italy

Corresponding author: Matthias Neuner, matthias.neuner@live.de
Speaker: Matthias Neuner, matthias.neuner@live.de

Abstract

Even nowadays, still some small huts in the alps need a helicopter to provide them with the necessities for living. This task is always connected with high costs and organisational effort. With the ongoing development in the field of unmanned aerial vehicles (UAVs) already a drone exist which can take over this task. Currently, this drone is limited to a payload of 100 kg. In this paper, the design and optimisation of a 19.5 kW 2-level inverter based on half-bridge configuration modules using Silicon Carbide (SiC) technology for this type of cargo drones is presented. The 1200 V, 16 mΩ SiC half-bridge modules of Wolfspeed provide the base for the power stage. With a switching frequency of 25 kHz and a power density of 8.29 $\frac{kW}{dm^3}$ and 20.06 $\frac{kW}{kg}$ the system achieves several advantages for this use-case, including a maximum payload of 200 kg. In addition, the layout of the Printed Circuit Boards (PCB's) is optimised for altitudes over 3500 m. In terms of control strategy, a Field Oriented Control (FOC) implementation on an ST microcontroller is performed. The system is optimised to work with a 19.5 kW Permanent Magnet Synchronous Motor (PMSM) of MAD Components. For validation of the functionality, several tests are performed including double-pulse tests, tests on a passive RL load as well as tests with a Motor test bench including the selected PMSM and a torque sensor. The measured system efficiency of the inverter and PMSM combination reaches 91.37 %. The inverter itself reaches a peak efficiency of 98.42 %. Further tests on the real drone are intended to be performed, and further optimisation will be done to allow a serial production of this design.

1   Introduction

UAV's, or more specific in this case cargo drones, can provide a cheap and efficient alternative to several tasks, e.g. delivering necessities to small huts high up in the mountains. Especially for these companies, it is of major interest to have a high performance, but also light weight motor drive. A previous work is done in article Design and Demonstration of High Power Density Inverter for Aircraft Applications ([1]). This paper describes a design methodology for such a high power density converter and considers different typologies and power devices. For verification purposes, a 50 kVA three-phase power inverter is designed and evaluated as an example. This implemented inverter has a switching frequency of 65 kHz and can reach an efficiency of 97.77 %. In the article Design of a Fast Switching 200 kVA Silicon-Carbide (SiC) Drive Inverter for Aviation Application ([2]) a 200 kVA SiC drive inverter is shown. This SiC drive inverter reaches a power density of 89 $\frac{kW}{dm^3}$ and 79 $\frac{kW}{kg}$ with a measured peak efficiency of 98.8 % at a switching frequency of 48 kHz. This design is intended to fulfil several requirements, including creepage distance of 12.5 mm for high voltage areas (IPC-2221B) for uncoated external conductors, weight under 1 kg CAN communication and own internal power supply for electronics, 3.3 V and 12 V. Table 1 shows the main parameters of interest for this design.

2   METHODS

2.1   MECHANICAL DESIGN

When mounted on the drone, the inverter will be exposed to vibrations, low temperatures, high humidity or even rain. To handle this conditions, a

Tab. 1: This table provides electrical parameters which the inverter should fulfil.

Parameter	Value
V_{bus}	400 V
I_{DC}	49.1 A
P_{out}	19.6 kW
I_{outp}	64 A(peak)
f_{sw}	25 kHz

well-designed sealed housing is needed. Figure 1 shows a first reference design of two inverters mounted on a drone arm. The placement of the inverter in this location provides forced air cooling for the system by the propellers.

Fig. 1: CAD model of the mechanical design mounted on one arm of the drone.

For handling humidity and temperature inside the housing, it is possible to install an additional PWM controlled fan and NTC inside the housing. Especially at low temperatures, the circulating air can be used to prevent condensation on the PCBs and components and a controlled environment is given. Furthermore, additional coating of the PCBs is in discussion for the 2nd generation of this inverter. A thermal simulation in Ansys Icepak for better understanding of the cooling concept is done in the thesis to this paper [3].

Figure 2 shows an exploded view of the design for first tests. Of special interest are the 3D-printed spacers located on the left and right, which guarantee the correct placement of the single PCBs and function as an additional electrical isolation in terms of air distance between the housing and the inverter. The SiC half-bridge modules are mounted on the heatsink with a defined pattern of heatpaste according to the application note of Wolfspeed to guarantee a good distribution over the whole surface. For additional mechanical stiffness and protection of the Press-Fit connectors, each half-bridge module

is screwed to the DC-Link PCB additionally on the top-side. [4]

Fig. 2: Exploded view of inverter housing.

2.2 PCB Design

The designed inverter consists of a stack-up of three PCBs. This allows a clear separation of different voltage levels. In addition, it fulfils the size requirements given by the customer.

2.2.1 DC-LINK PCB

This PCB works as the power stage of the inverter. On this the CAB016M12FM3 SiC half-bridge modules of Wolfspeed, Würth Electronics Redcube (7461101) connectors for power including DC input from the battery pack as well as AC output to the PMSM are located. Also, the DC-Link capacitance and differential hall sensors for current measurement of the three phases are placed on this PCB. The DC-Link capacitance consists of a combination of 11 5.6 µF C4AQLBU4560A1WK of Kemet as the main capacitance and three 1 µF CeraLink B58031I5105M062 of TDK located the nearest possible to the power terminals of the half-bridge modules. To keep a low profile, the TLI4971A120 differential hall sensors of Infineon are used to measure the phase currents. Figure 3 shows a 3D model of the designed PCB.

As can be seen, a strict separation between the different regions is performed. These regions are the DC input, the AC output as well as the signal part. This signal part includes the analogue output of the current measurements as well as the output of the in the modules, integrated NTCs. To handle the expected currents, of 64 A peak on the AC terminals and 49.1 A on the DC, a 4 layer design is used with a minimum copper thickness of 70 μm. Big planes and random vias as well as stitched vias around high power connections allow a sufficient current distribution throughout the regions.

In addition, cut-outs are located throughout the board to provide the necessary creepage distance for altitudes over 3500 m.

Fig. 3: Top view of DC-Link PCB.

2.2.2 DRIVER PCB

Due to limitations in dimension, the driver circuitry is located on a separated PCB, the Driver PCB. By using connectors, a short connection to the Gate and Kelvin Source terminals of the half-bridge modules can be guaranteed.

Figure 4 shows the top view of the 3D design done in Altium. On the top left, a connection of the DC-Link Voltage including the inductive part of an CLC filter is shown. This connection is needed for the voltage measurement as well as the internal supply, which will be discussed in the control PCB section.

On the lower part the Drivers are located with separation of low voltage and high voltage region. The low voltage region on the bottom is used for the PWM signals, driver enable signals, as well as the power supplies coming from the control PCB.
To guarantee again the required creepage distance, cutouts are located between the analogue part on the bottom and between high and low side of the gate driver.

Fig. 4: Top view of the Driver PCB.

2.2.3 CONTROL PCB

Figure 5 shows the top-view of the control PCB done in Altium. On the left top, the circuitry for the internal supply is located. By using a Viper26LN off-line converter, a buck converter is implemented [5]. This buck converter regulates down the DC-Link voltage from 400 V to 18 V DC, which then again is being regulated to 12 V and 3.3 V using additional DCDC converters and LDOs.

In this high voltage region also a Si923B-IS of Skyworks Solutions isolated amplifier for voltage measurement is located. By the use of a voltage divider, thin-film resistors with a tolerance of 1 % and a temperature coefficient of resistance (T.C.R) of ±100 ppm/°C, the nominal input for this amplifier can be achieved.

In the middle, operational amplifiers for signal conditioning of the current sensor signal and NTCs are placed. This operational amplifier circuitry provides the analogue signal for the ADCs of the MCU and is designed following the guidelines in [6].
To allow a flexible change of the MCU a SiBrain dotted board of Mikroe is used, currently with a STM32F746ZG mounted, which provides the CAN interface for communication with the flight controller of the drone as well as the PWM signals for the inverter and is the main control in the design. For communication and debugging, in addition, a JTAG and SWD interface is implemented. For testing ADC measurements, a DAC output is also provided, which can be used also for debugging and testing purposes.
On the right side, status LEDs and a thermocouple amplifier for possible measurement of Motor temperature is placed.

2.3 FIRST PROTOTYPE

Figure 6 shows the first assembled prototype of this series. For measurement and testing purposes, some modifications are done. This includes twisted

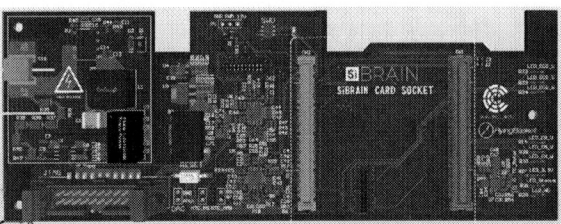

Fig. 5: Top view of the Control PCB.

cables for measurement of Gate and Source directly connected to the MOSFETs and lifted inductance of the buck converter for current measurement of supply. In addition, by means of dismounting of one DCDC-Converter, a separation of 12 V and 3.3 V auxiliary supply was done for testing purposes. For programming and debugging purposes, a STLINK V3SET debugger is connected via the JTAG interface.

Fig. 6: First fully assembled prototype.

2.4 SOFTWARE

In this section, a description of the implemented software is given. Figure 7 shows the main system architecture implemented on a STM32F746ZG of STMicroelectronics. The used control strategy for this implementation is sensor-less FOC. By the use of a Back EMF Observer, the motor speed as well as the rotor phase is estimated. For this purpose, a mathematical model of the selected M50C35 PRO EEE 9KV of MAD components is implemented. Two internal state machines are used for proper control of the system. The overall system is controlled by the global state machine. In this state machine four different states exist, fault state, stop state, ready state and run state. For the control of the inverter itself an additional

state machine, called inverter state machine, is implemented and works as a subsystem of the global state machine.

In this state machines, several functionalities are implemented [7]. Starting from the system initialisation where the timers get configured and variables are preset, e.g. level of under-voltage error. After this, an ADC calibration for the current measurement is done. Then the global state machine is started and is used to manage the overall system including enable and disable of PWM. With an internal global command generation the values of interest, including DC-Voltage, phase currents, temperature of the modules and heatsink and estimated RPM of the motor, get read-out in an 40 µs interval. Over-voltage as well as over-current are triggered separately with an interrupt function to provide a fast and reliable turn-off. The implemented software includes four different control strategies including torque control with FOC, speed control with FOC, Volts per Hertz control and Ampere per Hertz control. The main algorithm used is the torque control with FOC. By the use of command words, the space vector PWM generation can be started.

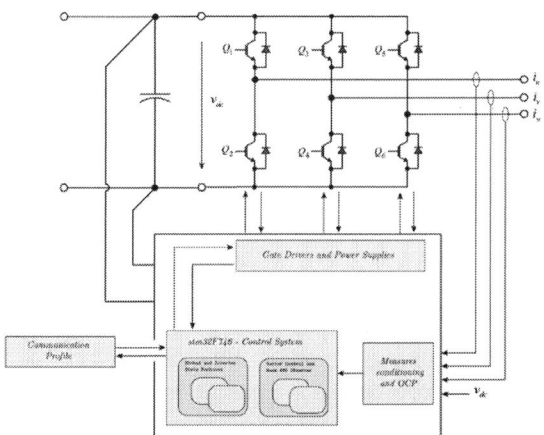

Fig. 7: Principle system architecture [8].

2.5 TEST SETUP

To provide a safe testing environment, especially at higher voltages and loads, different setups and considerations are made. Figure 8 shows the implemented testing box where a safe testing of the prototype can be performed. As in malfunction even an explosion of the half-bridge modules is possible, it consists of 10 mm thick Plexiglas plates with several cutouts for measurement and cooling purposes. It includes the possibilities for mounting fans to provide active cooling to the system, which is also given on the real drone.

Fig. 9: Motor test bench with torque sensor and motor mounted.

of Keysight is located. For supplying the fans and control electronics external during debugging, an additional laboratory power supply is used.

Fig. 8: Testing box for safety reasons.

To perform the necessary tests, especially on the real motor, a motor test bench is set up with the corresponding hardware, including two M50C35 PMSM of MAD components and a TS111 torque sensor of Magtrol. Figure 9 shows this motor test bench. The mounted motor on the left serves as the main motor for testing the inverter. To provide load to the system, a second motor can be mounted on the right and is then used in the generator load, where the phases will be connected to a resistive load. With the torque sensor located in the middle, a measurement of the applied torque is provided. Figure 10 shows the used power supplies. As a power supply, two GEN300-17-3P208 of TDK are used. This power supply can provide up to 300 V output voltage and 17 A output current each. With two power supplies in parallel, a maximum reachable power of 10.2 kW is given. By using them in series, the rated 400 V of the inverter can be reached with a maximum power of 6.8 kW. On top of these, a InfiniiVision DSOX2014A oscilloscope

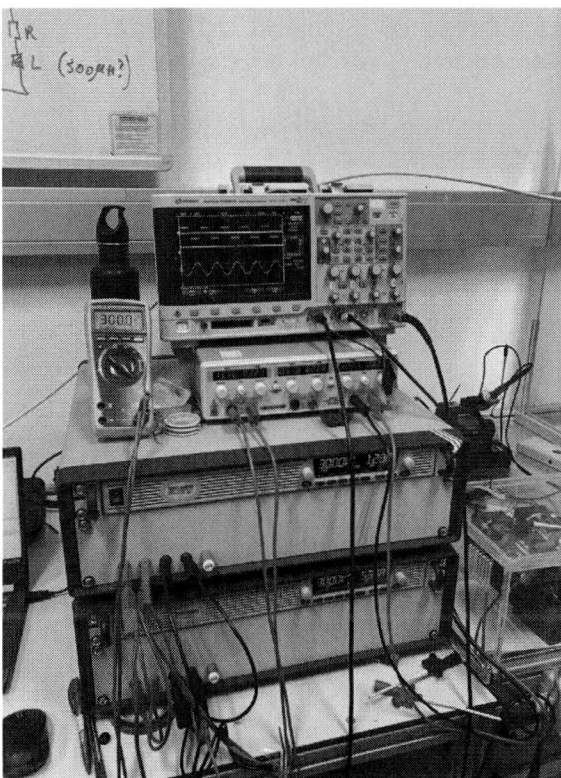

Fig. 10: GEN300-17-3P208 of TDK power supplies and InfiniiVision DSOX2014A oscilloscope of Keysight.

3 RESULTS

To guarantee an efficient measurement procedure, a test-plan is set up. This test-plan includes tests of the single PCBs, e.g. test of the Viper26, internal buck converter and check of driver functionality. The main tests including double-pulse tests, tests on a passive RL load and tests on the real motor

3.1 DOUBLE-PULSE TEST

For this test, the high-side of the half-bridge is being short-circuited with an 125 µH inductance and the low side is being switched with two pulses. The DC-Link voltage in this test is the nominal 400 V. To avoid too much current on the internal MOSFET diode, in addition also the high-side is being switched complimentary with a dead-time of 1 µs. This dead-time will be reduced to increase the system efficiency. The pulse time for the first pulse is chosen to be, 15 µs and for the second pulse, a pulse time of 5 µs is used. By using the known relationship between current and voltage inside a conductor, Equation 1, the maximum reached current can be calculated to be 64 A after the second pulse. For the measurement, differential probes of Ridley engineering with a bandwidth of 30 MHz and a Tektronix TCPA300 AC/DC current probe amplifier with a TC312A current probe is used with a InfiniiVision DSOX2014A oscilloscope of Keysight. Figure 11 shows this double-pulse test.

at the nominal current of 64 A. It corresponds to 13 ns and therefore an $\frac{dv}{dt}$ of 30.79 $\frac{V}{ns}$ and an $\frac{di}{dt}$ of 4.92 $\frac{A}{ns}$ is being calculated. Deeper investigation is done in the thesis to this paper [3].

$$V_L = L\frac{di}{dt} \tag{1}$$

3.2 Passive RL load

3.2.1 Thermal calibration

For better understanding of the generated losses inside the inverter, a thermal calibration is done. For this purpose, a reversed voltage is applied to the DC terminals. This allows that the structurally given body diode inside the MOSFETs are used and losses can be generated. With a temperature measurement using type K thermocouples with a TC-08 logger of Picotech. For this purpose, small holes are drilled in the heatsink directly underneath the three half-bridge modules. Figure 12 shows the result of this thermal calibration with forced air cooling. This results can now be used to validate the MOSFET losses in dependence of the heatsink temperature.

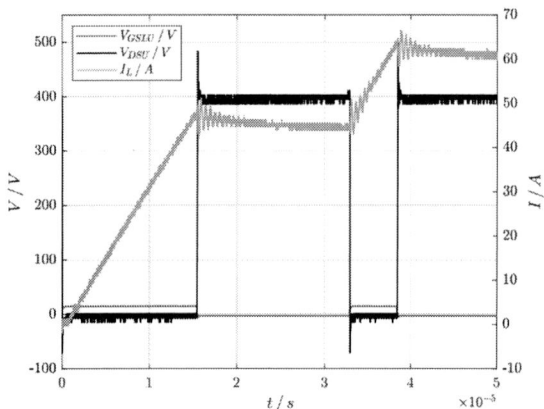

Fig. 11: 400 V double-pulse test with 125 µH inductance.

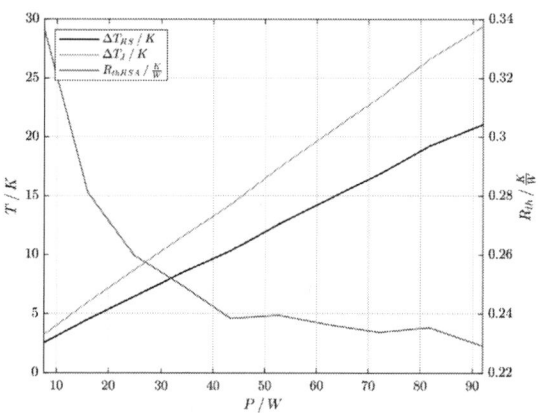

Fig. 12: Result of thermal calibration.

In the following, the results of this double-pulse test are investigated in deeper details. With this measurement set up a maximum overshoot of 99.52 V, 24.8 % of nominal voltage, is given after turn-off of the second pulse. The rise-time, from 90% nominal voltage to 10% of V_{DSU}, is measured to be 19 ns. This means that the $\frac{dv}{dt}$, the change of voltage over time, during turn-on is 21.05 $\frac{V}{ns}$ and the change of current over time $\frac{di}{dt}$ corresponds to 2.66 $\frac{A}{ns}$. The fall-time t_f is measured the same way as the rise-time, t_r with the difference that the turn-off occurs

3.2.2 Results

Here, the results of the inverter test with a passive RL Load is presented. For this purpose, a 4 Ohm 500 µH passive RL load is used. Figure 13 shows the phase to phase voltage, $V_{UVph-ph}$ which is measured using differential isolation probes of Ridley engineering. This test is performed using current control at a direct current reference i_{Sdref} of 33 % of the nominal current, or 21.12 A peak. Figure 14 shows the current waveform, measured directly on the phase output using current probes

PCIM Europe 2024, 11– 13 June 2024, Nuremberg   DOI: 10.30420/566262464

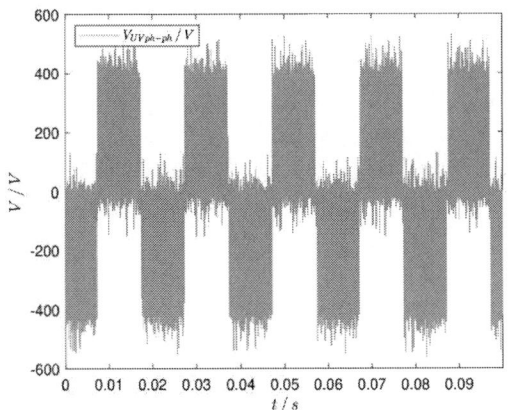

Fig. 13: Phase to phase voltage V_{Uph-ph} of phase U during passive RL load test in current control.

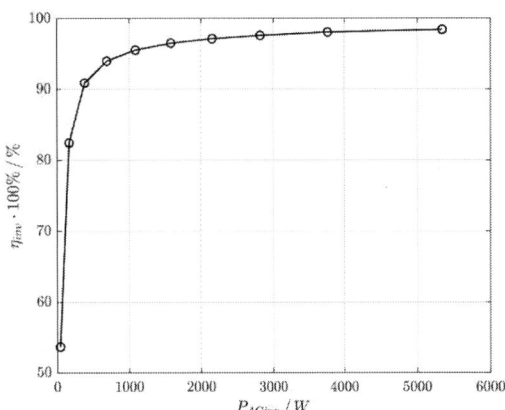

Fig. 15: Efficiency of different working points on the RL load.

of Tektronix, type TCP312A. As can be seen, the main frequency of the current is at 50 Hz, which is the set point for the current control. By the use of a PPA3500 power analyser of N4L and IT205-S current sensors of LEM the inverter efficiency is measured. The results of this measurement are shown in figure 15. A peak efficiency of 98.42 % is reached at an output power of 5.34 kW.

speed $\omega = 2185.15$ RPM.

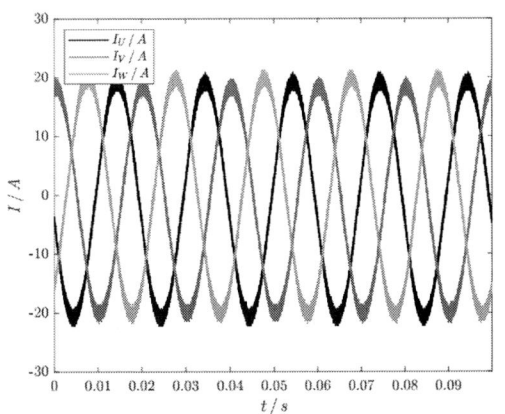

Fig. 14: Phase current I_U, I_V and I_W during passive RL load test in current control.

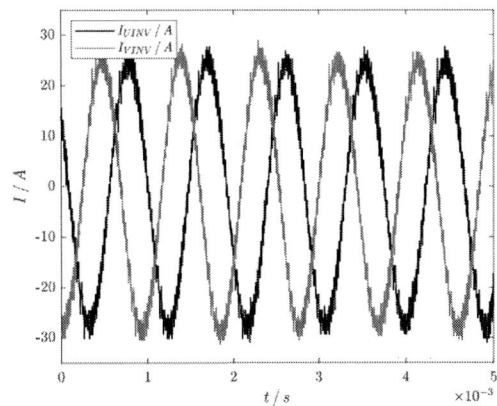

Fig. 16: Inverter phase currents I_{Uinv} and I_{Vinv} during real motor test at 30 % torque reference.

Figure 17 shows a contour plot of the system efficiency η_{sys}, including the inverter and PMSM motor, in dependence of the rotational speed w and the motor torque T. A peak system efficiency of 91.37 % is recorded.

3.4 Tests on real propeller

For verification of the functionality in addition, tests at the facility of the company FlyingBasket SRL in Bozen, Italy are performed using the selected motor in combination with a propeller and a 400 V battery. The build-up test stand also features a load cell for measuring the produced thrust and torque. Figure 18 shows the corresponding test stand. With this teststand the inverter is tested at the nominal power. Furthermore, influences of the propeller on the inverter are tests. This includes mechanical os-

3.3 Tests on real motor

In this section the results of the closed loop torque control at 30 % of nominal torque, $T = 19.2$ Nm, is shown. As a load in this test for the second motor used in generator mode, a $2\,\Omega$ resistance is chosen. Figure 16 shows the inverter output current I_{Uinv} of phase U and I_{Vinv} of phase V. The electrical frequency f_{elec} is equal to, f_{elec} = 1092.57 Hz which corresponds to a rotational

3302

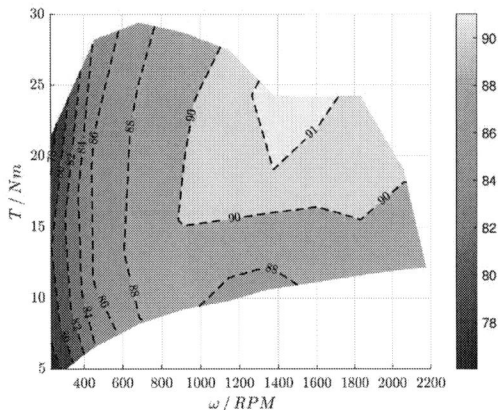

Fig. 17: 3D plot of system Efficiency versus torque T and rotational speed ω of the implemented inverter.

cillations as well as peak current operation. Figure 19 shows a phase current of the motor at a fundamental frequency of, 1200 Hz which corresponds to a rotational speed of 2322 RPM. A phase peak current of 64.12 A is reached at continuous operation. Furthermore, a thrust of 81.7 kg at 59.9 Nm is measured. As a control interface, the already existing user interface of FlyingBasket is used, which uses CAN as a communication interface. As a last test of functionality, several step-responses going from 10 % of nominal torque to 100 % are performed.

4 CONCLUSION AND OUTLOOK

In this paper, a 19.5 kW three-phase inverter design for a real UAV application is presented. A first prototype is designed, assembled and tested. First tests including start-up of the single PCB's as well as double-pulse tests on the whole inverter are done. The results of this are as expected. Further tests on a passive RL load and on a real motor are performed, and the nominal current is reached. In addition, the inverter and motor is tested on a propeller test stand, and it reaches the nominal power. A peak inverter efficiency of 98.42 % and a peak system efficiency of 91.37 % are measured.

As a next step further prototypes are produced and tested on the real drone. Employing the results of this test, further optimisations in software and hardware are performed.

Fig. 18: Test build-up including the selected Motor, a propeller and a battery used at FlyingBasket.

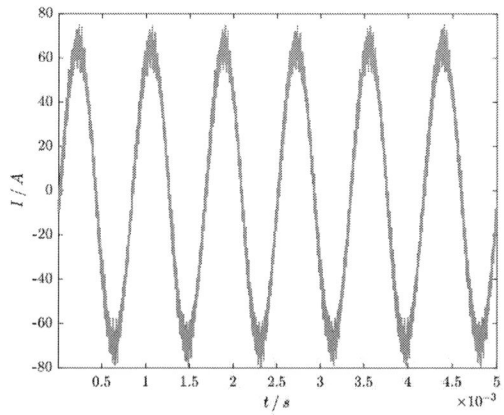

Fig. 19: Phase current of motor on propeller test stand at 1200 Hz.

References

[1] A. Nawawi, C. F. Tong, S. Yin, A. Sakanova, Y. Liu, *et al.*, "Design and demonstration of high power density inverter for aircraft applications," *IEEE Transactions on Industry Applications*, vol. 53, no. 2, pp. 1168–1176, 2016.

[2] C. A. Nicolas Vollmar Dennis Wöhrle and C. Schöner, "Design of a fast switching 200kva sic drive inverter for aviation application," in *PCIM Europe 2023*, VDE Verlag GmBh, 2023, pp. 965–972.

[3] Matthias Neuner, "Design of a high power density inverter and foc implementation for uavs," 2023.

[4] "Cpwr-an41 wolfpack mounting instructions and pcb requirement," Wolfspeed, Inc. (2021), [Online]. Available: https://cms.wolfspeed.com/download/21978.

[5] "Viper26," STMicroelectronics. (2020), [Online]. Available: https://www.st.com/resource/en/datasheet/viper26.pdf.

[6] "Analog engineer's circuit amplifiers temperature sensing with ntc circuit sboa323a," Texas Instruments Incorporated. (2018), [Online]. Available: https://www.ti.com/lit/an/sboa323a/sboa323a.pdf?ts=%20%5C%5C%201677502790693%20%5C&ref%5C_url=https%5C%253A%5C%252F%5C%252F%20%5C%5C%20www.google.com%5C%252F.

[7] Davide Bagnara, Maurizio Incurvati and Matthias Neuner, "Flying basket description of the firmware architecture and its implementation on stm32f746," 2023.

[8] Davide Bagnara, "Flying basket description of the firmware architecture and its implementation on stm32f746 (images)," 2023.

PCIM Europe 2024, 11– 13 June 2024, Nuremberg   DOI: 10.30420/566262465

Highly-Integrated, Flexible Power Solution for Aerospace 5kVA – 20 kVA Motor Drive Applications

Alain Calmels [1], Vincent Walsh [2], Laurence Egan [2]

[1] Microchip Technology Inc, France

[2] Microchip Technology Inc, Ireland

Corresponding author: Alain Calmels, Alain.Calmels@microchip.com

Abstract

As aerospace actuation systems become more electric, there is an increased demand for high-reliability, low-weight, cost-effective power electronic solutions to drive the electrical motors and reduce carbon emissions. This paper presents a highly integrated power solution (IPS) containing a power module and companion driver PCBA, which has been designed and manufactured in AS9100-approved facilities and qualified based on DO-160 specifications. The solution is extremely flexible with a variety of different product configurations and functionality options available. This paper presents technical information on all the functional elements and includes measurement, performance and trade-off details on silicon IGBT and mSiC™ silicon-carbide MOSFET solutions.

Introduction

The continued development of the 'More Electric Aircraft' and the emergence of platforms such as eVTOLs (electric vertical take-off and landing) and UAM (urban air mobility) is resulting in an increased demand for actuation solutions which combine high reliability with low weight, small size and cost effectiveness. This growth is driven by the aviation industry's goals to use the latest technology to deliver, more efficient low emission aircraft that contribute towards the decarbonization targets within the industry.

Microchip's new SP6HPD (Hybrid Power Drive) product range Fig. 1&2. offers the features required for these newer aircraft, while a companion driver PCBA is also available for a more complete, integrated solution. For increased flexibility and to suit a precise application, this high reliability solution is modular and configurable. One such example is the MSCSM120X10CTYZBNMG power module Fig. 1. which integrates mSiC™ Silicon Carbide MOSFETs and Schottky diodes from MICROCHIP, for increased ruggedness and reliability.

This paper presents the SP6HPD with companion driver PCBA product range with various options.

Performance measurements are presented with details and results of the reliability and qualification tests that were performed.

Fig. 1. SP6HPD Power Module

Fig. 2. Companion Driver PCBA + SP6HPD Module

1   Description and Characteristics

In the fully integrated device (power module and companion driver board), this all-in-one power solution integrates all the stages needed to drive an electrical motor and speeds up the design process for actuator integrators. A system block diagram / electrical diagram is shown in Fig. 3. This displays the functionality contained in, and the interfaces between, the SP6HPD (power module) and the companion PCBA (driver card).

This innovative, high performance and flexible power solution is particularly suitable for Electrohydrostatic actuators (EHA), Electrical back-up hydraulic actuators (EBHA), and Electromechanical actuators (EMA) in aviation power systems, when the criticality level is high (up to DAL-A/B). The Fig. 4. shows some typical applications.

The ultimate goal is to totally replace the central hydraulic system with full Electric power transmission (Power-by-Wire) and resulting in reducing weight, complexity and maintenance of the plane, while improving reliability.

The modularity of the Microchip solution associated with single package, envelope and design, supports the standardization, interchangeability and compatibility required by most of the aerospace Tier ones.

The efficiency gains, that are critical for this sort of high reliability application, can be achieved when using SiC semiconductors instead of IGBTs and Si diodes.

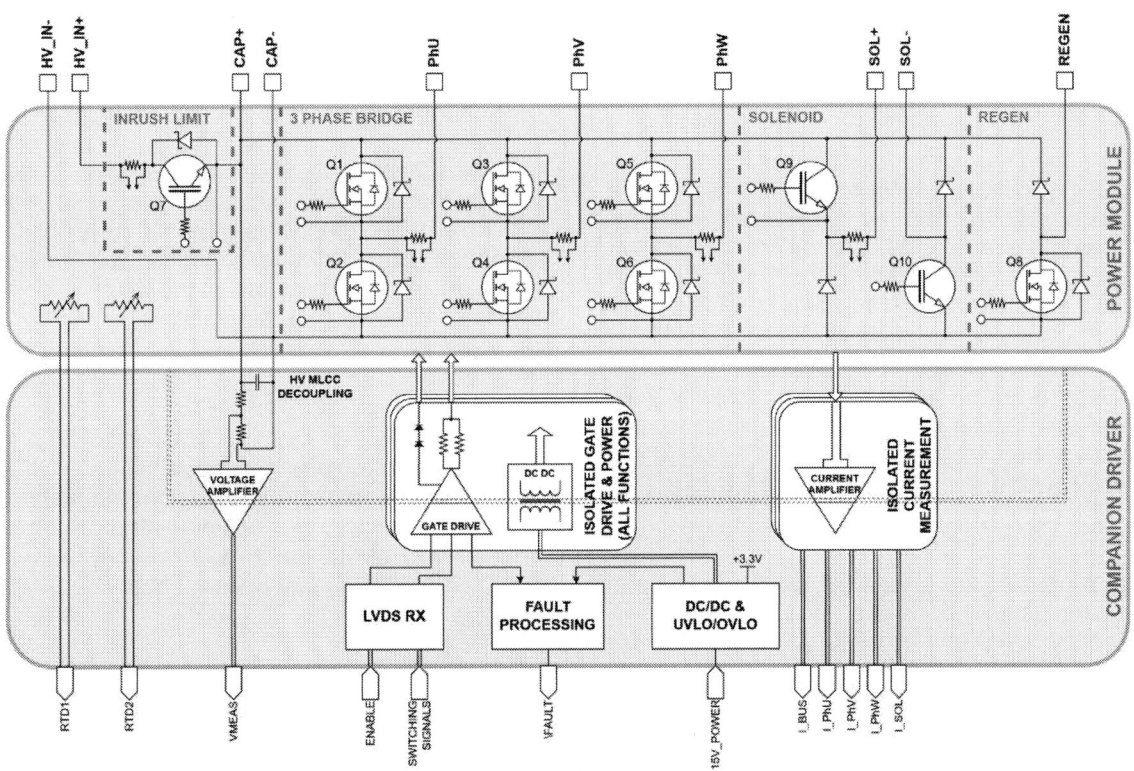

Fig. 3. System Block Diagram

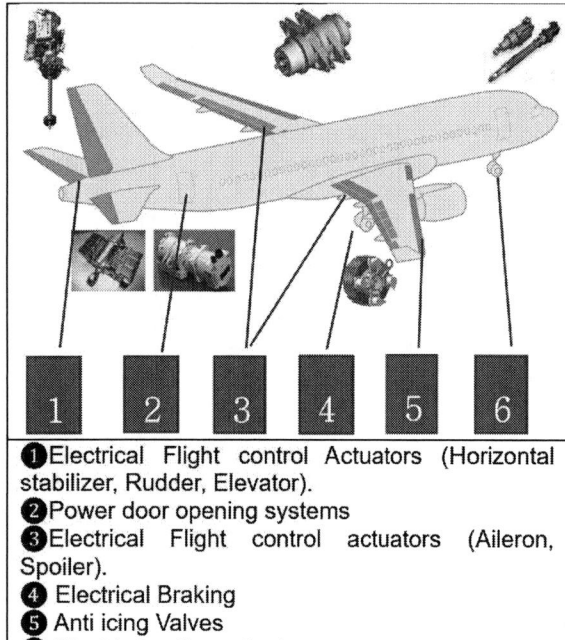

① Electrical Flight control Actuators (Horizontal stabilizer, Rudder, Elevator).
② Power door opening systems
③ Electrical Flight control actuators (Aileron, Spoiler).
④ Electrical Braking
⑤ Anti icing Valves
⑥ Electric landing actuator

Fig. 4. Some of the typical Aerospace applications

Switching losses associated with diode reverse-recovery effects can be eliminated with SiC Schottky diodes, while the difference in third quadrant ("reverse") current conduction between the two switch technologies yields conduction loss benefits when using SiC MOSFETs.

IGBT switches require anti-parallel diodes to facilitate reverse currents which typically have higher conduction losses than the power switch, but this is not required with their SiC alternatives, where reverse current is conducted through the MOSFET channel.

The products are available with a variety of semiconductor options and corresponding power capability. A summary of these is presented in Table 1.

1700V Power semiconductors are also available on demand.

Bus Voltage (V)	Power Level (kVA)	Semiconductor Technology
270	5 – 15	SiC MOSFET, IGBT
540	5 – 15	IGBT
540	5 – 20	SiC MOSFET

Table 1: Sample of available voltage and power options

1.1 SP6HPD Power Module Details

In this plug-and-play power module, the power switches (both SiC MOSFETs and IGBTs) are designed to provide the following functions: Three-Phase Inverter Bridge + Brake + Inrush current switch + solenoid as presented in Table 2. On request, additional topologies are possible, such as 3-phase rectifier bridge (400HzAC Input instead of HVDC).

1.1.1 Three-Phase Bridge

For better thermal performance and reliability, the full Low stray inductances 3 phase bridge is mounted on AMB (Active Metal Brazing) Si3N4 (Silicon Nitride) power substrate and associated AlSiC base plate. AlSiC is a matrix composite of Aluminum and Silicone Carbide with a low coefficient of thermal expansion which is better matched with the other components.

For ease of safe gate drive, Kelvin sources are offered for all switches. All the power semiconductors (SiC MOSFETs and IGBTs) are short circuit rated (both 1200V and 700V as standard). The module Integrates highly accurate phases current sensors and two thermal sensors for redundancy of power module base temperature information (PTC in standard, NTC is also possible on demand). See Table 2.

1.1.2 Brake Switch

This is an important function for safe braking operation. During regenerative braking, this switch integrated with an external power resistor will limit the voltage, avoiding over voltage on the HVDC Bus (High Voltage Direct Current DC bus).

3 Phase Bridge (with robust SiC MOSFETs or FST IGBTs power transistors).	Solenoid Stage (IGBTs in standard but SiC MOSFETs also possible)	Inrush current limiter switch (IGBTs or SiC MOSFETs).	Brake switch (robust SiC MOSFETs or IGBTs).

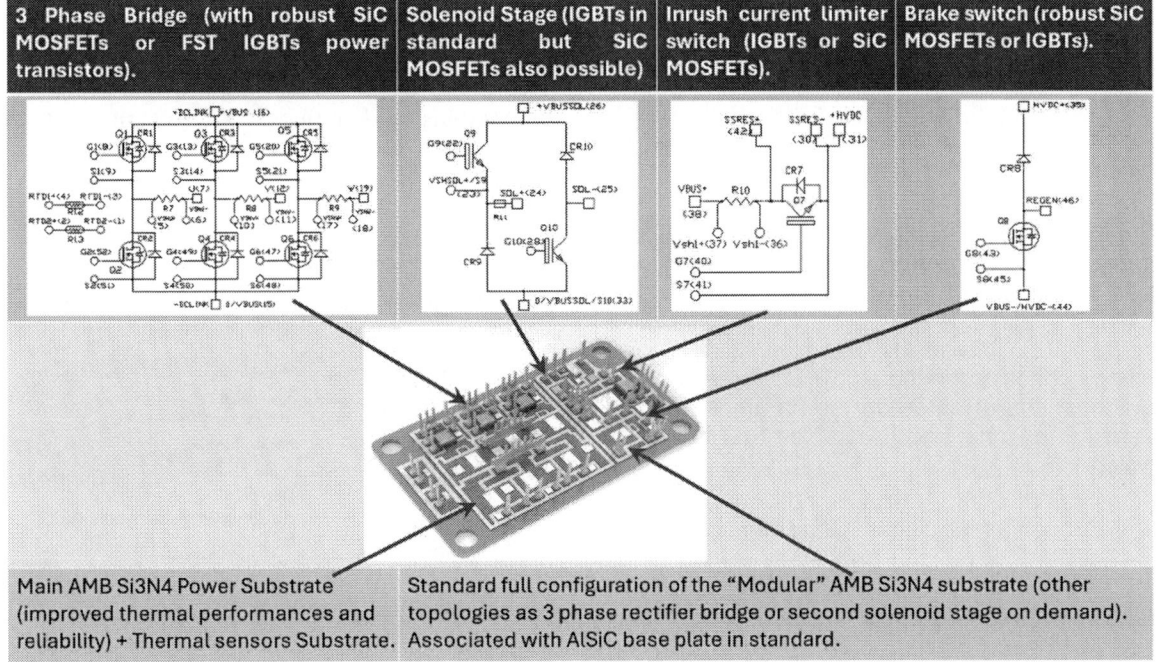

Main AMB Si3N4 Power Substrate (improved thermal performances and reliability) + Thermal sensors Substrate. Standard full configuration of the "Modular" AMB Si3N4 substrate (other topologies as 3 phase rectifier bridge or second solenoid stage on demand). Associated with AlSiC base plate in standard.

Table 2: SP6HPD integrated topologies (both 1200V and 700V in standard)

1.1.3 Inrush Current limiter Switch

Also named a soft starter switch. During starting operation this switch, integrated with an external high energy resistor, will limit the peak current due to the HVDC Capacitor bank charging. The current sensor provides the HVDC current information.

1.1.4 Solenoid Stage

Also named "Actuator Electromechanical function". For security reasons two switches guarantee the possibility to deactivate this function in any case. The current information is also provided thanks to a very accurate shunt resistor (two current sensors also possible on demand).

1.2 SP6HPD Companion Driver Details

The SP6HPD companion driver board provides isolated gate drive for all power module switches, with multiple protections such as short-circuit protection, Under Voltage Lock Out (UVLO), shoot-through protection, and active miller

clamping. It also provides isolated bus voltage measurement, isolated bus current measurement and isolated output current measurements.

1.2.1 Gate Drive

The companion driver board implements isolated gate drive for all switches.

Active miller clamping mitigates against false turn-on due to voltage disturbances on gate-source signals caused by current flow through the switch reverse transfer capacitance during switching. This is particularly relevant for SiC MOSFETs with high switching speeds.

Shoot-through protection is provided on the 3-phase bridge through interlocking of top and bottom switch PWM signals, preventing erroneous or corrupted drive signals creating a short on the bus.

Short circuit protection is provided through a 'desaturation' fault detection circuit which recognizes when the power switch is in an overcurrent/short-circuit condition and performs soft turn-off. Short-circuit response times and detection thresholds are tailored according to SiC

MOSFET/Si IGBT requirements. Short circuit testing is presented in section 2.1.

1.2.2 Under-Voltage Lock Out

UVLO is used to ensure that power switches are operated within a safe gate voltage range, minimizing the risk of overheating of the switches in case of a power supply malfunction or failure.

The gate driver board monitors the input low-voltage (15V) supply as well as each individual gate voltage supply. Hysteresis is provided to avoid oscillations when the input voltage is close to the UVLO threshold. Detection of an under-voltage condition on the input supply disables all switches, while detection of an under-voltage condition on one switch disables that individual switch.

Detection of UVLO faults is notified at system level through a fault signal at the signal connector.

1.2.3 Isolated Gate Bias Supplies

Isolated gate bias supplies are generated with on-board push pull converters. Custom transformers developed by Microchip minimize barrier capacitance and support good isolation barrier lifetime for aerospace applications, validated through partial discharge testing under low-pressure conditions. Partial discharge testing is presented in section 2.3.

1.2.4 Current Measurement

Shunt resistors integrated into the 3-phase bridge and inrush current limiter functions are complemented by isolated measurement amplifiers to provide accurate and high-bandwidth (>200kHz) analog output signals. This current measurement is ideally suited to high performance closed-loop current control for actuation applications.

1.2.5 Bus Voltage Measurement

Isolated bus voltage monitoring at the power-module level provides feedback on voltage seen by the power module.

2 Qualifications and Electrical Evaluations

This paper details some of the qualification tests completed by Microchip and summarizes the results in a table which illustrates that all tests were successfully passed (see Table 3).

Table 3: Qualification test plan

Test	Conditions
High Temperature Cycle	DO-160G, Section 4, Cat. D2 (100 °C)
Low Temperature Cycle	DO-160G, Section 4, Cat. D2 (-55 °C)
Cold Temperature Start-up	10 starts, -55 °C
Temperature Variation	DO-160G, Section 5, Cat. A (>10 °C/minute)
Altitude	DO-160G, Section 4, Cat. D2 (Unpressurized area) (50,000 feet)
Humidity	DO-160G, Section 6, Cat. C (55 °C, 95% RH)
Operational Vibration	2x DO-160G, Section 8, Cat. R, Curve E1 (22g)
Operational Shock and Crash Safety	DO-160G, Section 7, Cat. D (20g, 11 ms, saw-tooth)

In addition to the qualification testing performed at converter level, reliability testing is performed at semiconductor die and power module level.

2.1 Short Circuit Test

The objective of the short circuit tests is to verify that the short circuit protection feature provides protection to the switches to prevent failure or deterioration of function in the event of short circuit conditions.

The increased power density of SiC MOSFETs, compared to Si IGBT counterparts, typically demands faster protection response times to guarantee sufficient protection.

Short circuit testing performed on the SP6HPD module includes ten successive Hard Switching Fault (HSF / SC I) tests, in addition to fault under load and hard commutation at an output current just below the short-circuit detection threshold level to ensure protection under all conditions.

The Power Modules were screwed to a heat-sink, connected to a low-inductance bus capacitor. The phase output was shorted to bus with a short cable. (See Figure 5a).

Fig. 5a. Short circuit (HSF) test setup.

Waveforms captured during a HSF test with the MSCSM120XM19CTYZBNMG power module (SiC MOSFET, 1200V/17mΩ) are shown in Figure 5b. Current peaks at 1175A, before the short-circuit protection trips in <1.5µs, initiating a soft turn-off to limit dangerous over-voltages at power switch level. The duration of the short-circuit event is less than 2.5µs in total, giving good protection to the SiC MOSFET switch.

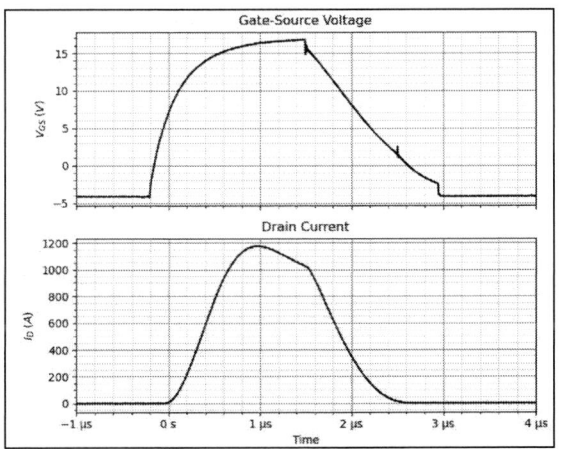

Fig. 5b. Short Circuit: Hard switching fault, SiC MOSFET, 1200V/17mΩ, V_{BUS}=800V

No degradation in power module performance is detected in testing following ten short circuit events on each switch.

2.2 Vibration and Shock Tests

The assembly was tested to 2x the DO-160 robustness vibration profile (E1) and the crash safety shock profile to ensure robustness in all the most extreme mechanical aviation environments.

An amplification of 2x is applied to the operational vibration profile to account for any potential resonances with a target system. For the duration of the testing, the assembly is operated so that any intermittent defects can be detected during the test.

See figure 6 which includes the DUT setup on the vibration shaker.

Fig. 6. Shock and vibration test setup.

2.3 Partial Discharge Test

The partial discharge test of aerospace modules is particularly important due to DC Bus voltage doubling (HVDC is 540V Typically instead of 270V and up to 900V) and the dv/dt generated by the new power semiconductors that will impact the components related to the insulation as the Power substrates used to isolate the power Electronics to the plane chassis ground.

Thanks to the capability to detect the local electrical discharge which result from defects or in homogeneous insulation material, this test will guarantee the reliability of all the components related to the insulation performances as the power substrates including in non-pressurized environment.

This is why the SP6HPD + PCB companions are qualified in PDT according with the EN 61287-1 standard. In addition, to simulate the non-pressurized conditions, the Microchip SP6HPD complete systems are tested at low pressure (116mbars to reproduce the conditions at 50000 feet).

All the modules passed the test before and after the qualifications according to the acceptance limits of 10pC max at 1200V AC. See Figure 7.

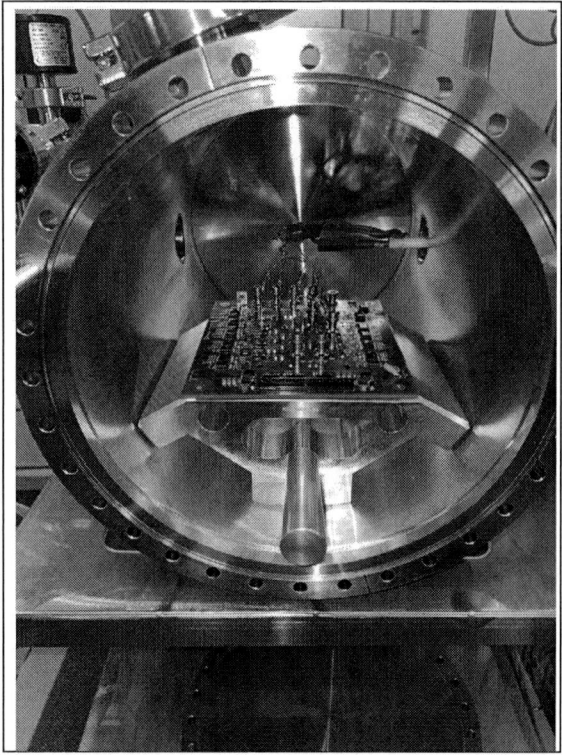

Fig. 7. Test fixture used during Partial Discharge

2.4 Reliability Analysis

The reliability evaluations are based on specific mission profiles, which allows for better understanding of the stresses to which the devices will be exposed. The mission profiles and the information mentioned below is critical to understanding the MTTF for a device. For the new-generation HPD, we are using a short-haul flight profile, which is considered a severe condition due to the higher frequency of temperature cycles. Both pressurized and non-pressurized mounting location profiles are considered. The mission profile is defined using the following stresses:

- Temperature cycles
- Minimum and maximum temperatures
- Number of cycles
- Cycle duration
- Ambient temperature
- Mechanical stresses
- Humidity

- Chemical stresses

Due to the criticality of the application, reliability prediction analysis is performed throughout the development life cycle. It is updated and re-evaluated at each of the major design phases.

Multiple reliability analysis tools are available to predict the MTTF of electrical systems, but the preferred option is FIDES (Latin for "trust"). The FIDES reliability prediction model accommodates complex mission profiles. It is adapted to account for rugged environments and allows for the inclusion of process factors, which provide a more accurate reliability prediction than alternative models. The analysis incorporates the component electrical stresses and temperature rises for more accurate Failure in Time (FIT) predictions. The results allow us to identify areas of the design where failure rates are higher, which can be subsequently addressed in the early design iterations. The FIDES library has a specific set of component failure models. If a model does not exist for a particular device, a representative alternative is selected.

Figure 8 shows an example of the breakdown of the component contributions to the overall FIT value for both a pressurized and non-pressurized mission profile produced from a FIDES analysis on an HPD device.

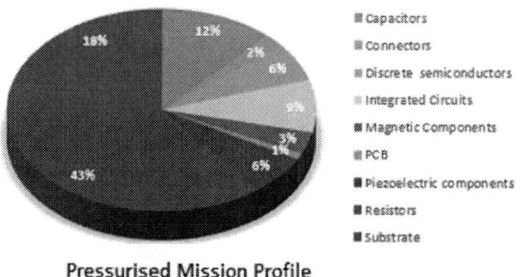

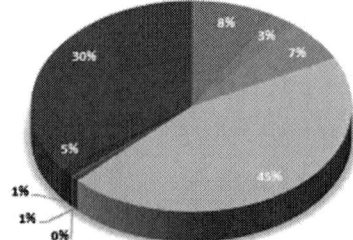

Fig. 8. Component contribution to FITS for pressurized and non-pressurized mission profiles

The difference in the contribution of each component type to the overall FIT figure is driven by the variation of environmental conditions in each of the mission profiles. The high contribution of resistors to the overall figure for the non-pressurized profile is mainly driven by the thermal cycling stresses associated with the non-pressurized environment, while the contribution from Integrated Circuits (ICs) in the pressurized profile is driven by a corresponding thermal stress in the pressurized zones.

Analysis like this allows for fine tuning of a design to improve reliability for specific applications.

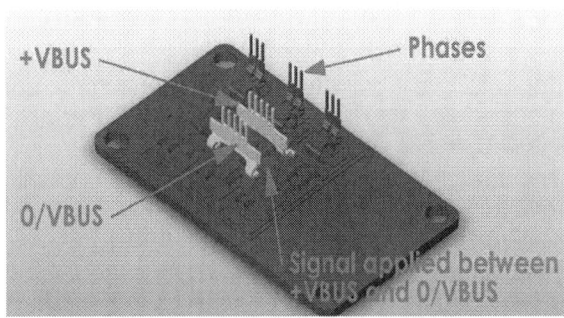

Fig. 9a. DC link stray inductance simulation (From +VBUS and 0/VBUS power connectors)

3 Stray Inductances (simulations and measurements)

Thanks to an optimized design of the Low profile Power module and also at PCB companion level (high performance MLCC ceramic capacitors mounted very close and between the HVDC power connectors), the SP6HPD power solution presents in standard a low stray inductance of 10nH in total (9nH for the Power module and 1nH for the PCB).

That allows a very good voltage safety margin including during worst operation cases conditions (VBUS max of about 900V) and very fast switching times.

These results are confirmed by both simulations and real switching measurements.

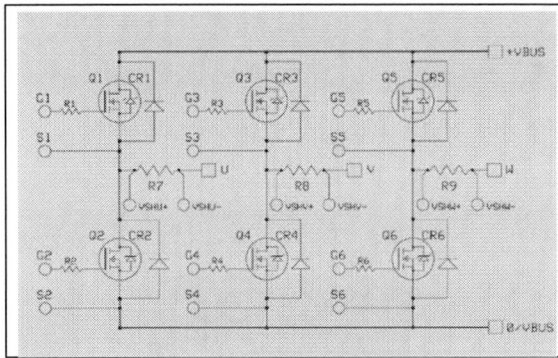

Fig. 9b. 3 Phases bridge worst Stray Inductance Path

3.1 SP6HPD Standard Power Module stray inductance simulation

For the simulations the signal is applied between the +HVDC and -HVDC power connectors as shown in Fig.9a. and on the Eletrical Diagram Fig.9b. that shows the worst case.

The inductance value is calculated by splitting the real part and the imaginary part as shown with the equation below:

$$L = imag\ (1/y) / 2 * Pi * f.$$

The parasitic inductance varies from 9.21 nH at 100 KHz to 8.14 nH at 1 MHz that is well below the 10nH target. See Fig. 9c.

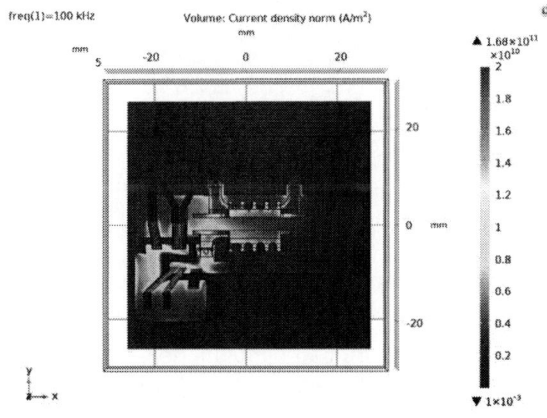

Fig. 9c. Simulated Loop inductance High Frequency current distribution (standard version).

3.2 SP6HPD Custom Power Module Stray Inductance Simulation

For this version the HVDC connectors are modified as Bus Bars to increase the STRIPLINE effect (see Fig. 10a.).

Thanks to those modifications the parasitic inductance is 6.69 nH at 1 MHz (see fig.10b.) and confirms improved values close to 6nH.

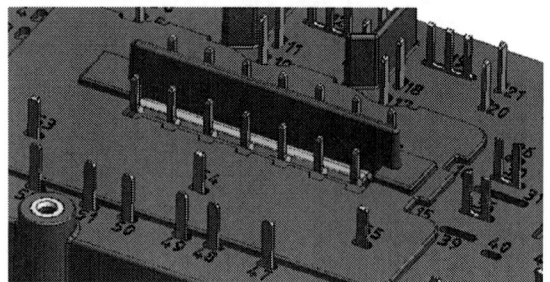

Fig. 10a. HVDC STRIP LINE optimized version.

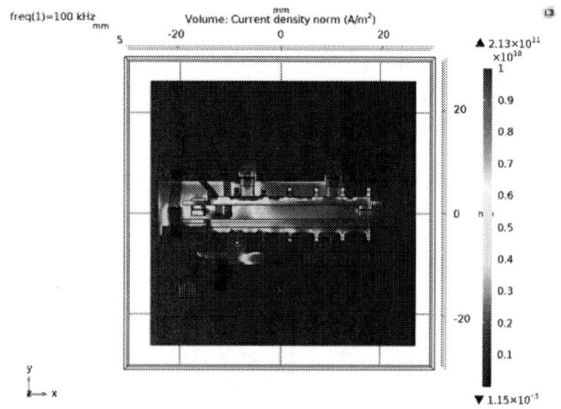

Fig. 10b. Simulated Loop inductance High Frequency current distribution (option available on demand)

3.3 Total solution switching measurement for parasitic inductances calculation.

Parasitic inductance simulations have been verified through measurement at a total solution level.

Measured chip voltages, close to the die, are compared to voltages measured at the Power Module VBUS terminals, as well as on the high-voltage MLCC capacitors used at PCB level to minimize the total loop inductance. See Figure 11a. By comparing voltage drops to switch current, measured through the VBUS- terminal, the inductance internal to the power module and at the total solution level are calculated.

The test was performed using an IGBT Power Module (MSCGLQ75X120CTYZBNMG), using switches Q1/Q2 in accordance with the simulation performed in Fig. 9c.

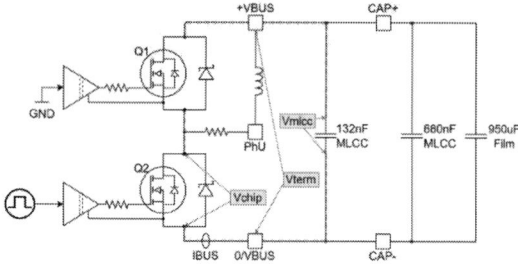

Fig. 11a. Parasitic inductance measurement setup

Measured voltage overshoot of $V_{CE(OS)}$=37.7V at the switch, compared to the power module terminals, corresponding to dI_C/dt=4.4kA/µs (see Figure 11b), indicates an internal power module inductance of 8.5nH, aligned with the simulated inductance range.

There is no significant voltage difference between voltage measured at the power module terminals and voltage at the MLCC endcaps, as a result of their close proximity. However, up to 1nH additional inductance may be considered internal to the MLCC, yielding a loop inductance of 9.5nH at total solution level.

This measurement considers parasitic inductance in the commutation loop, affecting switch over-voltage. Inductance related to the bulk capacitor connection, affecting low-frequency oscillations, is not considered.

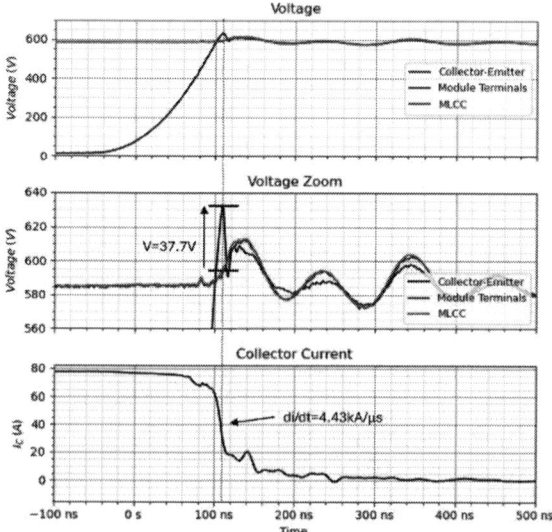

Fig. 11b. Parasitic inductance measurement

4 Thermal Simulations

To confirm the good thermal performances, thermal simulations were done on all power switches integrated into this All in one power module. The following are the results of one of the chips offered in the MSCSM120XM31CTYZBNMG power module.

4.1 Thermal Simulations

Description: The aim of the thermal simulation is to determine the SP6HPD modules thermal resistance and thermal impedance values. The chip used for the modules are the 1200V/25mΩ mSiC™ MOSFET die as shown in Figure 12a).

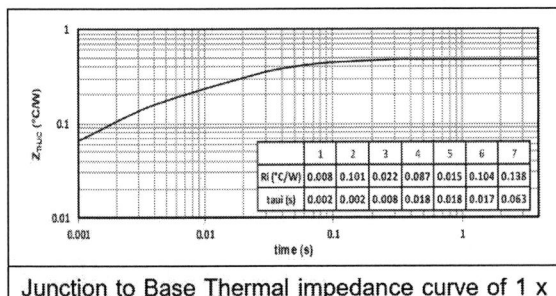

Junction to Base Thermal impedance curve of 1 x 1200V/25mR SiC MOSFET Die. R_{thJC}=0.48°C/W

Figure 12a. Thermal resistance and Impedance

5 Conclusions

Microchip has developed a new family of modular HPD (Hybrid Power driver) power modules and total solutions based on a compact standard package named SP6, containing SiC MOSFETs or IGBTs arranged in different topologies to complete this "All in one solution".

Based on the tests completed by Microchip, the technology is being qualified for aerospace applications and fully mature to serve civil aircraft and eVTOLs.

On request if additional qualifications are requested the Microchip Integrated Power Solutions (IPS) group has capability to perform complementary tests based on the conditions outlined in RTCA DO-160G (Environmental Conditions and Test Procedure for Airborne Equipment).

6 References

[1] Dennis Meyer, Steve Chenetz, Enhab Tarmoon, Kevin Speer, Failure Rate Calculation Due to Neutron Flux with SiC MOSFETs and Schottky Diodes. IEEE 2022.

[2] D. Othman, S. Lefebvre, M. Berkani, Z. Khatir, A.Ibrahim, A. Bouzourene, Thales SATIE and LTN, "Investigation of 1.2 kV Investigation of SiC MOSFETs for Aeronautics Applications" . EPE 2013.

[3] Avinash S. Kashyap, A. Gendron-Hansen, D. Sdrulla, B. Odekirk, D. Meyer, C. Hong, W. Brower, "Microsemi Exceptional Ruggedness of SiC devices" PCIM2018.

[4] Jacques Laeuffer, Towards_a_One_Nano-Henry_Power_Module_for_SiC_and_GaN, PCIM Europe 2021.

[5] Jean-Charles Maré, Review and Analysis of the Reasons Delaying the Entry into

Service of Power-by-Wire Actuators for High-Power Safety-Critical Applications. Actuators 2021.

[6] JENTINK Henk, Exploitation Of ACTUATION2015 Pre-Standardisation Activities On Power-By-Wire, R3ASC 2016.

[7] RTCA "DO-160G – Environmental Conditions and Test Procedures for Airborne Equipment", August 2010

[8] O'Donnell, Shane; Castellazzi, Alberto; Debauche, Jean-Louis; Wheeler, Pat; Silicon Carbide MOSFETs in More Electric Aircraft Power Converters: EPE'16 ECCE Europe,

PCIM Europe 2024, 11– 13 June 2024, Nuremberg   DOI: 10.30420/566262466

Database-supported Preliminary Design, Simulation and Evaluation of Power Converters in Electric Aircraft Propulsion Systems

Jeff Kugener, Ankit Pal, Stefan Kazula

German Aerospace Center (DLR), Institute of Electrified Aero Engines, Cottbus, Germany

Corresponding author: Jeff Kugener, jeff.kugener@dlr.de
Speaker: Jeff Kugener, jeff.kugener@dlr.de

H03 Power Electronics for Aerospace Applications
Preferred presentation form: Oral (poster presentation acceptable)

Abstract

In electric aircraft propulsion systems, the number and power level of power electronic components will increase significantly. Thereby, preliminary design tools of power converters are required to support system studies during early design stages. This paper assists those studies by introducing an automated approach to achieve a comprehensive preliminary design. Common inverter topologies are designed according to the challenging aircraft application requirements. Sub-components are selected from a database and used for a topology circuit parameterization via an XML-RPC interface. The topologies are simulated in PLECS and the results are returned to the implemented design tool for post-processing.

1 Introduction

The goal of limiting global warming to 1.5 °C set at the Paris Agreement can only be achieved by significantly reducing greenhouse gas emissions. The aerospace industry is therefore exploring the application of all-, turbo- and hybrid-electric aircraft propulsion systems [1]. In this matter electric powertrains rely on power converters conditioning the electric power.

A power converter is a highly optimized component of the electric powertrain. During the preliminary design, there is often insufficient knowledge of the propulsion system, which is crucial for any optimization task. Numerous different components and complex correlations complicate an accurate estimation of weight, volume and losses. Furthermore, the limited scalability of semiconductor devices and the dominant influence of their packaging is often over-simplified in system studies [2], where only a rough value for the power density of the converter is specified.

A detailed evaluation of converter key characteristics is a valuable contribution to system studies when the topology and sub-components are clearly specified [3]. This paper presents an approach for automated preliminary design and sizing of converters in Python including an SQL database. Analytical methods combined with empirical correlations are used to obtain an initial electrical design and a gravimetric and volumetric power density estimate. The aircraft application requirements are discussed and applied with an assessment of challenges due to high altitude operation. 2-Level (2L) and 3-Level Neutral-Point-Clamped (3L-NPC) inverter simulations in PLECS are automatically parameterized and evaluated in the proposed Python-PLECS-Framework to gain an extended insight into efficiency and power quality.

2 Methodology

Fig. 1 shows the principle workflow of the implemented approach. For a given topology, the preliminary electrical design is performed, including a first estimation of losses, voltage and current stress on the components. Based on the electrical design, the sizing and design of the various sub-components is carried out.

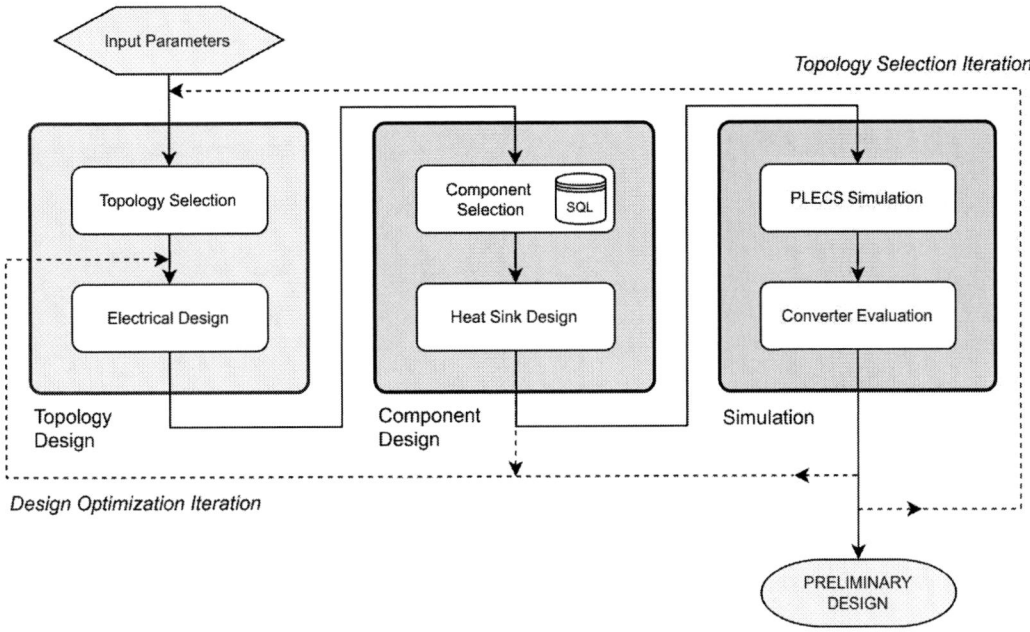

Fig. 1: Workflow of preliminary design framework

The volume of each component and its mass are summed up to determine a volumetric and gravimetric power density. To improve the efficiency and power quality assessment, the preliminary component design is implemented in an automated PLECS simulation model, including thermal models of semiconductors and the thermal management parameterized by an SQL database. The approach thus outperforms conventional preliminary design tools by providing a more accurate assessment of the losses with commercial semiconductors and an assessment of the power quality trough simulated waveform analysis.

3 Results and Conclusions

The study investigates a 500 kW inverter with a gravimetric power density of 42 kW/kg and a volumetric density of 59 kW/l. Figure 2 shows the output waveforms of the simulated 3L-NPC topology. The simulation enables an assessment of power quality and efficiency, while thermal models of SiC mosfets are selected through a SQL component database allowing an automated parameterization and simulation.

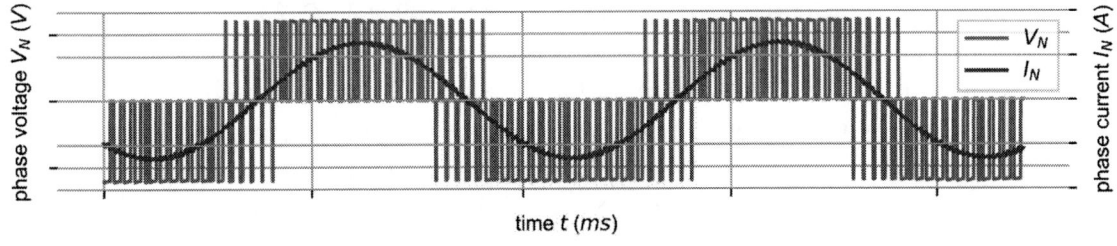

Fig. 2: PLECS simulation results - output phase voltage V_N and current I_N

The gravimetric power density assessment in Figure 3 (a) is in good agreement with various state-of-the-art inverter designs [4–8]. The mass breakdown in Fig. 3 (b) reveals that the total weight is not dominated by the semiconductor modules. Rather passive elements and cooling components account for more than half of the total weight. A comparative study of the sub-component weight distribution is carried out for a Wolfspeed 600kW reference design shows a similar distribution.

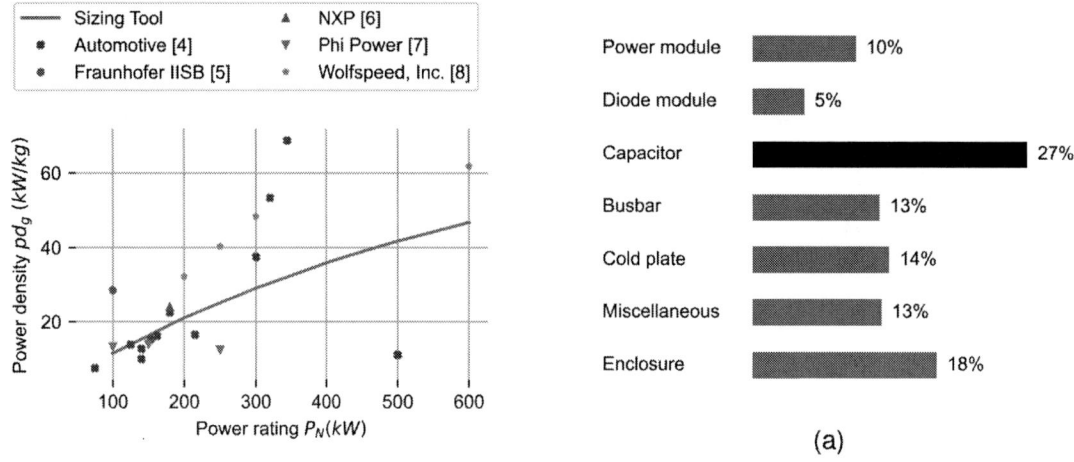

Fig. 3: Gravimetric power density (a) and mass breakdown (b) of the preliminary sizing tool results

The proposed design approach thus provides a basis for a comprehensive analysis of efficiency, power quality and power density for the early phase of the development of converters for electrical aircraft.

References

[1] C.K. Sain, S. Kazula, L. Enghardt, "Electric propulsion for regional aircraft - critical components and challenges," 2022. DOI: 10.25967/570151.

[2] J.-K. Mueller, A. Bensmann, B. Bensmann, T. Fischer, T. Kadyk, *et al.*, "Design considerations for the electrical power supply of future civil aircraft with active high-lift systems," 2018. DOI: 10.3390/en11010179.

[3] J. Kugener, P. Ankit, S. Kazula, "Preliminary design tool for power converters in electric aircraft propulsion systems," *Proceedings of DLRK23, Stuttgart*, 2023.

[4] C. S. Goli, S. Essakiappan, P. Sahu, M. Manjrekar, and N. Shah, "Review of recent trends in design of traction inverters for electric vehicle applications," pp. 1–6, 2019. DOI: 10.1109/PEDG51384.2021.9494164.

[5] Fraunhofer IISB, *100 kw sic-inverter for automotive application*, www.iisb.fraunhofer.de/content/dam/iisb2014/en/Documents/Research-Areas/vehicle_electronics/2017-05-08_FraunhoferIISB_Produktblatt_100kW-SiC-Inverter_WWW.pdf, Accessed: 12.09.2023, 2023.

[6] NXP, *Ev-inverterhd*, www.nxp.com/design/designs/ev-power-inverter-control-reference-platform-gen-1:RDPWRINVERTER, Accessed: 12.09.2023, 2023.

[7] Phi Power, *Pm100 - pm250*, www.phi-power.com/de/wechselrichter/, Accessed: 12.09.2023, 2023.

[8] Wolfspeed, *600 kw xm3 high performance dual three-phase inverter*, www.wolfspeed.com/products/power/reference-designs/crd600da12e-xm3/, Accessed: 12.09.2023, 2023.

PCIM Europe 2024, 11– 13 June 2024, Nuremberg   DOI: 10.30420/566262467

Preliminary Design and Evaluation of Inverter Power Density and Reliability in Electric Aircraft Propulsion Systems

Jeff Kugener, Stefan Kazula

German Aerospace Center (DLR), Institute of Electrified Aero Engines, Cottbus, Germany

Corresponding author: Jeff Kugener, jeff.kugener@dlr.de
Speaker: Jeff Kugener, jeff.kugener@dlr.de

Abstract

In electric aircraft propulsion systems, the number and power level of power electronic components increase significantly compared to conventional aircraft. Preliminary design tools for power converters are therefore required to support system studies in the early stages of design. This paper presents an automated approach to achieve a comprehensive converter evaluation. Common inverter topologies are designed according to the challenging aircraft application requirements. Commercial sub-components, including thermal models, are selected from a database and used for a preliminary sizing. A standardized methodology for the evaluation of failure rates is implemented. The resulting approach allows a comprehensive evaluation of power density, efficiency introduced and failure rates for preliminary aircraft concept studies.

1 Introduction

The mandate of the Paris Agreement to limit greenhouse gas emissions is driving the aviation industry to explore various electric aircraft concepts. Balancing the growing demand for air travel with the need to minimize environmental impact is a major challenge for system and component design. In this matter, the German Aerospace Center (DLR) conducts extensive concept studies to explore promising aircraft concepts and technologies on a large scale [1]. To increase the accuracy of economic and environmental analyses for holistic aircraft designs, each aspect of the aircraft design is based on an underlying component model.

In this paper, the approach for the preliminary modeling of power converters is introduced. Power electronics play a crucial role in electrified aircraft, as optimized power converters are essential for efficient control and reliable operation of the drivetrain. The increasing integration of power electronics in such engines requires a modelling approach to accurately represent the domain in aircraft studies and to communicate with peripheral systems such as the thermal management.

The design of a power converter is a highly optimized task considering details of the drivetrain architecture, mission profiles and parasitics. At this preliminary stage, with the according low degree of detail, an automated modelling approach should be able to represent general characteristics and challenges of power converter design in order to evaluate large sets of powertrain concepts and variations.

This work defines the key converter characteristics for preliminary aircraft studies. The reliability, power density and efficiency of inverter designs are evaluated using a commercial component database. Key input parameters such as switching frequency, voltage levels and heatsink temperature are varied to identify the main influencing factors on the evaluation criteria.

An approach for a comprehensive evaluation of a converters power density is presented in [2]. This paper extends the approach for a preliminary assessment of failure rates, the implementation of thermal networks and a wide parameter variation. The resulting tool allows an automated preliminary design and sizing of converters in Python based on an extendable component database

3318

implemented in SQLite. The aircraft application requirements are discussed and applied with an assessment of challenges due to high altitude operation. 2-Level (2L) and 3-Level Neutral-Point-Clamped (3L-NPC) inverter simulations in PLECS can be automatically parameterized via an XML-RPC interface for the most promising designs. The implemented approach provides an evaluation of power density, efficiency and reliability, including simplified thermal models and thus enables a comprehensive initial assessment for system studies.

2 Power converters in electrified aero engines

The potential electrified aircraft propulsion system topologies are divided into three categories according to the electrification of their components: all-electric, turbo-electric and hybrid-electric [3]. All electrified aircraft propulsion topologies rely heavily on power converters to efficiently control and distribute electrical power throughout the system [4].

2.1 Power Converters

Power converters are essential for conversion and conditioning of electrical power from energy sources such as batteries or fuel cells to drive electric motors, ensuring optimal performance and reliability.

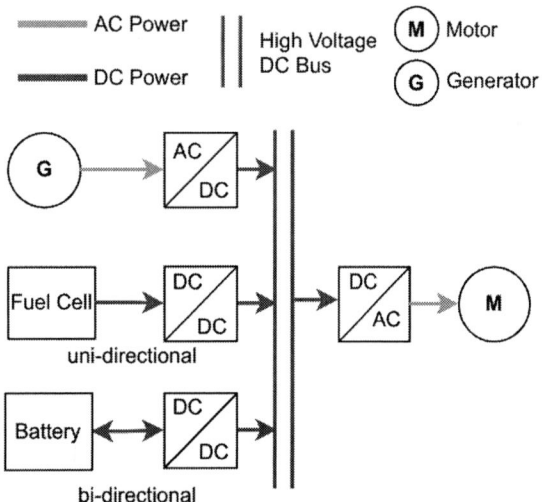

Fig. 1: Potential drive train components

Within an electrified aircraft, several power converters can be applied, including DC-DC, DC-AC,

AC-AC and AC-DC power conversion. Various use cases of these are illustrated in Fig. 1. This study focuses on the design aspects of 2L and 3L-NPC inverter (DC/AC) topologies, with the proposed methodology being adaptable to other converter configurations.

2.2 Requirements

The development of electrical components with high efficiency, power density and robustness is crucial for enabling commerical electric aircraft. With aircraft expected to operate for up to 100,000 hours in challenging environmental conditions [5], ensuring the reliable performance of the converter and its components is essential to ensure flight safety. Blaabjerg [6] estimates the active operating time of inverters in automotive environments to be around 12,000 hours. Extending the component lifetime and reducing the required maintenance poses significant challenges in electrified aviation.

At a preliminary design stage, consideration of failure rates can therefore be an important factor in making initial design decisions. Failure rates, expressed as failures in billion hours (FIT), are established in the Military Handbook - Reliability Prediction of Electronic Equipment [7] and IEC 61709 Electric components - Reliability [8]. The presented preliminary design approach uses these standardized methodologies to evaluate failure rate reference conditions and stress models applicable to electronic components and thereby facilitating early evaluation of reliability considerations.

Several challenges have been examined in literature, including reduced ambient pressure, cosmic radiation, humidity and heat dissipation during high altitude operation [5], [9]–[13]. Derating factors that affect the reliability of electrical components should therefore be considered in the preliminary design of the power converter. In addition, compliance with the mechanical and structural requirements outlined in the Environmental Conditions and Test Methods Standard for Airborne Equipment DO-160 [14] can have a significant impact on the resulting power density of a component, although these factors are often neglected in studies on power densities [15]–[17].

The SAE Aerospace Information Report AIR6127 for Managing Higher Voltages in Aerospace

Electrical Systems [18] discusses the transition to higher voltage levels in aerospace applications. At high altitudes, the dielectric strength of air decreases with air pressure according to Paschen's law. As a result, the thickness of the insulation layers and the clearance distance between components must be increased. According to the standard for insulation coordination for equipment in power supply systems IEC 60664 [19], an altitude correction factor is applied depending on the maximum altitude. This factor more than doubles the required clearance distance at 8000 metres above sea level.

In addition, power devices operating at high altitudes are more likely to be disrupted by high-energy particles, as increased neutron flux increases the risk of cosmic ray induced failures. High-energy particles colliding with the power module can cause electron-hole pairs or displacement of atoms within the semiconductor material. This risk of single event burnout (SEB) increases from possible localised charge build-up or structural damage to the semiconductors [20]. In SEB, the impact of a high-energy particle triggers a rapid voltage surge across the semiconductor device, exceeding its breakdown voltage and causing catastrophic failure. To reduce the risk of failure, Felgemacher [12] recommends operating silicon carbide (SiC) semiconductors at 70% and silicon (Si) devices at 50% of the maximum breakdown voltage.

2.3 Limitations

This concept study investigates promising topologies for future regional aircraft. It is important to understand the limitations of the state-of-the-art in order to provide a realistic outlook. This work focuses on commercially available components and aims to determine the achievable performance characteristics. Future developments can be realised by expanding the component database with new and potential future components.

The limiting factors include voltage, current and endurance. As state-of-the art power modules reach up to 3.3kV, the task of paralleling bare dies to reach high current ratings is a sensible task [21]–[23]. Low inductive design is a major requirement for reliable operation [24], which is compromised by increasing clearance distances in high altitude operation.

Based on currently available components, the implemented tool covers voltage levels up to 1700V and current capacities up to 800A. An extrapolation for modules with 3300V and 2000A is also performed. To achieve higher power levels, several power modules can be connected in series or in parallel according to Bolotnikov [25]. As the connection of multiple modules in one leg introduces further design challenges and unknowns for the behaviour in high altitude operation, this study chooses a conservative approach and does not connect multiple modules to achieve higher power levels. The limitations over the output power are therefore given by the characteristics of the included power modules.

This paper investigates inverter designs for an output power of 250 kW. To achieve higher power levels in a 2L topology, series and parallel connections of modules are necessary. Based on the analysis of state-of-the-art power modules, the study achieves potential power levels of up to 600 kW with a 3L-NPC topology without exceeding acceptable failure rates.

Effective thermal management is a challenge in both the design and operation of power converters within electric aircraft propulsion systems. The adequate dissipation of heat generated by power electronics components is critical to achieve the desired efficiencies and reliability, particularly in compact designs [26]. Irregular heat dissipation or high temperatures reduce the lifetime and introduce potential critical failures. Furthermore, the reduced air density at higher altitudes limits the effectiveness of conventional fan cooling techniques. As a result, traditional air cooling methods are replaced by liquid cooling solutions in inverter systems [27]. The thermal limitation in this tool is given by a maximum junction temperature of 175 °C.

3 Methodology

Figure 2 introduces the principle workflow of the implemented approach. For a given topology, the preliminary electrical design is performed, including a first estimation of losses, voltage, current and thermal stress on the components. Based on the electrical design, the sizing of the various sub-components is carried out.

A wide range of parameters is analysed to determine system correlations. The parameter configurations for each simulated converter are represented by a 6-dimensional parameter vector $\vec{\rho}$. The range of the paramter configurations is given in table 1. For each parameter configuration, the most promising power module components, including PLECS thermal models and capacitors, are selected from an SQL database. The resulting inverter variants are then evaluated for power density, efficiency and failure rates.

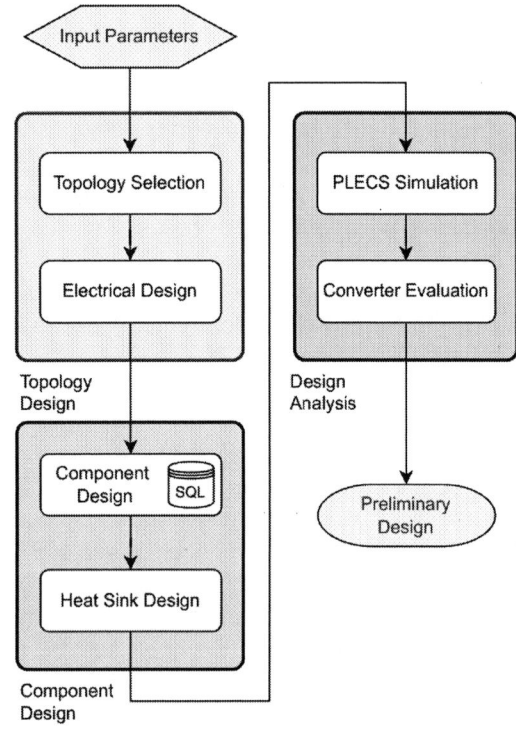

Fig. 2: Workflow of preliminary design framework

Tab. 1: Inverter input parameter variations ρ

Parameter	Range	Unit
Topology	$\rho_0 \in [2L, 3L\text{-}NPC]$	
Switching frequency	$5 \le \rho_1 \le 100$	kHz
DC-Link voltage	$800 \le \rho_2 \le 3300$	V
Modulation index	$0.5 \le \rho_3 \le 1$	
Heatsink temperature	$25 \le \rho_4 \le 100$	°C
Voltage derating	$0.5 \le \rho_5 \le 1$	

3.1 Power density

An approach for a rapid and comprehensive evaluation of an inverter power density is presented in earlier work [2]. The volume of each subcomponent and its mass are summed up to determine a volumetric and gravimetric power density.

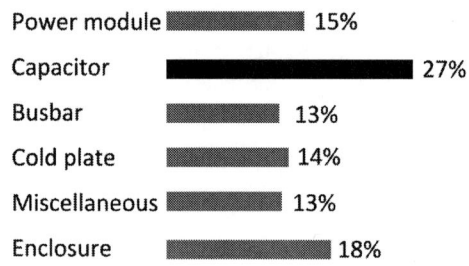

Fig. 3: Typical mass breakdown based on [2]

Figure 3 shows a typical mass breakdown resulting of a 500kW 3L-NPC inverter topology with a gravimetric power density of 42 kg/kW. The total weight is not dominated by the semiconductor modules. Rather passive elements and cooling components account for more than half of the total weight. The presented approach provides a volume- and mass-breakdown for each parameter configuration. The DC-link capacitance is considered in the sizing, additional output filters or snubber circuits are not taken into account yet.

3.2 Efficiency

The halfbridge modules included in the database are linked to a PLECS loss model. The model is parsed in Python to include the switching energy E_{sw} and the forward voltage V_f as a function of the blocking voltage V_{DS}, the switched current I_{DS} and the junction temperature T_j. The switching losses P_{sw} result from the switching frequency f_{sw} and the turn-on E_{on} and turn-off E_{off} energies.

$$E_{sw} = E_{on}(V_{DS}, I_{DS}, T_j) + E_{off}(V_{DS}, I_{DS}, T_j) \tag{1}$$

$$P_{sw} = f_{sw} \cdot E_{sw}(V_{DS}, I_{DS}, T_j) \tag{2}$$

The conduction losses P_{cond} result from the forward voltage V_f and conducted current I_{DS}.

$$P_{cond} = V_f(I_{DS}, T_j) \cdot I_{DS} \tag{3}$$

Figure 4 visualizes the implementation of the turn-on energy as a function of blocking voltage V_{DS}, switched current I_{DS} and junction temperature T_j.

The PLECS models for SiC modules include the thermal chain for a Cauer thermal model. This model uses thermal-electrical analogies, represented by a RC network of thermal resistors R_{th} and capacitors C_{th}, to describe a simplified thermal behaviour of the modules.

heat flow Q_{PM}.

$$\dot{\vec{T}}_{PM}(t) = A \cdot \vec{T}_{PM} + \vec{b}(t) \tag{4}$$

with

$$A = -C^{-1} \cdot G \quad \text{and} \quad \vec{b} = C^{-1} \cdot \vec{Q}_{PM}(t) \tag{5}$$

In this simplified approach, the steady-state junction temperatures T_j of the power modules are analyzed. An interface to thermal management design is provided by including a case-to-heatsink resistance $R_{th,c-hs}$ and a thermal capacitance for the heatsink $C_{th,hs}$. Initial assumptions about junction overtemperature detection and accurate heatsink temperatures can be thereby be implemented.

3.3 Failure Rates

A major challenge in qualifying power modules for aircraft application is to achieve the required failures rates while maximizing the power density. Nakashima [20] describes a required failure rate of 100 FIT per module for aircraft power modules. As research discusses higher bus voltages in electrical systems [29], the influence of voltage levels on failure rates should not be ignored. Derating the applied voltage V_{DS} compared to the maximum breakdown voltage V_B can reduce the risk of failure.

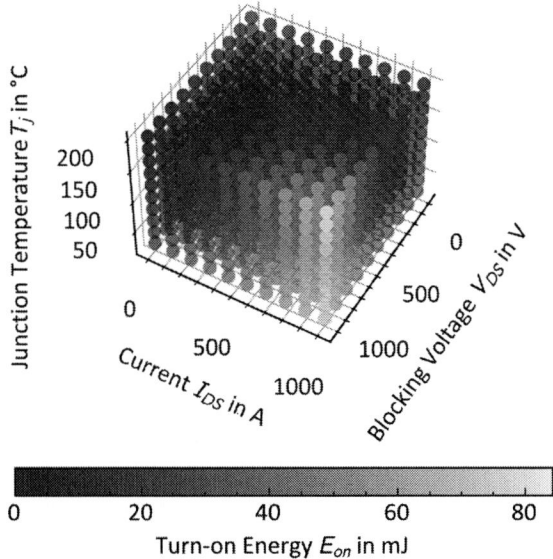

Fig. 4: Loss model visualization for a Wolfspeed CAB760M12HM3 SiC power module

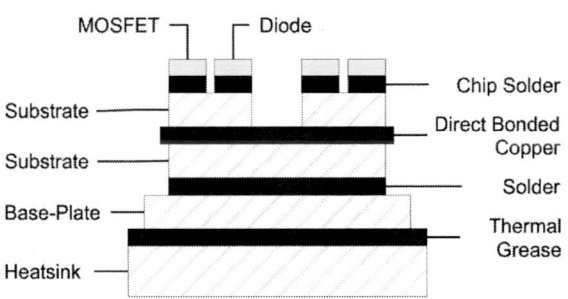

Fig. 5: Power module structure based on [28]

In a Cauer network, the potentials correspond to the physical layers within the component, represented in Fig. 5. The resulting thermal network describes the heat conduction with the thermal conductance matrix G and storage with the thermal capacitance matrix C within the module and allows analysis of temperature distribution T_{PM} and thermal behaviour. Solving the Cauer network according to Eq. (4) in Python allows fast computation to estimate the steady-state temperatures T for a

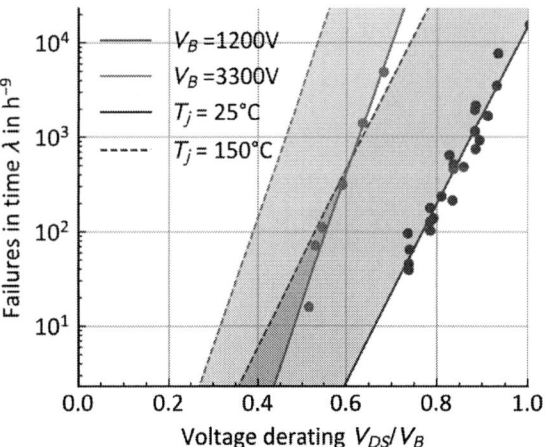

Fig. 6: Implementation of FIT rates based on [25]

The preliminary design in this work aims to balance the requirements for efficiency and power density with reliability. This involves selecting the appropriate voltage level for the power modules based on the DC bus voltage and considering the FIT rate λ of the inverter system, which includes

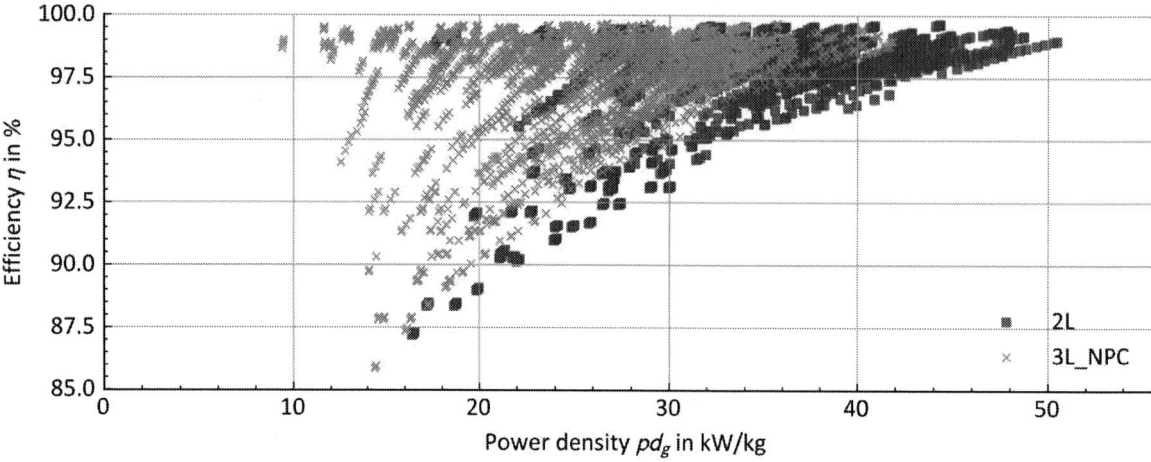

Fig. 7: Efficiency η over gravimetric power density pd_g for all inverter configurations with output power $P_N = 250\,\text{kW}$

multiple power modules and capacitors.

The IEC 61709 standard [8] describes reference conditions for failure rates and stress models for conversion for electronic components and provides a model for FIT calculation. Although, reference values for failure rates are not included. The methodology of the military handbook MIL-HDBK-217F [7] is implemented to evaluate failure rates for capacitors. Equation (6) calculates the capacitor failure rate λ_C based on a capacitance reference value $\lambda_{C,ref}$ and factors of voltage π_V, temperature π_T, environment π_E and power quality π_Q.

$$\lambda_C = \lambda_{C,ref} \cdot \pi_V \cdot \pi_T \cdot \pi_E \cdot \pi_Q \qquad (6)$$

Neither IEC 61709 nor MIL-HDBK-217F provide reference failure rates for state-of-the-art SiC power modules. However, Bolotnikov [25] presents failure rates for SiC power modules as a a function of of maximum blocking voltage and applied voltage. The presented approach models power module failure rates by using these state-of-the-art failure rates as reference values $\lambda_{PM,ref}$ and applying a temperature dependency factor π_T. Equation (8) extends the standarized model by a factor π_{CR} to account for cosmic radiation based on [20].

$$\lambda_{PM} = \lambda_{PM,ref}(V_B, V_{DS}) \cdot \pi_T \cdot \pi_{CR} \qquad (7)$$

Figure 6 illustrates the implemented failure rates with the reference values at sea level given by Bolotnikov in [25]. The curve shows that the module with the higher blocking voltage of 3300V

requires a significantly higher derating in order to achieve the same failure rates as the 1200V module.

With the amount power power modules N_{PM} the amount of capacitors N_C, the total failure rates of the components add up to the evaluated failure rate of the inverter according to Eq. (8).

$$\lambda = N_{PM} \cdot \lambda_{PM} + N_C \cdot \lambda_C \qquad (8)$$

According to Wang [13], the power modules and capacitors account for approximately 50% of the failures in power converters.

4 Results

The calculated inverter configurations are mapped in Figure 7. The exploration of a wide range of input parameters listed in table 1 allows the identification of correlations for the preliminary design of aircraft system studies. In table 2, reference values are defined for the following topology variations.

Tab. 2: Inverter reference topology

Parameter		Reference value
Output power	P_N	250 kW
Switching frequency	f_{sw}	20 kHz
DC-Link voltage	V_{DC}	1200 V
Modulation index	m_{idx}	1
Heatsink temperature	T_{hs}	25 °C
Voltage derating	V_{DS}/V_B	0.5

Figure 8 shows the gravimetric power density pd_g over the switching frequency f_{sw}. The power density peaks between 20 to 30 kHz and decreases at higher switching frequencies. As the frequency increases, the required DC-link capacitance C_{DC} decreases proportionally, but the switching losses also increase according to Eq. (2). At frequencies above 30 kHz, the effects of the increasing losses and the resulting size of the heatsink dominates the power density, which leads to a decrease.

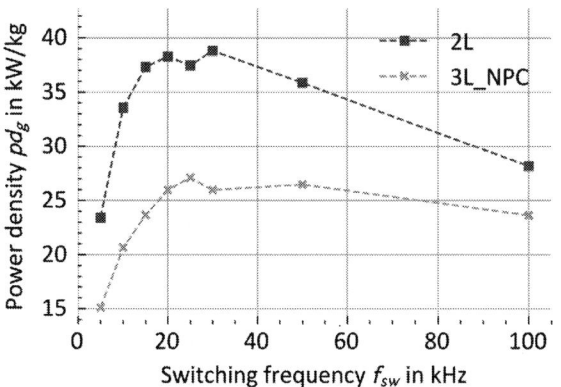

Fig. 8: Gravimetric power density pd_g over switching frequency f_{sw}

Figure 9 presents the influence of an increasing switching frequency on the steady-state junction temperature T_j. With increasing losses due to a higher switching frequency and a constant heatsink temperature, the junction temperature increases proportionally. It is important to consider the limiting design factor of the maximum junction temperature T_j at 175 °C during the preliminary design phase.

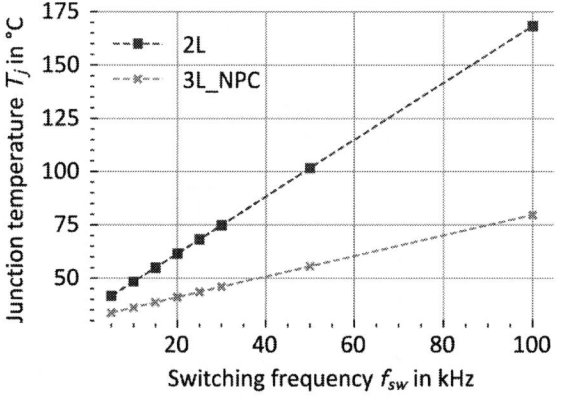

Fig. 9: Steady-state junction temperature T_j over constant heatsink temperature T_{hs}

Figure 10 shows an increase in efficiency up to 1.2 kV, then a drop in efficiency occurs for higher DC-link voltages. However, the current can be reduced by increasing the voltage level, the power loss of a module is a function of the junction temperature, the switched current and the switched blocking voltage, as defined in Eqs. (1) and (2). With higher voltages, the switching losses P_{sw} also increase.

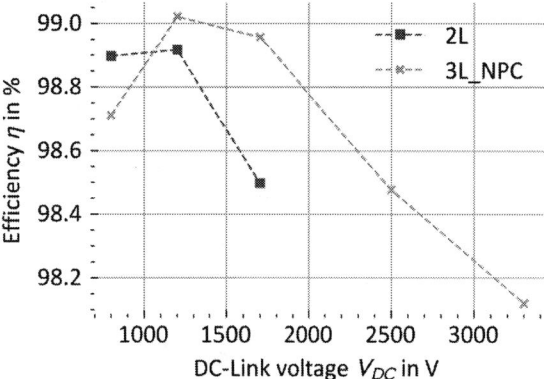

Fig. 10: Efficiency η over DC-link voltage V_{DC}

As indicated by the data in Fig. 6, power modules with higher breakdown voltages necessitate greater derating to maintain equivalent failure rates. Furthermore, Fig. 11 demonstrates the impact of a constant heatsink temperature T_{hs} on the inverter failure rate λ, revealing a notable increase of over 500% with a rise in heatsink temperature from 25 to 75 °C.

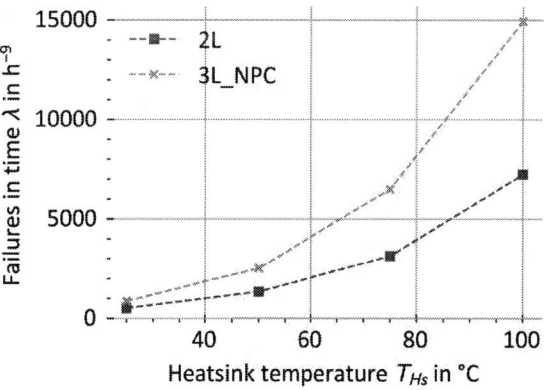

Fig. 11: Failure rate λ over heatsink temperature T_{hs}

To provide a comprehensive insight into the system behaviour, Pearson correlation coefficients are computed for all derived variants and visualized in Fig. 12. Primary influencers on the evaluation criteria, gravimetric power density pd_g, efficiency η, and failure rates λ are identified. The switching frequency f_{sw}, the DC-link voltage V_{DC}, the voltage derating $V_{applied}/V_{rated}$, the modulation index m_{idx}, and the heatsink temperature T_{hs} are varied according to table 1.

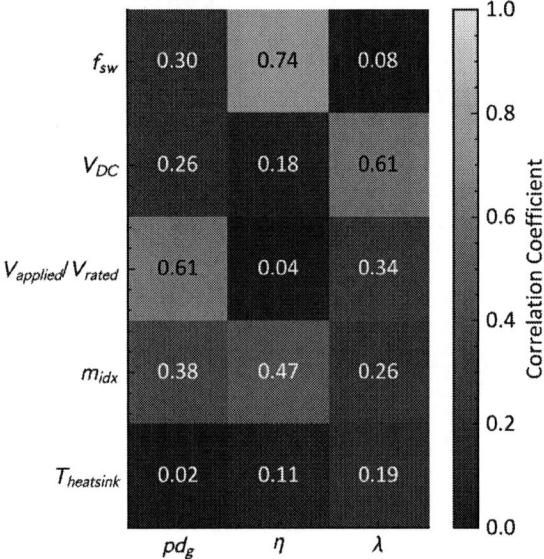

Fig. 12: Pearson correlation coefficients for 250 kW inverter variations

The correlation matrix highlights that voltage derating is the primary factor influencing gravimetric power density. With a decrease in the voltage applied to the components, the applied current increases. As shown in Fig. 3, the DC-link capacitance contributes a significant portion of the total mass of the inverter. The capacitance value C is determined by Eq. (9), which is dependent on the applied current I_C, voltage ripple ΔV_C, and switching frequency f_{sw}. Consequently, as the applied voltage decreases and the current increases, the required capacitance and thus the mass of the system also increase.

$$C = \frac{\sqrt{2} \cdot I_C}{2 \cdot \Delta V_C \cdot f_{sw}} \quad (9)$$

Furthermore, the system efficiency is strongly related to the applied switching frequency. As outlined in Eq. (2), the switching losses increase proportionally with the switching frequency and the total inverter efficiency decreases accordingly.

The DC-link voltage level has the most significant effect on the failure rates. As shown in Fig. 6,

the maximum breakdown voltage V_B and the switched voltage V_{DS} have a substantial impact on overall failure rates. In addition, higher voltage levels result in an increase in capacitance value which, according to [7], is associated with a higher susceptibility to failure.

It is to note that the modulation index m_{idx} heavily affects all three evaluation criteria. By defining the relationship between the input voltage V_{DC} and the output voltage V_{AC} of the inverter, the parameter influences the duty cycles of the modulation pulse, the applied currents and the voltage stresses on the components. In this steady-state open-loop study, the substantial influence of the modulation index highlights the significance of an optimized control strategy. Balancing loads and optimizing switching patterns have a considerable impact on converter design and should be included in ongoing studies.

Figure 13 shows the waveforms obtained from an automated PLECS simulation, displaying the phase voltage V_N and current I_L of the reference design described in table 2. These waveforms can provide an early stage insight into the power quality. For the 2L inverter topology, the total harmonic distortion (THD) of the phase current I_L is 2.2%, while the more sinusoidal wave form of the 3L-NPC topology reveals a THD of 1.0%.

5 Conclusions

The applied sizing approach, which was introduced in an early work publication [2], provides detailed information on each subcomponent. Despite potential inaccuracies in the preliminary sizing of individual subcomponents and the quantified failure rates, the system correlations and tendencies remain reliable and offer comprehensive analysis and interfaces to the preliminary design of the overall drivetrain.

The use of an advanced preliminary design approach that includes thermal network modelling and accurate loss estimation, together with the consideration of failure rates, provides extended insights for the concept studies. However, it is important to recognize the preliminary design stage, as transient switching characteristics and parasitics will have a significant impact on the final design. In addition, the integration of filters

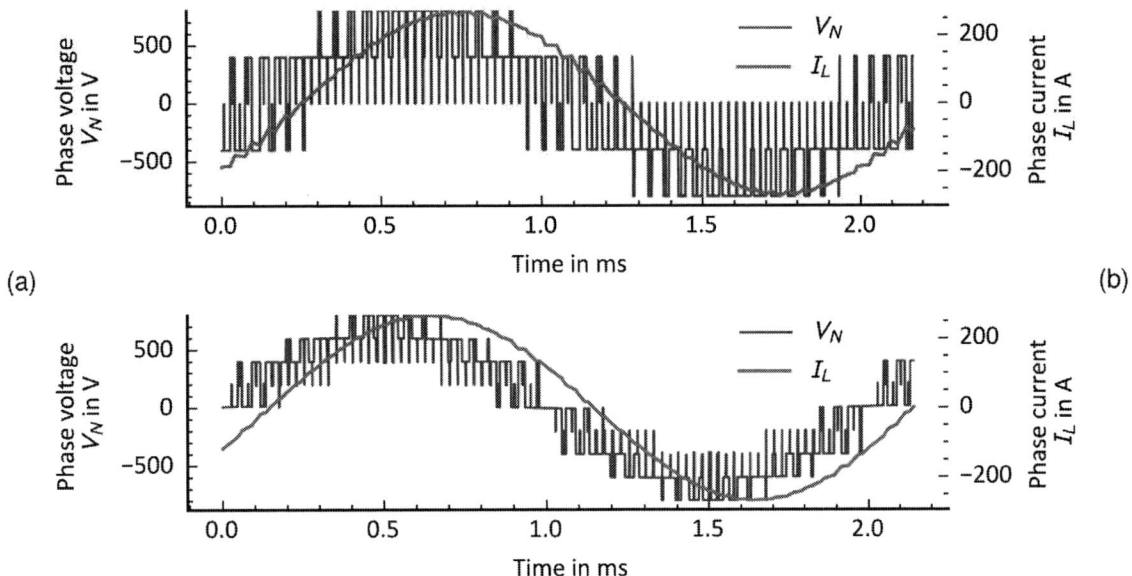

(a)

(b)

Fig. 13: PLECS scope for reference inverter (a) 2L & (b) 3L-NPC topology

and more passive components may have a strong influence and a selection of higher switching frequencies could be beneficial. By considering higher voltage and current ratings with power modules rated at 3300V and 2000A or even higher, the implementation allows the exploration of high power scenarios.

Extending the current steady-state design methods to include mission profiles would provide further insights in inverter performance for different operational scenarios. The PLECS simulations can be extended to include control strategies to provide insights into dynamic behaviour and optimization opportunities. Finally, a more sophisticated machine modelling including voltage stress on machine windings would allow a more accurate representation of the interactions in the inverter-machine system.

The results and outlook for further development demonstrate the potential to significantly improve the insight of the overall performance of electric propulsion systems in early-stage system studies for aircraft applications.

References

[1] J. Hartmann and B. Nagel, "Eliminating climate impact from aviation - a system level approach as applied in the framework of the dlr-internal project exact," *Proceedings of DLRK*, 2021.

[2] J. Kugener, A. Pal, and S. Kazula, "Preliminary design tool for power converters in electric aircraft propulsion systems," *Proceedings of DLRK*, 2023.

[3] C.K. Sain, S. Kazula, L. Enghardt, "Electric propulsion for regional aircraft - critical components and challenges," 2022. DOI: 10.25967/570151.

[4] J. Benzaquen, J. He, and B. Mirafzal, *Toward more electric powertrains in aircraft: Technical challenges and advancements*. 2021, vol. 5. DOI: 10.30941/CESTEMS.2021.00022.

[5] H. Schefer, L. Fauth, T. H. Kopp, R. Mallwitz, J. Friebe, and M. Kurrat, "Discussion on electric power supply systems for all electric aircraft," *IEEE Access*, vol. 8, pp. 84188–84216, 2020. DOI: 10.1109/ACCESS.2020.2991804.

[6] F. Blaabjerg, H. Wang, I. Vernica, B. Liu, and P. Davari, "Reliability of power electronic systems for ev/hev applications," *Proceedings of the IEEE*, vol. 109, no. 6, pp. 1060–1076, 2021. DOI: 10.1109/JPROC.2020.3031041.

[7] U.S. Department of Defense, *Military handbook - reliability prediction of electronic equipment: Mil-hdbk-217f*, 1991.

[8] IEC, *Electric components - reliability models for conversion: Iec 61709: Reference conditions for failure rates and stress*, 2017.

[9] M. Meindl, M. Maerz, F. Hilpert, and C. Bentheimer, "Power electronics design for a 50 pax hybrid-electric regional aircraft," 2023. DOI: 10.2514/6.2023-4545.

[10] Daniel J. Lichtenwalner, Akin Akturk, James McGarrity, Jim Richmond, Thomas Barbieri, et al., "Reliability of sic power devices against cosmic ray neutron single-event burnout," Materials Science Forum, no. 924, pp. 559–562, 2018. DOI: 10.4028/www.scientific.net/MSF.924.559.

[11] H. Wang, M. Liserre, F. Blaabjerg, P. de Place Rimmen, J. B. Jacobsen, et al., "Transitioning to physics-of-failure as a reliability driver in power electronics," IEEE Journal of Emerging and Selected Topics in Power Electronics, vol. 2, no. 1, pp. 97–114, 2014. DOI: 10.1109/JESTPE.2013.2290282.

[12] C. Felgemacher, S. V. Araujo, P. Zacharias, K. Nesemann, and A. Gruber, "Cosmic radiation ruggedness of si and sic power semiconductors," pp. 51–54, 2016. DOI: 10.1109/ISPSD.2016.7520775.

[13] H. Wang, M. Liserre, and F. Blaabjerg, "Toward reliable power electronics: Challenges, design tools, and opportunities," IEEE Industrial Electronics Magazine, vol. 7, no. 2, pp. 17–26, 2013. DOI: 10.1109/MIE.2013.2252958.

[14] RTCA, Environmental conditions and test procedures for airborne equipment: Do-160, 2010.

[15] S. Biser, G. Wortmann, S. Ruppert, M. Filipenko, M. Noe, and M. Boll, "Predesign considerations for the dc link voltage level of the centreline fuselage fan drive unit," Aerospace, vol. 6, no. 12, p. 126, 2019. DOI: 10.3390/aerospace6120126.

[16] J.-K. Mueller, A. Bensmann, B. Bensmann, T. Fischer, T. Kadyk, et al., "Design considerations for the electrical power supply of future civil aircraft with active high-lift systems," 2018. DOI: 10.3390/en11010179.

[17] Anton N. Varyukhin, Pavel S. Suntsov, Mikhail V. Gordin, Viktor S. Zakharchenko, and Daniil Ya. Rakhmankulov, "2019 international conference on electrotechnical complexes and systems (icoecs): Ufa state aviation technical university, ufa, russia, 22 -25 october 2019 : Proceedings," Energies, no. 1, p. 179, 2019. DOI: 10.1109/ICOECS46375.2019.

[18] SAE, Managing higher voltages in aerospace electrical systems: Air 6127, 2023.

[19] IEC, Insulation coordination for equipment within low-voltage supply systems: Iec 60664, 2020.

[20] J. Nakashima, T. Horiguchi, Y. Mukunoki, K. Hatori, R. Tsuda, et al., "Investigation of full sic power modules for more electric aircraft with focus on fit rate and high-frequency switching," IEEE Transactions on Industry Applications, vol. 58, no. 3, pp. 2978–2986, 2022. DOI: 10.1109/TIA.2022.3150624.

[21] Teresa Bertelshofer, Andreas Marz, and Mark.-M. Bakran, "Limits of sic mosfets & parameter deviations for safe parallel operation," 2018.

[22] S. Fukunaga, A. Castellazzi, and T. Funaki, "Development of reliable multi-chip power modules with parallel planar- and trench-gate sic mosfets," pp. 181–184, 2022. DOI: 10.1109/ISPSD49238.2022.9813624.

[23] H. Li, W. Zhou, X. Wang, S. Munk-Nielsen, D. Li, et al., "Influence of paralleling dies and paralleling half-bridges on transient current distribution in multichip power modules," IEEE Transactions on Power Electronics, vol. 33, no. 8, pp. 6483–6487, 2018. DOI: 10.1109/TPEL.2018.2797326.

[24] J. Schnack, V. Golev, J. P. Goerdes, U. Schuemann, R. Mallwitz, and S. Stahl, "Low-inductance dc-link design dedicated to sic-based highly integrated inverters," CIPS 2020; 11th International Conference on Integrated Power Electronics Systems, 2020.

[25] A. Bolotnikov, P. Losee, A. Permuy, G. Dunne, S. Kennerly, et al., "Overview of 1.2kv − 2.2kv sic mosfets targeted for industrial power conversion applications," pp. 2445–2452, 2015. DOI: 10.1109/APEC.2015.7104691.

[26] C.-W. Chang, X. Zhao, R. Phukan, D. Dong, R. Burgos, and A. PLAT, "Weight-minimizing optimization of microchannel cold plate for sic-based power inverters in more-electric aircraft," pp. 1–8, 2022. DOI: 10.1109/ECCE50734.2022.9947530.

[27] C. S. Goli, S. Essakiappan, P. Sahu, M. Manjrekar, and N. Shah, "Review of recent trends in design of traction inverters for electric vehicle applications," pp. 1–6, 2019. DOI: 10.1109/PEDG51384.2021.9494164.

[28] M. Shahjalal, H. Lu, and C. Bailey, "A review of the computer based simulation of electro-thermal design of power electronics devices," pp. 1–6, 2014. DOI: 10.1109/THERMINIC.2014.6972515.

[29] J. Ebersberger, M. Hagedorn, M. Lorenz, and A. Mertens, "Potentials and comparison of inverter topologies for future all-electric aircraft propulsion," IEEE Journal of Emerging and Selected Topics in Power Electronics, vol. 10, no. 5, pp. 5264–5279, 2022. DOI: 10.1109/JESTPE.2022.3164804.

PCIM Europe 2024, 11– 13 June 2024, Nuremberg   DOI: 10.30420/566262468

Addressing Testing Challenges for Power Modules and Three-Level Inverters

Oleg Fotteler[1]

[1] SPEA, Italy

Corresponding author: Oleg Fotteler, oleg.fotteler@spea.com
Speaker: Oleg Fotteler, oleg.fotteler@spea.com

Abstract

The need to improve efficiency and reliability of power modules is leading to the increasing diffusion of three-level IGBT inverters. Compared to the equivalent two-level modules, these products pose some specific challenges in regard to their production testing. In particular, they require to operate with multiple programmable and independent drivers to condition each section of the module under test properly, in order to measure all the working parameters. In addition to that, special attention must be paid to the design of a signal path with minimal stray inductance, in order to minimize voltage overshoots during signal commutation. This paper explores the unique challenges associated with production testing of power modules and three-level inverters and proposes solutions to overcome these hurdles.

Keywords: Power Modules, Three-Level Inverters, Production Testing, Efficiency, Reliability

1 The Power Surge: Driving Demand for Efficient and Compact Power Electronics

The global demand for electricity is experiencing an unprecedented surge, fueled by several key trends. The transportation sector is undergoing a paradigm shift with the mass adoption of electric vehicles (EVs). Renewable energy sources like solar and wind power are playing an increasingly prominent role in the energy mix, necessitating sophisticated power conversion solutions. Furthermore, the relentless march of industrial automation demands reliable and efficient power management across factory floors. These converging forces are creating a critical need for a new generation of power electronics devices – ones that are not only highly efficient but also remarkably compact.

1.1 Electrification Demands High-Performance Power Conversion

The transition to EVs represents a significant transformation in the automotive industry. Unlike traditional internal combustion engine vehicles, EVs rely heavily on power electronics to convert battery power into the precise form required to drive electric motors. These power electronics systems must handle significantly higher power densities while maintaining efficiency and minimizing size. Bulkier, less efficient devices would not only limit driving range but also compromise valuable cargo space within the vehicle.

1.2 Renewable Energy Integration Requires Efficient Power Management

The integration of renewable energy sources like solar and wind power presents unique challenges. These resources produce variable power output, requiring sophisticated power electronics systems to convert and condition the power for seamless integration into the grid. Additionally, maximizing energy capture and minimizing transmission losses hinge on the efficiency of the power electronics employed.

1.3 Industrial Automation Demands Compact and Reliable Power Solutions

The concept of Industry 4.0, characterized by high levels of automation and digitalization, is placing ever-increasing demands on power electronics. Factory robots, automated production lines, and advanced machine tools all require reliable and efficient power conversion to operate effectively. Compactness is also crucial, as space optimization is a key consideration in modern industrial settings. Large, bulky power electronics would hinder efficient design and potentially limit production capacity.

1.4 Efficiency and Compactness: A Synergistic Partnership

The emphasis on both efficiency and compactness in power electronics stems from a need for optimal resource utilization and system performance. High efficiency translates directly to maximizing energy use and minimizing heat generation. For EVs, this translates to longer driving ranges. In renewable energy systems, it translates to capturing more clean energy. Compactness is equally important, especially in space-constrained applications. Smaller power electronics devices allow for better packaging and design flexibility, leading to more streamlined and efficient systems.

1.5 The Increasing Diffusion of Three-Level Inverters

The need for improvements in efficiency and reliability is leading to the increasing diffusion of three-level IGBT inverters: featuring smaller output voltage steps than two-level topologies, they provide a cleaner output waveform, resulting in a more effective switching frequency.

Because the IGBTs are subjected to a lower bus voltage, lower-voltage modules can be used in their design. For this reason, three-level topologies are being widely used in various applications requiring high voltages, including photovoltaics and wind power inverters.

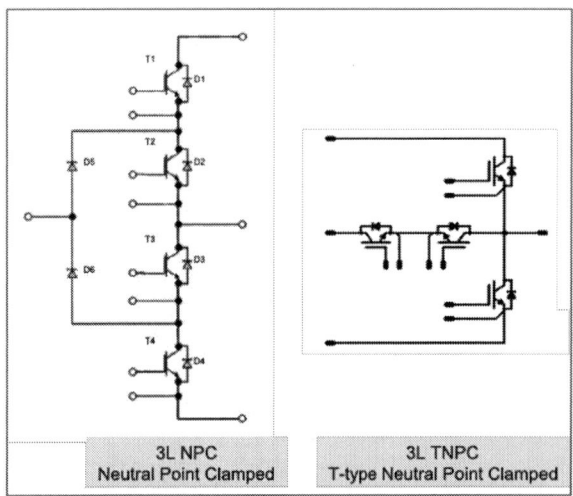

Fig. 1 Two examples of Neutral Point Clamped and T-type Neutral Point Clamped three-level inverters.

2 The Impact on Testing: Power Devices Test Challenges

Power modules, integrating multiple power semiconductor devices with associated circuitry, offer a compact and efficient solution for power conversion. However, ensuring the quality and reliability of these modules during production testing presents complex challenges.

They must be subject to several testing operations, including:

- **Static Parameter Testing**: Traditional methods like measuring DC resistance, leakage current, and voltage breakdown can be time-consuming.

- **Dynamic Performance Evaluation**: Evaluating switching characteristics, including turn-on and turn-off times, switching losses, and short-circuit withstand capability requires specialized equipment and expertise.

- **Isolation Test**: assessing the device dielectric strength, or ability to withstand high voltage without electrical breakdown. This ensures safe operation by verifying the integrity of the insulation between the high-voltage components and the heat sink or chassis.

By performing these tests, manufacturers can ensure the modules meet the required performance and safety standards and have sufficient electrical isolation for their intended application.

In a production environment, some specific challenges emerge in relation to the test processes.

2.1 "Test Early, Test Often": The Golden Principle of Production Testing

In the fast-paced world of electronics manufacturing, quality is paramount. A single faulty component can cripple an entire device, leading to costly rework, delays, and damaged reputations. That's why "Test Early, Test Often" is considered the golden principle of production testing.

This approach emphasizes the need to perform multiple tests along the manufacturing process, not just at the end. By screening parts at various stages - from raw silicon wafers to completed modules - defective components can be identified and removed as early as possible.

Here's why this philosophy is so crucial:

- **Cost Savings**: Catching defects early minimizes wasted resources. Fixing a faulty wafer is far cheaper than discovering a problem in a fully assembled module.

- **Improved Quality**: Early testing allows for adjustments in the manufacturing process, leading to a higher overall yield of functional parts.
- **Faster Time to Market**: Identifying and resolving issues early prevents delays caused by late-stage defect discovery and rework.

The "Test Early, Test Often" strategy involves testing at key points in the production line:

- **Silicon Wafers**: Here, tests ensure the material meets the required specifications for crystal structure and electrical properties.
- **Diced Dice**: Once the wafer is cut into individual die (chips), tests verify functionality and identify any physical defects.
- **DBC** (Direct Bond Copper): This stage involves attaching the die to a ceramic base. Tests ensure proper adhesion and electrical connectivity.
- **Packaged discretes**: The die is encased in a protective package. Tests confirm package integrity and functionality of the assembled unit.
- **Modules**: Finally, the completed modules undergo comprehensive testing to mimic real-world operation and identify any performance issues.

2.2 Specific Testing Challenges for Wide Band-Gap Devices

Silicon Carbide (SiC) and Gallium Nitride (GaN) are revolutionizing power electronics. These wide bandgap materials boast superior performance compared to traditional silicon, offering significant advantages like higher efficiency and operation at elevated temperatures. However, their exceptional properties pose a unique challenge for manufacturers: measuring their minuscule leakage currents.

Unlike silicon devices, which can exhibit leakage currents in the microamp (µA) range, the latest SiC and GaN components often have leakage currents well below 1 picoamp (pA). This incredibly low leakage translates to significant energy savings during operation, but it also presents a hurdle during production testing.

Here's why testing these next-generation devices requires a different approach:

- **Higher Test Voltage Sourcing**: Traditional testers might not be able to provide the high voltages needed to properly assess the leakage current of SiC and GaN devices. These materials can withstand significantly higher voltages compared to silicon, necessitating advanced testers capable of supplying the appropriate voltage levels.
- **Enhanced Current Measurement Sensitivity**: Since leakage currents in SiC and GaN devices are miniscule, testers need exceptional current measurement sensitivity. Standard testers might struggle to detect such low current flows accurately, potentially leading to undetected defects.

To address these challenges, manufacturers are turning to specialized test equipment designed specifically for SiC and GaN devices. These advanced testers incorporate features like:

- **High-Precision Voltage Sourcing**: The ability to generate the necessary high voltages to accurately assess leakage current behavior in these wide bandgap materials.
- **Femtoamp (fA) Current Measurement Capability**: Ultra-sensitive current measurement tools capable of detecting leakage currents in the picoamp (pA) range.
- **Advanced Filtering Techniques**: Advanced filtering algorithms to eliminate noise and ensure accurate measurement of the minuscule leakage currents.

2.3 Need to Minimize the Stray Inductance

During dynamic testing of power semiconductors, minimizing stray inductance along the entire signal path, from the test equipment to the device under test (DUT), is crucial. Stray inductance acts like an electrical "spring," storing and releasing energy during rapid switching events. This can lead to voltage overshoots, exceeding the intended voltage rating of the DUT and potentially causing damage.

When testing wide bandgap (WBG) power semiconductors like SiC and GaN, minimizing stray inductance becomes paramount. This is because WBG devices operate at high switching speeds, and any stray inductance acts like a tiny invisible coil, opposing rapid current changes. This opposition manifests as voltage spikes (overshoots) during dynamic testing, potentially exceeding the device's voltage rating and leading to inaccurate test results or even device damage. By minimizing stray inductance, we ensure clean, accurate test signals and reliable characterization of these next-generation power components.

2.4 Specific Testing Challenges for Three-Level Inverters

Compared to the equivalent two-level modules, these products pose some specific challenges in regard to their production testing. In particular, they require to operate with multiple programmable and independent drivers to condition each section of the module under test properly, in order to measure all the working parameters.

In addition to that, special attention must be paid to the design of a signal path with minimal stray inductance, in order to minimize voltage overshoots during signal commutation.

3 Answering the Testing Challenges

Overcoming the testing challenges presented above requires a strategic approach that leverages industry-standard test equipment specifically designed to handle the demands of power devices.

The key lies in versatility. Production test equipment needs to be adept at handling a variety of power module configurations. For instance, the ability to seamlessly test both three-level and two-level modules is crucial for manufacturers catering to a diverse range of applications.

Here's how industry-standard test equipment tackles the challenges of power device testing:

- **Comprehensive Test Capabilities**: Modern test equipment offers a comprehensive suite of test instrumentation to assess critical parameters of power modules, including ISO, AC and DC Testing.
- **Flexibility for Different Module Configurations**: The equipment should be adaptable to handle both three-level and two-level power module configurations. This caters to a broader range of power electronics applications.
- **Streamlined Test Process**: Standardized testing procedures and automation capabilities ensure a high-throughput testing environment, maximizing production efficiency without compromising quality.

By relying on industry-standard test equipment that addresses these critical factors, manufacturers can achieve a production test environment that is both performant and high-throughput. This translates to faster time-to-market, reduced production costs, and ultimately, the delivery of reliable power electronics that power our ever-evolving world.

3.1 Stray Inductance Control

Although ensuring the lowest value of parasitic inductance along the whole connection chain, from the tester to the device under test (including probe card, and sockets) is essential for testing power semiconductors, it can be a time consuming and costly process. A careful design of tester connection layout, sockets, and contactors, is mandatory to minimize the overall stray inductance, in order to avoid voltage overshoots during signal commutation. Validating the production test setup for these elements is crucial to ensure accurate and reliable testing. Incorporating an accurate software simulation of the contact unit and socket stray inductance into the design process proves to be a valuable tool. By modeling these elements, engineers can predict their inductive behavior and its impact on the test signals. This allows for optimizing the test fixture design to minimize stray inductance. The simulation can be used to virtually test different contact unit and socket configurations, identifying layouts that minimize inductance and ensure the integrity of the test signals reaching the device under test. This proactive approach using simulation translates to a robust and reliable production test setup, ultimately saving time and resources during the physical validation stages.

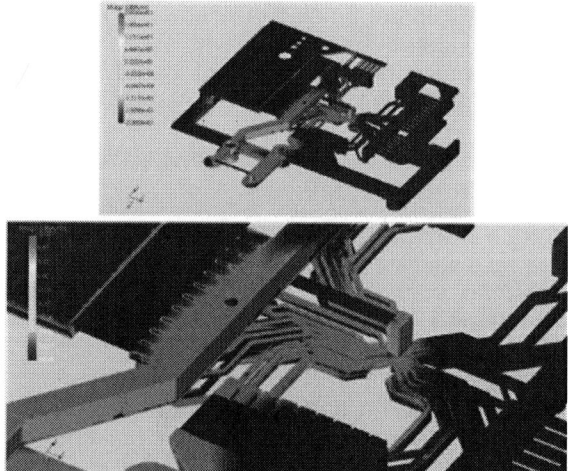

Fig. 2 Example of current density analysis on a software-modeled contact unit for power KGD testing.

3.2 Dynamic Test on Three-Level Inverters

Dynamic tests on three-level inverters require to operate with multiple programmable and independent drivers to condition each section of the module under test and measure its working parameters.

Figures below show how this can be performed on a Neutral Point Clamped module.

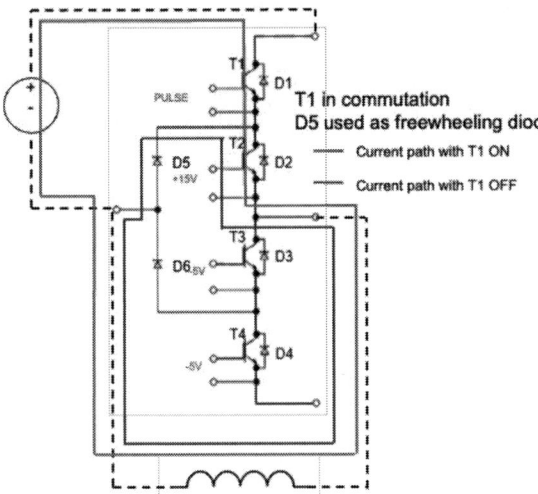

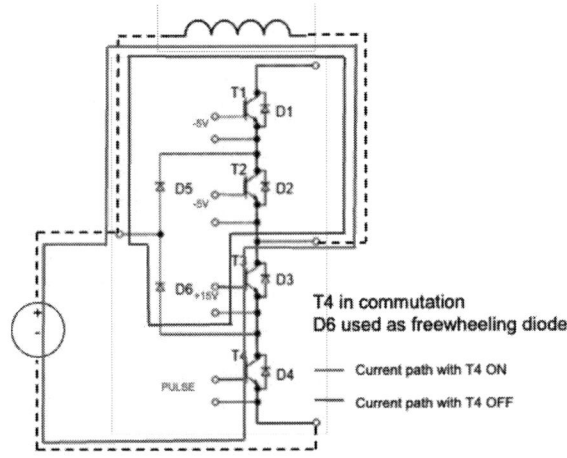

Fig. 3 Example of Dynamic Test scheme on Neutral Point Clamped three-level inverter.

4 Operator's Safety, Devices Protection

The very nature of power semiconductors, capable of handling significant currents and voltages, necessitates a paramount focus on safety throughout the testing process.

A robust suite of protection measures safeguards not only human operators but also the device under test (DUT) and the test equipment itself.

Test equipment shall incorporate a multi-layered approach to safety, including:

- **Overcurrent and Overvoltage Protection Circuits**: These act as a safety net, automatically shutting down the test if current or voltage levels exceed safe limits, protecting the DUT and preventing equipment damage.
- **Voltage Monitors on Output Relays Commands**: Constant monitoring of voltage at critical points, like output relays, ensures accurate command execution and prevents unexpected voltage spikes that could harm the DUT or operator.
- **Open/Short Check**: Verifying proper connection between the test equipment and DUT minimizes the risk of arcing due to open circuits or short circuits, safeguarding personnel and equipment.
- **Contact Needle Protection**: Physical barriers around contact needles prevent accidental contact during live testing, minimizing the risk of operator injury.
- **Embedded Alarms on HV and HI Modules**: Alarms triggered by high voltage (HV) or high current (HI) modules alert operators to potential issues, allowing for immediate intervention and preventing catastrophic failures.
- **Anti-arcing for HV KGD Test**: During Kelvin Guarding (KGD) tests on high voltage devices, specialized anti-arcing measures suppress potential arcing events, safeguarding both the DUT and the test equipment.

By implementing these comprehensive safety features, manufacturers of power semiconductor testing equipment can prioritize the well-being of personnel and users, while ensuring the integrity and accuracy of the testing process. This commitment to safety fosters a secure testing environment and ultimately contributes to the production of reliable power electronics.

5 Conclusions

This paper has explored the crucial role of industry-standard test equipment in addressing the key challenges associated with testing three-level power devices for high-throughput production. We have demonstrated that by leveraging these well-established testing solutions, manufacturers can achieve efficient and reliable characterization throughout the entire manufacturing process, from raw materials to final packaged modules.

Our exploration highlighted several key takeaways:

- **Standard ATE (Automated Test Equipment) proves its versatility**: Contrary to the notion that specialized equipment is necessary, standard ATE offers the capability to handle not only complex three-level device testing but also demanding two-level applications. This adaptability allows manufacturers to streamline their testing infrastructure and leverage existing expertise.

- **Software Modeling as a Powerful Ally**: The paper emphasized the value of software simulation in minimizing stray inductance and preventing voltage overshoots during commutation. By virtually prototyping the test setup, engineers can optimize hardware design and ensure the integrity of the test signals.

- **Safety Remains Paramount**: The final point underscored the unwavering importance of safeguarding operators, devices, and test equipment during the testing process. The integration of comprehensive protection measures within the testing system, encompassing both the tester and the contacting elements, guarantees a secure testing environment.

In conclusion, by embracing industry-standard test equipment and adopting a holistic approach that incorporates software simulation and prioritizes safety, manufacturers can navigate the challenges of power device testing and pave the way for high-performance, reliable production. This commitment to quality testing is fundamental to the advancement of power electronics technology and its diverse applications.

References

[1] Wang, Bo. "Review of Power Semiconductor Device Reliability for Power Converters." CPSS Transactions on Power Electronics and Applications 2, no. 2 (June 2017): 101–17.

[2] Lutz, Josef, Heinrich Schlangenotto, Uwe Scheuermann, and Rik De Doncker. Semiconductor Power Devices. Cham: Springer International Publishing, 2018.

[3] Rodrigues, Eduardo M. G., Radu Godina, and Edris Pouresmaeil. "Industrial Applications of Power Electronics." Electronics 9, no. 9 (September 19, 2020).

[4] Piumatti, Davide. "Reliability in Power Electronics and Power Systems." Doctoral thesis, Politecnico di Torino, 2021.

[5] Sadik, Diane-Perle. "On Reliability of SiC Power Devices in Power Electronics." Doctoral thesis, KTH, Elkraftteknik, 2017.

[6] Anvari-Moghaddam, Amjad. Power Semiconductors for An Energy-Wise Society. Aalborg University, 2023.

[7] Y. Zhang, "Compariosn between competing requirements of GaN and SiC family of power switching devices," IOP Conference Series: Materials Science and Engineering, vol. 738, 2020.

[8] Bo Wang; Jie Cai; Xiong Du; Luowei Zhou. Review of power semiconductor device reliability for power converters. CPSS Transactions on Power Electronics and Applications (Volume: 2, Issue: 2, 2017).

[9] Ruizhe Zhang, Yuhao Zhang. Power device breakdown mechanism and characterization: review and perspective. Japanese Journal of Applied Physics, 2023.

[10] Elvis Zeng; QingYang Zhang; ZhiPing Hu; Richard Whitcomb; Seok Hong; Simon Mei. Power Module Application Test System Setup in 3 Levels ANPC Solar Inverter. PCIM Asia 2019; International Exhibition and Conference for Power Electronics, Intelligent Motion, Renewable Energy and Energy Management, 2019.

[11] Dustert, Christoph, Volke, Andreas. Application of Gate Drivers for 3-Level NPC-2 Power Modules with Reverse Blocking IGBTs. PCIM Asia, 2014.

[12] T-type Advanced 3-level Inverter Module Power Dissipation and comparison tables. Fuji Electric, 2012.

[13] Marc Buschkühle. Highly Efficient 3-Level Solutions for Renewable Energy Applications. Power Electronics Europe, 2010.

PCIM Europe 2024, 11–13 June 2024, Nuremberg   DOI: 10.30420/566262469

Characterization of the Bonding Quality of Silver Sintered Compounds by Means of Laser-Induced Breakdown Spectroscopy

Yannick Bockholt[1], Knud Gripp[1], Aylin Bicakci[1], Ronald Eisele[1]

[1] Fachhochschule Kiel, Germany

Corresponding author: Yannick Bockholt, yannick.bockholt@fh-kiel.de

Abstract

In power electronics, the aim of bonding and joining technology is to continuously optimize the adhesion between the joining partners. The low-temperature sintering method has proven to be one of the most effective methods, especially under the aspect of constantly increasing requirements. In addition to the generation and optimization of the interconnection layers, this also leads to increasing demands on the analysis technology. Therefor the potential of laser-induced breakdown spectroscopy, a new analysis technique for power electronics, is investigated. In addition to the superficial measurement, a multiple measurement at one point is possible via an automatic refocusing, which allows depth profiles to be created. For example, a correlation was found between the extent of the fragmentation margins and the shear strength. The fragmentation margin is formed by the material evaporated for analysis and subsequently condensed, which forms around the crater of the measurement point.

1 Introduction and Motivation

Constantly increasing demands in the field of packaging technology also lead to higher requirements for analysis technology [1]. Currently, continuous process-accompanying quality assurance is essential for ensuring consistently high quality. Direct non-destructive testing of each component is the optimal and safest option. Consequently, only parts that do not meet the requirements would be sorted out and the yield loss is reduced. However, the transfer of individual test specimens to an entire production is still the standard in many areas. Here, non-destructive testing like X-Ray, scanning acoustic microscope and pulsed infrared thermography [2] [3] [4] already represents a potential solution. A low-destructive analysis for characterization of silver sintering compounds will be developed using laser-induced breakdown spectroscopy (LIBS) with automatic refocusing. Shear strength will be used as a reference method. Laser-induced breakdown spectroscopy leads to comparatively little damage to the component (Fig. 1), so that it can continue to be used. In addition, no sample preparation or special measuring environments are necessary and can therefore be used in production environments without any problems.

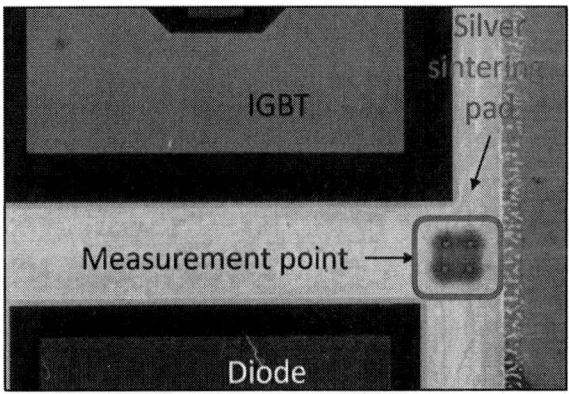

Fig. 1 Example of a LIBS measurement point on silver sinter pad between a diode and IGBT.

2 Measurement Methods

2.1 Shear-Strength Testing of Silver Sintering

Significant temperature differences occur during the operation of a power module. Due to the different CTEs of the materials used, shear stresses occur at the joint layers. In this analysis, the focus is on the silver sinter layer and connection surfaces. The shear strength is therefore used to determine the quality of the silver sintered connection. [1]

For the shear test standardized 2.3 x 2.3 x 1 mm cuboid shear bodies are used. The shear bodies are OF-Copper cores with a silver coating. The shear bodies are silver sintered on an OF-Copper substrate. The shear chisel is positioned parallel to one edge of the shear body at a height of 10%, see in Fig. 2. In the next step, the shearing process proceeds horizontally until the connection breaks. The maximum force determined corresponds to the shear strength.

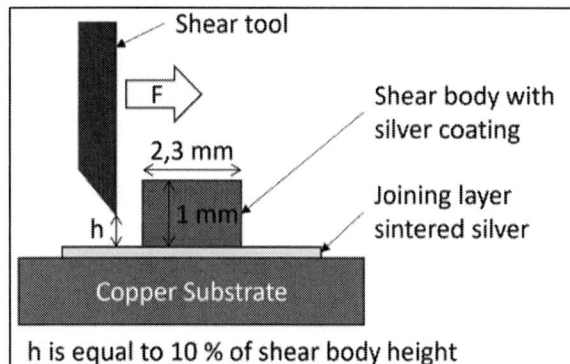

Fig. 2　Schematic sketch of the shear test

Determining the shear strength is an established method for qualifying the bonding quality of the sintering layer. It is therefore the reference method in this subdivision. The shear strength for this examination is tested with the Condor Sigma shear tester from XYZTEC.

2.2 LIBS – Laser-Induced Breakdown Spectroscopy

2.2.1 Measurement Principle

Laser-induced breakdown spectroscopy (LIBS) describes a method in the field of atomic emission spectroscopy. It is a rapid examination method that is carried out under normal ambient conditions without any special sample preparation. It enables the contactless analysis of the elemental compositions of a wide variety of samples using lasers. In general, laser-induced breakdown spectroscopy is a destructive method in which the damage can be classified in the low micrometer range. [5]

The system used is the add-on module EA300 for the microscope VHX-7000 from Keyence. The laser used is a class 1 Nd:YAG laser with a wavelength of 355 nm. The system has an automatic near focusing function, allowing multiple measurements to be taken automatically at one point. This enables depth profile scans as well as measurements on the surface. [6]

To analyze the material composition, material is vaporized with a laser and transferred to the plasma phase, as shown in Fig. 3. The plasma emits characteristic radiation (spectral lines), which is separated according to individual wavelengths in a spectrometer. The intensity of the individual wavelengths is then recorded. The material composition and concentrations can be derived from the spectrum obtained in this way. [7]

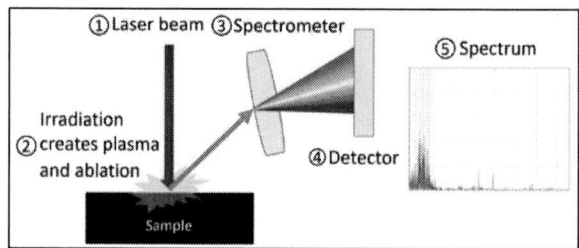

Fig. 3　Schematic sketch of LIBS measurement [8]

When measuring the laser-induced breakdown spectroscopy, another focus is on the deposition caused by the vaporized material. This can be seen in the more detailed sketch of the process in Fig. 4. In addition to the emission of the characteristic radiation, larger fragments that were vaporized by the laser pulse are deposited. The plasma then cools down and the previously vaporized material is deposited. Ultimately, an ablation crater with a fragmentation margin remains. The intensity of the fragmentation margin increases depending by different factors like the number and distance of the measurements or the vaporized material.

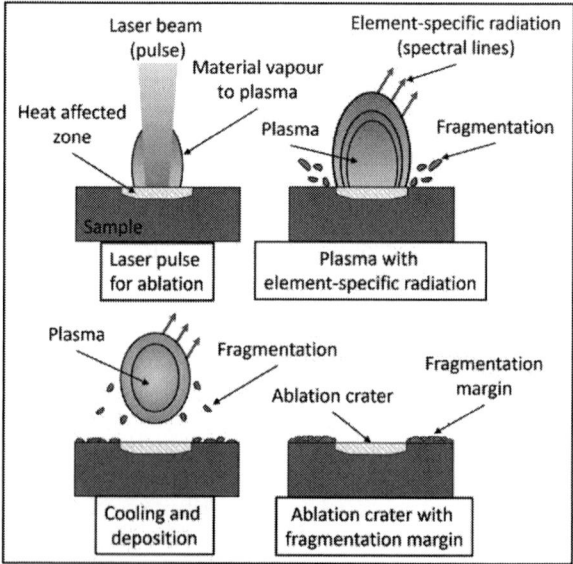

Fig. 4　Sketch of the different phases of the process of the laser-induced breakdown.

2.2.2 Characterization of Bonding Quality

For the characterization of the bonding quality of the silver sintering the focus is on the evaluation and analysis of the fragmentation margin. Particular attention is paid to the intensity of the discoloration of the edge. The analysis evaluates three approaches, which can be seen in Fig. 5.

The first method is the distance ramp. The main goal of this pattern is to determine the impact of the distance between the crates. Consequently, the overlay or interference of the fragmentation margins is analyzed. In this pattern the distance between the crates in one line is 340 µm. The distance between one pair of crates increases from 55 to 350 µm.

The second method is a grid of four crates with 340 µm between. The focus is on the direct comparison of the intensity of the fragmentation margin between the samples.

The third method is a single crate measurement as reference to analyze the fragmentation margin without any interference of other margins.

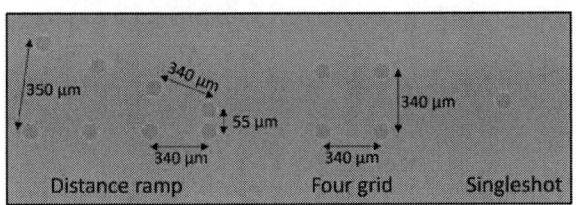

Fig. 5 Patterns of crates for different approaches.

The analysis is carried out for all samples with the same parameters. The laser intensity setting "strong" is used. For the analysis the automatic refocusing is used to make depth profiles. The number of repeat measurements for each measuring point is 15. Laser-induced breakdown spectroscopy (LIBS) analyses are made between the shear craters.

3 Sample Design and Buildup

The measurement of the fragmentation margins and the determination of the shear strength are carried out on the same samples to allow direct comparability. The samples with the layout shown in Fig. 6 are made automatically to reduce irregular variations due to manual work. As substrate a 42 x 25.5 x 1 mm OF-copper plate is used. In the first process step the samples are cleaned with isopropanol and deoxidized with citric acid and water. After that a silver sinter paste is applied as a stencil print using an Ekra X5 paste printer. The sintering pad measures 30.5 x 14.5 mm with a wet paste thickness of 80 µm. The paste is dried according to the manufacturer's instructions. After cooling down, the substrates are assembled with ten silver-coated copper shear bodies measuring 2.3 x 2.3 x 1 mm. Placement is automated using the Tresky T6000 die bonder machine.

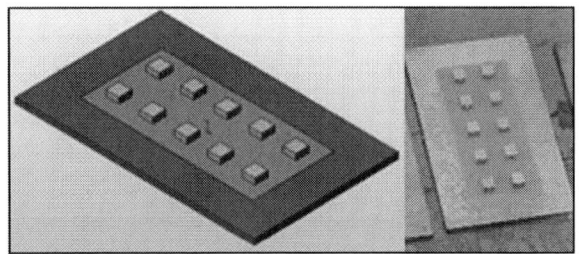

Fig. 6 Samples with shear bodies as CAD model (right) and the build sample (left).

In the next step, suitable sintering parameters are determined from preliminary examination with different parameters of sintering pressure and temperature. The chosen parameters can be taken from the Table 1. The aim of the selected parameters is to achieve significant differences in shear strength. Consequently, clearer differences in the evaluation of the fragmentation margins are also to be expected. Each selected parameter combination is determined in quadruplicate.

Adhesive strength	Low	Low	High	High
Sample Name	L10-220	L25-220	H25-250	H25-280
Pressure [MPa]	10	25	25	25
Temperature [°C]	220	220	250	280

Table 1 Example sintering parameters for comparison.

The sintering process takes place with a 150 °C preheated tool in an open sintering press in a quasi-hydrostatic process with a silicon stamp. The process atmosphere is air while the whole sintering process. For a comparison of oxygen content while sintering process a second closed two-chamber system sintering press is used. In addition, the tool used is inserted into the press at room temperature. It is then heated and compressed so that oxygen is sealed off before the sample has reached around 100 °C. As a result, the influence of oxygen to the fragmentation margins can be discussed.

4 Discussion of Measurement

4.1 Shear-Strength Testing

The boxplot representation of determined values is shown in the graph in Fig. 7. When analyzing the distribution of shear values, significant differences in shear strength can be observed depending on the sintering parameters.

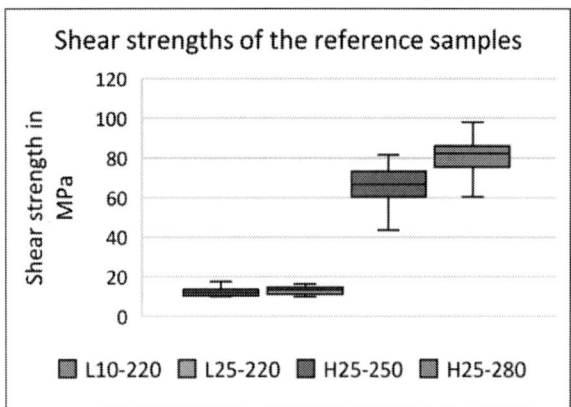

Fig. 7 Shear strength depending on the used sintering parameters.

When analyzing the shear strength, the values determined correspond to the expectations. The sintering temperature recommended by the manufacturer is 250 °C, which means that samples sintered at 220 °C are at least incompletely sintered. In addition, the influence of pressure on the joining strength is known, which explains the difference between the 220 °C samples. The dispersion of sample H25-250 is comparatively large, which should be considered when analyzing the fragmentation margins using LIBS. However, it should be noted that the scatter tends towards higher values. The values of sample H25-280 correspond to the experience that temperature increases improve the diffusion process and thus lead to higher shear strengths.

4.2 LIBS Analysis

4.2.1 Distance Ramp – Influence on Intensity

The analysis of the distance ramp Fig. 8 reveals several influencing factors. Firstly, a clear influence of the process parameters can be seen. Both samples are sintered in an air atmosphere. The upper part (Fig. 8a) shows a sample sintered at 10 MPa and 220 °C. A predominantly brownish coloration can be seen, which appears more intense with decreasing distance. The lower part (Fig. 8b) shows a distance ramp of a sample sintered at 25 MPa at 280 °C. A brown coloration can also be seen here, but this is mainly limited to the edge area and larger distances between the measuring points. The dominant coloration is bluish. The distance between the measuring points also has an impact. The intensity of the discoloration increases with decreasing distance. Even with the upper sample at a small distance, a blue coloration can be seen around the crater. As the distance increases, this can no longer be detected. The lower sample also shows a decrease in the intensity of the (bluish) discoloration as a function of the distance between the measuring points.

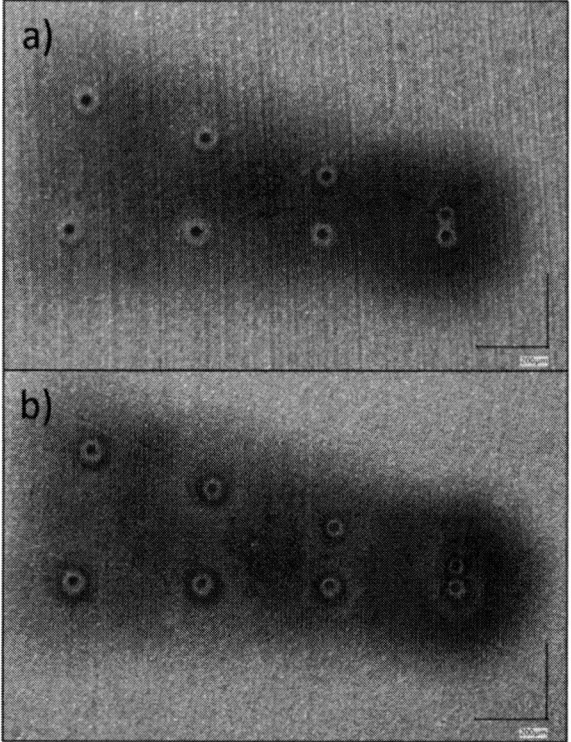

Fig. 8 Overview of the influence of the distance factor on the fragmentation margin (scale is 200 µm). **a)** Sample is sintered with 10 MPa and 220 °C. **b)** Sample is sintered with 25 MPa and 280 °C.

The influence of the distance can be explained by the size of the fragmentation margin itself. With the present layer structure and laser settings, the radius is around 275 µm. In addition, the amount of evaporated material deposited decreases with increasing distance from the center of the crater. Consequently, as the distance decreases, the fragmentation margins become increasingly overlapping. In particular, the areas in which more material is deposited are decisive here. This results in a thicker layer and thus an increasing color intensity.

4.2.2 Four Grid – Intensity Overview

As already shown in the comparison of the factors influencing the coloration of the fragmentation margin, the four-grid pattern also shows a dependence on the sintering parameters seen in Fig. 9. The intensity of the coloration also follows the shear values determined.

Sample L10-220 has the lowest intensity and shear strength in this comparison. In contrast to, both samples with good shear values show clear discoloration. Although sample L25-220 shows a more intense coloration than L10-220, there is a significant difference to the H25 samples.

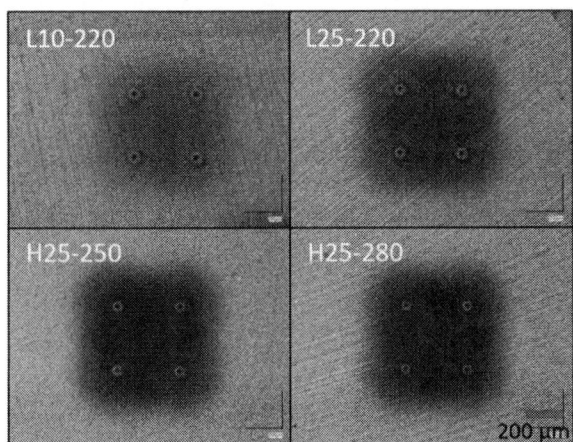

Fig. 9 Overview of four-grid pattern samples.

From the overview in Fig. 9, two factors influencing the shape of the fragmentation edge can be derived. Lower compression results in a larger pore volume, which is also determined in cross-sections. The larger pores consequently lead to a lower density and deviating thermal conductivity of the sintered layer formed. Consequently, less material is evaporated per laser pulse, which results in a lower coloration. In addition to compression, the sintering temperature is also decisive. If the temperature is too low, residues of the organic coating of the silver particles can remain in the paste, which also leads to a deviation in the layer properties. An increase in the sintering temperature generally results in an intensification of the sintering process, as can also be seen in the study.

Furthermore, the four-grid pattern shows the superposition of the fragmentation margins already described for the distance ramp. Although the

distance of 340 µm corresponds more to the end of the distance ramp, a significant coloration can be seen. In particular, the center shows an intense coloration. In addition, the inner margins of the crater have a more intense coloration towards the center, seen in Fig. 10.

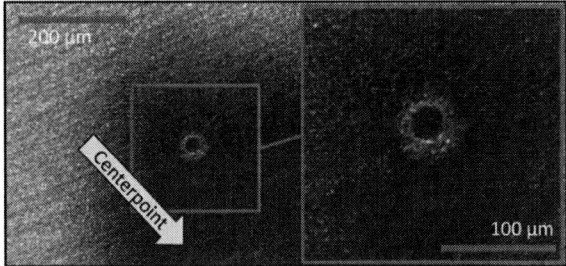

Fig. 10 Detailed view of four-grid of H25-280

4.2.3 Single Shot Reference

As already described, the single measurements serve as a reference for the fragmentation margins. They therefore show the extent of the overlapping of the fragmentation margins, which can be observed when creating different patterns such as the four-grid or the distance ramp. Consequently, it is to be expected that the intensity of the fragmentation margins is lower than in the previous evaluations, as can also be seen in Fig. 11. Nevertheless, the correlation between the intensity of the discoloration of the fragmentation edge and the shear strength is also clear in the individual measurement. As a result of the findings, a process-accompanying investigation using a depth profile with, for example, 15 individual measuring points would be possible.

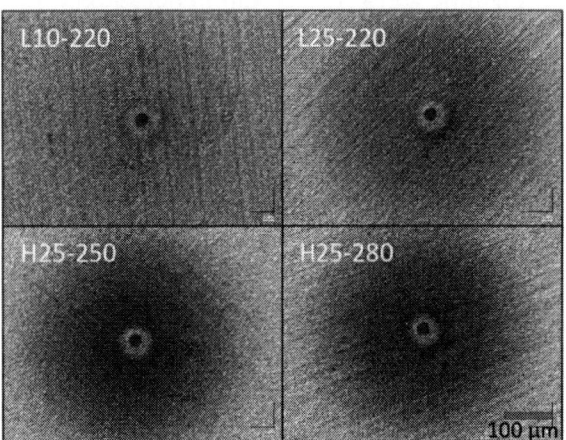

Fig. 11 Overview single shot measurements

4.2.4 Comparison of Oxygen Content

A comparison of the samples from the two sintering presses shows differences in the intensity of coloration. Two specimens sintered at 25 MPa and 280 °C are shown in Fig. 12. The sample on the left shows the sample from the closed chamber, the sample on the right from the open chamber. A direct comparison shows that the intensity of the coloration from the sample out of the closed chamber is less pronounced. Although a blue coloration can also be seen near the crater, the coloration is generally less pronounced.

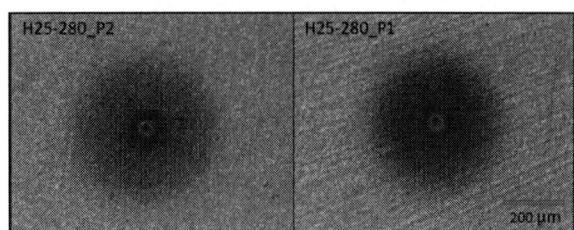

Fig. 12 Comparison of two samples made with 25 MPa and 280 °C in two different presses.

As a result of this finding, it can be deduced that the degree of oxidation during the whole sintering process could have an influence on the resulting fragmentation margins. Further investigations with samples sintered under vacuum and nitrogen atmosphere are necessary for this purpose.

5 Conclusion

The analysis shows that it is possible to draw conclusions about the quality of the sintered layer via the fragmentation margin of a LIBS analysis. It could therefore be possible to detect problems in the series production process at an early stage in the course of in-process quality control. These could be, for example, temperature fluctuations or pressure differences during the sintering process. Due to the existing influencing factors, a data basis for the corresponding process and suitable LIBS analysis parameters would first have to be determined.

In general, the relationship between fragmentation margin and sintering quality should be examined and expanded in further studies. This also includes the transfer to other sinter pastes and presses. In addition, optimization of the LIBS system for this measurement task would also be a further step.

Finally, the approach could be a suitable means for automated quality control during the process.

A test point of less than 1 mm² on the layout would be sufficient for the evaluation.

References

[1] Martin Becker, „Neue Technologien für hochzuverlässige Aufbau und Verbindungstechniken leistungselektronischer Bau-teile," Dissertation, University of Technology Chemnitz, 2015. Online available: http://nbn-resolving.de/urn:nbn:de:bsz:ch1-qucosa-183012 (27.04.2024)

[2] S. Brand et al., "Non-Destructive Assessment of Reliability and Quality related Properties of Power Electronic Devices for the In-Line Application of Scanning Acoustic Microscopy," CIPS 2016; 9th International Conference on Integrated Power Electronics Systems, Nuremberg, Germany, 2016, pp. 1-6.

[3] J. Rudzki, L. Jensen, M. Poech, L. Schmidt and F. Osterwald, "Quality evaluation for silver sintering layers in power electronic modules," 2012 7th International Conference on Integrated Power Electronics Systems (CIPS), Nuremberg, Germany, 2012, pp. 1-6.

[4] D. R. Wargulski et al., "Paving the way for the replacement of solder interconnections in power electronics by silver-sinter using pulsed infrared thermography," 2019 22nd European Microelectronics and Packaging Conference & Exhibition (EMPC), Pisa, Italy, 2019, pp. 1-8, doi: 10.23919/EMPC44848.2019.8951861.

[5] Prof. Dr. Georg Ankerhold, Prof. Dr. Peter Kohns, Christian Beresko, „Berührungsfreie Elementanalyse mit Lasern: Ein laserspektroskopisches Verfahren mit einem breiten Anwendungsspektrum," GIT Labor-Fachzeitschrift, p. 654–656.

[6] Keyence Deutschland GmbH. „Datenblatt EA-300: Laserbasierte Materialanalyse-Einheit." https://www.keyence.de/products/microscope/elemental-analyzer/ea-300/models/ea-300/ (27.03.2024)

[7] Cassian Gottlieb, „Einfluss der Korngröße auf die quantitative Elementanalyse heterogener, mineralischer Werkstoffe mittels der laserinduzierten Plasmaspektrokopie," Dissertation, Clausthal University of Technology, 2019, doi:10.21268/20190606-2

[8] Temmo-Frithjof Glaw, email communication, 11.01.2024

PCIM Europe 2024, 11– 13 June 2024, Nuremberg   DOI: 10.30420/566262470

Inverter-Integrated Measurement of the Frequency-Dependent Winding Impedance of Electric Machines

Christian Mühlfeld[1], Jens Onno Krah[1]

[1] TH Köln, Cologne, Germany

Corresponding author: Christian Mühlfeld, christian.muehlfeld@th-koeln.de
Speaker: Christian Mühlfeld

Abstract

This paper deals with resonance effects in windings of electrical machines, which can be excited by harmonic components of voltage source inverters. With increasing switching frequencies, the topic becomes more and more relevant. For standard automation drives, a method is presented that allows the frequency-dependent impedance to be measured with the inverter-integrated $\Sigma\Delta$-based current measurement. The frequency-dependent impedance can be measured up to about 200 kHz, even if the PWM of the inverter switches the power transistors at a lower frequency. No additional measurement equipment is required.

1 Introduction

Electric motors are ubiquitous in industrial applications ranging from logistics and manufacturing processes to robotic manipulators. The control of these motors is essential to achieve desired performance metrics such as speed, torque, and position accuracy, as well as good efficiency. Automation drives serve as the interface between the control system and the electric motor, enabling sophisticated control strategies to be effectively implemented.

The efficiency and performance of automation drives are highly influenced by the switching frequency and the resulting pulse-width modulation (PWM) used to operate them. Switching patterns determine how the DC link voltage is converted to the AC output voltage, affecting crucial aspects such as waveform quality, harmonic content, and electromagnetic compatibility (EMC).

A low switching frequency results in an imperfect current sine wave, which results in higher power losses in the motor compared to a perfect sine wave. Increasing the switching frequency would result in an output current that is closer to the perfect sine wave, which reduces losses in the motor but increases power electronic losses in the inverter due to higher switching losses of the semiconductors in the power stage.

With the availability and economic viability of wide bandgap devices such as silicon carbide (SiC) or gallium nitride (GaN) are increasingly used in the power stages of servo drives. Wide bandgap devices have higher voltage slew rates compared to their silicon counterparts, resulting in lower switching losses. In addition, this new feature enables higher switching frequencies. Because the technology is not widely used in the industry, there is little field experience that considers the interaction of such converters with their modulation scheme, the cable, and the winding of the electrical machine.

Each motor and cable manufacturer differs in the layout of the product and uses different shapes and materials for conductors and insulation. Therefore, each product and combination of products (e.g. motor with cable) has different characteristics. Great attention should be paid to the frequency-dependent impedance $|Z(f)|$ and any resonant frequencies. This impedance is composed of the desired inductance of the motor and the parasitic impedance (conductor resistance and inductance, strand-strand insulation capacitance, strand-case insulation capacitance, ...).

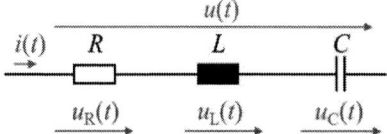

Fig. 1 Equivalent circuit for a series resonance

PCIM Europe 2024, 11– 13 June 2024, Nuremberg   DOI: 10.30420/566262470

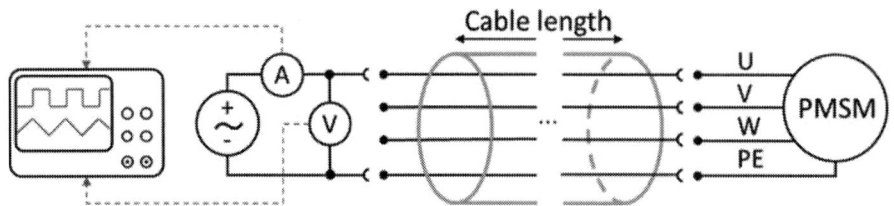

Fig. 2 Laboratory test setup to determine manually the frequency-dependent current of a PMSM-cable combination

When series resonance frequencies are reached by the inverter switching frequency or its harmonics, high voltage stress may be applied to the motor coil winding, possibly resulting in partial discharge or even winding failure [1-4]. Figure 1 shows an LCR series resonant circuit as an equivalent circuit for such resonances.

Such a potentially damaging voltage cannot be measured at the motor terminal. A workaround is to measure the frequency-dependent impedance $|Z(f)|$ and determine the series resonances for an estimation of the winding voltage. The frequency-dependent impedance $|Z(f)|$ of a motor winding can be measured for example with a network analyzer or manually with a signal generator in combination with a current probe. However, an individual measurement in the field, with its customized motor cable types and lengths, is time-consuming and costly.

The measurement suggested here determines any application-specific resonant frequencies of the motor-cable combination during commissioning. Once identified, the application-specific critical switching frequency ranges can be prohibited, for example.

2 Measurement of a sample motor with cable

Figure 2 shows the schematics of the laboratory test setup for determining the frequency-dependent current of an industrial 0.84 kW permanent magnet synchronous motor (PMSM) from a well-known manufacturer with a rated voltage of 480 V AC, a rated torque of 1 Nm, a rated speed of 8000 rpm, a phase-to-phase resistance of 8.5 Ω at 20°C, a phase-to-phase inductance of 20.8 mH at 1 kHz, and a rated continuous current of 2.2 A. The PMSM and the voltage source are connected with a 15 m shielded cable from the motor manufacturer. The cable has four 1 mm² cross-section wires for the three phases and PE, and five

0.38 mm² signal wires, which are not used in the test setup. A signal generator with 30 V_{PP} is connected to one phase (U) and PE (shield and motor housing). The resulting current is measured using a Tektronix TCP312 current probe with the amplifier TCPA300 in conjunction with a Rigol DS1054Z oscilloscope. The current signal is filtered using the oscilloscope's digital bandwidth limiting function with a cutoff frequency of 200 kHz. The resulting current magnitude due to a sinusoidal excitation voltage and a square-wave excitation voltage is plotted for excitation frequencies from 1 kHz to 40 kHz. The results are shown in Fig. 3.

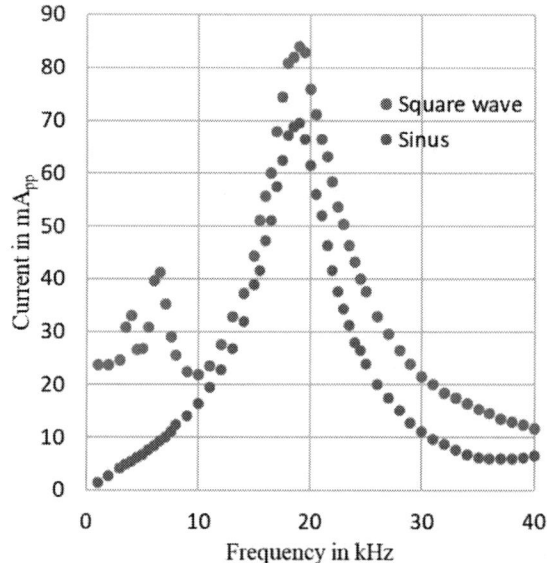

Fig. 3 Current-frequency curve of a 0.84 kW industrial servo motor connected with 15 m shielded cable and sinusoidal vs. square-wave excitation voltage

Both currents reach the global maximum at 19.24 kHz. The square-wave voltage produces two additional local maxima at 6.41 kHz and 3.85 kHz. Here, the 3rd and 5th harmonics of the square-wave voltage also excite the resonant frequency.

At 19.24 kHz the current due to the square-wave voltage is significantly higher than the current caused by the sinusoidal voltage. This is because the amplitude of the Fourier-decomposed fundamental frequency of the square-wave signal is $\frac{4}{\pi}$ times greater than that of a sinusoidal signal with the same peak-to-peak amplitude [5].

Only the sinusoidal signal component is used to calculate the frequency-dependent impedance using complex alternating current (AC) arithmetic.

This setup was used to measure the current amplitudes from 4 kHz to 100 kHz. The values measured are divided by the corresponding voltage amplitudes to obtain the frequency-dependent impedance $|Z(f)|$ shown in Fig. 4.

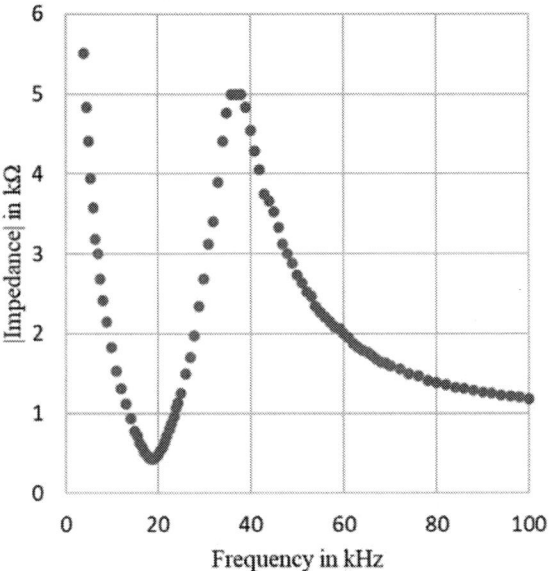

Fig. 4 Measured impedance curve $|Z(f)|$ of an industrial 0.84 kW servomotor with 15 m shielded cable

As described in [1], impedance minima due to resonances are critical. In the range of the minimum impedance around 20 kHz, the curve can be approximated by an equivalent LCR circuit in a series connection:

$$Z(f) = R + j(X_L - X_C)$$
$$= R + j\left(2\pi f L - \frac{1}{2\pi f C}\right) \quad (1)$$

The resonance frequency of an equivalent LCR circuit in a series connection is at the current maximum resp. at the impedance minimum at 19.24 kHz with a value of $R = 355\ \Omega$, which corresponds to the resistance of the equivalent LCR circuit. At lower frequencies, the reactance of the capacitor is dominant. At $f_C = 7$ kHz, the impedance is $|Z(f_C)| = 4$ kΩ. This results in an equivalent capacitance of $C = 5.7$ nF. At higher frequencies, the

reactance of the inductor tends to dominate. At $f_L = 30$ kHz the impedance is $|Z(f_L)| = 2.5$ kΩ. This results in an equivalent inductance of $L = 13$ mH.

With the parameters of the equivalent LCR circuit, the series resonance is 18 kHz, which is 5 % less than the measured minimum impedance. This can be caused by, among other things, a superposition of the parallel resonance.

When a sinusoidal voltage is applied at the resonant frequency, the current can be calculated using Ohm's law. According to the Fourier analysis, a sinusoidal voltage with resonance frequency can originate when for example a square-wave voltage with a frequency of $f_{PWM} = 19.24$ kHz is applied. With a DC link voltage of $U_{PP} = 600$ V, the fundamental Fourier oscillation (harmonic $n = 1$) would result in:

$$U_n = U\,\frac{4}{\pi}\frac{1}{n} = \frac{600\ \text{V}}{2}\,\frac{4}{\pi}\frac{1}{1} = 382\ \text{V} \quad (2)$$

$$I_n = \frac{U_n}{Z(f_n)} = \frac{382\ \text{V}}{355\ \Omega} = 1.08\ \text{A} \quad (3)$$

At the inductance (part of the motor winding) this would result in:

$$U_{n,L} = X_L(f_n)\,I_n = 1.57\ \text{k}\Omega \cdot 1.08\ \text{A}$$
$$= 1\,696\ \text{V} \quad (4)$$

As mentioned above this voltage is caused by the fundamental. To determine the maximum voltage, the voltages of all the harmonics must be superimposed on the fundamental voltage.

The resulting voltage, which can be several times higher than the excitation voltage, causes the high-voltage stress mentioned above. This resulting voltage is not measurable at the motor terminals by regular methods.

3 Fourier analysis for spectral analyzing

Typical impedance measurement methods apply a sinusoidal voltage of a given voltage and frequency to the device under test (DUT). The responding current (magnitude and phase shift) is used to calculate the impedance at the given frequency. By repeating this process for the desired frequency range, the frequency-dependent impedance curve can be determined incrementally. Conventional automation drives use B6 bridges to feed three square-wave voltages to the motor. Due to the low-pass behavior of the coil and the resistance of the motor, the current results in a

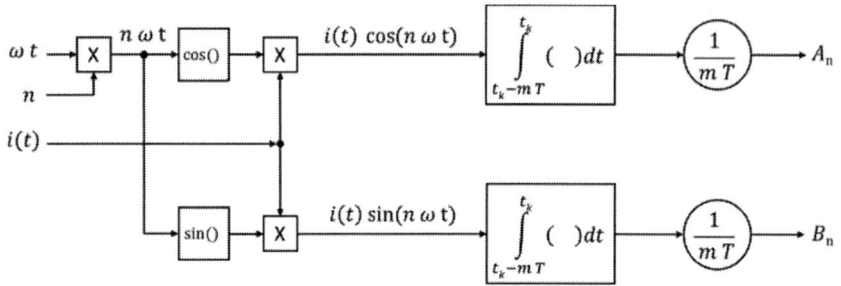

Fig. 5 Signal flow diagram of a motor phase current Fourier analysis

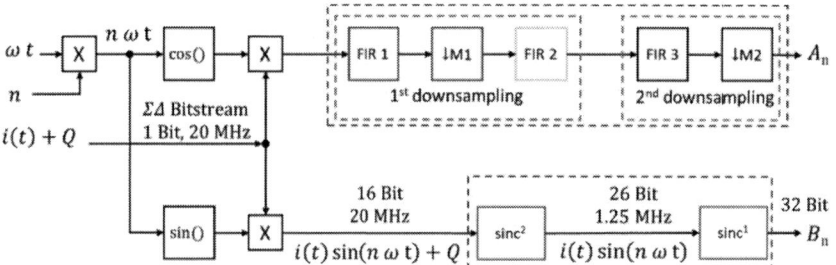

Fig. 6 Signal flow diagram of an FPGA implementation for motor phase current Fourier analysis [6]

formulas are obtained when m = 1. From a mathematical point of view, the extended representation does not make sense, because nothing changes with $m > 1$, but if $f(t)$ is a measured noisy signal, more accurate coefficients are obtained with $m > 1$.

Figure 5 shows a signal flow diagram of a motor phase current Fourier analysis, which determines the Fourier coefficients A_n and B_n for a given harmonic n. The integration is performed over the timespan $m\,T$, where T is the PWM period. The signal flow shown corresponds to equations (7) and (8).

Since the voltage is the excitation, the signal and its Fourier coefficients are known, so measuring the voltage is optional. To determine the coefficients of the resulting current, a high-bandwidth current measurement is required.

A PWM has the standard geometrical shape of a square-wave signal. If the PWM is without DC offset (placed symmetrically to the time axis) the coefficient $A_0 = 0$. This results in equation (9). [5]

Once the Fourier coefficients are determined, the amplitude of each signal harmonic can be calculated. Finally, the impedance for each harmonic can be calculated using equations (9-11). With this method, it is possible to measure a curve of a frequency-dependent impedance by repeating the algorithm for a range of frequencies of interest or to calculate the impedance only for a desired switching frequency and its harmonics.

triangular waveform. The square-wave voltage and the current response cannot be used directly to calculate the frequency-dependent impedance.

Both periodic signal waveforms, voltage and current, can be decomposed into their fundamental and harmonic components using Fourier analysis. Their waveforms can be formed by Fourier series according to the equation (5).

$$\hat{f}(t) = \frac{A_0}{2} + \sum_{n=1}^{\infty}[A_n \cos(n\,\omega\,t) + B_n \sin(n\,\omega\,t)] \tag{5}$$

$$\omega = \frac{2\pi}{T} = 2\pi f_{PWM} \tag{6}$$

$$A_n = \frac{1}{m\,T}\int_0^{m\,T} f(t)\,\cos(n\,\omega\,t)\,dt \tag{7}$$

$$B_n = \frac{1}{m\,T}\int_0^{m\,T} f(t)\,\sin(n\,\omega\,t)\,dt \tag{8}$$

Where $\hat{f}(t)$ is the resulting Fourier sequence, m is the number of periods over which the signal is integrated, T is the PWM period, n is the number of the harmonic of interest and A_n and B_n are the Fourier coefficients. Here the formulas for calculating the coefficients A_n and B_n do not integrate over just one PWM period as usual, but over m periods, where m is an integer number. The familiar

$$U_n = \sqrt{A_{n,U}{}^2 + B_{n,U}{}^2} = \frac{4\,\hat{U}}{\pi\,n} \tag{9}$$

$$I_n = \sqrt{A_{n,I}{}^2 + B_{n,I}{}^2} \tag{10}$$

$$Z(n\,f_{PWM}) = \frac{U(n\,f_{PWM})}{I(n\,f_{PWM})} = \frac{U_n}{I_n} \tag{11}$$

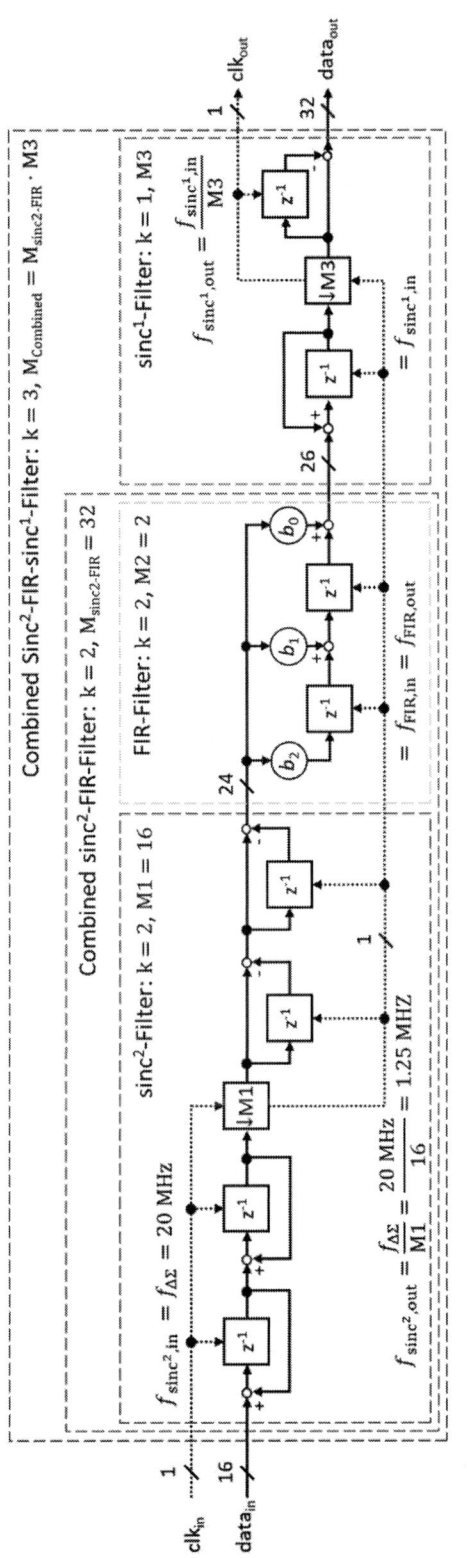

Fig. 7 Detailed signal flow diagram of the designed filter system [6]

4 FPGA-based implementation of the Fourier analysis

An Field Programmable Gate Array (FPGA) implementation with digital signal processing is shown in Fig. 6. The common phase current measurement for motor control is not shown in detail here, but it is based on $20\,\mathrm{MHz}$ bit streams ($i(t) + Q$) generated by $\Sigma\Delta$ modulators. These bit streams can be used for the Fourier analysis. For this purpose, each bit of the bit streams is multiplied by $\cos(n\,\omega\,t)$ and $\sin(n\,\omega\,t)$, both of which are represented by 16-bit fixed-point data words. This results in 16-bit fixed-point signal streams with significant high-frequency quantization noise from the $\Sigma\Delta$ modulation. This multiplication is intentionally performed before any filtering to avoid 16-bit times 16-bit multiplications to conserve digital signal processing (DSP) resources.

A 1st filter block consisting of a 1st FIR filter (dark red), a downsampler with M1 = 16, and a 2nd FIR filter (orange). This combined filter (red) behaves like a sinc² filter with twice the update rate. A classic sinc¹ filter with its signal-averaging behavior approximates the integration. Therefore, a regular sinc¹ is used in the 2nd filter block (blue) to calculate the Fourier coefficients A_n and B_n. [6]

A detailed overview of the filter system is shown in Fig. 7. The data signal flow and the current bit width are represented by the black lines. The clock signal is represented by the dashed black line.

The data enters the filter system with a width of 16 bits and an $\Sigma\Delta$ update frequency of $20\,\mathrm{MHz}$ and leaves the first filter with a data width of 24 bits and an update frequency of $1.25\,\mathrm{MHz}$. The second filter implemented as an FIR filter with the coefficients $b_0 = 1$, $b_1 = 2$, and $b_2 = 1$ produces a 26-bit wide data stream at the same update frequency.

The sinc¹ filter has a 32-bit wide output. The frequency depends on the downsampler M3, which depends on several factors:

$$M3 = \frac{f_{\Sigma\Delta}}{M1}\,m\,T_{\mathrm{PWM}} \qquad (12)$$

With $f_{\Sigma\Delta}$ as the $\Sigma\Delta$ modulation frequency, m is the number of periods to be integrated, and T_{PWM} is the selected PWM period of interest for the measurement. M3 is the number of samples taken to calculate the Fourier coefficients.

The measurement of the current amplitude at the desired frequency and the calculation of the frequency-dependent impedance curve is controlled by a microcontroller (µC). The µC configures the PWM period T_{PWM} and the harmonic order n successively for all f_n as the frequencies of interest, and the FPGA-based measurement of the corresponding Fourier coefficients is started.

$$f_n = \frac{n}{T_{PWM}} \qquad (13)$$

When the measurement is finished, the frequency-dependent impedance $|Z(f_n)|$ can be calculated by the µC.

5 Experimental Validation

To validate the presented concept, a servo drive with an Intel®/Altera® MAX® 10 FPGA with 25 000 logic cells (LC) is used for control. The softcore processor (Nios® II) from Intel/Altera is used as the microcontroller. The compiled configuration uses 6219 LCs and 8 DSP units. The V23990-K219-F40-PM module is used for the power stage. Texas Instruments AMC 1304M25 2nd order $\Sigma\Delta$ modulators are used for analog-to-digital conversion of the motor phase currents. A $100\ m\Omega$ CRA2512 R100E resistor is used as the current sense shunt. The voltage source is a $36.5\ V$ laboratory power supply connected to the DC link.

For validation, the measurement results (impedance and the corresponding frequency) are printed to the softcore processor's integrated development environment (IDE) console. Figure 8 shows the test setup for automation drive integrated measurements. The same motor-cable combination from Chapter 2 is used as the DUT.

The following parameters are used for the servo inverter integrated measurement: f_{start} = 4 kHz, f_{end} = 15 kHz, f_{step} = 200 Hz, m = 16. The fundamental, 3rd, 5th, and 7th-order harmonics are measured.

For comparison the measurements from chapter 2, only one phase (U) is used and the EMC capacitor C_{EMC} (which is usually more than $100\ nF$) is bypassed, therefore PE is connected directly to U_{DC}.

Distinct measurement points are taken from the console and compared to the results obtained with the signal generator and the current probe from chapter 2. Both data are plotted on the same graph and shown in Fig 9.

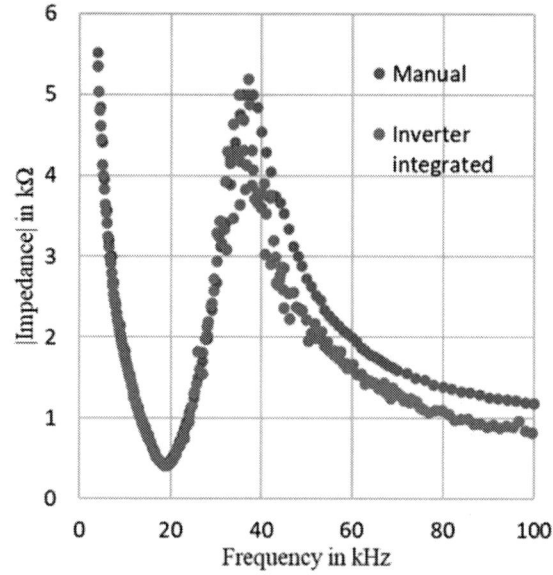

Fig. 9 Comparison of $|Z(f)|$ - blue curve: manual measured, red curve: inverter integrated measurement

The curves fit with minor differences for frequencies up to 30 kHz. Differences occur near the anti-resonance frequency in the range between 30 kHz and 50 kHz. At frequencies above 50 kHz, the inverter-integrated measurement is about $400\ \Omega$ lower than the manually measured curve.

Reasons for these differences could be:

1. At higher impedances near the anti-resonance frequency, there is less current, and therefore the current signal is noisier. This problem is reduced in real operation because inverters are usually operated with about $600\ V$ DC link voltage. Therefore, the current amplitude will be higher than in the laboratory test setup.

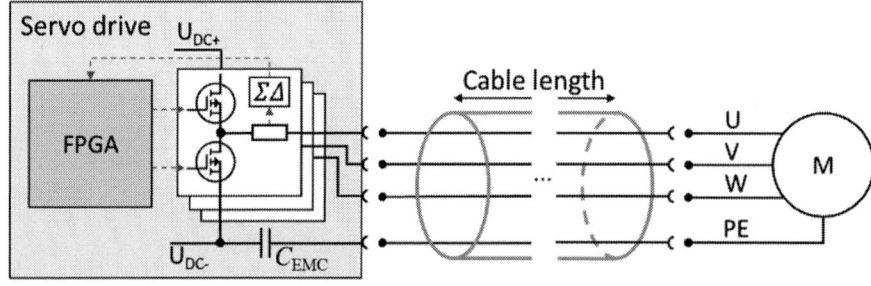

Fig. 8 Servo drive integrated measurement method to determine the frequency-dependent current of a PMSM-cable combination

2. The voltage generated by the PWM is due to limited voltage slew rates etc., not an ideal square-wave signal used to calculate the Fourier coefficients of the voltage. This has a non-negligible effect at higher frequencies, where the blocking and the slew times of the power stage take up a larger percentage of the period than at lower frequencies.

6 Additional use

This method allows the impedance of each phase to be measured online during operation. Impedance changes due to winding shortcuts can be detected during operation to implement a low-cost condition monitoring of the system.

Similar to [7], where the authors take a regular resolver for functional safe motion control applications, the current sensors can be monitored for diagnostics, and therefore only two current sensors are necessary for applications with safety-related current acquisition.

According to Kirchhoff, the sum of all three motor phase currents is zero. The absolute value of his sum is usually monitored to be less than a specified tolerance ε. With this common approach, three current sensors are needed for safety-related current acquisition.

7 Conclusion

It is possible to use off-the-shelf automation drive hardware to measure the frequency-dependent impedance $|Z(f)|$ of the motor-cable combination. The additional FPGA resources used are relatively small compared to a typical FPGA used in servo inverters. With a small additional step during commissioning, possible motor damage and the resulting downtime during the life cycle can be avoided.

In addition, this method can potentially replace some hardware with algorithms: It may be possible to replace one of the three phase current transducers in functionally safe servo drives with software.

The test setup used was not perfect and can be improved in several ways. However, in the critical resonance frequency range, where the fundamental oscillation can cause the greatest damage, the measurement error is negligible.

8 References:

[1] H. Bärnklau, M. Harnisch and J. Proske, "On Medium Frequency Differential Mode Resonance Effects in Electric Machines," in *IEEE Open Journal of Industry Applications*, vol. 2, pp. 310-319, 2021, doi: 10.1109/OJIA.2021.3121319.

[2] S. Sundeep, J. Wang, A. Griffo and F. Alvarez-Gonzalez, "Antiresonance Phenomenon and Peak Voltage Stress Within PWM Inverter Fed Stator Winding," in *IEEE Transactions on Industrial Electronics*, vol. 68, no. 12, pp. 11826-11836, Dec. 2021, doi: 10.1109/TIE.2020.3048286.

[3] A. Tessarolo, E. Ciceran and F. Luise, "Investigation into electrical resonance phenomena in the field circuit of wound-rotor synchronous machines," IECON 2016 - 42nd Annual Conference of the IEEE Industrial Electronics Society, Florence, Italy, 2016, pp. 1692-1697, doi: 10.1109/IECON.2016.7793686.

[4] J. Rodriguez et al., "Resonances and overvoltages in a medium-voltage fan motor drive with long cables in an underground mine," in *IEEE Transactions on Industry Applications*, vol. 42, no. 3, pp. 856-863, May-June 2006, doi: 10.1109/TIA.2006.872936.

[5] H. Ulrich and H. Weber, "Fourier-Reihen," in *Laplace-, Fourier- und z-Transformation*, 10th ed. Wiesbaden, Germany: Springer Vieweg Wiesbaden, 2017, ch. 1, sec. 2, pp. 2–8.

[6] A. Peters, "Entwicklung und Inbetriebnahme einer FPGA-basierten, umrichterintegrierten Messung der frequenzabhängigen Impedanz von Motorwicklungen", master thesis, control lab, TH Köln, Nov. 2023

[7] T. Schmidt, J. O. Krah and J. Holtz, "Mixed Critical Resolver-to-Digital Conversion for Safety-Related Servo Drive Applications," *PCIM Europe 2023; International Exhibition and Conference for Power Electronics, Intelligent Motion, Renewable Energy and Energy Management*, Nuremberg, Germany, 2023, pp. 1-8, doi: 10.30420/566091178.

PCIM Europe 2024, 11– 13 June 2024, Nuremberg   DOI: 10.30420/566262471

Compensation Techniques for Bandwidth-Distorted Measurements of Fast Transients in Double Pulse Tests

Christian Lottis [1], Sebastian Sprunck [2], Bikash Sah [1], Marco Jung [1,2]

[1] Bonn-Rhein-Sieg University of Applied Sciences, Germany
[2] Fraunhofer Institute for Energy Economics and Energy System Technology IEE, Germany

Corresponding author: Christian Lottis, christian.lottis@h-brs.de
Speaker: Christian Lottis, christian.lottis@h-brs.de

Abstract

While assessing the losses in Wide-Bandgap (WBG) devices in a Double-Pulse-Test (DPT), stringent requirements for high-fidelity probes have to be applied. The limited bandwidth of a probe has a significant influence on the measurement results due to its inherent delay and distortion effects that originate from its parasitic or discrete filter components. This paper proposes a technique to compensate for these influences by post-processing measured data based on the characteristics of the used probes. Instead of using a complex inverse transfer function of the probe, an average error compensation is developed and discussed for use in a DPT and validated based on circuit simulation.

1   Introduction

Wide-Bandgap (WBG) devices are becoming increasingly prevalent in applications that rely on power electronic converters. Above all the benefits which these components provide, such as the capability to operate at higher switching frequencies and better thermal performances, the dynamic losses of semiconductors are one of the main restraining factors in the design of high-power converters. To unlock the benefits of WBG devices, the switching frequency of the system must be increased to the physical limits, thereby reducing the size of the main passive components present within the power converters. However, increasing the switching frequency leads to higher switching loss energies in the WBG devices, which translates to joule heating that needs to be handled by an appropriate cooling system.

Manufacturers of power semiconductors often provide performance and characteristic data at a limited set of operating points in device datasheets. However, these operating points seldomly fall into the application-specific requirements and boundary conditions of their intended operation. Hence, it is critical for a power converter designer to have characteristic device data available at multiple operating points that are relevant for the given application, apart from those mentioned in the datasheets.

Better device characterization helps to perform accurate modelling, to optimize circuit performance and efficiency, and to determine limitations of the devices and circuits, such as maximum current and voltage ratings, switching speed limitations, and thermal behaviour. Further, a designer can design better power converters when accurate device models are available, allowing for early design measures that ensure safe and reliable operating conditions before the actual application hardware is assembled, saving valuable time and laboratory resources as well as possible design iterations. A well-established method of power semiconductor device characterizations is the Double Pulse Test (DPT). In DPT, measuring the high-frequency signals that emerge during switching is critical to characterize the losses. These signals consist of significant high-frequency content [1] and, hence, require high bandwidths for the applied probes. Further, the measurement results are strongly influenced by the insertion of the probe into the DPT circuit and by their transmission behaviour due to the presence of low-pass filtering. In addition to the high bandwidths, a high common mode rejection ratio (CMRR) of the used probes is also required for accurate measurements.

Literature on the measurement of switching waveforms of power semiconductor devices shows that even the use of high-quality probes leads to a signif-

3347

icant error between applications and DPT measurements [1, 2]. This error is caused by the inherent delay and distortion of the parasitic or intentional filters in the probes.

This manuscript takes a look at the simplified worst-case signals and analyzes different possibilities of post-processing the results to partly compensate for the low-pass filtering error introduced during measurements.

In recent literature, compensation techniques have been proposed that make use of the complex transfer function of the probe and try to improve it through suitable hardware [3] or software [4]. In contrast to these detailed approaches, this paper instead proposes a simple method to reduce the average error of the measurement, using a post-processing time shift. This post-processing time shift can be determined based on the bandwidth of the signal measurement probe used during experiments alone; no further pieces of information are required in addition. The proposed approach is supported by theoretical analysis and mathematical formulations, followed by validation using a simulation of a DPT.

1.1 Measuring switching losses of semiconductors

The most common procedure for characterizing switching losses of semiconductors is the well-known DPT [5, 6], whose basic setup is shown in Fig. 1. The main setup contains DC-Link capacitors C_{Link}, which will contain the required energy, and the inductor L_{DPT}, which will be used to control the desired switching current and also store energy during the off-period of the device under test (DUT) Q_2. The diode of the upper semiconductor Q_1 is used for freewheeling during the turn-off phase of the DUT. After a dead time to prevent cross conduction, also the channel of Q_1 can be used for this purpose. The DUT Q_2 controls the current through L_{DPT}. During the test, the DUT is first turned on, which will cause a rising current through L_{DPT}. When reaching the desired current, the Gate of the DUT is turned off, and the turn-off voltage and current at the DUT are recorded. After a short waiting time during which oscillations can settle, the DUT is turned on again, and the turn-on event is measured. After this event, the device remains in off-state to change the parameters (DC voltage, gate voltage/resistance or temperature) and/or perform the next measurement at different operating points. A multi-pulse test using an au-

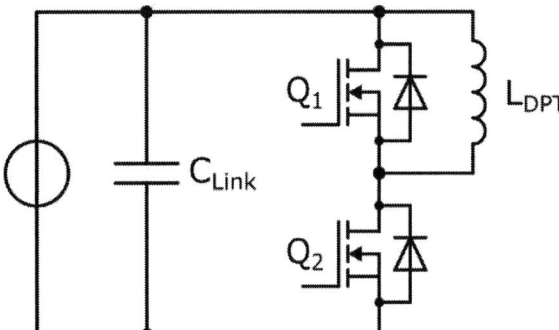

Fig. 1: Schematic of a common DPT setup.

tomated test bench can further ease the overall process, by chaining multiple load currents into a single measurement. The losses can be computed from the recorded measured quantities according to different standards, e.g. IEC 60747.

1.2 Requirements on measurements in DPT

The accuracy of the results of a DPT measurement, in which steep transients of current and voltage have to be recorded, highly depends on the bandwidth of the measurement devices and on the error that occurs due to the unavoidable insertion of parasitic elements in the power loop. For voltage measurements, the insertion error occurs due to the added capacitance of the probe in parallel to the drain and source terminals of the DUT [7]. For current measurements, the error is caused by the additional inductance in the power loop that is created due to necessary modifications in the layout to incorporate current sensors [8].

When using voltage and current signals in power measurements, it is essential to calibrate the transmission time of signals of the different channels. The transmission times of the signals will depend on different factors, such as the physical parameters of different probes, including bandwidth and the cable length used to connect these probes. Without this calibration, large errors will occur. These different transmission times can be aligned using the deskew function of the oscilloscope used, or, if the transmission times are known, post-processing can be performed as well. The deskew method is usually the preferred one, as it is required only once when the probe setup is changed, whereas for the post-processing approach, the same deskew data has to be recorded and included in the data evaluation for each individual measurement.

1.3 Compensation Techniques

The errors due to the addition of the parasitic components, which slow down the slopes or introduce additional ringing, are not easy or even possible to compensate most of the time. Hence, it is recommended that circuits and systems with parasitic component values be developed to be as small as possible. A lot of work is being put into the design of low inductive measurements. Besides the effect of parasitic components, the effect of the limited bandwidth is another important concern as well. This effect leads to distorted signals at the output of the probes. The bandwidth-related errors can partly be compensated either by hardware or software, e.g. using filters tuned to the specific probe properties [3,9] or through a characterization of the complex transfer function of the probe and a post-processing compensation by finding and applying the corresponding inverse transfer function [10].

2 Description of the Signal in a DPT

This section focuses on the post-processing of the waveforms in a DPT. Before discussing post-processing, the signals under examination need to be introduced. In [8], a worst-case estimation of the steepest signal contents is performed to estimate the worst possible error caused by the bandwidth limitations. A simplified description that can be applied to both current and voltage signals is provided as a linear ramp with a slope that is tangential to the steepest parts of each respective signal (Fig. 2). For simplified considerations, only areas A and C are taken into account, and the oscillations in area B are neglected. Figure 4 displays this simplified signal. To achieve generally valid results that can be used for both current and voltage measurements, normalized functions are used to calculate the low-pass error. The time t is normalized to the signal rise time T_R as (1). The signal amplitude is normalized to the maximum amplitude of the signal recorded during the DPT (2), i.e. the switching current I_{DPT} or the switching voltage U_{DPT}. A mathematical description of the simplified (orange) signal shown in Fig. 2 is then possible through (3), where $\epsilon(\tau)$ represents the unit step function.

$$\tau_r = \frac{t}{T_r} \tag{1}$$

$$S_{max} = \frac{i}{I_{DPT}} = \frac{u}{U_{DPT}} \tag{2}$$

$$S_{DPT}(\tau) = \tau \cdot \epsilon(\tau) - (\tau - 1) \cdot \epsilon(\tau - 1) \tag{3}$$

Waveform during switching event

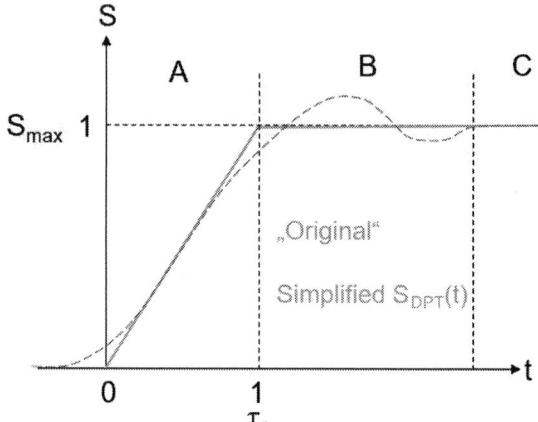

Fig. 2: Simplified, normalized signal of the semiconductor voltage during turn-off or current during turn-on.

3 Average Error Compensation

To elaborate on the method of compensating the bandwidth-related errors in a DPT, first, the ideal signal S_{DPT} and the filtered signal require a comparison. To receive a mathematical description of the filtered signal, S_{filt} is calculated based on the ideal signal (3) and the application of a first-order Butterworth low-pass filter H_1 (4), where $\tilde{\nu}$ is the normalized angular frequency (5). Figure 3 provides a visual representation of this process. A first-order Butterworth filter description is chosen based on its transmission characteristic over the frequency range, which provides a flat profile below the cutoff frequency and which should be applicable to most oscilloscope probes.

First, the simplified signal (3) is transformed into the Laplace domain. This transformed signal is then multiplied by the filter description (4) and the result is then transferred back to the time domain, yielding the filtered signal S_{filt} (6).

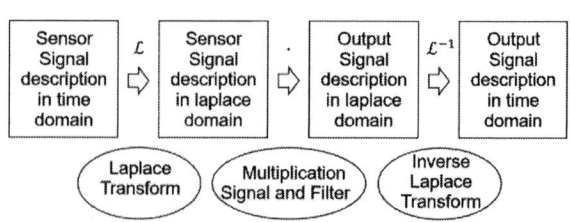

Fig. 3: Flow chart of the calculation of the filtered signal.

PCIM Europe 2024, 11– 13 June 2024, Nuremberg   DOI: 10.30420/566262471

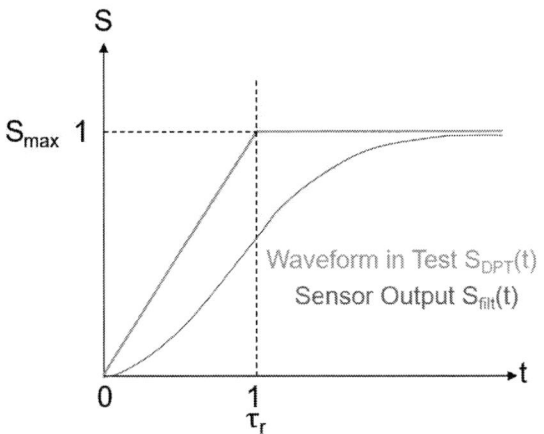

Fig. 4: Schematic representation of the waveform of the simplified normalized signals in a DPT S_{DPT} and the filtered, distorted output S_{filt}. For illustration purposes, the distortion effect of the filtered signal is strongly exaggerated.

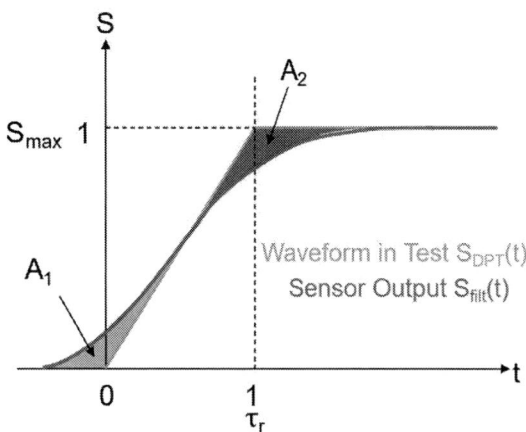

Fig. 5: Schematic representation of a time shift applied to the filtered signal of Fig. 4, producing the compensated signal $S_{filt,comp}$. For illustration purposes, the distortion effect of the filtered signal is strongly exaggerated.

$$H_1(s,\tilde{\nu}) = \frac{1}{\frac{s}{\tilde{\nu}}+1} \tag{4}$$

$$\tilde{\nu} = 2\pi\nu = 2\pi f_g T_r \tag{5}$$

$$S_{filt}(\tau,\tilde{\nu}) = \frac{(\tilde{\nu}\tau + e^{-\tilde{\nu}\tau} - 1)\cdot\epsilon(\tau)}{\tilde{\nu}}$$
$$\frac{(\tilde{\nu}\cdot(\tau-1) + e^{-\tilde{\nu}(\tau-1)} - 1)\cdot\epsilon(\tau-1)}{\tilde{\nu}} \tag{6}$$

Figure 2 shows that the filter applies a time delay and distortion to the measured signals in the DPT. To avoid the complex calculation of probe-specific inverse filters and its time-consuming application onto the recorded signals, this paper instead proposes to focus on the average error between the original and the filtered signal: By shifting the filtered signal in a post-process, the average error can be eliminated as sketched in Fig. 5. In this case, the positive and negative errors are cancelled due to the same area below and above the simplified signal.

For the average error to beome zero, the condition in (7) must be fulfilled, which is possible by finding a suitable time shift parameter z.

$$\int_0^\infty (S_{DPT}(\tau) - S_{filt}(\tau + z, \tilde{\nu}))d\tau = 0 \tag{7}$$

Solving (7) using the signal descriptions (3) and (6), this optimal time shift z can be calculated for non normalised values as (8).

$$z = \frac{1}{\tilde{\nu}}\cdot T_r = \frac{1}{2\pi\cdot\nu}\cdot T_r = \frac{1}{2\pi\cdot f_g} \tag{8}$$

This parameter only depends on the bandwidth of the probe, hence, it is generally valid for any DPT with any rise and fall times. Further, it does not require deeper knowledge of the physical structure and transmission behaviour of the probes. This compensation can be used as long as the Nyquist sampling theorem is fulfilled, meaning that the low-pass filter is not allowed to have an arbitrarily low cutoff frequency compared to the signal rise time.

4 Spice Simulation

To validate the previous considerations and calculations, a simulation of a DPT is performed in LTSpice. The simulation uses a Silicon Carbide (SiC) MOSFET C3M0060065K in a DPT environment, as shown in Fig. 6. The MOSFET used has a typical turn-on time specified with 11 ns. For validation purposes, the drain-to-source voltage over the DUT is taken directly out of the simulation as u_{DPT}. The voltage probe is represented through an RC-lowpass with a cutoff frequency of 500 MHz, which is a typical bandwidth of passive probes available in the market. A divider ratio of 1:10 is also used for this representation, yielding the low-pass filtered probe output voltage u_{filt}. Other parasitic components of the probe are not considered in this simulation, as it is meant to focus on the bandwidth. The simulated voltage waveform during the turn-off event of the SiC semiconductor can be seen in

3350

PCIM Europe 2024, 11–13 June 2024, Nuremberg   DOI: 10.30420/566262471

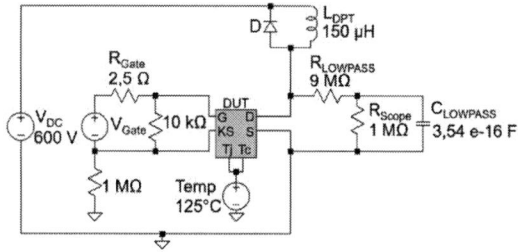

Fig. 6: Schematic of the LTSpice-Simulation to validate the compensation approach.

Fig. 7. As described earlier, the filter function of the probe causes a delay and distortion depending on its transfer function. According to (8), the probe bandwidth of 500 MHz leads to the following time shift for an average error compensation:

$$z = \frac{1}{2\pi \cdot 500\,\text{MHz}} \approx 318.3\,\text{ps}. \qquad (9)$$

Applying this time shift to the probe output signal u_{filt} provides the graphs shown in Fig. 8. Comparing Figs. 7 and 8 clearly shows a much better fit of the measured output signal to the original value. Without compensation, the filtered signal stays lower than the real signal until the first voltage peak is reached. On the other hand, when using the average error compensation, the filtered and compensated signal appears above the non-filtered signal in the first part of the rise time, and below it for the second half of the signal. This is caused by the conditions applied in (7).

This analysis proves that the measurement error can be reduced drastically with only the knowledge about the bandwidth of the probe. Obviously, this compensation does not affect the errors in the amplitude and the time-dependent distortions of the overall signal. However, the influence of the different transmission times of the different frequency-dependent signal contents can still be reduced without extensive characterization of the probes.

5 Conclusion

This paper presents an error compensation method that can be used when measuring fast transients, e.g. in DPT measurements, to reduce the influence of the limited bandwidth of probes and demonstrates its usability through a realistic LTSpice simulation. This method can be used in data postprocessing or can be added to the probe deskew right at the oscilloscope by adding the calculated

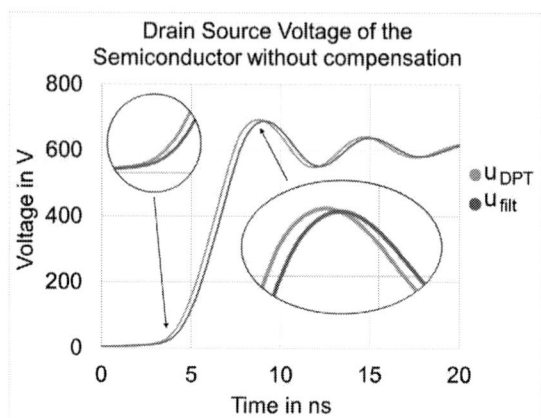

Fig. 7: Comparison of the ideal drain-to-source voltage of the DUT taken out of the simulation (u_{DPT}, orange) and the low-pass filtered output of a probe (u_{filt}, blue).

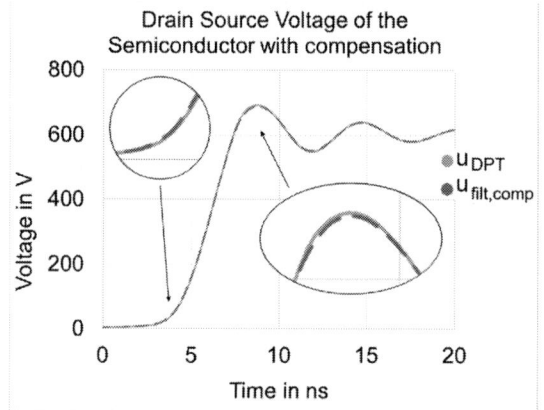

Fig. 8: Comparison of the ideal drain-to-source voltage of the DUT taken out of the simulation (u_{DPT}, blue) and the low-pass filtered signal with the applied average error compensation ($u_{filt,comp}$, blue).

time shift to the respective deskew timing of the probe channel. The post-processing approach is less complex compared to evaluating the exact high-frequency behaviour of the measurement probe to compensate for its behaviour with tuned filters or calculations of its inverse transfer function. The simulation of a DPT with a SiC semiconductor model as a DUT is used to evaluate the influence of the proposed method on the measurement of its Drain-to-Source voltage, proving that a substantial improvement of measurement fidelity is possible. As the method is developed for normalized signals, it is applicable to both voltage and current measurements.

3351

6 Acknowledgment

This work was supported by the German Federal Ministry for Economic Affairs and Climate Action BMWK under grants "GaN-Highpower" (FKZ 03EE1111A and -E) and "MarraKEsH" (FKZ 03EN5035C and -F). Only the authors are responsible for the contents of this publication.

References

[1] S. Sprunck, "Charakterisierung der Schaltverluste diskreter Wide Band Gap Leistungshalbleiter und Entwaermung kompakter Bauteile," Ph.D. dissertation, University of Kassel, 2021.

[2] H. Lutzen, J. Müller, and N. Kaminski, "A Review of Current Sensors in Power Electronics: Fundamentals, Measurement Techniques and Components to Measure the Fast Transients of Wide Bandgap Devices," in *2023 25th European Conference on Power Electronics and Applications (EPE'23 ECCE Europe)*. IEEE, 2023, pp. 1–11.

[3] L. Shillaber, Y. Jiang, L. Ran, and T. Long, "Ultrafast Current Shunt (UFCS): A Gigahertz Bandwidth Ultra-Low-Inductance Current Sensor," *IEEE Transactions on Power Electronics*, vol. 37, no. 12, pp. 15 493–15 504, 2022.

[4] M. Didat, C. D. New, S. Choi, and A. Lemmon, "Improved methodology for conducted EMI assessment of wide band-gap power electronics," *IEEE Open Journal of Power Electronics*, vol. 3, pp. 731–740, 2022.

[5] A. Ghosh, C. N. M. Ho, and J. Prendergast, "A cost-effective, compact, automatic testing system for dynamic characterization of power semiconductor devices," in *2019 IEEE Energy Conversion Congress and Exposition (ECCE)*. IEEE, 2019, pp. 2026–2032.

[6] D. Stracke, M. Klee, F. Schnabel, M. Jung, and A. Seibel, "Fully automated measurement of losses on wide bandgap power semiconductors," in *PCIM Europe 2019; International Exhibition and Conference for Power Electronics, Intelligent Motion, Renewable Energy and Energy Management*. VDE, 2019, pp. 1–6.

[7] S. Sprunck, M. Koch, C. Lottis, and M. Jung, "Suitability of Voltage Sensors for the Measurement of Switching Voltage Waveforms in Power Semiconductors," *IEEE Open Journal of Power Electronics*, vol. 3, pp. 587–598, 2022.

[8] S. Sprunck, C. Lottis, F. Schnabel, and M. Jung, "Suitability of Current Sensors for the Measurement of Switching Currents in Power Semiconductors," *IEEE Open Journal of Power Electronics*, vol. 2, pp. 570–581, 2021.

[9] P. Ziegler, F. Stjepandic, J. Ruthardt, P. Marx, M. Fischer, and J. Roth-Stielow, "Wide bandwidth current sensor for characterization of high current power semiconductor modules," in *2021 23rd European Conference on Power Electronics and Applications (EPE'21 ECCE Europe)*. IEEE, 2021, pp. 1–9.

[10] F. L. Vieira, M. S. Alamsyah, C. Siebauer, and H. Garbe, "Compensation of Time-Domain Waveforms by Applying the Complex Transfer Function of a Current Probe in the kHz-MHz Range," in *2022 Kleinheubach Conference*. IEEE, 2022, pp. 1–3.

PCIM Europe 2024, 11– 13 June 2024, Nuremberg

DOI: 10.30420/566262472

An Aerodynamic Load Measurement Technique for Autonomous Aerial Vehicles

Mehmet Oguz Girgin[1], Semih Cakiroglu[1]

[1] Roketsan Inc., Turkey

Corresponding author: Mehmet Oguz Girgin, oguz.girgin@roketsan.com.tr
Speaker: Mehmet Oguz Girgin, oguz.girgin@roketsan.com.tr

Abstract

In this study, an aerial load measurement technique is proposed specifically for autonomous aerial vehicles (AAVs). The approach aims to measure aerial load force through specific point of transmission within the fin actuation. Instead of measuring directly from fin surfaces, measuring from a particular transmission point of the fin actuation mechanism provides ability to gather load data when the AAV's fins are in motion. To collect accurate data, a fully strain gauge attached Wheatstone bridge hardware and a specialized signal conditioning circuit are located within AAV's fin actuation mechanism. To justify the measurements, an external torque sensor equipped fin loading test bench is used.

1 Introduction

Control actuation system design aims orientation of an aerial vehicle by manipulating aerodynamic flow. The load coming from aerial forces is an important design criterion in terms of performance, design components' selection and packaging. Related performance requirements, which affect auto-pilot stability, are defined by system engineers by the help of several computational fluid dynamics analysis. In design aspect, the more aerodynamic load range, the more heavyset mechanism, in terms of actuator and transmission in a limited space. By this reason, accuracy of computational analysis might cause over-safe design.

This study aims measuring aerial load forces for AAVs. Measurement with strain gauges in various cases for fixed wing aerial vehicles is in operation for many years [1-2]. The proposed approach aims to obtain load force from specific point of transmission within the fin actuation system instead of measuring directly from fin surfaces by strain gauges. Especially, when the rapid growth in autonomous aerial vehicle technology is considered, measurement of aerial load for an aerial vehicle that contains moving fins gains more importance for the development of autonomous aerial vehicle technology.

In addition to measuring load forces via strain gauges, more clear and accurate measurements can be gathered by fiber optic strain sensing method [3-4]. However, light beam modulation unit might not be applicable for fin mechanism that has limited space easily.

The proposed method is based on a fully strain gauge attached Wheatstone bridge located on the output shaft slot within the fin actuation system. The problems depending on motion are eliminated by well-located bridge hardware that provides differential electrical signals carried by flexible cables. In addition to that, a special signal conditioning circuit is attached for data acquisition.

On the other hand, strain gauge sensors bring along several drawbacks on the load monitoring systems. These drawbacks can be described in terms of applications. However, it is possible to sort essential ones. Strain gauges might require complex wirings, especially coming from fully attached strain gauge-based sensors which involve four strain gauges and eight cable for four strain gauge sensors [5]. The durability and reliability of strain gauges might be affected from bonding according to the used adhesive technique. In addition to that, strain gauges wires and materials can deteriorate depending on different environmental conditions and time period [6].

Realized sensor system and calculations are described in Section 2. Obtained results are justified by an external torque sensor attached test bench which is defined in Section 3. The results of the studies are discussed in Section 4.

2 Measurement Method

To understand this proposed approach, the actuation system should be well understood. In this work, a quite similar fin actuation system shown in Fig.1 is used. This actuation system is transferring rotational motion into the fin angular motion. The mechanism contains a dc motor, ball screw, gear and fork shaft. Load force measuring will be achieved by four strain gauges attached to the fork shaft as shown in Fig. 2. Bending at the fork shaft has been measured by strain gauges which give resistance value proportional to bending.

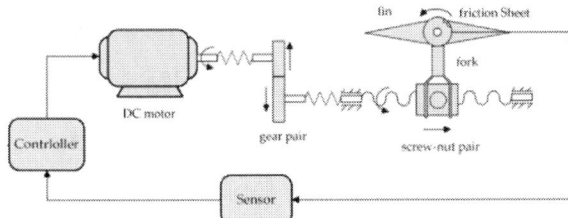

Fig. 1: Control Actuation System [7]

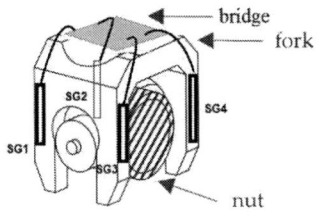

Fig. 2: Attached Strain Gauges to Representational Fork [7]

Four strain gauges are attached as shown in Fig.2 with respect to attachment method shown in Fig.3. They are connected as Wheatstone bridge as shown in Fig.4.

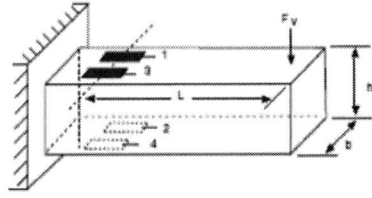

Fig. 3: Bending Strain Measurement [8]

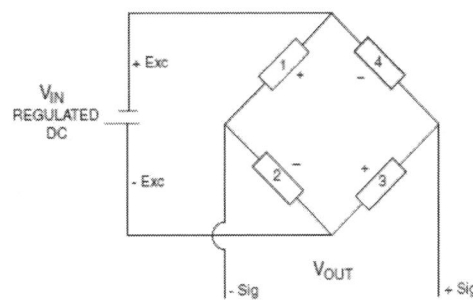

Fig. 4: Wheatstone Bridge for Strain Measurement [8]

Wheatstone bridge should be constructed as close as possible for the attached sensors to achieve minimum distortion depending on resistance changes as shown in Fig.2. Otherwise, eight strain gauge cables, each strain gauge has two cables, might be intertwisted as an improper condition because of the narrow fin mechanism space. Besides, it is observed that there are several pros of Wheatstone bridge differential output signals being twisted and cable shield grounded. On the other hand, material for adaptation is crucial for strain gauges. To attach strain gauges, the attachment area must be prepared for strain gauge bonding [9].

2.1 Signal Conditioning

Maximum differential output voltage is calculated in mV scale. When the load is small, the differential output voltage of the bridge is going to be small too. There might be distortion depending on environmental factors on such a small differential output signal. Consequently, the distortion on differential output voltages will be resulted with misleading measurement for straining. On the other hand, it is hard to digitize such a small voltage range with an analog to digital converter. Because of these reasons, a signal conditioning circuit has been designed. Fig. 5 shows the signal conditioning circuit design of Wheatstone bridge. The small level signal coming from Wheatstone bridge is amplified by a single supply instrumentation amplifier. After that, the amplified signal can be filtered with an anti-aliasing filter before digitizing according to the sampling frequency. Amplified and filtered signal can be digitized by the help of an analog to digital converter to be processed in control unit. To minimize measurement error, the amplifier and converter are powered by the same supply source of the bridge [10].

It is crucial that, strain gauges should be connected as Wheatstone bridge as close as possible to each sensor location at same distance as indicated before. Otherwise, the noise coming from improper cabling with several errors depending on external effects is going to be amplified too.

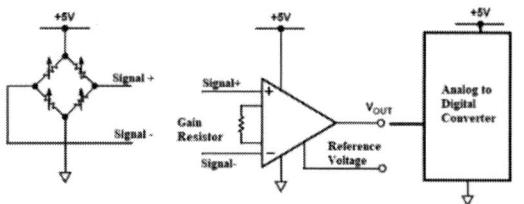

Fig. 5: Wheatstone Bridge Instrumentation Amplifier Circuit Diagram

2.2 Strain Data Computations

The output voltage data of sensor system should be calculated in a correlation with applied force. The voltage obtained by sensor system is too small and a few millivolts level. Because of this reason, this small output voltage is amplified and filtered. Amplified signal is digitized via an analog to digital converter, which is 16 bits. The resulting raw data coming from analog digital converter is obtained by Equation (1).

$$V_{out} = 5V \cdot \text{raw data} / 2^{15} \qquad (1)$$

After logging of these data, the resulting strain, electrical, due to load force is achieved by Equation (2). In Equation (2), V_{ref} is the amplifier circuit voltage reference and K is the constant amplifier circuit gain. These values should be obtained from amplifier circuit of signal conditioning section of sensor system.

$$\epsilon = (V_{out} - V_{ref}) / (K \cdot \text{"Gauge Factor"}) \qquad (2)$$

Where ϵ is elastic strain and gauge factor is the ratio of resistance change and length change as shown in Equation (3).

$$\text{"Gauge Factor"} = (\Delta R/R)/(\Delta l/l) \qquad (3)$$

Equation (4) describes the relation between stress σ, F, applied force and A, which is the modulus of fork cross sectional area. In addition to that, E is modulus of elasticity and Equation (5) describes the relation between modulus of elasticity, strain and stress.

$$\sigma = F/A \qquad (4)$$

$$E = \sigma/\epsilon \qquad (5)$$

By stress formulation Equation (4) and modulus of elasticity formulation Equation (5), the following equations, Equation (6) and Equation (7) are obtained.

$$\sigma = F/A = E. \epsilon \qquad (6)$$

$$F = E. \epsilon .A \qquad (7)$$

As a result, load torque, τ, can be calculated by applied force, F, and lever arm length, L, multiplication as shown in Equation (8) [11].

$$\tau = F .L \qquad (8)$$

3 Test Setup and Results

Sensor system is validated by an external load measurement sensor equipped test bench constructed for active aerodynamic load simulation as shown in Fig. 6.

The test bench generates load torque via an electromechanical servo actuator. The load command of the actuator is generated by the analog output module of a real time industrial computer. Generated reference is the result of torque control which is closed loop with torque sensor feedback gathered via analog input module. In addition to that, the real time test system transmits motion reference to fin mechanism electronics via UART protocol and RS485 standard. Strain gauge sensor system located within the fin mechanism sends analog amplified signals to an analog to digital converter attached on fin mechanism control electronic. By the help of this approach, digitized strain data are able to be sent to the real time computer via UART communication.

In order to test under different load levels, test mechanism has been configured with pre-defined load profile. When the load force is applied, the mechanism is kept stationary and on motion in different test scenarios. By the help of this approach, it is aimed to observe the distortion level, which might occur on the sensor measurements, depending on the motion.

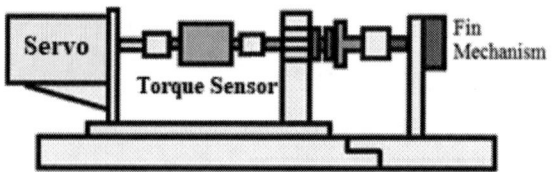

Fig. 6: Fin Mechanism Loading Test Bench

By using the test bench of fin loading mechanism that contains an external torque sensor as shown in Fig. 6, attached strain gauges sensor systems' results are given in Fig.7 and Fig. 8.

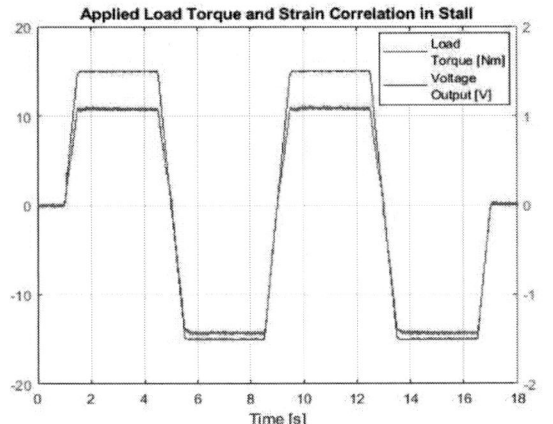

Fig. 7: Applied Load vs Sensor Voltage Comparison in Stall

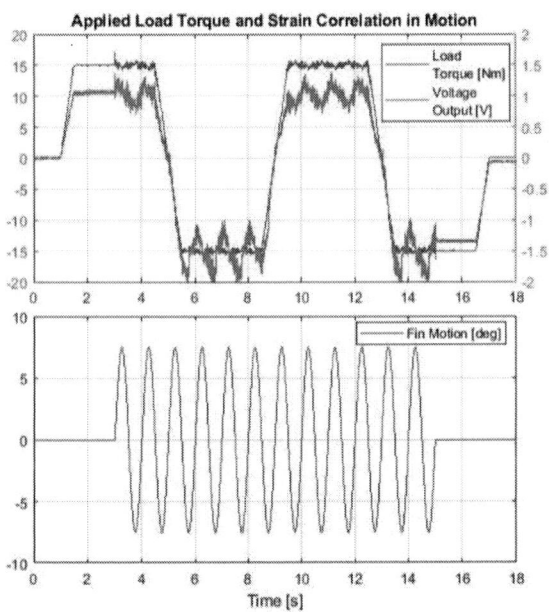

Fig. 8: Load vs Sensor Voltage Comparison Fins Under Dynamic Motion

Fig.7 shows the comparison for applied load on stationary fin by test bench and corresponding sensor output voltage that is measured by strain gauges and sensor electronics without any motion. It is obvious that the applied load and measured voltage are correlated. As shown in Fig.7, it is also observed that there is an asymmetrical behavior on the of sensor output voltage. On positive loading, sensor output voltage is around 1V as shown in Fig.7 while negative loading results with approximately -1.5V on the sensor output voltage results. This asymmetrical behavior can be explained by the exact position difference and surface quality where the strain gauges are bonded. Repeated tests resulted in each sensor system and fin mechanism pairs produce similar behavior. Therefore, the repeated error is decided to be handled with calibration process. To compute final load value, beside to computation in Section 2, there is a need to post-processing in terms of calibration of the sensor system.

Fig. 8 shows the similar comparison under motion. The fin movement characteristic is observed on the load but it is not dominant on load level computed from sensor system. The declared effect is decided to be clarified by including fin orientation for force calculation in trigonometric aspect.

4 Conclusion

Transmitted load measurement from fork shaft technique is applied on the fin actuation mechanism for an autonomous aerial vehicle to gather aerodynamic load data. The main subject was avoiding overdesign and optimizing mechanism according to the realistic aerodynamic load level. Therefore, getting distortion free data under motion is necessary for moving fins. Proposed method, measurement on transmission, is providing load information under motion as well. In addition to these, full bridge measurement discards thermal effects acting on each strain gauge. Fully attached strain gauge measurement technique is verified under dynamically changing load profile by an external loading setup that contains additional commercial of the shelf torque sensor product.

Currently, this study is executed in laboratory environment. It is aimed to measure under real flight loads on air, and there will be a correlation between real aerodynamic loads and current consumption of the actuator under load by a current sensor located in actuator driver circuit.

On future work, air tests and laboratory tests are going to be compared by all these measurement data.

5 Acknowledgment

The authors would like to thank Roketsan Inc. for their financial and laboratory support for this work.

References

[1]. M. J. Allen and R. P. Dibley, "Modeling Aircraft Wing Loads from Flight Data Using Neural Networks", NASA Dryden Flight Research Center, September 2003

[2]. A. M. Lizotte, W. A. Lokos, "Deflection-Based Aircraft Structural Loads Estimation from the Active Aeroelastic Wing F718 Aircraft", NASA Dryden Flight Research Center, May 2005

[3]. J. Bakalyar and C. Jutte, "Validation Tests of Fiber Optic Strain-Based Operational Shape and Load Measurements", NASA Dryden Flight Research Center

[4]. F.Pena, B. L. Martins, W. L. Rcihards, "Active In-flight Load Redistribution Utilizing Fiber-Optic Shape Sensing and Multiple Control Surfaces", February 2018

[5]. Wu, J., Yuan, S., Shang, Y. and Wang, Z., 2009, "Strain distribution monitoring wireless sensor network design and its evaluation research on aircraft wingbox", International Journal of Applied Electromagnetics and Mechanics, vol. 31, pp. 17–28, DOI 10.3233/JAE-2009-1042.

[6]. M. J. Martinez, B. Rocha, G. Shi, C. A. Beltempo, et.al, "Load monitoring of aerospace structures utilizing micro-electro-mechanical systems for static and quasi-static loading conditions" Smart Materials and Structures, September 2012, DOI: 10.1088/0964-1726/21/11/115001

[7]. J. Lu, Z. Wu, and C. Yang, "High-Fidelity Fin Actuator System Modeling and Aeroelastic Analysis Considering Friction Effect", MDPI, Appl.Sci.2021, 11, 3057, DOI:10.3390/app11073057

[8]. "Strain Gauge Installation How to Position Strain Gauges to Monitor Bending, Axial, Shear, And Torsional Loads", Omega

[9]. K. Hoffman, "Practical Hints for the Installation of Strain Gages", HBM Publication

[10]. C. D. Johnson, "Process Control Instrumentation Technology" 8th Edition, pp-54-140

[11]. C. D. Johnson, "Process Control Instrumentation Technology" 8th Edition, pp-232-246

PCIM Europe 2024, 11– 13 June 2024, Nuremberg   DOI: 10.30420/566262473

Compensation Techniques for Bandwidth-Distorted Measurements of Fast Transients in Double Pulse Tests

Christian Lottis [1], Sebastian Sprunck [2], Bikash Sah [1], Marco Jung [1,2]

[1] Bonn-Rhein-Sieg University of Applied Sciences, Germany
[2] Fraunhofer Institute for Energy Economics and Energy System Technology IEE, Germany

Corresponding author: Christian Lottis, christian.lottis@h-brs.de
Speaker: Christian Lottis, christian.lottis@h-brs.de

Abstract

While assessing the losses in Wide-Bandgap (WBG) devices in a Double-Pulse-Test (DPT), stringent requirements for high-fidelity probes have to be applied. The limited bandwidth of a probe has a significant influence on the measurement results due to its inherent delay and distortion effects that originate from its parasitic or discrete filter components. This paper proposes a technique to compensate for these influences by post-processing measured data based on the characteristics of the used probes. Instead of using a complex inverse transfer function of the probe, an average error compensation is developed and discussed for use in a DPT and validated based on circuit simulation.

1 Introduction

Wide-Bandgap (WBG) devices are becoming increasingly prevalent in applications that rely on power electronic converters. Above all the benefits which these components provide, such as the capability to operate at higher switching frequencies and better thermal performances, the dynamic losses of semiconductors are one of the main restraining factors in the design of high-power converters. To unlock the benefits of WBG devices, the switching frequency of the system must be increased to the physical limits, thereby reducing the size of the main passive components present within the power converters. However, increasing the switching frequency leads to higher switching loss energies in the WBG devices, which translates to joule heating that needs to be handled by an appropriate cooling system.

Manufacturers of power semiconductors often provide performance and characteristic data at a limited set of operating points in device datasheets. However, these operating points seldomly fall into the application-specific requirements and boundary conditions of their intended operation. Hence, it is critical for a power converter designer to have characteristic device data available at multiple operating points that are relevant for the given application, apart from those mentioned in the datasheets.

Better device characterization helps to perform accurate modelling, to optimize circuit performance and efficiency, and to determine limitations of the devices and circuits, such as maximum current and voltage ratings, switching speed limitations, and thermal behaviour. Further, a designer can design better power converters when accurate device models are available, allowing for early design measures that ensure safe and reliable operating conditions before the actual application hardware is assembled, saving valuable time and laboratory resources as well as possible design iterations. A well-established method of power semiconductor device characterizations is the Double Pulse Test (DPT). In DPT, measuring the high-frequency signals that emerge during switching is critical to characterize the losses. These signals consist of significant high-frequency content [1] and, hence, require high bandwidths for the applied probes. Further, the measurement results are strongly influenced by the insertion of the probe into the DPT circuit and by their transmission behaviour due to the presence of low-pass filtering. In addition to the high bandwidths, a high common mode rejection ratio (CMRR) of the used probes is also required for accurate measurements.

Literature on the measurement of switching waveforms of power semiconductor devices shows that even the use of high-quality probes leads to a signif-

3358

icant error between applications and DPT measurements [1, 2]. This error is caused by the inherent delay and distortion of the parasitic or intentional filters in the probes.

This manuscript takes a look at the simplified worst-case signals and analyzes different possibilities of post-processing the results to partly compensate for the low-pass filtering error introduced during measurements.

In recent literature, compensation techniques have been proposed that make use of the complex transfer function of the probe and try to improve it through suitable hardware [3] or software [4]. In contrast to these detailed approaches, this paper instead proposes a simple method to reduce the average error of the measurement, using a post-processing time shift. This post-processing time shift can be determined based on the bandwidth of the signal measurement probe used during experiments alone; no further pieces of information are required in addition. The proposed approach is supported by theoretical analysis and mathematical formulations, followed by validation using a simulation of a DPT.

1.1 Measuring switching losses of semi-conductors

The most common procedure for characterizing switching losses of semiconductors is the well-known DPT [5, 6], whose basic setup is shown in Fig. 1. The main setup contains DC-Link capacitors C_{Link}, which will contain the required energy, and the inductor L_{DPT}, which will be used to control the desired switching current and also store energy during the off-period of the device under test (DUT) Q_2. The diode of the upper semiconductor Q_1 is used for freewheeling during the turn-off phase of the DUT. After a dead time to prevent cross conduction, also the channel of Q_1 can be used for this purpose. The DUT Q_2 controls the current through L_{DPT}. During the test, the DUT is first turned on, which will cause a rising current through L_{DPT}. When reaching the desired current, the Gate of the DUT is turned off, and the turn-off voltage and current at the DUT are recorded. After a short waiting time during which oscillations can settle, the DUT is turned on again, and the turn-on event is measured. After this event, the device remains in off-state to change the parameters (DC voltage, gate voltage/resistance or temperature) and/or perform the next measurement at different operating points. A multi-pulse test using an au-

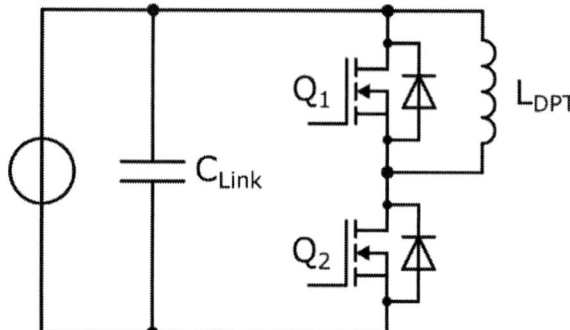

Fig. 1: Schematic of a common DPT setup.

tomated test bench can further ease the overall process, by chaining multiple load currents into a single measurement. The losses can be computed from the recorded measured quantities according to different standards, e.g. IEC 60747.

1.2 Requirements on measurements in DPT

The accuracy of the results of a DPT measurement, in which steep transients of current and voltage have to be recorded, highly depends on the bandwidth of the measurement devices and on the error that occurs due to the unavoidable insertion of parasitic elements in the power loop. For voltage measurements, the insertion error occurs due to the added capacitance of the probe in parallel to the drain and source terminals of the DUT [7]. For current measurements, the error is caused by the additional inductance in the power loop that is created due to necessary modifications in the layout to incorporate current sensors [8].

When using voltage and current signals in power measurements, it is essential to calibrate the transmission time of signals of the different channels. The transmission times of the signals will depend on different factors, such as the physical parameters of different probes, including bandwidth and the cable length used to connect these probes. Without this calibration, large errors will occur. These different transmission times can be aligned using the deskew function of the oscilloscope used, or, if the transmission times are known, post-processing can be performed as well. The deskew method is usually the preferred one, as it is required only once when the probe setup is changed, whereas for the post-processing approach, the same deskew data has to be recorded and included in the data evaluation for each individual measurement.

1.3 Compensation Techniques

The errors due to the addition of the parasitic components, which slow down the slopes or introduce additional ringing, are not easy or even possible to compensate most of the time. Hence, it is recommended that circuits and systems with parasitic component values be developed to be as small as possible. A lot of work is being put into the design of low inductive measurements. Besides the effect of parasitic components, the effect of the limited bandwidth is another important concern as well. This effect leads to distorted signals at the output of the probes. The bandwidth-related errors can partly be compensated either by hardware or software, e.g. using filters tuned to the specific probe properties [3,9] or through a characterization of the complex transfer function of the probe and a post-processing compensation by finding and applying the corresponding inverse transfer function [10].

2 Description of the Signal in a DPT

This section focuses on the post-processing of the waveforms in a DPT. Before discussing post-processing, the signals under examination need to be introduced. In [8], a worst-case estimation of the steepest signal contents is performed to estimate the worst possible error caused by the bandwidth limitations. A simplified description that can be applied to both current and voltage signals is provided as a linear ramp with a slope that is tangential to the steepest parts of each respective signal (Fig. 2). For simplified considerations, only areas A and C are taken into account, and the oscillations in area B are neglected. Figure 4 displays this simplified signal. To achieve generally valid results that can be used for both current and voltage measurements, normalized functions are used to calculate the low-pass error. The time t is normalized to the signal rise time T_R as (1). The signal amplitude is normalized to the maximum amplitude of the signal recorded during the DPT (2), i.e. the switching current I_{DPT} or the switching voltage U_{DPT}. A mathematical description of the simplified (orange) signal shown in Fig. 2 is then possible through (3), where $\epsilon(\tau)$ represents the unit step function.

$$\tau_r = \frac{t}{T_r} \tag{1}$$

$$S_{max} = \frac{i}{I_{DPT}} = \frac{u}{U_{DPT}} \tag{2}$$

$$S_{DPT}(\tau) = \tau \cdot \epsilon(\tau) - (\tau - 1) \cdot \epsilon(\tau - 1) \tag{3}$$

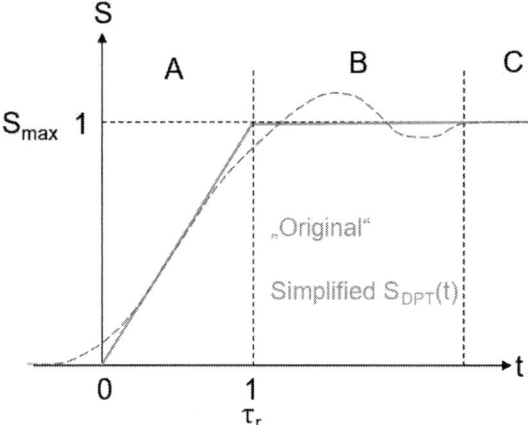

Fig. 2: Simplified, normalized signal of the semiconductor voltage during turn-off or current during turn-on.

3 Average Error Compensation

To elaborate on the method of compensating the bandwidth-related errors in a DPT, first, the ideal signal S_{DPT} and the filtered signal require a comparison. To receive a mathematical description of the filtered signal, S_{filt} is calculated based on the ideal signal (3) and the application of a first-order Butterworth low-pass filter H_1 (4), where $\tilde{\nu}$ is the normalized angular frequency (5). Figure 3 provides a visual representation of this process. A first-order Butterworth filter description is chosen based on its transmission characteristic over the frequency range, which provides a flat profile below the cutoff frequency and which should be applicable to most oscilloscope probes.

First, the simplified signal (3) is transformed into the Laplace domain. This transformed signal is then multiplied by the filter description (4) and the result is then transferred back to the time domain, yielding the filtered signal S_{filt} (6).

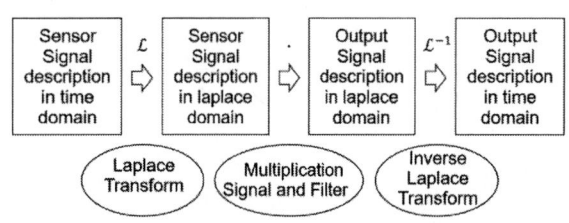

Fig. 3: Flow chart of the calculation of the filtered signal.

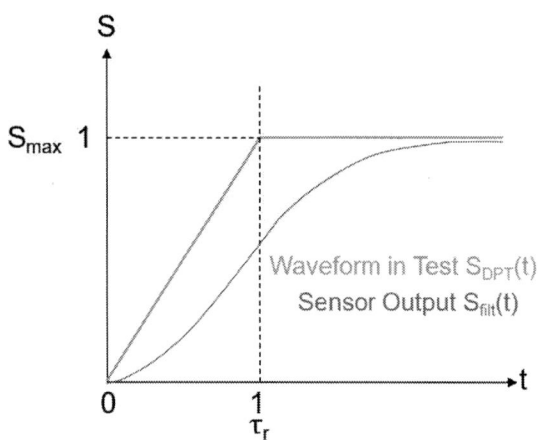

Fig. 4: Schematic representation of the waveform of the simplified normalized signals in a DPT S_{DPT} and the filtered, distorted output S_{filt}. For illustration purposes, the distortion effect of the filtered signal is strongly exaggerated.

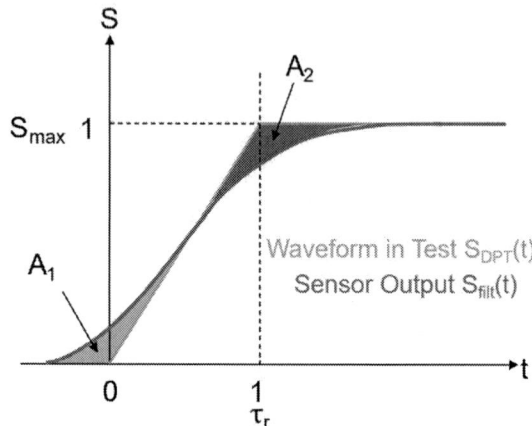

Fig. 5: Schematic representation of a time shift applied to the filtered signal of Fig. 4, producing the compensated signal $S_{filt,comp}$. For illustration purposes, the distortion effect of the filtered signal is strongly exaggerated.

$$H_1(s, \tilde{\nu}) = \frac{1}{\frac{s}{\tilde{\nu}}^2 + 1} \quad (4)$$

$$\tilde{\nu} = 2\pi\nu = 2\pi f_g T_r \quad (5)$$

$$S_{filt}(\tau, \tilde{\nu}) = \frac{(\tilde{\nu}\tau + e^{-\tilde{\nu}\tau} - 1) \cdot \epsilon(\tau)}{\tilde{\nu}}$$
$$\frac{(\tilde{\nu} \cdot (\tau - 1) + e^{-\tilde{\nu}(\tau-1)} - 1) \cdot \epsilon(\tau - 1)}{\tilde{\nu}} \quad (6)$$

Figure 2 shows that the filter applies a time delay and distortion to the measured signals in the DPT. To avoid the complex calculation of probe-specific inverse filters and its time-consuming application onto the recorded signals, this paper instead proposes to focus on the average error between the original and the filtered signal: By shifting the filtered signal in a post-process, the average error can be eliminated as sketched in Fig. 5. In this case, the positive and negative errors are cancelled due to the same area below and above the simplified signal.

For the average error to beome zero, the condition in (7) must be fulfilled, which is possible by finding a suitable time shift parameter z.

$$\int_0^\infty (S_{DPT}(\tau) - S_{filt}(\tau + z, \tilde{\nu}))d\tau = 0 \quad (7)$$

Solving (7) using the signal descriptions (3) and (6), this optimal time shift z can be calculated for non normalised values as (8).

$$z = \frac{1}{\tilde{\nu}} \cdot T_r = \frac{1}{2\pi \cdot \nu} \cdot T_r = \frac{1}{2\pi \cdot f_g} \quad (8)$$

This parameter only depends on the bandwidth of the probe, hence, it is generally valid for any DPT with any rise and fall times. Further, it does not require deeper knowledge of the physical structure and transmission behaviour of the probes. This compensation can be used as long as the Nyquist sampling theorem is fulfilled, meaning that the low-pass filter is not allowed to have an arbitrarily low cutoff frequency compared to the signal rise time.

4 Spice Simulation

To validate the previous considerations and calculations, a simulation of a DPT is performed in LTSpice. The simulation uses a Silicon Carbide (SiC) MOSFET C3M0060065K in a DPT environment, as shown in Fig. 6. The MOSFET used has a typical turn-on time specified with 11 ns. For validation purposes, the drain-to-source voltage over the DUT is taken directly out of the simulation as u_{DPT}. The voltage probe is represented through an RC-lowpass with a cutoff frequency of 500 MHz, which is a typical bandwidth of passive probes available in the market. A divider ratio of 1:10 is also used for this representation, yielding the low-pass filtered probe output voltage u_{filt}. Other parasitic components of the probe are not considered in this simulation, as it is meant to focus on the bandwidth. The simulated voltage waveform during the turn-off event of the SiC semiconductor can be seen in

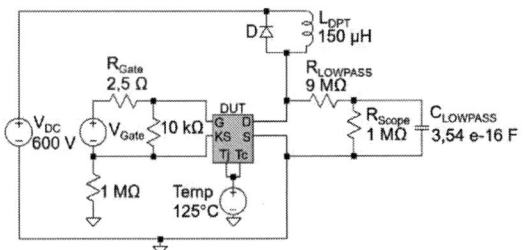

Fig. 6: Schematic of the LTSpice-Simulation to validate the compensation approach.

Fig. 7. As described earlier, the filter function of the probe causes a delay and distortion depending on its transfer function. According to (8), the probe bandwidth of 500 MHz leads to the following time shift for an average error compensation:

$$z = \frac{1}{2\pi \cdot 500\,\text{MHz}} \approx 318.3\,\text{ps}. \qquad (9)$$

Applying this time shift to the probe output signal u_{filt} provides the graphs shown in Fig. 8. Comparing Figs. 7 and 8 clearly shows a much better fit of the measured output signal to the original value. Without compensation, the filtered signal stays lower than the real signal until the first voltage peak is reached. On the other hand, when using the average error compensation, the filtered and compensated signal appears above the non-filtered signal in the first part of the rise time, and below it for the second half of the signal. This is caused by the conditions applied in (7).

This analysis proves that the measurement error can be reduced drastically with only the knowledge about the bandwidth of the probe. Obviously, this compensation does not affect the errors in the amplitude and the time-dependent distortions of the overall signal. However, the influence of the different transmission times of the different frequency-dependent signal contents can still be reduced without extensive characterization of the probes.

5 Conclusion

This paper presents an error compensation method that can be used when measuring fast transients, e.g. in DPT measurements, to reduce the influence of the limited bandwidth of probes and demonstrates its usability through a realistic LTSpice simulation. This method can be used in data post-processing or can be added to the probe deskew right at the oscilloscope by adding the calculated

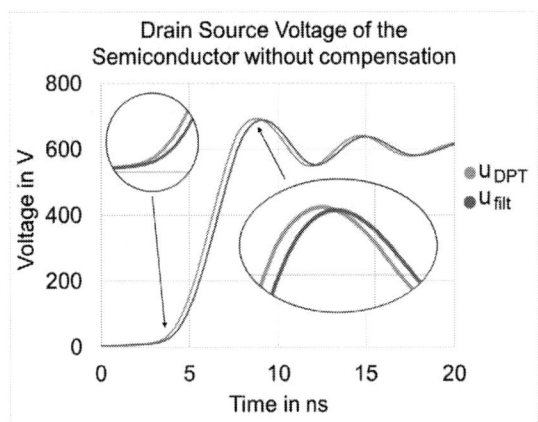

Fig. 7: Comparison of the ideal drain-to-source voltage of the DUT taken out of the simulation (u_{DPT}, orange) and the low-pass filtered output of a probe (u_{filt}, blue).

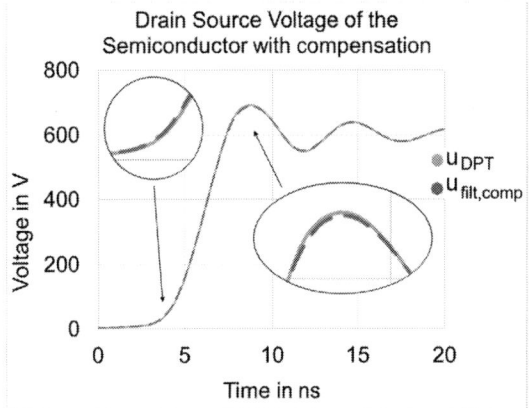

Fig. 8: Comparison of the ideal drain-to-source voltage of the DUT taken out of the simulation (u_{DPT}, blue) and the low-pass filtered signal with the applied average error compensation ($u_{filt,comp}$, blue).

time shift to the respective deskew timing of the probe channel. The post-processing approach is less complex compared to evaluating the exact high-frequency behaviour of the measurement probe to compensate for its behaviour with tuned filters or calculations of its inverse transfer function. The simulation of a DPT with a SiC semiconductor model as a DUT is used to evaluate the influence of the proposed method on the measurement of its Drain-to-Source voltage, proving that a substantial improvement of measurement fidelity is possible. As the method is developed for normalized signals, it is applicable to both voltage and current measurements.

6 Acknowledgment

This work was supported by the German Federal Ministry for Economic Affairs and Climate Action BMWK under grants "GaN-Highpower" (FKZ 03EE1111A and -E) and "MarraKEsH" (FKZ 03EN5035C and -F). Only the authors are responsible for the contents of this publication.

References

[1] S. Sprunck, "Charakterisierung der Schaltverluste diskreter Wide Band Gap Leistungshalbleiter und Entwaermung kompakter Bauteile," Ph.D. dissertation, University of Kassel, 2021.

[2] H. Lutzen, J. Müller, and N. Kaminski, "A Review of Current Sensors in Power Electronics: Fundamentals, Measurement Techniques and Components to Measure the Fast Transients of Wide Bandgap Devices," in *2023 25th European Conference on Power Electronics and Applications (EPE'23 ECCE Europe)*. IEEE, 2023, pp. 1–11.

[3] L. Shillaber, Y. Jiang, L. Ran, and T. Long, "Ultrafast Current Shunt (UFCS): A Gigahertz Bandwidth Ultra-Low-Inductance Current Sensor," *IEEE Transactions on Power Electronics*, vol. 37, no. 12, pp. 15 493–15 504, 2022.

[4] M. Didat, C. D. New, S. Choi, and A. Lemmon, "Improved methodology for conducted EMI assessment of wide band-gap power electronics," *IEEE Open Journal of Power Electronics*, vol. 3, pp. 731–740, 2022.

[5] A. Ghosh, C. N. M. Ho, and J. Prendergast, "A cost-effective, compact, automatic testing system for dynamic characterization of power semiconductor devices," in *2019 IEEE Energy Conversion Congress and Exposition (ECCE)*. IEEE, 2019, pp. 2026–2032.

[6] D. Stracke, M. Klee, F. Schnabel, M. Jung, and A. Seibel, "Fully automated measurement of losses on wide bandgap power semiconductors," in *PCIM Europe 2019; International Exhibition and Conference for Power Electronics, Intelligent Motion, Renewable Energy and Energy Management*. VDE, 2019, pp. 1–6.

[7] S. Sprunck, M. Koch, C. Lottis, and M. Jung, "Suitability of Voltage Sensors for the Measurement of Switching Voltage Waveforms in Power Semiconductors," *IEEE Open Journal of Power Electronics*, vol. 3, pp. 587–598, 2022.

[8] S. Sprunck, C. Lottis, F. Schnabel, and M. Jung, "Suitability of Current Sensors for the Measurement of Switching Currents in Power Semiconductors," *IEEE Open Journal of Power Electronics*, vol. 2, pp. 570–581, 2021.

[9] P. Ziegler, F. Stjepandic, J. Ruthardt, P. Marx, M. Fischer, and J. Roth-Stielow, "Wide bandwidth current sensor for characterization of high current power semiconductor modules," in *2021 23rd European Conference on Power Electronics and Applications (EPE'21 ECCE Europe)*. IEEE, 2021, pp. 1–9.

[10] F. L. Vieira, M. S. Alamsyah, C. Siebauer, and H. Garbe, "Compensation of Time-Domain Waveforms by Applying the Complex Transfer Function of a Current Probe in the kHz-MHz Range," in *2022 Kleinheubach Conference*. IEEE, 2022, pp. 1–3.

A High Bandwidth and Multilevel Counter Circuit for Bearing Current Evaluation

Felix Schulte[1], Moritz Seidel[1], Martin Pfost[1]

[1] Chair of Energy Conversion, TU Dortmund University, Germany

Corresponding author: Felix Schulte, felix.schulte@tu-dortmund.de
Speaker: Felix Schulte, felix.schulte@tu-dortmund.de

Abstract

Inverter-induced bearing currents are known to reduce bearing life in electrical machines. Several types of bearing currents can be distinguished, and the different types of bearing currents have different causes, but all these types occur in pulses. In general, the different types also differ in the pulse height. This paper proposes a circuit for live evaluation of bearing currents by counting their occurrence rate as a function of pulse height, while distinguishing between two different current levels. In addition, it is emphasized what needs to be considered when counting bearing current pulses. The occurrence rate and pulse height can be used as indicators for bearing current investigations.

1 Introduction

The mechanics of bearing currents in inverter-fed drives has been described many times. Bearing currents can cause damage to the bearings, resulting in increased machine vibration. To understand the occurrence of bearing currents, Fig. 1 is used for illustration purposes. It shows the common view of the motor capacities C_{wr} (winding to rotor), C_{wf} (winding to frame), C_{rf} (rotor to frame) and C_b (bearing). In [1] the authors use finite element method for the calculation of the motor capacities, while the authors of [2] present analytical and measurement methods.

The bearing current pulses appear with the breakthrough of the bearing capacity C_b, which leads to electric discharge machining (EDM), or with the gradient of the common-mode voltage v_{com}, which leads to high-frequency circulating bearing currents, dv/dt bearing currents or rotor ground currents. Since all of these types of bearing currents occur in pulses, counting them can provide important information about the bearing currents that are occurring.

In [3], the authors emphasize the bearing current density and the bearing current power as characteristic parameters for bearing damage, while these parameters depend on the bearing current pulse height. Consequently, monitoring the height

and rate of occurrence of the pulses is an important tool for assessing bearing current damage, as done in [4]. Typically, monitoring the rate of occurrence requires two oscilloscopes with real-time counting capability for each comparison level, which is not widely used and can result in multiple counts of ringing signals. This paper proposes a circuit that makes it easy to count these pulses using a microcontroller, since the signal preprocessing is done using an analog circuit. It takes a closer look at what needs to be considered when counting these pulses and determines the accuracy of the circuit using a reference measurement for various use cases.

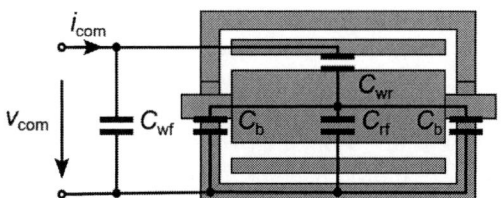

Fig. 1: The schematic representation of the motor capacities C_{wr} (winding to rotor), C_{wf} (winding to frame), C_{rf} (rotor to frame) and C_b (bearing)

2 Measurement Setup

To measure bearing currents in electrical machines, the machine must be prepared. In this work, the bearings of a 2.2 kW induction motor are insulated and bypassed to measure the bearing current through them with a current transformer. Fig. 2 shows two current probes measuring the bearing current through the bypass, one for the counting circuit and one for the reference measurement.

The induction motor runs without mechanical load. It is driven by a SiC frequency inverter running a 10 kHz PWM with a modulation index of 0.2 at a speed of 150 min^{-1} when not otherwise specified. The DC link voltage is varied to adapt the bearing current pulse heights and occurrence rates. Further details on motor preparation and measurement setup can be found in [5].

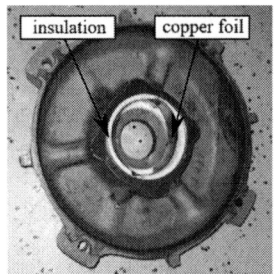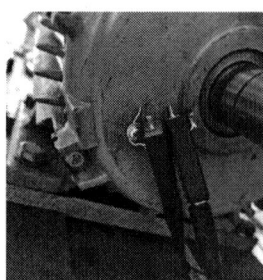

Fig. 2: The machine is prepared to measure bearing currents. The bearing is insulated with a 3D printed resin layer and the outer ring of the bearing is contacted with a copper foil.

3 Measurement Requirements

The most important aspects relevant to the design of the counting circuit and, more generally, to the measurement of bearing currents are discussed below. These include the required bandwidth, the expected pulse shape of the bearing currents, and the rate at which the pulses occur.

To examine these aspects, Fig. 3 shows the bearing current and the common-mode voltage, while Fig. 4 shows the FFT of the bearing current. The frequency domain represents the oscillation frequency of the appearing bearing current pulses that the bandwidth of the measurement must fulfill. In [6] it is shown that one reason for the oscillation is the measurement technique itself, since the bearing insulation adds a capacitance and the bypass adds an inductance to the bearing current path.

In the close-up of the EDM and dv/dt pulses in Fig. 5, the pulses cross the exemplary chosen comparison value for counting the pulse $\pm I_{comp,1}$ several times, resulting in incorrect counting if the counter is not blocked for a certain time.

Fig. 6 shows the minimum time Δt for which the counter must be blocked. The blocking circuit must be adapted to the specific pulse shapes.

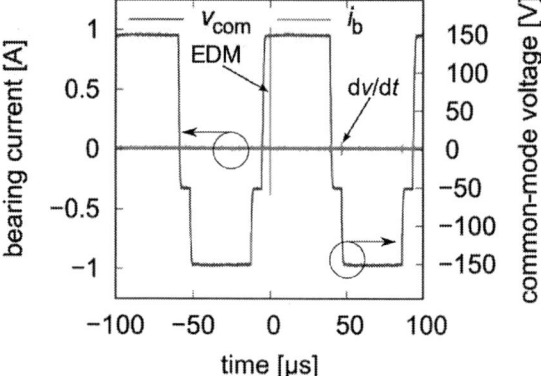

Fig. 3: The bearing current i_b occurs in pulses, while the EDM pulses are much larger than the dv/dt pulses.

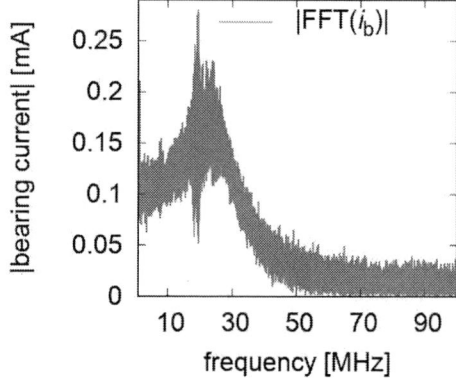

Fig. 4: The FFT of the bearing current in Fig. 3 shows the resonance frequency around 25 MHz.

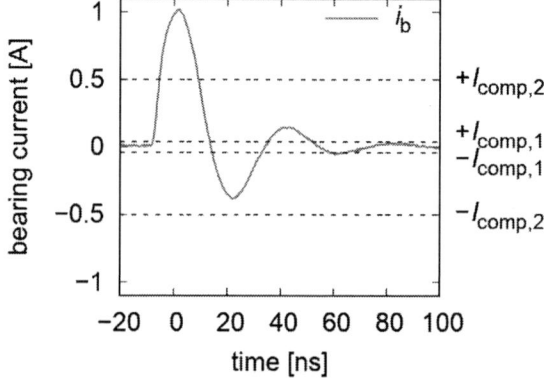

Fig. 5: This figure shows the close-up of the EDM pulse with exemplary comparison values $\pm I_{comp,1}$ and $\pm I_{comp,2}$ that must to be exceeded for counting.

PCIM Europe 2024, 11– 13 June 2024, Nuremberg   DOI: 10.30420/566262473

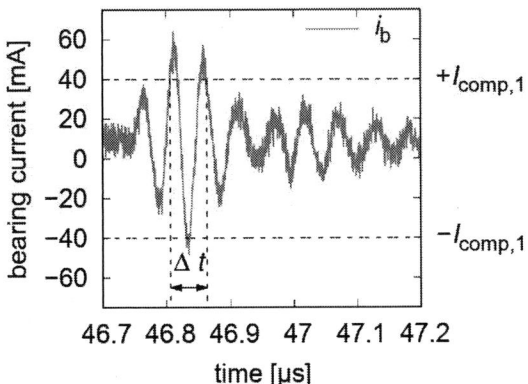

Fig. 6: Additional to the dv/dt bearing current pulse the minimum blocking time Δt for avoiding multiple counting is shown.

4 Proposed Circuit

The schematic block diagram of the proposed circuit is shown in Fig. 7. It detects two different current levels as well as the positive and the negative bearing current pulses that can occur equally. The current transformer (CT) measures the bearing current pulses and is used as the input to the circuit. An RC voltage divider is used to scale down the amplitude of the signal over a wide frequency range.

The circuit is divided into two sections to distinguish between small and large bearing currents. Operational amplifier OP1 is used as a non-inverting amplifier to amplify the voltage signal and focus the small bearing currents. Operational amplifier OP2 is used as a voltage follower to decouple the circuit

for the larger bearing currents. The voltage V_{comp} is the threshold voltage for selecting the counted bearing currents and can be set by the microcontroller (MCU). To avoid multiple counting of a pulse, cf. Fig. 5 and Fig. 6, two monostable multivibrators (monoflop) are combined with an OR gate for each level. This results in a square-wave signal for the lower and upper paths of the circuit, with each rising edge corresponding to a bearing current pulse. The microcontroller (MCU) increases a counter for each rising edge of the monoflop output.

Fig. 8 shows the bearing current i_b and the output voltage v_{edge} of the blocking circuit for positive and small pulses. It can be seen that exceeding the comparison threshold results in a pulse at the output of the monostable multivibrator, which is then combined with the pulse of the negative string in an OR gate.

To ensure an accurate conversion between the comparison current values I_{comp} and the set voltage V_{comp}, the circuit is calibrated with 25 MHz pulses from a signal generator, which functions as an emulated bearing current pulse. The linear relationship can be clearly seen in Fig. 9 for the positive and negative pulses, which is used to calculate the resulting voltage V_{comp} from a given current comparison value I_{comp}. The comparison voltage and current $V_{comp,1}$ and $I_{comp,1}$ is intended for small pulses and $V_{comp,2}$ and $I_{comp,2}$ for large pulses.

The frequency dependence of the circuit is also examined. Fig. 10 shows that the set value is largely independent of the frequency in the con-

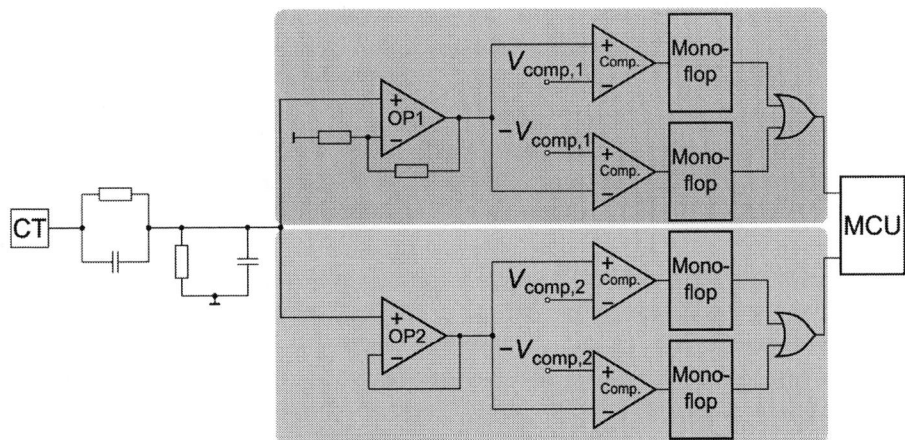

Fig. 7: Simplified schematic of the counting circuit that can count two different current levels. The blue part of the circuit counts the small pulses and the red part counts the large pulses. Each part contains two strings for positive and negative pulses, which are combined in an OR gate.

3366

PCIM Europe 2024, 11– 13 June 2024, Nuremberg   DOI: 10.30420/566262473

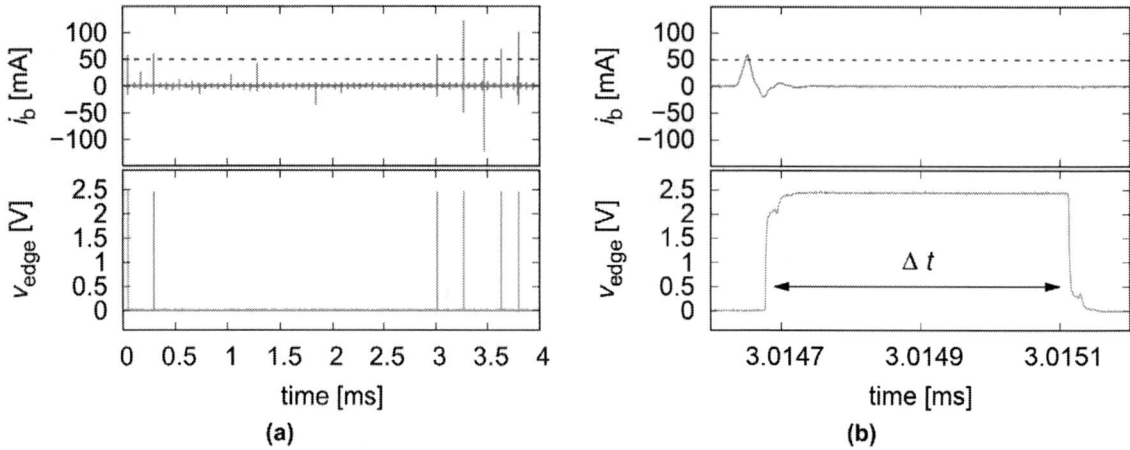

Fig. 8: In (a) the bearing current and the corresponding voltage from the blocking output are shown for small and positive pulses, while (b) shows a close-up view with the blocking time Δt.

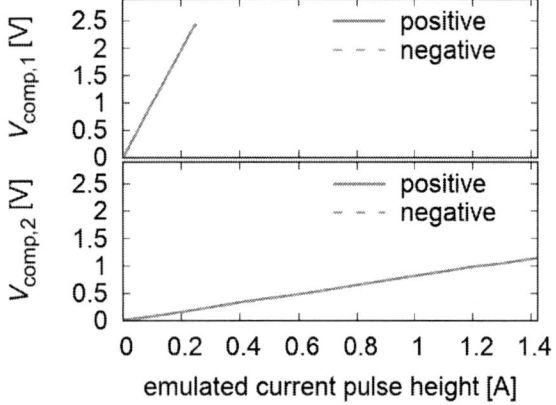

Fig. 9: The comparison voltages $V_{comp,1}$ and $V_{comp,2}$ and the input of the circuit that emulates a bearing current pulse show linear behavior.

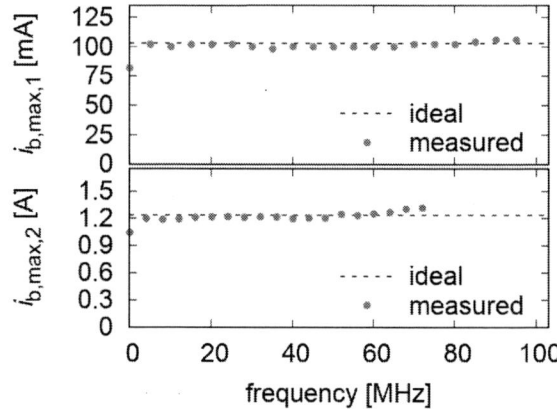

Fig. 10: The measured bearing current pulse height is largely constant over the frequency range considered.

sidered range, especially in the expected range around 25 MHz, which is the oscillation frequency of the bearing current pulses in the setup used.

The rate of occurrence of bearing current pulses can vary due to several mechanisms. One is the probability distribution of the EDM bearing currents, which results in different time intervals between two peaks (see Fig. 3). In addition, the time between an EDM pulse and a dv/dt pulse depends on the time interval between the edge of the common-mode voltage and the statistically occurring EDM pulses. Another reason is that the distance between the edges of the common-mode voltage varies with time and modulation index, while these edges result in dv/dt bearing currents, circulating bearing currents, or rotor ground bearing currents. In the laboratory test, the micro-

controller counts pulses with an interval of < 50 ns without error, which is below the blocking time Δt of 160 ns and therefore does not cause a problem.

5 Experimental Results

To evaluate the functionality of the proposed circuit, the bearing current is measured with an oscilloscope in parallel with the counting circuit, as shown in Fig. 11. Each frame of the oscilloscope data is post-processed and the reference count N_{ref} is determined in this process. Both methods are synchronized with a signal generator that triggers the measurement with the oscilloscope and the circuit in parallel. Due to the fact that the time range of the oscilloscope is limited, this method is useful for a comparison of several frames, but not for a continuous time course.

3367

To validate the accuracy of the circuit, the DC link voltage and the comparison value $I_{comp,1}$ for smaller pulses are varied. For each level and DC link voltage, 500 measurements are made with a time range of 900 µs, resulting in a total observed time of $500 \cdot 900$ µs = 450 ms for each parameter configuration. In Fig. 12 it can be seen that the failure rate $|N_{count} - N_{ref}|$/time with a DC link voltage of 300 V remains constant at the end of the observed time range, indicating a sufficient observation period. A higher comparison value reduces the error rate, which is due to a reduction in interference.

In Fig. 13 the accuracy of the count can be examined, where the ratio of the count value N_{count} to the reference value N_{ref} is shown for different comparison values and DC link voltages with a target value of 1. For large ranges, the ratio is close to or equal to this value, indicating that the pulses are being counted without major mistakes. For very small comparison voltages, too many pulses are counted because there is an increased susceptibility to interference from the drive system.

For the series of measurements with 300 V DC link voltage and 100 mA comparison value, the ratio is slightly below 1. All failed measurements of this point are shown in Fig. 14, where it can be seen that some pulses are very close to the decision threshold, which is interpreted differently by the reference measurement and the counting circuit. Since the difference is a few mA and the decision threshold is constant using the same circuit, it does not affect functionality.

Fig. 15 shows that small comparison values lead to huge failure rates. For very small comparison val-

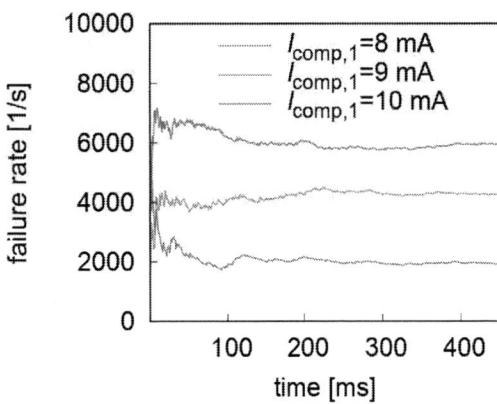

Fig. 12: The failure rate shows constant behavior at the end of the observation time for different comparison values with constant DC link voltage.

ues, a high DC link voltage tends to leads to a lower failure rate because the bearing current pulses are larger than the noise. As the comparison value increases, the failure rate decreases faster with lower DC link voltages due to less interference at lower voltages.

The statistical evaluation of the comparison value $I_{comp,2}$ for larger pulses is not possible with the present setup in such detail, since the large pulses occur much less frequently and the validation method used here only covers a measurement duration of 450 ms. Since this part of the circuit is designed for a wider range of pulses up to 3 A, both the resolution and the resistance to external interference are reduced. As shown in Fig. 9, a small distortion of $V_{comp,2}$ leads to a significantly increased shift of the measured pulse height com-

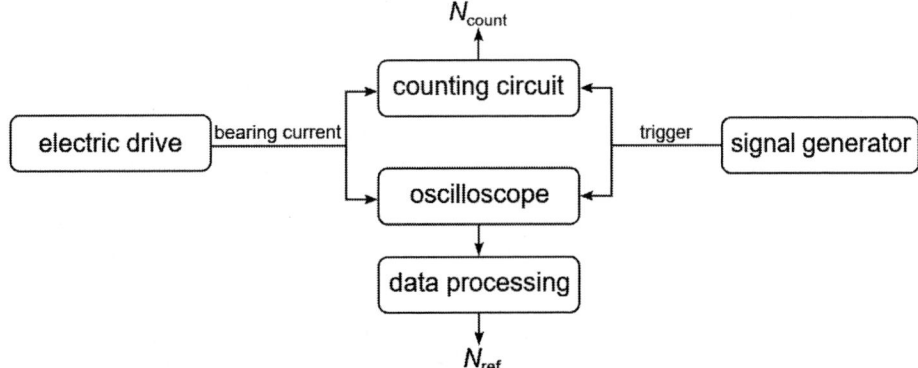

Fig. 11: The schematic diagram of the measurement setup shows how the data is generated. The bearing current is measured with two current probes, one connected to an oscilloscope and the other to the counting circuit. The measurements are synchronized by a signal generator. The counting circuit generates the value N_{count} and the reference value N_{ref} is generated by post-processing the oscilloscope data.

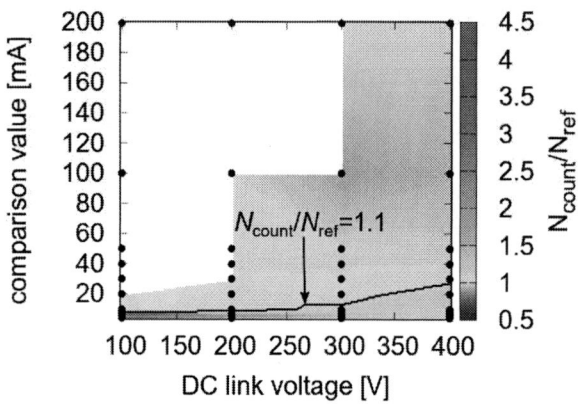

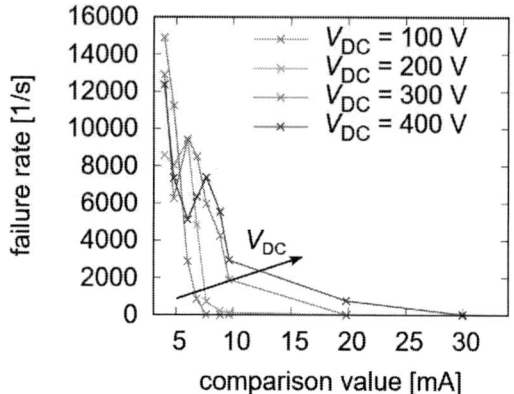

Fig. 13: The counting circuit is examined for different DC link voltages and comparison values. In the white area, no bearing current pulse could be detected. The black dots mark the measured values, while the remaining data is generated by linear interpolation. Above the line shown, the deviation is less than 10 percent.

Fig. 15: The failure rate decreases with increasing comparison values due to less interference. The main source of noise is the switched DC link voltage, which results in a faster decrease in failures at lower DC link voltages.

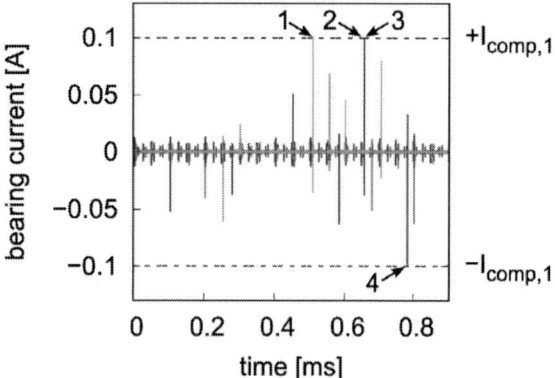

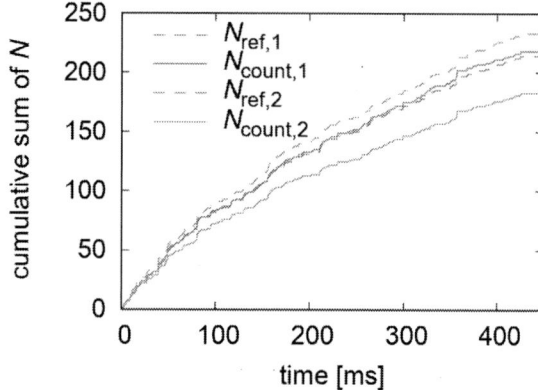

Fig. 14: All four failed measurements with V_{DC} = 300 V and $I_{comp,1}$ = 100 mA are pulses with an maximum very close to the comparison threshold.

Fig. 16: The cumulative sum of the counting values N for the reference measurement (N_{ref}) and the counting circuit (N_{count}) are shown for smaller (N_1) and larger (N_2) bearing currents.

pared to $V_{comp,1}$. To evaluate the functionality of the circuit part for large pulses Fig. 16 shows the cumulative sum of the counted pulses for a measurement series with a reference value of 100 mA and a DC link voltage of 400 V. The modulation index is set to 0.5 and the rational speed to 900 min^{-1} to aim for a high number of large pulses. The curves of the reference values $N_{ref,1}$ and $N_{ref,2}$ differ because they are adapted to different current resolutions of the comparison values. The measured and reference curves of the cumulative sum of the counting values differ more for larger pulses N_2 than for smaller pulses N_1. The increased inaccuracy can be seen when the two curves for comparison values $I_{comp,2}$ diverge fur-

ther than for $I_{comp,1}$ because many pulse heights are near to the comparison value and the deviation is due to the higher susceptibility to interference as explained above. In conclusion the maximum measurable values should be kept as low as possible to increase the interference resistance, because the motor test bench causes considerable interference.

6 Conclusion

This paper proposes a circuit for evaluating bearing current pulses, which can occur at rates of several kHz and oscillation frequencies of several MHz. Due to the presented analog pre-processing, a microcontroller with a simple edge counting algorithm is sufficient. The proposed circuit provides accurate live evaluation of bearing current pulses while taking into account the most important aspects of pulse counting.

Since the number of pulses is often used as a characteristic of bearing currents in electrical machines, this paper is intended to contribute to a better understanding of the bearing current measurement technique and to highlight the important aspects to be considered.

References

[1] C. S. Chaves, J. R. Camacho, H. de Paula, M. L. R. Chaves, and E. Saraiva, "Capacitances calculation using fem for transient overvoltage and common mode currents prediction in inverter-driven induction motors," in *2011 IEEE Trondheim PowerTech*, 2011, pp. 1–7.

[2] A. Muetze and A. Binder, "Calculation of motor capacitances for prediction of the voltage across the bearings in machines of inverter-based drive systems," *IEEE Transactions on Industry Applications*, vol. 43, no. 3, pp. 665–672, 2007.

[3] M. Weicker and A. Binder, "Characteristic parameters for electrical bearing damage," in *2022 International Symposium on Power Electronics, Electrical Drives, Automation and Motion (SPEEDAM)*, 22-24 June 2022, pp. 785–790.

[4] Y. Xu, Y. Liang, X. Yuan, X. Wu, and Y. Li, "Experimental assessment of high frequency bearing currents in an induction motor driven by a SiC inverter," *IEEE Access*, vol. 9, pp. 40 540–40 549, 2021.

[5] F. Schulte and M. Pfost, "Experimental investigation of the impact of soft switching on capacitive bearing currents in SiC-based motor drives," in *2023 IEEE International Electric Machines and Drives Conference (IEMDC)*, 15-18 May 2023, pp. 1–6.

[6] O. Magdun, Y. Gemeinder, A. Binder, and K. Reis, "Calculation of bearing and common-mode voltages for the prediction of bearing failures caused by EDM currents," in *8th IEEE Symposium on Diagnostics for Electrical Machines, Power Electronics & Drives*, 2011, pp. 462–467.

PCIM Europe 2024, 11– 13 June 2024, Nuremberg   DOI: 10.30420/566262474

Core Loss Model for considering Anisotropy and Temperature Effects on Electrical Steel under Power Electronic Conditions

Michael Owzareck[1], Sascha Langfermann[1], Lukas Fräger[1], Christian Kliesch[1]

[1]BLOCK Transformatoren-Elektronik GmbH, Germany,

Corresponding author: Michael Owzareck, Michael.Owzareck@block.eu
Speaker: Michael Owzareck, Michael.Owzareck@block.eu

Abstract

This paper presents a method for considering the influence of temperature and magnetic anisotropy on the core losses of grain-oriented electrical steel. To achieve this, the specific electrical resistance of grain-oriented electrical steel was measured as a function of temperature, considering the rolling direction, using a specially developed measurement setup in a climate chamber. The core losses were measured in the climate chamber using a 12.5 cm temperature-resistant Epstein frame custom-made for this purpose. The developed core loss model is based on the loss separation method and was tested for various typical magnetization scenarios in power electronics, with comparisons to measurements.

1   Introduction

Typically, manufacturer specifications for electrical steel are measured under laboratory conditions at an ambient temperature of 20°C. In practice, developers of passive magnetic components use these datasets to parameterize their loss models. Since the Curie temperature of silicon-iron is well above the usual insulation class limits, the influence of temperature dependence of the hysteresis loss is neglected [3,6]. In practice, magnetic cores of filter chokes often reach temperatures well above 100°C, leading to deviations in the calculation of specific core losses and measurements in the testing field. As the specific electrical resistance increases with rising temperature [5,7], this has a direct impact on eddy current losses, which are the dominant loss component, especially in power electronics applications.

2   Specific electric resistance of electric steel

Typically, it is argued that the temperature influence on core losses in typical electrical silicon steel sheets is negligible. However, the goal in the optimized development of passive magnetic components is to achieve as little deviation as possible from modeling to reality. Since all metals have electrical conductivity due to the polycrystalline structure of metals, the electrical conductivity influ-

ences the penetration depth and thus the skin effect in the plane of the sheet. This affects the specific eddy current losses [4]. This loss component plays a crucial role in power electronics applications at higher frequency components in the magnetization current of passive magnetic components.

2.1   Measurement setup for determining the electrical resistance

In the datasheets of manufacturers of silicon electrical steel, often only a single measurement value for the specific electrical resistance is provided for an ambient temperature of 20°C. To determine the temperature dependence of the specific resistance by themselves and to improve the quality of the calculation results, the measuring apparatus shown in Fig. 1 was developed.

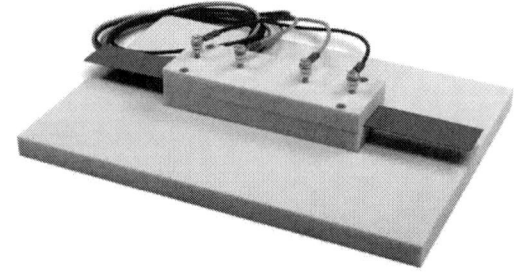

Fig. 1:   A specially developed test for measuring the specific electrical resistance for Epstein samples of 300 mm x 30 mm.

The measuring apparatus shown in Fig.1 is designed so that a typical Epstein sample with dimensions of 300 mm x 30 mm can be inserted into a body made of Durostone. The specific electrical resistance is determined technically by four measuring points. A constant direct current is applied at the outer measuring points. At the middle measuring points, the voltage drop is measured. From these two measured values, the specific electrical resistance can be determined as follows using Eq.1:

$$\rho_C = \frac{d_S \cdot b_S}{l_V} \cdot \frac{U}{I} \qquad (1)$$

Here, d_S is the sheet thickness and b_S is the sheet width. The distance between the measuring points for voltage measurement is l_V. Thus, the specific electrical resistance ρ_C can be measured indirectly through the constant imposed current I and the voltage drop U.

2.2 Measurement results for the electrical resistance of various electrical sheets

In Fig. 2, an excerpt from the measurements is shown. Here, for example, the specific electrical resistance for the electrical steel grades M165-35S, M400-50A, and N0-10 was measured as a function of temperature for various directions relative to the rolling direction.

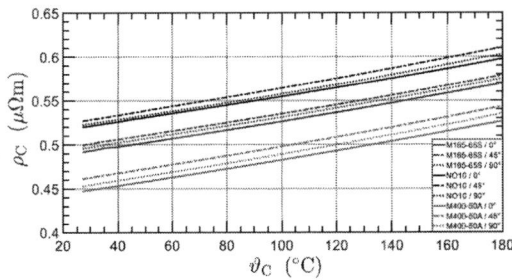

Fig. 2: Measured specific electrical resistances for the electrical steel grades M165-35S, M400-650A, and NO10 as a function of temperature for various directions relative to the rolling direction.

These measurement data will be used later in the paper to correct the penetration depth andalso the

eddy current losses accordingly. As can be seen from the figure, a slight anisotropy is also observed in the specific electrical resistance, which is due to the crystal anisotropy in polycrystalline solids. Based on these measurements, one would initially assume that the influence of temperature should not be significant. However, especially in grain-oriented electrical steel, where the magnetic anisotropy is strongly pronounced, this temperature influence becomes more pronounced in combination.

3 Measurement of the magnetic properties of electrical steel as a function of temperature

The typical manufacturer specifications for electrical steel, such as the normal magnetization curve and the specific core losses as a function of magnetic flux density or frequency, are measured at ambient temperature. These measurements are typically conducted using a 25 cm Epstein frame, as shown in Fig. 3.

Fig. 3: Standard 25 cm - Epstein frame according to IEC 60404-2 for measuring the magnetic properties of electrical steel.

To measure the magnetic properties of electrical steel as a function of temperature, a specially designed temperature-resistant 12.5 cm Epstein frame made of Durostone, as shown in Fig. 5, was developed. This makes it possible to conduct the measurements in a climatic chamber, which would typically not be possible with a classic Epstein frame, as the materials used there would not withstand the temperatures.

PCIM Europe 2024, 11– 13 June 2024, Nuremberg DOI: 10.30420/566262474

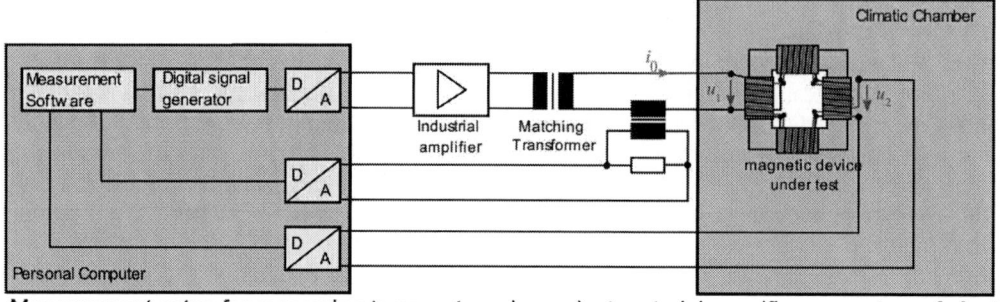

Fig. 4: Measurement setup for measuring temperature-dependent material-specific parameters of electrical steel in the climatic chamber.

This 17.5cm Epstein frame has a shortened magnetic field line length compared to the 25cm Epstein frame, resulting in a 32% reduction in the required apparent power demand of the magnet device under test. Due to the reduced apparent power demand and load adjustment using matching transformers, this enables measurements at medium frequencies up to 10 kHz and high magnetic flux densities with the measurement setup depicted in Fig. 4. Previous investigations, as in [8],[11],[14], use toroidal cores as test samples to examine the influence of temperature. Or the wattmetric method was not used, but rather argued through thermal images, as in [9]. Similar investigations in the climate chamber with an Epstein frame were also conducted by [10],[12]

Fig. 5. Optimize temperature-resistant 12.5 cm Epstein frame made of Durostone for low parasitic capacitances.

The apparent power from Eq.2 is directly proportional to the volume V_C of the test sample and thus directly proportional to the length of magnetic field lines. Now, if the magnetic properties are to be measured at higher frequency f and simultaneously at high magnetic field strength H or magnetic flux density B, the apparent power S_C demand of the magnetic device under test is very

high and must be provided by the industrial amplifier in Fig. 4. For this purpose, the Epstein frame is typically reduced in size. In this case, the apparent power demand decreases from the 25 cm Epstein frame in Fig. 2 to the 17.5 cm Epstein frame in Figure 5 by 32%, enabling the measurements.

$$S_C = \pi \cdot f \cdot \hat{B} \cdot \hat{H} \cdot V_C = \pi \cdot f \cdot \hat{B} \cdot \hat{H} \cdot A_C \cdot l_C \quad (2)$$

$$S_{17,5} = \frac{l_{C17,5}}{l_{C25}} \cdot S_{25} = 0{,}68 \cdot S_{25} \quad (3)$$

4 Theoretical discussion of the temperature dependence and magnetic anisotropy of classical eddy current losses.

The classical eddy current losses p_E can be calculated according to Eq. 4 as follows. Here, the skin effect in the sheet metal is considered:

$$p_E(\vartheta, \alpha) = \frac{\rho_C(\vartheta) \cdot \hat{B}^2}{d_S^{~2} \cdot \rho_V} \cdot \gamma \cdot \frac{\sinh(\gamma) - \sin(\gamma)}{\cosh(\gamma) - \cos(\gamma)} \quad (4)$$

Here, ρ_V is the specific mass-related density and d_S is the sheet thickness. The specific electrical resistance ρ_C is a function of temperature ϑ of the core, and thus the specific classical eddy current loss $p_E(\vartheta)$ is also temperature dependent.
Here γ is a quotient of the sheet thickness d_S and the penetration depth δ of the magnetic field:

$$\gamma(\vartheta, \alpha) = \frac{d_S}{\delta} = \frac{d_S}{\sqrt{\dfrac{\rho_C(\vartheta)}{\pi \cdot \mu_0 \cdot \mu_r(\alpha)}}} \quad (5)$$

Now, the penetration depth δ is a function of the specific electrical resistance $\rho_C(\vartheta)$ and of the

3373

magnetic relative permeability $\mu_r(\alpha)$. The magnetic relative permeability depends on the magnetization direction α in relation to the rolling direction. Thus, the specific classical eddy current losses $p_E(\vartheta, \alpha)$ depend both on the temperature and on the magnetization direction. Typical developers of passive magnetic components usually do not have their own measuring equipment and the time to conduct their own magnetic material measurements, and therefore rely on the data sheets provided by the material manufacturers. However, these data sheets for electrical steels used as core materials only contain measurement data for the 0° magnetization direction α relative to the rolling direction, which leads to significant calculation errors, especially in grain-oriented electrical steels, due to insufficient data sets. In Fig. 6, measured data sets with the measurement setup from Fig. 4 are presented. It is clearly visible that when magnetized at 45° or 90° relative to the rolling direction, the relative amplitude permeability $\mu_r(\alpha)$ is significantly lower than at 0° magnetization direction relative to the rolling direction.

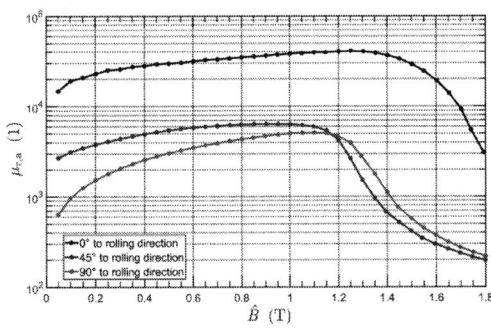

Fig. 6: Relative amplitude permeability as a function of the magnetic flux density B for various magnetization directions at 20°C ambient temperature of M165-35S grain-oriented electrical steel.

When the insights gathered from measurements of $\mu_r = f(\alpha, \hat{B})$ and $\rho_C(\vartheta)$ are inserted into Eq. 4 and 5, and calculated for a constant magnetic flux density of 1 Tesla for the electrical steel M165-35S, the results for the magnetization directions α of 0°, 45°, and 90° are shown in Fig. 7 to 9 as the specific classical eddy current losses $p_E(\vartheta, \alpha)$. Comparatively, it is clearly visible that the influence of temperature only becomes apparent with the onset of the skin effect above the eddy current limit frequency. Fig. 10 shows the error that would typically be made in the development of passive magnetic components if only the data sheet specifications were used. These specify, for example, only

the 0° magnetization direction and the specific electrical resistance of grain-oriented electrical steel at an ambient temperature of 20°C. Fig. 10 illustrates the ratio of specific classical eddy current losses at a 45° magnetization direction at any temperature to specific classical eddy current losses at a 0° magnetization direction and 20°C. Above the eddy current limit frequency, a factor of about 16 must be considered if the magnetic anisotropy and temperature dependence are not taken into account.

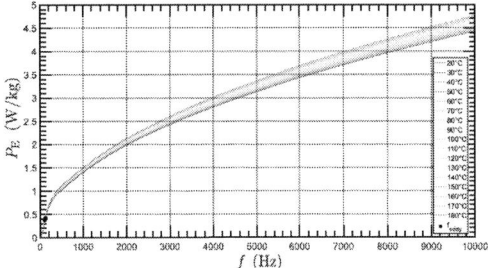

Fig. 7: Calculated specific classical eddy current loss for M165-35S grain-oriented electrical steel with d_S= 0.35 mm and α = 0° magnetization direction at $\hat{B}$ = 1 T = const.

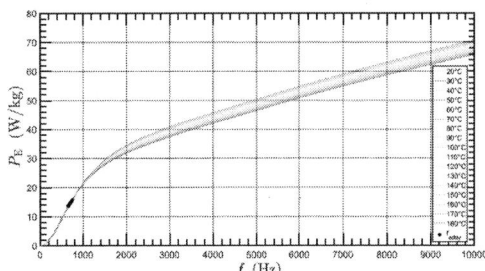

Fig. 8: Calculated specific classical eddy current loss for M165-35S grain-oriented electrical steel with d_S= 0.35 mm and α = 45° magnetization direction at $\hat{B}$ = 1 T = const.

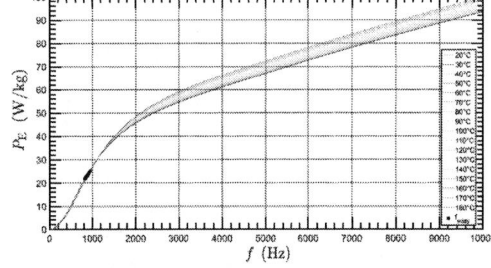

Fig. 9: Calculated specific classical eddy current loss for M165-35S grain-oriented electrical steel with d_S= 0.35 mm and α = 90° magnetization direction at $\hat{B}$ = 1 T=const.

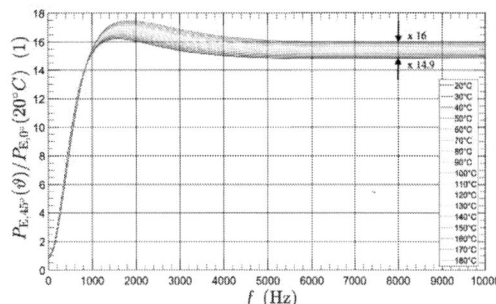

Fig. 10: The ratio of specific classical eddy current losses at a 45° magnetization direction to the specific classical eddy current losses at a 0° magnetization direction at 20°C

Fig. 11 illustrates the error that would occur if a developer relies on manufacturer specifications and works in application areas where the magnetization direction in a grain-oriented electrical steel would be 90°.

It also shows the ratio of specific classical eddy current losses at a 90° magnetization direction at any temperature to specific classical eddy current losses at 0° magnetization direction and 20°C. Above the eddy current limit frequency, a factor of about 22 must be considered if the magnetic anisotropy and temperature influence are not considered.

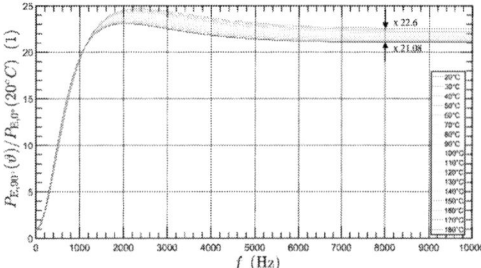

Fig. 11: The ratio of specific classical eddy current losses at a 90° magnetization direction to the specific classical eddy current losses at a 0° magnetization direction at 20°C

5 Core loss model based on the separation of losses considering temperature and anisotropy.

The aim of this paper was not only to empirically substantiate the errors caused by magnetic anisotropy under the influence of temperature on the specific electrical resistance but also to develop an efficient method to address these influences de-

scribed in Chapter 4 using as few pre-known datasets as possible. "There are models that already take temperature influences into account, such as [13],[15], but they require many datasets to parameterize them. The chosen model approach is based on the loss separation by Bertotti [1]. Here, the specific core loss should now be a function of temperature and the angle of magnetization. The specific core loss $p_C(\vartheta, \alpha)$ is the sum of specific hysteresis loss p_H specific classical eddy current loss $p_E(\vartheta, \alpha)$, and specific anomalous eddy current loss $p_A(\vartheta, \alpha)$, which arises from eddy current losses at the Bloch walls caused by their movement through the material.

$$p_C(\vartheta, \alpha) = p_H(\alpha) + p_E(\vartheta, \alpha) + p_A(\vartheta, \alpha) \quad (6)$$

The classical eddy current loss can be calculated for sinusoidal magnetic flux density profiles according to [1].

$$p_E(\vartheta, \alpha) = \frac{\pi^2 \cdot d_S^2}{\rho_C(\vartheta, \alpha) \cdot 6 \cdot \rho_V} \cdot \hat{B}^2 \cdot f^2 \quad (7)$$

The anomalous eddy current results from [2] for sinusoidal magnetic flux densities:

$$p_A(\vartheta, \alpha) = 8{,}76 \sqrt{\frac{G_{Bw} A_e V_0}{\rho_C(\vartheta, \alpha)}} \cdot (\hat{B} \cdot f)^{1{,}5} \quad (8)$$

Typically, Eq.(9) is used as a simplification of Eq. (6), (7), and (8), where k_E is the classical eddy current coefficient and k_A is the anomalous eddy current coefficient. Both k_E and k_A are typically determined by curve fitting to measurement data of specific core losses.

$$p_C = p_H + k_E \cdot \hat{B}^2 f^2 + k_A \cdot \hat{B}^{1{,}5} f^{1{,}5} \quad (9)$$

According to Fiorillo [2], the classical eddy current losses can be calculated for arbitrary waveform shapes of the magnetic flux density:

$$p_E(\vartheta, \alpha) = \frac{d_S^2}{\rho_C(\vartheta, \alpha) \cdot 12 \cdot \rho_V} \frac{1}{T} \int_0^T \left(\frac{dB}{dt}\right)^2 dt \quad (10)$$

We have made the additional assumption here that the loss depends both on the temperature and on the magnetization angle.

The specific anomalous eddy current loss for arbitrary waveform shapes of the magnetic flux density is derived as follows:

$$p_A(\vartheta, \alpha) = \sqrt{\frac{G_{Bw} A_e V_0}{\rho_C(\vartheta, \alpha)}} \cdot \frac{1}{T} \int_0^T \left|\frac{dB}{dt}\right|^{1,5} \cdot dt \quad (11)$$

We also assume a dependence on temperature and anisotropy. Since typically only the specific core losses, measured with a sinusoidal waveform, are available in the manufacturers' datasheets, and thus only $k_E(\vartheta, \alpha)$ and $k_A(\vartheta, \alpha)$ are known from Equation (9), Equation (13) was developed from Equation (12)

$$\frac{k_{E,G}(\vartheta, \alpha)}{k_E(\vartheta, \alpha)} = \frac{\dfrac{d_S^2}{\rho_C(\vartheta, \alpha) \cdot 12 \cdot \rho_V}}{\dfrac{\pi^2 \cdot d_S^2}{\rho_C(\vartheta, \alpha) \cdot 6 \cdot \rho_V}} = \frac{1}{2 \cdot \pi^2} \quad (12)$$

With Eq. 13, it is now possible to calculate the classical eddy current loss for arbitrary waveform shapes using the classical eddy current coefficients $k_E(\vartheta, \alpha)$ extracted from curve fitting under sinusoidal conditions.

$$p_E(\vartheta, \alpha) = k_E(\vartheta, \alpha) \frac{1}{2 \cdot \pi^2} \cdot \frac{1}{T} \cdot \int_0^T \left(\frac{dB}{dt}\right)^2 dt \quad (13)$$

By a clever comparison of coefficients in Equation (14), the anomalous eddy current coefficient $k_A(\vartheta, \alpha)$ can also be converted into a general formulation:

$$\frac{k_{A,G}(\vartheta, \alpha)}{k_A(\vartheta, \alpha)} = \frac{\sqrt{\dfrac{G_{Bw} A_e V_0}{\rho_C(\vartheta, \alpha)}}}{8,76 \cdot \sqrt{\dfrac{G_{Bw} A_e V_0}{\rho_C(\vartheta, \alpha)}}} = \frac{1}{8,76} \quad (14)$$

With Equation (15), the anomalous eddy current coefficient, which was extracted from curve fitting under sinusoidal conditions, can now be used to calculate the anomalous eddy current losses for arbitrary waveform shapes as follows:

$$p_A(\vartheta, \alpha) = \frac{k_A(\vartheta, \alpha)}{8,76} \cdot \frac{1}{T} \cdot \int_0^T \left|\frac{dB}{dt}\right|^{1,5} \cdot dt \quad (15)$$

Thus, the specific core loss for arbitrary waveform shapes can be calculated from the loss coefficients $k_E(\vartheta, \alpha)$ and $k_A(\vartheta, \alpha)$ based on manufacturer specifications or one's own measurements under sinusoidal conditions using Eq. 16.

$$p_C(\vartheta, \alpha) = p_H + \frac{k_E(\vartheta, \alpha)}{2\pi^2} \frac{1}{T} \cdot \int_0^T \left(\frac{dB}{dt}\right)^2 dt$$
$$+ \frac{k_A(\vartheta, \alpha)}{8,76} \frac{1}{T} \cdot \int_0^T \left|\frac{dB}{dt}\right|^{1,5} dt \quad (16)$$

Here, the loss coefficients $k_E(\vartheta, \alpha)$ and $k_A(\vartheta, \alpha)$ inherently have a temperature dependence since they were derived through curve fitting at 20°C; therefore, this equation is only valid at this temperature. Here, the measurement data from Figure 2, which were taken with the measuring apparatus in Figure 1, come into play. To minimize the number of datasets needed by developers of passive magnetic components, a model was aimed for that relies solely on the loss coefficients $k_E(\vartheta, \alpha)$ and $k_A(\vartheta, \alpha)$, which were extracted from sinusoidal datasets, and the dataset from Figure 2, the specific electrical resistance as a function of temperature $\rho_C = f(\vartheta)$. It was recognized that the loss coefficient $k_A(\vartheta, \alpha)$ is proportional to the reciprocal value of the specific electrical resistance.

$$k_E(\vartheta, \alpha) \sim \frac{1}{\rho_C(\vartheta, \alpha)} \quad (17)$$

This implies for arbitrary temperatures:

$$\frac{k_E(\vartheta, \alpha)}{k_E(20°C, \alpha)} = \frac{\rho_C(20°C, \alpha)}{\rho_C(\vartheta, \alpha)} \quad (18)$$

It was also recognized that the loss coefficient $k_E(\vartheta, \alpha)$ is proportional to the reciprocal value of the square root of the specific electrical resistance:

$$k_A(\vartheta, \alpha) \sim \sqrt{\frac{1}{\rho_C(\vartheta, \alpha)}} \quad (19)$$

This implies for arbitrary temperatures:

$$\frac{k_A(\vartheta, \alpha)}{k_A(20°C, \alpha)} = \sqrt{\frac{\rho_C(20°C, \alpha)}{\rho_C(\vartheta, \alpha)}} \quad (20)$$

So, equation (16) is reformulated with the loss coefficients $k_E(20°, \alpha)$ and $k_E(20°, \alpha)$ and the ratios of the specific electrical resistance at any temperature to the specific electrical resistance at which the loss coefficients were determined. This equation now represents a solution to the challenges presented in Chapter 4.

$$p_{\mathrm{C}}(\vartheta, \alpha) = p_{\mathrm{H}} +$$

$$\frac{k_{\mathrm{E}}(20°C, \alpha)}{2\pi^2} \frac{\rho_{\mathrm{C}}(20°C, \alpha)}{\rho_{\mathrm{C}}(\vartheta, \alpha)} \frac{1}{T} \int_0^T \left(\frac{dB}{dt}\right)^2 dt$$

$$+ \frac{k_{\mathrm{A}}(20°C, \alpha)}{8{,}76} \sqrt{\frac{\rho_{\mathrm{C}}(20°C, \alpha)}{\rho_{\mathrm{C}}(\vartheta, \alpha)}} \frac{1}{T} \int_0^T \left|\frac{dB}{dt}\right|^{1{,}5} dt \quad (21)$$

The total specific core loss in Eq.(22) thus results from the sum of the sub-volumes of the core, which are magnetized in different magnetization directions. Here, the individual terms are calculated using Eq. (21):

$$p_{\mathrm{C}}(\vartheta, \alpha) = p_{\mathrm{C}}(\vartheta, 0°) + p_{\mathrm{C}}(\vartheta, 45°) + p_{\mathrm{C}}(\vartheta, 90°) \quad (22)$$

6 Model Verification

Eq.(21) was compared with measurement data obtained using the temperature-resistant 12.5 cm Epstein frame in the experimental setup shown in Fig. 5 in the climate chamber. The sample used was the grain-oriented electrical steel M165-35S. The fig.12 to14 illustrates comparisons at a core temperature of 20°C for the magnetization angles of 0°, 45°, and 90° with a sinusoidal waveform across a frequency range from 50 Hz to 10 kHz. Eq. (21) shows good agreement.

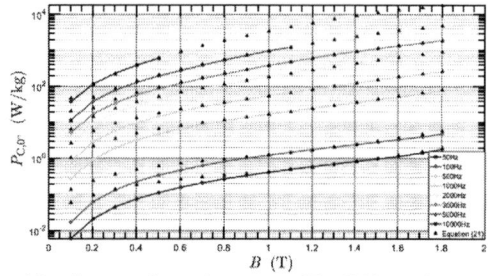

Fig. 12: Comparison between Eq.(21) and measurement values of an M165-35S sample with 0° magnetization direction at 20°C and sinusoidal waveform.

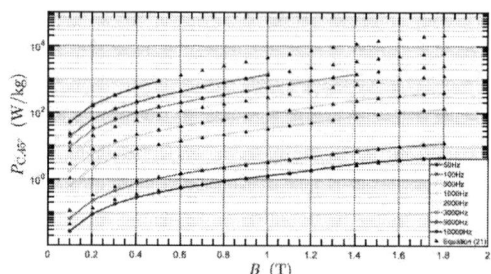

Fig. 13: Comparison between Eq.(21) and measurement values of an M165-35S sample with 45° magnetization direction at 20°C and sinusoidal waveform.

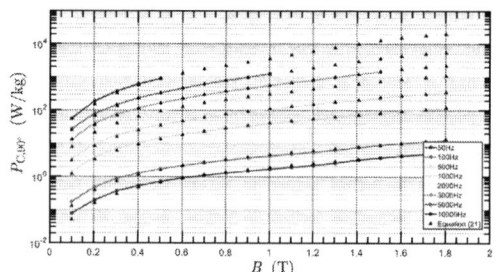

Fig. 14: Comparison between Eq.(21) and measurement values of an M165-35S sample with 90° magnetization direction at 20°C and sinusoidal waveform

Fig. 15 shows a comparison between measured and calculated specific core losses at a core temperature of 150°C and a magnetization frequency of 1000 Hz. Measurements and calculations were selectively conducted in the 0° magnetization direction. The temperature factor from Eq.(21) demonstrates applicability with respect to the measurement data.

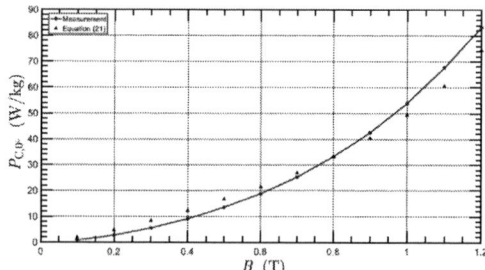

Fig. 15: Comparison between measurement and calculation using Eq.(21) at a frequency of 1000 Hz and a core temperature of 150 °C for an M165-35S sheet sample in 0° magnetization direction and sinusoidal waveform

Fig. 16 once again partially presents the correction factor for the temperature from Eq.(21) at a constant magnetic flux density of 1 Tesla and a core temperature of 100°C. Here too, the results are satisfactory.

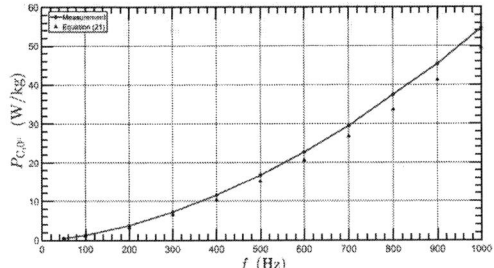

Fig. 16: Comparison between Eq.(21) and measurement data at a constant 1 Tesla and a core temperature of 100°C for an M1565-35S sample in 0° magnetization direction.

Lastly, it should also be mentioned that the eq.(11) withstands power electronic waveforms. For example, the waveform of a converter output filter or a sinusoidal filter was chosen. Here, a 10% ripple current is superimposed on the fundamental frequency.

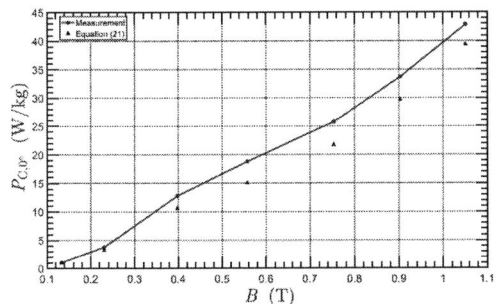

Fig. 17: Comparison between measurement and calculation at 20°C and a waveform of a sinusoidal filter or an inverter output filter with 10 ripples at 4 kHz for an M1565-35S sample in 0° magnetization direction

7 Conclusion

This paper demonstrates the dependency of core losses in electrical steel sheets on magnetic anisotropy and temperature. For this purpose, a special measurement apparatus was developed to measure the specific electrical resistance of the electrical steel sheets depending on the rolling direction. To measure the magnetic properties, such as the normal magnetization curve, and the specific core losses as a function of temperature, frequency, and magnetic flux density, a special 12.5

Epstein frame made of temperature-resistant material was developed to conduct measurements in the climatic chamber. Furthermore, a core loss model was developed, which takes into account the magnetic anisotropy and the temperature dependence of eddy current losses. This model utilizes loss coefficients measured under sinusoidal conditions and at a temperature of 20°C. With this model, specific losses can now be predicted for arbitrary waveforms, magnetization directions, and core material temperatures. Comparisons with measurements demonstrate a good agreement between theory and practical measurement data

8 References

[1] G. Bertotti: Hysteresis in magnetism 1998, Academic Press, p. 558

[2] E. Barbisio, F. Fiorillo and C. Ragusa, "Predicting loss in magnetic steels under arbitrary induction waveform and with minor hysteresis loops," in IEEE Transactions on Magnetics, vol. 40, no. 4, pp. 1810-1819, July 2004

[3] M. Maclaren, "The effect of temperature upon the hysteresis loss in sheet steel," in Proceedings of the American Institute of Electrical Engineers, vol. 31, no. 10, pp. 1895-1905, Oct. 1912, doi: 10.1109/PAIEE.1912.6660097.

[4] K. Foster, "Temperature dependence of loss separation measurements for oriented silicon steels," in IEEE Transactions on Magnetics, vol. 22, no. 1, pp. 49-53, January 1986,

[5] A. Mouillet, J.Ille,M. Dami,M. Akroune: Magnetic and loss characteristics of nonoriented silicon-iron under unconventional conditions, IEE Proceedings - Science, Measurement and Technology , Vol. 141, No. 1 Institution of Engineering and Technology (IET), p. 75-78

[6] V. T. Morgan, B. Zhang and D. Findlay, "Effects of temperature and tensile stress on the magnetic properties of a steel core from an ACSR conductor," in IEEE Transactions on Power Delivery, vol. 11, no. 4, pp. 1907-1913, Oct. 1996, doi: 10.1109/61.544275.

[7] T. Saito, S. Takemoto and T. Iriyama, "Resistivity and core size dependencies of eddy current loss for Fe-Si compressed cores," in IEEE Transactions on Magnetics, vol. 41, no. 10, pp. 3301-3303, Oct. 2005

[8] N. Takahashi, M. Morishita, D. Miyagi and M. Nakano, "Examination of Magnetic Properties of Magnetic Materials at High Temperature

Using a Ring Specimen," in IEEE Transactions on Magnetics, vol. 46, no. 2, pp. 548-551, Feb. 2010,

[9] H. Shimoji, B. E. Borkowski, T. Todaka and M. Enokizono, "Measurement of Core-Loss Distribution Using Thermography," in IEEE Transactions on Magnetics, vol. 47, no. 10, pp. 4372-4375, Oct. 2011

[10] Z. Cheng et al., "Modeling of Magnetic Properties of GO Electrical Steel Based on Epstein Combination and Loss Data Weighted Processing," in IEEE Transactions on Magnetics, vol. 50, no. 1, pp. 1-9, Jan. 2014, Art no. 6300209,

[11] N. Yogal, C. Lehrmann, M. Henke and H. Zheng, "Measurement, simulation and calculation using fourier transformation of iron losses for electrical steel sheets with higher frequency and temperature effects," 2016 XXII International Conference on Electrical Machines (ICEM), Lausanne, Switzerland, 2016, pp. 2655-2661

[12] M. L. Ababsa, O. Ninet, G. Velu and J. P. Lecointe, "High-Temperature Magnetic Characterization Using an Adapted Epstein Frame," in *IEEE Transactions on Magnetics*, vol. 54, no. 6, pp. 1-6, June 2018, Art no. 6201406,

[13] G. Novak, J. Kokošar, M. Bricelj, M. Bizjak, D. Steiner Petrovič and A. Nagode, "Improved Model Based on the Modified Steinmetz Equation for Predicting the Magnetic Losses in Non-Oriented Electrical Steels That is Valid for Elevated Temperatures and Frequencies," in *IEEE Transactions on Magnetics*, vol. 53, no. 10, pp. 1-5, Oct. 2017, Art no. 2001805

[14] A. Yao, A. Adachi and K. Fujisaki, "Iron loss characteristics of electric motor in high-temperature environment," *2017 IEEE International Electric Machines and Drives Conference (IEMDC)*, Miami, FL, USA, 2017

[15] S. Xue, J. Feng, S. Guo, J. Peng, W. Q. Chu and Z. Q. Zhu, "A New Iron Loss Model for Temperature Dependencies of Hysteresis and Eddy Current Losses in Electrical Machines," in *IEEE Transactions on Magnetics*, vol. 54, no. 1, pp. 1-10, Jan. 2018, Art no. 8100310,

PCIM Europe 2024, 11– 13 June 2024, Nuremberg   DOI: 10.30420/566262476

Circular Economy Oriented and Reconfigurable Planar Transformer Design for Isolated DC/DC Converters

Fabian Groon[1], Hamzeh Beiranvand[1,2], Görkem Can[1], Marco Liserre[1,2,3]

[1] Kiel University, Germany
[2] Kiel Nano, Surface and Interface Science (KiNSIS), Germany
[3] Fraunhofer Institute for Silicon Technology ISIT, Germany

Corresponding author: Fabian Groon, fagr@tf.uni-kiel.de
Speaker: Fabian Groon, fagr@tf.uni-kiel.de

Abstract

The need for high power density, efficiency and predictable parasitic parameters is driving the use of high-frequency planar transformers (PTs). The scarcity of resources and the generated electronic (e)-waste impose new constraints on the PT design. In circular economy, hardware design should be adapted such that components are reused, repaired, maintained, refurbished and upgraded. Therefore, this paper introduces modularity to the printed circuit board (PCB) layer design for PTs as well as proposes a reconfigurable design methodology to fulfill the circular economy implications on PTs. The efficiency (η) and power density (ρ) of the proposed PT are optimized employing $\eta\rho$-Pareto optimization and the reconfigurable and modular PTs are selected from the optimal solutions. Four different PTs are built and extensively tested. The theoretical, analytical, and experimental results demonstrate that the objectives of circular economy can be addressed without violating the $\eta\rho$-Pareto optimal solutions.

1 Introduction

Transformers enable isolated DC/DC and AC/AC conversion in a wide range of products covering applications such as home appliances, grid and e-mobility. Planar Transformers (PTs) with high potential for weight and volume reduction have been widely used in on-board chargers for electric vehicles. This has motivated the design formulations to maximize efficiency and power density without considering the contribution of PT design to electronic (e)-waste and its environmental impacts [1]. Thereby, it is sensible for a study to incorporate the constraints that minimize e-waste and enable reuse of PT parts.

A number of methods have been realized to design both conventional transformers and PTs. Regarding conventional transformer designs, genetic algorithm was applied to multi-objective transformer design optimization with goals optimizing weight, efficiency, and inductance [2]. A comprehensive study was conducted in [3], including control variables such as number of turns, frequency, core and conductor material and their constraints to

maximize power density and efficiency. Similarly, the scaling laws for medium-frequency transformers are derived by setting constraints and optimizing for frequency and number of turns [4]. Even more variables were considered in [5], where a 100 kW 10 kHz transformer was optimized based on database input and direct user input. The constraint imposed by the converter voltage and frequency on transformer design were addressed in [6]. Similar approaches have been applied to design optimization of PTs [7, 8, 9]. So far, it has not been shown how a design can enable reuse of transformer parts before complete wear out and recycling of transformers.

The concept of circular economy was introduced to use resources for the longest time possible [10, 11, 12]. It is of a great advantage to consider it early in the design process, as a large part of the sustainability is already determined at this stage [13]. While circular economy has already been described in power electronics and batteries [14], it has never been introduced to PTs. In power electronics, the energy consumption of inverters in standby mode, which reduces system efficiency,

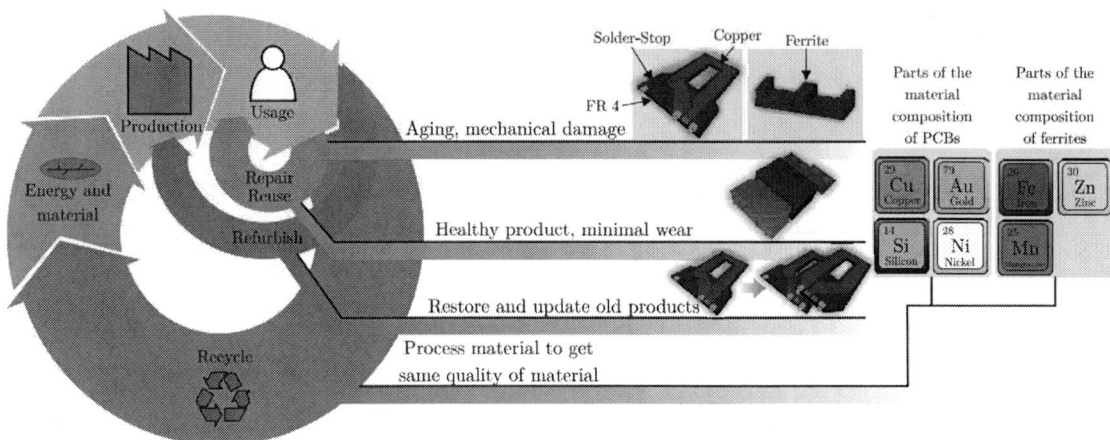

Fig. 1: Circular economy applied to PTs for e-waste prevention.

has a significant impact on the environmental footprint [15]. In case of PTs, high electrical stress and/or slow aging over time might lead to damage to the insulation, copper and core [16]. However, a part of the failed PTs can still be reused. A modular structure enables repairs, reuse of modules or the possibility of upgrading a product [10, 15, 17]. The upgrade option of a product is particularly interesting, as this allows a product to continue to be used despite increasing needs, production time can be reduced and scalability increased [18, 19]. Nevertheless, PTs also have an impact on e-waste and shall be included in the circular economy.

The lack of transformer designs compatible with circular economy constrained optimization motivates this paper to develop a PT structure that meets the requirements of the circular economy. A PCB layout is proposed that can be easily repaired, reused, upgraded and reconfigured. A design optimization including circular economy constraints is presented and it is shown that a Pareto-optimal solution can be achieved despite the circular economy constraints. Four optimal PTs solutions are built that scale the voltage between 400 V and 800 V and power between 5 to 7.5 kW. Experiments are carried out on the designed PTs to demonstrate the feasibility of the PCB layout and the Pareto optimal solutions.

2 Proposed Reconfigurable PCB-Design

The economic cycle is conceptualized for PTs in Fig. 1. Replacing damaged or worn PCBs or ferrites puts the product directly back into use and creates a small economic loop, which is very beneficial. Refurbishing used PTs helps to resell an old product by upgrading it and bringing it to new customers. The upgradability and reconfigurability will be a great advantage, making it easier to give the product a new life. If none of the previous steps can be achieved, the last step should be to recycle the transformers.

A conventional PCB layout design for a PT is shown in Fig. 2(a), where 3 different PCBs are used to create the primary side of the transformer. As it can be seen, the layers have different structures and each new PT should be designed from the beginning. The proposed PCB design is shown in Fig. 2(b). The top side features four connection terminals, while the bottom side has two. The particular example shows a 4-layer PCB with 2 turns per layer, where the bottom and top layers are kept free of turns and only have terminals for connection. This PCB design enables universal connection of the PCBs using the four connection terminals. The turns of the winding start at terminal 1 and finish at terminal 2 on the top of the PCB. The turns are connected through buried vias in the inner layers of the PCB.

Different configurations are easily introduced using this layout without modifying the PCB: interleaved, non-interleaved, and paralleled layers. The first configuration, shown in Fig. 3, represents the standard connection of the PCBs. The PCBs are connected in series and interleaved by design according to the number of turns in each PCB. In this example, the configuration is P-P-P-P-S-S-S-S-P-P-P-P and will continue in that way. In order to achieve this configuration, the second PCB is rotated 180° and terminal 4 is connected to terminal 2 of the pre-

PCIM Europe 2024, 11– 13 June 2024, Nuremberg   DOI: 10.30420/566262476

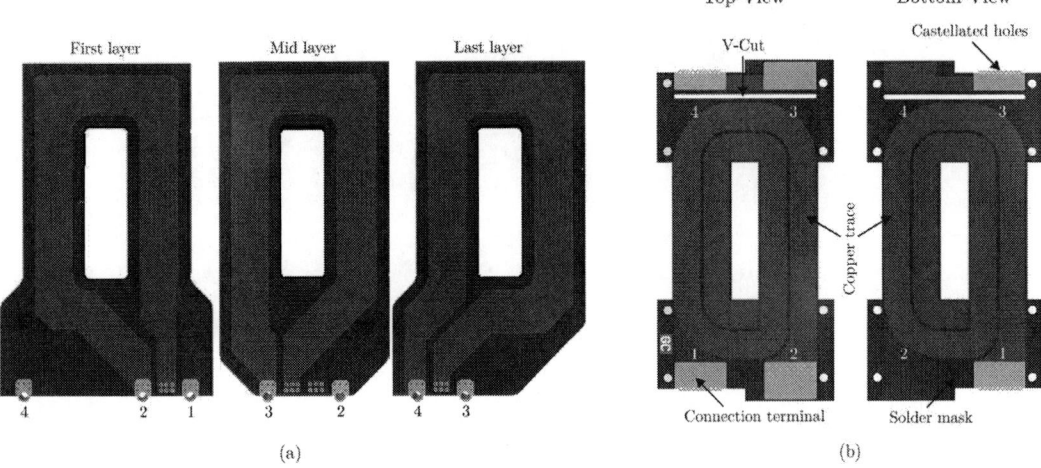

(a)   (b)

Fig. 2: Transformer PCB layout for (a) conventional design with different layers and (b) proposed PCB design top view and bottom view.

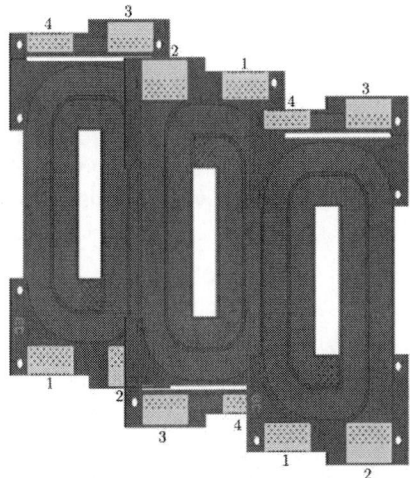

Fig. 3: Assembly of interleaved design.

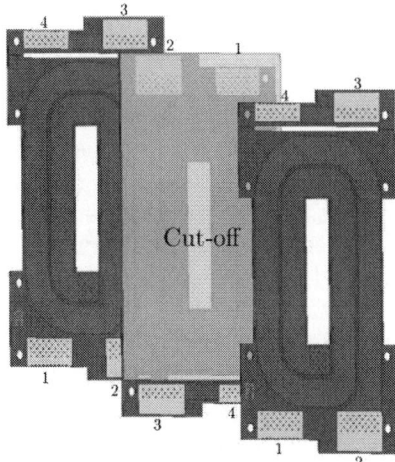

Fig. 4: Assembly of non-interleaved design.

vious PCB. By connecting terminals 3 and 4, the connection is moved back to the left side to start the next PCB for the primary on top of the PCB for the secondary. Then terminal 1 of the third PCB is connected to terminal 4 of the second PCB. This arrangement can be continued indefinitely until the desired number of turns is achieved. The design enables the product to be upgraded for higher voltage requirements in the future. Furthermore, it allows the PCBs to be easily reused and reconfigured for other applications.

The option to change the interleaving pattern is important and depends on the requirements of the application. Interleaving the windings results in a trade-off between low leakage inductance, low copper losses and high winding capacitance [9, 20]. It is clear that low leakage, low copper losses and

low winding capacitance cannot be achieved at the same time, and the interleaving pattern must be selected depending on the application. The proposed PCB design enables the degree of interleaving by controlling the number of turns inside each PCB. The second configuration is shown in Fig. 4 and implements a non-interleaved design. The V-cut of the proposed PCB design is used to achieve the non-interleaved design. Only the small PCB area that will be separated is used and connected as shown in Fig. 4. A filler material can be used to replace the missing part of the PCB. The third PCB is then connected to this terminal as before. This process continues until all the primary layers are connected. The next step is to connect the secondary side in exactly the same configuration as the primary side, with a 180° rotation on the z-

3382

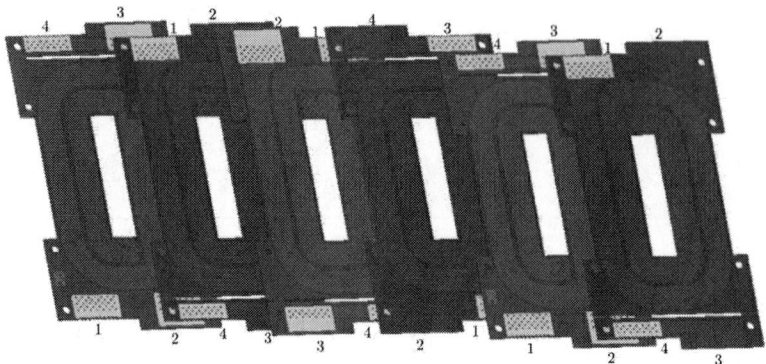

Fig. 5: Assembly of parallel layers.

axis. This configuration has the drawback of lower power density, as one layer of PCB thickness is used to achieve the non-interleaved connection. This configuration is not recommended where high power density is critical. However, this configuration can help to achieve better circular economy, as PCBs can be reused to achieve non-interleaved patterns.

The third and final possible configuration is shown in Fig. 5. This configuration demonstrates the use of parallel layers in addition to the series connections. It allows 2 layers to be connected in parallel before continuing with the series connection. In order to achieve this configuration, the second PCB is flipped and terminal 1 of the first PCB is connected to terminal 2 of the second PCB and terminal 2 of the first PCB is connected to terminal 2 of the second PCB. This results in a parallel connection of the layers. The first turns of the secondary side are then placed on top of this PCB stack and terminal 4 of the third PCB is connected to terminal 1 of the second PCB. The fourth board is then connected to the third PCB in the same way as the second PCB is connected to the first, with the addition of connecting terminal 4 of the fourth PCB to terminal 3 of the third PCB. This moves the primary connection back to the top left to connect the next PCBs for a continuous series connection of two parallel layers.

The proposed modular and reconfigurable PCB design allows for easy reuse for other applications, as the PCB can be reconfigured and modified to suit the new requirements. In addition, repair and maintenance of the PT is straightforward with this PCB design, as a failed PCB can be easily replaced with a new one.

3 Analysis of Planar Transformer Designs and Upgrade Potential

This section provides the design optimization methodology for the proposed PCB layout that meets the requirements of circular economy. Pareto optimal solutions are derived with the aim of maximizing efficiency (η) and power density (ρ) as in [6]. Four $\eta\rho$-Pareto optimal solutions are selected to maintain the modularity and reconfigurability constraints. Finally, finite element method (FEM) simulations are performed to validate the proposed PT designs.

3.1 Design Methodology

Flux density and current density are the main drivers of power losses in the transformers core and windings, respectively. The flux density of a transformer with applied square wave depends on the core area A_{E}, the number of turns N_{T}, the frequency f_{s} and the input voltage V_{in} and can be calculated using the following equation:

$$B_{\max} = \frac{V_{\mathrm{in}}}{4 f_{\mathrm{s}} N_{\mathrm{T}} A_{\mathrm{E}}}. \tag{1}$$

Based on this, the required core cross-section and the number of cores in parallel can be calculated. The core losses are calculated by the steinmetz equation [21, 22]:

$$P_{\mathrm{V,vol}} = C_{\mathrm{m}} f_{\mathrm{s}}^{\alpha} B_{\max}^{\beta}. \tag{2}$$

The current density in the windings is dependent on the current I, the copper width L_{Cu} and the copper thickness H_{Cu}:

$$J = \frac{I}{L_{\mathrm{Cu}} H_{\mathrm{Cu}}}. \tag{3}$$

This can be used to calculate the required copper area with copper width L_{Cu} and thickness H_{Cu}.

The winding losses of the PT can be calculated using the equation from [23]:

$$\frac{R_{AC}}{R_{DC}} = \frac{\xi}{2} \left[(H_{ext} - H_{int})^2 \frac{\sinh(2\xi) + \sin(2\xi)}{\cosh(2\xi) - \cos(2\xi)} \right. \\ \left. + 2H_{ext}H_{int} \frac{\sinh(\xi) - \sin(\xi)}{\cosh(\xi) + \cos(\xi)} \right], \quad (4)$$

where $\xi = \frac{H_{Cu}}{\delta}$ is the ratio between copper thickness H_{Cu} and skin depth δ, H_{ext} and H_{int} represents the magnetic field strength at the bottom and on top of the layer. The DC resistance of the winding for PTs can be estimated by:

$$R_{DC} = \frac{\rho_{Cu} \cdot MLT \cdot N_T}{L_{Cu}H_{Cu}}, \quad (5)$$

where ρ_{Cu} is the resistivity of copper, MLT the mean length track of one turn and L_{Cu} the width of one track. To calculate power density, the volume of the PT is estimated by

$$V_{PT} = (D_C N_C + 2L_{Cu}N_L)L_C H_C \quad (6)$$

where D_C is the length of the core, L_C the width of the core, H_C the height of the core, N_C the number of cores, L_{Cu} the width of the tracks per layer, and N_L is the number of tracks per layer.
A simplified temperature assumption from [24] is used for the calculation of the overall system temperature

$$T_{max} = \left(\frac{24}{V_C}\right)^{0.5} P_v + T_A, \quad (7)$$

where V_C is the core volume in cm³, P_v the total losses including core and winding losses and T_A the ambient temperature. The temperature limit is set to 125 °C, as above this value the core losses increase and the PCB starts to go into glass transition.

3.2 Pareto Optimization

The objectives are to maximize the efficiency (η) and power density (ρ) with consideration of the circular economy modularity. Control variables are B_{max}, J, H_{Cu}, N_T, N_L, the number of parallel layers N_P, and core configuration. The copper thickness is limited between 35 µm and 70 µm. All commercially available E and I core combinations are considered for optimization. The current density J is varied between 2 - 12 A/mm² and the flux density between 0.05 - 0.3 T. The constraints are defined by the core geometry and maximum temperature. The design specifications to carry out the

Tab. 1: Specifications of the PTs.

Parameter	
Output Power P_{out}	5 kW
Nominal input voltage $V_{in,nom}$	400 V
Nominal Output voltage $V_{out,nom}$	400 V
Nominal switching frequency f_s	50 kHz
Transformer turns ratio	1:1

$\eta\rho$-Pareto optimization of the initial Design A are given in Tab. 1.

The obtained results are shown in Fig. 6(a) for Design A. It results in a lower front constrained by the thermal boundary and an upper front constrained by the geometric boundary. Any solution that is located on the upper front of the geometric boundary is suitable for the selection of an optimal design. A trade-off is made between efficiency and power density. The final solution for the design is marked in Fig. 6(a) and the values for this solution are also given. A 3D render of the final solution is shown and is referred to as Design A. The frontmost values have not been selected as a design with a lower temperature should be selected. In addition, the circular economy should also be considered here, as the upgradability of the design is a vital point. Since the solution selected is a combination of an EI core, it can be upgraded to an EE core, which means that additional PCBs can fit into the core. In order to validate this, one of the frontmost solutions is also shown and will be referred to as Design E. The full list of parameters for both designs can be found in Tab. 2.

As shown in section 2, additional windings can be added in series or in parallel. For this reason, the next case to be considered is one in which a higher output power is required. The system power is to be increased from 5 kW to 7 kW at the same voltage. Again, a $\eta\rho$-Pareto optimization is performed, as shown in Fig. 6(b). This reveals that a possible solution at the upper boundary is the selected design for 5 kW, but now with two PCB layers in parallel and an EE core combination to accommodate the additional PCBs. The full list of parameters for the design is given in Tab. 2 and the design is referred to as Design B. The current density of 8.4 A/mm² is changed proportional to the previous current density of 12 A/mm², as the current density has increased by a factor of 1.4 and the copper thickness has doubled due to the parallel layers. This shows that the upgrade option for the circu-

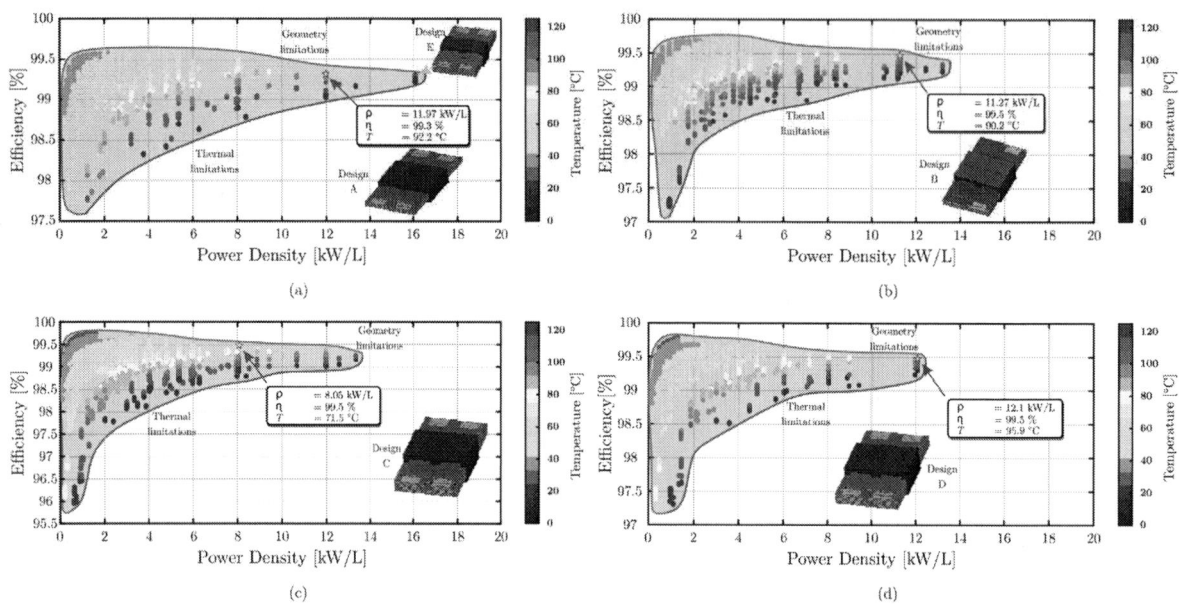

Fig. 6: $\eta\rho$-pareto-fronts for (a) 400 V 5 kW, (b) 400 V 7 kW, (c) 800 V 5 kW and (d) 800 V 7.5 kW.

lar economy is provided by the design and that an optimized design can still be achieved.

Another case considered is an increase in the required voltage from 400 V to 800 V. The final $\eta\rho$-Pareto-front is shown in Fig. 6(c). The optimally selected solution at the upper boundary of the Pareto front shows the solution of the previous transformer Design A but now with EE core combination to accommodate the 24 turns. The selected design referred to as Design C is shown in Fig. 6(c) and the parameters are given in Tab. 2. The current density of 6 A/mm² is changed proportional to the previous design as the required current has been halved. This again shows that an optimal solution can be found at the upper boundary of the Pareto front and the original selected design can be upgraded by adding another 3 PCBs in series and changing to the EE core.

The last case considered is an increase in voltage from 400 V to 800 V with an increase in power from 5 kW to 7.5 kW. The final $\eta\rho$-pareto-front is shown in Fig. 6(d). The selected solution represents the same system as Design C, since 24 turns are required and the current density of 9 A/mm² in Tab. 2 is altered proportional to 6 A/mm² due to the 50 % increase in current.

In summary, it can be shown that an upgrade option with the proposed PCB layout is possible by selecting the appropriate initial system. It has been shown that an optimal design can always be found for the selected examples, despite the consideration of circularity. The only limitation is that the upgrade option does not allow a completely free trade-off between efficiency and power density.

Tab. 2: Specifications of the Designed Planar Transformers.

	Parameter	Design A	Design B	Design C/D	Design E
Specifications	Output Power P_{out}	5 kW	7 kW	5/7.5 kW	5 kW
	Input voltage V_{in}	400 V	400 V	800 V	400 V
	Output voltage V_{out}	400 V	400 V	800 V	400 V
Control Variables	Current Density J	12 A/mm²	8.4 A/mm²	6/9 A/mm²	12 A/mm²
	Copper Thickness H_{cu}	70 µm	70 µm	70 µm	70 µm
	Core	EI-Core	EE-Core	EE-Core	EE-Core
	Number of Cores Parallel N_C	2	2	2	1
	Number of Turns N_T	12	12	24	24
	Number of Parallel Layers N_P	1	2	1	1

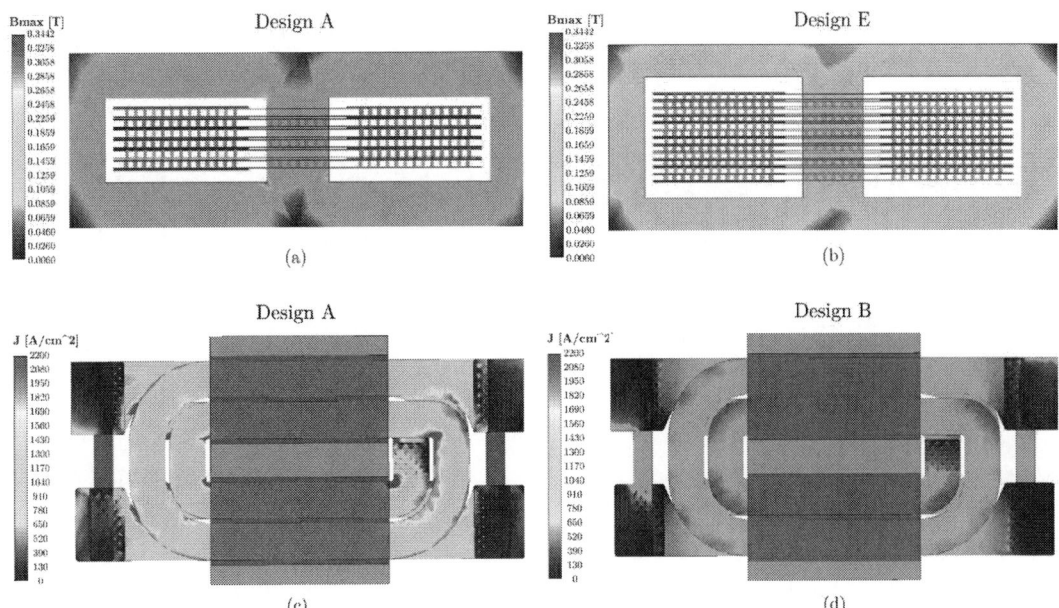

Fig. 7: Simulation results for (a) flux density of Design A, (b) flux density of Design E, (c) current density of Design A and (d) current density of Design B.

3.3 FEM Verification

To confirm the different configurations of the proposed PCB design and the previous analysis, some of them are built in Ansys FEM simulation software. A 3D model of Design A is implemented with interleaved windings and 12 turns connected in series. A 2D flux density distribution in the core center is shown in Fig. 7(a). The final design has a maximum flux density of approximately $B = 0.15$ T. This flux density can be confirmed by (1) and shows that the transformer design is correct.

Next, a PT design is selected to demonstrate the correctness of the PCB layout for different winding geometries. Therefore, Design E has been selected as an example and its flux density is illustrated in Fig. 7(b). As expected, Design E has a maximum flux density of $B = 0.15$ T as in Design A. With increasing number of turns in series, the DC resistance increases and so does the temperature, as shown in Fig. 6(a). This also confirms the upgrade potential of Design E is very limited compared to Design A, which is the best solution.

The last simulated PT is used to confirm the parallel configuration of the layers correspond to Design B. For ease of comparison, the power for the simulated transformer with parallel layers remains at 5 kW. To validate the parallel connection of the layers, the current density of the turns can be used. Paralleling the layers effectively increases the copper thickness H_{Cu} of the individual turns in (3). It is expected to have half the current density. Fig. 7(c) and Fig. 7(d) show the current density of the interleaved design of Design A and the parallel design of Design B, respectively. The interleaved design shows a current density of $J = 12 \text{ A/mm}^2$ and the parallel design shows a current density of $J = 6 \text{ A/mm}^2$. This shows the correctness of the selected Design B.

4 Experimental Results

The built PTs and their path in the economic cycle are shown in Fig. 8. A DAB converter as shown in Fig. 9(a) with SiC Mosfets is used to validate the PT designs. A FPGA is used to generate the PWM signals and control the phase shift, and an additional inductor is placed in series to the PT on the primary side. The experiments are performed in open-loop mode and a thermal camera is used to measure the temperature of the PTs. All PTs are tested for 15 minutes at full power to reach near steady-state temperature in order to validate the design. The full-bridge converter used for both sides is shown in Fig. 9(b) and waveforms for the experiments of 400 V 5 kW and 800 V 7.5 kW are shown in Fig. 9(c) and (d) respectively.

The first PT to be validated is Design A at 400 V and 5 kW power. From Fig. 10(a) it can be seen that the maximum surface temperature of the transformer is 89 ℃. This temperature is reached for the windings

PCIM Europe 2024, 11– 13 June 2024, Nuremberg   DOI: 10.30420/566262476

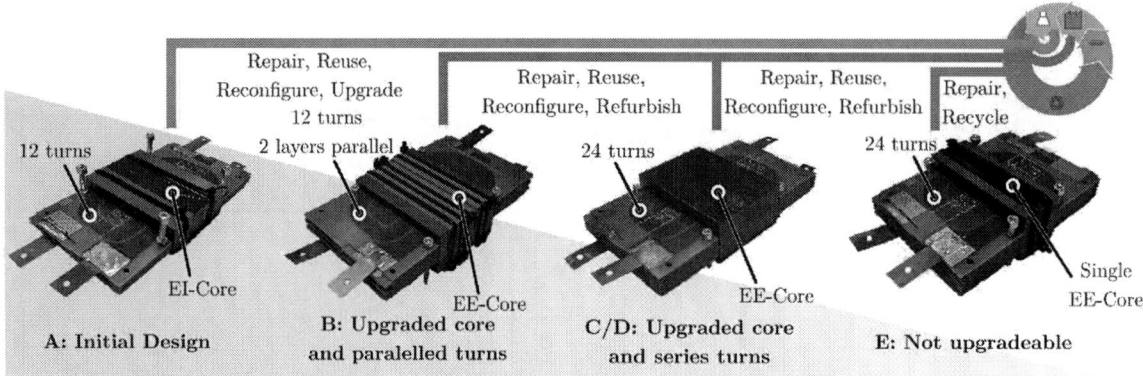

Fig. 8: Built PTs and their relation to the concept of circular economy.

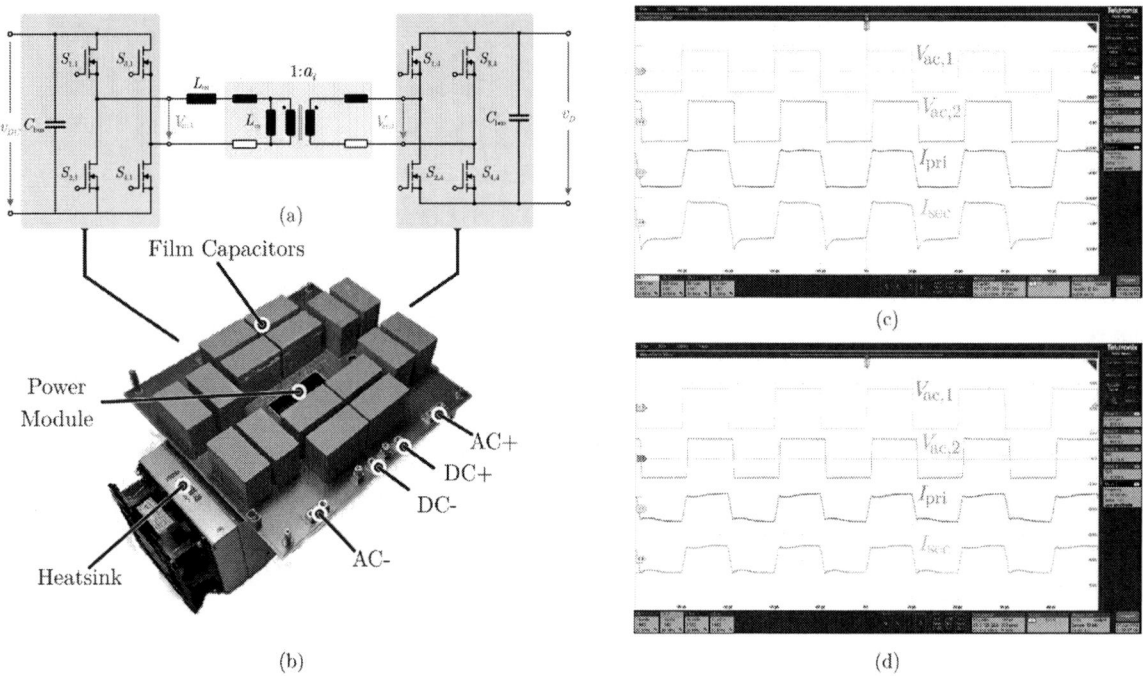

Fig. 9: Overview of the setup showing (a) the electrical schematic, (b) the full-bridge converter used for primary and secondary side, (c) the waveforms for 400 V 5 kW and (d) the waveforms for 800 V 7.5 kW.

due to their high current density J of 12 A/mm². The core temperature reaches approximately 72 °C. The temperature of the inner layers is expected to be slightly higher than the 89 °C of the top PCB. The final recorded temperature is close to the estimated temperature and confirms Design A.

Next, Design B with parallel layers is validated at 400 V and 7 kW power. The temperature after 15 minutes is shown in Fig. 10(b). The core temperature is approximately the same as for Design A and reaches 67 °C. The maximum surface temperature measured for the windings reached 103 °C. This is higher than calculated, as the same losses and therefore the same temperature would be expected

from the current density. The higher temperature can be explained by the fact that a simple temperature estimation method is used. Nevertheless, the temperatures are within a reasonable range to verify the upgradability with parallel layers.

In order to validate the additional windings in series, Design C is validated at 800 V and 5 kW power. The final temperature is shown in Fig. 10(c) and reaches around 60 °C for the core. This is lower than expected as the temperature should be similar to Design A and B given that the flux density is the same when the number of turns is increased to 24. The winding temperature of around 47 °C is lower than the other designs because the current density

3387

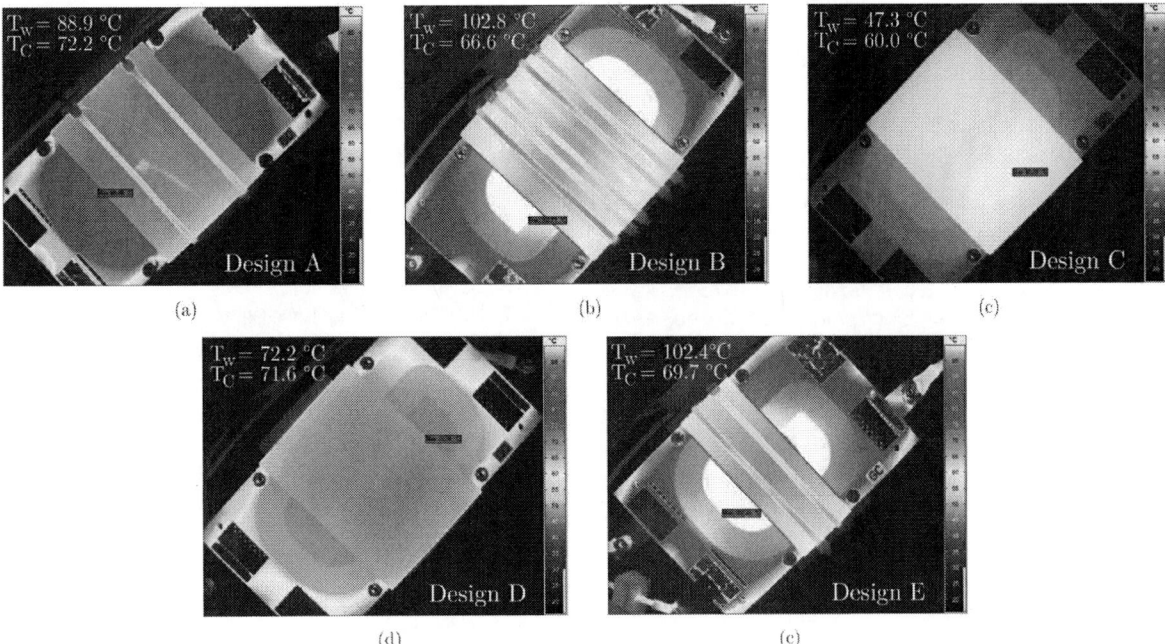

Fig. 10: Temperature of the designed transformers after 15 minutes for (a) Design A at 400 V 5 kW, (b) Design B at 400 V 7 kW, (c) Design C at 800 V 5 kW and (d) Design D at 800 V 7.5 kW.

in this design is very low compared to the other designs. Therefore, the upgradability with more turns in series is also validated.

Further, Design D is tested at 800 V with 7.5 kW power. The maximum temperature for the windings and the core is around 72 °C as shown in Fig. 10(d). The overall maximum temperature is lower than estimated, which is due to the fact that the estimation assumes combined losses and does not separate winding and core losses. In the experiment, the winding and core losses do not affect each other significantly, resulting in a lower temperature.

Finally, Design E can be validated at 400 V and 5 kW power. It is noticeable that the winding temperature rises to 102 °C and the core temperature reaches around 70 °C. This is as expected as this is the frontmost solution in the first pareto front in Fig. 6(a) and therefore has higher losses and temperature. This design confirms that different winding geometries are achievable and shows that the upgradability of this design is limited as it already uses an EE core combination and a large part of the core window.

The experiments confirm the analysis and the achievement of upgradability. It shows that the design can help to achieve circularity and reduce the e-waste produced by PT.

5 Conclusion

While transformers are conventionally designed to maximize the efficiency and power density, the environmental impact has never been addressed at this stage. This paper proposes a reconfigurable and modular PCB design for PTs that introduces the circular economy constraints at early design stages. The PCB design is shown and different configurations that can be achieved using this design are explained. Extensive $\eta\rho$-Pareto optimization was carried out and four designs complying with circularity were selected from the optimal Pareto-front. FEM simulations were carried out to validate the analysis and the derived configurations. Finally, the selected solutions were built and experiments were performed on the PTs at voltage levels of 400 V and 800 V and power levels of 5 kW and 7.5 kW. The theoretical, simulation, and experimental results confirm that the proposed PCB design meets the circular economy objectives and contributes to the reduction of e-waste for PTs.

Acknowledgment

This research has received funding from the European Innovation Council (EIC) under the European Union's Horizon 2020 research and innovation programme Grant agreement No 101057679, project Super-HEART.

References

[1] J. Huber, L. Imperiali, D. Menzi, F. Musil, and J. W. Kolar, "Energy efficiency is not enough!" *IEEE Power Electronics Magazine*, vol. 11, no. 1, pp. 18–31, 2024.

[2] A. Garcia-Bediaga, I. Villar, A. Rujas, L. Mir, and A. Rufer, "Multiobjective optimization of medium-frequency transformers for isolated soft-switching converters using a genetic algorithm," *IEEE Transactions on Power Electronics*, vol. 32, no. 4, pp. 2995–3006.

[3] M. Leibl, G. Ortiz, and J. W. Kolar, "Design and experimental analysis of a medium-frequency transformer for solid-state transformer applications," *IEEE Journal of Emerging and Selected Topics in Power Electronics*, vol. 5, no. 1, pp. 110–123.

[4] T. Guillod and J. W. Kolar, "Medium-frequency transformer scaling laws: Derivation, verification, and critical analysis," *CPSS Transactions on Power Electronics and Applications*, vol. 5, no. 1, pp. 18–33.

[5] M. Mogorovic and D. Dujic, "100 kW, 10 kHz medium-frequency transformer design optimization and experimental verification," *IEEE Transactions on Power Electronics*, vol. 34, no. 2, pp. 1696–1708.

[6] H. Beiranvand, E. Rokrok, and M. Liserre, "Vf-constrained $\eta \rho$-pareto optimisation of medium frequency transformers in ISOP-DAB converters," *IET Power Electronics*, vol. 13, no. 10, pp. 1984–1994.

[7] F. Groon, H. Beiranvand, T. Pereira, G. Can, and M. Liserre, "PCB layer optimization of planar medium frequency transformer for on-board EV chargers," in *2022 24th European Conference on Power Electronics and Applications (EPE'22 ECCE Europe)*, pp. P.1–P.9.

[8] F. Groon, T. Pereira, H. Beiranvand, S. Schikowski, D. Metschies, and M. Liserre, "GaN-based multiport resonant converter for automotive applications," in *2023 IEEE Applied Power Electronics Conference and Exposition (APEC)*, pp. 892–899.

[9] Z. Ouyang, O. C. Thomsen, and M. A. Andersen, "Optimal design and tradeoff analysis of planar transformer in high-power DC–DC converters," *IEEE Transactions on Industrial Electronics*, vol. 59, no. 7, pp. 2800–2810.

[10] W. R. Stahel, "The circular economy," *Nature*, vol. 531, no. 7595, pp. 435–438.

[11] P. Morseletto, "Targets for a circular economy," *Resources, conservation and recycling*, vol. 153, p. 104553.

[12] J. Potting, M. P. Hekkert, E. Worrell, and A. Hanemaaijer, "Circular economy: measuring innovation in the product chain," *Planbureau voor de Leefomgeving*, no. 2544.

[13] J. W. Kolar, "Net-zero-CO 2 by 2050 is NOT enough!" in *2023 25th European Conference on Power Electronics and Applications (EPE'23 ECCE Europe)*, pp. 1–2.

[14] A. Sangwongwanich, D.-I. Stroe, C. Mi, and F. Blaabjerg, "Sustainability of power electronics and batteries: A circular economy approach," *IEEE Power Electronics Magazine*, vol. 11, no. 1, pp. 39–46.

[15] F. Musil, C. Harringer, A. Hiesmayr, and D. Schoenmayr, "How life cycle analyses are influencing power electronics converter design," in *PCIM Europe 2023*, pp. 1–9.

[16] Z. Shen, Q. Wang, and H. Wang, "Degradation analysis of planar magnetics," in *2020 IEEE Applied Power Electronics Conference and Exposition (APEC)*, pp. 2687–2693.

[17] K. Parajuly, C. Fitzpatrick, O. Muldoon, and R. Kuehr, "Behavioral change for the circular economy: A review with focus on electronic waste management in the EU," *Resources, Conservation & Recycling: X*, vol. 6, p. 100035.

[18] T. T. Romano, T. Alix, Y. Lembeye, N. Perry, and J.-C. Crébier, "Towards circular power electronics in the perspective of modularity," *Procedia CIRP*, vol. 116, pp. 588–593.

[19] Y. Umeda, S. Fukushige, K. Tonoike, and S. Kondoh, "Product modularity for life cycle design," *CIRP annals*, vol. 57, no. 1, pp. 13–16.

[20] M. A. Saket, N. Shafiei, and M. Ordonez, "LLC converters with planar transformers: Issues and mitigation," *IEEE Transactions on power electronics*, vol. 32, no. 6, pp. 4524–4542.

[21] C. P. Steinmetz, "On the law of hysteresis," *Proceedings of the IEEE*, vol. 72, no. 2, pp. 197–221.

[22] K. Venkatachalam, C. R. Sullivan, T. Abdallah, and H. Tacca, "Accurate prediction of ferrite core loss with nonsinusoidal waveforms using only steinmetz parameters," in *2002 IEEE Workshop on Computers in Power Electronics, 2002. Proceedings.*, 2002, pp. 36–41.

[23] I. Villar, "Multiphysical characterization of medium-frequency power electronic transformers," *Ph.D dissertation, EPFL*, p. 234, 2010.

[24] R. Bakri, G. Corgne, and X. Margueron, "Thermal modeling of planar magnetics: Fundamentals, review and key points," *IEEE Access*, vol. 11, pp. 41 654–41 679, 2023.

PCIM Europe 2024, 11– 13 June 2024, Nuremberg   DOI: 10.30420/566262477

Controllable Magnetics: Variable Transformers and Variable Inductors, Theory – Production – Application

Florian Fenske, Paul Ott

mdexx inductive electronics GmbH, Zeppelinstraße 30, 28844 Weyhe, Germany

Corresponding author: Florian Fenske, Florian.Fenske@mdexx.com
Speaker: Florian Fenske, Florian.Fenske@mdexx.com

Abstract

This article presents controllable magnetics with linear characteristics under electrical pre-magnetization. The key to linearity is the geometric implementation of the pure dc pre-magnetization in the core, which is represented by the Theory of superposition and coupling of electromagnetic fields. This achieves linear performance in the large signal characteristics, in reducing the inductance of inductors; in transformers, the voltage transformation ratio, the leakage or the main inductance can be varied. This is experimentally demonstrated, verified and validated using functionally controllable samples. Finally, the application possibilities are identified and the potentials are shown.

1   Introduction

The demands on modern power electronics will continue to increase in terms of size, cost and regulatory requirements. Controllable magnetics will play a key role in solving these challenges.

System performance is increased by wide band semiconductors, which opens completely new possibilities. This means that very efficient systems are possible - high power over more than 100kVA at the low switching frequencies around 10kHz and moderate power of several 10kVA at the switching frequencies around 100kHz. The key driver of these systems is the energy transition and sustainability. Electrical pre-magnetization has been put on the back burner by the advent of semiconductors - can bring now significant benefits to the energy system.

1.1   Theory of controllable magnetics

The theory of controllable magnetics is based on a mechanical or electrical effect. For reasons of robustness, durability and size, mechanical systems are not considered. In this publication, the electrically controllable magnetics are used, which is based on the non-linearity of the core materials, the B-H curve.

1.2   State of Art

In the state of the art, controllable magnetics are represented by the "magnetic amplifier", a good overview is available in [1] and the possibilities for power electronics is already known in [2, 3]. The types of pre-magnetizations can be divided into 3 operating principles: parallel, local and orthogonal. The main disadvantage of parallel and local pre-magnetization (such as **V**irtual **A**ir **G**ap) is the non-linearity of the materials, i.e. there is no linear variation of the reluctance. The implementation of local pre-magnetization based on the VAG concept is shown in Fig. 1 for a controllable inductor.

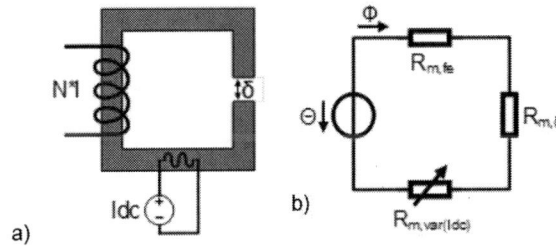

Fig. 1 Controllable device, example of an inductor: a) construction component b) magnetic circuit

The orthogonal pre-magnetization requires a complex field simulation to overlay the magnetic ac and dc fields and determine the change in reluctance. Newer designs than the classic magnetic amplifier is shown in [4, 5], but the linearity the characteristic curve under pre-magnetization is still a problem [6]. If the ac and dc fields are not parallel to each other, but only locally in the core material, they cannot be described analytically in the design, as is the case for the magnetic amplifiers. For this purpose, the pre-magnetized field area must be characterized using an FEM analysis.

From this, analytical models can be also developed and generated for the local pre-magnetization, if the characteristic results in the same response. It is crucial that the pre-magnetization creates a point symmetry of Ψ-I resp. Φ-V characteristic and the ac pre-magnetization is scaled in the reaction to the ac flow/current, or described by a model. This is for different implementations of the pre-magnetization in a core-section in the Fig. 2 shown.

If the ac and dc fields are not parallel to each other, but only locally in the core material, they cannot be described analytically in the design, as is the case for the magnetic amplifiers. For this purpose, the pre-magnetized field area must be characterized using an FEM analysis. It is crucial that the pre-magnetization creates a point symmetry of Ψ-I resp. Φ-V characteristic and the ac pre-magnetization is scaled in the reaction to the ac flow/current, or described by a model. This is for different implementations of the pre-magnetization in a core-section in the Fig. 2 shown.

1.3 New approach

To achieve a linear change in reluctance, the local pre-magnetization concept can be used. This involves arranging the holes of the VAG-concept in such a way that there are no permanently saturated areas above the core cross-section. This concept has been developed and presented in [9]. Because of the local effect of the control, the area of the pre-magnetization can be considered independently of the standard design of the components. This allows, for example, serial or parallel points/paths to be implemented/influenced in the component to achieve different operating modes.

The holes of the VAG-concept for dc bias are shown in Fig. 3a), below which the Φ-V characteristic of the core-section is shown in Fig. 3b). The effect corresponds to a linear change in reluctance proportional to the dc bias.

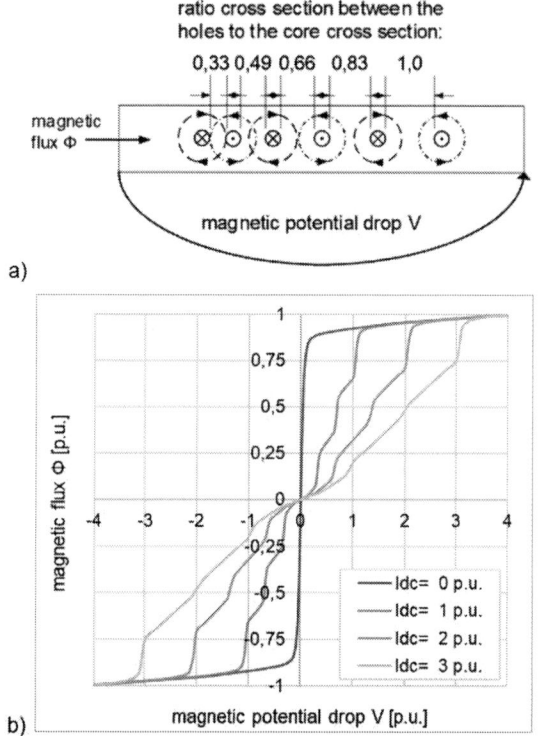

Fig. 3 Local pre-magnetization: a) controllable core-section b) FEM-simulated Φ-V characteristic curves

The superposition of the magnetic fluxes of the ac and dc bias is shown in Fig. 4. It can be seen that the flux interacts in an anti-parallel direction, proportional to the bias field strength. Another advantage, besides the linearity, is the analytical design of the concept, which can be scaled using the geometric implementation in the core under the ratio of ac to dc bias magnetic voltage drops on the windings.

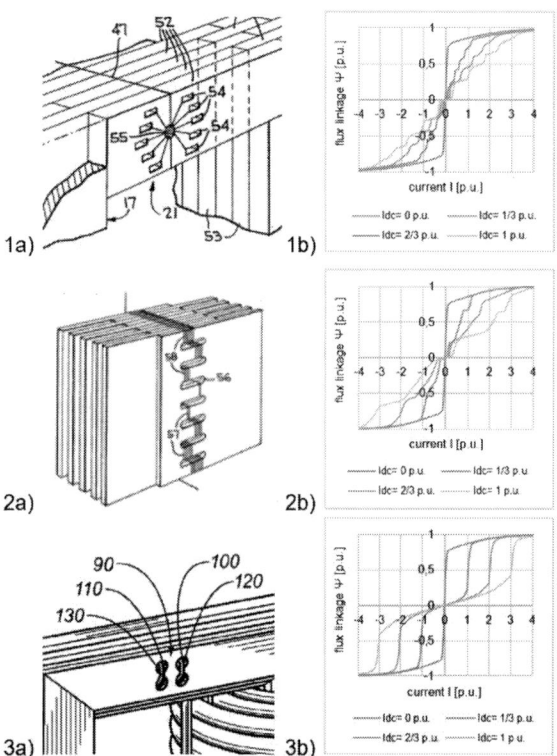

Fig. 2 Investigation of local pre-magnetization: a) construction method b) FEM Ψ-I characteristic curves 1) star layout [7, FIG. 2] 2) parallel layout [7, FIG. 3] 3) serial/parallel layout [8, FIG. 3]

The results of Fig. 2 make it clear that an FEM analysis of the characteristics of the implementation type of pre-magnetization is essential.

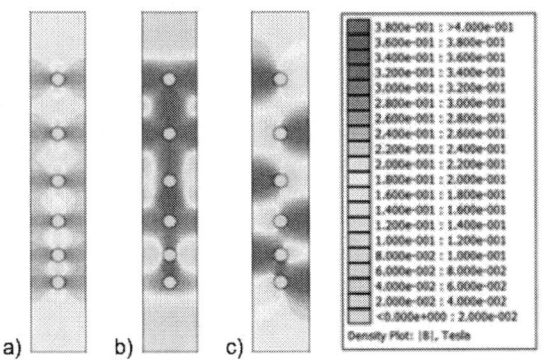

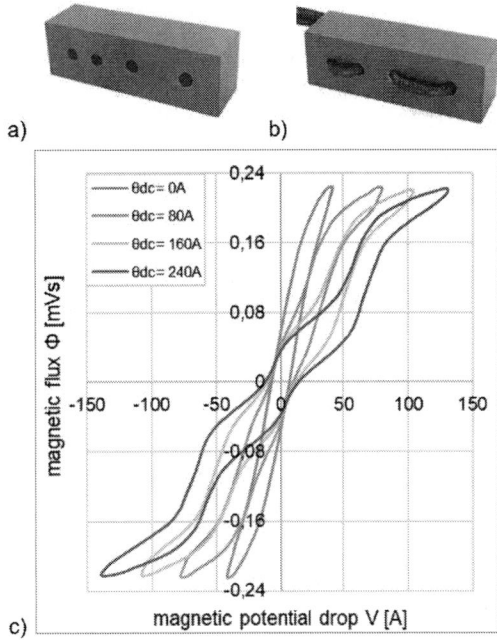

Φ = 0.5 p.u. Φ = 0 p.u. Φ = 0.5 p.u.

Idc = 0 p.u. Idc = 2 p.u. Idc = 2 p.u.

Fig. 4 FEM-results of the field images for the distribution of magnetic induction in the ferrite, equivalent to Fig. 3: a) ac flux b) dc flux c) ac/dc flux

The expression of the pre-magnetized area can be controlled by the distance of two opposite pre-magnetization fields or a pre-magnetization hole to the edge of the core. As a result of the flux leakage between the core and winding, the characteristics are further linearized under the pre-magnetization, which consequently corresponds to the reluctance of an adjustable air gap.

This effect will be used to produce very linear systems that do not generate harmonics or require complex controls to suppress it. A higher linearization of the pre-magnetization can be achieved by increasing the VAG-holes and an optimized spacing of the holes/limitation of the pre-magnetization field.

2 Production

The new approach of linear pre-magnetization can be used to obtain controllable inductive components such as inductors and transformers. An overview of the design methods and possibilities is given in [10, 11]. The inductance of inductors can be reduced, the coupling of the windings of coupled inductors can be changed and the main and leakage inductance and voltage transformation ratio of transformers can be varied.

The implementation of the pre-magnetization is shown in Fig. 5, where one or more conductors (winding) can be used for pre-magnetization. A winding is used to reduce the bias current, as the dc current flux is $\theta dc= N*Idc$. The advantage of local pre-magnetization is that the magnetic circuit is only influenced in a selected area. All known production methods and processes remain the same, which allows it to be used for a wide variety of components.

Fig. 5 Implementation of the new pre-magnetization concept in ferrit: a) core-section VAG-holes with different spacing ratio b) core-section with pre-magnetization conductor c) measured Φ-V characteristic curves

3 Application and potential

The greatest potential for controllable magnetics lies in isolated and resonant power converters. Resonant converters can operate at higher switching frequencies and allow wider bandwidths. The result is an optimal switching range of the semiconductors, which extends the soft switching zones and voltage operating ranges, benefiting zero voltage switching and zero current switching. Compared to the conventional solution, high efficiencies can be achieved over wider load ranges. Stability is improved and power electronics performance can be increased compared to the conventional solution.

High efficiencies can be increased over wider load ranges and another control option for switching frequency variation is available. The size of the magnetic core in the converter can be reduced, resulting in an optimization of the overall system size. With the availability of controllable magnetics, power electronics would need to focus on innovative topologies to better understand the full potential of this new family of devices. This is expected to lead to dramatic advances in the design of the next generation of power electronics. [12-19]

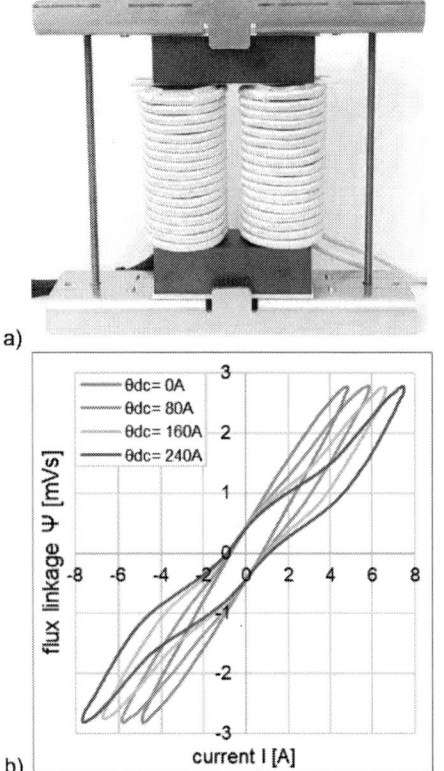

a)

b)

Fig. 6 electrical controllable variable inductor: a) structure b) measured Ψ-I characteristic

The controllable inductor in Fig. 6 can be used as a control element for a resonant converter. This largely or completely eliminates to need the change switching frequency of the semiconductors to control and adapt the power flow [20]. With this concept of linear reluctance control, variable transformers can be obtained through the design of the magnetic circuit. The parameters of the transformer, such as leakage inductance, winding coupling or main inductance, can be varied individually or collectively by the pre-magnetization.

4 Conclusions

Controllable magnetics allow an improvement of power electronics and bring advantages to the systems. The presented implementation of pre-magnetization in the core allows linearity of the characteristic in the large signal range and solves the disadvantages of the classical magnetic amplifier. A further main advantage of this type of pre-magnetization is the electrical dc control system, which controls the mag. reluctance of a core-section. No mechanical actuators are required, which maximizes the lifetime of the components.

This allows variable inductors and variable transformers to be obtained in power electronics, and this concept is also used for laminated components at grid frequency.

References

[1] [1] C. W. Lufcy, "A Survey of Magnetic Amplifiers," in Proceedings of the IRE, vol. 43, no. 4, pp. 404-413, April 1955, doi: 10.1109/JRPROC.1955.278187.

[2] J. Pfeiffer, P. Küster, I. E. M. Schulz, J. Friebe and P. Zacharias, "Review of Flux Interaction of Differently Aligned Magnetic Fields in Inductors and Transformers," in IEEE Access, vol. 9, pp. 2357-2381, 2021, doi: 10.1109/ACCESS.2020.3047156.

[4] P. Zacharias, T. Kleeb, F. Fenske, J. Wende and J. Pfeiffer, "Controlled magnetic devices in power electronic applications," 2017 19th European Conference on Power Electronics and Applications (EPE'17 ECCE Europe), Warsaw, Poland, 2017, pp. P.1-P.10, doi: 10.23919/EPE17ECCEEurope.2017.8099004.

[5] M. S. Perdigão, M. F. Menke, Á. R. Seidel, R. A. Pinto and J. M. Alonso, "A Review on Variable Inductors and Variable Transformers: Applications to Lighting Drivers," in IEEE Transactions on Industry Applications, vol. 52, no. 1, pp. 531-547, Jan.-Feb. 2016, doi: 10.1109/TIA.2015.2483580.

[6] F. Fenske, dissertation, "Nutzung nichtlinearer Effekte zur elektrischen Steuerung von Netzdrosseln im Impedanzverhalten", kassel university press, 2023, doi:10.17170/kobra-202303207662.

[7] N. Moerman, "CONVERSION AND CONTROL OF ELECTRICAL ENERGY BY ELECTROMAGNETIC INDUCTION", US patent 4020440, April 26, 1977.

[8] J. Montes and R Aguirre, "ELECTIC REACTOR OF CONTROLLES REACTIVE POWER AND METHOD TO ADJUST THE REACTIVE POWER", US patent 764288 B2, January 5, 2010.

[9] S. Saeed, J. Garcia, M. S. Perdigão, V. S. Costa, B. Baptista and A. M. S. Mendes, "Improved Inductance Calculation in Variable Power Inductors by Adjustment of the Reluctance Model Through Magnetic Path Analysis," in IEEE Transactions on Industry Applications, vol. 57, no. 2, pp. 1572-1587, March-April 2021, doi: 10.1109/TIA.2020.3047593.

[10] F. Fenske, "Controllable inductor coil and method for restricting electric current", WO patent disclosure 002023006730 (A1), Juli 26, 2022.

[11] F. Fenske, "Steuerbarer Transformator und Verfahren zur Steuerung eines Transformators", DE patent disclosure 102023101986 (A1), Januar 27, 2023.

[12] C. S. Buitrago, D. B. Cobaleda and W. Martinez, "Magnetically Controlled Transformer With Variable Turns Ratio and Low Series Inductance: Analysis and Implementation Toward Its Application in SMPS," in IEEE Transactions on Power Electronics, vol. 38, no. 11, pp. 14360-14374, Nov. 2023, doi: 10.1109/TPEL.2023.3300583.

[13] M. Liserre et al., "Voltage Controlled Magnetic Components for Power Electronics," in IEEE Power Electronics Magazine, vol. 10, no. 2, pp. 40-48, June 2023, doi: 10.1109/MPEL.2023.3273892.

[14] S. Brandt, M. Meissner, N. Polap, G. Schierle and K. F. Hoffmann, "A Survey on Adjustable Inductances for Power Electronic Circuits," PCIM Europe 2022; International Exhibition and Conference for Power Electronics, Intelligent Motion, Renewable Energy and Energy Management, Nuremberg, Germany, 2022, pp. 1-9, doi: 10.30420/565822219.

[15] T. Pereira, Y. Pascal, M. Liserre, Y. Wei and H. A. Mantooth, "Multiport Resonant DC-DC Converter using Actively-Controlled Inductors for Hybrid Energy Storage System Integration," 2022 IEEE Applied Power Electronics Conference and Exposition (APEC), Houston, TX, USA, 2022, pp. 1154-1161, doi: 10.1109/APEC43599.2022.9773497.

[16] S. Saeed, J. Garcia and R. Georgious, "Dual-Active-Bridge Isolated DC–DC Converter With Variable Inductor for Wide Load Range Operation," in IEEE Transactions on Power Electronics, vol. 36, no. 7, pp. 8028-8043, July 2021, doi: 10.1109/TPEL.2020.3048928.

[17] Y. Wei, Q. Luo, X. Du, N. Altin, A. Nasiri and J. M. Alonso, "A Dual Half-Bridge LLC Resonant Converter With Magnetic Control for Battery Charger Application," in IEEE Transactions on Power Electronics, vol. 35, no. 2, pp. 2196-2207, Feb. 2020, doi: 10.1109/TPEL.2019.2922991.

[18] L. Zhang, W. G. Hurley and W. H. Wölfle, "A New Approach to Achieve Maximum Power Point Tracking for PV System With a Variable Inductor," in IEEE Transactions on Power Electronics, vol. 26, no. 4, pp. 1031-1037, April 2011, doi: 10.1109/TPEL.2010.2089644.

[19] J. Guo, H. Wang, G. Xu, X. Li, Y. Sun and M. Su, "Dual Coupled Inductors With Controllable Integrated Leakage Inductance and CM Noise Suppression for CF-DAB Converter," in IEEE Transactions on Power Electronics, vol. 38, no. 7, pp. 8033-8038, July 2023, doi: 10.1109/TPEL.2023.3267728.

[20] H. Edel, "Hoch effizienter und optimal regelbarer Resonanzwandler mit spannungsunabhängigem bidirektionalem Betrieb", DE patent DE 10 2021 002 626 B4, March 2, 2023.

[21] S. B. Camilo, PhD thesis, "Magnetically Controlled Transformers: Analysis and Implementation in a Dual Active Bridge Converter", KU Leuven, 16.02.2024.

A Three-phase Interleaved LLC Integrated Transformer Using PCB Windings for Fuel Cell DCDC Converters

Jiajia Guan[1] , Jin Wen[1] , Shuangxi Zhu[1] , Zongheng Wu[1] , Cai Chen[1] , Yong Kang[1]
Yue Wu[2], Zhipeng He[2]

[1] School of Electrical and Electronic Engineering, Huazhong University of Science and Technology, China

[2] State Key Laboratory of HVDC, Electric Power Research Institute, CSG, China

Corresponding author: Cai Chen, caichen@hust.edu.cn
Speaker: Jiajia Guan, jiajiaguan@hust.edu.cn

Abstract

Compared with single-phase LLC, three-phase interleaved LLC can increase the power capacity and reduce current ripple. However, the number of resonant components has also tripled. Without magnetic components integrated design will affect the power density and efficiency. This paper proposes a new integration method for three-phase transformer based on the reluctance model and magnetic flux cancellation principle. It has three magnetic columns, has better magnetic flux distribution characteristics, and can reduce the volume of the magnetic core compared with existing integrated methods. Finally, a prototype with 98% peak efficiency was designed to verify the performance of the proposed transformer.

1 Induction

With the development of automotive electronics and data centers, the power level of converters is getting higher and higher. The LLC topology is widely used because it can achieve zero-voltage turn-on of the primary side and zero-current turn-off of the secondary side [1] - [4]. However, the output current of single-phase LLC is a half-sine wave and the current ripple is large. On the one hand, this requires increasing the volume of the output capacitor, which affects the power density. On the other hand, it increases the AC loss of copper.

The three-phase interleaved LLC converter can reduce current ripple, which is beneficial to reducing the volume of filter capacitors and improving power capacity [5] - [9]. However, the three-phase interleaved LLC topology contains six magnetic components: three resonant inductors and three transformers (as shown in Fig. 1). The increase in the number of magnetic components will affect the system power density and cost. Therefore, integration and optimized design are needed to reduce the impact of the increase in magnetic components [7].

As shown in Fig. 2, currently three-phase integrated transformers can be divided into two types: rectangular type and triangular type [8] – [10]. For the rectangular structure, the winding outlet is more convenient, and resonant inductor magnetic columns are usually added next to the transformer magnetic columns to integrate all magnetic components. However, the magnetic flux distribution of the rectangular structure is uneven. As shown in Fig. 2(a), when the A-phase magnetic flux is maximum, the magnetic flux of the cover plate between phase A and phase B will be greater than that between phase B and phase C. For the triangular structure, as shown in Fig. 2(b), the fourth magnetic column is located at the center of the three-

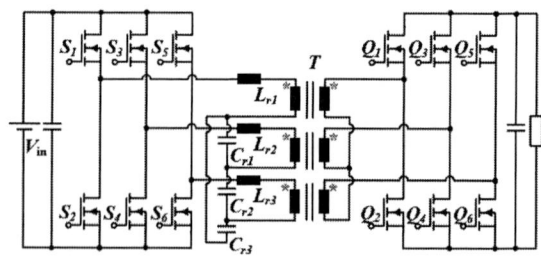

Fig. 1 Topology of three-phase interleaved LLC.

PCIM Europe 2024, 11– 13 June 2024, Nuremberg   DOI: 10.30420/566262478

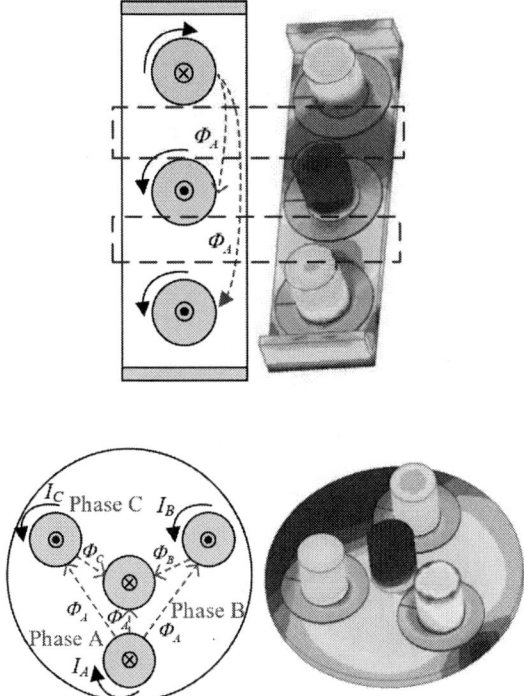

Fig. 2 Two core structures and magnetic flux distribution. (a) Rectangular. (b) Triangular.

phase magnetic column, and the positions of the three-phase magnetic columns are symmetrical. Therefore, the magnetic flux on the cover plate is evenly distributed and the core loss is smaller than that of the rectangular structure. Different from the rectangular structure, the resonant inductor of the triangular structure can be placed on the top of the transformer or completely replaced by leakage inductance.

This paper proposes a new three-phase transformer integration method. By analyzing its magnetoresistance model, the middle magnetic column of the traditional triangular structure is eliminated, thereby further reducing the volume of the magnetic components.

2 Three-phase Transformer Reluctance Model and Optimization

Based on the above analysis, to reduce transformer losses, this article adopts a triangular magnetic core structure, and the primary and secondary windings of the same phase are placed on the

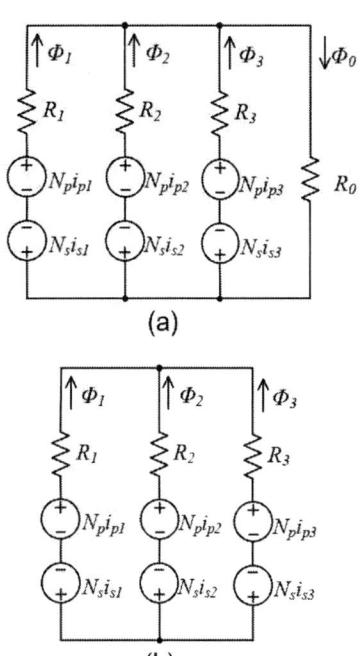

Fig. 3 Two reluctance models. (a) Four magnetic columns. (b) Three magnetic columns.

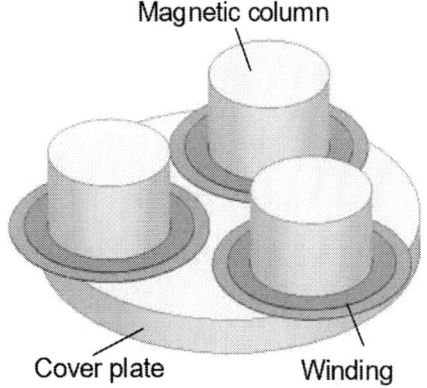

Fig. 4 Magnetic core structure with three magnetic columns.

same magnetic column. As shown in Fig. 3(a), an equivalent reluctance model is established based on the core structure and winding structure.
According to the reluctance model and Kirchhoff's law, when the reluctance of each phase is equal ($R_1 = R_2 = R_3 = R$), assuming $R_0 = \alpha R$, the magnetic flux of the three side magnetic columns can be obtained (1) (2). The peak value of the three-phase current of the transformer is equal and the phases are 120° different from each other. The magnetic flux of the central magnetic column is the sum of the three-phase magnetic fluxes, so the magnetic flux of the central magnetic column is 0

3396

(3). It can also be seen from the simulation results in Fig. 2(b) that the magnetic flux of the central magnetic column is 0, which is consistent with the calculation results. Therefore, in high-power applications, to reduce the size of the transformer, the central magnetic column can be eliminated and a three magnetic columns core structure as shown in Fig. 4 can be adopted. The optimized transformer reluctance model is shown in Fig.3(b). The air gap of the transformer is between the magnetic column and the upper cover plate. The magnetizing inductance can be changed by adjusting the length of the air gap (4).

$$\begin{cases} \Phi_1 = \left(\dfrac{N_p i_{p1} - N_s i_{s1}}{R}\right) - \dfrac{\alpha}{1 + 3\alpha}\Phi_k \\[2mm] \Phi_2 = \left(\dfrac{N_p i_{p2} - N_s i_{s2}}{R}\right) - \dfrac{\alpha}{1 + 3\alpha}\Phi_k \\[2mm] \Phi_3 = \left(\dfrac{N_p i_{p3} - N_s i_{s3}}{R}\right) - \dfrac{\alpha}{1 + 3\alpha}\Phi_k \end{cases} \quad (1)$$

$$\Phi_k = \dfrac{\left(\dfrac{N_p i_{p1} + N_p i_{p2} + N_p i_{p3}}{R}\right)}{-\left(\dfrac{N_s i_{s1} + N_s i_{s2} + N_s i_{s3}}{R}\right)} \quad (2)$$

$$\Phi_0 = \Phi_1 + \Phi_2 + \Phi_3 \quad (3)$$

$$L_m = N_p \Phi_1 / i_p = 2N_p{}^2 \mu_0 A_g / (3l_g) \quad (4)$$

Where l_g is the length of the air gap, A_g is the cross-sectional area, μ_0 is the vacuum permeability, and L_m is the magnetizing inductance.

Resonant frequency	300 kHz
Transformer ratio	2:3
Magnetizing inductor	10 μH
Resonant inductor	0.5 μH
Resonant capacitor	188 nF
Primary device	SCT3030ALHR
Secondary device	SCT3040KLHR
Gate drive	UCC21750

Table 1 Parameters of the prototype.

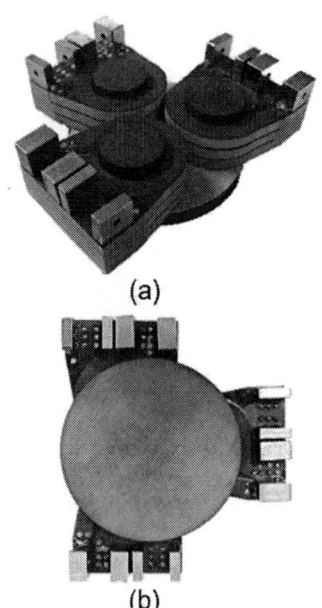

(a)

(b)

Fig. 5 Optimized transformer prototype. (a) Without cover plate. (b) With cover plate.

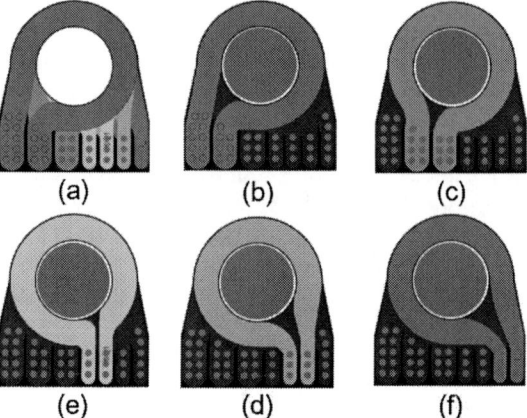

(a)   (b)   (c)

(e)   (d)   (f)

Fig. 6 The structure of PCB winding. (a) The overall structure. (b) The first layer. (c) The second layer. (d) The third layer. (d) The fourth layer. (e) The fifth layer. (f) The sixth layer.

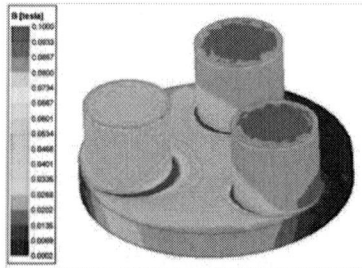

Fig. 7 Simulation result of the magnetic core.

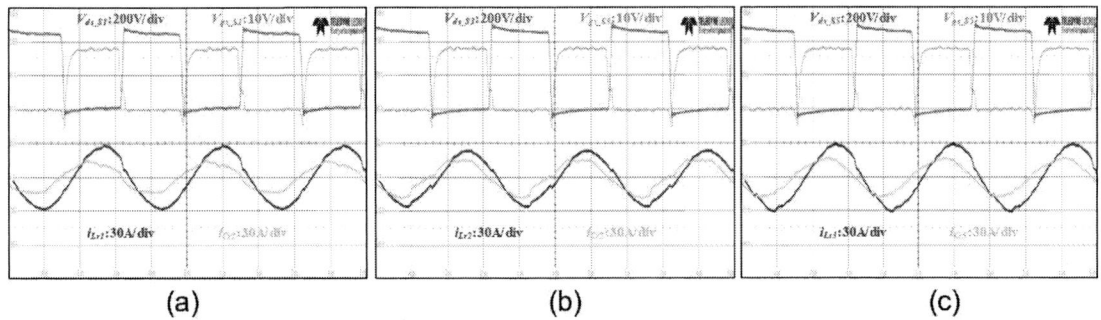

Fig. 8 Voltage and current waveforms of three-phase circuits @V_i=250V. (a) Phase A. (b) Phase B. (c) Phase C.

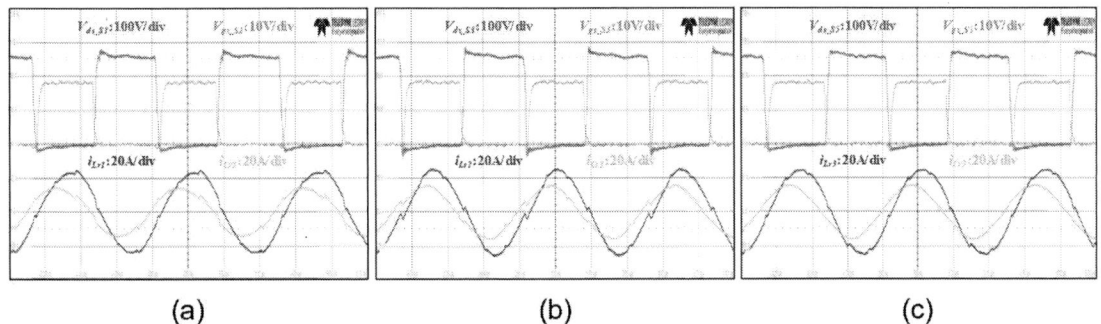

Fig. 9 Voltage and current waveforms of three-phase circuits @V_i=450V. (a) Phase A. (b) Phase B. (c) Phase C.

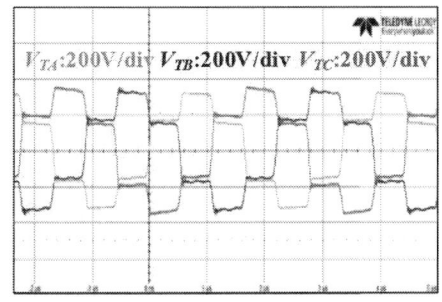

Fig. 10 Transformer winding voltage waveform.

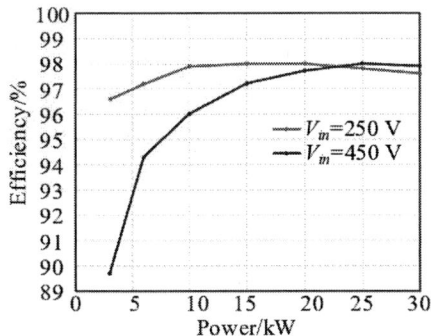

Fig. 11 Efficiency curve.

3 Prototype and Experimental Results

Based on the size model and loss model, the transformer can also be optimized [7]. According to the optimization results, the designed integrated transformer is shown in Fig. 5. The radius of the magnetic column is 16 mm, the width of the winding is 10.5 mm, the material of the magnetic core is DMR96, and the winding is made of the printed circuit board (PCB). The transformation ratio is 2:3, and the windings are made of 10-layer PCB. The windings are placed interleaved to reduce the peak magnetomotive force and AC losses. The structure is PPSSSPPSSS (P is the primary winding, and S is the secondary winding). The five-layer PCB structure is shown in Fig. 6. The simulation result of the magnetic core is shown in Fig. 7.

Taking the fuel cell DCDC converter with 250V-450V input as an example, the experimental waveforms of the designed three-phase interleaved LLC prototype is shown in Fig. 8 - Fig. 10. The prototype parameters are shown in Table 1. Fig. 8 is the waveform of the drive voltage, switching node, current of resonant inductor and capacitor under the 250V input. Fig. 9 is the waveform of the drive voltage, switching node, current of resonant inductor and capacitor under the 450V input. It can be seen that the zero-voltage switching-on can be achieved under the full input voltage range. And thanks to the symmetrical design of the three phases, the resonant current of the three phase is balanced. Fig.10 is the voltage waveform of the transformer, which is consistent with the theoretical analysis.

Fig. 11 is the efficiency curve of the designed prototype. It is worth noting that the lower efficiency at light load at 450V input is due to larger core losses. The prototype can achieve peak and full load efficiency of 98%. It can be seen that the designed integrated transformer can achieve high efficiency and reliable power conversion, which verifies its feasibility.

4 Conclusion

This paper proposes a three-phase integrated transformer structure with three magnetic columns. Compared with the rectangular structure, it has a more balanced magnetic flux distribution. Compared with the traditional four magnetic column structure, it can reduce the volume and cost of the transformer. Simulation and experiments have verified that the proposed integrated transformer structure has high-efficiency characteristics. Using the proposed transformer structure, the three-phase interleaved LLC prototype can achieve a peak efficiency of 98%.

References

[1] J. Wen et al., "An Iterative-Based Dead-Time Compensation Method for Integrated Interleaved Boost-LLC Converter," 2023 IEEE Energy Conversion Congress and Exposition (ECCE), Nashville, TN, USA, 2023, pp. 3341-3348.

[2] Z. Wu, Z. Wang, T. Liu, W. Xu, C. Chen and Y. Kang, "High Efficiency and High Power Density Partial Power Regulation Topology With Wide Input Range," in IEEE Transactions on Power Electronics, vol. 38, no. 2, pp. 2074-2091, Feb. 2023.

[3] Bo Yang, F. C. Lee, A. J. Zhang and Guisong Huang, "LLC resonant converter for front end DC/DC conversion," APEC. Seventeenth An-

nual IEEE Applied Power Electronics Conference and Exposition (Cat.No.02CH37335), 2002, pp. 1108-1112 vol.2.

[4] S. Luan, Z. Wu, Z. Wang, X. Liu, C. Chen and Y. Kang, "A High Power Density Two-Stage GaN-Based Isolated Bi-Directional DC-DC Converter," 2019 IEEE Energy Conversion Congress and Exposition (ECCE), 2019, pp. 3240-3244.

[5] J. Guan et al., "A High Efficiency Δ-Lr-Y Type Three-Phase Interleaved LLC Converter with Less Transformer Loss," in IEEE Transactions on Power Electronics, vol. 38, no. 9, pp. 11152-11168, Sept. 2023.

[6] J. Guan, J. Wen, S. Zhu, Z. Wu, C. Chen and Y. Kang, "Current Sharing Analysis of Three-Phase Interleaved LLC and Optimization Method to Reduce the Influence of Stray Inductance," in IEEE Transactions on Power Electronics, vol. 39, no. 3, pp. 3422-3437, March 2024.

[7] C. Fei, R. Gadelrab, Q. Li and F. C. Lee, "High-Frequency Three-Phase Interleaved LLC Resonant Converter with GaN Devices and Integrated Planar Magnetics," in IEEE Journal of Emerging and Selected Topics in Power Electronics, vol. 7, no. 2, pp. 653-663, June 2019.

[8] R. Gadelrab, F. C. Lee and Q. Li, "Three-Phase Interleaved LLC Resonant Converter with Integrated Planar Magnetics for Telecom and Server Application," 2020 IEEE Applied Power Electronics Conference and Exposition (APEC), New Orleans, LA, USA, 2020, pp. 512-519.

[9] M. Noah et al., "A Current Sharing Method Utilizing Single Balancing Transformer for a Multiphase LLC Resonant Converter with Integrated Magnetics," in IEEE Journal of Emerging and Selected Topics in Power Electronics, vol. 6, no. 2, pp. 977-992, June 2018.

[10] J. Huang, Z. Zhang, Y. Xiao, B. Sun and M. A. E. Andersen, "An Integrated Three-phase Transformer for Partial Parallel Dual Active Bridge Converter," 2019 10th International Conference on Power Electronics and ECCE Asia (ICPE 2019 - ECCE Asia), Busan, Korea (South), 2019, pp. 1810-1816.

PCIM Europe 2024, 11– 13 June 2024, Nuremberg   DOI: 10.30420/566262479

Testing the Primary-Secondary Coil Coupling of High-Frequency Transformer Implemented on ETD and Toroidal Cores

Alexis Gioda[1] , Daniel Chatroux[1]

[1] Univ. Grenoble Alpes, CEA, Liten, DEHT, 38000 Grenoble, France

Corresponding author: Daniel Chatroux, daniel.chatroux@cea.fr
Speaker: Alexis Gioda, alexis.gioda@cea.fr

Abstract

In conventional literature, ideal transformers are depicted with primary and secondary windings tightly linked to the magnetic circuit, yet uncoupled from each other, leading to substantial energy storage in leakage inductors. This study challenges this understanding by exploring coupling configurations' implications on high-frequency transformer performance. Meticulous testing on toroidal and ETD setups reveals the criticality of effective coupling for an optimal transformer. The results emphasize the drawbacks of insufficient coupling in toroidal setups and a very efficient coupling realized in ETD configurations. This underscores the vital role of precise coil coupling strategies in improving high-frequency transformers performances.

1   Magnetism applied to transformers

Magnetism is a discipline of physics that studies phenomena related to magnetic fields and their interactions with materials. Two fundamental quantities are the vectors of magnetic induction B and magnetic field H.

B represents the magnetic induction measured in tesla (T) and results from the influences of the magnetic field produced by electric currents and magnetic materials. The fundamental relationship between B and H is expressed by the Eq. (1).

$$B = \mu \times H \tag{1}$$

Where μ is the magnetic permeability of the material.

The magnetic permeability μ can be broken down into two quantities: μ_0 the magnetic permeability in the vacuum and μ_r the relative magnetic permeability of the material as in Eq. (2).

$$\mu = \mu_0 \times \mu_r \tag{2}$$

The magnetic field H is measured in amperes per meter (A/m) and is proportional to the electric current flowing through the conductors.

It is related to the current I by the Eq. (3).

$$\oint H dl = N \times I \tag{3}$$

With I the current flowing through the conductor (A) and N the number of turns in a winding.

An electrical transformer is a fundamental component in the distribution and transmission of electrical energy, enabling voltage and current levels to be modified while ensuring electrical insulation.
It is made up of at least two main coils, namely the primary coil and the secondary coil, which are essential for the efficient transmission of electrical energy from one voltage level to another, while maintaining electrical insulation between the two coils.

More precisely, an electrical transformer works by using the principles of electromagnetic induction to transfer electrical energy from one circuit to another. It usually consists of at least two separate wire windings. When an alternating current flows through the primary, it creates a variable magnetic field, which induces a current in the secondary via electromagnetic induction.
The ratio of the number of turns of the primary (N_p) and secondary windings (N_s) determines the ratio of electrical voltages (V_p/V_s) between the two circuits, enabling efficient voltage matching for different applications.

3400

Electrical transformers are useful in many applications, such as power electronics associated with energy storage systems, emergency power supplies and electric vehicles, where it is necessary to be able to change voltage or current levels. The Fig. 1 shows some audio transformers.

In addition, electrical insulation between primary and secondary voltages is essential to guarantee the safety of people and equipment.

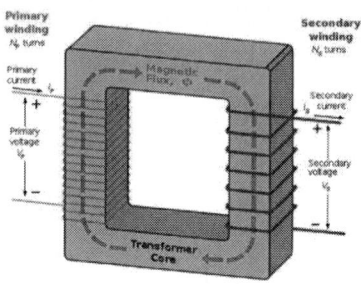

Fig. 2 Representation of an ideal transformer [2]

The Fig. 2 illustrates the dominant representation in the literature that focuses on optimizing the interaction of magnetic materials in relation to the coupling between primary and secondary windings, often neglecting the importance of their mutual interaction[4].

In this study, we seek to challenge this conventional understanding by exploring the complex relationship between coupling configurations and their implications for the performance and energy efficiency of high-frequency transformers.

The Fig. 3 is a simplified representation of the main magnetic field line zones, showing the common field lines running through both windings, corresponding to the magnetizing inductor and coupling between the two coils, and the field line zones specific to the primary winding and those specific to the secondary winding. These two field line zones correspond to the two leakage inductors of each of these windings.

Fig. 1 Audio transformers [1]

The primary and secondary windings can be considered as two coupled coils. Coils coupling refers to the extent to which magnetic fields generated by adjacent or nearby coils interact with each other. Strong coupling indicates significant interaction between the fields, while weak coupling corresponds to limited interaction.

2 State of the art

Typically, books and manuals describe ideal transformers with a primary winding tightly coupled to the magnetic material, a secondary winding tightly coupled to the magnetic material, and the two windings spaced apart so not or loosely coupled to each other. [2]

"An ideal transformer is linear, lossless and perfectly coupled. Perfect coupling implies infinitely high core magnetic permeability and winding inductance and zero net magnetomotive force." [3]

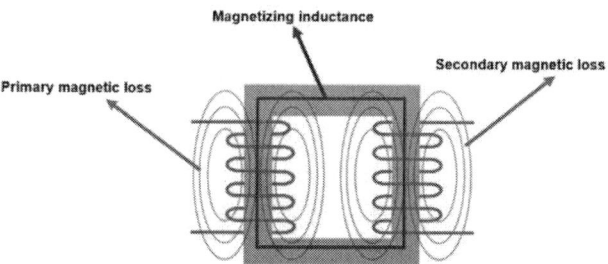

Fig. 3 Magnetic field distribution of primary and secondary windings

As shown in Fig. 3, with this embodiment, one would rather expect to obtain a transformer with poor coupling, resulting in significant energy storage in the primary and secondary leakage inductors. Thus, it is important to note that taking inspiration from this representation can unintentionally lead to the design of transformers with inefficient coupling, resulting in sub-optimal energy transfer and potential operational inefficiencies. [5]

The effects of excessive magnetic energy storage in the leakage inductors can include:

- Reduced secondary voltage and power due to the series impedance of the two leakage inductors.

- Increased magnetic losses: high-energy storage increases transformer losses, reducing overall system efficiency.

- Non-uniform magnetic flux distribution: excessive energy storage disturbs the uniform distribution of flux, affecting electrical performance.

- Increased risk of core saturation: excessive energy storage increases the risk of core saturation, leading to reduced performance and oversized transformers. Locally, common and winding-specific field lines can lead to local saturation of the magnetic material.

In practice, the coupling efficiency of a transformer depends mainly on the optimum level of coupling between its two windings, rather than on the coupling between each of them and the magnetic circuit.

In order to address this issue, this paper proposes a simple experimental method to study the optimal coupling of different transformer core and winding configurations.

3 Experimental method

3.1 Test on Toroidal core

To assess the validity of the hypothesis, a transformer was built using a toroidal ferrite core and copper windings with a diameter of 1 mm, wound on 10 turns for both primary and secondary windings, as shown in Fig. 4. The core had an external diameter of 23.8 mm, an internal diameter of 13.2 mm, a width of 7.7 mm and a measured relative permeability μ_r of 1360. The relative permeability of the magnetic material is probably a little higher due to the external insulation of unknown thickness, so the actual ferrite cross-section is a little lower by 5 % than the measurable cross-section assuming a coating difference of 500µm

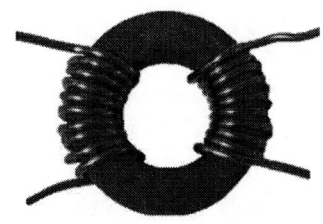

Fig. 4 Toroidal transformer wound for testing

Then, six different winding coupling configurations were tested, as shown in Fig. 5. These configurations were chosen to bring the primary and secondary windings closer, with the aim to improve the magnetic couplings. The chosen configuration are side-by-side (case 2) or superimposed (case 3), or twisted together (case 4), or twisted together but poorly coupled with the toroid (case 5). The final configuration consists of having two primary windings in parallel to sandwich the secondary between these two primary layers (case 6).

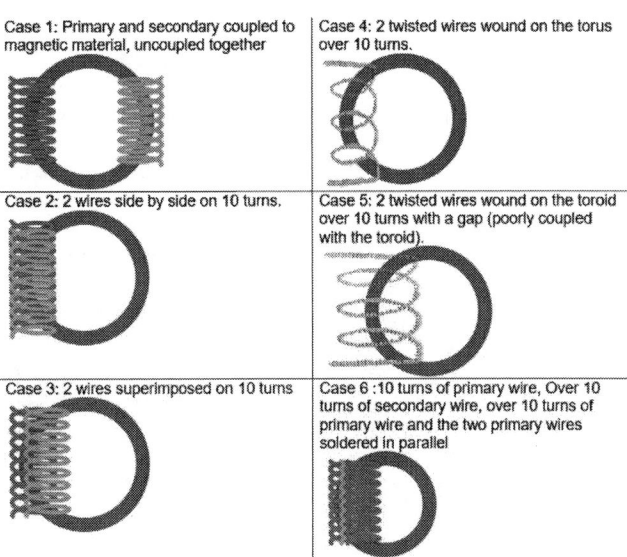

Fig. 5 The six coupling configurations used for testing on toroidal core

3.2 Test on ETD

Economical Transformer Design (ETD), are the second magnetic core tested. For ETD, a unique transformer design is used. Instead of separate cores for each winding, all the primary and secondary windings are wound on the same ETD core. The secondary side is winded between two primary windings as shown in Fig.6. Each colors on the figure translates to a winding [6].

The key concept here is achieving multiple transformer configurations with just one physical transformer. This is achieved by welding specific combinations of primary and secondary windings together. This approach aims to reduce the number of transformers needed to be built and allows for greater flexibility in achieving different test.

3402

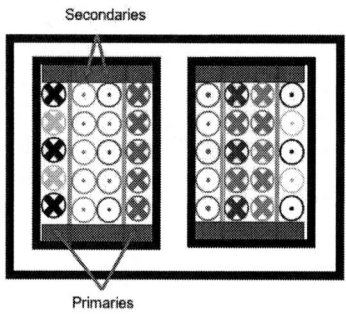

Fig. 6 ETD34 Transformer configuration

The transformer is constructed on a cylindrical ETD core measuring 34 mm by 11.1 mm by 17.5 mm. It utilizes 0.5 mm diameter wires for both the primary and secondary windings, which are mirrored. This means the primary and secondary coils have the same number of turns in a mirrored configuration. To ensure proper separation, winding spacers are used, and polyimide insulating tape is applied between both windings. Fig. 7 presents the realization of this transformer.

Each winding exhibits an inductance of 376 µH with 12 turns for a frequency of 100 kHz. This translates to a specific inductance of 2626 nH/turn².

Fig. 7 Picture of the ETD34 Transformer

Following the same methodology as the toroidal core tests, the transformer with the ETD34 core undergoes examination across six configurations. Fig.8 illustrates the six configurations used for the tests. The initial scenario as two coils in parallel, while the second scenario mirrors two half-coils in series. In the third configuration, primary coils 1 and 2, along with coils 3 and 4, are serially connected, followed by parallel connection of these two assemblies. On the secondary side, coils 1 and 2, as well as coils 3 and 4, are connected in parallel, then in series. In the fourth configuration, primary coils 1 and 2, and coils 3 and 4, are paralleled before being serially connected. Similarly, on the secondary side, coils 1 and 2, and coils 3 and 4, are connected in series, then in parallel.

The fifth configuration features serial connection of primary coils 1 and 2, and coils 3 and 4, followed by serial connection of these two assemblies. Likewise, on the secondary side, coils 1 and 2, and coils 3 and 4, are connected in series, maintaining the series connection.

Finally, in the sixth configuration, primary coils 1 and 2, and coils 3 and 4, are connected in parallel, then in series. On the secondary side, coils 1 and 2, and coils 3 and 4, are connected in parallel, maintaining the parallel connection.

ETD34 transformer configurations are not the same as for toroidal transformers.

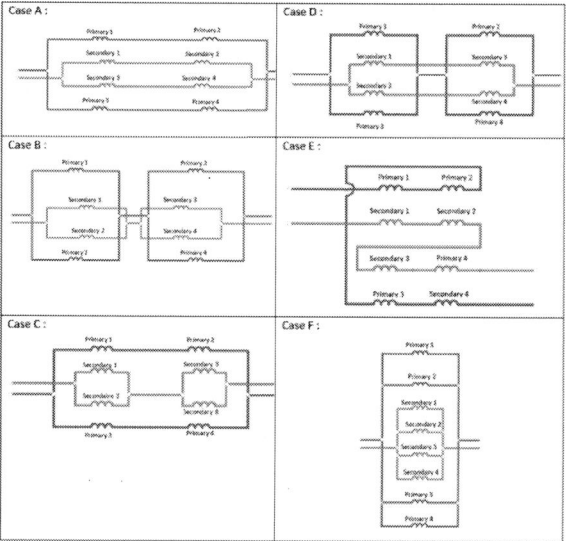

Fig. 8 The six coupling configurations used for testing on ETD34 core

4 Test procedure

The Fig. 9 show the standard model of a transformer. It is composed of three virtual inductors: leakage inductors on the primary and secondary winding and the magnetizing inductor.

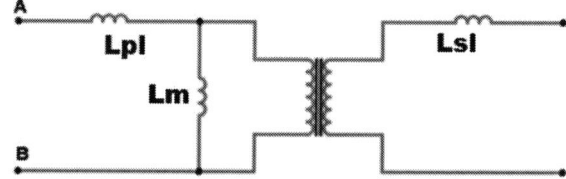

Fig. 9 Model of a Transformer Circuit with the primary and secondary leakage inductors

The magnetic flux of the primary winding is a function of the current at the primary and stores energy according to Eq. (4).

$$\frac{1}{2}L_{pl} \times I_p{}^2 \qquad (4)$$

The same principle applies to secondary-specific magnetic flux. Finally, the shared flow represents the magnetization inductor. It should be noted that it is the voltage at the terminals of the windings that gives rise to the voltage at the terminals of the magnetizing inductor that generates the magnetizing current and therefore the magnetizing energy. The magnetization current does not depend on the incoming current, nor on the transmitted power, only on the voltages at the terminals of the winding.

To tend towards a perfect transformer, consists firstly to minimize the magnetization energy, therefore the magnetization current and for this the magnetization inductor must be as high as possible, and secondly to minimize the leakage energy at primary and secondary, so to minimize the two leakage inductors.

As indicated above, Fig. 9 illustrates the transformer model used in the tests, with three main components: magnetizing inductor (Lm), primary winding leakage inductor (Lpl), and secondary winding leakage inductor (Lsl).
The comparison criterion we chose is based on the ration between these two quantities to determine the quality of the transformer coupling.
For each of the six configurations, the inductor at the primary level is measured at points A and B while leaving the secondary winding free. This measure allows to know the sum of leakage inductors and magnetizing inductor (Lm + Lpl) that we will approximate to the value of Lm, since Lpl is supposed negligible, which we can verify. Then, a second measurement is made of the inductor at the primary level, still at points A and B, this time by short-circuiting the secondary winding.

This measurement makes it possible to determine the sum of the leakage inductors (Lpl + Lsl). Assuming Lm very large in front of Lsl. These measurements are performed by a scan in frequency ranging from 100 Hz to 1 MHz for the toroidal transformer and 100 Hz to 300 kHz for the ETD34 transformer to guarantee the accuracy of the results and to avoid the measurement artifacts that can generate the parasitic winding capacitance [7].

The tests were performed using a LCR meter and 4-points measurement probes as shown in Fig.10.

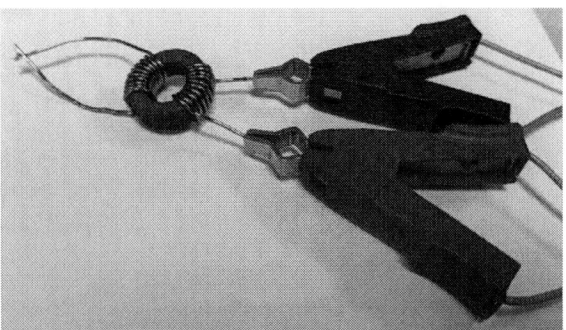

Fig. 10 Leakage Inductors Measurement on Configuration 1

5 Results

5.1 Results of toroidal core tests

After conducting the experimental tests on the six coupling configurations, in accordance with the methodology previously described, we do the measurements.

The Fig. 11 illustrates the ratio between the combined leakage inductances and the magnetizing inductance obtained from measurements
For the case 1 (presented in Fig. 5) configuration, two coils were well coupled to the magnetic material, but separated from each other. This configuration presented a sum of the leakage inductors (Lpl + Lsl) measured at 5,3 µH, equivalent to a 5 % ratio between this sum and the magnetizing inductor.

For the case 2 configuration, where two wires are placed side by side, the ratio of leakage inductors to magnetizing inductor is 1,37%. This configuration presents a clear improvement compared to the previous case.
The case 3 configuration, where the two windings are superimposed, has a ratio of leakage inductors to magnetizing inductor of 0,62 %.

The case 4 scenario involves two twisted wires coiled on the torus, displaying a ratio of leakage inductors to magnetizing inductor of 0,24 %.

In the case 6, we have the same two twisted wires coiled on the torus but with a few centimeters gap to the torus to ensure a weak coupling with the torus, with the magnetic material. The ratio of leakage inductors to magnetizing inductor is 0,34%. In fact, the increase in leakage inductor is due to the greater length of both primary and secondary wires. The magnetization inductor is not changed, since the number of turns and the magnetic section are unchanged.

Finally, the case 6 whose presents 10 turns of primary wires, then 10 turns of secondary wires above, and finally 10 turns of primary wires; with the two primary welded in parallel. This case shows an inductor of leakage inductors to magnetizing inductor of 0,4%.

All these results are summarized in Table 1.

The measurements of the leakage inductors and the magnetizing inductor allowed us to validate our hypothesis that optimal coupling is obtained by a close coupling between the windings by proximity, superimposition, twisting or sandwich of the wires of the primary and secondary windings. The tests of case 5 show that even a very bad coupling with the magnetic material by voluntary distancing still allows realizing a transformer with the primary and secondary leakage inductors minimized. They only increase because of the extra length of the winding imposed by the distance to the core.

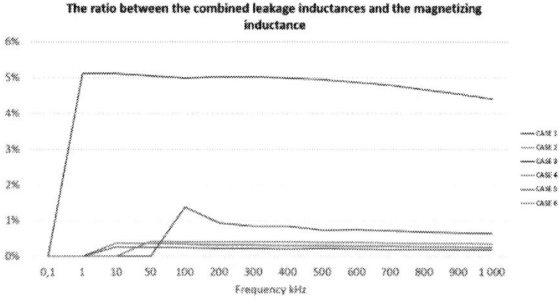

Fig. 11 Results of transformer coupling tests on torus

Table 1: Results of leakage inductors ratios on magnetizing inductors for each case

Case	Description	Ratio of leakage inductors to magnetization inductor
1	Windings apart	4,98 %
2	Rib-to-rib wires	1,37 %
3	Superimposed windings	0,62 %
4	Twisted wires	0,24 %
5	Twisted wires with a gap	0,34 %
6	Sandwich winding	0,4 %

5.2 Results of ETD34's tests

Table 2 shows the ratio between the sum of the inductors and the magnetizing inductor for six different configurations (cases A to F) as a function of frequency.

Table 2: Results of transformer coupling tests on ETD34 core

	20	50	75	100	150	200	300
CASE A	0,1504%	0,1098%	0,1023%	0,0964%	0,0836%	0,0702%	0,0385%
CASE B	0,1462%	0,1112%	0,1053%	0,1005%	0,0918%	0,0820%	0,0598%
CASE C	0,1470%	0,1129%	0,1064%	0,1012%	0,0899%	0,0764%	0,0444%
CASE D	0,1553%	0,1137%	0,1072%	0,1024%	0,0931%	0,0841%	0,0619%
CASE E	0,1018%	0,0983%	0,0914%	0,0820%	0,0564%	0,0237%	
CASE F	0,1220%	0,1212%	0,1192%	0,1169%	0,1138%	0,1126%	0,1032%

Frequency kHz

The overarching trend observed in the experimental findings indicates a consistent decline in the ratio as the frequency increases across all examined cases. Among these cases, Case F notably stands out with the highest ratio, averaging at 0,115 %, closely trailed by Case D, which demonstrates a ratio averaging at 0,102 %. It is noteworthy that Cases D and B display strikingly similar ratios, indicating comparable performance under the tested conditions. On the other end of the spectrum, Case E presents the lowest ratio among the configurations, with an average of 0,075 %.

Moreover, an important observation pertains to the discernible variation in ratios between cases, particularly accentuated at higher frequencies. This suggests that the behavior of the transformer varies significantly depending on the frequency range. For instance, the ratio for Case E remains relatively steady up to approximately 100 kHz before undergoing a more pronounced decline.

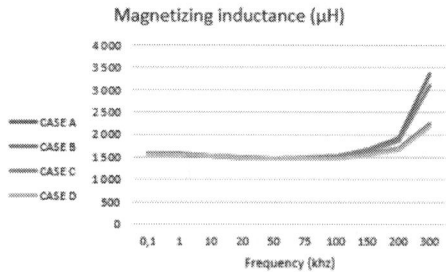

Fig. 12 : Magnetizing inductance measurement on ETD34

Fig. 12 shows that the magnetizing inductance measurement changes significantly with frequency. However, it should be noted that this change is not due to this inductance, but rather to limitations of the measuring device itself.

Indeed, the LCR meter operates by measuring impedance using a parallel LC circuit model.

Due to the resonant parallel circuit, the impedance of the resonant circuit increases when the frequency is closer of the resonance frequency, so the LCR meter analyzes this information as a higher value than the real one. It is the reason why a frequency scanning is mandatory for inductance and capacitance measurement to detect a possible resonance with the parasitic component: capacitance or inductance.

Leakage inductance, on the other hand, varies only slightly with frequency.

In the case A, B, C and D, the number of turns is 24, due to the series connection of the windings, so, $Lm = 1,5 \, mH$.

In the case E, four windings are in series, so the number of turns is 48, and so, $Lm = 6,2 \, mH$.

In the case F, the windings are only in parallel so the number of turns is 12 and $Lm = 380 \, \mu H$.

The higher the number of turns, the higher the magnetic inductance, and the lower the resonance frequency.

For all configuration, due to the close coupling in mirror between each primary turn and each secondary turn, the leakage inductances are minimized, these values are summarized in Table 3.

Table 3: Average ratio of leakage inductor to magnetizing inductor across the six configurations

Case	Average ratio of leakage inductor to magnetizing inductor
A	0,093%
B	0,099%
C	0,096%
D	0,102%
E	0,075%
F	0,115%

6 Conclusion

In summary, optimal coupling between primary and secondary windings is crucial to minimize the leakage inductances of a transformer.
Experimental tests conducted on various toroidal transformer configurations revealed that the ratio of leakage inductance to magnetizing inductance is typically several percent (e.g., 5 % in our implementation) if the two windings are well coupled with the magnetic material but are distant from each other, and decreases with tighter coupling between primary and secondary windings. The leakage inductances decrease more and more (reaching the minimum of 0,24%) if the primary and secondary wires by placing them side-by-side or superimposed, or twisted together. One test shows even that a very bad coupling with the magnetic material by voluntary distancing still allows realizing a transformer with the primary and secondary leakage inductors minimized. The leakage inductors only increase because of the extra length of the winding imposed by the distance to the core.
Another series of test on an ETD34 magnetic circuit prove very good coupling if each primary and secondary windings are designed in mirror with each primary turn in mirror of a secondary turn. On this ETD34 magnetic circuit, our tests also reveal parasitic capacitance due to the stray capacitance of the windings.
A frequency scanning is mandatory for inductance and capacitance measurement to detect a possible resonance with the parasitic component (i.e. capacitance or inductance).

The limitation of this study is that the tests were performed only on a specific transformer manufactured using a ferrite core and copper windings. The relative permeability of the core is slightly higher than calculated, as it is based on an overall measurement of the torus cross-section and is therefore influenced by the thickness of the outer insulation which is unknown. Six coupling configurations have been tested, other configurations exist. This study did not explore all possible variations in transformer design parameters.
For this study, we limited ourselves to a configuration of ten turns. We have not studied the impact of the number of turns and thus the length of the winding on the coupling of the transformers.

References

[1] R. Teja, 'Different Types Of Transformers', ElectronicsHub USA. Accessed: Apr. 09, 2024. [Online]. Available: https://www.electronicshub.org/transformer-types/

[2] V. Lebedev, 'Transformer basics', in *2007 Electrical Insulation Conference and Electrical Manufacturing Expo*, Oct. 2007, pp. 356–359. doi: 10.1109/EEIC.2007.4562642.

[3] 'Transformer', *Wikipedia*. Oct. 08, 2023. Accessed: Oct. 24, 2023. [Online]. Available: https://en.wikipedia.org/w/index.php?title=Transformer&oldid=1179235723

[4] 'Basic Inductance Principles in Transformers - Technical Articles'. Accessed: Oct. 24, 2023. [Online]. Available: https://eepower.com/technical-articles/transformer-operating-principles/

[5] 'Electrical Engineering Handbook'. Accessed: Oct. 26, 2023. [Online]. Available: https://www.docdroid.net/OeGMLxF/electrical-engineering-handbook-pdf

[6] L. Dixon, 'Transformer and Inductor Design for Optimum Circuit Performance', 2003. Accessed: Mar. 14, 2024. [Online]. Available: https://www.semanticscholar.org/paper/Transformer-and-Inductor-Design-for-Optimum-Circuit-Dixon/1d0e2ba266b2ca222ed0499cc605e538d956b8dd

[7] Abdi, B., Nasiri, A., Aslinezhad, M., & Abroshan, M. (2011). Winding Considerations on the High Frequency Transformers. Energy Procedia, 12, 656 661. https://doi.org/10.1016/j.egypro.2011.10.089

AUTHOR INDEX

Abbas, Khizra 764
Ackermann, Martin 1336
Aiello, Giuseppe 1217
Akbari, Saeed 2094
Akturk, Akin 739
Alauzet, Louis 2811
Albert, Tianlong 1759
Alfonso, Irene Maria Torres 2503
Alfonzetti, Emanuela 1844
Allioua, Abdelmoumin 2128
Ammar, Ahmed 1087
Appleby, Matthew 3276
Arai, Nobuhide 298
Araujo, Lucas 1673
Arnaudov, Dimitar 2268
Askan, Kenan 1545
Aspalter, Paul 2258
Augustin, Tim 3086
Aunon, Fernando 1467
Ausseresse, Pierrick 1082
Austrup, Isabel 2956
Babaki, Amir 1227
Bagheribavaryani, Mohammadreza 1418
Baharizadeh, Mehdi 378
Bai, Yeriel 1804
Baker, Nick 1923
Bándy, Kristóf 403, 2566
Barcelos, Renan Pillon 264
Barón, Kevin Muñoz 1978
Barth, Henry 2838
Basso, Christophe 3096
Bastawros, Adel 440, 1951
Batista, Emmanuel 2394
Baudais, Briac 3187
Behrendt, Stefan 361
Beiranvand, Hamzeh 1105
Beyerle, Raphael 958
Bhatia, Tamanna 1259
Bicer, Ekin Alp 40
Bimmel, Luc 3206
Blechinger, Christoph 1717
Block, Marius 2217
Bockholt, Yannick 3334
Böhning, Lukas 2208
Boldyrjew-Mast, Roman 723
Bosnjic, Zlatko 1788
Boutry, Arthur 1878
Bouzerd, Souhila 581

Branas, Christian 2286
Brandl, Anja Katerina 1613
Breidenstein, Daniel 1634
Bürger, Matthias 863
Cairnie, Mark 599
Calmels, Alain 3305
Cammarata, Federica 1289
Campos, Adriana 2663
Cannone, Marco 502
Capobianco, Thomas Anthony 1168
Çay, Yunus 3247
Cepin, Simon 1051
Chaisakdanugull, Chanuch 3067
Chatroux, Daniel 2278
Chatterjee, Bhaskar 774
Chen, Mengxing 424
Cherief, Wahid 1910
Cho, Wonjin Dylan 1046
Choo, Vin Loong 1775
Chorfi, Ilias 2175
Cinik, Sadik 2453
Colak, Baris 490
Colomer, Pau 456
Conilh, Christophe 2227
Corbitt, Anna 135, 1123, 1821
Croston, Jose Andres Aguilar 3150
Curbow, Austin 1475
Cusumano, Andrea 1627
Czerwenka, Philipp 1139, 3034
Daire, Baptiste 3110
Dasch, Michael 1907
Davoodi, Hossein 1013
Debbadi, Karthik 2963
Deboy, Gerald 15
Dedew, Mohamed Lemine 34
Delaforge, Timothé 1797
Denk, Marco 1192
Despesse, Ghislain 797
Diz, Sergio De Lopez 411
Do, Nguyen Nghia 1428
Dresel, Lars 2737
Du, Xinyuan 1987
Duijsen, Peter Van 1658, 2248, 2657, 3213
Dumollard, Yannick 1751
Dupont, Max 93
Dusmez, Serkan 383, 2334, 3060
Eichler, Felix 3020
Eyama, Takaaki 56

Fabian, Benjamin 190
Fenske, Florian 3390
Fey, Justin 1902
Fleck, Soenke 338
Förster, Nikolas 3237
Fotteler, Oleg 3328
Fräger, Lukas 926
Frank, Michael 754
Frank, Wolfgang 1770
Frei, Steffen 2478, 3007
Fuchs-Gade, Jannik 2632
Fuhrmann, Jan 1315
Gackowski, Bartosz 1504
Gandluru, Veera Bharath Chandra Reddy 2167
Gavin, Serge 1101
Gebhard, Thomas 1128
Gebhardt, Mathias 2769
Gellman, Ziv 608
Gendrin, Martin 909
Ghanbari, Alireza Ramezan 3175
Ghosh, Priyanka 1523
Gick, Sebastian 1264
Gioda, Alexis 3400
Girgin, Mehmet Oguz 3353
Giuffrida, Simone 248
Giuffrida, Vittorio 1065
Gleissner, Michael 2803
Goff, Gregoire Le 3160
Gomez, Antonio Miguel Munoz 625
Gottardo, Davide 2461
Gragger, Johannes 2104
Graham, Robert 1410
Groon, Fabian 3380
Groos, Gerhard 986
Guan, Jiajia 2591, 3395
Gudala, Bhavana 2524
Guiot, Eric 1604
Gunes, Ekrem R. 3221
Gupta, Gaurav 534
Gürlek, Yavuz 745
Haake, Daniel 2538
Haas, Tobias 2326, 3017
Haehre, Karsten 214
Haensel, Stefan 230
Hanf, Michael 351, 571
Harmand, Thomas 2138
Hasegawa, Kazunori 3002
Hauenschild, Philipp 1969
Hegarty, Timothy 1092
Hegde, Niranjan 1374
Heimler, Patrick 1955
Hellinger, Rolf 1

Hepp, Maximilian 3045
Herrera, Adolfo 1057
Herrmann, Clemens 731
Hertline, Joseph 1886
Herzog, Fabian 3136
Hirao, Takashi 1007
Hironaka, Yoichi 699
Hoffmann, Lennart 3264
Horat, Andreas 480
Hornbuckle, Malachi 2724
Hosseinzadehlish, Mana 1402, 1610
Hu, Jhih-Cheng 791
Huber, Jonas 254
Huerner, Andreas 681
Huselstein, Jean-Jacques 2547
Husev, Oleksandr 893
Igartuburu, Daniel San Laureano 2303
Imai, Ayano 180
Ippisch, Matthias 2638
Irifune, Hiroyuki 2028
Jahn, Simon 883
Jamal, Adeel 1346
Jappe, Tiago 2843
Jegal, Junhyeok 1590
Jha, Kunal 2930
Jia, Minli 2730
Jo, David 1732
Jones, Jeremy 1031
Kaiser, Jeremias 1538
Kampert, Erik 2342
Kanatzar, Paul 1361
Kangjia, He 62
Karout, Mohammed Amer 1835
Kasko, Igor 1991
Kato, Koji 1368
Kaufmann-Bühler, Marius 2400
Kawabata, Junya 2049
Keilmann, Robert 2972
Kempitiya, Asantha 497
Klever, Severin 1561
Knappstein, Lukas 1745
Knecht, Martin 3142
Koch, Jan-Niklas 2240
Koczy, Dawid 1651
Kohlhepp, Benedikt 2316
Koi, Kenichi 67
Kono, Hiroshi 2022
Kopischke, Ruben 2796
Körner, Patrick 615
Kragl, Robert 1385
Kreppel, Thomas 2416
Krigar, Tim 174

Kroics, Kaspars	510
Kugener, Jeff	3315, 3318
Kurukuru, Varaha Satya Bharath	875
Kuzmanoska, Sara	2745
Ladentin, Kevin	1964
Lambert, Adrien	1574
Langfermann, Sascha	1516
Lavery, Melanie	1485
Lee, Chih Hui	1152
Lee, Jongmu	1712
Lee, Kihyun	1724, 1737
Lemaitre, Damien	2596
Lenz, Travis	1352
Lenzen, Patrick	903
Leung, Wing Tai	74
Liao, Xinyuan	322
Lim, Alex	2937
Lindner, Lars	2370
Lippold, Florian	2981
Liu, Baihan	1072
Liu, Iris	1222
Liu, Yusi	3181
Lottis, Christian	3347, 3358
Lotz, Marc René	1457
Lu, Juncheng	19, 837
Lucia, Oscar	2448, 2513
Lutzen, Hauke	976
Lv, Jianwei	1872
Ma, Kwokwai	2778
Machtinger, Katharina	2119
Madloch, Sonja	369
Maheshwari, Ramkrishan	3118
Mai, Annette	284
Maier, Jannik	1642
Mandrioli, Riccardo	2576
Mannen, Tomoyuki	831
Mari, Jorge	843
Marie, Alexandre	2819
Martano, Emanuele	2874
Martínez, Alfonso	2359
Masuda, Akiyoshi	1018
Mauromicale, Giuseppe	2751
Mazzer, Simone	2162, 2532
McRae, Tim	2190, 2364, 3042
Medina-Garcia, Alfredo	1207
Meligy, Ahmed	1495
Menzel, Steffen	933
Merrouche, Abdennour	1113
Minamisawa, Renato Amaral	2036
Mirkovic, Nikola	2488
Mo, Xianghao	2386
Mochizuki, Yo	870

Mönch, Stefan	167
Mueller, Lukas	2425
Mühlfeld, Christian	3340
Muralikrishna, Ajay Krishna Voppu	2886
Nachete, Idriss	2408
Nakako, Hideo	49
Nawaz, Muhammad	2013
Nehmer, Dominik	24
Neira, Sebastian	2700
Neuner, Matthias	3296
Nikiforidis, Ioannis	916, 2718
Nkembi, Armel Asongu	2469
Oberdieck, Karl	1828
O'Keeffe, Rosemary	1249
Olalla, David	1814
Ong, Shu Ee	2942
Orlando, Stefano	1434, 2673
Otori, Daichi	518
Otte, Raphael	2234
Ouhab, Merouane	589, 2948
Owzareck, Michael	3371
Palma, Marco	1568
Panchal, Pranav	315
Paradkar, Sachin Shridhar	2627
Patterson, Andrew	1330
Paul, Indrajit	1133
Peng, Hujun	665
Petzold, Tom	1158
Pham, Thanh-Toan	2786
Philippe, Antoine	1441, 1449
Phung, Thanh Hai	1555, 2612
Piccioni, Andrea	2680
Piepenbrock, Till	463
Poller, Tilo	2914
Porpora, Francesco	222
Pouresmaeil, Mobina	419
Prince, Aswathy M.	2850
Rabay, Battist	2082
Radix, Bryan	2112
Radomsky, Lukas	3286
Randerath, Joschka	3256
Raßmann, Rando	2350
Rauh, Michael	690
Rebenklau, Lars	2088
Reddy, Niranjan Suravarapu	2273
Rehlaender, Philipp	2686
Reimann, René	2377
Reiner, Richard	557
Reißenweber, Lukas	447, 525
Reitz, Niclas	1393
Ren, Linhao	1917
Ren, Xufu	803

Rendek, Karol	2831
Reymond-Laruina, Frédéric	103
Rezaeizadeh, Amin	1686
Ribarich, Tom	1254
Ribeiro, Kelly	2294
Rillo, Oriol Subirats	290
Ringelmann, Tim	2708
Rodrigues, Luis Alves	635
Rodriguez, Manuel Escudero	812
Rodruigez, Manuel Escudero	3077
Rosensaft, Boris	2620
Rudzki, Jacek	1942
Ruoff, Dominik	1277
Ruppert, Lukas	1703
Sakai, Junya	2006
Salomez, Florentin	2431
Samura, Koki	2764
Sankari, Rasched	197
Sawada, Takashi	161
Schindler, Stefanie	1147
Schindler, Tobias	140
Schmidhuber, Michael	2438
Schmidt, Matthias	397
Schmidt, Paul	1999
Schmitz, Laurids	2995
Schnell, Raffael	855
Schnitzler, Ruben	1851
Schulte, Felix	3364
Schulz, Martin	1077
Schwab, Stefan	343
Schwarz, Niklas	1200
Scuto, Alfio	2921
Seber, Elizabeth	888
Sekar, Ajith Kumar	2041
Sen, Gokhan	784
Seo, Hansol	1896
Sheikhan, Alireza	549, 2606
Shi, Sanbao	1212, 3029
Sifoune, Sarah	390
Singer, Mehyeddine	2309
Solomakha, Oleksandr	1174
Somarin, Hasan Mousavi	2494
Sos, Carlos Costas	205
Sousa, Gean	2557
Srikrishna, N. H	3269
Steenbock, Liska	3169
Steiner, Felix	1891
Stone, David A.	949
Subotic, Stefan	274
Sugie, Hisashi	1765
Sun, Qing	2791
Suzuki, Keita	2053

Syed, Hadiuzzaman	2758
Talits, Kevin	997
Tan, John Emmanuel	150
Tanikawa, Kohei	1272
Tarmoom, Ehab	942
Tekir, Bünyamin	672
Tengvall, Sebastian	1380
Thamm, Merlin	1532
Thekemuriyil, Tanya	645, 2986
Thirukoluri, Rajani Kumar	1865
Thomas, Mark	564
Thönnessen, André	1584
Tigira, Sandu	3130
To, Pham Ha Trieu	707
Tobler, Stefan	472
Tokorozuki, Takeshi	849
Torrisi, Marco	822
Tranchero, Maurizio	1322
Troudi, Rami	1185
Tuncay, Sebnem	1858
Uemura, Hirofumi	433
Ueno, Masaki	2909
Ugur, Abdulkerim	3229
Uhlemann, Andre	1283
Urbaneck, Daniel	2152
Varadarajan, Kamal	1598
Vemulapati, Umamaheswara Reddy	1025
Vinciguerra, Vincenzo	1039
Vobecky, Jan	1002
Vogelsberger, Markus	1307
Vogt, Michael	2866
Vuletic, Radovan	305
Walter, Michael	1297
Wang, Hamlin	2185
Wang, Hao	2693
Wang, Lei	1242
Wang, Lisheng	2060
Wang, Qilei	113
Wang, Rui	84
Wang, Yushi	966
Watanabe, Hiroki	3125
Weckbrodt, Julien	121
Wei, Frank	2146
Wei, Suhang	2074
Weihe, Sven	330
Wen, Jin	2182, 3103
Wessel, Wilfried	655
Wietschel, Martin	7
Wille, Christopher	128
Winkler, Paul	1511
Xie, Dong	2880
Xie, Luhong	1930

Yadav, Sachin 2583
Yan, Xingda 1680
Yan, Yiyang 1809
Ye, Yijun 2826
Ye, Zhong 1621, 2646
Yoshida, Satoshi 543
Yoshioka, Kentaro 2067
Yu, Renze 717
Yu, Sean 1180, 2518
Yu, Sheng-Yang 2200
Zeng, Chenhang 1693
Zhang, Chi 1781
Zhang, Hongpeng 238
Zhang, Huaiyuan 2901
Zhang, Yi 1936
Zhao, Yue 3197
Zheng, Zexiang 2860
Zhu, Shiwu 2893
Zipperstein, David 2651
Zipprich, Robert 1664
Zocher, Markus 1233